Lecture Notes in Computer Science 5554

Commenced Publication in 1973
Founding and Former Series Editors:
Gerhard Goos, Juris Hartmanis, and Jan van Leeuwen

Lora Aroyo Paolo Traverso
Fabio Ciravegna Philipp Cimiano
Tom Heath Eero Hyvönen
Riichiro Mizoguchi Eyal Oren
Marta Sabou Elena Simperl (Eds.)

The Semantic Web: Research and Applications

6th European Semantic Web Conference, ESWC 2009
Heraklion, Crete, Greece, May 31–June 4, 2009
Proceedings

Springer

Volume Editors

Lora Aroyo
Eyal Oren
VU Amsterdam, The Netherlands, E-mail: {l.m.aroyo, eyal}@cs.vu.nl

Paolo Traverso
FBK IRST, Trento, Italy, E-mail: traverso@fbk.eu

Fabio Ciravegna
University of Sheffield, UK, E-mail: Fabio.Ciravegna@shef.ac.uk

Philipp Cimiano
TU Delft, The Netherlands, E-mail: p.cimiano@tudelft.nl

Tom Heath
Talis Information Ltd., Birmingham, UK, E-mail: Tom.Heath@talis.com

Eero Hyvönen
Helsinki University of Technology, Finland, E-mail: Eero.Hyvonen@tkk.fi

Riichiro Mizoguchi
Osaka University, Japan, E-mail: miz@ei.sanken.osaka-u.ac.jp

Marta Sabou
The Open University, Milton Keynes, UK, E-mail: r.m.sabou@open.ac.uk

Elena Simperl
STI Innsbruck, Austria, E-mail: elena.simperl@sti2.at

Library of Congress Control Number: Applied for

CR Subject Classification (1998): H.4, H.3, C.2, H.5, I.2, K.4, D.2

LNCS Sublibrary: SL 3 – Information Systems and Application, incl. Internet/Web and HCI

ISSN 0302-9743
ISBN-10 3-642-02120-4 Springer Berlin Heidelberg New York
ISBN-13 978-3-642-02120-6 Springer Berlin Heidelberg New York

springer.com

Printed in Germany

Typesetting: Camera-ready by author, data conversion by Scientific Publishing Services, Chennai, India
Printed on acid-free paper SPIN: 12690930 06/3180 5 4 3 2 1 0

Preface

This volume contains papers from the technical program of the 6th European Semantic Web Conference (ESWC 2009), held from May 31 to June 4, 2009, in Heraklion, Greece. ESWC 2009 presented the latest results in research and applications of Semantic Web technologies. In addition to the technical research track, ESWC 2009 featured a tutorial program, a PhD symposium, a system demo track, a poster track, a number of collocated workshops, and for the first time in the series a Semantic Web in-use track exploring the benefits of applying Semantic Web technology in real-life applications and contexts.

The technical research paper track received over 250 submissions. The review process was organized using a two-tiered system, where each submission was reviewed by at least three members of the Program Committee. Vice Program Committee Chairs organized a discussion between reviewers, collected additional reviews when necessary and provided a metareview for each submission. During a physical Program Committee meeting, the Vice Program Committee Chairs together with the Program Chairs selected 45 research papers to be presented at the conference.

In addition to the technical research paper track, this year a refereed Semantic Web in-use track was introduced. This track complements the more theoretical aspects of the technical research paper track by providing accounts of rigorous, principled and original work in the Semantic Web field that has a more practical flavor. The Semantic Web in-use track received 27 submissions, which were thoroughly peer-reviewed by a separate Program Committee of Semantic Web researchers and practitioners from industry and academia. Eight papers were selected for publication in the ESWC 2009 proceedings, and for presentation at the conference in two in-use sessions, titled "Bringing Legacy Content to the Semantic Web" and "Linked Data."

The ESWC 2009 proceeding also include eight PhD symposium papers presented at a separate track preceeding the main conference and 24 demo papers giving a brief description of the system demos that were accepted for presentation in a dedicated session during the conference.

ESWC 2009 had four keynote speakers. "Tonight's Dessert: Semantic Web Layer Cakes" was given during the conference dinner by Jim Hendler, Tetherless World Professor of Computer and Cognitive Science, and the Assistant Dean for Information Technology, at Rennselaer. He gave his vision of the future work on the Semantic Web by looking at the layer cake evolution over time. Craig Knoblock, Senior Project Leader at the Information Sciences Institute and a Research Professor in Computer Science at the University of Southern California (USC), discussed work on discovering and building semantic models of Web sources. Alan Smeaton, Professor of Computing and Deputy Director of CLARITY: Centre for Sensor Web Technologies, at Dublin City University,

Ireland, discussed video semantics and the Sensor Web, in the context of content-based video retrieval. "Keys, Money and Mobile Phone" presented by Matthias Wagner, Senior Manager of the Smart and Secure Services Research Group at DOCOMO Euro-Labs, the European research laboratories of the leading Japanese mobile operator NTT DOCOMO, discussed how semantic technologies can add value to mobile services.

To encourage deployment of Semantic Web technologies and to gather and share semantic data, in the tradition of "eating our own dog food," the metadata co-chairs this year extended the usual publication of a rich set of conference semantic metadata with the launch of an ambitious social experiment where the Semantic Web, the Social Web, and the Physical World came together to create a rich and integrated network of information.

Special thanks go to all the members of the Program Committee of the different refereed tracks and the additional reviewers who helped with the review process and contributed to ensuring the scientific quality of ESWC 2009, as well as the members of the Organizing Committee for their hard work in selecting outstanding tutorials, workshops, panels, lightning talks, demos, posters and PhD symposium papers. We would like to also thank the Sponsorship Chair for reaching out to industry and various organizations supporting the 2009 edition of the conference, as well as the local organization, website and conference administration team put together by STI International for their excellent coordination during the conference preparation.

Finally, we would like to thank the Proceedings Chair for the hard work in preparing these proceedings, to Springer for the support with the preparation of the proceedings and to the developers of the EasyChair conference management system, which was used to manage the submission and review of papers and the production of this volume.

June 2009

Lora Aroyo
Paolo Traverso
Fabio Ciravegna
Philip Cimiano
Tom Heath
Eero Hyvönen
Riichiro Mizoguchi
Eyal Oren
Marta Sabou
Elena Simperl

Organization

General Chair

Fabio Ciravegna	University of Sheffield, UK

Program Chairs

Lora Aroyo	VU University Amsterdam, The Netherlands
Paolo Traverso	FBK IRST, Italy

In-Use Chairs

Tom Heath	Talis Information Ltd., UK
Eero Hyvönen	Helsinki University of Technology, Finland

Tutorial Chairs

Philippe Cudré-Mauroux	Massachusetts Institute of Technology, USA
Chiara Ghidini	FBK IRST, Italy

Workshop Chairs

Jérôme Euzenat	INRIA and LIG, Grenoble, France
Valentina Presutti	ISTC-CNR, Italy

Demo and Poster Chairs

Philipp Cimiano	TU Delft, The Netherlands
Elena Simperl	STI Innsbruck, Austria

Sponsorship Chair

Pascal Hitzler	AIFB, University of Karlsruhe, Germany

Panel Chair

Fausto Giunchiglia	University of Trento, Italy

Proceedings Chair

Eyal Oren	VU University Amsterdam, The Netherlands

PhD Symposium Chairs

Riichiro Mizoguchi	Osaka University, Japan
Marta Sabou	The Open University, UK

Semantic Web Technology Coordinators

Harith Alani	University of Southampton, UK
Michael Hausenblas	DERI, National University of Ireland, Galway

Treasurer

Alexander Wahler	STI International, Austria

Local Organizer

Lejla Ibralic-Halilovic	STI International, Austria

Publicity Chair

Charlotte Finkbeiner	STI International, Austria

Vice-Chairs – Research Track

Karl Aberer
Sean Bechhofer
Isabel Cruz
David De Roure
John Davies
John Domingue
Thomas Eiter
Asunción Gómez-Pérez
Nicola Guarino
Martin Hepp
Masahiro Hori
Ian Horrocks
Matthias Klusch
Manolis Koubarakis
Natasha Noy
Massimo Paolucci
Luciano Serafini
Frank van Harmelen
Chris Welty

Program Committee – Research Track

José Júlio Alferes
Yuan An
Ion Androutsopoulos
Galia Angelova
Anupriya Ankolekar
Marcelo Arenas
Alessandro Artale
Yannis Avrithis
Danny Ayers
Christopher Baker
Cristina Baroglio
Nick Bassiliades

Zohra Bellahsene
Abraham Bernstein
Leopoldo Bertossi
Walter Binder
Kalina Bontcheva
Alex Borgida
Paolo Bouquet
John Breslin
Dan Brickley
Jos de Bruijn
François Bry
Paul Buitelaar
Christoph Bussler
Andrea Cali
Diego Calvanese
Mario Cannataro
Silvana Castano
Vinay Chaudhri
Paul-Alexandru Chirita
Fabio Ciravegna
Oscar Corcho
Bernardo Cuenca Grau
Mike Dean
Thierry Declerck
Vania Dimitrova
Ying Ding
Wlodek Drabent
Martin Dzbor
Boi Faltings
Giorgos Flouris
Gerhard Friedrich
Fabien Gandon
Aldo Gangemi
Nicholas Gibbins
Christine Golbreich
John Goodwin
Guido Governatori
Mark Greaves
Gianluigi Greco
Peter Haase
Mohand-Said Hacid
Axel Hahn
Harry Halpin
Herman ter Horst
Laura Hollink
Masahiro Hori
Andreas Hotho
Geert-Jan Houben
Jane Hunter
Carlos Hurtado
Arantza Illarramendi
Antoine Isaac
Epaminondas Kapetanios
David Karger
Vangelis Karkaletsis
Yevgeny Kazakov
Anastasios Kementsietsidis
Hak Lae Kim
Mieczyslaw Kokar
Yiannis Kompatsiaris
Georgia Koutrika
Rubén Lara
Georg Lausen
Nicola Leone
Yaoyong Li
Francesca Alessandra Lisi
Joanne Luciano
Bertram Ludaescher
Alexander Löser
Véronique Malaisé
Jan Maluszynski
Zlatina Marinova
M. Scott Marshall
Mihhail Matskin
Wolfgang May
Diana Maynard
Peter Mika
Simon Miles
Dunja Mladenic
Ralf Moeller
Boris Motik
Saikat Mukherjee
Wolfgang Nejdl
Lyndon Nixon
Daniel Oberle
Leo Obrst
Daniel Olmedilla
Jacco van Ossenbruggen
Sascha Ossowski
Georgios Paliouras

Jeff Z. Pan
Bijan Parsia
Terry Payne
Paulo Pinheiro da Silva
Marco Pistore
Axel Polleres
Line Pouchard
Jinghai Rao
Alan Rector
Gerald Reif
Ulrich Reimer
Mark Roantree
Riccardo Rosati
Marie-Christine Rousset
Alan Ruttenberg
Vasilis Samoladas
Kai-Uwe Sattler
Sebastian Schaffert
Stefan Schlobach
Michael Schumacher
Pavel Shvaiko
Wolf Siberski
Alkis Simitsis
Munindar Singh
Michael Sintek
Sergej Sizov
Spiros Skiadopoulos
Jin Song Dong
Kavitha Srinivas
Giorgos Stamou
Heiko Stoermer
Giorgos Stoilos
Ljiljana Stojanovic
Michael Stollberg
Umberto Straccia
Heiner Stuckenschmidt
Rudi Studer
Gerd Stumme
York Sure
Vojtech Svatek
Hideaki Takeda
Andrei Tamilin
Bernhard Thalheim
Robert Tolksdorf
Christos Tryfonopoulos
Victoria Uren
Costas Vassilakis
Laure Vieu
Ubbo Visser
Tomas Vitvar
Raphael Volz
George Vouros
Holger Wache
Chris Welty
Graham Wilcock
Peter Wood
Takahira Yamaguchi
Milena Yankova
Yiyu Yao
Djamel Zighed
Michal Zaremba

Program Committee – In-Use Track

Harith Alani
Sören Auer
Mathieu d'Aquin
Chris Bizer
Andreas Blumauer
Christopher Brewster
Òscar Celma
Sam Chapman
Huajun Chen
Mariano Consens
John Davies
Stefan Decker
Catherine Dolbear
Robert Engels
Lee Feigenbaum
Hugh Glaser
Mark Greaves
Siegfried Handschuh
Martin Hepp
Ivan Herman
Andreas Hotho
Anthony Jameson

Brian McBride
Eetu Mäkelä
Knud Möller
Lyndon Nixon
Daniel Olmedilla
Eyal Oren
Yuzhong Qu
Yves Raimond
Dave Reynolds
Alan Ruttenberg
Marta Sabou
Leo Sauermann
Sebastian Schaffert
Daniel Schwabe
Andy Seaborne
John Sheridan
Susie Stephens
York Sure
Klaus Tochtermann
Robert Tolksdorf
Raphaël Troncy
Denny Vrandečić
Jan Wielemaker
Milena Yankova

Program Committee – Demo and Poster Track

Harith Alani
Anupriya Ankolekar
Mathieu d'Aquin
Sören Auer
Christopher Brewster
Paul Buitelaar
Klaas Dellschaft
Alfio Ferrara
Agata Filipowska
Florian Fischer
Blaz Fortuna
Dragan Gašević
Birte Glimm
Peter Haase
Willem van Hage
Siegfried Handschuh
Tom Heath
Martin Hepp
Stijn Heymans
Nicholas J. Kings
Reto Krummenacher
Markus Krötzsch
Jens Lehmann
Francesca Alessandra Lisi
Diana Maynard
Adrian Mocan
Malgorzata Mochol
Lyndon Nixon
Eyal Oren
Raul Palma
Viktoria Pammer
Jeff Z. Pan
Sofia Pinto
Angus Roberts
Dumitru Roman
Sebastian Rudolph
Marta Sabou
Leo Sauermann
Arno Scharl
Simon Schenk
Michael Sintek
Katharina Siorpaes
Giorgos Stoilos
Michael Stollberg
Vojtech Svatek
Vlad Tanasescu
Raphaël Troncy
Tania Tudorache
Giovanni Tummarello
Victoria Uren
Denny Vrandečić
Johanna Völker
Haofen Wang
Ingo Weber
René Witte
Maciej Zaremba
Anna Zhdanova

Program Committee – PhD Symposium

Mathieu d'Aquin
Paolo Bouquet
Paul Buitelaar
Martin Dzbor
Jérôme Euzenat
Adam Funk
Stefania Galizia
Carole Goble
Alessio Gugliotta
Peter Haase
Frank van Harmelen
Tom Heath
Martin Hepp
Pascal Hitzler
Diana Maynard
Eduardo Mena
Peter Mika
Dunja Mladenic
Natasha Noy
Daniel Olmedilla
Jeff Z. Pan
Terry Payne
Valentina Presutti
Guus Schreiber
Pavel Shvaiko
Elena Simperl
Steffen Staab
Heiner Stuckenschmidt
Hideaki Takeda
Valentina Tamma
Johanna Voelker
Holger Wache

External Reviewers

Alessandro Adamou
Sudhir Agarwal
Ricardo Amador
Claudia d'Amato
Georgios Apostolikas
David Aumueller
Ulrike Bauernfeind
Thorsten Berger
Ravish Bhagdev
Holger Billhardt
Sebastian Blohm
Eva Blomqvist
Stefano Borgo
Stefano Bortoli
Amancio Bouza
Shawn Bowers
Carlos Buil
Paul Buitelaar
Michele Catasta
Òscar Celma
Huajun Chen
Philipp Cimiano
Remi Coletta
Gianluc Colombo
Oscar Corcho
Ilaria Corda
Ronald Denaux
Sebastian Dietzold
Nima Dokoohaki
Alistair Duke
Paul El Khoury
Sofia Espinosa
Nicola Fanizzi
Alexander Felfernig
Yuzhang Feng
Alberto Fernandez Gil
Roberta Ferrario
Angela Fogarolli
Avigdor Gal
Leyla Jael García Castro
Anna Lisa Gentile
Chiara Ghidini
Guido Governatori
Gisela Granitzer
Giovanni Grasso
Laurian Gridinoc
Gunnar Grimnes
Benjamin Grosof

Tudor Groza
Pietro Hiram Guzzi
Peter Haase
Willem van Hage
Harry Halpin
Michael Hausenblas
Tom Heath
Sebastian Hellmann
Jan Hidders
Patrick Hoefler
Thomas Hornung
Geert-Jan Houben
Wei Hu
Zhisheng Huang
Fabian Hueske
Seylan Inanc
Antoine Isaac
Robert Jäschke
Nophadol Jekjantuk
Philipp Kärger
Atila Kaya
C. Maria Keet
Shahan Khatchadourian
Malte Kiesel
Nick Kings
Styliani Kleanthous
Matthias Knorr
Mieczyslaw Kokar
Stasinos Konstantopoulos
Efstratios Kontopoulos
Markus Krötzsch
Anastasia Krithara
Vitaveska Lanfranchi
Yaoyong Li
Frank Loebe
Nuno Lopes
Davide Lorusso
Marco Manna
Rainer Marrone
Alessandra Martello
Diana Maynard
Georgios Meditskos
Christina Meilicke
Eduardo Mena
Fergal Monaghan
Stefano Montanelli
Preslav Nakov
Jan Nemrava
David Newman
Andriy Nikolov
Barry Norton
Vit Novacek
Tun Nwe Ni
Alessandro Oltramari
Eyal Oren
Jacco van Ossenbruggen
Viktoria Pammer
Jorge Perez
Georgios Petasis
Thanh Pham
Dimitris Pierrakos
Valentina Presutti
Alan Rector
Katharina Reinecke
Yuan Ren
Francesco Ricca
Mariano Rodriguez-Muro
Leonardo Salayandia
Yacine Sam
Simone Santini
Simon Scerri
Thomas Scharrenbach
Florian Schmedding
Michael Schmidt
Guus Schreiber
Alexander Schutz
Lei Shu
Kamil Sokolski
Nenad Stojanovic
Heiner Stuckenschmidt
Yehia Taher
Diana Tanase
Jonas Tappolet
Stuart Taylor
VinhTuan Thai
David Thau
Thanh Tran
Yiorgos Trimponias
Georgios Tsatsaronis
Gaia Varese

Luis Manuel Vilches
Alex Villazon
Shenghui Wang
Cathrin Weiss
Stefan Woltran
Huiyong Xiao
Ondrej Zamazal
Markus Zanker
Ilias Zavitsanos
Yuting Zhao
Dimitri Zibold
Juri Luca de Coi

Sponsors

Table of Contents

Invited Talks

Research Track

Applications

Evaluation and Benchmarking

Ontologies and Natural Language

Ontology Alignment

Ontology Engineering

Query Processing

Reasoning

Search and Identities

Semantic Web Architectures

Semantic Web Services

Tagging and Annotation

Semantic Web In-Use Track

Demo Track

PhD Symposium

Tonight's Dessert: Semantic Web Layer Cakes

James A. Hendler

Tetherless World Chair of Computer and Cognitive Science
Rensselaer Polytechnic Institute
hendler@cs.rpi.edu

Abstract. Tim Berners-Lee's "Semantic Web Layer Cake" is one of the most used figures in our field. It has changed numerous times over the years, as new technologies have come along, and as the field has changed. Despite this, it appears to be incomplete – lots happening in the field is not included. In this talk, we take a look at the layer cake as it has changed over time and see what it can tell us about current and future Semantic Web work.

L. Aroyo et al. (Eds.): ESWC 2009, LNCS 5554, p. 1, 2009.

Discovering and Building Semantic Models of Web Sources*

Craig A. Knoblock

University of Southern California
Information Sciences Institute
4676 Admiralty Way
Marina del Rey, CA 90292 USA
knoblock@isi.edu

Abstract. To achieve widespread use of the Semantic Web depends on having a critical mass of Web data available with semantic annotations. Since there are a huge number of sources available today without any such annotations, the challenge is how to find and build semantic models for these sources. In this talk I will describe an integrated end-to-end approach that automatically discovers information-producing web sources, invokes and extracts the data from these sources, builds semantic models of the sources, and validates the results by comparing the data produced by the source with the model of the source. These techniques are implemented in a system called DEIMOS, which integrates a diverse set of technologies to completely automate this task. DEIMOS starts with a "seed" source and finds other similar sources online using data from a social networking web site. Next the system learns how to invoke these sources through experimentation and then extracts data from these sources with automatic wrapping techniques. Finally, DEIMOS learns a semantic model of a source, which identifies the semantic types of the data produced by a source as well as the function that maps the inputs to the outputs. I will describe the challenges in integrating the component technologies into a unified approach to discovering, extracting and modeling new online sources. I will also present an evaluation of the integrated system on three different domains to demonstrate that it can automatically discover and model new Web sources.

* This talk is based on joint work with Jose Luis Ambite, Mark Carman, Cenk Gazen, Kristina Lerman, Steven Minton, Anon Plangprasopchok, and Tom Russ. This research is based upon work supported in part by the National Science Foundation under award number IIS-0535182, in part by the Air Force Office of Scientific Research under grant number FA9550-07-1-0416, and in part by the Defense Advanced Research Projects Agency (DARPA) under Contract No. FA8750-07-D-0185/0004.

L. Aroyo et al. (Eds.): ESWC 2009, LNCS 5554, p. 2, 2009.

Video Semantics and the Sensor Web

Alan F. Smeaton*

CLARITY: Centre for Sensor Web Technologies
Dublin City University
Glasnevin, Dublin 9, Ireland
alan.smeaton@dcu.ie

The most widespread way in which content-based access to video information is supported is through using a combination of video metadata (date, time, format, etc.) and user-generated description (user tags, ratings, reviews, etc.). This has had widespread usage and is the basis for navigation through video archives in systems such as YouTube, Open Video and the Internet Archive. However, there are limitations with this such as vocabulary issues, and authentication across the users who annotate content.

Our work has concentrated on addressing *content-based* video retrieval, the kind of direct retrieval we routinely perform on text, retrieving web pages or blog posts based primarily on actual content. One premise for our work is that we can extract descriptions of content reliably.

We can easily extract low-level features such as colour, texture and shapes from the visual aspects of video (the picture) and from these we can compute visual similarity. In a video retrieval system this manifests as being able to find video shots based on the visual similarity between one or more query images and each of the keyframes in the video. Once again, however, this also has limits, such as having to locate query images to start the query, and the fact that a keyframe is just one moment of a video shot which may not be representative.

More relevant to this talk is the challenge of automatically identifying semantic features from video and the talk will present a snapshot of the current approaches taken to identifying the presence or absence of groups of semantic features, in video. The annual TRECVid activity has been benchmarking the effectiveness of various approaches since 2001 and we will examine what is the performance of these detectors, what are the trends in this area, and what is the state of the art in this. We will discover that the performance of individual detectors varies widely depending on the nature of the semantic feature, the quality of training data and its dependence on other detectors. Current work addresses the inter-dependence among semantic features, supported by the use of ontologies, and also how the relatively poor quality of detection of these semantic features when taken in isolation, can be ameliorated by orchestrating them to work together. There is a strong analogy between this and the way that sensors (environmental, physiological, etc.) which make up the sensor web, can also have poor accuracy levels when used in isolation but whose individual performances can be improved when used in combination.

* The author acknowledges the support of Science Foundation Ireland under grant number 07/CE/I1147.

L. Aroyo et al. (Eds.): ESWC 2009, LNCS 5554, p. 3, 2009.

Keys, Money and Mobile Phone

Matthias Wagner

DOCOMO Euro-Labs

Abstract. Across cultures, genders and generations, the mobile phone has become one of the most essential objects in peoples everyday life. In this talk we will go through the evidence on how, next to money and keys, the mobile phone is the third-most thing to be carried around by almost everybody, nearly all the time. By looking at the demands and needs of todays mobile users, we discuss how Semantic technologies can add value to phones as well as to mobile services.

Walking through selected projects of DOCOMO Euro-Labs, we advocate that the success of future mobile services will largely depend on their ability to maximize their value in varying context. Contextual Intelligence in devices, mobile applications and service platforms will be needed to manage different mobile terminals, personalize content and services or to narrow down possibly very large sets of applicable services in a given situation. With this vision of Contextual Intelligence in mind, we are exploiting technologies from the Semantic Web for the mobile domain. For instance, to extend location-based service with knowledge and reasoning on places, people and things.

L. Aroyo et al. (Eds.): ESWC 2009, LNCS 5554, p. 4, 2009.

Querying Trust in RDF Data with tSPARQL

Olaf Hartig

Humboldt-Universität zu Berlin
hartig@informatik.hu-berlin.de

Abstract. Today a large amount of RDF data is published on the Web. However, the openness of the Web and the ease to combine RDF data from different sources creates new challenges. The Web of data is missing a uniform way to assess and to query the trustworthiness of information. In this paper we present tSPARQL, a trust-aware extension to SPARQL. Two additional keywords enable users to describe trust requirements and to query the trustworthiness of RDF data. Hence, tSPARQL allows adding trust to RDF-based applications in an easy manner. As the foundation we propose a trust model that associates RDF statements with trust values and we extend the SPARQL semantics to access these trust values in tSPARQL. Furthermore, we discuss opportunities to optimize the execution of tSPARQL queries.

1 Introduction

During recent years a large amount of data described by RDF has been published on the Web; large datasets are interlinked; new applications emerge which utilize this data in novel and innovative ways. However, the openness of the Web and the ease to combine RDF data from different sources creates new challenges for applications. Unreliable data could dominate results of queries, taint inferred data, affect knowledge bases, and have negative or misleading impact on software agents. Hence, questions of reliability and trustworthiness must be addressed. While several approaches consider trustworthiness of potential sources of data (e.g. [1,2,3]), little has been done considering the actual data itself.

What is missing for is a uniform way to rate the trustworthiness of the data on the Web and standardized mechanisms to access and to use these ratings. Users as well as software agents have to be able to utilize trust ratings and base their decisions upon them. They have to be enabled to ask queries such as:

Q_1: Return a list of garages close to a specific location ordered by the trustworthiness of the data.
Q_2: Return trustworthy reviews for a specific restaurant.
Q_3: Return the most trustworthy review for each hotel in the city of Heraklion.

To ask queries like these this paper presents appropriate extensions for RDF and its query language SPARQL. Our main contributions are:

- a trust model for RDF data which associates triples with trust values, and
- a trust-aware query language, tSPARQL, which extends SPARQL to describe trust requirements and to access the trustworthiness of query solutions.

L. Aroyo et al. (Eds.): ESWC 2009, LNCS 5554, pp. 5–20, 2009.

This paper is structered as follows. First, Section 2 outlines our trust model for RDF. In Section 3 we informally present tSPARQL by introducting two new query clauses that enable users to ask queries such as Q_1 to Q_3. In the remainder we discuss the necessary extensions for tSPARQL in detail. We present our approach for trust-aware processing of queries (Section 4); followed by the extensions for the new clauses (Section 5) and a discussion of opportunities to optimize the execution of tSPARQL queries (Section 6). An evaluation of our approach is given in Section 7. Finally, we review related work in Section 8 and conclude in Section 9.

2 A Trust Model for RDF

As the foundation of a trust infrastructure that considers the trustworthiness of RDF data we introduce a trust model for RDF. The goal of our trust model is to rate information expressed in RDF according to trustworthiness. Since information expressed in RDF is represented by triples our model represents the trustworthiness of triples. Our fundamental understanding of the *trustworthiness of RDF triples* is the subjective belief or disbelief in the truth of the information represented by the triples. Notice, disbelief is a negative form of belief: disbelief in the truth of information is belief in the untruth of this information; i.e., disbelief is the belief that the information is false. Since belief is a personal attitude the trustworthiness of triples depends on the information consumer.

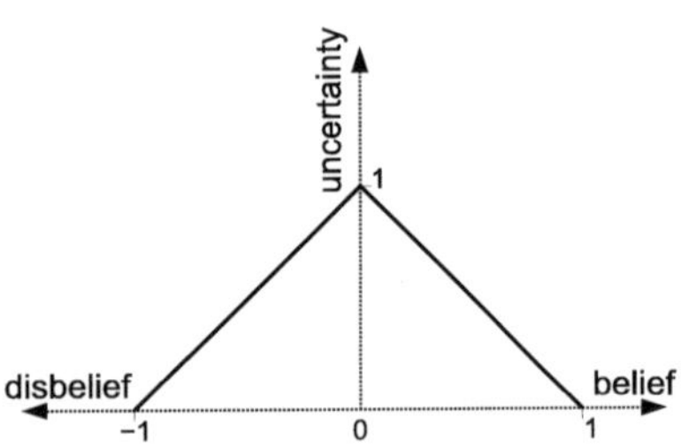

Fig. 1. Meaning of trust values

To enable machine-based processing we introduce a quantifiable measure; we represent the trustworthiness of RDF triples by a *trust value* which is either unknown or a value in the interval [-1,1]. We define the meaning of these values by a specification of the interval boundaries: a trust value of 1 represents absolute belief in the information represented by the corresponding triples; -1 represents absolute disbelief; intermediary values represent degrees of belief/disbelief. We understand the difference between trust values and the extrema 1 and -1 as uncertainty (cf. Figure 1). For a value of 1 the consumer is absolutely sure about the truth of the corresponding triples; a positive value less than 1 still represents belief in the truth; however, to a certain degree the consumer is unsure regarding the assessment. Hence, the lower the trust value, the greater the uncertainty. A value of 0, finally, represents absolute uncertainty. The same holds for negative trust values with respect to disbelief: the higher the negative trust value, the greater the uncertainty. Hence, absolute uncertainty regarding a truth assessment, i.e. the value 0, is equal to the lack of belief as well as the lack of disbelief. Furthermore, we permit unknown trust values, denoted by $\emptyset$, for cases where it is impossible to determine the trustworthiness of triples. Please note the significant difference between a trust value of 0 and an unknown trust value; while the latter denotes the trust management system has no information, a value of 0 represents the expressed lack of belief/disbelief. To determine trust values we define a *trust function.*

Definition 1. *Let T be the set of all RDF triples. A* **trust function** tv^C *for RDF triples is a mapping* $\text{tv}^C : T \to \{tv \mid tv \in [\text{-}1, 1]\} \cup \{\emptyset\}$ *which assigns every triple in T a subjective trust value that represents the trustworthiness of the triple specific to an information consumer C.*

Our trust model does not prescribe a specific implementation of determining trust values. Instead, we allow each system to provide its own, application-specific trust function. Determining trust values may be based on provenance information [4] as in the Inference Web trust component [5]; the TRELLIS system [6] additionally considers related data; the FilmTrust application [7] combines provenance information and recommendations from other consumers.

We refer to an RDF graph with triples that are all associated with trust values for a specific information consumer as a *trust weighted RDF graph.*

Definition 2. *A* **trust weighted RDF graph** $\widetilde{G}^C$ *for information consumer C is a pair (G, tv^C) consisting of an RDF graph G and a trust function* tv^C.

Example 1. Figure 2 depicts a trust weighted RDF graph; The edges represent the predicates of triples. They are annotated with the predicate identifier as usual and with an additional label for the consumer-specific trust value of the corresponding triple. One of the triples in the graph asserts that resource `ex1:Kastro` is a hotel building; this triple is associated with a trust value of 0.95. □

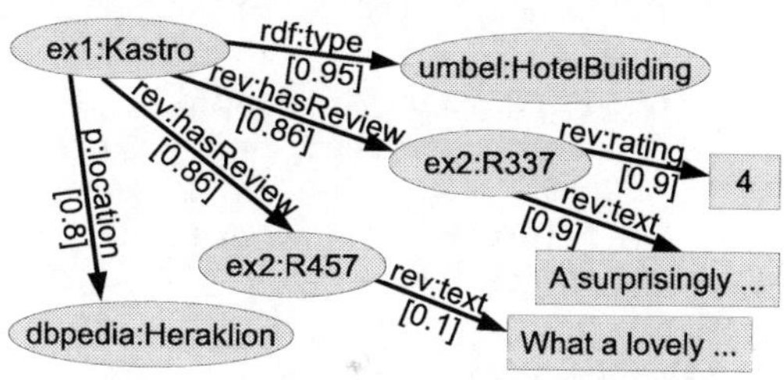

Fig. 2. A trust weighted RDF graph

In addition to the trustworthiness of single triples we represent the trustworthiness of sets of triples, i.e. of whole RDF graphs. The trustworthiness of an RDF graph is an aggregation of the trustworthiness of its triples. Hence, we introduce a *trust aggregation function* to calculate the trust value for RDF graphs.

Definition 3. *A* **trust aggregation function** *for trust weighted RDF graphs is a function* ta *which assigns each trust weighted RDF graph $\widetilde{G}^C$ an aggregated trust value* $\text{ta}(\widetilde{G}^C)$ *that represents the trustworthiness of $\widetilde{G}^C$.*

As with the trust function we do not prescribe a specific aggregation function. Applications have the freedom to choose a function that fits their use cases. The minimum, for instance, is a cautious choice; it assumes the trustworthiness of a set of triples is only as trustworthy as the least trusted triple. The median, a more optimistic choice, is another reasonable trust aggregation function. However, each trust aggregation function must process unknown trust values in a meaningful way. A possible approach is to return an unknown value if an input value is unknown. We will investigate aggregation of trust values more closely in the future.

Please notice, we see the primary usage of trust values for triples within the trust component of a system; trust values are parts of data structures that represent trust weighted RDF graphs. However, to explicitly assert the trustworthiness of RDF data we provide a trust vocabulary[1].

3 Accessing Trust Values in Queries

Trust ratings for RDF data add little value if they cannot be utilized while accessing the data. Therefore, we propose tSPARQL as an extension to the RDF query language SPARQL [8]. SPARQL is of declarative nature; evaluation is based on graph patterns that must match subgraphs in the queried RDF graphs. With tSPARQL users can additionally access trust values that represent the trustworthiness of matching subgraphs. This additional expressivity allows for queries such as Q_1 to Q_3. To express these queries we add a `TRUST AS` clause to the query language. Consider the query in Figure 3 which expresses Q_1 in tSPARQL. The query contains a `TRUST AS` clause with a new variable `?t` which allows access to the trust values of the subgraphs that match the query pattern.

```
SELECT ?garage ?t WHERE {
  ?garage a <http://umbel.org/umbel/sc/AutoRepairShop>;
            geo:lat ?lat; geo:long ?long .
  FILTER ( ex:distance(?lat,?long,35.33,25.13) < 10.0 )
  TRUST AS ?t
}
ORDER BY ?t
```

Fig. 3. tSPARQL representation of query Q_1 (prefix declarations omitted)

The `TRUST AS` clause offers the following novel features: i) the new variable can become part of the query result, ii) it can be used for sorting the results, iii) it can be associated with parts of the query pattern, and iv) two variables that represent trust values of different query pattern parts can be compared. This approach can even be used for expressing requirements regarding the trustworthiness of query results as in query Q_2: in addition to the `TRUST AS` clause users simply add a `FILTER` clause which restricts the new variable. However, for convenience we propose another new clause for these cases, namely the `ENSURE TRUST` clause which includes a pair of numbers that denote a lower bound and an upper bound, respectively. Figure 4 depicts the tSPARQL representation of Q_2; due to the `ENSURE TRUST` clause only those solutions become part of the result where the trust value of the matching subgraph is at least 0.5 and at most 1.0.

Queries Q_1 and Q_2 access the trust values of the subgraphs that match the whole query pattern. Query Q_3, in contrast, compares the trustworthiness of different parts of each query result. Hence, for Q_3 we must limit the scope of

[1] http://trdf.sourceforge.net/trustvocab

```
SELECT ?r ?text WHERE {
  <http://dbpedia.org/resource/The_Cherry_Street_Tavern>
      rev:hasReview ?r .
  ?r rev:text ?text .
  ENSURE TRUST ( 0.5, 1.0 )
}
```

Fig. 4. tSPARQL representation of query Q_2 (prefix declarations omitted)

```
1 SELECT ?h ?txt1 WHERE {
2   ?h rdf:type <http://umbel.org/umbel/sc/HotelBuilding> ;
3      p:location <http://dbpedia.org/resource/Heraklion> .
4   { ?hotel rev:hasReview [ rev:text ?txt1 ]
5      TRUST AS ?t1 }
6   OPTIONAL { ?h rev:hasReview [ rev:text ?txt2 ]
7               TRUST AS ?t2 FILTER ( ?t2 > ?t1 ) }
8   FILTER ( ! BOUND (?txt2) )
9 }
```

Fig. 5. tSPARQL representation of query Q_3 (prefix declarations omitted)

`TRUST AS` clauses to subgraphs that match only parts of the whole query pattern. This limitation can easily be expressed by associating the clause with a specific part of the query pattern as the tSPARQL representation of Q_3 in Figure 5 illustrates. The `TRUST AS` clause in line 6, for instance, solely binds variable ?t1 with trust values of subgraphs that match the graph pattern in line 5. Hence, the position of `TRUST AS` clauses in the query pattern matters; the same holds for `ENSURE TRUST`.

To enable the proposed extensions we must enhance SPARQL in two ways. First, we extend the query language with the new clauses and define operations for them. Second, since trust values are currently not part of SPARQL we also extend the processing of SPARQL queries to consider trust values. In the remainder of this paper we describe the details of our extensions.

4 Trust in SPARQL Query Processing

The semantics of SPARQL do not consider trust values. To implement trust-aware query processing we redefine parts of the semantics for tSPARQL. In this section, we first take a brief look at SPARQL query processing and query evaluation; afterwards we present our adaptations.

The SPARQL specification [8] gives an operational definition of the semantics of SPARQL. In brief, the specification defines a grammar for the query language, a translation from a parse tree to an abstract syntax tree (AST), a transformation from an AST to an abstract query with an algebra expression, and an operation to evaluate abstract queries based on algebra operators. The algebra is defined to calculate query solutions and operate on them. Finally, a result form definition specifies how to create the query result from the solutions. This operational definition of the semantics forms the basis for query processing.

To consider trustworthiness during query processing and to enable operators to access trust values we extend query evaluation for tSPARQL. First, we redefine the notion of solutions because the trust values have to become part of the solutions. Second, we specify how the algebra operates on the extended solutions.

4.1 Trust-Aware Basic Graph Pattern Matching

SPARQL is based on graph patterns and subgraph matching. The elementary graph pattern is called a *basic graph pattern* (BGP); it is a set of *triple patterns* which are RDF triples that may contain variables at the subject, predicate, and object position. During evaluation values are bound to the variables according to the subgraphs that match BGPs. These variable bindings are called *solution mapping*. Besides variables SPARQL permits blank nodes in triple patterns. Blank nodes have to be associated with values during pattern matching similiar to variables. Hence, the SPARQL specification additionally introduces *RDF instance mappings* that map blank nodes to values. A *solution* for a BGP is each solution mapping which, in combination with an RDF instance mapping, maps the BGP to a subgraph of the queried RDF graph. The result of BGP matching is a multiset (or bag) of solution mappings; a solution mapping can be combined with different RDF instance mappings to map the BGP to different subgraphs.

We adapt the definitions for tSPARQL. However, we associate each solution mapping with a trust value.

Definition 4. *A* ***trust weighted solution mapping*** $\widetilde{\mu}$ *is a pair* (μ, t) *consisting of a solution mapping* μ *and a trust value* t*. We denote the cardinality of* $\widetilde{\mu}$ *in a multiset* $\widetilde{\Omega}$ *of trust weighted solution mappings with* $\mathrm{card}_{\widetilde{\Omega}}(\widetilde{\mu})$.

Following the definition from the SPARQL specification, we define solutions for a BGP over a trust weighted RDF graph. Every solution mapping that is a solution to a BGP represents one matching subgraph; the trust value of this solution mapping must represent the trustworthiness of the subgraph; hence, the trust value can be calculated by a trust aggregation function (cf. Definition 3):

Definition 5. *Let* b *be a BGP; let* $\widetilde{G}^C = (G, tv^C)$ *be a trust weighted RDF graph. The trust weighted solution mapping* (μ, t) *is a* **solution** *for* b *in* $\widetilde{G}^C$ *if there is an RDF instance mapping* σ *such that i)* $\mu(\sigma(b))$ *is a subgraph of* G*, ii)* μ *is a mapping for the query variables in* b*, and iii)* $t = \mathrm{ta}(\widetilde{S}^C)$ *is the aggregated trust value of the trust weighted RDF graph* $\widetilde{S}^C = (\mu(\sigma(b)), tv^C)$ *calculated by trust aggregation function* ta*. For each solution* μ *for* b $\mathrm{card}_{\widetilde{\Omega}}(\widetilde{\mu})$ *is the number of distinct RDF instance mappings* σ *such that* $\mu(\sigma(b))$ *is a subgraph of* G*.*

With our definition of solution the result of BGP matching is a multiset of trust weighted solution mappings. Since the solutions are calculated for a trust weighted RDF graph they are calculated in the context of a specific information consumer. BGP matching in the context of another consumer may yield solutions with different trust values because assessing the trustworthiness of matching subgraphs is subjective; hence, the trust values associated with the triples of matching subgraphs are consumer-specific.

Example 2. When we apply the BGP in line 5 of Figure 5 to our sample trust weighted RDF graph in Figure 2 we find two matching subgraphs resulting in the two solutions shown in Figure 6. μ_1 maps `?h` to `ex1:Kastro` and `?txt1` to the literal "A surprisingly ..."; μ_2 maps `?h` to `ex1:Kastro` and `?txt1` to "What a lovely ... ." To determine the trust values for both, μ_1 and μ_2, we choose the minimum as our application-specific trust aggregation function. The subgraph for μ_1 consists of two triples with trust values 0.86 and 0.9. Hence, our first solution is the trust weighted solution mapping $\widetilde{\mu_1} = (\mu_1, 0.86)$. For μ_2 we have the two trust values 0.86 and 0.1; our second solution is $\widetilde{\mu_2} = (\mu_2, 0.1)$. □

μ		t
?h	?txt1	
ex1:Kastro	"A surprisingly ..."	0.86
ex1:Kastro	"What a lovely ..."	0.1

Fig. 6. Trust weighted solution

4.2 Enhanced SPARQL Algebra

After defining trust weighted solution mappings we now explain how these mappings are combined in more complex queries. Besides BGPs, the SPARQL specification introduces other graph patterns. During query evaluation they are represented by algebra operators which operate on multisets of solution mappings. For our new clauses (cf. Section 3) we need new types of operators. To enable these new operators to access the trust values in solutions all operators have to consider the trust values. Hence, for tSPARQL we redefine the conventional SPARQL algebra operators to operate on multisets of trust weighted solution mappings. In the following we exemplarily present the redefined join operator.

The conventional join operator represents a group graph pattern. The two operands of the operator are multisets of solution mappings. Every mapping from one operand is merged with every mapping from the other if they are compatible. Solution mappings are *compatible* if all variables specified in both mappings are bound to the same values. *Merging* two solution mappings combines all variable bindings from both mappings in a new one.

A join operator that operates on trust weighted solution mappings has to consider the trust values while merging solutions. The trust value of a merged solution mapping is an aggregation of the trust values associated with the individual mappings that has been used for merging. For this purpose we introduce another aggregation function which we call *trust merge function*.

Definition 6. *A* ***trust merge function*** *for two trust weighted solution mappings* $\widetilde{\mu_1}$ *and* $\widetilde{\mu_2}$ *is a commutative and associative function* tm *that determines a merged trust value* $\mathrm{tm}(\widetilde{\mu_1}, \widetilde{\mu_2})$.

We notice that trust merge functions must be commutative and associative because the join operator is a commutative and associative operation. Using trust merge functions we redefine the join operator.

Definition 7. *Let $\widetilde{\Omega_1}$ and $\widetilde{\Omega_2}$ be multisets of trust weighted solution mappings; let* merge *be the merge operation for solution mappings [8]. The result of a* ***join operator*** *is a multiset of trust weighted solution mappings which is defined as*

$$Join(\widetilde{\Omega_1}, \widetilde{\Omega_2}) = \big\{ \big(\mathrm{merge}(\mu_1, \mu_2), \mathrm{tm}(\widetilde{\mu_1}, \widetilde{\mu_2})\big) \;\big|\; \widetilde{\mu_1} = (\mu_1, t_1) \in \widetilde{\Omega_1} \wedge \widetilde{\mu_2} = (\mu_2, t_2) \in \widetilde{\Omega_2} \wedge \mu_1 \text{ and } \mu_2 \text{ are compatible} \big\}$$

with

$$\mathrm{card}_{Join\left(\widetilde{\Omega_1}, \widetilde{\Omega_2}\right)}(\widetilde{\mu}) = \sum_{\substack{\widetilde{\mu_1} \in \widetilde{\Omega_1} \\ \widetilde{\mu_2} \in \widetilde{\Omega_2}}} \begin{cases} \mathrm{card}_{\widetilde{\Omega_1}}(\widetilde{\mu_1}) \cdot \mathrm{card}_{\widetilde{\Omega_2}}(\widetilde{\mu_2}) & \text{if } \widetilde{\mu} = (\mu, t) \text{ with } \\ & \quad t = \mathrm{tm}\left(\widetilde{\mu_1}, \widetilde{\mu_2}\right) \text{ and } \\ & \quad \mu = \mathrm{merge}(\mu_1, \mu_2) \\ & \quad \text{where } \widetilde{\mu_i} = (\mu_i, t_i) \\ 0 & \text{else} \end{cases}$$

where tm *is an application-specific trust merge function.*

Again, the definition does not prescribe a specific trust merge function; thus giving applications a choice (insofar as the function is commutative and associative). Possible choices are the minimum or the arithmetic mean.

Example 3. The group graph pattern

{
{ ?h rdf:type umbel:HotelBuilding;
p:location dbpedia:Heraklion. }
{ ?h rev:hasReview [rev:text ?txt1] }
}

groups two BGPs. Figure 7 depicts the corresponding algebra expression as an operator tree. The tree is annotated with trust weighted solution mappings that are solutions for our sample graph in Figure 2. The two tables near the bottom of Figure 7 contain the solutions for the BGPs. Joining the solutions from both multisets results in the solutions represented by the upper table. The trust merge function applied is the minimum function. □

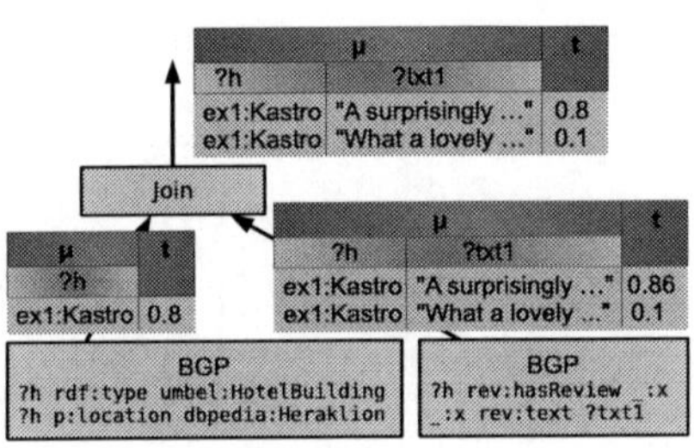

Fig. 7. Operator tree with solutions

Similiar to the join operator we adapted all algebra operators for tSPARQL [9].

5 SPARQL Extension for Trust Requirements

Section 3 gives a high-level overview of tSPARQL and the new clauses `TRUST AS` and `ENSURE TRUST`. In this section we provide a more formal description.

5.1 Accessing Trust Values

The `TRUST AS` clause permits access to the trust values associated with solutions. This is impossible with conventional SPARQL queries since conventional query evaluation does not consider trust values. To process `TRUST AS` clauses we extend the grammar of the query language, adapt the translation to an abstract syntax tree (AST) as well as the transformation from an AST to an abstract query; we define a new algebra operator and we extend the operation to evaluate abstract queries. Due to the limited space we only give a brief informal description of the grammar extension and present the algebra operator here. The tSPARQL specification [9] covers all necessary extensions in detail.

The `TRUST AS` clause is denoted by the keywords `TRUST AS` which are followed by a query variable. This variable must not be contained in any other pattern of the query. A `TRUST AS` clause can occur at any position in a query where `FILTER` clauses are permitted. The corresponding algebra operator, called *project trust operator*, operates on a multiset of trust weighted solution mappings. For every mapping it accesses the trust value, creates a new variable binding which maps the specified variable to an RDF literal that represents the trust value, and adds the new binding to the mapping.

Definition 8. *Let $\widetilde{\Omega}$ be a multiset of trust weighted solution mappings; let v be a query variable which is not bound in any $\widetilde{\mu} \in \widetilde{\Omega}$. The result of a* ***project trust operator*** *is a multiset of trust weighted solution mappings which is defined as*

$$PT\left(v, \widetilde{\Omega}\right) = \left\{ (\mu', t) \mid (\mu, t) \in \widetilde{\Omega} \wedge \mu' = \mu \cup \{(v, t)\} \right\}$$

with $\mathrm{card}_{PT\left(v,\widetilde{\Omega}\right)}(\widetilde{\mu}) = \mathrm{card}_{\widetilde{\Omega}}(\widetilde{\mu})$.

The following example illustrates query evaluation with a project trust operator.

Example 4. Consider a group graph pattern similar to the pattern in Example 3 where the second BGP is associated with a `TRUST AS` clause. Figure 8(a) depicts an operator tree, annotated with sample solutions, for this pattern. Compare the solutions consumed and provided by the project trust operator. Every solution provided by this operator contains an additional binding for variable `?t1`. This binding maps `?t1` to a value that corresponds to the trust value associated with the respective solution when the project trust operator is evaluated (e.g. 0.86 for the first solution). Note, we used the trust merge function $tm_{\min}$ for the join operation. Thus, the trust value of the first solution after the join is 0.8. However, the value bound to variable `?t1` has not changed; it is still 0.86. This can be attributed to the limited scope of the trust projection and reflects the intention of the `TRUST AS` clause and its position in the query. To illustrate the role of the limited scope consider a slight variation of the query where the `TRUST AS` clause has been defined for the whole group graph pattern (i.e. before the last closing brace in the pattern of Example 3). Figure 8(b) depicts the corresponding operator tree with sample solutions. Notice, the solutions from BGP matching are the same as in Figure 8(a). Even so, the first of the overall resulting solutions differ for `?t1` because the project trust operator is applied after joining the solutions. Obviously, the position of a `TRUST AS` clause in a query pattern matters. □

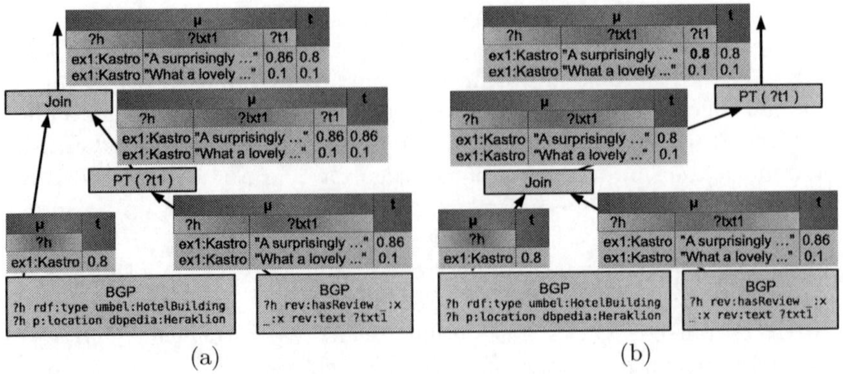

Fig. 8. Project trust operators in an operator tree with sample solutions

5.2 Expressing Trust Requirements

To express trust requirements as in query Q_2 we propose the ENSURE TRUST clause. Conventional FILTER clauses are not defined for trust weighted solution mappings and, thus, are inapplicable to filter solutions by restricting trust values directly. This holds for application-specific extension functions too. Instead of attempting to redefine FILTER we introduce the new clause as a clear separation between restricting with respect to variable bindings and associated trust values.

To process ENSURE TRUST clauses we extend the query language similar to our extensions for the TRUST AS clause. We refer to the tSPARQL specification [9] for the details and confine ourself here to present the new algebra operator.

Definition 9. *Let $l, u \in [-1,1]$ be lower and upper bound values, respectively; let $\widetilde{\Omega}$ a multiset of trust weighted solution mappings. The result of an* ***ensure trust operator*** *is a multiset of trust weighted solution mappings which is defined as*

$$ET\left(l, u, \widetilde{\Omega}\right) = \left\{(\mu, t) \mid (\mu, t) \in \widetilde{\Omega} \wedge l \leq t \leq u\right\}$$

with $\mathrm{card}_{ET(l,u,\widetilde{\Omega})}(\widetilde{\mu}) = \mathrm{card}_{\widetilde{\Omega}}(\widetilde{\mu})$.

The ensure trust operator accepts only those trust weighted solution mappings that have a trust value within a specified interval, i.e., it eliminates any solutions with trust values lesser than the lower bound or larger than the upper bound. As for the TRUST AS clause the position of an ENSURE TRUST clause in a query pattern matters.

Please notice, the results of a tSPARQL query may differ for different users. As discussed in Section 2, trustworthiness of triples is a subjective judgment. Hence, the trust values associated with query solutions depend on the information consumer. For this reason, each tSPARQL query must be executed in the context of a specific consumer in order to determine consumer-specific results.

6 Optimization of tSPARQL Query Execution

A well-known heuristic to optimize query execution in relational database systems is selection push-down. Algebra expressions are being rewritten to push down selections in the operator tree in order to reduce intermediary solutions and, thus, evaluate queries more efficiently. We adapt this heuristic to tSPARQL. In this section we present rewrite rules to push down trust constraints.

Enforcing trust constraints as early as possible may reduce query execution costs by reducing the number of trustweighted solution mappings that have to be processed. However, pre-drawing the evaluation of trust constraints is not as simple as pushing down ensure trust operators: this transformation may modify the semantics of the query unintentionally. In particular, pushing trust constraints in join operations may result in algebra expressions not equivalent to the original expressions. The soundness of rewrite rules that incorporate join operators depends on the trust merge function employed for joins. In the following we focus on rewrite rules that are only valid for the minimum trust merge function $tm_{\min}$.

Let $\widetilde{\Omega_1}$ and $\widetilde{\Omega_2}$ be multisets of trust weighted solution mappings. For join operators that employ $tm_{\min}$ the following equivalence of algebra terms holds:

$$ET\Big(l, u, Join(\widetilde{\Omega_1}, \widetilde{\Omega_2})\Big) \equiv ET\Big(l, u, Join\big(ET(l, 1, \widetilde{\Omega_1}), ET(l, 1, \widetilde{\Omega_2})\big)\Big) \tag{1}$$

Due space limitation we do not proof the equivalence. Instead, we refer to the tSPARQL specification [9] which contains proofs for all equivalences presented here. Based on (1) we propose to rewrite algebra expressions by replacing terms of the form on the left hand side of (1) by the corresponding term of the form on the right hand side of (1). Furthermore, for left-join operators that employ $tm_{\min}$ we propose a similar rewrite rule based on the following equivalence:

$$ET\Big(l, u, LJoin(\widetilde{\Omega_1}, \widetilde{\Omega_2}, ex)\Big) \equiv ET\Big(l, u, LJoin\big(ET(l, 1, \widetilde{\Omega_1}), \widetilde{\Omega_2}, ex\big)\Big) \tag{2}$$

To enable an even more extensive push-down of trust constraints we introduce the following equivalences and propose to apply the corresponding rewrite rules.

$$ET\left(l, u, Filter(ex, \widetilde{\Omega})\right) \equiv Filter\left(ex, ET(l, u, \widetilde{\Omega})\right) \tag{3}$$

$$ET\left(l, u, PT(v, \widetilde{\Omega})\right) \equiv PT\left(v, ET(l, u, \widetilde{\Omega})\right) \tag{4}$$

$$ET\left(l_1, u_1, ET(l_2, u_2, \widetilde{\Omega})\right) \equiv ET\left(\max(l_1, l_2), \min(u_1, u_2), \widetilde{\Omega}\right) \tag{5}$$

In contrast to (1) and (2), the equivalences (3) to (5) hold for all trust merge functions. Applying all the proposed rewrite rules during the optimization of tSPARQL queries reduces query execution times significantly (cf. Section 7.3).

7 Evaluation

In this section we evaluate the impact of our trust extension on query execution times. We implemented a tSPARQL query engine which extends the SPARQL query engine ARQ[2]. Our engine is available as Free Software from our project website[3]. For the evaluation we use a simple provenance-based trust function that

[2] http://jena.sourceforge.net/ARQ

[3] http://trdf.sourceforge.net/tsparql

assumes the existence of trust assessments for RDF graphs; these assessments associate a consumer-specific trust value with each graph. The trust function simply adopts these trust values for all triples in a graph. For our tests we use an extended version of the Berlin SPARQL Benchmark (BSBM) [10]. The BSBM executes a mix of 12 SPARQL queries over generated sets of RDF data; the datasets are scalable to different sizes based on a scaling factor. The generated data is created as a set of named graphs [11]. We extend the BSBM by trust assessments. Our extension, available from the project website, reads the named graphs-based datasets, generates a consumer-specific trust value for each named graph, and creates an assessments graph. The *assessments graph* is an additional RDF graph with statements that assign the generated trust values to the named graphs; for these statements we use our trust vocabulary (cf. Section 2). The proposed trust function determines the trust value for a triple by querying the assessments graph for the trust value associated with the graph that contains the triple. For all tests we use the minimum function to aggregate and to merge trust values. We conduct our experiments on a Intel Core 2 Duo processor with 2 GHz and 2 GB main memory. Our test system runs a recent 32 bit version of Gentoo Linux with Sun Java 1.6.0. In the following we, first, investigate how processing trust values impacts query execution time; we analyze the impact of determining trust values during query execution; and, finally, we evaluate the benefits of query rewriting as presented in the previous section.

7.1 The Impact of Processing Trust Values

As presented in Section 4.2 tSPARQL redefines the algebra of SPARQL in order to consider trust values during query execution. To measure the impact of this extension on query execution times we compare the results of executing our extended version of the BSBM with ARQ and with our tSPARQL query engine. To eliminate the effects of determining trust values in our engine we precompute the trust values for all triples in the queried dataset and store them in a cache. We execute the usual BSBM query mix for datasets generated with a scaling factor of 100, 200, 300, and 400; these datasets have the size of 31800, 60424, 92337, and 124305 triples, respectively. For each dataset we run the query mix 10 times for warm up and 50 times for the actual test. Figure 9(a) depicts the average times to execute the query mix with ARQ and with our engine, respectively. The main additional tasks performed by our engine, in contrast to ARQ, are accessing the trust value cache and aggregating trust values during BGP matching as well as merging trust values during join operations. Naturally, this additional functionality comes not for free. Nonetheless, the processing of trust values does not increase query execution times to a significant extent, especially for larger datasets, as can be seen in Figure 9(a).

7.2 The Impact of Determining Trust Values

While we analyze the processing of trust values in the previous experiment we focus on determining trust values during the execution of queries in the following.

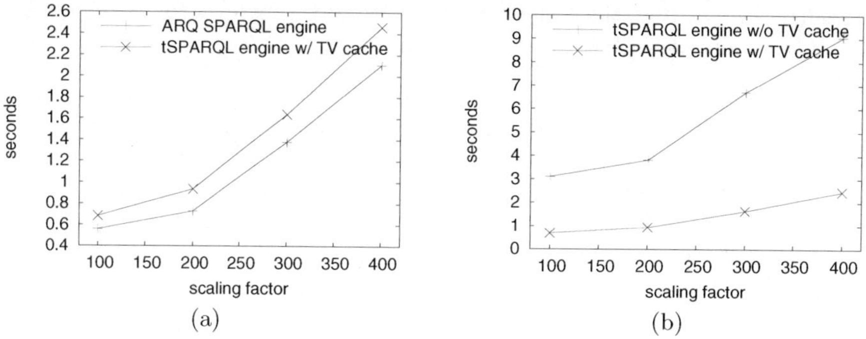

Fig. 9. Average times to execute the BSBM query mix for datasets of different sizes measured for ARQ and for the tSPARQL engine with and without a trust value cache

To measure how determining trust values may impact query execution times we use our tSPARQL query engine with a disabled trust value cache to execute the extended BSBM. During query execution the engine determines trust values with the simple, provenance-based trust function introduced before. For this setting, the chart in Figure 9(b) depicts the average times to execute the BSBM query mix; furthermore, the chart puts the measures in relation to the times obtained for our engine with a trust value cache in the previous experiment. As the chart illustrates, determing trust values during query execution dominates the query execution time. Thus, we strongly suggest to make use of trust value caches in a tSPARQL engine. As a future work we will look deeper into the topic of caching trust values.

7.3 The Impact of Query Rewriting

In Section 6 we propose rewrite rules that push down trust constraints in order to execute tSPARQL queries more efficiently. To evaluate the potential impact of the application of these rules we implemented them in our tSPARQL query engine. Furthermore, we developed a mix of six tSPARQL queries. The queries are in line with the BSBM mix of SPARQL queries and with the BSBM e-commerce use case that considers products as well as offers and reviews for these products. Due to space limitations, we do not present our queries in detail; we refer the reader to the tSPARQL specification instead. However, to give an impression the following list shows two of the queries in their textual form:

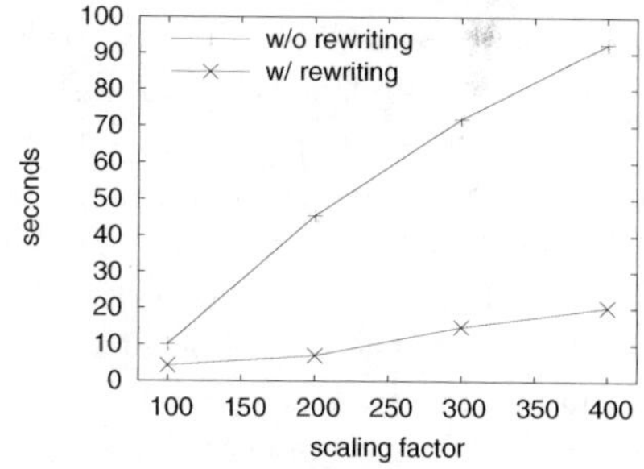

Fig. 10. The impact of rewrite rules

– For all products of a specific type return the cheapest offer, respectively; ensure all information can at least moderately be trusted.

– Return reviews and the trustworthiness of these reviews for products of a specific type; ensure trustworthy information.

For our test we enable the trust value cache in our tSPARQL engine and compare query execution times when query rewriting is enabled and disabled. For both cases we run our query mix 10 times for warm up and 50 times for the actual test; we use four datasets (scaling factor: 100, 200, 300, and 400). As can be seen in Figure 10 the average execution times for a query mix differ drastically depending on the application of our rewrite rules. Hence, rewriting tSPARQL queries may reduce the execution time significantly.

8 Related Work

Formalizing trust and trust in the Web is a topic of research since several years. For instance, Marsh [12] analyzes trust as a computational concept. This work influenced our approach to measure the trustworthiness of triples by a trust value in the interval [-1,1].

The basic idea of expressing trust requirements in a query language, as in our proposal, has also been presented by Bizer et al. [13]. They propose an extension to a query language for RDF, namely TriQL.P, to express trust-policies within queries. This approach is based on additional RDF statements that explicitly describe provenance and trustworthiness of information or information sources. Additionally TriQL.P permits a `METRIC` clause as "an open interface for different trust metrics and algorithms" which "are implemented as plug-ins for the query engine." In contrast to our approach, Bizer et al. do not explicitly provide a trust model for RDF data and their approach requires annotations regarding provenance and trustworthiness.

The most common approach to address trustworthiness on the Web are trust infrastructures based on a Web of trust (e.g. [1,2,3]). These Web of trust approaches consider the trustworthiness of members of the web. In contrast to these approaches, we focus on the trustworthiness of the data published on the Web, instead of the publishers. To the best of our knowledge, only Mazzieri and Richardson et al. propose trust models with an intention similar to ours. Mazzieri [14] introduces fuzzy RDF; a *membership value* associated with each triple represents the likelyhood the triple belongs to the RDF graph. By equating those membership values with trustworthiness of triples Mazzieri inappropriately mixes two different concepts; trustworthiness is not the same as a fuzzy notion of truth nor is trustworthiness of RDF triples tied to a specific RDF graph. Richardson et al.'s [15] approach is very close to ours; they represent a user's personal belief in a statement by a value in the interval [0,1]. What is missing in their approach is a possibility to finally utilize the ratings, e.g. in queries.

Other systems that consider trust in content are TRELLIS, IWTrust, and FilmTrust. The TRELLIS [6] system assesses the truth of statements by considering their provenance and related statements. Users can rate information sources and follow the assessments that are presented with the corresponding analysis and the influencing facts. The information assessed in TRELLIS is not as granular as the single triples in our approach and there is no trust model that rates the information directly. However, the presented possibility to assess information based on corresponding statements can be used to determine trust ratings of triples or sets of them. IWTrust [5], the trust component of the Inference Web answering engine, understands trust in answers as the trust in sources and in users. Similarily, FilmTrust [7] represents the trustworthiness of movie reviews by a user's trust in the reviewer and in other users' competence to recommend movies. These provenance-based notions of content trust may guide the development of trust functions.

9 Conclusion

In this paper we present a trust model for RDF data and tSPARQL, a trust-aware extension of the query language SPARQL. Our model associates every triple with a trust value. To access and use the trust values and to describe trust requirements we propose the `TRUST AS` and `ENSURE TRUST` clauses as extensions for tSPARQL. To enable tSPARQL we developed concepts for a trust-aware query processing. Conceptually, our solution is entirely independent from the applied methods to determine trust values; e.g. by not prescribing a specific trust function the two tasks, determining trust values and BGP matching, are clearly separated. Our approach can even be adapted for other trust models with a different definition of the trust values for triples; users merely have to provide trust functions and aggregation functions for their settings.

As future work, we plan to integrate tSPARQL in applications that process RDF data from the Web. Today, a majority of these applications do not consider the trustworthiness of the data. With tSPARQL we provide an easy tool to enhance these applications and make them more trust-aware. One of the requirements for integrating our concepts in applications is the existence of application-specific trust functions. For this reason, we will research different possibilities to determine trust values and we will integrate them in the tSPARQL query engine. In addition to trust functions applications require a method to aggregate trust values. We will develop trust aggregation functions and analyze their suitability for different applications and scenarios. Furthermore, we will develop concepts to enhance the trust value cache in our engine because efficient query execution benefits from caching trust values as our evaluation illustrates.

References

1. Golbeck, J., Parsia, B., Hendler, J.A.: Trust networks on the semantic web. In: Klusch, M., Omicini, A., Ossowski, S., Laamanen, H. (eds.) CIA 2003. LNCS, vol. 2782, pp. 238–249. Springer, Heidelberg (2003)
2. Ziegler, C.N., Lausen, G.: Spreading activation models for trust propagation. In: Proc. of EEE 2004 (March 2004)
3. Brondsema, D., Schamp, A.: Konfidi: Trust networks using PGP and RDF. In: Proc. of the Workshop on Models of Trust for the Web at WWW 2006 (May 2006)

4. Hartig, O.: Provenance information in the web of data. In: Proc. of the Linked Data on the Web Workshop at WWW 2009 (April 2009)
5. Zaihrayeu, I., da Silva, P.P., McGuinness, D.L.: IWTrust: Improving user trust in answers from the web. In: Herrmann, P., Issarny, V., Shiu, S.C.K. (eds.) iTrust 2005. LNCS, vol. 3477, pp. 384–392. Springer, Heidelberg (2005)
6. Gil, Y., Ratnakar, V.: Trusting information sources one citizen at a time. In: Horrocks, I., Hendler, J. (eds.) ISWC 2002. LNCS, vol. 2342, pp. 162–176. Springer, Heidelberg (2002)
7. Golbeck, J., Hendler, J.: FilmTrust: Movie recommendations using trust in web-based social networks. In: Proc. of CCNC 2006 (January 2006)
8. Prud'hommeaux, E., Seaborne, A.: SPARQL query language for RDF. W3C Recommendation (January 2008)
9. Hartig, O.: Specification for tSPARQL (December 2008), `http://trdf.sf.net/documents/tsparql.pdf`
10. Bizer, C., Schultz, A.: Benchmarking the performance of storage systems that expose SPARQL endpoints. In: Proc. of the Workshop on Scalable Semantic Web Knowledge Base Systems at ISWC 2008 (October 2008)
11. Carroll, J.J., Bizer, C., Hayes, P., Stickler, P.: Named graphs, provenance and trust. In: Proc. of WWW 2005 (May 2005)
12. Marsh, S.P.: Formalising Trust as a Computational Concept. PhD thesis, University of Stirling, Department of Mathematics and Computer Science (April 1994)
13. Bizer, C., Cyganiak, R., Gauss, T., Maresch, O.: The TriQL.P Browser: Filtering information using context-, content- and rating-based trust policies. In: Proc. of the Semantic Web and Policy Workshop at ISWC 2005 (November 2005)
14. Mazzieri, M.: A fuzzy RDF semantics to represent trust metadata. In: Proc. of Italian Workshop on Sem. Web Applications and Perspectives (December 2004)
15. Richardson, M., Agrawal, R., Domingos, P.: Trust management for the semantic web. In: Fensel, D., Sycara, K.P., Mylopoulos, J. (eds.) ISWC 2003. LNCS, vol. 2870, pp. 351–368. Springer, Heidelberg (2003)

RadSem: Semantic Annotation and Retrieval for Medical Images

Manuel Möller[1], Sven Regel[2], and Michael Sintek[1]

[1] German Research Center for Artificial Intelligence (DFKI) GmbH
Kaiserslautern, Germany
manuel.moeller@dfki.de, michael.sintek@dfki.de
[2] Chemnitz University of Technology, Chemnitz, Germany
sven.regel@cs.tu-chemnitz.de

Abstract. We present a tool for semantic medical image annotation and retrieval. It leverages the MEDICO ontology which covers formal background information from various biomedical ontologies such as the Foundational Model of Anatomy (FMA), terminologies like ICD-10 and RadLex and covers various aspects of clinical procedures. This ontology is used during several steps of annotation and retrieval: (1) We developed an ontology-driven metadata extractor for the medical image format DICOM. Its output contains, *e. g.*, person name, age, image acquisition parameters, body region, *etc.* (2) The output from (1) is used to simplify the manual annotation by providing intuitive visualizations and to provide a preselected subset of annotation concepts. Furthermore, the extracted metadata is linked together with anatomical annotations and clinical findings to generate a unified view of a patient's medical history. (3) On the search side we perform query expansion based on the structure of the medical ontologies. (4) Our ontology for clinical data management allows us to link and combine patients, medical images and annotations together in a comprehensive result list. (5) The medical annotations are further extended by links to external sources like Wikipedia to provide additional information.[1]

1 Introduction

Advances in medical imaging have enormously increased the volume of digital images produced in clinical practice. At the same time, modern hospital information systems have also become more complex. Radiological findings are kept separately from images which, in turn, are kept separately from patient accounting and billing information.

Currently, these systems are more or less isolated from each other and do not allow queries to span across these systems. Thus it has become challenging for clinicians to query for and retrieve relevant historical data due to the volume

[1] This research has been supported in part by the THESEUS Program in the MEDICO Project, which is funded by the German Federal Ministry of Economics and Technology under the grant number 01MQ07016. The responsibility for this publication lies with the authors.

L. Aroyo et al. (Eds.): ESWC 2009, LNCS 5554, pp. 21–35, 2009.

of information as well as the complexity of information systems. In particular, historical patient images would be useful for analyzing images of a current examination since they help in understanding any progression of pathologies or development of recent abnormalities. Today, such images can only be retrieved by attributes stored in the DICOM[2] headers of the images such as patient name, age or gender. However, these attributes do not contain any information about the anatomy or disease associated with the image. Hence, radiologists are often overwhelmed with irrelevant images not connected with the current examination or—the other extreme—are unable to retrieve any similar case.

To overcome these limitations our system uses the Semantic Web standards OWL [9] and RDF [7] as a common umbrella to represent domain knowledge and annotations in the same formalism. We have developed the MEDICO Ontology Hierarchy (see Fig. 1) which models various aspects of clinical data management and medical background knowledge. Our rationale was to reuse background knowledge represented in formal ontologies such as the Foundational Model of Anatomy ontology (FMA) [19] and terminologies like RadLex [8] and the International Classification of Diseases version 10 (ICD-10).[3] Wherever possible we used existing translations to OWL. For the ICD-10 we implemented our own technique which is described in Sect. 3.2. These ontologies were integrated into our hierarchy to provide background knowledge about the medical domain during annotation and retrieval.

On the search side we leverage the structural information in the ontologies to allow multilingual search and to perform query expansion to retrieve images which are annotated with semantically similar concepts.

Semantic Web technologies allows us to reuse highly qualified medical domain knowledge for annotation and query expansion without being medical experts ourselves. Storing the medical annotations as instances of well defined OWL classes—rather than in a proprietary relational database—fosters an open interchange of this data with other applications and makes them easily available for other research goals, such as clinical data mining. The fusion of medical annotations with automatically extracted patient and image metadata in the same representation formalism provides the basis for new intuitive visualizations.

The research on semantic annotation and retrieval described in this paper is part of a broader effort to understand the semantic of medical images in the THESEUS MEDICO [12] project.

2 Related Work

The need for representing high-level annotations of medical images on an abstract level has been emphasized in various publications in recent years, *e. g.*, in [13, 23, 14].

Biomedical ontologies and terminologies received high attention in the last decade and provide promising technologies. [3] evaluated popular large scale

[2] Digital Imaging and Communication standard, `http://medical.nema.org/`

[3] `http://www.who.int/classifications/apps/icd/icd10online`

ontologies such as SNOMED, FMA, and Gene Ontology and stated that "ontologies play an important role in biomedical research through a variety of applications". Only recently, [10] presented an approach for integrating FMA and RadLex as FMA-RadLex to compile a robust application ontology. Besides efforts combining ontologies and radiological reports (*e. g.*, [15]), other approaches using ontologies in medical image processing have been proposed [18, 22].

The work in [20] is similar to our research in annotating medical images with ontology-based semantics and the use of context for faster annotation. However, their notion of context is primarily oriented on anatomy than *diseases*.

3 Ontological Modeling

Fig. 1 gives an overview of the general structure of the MEDICO Ontology Hierarchy. The following paragraphs describe the different parts of the hierarchy in detail as well as their interrelations and the design rationales. We follow Gruber's definition [5] for the term *ontology* that "an ontology is a formal specification of a (shared) conceptualization." An OWL model of this ontology hierarchy can be browsed online.[4]

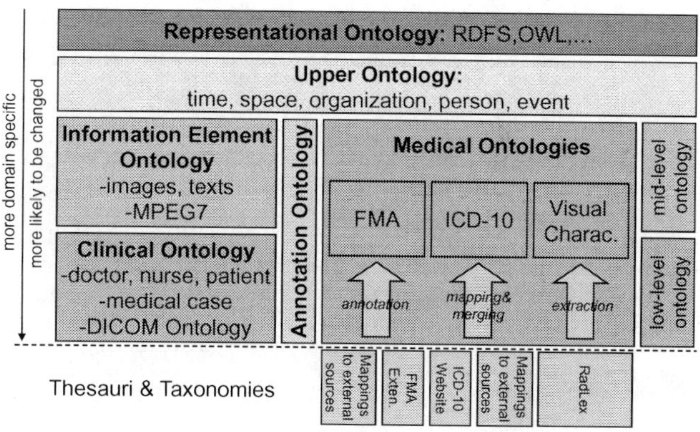

Fig. 1. MEDICO Ontology Hierarchy

Ontologies are usually structured in various layers or levels, based on the assumption that those at higher levels are more stable, shared among more people, and thus change less often than those at lower levels. Following [21], we distinguish representational ontologies, upper-level ontologies, mid-level ontologies, and low-level or domain ontologies.

3.1 Medical Ontologies

In the medical domain, large amounts of knowledge are already formulated in ontologies like the FMA. Therefore we decided to reuse those sources and opted for a mapping to/merging of those ontologies and thesauri each covering different dimensions of image annotation.

We differentiate between three different aspects or *dimensions* of medical annotation. For *anatomy* we use the FMA. The concepts for the *visual manifestation* of an anatomical entity on an image is derived from the *modifier* and

[4] http://www.dfki.uni-kl.de/~moeller/ontologies/medico-browser

imaging observation characteristic sub-trees of RadLex. We consider the *disease* aspect as the interpretation of the combination of the previous two. Here we use the ICD-10.

3.2 ICD-10

At the time of writing, ICD-10 was not available in any format following the Semantic Web standards. Due to our decision to use OWL and RDF in MEDICO, we generated a light-weight ICD-10 model in OWL which mainly reflects the terminological structure.

From the semistructured online version of the terminology, we have created a simple model using the OWL API [2]. It reflects the hierarchy as defined in the ICD-10 manual [4]. We modeled this hierarchy on the class level while instances are considered to reflect concrete diseases of concrete patients. The whole OWL model contains 11,290 classes and has a maximum depth of 5.

3.3 Annotation Ontology

The Annotation Ontology is used to connect a given medical document—be it image or text—with the concepts in the ontologies described in the previous sections. The diagram in Fig. 2 illustrates the ontology we created to generate annotations. (Ovals depict properties, rectangles classes.)

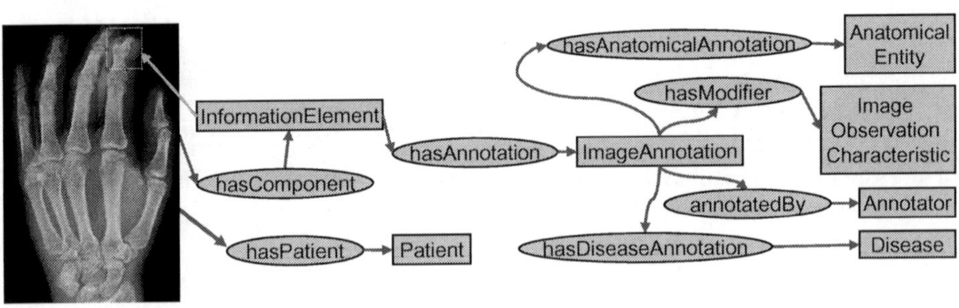

Fig. 2. Annotation Ontology

The image is connected to the patient via the `hasPatient` property. Additionally, we add annotations to the study and series data the image belongs to. For the sake of clarity we omitted these links in the illustration.

The image itself is decomposed into `InformationElement` instances with each of these instances being connected via the property `hasAnnotation` to an `ImageAnnotation` instance. `InformationElement` subsumes arbitrary documents (text, images, 3D volumes, ...) as well as parts or regions of them.

For each `ImageAnnotation` we store information on the annotation provenance. `Annotator` can either be a `Physician` instance for manual annotations by a medical expert or an instance of `Classifier` for automatically generated image annotations which we plan to integrate in the future.

Finally, ImageAnnotations can be further qualified by adding instances of VisualModifier. Here we reuse the existing RadLex vocabulary enriched by terms from our own studies on radiology reports to identify further terms that are common in the clinical practice but not yet part of RadLex. These qualifiers comprise attributes to describe alterations from normal and healthy features like bulky or compressed.

3.4 Mapping from DICOM Body Regions to FMA Body Regions

Standardized formats such as DICOM have been developed for digitally representing the acquired images from the various modalities. Apart from the image pixels, a DICOM image also contains a header, which is used to store certain patient information such as name, gender, demographics, etc.

In order to create a deep interconnection between data from the DICOM metadata and the medical domain knowledge, we manually defined mappings for the 20 allowed values of the DICOM tag Body Part Examined to FMA concepts. We are aware of former studies like [6] which showed that the value of this field is in practice neither mandatory nor always correct. But in our opinion it would be wrong to completely ignore this information since it is correct in most of the cases. Instead we propose to put a bias on their value, signifying that their value is not absolutely reliable. In fact this is also the case with all results from automatic object recognition algorithms we plan to incorporate.

If there is no body region set in the DICOM header, we offer all concepts from the FMA for the anatomical annotation. If the wrong body region was saved to the DICOM header, this becomes apparent by comparing the body region visualization and the anatomical structures on the image. In both cases the user can set the correct body region manually in the application.

In the manual annotation tool we use the body region information from the DICOM header to select only those concepts from the FMA which belong to the body region. This reduces the number of potential anatomical annotation concepts from approx. 80,000 down to a few hundred to a few thousand.

Another application in the area of visual object recognition of the body region annotations is described in [11]: A *Metadata Extraction Module* retrieves the BodyRegion from the DICOM header. Based on the mapping, the FMA is used to retrieve a list of anatomical entities that could possibly be detected in this body region. This narrows down the search space of potential objects on the image significantly.

3.5 Body Substance Annotations for FMA

Each body substance has its own characteristic radiation attenuation spectrum which is typically measured in *Hounsfield Units*.[5] On CT and X-ray images this manifests in specific pixel contrast ranges for the tissue types. We have modeled the class TissueType which currently subsumes classes for 13 different body

[5] The Hounsfield Scale is a quantitative scale describing radiodensity.

substances like Bone, Blood, *etc.* each annotated with the respective Hounsfield range.

In a second step they were mapped manually to FMA concepts. We considered implementing generic heuristics which traverse particular relationships in the FMA to retrieve all related concepts belonging to a specific tissue type. But so far we did not discover any heuristic generic enough to treat organs and their sub-components the same way as blood vessels which belong to organs.

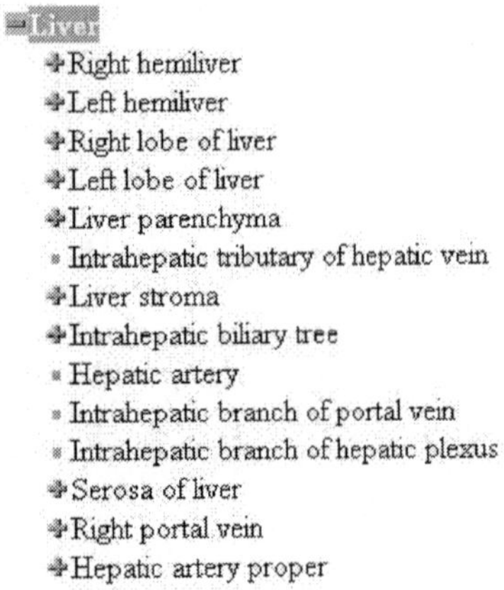

Fig. 3. part_of sub-tree of the Liver in the FMA

The snapshot in Fig. 3 of the FMA shows a part of the part_of sub-tree of the concept Liver. It clearly shows, that veins and arteries are also considered to be part of the liver and not the artery system. This also holds for the more specific properties systemic_part_of, regional_part_of and constitutional_part_of. We decided against the annotation of all constitutional parts of an organ with the same tissue type as their super class. The reason was that constitutional parts also comprise blood supply which has a different radiodensity.

3.6 Mappings to External Biomedical Data Sources

Here, our goal was to augment the annotations of search results with mappings to external data sources such as Wikipedia to provide additional information explaining the annotation concepts. The first step was to extract a list of all class labels and synonyms from the ontologies. Next, we iterated over this list and tried to identify the most relevant document in a corpus of potentially relevant web pages by performing a statistical keyword/phrase search. The corpus was indexed offline using the DynaQ API [1].

We searched for English labels and synonyms of the FMA in Wikipedia. In our first attempt we did a plain full text keyword search for labels and synonyms and created one mapping for the best match if there was one. This resulted in mappings for 86% (84 685 absolute, column keyword in Tab. 1) of all FMA labels and synonyms. Prima facie this number is surprisingly high since most of the FMA concepts are very specific and we did not expect Wikipedia to cover them to such a high extent.

However, the quality of these mappings turned out to be very limited. We imposed a systematic evaluation by picking a random set of 100 mappings and checking each of them manually. We distinguished between (1) *perfect match* where the Wikipedia article is covering exactly the FMA concept; (2) a *"useful" match* meaning that the Wikipedia article covers (a) an anatomical part of the search label, (b) the FMA label is a part of the concept covered in the Wikipedia article, or (c) the Wikipedia article deals with a direct sub- or

Table 1. Precision for different mapping types

search method	**keyword**	**phrase**	**phrase AND "Anatomy"**	**phrase in title**
perfect match	33	30	26.7	64.7
"useful" match	33	21	18.7	15.7
non-match	34	49	54.6	19.6

superclass of the FMA concept; and (3) the rest is regarded as *non-match*. It turned out that only 33% of the mappings were perfect matches. Subsequently, we switched to phrase-based search which lowered the number of mappings to 6.7% (6 439 absolute, column phrase in Table 1). To overcome the problem of obviously irrelevant mappings we confined the results to articles containing the FMA label as a phrase *and* the keyword "Anatomy". This resulted in almost 50% wrong mappings (column phrase AND "Anatomy" in Tab. 1). By further restricting the search to pages which also contain the keyword "Anatomy" we tried to limit the results to articles of the target domain. However, this dropped the number of matches because "Anatomy" does not necessarily appear in all articles dealing with anatomical topics. Still this did not lower the number of irrelevant articles significantly. Finally we confined the search to the article titles which reduced the ratio of non-matches to less than 20% (column phrase in title in Table 1).

The additional restrictions cut down the overall number of mappings from 6 439 to 4 104. We decided to trade precision for recall, provide less mappings and in turn avoid dissatisfaction due to irrelevant mappings.

3.7 DICOM Ontology

Since one of our main goals is to incorporate existing data based on established standards, the design of the clinical ontologies is to quite some extent aligned with the DICOM standard. This standard is a world wide established format used by hospitals to store the results of visual examinations (*e. g.*, X-Ray, CT, PET, *etc.*) as well as patient and image metadata.

There are other efforts towards creating a DICOM ontology[6] but to our knowledge no OWL model has been released so far. Therefore we decided to model a MEDICO DICOM ontology. Our rationale was to keep closely with the DICOM standard in its version 3 from 2008. The reason was that many processes of clinical data management are use this structure and our intention is to support these processes instead of replacing them.

As discussed above (Sect. 3.4), the imaging equipment manufacturers do not always follow standards when it comes to storing imaging metadata. We also see this in our corpus. To cope with this fact, we decided on a hybrid methodology for our modeling. We combined a data-driven bottom-up with a top-down approach driven by the DICOM standard. We started by looking at the metadata available in the images of our corpus (see Sect. 5.1) and checked the standard for regulations regarding format, value ranges and meaning.

[6] For example within the caBIG project, `http://cabig.nci.nih.gov/`

The DICOM standard organizes data elements in a simple hierarchy. In a clinical data management system each patient's images arranged in one or more *Studies*. A *Study* contains one or more *Series* which in turn contains one or more *Image(s)*. Our modeling reflects this hierarchy and assigns properties only once depending on their domain. Each DICOM file has a comprehensive DICOM header which contains information about the patient, study and series it belongs to, date and time and several image acquisition parameters. In the following we introduce our approach by example.

Each metadata element in the DICOM standard has a fixed so called *DICOM tag* identifying it in the DICOM file header. To control the mapping process from DICOM metadata fields to properties in our ontology we annotated those properties with the respective tag using the annotation property `dicomTag`.

For example, the DICOM tag `0010,1010` always identifies the patients age for the current image/series, which we store in the datatype property `age`. Since DICOM specifies date and time formats differently from the XSD-based datatype properties in our ontology we have added the annotation property `converter` in order to control the conversion of each datatype property separately.

4 Application

Fig. 4 shows the annotation frontend of our application. Images are loaded and displayed using the API of the image processing program ImageJ[7]. Using this library images are segmented into *regions of interest* (ROI). Currently we support rectangular, ellipsoid and arbitrary polygons to confine these regions. These three types are modeled as different classes in our ontology. As discussed in Sect. 3.1, we distinguish between *anatomy*, *image observation characteristic* and *disease* which can be annotated separately for each ROI. To ease the task of finding appropriate annotations we use *auto-completing* combo-boxes. While typing in a search term, concept names with matching prefixes are shown in a drop down box and can be selected.

4.1 DICOM Metadata Extractor

Based on our modeling of the DICOM ontology we have extended the metadata extraction framework Aperture[8] with a metadata extractor for DICOM images. Controlled by the special annotation properties described in Sect. 3.7, it crawls file systems, web servers and WebDAV repositories and converts all available DICOM metadata to instances of the classes modeled in our ontology.

In fact, each DICOM image contains all information about patient, series and study which leads to a huge data redundancy considering volume datasets (*i. e.*, a stack of images generated by a CT scan) which often contain more than 100 images. In such cases all images of a series are linked to the same instance of a patient in contrast to the redundant storage in the DICOM headers.

[7] `http://rsbweb.nih.gov/ij/`

[8] `http://aperture.sourceforge.net/`

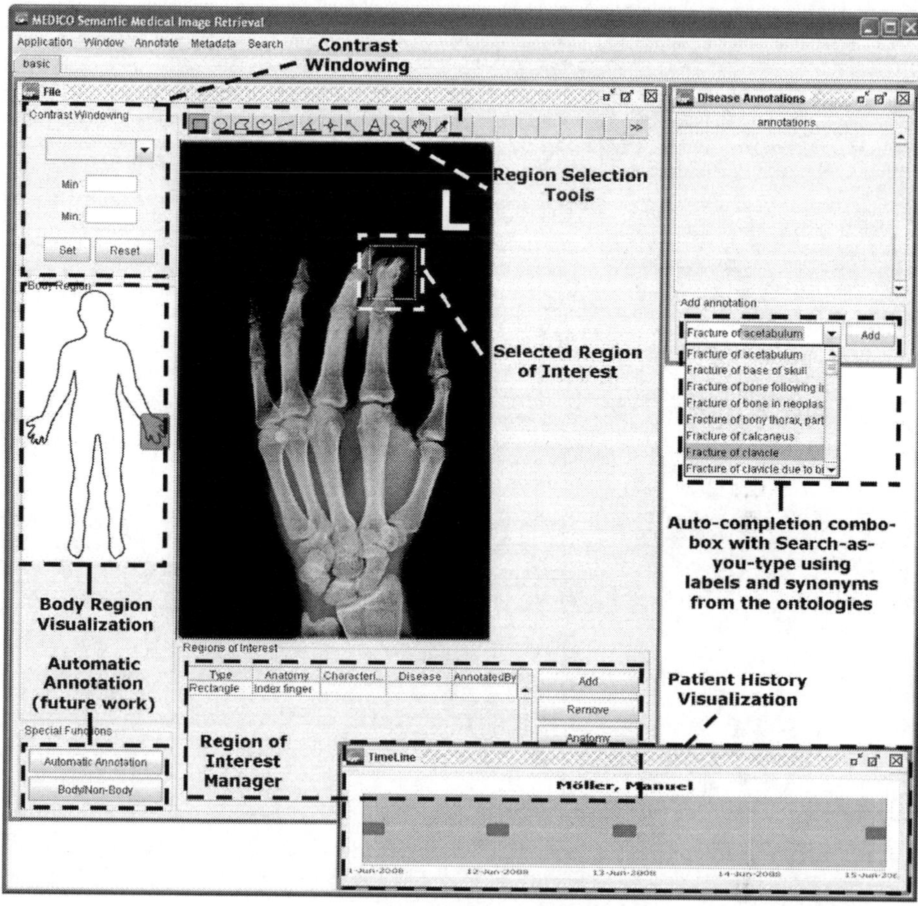

Fig. 4. Annotation Frontend

In our annotation application this metadata can be accessed during annotation and retrieval and is rendered using the presentation vocabulary Fresnel [16]. Additionally, we have set up a separate faceted browsing application[9] using MIT's Longwell.[10]

4.2 Annotation Interface

Contrast Windowing: The medical background knowledge tells a radiologist that, *e. g.*, he has to adjust the contrast range to 80-1000 if he wants to highlight bones in a CT scan. Likewise we offer the `TissueType` annotations introduced in Sect. 3.5 in the annotation tool to dynamically adjust the contrast range of the current image (see Fig. 4). This feature is included in many medical imaging

[9] `http://www.dfki.uni-kl.de/medico-longwell-2.5.5`
[10] `http://simile.mit.edu/wiki/Longwell`

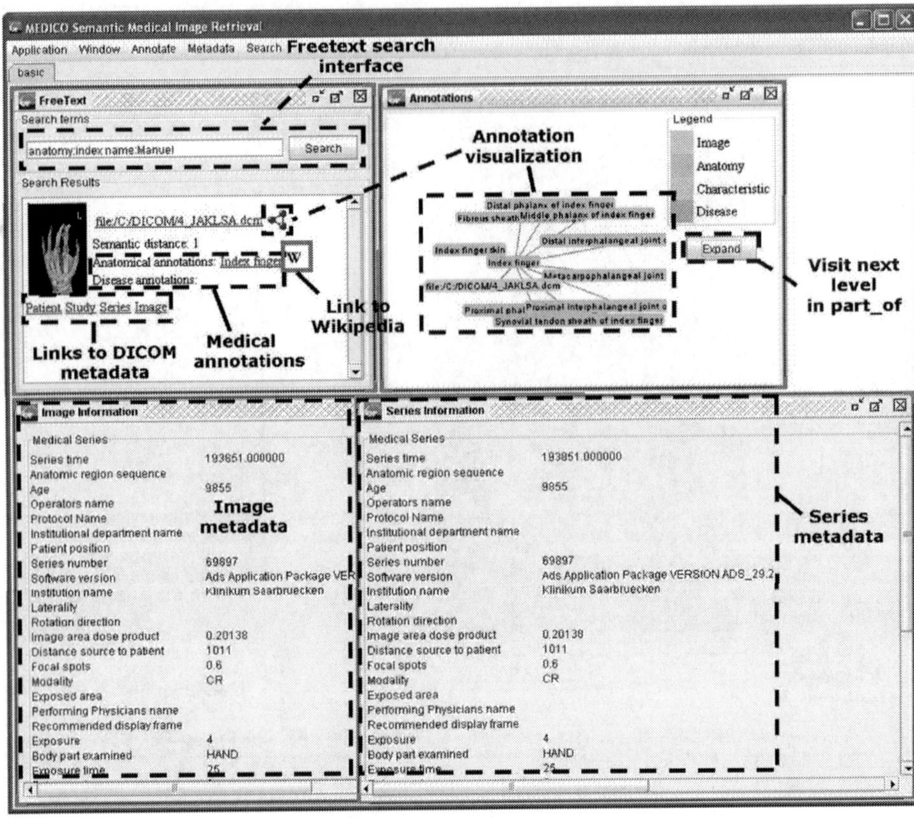

Fig. 5. Search Frontend

applications. But here we use one single abstract modeling of tissue types both for manual annotation as well as to control the preprocessing in automatic image annotation workflows, which is beyond the scope of this publication and further described in [11].

Body Region Visualization: The mapping between DICOM body region and FMA body region annotations as introduced in Sect. 3.4 is also used to provide a visualization of the body region from which the current image is taken. Upon loading an image, the body region metadata tag of the DICOM image is used to visualize the body region of the current image in the silhouette of a prototypical human (see Fig. 4).

Patient History Visualization: Medical diagnosis is largely dependent on the medical history of a patient. On this account we offer a visual timeline of previous radiological examinations. Each available image series is displayed as a box on a time axis in chronological order (see Fig. 4). This timeline is generated on the fly from the automatically extracted information from the

DICOM headers and former manual annotations from the radiologists. Even without any manual annotations, the user immediately gets an overview about former visual examinations of the patient. Clicking on a series opens a window with all available information about this series and links to corresponding images.

4.3 Search Interface

In the Lucene[11]-like free text search, arbitrary queries can be created by specifying Lucene labels such as "anatomy:" and corresponding values like "Index".

This allows the creation of hybrid queries searching for manual medical annotations as well as specific metadata values as extracted from the DICOM metadata extractor (Sect. 4.1). For example, you can specify that the name of the patient has to match a certain substring, *e. g.*, by adding "name:Manuel" to the query. The values of search labels such as "anatomy" and "disease" are mapped in the background to concepts in the medical ontologies.

The results contain a thumbnail, the full path of the image and all available annotations. By clicking on the RDF icon a separate frame is opened which visualizes all annotations. A click on the "Expand" button retrieves all child concepts of all annotated concepts and thus allows the exploration of the semantic "neighborhood".

Query Expansion: For searching we leverage the structural information in the medical ontologies. In the *part-of* hierarchy of the FMA there is a path from the concept *Index finger* (the concept used originally to annotate the image) to *Index* which was the search concept in the example in Fig. 5. Internally we use this information to compute a query expansion and translate it into a SPARQL [17] query. Using the FMA, a search for concept X searches for the concept itself and all concepts which have a (transitive) link via the properties *regional_part_of* and *constitutional_part_of* to X. To rank the results, we compute the path length from the search concept to the concept used for annotation of the search results. Using ICD-10, we traverse the *subClassOf* property downwards.

Traversing the FMA recursively with SPARQL queries at runtime turned out to be way too slow. The FMA uses the equivalent of "multiple inheritance" in the *part_of* hierarchy. *E. g.*, *Heart* is both part of *Set of thoracic viscera* and of *Content of middle mediastinum*. Thus we decided to materialize a spanning tree for the properties *regional_part_of* and *constitutional_part_of* starting from the concept *Human body*. By materializing the shortest distances to the root node of this closure, we are able to perform a query expansion with arbitrary search depth limit in two retrieval operations: The first query is used to retrieve the list of classes in the *part_of* spanning tree down to a certain depth limit and the second to retrieve the annotated images.

Of course we would prefer to find a solution that did not require to materialize closure and distance information, since this would save us a lot of memory. But at the time of writing, our system setup with SwiftOWLIM[12] version 3 on a

[11] http://lucene.apache.org/

[12] http://www.ontotext.com/owlim/

quadcore machine with 32 GB RAM did not allow reasonable retrieval times without materialization.

Multilingual Annotation and Search: Another advantage of using existing ontologies such as the FMA is the availability of multilingual synonyms. Currently, we allow annotation and search using any synonym available in the FMA. English, Latin, French, German and Filipino are covered and the preferred[13] annotation language can be changed at runtime.

5 Evaluation

5.1 Our Corpus

Acquiring medical images for testing purposes is a big problem. Legal regulations enforce a very strict anonymization both for clinical patient records as well as any personal information in the DICOM headers. Because we wanted to test our extractor with real data and avoid systematic errors by testing only with values generated by anonymization algorithms, we added non-anonymized personal images to the corpus. All images and metadata shown on screenshots in this publication belong to examinations performed on one of the authors during the summer of 2008.

5.2 Qualitative Evaluation

When it comes to the evaluation of systems using technologies from the Semantic Web, evaluation turns out to be problematic. Measuring only *quantitative* features of such systems is certainly possible but in many cases does not reflect a measurement of the claimed scientific contributions, since they are first and foremost of a *qualitative* nature.

To our knowledge, there is no clinical imaging database system that fuses image metadata (patient name, age, gender, *etc.*) and medical annotations (anatomy, imaging characteristic, disease) in the same formalism and allows searching for both in a unified manner. Thus we decided against presenting tables of retrieval times of our system.[14] Firstly, because they do not really reflect the important features of this system and secondly because they are not comparable to existing systems in clinical practice. Instead we would like to discuss three query scenarios which contrast the features of RadSem with existing systems.

Scenario 1: "Give me all images of patient X." This can be handled directly by existing clinical image data base systems (PACS[15]) since this query can be

[13] The multilingual synonyms are not available for all FMA classes.

[14] Our efforts, *e. g.*, for materialization of distances as described above, result in retrieval times always below one second even for large query expansions.

[15] Picture Archiving and Communications Systems.

reduced to a simple keyword search across metadata elements. RadSem supports this type of query in the same manner.

Scenario 2: "Give me all images from patients with a broken index finger." Existing systems require two tasks to be performed for the retrieval of this type of query. In the first task, the radiology information system (RIS) which stores all radiological findings has to be searched for textual descriptions of "broken index fingers". The facts that radiologists tend to use abbreviations in their reports and that the same medical incident could also be described using different words (*e. g.*, fracture of index finger) or a different language limit the recall significantly. The second task for the radiologist is then to choose to retrieve a selection of images referenced by the radiology reports. Not all of these images necessarily have to deal with a "broken index finger" but might also deal with fractures of other bones which the patient might have had during the same time.

In RadSem both tasks can be performed in a single step. In the free text search, the query would simply be "anatomy:index disease:fracture".

Scenario 3: "Give me all pictures dealing with enlarged lymph nodes." Patients potentially suffering from lymph node cancer (lymphoma) are usually scanned using Computed Tomography, treated if cancerous lymph nodes are detected and called in for a number of follow-up examinations after the treatment over a period of several months. During these follow-up examinations the radiologist searches for lymph nodes which have changed in size compared to previous examinations. This requires access to previous images. For the same patient, this query can be handled by existing PACS. Their interfaces usually allow opening of any image series available for the current patient.

One important advantage of RadSem is that the radiologist here can also directly search for images from other patients with the clinical finding of having enlarged lymph nodes. Additionally, the query is not limited to a specific lymph node. The query "anatomy:lymph_node visual:enlarged" retrieves all images which are annotated with the anatomical concept "Lymph node" or any part of it and the visual characteristic enlarged. One could also think of refining this query by adding date ranges, patient gender, etc.

Summary: These examples clearly show that there are severe limitations for searching existing medical image database systems. Despite the fact that there are standardized vocabularies such as ICD-10 and well developed ontologies like the FMA, these technical developments have not yet found adaptation in clinical practice. This section pointed out these limitations and described our approach to overcome them by using a common representation for medical background knowledge and annotations, patient metadata and image acquisition parameters using the Semantic Web standards RDF and OWL.

6 Conclusion and Future Work

We have presented a semantic annotation and retrieval tool RadSem for medical images. It leverages on background information about the medical domain from formal ontologies and also covers clinical data management at several steps. Ontology based metadata extraction and medical annotations are represented using RDF and OWL and lay the basis for clearly arranged information visualizations of patient history, image annotations and search results. We presented different approaches to minimize the additional effort for adding medical annotations.

We have described our approach for formalizing the structure of the ICD-10 in OWL and presented several augmentations and mappings for FMA with further medical knowledge (*e. g.*, about tissue types) and external data sources.

Articles about medicine in Wikipedia are in the majority of the cases understandable for a general audience but at the same time the scientific level of the articles is rather low compared to peer reviewed scientific publications as for example in PubMed.[16] Therefore we will apply the technique described in Sect. 3.6 to generate links from ICD-10 concepts to relevant PubMed articles.

On the imaging side, we plan to integrate low-level content-based image retrieval techniques to additionally support query by image in conjunction with the presented search across medical annotations and patient metadata. We also aim to integrate fully-automatic semantic image parsing techniques to provide facilities for automatic image segmentation and the automatic suggestion of semantic image annotations.

References

1. Agne, S., Reuschling, C., Dengel, A.: DynaQ – dynamic queries for electronic document management. In: 10th IEEE International Enterprise Distributed Object Computing Conference Workshops (EDOCW 2006) (2006)
2. Bechhofer, S., Volz, R., Lord, P.W.: Cooking the semantic web with the OWL API. In: Fensel, D., Sycara, K.P., Mylopoulos, J. (eds.) ISWC 2003. LNCS, vol. 2870, pp. 659–675. Springer, Heidelberg (2003)
3. Bodenreider, O.: Biomedical ontologies in action: role in knowledge management, data integration and decision support. In: International Medical Informatics Association (IMIA) (ed.) IMIA Yearbook 2008, pp. 67–79. Schattauer (2008)
4. World Health Organization Geneva. International statistical classification of diseases and related health problems (2004)
5. Gruber, T.R.: Toward principles for the design of ontologies used for knowledge sharing. International Journal of Human-Computer Studies 43, 907–928 (1995)
6. Güld, M.O., Kohnen, M., Keysers, D., Schubert, H., Wein, B.B., Bredno, J., Lehmann, T.M.: Quality of DICOM header information for image categorization. Proc. SPIE 4685(39), 280–287 (2002)
7. Hayes, P.: RDF semantics. W3C Recommendation (February 2004)
8. Langlotz, C.P.: Radlex: A new method for indexing online educational materials. RadioGraphics 26, 1595–1597 (2006)
9. McGuinness, D.L., van Harmelen, F.: OWL Web Ontology Language overview. W3C recommendation, World Wide Web Consortium (February 2004)
10. Mejino, J.L., Rubin, D.L., Brinkley, J.F.: FMA-RadLex: An application ontology of radiological anatomy derived from the foundational model of anatomy reference ontology. In: AMIA Annual Symposium Proceedings, p. 465 (2008)

[16] `http://www.ncbi.nlm.nih.gov/pubmed`

11. Möller, M., Sintek, M.: A scalable architecture for cross-modal semantic annotation and retrieval. In: Dengel, A.R., Berns, K., Breuel, T.M. (eds.) KI 2008. LNCS, vol. 5243. Springer, Heidelberg (2008)
12. Möller, M., Sintek, M., Buitelaar, P., Mukherjee, S., Zhou, X.S., Freund, J.: Medical image understanding through the integration of cross-modal object recognition with formal domain knowledge. In: Proc. of HEALTHINF 2008, Funchal, Madeira, Portugal, vol. 1, pp. 134–141 (2008)
13. Mojsilovic, A., Gomes, J., Rogowitz, B.E.: Semantic-friendly indexing and querying of images based on the extraction of the objective semantic cues. International Journal of Computer Vision 56, 79–107 (2004)
14. Mueen, A., Zainuddin, R., Baba, M.S.: Automatic multilevel medical image annotation and retrieval. Journal of Digital Imaging 21(3), 290–295 (2008)
15. Mykowiecka, A., Marciniak, M., Podsiadly-Marczykowska, T.: Data-driven ontologies for an information extraction system from polish mammography reports. In: Proc. of the 10th International Protégé Conference, Budapest, Hungary (July 2007)
16. Pietriga, E., Bizer, C., Karger, D.R., Lee, R.: Fresnel: A browser-independent presentation vocabulary for RDF. In: Cruz, I., Decker, S., Allemang, D., Preist, C., Schwabe, D., Mika, P., Uschold, M., Aroyo, L.M. (eds.) ISWC 2006. LNCS, vol. 4273, pp. 158–171. Springer, Heidelberg (2006)
17. Prud'hommeaux, E., Seaborne, A.: SPARQL query language for RDF. Technical report, W3C (March 2007)
18. Raicu, D.S., Varutbangkul, E., Furst, J.D., Armato III, S.G.: Modeling Semantics from Image Data: Opportunities from LIDC. International Journal of Biomedical Engineering and Technology (2007)
19. Rosse, C., Mejino, R.L.V.: A reference ontology for bioinformatics: The foundational model of anatomy. Journal of Biomedical Informatics 36, 478–500 (2003)
20. Rubin, D.L., Mongkolwat, P., Kleper, V., Supekar, K., Channin, D.S.: Medical imaging on the semantic web: Annotation and image markup. In: AAAI Spring Symposium Series, Stanford University (2008)
21. Semy, S.K., Pulvermacher, M.K., Obrst, L.J.: Toward the use of an upper ontology for U.S. government and U.S. military domains: An evaluation. Technical report, MITRE Corporation (September 2004)
22. Temal, L., Dojat, M., Kassel, G., Gibaud, B.: Towards an ontology for sharing medical images and regions of interest in neuroimaging. Journal of Biomedical Informatics (2008)
23. Yao, J., Antani, S., Long, R., Thoma, G., Zhang, Z.: Automatic medical image annotation and retrieval using SECC. In: Proc. of 19th International Symposium on Computer-Based Medical Systems (CBMS 2006), Salt Lake City (June 2006)

Semanta – Semantic Email Made Easy

Simon Scerri, Brian Davis, Siegfried Handschuh, and Manfred Hauswirth

Digitial Enterprise Research Institute,
National University of Ireland Galway, IDA Business Park, Galway, Ireland
{simon.scerri,brian.davis,siegfried.handschuh,
manfred.hauswirth}@deri.org

Abstract. In this paper we present Semanta – a fully-implemented system supporting Semantic Email Processes, integrated into the existing technical landscape and using existing email transport technology. By applying Speech Act Theory, knowledge about these processes can be made explicit, enabling machines to support email users with correctly interpreting, handling and keeping track of email messages, visualizing email threads and workflows, and extracting tasks and appointments from email messages. Whereas complex theoretical models and semantics are hidden beneath a simple user interface, the enabled functionalities are clear for the users to see and take advantage of. The system's evaluation proved that our experiment with Semanta has indeed been successful and that semantic technology can be applied as an extra layer to existing technology, thus bringing its benefits into everyday computer usage.

Keywords: Semantic Email, Speech Act Theory, Semantic Web Applications.

1 Introduction

In earlier work [1] we discussed the notion of *Semantic Email* – whereby the email process is semantically-enabled so that intelligent mail-user agents can support the user with their daily email chores like writing, reading, tracking emails and action items within, as well as managing personal information generated within. We all use email subconsciously to organise our work, but little do we realize that email is just the transportation layer supporting what effectively is a number of distributed, well-defined ad-hoc workflow processes. As a consequence of the inability to manage these implicit workflows, the productivity of knowledge workers that rely on email communication suffers. By adding semantics to the email process we can add new functionality, whereby machines can support the user in managing and keeping track of the ongoing workflows. If an email system is aware of the executing workflows, it can provide semantic ad-hoc process support.

In this paper, we will present *Semanta*, a fully-implemented system supporting Semantic Email, whereby we have lifted email processes to a semantic level via speech act theory and a formally-defined ad-hoc email workflow model. There is a lot of talk about semantic software, but these are seldom fully-implemented. In the cases when they are, a frequently forgotten golden rule is that semantics should be the enabler of further functionality and it is the latter not the former which the user should be aware of. Our achievements with regards to Semanta are that:

L. Aroyo et al. (Eds.): ESWC 2009, LNCS 5554, pp. 36–50, 2009.

- we did not change the transport layer for email processes
- we did not create new Mail User Agents (MUAs) but provided lightweight extension to existing email applications
- we abstracted a complex theoretical model beneath a very simple Graphical User Interface (GUI)
- the extensions will be immediately useful by everyone who uses email

We would like to stress the final point, because we believe that such a viral approach is what is really required by the Semantic Web community to get semantics into everyday computer usage. The main results of the evaluation of the implemented system have confirmed that its support for semantic email processes is applicable to everyday email usage and that the chances for Semanta's take-up are very high.

2 Semantic Email

The notion of Semantic Email was originally introduced by McDowell et. al. [2] to refer to an email message consisting of a structured query (or an update to the query) coupled with a corresponding explanatory text. Their approach was based on the provision of a broad class of *Semantic Email Processes* that represent commonly occurring workflows within email (e.g. collecting RSVPs, coordinating group meetings). Implemented within Mangrove [3] the system provided templates which exposed structured knowledge about these scenarios to both humans and machines. The ultimate goal was to support the user with common email related tasks such as collecting information from a group of people, handling schedule/event information and reminding others about previous unanswered emails etc.

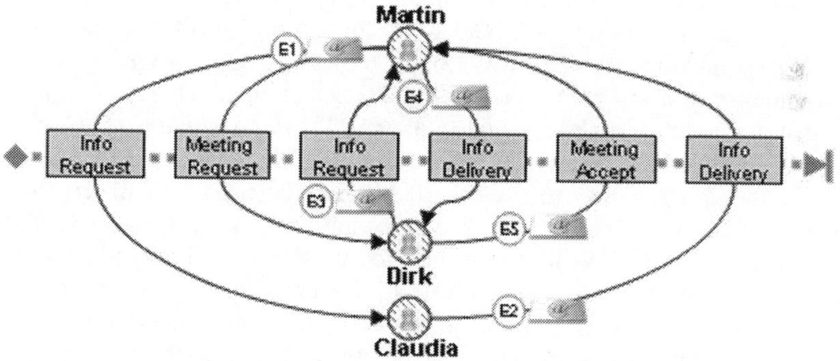

Fig. 1. An Email Workflow between Martin, Dirk and Claudia

Although we also handle commonly occurring email workflows, we take a different approach [1] whereby the only real similarity is that we also expose all the available knowledge to machines for querying, reasoning and eventually integration. For this purpose, we employ the Resource Description (RDF)[1] to represent information extracted from email messages as machine-understandable metadata. In

[1] http://www.w3.org/RDF/

our vision we consider the fact that an email has one or more purposes, or *Action Items*. The content of an email message can be summarized into a number of such items (e.g. *Meeting Request*, *Task Assignment*, *File Request* etc.). Once exchanged, every single action item can be seen as the start, or continuation of a separate workflow. The *sMail Conceptual Framework* [4] applies *Speech Act Theory* [5] to the email communication process, in order to provide a formal structure and semantics for these action items and their workflows. Email action items like the ones above can be represented by a number of speech act instances provided in the sMail ontology[2] (in this paper we use the terms speech acts and action items interchangeably).

At the design stage, we learned a lot from a debate over the use of technology for the coordination of individual actions in organisations, sparked off by the Coordinator [6] – a heavily criticised system based on speech act theory. This work highlighted the need for a more sophisticated coordination of interactions among individuals, so as to avoid breakdowns in an organisation's efficiency. As documented in [7], a major concern was that shaping conversation to constrain behaviour reduces the potential of electronic communication to support ubiquitous interaction. The system we designed is based on an acceptable compromise between *Computability* and *Communication Efficiency*. We are aware that email workflows are flexible, evolve unpredictably with time, and that one can only predict what is *likely* to occur after certain action items are received or sent. For example, on receiving a *Meeting Request*, the user tends to accept the request, decline it or propose changes to the requested meeting. However, the user might also want to pursue other paths spontaneously. This is demonstrated in Fig.1. Martin sends an email to Claudia and Dirk (E1). The email contains a *Meeting Request* addressed to Dirk, and an *Information Request* addressed to Claudia. When Claudia receives the information request from Martin, she immediately sends an email (E2) back with the required information. On the other hand, when Dirk receives the meeting request, he decides to send an email (E3) with a request for more information regarding the meeting, and waits for a reply (E4) before also approving the meeting via another email (E5). We believe that this flexibility is an advantage of email communication and consequently should be retained. Therefore, although we believe that the option of fixed templates taken in [2] can in some cases be useful, our approach is more oriented towards the handling of what we termed *Email Ad-hoc Workflows*. In earlier work we presented the *Email Speech Act Workflow* model, explicitly modeled via standardised Workflow Patterns [8], to support exactly these kinds of email workflows. While it outlines the most common reactions to incoming email action items, it simultaneously allows for the characteristic spontaneity of email as demonstrated in our example. In this paper we will not go again into the details and complexity of the workflow model. Instead, we will demonstrate its usefulness in supporting semantic email processes as experienced by the user when using Semanta.

3 Semanta

Semanta is a system that supports the handling and tracking of common email processes (e.g. *Meeting Scheduling*, *Task Delegation*, *Information Request*) (co)executing in email

---

[2] http://ontologies.smile.deri.ie/smail

threads, and manages resulting artefacts (e.g. events, tasks). This implementation follows a limited prototype presented earlier [9]. Additionally, Semanta is now fully-integrated within the Social Semantic Desktop [10]. After an overview of the implemented system's architecture we will demonstrate its features.

3.1 Architecture

Fig. 2 illustrates the general system architecture. A simple intuitive user interface is crucial for the take-up of a semantic email application. Hiding the underlying complex workflow model beneath such a UI was not straightforward, but as we demonstrate in the next section the user is exposed only to the additional functionalities provided by the hidden semantics, and not the semantics themselves. Semanta's interface has been implemented as an add-in/extension to two popular Mail User Agents – Microsoft Outlook 2003 and Mozilla Thunderbird. The MUAs are still responsible for creating and sending email messages using the existing, time-tested transport technologies. Semanta is integrated within the Social Semantic Desktop (SSD) and since it employs Semantic Web technologies to represent generated data, not only is this data exposed to semantic operations such as queries and inference, but it is also immediately available for re-use by other intelligent SSD applications, and vice versa. On the SSD, personal information such as address book data, calendar data, email data, folder structures, etc. is lifted onto an RDF representation. These information items are treated as Semantic Web resources and ontologies allow the user to express such desktop data formally. The NEPOMUK Project[3] developed an infrastructure of ontologies[4] to handle different aspects of the SSD - from the Personal Information Model Ontology (PIMO) [11] to represent the structures and concepts within a knowledge workers mental model; to the Information Element Ontologies (NIE) to represent files (NFO), contacts (NCO), messages (NMO) etc. At the semantic level, integrating Semanta on the SSD required the modification of the sMail Ontology to extend existing concepts in the SSD ontologies (e.g. nmo:Email). At the

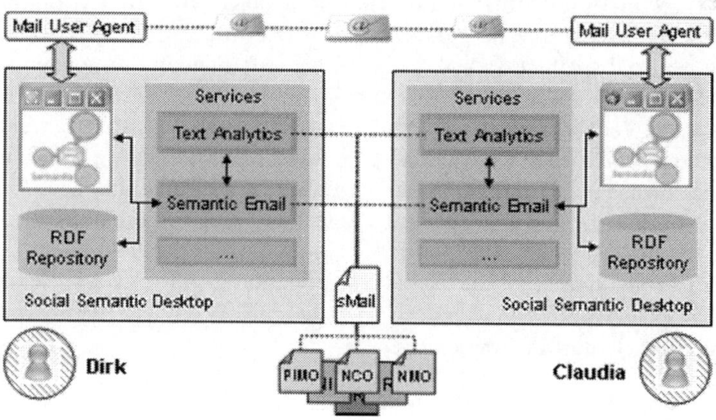

Fig. 2. Semanta within the Social Semantic Desktop

[3] http://www.nepomuk.semanticdesktop.org/
[4] http://www.semanticdesktop.org/ontologies/

practical level, Semanta was integrated by enabling the access and storage of generated metadata on the desktop's RDF repository.

The SSD provides a number of underlying services serving specific functions. Two crucial architectural components are the *Semantic Email* and the *Text Analytics* services. Both access the knowledge of the sMail Framework via the sMail Ontology, providing the Business Logic, which is separate from the Graphical User Interface (GUI). Thus only the GUI is dependent on the targeted MUA, i.e. Semantic Email can be exchanged between different users using different MUAs on different platforms. The semantic email service acts as an invisible layer beneath the GUI performing:

- Semi-automatic content annotation (via the use of the text analytics service)
- Writing/reading RDF statements into/from a specific 'x-smail' email header
- Reasoning over which options a user is given when reacting to action items (given the sMail Email Speech Act Workflow model)
- Detecting tasks or events generated within email
- Storing metadata in the SSD's central RDF Repository
- Querying data in the repository to provide the user with information regarding action items, emails, tasks, events, people and their relationships.

The semantic email service utilises the text analytics service to provide semi-automatic annotation of email content. This service uses Ontology-Based Information Extraction (OBIE) techniques to elicit speech acts (action items) in email bodies. The information extraction is based on a declarative model which classifies text into speech acts based on a number of linguistic features like sentence form, tense, modality and the semantic roles of verbs. The service deploys a GATE [12] corpus pipeline consisting of a tokeniser, modified sentence splitter, POS tagger, keyphrase lookup via Finite State gazetteers and several JAPE [13] grammars. Some of these grammars are conditionally run based on the outcome of previous JAPE annotations and are ordered in priority to consume the longest matching annotation. Earlier work [14] implemented similar Knowledge Based (KB) approaches using earlier versions of GATE. The current service is at the beta testing stage whereby JAPE grammars are iteratively tuned based on the outcome of each test cycle. The results of this technology's initial evaluation, where a corpus of manually annotated email was compared against automatic annotation, gave an f-measure of 0.648. However, when these were averaged by speech act category, the average f-measure went down to 0.468. This low score is due to a deficit of dictionary entries and we intend to improve upon our performance by extending our gazetteer lists through the use of the GATE *Annotation Diff* tool, which also allows us to test and tune rules that do not fire. At the current development stage, the service is suitable for providing semi-automatic classification, given that fully automated classification is not yet very reliable.

3.2 Supporting Semantic Email Processes

We will now demonstrate the added functionalities of Semantic Email as provided by Semanta by going through the example provided in Section 2 and simultaneously showing the support provided via Semanta's GUI (as implemented for Outlook). The email workflow demonstrated in Fig. 1 can be represented as a timeline depicting *Action Item Threads* within an *Email Thread*. Fig. 3 shows how an email (E1)

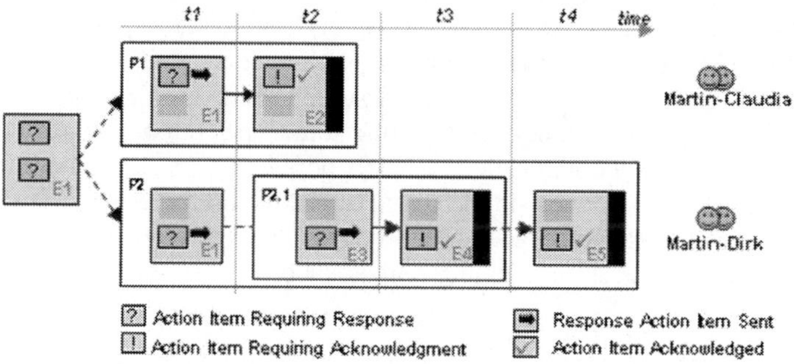

Fig. 3. Email breakdown into two separate workflows

containing two action items (marked '[?]') is conceptually broken down into two 1-1 transactions between Martin-Dirk (P1) and Martin-Claudia (P2). In the remainder of this section, we trace the paths undertaken by these workflows while demonstrating how Semanta supports Martin, Dirk and Claudia with their semantic email processes.

3.2.1 Initiating Email Processes

After writing email E1, Martin annotates it via the Semanta toolbar. Specific action items in the content are annotated, rather than the email as a whole. After reviewing the automatically annotated action items, he can remove, change or create new annotations via a right click which brings up Semanta's *Annotation Wizard* (Fig. 4). Rather than providing a list of predefined action items to the user, the wizard supports the user with building annotations by selecting the:

1. *Action* that represents the textual statement (e.g. Request, Assign, Suggest)
2. *Object* of the action (e.g. Event, Task, Information, Resource)
3. *Subject(s)* of the action, if applicable (e.g. who is implicated in a Task/Event)
4. *Target(s)* of the action – i.e. to whom the action is directed to.

These are the four defining parameters for a speech act in the sMail Speech Act Model. This process takes up between 3 clicks for simpler speech acts, e.g. the first annotation in E1 – "Request

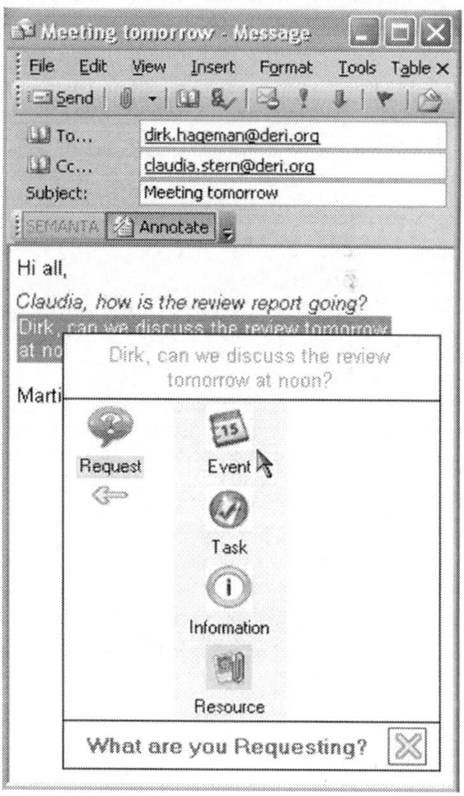

Fig. 4. Modifying and creating annotations

(action) Information (object) from Claudia (target)"; up to 4 or more clicks, depending on the amount of recipients, for more complex speech acts e.g. the second speech act in E1 - "Request (action) a Meeting (object) between Yourself and Dirk (subjects) and direct the request to Dirk (target)". A dynamic sentence (bottom of the wizard) guides the user in constructing the annotation. Objects populating the wizard are dynamically loaded from the sMail ontology via the semantic email service. When Martin sends the email, these annotations together with other metadata are invisibly transported alongside the email content in the custom 'x-smail' email header.

3.2.2 Executing Email Workflows

At time *t1* both Dirk and Claudia receive Martin's email. Only one of the action items is addressed to Claudia – the *Information Request*. The other action item – the *Meeting Request*, is addressed to Dirk. Semanta flags incoming email if they possess action items addressed to the current user. Although the email here is flagged as a whole, the flagging is based on how many, if any, action items in the message are addressed to the user. When viewing email with Semanta, users can react to *each* individual action item separately. Depending on its type, they are given a number of appropriate options, as determined by the sMail Email Speech Act Workflow. In Claudia's case, when she right-clicks the *Information Request*, a number of appropriate options are given whereby she can choose to deliver or decline the requested information. She can also ignore the action item (can be undone). More importantly Claudia is allowed to react in any other way. Claudia selects the 'Deliver Information' option, upon which she is prompted to input the required information - "So far, so good". This is automatically annotated as an *Information Delivery* and sent back in a reply email E2 to Martin, who receives and acknowledges it at time *t2*. Below we produce snippets from the RDF generated for this email reply, as transported within the email headers and stored on Claudia's SSD (some URI's have been simplified).

```
<EMAIL2> a <nmo#Email>;
 nmo:from :ClaudiaStern;
 smail:hasPrecedingEmail :EMAIL1;
 smail:hasSpeechAct :EMAIL2SA0;
 pimo:isDefinedBy
  <http://nepomuk.semanticdesktop.org/users/claudias/pimo>;
 nmo:messageId "0000000094A633229025F385F0";
 nmo:sentDate "2008-10-03T13:33:53";
 nmo:to :MartinWilliams .

<EMAIL2SA0> a <smail#SpeechAct>;
 smail:hasAction <smail#Deliver>;
 smail:hasNoun <smail#Information>;
 smail:hasPrecedingSpeechAct :EMAIL1SA0;
 smail:hasRole <smail#Completive>;
 smail:hasSpeechLength "16";
 smail:hasSpeechStart "46";
 smail:hasStatus <smail#Completed>;
 smail:hasTarget :MartinWilliams;
 smail:recipientExpectation <smail#Acknowledge>;
 smail:senderExpectation <smail#None> .
```

```
<MartinWilliams> a <nco#PersonContact>;
 nco:hasEmailAddress :MartinWilliamsEmail;
 <rdfs:label> "Martin Williams" .
<MartinWilliamsEmail> a <nco#PersonContact>;
 nco:emailAddress "martin.williams@deri.org" .
```

The first definition describes the email as a whole, providing an RDF representation of customary email metadata, e.g. sender, recipients, etc.; as well as information specific to Semanta, e.g. references to the speech act within, to the SSD PIMO where this email is relevant (i.e. Claudia), the system (Outlook) message ID as well as a reference to the URI of the preceding email message. The second definition represents the annotation of an information delivery for the text "So far, so good". This annotation consists of an *Information* (noun) *Delivery* (action) addressed to Martin (target). This speech act is in response to the preceding *Information Request* from Martin ("Claudia, how is the review report going?", represented by '#EMAIL1SA0') in the preceding email ('#EMAIL1'). The following properties are defined on the fly by the semantic email service. Given the nature of this action item and of the one preceding it, the role of the former is set to *Completive* – since it is completing the *Information Request* process. Since Claudia is expected to do nothing else on sending this action item (senderExpection), the status of this action item within its representation on her SSD is set to *Completed* right after the email is sent (as shown below). However, since the recipient is expected to acknowledge this action item, the status within its representation in the email sent to Martin will be set to *Pending*, and will remain so until it is acknowledged. The last two definitions describe an email participant. They make sole use of the NCO ontology (and RDF/S) whereas the previous use PIMO, NMO and the sMail ontologies (see 3.1).

Fig. 5 shows Dirk considering the *Meeting Request* he receives from Martin at *t1* - "Dirk, can we discuss the review tomorrow at noon?". On right click, Dirk is given the option to approve, decline or amend the proposal, ignore it or do something else. Dirk chooses the latter option ("Other..") since he decides that he wants to request more information before making a decision on the meeting request. The annotation wizard is shown right away and Dirk inputs the desired text - "What is there to be discussed?", and annotates it right-away as an *Information Request*. Once Dirk has finished, a reply (E3) with this information request is automatically sent back to Martin. This initiates the sub-workflow P2.1 shown in Fig.3. Martin receives E3 at time *t2* and promptly provides the required information in E4. When Dirk receives E4 and acknowledges this action item at time *t3*, P2.1 is terminated.

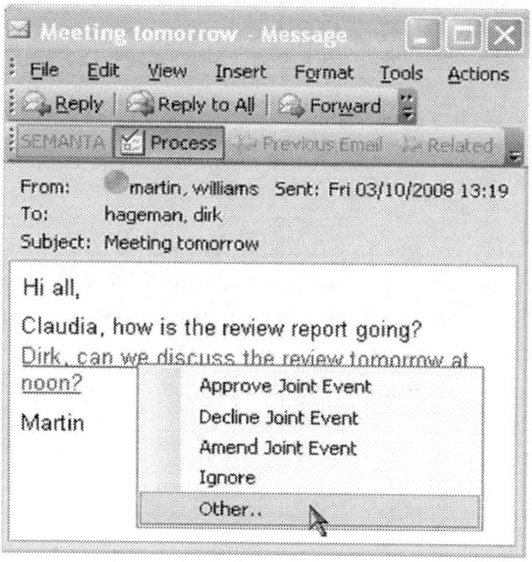

Fig. 5. Processing Incoming Email

3.2.3 Tracking Email Workflows

So far we have shown how Semanta can support the user with finding and exchanging action items within emails. We have also shown how by reacting to individual action items within email messages the user is effectively executing implicit ad-hoc workflows with the support of Semanta. We will now demonstrate how Semanta can visualise these implicit email workflows. In the Outlook main window, Semanta's toolbar provides the *Action Items* button for viewing and tracking email workflows. The action item tracker has three main views:

1. Pending Incoming – this view shows all action items (e.g. requests, assignments, suggestions) which require action from the user.
2. Pending Outgoing – this view shows all outgoing action items (e.g. requests) for which the user is still awaiting a reply.
3. All Items – this view shows all incoming and outgoing action items, regardless whether they have been tackled by the user or the user's contacts accordingly.

By viewing the pending incoming action items, the users can tackle those email workflows they are personally stalling. By viewing the pending outgoing action items, the user can view those workflows which have been stalled by someone else and decide whether to send a reminder or otherwise. The all items view shows all exchanged action items within ongoing and terminated workflows.

Fig. 6. Tracking Email Action Item Threads

For items in all three views, the user can expand the window to view the whole thread of action items exchanged for that workflow. This is presented to the user as the *context* of that action item within its thread. The options for visualising email action items threads are limitless and can be improved. The current design mimics the one for message folders in the host application, without being too similar so as not to confuse the user. Back to our example Dirk chooses to view more information about one of his four pending incoming action item (Fig. 6). He can see that this action item represents the event requested by his manager ("Dirk, can we discuss the review tomorrow at noon?") in email E1. In the context panel Dirk can see that he reacted to this request by asking for more information regarding the event. He can also see that the manager already replied to this request ("The way forward!"). Dirk therefore decides to reconsider the *Meeting Request* and on double-clicking this item he is again shown email E1. He now selects the 'Approve Joint Event' option upon which an automatic reply ("Yes") is embedded as reply text (the text can be changed). This text is automatically annotated as an *Event Assignment* where the event involves both Martin and Dirk, thus binding both of them to the now scheduled meeting.

3.2.4 Handling Workflow Artefacts

When Dirk is done viewing email E4, the task assignment is automatically sent in E5 to Martin. Semanta detects that an event involving Dirk and Martin has been generated and presents the item to Dirk for review (Fig. 7), where he can either dismiss it or add it to his Outlook Calendar (an analogous process takes places for detected tasks which can be added to the Outlook Tasklist). On choosing to add the event to his calendar, the native Outlook Appointment Item window is brought up. Semanta populates this item up to a certain degree, and presents it to Dirk for review. Currently Semanta can provide the item subject as extracted from the action item requesting the activity and the one approving

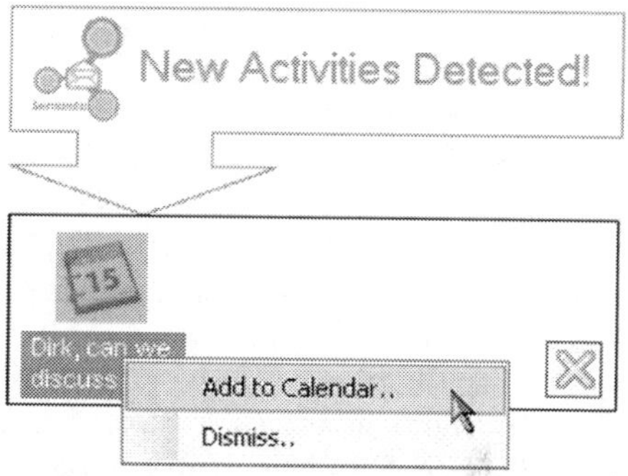

Fig. 7. Detecting new events/tasks

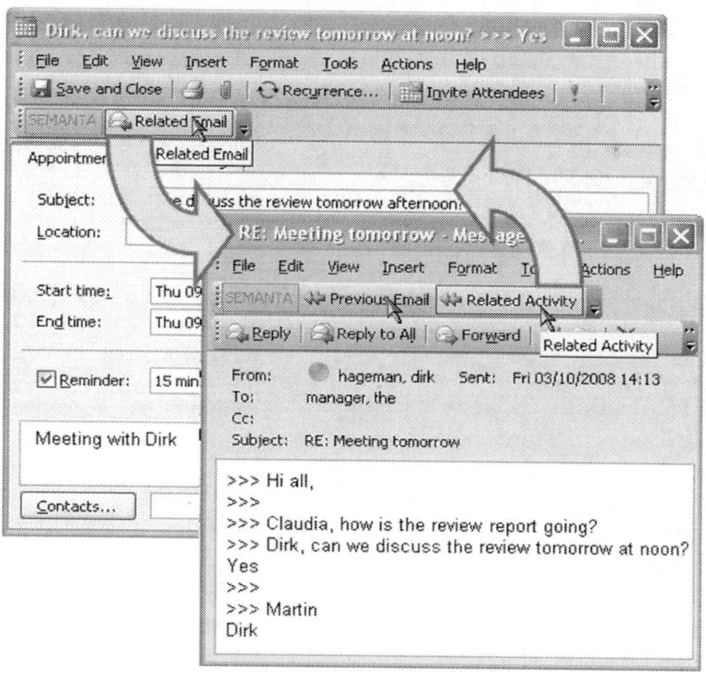

Fig. 8. Linking related Email, Task or Event items

it ("Dirk, can we discuss the review tomorrow afternoon? >>> Yes"). Semanta is also aware of the meeting participants and lists them as the appointment attendees. At *t4*, Martin acknowledges the *Event Assignment* in the automatic reply (E5). Since the event assigned also involves Martin, he is similarly prompted to review and/or add the generated event to his Calendar. At this point all the email workflows generated by the first email in our example (E1) are terminated.

Links between event/tasks created via Semanta and the email messages wherein they were generated are stored in the users' SSD RDF Repositories. Figure 8 shows two buttons provided in the Semanta command bar for this purpose – the 'Related Email' button loads the email source for stored tasks and events; whereas the 'Related Activity' button points back to any generated tasks/events from an email message. The same linking functionality provided between emails and the activities generated within is also provided for email messages within one email thread. In Fig. 8 the reader can also note that alongside the 'Related Activity' button the user is given the possibility to traverse up the email thread via the 'Previous Email' button. All these links are stored and retrieved from the SSD's RDF Repository by Semanta's semantic email service. Semantic interoperability between the various items is thus enabled.

3.2.5 Semantic Email on the Social Semantic Desktop

Since Semanta is integrated within the SSD, metadata generated by Semanta is also immediately available to other SSD applications and not just to itself. Fig. 9 shows the event generated in our example at time *t4*, as viewed from Martin's desktop via the P2P Semantic Eclipse Work--bench (PSEW)[5] - an integrated environment based on the SSD architecture allowing the users to browse, query, share and annotate resources to which they have access. Other applications like SSD calendar and task management tools will also have access to these resources. Likewise it must be

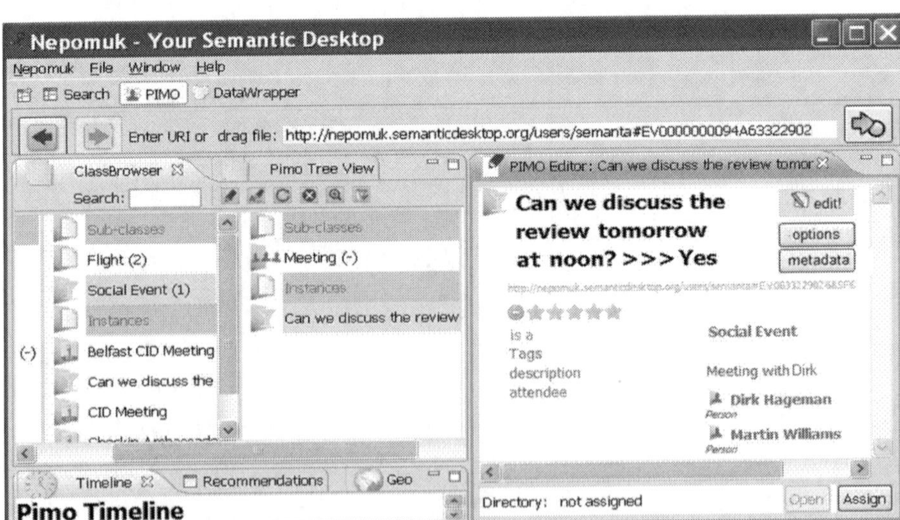

Fig. 9. Sharing metadata generated by Semanta across the SSD

[5] http://nepomuk-eclipse.semanticdesktop.org/xwiki/bin/view/Main/PSEW

pointed out that metadata generated by these applications is available to Semanta. In the future we would like to have a system in place where any proposed meetings (for example) prompt Semanta to check for any time clashes with other user commitments as know by the SSD, irrespective of the application wherein they were generated.

4 Evaluation

The evaluation process involved 6 computer science researchers. Although this is not large number, previous experiments [15] concluded that 5-12 users are considered as an acceptable number for a system-usability study. In particular Nielsen et. al. found that 5 users can already find 75% of existing problems [16]. Once we improve our system we intend to test Semanta against an even wider audience.

4.1 Experimental Setup

The evaluators averaged 32.3 years of age and they had above average familiarity with the MUA used for the evaluation – Outlook. Five typical email situations (workflows) were given to the user for consideration - Task Delegation, Task Acceptance, Data Request, Appointment Scheduling and Event Announcement. To carry out these tasks, the users had to write emails to a number of people with the assistance of Semanta. Whereas some scenarios required only one email to be sent, other scenarios included multiple exchanges, i.e. writing and reading email replies. Other scenarios included the generation of tasks and events from email. Throughout the experiment, the time taken by the user to complete different tasks was recorded. The users' behaviour was also recorded (filmed) during the evaluation and comments, suggestions, problems and errors were noted down. Finally the users were provided with a questionnaire where they could rate different aspects of Semanta with respect to consistency, functionality and design. The evaluation process took an average of 01:23 hours. All material used for the evaluation, including the presentation, scenario outlines, questionnaire, time sheets, videos and the results are available online[6].

4.2 Main Findings

The users were quite satisfied with both the automatic and semi-automatic annotation functionalities provided by the system. Of the action items detected by the text analytics service, 54% were approved by the users without any changes, 32% were improved (i.e. although the automatic annotations made sense, the user felt that the level of sophistication of the annotations could be improved, e.g. a *Request Task* rather than the more generic *Request Information*) via the annotation wizard, and only 13% were either removed or changed completely (i.e. wrong annotations). Of the final number of exchanged annotations, 27% where manually created by the users (i.e. not detected by the service). Echoing the latter was the general feeling amongst users that automatic annotation had higher precision than recall. This reflected a decision taken at the implementation stage; whereby we acknowledged that it is better for action items to pass undetected, rather than having Semanta trying to support the users with

[6] http://smile.deri.ie/projects/semanta/evaluation/

inexistent workflows bound to incorrectly annotated action items. With regards to the Annotation Wizard, although users demanded more meaningful tooltips for items within, they found the semi-automatic annotation experience easy and intuitive. Thus we feel that the effort to abstract the sMail Speech Act Model and semantic annotation beneath a simple, user-friendly interface was successful. On the down side, the ratio of time the users required to review and/or add annotations, against the total time required to create and send the email only went down to just above 50% by the end of the experiment. This is not satisfactory given that this means that users were still taking around as much time to annotate the email as to write it. However, it is a fact that throughout the evaluation many users were giving suggestions and 'thinking-aloud', thus considerably increasing the time spent on each annotation task.

With respect to reading and processing incoming email, the evaluators were satisfied with the options given for different incoming action items, i.e. Semanta's support for email processes, and there were no cases when the users were unsure of which option to select. The users liked the event/task detection and were satisfied with the partial automatic population provided when adding these items to their task list/calendar. They also appreciated the links generated between detected tasks/events and their source email messages as well as the fact that one can traverse up the email thread from the email message window.

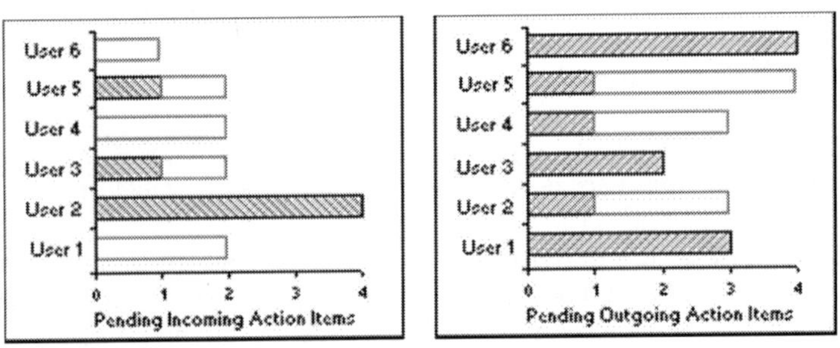

Fig. 10. Tracked action items with (outline) and without (shaded) Semanta

After the users carried out the scenarios, they were asked to enumerate all exchanged pending outgoing action items, i.e. action items they have sent for which they were still awaiting a reply; and pending incoming action items, i.e. action items they have received and still needed to consider; that they could remember. The users were allowed the use of Outlook for this task, i.e. looking at exchanged emails in the inbox and sent items folders. Although this process took the users an average of 6.66 minutes, they still could not point them all out. The same information can be instantly and fully retrieved by Semanta. On average, after browsing exchanged messages, the users could recall just 33% of total pending incoming email action items (Fig. 10). The recall rate for the outgoing email action items amounted to 65%. This suggests that users tend to remember pending action items that they have sent to their contacts with twice the accuracy that they remember pending action items addressed to them. In conclusion, the users were highly appreciative of Semanta's action item tracking function – which effectively provides visualisation for ad-hoc email workflows.

4.3 Improving Semanta

After the evaluation we made a number of improvements, both in terms of functionality as well as the user interface. With regards to the text analytics service we are still improving the recognition of persons involved in the email (taking advantage of existing structured sources such as Address Books) but have since included simple co-reference resolution. We will investigate the use of ML techniques to improve both precision and recall of automatic annotation. Eventually we want to bring the ratio of time required to annotate an email against the time required to write, annotate and send it to at most 25%. We also plan to extend the text analytics service in order to capture date and time information for detected tasks and events. Although we criticised the system implemented in Mangrove [3], we acknowledge that in some cases the option of using fixed templates can save the user's time. We therefore added QuickShot email function, offering a selection of email templates where the purpose is predefined and the user is required only to input the content.

We will provide the following additions to Semanta's email action item tracking. When viewing pending outgoing action items, users will be able to quickly send reminders to their contacts to inform them that they still await a response. Reminders will also be incorporated in the sMail Workflow Model, as special kinds of email messages. Action item thread visualisation will be extended such that, when a workflow ends with a generated event/task, these are also incorporated in this view and in the case of tasks, their status will be dynamically updated. For example, after Claudia accepts a *Task Request* from Dirk, it will be shown at the end of the workflow in both Claudia's and Dirk's action item tracker. When Claudia eventually ticks the task done, the task will be marked completed in both systems.

5 Conclusion

In this paper we have presented a fully-implemented system supporting Semantic Email Processes – Semanta. We have shown the reader how Semanta can assist the user with the execution of previously implicit ad-hoc email workflows. The main contribution of this paper is the successful implementation of a system whereby we have seamlessly integrated semantic technology supporting semantic process management into the existing technical landscape, used existing transport technology, and hidden a complex workflow model and semantics under an intuitive user interface that can be used by anyone on their own computer. In contrast, the additional functionalities enabled by these semantics are clearly visible to the end-user. We believe that this is exactly what the Semantic Web needs to be taken-up by the computer-literate society at large. We verified the acceptance and applicability of the user interface by an experimental case-study, where one of the goals was to find out how well we can actually hide the semantic technology. The results of the evaluation have confirmed our success in these respects.

Acknowledgments. The work presented in this paper was supported (in part) by the Lion project supported by Science Foundation Ireland under Grant No. SFI/02/CE1/I131 and (in part) by the European project NEPOMUK No FP6-027705.

References

1. Scerri, S., Handschuh, S., Decker, S.: Semantic Email as a Communication Medium for the Social Semantic Desktop. In: Bechhofer, S., Hauswirth, M., Hoffmann, J., Koubarakis, M. (eds.) ESWC 2008. LNCS, vol. 5021, pp. 124–138. Springer, Heidelberg (2008)
2. McDowell, L., Etzioni, O., Halevey, A., Levy, H.: Semantic Email. In: Proc. World Wide Web (2004)
3. McDowell, L., Etzioni, O., Gribble, S., Halevey, A., Levy, H., Pentney, W., Verma, D., Vlasseva, S.: Evolving the Semantic Web with Mangrove. UW Tech Report (2003)
4. Scerri, S., Mencke, M., Davis, B., Handschuh, S.: Evaluating the Ontology underlying sMail – the Conceptual Framework for Semantic Email Communication. In: Proc. Language Resources and Evaluation 2008, Marrakech, Morocco (2008)
5. Searle, J.: Speech Acts. Cambridge University Press, Cambridge (1969)
6. Winograd, T., Flores, F.: Understanding Computers and Cognition: A New Foundation for Design, p. 220. Ablex, Norwood (1986)
7. Curtis, B.: Can speech acts walk the talk? Proc. of Comput. Supported Coop. Work 3(1), 61–64 (1994)
8. Voorhoeve, M., Van der Aalst, W.: Ad-hoc workflow: problems and solutions. In: 8th International Workshop on Database and Expert Systems Applications (DEXA 2007) (1997)
9. Scerri, S.: Semanta – Your Personal Email Semantic Assistant. In: Demo Session at the Intelligent User Interface Conference 2009, Gran Canaria, Spain (2008)
10. Decker, S., Martin, F.: The Social Semantic Desktop. Technical Report (2004)
11. Sauermann, L., van Elst, L., Dengel, A.: PIMO - a Framework for Representing Personal Information Models. In: Proc. I-Semantics 2007, pp. 270–277 (2007)
12. Cunningham, H., Maynard, D., Bontcheva, K., Tablan, V.: GATE: A Framework and Graphical Development Environment for Robust NLP Tools and Applications. In: Proc. 40th Anniversary Meeting of the Association for Computational Linguistics (ACL 2002), Philadelphia (2002)
13. Cunningham, H., Maynard, D., Tablan, K.: JAPE: a Java Annotation Patterns Engine (Second edn.). Technical report CS–00–10, University of Sheffield, Department of Computer Science (2000)
14. Khosravi, H., Wilks, Y.: Routing email automatically by purpose not topic. Natural Language Engineering 5, 237–250 (1999)
15. Dumas, J.S., Redish, J.C.: A practical guide to Usability Testing (revised edn.). Intellect, Exeter (1999)
16. Nielsen, J.: Heuristic Evaluation. In: Nielsen, J., Mack, R.K. (eds.) Usability Inspection Methods. John Wiley & Sons, New York (1994)

The Sile Model — A Semantic File System Infrastructure for the Desktop

Bernhard Schandl and Bernhard Haslhofer

University of Vienna, Department of Distributed and Multimedia Systems
{bernhard.schandl,bernhard.haslhofer}@univie.ac.at

Abstract. With the increasing storage capacity of personal computing devices, the problems of information overload and information fragmentation become apparent on users' desktops. For the Web, semantic technologies aim at solving this problem by adding a machine-interpretable information layer on top of existing resources, and it has been shown that the application of these technologies to desktop environments is helpful for end users. Certain characteristics of the Semantic Web architecture that are commonly accepted in the Web context, however, are not desirable for desktops; e.g., incomplete information, broken links, or disruption of content and annotations. To overcome these limitations, we propose the sile model, an intermediate data model that combines attributes of the Semantic Web and file systems. This model is intended to be the conceptual foundation of the Semantic Desktop, and to serve as underlying infrastructure on which applications and further services, e.g., virtual file systems, can be built. In this paper, we present the sile model, discuss Semantic Web vocabularies that can be used in the context of this model to annotate desktop data, and analyze the performance of typical operations on a virtual file system implementation that is based on this model.

1 Introduction

A large amount of information is stored on personal desktops. We use our personal computing devices—both mobile and stationary—to communicate, to write documents, to organize multimedia content, to search for and retrieve information, and much more. With the increasing computing and storage power of such devices, we face the problem of *information overload*: the amount of data we generate and consume is increasing constantly, and because of the availability of cheap storage space, each and every bit of information is stored. Another problem is even more prevalent on the desktop than on the Web: *information fragmentation*. Data of different kinds are stored in silos, and—contrary to the Web, where hyperlinks between documents, and across site boundaries, can be defined—there exist only limited means to define and retrieve relationships between different resources on the desktop.

The Semantic Web tries to deal with the problems mentioned before by adding a layer on top of the existing Web infrastructure, in which descriptions about

L. Aroyo et al. (Eds.): ESWC 2009, LNCS 5554, pp. 51–65, 2009.

web resources are expressed in RDF using commonly accepted vocabularies or ontologies. This allows machines to interpret the published data and thus helps end users to find information more efficiently. A large number of data sets[1] and vocabularies[2] have already been published on the Semantic Web, forming a solid data corpus that can be crawled by (semantic) search engines, or serve as foundation for applications.

Recent research in the field of the Semantic Desktop [2,10,12] has shown that a number of characteristics of Semantic Web technologies are also suitable for the problem of information management on the desktop; especially, the provision of unified identifiers, the ability to represent data in an application-independent generic format, and the flexible usage of formalized vocabularies to describe resources. It has been shown [15] that the inclusion of semantic technologies on the desktop can significantly improve the user's perceived quality of personal information management, especially when they are applied during a longer time period.

However there exist some significant differences between the Web and the desktop contexts. First, in contrast to the World Wide Web, the desktop already has a *de facto* organization metaphor for data: hierarchical file systems. File hierarchies have been in use for decades, and the vast majority of personal information is stored in these structures. Therefore it is crucial for the Semantic Desktop to smoothly integrate with file systems in a way that allows for the annotation of files without breaking the behavior of existing desktop applications. As a second difference, while it is accepted on the Web that URLs may be broken and web resources appear and disappear, this is not the case for the desktop. Users rightfully expect their data to remain consistent and complete at all times.

Since the RDF data model has by design a number of shortcomings that may cause problems when a Semantic Desktop is to be efficiently implemented (see Section 2), we propose in this paper the *sile model*, a data model that acts as an intermediate layer between file systems and Semantic Web technology. This model allows users and applications to annotate and interrelate desktop resources, but preserves the possibility to be used as a hierarchical file system. Thus, it maintains full backwards-compatibility to existing systems and applications.

Despite the number of existing differences between the sile model and RDF, the former has been designed to be interoperable with vocabularies from the Semantic Web. Interoperability between the desktop and the Web is desirable in order to provide a unified data space to the user. Tools and applications are enabled to operate on data both locally and globally, and data can be seamlessly exchanged between these two worlds. Hence we propose to use Web vocabularies for data representation wherever possible, and we support this requirement by

---

[1] `http://esw.w3.org/topic/TaskForces/CommunityProjects/LinkingOpenData/DataSets`

[2] `http://esw.w3.org/topic/TaskForces/CommunityProjects/LinkingOpenData/CommonVocabularies`

using URIs, which can refer to Semantic Web resources, as identifiers for all annotations in our data model.

This paper discusses the issues that arise when file systems and semantic technologies are integrated in Section 2, and presents the design of the sile model (Section 3) as a data abstraction layer for the desktop. We discuss how the sile model can be used in applications, and which already existing Semantic Web vocabularies can be used for the annotation of desktop data (Section 4). Finally, we present our prototypical implementation (Section 5), which we have evaluated under the assumption of realistic amounts of data (Section 6).

2 Integrating File Systems and RDF Descriptions

File Systems as Data Organization Metaphor. The majority of personal information is stored within file systems, where we can observe three usage patterns. We denote the first one as *user-driven file structures*: a large number of applications do not use internal directory structures or file name conventions, but allow the users to organize files in directories named and nested according to their needs. Examples for this group include applications like word processors or spreadsheet tools. A second group of applications establish *application-driven file structures*, where directory hierarchies and files are managed entirely by applications, but still expose a certain meaning to end users. Examples of such applications are e-mail clients like Mozilla Thunderbird or media players like iTunes. Typically, these applications do not operate on single files but on file collections, and interpret the directory hierarchies and file names in a schematic manner. A third group of applications do not expose a file structure to the user, but rather operate on continuous data corpora; examples for this group include calendar applications that store appointments in one large file, or database-driven applications. We denote the data structures of such applications as *hidden file structures*, because in the end even these data are stored in file systems.

For all three classes of applications it is desirable to make information accessible in a semantically meaningful, application-independent way. This has two main benefits: first, it makes it possible to interrelate objects from different sources, which otherwise form disjunct structures in one's personal information space [3]; and second, it extends the possibilities provided by file systems with respect to search and retrieval [15]. All three classes of applications, as described before, could benefit from such an integration: user-driven file structures could be extended by more powerful and non-hierarchical annotation mechanisms, like tagging and typed relationships. Application-driven file structures could be disbanded in favor of explicit semantic annotations that ideally adhere to open vocabularies and therefore are interpretable by external entities. Finally, hidden file structures could be revealed and hence be integrated with other information sources.

The Gap to RDF. The RDF data model can be used to identify information units (*resources*), and to describe these units and the relationships between them using subject-predicate-object triples. Its successful usage in the context of Linked Open Data [1] indicates that it is sufficiently expressive for a large number of applications, ranging from statistical data to geo information and multimedia content. Certain RDF characteristics, however, are problematic in the context of the desktop: for instance, RDF does not consider the actual *content* of information items. The relationship between a resource identifier (a URI) and the resource itself has been explicitly left out of the scope of the RDF specification (cf. [11], Section 1.2). In the open Web environment, it is acceptable for users and applications that resources may become unavailable and links may become broken. However, on the desktop, integrity of content objects and their annotations must always be ensured.

Another potentially problematic aspect of RDF is the Open World Assumption. Since RDF is designed for the Web, incomplete or unknown information is accepted by design. Again, for the closed environment of the desktop, incomplete information is not desirable, and hence recent approaches to Semantic Desktop information modelling have restricted the interpretation of RDF data to a closed world semantics (e.g., [18]).

As a third problematic issue, the application of certain RDF language elements (in particular collections, blank nodes, and reification), has been discouraged both in the context of the Web (cf. [1], Section 2.2) and the desktop (cf. [19], Section 2.3.2). These features significantly increase the complexity of RDF processing, as well as representing RDF in user interfaces [13,17], but expose a very light semantics [13]. Hence it is doubtful whether these elements are useful in the context of the desktop, where end users have only limited knowledge about the characteristics of the underlying data structures, and the available computing power is limited in comparison to large-scale database servers.

Recent Semantic Desktop projects (e.g., [2,10,12]) add an RDF layer on top of existing desktop infrastructures and refer to the actual content representation using local URIs. These URIs are often minted based on the characteristics of the underlying system. For instance, a URI that refers to a file in the user's home directory could be `file:////home/bernhard/work/papers/eswc2009-submission.pdf`. This URI, however, becomes invalid when a file is renamed, moved to a different directory, or deleted. Consequently, synchronization mechanisms are needed that propagate modifications to the semantic layer; however often it is difficult to track such modifications accurately. In the worst case, where such propagation is not possible, the applicability of the entire system is heavily limited.

In the following, we present an alternative approach that aims to solve the problems described before: we consider an object's content and its annotations as integral components, which are always processed together. Instead of adding a semantic layer on top of existing structures, we inject the Semantic Web into the core of the data representation structures—in our case, the file system—, and provide a virtual, file system-like representation of the data stored therein which can be accessed by existing desktop applications.

3 The Sile Model

Siles: Adding Semantics to Files. *Siles* (*S*emantic f*iles*) can be regarded as combinations of files and semantic annotations. A sile is always identified by a globally unique URI and consists of a (binary) string of arbitrary *content*, as well as an arbitrary number of *annotations*. In the context of siles, URIs are not used as URLs: while in the Semantic Web it is recommended that the URI of a resource is at the same time a URL that can be used to retrieve the resource's representation (cf. Figure 1(a)), this is not the case for siles. In our model, URIs are solely used for identification purposes, and the sile identifier, the content, and the associated annotations are integral parts of the sile, as depicted in Figure 1(b).

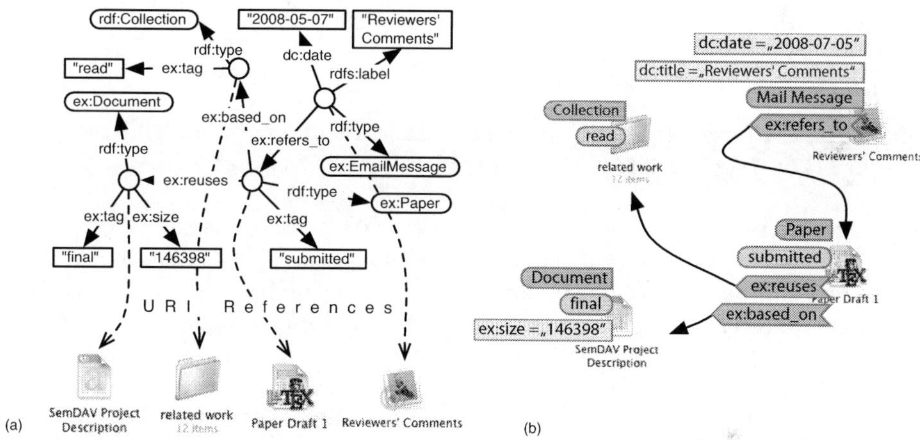

Fig. 1. (a) RDF model: URIs refer to actual content; (b) Sile model: integrated view on content and annotations

In our model we distinguish between four types of annotations, which from our experience are able to cover a large share of annotation needs in the desktop domain: *tags*, which are plain strings; *categories*, which refer to entities with machine-processable semantic interpretation (e.g., classes from an ontology); *attributes* in the form of typed name/value pairs; and *slinks* (semantic links), i.e., directed typed relationships between siles. Categories, attributes, and slinks are identified by URIs, which allow for a non-ambiguous interpretation of their semantics; however the formalism used for this purpose is out of the scope of this definition. A category annotation, for instance, may refer to an OWL ontology class as well as to a table within a relational database schema. Figure 1 depicts a number of siles and their associated annotations, whereas different shapes and colors are used to indicate different annotation types.

Because our model relies on URIs to identify siles and annotations, it is straightforward to represent sile annotations in RDF. Figure 2 depicts an example of a sile that represents a project report. This sile is annotated with a

```
<urn:uuid:57207370-6880-11dd-ad8b-0800200c9a66>
  a sile:Sile ;
  sile:label "SemDAV Project Description"^^xsd:string ;
  sile:cat-type nfo:PaginatedTextDocument ;
  sile:tag-type sile-tag:final ;
  sile:creation-date "2008-07-11T16:21:14"^^xsd:dateTime ;
  sile:update-date "2008-07-11T17:14:21"^^xsd:dateTime ,
                   "2008-07-11T17:32:02"^^xsd:dateTime ;
  sile:content-type "application/msword"^^xsd:string ;
  sile:content-length "146398"^^xsd:long ;
  dc:subject <http://www.semdav.org> ;
  sile:cat-type silefs:File ;
  silefs:path "/home/projects/semdav/docs/semdav_description.doc"
    ^^xsd:string ;
  silefs:parent <urn:uuid:60ad6a73-1b60-4553-9436-d09d395fc29c> .
```

Fig. 2. RDF representation of a sile (N3 notation)

number of attributes (like label, creation date, content length, and subject), one category (`nfo:PaginatedTextDocument`), and one tag (`final`).

Representation of Directory Hierarchies. We can now use our data model to represent all elements of a file system without loss of information. A basic file system can be represented by an unordered tree having two types of nodes: directories and files. Directory nodes may have children, whereas file nodes are always leaf nodes. Each node is labelled with a local name (i.e., the file or directory name), which is unique within the scope of its parent.

We map each node in the directory tree to a sile and represent its type by a category. We further represent the node name as an attribute annotation. By concatenating the labels of the nodes along a path, a unique identifier for each element (directory or file) of the file system can be constructed. However, when a path expression consists of many elements, traversal across a large fraction of the graph is required in order to locate the described file. To enable direct access to a file given its path, we additionally store the file's absolute path as a sile attribute. This requires processing overhead when modifications to the file tree are made, but significantly increases the performance of read operations.

The RDF representation of these additional sile annotations is shown in the last three lines of Figure 2. These annotations identify the resource as a file, describe its file name and absolute path, and relate the resource to its parent directory. This representation allows for efficient execution of typical file system operations. For most such operations the full paths of involved files are given, which allows us to retrieve the corresponding sile with a simple query for the `silefs:path` attribute. Operations that involve a directory and its children (e.g., a directory listing) can be resolved by querying for `silefs:parent` relationships between siles. When a file or directory is renamed or moved[3], the steps to be taken depend on whether the file remains within its parent directory or not. In

[3] Moving and renaming of files are implemented identically in many operating systems.

the former case, it is sufficient to update the sile's `silefs:path` attribute, while in the latter case the `silefs:parent` slink must also be updated to point to the sile's new parent directory. However, the identity of this sile can be efficiently retrieved by querying for the directory sile's `silefs:path` attribute.

The representation of files using a graph-based model enables us also to represent hard links by storing additional `silefs:path` attributes and `silefs:parent` relationships. The semantics of the deletion of a link can be simulated by checking whether a sile has more than one `silefs:path` attribute before deletion, and by only deleting the entire sile if it has only one such attribute. In the case of multiple `silefs:path` attributes, only the attribute and the `silefs:parent` slink are deleted.

4 Annotating Files Using the Sile Model

Accessing Semantic Files. Using the mechanisms described in the previous section, we can now implement a common hierarchical file system that can be used by any application without modifications. To read, write, and search for semantic annotations of files that are represented as siles we have developed an Application Programming Interface (API). A detailed discussion of this API is outside the scope of this paper; however to give the reader an impression of the structure and usage of the API, a code snippet is depicted in Figure 3. This example reproduces the annotations depicted in Figure 1 and Figure 2: a category and a tag are attached to a file (`"semdav_description.doc"`). Later, the application (or the user) could easily retrieve the object by searching for these annotations.

The goal of this API is to allow desktop application developers to easily integrate semantic annotations into their code. Assuming that files are stored on a sile-based virtual file system, one can retrieve the sile that represents a certain file by a single API call, and access or manipulate annotations. Additionally, the API allows an application to search for siles that match certain criteria and to retrieve the corresponding file paths. For instance, a word processing application could be extended so that it stores metadata about created documents (e.g., author and title) as sile attributes, or a search operation could be used to retrieve files that are associated with a certain project.

We believe that our API, which tightly integrates files and semantic annotations, has the potential to significantly increase the proliferation of semantic

```
Sile docSile = repo.getSileForFile("/home/projects/semdav/docs/
   semdav_description.doc");
docSile.addAnnotation(repo.getCategory("nfo:PaginatedTextDocument"));
docSile.addAnnotation(repo.getTag("final"));
Sile paperSile = repo.getSileForFile("/home/papers/eswc2009/paper.tex");
paperSile.addAnnotation(repo.getSlink("ex:based_on", docSile));
```

Fig. 3. Sile API Code Example

technologies on the desktop. As described in Section 2, the majority of desktop applications operate directly on the file system. By transferring existing file hierarchies to a virtual, sile-based file system and by integrating semantic annotations into applications, the desktop can be extended using semantic technology while avoiding the need for fundamental architecture changes, either for single applications or operating systems.

Vocabularies for Desktop Data. Since the sile model is closely related to the RDF model and uses URIs to identify annotations, it is obvious to reuse existing vocabularies and ontologies from the Semantic Web for interoperability purposes. Although, due to the open design of the sile model, an application can freely define terms, the usage of shared and commonly accepted vocabularies is preferred. Shared vocabularies enable other applications—either on the same machine, or when siles are exchanged across systems—to interpret annotations in a semantically correct way. To establish that kind of interoperability, we propose the following strategy for sile annotation vocabularies:

1. Whenever possible, use terms taken from widely used vocabularies that are published on the Web in a structured, machine-readable format, i.e., RDFS or OWL.
2. If there is no such term that reflects the required semantics, reuse a semantically broader term by establishing e.g., an `rdfs:subPropertyOf` relationship, refine it for the purpose of the target application within a new namespace, and publish it on the Web.
3. If (1) and (2) are not feasible, create a suitable vocabulary and publish it on the Web in order to make it accessible also for other users and applications.

There already exist a number of widely used vocabularies, many of which are applicable for desktop data. Semantic search engines, such as Sindice [14] and Swoogle [5], or index sites for the Semantic Web[4] are good starting points to search for existing vocabularies. In Figure 4 we present a representative set of Semantic Web vocabularies that are relevant for the desktop, grouped by their application domain. For each vocabulary we also indicate their base language as well as the number of concepts and properties they define.

Our analysis indicates that the Semantic Web already provides a large number of vocabularies, which cover a large share of the data we find on typical desktops. Many of the vocabularies we have included in our analysis are compact in terms of the number of classes and properties, and hence are relatively easy to understand and to implement. Moreover, a number of these vocabularies have been defined by re-using terms from other vocabularies. For instance, the *Description of a Project (DOAP)* vocabulary is based on *Friend-of-a-Friend (FOAF)* and therefore each application that understands FOAF is also enabled to interpret DOAP-based data to a certain extent.

[4] E.g., `http://pingthesemanticweb.com/stats/namespaces.php`

Vocabulary Name	Base	Concepts	Props.
General, Documents			
Dublin Core (DC)[a]	RDFS	22	55
NEPOMUK Annotation Ontology (NAO)[b]	RDFS	4	31
NEPOMUK File Ontology (NFO)[c]	RDFS	47	60
Contacts, Communication			
Friend of a Friend (FOAF)[d]	OWL	12	54
Semantically Interlinked Online Communities (SIOC)[e]	OWL	11	53
NEPOMUK Contact Ontology (NCO)[f]	RDFS	30	55
NEPOMUK Message Ontology (NMO)[g]	RDFS	7	23
VCard Ontology[h]	OWL	5	54
Calendar and Events, Project Management			
Description of a Project (DOAP)[i]	RDFS	7	30
RDF Calendar Schema[j]	OWL	14	48
NEPOMUK Calendar Ontology (CAL)[k]	RDFS	51	107
Location			
WGS84 Geo Positioning[l]	RDFS	2	4
GeoNames Ontology[m]	OWL	7	18
Multimedia			
Music Ontology[n]	OWL	53	131

[a] `http://purl.org/dc/terms/`
[b] `http://www.semanticdesktop.org/ontologies/2007/08/15/nao#`
[c] `http://www.semanticdesktop.org/ontologies/2007/03/22/nfo#`
[d] `http://xmlns.com/foaf/0.1/`
[e] `http://rdfs.org/sioc/ns#`
[f] `http://www.semanticdesktop.org/ontologies/2007/03/22/nco#`
[g] `http://www.semanticdesktop.org/ontologies/2007/03/22/nmo#`
[h] `http://www.w3.org/2006/vcard/ns#`
[i] `http://usefulinc.com/ns/doap#`
[j] `http://www.w3.org/2002/12/cal/ical#`
[k] `http://www.semanticdesktop.org/ontologies/2007/04/02/ncal#`
[l] `http://www.w3.org/2003/01/geo/wgs84_pos#`
[m] `http://www.geonames.org/ontology#`
[n] `http://purl.org/ontology/mo/`

Fig. 4. Relevant Semantic Web Vocabularies for the Desktop

5 Implementation Case Study

We have realized a virtual file system on top of our SemDAV semantic repository[5], which implements the sile model and uses a combination of an RDF store (Jena, SDB, PostgreSQL) and plain files to persist siles and their annotations. It exposes the stored siles via a variety of protocols and interfaces (including XML-RPC, RMI, WebDAV, and HTTP).

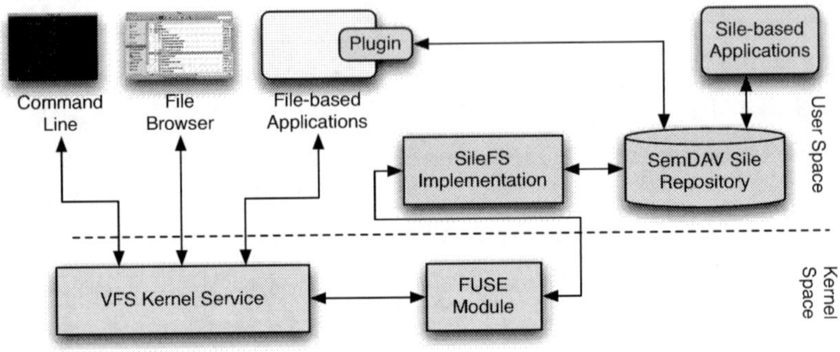

Fig. 5. Virtual File System Architecture

In addition to this repository implementation, we have developed a component that handles file system calls that are forwarded from the local machine's kernel and translates them to corresponding queries and operations on the sile model, according to the file system representation described in Section 3. To integrate the system with the local file system we have used the FUSE framework[6] and its Java binding FUSE-J[7]. FUSE defines a set of interfaces and data structures that describe files, their metadata structures, and operations thereon. Based on these frameworks we simulate a local file system which can be accessed by applications and users as if it were a common file system; but the files can also be annotated and queried through the sile API. Since all files in this virtual file system are persisted as siles in our repository and hence can be accessed only through the sile API or through the virtual file system, data consistency and completeness is ensured at all times. The architecture of this implementation is depicted in Figure 5.

Since file paths are represented as explicit sile annotations, it is straightforward to implement extensions or plug-ins for existing file-based applications: whenever an application operates on a file that is stored on a sile-backed file system, the corresponding sile can be easily retrieved, and vice versa.

6 Performance Evaluation

To evaluate the performance of our approach, we have analyzed the execution times of typical file system operations. To estimate a realistic amount of data, we

[5] `http://www.semdav.org`
[6] `http://fuse.sourceforge.net`
[7] `http://fuse-j.sourceforge.net`

Dataset #	**1**	**2**	**3**
Hierarchy depth	2	3	4
Average no. of sub-directories per directory	5	6	7
Average no. of files per directory	12	15	15
Total number of siles (directories and files)	403	4,144	44,816
Total number of RDF triples	3,626	37,295	403,343
Total number of RDF triples incl. ontologies	4,361	38,030	404,078

Fig. 6. Datasets for Performance Evaluation

crawled the home directories of our department's members, which includes scientific staff (7 persons) as well as technical and administrative staff (3 persons). We used only home directories in favor of scanning entire hard disks because personal data will be the target domain for a semantic file system, and there is little need to semantically annotate system- and application-internal file structures. We discarded files that were on a black list of files and directories that usually are present in users' home directories but are not directly accessed by end users; e.g., `.svn`, `desktop.ini`, and `*.tmp`. The resulting average size of the home directory was 38,000 files stored within 5,150 directories. We view these numbers as upper limits, since we assume that the home directories of computer scientists will typically contain more files (e.g., source code trees) than those of average end users.

To estimate the influence of the size of home directories on our system's performance, we artificially created three test data sets, which are described in Figure 6. To represent basic data about files and directories (cf. Section 2) nine triples per object were created. Note that this does not include any additional descriptive triples (i.e., semantic annotations); these were not considered in our performance evaluation. Our implementation also requires loading a set of core ontologies, which add another $\approx$700 triples to the database.

We have analyzed the runtime performance of typical access patterns to file systems: navigation between directories, listing of directory contents, deletion, moving, and renaming of files. We have carried out the experiments on a high-end consumer notebook (MacBook Pro, Core 2 Duo, with 2 GB RAM) running Mac OS X 10.5 and JVM 1.5. We have used the command shell (`/bin/bash`) to perform our measurements and used only standard commands (`cd`, `ls`, `rm`, and `mv`). Because of our implementation architecture, each operation is processed by a number of external components (e.g., the FUSE kernel module; see also Figure 5) which are not under our direct control. Hence we do not have influence on how shell commands are translated to file system driver calls; for instance, issuing a directory listing command (`ls`) causes the execution of four FUSE calls being passed to our implementation. Nevertheless, our goal was to measure the execution time as experienced by the end user, hence we tracked the total processing time of commands, including overhead caused by the operating system and the FUSE kernel module.

The operations we have evaluated involve read-only access (directory navigation and directory listing) and read+write operations (deletion, moving, renaming). For the latter, the complexity of read and write operations differs: for a sile deletion, (*1*) the triples within the store that describe the object to be deleted have to be identified (read), and (*2*) these triples have to be removed from the store (write). Move and rename operations require in principle the same access operations, whereas a move across directories requires an additional read and write operation, namely the update of the relationship between the file and its parent directory. For our experiment, we have executed each of these operations 10 times in random order, and the entire experiment was repeated five times.

Dataset #	**1**	**2**	**3**
Total number of siles	403	4,144	44,816
`cd`	0.029	0.048	0.107
`rm`	0.063	0.142	0.879
`ls`	0.258	0.464	1.547
`mv` within directory	0.254	0.488	2.488
`mv` across directories	0.296	0.688	3.238

Fig. 7. Evaluation Results: Average Execution Time in Seconds

The results of our experiments are depicted in Figure 7. For the first two datasets ($\approx$400 and $\approx$4,000 siles) we can observe very low execution times, which allow for uninterrupted interactive work with virtual file systems. For a dataset consisting of $\approx$40,000 siles, the response times for simple operations (change directory, remove file) are still in a reasonable range, and even operations that involve multiple, complex queries (directory listing, moving) are within a range comparable to accessing remote file systems via the Web. We did not evaluate the performance of actual read and write operations on the file content: the modifications to metadata caused by these actions are comparable to those of a move operation (i.e., an update of the `content-length` and `update-time` properties), and the actual file content is provided by the underlying file system and hence is out of the scope of our performance measurements.

These numbers indicate that even a prototypical implementation of a virtual file system, based on our data model and built using an off-the-shelf RDF triple store, has acceptable performance for everyday usage on a typical consumer machine. A semantic file system and a more efficient triple store, more tightly integrated into the operating system, could achieve even better performance, since this would allow us to circumvent the rather inefficient architecture that we have chosen for the sake of implementation simplicity.

7 Related Work

The idea of adding semantic technologies to file systems has been widely researched. The benefits of virtual file systems based on semantically enriched data have been discussed and demonstrated in a number of works [4,6,7,8,9,16]. Usually, semantic annotations of files (e.g., tags, ontology classes, or property/value pairs) are mapped to virtual directories which contain the resources that match the criteria. This often implies however that a resource may be accessible through multiple paths, and also may induce problems when files are written to virtual directories, as the mapping between directory names and semantic annotations is often not bijective (e.g., when disjunctions are included in the underlying query). In the case of large datasets and extensive annotations, virtual hierarchies may become very large and therefore difficult to understand and browse by the user. Our approach does not attempt to perform such a mapping; instead we use explicit annotations for directory structures. Hence we preserve full compatibility with existing applications and tools that are implemented under basic assumptions about the structure of file systems (e.g., each resource exists at exactly one place), which is a crucial requirement for applications built on top of file systems; e.g., desktop search engines.

A number of Semantic Desktop projects [2,10,12] aim to provide a semantic infrastructure that covers all applications and information needs that users operate on. These approaches often use file system crawlers to incorporate file systems into the personal information space. As we have described in Section 2, this may cause problems; therefore we consider our work a complementary virtual file system that may be integrated into a Semantic Desktop.

8 Conclusions

We have discussed a number of issues regarding the integration of semantic technologies with file systems, which is a crucial requirement for a successful deployment of Semantic Desktop solutions. First, we showed that the RDF data model exposes a number of characteristics that may cause problems when used in the context of information management on the desktop. To overcome these limitations, we have proposed the sile model, which combines characteristics from both the Semantic Web and file systems. This model provides an integrated view on desktop resources and associated semantic annotations, and it is intended to serve as an intermediate layer between applications and actual storage infrastructure. Second, we have analyzed a representative set of Semantic Web vocabularies and showed that the Semantic Web already provides a sufficient number of suitable vocabularies to be used in the desktop context. Third, we presented our RDF-based implementation of the sile model and a virtual file system that is backed by our system. We have analyzed the performance of typical file system operations under the consideration of realistic amounts of data that can be found on typical users' desktops, and demonstrated that the performance of such a virtual file system is acceptable for interactive usage. We aim to more tightly integrate our approach with common desktop operating systems, from which we expect significant performance improvements.

Acknowledgements. Parts of this work have been funded by FIT-IT grants 812513 and 815133 from the Austrian Federal Ministry of Transport, Innovation, and Technology. The authors thank Stefan Pomajbik, Diman Todorov, and Arash Amiri for support in the implementation of our system, and Stefan Zander and Wolfgang Jochum for their valuable comments on this paper.

References

1. Bizer, C., Cyganiak, R., Heath, T.: How to Publish Linked Data on the Web (2007), `http://www4.wiwiss.fu-berlin.de/bizer/pub/LinkedDataTutorial/` (retrieved December 2, 2008)
2. Blunschi, L., Dittrich, J.-P., Girard, O.R., Karakashian, S.K., Salles, M.A.V.: A Dataspace Odyssey: The iMeMex Personal Dataspace Management System. In: CIDR, pp. 114–119 (2007), `www.crdrdb.org`
3. Boardman, R.: Multiple Hierarchies in User Workspace. In: CHI 2001: CHI 2001 Extended Abstracts on Human Factors in Computing Systems, pp. 403–404. ACM Press, New York (2001)
4. Bowman, C.M., Dharap, C., Baruah, M., Camargo, B., Potti, S.: A File System for Information Management. In: Proceedings of the ISMM International Conference on Intelligent Information Management Systems (March 1994)
5. Ding, L., Finin, T., Joshi, A., Pan, R., Cost, R.S., Peng, Y., Reddivari, P., Doshi, V., Sachs, J.: Swoogle: A Search and Metadata Engine for the Semantic Web. In: CIKM 2004: Proceedings of the Thirteenth ACM International conference on Information and Knowledge Management, pp. 652–659. ACM, New York (2004)
6. Dourish, P., Edwards, W.K., LaMarca, A., Salisbury, M.: Using Properties for Uniform Interaction in the Presto Document System. In: UIST 1999: Proceedings of the 12th Annual ACM symposium on User Interface Software and Technology, pp. 55–64. ACM, New York (1999)
7. Faubel, S., Kuschel, C.: Towards Semantic File System Interfaces. In: Bizer, C., Joshi, A. (eds.) Proceedings of the Poster and Demonstration Session at the 7th International Semantic Web Conference (ISWC 2008), vol. 401, CEUR Workshop Proceedings (2008)
8. Gifford, D.K., Jouvelot, P., Sheldon, M.A., O'Toole Jr., J.W.: Semantic File Systems. In: SOSP 1991: Proceedings of the 13th ACM Symposium on Operating Systems Principles, pp. 16–25. ACM Press, New York (1991)
9. Gopal, B., Manber, U.: Integrating Content-based Access Mechanisms with Hierarchical File Systems. In: OSDI 1999: Proceedings of the 3rd Symposium on Operating Systems Design and Implementation, pp. 265–278. USENIX Association (1999)
10. Groza, T., Handschuh, S., Moeller, K., Grimnes, G., Sauermann, L., Minack, E., Mesnage, C., Jazayeri, M., Reif, G., Gudjonsdottir, R.: The NEPOMUK Project — On the Way to the Social Semantic Desktop. In: Pellegrini, T., Schaffert, S. (eds.) Proceedings of I-Semantics 2007, pp. 201–211. JUCS (2007)
11. Hayes, P.: RDF Semantics (W3C Recommendation 10 February 2004). World Wide Web Consortium (2004)

12. Karger, D.R.: Haystack: Per-User Information Environments Based on Semistructured Data. In: Kaptelinin, V., Czerwinski, M. (eds.) Beyond the Desktop Metaphor, pp. 49–100. Massachusetts Institute of Technology (2007)
13. Muñoz, S., Pérez, J., Gutierrez, C.: Minimal deductive systems for RDF. In: Franconi, E., Kifer, M., May, W. (eds.) ESWC 2007. LNCS, vol. 4519, pp. 53–67. Springer, Heidelberg (2007)
14. Oren, E., Delbru, R., Catasta, M., Cyganiak, R., Stenzhorn, H., Tummarello, G.: Sindice.com — A Document-oriented Lookup Index for Open Linked Data. Int. J. Metadata Semant. Ontologies 3(1), 37–52 (2008)
15. Sauermann, L., Heim, D.: Evaluating Long-Term Use of the Gnowsis Semantic Desktop for PIM. In: The Semantic Web — ISWC 2008. LNCS, vol. 5318, pp. 467–482. Springer, Heidelberg (2008)
16. Schenk, S., Görlitz, O., Staab, S.: TagFS: Bringing Semantic Metadata to the Filesystem. In: 3rd European Semantic Web Conference (ESWC). (2006)
17. Schmedding, F., Hanke, C., Hornung, T.: RDF Authoring in Wikis. In: Lange, C., Schaffert, S., Skaf-Molli, H., Völkel, M. (eds.) Proceedings of the 3rd Semantic Wiki Workshop (SemWiki 2008). CEUR Workshop Proceedings, vol. 360. CEUR-WS.org (2008)
18. Sintek, M., van Elst, L., Scerri, S., Handschuh, S.: Distributed Knowledge Representation on the Social Semantic Desktop: Named Graphs, Views and Roles in NRL. In: Franconi, E., Kifer, M., May, W. (eds.) ESWC 2007. LNCS, vol. 4519, pp. 594–608. Springer, Heidelberg (2007)
19. Sintek, M., van Elst, L., Scerri, S., Handschuh, S.: NEPOMUK Representational Language Specification. Technical report, NEPOMUK Project Consortium (2007), `http://www.semanticdesktop.org/ontologies/nrl` (retrieved December 2, 2008)

Who the Heck Is the Father of Bob?

A Survey of the OWL Reasoning Infrastructure for Expressive Real-World Applications

Marko Luther[1], Thorsten Liebig[2], Sebastian Böhm[1], and Olaf Noppens[2]

[1] DOCOMO Communications Laboratory Europe GmbH, Munich, Germany
lastname@docomolab-euro.com
[2] Inst. of AI, Ulm University, Ulm, Germany
firstname.lastname@uni-ulm.de

Abstract. Finding the optimal selection of an OWL reasoner and service interface for a specific ontology-based application is challenging. Over time it has become more and more difficult to match application requirements with service offerings from available reasoning engines, in particular with recent optimizations for certain reasoning services and new reasoning algorithms for different fragments of OWL. This work is motivated by real-world experiences and reports about interesting findings in the course of developing an ontology-based application. Benchmarking outcomes of several reasoning engines are discussed – especially with respect to accompanying sound and completeness tests. We compare the performance of various service and communication protocols in different computing environments. Hereby, it becomes apparent that these largely underrated components may have an enormous impact on the overall performance.

1 Introduction

In the recent past, the application of Semantic Web technologies has seen a steady increase in a broad variety of services and applications. On the foundation of W3C's Web Ontology Language (OWL) [1], Semantic Technologies are meant to be soon adopted by the industry. In the course of developing applications that rely on ontology based reasoning mechanisms, not only an appropriate reasoning engine needs to be selected. In fact, finding the right combination of reasoner and communication interface for expressive ontologies with the required level of expressivity is crucial for the overall reasoning performance. Existing benchmarks promise to shed some light on the performance of popular inference engines, at least for some selected ontologies and thus particular reasoning tasks. Our own experience has shown that most of the benchmark results are not as helpful as desired. When building an ontology based application, designers typically have to deal with many different requirements and constraints on the query interface, ontology updates, reliability of results, etc. To utilize ontology reasoning for applications in the mobile space, we first tried to identify an appropriate reasoning infrastructure by applying standard benchmarks as well as spot tests

L. Aroyo et al. (Eds.): ESWC 2009, LNCS 5554, pp. 66–80, 2009.

for different ontologies. In a second step, concrete reasoning scenarios for our application with specifically designed ontologies have been evaluated. In this paper, our experiences in the development are discussed to help others in realizing an appropriate reasoning infrastructure.

The remainder of this work is organized as follows: Section 2 gives an overview of existing reasoning engines and discusses some benchmarking results of the former. In the subsequent Section 3, a set of tests gained from developing real-world applications are performed. Concluding remarks can be found in Section 4.

2 OWL Reasoning Infrastructure

The core element of any OWL infrastructure consists of a reasoning engine capable of processing OWL ontologies at a certain level of expressivity. The typical processing services cover consistency and entailment checks as well as any kind of query answering. Important criteria of these services are soundness resp. completeness and practical efficiency. It is worth noting that both criteria are of equal importance and inherently depend on each other. For instance, efficiency can easily be achieved by giving up sound- and completeness. On the other hand, practical efficiency in a real-world usage scenario is without doubt an important requirement for semantic applications, even if reasoning with OWL is known to be unfeasible in the worst case. Nevertheless, highly optimized implementations of well-chosen calculi have empirically proven that real-world applications are possible. Besides, the variety of profiles of OWL 2 provides a selection of language fragments with different characteristics for various needs and ontology sizes.

Finding the appropriate reasoning approach resp. system for a given expressivity and problem size is all but obvious. Certain approaches perform well only with specific language fragments and the performance of reasoning engines may vary significantly from case to case because of subtle optimization techniques. However, when it comes to large volumes of ontology data, incremental reasoning or answer caching may be more important than any other optimization technique. Besides all these partly conflicting requirements, an application designer needs to know whether or not the chosen reasoning engine is mature and reliable enough for a particular setting. In other words, if a system is not sound (for whatever reason), it is practically useless for the vast majority of applications.

This sections summarizes empirical results from currently available OWL reasoning systems with respect to the above mentioned issues. Our findings only draw a very fragmentary picture, even though valuable insights can still be given. Our survey is based on experiences with a benchmark suite for large ABox data on the one hand, and a selection of small but difficult T- and ABox test cases on the other hand. This work updates and extends previous analyses with respect to hard spot tests [2] as well as huge volumes of instance data [3].

Our evaluation shows that even with only one benchmark suite, it is hard to draw a reliable conclusion. The runtime of systems heavily varies from case to case and is utterly unpredictable. We even found out that the benchmark suite itself was incorrect with respect to the official number of results. On the opposite end of testing, namely the small but hard test cases, the results are likewise disappointing. Almost all reasoning systems failed at least for one of the cases.

2.1 Benchmarking Scalability

In order to measure the performance of the given systems with respect to large volumes of instance data we have chosen the University Ontology Benchmark (UOBM) [4]. The UOBM extends the well-known Lehigh University Benchmark (LUBM) [5] by adding extra TBox axioms making use of all of OWL Lite (UOBM Lite) and OWL DL (UOBM DL). The ABox is enriched by interrelations between individuals of formerly separated units, which makes it less artificial and more realistic. Please note that our intention is not to simply run this benchmark again. Instead, our main interest is to analyze different settings and outcomes.

The set of reasoning engines within our survey consists of FaCT++ (v1.2.3)[1] [6], KAON2 (2008-06-29)[2], Pellet (v2.0.0RC5)[3] [7] and RacerPro (1.9.3b)[4] [8]. We have used the latest available versions of these systems except for RacerPro, for which we had access to a soon to be released beta version. Hermit (v0.9.3)[5] [9] was initially also within the set of potential candidates but repeatedly produced exceptions while processing the UOBM. A feature matrix of selected system characteristics can be found in Table 2. For all tests in this section we used 64bit versions of the systems, which were running on a 16GB quad-core Linux machine.

Since FaCT++ does not come with a SPARQL or any other conjunctive query language, the set of test systems with respect to query answering only consists of Pellet, RacerPro and KAON2. However, KAON2 is known for its performance decrease in the presence of cardinality restrictions. In fact, the KAON2 engine was two and more magnitudes slower than the other systems. In addition more than 10 GB of main memory were allocated even for the smallest UOBM data set (lite-1). A comparison of KAON2 with the SHER[6] system for the UOBM without cardinality restrictions can be found in [10].

We tested the systems with their standard settings for optimized but complete query answering (e.g. nRQL mode 3) and a warm-up phase of three queries disjoint from the set of benchmark queries. In our comparison, we just computed the number of answers without retrieving them to get measures which are independent of the utilized interface. The systems do not vary dramatically in loading the data. The smallest data set (lite-1) was loaded into KAON2 in 12 seconds, in Pellet in 20 seconds, whereas RacerPro required about 30 seconds. The setup

[1] `http://owl.man.ac.uk/factplusplus`
[2] `http://kaon2.semanticweb.org`
[3] `http://clarkparsia.com/pellet`
[4] `http://www.racer-systems.com`
[5] `http://www.hermit-reasoner.com`
[6] `http://www.alphaworks.ibm.com/tech/sher`

and query preparation time for RacerPro was about 85 seconds compared to Pellet with 9 seconds. Computing the query results for lite-1 was accomplished in less than a second for both Pellet and RacerPro, except for two queries. Pellet needed more than 100 resp. 500 seconds for the queries number 8 and 13 whereas RacerPro required more than 4 resp. 8 seconds for the queries number 2 and 9. Altogether RacerPro completed the whole lite-1 test in roughly 100, Pellet in 680 seconds.

Overall memory consumption was lowest for RacerPro with not more than 3GB for lite-1. Pellet requires up to 4GB of main memory for the same test set. When classifying the TBox before querying, the timings do not vary significantly. The only exception is query number 13 with Pellet only requiring 270 (instead of 550) seconds.

In case of realizing the ABox (as well as classifying the TBox) before querying, it turned out that RacerPro performs slightly better for almost all queries but required more than 16 instead of 8 seconds for query 9. Likewise, Pellet performed dramatically better for three queries (query 2, 8, and 13) with a slight performance decrease for query 4. Realizing the ABox took about 10 minutes for RacerPro and more than one hour for Pellet. We did not succeed in realizing the smallest UOBM lite data set with FaCT++. The system aborted after several minutes, with having allocated all of the 16GB main memory. Interestingly, we were able to realize the small and mid-size UOBM dl data sets (realizing the mid-size data set consumed less than 2.4GB main memory). It seems that the reasoning procedure of FaCT++ for less expressive languages is not as optimized as for the expressive ones. In conclusion, realizing the ABox helps to increase performance for some queries. However, the influence on query answering is somehow unpredictable and due to the computational overhead it is not feasible for larger data sets and was only performed here for comparison.

Interestingly, the number of results for two out of 13 UOBM queries for lite-1 with RacerPro (nRQL mode 3) were different compared to the number of results returned by Pellet as well as the official numbers. For query 10, RacerPro came up with fewer, for query 9 with more answers. When using nRQL's incomplete mode 1 or 2 even eight answers were different, but the performance increased (all queries were processed in less than a second except for query 9).

In a different case, namely query 11 of the dl-5 test set, the official number of query results is obviously wrong. Pellet as well as RacerPro provide 6230 answers here, whereas the UOBM only states 6225 correct results. However, when analyzing the five additional answers it can be easily seen that they meet the query condition. More precisely, all results are correct because of their equivalence to other individuals in the result set. This is due to an individual merge caused by the functional object property "isTaughtBy". Pellet and RacerPro also disagree with the official result on query 15 (also dl-5) with 72 individuals. Both reasoner, however, return no answer at all.

Based on our findings we conclude that it is not only difficult to write expressive and scalable reasoning engines but also to generate sophisticated benchmark suites. More than that, conducting benchmarks requires a careful interpretation of the results from different view points and for different requirements.

Table 1. Correctness Spot Tests

no.	expressivity	C	P	D	I	FaCT	Racer	Pellet	HermiT	KAON2
1b	$\mathcal{SHIN}$	6	7	0	0	+	time	time	+	time
2a	$\mathcal{SHN}$	8	9	0	0	+	+	+	time	time
2b	$\mathcal{SHN}$	8	9	0	0	+	+	time	time	time
9	$\mathcal{SHF}$	7	7	0	0	-	+	+	+	+
28	$\mathcal{ALC}$	33	7	0	0	+	+	+	time	+
29b	$\mathcal{SHIF}$	12	6	0	0	-	+	+	+	+
30	$\mathcal{SHOIF}$	23	6	0	41	+	(+)	error[7]	+	n.a.
35	$\mathcal{ALCOI}$	17	7	0	9	memory	(+)	+	+	n.a.

2.2 Sound- and Completeness Spot Tests

Reasoning with OWL Lite as well as OWL DL is known to be of high worst-case complexity. However, trading soundness/completeness for efficiency is not a solution, in our opinion. Instead, application designers should carefully choose their language fragment and size of ontology in order to be able to pick the right approach resp. system. Implementors of reasoning engines should, on the other hand, take special care about creating a reliable system in terms of soundness and completeness (which is, for sure, a formidable challenge). Otherwise, performance results of different systems are not comparable at all. Even if a bug-free reasoning system may never be available, our experiences in intensively using various engines have revealed a number of disappointing results. Some systems fail to answer even small or trivial test cases, some failures tend to re-appear in subsequent system versions, or termination varies not only from system to system but also from release to release.

As a consequence, we tested our systems with an empirical evaluation using spot tests which are intentionally designed to be hard to solve but small in size. They try to meter the correctness of the reasoning engines with respect to inference problems of selected language features and were published two years ago [2]. The tests were conducted with the systems listed in Table 2. We used FaCT, Pellet, Hermit via the OWL-API, whereas KAON2 and RacerPro were tested via their native interfaces for loading OWL files. Surprisingly, almost all reasoners failed in some test cases by providing either wrong results or were not able to find a solution within 10 minutes. Table 1 shows a selection of these test cases, all ran on a Linux maschine (ubuntu 2.6.17, 16GB ram, Java 1.6, gcc 4.0.3):

In the first column, each test case has been assigned an identifier.[8] The subsequent columns list the expressivity, the number of classes (C), object properties (P), datatype properties (P), and individuals (I) of the respective test case. A

[7] Varies with respect to platform (Linux resp. MacOS) and Java version (1.5 resp 1.6) from time-out to an error caused by an exception.

[8] The corresponding OWL file can be found at `http://www.informatik.uni-ulm.de/ki/Liebig/reasoner-eval/no.owl`

Table 2. Reasoner Feature Matrix

	OWL-API	Jena	DIG	OWLlink	native API	language support	retraction	incr. reas.	SWRL supp.	query language	query entailm.	available license	impl. lang.
FaCT++ 1.2.3	X		X		KRSS like	OWL 2 $\mathcal{SROIQ}$(D)						LGPL v2.1	C++
Pellet 2.0.0RC5	X	X	X			OWL 2 $\mathcal{SROIQ}$(D)	X	X	X	SPARQL		AGPL v3 & Commercial	Java
RacerPro 1.9.3b	X		X	X	KRSS like	OWL 1 $\mathcal{SHIQ(D)}$	X		X	nRQL	X	Academic & Commercial	Lisp
KAON2 2008-6-29			X		KAON	OWL 1 $\mathcal{SHIQ(D)}$			X	SPARQL		Academic & Commercial	Java
HermiT 0.9.3	X				KAON	OWL 2 $\mathcal{SHOIQ(D)}$						GPL	Java

"+" for a specific system means that the system could provide the correct answer (typically satisfiability/unsatisfiability of a class), whereas round brackets denote that the result was computed on the basis of approximate reasoning. Furthermore, "time" indicates that the system could not provide an answer within ten minutes. Since our test environment provided more than 16GB of main memory an "out of memory" error did only occur once within our time frame of 10 minutes. A "-" signals that the system returned the wrong result. To identify potential modeling patterns that are supposed to slow down reasoning, we fed all the test cases into Pellint[9] v0.2 [11]. Hereby, no performance barriers have been reported.

It is worth noting that the results of our spot test often change from version to version for FaCT++ and Pellet and even from platform to platform resp. environment (operating system and/or Java version). For instance, previous versions were able to solve 1b (Pellet) or 2b (FaCT++) but did not terminate or failed with others. In addition, FaCT++ v. 1.2.1 on Mac OS X (gcc 4.0.1) succeeds for test case 2b but fails for 1b. With version 1.2.0 it is the other way round. The newest Pellet release candidate (2.0.0RC5) is able to solve test case 28 within seconds whereas previous versions did not. Interestingly, various versions of Pellet randomly abort with an exception on test case 30 (whereas Pellet 1.4 produces a time-out). Test case 30 is the only one using all of OWL DL and cannot be handled by KAON2 and only in an approximative way by RacerPro. Again, the MacOS X version of FaCT++ does not terminate on test case 30.

Please note that all test cases above are within OWL 1 and even only a fraction of OWL Lite in some cases. The upcoming OWL 2 adds further language constructs which makes reasoning even more difficult. Extrapolating the given

[9] `http://pellet.owldl.com/pellint`

results indicates that reliable OWL 2 reasoning engines might still need some more time to mature.

3 Reasoning in Practice

IYOUIT[10] [12], a mobile community service in the field of context-awareness makes use of formal representations to reason about gathered data. IYOUIT allows people equipped with an ordinary mobile handset like the gPhone to instantly share personal experiences with others while on the go. The mobile application is connected to a network of components on the Internet. All data gathered by the mobile handset, such as the current location or nearby Bluetooth beacons, are sent to the appropriate component, which stores and further processes this information. One of the main tasks of each component is to abstract from low-level sensor data (e.g., location traces) to corresponding qualitative representations (e.g., important places of a user). Once theses abstractions are linked to concepts formalized in OWL ontologies, reasoning is applied to derive additional information. In the following, two of these reasoning use cases are highlighted, both with different requirements on the expressiveness of the underlying ontology and temporal constraints on the overall reasoning process.

All relationships in the IYOUIT user community are represented within a social ontology to allow for expressing social relationships between users. Ontology based reasoning is applied to maintain the consistency of the social network and to make implicit relations explicitly available. The structure of the complemented social network is in turn used for privacy control.

We performed several tests to find the appropriate combination of reasoning engine and ontology modeling style that complies with the requirements on correctness as well as performance. To ensure comparable results, we connected all reasoning engines via the standard Java-based OWL API[11] [13] using the corresponding connector.[12] In addition, we analyzed the performance of the HTTP-based DIG protocol via the OWL API DIG connector, OWLlink[13] [14] and RacerPro's native TCP interface. All tests have been performed with an Apple MacBook Pro 2,33 GHz Intel Core 2 Duo notebook computer with 3GB of memory running Mac OS X 10.5.5 with Java 5.0 32bit and Java 6.0 64bit, respectively. The reasoning engines were initialized with their default configuration and the Java Virtual Machine has been configured as follows: The maximum stack size for each thread has been set to 4MB, the minimum heap space to 30MB and the maximum heap space has been limited to 200MB. Before the actual measurements have been recorded, five consecutive runs were applied to warmup the Java hotspot compiler and to initialize the reasoner. In total, 100

[10] `http://www.iyouit.eu`

[11] We used the SVN version of the OWL API from `http://owlapi.sourceforge.net` as of 9.12.2008 for all experiments.

[12] Reasoners analyzed in Section 2 that do not implement all required functions of the OWL API interface yet, were not considered in the following tests.

[13] `http://www.owllink.org`

runs were performed to average out minimum and maximum measurements and to simulate the actual real-world setting as part of a server infrastructure.

3.1 Social Ontology

To model qualitative social relationships we developed a social ontology for the IYOUIT system. Formalized in OWL-Lite, a fragment of OWL 1, it fulfills the requirements on expressivity (in modeling the social network) and decidability. Complying with the Description Logic $\mathcal{SHIF}$, the social ontology makes use of an object-property hierarchy with transitive, inverse and functional properties (cf. Figure 1). Social relationships are represented as instantiations of those object-properties, whereas entities that represent IYOUIT users are formulated as instances of the class `Person`. In total, the social ontology contains, 6 primitive concepts, 1 disjoint class axiom, 20 object properties and 110 individuals with 280 object property assertion axioms.

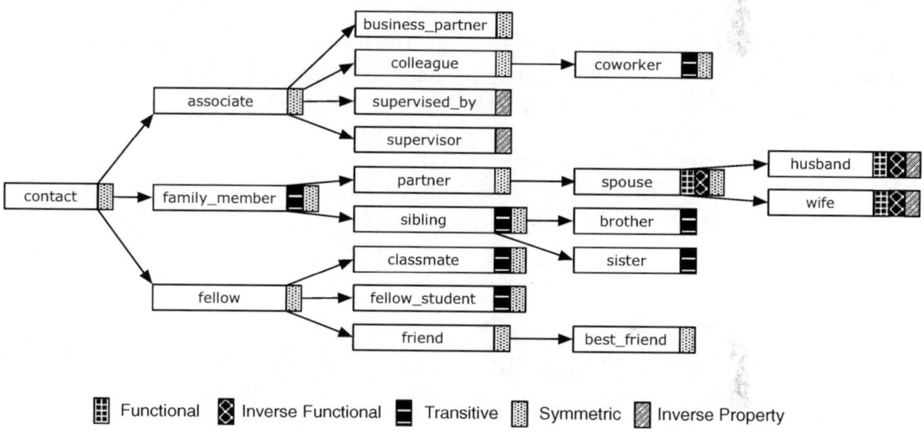

Fig. 1. Social Object Properties

We initially defined a set of test cases to verify that our social ontology captures the intended semantics.[14] The first two queries, encoded in Test A and Test B, analyze the inferred object property structure by requesting the direct and the full set of sub-properties of the property `contact`, respectively. Please note that for Test A and B an earlier and slightly larger version of the social ontology with in total 74 object properties has been used. The last query, Test C, requests the total set of object property relationships of one specific individual. Here, the optimized social ontology with its 20 object properties has been used (cf. Figure 1). This ontology and the respective query can also be found in the actual real-world application within IYOUIT.

[14] The ontologies for all tests in this section can be found at `http://www.informatik.uni-ulm.de/ki/Liebig/reasoner-eval/IYOUIT.zip`

Interestingly, different reasoning engines returned different results for our queries. Table 3 lists the number of results returned by different engines for all 3 test cases. To the best of our knowledge, only the results returned by RacerPro v1.9.3b are correct, whereas all other results are wrong, which has also been confirmed by the respective developers. Earlier versions of RacerPro returned wrong results as well, similar to FaCT++ for versions in between 1.1.9–1.1.11 that returned the same wrong results as the latest FaCT++ version 1.2.1. Version 1.2.0 is the only version of FaCT++ that provided the correct results for case A and B (case C could not be tested via the OWL API since the required interface function is only supported in v1.2.1). Pellet version 1.5.0 and 1.5.1 suffer from a typo in the OWLAPI interface implementation and return only the direct sub-properties in Test case B. For Pellet 2.0.0RC1, the result set for Test A and B actually depends on the JavaVM version. Here, the number of results indicated before the slash are obtained with Java v1.5, those after the slash with v1.6.

In summary, we were astound by the number of issues (sometimes reoccurring in later versions after having been fixed before) produced by our relatively simple ontology. It seems that the use of inverse properties accounts for the complications in the reasoning kernel in most cases, while other problems were just caused by implementation errors in the corresponding interface functions.

Table 3. Object Property Inferences

	Test A	Test B	Test C
FaCT++ 1.2.1	12	50	53
Pellet 1.4	10	10	38
Pellet 2.0.0RC1	10/9	44/46	71
Pellet 1.5.2 & 2.0.0RC3	10	50	71
RacerPro 1.9.3b	10	50	72

3.2 Object Property Reasoning

Social reasoning mechanisms are applied to complement the social network of IYOUIT users and to maintain the consistency of explicitly given social relations. To find the most appropriate reasoner for the social network reasoning, four different reasoning engines, namely RacerPro 1.9.3 beta, Pellet 1.5.2, the latest Pellet release candidate version 2.0.0RC3 and FaCT++ 1.2.1, were compared via the OWL API. To measure the practical reasoning performance and to trigger the actual classification process, an additional relationship property axiom has been added to the already defined 280 object property assertion axioms of the social ontology. Next, all relationships were retrieved from the reasoner and finally, the initially added relationship has been removed again. While adding and removing axioms has been performed very fast by all reasoning engines, significant differences were observed in retrieving the actual inference result. On the left-hand side of Figure 2, the results of this retrieval process are shown in more detail. With an average of 9,3 milliseconds, FaCT++ has been the

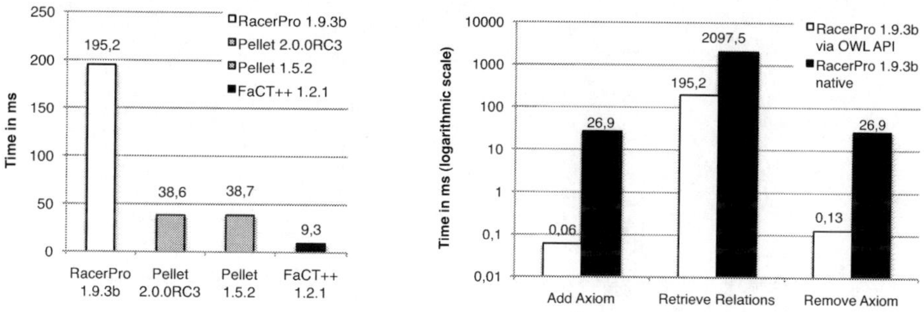

Fig. 2. Object Property Reasoning Performance

fastest engine, followed by Pellet with 38,6 milliseconds. Here, no significant performance difference was found when comparing Pellet 1.5.2 with the latest version 2.0.0RC3. With 195,2 milliseconds, RacerPro 1.9.3b took considerably longer to answer the given query. Performance aside, RacerPro has been the only system that returned the correct result (1620 relationships). Pellet was missing exactly one relationship and FaCT++ only returned 1440 relationships (cf. Section 3.1).[15] As a result, a concluding performance assessment cannot be made as long as not all reasoning engines do return the correct result.

In a second test, we compared two currently available interfaces for RacerPro, the OWL API and its native JRacer interface, both connecting the reasoner and the Java application via TCP. Again, adding and removing axioms took considerably less time than retrieving the inference result. To our surprise, the OWL API connection performed dramatically better than the native connection for both, the inference retrieval as well as the axiom retraction. As summarized on the right-hand side of Figure 2, transferring all relationships from RacerPro was finished within 195,2 milliseconds in case of the OWL API and 2097,5 milliseconds in case of the native JRacer connection. One feasible explanation, which has also been confirmed by the implementors of RacerPro, could be the fact that the RacerPro OWL API connector is based on an optimized TCP protocol, while the JRacer connection still relies on an older and thus un-optimized version of the latter.

3.3 Dynamic Classification

The situation estimation component [15] of the IYOUIT system constantly retrieves abstractions of the latest sensor data produced by the IYOUIT mobile client such as the actual place of a user, a qualitative description of the current time and the people in proximity. Its task is to compute an abstract description of a user's situation based on live context data and static profile information, including all established qualitative social relationships.

[15] In its latest release candidates RC4 and RC5, Pellet now also returns the correct result. Two newer versions of Fact++, however, do not include a respective fix.

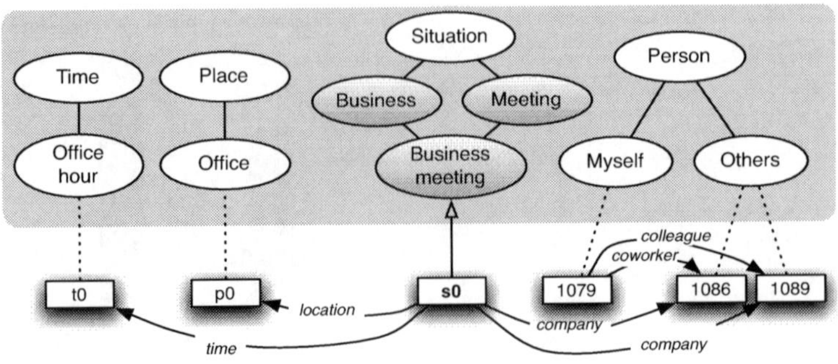

Fig. 3. Situation Description

The set of OWL DL ontologies used to realize this situational reasoning has been refined step by step. The initial version, in the following referred to as Ontology C, is within $\mathcal{SHOIN}$ and consists of 270 (cyclic) classes, 7 CGIs, 135 object properties, 65 datatype properties and 333 individuals. It is distributed over 9 component ontologies including the social ontology (without its individuals and object property assertion axioms) discussed above and the standard time-entry ontology [16]. The second version of the ontology, Ontology B, was derived from Ontology C by removing all remaining individuals and nominals, breaking the cycles between classes, reducing the number of domain/range restrictions, datatype properties and inverse functional declarations, adding additional classes and disjoint class axioms, replacing the time-entry ontology by a simpler time ontology and reducing the number of components. This results in 324 classes, 0 CGIs, 116 object properties, 43 datatype properties and 0 individuals distributed over 4 component ontologies, reducing the complexity to $\mathcal{SHIN}$. The latest ontology, Ontology A, was derived from Ontology B by reducing the number of object properties, classes and disjointness axioms, by removing all datatype properties, most of the remaining domain/range restrictions and all annotations, resulting in 137 classes, 0 CGIs, 24 object properties, 0 datatype properties and 0 individuals, distributed over 4 component ontologies in $\mathcal{SHIN}$.

Once the ontologies are loaded into the reasoner and all qualitative data used to characterize a users situation is gathered from the IYOUIT network, the data is translated into an OWL situation description, represented as individuals of the corresponding classes within the ontology (cf. Figure 3). A situation individual `s` of class `Situation` is associated with individuals representing the abstract time and location of the user, as well as the buddies detected in proximity using object property axioms. Additionally, object property axioms are defined between the considered user (a member of the `Myself` class) and the detected nearby buddies corresponding to the established social relations as discussed above. Based on the axiomatization of the subclasses of `Situation`, the derived situation of `s` can be retrieved by asking the reasoning engine for its direct types. In a final clean-up step, the situation individual `s` is again removed from the ontology, together with all associated individuals and object property axioms.

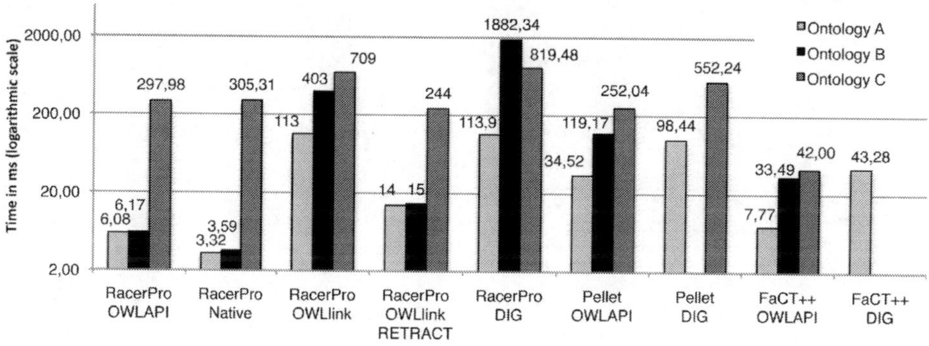

Fig. 4. Interface Performance

In case of the situational reasoning, our main interest is to compare the effect of the utilized interface between the reasoner and the application, since the actual inference problem is rather simple. Comparing the performance of one classification task formulated with ontologies of varying size and complexity, we examine the influence of the ontology modeling and the effect of the potential use of (large) standard ontologies. Furthermore, we analyze the effect of retraction and incremental reasoning by defining one distinct classification task in the following three steps: a) establishing a situation description by adding the corresponding axioms; b) requesting the classification result; c) removing the axioms describing the situation again.

Figure 4 summaries the results we obtained with the latest available versions of RacerPro (v1.9.3b), FaCT++ (v1.2.1) and Pellet (v2.0.0RC3). The reasoners were connected via the OWL API, via DIG over HTTP using the corresponding DIG OWL API connector, and, for RacerPro, additionally via the native JRacer connection over TCP and the OWLlink interface over HTTP. While Pellet and FaCT++ run within the same memory partition with the OWL API interface being directly connected via the corresponding connector (in case of FaCT++ using its JNI interface), the RacerPro OWL API connector accesses RacerPro via its native TCP interface.

Not surprisingly, the actual performance largely depends on the complexity of the underlying ontology. However, the effect of the selected ontology seems to influence the performance of RacerPro differently than the other reasoners. RacerPro, connected via TCP and with support for retraction, outperformed the in-memory connected reasoners Pellet and FaCT++ for the less complex and non-cyclic ontologies. The native RacerPro connection, without the additional overhead caused by the OWL API, showed an even better performance. However, in earlier experiments we discovered that predecessors of Racer v1.9.1 performed slower from run to run, an issue that seems to have been solved in later versions. The DIG interface fails to properly transmit the ontology to the reasoner as some constructs (like cardinality restrictions on datatype properties) are not supported

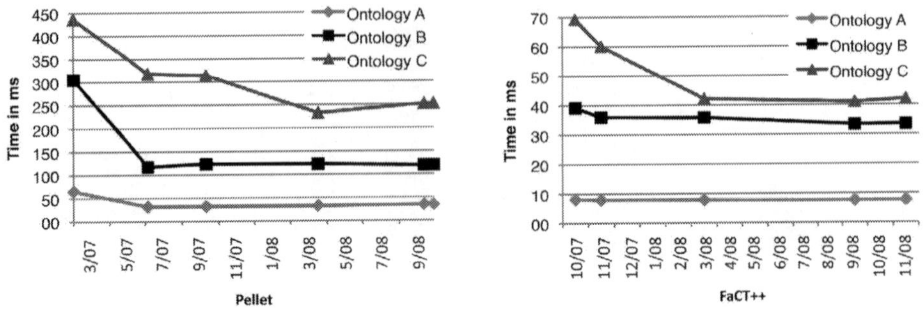

Fig. 5. Reasoner Development

by this protocol.[16] This results in cyclic knowledge bases for all three variants of the ontologies, a bad overall performance, in OutOfMemory errors for Pellet in some cases and DIGReasonerExceptions for FaCT++. OWLlink, the successor of DIG, allows to transmit all of OWL 2 and supports retraction. OWLlink via HTTP using retraction outperforms even Pellet's in-memory connection. On the positive side, different versions of Pellet and FaCT++ connected via the OWL API (cf. Figure 5) show a steady improvement of performance during the last 1.5 years, especially for the more complex versions of the ontology.

4 Conclusions

By applying Semantic Technologies in real-world applications, complex facts can be described by a well-defined model instead of writing application specific code. A major advantage of this approach is the clear separation of the application logic from the underlying model, which (in principle) reduces the risk of system failures. However, choosing the appropriate combination of a reasoning engine, a communication interface and expressivity of the utilized ontology is an underestimated complex and time-consuming task. To evaluate the performance of the selected components, appropriate test cases seem to be inevitable. However, what is known for being a good solution in one case, might as well be an inappropriate solution in another case. Standard test environments such as the one introduced by Gardiner et al. [17] point in the right direction as they allow for specifying and verifying the expected results. However, this approach is based on DIG, and therefore fails to transfer the entire set of OWL axioms to the reasoning engine. Isolating condensed test cases when either dealing with huge knowledge bases or in case the correct result is not known beforehand is a challenging issue in practice. Here, the recent advancement of the OWL API is most welcome, since it makes comparing different reasoning engines and ontologies far easier without having to adapt the core application logic. Likewise, OWLlink,

[16] Furthermore, the latest OWL API version translated exact cardinality restrictions into invalid DIG syntax. An issue we fixed before running the experiments.

the recently introduced successor of DIG, offers an implementation-neutral and extensible protocol for the communication with diverse OWL reasoning systems (also supporting retraction). Yet, deploying applications as part of a Web-service based server infrastructure, native reasoner connections via TCP can still have an advantage compared to in-memory solutions. Here, the actual application is not affected in case the reasoning engine might not be responsive any more.

Another noticeable trend in recent years is concerned with the improvement of implemented optimization strategies of most reasoning engines. However, we were wondering why the addition of incremental consistency checking under syntactic Abox updates in Pellet v1.5.0 [18] and the newly introduced incremental classifier of Pellet 2.0.0RC1 [19] did no have a larger impact on the performance of our incremental situation classification task. At least for some cases, these optimizations result in more efficient reasoning tasks. Yet, we found that for relatively simple reasoning use cases in which the actual classification takes only a minimal amount of time, the inference retrieval is by far the most expensive task in the whole process. Part of the problem lies in the fact that there is no appropriate interface that would allow for retrieving only inferred axioms from the reasoning engine. As a result, the application itself needs to retrieve the entire inference result, including all told axioms, to compute only newly inferred axioms. To maintain correct inference results and to ensure efficient reasoning tasks, only lightweight, specifically designed ontologies were considered. Utilizing standard ontologies such as DOLCE [20] or OpenCyc[17] to profit from a well-defined cross domain vocabulary concurrently increases complexity with respect to the inference traceability and might as well have a negative impact on the overall reasoning performance since reasoning in OWL is not modular.

References

1. Grau, B.C., Horrocks, I., Motik, B., Parsia, B., Patel-Schneider, P., Sattler, U.: OWL 2: The next step for OWL. In: Web Semantics: Science, Services and Agents on the World Wide Web, vol. 6, pp. 309–322 (2008)
2. Liebig, T.: Reasoning with OWL: System Support and Insights. Technical Report TR-2006-04, Ulm University, Germany (2006)
3. Weithöner, T., Liebig, T., Luther, M., Böhm, S., von Henke, F.W., Noppens, O.: Real-world reasoning with OWL. In: Franconi, E., Kifer, M., May, W. (eds.) ESWC 2007. LNCS, vol. 4519, pp. 296–310. Springer, Heidelberg (2007)
4. Ma, L., Yang, Y., Qiu, Z., Xie, G., Pan, Y., Liu, S.: Towards a Complete OWL Ontology Benchmark. In: Sure, Y., Domingue, J. (eds.) ESWC 2006. LNCS, vol. 4011, pp. 125–139. Springer, Heidelberg (2006)
5. Guo, Y., Pan, Z., Heflin, J.: An Evaluation of Knowledge Base Systems for Large OWL Datasets. In: McIlraith, S.A., Plexousakis, D., van Harmelen, F. (eds.) ISWC 2004. LNCS, vol. 3298, pp. 274–288. Springer, Heidelberg (2004)
6. Horrocks, I.: Using an expressive description logic: FaCT or fiction? In: Proc. of the 6th Int. Conf. on Principles of Knowledge Representation and Reasoning (KR 1998), June 1998, pp. 636–647 (1998)

[17] `http://www.opencyc.org`

7. Sirin, E., Parsia, B., Grau, B.C., Kalyanpur, A., Katz, Y.: Pellet: A practical OWL-DL reasoner. Journal of Web Semantics 5, 51–53 (2007)
8. Haarslev, V., Möller, R.: Description of the RACER system and its applications. In: Proc. of the Int. Workshop on Description Logics (2001)
9. Shearer, R., Motik, B., Horrocks, I.: HermiT: A Highly-Efficient OWL Reasoner. In: Proc. of the OWL Experiences and Directions Workshop at the ISWC 2008 (2008)
10. Dolby, J., Fokoue, A., Kalyanpur, A., Ma, L., Schonberg, E., Srinivas, K., Sun, X.: Scalable conjunctive query evaluation over large and expressive knowledge bases. Technical report, IBM Watson Research Center (2008)
11. Lin, H., Sirin, E.: Pellint – A Performance Lint Tool for Pellet. In: Proc. of the OWL Experiences and Directions Workshop at the ISWC 2008 (2008)
12. Böhm, S., Koolwaaij, J., Luther, M., Souville, B., Wagner, M., Wibbels, M.: Introducing IYOUIT. In: Int. Semantic Web Conf. (ISWC 2008). LNCS, vol. 5318, pp. 804–817. Springer, Heidelberg (2008)
13. Horridge, M., Bechhofer, S., Noppens, O.: The OWL API. In: Proc. of the 3rd OWL Experiences and Directions Workshop at the ESWC 2007 (2007)
14. Liebig, T., Luther, M., Noppens, O., Rodriguez, M., Calvanese, D., Wessel, M., Horridge, M., Bechhofer, S., Tsarkov, D., Sirin, E.: OWLlink: DIG for OWL 2. In: Proc. of the OWL Experiences and Directions Workshop at the ISWC 2008 (2008)
15. Luther, M., Fukazawa, Y., Wagner, M., Kurakake, S.: Situational reasoning for task-oriented mobile service recommendation. The Knowledge Engineering Review 23, 7–19 (2008)
16. Pan, F., Hobbs, J.: Time in OWL-S. In: Proceedings of the AAAI Spring Symposium on Semantic Web Services, California, Stanford University, pp. 29–36 (2004)
17. Gardiner, T., Horrocks, I., Tsarkov, D.: Automated Benchmarking of Description Logic Reasoners. In: Parsia, B., Sattler, U., Toman, D. (eds.) Proc. of the Int. Workshop on Description Logics (DL 2006), vol. 189, pp. 167–174. CEUR.org (2006)
18. Halaschek-Wiener, C., Parsia, B., Sirin, E.: Description logic reasoning with syntactic updates. In: Meersman, R., Tari, Z. (eds.) OTM 2006. LNCS, vol. 4275, pp. 722–737. Springer, Heidelberg (2006)
19. Grau, B.C., Halaschek-Wiener, C., Kazakov, Y.: History matters: Incremental ontology reasoning using modules. In: Aberer, K., Choi, K.-S., Noy, N., Allemang, D., Lee, K.-I., Nixon, L., Golbeck, J., Mika, P., Maynard, D., Mizoguchi, R., Schreiber, G., Cudré-Mauroux, P. (eds.) ASWC 2007 and ISWC 2007. LNCS, vol. 4825, pp. 183–196. Springer, Heidelberg (2007)
20. Gangemi, A., Guarino, N., Masolo, C., Oltramari, A., Schneider, L.: Sweetening ontologies with DOLCE. In: Gómez-Péréz, A., Benjamins, V.R. (eds.) EKAW 2002. LNCS, vol. 2473, pp. 166–181. Springer, Heidelberg (2002)

Benchmarking Fulltext Search Performance of RDF Stores

Enrico Minack, Wolf Siberski, and Wolfgang Nejdl

L3S Research Center
Leibniz Universität Hannover
30167 Hannover, Germany
lastname@L3S.de

Abstract. More and more applications use the RDF framework as their data model and RDF stores to index and retrieve their data. Many of these applications require both structured queries as well as fulltext search. SPARQL addresses the first requirement in a standardized way, while fulltext search is provided by store-specific implementations. RDF benchmarks enable developers to compare structured query performance of different stores, but for fulltext search on RDF data no such benchmarks and comparisons exist so far. In this paper, we extend the LUBM benchmark with synthetic scalable fulltext data and corresponding queries for fulltext-related query performance evaluation. Based on the extended benchmark, we provide a detailed comparison of fulltext search features and performance of the most widely used RDF stores. Results show interesting RDF store insights for basic fulltext queries (classic IR queries) as well as hybrid queries (structured and fulltext queries). Our results are not only valuable for selecting the right RDF store for specific applications, but also reveal the need for performance improvements for certain kinds of queries.

1 Introduction

Since the emergence of the Semantic Web [1], more and more data is stored in semi-structured form, using the RDF framework[1]. This annotated data becomes easier processable, or even "understandable" by machines. However, the data on the Semantic Web is frequently related to textual human-readable information (e.g., publication metadata). Therefore, most applications that provide their users access to RDF data also need support for fulltext search. This was confirmed in the Semantic Web Survey [2], where all 35 studied semantic search applications index textual data, and most of them provide fulltext search.

Whereas structured query languages such as SQL or SPARQL [3] provide a high expressivity regarding the *structure* of the data, fulltext queries target at the *content*. Integrating fulltext search (IR) [4] with structured search (DB) [5] is an actively discussed undertaking [6,7]. A lot of research has been

[1] Resource Description Framework: http://www.w3.org/RDF/

L. Aroyo et al. (Eds.): ESWC 2009, LNCS 5554, pp. 81–95, 2009.

conducted in DB fields such as relational DBMS [8,9], XML databases [10], and RDF databases [11,12], to name only some. Fulltext search clearly adds a large set of rich features to SPARQL, *e.g.* boolean, phrase, wildcard, and proximity queries [4]. A score that represents the relative relevance of matching resources can also be obtained. Further, the so called *snippet generation* provides a small summary of the text around the matching keywords.

While SPARQL [3] offers a high expressivity with respect to structured queries, fulltext-related queries are not well supported. The only feature being targeted on fulltext search is regular expression evaluation against string values. Interestingly, none of the widely used RDF stores support the standardized regular expressions efficiently, *i.e.,* with the help of fast index structures [13,14]. The reason is probably that regular expressions are much more expressive than required, and thus much more difficult to implement in an efficient manner, than the usual keyword search. In contrast, all RDF store implementations support efficient keyword search, putting significant effort in a feature that is *not* part of the W3C Recommendation. Some RDF indexing systems even consider the combination of inverted keyword indices and statement indices as fundamental building blocks for efficient RDF indexing [15]. This investment signifies the high demand of Semantic Web applications for support of fulltext search or *hybrid queries* which combine structured and fulltext queries [11].

For benchmarking RDF stores, a number of benchmarks have been proposed which evaluate performance regarding structured queries [16,17,18]. However, none of them addresses fulltext search capabilities. Therefore, RDF store developers test their fulltext search implementations with their own ad-hoc benchmarks [15,12,19], rendering efficiency evaluations difficult to be repeated or compared. A commonly available RDF fulltext benchmark is still missing.

In this paper, we propose such a benchmark that measures performance and feature richness of fulltext search facilities of RDF stores[2]. More precisely, we extend the well-known and widely used *Lehigh University Benchmark* (LUBM) [16] with fulltext content and hybrid queries. Using this new benchmark, we evaluate the most widely used open-source RDF stores and discuss the results. In particular, our work makes the following contributions:

1. We identify properties of datasets needed for an RDF fulltext benchmark, taking into account real fulltext characteristics such as term distribution.
2. We extend the LUBM benchmark dataset accordingly, and propose an algorithm for scalable generation of synthetic fulltext literals.
3. We design RDF fulltext queries to investigate different aspects of fulltext search on RDF, and evaluate well-known open-source RDF stores with our benchmark (in alphabetical order): *Jena*, *Sesame2*, *Virtuoso*, and *YARS*.

The paper is structured as follows. In section 2, we define objectives and desired features of an RDF fulltext benchmark, and describe our extension of the LUBM benchmark. Section 3 presents our evaluation of popular RDF stores against our proposed RDF fulltext benchmark. Relevant related work is discussed in section 4. We close with a conclusion.

[2] LUBMft: http://www.l3s.de/%7eminack/rdf-fulltext-benchmark/

2 A Fulltext Extension for LUBM

2.1 LUBM Overview and Discussion

The *Lehigh University Benchmark*[3] (LUBM) is a very frequently used synthetic benchmark for RDF stores [20,15,12]. It provides the *Univ-Bench Artificial data generator* (UBA) and a set of 14 test queries. Its dataset complies to the *Univ-Bench ontology*[4] describing universities. It contains 43 classes, such as `ub:University`, `ub:Department`, `ub:Professor`, and `ub:Publication`. It further contains 25 object properties (pointing to resources), as well as 7 datatype properties (pointing to literals), such as `ub:name` and `ub:publicationAuthor`. Datasets are generated using a scaling factor N which determines the size of the generated dataset and serves to investigate RDF store scalability. In addition, a seed M for the random number generator can be specified to allow repeated regeneration of the same synthetic dataset. These generated datasets are referred to as *LUBM(N, M)*. The notation *LUBM(N)* indicates that M is set to 0.

Even though this benchmark was primarily designed to evaluate OWL and DAML+OIL capabilities of RDF stores, it was also very often used to benchmark RDF stores and RDF related systems without OWL or DAML+OIL support [20,15,12]. This may be due to the very easy generation of arbitrary size RDF datasets and the familiar ontology domain. As LUBM, our benchmark works perfectly on stores without any kind of reasoning.

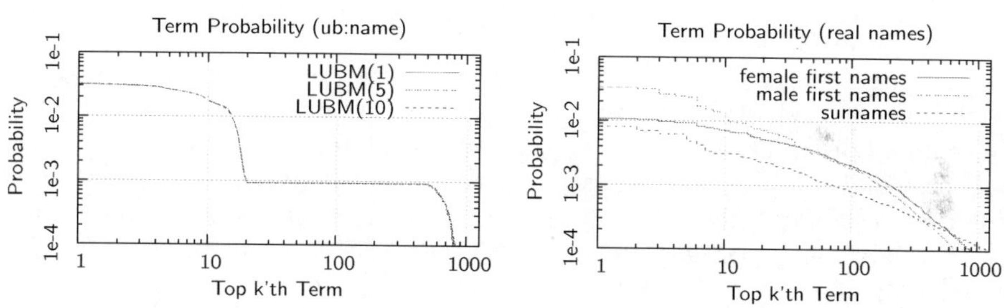

Fig. 1. (Left) Term distribution of the `ub:name` predicate. (Right) Term distribution of first names and surnames provided by the 1990 U.S. Census.

During the generation process, instances and relations are created on the basis of uniform distributions. For instance, a university contains 15 to 25 departments, each employing 7 to 10 full professors, whereas each authored 15 to 20 publications. This gives the LUBM datasets quite a complex structure with well defined connectivity properties. However, the literal values lack such sophisticated characteristics. The name of a person, for instance, is a one term literal like `"FullProfessor0"`. A fulltext search for this term would have exactly one

[3] Lehigh University Benchmark (LUBM): http://swat.cse.lehigh.edu/projects/lubm/

[4] Throughout this paper we refer to the Univ-Bench namespace using `ub`, which resolves to http://www.lehigh.edu/%7Ezhp2/2004/0401/univ-bench.owl

match for each department of each university. In addition, the `ub:name` property unfortunately is used to label all kinds of resources. Therefore, the terms of this predicate follow a mixture of uniform distributions, a highly unrealistic setting. Figure 1 shows the distribution of the `ub:name` terms (left-hand side), and, as an example, a real world distribution of first names and surnames (right-hand side). One can see that surnames and male first names are even power law distributions, as most real-world term distributions. We provide further details on first name and surname distributions in section 2.2.1.

2.2 Fulltext Content Generation

The proposed fulltext extensions modifies the LUBM dataset in two places. First, names are generated such that they follow realistic term distributions. Second, to measure performance on large text literals, fulltext content is added to the publication instances created by LUBM. To realize this extension, we add two sources of literals to the LUBM data generator, 1) the *Name Generator* adding real person names and 2) the *Document Generator* providing publication content. The enhanced datasets contain the same statements as the original LUBM datasets, extended by the `ub:firstname`, `ub:surname`, `ub:fullname`, as well as the `ub:publicationText` predicate, containing the respective content. Whereas the original LUBM dataset is referred to as LUBM(N), the fulltext extended dataset is denoted as LUBMft(N).

2.2.1 Name Generatator

The first new source of literals we add is the *Name Generator* which produces random names consisting of a first name and a surname. The generated terms follow a real distribution of names. As input for the term frequencies we use the data provided by the U.S. Census Bureau from the 1990 census[5]. This dataset provides probabilities for the top $\approx 1,200$ male, and for the top $\approx 4,300$ female first names. It also contains the top $\approx 89,000$ surnames, but due to rounding errors, only the top $\approx 19,000$ surnames have sufficiently accurate frequency information.

2.2.2 Document Generator

The second source of literals with realistic term distributions is the *Document Generator*. Since LUBM is a synthetic benchmark that produces data of any specified size, the content generation also needs to scale. Therefore, we can't directly use a real-world document collection as content. Instead, we use a Probabilistic Topic Model [21] to learn features that characterize such a collection. We then generate a synthetic set of documents which exhibit the learned real-world characteristics. This approach preserves the desired scalability feature of LUBM and at the same time ensures realistic term frequency distributions of the generated collection.

[5] Frequently Occurring First Names and Surnames From the 1990 Census: http://www.census.gov/genealogy/names/

To model a realistic topic distribution of documents, we apply *Latent Dirichlet Allocation* (LDA) [21] where a document is the result of a mixture of topics, and a topic is a distribution over words. We reuse the Matlab® implementation of the LDA model provided by the Steyvers and Griffiths [21]. As training set we employ the NIPS document collection[6], consisting of 1,740 papers, written by 2,037 authors with a total of 2,301,375 terms, out of 13,649 unique terms. From the topic distribution of documents, we derive a topic cooccurrence probability: when a topic has a high probability for a document, then it also gets a higher cooccurrence probability with all other topics of this document.

Given the topics and their cooccurrence, we first assign a dedicated topic to each department. Then, we distribute topics cooccurring with the department topic among the faculty staff of the department, according to their probabilities. Together with the department topic, the topics of the authors determine the topic mixture of a publication. This topic assignment algorithm leads to publications that cluster around departments by means of used terms: each department and author has its own specific vocabulary. This better reflects term usage among universities, departments, and authors than one global term distribution. Table 1 gives an overview of the resulting LUBMft datasets.

Table 1. Statistics of the LUBM fulltext datasets

	LUBMft(1)	LUBMft(5)	LUBMft(10)	LUBMft(50)	LUBMft(100)
Universities	1	5	10	50	100
Person Names	8,330	51,955	106,409	555,815	1,120,834
Publications	5,999	37,854	76,529	402,142	808,741
Unique Terms	17,611	28,171	32,438	36,456	36,518
Terms	6,032,320	38,061,820	76,954,636	404,365,260	813,224,336
Statements	134 k	840 k	1.7 M	8.96 M	18 M
Size (XML)	56 MB	356 MB	719 MB	3.8 GB	7.6 GB

2.3 Fulltext Test Queries

The fulltext dataset generated needs to be complemented with related queries to form a complete benchmark. The goal is to design these queries that evaluating them against the LUBMft datasets provides insights in the strengths and weaknesses of RDF stores regarding fulltext search. We deem query plan optimization an integral part of an efficient query evaluation. Therefore, we follow the same principle as LUBM where query patterns are stated in descending order, *w.r.t.* their cardinality. This gives RDF stores the maximum possible opportunity to optimize the query plan, and reduces the probability that a store just by chance executes a better query plan than another store.

Our benchmark queries are subdivided into three sets, each targeting a different area:

[6] Author-Topic Model: NIPS: http://www.datalab.uci.edu/author-topic/NIPs.htm

1. *Basic IR performance*: These queries are equivalent to classic IR queries, *i.e.,* keyword queries without any semantic relations.
2. *Semantic IR performance*: This set of queries targets at the performance of the combination of keyword queries and semantic relations. These are hybrid queries. Furthermore, the quality of integration of fulltext search into the SPARQL evaluation process is investigated.
3. *Advanced IR features*: Finally, this set of queries explores advanced IR features like boolean, phrase, or wildcard queries, score or snippet retrieval.

Queries. In the following, we present the three query sets, in which the queries get more and more complex, and where subsequent queries build on the results of previous ones. Note that some queries have variants, *e.g.* Query 1 is evaluated with different keywords to investigate the impact of those different keywords for the same dataset size. The two variants are denoted as Query 1.1 and 1.2.

All queries focus on fulltext performance, their *combination* with structured queries and their *integration* into the query evaluation process. We do not investigate the performance of the structural part of the queries. The performance of the structured query parts is sufficiently investigated by RDF benchmarks discussed in section 4.

2.3.1 Basic IR Performance

The first set of queries that we design will only contain queries that can equivalently be expressed in pure IR queries as well. Further, only those IR features are exploited that are expected to be the minimum set of features all RDF stores with fulltext search do support, namely keyword and phrase queries.

Table 2. Basic IR Queries

Q1: "All resources matching the keyword k". $k \in \{$'engineer', 'network'$\}$
Q2: "All resources matching k_i in `ub:publicationText`". $k \in \{$'engineer', 'network'$\}$
Q3: "All resources matching 'network' or 'engineer' in `ub:publicationText`".
Q4: "All resources matching the phrase 'network engineer' in `ub:publicationText`".
Q5: "All resources matching keyword 'smith' / having literal "Smith" in `ub:surname`".

We start with a simple one-keyword query. Its performance is of interest, since in the more complex benchmark queries the evaluation of keyword query conditions produces equivalent intermediate result sets with a proportional performance and scalability impact. We use one infrequent term, 'engineer', and one very frequent term, 'network', to investigate influence of result set cardinality. For LUBMft(1), the result set cardinality for these terms is 40 and 1,013, respectively.

In Query 2, we restrict the terms to match in the `ub:publicationText` predicate. The result set is the same, so the difference in performance compared to Query 1 is caused by this additional constraint. This query is equivalent to fielded keyword search in IR. Result sets of two keywords have to be looked up and merged to evaluate Query 3, which is the union of both result sets. To evaluate a phrase query of the two keywords, the position information of each occurrence of both keywords have to be processed by the fulltext search engine. Query 5 finally let us compare keyword lookup performances with literal lookup, since both queries are semantically equivalent and have the same result set.

2.3.2 Semantic IR Performance

The following queries contain structural conditions to investigate their impact on the query evaluation performance. These queries show the quality of integration of IR features with general SPARQL query processing.

Table 3. Semantic IR Queries

Q6:	"All `ub:Publications` matching 'engineer' in `ub:publicationText`".
Q7:	"All `ub:Publications` and their `ub:title` matching 'engineer' in predicate `ub:publicationText`".
Q8:	"All `ub:Publications`, their `ub:titles`, and the `ub:FullProfessor` author's `ub:fullname`, matching 'engineer' in `ub:publicationText`".
Q9:	"All resources matching 'engineer' in `ub:publicationText` and all resources matching 'smith' in `ub:fullname`, being connected via the predicate `ub:publicationAuthor`".
Q10:	"All resources matching 'network' in `ub:publicationText` and all resources matching 'engineer' in `ub:publicationText`, that are both connected via `ub:publicationAuthor` to the same `ub:FullProfessor`".
Q11:	"All distinct `ub:FullProfessors` matching 'smith' in `ub:fullname` that authored both, resources matching 'network' and resources matching 'engineer' in the `ub:publicationText` property".

From Query 6 to 8, we increase the number of structural constraints of the query. The result set of the keyword condition of the query has to be joined with more and more triple patterns. This allows us to evaluate the impact of structural parts of the query. The next step is the extension to two-keyword queries, where each keyword matches a different resource in the query graph. In Query 9, both matching resources are tested to be connected via a `ub:publicationAuthor` predicate. Query 10 interconnects both matching resources via a `ub:FullProfessor` resource. Finally, we extend our query to three keywords, where the intermediate full professor additionally has to match the third keyword. This requires three result sets of the fulltext search to be joined via structural restrictions.

2.3.3 Advanced IR Features

In our final query set we investigate the existence of certain advanced IR features. In contrast to the feature evaluated by Query 1 to 5, the advanced features are not expected to be supported by all RDF stores offering fulltext search.

Table 4. Advanced IR Queries

Q12:	"All resources matching 'network' **and** 'engineer' in `ub:publicationText`".
Q13:	"All resources matching 'network', **but not** 'engineer' in `ub:publicationText`".
Q14:	"All resources matching 'network' and 'engineer' in `ub:publicationText`, both keywords appearing within a distance of at most 10 words of each other".
Q15:	"All resources matching wildcard keyword 'engineer*' in `ub:publicationText`".
Q16:	"All resources matching wildcard keyword 'engineer?' in `ub:publicationText`".
Q17:	"All resources matching fuzzy keyword 'engineer 0.8' in `ub:publicationText`".
Q18:	"All resources and their relevance to the keyword 'engineer' in predicate `ub:publicationText`".
Q19:	"All resources and a snippet matching keyword 'engineer' in predicate `ub:publicationText`".
Q20:	"The top-10 resources with their relevance to the keyword 'network' in `ub:publicationText`".
Q21:	"The resources and their relevance to 'network' in predicate `ub:publicationText`, that have a relevance of > 0.75".

We test for conjunction and negation of fulltext related conditions, for a proximity query, where both keywords have to appear in a distance of at most 10 words, and for wildcard queries. Query 17 matches all terms that have a similarity of 0.8 to the given keyword. The similarity measure is application dependent, so for different implementations, different similarity values may be used to retrieve a comparable result set size. The relevance score and a snippet is retrieved by Query 18 and 19, respectively. Some RDF stores allow to limit the results of a fulltext search to the top-k, or to all results that exceed a certain relevance score threshold. These features are tested by Query 20 and 21. If these features are not supported, then they can be mimicked by a SPARQL LIMIT or FILTER operation. In that case, a similar performance to Query 2.2 is expected.

3 Evaluation

After generating the enhanced LUBMft dataset and designing 21 queries, we now evaluate a number of well-known open-source RDF stores using our RDF fulltext benchmark. We decided for the following RDF stores (in alphabetical order): Jena [22], Sesame2 [23], Virtuoso[7], YARS [20]. Since Jena provides a large number of different backend stores, we performed preliminary tests to identify the fastest configuration, where Jena+TDB performed best. We used the following configurations, which we will further refer to using these abbreviations:

Jena: Jena 2.5.6, ARQ 2.5.0, Lucene 2.3.1, TDB 0.6.0
Sesame2: Sesame 2.2.1, NativeStore, LuceneSail (Hits-Set) 1.3.0, Lucene 2.3.2
Virtuoso: Virtuoso 5.0.9
YARS: YARS post beta 3, Lucene 1.9.1

[7] Virtuoso Open-Source Edition: http://virtuoso.openlinksw.com/wiki/main/Main/

All Java-based RDF stores employ the Lucene[8] full-featured text search engine Java library, the state-of-the-art Java implementation for Information Retrieval.

3.1 Evaluation Methodology

The evaluation is conducted over LUBMft(N) with $N \in \{1, 5, 10, 50\}$. We reused the *LUBM benchmark test tool* (UBT) in version 1.1 to perform the tests. We modified its program sequence towards the following behavior:

Since we have designed our queries in an incremental manner, subsequent queries build on findings of previous queries. This means, that the result set of an early query is sometimes the intermediate result set of a later query. Due to RDF store specific caching mechanisms and filesystem cache managed by the operating system, subsequent executions of the same query, as well as similar queries (as described above), will benefit from those "warmed up" caches. In order to surpress these side-effects, we clear the RDF store caches by restarting it for each query, and clear the filesystem cache. We then evaluate each query six times, where the first duration is considered as "uncached", while the subsequent durations are considered as "cached". This will provide us insights in the performance of each store in the two cases of insufficient and sufficient memory for caching. Queries that exceed a limit of 1,000 s are not evaluated any further. As a duration of a query we define the time that passes from parsing the query until no more results are provided by the RDF store.

For each RDF store and each LUBMft dataset, all defined 21 queries are evaluated in this manner. This whole process is repeated 5 times. In the end we have 5 evaluation times for each query referring to the "uncached" case, as well as 25 times under the presence of "warmed up" caches.

For the evaluation we use a GNU/Linux machine with a 2 GHz AMD Athlon™ 64bit Dual Core Processor, 3 GByte RAM, and a RAID 5 array. The UBT Tester runs with Java™ SE Runtime Environment 1.6.0_10 and 2 GB Memory.

3.2 Evaluation Results

In the following, we present the results of our benchmark evaluation. For easier comparison, all figures will share the same layout. Each figure depicts one benchmark query and aggregates all RDF stores. Along the x-axis the different datasets are aligned by increasing scaling factor N. The logarithmic y-axis depicts the evaluation time of a query in *milliseconds*. The bars of the RDF stores are subdivided into two parts: the whole bar represents the evaluation time with cleared caches, whereas the lower part of the bar illustrates the evaluation performance under the existence of "warmed up" caches.

3.2.1 Basic IR Performance

Figure 2 depicts the one-keyword Queries 1.1 and 1.2, which represent the basis of fulltext search. For Query 1.1, Jena is slower than Sesame2, but Jena can better

[8] Apache Lucene: http://lucene.apache.org/

deal with the larger result set of Query 1.2, and can therefore exhibit a little performance advantage. Virtuoso, which is very fast for the smallest dataset, performs worse than Jena and Sesame2 with increasing dataset size for small result sets (Query 1.1), but performs better for larger result sets (Query 1.2). For both queries, YARS reveals the worst performance. For all queries, Virtuoso has an impressive performance with "warmed up" caches. The results of Query 2.1 and 2.2 exhibit for all stores the same performance as for the Queries 1.1 and 1.2. We therefore omitted the figures of Query 2.1 and 2.2 due to space limitations.

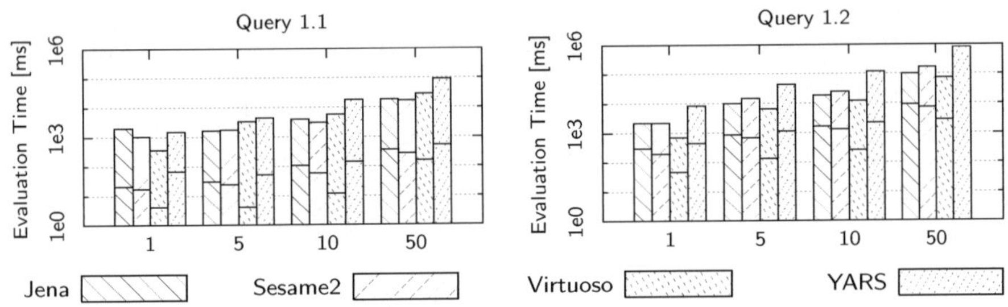

Fig. 2. Single keyword queries without predicate binding

With Query 3 we did not significantly increase result set size compared to Query 1.2 and 2.2, but we use multiple keywords in a boolean *OR* semantic. As we can see in Figure 3, all three stores that use Lucene exhibit exactly the same performance for Query 3 and 1.2. Note that Virtuoso does not support this query. Compared to Sesame2, Jena also yields a better performance for Query 4. Here, Virtuoso is even slower than Sesame, but it shows impressive response times in case of "warmed up" caches. YARS does not allow for phrase query articulation. Looking at Query 5.1 and 5.2, all RDF stores have a better performance evaluating the latter one.

Discussion. From Query 1 and 2 we can see that none of the stores has a problem with missing predicate bindings for the fultext search, nor does it gain performance when it is given. The implementations are robust in both situations.

Regarding basic IR queries, Jena is slightly faster than Sesame2. Both use Lucene as their fulltext search engine, so the overhead of the integration of the IR engine into the RDF evaluation might be a little higher in Sesame2. YARS also uses Lucene, but is much slower. This may be due to the outdated lucene version they use. Virtuoso is for all but one cases the slowest RDF store. Further, it does only support a subset of the basic IR features.

3.2.2 Semantic IR Performance

With our semantic IR queries depicted in Figure 4, we tested for the performance of fulltext search combined with structural queries. From Query 6 to 8

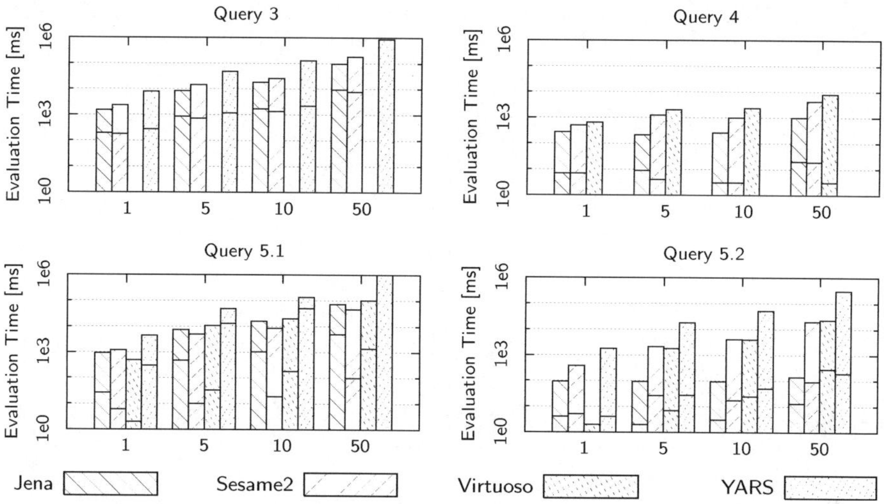

Fig. 3. Multiple-keyword, phrase, and keyword vs. literal queries

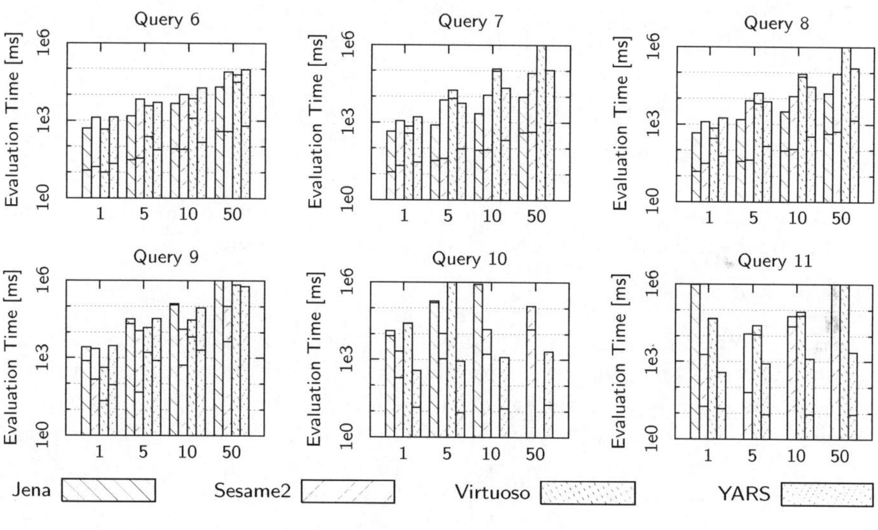

Fig. 4. Semantic IR queries combining fulltext search with structured queries

an increasing amount of structural information around the resources that match the fulltext search are retrieved. For Query 6, all stores have similar performance and scaling properties, but Jena always performs best. In Query 7 and 8, Virtuoso has significant scaling problems, while all other stores exhibit quite the same performance as for Query 6.

With multiple keywords (Query 9 to 11), Jena is completely overloaded. Virtuoso exceeds our query evaluation limit for the second smallest dataset already. Sesame2 scales up best with multiple keywords. Note that YARS is not capable of retrieving the correct results for queries with multiple fulltext searches.

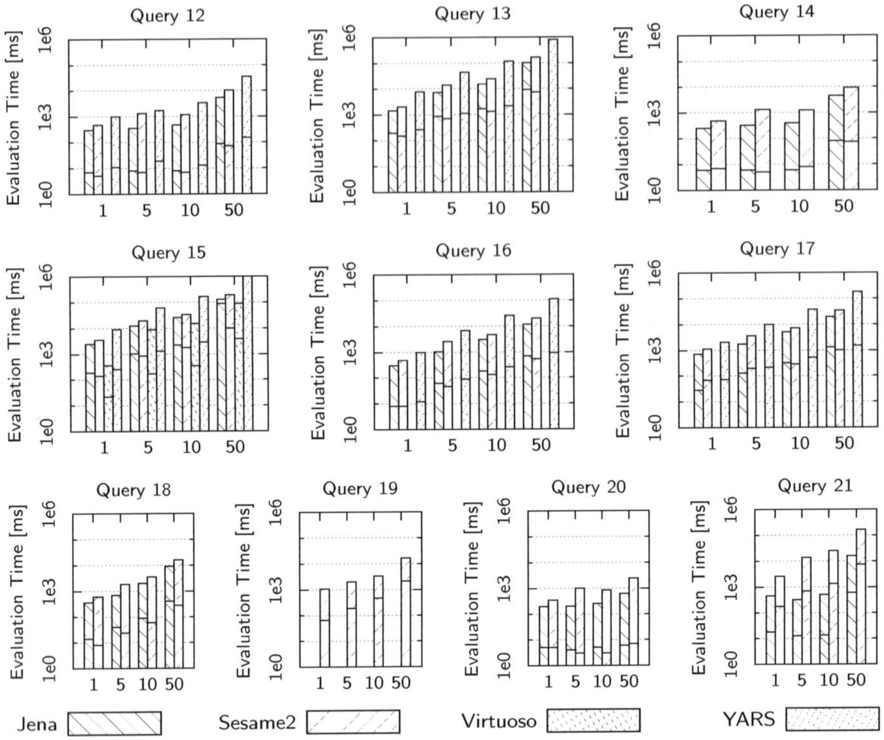

Fig. 5. Advanced IR feature queries: boolean, proximity, wildcard, and fuzzy queries, score and snippet retrieval, result and relevance score limitation queries

Discussion. By evaluating the semantic IR queries, we could recognize that Jena and Virtuoso have scalibility problems for hybrid queries in general, and in particular with those containing multiple fulltext queries assigned to different resources in the query. YARS could also not properly evaluate these queries. Further investigations on Jena showed that this is highly influenced by the very simple query optimizer provided by TDB. With an optimized query planning, Jena could potentially outperform Sesame2. This class of queries seems not to be considered by the evaluated RDF stores (except by Sesame2), however, it is an important class of queries in the field of semantic keyword queries.

3.2.3 Advanced IR Features

Finally, we tested for the advanced IR features. Virtuoso only supports one out of ten queries, but this one with the smallest response time of all RDF stores. YARS only supports a subset of five queries. Jena and Sesame2 both have similar performance, but Jena is always faster. For Query 20 one can see that the response time is quite constant for all datasets, whereas for Query 21, this only holds for Jena.

Discussion. Looking at advanced IR features, Sesame2 stands out by supporting all features. For those supported by Jena, its overall performance is better. The result set limitation feature of Query 20 perfectly works for both. The evaluation time neither grows with the dataset size nor with the correlating complete result set size. Limiting the results yields to an almost constant response time. Due to the fact that Sesame2 does not support the relevance score limitation feature of Query 21, this feature has to be achieved using a SPARQL FILTER. This, first evaluates the query completely and then filters all results that match the score limit. Jena provides this feature with its fulltext querying facility and therefore performs as expected.

Summary. For basic or advanced IR queries, Jena and Sesame2 are the best choices. Considering complex semantic IR queries, Sesame2 is the only RDF store sufficiently solving the tasks of fulltext search on RDF data. With "warmed up" caches, Virtuoso yields an impressive performance, but does only supported a small subset of queries.

4 Related Work

In software engineering domains, benchmarks are used to assess the relative performance and absolute feature-richness of a target system [24]. For this, well designed standard tasks mimic a particular type of workload. The *Full-Text Document Retrieval Benchmark* [24, ch. 8] (FTDR) generates as workload a mixture of searching and retrieving documents from the system, as well as adding documents. In contrast, our benchmark only investigates search.

In the area of relational databases, the *TEXTURE* benchmark [25] investigates performance of combining text processing with relational workload. As the FTDR Benchmark, it randomly generates queries based on certain term occurrence characteristics, whereas we intend to provide fixed queries that can be repeatably evaluated and studied for different RDF stores and hardware configurations. Similarily to our benchmark, they synthetically generate fulltext, but they only maintain one global word distribution.

In the area of RDF stores, a number of benchmarks are available. The *Berlin SPARQL Benchmark* [17] (BSBM) also generates fulltext content and person names. The SP^2Bench [18] uses a dataset that refers to the structure of the DBLP Computer Science Bibliography[9]. Both benchmarks pick terms from dictionaries with uniform distribution. The SP^2Bench and BSBM were not considered for our RDF fulltext benchmark simply due to the fact of their very recent publication. Enriching these benchmarks with real world fulltext content and fulltext queries is very much in our favor. To the best of our knowledge, the presented fulltext extension to the LUBM is the first fulltext Benchmark for Semantic Web systems.

Due to the lack of RDF fulltext benchmarks, the authors of [19] employed the RDF version of Wordnet[10] and queried for randomly selected terms. However, this dataset does not simply scale up to an arbitrary size. Similarily to

[9] DBLP Computer Science Bibliography: http://dblp.uni-trier.de/
[10] Wordnet RDF: http://www.semanticweb.org/library/

our approach, the authors of [15] generated a LUBM(50,000) dataset, where literals were generated by keywords randomly selected from a dictionary. Unfortunately, the authors did neither sufficiently report on this benchmark in order to reproduce results or reuse the benchmark, nor did they apply that benchmark on other RDF stores. Further, our benchmark provides more realistic content characteristics.

5 Conclusion

The work presented in this paper fulfills the strong need of application and RDF store developers for an RDF fulltext benchmark. With the provided dataset and hybrid queries, fulltext capabilities of any RDF store can be evaluated and compared. Our evaluation shows that on the one hand basic as well as advanced IR features are already sufficiently supported by today's RDF stores. On the other hand, it also uncovered several areas where performance needs to be improved. This holds primarily for complex hybrid queries that contain more than one fulltext search condition. The evaluation results also show that a good query plan optimization is crucial for competitive performance, as well as a tight integration of fulltext search features into this planning phase. We are convinced that the availability of a standard fulltext benchmark for RDF data stores will foster the development of more efficient hybrid query processing algorithms.

Acknowledgement

This work was supported by the European Union IST fund (Grant FP6-027750, Project NEPOMUK).

References

1. Berners-Lee, T., et al.: The Semantic Web. Scientific American 279(5) (May 2001)
2. Hildebrand, M., et al.: An analysis of search-based user interaction on the Semantic Web. Report, Centrum Wiskunde & Informatica (2007) INS-E0706, ISSN 1386-3681
3. Prud'hommeaux, E., Seaborne, A.: SPARQL Query Language for RDF (January 2008)
4. Baeza-Yates, R., Ribeiro-Neto, B.: Modern Information Retrieval. ACM Press / Addison-Wesley (1999)
5. Garcia-Molina, H., et al.: Database Systems: The Complete Book. Prentice Hall, Englewood Cliffs (2008)
6. Chaudhuri, S., et al.: Integrating DB and IR Technologies: What is the Sound of One Hand Clapping? In: CIDR, pp. 1–12 (2005)
7. Amer-Yahia, S., et al.: Report on the db/ir panel at sigmod 2005. SIGMOD Rec. 34(4), 71–74 (2005)
8. Hristidis, V., et al.: Efficient IR-Style Keyword Search over Relational Databases. In: VLDB, pp. 850–861 (2003)

9. DeFazio, S., et al.: Integrating IR and RDBMS Using Cooperative Indexing. In: SIGIR, Seattle, Washington, USA, July 9–13, 1995, pp. 84–92. ACM Press, New York (1995)
10. Consens, M.P., et al.: XML Retrieval: DB/IR in Theory, Web in Practice. In: VLDB, University of Vienna, Austria, September 23–27, 2007, pp. 1437–1438. ACM, New York (2007)
11. Bhagdev, R., et al.: Hybrid Search: Effectively Combining Keywords and Semantic Searches. In: ESWC, Tenerife, Canary Islands, Spain, June 1-5, 2008, pp. 554–568 (2008)
12. Zhang, L., et al.: Semplore: An IR Approach to Scalable Hybrid Query of Semantic Web Data. In: Aberer, K., Choi, K.-S., Noy, N., Allemang, D., Lee, K.-I., Nixon, L., Golbeck, J., Mika, P., Maynard, D., Mizoguchi, R., Schreiber, G., Cudré-Mauroux, P. (eds.) ASWC 2007 and ISWC 2007. LNCS, vol. 4825. Springer, Heidelberg (2007)
13. Baeza-Yates, R.A., Gonnet, G.H.: Fast Text Searching for Regular Expressions or Automaton Searching on Tries. Journal of the ACM 43(6), 915–936 (1996)
14. Cho, J.: A fast regular expression indexing engine. In: Proceedings of the 18th International Conference on Data Engineering (2002)
15. Harth, A., et al.: YARS2: A Federated Repository for Querying Graph Structured Data from the Web. In: Aberer, K., Choi, K.-S., Noy, N., Allemang, D., Lee, K.-I., Nixon, L., Golbeck, J., Mika, P., Maynard, D., Mizoguchi, R., Schreiber, G., Cudré-Mauroux, P. (eds.) ASWC 2007 and ISWC 2007. LNCS, vol. 4825, pp. 211–224. Springer, Heidelberg (2007)
16. Guo, Y., Pan, Z., Heflin, J.: LUBM: A Benchmark for OWL Knowledge Base Systems. Journal of Web Semantics 3(2), 158–182 (2005)
17. Bizer, C., Schultz, A.: Benchmarking the Performance of Storage Systems that expose SPARQL Endpoints. In: Proceedings of the 4th International Workshop on Scalable Semantic Web knowledge Base Systems (SSWS 2008) (2008)
18. Schmidt, M., et al.: An Experimental Comparison of RDF Data Management Approaches in a SPARQL Benchmark Scenario. In: ISWC, October 26–30 (2008)
19. Minack, E., et al.: The Sesame LuceneSail: RDF Queries with Full-text Search. Technical Report 2008-1, NEPOMUK (February 2008)
20. Harth, A., Decker, S.: Optimized Index Structures for Querying RDF from the Web. In: Proceedings of the 3rd Latin American Web Congress. IEEE Press, Los Alamitos (2005)
21. Steyvers, M., Griffiths, T.: Probabilistic Topic Models. Lawrence Erlbaum, Mahwah (2006)
22. Carroll, J.J., et al.: Jena: Implementing the Semantic Web Recommendations. In: WWW Alternate track papers & posters, pp. 74–83. ACM, New York (2004)
23. Broekstra, J., et al.: Sesame: A Generic Architecture for Storing and Querying RDF and RDF Schema. In: Horrocks, I., Hendler, J. (eds.) ISWC 2002. LNCS, vol. 2342, pp. 54–68. Springer, Heidelberg (2002)
24. Gray, J. (ed.): The Benchmark Handbook For Database and Transaction Processing Systems. Morgan Kaufmann, San Francisco (1993)
25. Ercegovac, V., et al.: The TEXTURE Benchmark: Measuring Performance of Text Queries on a Relational DBMS. In: VLDB, Trondheim, Norway, pp. 313–324 (2005)

A Heuristics Framework for Semantic Subscription Processing

Martin Murth and eva Kühn

Vienna University of Technology
Institute of Computer Languages,
Space Based Computing Group
Argentinierstraße 8, 1040 Vienna, Austria
{mm,eva}@complang.tuwien.ac.at

Abstract. The increasing adoption of semantic web technology in application scenarios with frequently changing data has imposed new requirements on the underlying tools. Reasoning algorithms need to be optimized for the processing of dynamic knowledge bases and semantic frameworks have to provide novel mechanisms for detecting changes of knowledge. Today, the latter is mostly realized by implementing simple polling mechanisms. However, this implies client-side post-processing of the received results, causes high response times and limits the overall throughput of the system. In this paper, we present a heuristics framework for realizing a subscription mechanism for dynamic knowledge bases. By analyzing similarities between published information and resulting notifications, heuristics can be employed to "guess" subsequent notifications. As testing the correctness of guessed notifications can be implemented efficiently, notifications can be delivered to the subscribers in an earlier processing phase and the system throughput can be increased. We experimentally evaluate our approach based on a concrete application scenario.

Keywords: Semantic subscription processing, heuristics framework, continuous queries, similarity heuristics, incremental result set updates.

1 Introduction

The employment of semantic web technology in an ever increasing number of different application areas has imposed new requirements on the underlying tools and frameworks. Besides the traditionally high requirements on storing, querying and reasoning about large static data sets, efficient processing of more dynamic data has become a central issue in many application scenarios, e.g., knowledge distribution for synchronization or clustering [1], semantically-enabled data integration [2], knowledge monitoring [3], semantic message routing [4-5], and knowledge-driven coordination [6-7].

The new requirements can generally be divided into two groups. First, reasoning algorithms and engines need to be optimized for the processing of frequently changing data. Different approaches have already been investigated in this field of

L. Aroyo et al. (Eds.): ESWC 2009, LNCS 5554, pp. 96–110, 2009.

research. In [8], for instance, an incremental reasoning technique for dynamic ABoxes is described, which allows for improvements of reasoning about $\mathcal{SHOQ}$ and $\mathcal{SHIQ}$ description logics [9] of "up to three orders of magnitude".

Second, semantic frameworks need to provide novel mechanisms for detecting changes of knowledge in a knowledge base. Today, this is often realized by implementing a simple client-side polling mechanism. The client system executes the query, retrieves the full result set and compares it with previously received result sets. Although the client is actually only interested in relevant updates, it needs to store the entire query result sets and analyze potentially large amounts of data.

Accounting for this, we developed a lightweight framework that implements a simple subscription mechanism for semantic repositories. Whenever data is added to the repository, subscribers are notified about new results of the registered queries. With this approach, the client does not need to store and post-process the received results and the amount of data that needs to be transferred between the semantic repository and the client is minimized. However, it is still necessary to execute all registered queries whenever new data is added. This often causes high response times and limits the overall throughput of the system (cf. [6]).

In this paper, we present a mechanism for reducing the time required to notify subscribers by applying heuristics on the published data for finding new query results. The core idea is based on the following observation: While adding new data to the knowledge base often entails a costly reasoning and query execution task to find new query results, the contents of the published graphs and the new query results can frequently be correlated following simple patterns. Hence, finding such patterns would allow the system to "guess" possible new results and to test whether they are valid in the new system state. Testing of possible results can be implemented efficiently and the valid results can be sent to the subscriber already before the usually much more expensive query is executed. Furthermore, if the heuristic is able to find the majority of new results, query execution can even be skipped for several processing cycles. This decreases the average time required to process a publication request and therefore increases the throughput of the semantic subscription system.

The rest of this paper is organized as follows: The remainder of Section 1 discusses related work in this field. Section 2 introduces SENS (Semantic Event Notification Service), our subscription processing framework for semantic repositories, and describes an extension for heuristic result set update algorithms. A simple heuristic for this framework is developed in Section 3. Section 4 evaluates the approach based on a concrete application scenario. Conclusions and future work are presented in Section 5.

1.1 Related Work

The work presented in this paper was motivated by experiences with application development using semantic tuplespace technology. Semantic tuplespaces (e.g., sTuples [10], TSC [11], Semantic Web Spaces [12], TripCom [13]) are usually built upon a semantic repository and implement a Linda-based [14] or a publish/subscribe-based interface to access the stored knowledge. Clients interact with each other by adding knowledge to the knowledge base and by retrieving notifications about events they have subscribed for. Although the proposed systems allow for elegant solutions of many complex coordination problems (e.g. [15]), the implementation of real world applications showed that the existing systems do not perform well for bigger knowledge bases as the continuous execution of queries causes a too high load on the server. In the TripCom project, mechanisms are being developed for clustering and distributing the data to multiple server instances. While this approach aims at improving scalability by intelligently distributing the load, our approach focuses on improving the performance of a single server instance. Thus, both approaches complement each other.

Many semantic repositories and frameworks implement a simple subscription mechanism. Jena [16] and Sesame [17], for example, implement listener interfaces that generate events whenever new triples are added to the repository. However, they do not support expressive subscription languages and notify listeners only about explicitly inserted triples. Consequently, subscribers are not notified about inferred information which is one of the crucial advantages of communicating via a shared knowledge base.

Some RDF query engines (e.g. [18-19]) employ caches for improving the performance of query answering. While this technique leads to good results for querying static data, it is hardly applicable for highly dynamic knowledge bases as the caches are invalidated each time new data is published.

Several algorithms for continuous queries on RDF data have been proposed in [4-5,20-21]. However, these algorithms require RDF triples as input data, which requires to instantiate all triples at least in memory. For applications that employ ontologies with extensive use of transitive properties or classification conditions, this might cause the instantiation of a huge, possibly exponentially growing number of RDF triples. In contrast, our approach builds upon repeated query executions but tries to find the optimal intervals and to reduce the number of execution cycles. Thus, triples only need to be instantiated if they are requested by the reasoning or the query engine. Moreover, the heuristics work independently of the employed query engine and reasoning technology.

2 A Heuristics Framework for Semantic Subscriptions

In this section we briefly introduce the subscription framework SENS and present a framework extension for heuristic result set update algorithms.

2.1 SENS – Semantic Event Notification Service

SENS (Semantic Event Notification Service) [6] is a framework for implementing and controlling complex and heterogeneous application scenarios by leveraging the reasoning capabilities of logic-based reasoning systems. Software clients can publish data at SENS and subscribe to changes of particular parts of the inferable knowledge. In contrast to conventional publish/subscribe systems, the published data is not directly forwarded to the subscribers but is, in a first step, added to the system's knowledge base. Afterwards, the system evaluates the new state of the knowledge base and notifies the subscribers about the according updates.

SENS is accessed via the API shown in Listing 1.

Listing 1. SENS API (Java)

```
public interface SENS {

    // publishes knowledge in the form of RDF triples to the knowledge base
    void publish(Graph graph);

    // deletes all triples that match the given statement pattern
    void delete(StatementPattern pattern);

    // creates an event channel by providing a channel name and defining a graph
    // pattern based query
    void createChannel(String channelQN, GraphPattern query);

    // registers a subscription at the given channel
    void subscribe(String channelQN, Subscriber subscriber);

    // removes a subscription
    void unsubscribe(String channelQN, Subscriber subscriber);
}
```

Clients can add knowledge to SENS in form of an RDF graph data structure (*publish*). The system tries to infer new knowledge right after the publication operation.

RDF triples can also be removed from the knowledge base by providing a statement pattern matching the triples to be deleted (*delete*). The deletion of triples also removes the knowledge that has been inferred from these triples. Previously inferred knowledge is thus not invalidated, but it is no longer guaranteed to be true.

For being notified about changes of knowledge, clients have to subscribe to a semantic event channel via its qualified name (*subscribe/unsubscribe*). A semantic event channel generates notifications about changes of a part of a knowledge base that has to be defined by means of a query statement (*createChannel*). As state-of-the-art semantic repositories implement a variety of different query languages, the current version of SENS only accepts query statements formulated as simple graph patterns as these can be expressed with most existing RDF query languages. However, it is planned to add support for SPARQL [22] queries in the next version. Whenever the knowledge base is updated, SENS executes the channels' queries and tags the returned bindings. If it finds bindings that were not contained in the previous result set of a particular channel (i.e., are not tagged yet), the subscribers are notified about these new bindings. Notifications are sent to the subscriber as variable bindings for the channel's query and can be accessed in tabular form or as variable substitutions of the query's graph pattern (analogous to SPARQL *select* and *construct* forms).

One of the main advantages of employing a semantic subscription framework like SENS is the high expressivity of the subscription mechanism. The use of logic-based reasoning allows detecting and reacting to complex dependencies between distributed clients which are not explicitly modeled in the application scenario but which can be inferred from the provided data. For describing the same dependencies using a query on a database or an event stream [23], one would require, e.g., nested subqueries, constraint evaluation, special treatment of null values, explicit case differentiations, and client-side post-processing of the received results. Besides this, the semantic publish/subscribe interaction pattern facilitates strong decoupling of the communicating clients with respect to *time* (clients do not need to be connected to the system at the same time), *reference* (clients do not need to know each other explicitly), and *data schema* (clients do not rely on homogenous data structures and semantics). This increases flexibility in application design and allows for more autonomous system architectures.

2.2 An Extension for Heuristic Query Result Update Algorithms

For the improvement of notification times and throughput of SENS, we developed a framework extension for heuristic query result set update algorithms. These algorithms try to find new query results by analyzing the history of published graphs and found notifications, but they do not execute the actual query itself.

As *notification time* $\delta(notify_{i.j})$ of a notification[1] $notify_{i.j}$, we define the time between the client-side invocation of the publish operation pub_i that triggered the notification and the client-side reception of the notification. Accordingly, the *average notification time* $\delta(C)$ for a channel C is defined as the average of all notification times of those notifications that were delivered to the subscribers of a channel during their registration time.

Generally, the performance of the system is determined by the performance of all supported operations (*publish, delete, createChannel, un/subscribe*). However, for the applications we investigated, only the efficiency of publication processing had a significant impact on the overall system performance. Consequently, we define the *throughput* $\mu(S)$ of a semantic subscription system S as the number of publication requests that can be processed (i.e., publication and notification of all concerned subscribers) within a certain time unit.

For the development of the framework extension, we had to adapt the processing model of SENS. The original processing phases executed after each publication operation are illustrated in Figure 1.

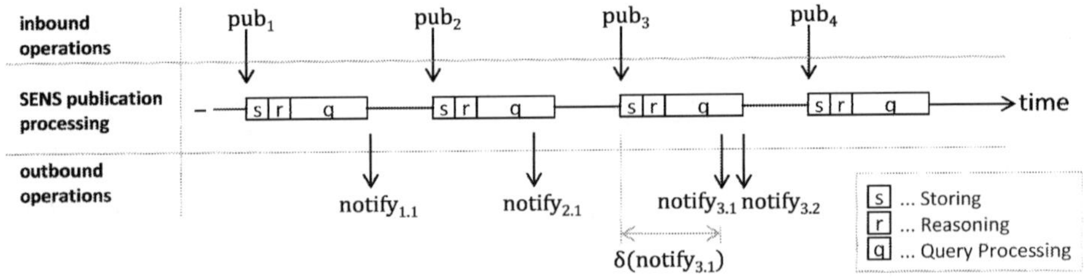

Fig. 1. SENS publication processing

When a publication request is issued to SENS (e.g., pub_3), the contained graph is first stored (**s**) in the connected RDF repository. Then a reasoning engine is employed to infer new knowledge (**r**) and finally, SENS executes the queries (**q**) of the subscriptions of all defined channels. Depending on the query and the employed RDF repository, the execution of the query may involve additional reasoning on the knowledge base. Whenever a new binding is found, a notification is immediately sent to the concerned subscribers (e.g., $notify_{3.1}$).

[1] Notify operations are given with two subscript numbers. The first one refers to the publication operation that triggered the notification and the second one is used to distinguish notifications that were triggered by the same publication operation.

The main benefit of employing a heuristic algorithm for finding potential bindings of subscriptions is that testing single bindings takes considerably less time than the complete execution of a query. The system only has to verify whether the bound triples of the graph pattern also exist in the new state of the RDF repository.

Figure 2 illustrates the additional phases of the extended processing model.

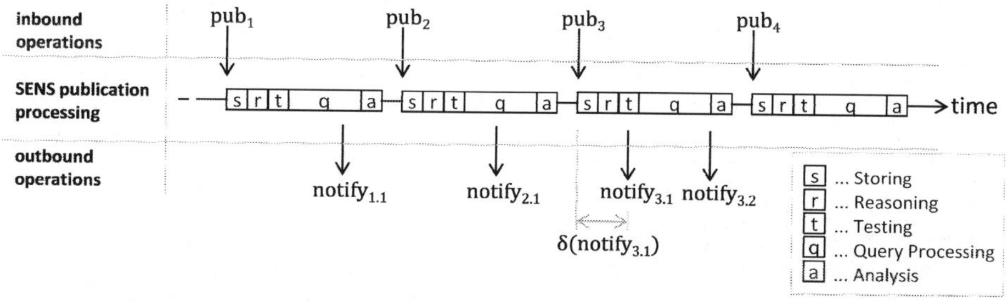

Fig. 2. SENS publication processing extension for result set update algorithms

Like in the original processing model, the content of a publication request is first stored (**s**) in the RDF repository and the reasoning engine tries to infer new knowledge (**r**). Afterwards, the heuristic analyzer component of SENS is provided with the new graph and returns a list of candidate bindings for each defined channel. SENS tests (**t**) for each candidate binding whether it represents a new notification event, i.e., whether it is a valid binding for the defined query and has not yet been delivered to the subscribers. If this is the case, subscribers are immediately notified about the new binding. In the next step, the standard query processing task (**q**) is executed in order to reliably find all new bindings that have not been discovered in the testing phase. Finally, the heuristic analyzer component tries to adjust and improve its heuristic method for subsequent requests. It therefore searches for further correlations between published graphs and resulting notifications (**a**) based on all found notifications and the history of preceding publication operations.

3 Simple Similarity Heuristic

In the following, we present a result set update heuristic for the previously described framework. The algorithm should serve as a proof of concept rather than a final solution and should motivate the development of more advanced algorithms. We will illustrate the algorithm's functionality by means of a simplified example of a data integration application. In the application scenario, a semantic repository is used to manage organizational and reference data of a university, e.g., name, address, telephone number, courses, publications, etc. of students and employees. A number of (heterogeneous) systems are storing and retrieving data from the shared repository. As several different terminologies are being used, an ontology is employed that maps concepts of one terminology to semantically equivalent concepts of the others. By implementing client-to-client communication via publishing and subscribing to knowledge at SENS, data of different data schemas is translated automatically and transparently for the client.

In the example, a client registers the following subscription in order to be notified whenever new information about a *Course* of a *Lecturer* becomes available:

Statement Patterns of Subscription Query
?L http://www.w3.org/1999/02/22-rdf-syntax-ns#type http://www.uni.org./onto/Lecturer
?L http://www.uni.org/onto/holdsCourse ?C

Each time a new binding for this query is found, the binding and the previously published graph are assigned to a particular *binding* and *graph category*. The following two tables show a published graph consisting of one single triple and a new binding resulting from the according publication operation. Note that in the ontology of the example, an individual is classified as *Lecturer*, if it teaches at least one course. The role *teaches* is defined to be the same as *holdsCourse*.

Published Graph[2]	Number of Elements
http://www.uni 1.org/dep 1/Professor 137	9
http://www.uni.org/onto/teaches	6
http://www.uni 1.org/courses/Course 148	8
	Graph Category = 23

Binding	Number of Elements
?L = http://www.uni 1.org/dep 1/Professor 137	9
?C = http://www.uni 1.org/courses/Course 148	8
	Binding Category = 17

Both categories are determined by the number of elements of the object's (i.e., graph or binding) textual representation. *Elements* (underlined text blocks) are character sequences that are separated by either a delimiter character (e.g., ":", "/", ".", "#", etc.) or by a digit following a letter character (e.g., "Course-1", "Professor-137"). Accordingly, the graph of the above example is assigned to the graph category 23 (9+6+8) and the binding is assigned to the binding category 17 (9+8).

Whenever a similar binding is found, it is checked whether the graphs that were published before are similar as well. *Similarity* is defined by means of a relation between two bindings or graphs. In our heuristic, we simply define similarity via an object's category: If and only if two objects belong to the same category, they are considered to be similar. While more complex similarity relations would obviously allow for a more accurate detection of similar objects, simple relations can usually be evaluated more efficiently.

The following two tables show a second graph that is published after the first one and the new binding that results from this publication.

[2] We added white space characters in the URIs for a better illustration of single elements.

Published Graph	Number of Elements
http^1://www^1.uni^1 1^1.org^1/dep^1 1^1/FullProfessor^1 [137]^2	9
http^1://www^1.uni^1.org^1/onto^1/teaches^1	6
http^1://www^1.uni^1 1^1.org^1/courses^1/Course^1 [259]^2	8
	Graph Category = 23

Binding	Number of Elements
?L = http^1://www^1.uni^1 1^1.org^1/dep^1 1^1/Professor^1 [137]^2	9
?C = http^1://www^1.uni^1 1^1.org^1/courses^1/Course^1 [259]^2	8
	Binding Category = 17

According to our heuristic, both bindings as well as both published graphs are similar. The binding's and graph's elements are now further separated into two groups:

- **Structure elements:** These elements (almost) always remain the same for all members of a category. They represent a common structure of members of a category.
- **Key elements:** These elements exhibit high diversity. They represent the distinguishing parts of members of a category.

In order to determine the group of an element, a history of added elements is kept for each category. If the n-th element of a graph or binding has not yet been added at the same position before, then the *diversity index* of this position is increased by one (shown as superscript number). As long as the diversity index of an element does not exceed a certain threshold (here < 2), it is considered to be a structure element. As soon as it exceeds the threshold, it is considered to be a key element (shown as framed text block).

If all key elements of a binding can also be found in the published graph, then a new mapping entry is generated. The following mapping entry defines that the publication of a graph of graph category 23 has caused the detection of a new binding of binding category 17 for the query of channel "NewCourseChannel". The key elements of the binding at positions 9 and 17 were found in the published graph at the positions 9 and 23, respectively.

Graph Category	Channel (Query)	Binding Category	Key Element Mappings
...	...	...	...
23	NewCourseChannel	17	BC(9)=GC(9), BC(17)=GC(23)

When the next graph is published, it is looked up in the mapping table whether it already contains entries for similar graphs, i.e. for the respective graph category. If this is the case, candidate bindings are generated by taking the structure elements of the binding category and the key elements of the published graph.

The described heuristic will, obviously, not always find all new bindings. The hit ratio highly depends on particular characteristics of the registered channel's query. The heuristic will probably find most new bindings for *queries with high structural similarities* between published graphs and found new bindings (independently of the complexity of the required querying and reasoning task), but it may not be able to find any new bindings for *queries with indirect relations* between the key elements of the published graph and the new bindings. For example, it will be difficult to predict new bindings of a query that reports the name of the lecturer who held the same course in the previous semester, since the newly published graph does not contain any information about this lecturer. In the worst case, a certain (configurable) amount of time is spent for testing and analysis without finding any new bindings. However, we saw that in many applications that build upon semantic technologies, a large number of queries exhibit high similarity between published graphs and new bindings and our approach can therefore help to significantly improve the performance of these applications.

4 Evaluation

In this section, we present an experimental evaluation of the proposed heuristic result set update algorithm. We therefore implemented an application scenario based on the LUBM benchmark [24]. We employed SENS to manage organizational and reference data of universities and realized client-to-client interaction by publishing knowledge and subscribing to knowledge updates. For populating the knowledge base, we took the data set of LUBM(10) (~10^6 explicit statements) and split it into n = 114789 separate graphs, each of which containing 2-20 semantically related statements. Graphs describing resources of the same type differ in size, structure and order of statements. This allows us to investigate whether the heuristic is capable to handle a certain level of "noise" in the published data. In the test scenario, we published n – 1000 graphs to SENS in random order. We then created a semantic event channel for each of the 14 LUBM benchmark queries and registered one subscriber at each channel. Finally, we published the remaining 1000 graphs and measured the average notification time of the resulting notifications and the publication throughput of the system.

In the test setting, SENS was run with SwiftOWLIM 2.9 [25], a semantic repository with a native full materialization[3] rule entailment engine that supports RDFS [26], OWL DPL [27], OWL Horst and most features of OWL Lite (cf. [25]). All tests were run on a Pentium IV HT 3,2GHz, 4GB RAM Windows Vista server.

4.1 Notification Time

For our measurements, we implemented a benchmark framework that executes all tests in two runs. In the first run, SENS is started without heuristic result set updates and with a query execution phase for each publication operation. The framework records which publication operation preceded a notification, assigns each operation a unique ID and stores benchmark configuration and reference data to an XML file. In the second run, the framework restarts the Java virtual machine, loads the configuration and runs the benchmark using the heuristic result set update algorithm. By comparing the results with the results of the first run, notifications can be traced back to the publication operations they resulted from, even if multiple graphs are published at once.

The notification time benchmark results are shown in Table 1.

[3] Materialization can be switched off for transitive closure to reduce memory consumption.

Table 1. LUBM(10)/SwiftOWLIM 2.9 Notification Times

Channel	Num. of Notif's	Tests Positives	Tests Negatives	Pos. Hitrate (Pos./Notifs#)	Mapping Entries	Avg. Notification Time Standard	Avg. Notification Time Heuristic	Heur. vs. Std. (-1+Heur./Std.)	Heuristic Overhead
Q1	1	0	0	0%	0	11 ms	12 ms	9%	2 ms
Q2	0	0	0	0%	0	0 ms	0 ms	0%	4 ms
Q3	1	0	0	0%	0	10 ms	10 ms	0%	3 ms
Q4	2	0	0	0%	0	12 ms	12 ms	0%	1 ms
Q5	74	51	13	69%	9	18 ms	14 ms	-22%	1 ms
Q6	454	423	0	93%	13	526 ms	57 ms	-89%	64 ms
Q7	11	2	123	18%	2	12 ms	12 ms	0%	1 ms
Q8	487	458	0	94%	12	148 ms	19 ms	-87%	1 ms
Q9	14	0	48	0%	3	381 ms	394 ms	3%	23 ms
Q10	1	0	0	0%	0	11 ms	12 ms	9%	1 ms
Q11	18	16	0	89%	1	19 ms	12 ms	-37%	1 ms
Q12	1	0	0	0%	0	13 ms	14 ms	8%	1 ms
Q13	0	0	0	0%	0	0 ms	0 ms	0%	1 ms
Q14	359	344	0	96%	7	447 ms	31 ms	-93%	13 ms
Q1-14	1423	1294	184	91%	47	813 ms	95 ms	-88%	21 ms

The channels with the highest average notification time in the SENS standard processing mode are Q6, Q8, Q9, and Q14 (marked rows). In this configuration, the average notification time directly corresponds to the query execution time. Hence, a notification time of ~500ms not only causes recognizable delays for the client but also imposes a high load on the server.

The update heuristic works best for those channels with a high number of notifications (Q5, Q6, Q8, and Q14). Since more statistical information about published graphs and found bindings can be collected, also the hit rate is improved. The resulting average notification times are reduced by 22%, 89%, 87%, and 93% of the original notification times.

The tests showed that the notification times for three of the four channels with particularly long notification times in the standard mode could be significantly reduced in the heuristics mode (Q6, Q8, and Q14). The reason for this lies in the high number of notifications of these channels. On the one hand, a high number of notifications provides for better hit rates during the test phase. On the other hand, a

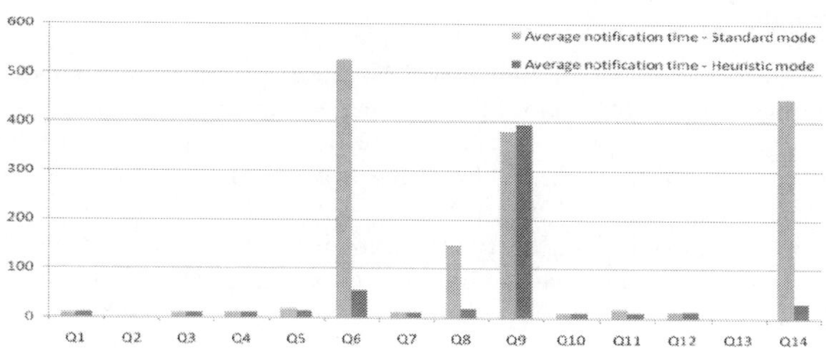

Fig. 3. Average notification times (ms)

high number of notifications is also often the cause for high response times. The result sets of queries with many result updates grow steadily (74, 454, 487, and 359 notifications) and even if the size of a result set is decreased by deleting knowledge from the knowledge base, the average amount of contained bindings is typically high. Since the execution time of a query primarily depends on the complexity of the query and on the size of the result set, the execution time of these queries is usually higher than the execution time of those with a low number of notifications.

A comparison of response times of all 14 channels is provided in Figure 3.

Employing the result set update algorithm led to response times of less than 60ms for all channels except of Q9. For Q9, the high response time of originally 381ms is primarily caused by the structure and complexity of the registered query but not by a large result set (only 14 notifications). The query is characterized by the most classes and properties of all 14 LUBM queries and includes a triangular relationship between its variables. Moreover, new query results frequently contain key elements which are not contained in the published graph and this prevents the result set update heuristic from predicting correct new bindings. Consequently, the high response time in the heuristics mode results from the original processing time plus a small overhead generated by the test and analysis phases.

While the tests showed promising results with respect to the achieved notification times, it also needs to be taken into account that the test and analysis phases of the publication operations cause a certain processing overhead. Especially for those channels with a high number of mapping entries, these two phases could even last longer than the query execution phase. If this is the case, the heuristic update mechanism should either be deactivated (for the particular channel) or the number of mapping entries should be reduced, e.g., by dropping those with comparably low hit rates.

In summary, the tests showed that the most significant improvements are obtained for channels with a high number of notifications. However, the effectiveness of the heuristic also highly depends on particular characteristics of the published graphs and the registered query. Generally, the highest hit rates will be achieved when (i) key elements of bindings are typically contained in the published graphs and (ii) there is high similarity between the published graphs and the new bindings.

4.2 Throughput

In the previous section we showed that the proposed query result update heuristic reduces response times of certain kinds of queries. However, employing the update heuristic does not yet improve the throughput of the system. For this purpose, the collected statistics can be leveraged in another way: If most of the new bindings are found by the update heuristic, then the query processing phase could be skipped in certain intervals. In many cases, query processing is the most costly task of publication processing and skipping this phase would save the according processing time.

Figure 4 and Figure 5 show how many of the previous 10 notifications were found by the update heuristic for the channels Q5 and Q6, respectively. Both figures show the last 1000 publication operations.

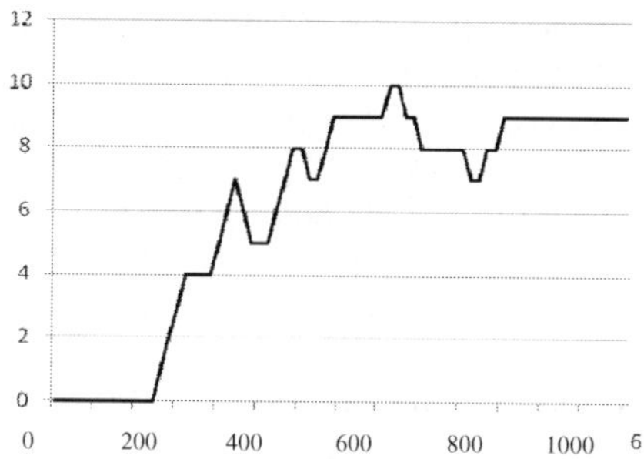

Fig. 4. Hits per 10 Notifications (Q5)

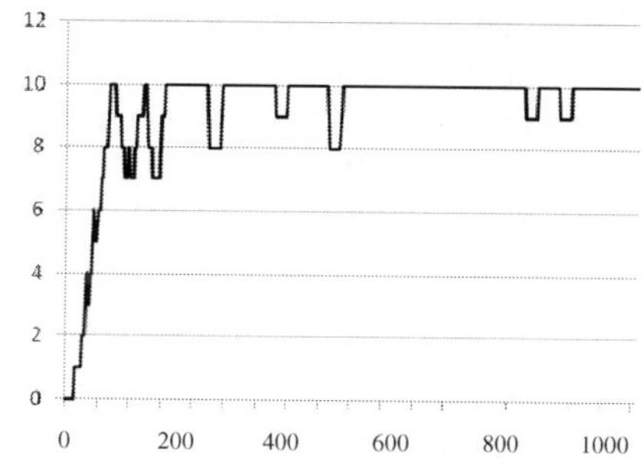

Fig. 5. Hits per 10 Notifications (Q6)

For channel Q5, the number of found bindings increases until the 400th publication operation and then remains between 7 and 10. If the query processing would only be executed in certain greater intervals, an average of ~20% of the notifications would not be found immediately. These notifications would be sent to the subscribers after the next query processing phase and exhibit essentially increased notification times. In contrast, the number of found bindings for channel Q6 increases until publication number 100 and after about 200 publication operations, the update heuristic finds almost all new bindings. Consequently, skipping the query processing for this channel introduces only a low risk of missing new bindings.

We implemented a simple extension for SENS that records the rate of notifications per publication as well as the rate of notifications found by the update heuristic. As an experimental configuration, we defined that if a channel exhibits an average of less than 0.1 notifications per publication or the update heuristic finds more than 80% of the new bindings, then the query processing is skipped for up to 10 publication processing cycles. Table 2 shows a comparison of average notification time, average time required for processing a publication, and the publication throughput (processed publication operations per second) for all three configurations: SENS standard mode, SENS with heuristic result set updates, and SENS with heuristic result set updates and applying the throughput optimization configuration.

Table 2. LUBM(10)/SwiftOWLIM 2.9 Throughput

	Standard			Heuristic			Heuristic - Optim. Query. Proc.		
Channel	Avg. Notif. Time	Avg. Pub. Proc. Time	Pub./sec	Avg. Notif. Time	Avg. Pub. Proc. Time	Pub./sec	Avg. Notif. Time	Avg. Pub. Proc. Time	Pub./sec
Q1	11 ms	10 ms	100.0	12 ms	12 ms	83.3	163 ms	10 ms	100.0
Q2	0 ms	32 ms	31.3	0 ms	36 ms	27.8	0 ms	32 ms	31.3
Q3	10 ms	10 ms	100.0	10 ms	13 ms	76.9	164 ms	10 ms	100.0
Q4	12 ms	12 ms	83.3	12 ms	13 ms	76.9	149 ms	12 ms	83.3
Q5	18 ms	19 ms	52.6	14 ms	20 ms	50.0	122 ms	14 ms	71.4
Q6	526 ms	556 ms	1.8	57 ms	620 ms	1.6	66 ms	145 ms	6.9
Q7	12 ms	11 ms	90.9	12 ms	12 ms	83.3	188 ms	11 ms	90.9
Q8	148 ms	150 ms	6.7	19 ms	151 ms	6.6	53 ms	41 ms	24.4
Q9	381 ms	394 ms	2.5	394 ms	417 ms	2.4	510 ms	77 ms	13.0
Q10	11 ms	11 ms	90.9	12 ms	12 ms	83.3	163 ms	11 ms	90.9
Q11	19 ms	20 ms	50.0	12 ms	21 ms	47.6	60 ms	13 ms	76.9
Q12	13 ms	13 ms	76.9	14 ms	14 ms	71.4	104 ms	11 ms	90.9
Q13	0 ms	11 ms	90.9	0 ms	12 ms	83.3	0 ms	11 ms	90.9
Q14	447 ms	462 ms	2.2	31 ms	475 ms	2.1	47 ms	93 ms	10.8
Q1-14	813 ms	1425 ms	0.7	95 ms	1446 ms	0.7	171 ms	299 ms	3.3

The channels with the highest average publication processing times in both SENS standard and heuristics mode are Q6, Q8, Q9, and Q14 (marked rows).

As expected, the throughput optimization extension works best for channels with a hit rate greater than 80% (see Table 1), namely Q6, Q8, and Q14. For these three channels, the publication rate increased by a factor of 3.7-5.0. Since the extension also skips query processing cycles for channels with a low notification frequency, the publication throughput of Q9 could also be increased from 2.5 to 13.0 publications per second (factor 5.1).

For the other channels, only small throughput improvements could be achieved. As query processing takes only a view milliseconds for these channels, the overhead introduced by the testing and analysis phase is almost as high as the time saved by skipping a query processing cycle. Furthermore, these channels exhibit essentially worse notification times, as SENS requires up to 10 processing cycles until it finds a new binding.

In summary, we saw that there is a trade-off between notification time and system throughput for all 14 channels. By skipping query processing cycles, more operations can be processed, but the probability of missing notifications increases accordingly. Concluding from the implemented tests, skipping query processing cycles is thus best applied for channels with (i) high query processing times and (ii) either low notification frequency or a high percentage of notifications found by the update heuristic.

5 Conclusion

In this paper, we introduced a heuristics framework for incremental result set updates for continuous queries on highly dynamic knowledge bases. We presented a simple heuristic for this framework and described how the collected statistical information can be employed to improve responsiveness and throughput of the semantic subscription system. We conducted an experimental evaluation based on a concrete application scenario, which demonstrated that the approach can successfully be employed to reduce response times of continuous queries by up to 90% of the original response times. Moreover, by leveraging the collected statistics for optimizing the query execution intervals, throughput could be improved by a factor of 3.7-5.1 for the most heavyweight publication operations.

Nonetheless, we also saw that the performance improvement highly depends on the concrete query. Consequently, we plan to further investigate the behavior of the heuristic in other benchmark environments as well as in real world applications in order to derive more detailed characteristics of applications for which the heuristic approach is best applied. Moreover, future work will include the improvement of the heuristic algorithm and the development of a framework extension that automatically selects the most suitable out of a collection of heuristics based on previously collected statistics.

Acknowledgements. We would like to thank Vesna Sesum-Cavic and Laszlo Keszthelyi for their valuable suggestions and numerous comments on this work.

References

1. Nixon, L.J.B., Teymourian, K., Krummenacher, R., Moritsch, H., Momtchev, V., Ghioni, A., Schütz, A.: Semantic Clustering and Self-Organisation in Triple Space, TripCom Research Deliverable (2008), http://www.tripcom.org/docs/del/D2.4.pdf
2. Catarci, T., Lenzerini, M.: Representing and Using Interschema Knowledge in Cooperative Information Systems. Journal of Intelligent and Cooperative Information Systems 2(4), 375–398 (1993)
3. Murth, M., Kühn, e.: A Semantic Event Notification Service for Knowledge-Driven Coordination. In: Proc. of 1st Int'l. workshop on emergent semantics and cooperation in open systems (ESTEEM), cooperation with the 2nd Int'l. Conf. on Distributed Event-Based Systems (DEBS 2008), Rome, Italy, July 1 (2008)
4. Petrovic, M., Liu, H., Jacobsen, H.: G-ToPSS: Fast Filtering of Graph-based Metadata. In: Proceedings of the 14th International Conference on World Wide Web (WWW 2005). ACM Press, New York (2005)
5. Wang, J., Jin, B., Li, J.: An ontology-based publish/Subscribe system. In: Jacobsen, H.-A. (ed.) Middleware 2004. LNCS, vol. 3231, pp. 232–253. Springer, Heidelberg (2004)
6. Murth, M., Kühn, e.: Knowledge-based Coordination with a Reliable Semantic Subscription Mechanism. In: Proc. of 24th ACM Symposium of Applied Computing (SAC 2009) – Special Track on Coordination Models, Languages and Applications. ACM Press, Honolulu (2009)
7. Nixon, L.J.B., Simperl, E., Krummenacher, R., Martin-Recuerda, F.: Tuplespace-based Computing for the Semantic Web: A Survey of the State-of-the-Art. The Knowledge Engineering Review 23(2) (2008)
8. Halaschek-Wiener, C., Parsia, B., Sirin, E.: Description logic reasoning with syntactic updates. In: Meersman, R., Tari, Z. (eds.) OTM 2006. LNCS, vol. 4275, pp. 722–737. Springer, Heidelberg (2006)
9. Baader, F., Calvanese, D., McGuinness, D.L., Nardi, D., Patel-Schneider, P.F.: The Description Logic Handbook: Theory, Implementation, Applications. Cambridge University Press, Cambridge (2003)
10. Khushraj, D., Lassila, O., Finin, T.W.: sTuples: Semantic Tuple Spaces. In: 1st Ann. Int'l. Conf. on Mobile and Ubiquitous Systems (2004)
11. Fensel, D., Krummenacher, R., Shafiq, O., Kühn, e., Riemer, J., Ding, Y., Draxler, B.: TSC - Triple Space Computing, In Special issue on ICT research in Austria. Journal of Electronics & Information Technology (e&i Elektrotechnik & Informationstechnik) (January-February 2007)
12. Nixon, L.J.B., Paslaru Bontas Simperl, E., Antonenko, O., Tolksdorf, R.: Towards Semantic Tuplespace Computing: The Semantic Web Spaces System. In: 22nd Ann. ACM Symposium on Applied Computing (2007)
13. Simperl, E., Krummenacher, R., Nixon, L.: A Coordination Model for Triplespace Computing. In: Murphy, A.L., Vitek, J. (eds.) COORDINATION 2007. LNCS, vol. 4467, pp. 1–18. Springer, Heidelberg (2007)
14. Gelernter, D.: Generative Communication in Linda. ACM Transactions in Programming Languages and Systems (TOPLAS) 7(1), 80–112 (1985)
15. Cerzza, D., Della Valle, E., Foxvog, D., Krummenacher, R., Murth, M.: Towards European Patient Summaries based on Triple Space Computing. In: Proc. of 1st European Conf. on eHealth, Fribourg, Switzerland, October 12-13 (2006)
16. Jena. Jena – A Semantic Web Framework for Java, http://jena.sourceforge.net/ (last accessed: September 2007)

17. Broekstra, J., Kampman, A., van Harmelen, F.: Sesame: A Generic Architecture for Storing and Querying RDF and RDF Schema. In: Horrocks, I., Hendler, J. (eds.) ISWC 2002. LNCS, vol. 2342, pp. 54–68. Springer, Heidelberg (2002)
18. ARQ – A SPARQL Processor for Jena, `http://jena.sourceforge.net/ARQ/`
19. Kaon2. Ontology management for the semantic web, `http://kaon2.semanticweb.org/`
20. Cai, M., Frank, M.: RDFPeers: A Scalable Distributed RDF Repository Based on a Structured Peer-to-Peer Network. In: 13th Int'l. Conference on World Wide Web (2004)
21. Liarou, E., Idreos, S., Koubarakis, M.: Continuous RDF Query Processing over DHTs. In: Aberer, K., Choi, K.-S., Noy, N., Allemang, D., Lee, K.-I., Nixon, L., Golbeck, J., Mika, P., Maynard, D., Mizoguchi, R., Schreiber, G., Cudré-Mauroux, P. (eds.) ASWC 2007 and ISWC 2007. LNCS, vol. 4825, pp. 324–339. Springer, Heidelberg (2007)
22. Prud'hommeaux, E., Seaborne, A.: SPARQL Query Language for RDF. W3C Recommendation (2008), `http://www.w3.org/TR/rdf-sparql-query/`
23. Luckham, D.: The Power of Events: An Introduction to Complex Event Processing in Distributed Enterprise Systems. Addison-Wesley, Reading (2002)
24. Guo, Y., Pan, Z., Heflin, J.: LUBM: A Benchmark for OWL Knowledge Base Systems. Journal of Web Semantics 3(2), 158–182 (2005)
25. Ontotext. OWLIM Fact Sheet, `http://www.ontotext.com/owlim/OWLIMFactSheet.pdf`
26. Manola, F., Miller, E.: RDF Primer W3C-Recommend (2004), `http://www.w3.org/TR/rdf-primer/`
27. McGuinness, D.L., van Harmelen, F.: OWL Web Ontology Language, W3C Recommendation (2004), `http://www.w3.org/TR/owl-features/`

Towards Linguistically Grounded Ontologies

Paul Buitelaar[1], Philipp Cimiano[2], Peter Haase[3], and Michael Sintek[4]

[1] DERI, Unit for Natural Language Processing
National University of Ireland, Galway
paul.buitelaar@deri.org
[2] Web Information Systems Group, TU Delft, The Netherlands
p.cimiano@tudelft.nl
[3] Institute AIFB, Universität Karlsruhe (TH), Germany
haase@aifb.uni-karlsruhe.de
[4] Knowledge Management Dept. & Competence Center Semantic Web
DFKI GmbH, Kaiserslautern, Germany
sintek@dfki.de

Abstract. In this paper we argue why it is necessary to associate linguistic information with ontologies and why more expressive models, beyond RDFS, OWL and SKOS, are needed to capture the relation between natural language constructs on the one hand and ontological entities on the other. We argue that in the light of tasks such as ontology-based information extraction, ontology learning and population from text and natural language generation from ontologies, currently available data-models are not sufficient as they only allow to associate atomic terms without linguistic grounding or structure to ontology elements. Towards realizing a more expressive model for associating linguistic information to ontology elements, we base our work presented here on previously developed models (*LingInfo, LexOnto, LMF*) and present a new joint model for linguistic grounding of ontologies called *LexInfo*. LexInfo combines essential design aspects of LingInfo and LexOnto and builds on a sound model for representing computational lexica called LMF which has been recently approved as a standard under ISO.

1 Introduction

Standards for representing ontologies have been developed in the last decade, in particular RDF Schema (RDFS [2]) and OWL [1,10]. Whereas ontologies are logical theories and independent of natural language,[1] a grounding in natural language is needed for several reasons:

- When engineering an ontology, human developers will be able to better understand and manipulate ontologies. Associating linguistic information to ontologies (in the simplest form by labels) allows people to ground concepts

[1] We are aware that some people might disagree with this statement (see [16]). Independently of the position one adopts here, the need for associating rich linguistic information to ontologies remains unaffected.

L. Aroyo et al. (Eds.): ESWC 2009, LNCS 5554, pp. 111–125, 2009.

and relations defined in the ontology with their own linguistic and cognitive systems.
- In ontology population, automatic procedures for ontology-based information extraction from text will be better equipped to link textual data with ontology elements when these are associated with information on their linguistic realization.
- In verbalizing an ontology, i.e., in generating natural language text descriptions, richer models that capture how concepts and relations are realized linguistically will be needed.

However, the development of models for associating linguistic information (part-of-speech, morphological decomposition, subcategorization frames etc.) with ontology elements (concepts, relations, individuals, etc.) is not as advanced as the corresponding ontology representation languages. While RDFS/OWL allow to associate labels to ontology elements, we argue that this is not enough for actual use of ontologies in connection with human users and textual data as described above. Even the datamodel behind SKOS [11] does not suffice for these purposes, as it only allows for the representation of atomic terms (in addition to plain labels as with RDFS/OWL) without a possibility for representing their linguistic (sub-)structure. However, SKOS was not developed with the aim of associating lexical and linguistic information to arbitrary (domain) ontologies, but with the goal of producing a datamodel (building on RDFS/OWL) for representing classification schemas. Thus, by definition, SKOS does not fulfill the requirements for an expressive model that allows for the association of linguistic information to arbitrary (domain) ontologies.

In this paper we present a unified model for associating linguistic information to ontologies. We call this model *LexInfo* as it emerged out of efforts of aligning the *LingInfo* ([4,5]) and *LexOnto* ([6]) models that were developed independently from each other but with similar goals and motivations. The LingInfo model provides a mechanism for modeling label-internal linguistic structure: inflection and morphosyntactic decomposition of complex ontology labels (i.e., of complex terms). The LexOnto model on the other hand enables the representation of label-external linguistic structure: predicate-argument structure that can be projected by lexical heads of ontology labels (terms) and their mapping to corresponding ontology elements. While the two models have the same aim of providing more expressive lexicon models for ontologies, they have focused on rather complementary aspects. In addition, LexInfo builds on the *Lexical Markup Framework* or LMF[2] ([7,8]), a metamodel for describing computational lexica that has been recently approved as an ISO standard under number ISO-24613:2008.

Our main contribution in this paper is the LexInfo model itself, besides providing a clear motivation why more expressive models for associating linguistic knowledge to ontologies are necessary. In aligning LingInfo, LexOnto and LMF, we also introduce machinery that allows for describing the semantic aspects of a computational lexicon with respect to a given ontology. A principled knowledge

[2] `http://www.lexicalmarkupframework.org/`

representation approach in both directions is necessary to enable reuse of lexical knowledge for particular application domains, i.e., to prevent ad-hoc integration of linguistic and domain knowledge for every new application. We therefore hope that the work presented here can provide a solid basis for any future discussion on standardization of lexicon models for OWL (and also RDFS) ontologies. Further development of the Semantic Web will build crucially on the analysis of unstructured and in particular text data, which requires linguistic knowledge to be associated with ontologies in order to enable automatic extraction of semantic knowledge from text. The LexInfo vision as outlined in this paper will support this by providing a detailed proposal for a standardized approach to linguistically grounded ontologies. LexInfo can be employed by use of a first API for this model available at `http://ontoware.org/projects/lexonto/`

The paper is structured as follows. In Sect. 2 we provide an extensive motivation for the work discussed here as well as a comparison with related work. In Sect. 3 we briefly discuss the LingInfo and LexOnto models, as well as the LMF model, thus providing the basis for understanding our newly proposed model LexInfo that we describe in more detail also in this section. Finally, in Sect. 4 we draw some conclusions from the work presented and discuss ideas for future work.

2 Motivation

In this section, we first argue why a separation between the linguistic and ontological levels is needed. Second, we motivate why a more flexible coupling between linguistic structure and ontology classes is needed, beyond what is provided by the labeling system of RDFS. Third, we motivate why subcategorization frames and predicate-argument structures should be an integral part of any proposal for linguistic grounding of ontologies. Finally, we discuss why previous work fails to address the requirements of an expressive model that allows us to associate linguistic information to ontologies.

2.1 Separation between Linguistic and Ontological Level

RDFS and OWL allow for the representation of a 'concept handler' through specification of the `rdfs:label` property, which is defined for `Resource` as domain and `Literal` as range. We could use this to specify that the class `cat` is typically expressed in natural language by words such as *'cat'* (in English), *'Katze'* (in German), etc. If we additionally want to represent linguistic variants of *'cat'*, e.g., its plural *'cats'*, the RDFS data model gives us only one choice, namely to add an additional and independent label, yielding something like:

```
<rdfs:Class about="#Cat">
  <rdfs:label xml:lang="en">cat</rdfs:label>
  <rdfs:label xml:lang="en">cats</rdfs:label>
  <rdfs:label xml:lang="de">Katze</rdfs:label>
  <rdfs:label xml:lang="de">Katzen</rdfs:label>
</rdfs:Class>
```

Although RDFS thus allows us to represent variants (different labels for the same concept), this is unsatisfactory as it fails to capture the linguistic relation between *'cat'* and *'cats'*, i.e., that the latter is the plural of the former. Such linguistic properties of ontology labels however should have no place in the domain ontology and should be captured in a separate linguistic model (i.e., *'lexicon'*) with appropriate pointers to the domain ontology. Along these lines, much of the lexical modeling is outsourced to the lexicon ontology, producing additional modeling "overhead" at the lexicon side, but clearly separating the two representation levels.

2.2 Flexible Coupling of the Ontological and Language Systems

The label property for RDFS and OWL in essence specifies an $n : m$ relation between a class or property and one or more labels, without allowing for a more complex correspondence between a class or property on one side and a "syntagmatic"[3] composition of several labels on the other. Why a more complex correspondence is needed may be explained with the following example. Let us consider a composite term like the German *'Schweineschnitzel'* (*pork cutlet*). We have the following possibilities to associate this term with ontological elements:

- There might be a class `Schweineschnitzel` to which *'Schweineschnitzel'* can refer to.
- There might be a composite class $Schnitzel \sqcap \exists madeOf.Pork$ to which *'Schweineschnitzel'* can point to.
- There might be simply the general class `schnitzel`. In which case we want to specify that only the second part of the composite term *'Schweineschnitzel'*, i.e., *'schnitzel'* refers to the class `schnitzel`.
- There might be both classes `pork` and `schnitzel` represented. In which case we want to specify that the second part of the composite term refers to the class `schnitzel` and the first part to the class `pork`.

It thus seems that we require a flexible system for associating terms to concepts that is sensitive to the way concepts or properties have been modeled and allowing to assign them to the whole term or to individual parts of it. Further, we see it as a requirement that this model does not assume that the linguistic and ontological levels are "fully synchronized".[4] For this we need appropriate means to represent the decomposition of terms and for associating ontological entities to terms and their sub-structure. Obviously, this is out of the scope of the RDFS label system, as it does not allow to model any of the semantic implications of the morphosyntactic, internal structure of complex labels (i.e., composite terms).

[3] Syntagmatic relations are between words in a sentence in sequence, whereas "paradigmatic" relations are between words according to meaning, i.e., between synonyms.

[4] Hirst [9] even argues that they cannot be synchronized as there are ontological distinctions that are never lexicalized and linguistic distinctions that are ontologically irrelevant.

2.3 Subcategorization and Predicate-Argument Structure

Further motivation for the definition of more expressive, linguistically grounded ontologies is provided by properties such as in the following examples:

```
<rdf:Property about="#capital">
  <rdfs:domain rdf:resource="#Country"/>
  <rdfs:range rdf:resource="#City"/>
  <rdfs:label xml:lang="en">capital</rdfs:label>
</rdf:Property>

<rdf:Property about="#flowThrough">
  <rdfs:domain rdf:resource="#River"/>
  <rdfs:range rdf:resource="#City"/>
  <rdfs:label xml:lang="en">flow through</rdfs:label>
</rdf:Property>

<rdf:Property about="#locatedAt">
  <rdfs:domain rdf:resource="#City"/>
  <rdfs:range rdf:resource="#Highway"/>
  <rdfs:label xml:lang="en">located at/rdfs:label>
</rdf:Property>
```

Although each property in these examples has been associated with meaningful labels (*'capital'*, *'flow through'*, *'located at'*) this is not sufficient for various reasons:

- Lack of linguistic information about part-of-speech of the lexical item expressed by the label. Consider for example the `capital` property and assume we want to generate a natural language description for the triple $(Germany, capital, Berlin)$. To prevent a system from generating a sentence like *"Germany capitals Berlin."*, it needs to know that *capital* is a noun and cannot be used as a verb. Capturing part-of-speech information (defining if it expresses a noun, verb, etc.) for labels is thus essential.
- Lack of deeper linguistic knowledge on subcategorization frames[5] that constrain the linguistic constructions in which such labels may appear. Consider for example the `flowThrough` relation in generating a natural language description for the triple $(Rhein, flowThrough, Karlsruhe)$. Here we need to know that *flow* is an intransitive verb[6] that requires a prepositional phrase introduced by the preposition *'through'* in order to generate an appropriate sentence like *"The Rhein flows through Karlsruhe"* (provided we also specify morphological information about the verb *'flow'*, in particular that the 3rd person singular is *'flows'*, see the discussion on inflection above).

[5] A subcategorization frame of a word is the number and types of syntactic arguments (subject, direct object, prepositional object, etc.) as well as their linguistic structure (nominal phrase, prepositional phrase, relative clause, etc.) that it can possibly co-occur with in a sentence.

[6] Transitive verbs (e.g., *'love'*) require both a subject and a (direct) object, while intransitive verbs do require only a subject but no direct object (e.g., *'sleeps'*).

- Lack of ways for capturing the variation in relation expression, as there are many ways in which a certain relation or property can be expressed in language. Consider for example the `locatedAt` relation, which can be expressed by *"The A8 passes by Karlsruhe"*, *"The A8 connects Karlsruhe"*, *"The A8 goes through Karlsruhe"*, etc. Although we would not necessarily want to add *'pass'*, *'connect'* and *'go'* as labels to the *locatedAt* property, we may want to express that all of the corresponding verbal forms are valid ways of expressing the `locatedAt` property.
- Lack of ways for expressing how and in which order linguistic arguments of a certain verb map to corresponding semantic arguments of a predicate as modeled in the ontology. For example, given a transitive verb such as *connects*, we may want to specify that its linguistic subject maps to the range of the `locatedAt` property and its direct object to the domain, as in *[The A8: subject] connects [Karlsruhe: direct object]*, which would map to the triple `(Karlsruhe,locatedAt,A8)`.

2.4 Why Related Work Is Not Enough

Given these explanations, it is clear that more expressive models than those currently available are needed to associate linguistic information with ontology elements. In particular, we derive from the discussion above at least the following specific requirements on a richer model for grounding linguistic information in ontologies, i.e., the model should allow to:

1. capture **morphological relations** between terms, e.g., through inflection (*cat, cats*), separately from the domain ontology;
2. represent the **morphological or syntactic decomposition** of composite terms and the linking of the components to the ontology;
3. model **complex linguistic patterns**, such as subcategorization frames for specific verbs together with their mapping to arbitrary ontological structures;
4. specify the meaning of linguistic constructions with respect to an **arbitrary (domain) ontology**, and
5. clearly **separate** the linguistic and semantic (ontological) representation levels.

The definition of more expressive data models for the representation of multilingual terms and/or linguistic information with ontologies has been addressed by a number of initiatives. However, we will argue that none of these address the issue in a completely satisfactory way.

The Simple Knowledge Organization System format (SKOS [11]) is a model which allows us to represent classification schemas using the datamodels of RDFS and OWL. Thus, it has completely orthogonal goals to our proposal and fulfills none of the requirements 1–5.

The Lexical Markup Framework (LMF [7,8]) is a proposal for an interoperable metamodel to represent computational lexica. It fulfills requirements 1–3 and 5 but does not allow to specify the semantics of linguistic constructs with respect to an arbitrary domain ontology (req. 4). Though LMF is not incompatible with this requirement (as we will see later), this has not been worked out so far.

The Linguistic Information Repository (LIR [14]) represents a metamodel for associating multilingual terms to ontology elements. As the LexOnto model [6], it builds on the OWL metamodel. It fails to fulfill requirements 1–3, whereas it clearly fulfills requirements 4 and 5.

Additionally, some natural language processing frameworks rely on ontologies as representations of linguistic meaning in their lexicon models. However, these approaches are typically restricted to specific ontologies, i.e., the Mikrokosmos ontology in Ontological Semantics [12]. Such approaches clearly fail on requirement 4 as they typically build on proprietary or general (top-level) ontologies. Arguably, it would be possible to generalize the machinery used in these projects to accommodate arbitrary ontologies, but this seems not to have been a focus so far.

Finally, the Linguistic Watermark Suite [13] also provides a metamodel for representing lexical resources, which makes it similar to LMF, but the focus seems on the software framework allowing to import different lexical resources in various formats into ontologies.

For a more detailed description of these related approaches and the main differences to our approach see our technical report on LexInfo [3].[7]

3 Towards an Ontological and Linguistic Joint Model

3.1 Previous Work

In this section we briefly introduce the basic ideas behind the models that LexInfo builds on: LingInfo, LexOnto and LMF.

LingInfo: To allow for a direct connection of linguistic information with corresponding classes and properties in a domain ontology, Buitelaar et al. ([5,4]) developed an RDFS-based lexicon model (LingInfo) that enables the association of linguistic information with ontology elements through the definition of 'LingInfo' objects, i.e., terms with their linguistic (morphosyntactic: inflection, decomposition) information. To accomplish this, these LingInfo objects (modeled as instances of a class `LingInfo`) are attached to classes and properties with a property `linginfo`, which is defined on these classes and properties with the help of a meta-class[8] (`ClassWithLingInfo`) and a meta-property (`PropertyWithLingInfo`). We refer to [5] and [4] and the LingInfo website[9] for details.

LexOnto: The LexOnto model [6] has been developed in order to specify the meaning of complex linguistic structures (in particular subcategorization frames)

[7] Available online at
`http://www.aifb.uni-karlsruhe.de/WBS/pci/lexinfo_tech_report_08.pdf`

[8] A meta-class is a class the instances of which are again classes.

[9] `http://olp.dfki.de/LingInfo/`

with respect to ontology elements (in particular properties). The main class of the LexOnto model is the class `LexicalElement`, which has the subclasses `PredicativeLexicalElement` (PLE) and `WordForm`. `WordForms` correspond to nouns, verbs and adjectives that project a predicate-argument structure. PLEs correspond to predicative lexical elements, i.e., subcategorization structures for verbs and nouns as well as adjectives seen as predicates. LexOnto focuses in particular on the mapping of subcategorization frames for verbs (and nouns) to predicate-argument structures, allowing to clearly specify how and in which order the linguistic arguments of a verb map to the semantic arguments (domain and range) of a corresponding property. We refer the interested reader to [6] and [3] and the LexOnto project site[10] for details.

Lexical Markup Framework (LMF): The Lexical Markup Framework is a meta-model that provides a standardized framework for the creation and use of computational lexica, allowing for interoperability and reuseability across applications and tasks [8]. As the lexicon for an ontology is a special type of computational lexicon, we build on LMF to describe lexica for ontologies. The LMF meta-model is organized in a number of packages of which the following are relevant to our work (see the LMF specification [8] and our technical report [3] for further details):

1. **core package:** containing the basic classes `Lexical Resource`, `Lexicon`, `Global Information`, `LexicalEntry` etc.
2. **morphology extension:** providing a mechanism for describing the morphological structure of lexical entries (extensionally)
3. **NLP syntax extension:** allowing to describe the syntactic behavior and properties of a lexical entry, in particular the subcategorization frame structure for predicative elements such as verbs etc.
4. **NLP semantics extension:** providing a way to associate semantic representation structures to syntactic structures, which has clearly a strong relation with the syntax package, allowing to define semantic predicates and associate their semantic arguments to syntactic arguments of a subcategorization frame.

We will see in the next section how we build on the syntax and semantics extensions of LMF to capture the modeling of subcategorization frames and their mapping to ontological structures as well as the morphology extension to capture the inflection and decomposition aspects.

3.2 The LexInfo Model

Our starting point for defining the LexInfo model was to build on the previously discussed LingInfo, LexOnto and LMF models, with the latter providing the glue for integrating these three frameworks. We proceeded as follows:

---
[10] `http://ontoware.org/projects/lexonto/`

- We downloaded the OWL version of the LMF model from the LMF website.[11]
- Unfortunately this was not a valid OWL ontology, so we fixed the errors of this ontology, thus yielding a syntactically valid OWL ontology. We also sent the error list to the LMF Working Group to allow for the correction of the LMF metamodel serialization in OWL.
- We commented most of the ontology classes on the basis of the descriptions of the LMF Specification model. The resulting corrected and documented LMF ontology is available for download.[12]
- As the ontology has been originally created starting from an UML model and only uses the property `AssociatedTo`, we have introduced appropriate subproperties of `AssociatedTo` for most of the associations between entities described in the LMF Specification.
- Then, we created a new ontology `LexInfo` importing the corrected LMF ontology, introducing our monotonic extensions on top of it. The LexInfo ontology can be downloaded as well.[13]

While we gloss over the aspects related to multilinguality in this paper, it is important to note that according to the LMF model, language information is attached to the object representing the whole `Lexicon` object. As a consequence, we need different lexica for each of the languages we consider and the language information is inherited to all the elements contained in the lexicon (in particular to the `LexicalEntry` objects). Interesting questions are here for example if lexica for different languages can share lexical entries, which would foster conciseness. We leave such questions as well as the exact specification of the semantics of the inheritance aside for future work.

The LexInfo model meets our requirements 1–5 as follows:

Req. 1: Capturing Morphological Relations between Terms. Morphological relations between terms are already directly captured by the LMF machinery. For example, inflected forms can be attached to a base `LexicalEntry` for *'Schwein'* for which alternative `writtenForms` can be specified (e.g., the plural *'Schweine'*, the genitive *'Schweins'* etc.). No crucial extensions of the LMF model were needed for this other than the labeling of the relations which were all named `AssociatedTo` in the original LMF model (see discussion above).

Req. 2: Morpho-syntactic Decomposition of Complex Terms. The morphological decomposition of terms is done in LexInfo by building on the morphological extension package of LMF, which essentially allows us to associate a `ListOfComponents` with a `LexicalEntry`, which has an ordered list of `Components` (with a minimum of 2) (see [8]). We have modeled this in OWL by introducing an additional datatype property `order` specifying the absolute order of a `Component` within a `ListOfComponents`. Components then in

[11] http://www.lexicalmarkupframework.org/

[12] http://lexonto.ontoware.org/lmf

[13] http://lexonto.ontoware.org/lexinfo

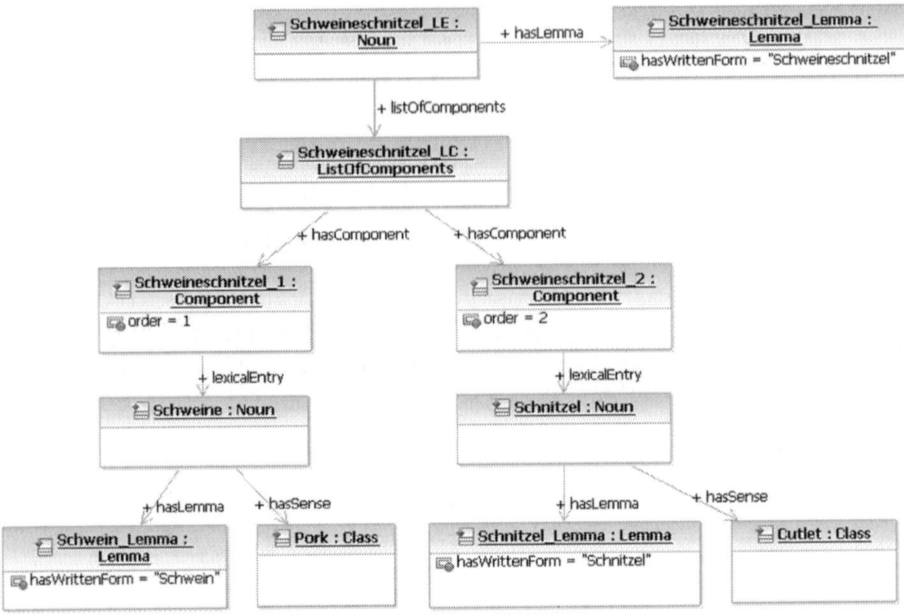

Fig. 1. Example of decomposition with linking to ontology concepts

essence point to `LexicalEntries` which can again be composite, thus allowing for recursion. In order to capture how the parts of a compound are associated to the ontology, we build on the general mechanism of LMF allowing to associate `LexicalEntry` objects with a `Sense`, for which we define subclasses `owl:Property` and `owl:Class`[14] (reusing the OWL 2 metamodel[15]). In this way, we are able to state that *'Schweineschnitzel'* is composed of two `LexicalEntry` objects where the first refers to the class `pork` and the second to the class `cutlet` (see Fig. 1). The crucial extension of LMF was here the fact that `Entity` (in the OWL 2 metamodel) is specified as a subclass of `lmf:Sense` in the LexInfo ontology.

Req. 3: Representing Subcategorization Frames. Figure 2 depicts the modeling of the mapping of subcategorization frames to ontological structures at the schema level. Figure 3 in particular shows how the *'flows(subj,(through) pobj)'* subcategorization frame is mapped to the `lexinfo:PropertyPredicate` standing proxy for the `flowsThrough` property (and linked through the `lexinfo:property` to an instance of the OWL meta-ontology representing the property in question). At the syntactic level, we model the intransitive verb *'flow'* by use of an instance of `lexinfo:IntransitivePP` (subclass

[14] Note that it is debatable whether classes and properties actually *are* senses or whether they just *convey* senses; for pragmatic reasons, we decided to go for the first interpretation.

[15] `http://owlodm.ontoware.org/OWL2`

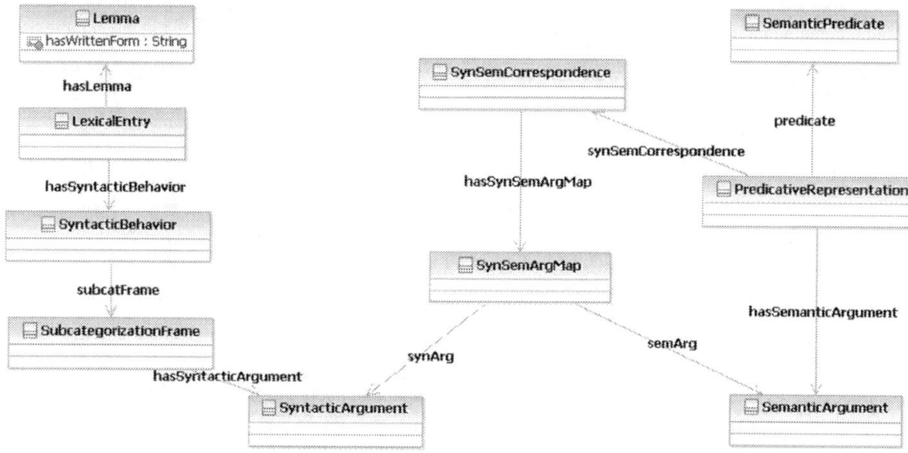

Fig. 2. Modeling the Syntax/Semantics Correspondence

of `lmf:SubcategorizationFrame`), subcategorizing for a `lexinfo:Subject` and prepositional object (`lexinfo:PObject`) as arguments. We follow here the LMF modeling in terms of an `lmf:SyntacticBehaviour` object linking to the corresponding subcategorization frame. At the semantic level, we instantiate a `lmf:PredicativeRepresentation`, in this case in particular a `lexinfo:PropertyPredicativeRepresentation` linking to the property `flowsThrough` (`PropertyPredicate`) as a subclass of `SemanticPredicate`. The semantic arguments (domain and range) of the `flowsThrough` property are attached to the `PropertyPredicate` instance. The crucial class establishing the connection between the syntax and semantic (ontological) levels is the `lmf:SynSemCorrespondence` class, which is associated to various `lmf:SynSemArgMaps` mapping a certain syntactic position, `lexinfo:Subject` and `lexinfo:PObject` in this case, to semantic arguments of an ontological predicate, in this case to the domain and range (as an instance of `lmf:SemanticArgument`) of the `flowsThrough` property, respectively.

In order to accomplish the above, we have introduced the following extensions on top of LMF in the LexInfo ontology:

1. Subclasses of `lmf:LexicalEntry`, i.e., `lexinfo:Verb`, `lexinfo:Noun` etc., which are distinguished by way of attributes in the LMF model (see Fig. 4).
2. Subclasses of `lmf:SubcategorizationFrame`, i.e., `lexinfo:Transitive`, `lexinfo:IntransitivePP`, etc. (see Fig. 5).
3. Subclasses of `lmf:SyntacticArgument`, i.e., `lexinfo:Subject`, `lexinfo:Object`, `lexinfo:PObject`), etc.
4. Subclasses of the `lmf:PredicateRepresentation` and `lmf:SemanticPredicate` classes, e.g., the classes `lexinfo:ClassPredicativeRepresentation` and `lexinfo:ClassPredicate` as well as `lexinfo:PropertyPredicativeRepresentation` and `lexinfo:PropertyPredicate` allowing to refer to a class or property (as predicate), respectively (see Fig. 6).

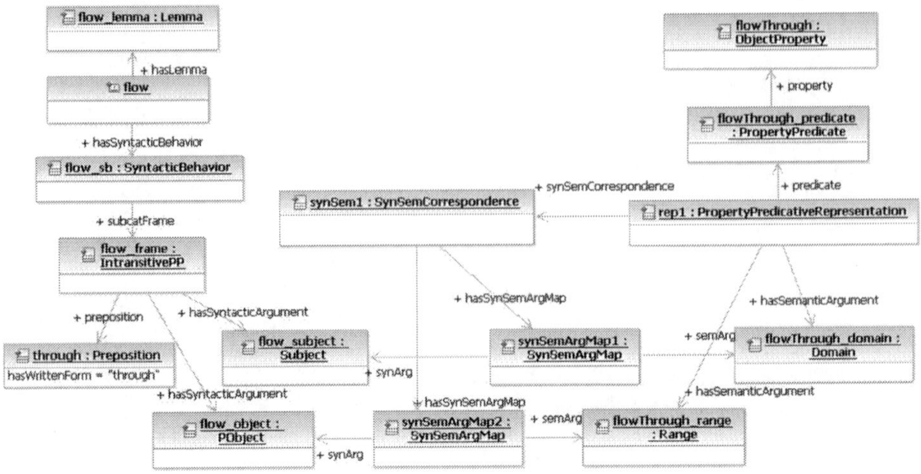

Fig. 3. Modeling the subcategorization frame *flow (through)* and its mapping to the `flowsThrough` property in LexInfo

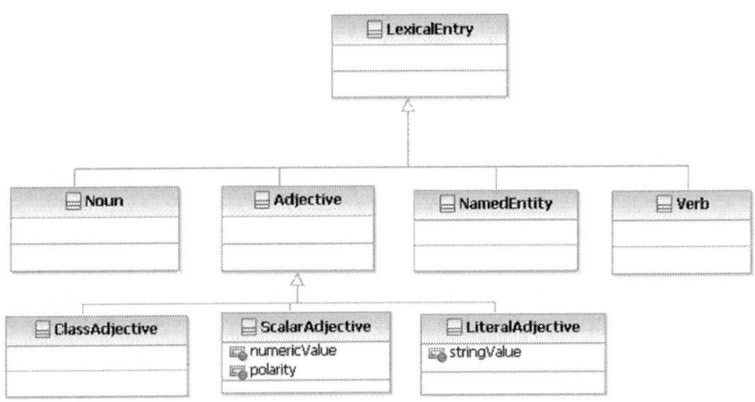

Fig. 4. Subclasses of LexicalEntry

5. Subclasses of the `lmf:SemanticArgument` class, i.e. `lexinfo:Domain`, `lexinfo:Range` etc., as well as appropriate subclasses allowing to specify the semantic arguments of a class (where properties are understood as slots of the frame represented by the class).

It is important to note that LMF also distinguishes between different types of subcategorization frames. However, the distinction is encoded as an attribute, i.e., *"regularSVO"* for a transitive verb, for instance. The advantage of modeling the different subcategorization frames as subclasses (as we have done) is that this allows us to formulate additional axioms, requiring for example that a **Transitive** subcategorization frame has exactly two syntactic arguments: a

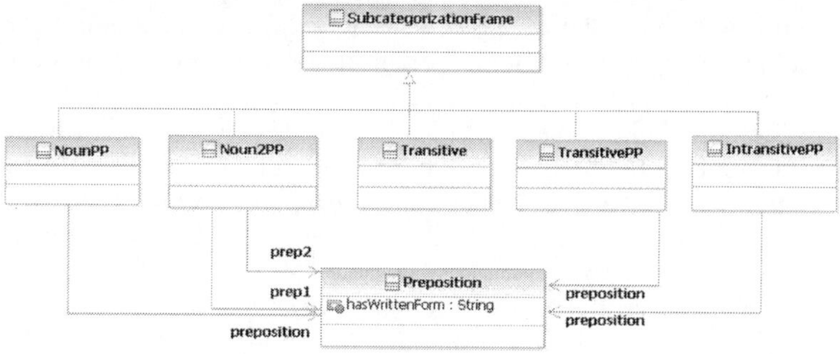

Fig. 5. Subclasses of SubcategorizationFrame

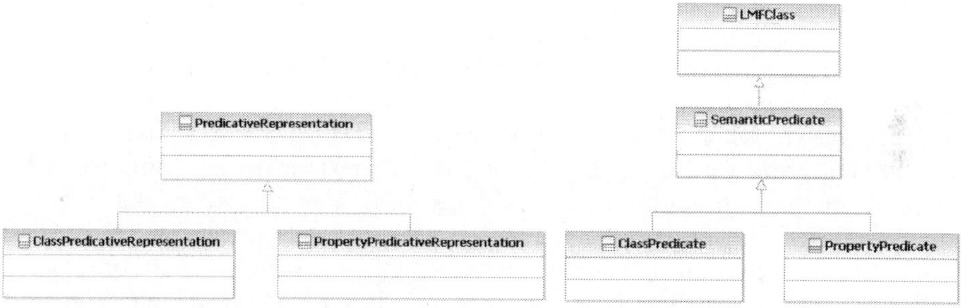

Fig. 6. Subclasses of PredicateRepresentation, SemanticPredicate

subject and an object. Such general axioms that allow to check the lexicons for consistency are clearly not possible in the original LMF model.

Req. 4: Specification of Meaning w.r.t. an Arbitrary (Domain) Ontology. The fulfillment of this requirement trivially follows from the way we have conceived our model as described above on how the semantics of terms and compounds are specified with respect to a domain ontology and how the ontological meaning of subcategorization frames is represented.

Req. 5: Clear Separation Between Linguistic and Semantic Levels. This requirement is also fulfilled, as we keep the linguistic and ontology levels fully separate, establishing the connection between them by use of the OWL 2 meta-ontology to point to classes and properties of the (domain) ontology. To some extent this requirement is already fulfilled by the LMF model in the sense that the syntactic and the semantic level are clearly separated, albeit interlinked with each other.

4 Conclusions

The interface between language and knowledge as captured by ontologies is much richer and more complex than can be expressed by current ontology models, such as the labeling system of RDFS, OWL, and SKOS. Enhanced models that couple linguistic information with ontological structure are certainly required for tasks such as ontology learning and population from text, natural language generation from ontologies, etc. While there is no doubt that ad-hoc models for representing linguistic information might be suitable for individual systems and solutions to these problems, we here propose a sound and principled model that can be used to exchange lexica across systems. While there are many lexica that can be used for the tasks we envision (population, generation etc.), we want to move towards avoiding that each application needs to make the connection between lexica and ontologies again and in an ad-hoc fashion. Our model allows to publish the link between lexica and ontologies in a declarative way on the Web together with the ontologies themselves, such that other systems can simply reuse them. Towards this goal we have clearly spelled out the requirements for such models and argued that many related proposals fall short of fulfilling these.

In this paper we have instead presented a model, *LexInfo*, that does fulfill the requirements stated and clearly spells out details of how the mapping from language to ontologies might work. In order to construct LexInfo, we have built on two previously developed models (LingInfo and LexOnto) with complementary focus. In our current proposal we used the LMF metamodel to glue together the crucial ingredients of these models. LMF represents a solid and principled framework for representing computational lexica, of which we regard ontological lexicons a special case. The LingInfo, LexOnto, LexInfo and LMF ontologies are available from the project website, as well as a corresponding Java API with an implementation for the commonly used OWL API.[16]

In future work we intend to further develop this API in building up services for the automatic generation of lexical knowledge bases on the basis of and in conjunction with domain ontologies by integrating existing LMF-conform computational lexicons and other resources such as Wikipedia etc. (see [15] for initial experiments in this direction). Additionally, we intend to continue our dialog with the LMF working group, aimed at the incorporation of aspects related to ontological meaning representation in the LMF model. Finally, we do hope to have provided a solid basis for any future discussion on standardization of lexicon models for OWL ontologies.

Acknowledgements. This research has been supported in part by the THESEUS Program in the MEDICO Project, funded by the German Federal Ministry of Economics and Technology under grant number 01MQ07016, by the MULTIPLA project, funded by the Deutsche Forschungsgemeinschaft (DFG) under grant number 38457858, and by the EU funded projects NeOn and K-Space under grant number IST-2005-027595 and FP6-027026. We thank Matthias Mantel and Thomas Wangler for developing the LexInfo API and Elena Montiel for very helpful comments on a draft versions of this paper. The responsibility for this publication lies entirely with the authors.

[16] `http://owlapi.sourceforge.net/`

References

1. Bechhofer, S., van Harmelen, F., Hendler, J., Horrocks, I., McGuinees, D.L., Patel-Schneider, P.F., Stein, L.A.: OWL Web Ontology Language Reference (2004)
2. Brickley, D., Guha, R.V.: RDF Vocabulary Description Language 1.0: RDF Schema. Technical report, W3C (2002)
3. Buitelaar, P., Cimiano, P., Haase, P., Sintek, M.: Towards linguistically grounded ontologies. Technical report, Institute AIFB, Universität Karlsruhe, TH (2008)
4. Buitelaar, P., Declerck, T., Frank, A., Racioppa, S., Kiesel, M., Sintek, M., Engel, R., Romanelli, M., Sonntag, D., Loos, B., Micelli, V., Porzel, R., Cimiano, P.: Linginfo: Design and applications of a model for the integration of linguistic information in ontologies. In: Proceedings of the LREC OntoLex 2006 Workshop (2006)
5. Buitelaar, P., Sintek, M., Kiesel, M.: A lexicon model for multilingual/multimedia ontologies. In: Sure, Y., Domingue, J. (eds.) ESWC 2006. LNCS, vol. 4011, pp. 502–513. Springer, Heidelberg (2006)
6. Cimiano, P., Haase, P., Herold, M., Mantel, M., Buitelaar, P.: LexOnto: A model for ontology lexicons for ontology-based NLP. In: Proceedings of the ISWC 2007 OntoLex Workshop (2007)
7. Francopoulo, G., Bel, N., Georg, M., Calzolari, N., Monachini, M., Pet, M., Soria, C.: Lexical markup framework: ISO standard for semantic information in NLP lexicons. In: Proceedings of the Workshop of the GLDV Working Group on Lexicography at the Biennial Spring Conference of the GLDV (2007)
8. The LMF Working Group. Language resource management – lexical markup framework (LMF). Technical Report ISO/TC 37/SC 4 N453 (N330 Rev.16), ISO (2008)
9. Hirst, G.: Ontology and the lexicon. In: Staab, S., Studer, R. (eds.) Handbook on Ontologies. Springer, Heidelberg (2004)
10. McGuiness, D.L., van Harmelen, F.: OWL Web Ontology Language Overview. W3C Recommendation (February 2004)
11. Miles, A., Matthews, B., Beckett, D., Brickley, D., Wilson, M., Rogers, N.: Skos: A language to describe simple knowledge structures for the web. In: Proceedings of the XTech Conference (2005)
12. Nirenburg, S., Raskin, V.: Ontological Semantics. MIT Press, Cambridge (2004)
13. Oltramari, A., Stellato, A.: Enriching ontologies with linguistic content: An evaluation framework. In: Proceedings of the LREC OntoLex 2008 Workshop (2008)
14. Peters, W., Montiel-Ponsoda, E., Aguado de Cea, G., Gómez-Pérez, A.: Localizing ontologies in OWL. In: Proceedings of the ISWC OntoLex 2007 Workshop (2007)
15. Reiter, N., Buitelaar, P.: Lexical enrichment of biomedical ontologies. In: Prince, V., Roche, M. (eds.) Information Retrieval in Biomedicine: Natural Language Processing for Knowledge Integration. IGI Global (2009)
16. Wilks, Y.: The Semantic Web: Apotheosis of annotation, but what are its semantics? IEEE Intelligent Systems 23(3), 41–49 (2008)

Frame Detection over the Semantic Web

Bonaventura Coppola[2], Aldo Gangemi[1], Alfio Gliozzo[1], Davide Picca[3], and Valentina Presutti[1]

[1] STLab, ISTC-CNR, Roma, Italy
{aldo.gangemi,alfio.gliozzo,valentina.presutti}@istc.cnr.it
[2] Department of Computer Science and Engineering, University of Trento, Italy
coppola@disi.unitn.it
[3] Université de Lausanne, Switzerland
davide.picca@unil.ch

Abstract. In the past, research in ontology learning from text has mainly focused on entity recognition, taxonomy induction and relation extraction. In this work we approach a challenging research issue: detecting *semantic frames* from texts and using them to encode web ontologies. We exploit a new generation Natural Language Processing technology for frame detection, and we enrich the frames acquired so far with argument restrictions provided by a *super-sense* tagger and domain specializations. The results are encoded according to a Linguistic MetaModel, which allows a complete translation of lexical resources and data acquired from text, enabling custom transformations of the enriched frames into modular ontology components.

1 Introduction

Ontology learning from text has become an important functionality for ontology design, due to the demand for domain-specific knowledge that supports semantic solutions scaling from personal knowledge management to large organizations, without requiring substantial competence from domain experts or expensive assistance from knowledge engineers. This is specially true if we consider the large number of different domains involved in modelling knowledge at an organization or even at a Web scale.

The current state of art in ontology learning deals with taxonomical relations and simple binary domain relations, but research has just started dealing with more complex domain relations that can be able to answer realistic competency questions [20], such as *who is communicating what to whom and for what reason?*. Complex relations are called *semantic frames* in linguistics, and show interesting analogies with other data structures that are known in AI and the Semantic Web, such as *frames*, *microformats*, *infoboxes*, etc.

In this paper we present a methodology for detecting and learning frames and situations from textual domain corpora, for using them to generate modules of a domain ontology, and possibly for populating the ontology with named entities and facts extracted from the same corpora. The methodology aims at specializing and/or instantiating a set of frames from the FrameNet lexicon [3], by extracting and filtering configurations of data from text corpora. We want to use the

L. Aroyo et al. (Eds.): ESWC 2009, LNCS 5554, pp. 126–142, 2009.

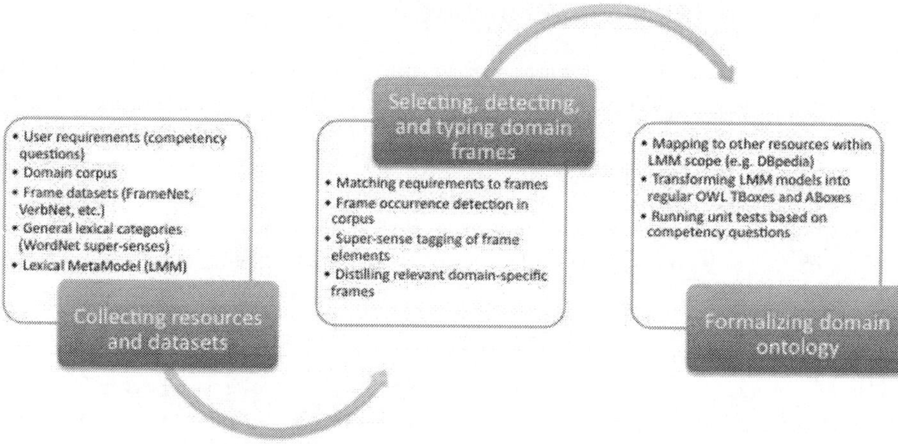

Fig. 1. A summary of the datasets, components, functionalities, and phases of the frame-based ontology learning method

extracted data in order to build domain-specific ontologies that show desirable characteristics such as cognitive and linguistic relevance, density, right granularity level, and inherent capability of being automatically evaluated based on expert requirements (in the form of competency questions). A previous attempt to achieve a similar objective has been reported by [4]. In contrast to that work we have used different text processing technology and knowledge representation standards.

The natural language processing technology for detecting occurrences of frames in corpora is presented in section 3.1), and their representation according to the FrameNet ontology is presented in section 2). Frame detection learns typical patterns from the occurrences in a corpus, providing raw material to start a domain adaptation process of frames by means of a WordNet Supersense tagging component (see section 3.2). The acquired information is then connected to a pile of reusable ontologies: the lexical ontologies OntoFrameNet [16] and OntoWordNet [18], DBpedia [1], and LMM [24,16], which allows a general representation model for lexical ontologies. The LMM representation is eventually converted into a regular ontology by means of *transformation patterns* (see section 4).

Results, described in the evaluation section (4) show that our workflow is sustainable and cost effective, being the speed of the adopted technology fast enough to work with mid-size corpora in reasonable time, and the accuracy of the data distilled at the end of the acquisition process good enough to be easily checked by an expert (or even automatically) with reference to competency questions. This turns out in a substantial reduction of the manual effort required for designing domain-specific ontologies that are design-ready to be evaluated against user requirements. Figure 1 summarizes the main components and functionalities involved in our workflow.

2 Representing Frames in LMM

Frame Semantics [15] is a linguistic theory that attempts to represent the conceptual interface between linguistic and real-world knowledge in the form of *semantic frames*, which are structures that describe particular types of situations, objects, or events along with the roles of their participating elements. Based on Frame Semantics, FrameNet [3] is a lexical knowledge base, consisting of a set of *frames*, which have proper *frame elements* and *lexical units*. Frame elements are unique to their frame, and can be optional ('non-core'). An occurrence of a frame consists in some piece of text whose words can be normalized as *lexemes*, and which have semantic roles dictated by the elements of that frame. A frame usually has only some of its roles actually lexicalized in texts. Types of linguistic realizations of a frame are called *lexical units*. Frames respectively frame elements are related between them, e.g. through the *subframe* relation. For example, a short definition of **CommerceScenario** in FrameNet is reported below:

CommerceScenario
Core Elements: **Buyer**, **Goods**, **Money**, **Seller**
non-Core Elements: **Manner**, **Means**, **Purpose**, **Rate**
Subframes: **CommercialTransaction**

The intuition underlying Frame Semantics is not very dissimilar from that described in [22], but the latter developed in the AI context, and eventually acquired a formal semantics, as in [7]. [16] is a reconstruction of different semantics of frames, and how they can be reconciled. As an assumption that can be adopted for all different approaches to frame semantics, [16] suggests that frames can be represented as (intensional) n-ary relations, with typed arguments (either mandatory or optional). For example,

$$\texttt{Desiring(x,y,e)} \rightarrow \texttt{Agent(x)} \wedge \texttt{Agent(y)} \wedge \texttt{Event(e)} \tag{1}$$

An occurrence of a frame can be straightforwardly treated as an instance of an n-ary relation, e.g.:

$$\texttt{Desiring(Susan, Marko, ListeningToHer)} \tag{2}$$

The logical representation of frames as n-ary relations is easily generalizable and clear, but hardly manageable by automated reasoners on large knowledge bases. A hard design problem is constituted by the polymorphism of many frames/n-ary relations, which can vary in number of the arguments that can be taken by the relation. For example, the same frame **Desiring** can be assumed with four arguments:

$$\texttt{Desiring(x,y,e,t)} \rightarrow \texttt{Agent(x)} \wedge \texttt{Agent(y)} \wedge \texttt{Event(e)} \wedge \texttt{Time(t)} \tag{3}$$

This problem was originally evidenced by Davidson with reference to a logic of events [12]. A neo-davidsonian approach is actually taken by [6], where an attempt is made to reconstruct FrameNet in terms of DRT (Discourse Representation Theory), with *events* being the DRT correlate of frames, and *semantic roles* being generic correlates for equivalence classes of frame elements.

A limitation found by [6] lies in the fact that DRT semantic primitives are tightly associated with elements having a given syntactic category (e.g. it assumes that an event cannot be expressed by an adverb), while FrameNet semantics only requires that any grammatical construction expresses a frame, independently from its syntactic category. The DRT approach has anyway the advantage of simplifying some assumptions of Frame Semantics, for example the uniqueness of frame elements to their frame. A formal semantics for frames that is also computationally manageable has been provided by Description Logics (DL) [2], such as OWL(DL). Due to their limited expressive power, description logics represent frames as classes, with roles (binary relations) that link a class to the types of the arguments of the original n-ary relation. Those types are classes as well, so that a graph of frames emerges out of this semantics. The example in (3) can be reengineered in DL as follows:

$$\mathtt{T} \sqsubseteq \forall\ \mathtt{R_1.Agent},\ \mathtt{T} \sqsubseteq \forall\ \mathtt{R_1^-.Desiring} \tag{4}$$

$$\mathtt{T} \sqsubseteq \forall\ \mathtt{R_2.Agent},\ \mathtt{T} \sqsubseteq \forall\ \mathtt{R_2^-.Desiring} \tag{5}$$

$$\mathtt{T} \sqsubseteq \forall\ \mathtt{R_3.Event},\ \mathtt{T} \sqsubseteq \forall\ \mathtt{R_3^-.Desiring} \tag{6}$$

$$\mathtt{T} \sqsubseteq \forall\ \mathtt{R_4.Time},\ \mathtt{T} \sqsubseteq \forall\ \mathtt{R_4^-.Desiring} \tag{7}$$

$$\mathtt{Desiring} \sqsubseteq (=\mathtt{1R_1} \sqcap =\mathtt{1R_2} \sqcap =\mathtt{1R_3} \sqcap =\mathtt{1R_4}) \tag{8}$$

The computational features of description logics make them a reasonable choice to formally represent linguistic frames, and this is the approach followed e.g. by [27]. On the other hand, description logics cannot be used to formalize the meta-level part of the intended semantics of frames: relations between frames, between frame elements, and between frames and frame elements, lexical units, and lexemes can be hardly represented in a DL.[1] The Lexical Meta-Model (LMM) [24] contains a viable solution to keep both the (reified) meta-level constructs of FrameNet in an OWL(DL) ABox, and to deliver custom views of frames in an OWL(DL) TBox when needed. LMM provides a formal semiotic representation of lexical knowledge. It is implemented in OWL(DL)[2]. LMM facilitates the integration of heterogeneous lexical knowledge resources, such as WordNet [14] and FrameNet[3], as well as of other semi-formal resources (thesauri, subject directories, etc.), with regular ontologies expressed in a formal language like OWL. LMM can be considered as a *semiotic façade* enabling the reengineering of any lexically-based resource to an ontology resource [16]. In the methodology reported in this paper, it is used to make different lexical data interoperable, and to transform the output of the three learning components *frame detector*, *super-sense tagger*, and *distiller* (see next sections) into an OWL ontology. The core LMM elements are `lmm:Expression`, which can `lmm:express` a `lmm:Meaning` (that is `lmm:conceptualizedBy lmm:Agent`), and which can

[1] F-logic [21] is another alternative that however has no direct correspondence to current standard semantic web languages.

[2] `http://www.ontologydesignpatterns.org/ont/lmm/lmm1.owl`

[3] The OWL ontology that encodes the alignment of OntoFrameNet to LMM can be downloaded at `http://www.ontologydesignpatterns.org/ont/lmm/ofn2lmm.owl`

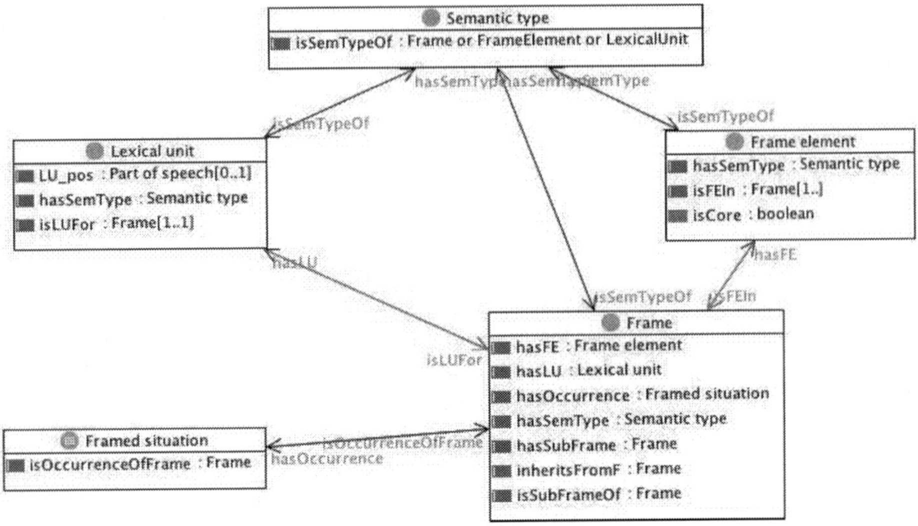

Fig. 2. OntoFrameNet: A full-fledged metamodel for FrameNet

`lmm:denote` a `lmm:Reference`; in addition, any entity can have a `lmm:Context` and can be `lmm:representedBy` a `lmm:FormalExpression`. Expressions can be any information object, and lexical items in particular; references can be any entity; meanings can be any information or conceptual entity that is used to describe or encode the (informal, natural) semantics of an expression, i.e. other expressions, concepts from a thesaurus, frames, topics, etc. A rich set of relations is defined in LMM to associate meanings between them, as well as to associate expressions, meanings, and references to formal constructs. A complete reengineering of FrameNet as a plugin to LMM can be found in the OWL version of OntoFrameNet[4][16] (Fig. 2). A `ofn:Frame` has some `ofn:FrameElement`, is lexically realized by some `ofn:LexicalUnit`, and all these can be evoked by some `ofn:Lexeme`, and can have some `ofn:SemanticType`. Since semantic types do not cover all frame elements and lexical units,[5] we use the super-sense tagger component (section 3.2) to infer WordNet `wn:SuperSenses` as additional types for frame elements. All mentioned classes and relations are aligned to LMM; an excerpt of alignments is included here:[6]

`ofn:Lexeme rdfs:subClassOf lmm:Expression` (9)

`ofn:Frame rdfs:subClassOf lmm:Meaning` (10)

`ofn:FrameElement rdfs:subClassOf lmm:Meaning` (11)

---

[4] `http://www.ontologydesignpattern.org/ont/ofn/ofntb.owl`

[5] Currently the majority of frames (719 out of 795) has types for one or more of their frame elements (4187 out of 7124), while no lexical unit (about 10000) has semantic types assigned to the application of frame elements in their context.

[6] `http://www.ontologydesignpattern.org/ont/lmm/ofn2lmm.owl`

$$\texttt{ofn:evokes rdfs:subClassOf lmm:expresses} \tag{12}$$

$$\texttt{ofn:FramedSituation rdfs:subClassOf lmm:Reference} \tag{13}$$

A thorough discussion on alternative formal translations of FrameNet is contained in [16].

3 The Frame Learning Components

In this section, we propose an ontology learning method aimed at acquiring frames and situations from text. This is a novel method involving challenging parsing capabilities. Ontology learning from text is usually performed with three components: a text corpus of an arbitrary size, a set of lexical resources, and some algorithms to extract, rank, and reengineer selected invariances from texts. Preliminary text classification may be also performed [5]. In the literature, most of the effort has concentrated in learning the terminology for a specific domain, taxonomies of concepts and their instances, relations between instances, trying to augment and adapt the information contained in e.g. WordNet with domain specific data. Unfortunately, this is not enough for our purposes, since we are trying to learn structures with multiple related elements in the context of a frame, as occurring in a particular situation. Therefore, we need text processing components that are able to identify occurrences of frames in text, and to recognize the corresponding frame elements and their types, also allowing for the learning of domain-specific specializations and instantiations.

To this aim, we have integrated two advanced text processing components: a frame detector and a super-sense tagger, described in the following subsections. As a result, portions of text corresponding to lexical units and frame elements are collected for each frame, which describe particular *framed situations* within a text.

As an example of the workflow, we use here the **JudgmentCommunication** frame. A sample annotated sentence follows:

`(S383) Public opinion and the press nowadays accuse us of being unavailable to give a response.`

`FRAME:` **JudgmentCommunication**

Communicator: `"Public opinion"` noun.cognition Concept
`"and the press"` noun.group-Concept"

Target: `"accuse"`_verb.communication

Evaluee: `"us"`

Reason: `"of being"` verb.stative `"unavailable"`
`"to give"` verb.possession `"a response"` noun.phenomenon-Concept"

The **Target** frame element is present in all FrameNet frames, and refers to the lexical unit (e.g. *accuse.v* in the previous example) that abstracts from the words that realize the frame in a particular text occurrence. This is actually used to assign that particular occurrence to one of the frames in FrameNet.

The remaining slots describe the frame elements identified in a text, and their corresponding types, as recognized by the super-sense tagger. We eventually learn domain frames by matching the frame element names to the corresponding super-senses, and by filtering out low-ranked results. This process is described in subsection 3.3.

3.1 The Frame Detection System

The Frame Detection component takes care of analyzing plain English input text, locating possible candidate frames invoked along by the terms in a text, and then tagging individual Frame Element instances of such frames. Tagging involves both locating the exact text boundary of each Frame Element, and assigning its correct semantic label.

The Berkeley FrameNet Project currently includes the definitions of nearly 800 Frames, 4,000 Frame Elements, and 135,000 annotated English sentences. An example of sentence annotation for the **CommerceScenario** is reported hereafter:

Ralemberg said [*he*]$_{\text{SELLER}}$ *already had a* [<u>*buyer*</u>]$_{\text{BUYER}}$ [*for the wine*]$_{\text{GOODS}}$

where the underlined word *buyer* is the target word (or *lexical unit* or predicate) which plays the role of *evoker* for this particular Frame.

To implement a FrameNet parsing system we adopted a multi-stage classification approach over natural language. Foundational studies in this area generally refer to Semantic Role Labeling [19]. In particular, our strategy extends the approaches in [23] to FrameNet and it is thoroughly described in [11]. Shortly, moving from previous pattern-based approaches [28], we exploit Support Vector Machines (SVMs) as a general statistical machine learning framework, in which we plug in Tree Kernels to obtain a similarity measure capable of working over structured objects as syntactic trees. Our strategy includes (1) **Target Word Detection**, where the semantically relevant words bringing predicative information are detected; (2) **Frame Disambiguation**, where the correct Frame for any target word has to be selected among several candidates; (3) **Boundary Detection**, where the sequences of words constituting the Frame Elements (arguments) are detected; and (4) **Role Classification**, which assigns semantic labels to the Frame Elements detected in the previous stage.

3.2 The Super-Sense Tagger

In this work, we exploit the Super-Sense Tagger (SST) described in [9] for a deeper semantic interpretation of texts finalized to the acquisition of structured knowledge in a frame-based representation. Supersenses are lexicographer's categories, used to provide an initial broad classification for the lexicon entries in WordNet [14]. Although simplistic in many ways, the supersense ontology has several attractive features for ontology engineering. In fact, supersenses correspond to the typical high level classes of most ontologies, reflecting distinctions between persons, locations, organizations, acts, etc. In this work we use supersenses to characterize the type of Frame Elements. Exploiting SST, we identify types in frame occurrences by filling the frame slots that are identified by the frame detection component.

SST[7] has been implemented using the SemCor corpus (a fraction of the Brown corpus annotated with WordNet word senses) and can be used for annotating large collections of English texts [8] The tagger implements a Hidden Markov Model, trained with the perceptron algorithm introduced in [10]. The tagset used by the tagger defines 26 supersense labels for nouns and 15 supersense labels for verbs. The tagger outputs named entity information, but also covers other relevant categories and attempts lexical disambiguation at the supersense level. In recent work, we have extended the tagset by subdividing each category into two sub-categories, *Instance* and *Concept* so that now the term "president" is tagged as *noun.person_Concept* and the term "Bill Clinton" as *noun.person_Instance*. An example of the output of this tagger is reported below.

$\text{Guns}_{B-noun.group\_instance}$ $\text{and}_{I-noun.group\_instance}$ $\text{Roses}_{I-noun.group\_instance}$ $\text{plays}_{B-verb.communication}$ at_{O} the_{O} $\text{stadium}_{B-noun.location\_concept}$

3.3 Distilling Knowledge from Tagged Frames

The output of the frame detector component provides a shallow semantic parsing of text, where text fragments are associated with frame elements and the lexical units are identified. After joining these data along with the tagging provided by the SST, the final output of the text processing components is illustrated in the example below:

`FRAME:` **CommerceScenario**
Buyer:`"People_noun.group_Instance who read_verb.cognition this warning_noun.communication"`
Target:`"buy_verb.possession"`
Goods:`"these toys_noun.artifactInstance"`
Recipient:`"for their children_noun.person_Concept or grandchildren_noun.person_Concept"`

The next step consists in converting this format into a formal representation. To this aim, the *distiller* component involves both semantic and statistical criteria:

Syntactic constraints. Occurrences of frame elements should only be nominals or noun phrases, while target lexemes can be verbs, nouns or adjectives

Type constraints. Most of the occurring elements that are recognized by the frame detector belong to a rather small number of super-senses (e.g. in a given domain buyers are either persons or groups, less likely they could be animals)

Redundancy constraints. Redundancy helps in recognizing fine-grained restrictions for frame elements with very high accuracy. Only those concepts occurring in the context of a particular frame with a certain degree of redundancy can be regarded as "valid" subtypes.

[7] The tagger is publicly available at:
`http://sourceforge.net/projects/supersensetag/`

At this stage, we have implemented a simple thresholding mechanism on frequency-based countings in order to model those assumptions, and in particular:

- The implementation of the syntactic constraints directly comes from the SST annotation, which distinguishes nouns, verbs and adjectives, and recognizes compound terms.
- Concerning type constraints, for each occurrence of a frame element, we count the occurrences of all different super-senses associated with the matched arguments, and we filter out those having less that 10 occurrences. For example, for the frame KILLING the possible types of the element VICTIM are as follows: group_concept: 90, person_concept: 79, person_instance: 41.
- Concerning redundancy constraints, for any occurrence of the frame elements, we keep the terms belonging to the super-senses selected at the first step. Since the SST distinguishes between concepts and instances, we select only those terms belonging to the concept types having more than three occurrences. For example, we have found that occurrences of the frame element VICTIM belonging to the **person_concept** super-sense include notions like *child, woman* and *civilian*, while occurrences categorized by the **noun.state** super-sense, e.g. *recognition*, have a sensibly lower frequency, and are most likely to be errors.

The result of the three steps above is a list of "qualified", domain-oriented semantic types for domain-specific frame element that can be easily leveraged to improve the quality of the frame representation. As a final step, we are interested in detecting specializations of frames. To this aim, we can adopt the results of the frame detector component, applying each of the cleaning steps described above. Among all possible occurrences of a certain frame in the corpus, we select only those having at least three of their element occurrences satisfying the constraints above, and we regard them as candidate *domain-specific* frames, like in the following case, accompanied by its generalization:

FRAME: **Killing** : **Victim**:`smoker` ; **Killer**:`tobacco`
FRAME: **Killing** : **Victim**:`person` ; **Killer**:`artifact`

The described procedure is fully automatic, and we have adopted the same threshold uniformly for three case study frames under analysis in this paper (Section 4.2). The results of this method include a set of domain-specific frames that can be candidate data for answering a competency question, and therefore for implementing an ontology that fits the user requirements.

4 Empirical Evaluation

4.1 NLP evaluation

Frame Detection Evaluation. We have tested the Frame Detection Component in the natural setting of the FrameNet Corpus. Version 1.3 of FrameNet was used for both learning and test. After preprocessing and parsing with the Charniak's constituency-based syntactic parser we obtained 135,293 annotated and parsed sentences. We split the data considering the part of speech of predicates, ending up with 782 different frames. The overall dataset was then partitioned into 90% training set and 10% test set.

We tested the plain polynomial kernel (*poly*, with a degree of 3) and the SubSet Tree Kernel (SST) respectively over standard [19] and structured features [23]. Also, their combination was tested. Please see [11] for full details.

Table 1 reports Precision, Recall and F_1 measure of our classifiers over different tasks. The four rows in the table show in turn: (1) the "pure" performance of the Boundary Detection (BD) classifier, i.e. considering correct the classification decisions also when a correctly classified tree *node* does not exactly correspond to a valid sentence constituent. Such mismatch frequently happens when the parse tree (which is automatically generated) includes incorrect nodes and attachments; (2) the performance of the BD classification "projected" on the tree leaves, i.e. when matching not only the constituent node as in 1, but also the selected *words* (leaves) with those in the FrameNet gold standard. This implies an exact syntactic analysis being realized in the subtree; (3) the same as 1, with the argument role classification (RC) also performed (i.e. Frame Element labels must match as well); (4) the same as 2, with RC also performed.

As shown in the table, the SST+*poly* kernel achieves 1.0 Precision, 0.732 Recall and 0.847 F_1 on NS. These figures can be compared to 0.855 Precision, 0.669 Recall and 0.751 F_1 of the system described in [13], achieved with the same amount of training data. In conclusion, our best learning scheme is currently capable of tagging FrameNet data from noisy syntax with exact boundaries and role labels at 0.63 F_1.

Table 1. Results for Frame Detection on the FrameNet dataset. SST+poly with 90% training and 10% test data.

Enhanced SST + poly			
Eval Setting	P	R	F_1
BD (nodes)	1.0	.732	.847
BD (words)	.963	.702	.813
BD+RC (nodes)	.784	.571	.661
BD+RC (words)	.747	.545	.630

Super-Sense Tagger Evaluation. We evaluated the performances of SST by adopting a n-fold cross validation strategy on the SemCor corpus exploited for training. Results for the chosen categories are illustrated in Table 2, reporting Precision, Recall and F_1 for any Supersense. Looking at the table we can notice that for some categories the F_1 is exceptionally high. Some of those best classified categories are really essential for ontology learning. For example, important labels such as *noun.person*, *noun.group* or *noun.location* achieve results higher than 0.70. We obtained a general F_1 of 0.76. For some categories we got a F_1 over 0.80: e.g. *noun.person_Instance*: F_1 0.90) or *noun.person_Concept*: F_1 0.81, etc.

Table 2. Super-Sense Tagger Evaluation Results: Recall, precision and F1 for each classification category

Category	Recall	Precision	F_1
noun.animal_Concept	0.712	0.766	0.738
noun.animal_Instance	0.416	0.793	0.545
noun.artifact_Concept	0.726	0.737	**0.731**
noun.artifact_Instance	0.596	0.646	0.620
noun.food_Concept	0.687	0.720	0.703
noun.food_Instance	0.444	0.500	0.457
noun.group_Concept	0.729	0.731	**0.730**
noun.group_Instance	0.683	0.703	0.693
noun.location_Concept	0.682	0.653	0.667
noun.location_Instance	0.752	0.800	**0.775**
noun.person_Concept	0.838	0.804	0.821
noun.person_Instance	0.927	0.881	**0.904**

4.2 Knowledge Engineering Evaluation

We have tested our methods on the Europarl corpus[8], comprising about 30 million documents, extracted from the proceedings of the European Parliament. It includes 11 official languages of the European Union. For our experiments, we have used the English part, including 1,461,429 sentences and 39,618,240 words. In step 1, sentences from the text corpus are annotated with super-senses by the SST, and with frames, frame elements, and lexical units by the Frame Detector component. For the sake of this experiment, we focused on the annotation of three frames: **Killing**, **JudgmentCommunication**, and **Commerce**. Results for each frame have been collected in tables reporting, for each frame occurrence recognized in the corpus, all the frame elements and their types, as recognized by SST. Occurrences of core frame elements show meaningful frequency in the corpus, so confirming the hypothesis that some elements are more central than others in the conceptualization of frames. About 1000 occurrences for each frame have been collected after step 1.

In step 2, the raw frame occurrence data have been filtered by applying the procedure described in section 3.3, consisting of identifying the most frequent super-senses for each frame element, in order to learn more specific types (e.g. `Person` or `Group`, and to filter out instances having different types (all super-senses can be assumed as disjoint classes of synsets). After applying the automatic distiller procedure based on combined frequency thresholds, data from only about 100 frame occurrences have been retained as candidate domain-specific frames for each generic frame. In addition to types, step 2 allows to learn sensible lexemes for the domain specialization of frame elements (e.g. *people*), and sensible lexical units (e.g. `Accuse` for the candidate frame specializations. An example of the outcome of the distiller after evaluation is shown in Table 3. Each distilled

[8] `http://www.iccs.inf.ed.ac.uk/~pkoehn/publications/europarl-mtsummit05.pdf`

Table 3. A sample output of the distiller with manual evaluation

Communicator	TargetVerb	Evaluee	Reason	*Validity*
people	accuse	european union	-	*ok*
-	accused	security	violations	*ok*
member states	blame	problems	-	*ok*
-	blamed	fishermen	negotiations	*ok*
companies	charge	price	-	*no*
amnesty international	cited	-	torture	*ok*
commission	charged	-	failure	*ok*
-	charged	authorities	crime	*ok*
-	charged	group	report	*ok*
employment	charged	-	-	*no*

domain frame is formalized as an `owl:Individual rdf:type ofn:Frame`, thus resulting to be also `rdf:type lmm:Meaning`, according to the guidelines provided in section 2. For example, in one filtered occurrence we first specialize the generic frame (14), and then add a new super-sense typing over its frame elements (15).

`europarl:AccuseOfViolation lmm:specializes`
`ofnabox:JudgmentCommunication` (14)

`ofnabox:Communicator ofn:hasSemType europarl:People)` (15)

After a manual evaluation on the domain-specific frames detected for the three generic frames chosen, 62% of distilled domain-specific frames resulted to be valid, where validity is assumed as respecting the conceptualization of the generic frame, in terms of extensional interpretation of the classes derived from target verbs, and of properties derived from frame elements. In this section we explain the workflow used to make this interpretation viable.

In this rapid ontology design process, an ontology designer is only asked to validate the detected frames that present a desirable design property, *density* i.e. the resulting ontology corresponding to each domain-specific frame is a labeled connected graph with nodes labeled by owl:Class names and edges labeled by owl:ObjectProperty names. Density helps preserving the conceptual structure of the generic frame, which is known to feature cognitive relevance and linguistic appropriateness. Notice that typical ontologies produced by traditional learning methods result to lack density and do not feature the expected formal structure implicit in experts' conceptualization of their domain.

Besides desirable design properties, we are interested in designing ontologies that *work*. Quality is therefore evaluated in terms of fitness to *competency question*-based (or task-based) ontology design [20], as assessed in the ontology evaluation framework described in [17]. We notice that frame-based ontology learning has also this very desirable property. Since frames are formally equivalent to n-ary relations at design time (section 2), and competency questions can be formalized as n-ary relations at query time [25], it is straightforward to match a linguistic competency question like: *what are the reasons for those events?* to the knowledge base of domain-specific frames.[9]

Domain frames can be used for two different design tasks: the first one is producing a traditional TBox domain ontology, e.g. consisting of OWL classes like `AccuseOfViolation`; the second one is to classify named entities that instantiate the domain frame (*framed situations*), which are formalized as OWL individuals.

The generation of a TBox from the LMM-encoded domain frame and framed situations is performed by means of customizable *transformation patterns* [25], which contain rules for translating each instance of an LMM type (Meaning, Expression, etc.) into an OWL element.

An additional advantage of frame-based ontology learning is the ability to generate small ontologies that respect the characteristics of highly-reusable *content ontology design patterns* [26,25]: small size, density, inherent task-based evaluability, cognitive and linguistic relevance.

As a concrete example, we show a diagram depicting a proposed design patterns that has been derived from the domain-specific frame **AccuseOfViolation** (Fig. 3). The transformation patterns adopted are exemplified here as a set of OWL(Full) axioms over the OntoFrameNet TBox, and the OWL metamodel:[10]

```
ofn:Frame rdfs:subClassOf
(owl:Restriction(owl:onProperty trans:transformableTo)
(owl:allValuesFrom owl:Class))                        (16)
ofn:FrameElement rdfs:subClassOf
(owl:Restriction (owl:onProperty trans:transformableTo)
(owl:allValuesFrom owl:ObjectProperty))               (17)
ofn:SemanticType rdfs:subClassOf
(owl:Restriction (owl:onProperty trans:transformableTo)
(owl:allValuesFrom owl:Class))                        (18)
ofn:hasFE rdfs:subClassOf
(owl:Restriction (owl:onProperty trans:transformableTo)
(owl:allValuesFrom owl:Restriction))                  (19)
ofn:hasSemType rdfs:subClassOf
(owl:Restriction (owl:onProperty trans:transformableTo)
(owl:allValuesFrom rdfs:range))                       (20)
```

[9] Several domain-specific frames can be retrieved which have evidence in the corpus: **AccuseOfViolation**, **BlameNegotiation**, **ChargeForFailure**, **ChargeForCrime**, etc.

[10] Transformation patterns are based on alignments between OntoFrameNet and LMM (cf. section 2), and between LMM and the FormalSemantics vocabulary: `http://ontologydesignpatterns.org/ont/dul/FormalSemantics.owl`

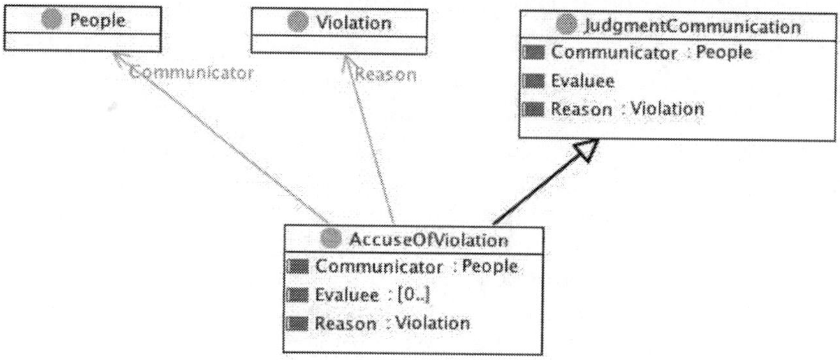

Fig. 3. A content ontology design pattern derived from the domain-specific frame **AccuseOfViolation**, based on the transformation patterns

For example, given the previous axioms 14 and 15, and the following axioms extending OntoFrameNet ABox,[11]:

europarl:AccuseOfViolation ofn:hasFE ofnabox:Communicator (21)

europarl:AccuseOfViolation ofn:hasFE ofnabox:Reason (22)

ofnabox:Reason ofn:hasSemType europarl:Violation (23)

we obtain the following TBox axioms,[12] after applying the transformation patterns:

europarl:AccuseOfViolation rdfs:subClassOf
(owl:Restriction (owl:onProperty ofnowl:Communicator)
(owl:allValuesFrom europarl:People)) (24)

europarl:AccuseOfViolation rdfs:subClassOf
(owl:Restriction (owl:onProperty ofnowl:Reason)
(owl:allValuesFrom europarl:Violation)) (25)

Task-based fitness can be proved by launching a unit-test that contains data about the europarl:AccuseOfViolation competency question: *what accuses of violation any people has made for what reason?*. From the n-ary relation hypothesis, we know that a frame has direct translations into competency questions, queries, and ontologies: if the TBox produced is able to support the proper queries, then the TBox fits the task. This is the case actually, since the classes and properties that have been built (Fig. 3) are able to encode the query derived from the competency question:

SELECT DISTINCT ?x ?y ?z WHERE {?x a Accuse.
?y a People . ?z a Violation . ?x :Communicator ?y . ?x :Reason ?z} (26)

[11] http://www.ontologydesignpatterns.org/ofnframes/accuseofviolation.owl

[12] http://www.ontologydesignpatterns.org/owlframes/accuseofviolation.owl

5 Conclusions

In this paper we have presented an implemented method for frame-based ontology learning by using components for frame detection and super-sense tagging on a text corpus, the FrameNet and WordNet lexical knowledge bases, and the LMM metamodelling framework.

The method can be used to generate candidate *domain-specific* frames suggested by qualified corpus evidence, and to populate an ontology with complex facts extracted from the corpus. The method fits a task-based analysis and evaluation of ontologies, as it is being developed in the NeOn project, where we are implementing novel functionalities for pattern-based design.

The method presented shows some significant improvements in terms of functionalities and technology with respect to existing ontology learning components, not only because are we acquiring more advanced knowledge structures (i.e. frames, instead of simple taxonomies or binary relations) but also do we provide an easier alignment to existing semantic web resources: the use of LMM makes the results more easily mappable to different kinds of data, such as datasets from Linking Open Data.

Our experiment proves that we can build a resource of reusable design patterns also by learning them from domain text corpora. There are challenging research issues in this area of ontology design. An important one is: since the amount of domain-oriented patterns can be huge, and FrameNet alone provides less than 1000 generic frames, are they complete, so that all domain patterns specialize them? Should we consider a procedure to learn patterns without previous knowledge of generic ones?

Acknowledgements

We thank Daniele Pighin and Alessandro Moschitti at University of Trento for their critical support in developing the Frame Detection module. This work has been partly supported by the EU projects NeOn, funded within the 6th IST Framework Programme, and BONy, funded within the Lifelong Learning Programme 2007.

References

1. Auer, S., Bizer, C., Lehmann, J., Kobilarov, G., Cyganiak, R., Ives, Z.: DBpedia: A Nucleus for a Web of Open Data. In: Aberer, K., Choi, K.-S., Noy, N., Allemang, D., Lee, K.-I., Nixon, L., Golbeck, J., Mika, P., Maynard, D., Mizoguchi, R., Schreiber, G., Cudré-Mauroux, P. (eds.) ASWC 2007 and ISWC 2007. LNCS, vol. 4825, pp. 722–735. Springer, Heidelberg (2007)
2. Baader, F. (ed.): The Description Logic Handbook: theory, implementation, and applications. Cambridge University Press, Cambridge (2003)
3. Baker, C.F., Fillmore, C.J., Lowe, J.B.: The Berkeley FrameNet Project. In: Proceedings of the 17th International Conference on Computational Linguistics, Morristown, NJ, USA, pp. 86–90 (1998)

4. Basili, R., Giannone, C., De Cao, D.: Learning domain-specific framenets from texts. In: ECAI Workshop on Ontology Learning and Population (2008)
5. Basili, R., Moschitti, A., Pazienza, M.T.: A text classifier based on linguistic processing. In: Proceedings of the IJCAI 1999 Workshop on Machine Learning for Information Filtering (1999)
6. Bos, J., Nissim, M.: Combining Discourse Representation Theory with FrameNet. In: Favretti, R.R. (ed.) Frames, Corpora, and Knowledge Representation, pp. 169–183. Bononia University Press (2008)
7. Brachman, R.J.: A Structural Paradigm for Representing Knowledge. Ph.d. thesis, Harvard University, USA (1977)
8. Ciaramita, M., Altun, Y.: Broad-coverage sense disambiguation and information extraction with a supersense sequence tagger. In: Proceedings of EMNLP 2006, Sydney, Australia, pp. 594–602 (2006)
9. Ciaramita, M., Johnson, M.: Supersense tagging of unknown nouns in wordnet. In: Proceedings of EMNLP 2003, Sapporo, Japan, pp. 168–175 (2003)
10. Collins, M.: Discriminative training methods for hidden markov models: Theory and experiments with perceptron algorithms. In: Proceedings of EMNLP 2002 (2002)
11. Coppola, B., Moschitti, A., Pighin, D.: Generalized framework for syntax-based relation mining. In: Proceedings of the IEEE International Conference on Data Mining (ICDM 2008), Pisa, Italy (2008)
12. Davidson, D.: The Logical Form of Action Sentences. In: The Logic of Decision and Action, 2nd edn. University of Pittsburgh Press, Pittsburgh (1967)
13. Erk, K., Pado, S.: Shalmaneser - a flexible toolbox for semantic role assignment. In: Proceedings of LREC 2006, Genoa, Italy (2006)
14. Fellbaum, C. (ed.): WordNet. An Electronic Lexical Database. MIT Press, Cambridge (1998)
15. Fillmore, C.J.: The Case for Case. In: Bach, E., Harms, R.T. (eds.) Universals in Linguistic Theory, pp. 1–210. Holt, Rinehart, and Winston, New York (1968)
16. Gangemi, A.: What's in a Schema?. In: Huang, C.R., Calzolari, N., Gangemi, A., Lenci, A., Oltramari, A., Prevot, L. (eds.) Ontologies and the Lexicon. Cambridge University Press, Cambridge (2009)
17. Gangemi, A., Catenacci, C., Ciaramita, M., Lehmann, J.: Modelling Ontology Evaluation and Validation. In: Sure, Y., Domingue, J. (eds.) ESWC 2006. LNCS, vol. 4011, pp. 140–154. Springer, Heidelberg (2006)
18. Gangemi, A., Navigli, R., Velardi, P.: The OntoWordNet Project: Extension and Axiomatization of Conceptual Relations in WordNet. In: Meersman, R., Tari, Z., Schmidt, D.C. (eds.) CoopIS 2003, DOA 2003, and ODBASE 2003. LNCS, vol. 2888, pp. 820–838. Springer, Heidelberg (2003)
19. Gildea, D., Jurafsky, D.: Automatic Labeling of Semantic Roles. Computational Linguistics 28(3), 245–288 (2002)
20. Gruninger, M., Fox, M.: The role of competency questions in enterprise engineering (1994)
21. Kifer, M., Lausen, G., Wu, J.: Logical Foundations of Object-Oriented and Frame-Based Languages. Journal of Association for Computing Machinery 42(4), 741–843 (1995)
22. Minsky, M.: A Framework for Representing Knowledge. In: Winston, P. (ed.) The Psychology of Computer Vision. McGraw-Hill, New York (1975)
23. Moschitti, A., Pighin, D., Basili, R.: Tree kernels for semantic role labeling. Computational Linguistics 34(2), 193–224 (2008)

24. Picca, D., Gangemi, A., Gliozzo, A.: LMM: an OWL Metamodel to Represent Heterogeneous Lexical Knowledge. In: Proc. of the International Conference on Language Resources and Evaluation (LREC), Marrakech, Morocco (2008)
25. Presutti, V., Gangemi, A., et al.: Library of design patterns for collaborative development of networked ontologies. Deliverable D2.5.1, NeOn project (2008)
26. Presutti, V., Gangemi, A.: Content Ontology Design Patterns as Practical Building Blocks for Web Ontologies. In: Li, Q., Spaccapietra, S., Yu, E., Olivé, A. (eds.) ER 2008. LNCS, vol. 5231. Springer, Heidelberg (2008)
27. Scheffczyk, J., Baker, C.F., Narayanan, S.: Reasoning over Natural Language Text by means of FrameNet and Ontologies. In: Ontologies and the Lexicon. Cambridge University Press, Cambridge
28. Tanev, H., Kouylekov, M., Negri, M., Coppola, B., Magnini, B.: Multilingual pattern libraries for question answering: a case study for definition questions. In: Proceedings of LREC 2004, Lisbon, Portugal (2004)

Word Sense Disambiguation for XML Structure Feature Generation

Andrea Tagarelli, Mario Longo, and Sergio Greco

Dept. of Electronics, Computer and Systems Sciences, University of Calabria, Italy
tagarelli@deis.unical.it

Abstract. A common limit of most existing methods that manage XML structure information is that they do not handle the semantic meanings that might be associated to the markup tags. In this paper, we study how to map structure information available from XML elements into semantically related concepts in order to support the generation of XML semantic features of XML structural type. For this purpose, we define an unsupervised *word sense disambiguation* method to select the most appropriate meaning for each element contextually to its respective XML path. The proposed approach exploits conceptual relations provided by a lexical ontology such as WordNet and employs different notions of *sense relatedness*. Experiments with data from various application domains are discussed, showing that our approach can be effectively used to generate structural semantic features.

1 Introduction

Structure information is essential in many problems of semistructured data management, such as indexing, query processing, change detection and schema matching. There is also a variety of tasks of learning and understanding from XML data (e.g., similarity detection, summarization, classification and clustering) focusing on the structural characteristics of such data.

As XML becomes increasingly widespread, handling the semantics of XML data arises as one of the most difficult challenges in data management and knowledge discovery. Indeed, XML data exhibiting proper structures and text contents may in principle encode related semantics due to a subjective definition of markup tags. Within this view, a common limit of most existing approaches is that they represent XML data according to syntactic structural information, whereas they do not consider an important characteristic of XML data: the semantic meanings that might be associated to the markup tags.

In this paper, we address the problem of generating semantic features for XML data focusing on structure information only. A major assumption of our work is that a *lexical ontology*, such as WordNet [4], can play a central role in identifying semantic relationships among the concepts underlying the constituents of structure information. A concept is represented by a lexical meaning (sense) which can be assigned to one or more terms. In the proposed structural analysis, these terms correspond to the tag names in XML paths and the objective is to

L. Aroyo et al. (Eds.): ESWC 2009, LNCS 5554, pp. 143–157, 2009.

couple syntactic information (the tag name) with a semantic one (the concept associated to the tag name).

To address the inherent ambiguity of the meaning of tag names, we have devised a *word sense disambiguation* method to select the most appropriate sense for each tag name in the context of an XML path. This method is based on different notions of *sense relatedness*, which can exploit scoring functions for overlaps between dictionary glosses and distance measures for ontology paths.

We have experimentally evaluated our approach on various data belonging to different application domains. Results have raised the significance of our approach and provided important indications on the relative roles of the devised notions and methods.

2 Related Work

Representing semistructured data is typically addressed by labeled rooted trees. Consequently, in the past few years, handling such data has leveraged results from research on tree matching, including a number of algorithms for computing tree edit distances (e.g., [9]). More recently, attention has also been drawn toward using simple Vector-space models to represent XML data, which substantially differ in the definition of feature space (e.g., [3,16,2]). However, in order to represent the structural characteristics of XML data, all the aforementioned approaches use syntactic information only, while ignoring semantic information.

The key element in many tasks involving a notion of semantics has been identified in *ontologies*. In particular, *WordNet* [4] is widely known as the most important publicly available large-scale lexical ontology. In WordNet, related concepts are grouped into equivalence classes, called *synsets* (sets of synonyms). Each synset represents one underlying lexical concept and is described by a short text (*gloss*). Synsets are explicitly connected to each other in the form of ontologies through different relations, such as *is-a* relations (hypernymy/hyponymy), *part-of* relations (meronymy/holonymy), etc. In this context, an early study is presented in [15], where texts and structure information are combined together to generate XML structure and content features, although ontological knowledge is marginally used. In the XML structural clustering approach proposed in [8], an element name is associated with a set containing its constituent words and semantically related words (e.g., the synonyms, hyponyms and hypernyms for each of the constituent words). Given any two element names, the similarity is computed by considering the intersection between their respective word sets. In the case that there are no common words between such sets, the similarity is computed by averaging the pair-wise syntactic matchings based on a combination of edit distances and n-gram functions.

In our previous works [14,13], we have addressed the problem of semantic relatedness between XML documents in a transactional clustering framework, focusing on the generation of semantically-enriched structure and content features.

In particular, we originally involved an unsupervised analysis of the meanings of the tag names into the structural feature generation.

Semantic analysis of XML structural characteristics naturally involves the possible senses of the tag names. This may eventually require a task of *word sense disambiguation* (WSD) in order to assign each tag with the most appropriate sense in a given context. In dictionary-based WSD the assumption is that the most plausible sense to assign to multiple co-occurring words is the one that maximizes the relatedness among the chosen senses. Within this view, the pioneer Lesk method [5] disambiguates a target word by choosing the meaning whose gloss shares the largest number of words with the glosses of the neighboring words. However, using a lexical ontology like WordNet allows for capturing semantic relationships based on concepts by also exploiting hierarchies of concepts besides dictionary glosses. The basic Lesk algorithm can be enhanced in this way to take advantage of the network of relations provided in WordNet. This idea has been formalized in a measure of semantic relatedness between word senses based on the notion of *extended gloss overlap* [10,1], which has the merit of considering phrasal matches and weighting them more heavily than single word matches. The extended gloss overlap measure takes as input two concepts (i.e., WordNet synsets) and computes a gloss overlap score, hereinafter denoted as *go-score*, as the sum of the squared sizes of the distinct overlaps between the glosses, where any *overlap* is detected whenever a shared maximal sequence of words occurs. Other functions of semantic relatedness/similarity for WSD problems have recently been proposed in, e.g., [17,11].

The study in [7] describes a structural approach to sense classification. A WSD algorithm based on a context-free grammar is developed to find structural semantic interconnections, i.e., structural specifications of the possible senses for each word in a context. The graph representation of word senses is built from several sources, including WordNet but also annotated corpora and glossaries. This makes the disambiguation approach semi-supervised. A recent attempt to the disambiguation of tag names in XML trees has been proposed in [6]. The disambiguation of any given tag is accomplished in an unsupervised way by analyzing the relative structural context with the support of an external knowledge source such as WordNet. The structural context of each word appearing in the target tag is a graph built on the ancestors, descendants and siblings of its corresponding node. Each word in the target tag is compared to all the words derived from this graph-context in order to calculate a vector of similarity scores associated to the possible senses. This vector is seen as a ranking of the plausible senses for each term, which can further be refined by considering the gloss definitions and linear ordering of the noun synsets.

Like [6], our work proposes a method of unsupervised disambiguation of XML elements, which uses WordNet as ontological knowledge. However, the use of WordNet in [6] is limited to synonymies and is-a relations, whereas our method involves other concept hierarchies (e.g., part-of relations). Also, we devise different notions of sense relatedness and various strategies of search through the WordNet hierarchies.

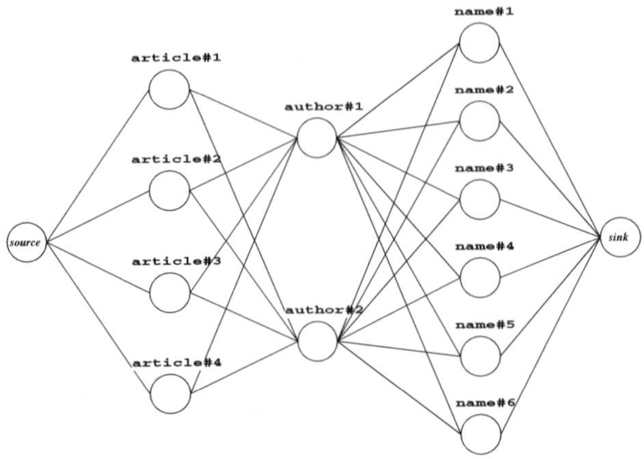

Fig. 1. The synset graph for the path `article.author.name` (Note that edge weights are not shown to avoid cluttering for clear presentation)

3 Tag Sense Disambiguation

Tag paths represent the natural basis for extracting structure features from XML data. Since we desire to go beyond a context-free use of syntactic terms (tag names), we aim to map them into semantically related concepts.

For any given tag path p, the objective is to provide each tag in p with a unique sense chosen from the reference lexical ontology; we assume that a fictitious sense is assigned to any term (tag name) for which no matches are found in the reference lexical ontology. Since selected senses are to be appropriate contextually to p, we need a task of WSD to handle the various senses of any given tag name and to finally select the most appropriate one with respect to a *path context*.

3.1 Building the Synset Graph

We conceive a path context as a "semantic network" which is built on the senses of the tags of a given path. More precisely, this network is modeled as a weighted graph of layered form, called *synset graph*, such that: *i)* each layer consists of all the possible senses of any given tag in the path, *ii)* the layers are connected according to the order of the tags in the path, and *iii)* the edge weights are computed in such a way that they measure the relatedness between nodes (synsets) belonging to adjacent layers. Figure 1 shows an example synset graph.

3.2 Computing the Edge Weights in a Synset Graph

The crucial aspect in building a synset graph is how to compute the edge weights. For this purpose, we pursue two ideas.

The first idea is to exploit the mechanism of *gloss overlap scoring* (i.e., the *go-score* computation discussed in Section 2) which is possibly enhanced by using the WordNet concept hierarchies.

The second idea is to define a notion of sense relatedness which is evaluated with respect to WordNet by applying an *ontology-path-based similarity* measure (e.g., [12]). The degree to which two senses are related is seen as a function of their respective locations in the lexical ontology, in such a way that the higher a sense in a hierarchy the more general it is.

Gloss-overlap-based sense relatedness. Let $\mathcal{WSR}$ denote a selected set of WordNet synset relations, namely *hypernymy*, *hyponymy*, *meronymy*, and *holonymy*. Given a relation $r \in \mathcal{WSR}$ and a synset σ, we define a function ω such that, when applied to r and σ, it yields the set $\omega(r, \sigma)$ of synsets directly connected to σ through r.

Function ω can be enhanced to include synsets that are indirectly connected to a target synset to a given distance in the WordNet taxonomy. Given a relation $r \in \mathcal{WSR}$, a synset σ and an integer $d \geq 1$, the set of synsets σ' connected to σ, through r, across a path of length d is defined as $\omega^*(r, \sigma, d) = \bigcup_{\sigma' \in \omega(r,\sigma)} \omega^*(r, \sigma', d-1)$, if $d > 1$; otherwise, $\omega^*(r, \sigma, d) = \omega(r, \sigma)$.

Given any two synsets σ and ρ, a non-negative integer $\overline{d}$, and a function $\mathrm{f}(d)$ monotonically decreasing for increasing values of d, we define the gloss-overlap-based sense relatedness as:

$$\begin{aligned} \textit{go-relatedness}(\sigma, \rho, \overline{d}) &= \textit{go-score}(\sigma, \rho) \\ &+ \sum_{d=1}^{\overline{d}} \Bigg[\sum_{\substack{\rho' \in \omega^*(\textit{hypernymy}, \rho, d) \vee \\ \rho' \in \omega^*(\textit{holonymy}, \rho, d)}} \textit{go-score}(\sigma, \rho') \times \mathrm{f}(d) \Bigg] \\ &+ \sum_{d=1}^{\overline{d}} \Bigg[\sum_{\substack{\sigma' \in \omega^*(\textit{hyponymy}, \sigma, d) \vee \\ \sigma' \in \omega^*(\textit{meronymy}, \sigma, d)}} \textit{go-score}(\sigma', \rho) \times \mathrm{f}(d) \Bigg] \end{aligned}$$

A simple enhancement to the above measure can be made by introducing a variant in the *go-score* computation: for any synset σ, the set of synonyms associated with σ can be included in the set of terms defining the σ's gloss. In this way, a gloss can be enriched with more, semantically equivalent terms.

Ontology-path-based sense relatedness. We believe that searching through the (WordNet) is-a hierarchy upwardly and/or through the part-of hierarchy is useful to capture relatedness among synsets.

Given any two synsets σ and ρ, we define the ontology-path-based relatedness between σ and ρ as:

$$\textit{p-relatedness}(\sigma, \rho) = \max_r \left\{ \frac{2 \times \textit{depth}_r(\textit{lcs}_r(\sigma, \rho))}{\textit{depth}_r(\sigma) + \textit{depth}_r(\rho) + |\textit{lcs}_r(\sigma, \rho)| - 1} \right\}$$

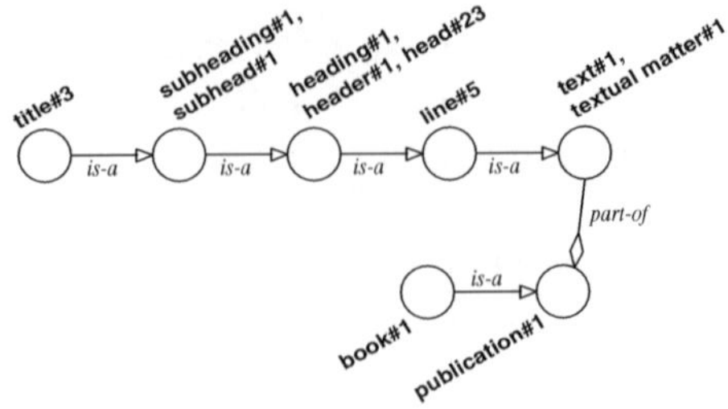

Fig. 2. The portion of WordNet explored to compute the ontology-path-based relatedness between the 1th synset of `book` and the 3rd synset of `title`

where

- $r \in \{hypernymy, holonymy\}$.
- $depth_r()$ computes the maximum distance (path length) from a given synset to a top hypernym (if $r = hypernymy$) or top holonym (if $r = holonymy$). The depth of a top synset is set to 1, whereas the depth of a synset whose sense is not found in the hierarchy is assumed to be 0.
- $lcs_r()$ computes the set of lowest common synsets between two given synsets with respect to r. This set contains a unique synset if a common hypernym ($r = hypernymy$) or holonym ($r = holonymy$) exists. Otherwise, if $r = hypernymy$, the nearest hypernym of σ and the nearest hypernym of ρ are identified in such a way that they are connected through a minimum-length chain of holonyms/meronyms; the lcs hence contains these hypernyms and the related holonyms/meronyms. Analogously, if $r = holonymy$, the lcs contains the nearest holonyms and the related hypernyms/hyponyms.

It easy to see that the *p-relatedness* function is defined to capture not only relationships involving the is-a and part-of hierarchies distinctly. Indeed, semantic relatedness between senses may be found through both the is-a and part-of hierarchies.

Figure 2 shows the portion of WordNet which is explored to compute the semantic relatedness between the synsets of `book` and `title`. Note that no is-a or part-of direct relationship holds for the two target synsets. Therefore, part-of relationships are to be searched among `book`#1's hypernyms and `title`#3's hypernyms. In this way, a 4-level hypernym of `title`#3, that is `text`#1 `textual matter`#1, is found as a meronym of a hypernym of `book`#1, that is `publication`#1. Thus, `text`#1 `textual matter`#1 and `publication`#1 represent the lcs for the two target synsets.

Definitions of synset graph edge weight. Let $a = \langle t_i, \sigma \rangle$ and $b = \langle t_{i+1}, \rho \rangle$ be synsets corresponding to connected nodes in a synset graph. The weight on the edge (a, b) can be defined in various ways according to the different settings of the two approaches previously discussed. In the following, for each definition we fix the corresponding notation within brackets, which will be used in the presentation of the experiments.

- Gloss-overlap-based definitions — various forms of the function $f(d)$ can be chosen, such as exponential functions. Also, the parameter $\overline{d}$ will be tuned experimentally. We provide here the following definitions:
  - [G] $weight(a,b) = go\text{-}relatedness(\sigma,\rho,0) \equiv go\text{-}score(\sigma,\rho)$.
  - [G_Exp] $weight(a,b) = go\text{-}relatedness(\sigma,\rho,\overline{d})$, with $f(d) = e^{-d}$.
  - [G_SqExp] $weight(a,b) = go\text{-}relatedness(\sigma,\rho,\overline{d})$, with $f(d) = e^{-d^2}$.
  - [G_InvLin] $weight(a,b) = go\text{-}relatedness(\sigma,\rho,\overline{d})$, with $f(d) = 1/(1+d)$

  Moreover, for each of the definitions above, we need to decide whether a gloss is to be enhanced with the corresponding set of synonyms; if this is the case, we will use the notation suffix /+S.
- Ontology-path-based definition — this setting corresponds to the definition [P] $weight(a,b) = p\text{-}relatedness(\sigma,\rho)$.
- Composite definitions — the *go-relatedness* and *p-relatedness* measures can be combined together in various ways. We provide here the following definitions:
  - [P_cnd] $weight(a,b) = go\text{-}relatedness(\sigma,\rho)$, if $p\text{-}relatedness(\sigma,\rho) > 0$; otherwise, $weight(a,b) = 0$. Here we are assuming that the sense relatedness is computed on the basis of the gloss overlap method conditionally to the path-based relatedness.
  - [GxP] $weight(a,b) = go\text{-}relatedness(\sigma,\rho) \times p\text{-}relatedness(\sigma,\rho)$. In this function the path-based term acts as a damping factor for the gloss-based term, since *p-relatedness* assumes values in $[0..1]$.

Disambiguating the path tags. Once the synset graph for a given XML path p has been established, the disambiguation of all the tag names in p is accomplished by finding the "best" path in the graph. The senses corresponding to this graph path are recognized as the most appropriate senses for the tag names in p. We devise two ways to find the "best" path in a synset graph:

- *Maximum-weight path* [MWP]: the best path is such that the sum of the weights over its edges is maximum. In the case of multiple best paths in the synset graph, the preferred path can be computed by exploiting the dictionary-supplied linear order of synsets associated with the tag names.
- *Direction-driven path*: the best path is detected as the one resulting from a search of the maximum weight layer-by-layer in the graph; precisely, once the maximum weight has been found in a certain edge of the first layer visited, this edge fixes a portion of the best path being detected, and the search analogously continues on the next layer. Clearly, this search depends on the direction the path is visited, namely "from left" [DDP_sx] (i.e., top-down in the XML tree) or "from right" [DDP_dx] (i.e., bottom-up in the XML tree).

The maximum-weight method appears to be more general than the direction-driven method; however, the latter could in principle perform better in some cases, since it takes into account the specific order of the tags in an XML path, either in a top-down or bottom-up fashion.

Note also that in the case of weights equal to zero for all the edges between any two adjacent tags, a maximum weight is set for the edge between the first senses of the given tags—which is in agreement with the dictionary-supplied linear order of the synsets.

4 Experimental Evaluation

We collected various XML data belonging to different application domains. In this section, we present experimental results obtained on XML structures whose schemas are shown in Fig. 3:

- *DBLP*: data concerning scientific bibliography with a DBLP-like structure.[1]
- *Reuters*: news headlines from the Reuters RSS news channel.[2]
- *People*: customized data structure for recording personal information.
- *Wikipedia*: data representing encyclopedia articles with a Wikipedia-like structure.
- *Shakespeare*: plays from the Shakespeare 2.00 collection.[3]

In principle, tag names should be subject to a number of text processing operations. Removal of stopwords might be performed, as long as such stopwords do not represent lexical constituents of a compound term; as an example, given the tag `state-of-the-art`, the stopwords "of" and "the" should not be removed from that tag. Word splitting can also be useful; in this work, we assume to perform a semi-automatic operation of word splitting, which is driven by both the tag tokenization based on delimiters (e.g., hyphen, underscore, or variation in the letter case) and ad-hoc lists of domain-specific compound terms.

We evaluated the results of tag sense disambiguation by comparing them with a human disambiguation based on the latest version of WordNet. More precisely, for each dataset, we manually selected the appropriate senses for the tag terms according to our comprehension of the meanings of the synset glosses. These senses were used as reference for the evaluation, as shown in Fig. 4. It should be noted that we selected more than one "ideal" sense for some tag terms in a specific data. This was mainly due to our difficulty in capturing the underlying relationships between the tag terms which may depend on a subjective interpretation of the synset descriptions.

To assess the accuracy in the disambiguation, we define a measure that considers the particular form of the disambiguation context (i.e., the synset graph). This measure, called *tag-pair-accuracy*, computes for each path the fraction of pairs of adjacent tags in the path which have been correctly disambiguated. The overall tag-pair-accuracy is finally obtained by averaging the local accuracy values over all the distinct paths in an XML document.

---

[1] http://www.informatik.uni-trier.de/~ley/db/
[2] http://www.reuters.com/tools/rss
[3] http://metalab.unc.edu/bosak/xml/eg/shaks200.zip

```
<!ELEMENT bibliography (article | inproceedings | book)+>
<!ELEMENT article (author+, title, journal, volume, year, publisher)>
<!ELEMENT inproceedings (author+, title, pages, year, booktitle)>
<!ELEMENT book (author+, title, publisher, year, editor?, series?, volume?)>
<!ELEMENT author (#PCDATA)>
<!ELEMENT booktitle (#PCDATA)>
<!ELEMENT editor (#PCDATA)>
<!ELEMENT journal (#PCDATA)>
<!ELEMENT pages (#PCDATA)>
<!ELEMENT publisher (#PCDATA)>
<!ELEMENT series (#PCDATA)>
<!ELEMENT title (#PCDATA)>
<!ELEMENT volume (#PCDATA)>
<!ELEMENT year (#PCDATA)>
```

(a) *DBLP*

```
<!ELEMENT news (channel+)>
<!ELEMENT channel (title, link?, description, image?, item+)>
<!ELEMENT image (title, width, height, link, url)>
<!ELEMENT item (title, description)>
<!ELEMENT description (#PCDATA)>
<!ELEMENT height (#PCDATA)>
<!ELEMENT link (#PCDATA)>
<!ELEMENT title (#PCDATA)>
<!ELEMENT url (#PCDATA)>
<!ELEMENT width (#PCDATA)>
```

(b) *Reuters RSS*

```
<!ELEMENT Shakespeare (play+)>
<!ELEMENT play (title, prologue?, act*)>
<!ELEMENT act (scene+, epilogue?)>
<!ELEMENT prologue (title, speech)+>
<!ELEMENT epilogue (title, speech)+>
<!ELEMENT scene (title, speech)+>
<!ELEMENT speaker (#PCDATA)>
<!ELEMENT speech (speaker)>
<!ELEMENT title (#PCDATA)>
```

(c) *Shakespeare*

```
<!ELEMENT encyclopedia (article+)>
<!ELEMENT article (header, body)>
<!ELEMENT header (title, revision, categories)>
<!ELEMENT body (section+)>
<!ELEMENT section (title, link?, list*, figure*)>
<!ELEMENT categories (category+)>
<!ELEMENT figure (image, caption)>
<!ELEMENT caption (#PCDATA)>
<!ELEMENT category (#PCDATA)>
<!ELEMENT entry (#PCDATA)>
<!ELEMENT image (#PCDATA)>
<!ELEMENT link (#PCDATA)>
<!ELEMENT list (entry+)>
<!ELEMENT revision (#PCDATA)>
<!ELEMENT title (#PCDATA)>
```

(d) *Wikipedia*

```
<!ELEMENT people (person+)>
<!ELEMENT person (name, address, email, phonenumber, creditcard?, profile)>
<!ELEMENT address (street, city, country, province, zipcode)>
<!ELEMENT profile (gender, age, income, education, business)>
<!ELEMENT age (#PCDATA)>
<!ELEMENT business (#PCDATA)>
<!ELEMENT city (#PCDATA)>
<!ELEMENT country (#PCDATA)>
<!ELEMENT creditcard (#PCDATA)>
<!ELEMENT education (#PCDATA)>
<!ELEMENT email (#PCDATA)>
<!ELEMENT gender (#PCDATA)>
<!ELEMENT income (#PCDATA)>
<!ELEMENT name (#PCDATA)>
<!ATTLIST person id NMTOKEN #REQUIRED>
<!ELEMENT phonenumber (#PCDATA)>
<!ELEMENT province (#PCDATA)>
<!ELEMENT street (#PCDATA)>
<!ELEMENT zipcode (#PCDATA)>
```

(e) *People*

Fig. 3. DTDs associated with the XML data used in the experiments

tag	*data*	*senses*
act	*Shakespeare*	3
address	*People*	2, 6
age	*People*	3
article	*DBLP*	1
article	*Wikipedia*	1, 3
author	*DBLP*	1
bibliography	*DBLP*	1
body	*Wikipedia*	9
book	*DBLP*	1
business	*People*	6
caption	*Wikipedia*	3
categories	*Wikipedia*	1, 2
category	*Wikipedia*	1, 2
channel	*Reuters*	5, 7, 8
city	*People*	1
country	*People*	1
credit_card	*People*	1

tag	*data*	*senses*
description	*Reuters*	1, 3
editor	*DBLP*	1
education	*People*	1, 3
email	*People*	1
encyclopedia	*Wikipedia*	1
entry	*Wikipedia*	1
epilogue	*Shakespeare*	1, 2
figure	*Wikipedia*	1
gender	*People*	2
header	*Wikipedia*	1
height	*Reuters*	1, 3
id	*People*	2
image	*Reuters*	2
image	*Wikipedia*	2
income	*People*	1
item	*Reuters*	1, 4
journal	*DBLP*	2

tag	*data*	*senses*
link	*Reuters*	1, 7
link	*Wikipedia*	6, 7
list	*Wikipedia*	1
name	*People*	1
news	*Reuters*	3, 4
pages	*DBLP*	1
people	*People*	1
person	*People*	1
phone_number	*People*	1
play	*Shakespeare*	1
proceedings	*DBLP*	2
profile	*People*	3
prologue	*Shakespeare*	1
province	*People*	1
publisher	*DBLP*	2
revision	*Wikipedia*	3

tag	*data*	*senses*
scene	*Shakespeare*	6, 1
section	*Wikipedia*	1
series	*DBLP*	3
Shakespeare	*Shakespeare*	1
speaker	*Shakespeare*	1
speech	*Shakespeare*	7, 1
street	*People*	1, 2
title	*DBLP*	3
title	*Reuters*	3, 1
title	*Shakespeare*	3, 2
title	*Wikipedia*	3
url	*Reuters*	1
volume	*DBLP*	4, 3
width	*Reuters*	1
year	*DBLP*	1
zip_code	*People*	1

Fig. 4. The senses selected for the evaluation

We would like to point out that we have chosen a single assessment criterion in this work for the sake of brevity. However, the notion of tag-pair-accuracy has been preferred to other, more standard criteria such as the classic precision, which evaluate the contingencies of single instances (i.e., the tag names) rather than pairs of instances. This results in an assessment criterion which is in principle tougher than precision or other similar measures, thus causing a relative underestimation of the disambiguation results.

4.1 Evaluation of Gloss-Overlap-Based Disambiguation

Figure 5 shows the accuracy results obtained by using the gloss-overlap-based disambiguation. In general, the form of the $f(d)$ function did not affect the accuracy substantially. The option /+S turned out to be beneficial to the disambiguation in nearly all the datasets. Regardless of $f(d)$, $\overline{d}$ and the strategy of search in the synset-graph, the average improvement was up to 11% on *Shakespeare*, 6% on *People*, 0.7% on *Reuters*, 0.6% on *DBLP*. However, the accuracy

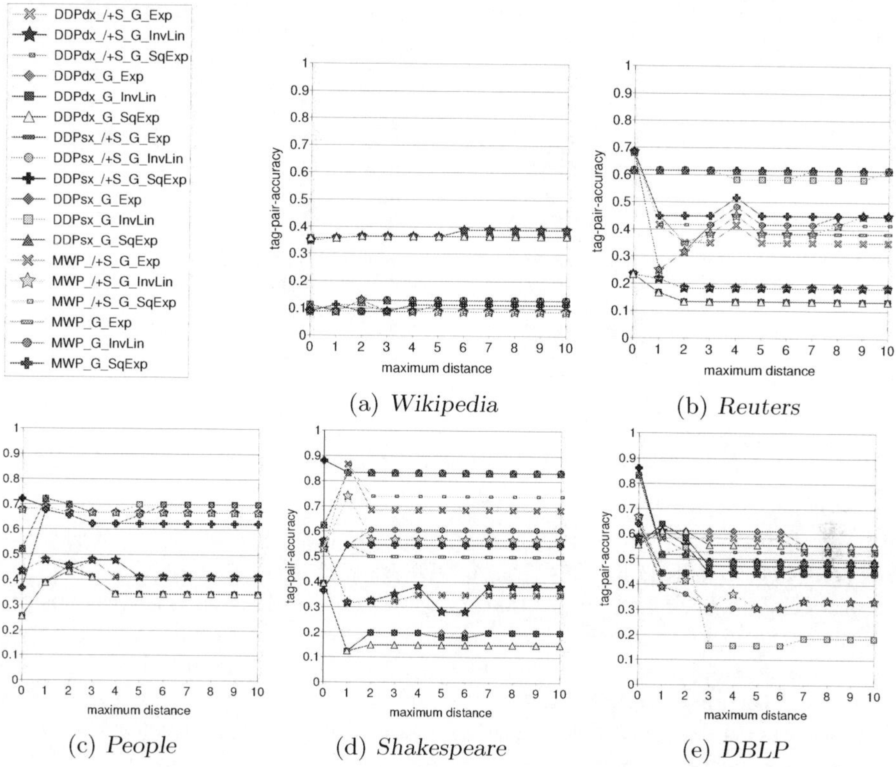

Fig. 5. Accuracy results using gloss-overlap-based methods

improvement was even higher when the best search strategy was used (e.g., 17% on *Shakespeare* for DDP_dx, 8% on *People* for DDP_dx, 4% on *Reuters* for DDP_dx). For example, the expected disambiguation for `speech` and `speaker` (i.e., synset #7 and #1, respectively) was captured thanks to the enhancement of the `speech`#7's gloss description, which leads to the overlap 'speech' with the `speaker`#1's gloss description.

The accuracy trends mainly depended on the value of the maximum distance $\overline{d}$ and the strategy of search in the synset-graph. In general, the accuracy tended to stabilize as $\overline{d}$ increases, especially for $\overline{d} \geq 2$. However, in some cases, the accuracy may decrease as $\overline{d}$ increases depending on the search strategy. For example, on *DBLP*, regardless of $\mathrm{f}(d)$ and /+S the difference between the maximum and the minimum accuracy was around 8% for DDP_dx, 14% for MWP and 31% for DDP_sx. Considering an average over $\mathrm{f}(d)$ and /+S, the best accuracy was achieved for small $\overline{d}$, in particular: 0.85 on *DBLP* for $\overline{d} = 0$, 0.88 on *Shakespeare* for $\overline{d} = 1$, 0.74 on *People* for $\overline{d} = 1$, 0.68 on *Reuters* for $\overline{d} = 0$.

As concerns the strategy of search in the synset-graph, DDP_sx behaved better than DDP_dx and MWP in all the datasets (except *Wikipedia*) considering the accuracy values averaged over $\overline{d}$, $\mathrm{f}(d)$ and /+S. In particular, comparing DDP_sx to DDP_dx, the improvement was up to 51% on *Shakespeare*, 44% on *Reuters*,

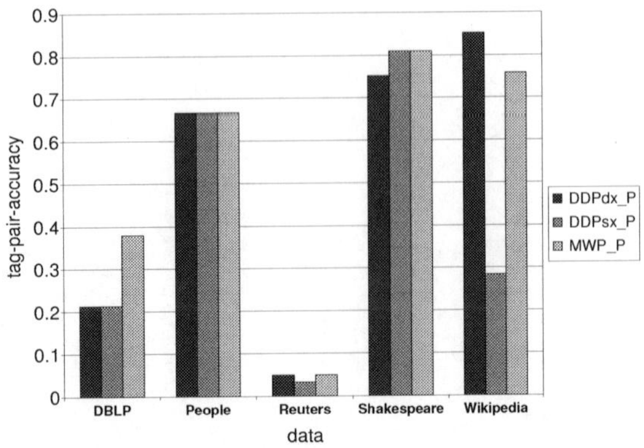

Fig. 6. Accuracy results using ontology-path-based methods

30% on *People*, 11% on *DBLP*. However, it should be noted that DDP_dx was less sensitive to $\overline{d}$, f(d) and /+S than DDP_sx and MWP.

A special remark has to be made on *Wikipedia*. In this data we observed a failure of the gloss-overlap-based disambiguation due to the lack of "significant" overlaps, that is overlaps consisting of false content-bearing words. For example, synsets `encyclopedia`#1 and `article`#4 share the term 'reference', and `article`#4 and `title`#1 share the term 'may', which mislead to disambiguate the meanings of such tags in their context. The accuracy was further negatively affected (especially when MWP and DDP_sx were used) by the fact that `encyclopedia`, `article` and `title` form a common prefix of XML paths in *Wikipedia*.

4.2 Evaluation of Ontology-Path-Based Disambiguation

The ontology-path-based disambiguation may take advantage of some characteristics of certain data. Comparing the results in Fig. 6 with Fig. 5, significant improvements were observed in some cases, in particular: on *Wikipedia*, up to 60% for MWP, 50% for DDP_dx and 10% for DDP_sx; on *Shakespeare*, up to 30% for DDP_dx and 15% for MWP; on *People* up to 20% for DDP_dx. By contrast, the accuracy drastically decreased on *Reuters* and *DBLP* regardless of the search strategy used.

We tend to justify these different behaviors by observing the cohesiveness of the path tags with respect to a specific application domain. Indeed, most tags in *Wikipedia* (e.g., `encyclopedia.article`), *Shakespeare* (e.g., `act.scene.speech`) and, partly, *People* (e.g., `zipcode`, `address`, `phonenumber`) show high pertinence to their respective domains. By contrast, some tag names in *Reuters* typically occur in different contexts with multiple meanings; in particular, when these tag names are adjacent and form prefixes of path (e.g., `channel` and `item`), the accuracy of disambiguation will result very low.

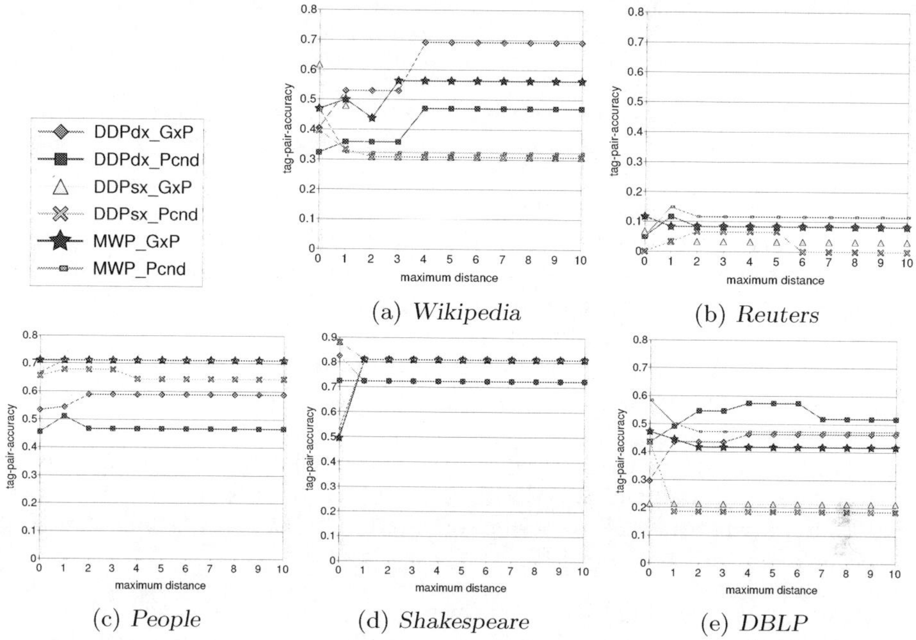

Fig. 7. Accuracy results combining gloss-overlap- and ontology-path-based methods

4.3 Evaluation of Combined Relatedness

Figure 7 shows the results obtained by combining weighting methods of the approaches previously discussed—we present here results referring to the case $f(d) = e^{-d}$ and gloss enhancement (option /+S). In general, using *p-relatedness* was relevant for the disambiguation, since the accuracy achieved average values similar to those observed with the pure ontology-path-based approach.

Compared to the results obtained by using gloss-overlap-based methods, the accuracy significantly improved in *Wikipedia* (from 40% to 70%), while substantially did not vary in *Shakespeare* and *People*. A drastic decrease in accuracy occurred in *Reuters* (which was due to the lack of *p-relatedness* as previously observed in Fig. 6), while the performance of *DBLP* tended to decrease depending on the search strategy and the composite methods.

Focusing the attention on the composite methods, we observed there was no particular setting in which one method prevailed against the other one, averaging over $\overline{d}$. For instance, on *DBLP*, P_cnd led to better average results when DDP_dx or MWP was used, whereas GxP slightly prevailed for DDP_sx and higher $\overline{d}$. On *Reuters*, P_cnd performed as good as or better than GxP. On *Wikipedia* and *People*, GxP performed as good as or better than P_cnd; in particular, on *Wikipedia*, a good value of *p-relatedness* lowered the impact of "false" overlaps, leading to relatively high performance for DDP_dx and GxP. In general, P_cnd seemed to be less sensitive to $\overline{d}$ than GxP.

4.4 Lessons Learned

Facing with the above results, we can summarize the following main remarks:

- Using the set of synonyms to enhance the text of a gloss definition is effective since it increases the probability of finding (semantically useful) overlaps between glosses.
- The maximum distance by which a synset is compared with relating indirect synsets may affect the disambiguation accuracy in different ways depending on the conceptual relatedness (*p-relatedness*) among the path tags; however, in general, accuracy tends to stabilize with $\overline{d} \geq 3$, and may have maximum peaks with smaller $\overline{d}$ in some cases.
- The form of the damping function $\mathrm{f}(d)$ is not crucial to the disambiguation, which is probably due to a small contribution of the gloss overlaps using indirect synsets.
- Different choices of the strategy of search in the synset graph may result in different disambiguations. DDP_sx and MWP tend to lead to better performance, whereas DDP_dx seems to be more robust to variations that may occur in the other parameters.

It is worth emphasizing that measuring the tag sense relatedness by using the gloss-overlap-based methods has proved to be effective, although it may lead to scarcely or not significant scorings; indeed, gloss descriptions may be not enough to capture relatedness between close synsets, and may also contain false content-bearing words. On the other hand, using only the WordNet concept hierarchies may have a very different impact on the disambiguation accuracy: in general, the higher the domain homogeneity of the path tags, the more significant the contribution of the WordNet hierarchies (especially the *is-a* hierarchy) to achieve better disambiguation.

5 Conclusion

We investigated how to semantically-enrich structure information available from XML data in order to support the generation of XML semantic features. In our approach, lexical ontology (WordNet) plays a central role in identifying semantic relationships among the concepts underlying the constituents of structure information. A method of word sense disambiguation is defined to couple each tag name with its appropriate meaning in the context of an XML path. We presented an experimental evaluation of the proposed approach and discussed some lessons learned.

We plan to extend our approach in several directions. One possible concerns a simultaneous analysis of all the distinct paths in an XML document, which could aid to adaptively disambiguate the tag senses with respect to the context of the document. We also intend to enhance the notions of sense relatedness with information on the tag-concept domains available from WordNet or other ontologies.

References

1. Banerjee, S., Pedersen, T.: Extended Gloss Overlaps as a Measure of Semantic Relatedness. In: Proc. IJCAI, pp. 805–810 (2003)
2. Candillier, L., Tellier, I., Torre, F.: Transforming XML Trees for Efficient Classification and Clustering. In: Fuhr, N., Lalmas, M., Malik, S., Kazai, G. (eds.) INEX 2005. LNCS, vol. 3977, pp. 469–480. Springer, Heidelberg (2006)
3. Doucet, A., Lehtonen, M.: Unsupervised Classification of Text-Centric XML Document Collections. In: Fuhr, N., Lalmas, M., Trotman, A. (eds.) INEX 2006. LNCS, vol. 4518, pp. 497–509. Springer, Heidelberg (2007)
4. Fellbaum, C.: WordNet: An Electronic Lexical Database. MIT Press, Cambridge (1998)
5. Lesk, M.: Automatic Sense Disambiguation Using Machine Readable Dictionaries: How to tell a pine cone from a ice cream cone. In: Proc. ACM SIGDOC Int. Conf. on Systems Documentation, pp. 24–26 (1986)
6. Mandreoli, F., Martoglia, R., Ronchetti, E.: Versatile Structural Disambiguation for Semantic-aware Applications. In: Proc. CIKM (2005)
7. Navigli, R., Velardi, P.: Structural Semantic Interconnections: A Knowledge-Based Approach to Word Sense Disambiguation. IEEE Transactions on Pattern Analysis and Machine Intelligence, 1075–1086 (2005)
8. Nayak, R.: Fast and effective clustering of XML data using structural information. Knowledge and Information Systems 14, 197–215 (2008)
9. Nierman, A., Jagadish, H.V.: Evaluating Structural Similarity in XML Documents. In: ACM SIGMOD WebDB Workshop, pp. 61–66 (2002)
10. Patwardhan, S., Banerjee, S., Pedersen, T.: Using Measures of Semantic Relatedness for Word Sense Disambiguation. In: Gelbukh, A. (ed.) CICLing 2003. LNCS, vol. 2588, pp. 241–257. Springer, Heidelberg (2003)
11. Pedersen, T., Banerjee, S., Patwardhan, S.: Maximizing Semantic Relatedness to Perform Word Sense Disambiguation. Tech. rep. UMSI 2005/25, Supercomputing Institute Research at University of Minnesota, pp. 7–9 (2005)
12. Resnik, P.: Semantic Similarity in a Taxonomy: An Information-based Measure and its Application to Problems of Ambiguity in Natural Language. Journal of Artificial Intelligence Research 11, 95–130 (1999)
13. Tagarelli, A., Greco, S.: Clustering Transactional XML Data with Semantically-Enriched Content and Structural Features. In: Zhou, X., Su, S., Papazoglou, M.P., Orlowska, M.E., Jeffery, K. (eds.) WISE 2004. LNCS, vol. 3306, pp. 266–278. Springer, Heidelberg (2004)
14. Tagarelli, A., Greco, S.: Toward Semantic XML Clustering. In: Proc. SIAM Data Mining, pp. 188–199 (2006)
15. Theobald, M., Schenkel, R., Weikum, G.: Exploiting Structure, Annotation, and Ontological Knowledge for Automatic Classification of XML Data. In: ACM SIGMOD WebDB Workshop, pp. 1–6 (2003)
16. Vercoustre, A.M., Fegas, M., Gul, S., Lechevallier, Y.: A Flexible Structured-based Representation for XML Document Mining. In: Fuhr, N., Lalmas, M., Malik, S., Kazai, G. (eds.) INEX 2005. LNCS, vol. 3977, pp. 443–457. Springer, Heidelberg (2006)
17. Yang, D., Powers, D.M.W.: Measuring semantic similarity in the taxonomy of WordNet. In: Proc. Australasian Conf. on Computer Science, pp. 315–322 (2005)

Improving Ontology Matching Using Meta-level Learning

Kai Eckert, Christian Meilicke, and Heiner Stuckenschmidt

KR&KM Research Group
University of Mannheim, Germany
{kai,christian,heiner}@informatik.uni-mannheim.de

Abstract. Despite serious research efforts, automatic ontology matching still suffers from severe problems with respect to the quality of matching results. Existing matching systems trade-off precision and recall and have their specific strengths and weaknesses. This leads to problems when the right matcher for a given task has to be selected. In this paper, we present a method for improving matching results by not choosing a specific matcher but applying machine learning techniques on an ensemble of matchers. Hereby we learn rules for the correctness of a correspondence based on the output of different matchers and additional information about the nature of the elements to be matched, thus leveraging the weaknesses of an individual matcher. We show that our method always performs significantly better than the median of the matchers used and in most cases outperforms the best matcher with an optimal threshold for a given pair of ontologies. As a side product of our experiments, we discovered that the majority vote is a simple but powerful heuristic for combining matchers that almost reaches the quality of our learning results.

1 Motivation

Despite significant research efforts automatic ontology matching is one of the unfulfilled promises of semantic web technologies and might turn out to become the Achilles' heel for large scale applications of ontologies on the web. So far, a significant number of automatic matching systems have been developed that address the matching problem by applying different heuristics, most of which are based on the similarity of representations. Depending on the kind of heuristics used, these matchers show a varying quality on different matching problems. This problem is typically addressed by approaches for selecting the optimal matcher based on the nature of the matching task and the known characteristics of the different matching systems. Such an approach that has been based on extensive interviews and tests is described in [1]. Another typical approach for dealing with the problem of adapting to a given matching task is to apply machine learning techniques for learning the optimal configuration of a matcher for a given dataset [2]. This approach amounts to determining the right heuristics and the appropriate parameters to be used in order to achieve the best result.

The approach proposed in this paper differs from the above mentioned ones by not signing up to a specific matcher but trying to exploit the results of available matchers which are treated as a black-box. This has the advantage that the weaknesses of individual matchers are compensated. Further, the approach settles on the idea that by using

L. Aroyo et al. (Eds.): ESWC 2009, LNCS 5554, pp. 158–172, 2009.

multiple matchers we can benefit from the high degree of precision of some matchers and at the same time the broader coverage of other matchers to complete the picture where highly precise matchers did not produce results.

Our approach is verified in a realistic setting on different standardized and open available datasets - details are discussed in Section 3.2. In particular, we use the results of the past OAEI campaigns that provide us with a gold standard mapping as well as a large number of mappings created by different matching systems. Thus, we can train a classifier on the outcome of different matching systems and learn what combination of results from different matchers provides the best indication of a correct correspondence. This proves to outperform previous attempts of combining matchers which have often been based on ad hoc methods or had to be customized manually.

Related Work. An approach to solve the problem of selecting correspondences from a set of matcher-generated mappings within the context of argumentation frameworks is presented in [3]. While it is based on theoretically well founded principles, the authors describe first experimental results as inconclusive and point out the necessity of further research. The general idea of combining individual matchers into a combined matching system constitutes also a principle used inside many existing matchers (e.g. [4,5]). The difference, however, is in the way the results of the individual matchers are combined. In contrast to existing approaches that combine a number of specific predefined classifiers, our approach is a more general one, as we do not make any assumptions about the individual matchers to be combined apart from the fact that they provide their results in a standardized format.

An approach that is very similar to ours is implemented in the GLUE System [6] that applies a meta-learning approach for generating matching hypotheses on the basis of multiple local classifiers that are trained on different aspects of the models to be matched. This approach, however requires that the input for meta learning is generated by specific probabilistic learning methods, in that case naive Bayes classifiers, that are integrated using a linear combination at the meta level. In our case, we do not make any assumptions about the matchers used and apply different machine learning techniques at the meta level. This makes our approach more widely applicable. Further, the evaluation in [6] is performed on a rather limited set of ontologies without the existence of commonly agreed reference alignments. We use two widely used benchmark datasets each containing multiple ontologies that have to be aligned. Looking at the results of [6] reveals that the meta-learning approach is dominated by the so-called content learner, which always performs almost as good as the integrated learning approach making the meta-learning step less important. As reported in section 3.3 our approach is in many cases significantly better than any of the local matchers which underlines the usefulness of our approach to meta-level learning.

Three other approaches in line with our ideas are described in [7], [8] and [9]. In [7] support vector machines are used to learn a classifier for mapping correctness based on a set of simple similarity measures. The classifier is evaluated on the benchmark datasets of the ontology alignment evaluation initiative and outperforms existing matching systems. The setting of the experiments, however, is a rather unrealistic one as all existing reference mappings are thrown together in one large training set and 10-fold cross-validation is used for computing the accuracy of the classifier. The results are only a very unreliable estimation of the behavior that can be expected from the classifier for a realistic matching problem. To avoid this problem, we carefully design the evaluation scenario and test our method in a setting that can be expected for a real integration task (compare section 3.2). Further, the approach uses only similarity values as input for learning.

In [8] the authors focus on the use of different bayesian classifiers. Again their feature set is based on string distance measures typically used in matching systems to derive a syntactic similarity. The experimental results show that there is a strong correlation between different measures and the machine learning approach cannot significantly improve the results of the best individual measure. Contrary to this, we implemented a rich set of features describing different aspects of ontologies and correspondences. We show that using the additional feature set can significantly improve the classification result. In fact, we even show that confidence values of individual matchers, which are normally the result of a similarity estimation are not significant and do not contribute to the learning result.

The approach reported in [9] is probably the most similar to our work. Here decision trees and rule learners are used to learn rules for integrating the results of different matching systems, which is more or less the same we do. Similar to [7], the difference to our work is the restriction to confidence values respectively measured similarities as the basis for learning. The approach has been evaluated on a subset of the OAEI benchmark dataset, but no detailed results of the evaluation are provided making it hard to judge the quality of the proposed method. Our results suggest that basing the learning step just on confidence values is not a good choice (compare section 3.3).

Contributions. The contributions of this paper are the following:

- We present a new approach for combining different matching systems using machine learning techniques.
- We evaluate the approach in a realistic setting using well established benchmark datasets.
- We show that our approach systematically outperforms existing matching tools in the sense that it not only produces better results than the median of the matchers but also outperforms or measures up to the best matching system for every matching task we investigate.
- We identify a simple but very powerful heuristic for combining matching results that outperforms the best matching system and almost reaches the performance of the machine learning approach.

The paper is structured as follows: we first briefly discuss the problem of combining the result of different matching systems and present our approach for solving this problem in more detail. The major part of the paper is concerned with describing the setting and the results of the matching experiments we performed. We motivate the choice of the datasets and the setting of the experiments and compare our results with three different baselines. We conclude with a discussion of the results and their implication for ontology matching on the semantic web.

2 A Meta-level Learning Approach

As indicated in the introduction, we approach the problem of combining different matching systems by applying machine learning techniques on top of the results produced by different state of the art matching systems. Our approach is presented in more details in this section.

2.1 The Problem of Combining Matchers

The ontology matching can be defined as follows [10]: Given two ontologies O_1 and O_2, establish semantic relations - also referred to as correspondences - between certain matchable elements. In this work, we restrict ourselves to the detection of equivalence correspondences between named concepts and properties. This restriction is motivated by the state of the art in ontology matching as it is documented in the annual benchmarking activities of the ontology alignment evaluation initiative (cf. [11] and [12]). Most of the systems that have successfully participated in the benchmarking activities are limited to one-to-one mappings between named concepts and properties; only very few systems produce relations other than equivalence. Another reason for only considering equivalence is the absence of commonly accepted reference mappings that contain non-equivalence correspondences.

Besides a set of possible equivalence correspondences, matching systems often provide information about a level of confidence in the correctness of the correspondence. The meaning of this level of confidence has been subject of many discussions. According to Bouquet et al. [13] the confidence is

> ...a degree of trust (confidence) in that mapping (notice, this degree does not refer to the relation R, it is rather a measure of the trust in the fact that the mapping is appropriate ("I trust 70% the fact that the mapping is correct/reliable/...")). The trust degree can be computed in many ways, including users' feedback or log analysis.

The only formal requirement on the level of confidence mentioned in [13] is the existence of some partial order over the confidence values that allows the matcher to order its results according to its own belief in the correctness of the relation represented. According to the definition of Euzenat and Shvaiko, the result of a matching system can be described in terms of a set of correspondences of the form $O_1 : E = O_2 : E' \ (n)$ where E and E' are either both named concepts or properties and n is a degree of confidence, normally a real number from the interval $[0, 1]$. An example could be $O_1 : \mathit{Hotel} = O_2 : \mathit{Accomodation}\ (0.6)$. This means that the matcher has a confidence of 0.6 that concept *Hotel* in ontology O_1 describes the same objects as concept *Accomodation* in ontology O_2. In this definition, the rather vague notion of confidence causes serious problems when trying to combine the result of different matching systems as it cannot be guaranteed that they use the same notion of confidence, and in many cases the confidence values are not comparable because they have been computed based on fundamentally different principles. This means that if there are two other matchers that produce the correspondences $O_1 : \mathit{Hotel} = O_2 : \mathit{Accomodation}\ (0.4)$ and $O_1 : \mathit{Hotel} = O_2 : \mathit{Hotel}\ (0.7)$ it is not clear how to combine the results. On the one hand, one might argue as follows: two matchers think that *Hotel* and *Accomodation* are equivalent concepts while only one does not agree. This means that *Hotel* = *Accomodation* should be accepted. On the other hand, one might raise an objection: the confidence in the truth of this statement is rather low compared to the confidence of the matcher which thinks that *Hotel* should be matched on *Hotel*. As different matchers can use very different methods for computing the confidence, there is no standard way of combining respectively comparing confidences to come to a conclusion.[1]

2.2 Combining Matchers Using Machine Learning

Our solution to the problem of combining different matchers is not to try to directly combine their output but to use machine learning techniques to train a classifier that decides whether two elements from different ontologies should be linked by an equivalence relation based on the output of different matching systems. As described above, this idea is similar to existing work in the area of ontology mapping. A major difference of our approach is that we do not restrict the learning approach to the output of the matchers, but add additional features as input for the learning approach.

The rationale for using these additional features is the observation that matchers more or less heavily rely on the existence of certain structures and information in the ontologies to be matched. Examples are the existence of additional concept descriptions encoded in labels or the existence of a WordNet synset the description can be linked to. In the first case, a matcher that uses information retrieval techniques to compare concept descriptions will outperform a matcher that just uses string matching on the names of concepts or properties. In the latter case a matcher that uses WordNet as background knowledge will detect more correspondences than a matcher that does not. Approaches like [1] make this information explicit for a limited set of known matching systems. By including certain properties of the ontologies and the elements to be matched into the learning process, we are able to take the strengths and weaknesses of matchers into account without having to know the concrete matchers. The fact that a certain matcher performs better if concepts can be linked to WordNet is detected in the learning step. This means that we do not need to know the characteristics of involved matchers as required precondition.

Figure 1 illustrates the generation of training data for the machine learning step with respect to the simple case of two matchers. In a first step, the matchers are run on the ontologies and generate two mappings. Then a training set is created that contains the union of both mappings. Correspondences contained in the reference mapping are positive examples, those that are not contained are negative examples. Besides the distinction between positive and negative examples indicated by the 'target' attribute, attributes include the results of the individual matchers in terms of an attribute indicating whether the correspondence was found by a certain matcher and what confidence was

[1] As described above, the GLUE approach solves this problem by only allowing individual matchers that use the same notion of confidence (in this case Bayesian probabilities). Their approach, however, does not solve the general problem of integrating arbitrary matching systems in an optimal way.

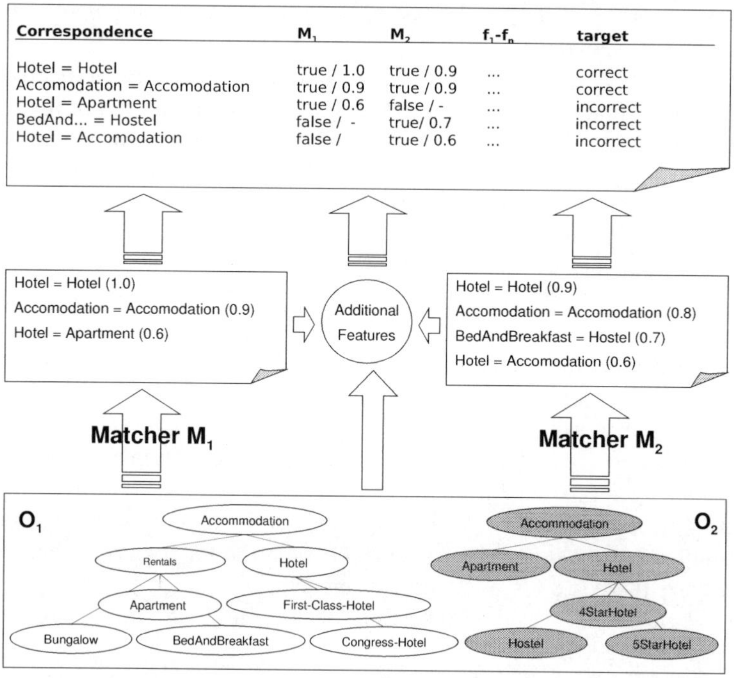

Fig. 1. Illustration of the Meta-level Approach to matcher combination

assigned by that matcher. A number of other features $f_1, \ldots, f_n$ that are extracted from the results of the matchers and the ontologies themselves are included. These features are motivated and described in detail in the following section.

The result of the learning step is a classifier modeling the training data set. Given a new matching problem, this classifier can be used to predict the correctness of correspondences. For this purpose, the same set of matchers used in the training phase is executed on the new matching problem. The resulting correspondences are used as input for the classifier together with the additional features that have to be computed for the new situation as well. The learned classifier can now be applied to the resulting descriptions of correspondences. In doing so, the correctness of a correspondence is decided based on matcher results and additional features.

2.3 Description of Features

The following set of additional features were used as input for our learning experiments. These features can roughly be separated into four groups:

1. **Matcher Features:** This group contains features that reflect the results of the involved matchers.

   Matcher Found: a boolean feature for each matcher that is *true*, if the corresponding matcher returned the correspondence in question as part of its mapping and *false* otherwise.

Matcher Confidenc: the confidence value specified by a particular matcher for a given correspondence.

Matcher Vote: a numerical feature with values between 0 and 1 reflecting the percentage of matchers that returned the correspondence as part of their mapping. Matcher Vote is a feature that follows the idea of a voting model to decide wether a given correspondence should be taken as correct. By modeling the feature this way we allow the classifier to learn how many individual votes are needed in combination with other feature values to decide the correctness of a correspondence.

2. **Ontology Features:** These features describe global characterstics of the two ontologies.

   Ontology Ratios: three features representing the ratio of the number of concepts, object properties and datatype properties between the two ontologies. We use the reciprocal value of the ratio, if the original value is above 1, thus the ratios are always between 0 and 1 with smaller values reflecting bigger differences in the number of matchable elements.

3. **Lexical Features:** As the lexical comparison of the texts belonging to the matching candidates are a common source of information for some matchers, we try to characterize these texts in this group of features.

   String Equivalence: a boolean feature that is *true*, if the normalized id of both elements are identical.

   Number of Token Ratios: These features are calculated in the same way as the Ontology Ratios and compare the number of tokens in id, label and comment. The rationale behind these features is the idea that lexical comparisons on these strings can only be effective if there is a similar amount of text available for both concepts. We decided for the number of tokens as a measure for the length of a text, as every token is a potential source for further analysis techniques.

   Mean Significance: We use the well known TF/IDF measure to calculate the mean significance S of tokens in a given text T against a text base D containing some other texts. With tf_t for the number of occurrences of token t in T and df_t for the number of texts containing token t in D, we get

$$S = \frac{1}{|T|} \cdot \sum_{t \in T} \left((1 + \log(tf_t)) \cdot \log\left(|D| \cdot df_t^{-1}\right) \right). \quad (1)$$

   We use Equation 1 to calculate the mean significance of the comment against the id and the label, as well as the mean significance of the label against the id. Both features are calculated for each element in the correspondence in question. The mean significance is a measure for the additional information that can be obtained from a text (like the comment), if other texts (like id and label) are already known.

   Wordnet Coverage. As Wordnet is commonly used to get further information about common terms that might occur in the id or label of an element, we introduced the Wordnet Coverage as a feature that returns a nominal value that says whether *none*, *one* or *both* of the elements contain terms that can be found in Wordnet. The Wordnet Coverage is calculated separately for id and label.

4. **Structural Features:** These features reflect the structural environment of the two elements in question.
   Type of Correspondence: a nominal feature that characterizes the type of elements involved in a given correspondence. The result may be one of four values: *Concept - Concept*, *Object Property - Object Property*, *Datatype Property - Datatype Property* and *Object Property - Datatype Property*. This feature has been introduced to enable the classifier to learn in how far different matching systems might perform better with respect to matching certain types of ontological elements.
   Node position. The Node Position is used to get information about two special cases, namely, if *none*, *one* or *both* of the elements linked by a correspondence are root or leaf nodes in their respective ontologies. Obviously, techniques like e.g. similarity flooding will perform differently depending on the position of an element within the ontological hierarchy.

3 Experiments

We tested our approach using systematic experiments on the benchmark datasets provided by the ontology alignment evaluation initiative. The experiments, which are described in more detail in the following, show that our approach systematically outperforms state of the art matching systems.

3.1 Datasets

The ontologies used in our experiments have been part of the Ontology Alignment Evaluation Initiative (OAEI) over the previous years. The OAEI offers several tracks and sub tracks concerned with different types of matching problems.[2] We can distinguish between those tracks where automatically generated mappings are compared against a reference mapping (also referred to as gold standard) and tracks using other evaluation techniques. Only the matching problems of the former tracks can be used as both input and basis for evaluating our approach and are described in the following.

The test set of the *benchmark track* is based on one particular ontology #101 dedicated to the very narrow domain of bibliography and a number of alternative ontologies of the same domain. It consists of a series of synthetic ontologies (#1xx and #2xx series) and four real world ontologies #301 to #304 which have to be matched on ontology #101. In our experiments we only consider test cases #301 to #304 because we are in particular interested in how far our approach can successfully be applied to realistic matching problems. The test set of the *conference track* consists of 15 ontologies where each pair of ontologies constitutes a matching problem. These ontologies have been developed as part of the OntoFarm project [14] and describe the domain of conference organization covering the structure of a conference, involved actors, as well as issues related to submission and review process. Since october 2008 reference mappings for all possible combinations between five of these ontologies are available which we used in

[2] Detailed information available at `http://oaei.ontologymatching.org/2008/`

our experiments. These ontologies are CMT, ConfTool, Ekaw, Iasted and Sigkdd. The conference dataset can be seen as a much harder testcase compared to the benchmark dataset that is less heterogeneous and has been extensively studied over the past years. This claim is also supported by our experimental results reported in Section 3.3.

Since we base our experiments on relevant OAEI datasets, we can make use of a rich set of diverse matching systems. In 2008 thirteen matching systems submitted mappings to the OAEI. We only included those matching systems generating a non boolean confidence value in the range $[0, 1]$ because we expected the confidence value to be an important feature for our learning approach. Matchers considered are *Aroma* [15], *ASMOV* [5], *DSSim* [16], *Lily* [17], *RiMOM* [18], *SAMBOdtf* and *SAMBO* [19][3]. While all of these systems submitted mappings for the benchmark track, only *ASMOV*, *DSSim* and *Lily* participated in the conference track. Thus, for each matching problem of the benchmark track there are mappings available generated by seven state of the art matching systems implementing diverse matching approaches, while for the conference track we can access the mappings of three matching systems.

3.2 Experimental Setup

Figure 2 shows the structure of the data sets and the derived experimental setting. On the left side the structure of the benchmark dataset is depicted as tree. Ontology #101 can be seen as a central knowledge base with several knowledge sources linked to it via mappings. Contrary to this, the conference data set on the right side is a full mesh where no ontology plays an outstanding role. We assume a setting, where a system of linked ontologies, e.g. #101, #301, #302 and #303 (colored white) exists. These ontologies have been linked in the depicted way and the mappings have been set up and verified by domain experts. We further assume that another knowledge source has to be added to this structure, in this case ontology #304. Given a set of matching systems respectively automatically generated alignments, we can profit from the reference mappings available in the way described above and learn a classifier C.

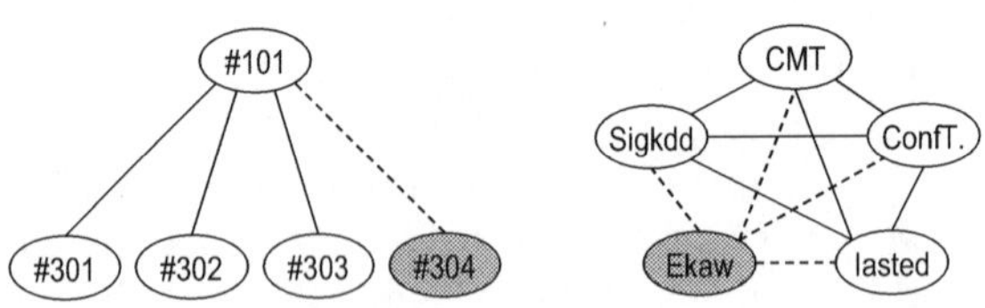

Fig. 2. Mapping structure of the benchmark (tree on the left) and conference data set (full mesh on the right) and experimental setting

Once we learned a classifier C, we apply it on the set of automatically generated mappings $M_{Aroma}, \ldots, M_{SAMBOdtf}$. We compare the mapping M_C, which consists of all those correspondences classified as correct by applying C, against reference mapping

[3] We did not include results of matching system SPIDER [20] due to its focus on generating complex non-equivalence correspondences not included in reference mappings.

$R_{\#304}$ to compute precision p, recall r and f-measure f, the standard measures for evaluating mappings, defined as follows:

$$p(M,R) = \frac{|M \cap R|}{|M|} \quad r(M,R) = \frac{|M \cap R|}{|R|} \quad f(M,R) = \frac{2 \cdot p(M,R) \cdot r(M,R)}{p(M,R) + r(M,R)}$$

This way we evaluate the performance of our approach with respect to sub testcase #304. We apply the same procedure for the other three sub testcases of the benchmark dataset.

For the conference dataset we apply a similar approach: We assume that a system of four pairwise mapped ontologies (Sigkdd, CMT, ConfTool and Iasted) exists and that another ontology (in this case EKAW) has to be linked with all of these ontologies. Here we use the pre-existing mappings between the four ontologies as training data and evaluate it on the mappings between the new ontology and the existing ones (dashed lines in Figure 2). We use each 4:1 split of the ontologies as a test case and aggregate the results.

Baselines. We used a number of baselines to compare our results to the state of the art in ontology matching. These baselines are described in the following. The first one is motivated by the usual approach to select one matcher from the set of available systems for a given task. This corresponds to selecting a mapping from $M_{Aroma}, \ldots, M_{SAMBOdtf}$. Without any additional knowledge this choice will be at random. We computed the f-measure for all mappings $M_{Aroma}, \ldots, M_{SAMBOdtf}$ and picked out the median of all measured values as first baseline to compare our results with. This baseline is referred to as *median matcher baseline* in the following.

In [1] the authors argue that it is possible to choose a well-suited matcher for a particular matching task given appropriate metadata describing the set of available matching systems. Although this metadata will often not be available and the approach does not ensure to choose the best system, we included a second baseline in our experiments by comparing our approach with the best mapping available for each particular matching task. This baseline will be referred to as *best matcher baseline*.

A more detailed analysis revealed that in many cases the threshold used for generating the mappings was not optimal with respect to the f-measure. Therefore, we included a third baseline where we successively increased the threshold to find the optimal threshold for each system with respect to each particular matching problem. This baseline is referred to as *optimal threshold baseline* and models the situation where the matcher is successfully adapted to the given matching task as suggested in [2].

Notice that it will not be easy to exceed any of these three baselines. In particular a comparison with the optimal threshold baseline will be a hard criterion for our approach. Another obvious baseline would be to compute the union of all mappings $M_1 \cup \ldots \cup M_n$. This baseline was used in [3]. Our experiments revealed that the union baseline cannot even reach the *median matcher baseline* for any of the described testcases. Therefore we resigned to include it in our experiments.

Implementation. Our experiments were conducted by means of the Weka Toolkit [21]. We used two different machine learning algorithms, on the one hand the decision tree algorithm J48, the Weka implementation of the last publicly available version of C4.5 [22], and on the other hand the Weka implementation for a Bayesian network Learner, BayesNet, which, for our data turned out to be equivalent to the Naive Bayes approach in principle[4]. All experiments were conducted with the default settings (2005 book version) making our experiments easily reproducible.

3.3 Results

As described above, [8], [9] and [7] propose approaches where machine learning techniques have been applied to a feature set only consisting of confidence values respectively measured similarities. Therefore, in a first set of experiments we focussed on a comparable setting based on a reduced feature set which consist of a single group of features, namely the generated confidence values. Table 1 lists detailed results for each subtestcase as well as aggregated values for the benchmark and conference dataset.

With respect to the benchmark testcases we could beat the median baseline with the use of both decision trees and Naive Bayes by approximately 3%. In particular, we outperformed the median baseline for each individual testcase. Although this baseline seems to be the most realistic one, the improvements are only minor and thus the best matcher baseline could not be reached.

With respect to the conference testcase we observe a different behavior. While decision trees perform very well (we measured an improvement of 13.3% in average) no significant difference between Naive Bayes and the median baseline can be observed. A look at detailed results reveals, that this has been mainly caused by the bad performance for subtestcase ConfTool. A posteriori, this outlier can be explained by an unfortunate choice in the discretization of the numerical attributes. Overall, these results partially coincide with our expectations based on existing studies cited above. A feature set restricted to confidences (or similarities) slightly improves the results, but cannot optimize the results to a significant degree in general.

In a second series of experiments we used the complete set of features listed in section 2.3. The results are presented in Table 2. For the benchmark dataset we observe only minor changes compared to our first experiments. Again, we measure only slight improvements with respect to the median baseline. This might be caused by a ceiling effect, since most of the input mappings were highly precise contrary to the conference dataset. Since testcase #301 to #304 have been part of the OAEI evaluation and with reference mappings open available, an overfitting of some matching systems cannot be excluded. Thus, it is extremely hard to improve the input by a significant degree.

With respect to the conference dataset we observe a dramatic change to the positive. Adding the complete feature set we measured an enhancement of the average f-value from 55.5% to 65.7% for decision trees and from 41.4% to 63.1% for Naive Bayes classifiers. This improvement topped our expectations. Obviously, our meta-level learning approach detects important interdependencies between those aspects described by the set of chosen features. Thus, we not only outperformed the median baseline but also the

[4] The results differ slightly to the default settings of the Weka NaiveBayes implementation, as the BayesNet implementation uses a supervised discretization step, which is not the default for NaiveBayes.

Table 1. Results based on features restricted to confidences

	Baselines			Decision Tree				Naive Bayes			
Ontology	M(edian)	B(est)	O(ptimal)	results	Δ-M	Δ-B	Δ-O	results	Δ-M	Δ-B	Δ-O
#301	0.825	0.877	0.877	0.855	+0.030	-0.022	-0.022	0.863	+0.038	-0.014	-0.014
#302	0.709	0.753	0.753	0.753	+0.044	+0.000	+0.000	0.753	+0.044	+0.000	+0.000
#303	0.804	0.860	0.891	0.816	+0.012	-0.044	-0.075	0.860	+0.056	+0.000	-0.031
#304	0.940	0.961	0.961	0.967	+0.027	+0.006	+0.006	0.954	+0.014	-0.007	-0.007
Average	**0.820**	**0.863**	**0.871**	**0.848**	**+0.028**	**-0.015**	**-0.023**	**0.857**	**+0.038**	**-0.005**	**-0.013**
CMT	0.435	0.512	0.512	0.500	+0.065	-0.012	-0.012	0.400	-0.035	-0.112	-0.112
ConfTool	0.471	0.484	0.484	0.474	+0.003	-0.010	-0.010	0.203	-0.268	-0.281	-0.281
Ekaw	0.411	0.471	0.516	0.593	+0.182	+0.122	+0.077	0.441	+0.030	-0.030	-0.075
Iasted	0.403	0.478	0.489	0.649	+0.246	+0.171	+0.160	0.453	+0.050	-0.025	-0.036
Sigkdd	0.390	0.462	0.475	0.560	+0.170	+0.098	+0.085	0.575	+0.185	+0.113	+0.100
Average	**0.422**	**0.481**	**0.495**	**0.555**	**+0.133**	**+0.074**	**+0.060**	**0.414**	**-0.008**	**-0.067**	**-0.081**

Table 2. Results based on a complete feature set

	Baselines			Decision Tree				Naive Bayes			
Ontology	M(edian)	B(est)	O(ptimal)	results	Δ-M	Δ-B	Δ-O	results	Δ-M	Δ-B	Δ-O
#301	0.825	0.877	0.877	0.883	+0.058	+0.006	+0.006	0.830	+0.005	-0.047	-0.047
#302	0.709	0.753	0.753	0.759	+0.050	+0.006	+0.006	0.753	+0.044	+0.000	+0.000
#303	0.804	0.860	0.891	0.816	+0.012	-0.044	-0.075	0.851	+0.047	-0.009	-0.040
#304	0.940	0.961	0.961	0.960	+0.020	-0.001	-0.001	0.966	+0.026	+0.005	+0.005
Average	**0.820**	**0.863**	**0.871**	**0.855**	**+0.035**	**-0.008**	**-0.016**	**0.850**	**+0.031**	**-0.012**	**-0.020**
CMT	0.435	0.512	0.512	0.580	+0.145	+0.068	+0.068	0.546	+0.111	+0.034	+0.034
ConfTool	0.471	0.484	0.484	0.572	+0.101	+0.088	+0.088	0.480	+0.009	-0.004	-0.004
Ekaw	0.411	0.471	0.516	0.621	+0.210	+0.150	+0.105	0.659	+0.248	+0.188	+0.143
Iasted	0.403	0.478	0.489	0.746	+0.343	+0.268	+0.257	0.750	+0.347	+0.272	+0.261
Sigkdd	0.390	0.462	0.475	0.766	+0.376	+0.304	+0.291	0.723	+0.333	+0.261	+0.248
Average	**0.422**	**0.481**	**0.495**	**0.657**	**+0.235**	**+0.175**	**+0.162**	**0.631**	**+0.209**	**+0.150**	**+0.136**

best matcher baseline. Even the best matcher with an optimal threshold obtains 16.2% respectively 13.6% worse results for the conference dataset.

The significant difference caused by taking into account the full feature set raises doubts about the importance of confidence values. Thus, in a third series of experiments we removed the confidence values from the complete feature set. Results presented in Table 3 confirm our scepticism about the validity of confidence values. Decision trees performed as good or even better for all of our nine sub testcases. With suppressed confidences a Naive Bayes classifier performs in six sub testcases better, in two cases we observed no changes, and only in one case we measured a slight decline. In average for all of our four combinations of dataset and classifier we nearly measure up or even clearly exceed the optimal threshold baseline. Notice again, that this baseline requires a perfect knowledge about the best matcher and its optimal threshold with respect to a certain matching task that is not available in nearly all realistic scenarios.

Table 3. Results based on a feature set where confidences have not been included

	Baselines			Decision Tree				Naive Bayes			
Ontology	M(edian)	B(est)	O(ptimal)	results	Δ-M	Δ-B	Δ-O	results	Δ-M	Δ-B	Δ-O
#301	0.825	0.877	0.877	0.883	+0.058	+0.006	+0.006	0.841	+0.016	-0.036	-0.036
#302	0.709	0.753	0.753	0.759	+0.050	+0.006	+0.006	0.753	+0.044	+0.000	+0.000
#303	0.804	0.860	0.891	0.816	+0.012	-0.044	-0.075	0.860	+0.056	+0.000	-0.031
#304	0.940	0.961	0.961	0.960	+0.020	-0.001	-0.001	0.966	+0.026	+0.005	+0.005
Average	**0.820**	**0.863**	**0.871**	**0.855**	**+0.035**	**-0.008**	**-0.016**	**0.855**	**+0.035**	**-0.008**	**-0.015**
CMT	0.435	0.512	0.512	0.625	+0.190	+0.113	+0.113	0.597	+0.162	+0.085	+0.085
ConfTool	0.471	0.484	0.484	0.572	+0.101	+0.088	+0.088	0.526	+0.055	+0.042	+0.042
Ekaw	0.411	0.471	0.516	0.621	+0.210	+0.150	+0.105	0.667	+0.256	+0.196	+0.151
Iasted	0.403	0.478	0.489	0.746	+0.343	+0.268	+0.257	0.740	+0.337	+0.262	+0.251
Sigkdd	0.390	0.462	0.475	0.766	+0.376	+0.304	+0.291	0.732	+0.342	+0.270	+0.257
Average	**0.422**	**0.481**	**0.495**	**0.666**	**+0.244**	**+0.184**	**+0.171**	**0.652**	**+0.230**	**+0.171**	**+0.157**

One major result of our experiments is concerned with the role of confidence values. By comparing the results based on three feature configurations we conclude that confidence values without additional features cannot be exploited successfully by a meta-level learning approach. Moreover, given a well designed comprehensive set of additional features, confidence values have even negative effects on the performance of the overall approach. Notice that our feature set still contains matcher specific information. These are the features describing wether a correspondence has been found by a particular matcher and the aggregated feature reflecting how many systems found the correspondence. These features are obviously sufficient to model interdependencies between certain characteristics of a particular correspondence and its correctness.

4 Conclusions

There are a couple of implications resulting from our work. Some of them have already been discussed in connection with the motivation of this work and the contribution to the state of the art. Others where hidden in the results of the matching experiments. We now briefly recall the major implications and point to some of the hidden aspects.

First of all, we have shown that machine learning is an adequate - maybe the best - way of combining the results of different heterogeneous matching systems. At the same time, we have shown that it is not enough to base the learning step on the results of the matching systems alone, but that additional features representing and aggregating information about the mapping and the mapped ontology have to be taken into account. These features enable us to put matcher results into context and get a better basis for deciding when to trust a certain matcher. We have also shown, that the confidence measures produced by matching systems are almost meaningless if these contextual features are known as they do not improve the result. A hidden implication of this observation is, that we can also apply our method on the result of matchers that do not return a degree of confidence. This significantly broadens the applicability of our approach.

Another very interesting and important result of our work is hidden in the classifiers learned. Analyzing the decision trees generated for the two datasets we discovered that the most significant feature was the fraction of matchers that found a correspondence. From this observation, we can derive a straightforward heuristic for combining matchers. In particular, it turned out that just following the majority of the matchers involved produces results that are almost as good as the results reported in the last section. After making this observation, we compared the result of the majority vote heuristic with our baselines. While on the benchmark dataset the result of the majority vote was comparable with the best matcher, it significantly outperformed all baselines on the conference data set: While the baselines ranged from 0.422 to 0.495 on average, the majority vote reached an average of 0.633, which is more than 25% better than the best baseline. For comparison the averages we reached on this dataset with the two machine learning methods where 0.666 and 0.652 which is only about 5% better than the result of the majority vote.

So far, we have shown that our approach works very well in a setting where an existing system of ontologies is to be extended with links to additional ontologies about the same domain. This is a very realistic setting for concrete applications as there is often a central ontology, i.e. the Gene Ontology many other ontologies are connected to. From a theoretical point of view, it would be interesting to see how far we can generalize this to arbitrary matching problems, i.e. it would be interesting to test whether the classifier learned for the benchmarking dataset can be applied to the conference dataset and vice versa. This would further enhance the usefulness of your method as there is no need to have reference mappings for new application domains, but classifiers could be trained on existing mapping sets like the ones we used in our experiments or the ones available in the area of medicine.[5] Investigating this question in detail will be subject of future work.

References

1. Mochol, M., Jentzsch, A.: Towards a rule-based matcher selection. In: Proc. of the 16th international conference on Knowledge Engineering, Acitrezza, Italy (2008)
2. Ehrig, M., Staab, S., Sure, Y.: Bootstrapping ontology alignment methods with APFEL. In: Gil, Y., Motta, E., Benjamins, V.R., Musen, M.A. (eds.) ISWC 2005. LNCS, vol. 3729, pp. 186–200. Springer, Heidelberg (2005)
3. Isaac, A., Trojahn, C., Wang, S., Quaresma, P.: Using quantitative aspects of alignment generation for argumentation on mappings. In: Proceedings of the ISWC 2008 Workshop on Ontology Matching, Karlsruhe, Germany (2008)
4. Aumueller, D., Do, H.H., Massmann, S., Rahm, E.: Schema and ontology matching with COMA++. In: Proc. of the ACM SIGMOD International Conference on Management of Data, Baltimore, Maryland, USA (2005)
5. Jean-Mary, Y.R., Kabuka, M.R.: Asmov: results for OAEI 2008. In: Proc. of the ISWC 2008 Workshop on Ontology Matching, Karlsruhe, Germany (2008)
6. Doan, A., Madhavan, J., Dhamankar, R., Domingos, P., Halevy, A.: Learning to match ontologies on the semantic web. VLDB Journal (2003)
7. Ichise, R.: Machine learning approach for ontology mapping using multiple concept similarity measures. In: Seventh IEEE/ACIS International Conference on Computer and Information Science, ICIS 2008, Portland, Oregon, USA (2008)

[5] http://www.obofoundry.org/index.cgi?sort=type&show=mappings

8. Svab, O., Svatek, V.: Combining ontology mapping methods using bayesian networks. In: Proc. of the ISWC 2006 Workshop on Ontology Matching, Athens, USA (2006)
9. Maio, P., Bettencourt, N., Silva, N., Rocha, J.: Evaluating a confidence value for ontology alignment. In: Proc. of the ISWC 2007 Workshop on Ontology Matching, Busan, Korea (2007)
10. Euzenat, J., Shvaiko, P.: Ontology Matching. Springer, Heidelberg (2007)
11. Euzenat, J., Isaac, A., Meilicke, C., Shvaiko, P., Stuckenschmidt, H., Svab, O., Svatek, V., van Hage, W., Yatskevich, M.: Results of the ontology alignment evaluation initiative 2007. In: Proc. of the ISWC 2007 Workshop on Ontology Matching, Busan, Korea (2007)
12. Caracciolo, C., Euzenat, J., Hollink, L., Ichise, R., Isaac, A., Malaise, V., Meilicke, C., Pane, J., Shvaiko, P., Stuckenschmidt, H., Svab-Zamazal, O., Svatek, V.: First results of the ontology alignment evaluation initiative 2008. In: Proc. of the ISWC 2008 Workshop on Ontology Matching, Karlsruhe, Germany (2008)
13. Bouquet, P., Ehrig, M., Euzenat, J., Franconi, E., Hitzler, P., Krötzsch, M., Serafini, L., Stamou, G., Sure, Y., Tessaris, S.: Specification of a common framework for characterizing alignment. Deliverable D2.2.1 (version 2), KnowledgeWeb Network of Excellence (February 2005), `http://knowledgeweb.semanticweb.org/semanticportal/deliverables/D2.2.1v2.pdf`
14. Svab, O., Svatek, V., Berka, P., Rak, D., Tomasek, P.: Ontofarm: Towards an experimental collection of parallel ontologies. In: Poster Proceedings of the International Semantic Web Conference (2005)
15. David, J.: AROMA results for OAEI 2008. In: Proc. of the ISWC 2008 Workshop on Ontology Matching, Karlsruhe, Germany (2008)
16. Nagy, M., Vargas-Vera, M., Stolarski, P., Motta, E.: DSSim results for OAEI 2008. In: Proc. of the ISWC 2008 Workshop on Ontology Matching, Karlsruhe, Germany (2008)
17. Wang, P., Xu, B.: Lily: ontology alignment results for OAEI 2008. In: Proc. of the ISWC 2008 Workshop on Ontology Matching, Karlsruhe, Germany (2008)
18. Zhang, X., Zhong, Q., Li, J., Tang, J.: RiMOM results for OAEI 2008. In: Proc. of the ISWC 2008 Workshop on Ontology Matching, Karlsruhe, Germany (2008)
19. Lambrix, P., Tan, H., Liu, Q.: SAMBO and SAMBOdtf results for the ontology alignment evaluation initiative 2008. In: Proceedings of the ISWC 2008 Workshop on Ontology Matching, Karlsruhe, Germany (2008)
20. Sabou, M., Gracia, J.: Spider: Bringing non-equivalence mappings to OAEI. In: Proc. of the ISWC 2008 Workshop on Ontology Matching, Karlsruhe, Germany (2008)
21. Witten, I.H., Frank, E.: Data Mining: Practical Machine Learning Tools and Techniques. Morgan Kaufmann, San Francisco (2005)
22. Quinlan, R.: C4.5: Programs for Machine Learning (1993)

Ontology Integration Using Mappings: Towards Getting the Right Logical Consequences

Ernesto Jiménez-Ruiz[1,*], Bernardo Cuenca Grau[2], Ian Horrocks[2], and Rafael Berlanga[1]

[1] Universitat Jaume I, Spain
{ejimenez,berlanga}@uji.es
[2] University of Oxford, UK
{berg,ian.horrocks}@comlab.ox.ac.uk

Abstract. We propose a general method and novel algorithmic techniques to facilitate the integration of independently developed ontologies using mappings. Our method and techniques aim at helping users understand and evaluate the semantic consequences of the integration, as well as to detect and fix potential errors. We also present ContentMap, a system that implements our approach, and a preliminary evaluation which suggests that our approach is both useful and feasible in practice.

1 Introduction

Effective ontology integration techniques are often needed both during ontology development (e.g., when an ontology being developed reuses one or more external ontologies), and when ontologies are used in conjunction with data (e.g. when integrating and querying data sources annotated using different ontologies).

When the ontologies to be integrated have been independently developed, their vocabularies will most likely diverge, either because they use different namespaces, or because they use different names or naming conventions to refer to their entities. As a consequence, these ontologies will most likely be unrelated from a logical point of view, even if they intuitively overlap.

For example, suppose that Peter wants to develop an ontology about juvenile forms of arthritis.[1] He finds on the Web two independently developed ontologies $\mathcal{O}_1$ and $\mathcal{O}_2$ with different vocabularies, which describe different types of arthritis and juvenile diseases respectively, and decides that an integration of these two ontologies would make a good starting point. Although largely independent, the two ontologies do overlap; for example, both describe a particular form of juvenile arthritis known as JRA (Juvenile Rheumatoid Arthritis), although they use different vocabulary in their descriptions.

As a first step in the integration of such independently developed ontologies it is usually necessary to establish appropriate correspondences (or *mappings*) between the terms used in the various ontologies. For example, Peter may need to establish that Rheumatoid_Arthritis in $\mathcal{O}_1$ has the same intended

* Ernesto was supported by his PhD Fellowship of the *Generalitat Valenciana*.
[1] The development of an ontology describing types of juvenile arthritis is one of the goals of the Health-e-Child project (`http://www.health-e-child.org/`).

L. Aroyo et al. (Eds.): ESWC 2009, LNCS 5554, pp. 173–187, 2009.

meaning as Rheum_Arthritis in $\mathcal{O}_2$. Ontologies are, however, often too large and complex for ontology developers to manually establish these correspondences. To address this problem, a number of sophisticated tools have been developed (see `http://www.ontologymatching.org/` for a comprehensive list), the purpose of which is to (semi-) automatically establish appropriate correspondences.

When reasoning with the ontologies to be integrated together with the automatically generated mappings, however, errors are likely to occur. There are two main causes for these errors. On the one hand, mappings suggested by automated tools are likely to include some errors. On the other hand, even if the correct mappings have been found, the ontologies may contain conflicting descriptions of the overlapping entities. These errors manifest themselves as unintended logical consequences (e.g., unsatisfiable concepts or unintended subsumptions), and they can be difficult to detect, understand and repair. However, existing tools provide little or no support for the user in trying to obtain the right logical consequences when integrating ontologies using mappings—as discussed in Section 5, research in this area is still at an early stage [1, 2].

In this paper, we present a new general method and novel algorithmic techniques designed to provide just such support. We also present ContentMap,[2] a system that implements our approach, and a preliminary evaluation which suggests that our approach is both useful and feasible in practice.

We will assume some basic knowledge of OWL and Description Logics, and refer to [3, 4] for comprehensive overviews. Moreover, due to lack of space, proofs and some details concerning the empirical evaluation are included in an online technical report [5].

2 Problems with Ontology Integration Using Mappings

In this section, we discuss some of the problems that arise when integrating ontologies using mappings, and analyze the requirements for a suitable tool.

Let us consider the ontologies $\mathcal{O}_1$ and $\mathcal{O}_2$ from Section 1 about arthritis and juvenile diseases respectively. As already mentioned, $\mathcal{O}_1$ and $\mathcal{O}_2$ overlap conceptually in the description of different types of juvenile rheumatoid arthritis. The overlapping parts of $\mathcal{O}_1$ and $\mathcal{O}_2$ are shown in Table 1.

Suppose that Peter—the ontology developer—uses an automatic mapping tool to find correspondences between entities in $\mathcal{O}_1$ and $\mathcal{O}_2$. Mappings are often represented as a tuple $\langle id, e_1, e_2, n, \rho \rangle$ [6], where id is a unique identifier for the mapping, e_1, e_2 are entity names in the vocabulary of $\mathcal{O}_1$ and $\mathcal{O}_2$ respectively, n is a numeric confidence measure between 0 and 1, and ρ is a relation between e_1 and e_2 (typically subsumption ($\sqsubseteq$), equivalence ($\equiv$), or disjointness ($\perp$)).

Table 2 contains the mappings obtained by Peter using the mapping tool OLA (see `http://ola.gforge.inria.fr/`). The prefixes '1:' and '2:' for the entity names refer to the namespaces (omitted in Table 1) of $\mathcal{O}_1$ and $\mathcal{O}_2$. Note that the mappings and confidence values suggested by different tools can vary enormously, so experience and/or experimentation may be needed in order to select the most suitable tool (or combination of tools) for the problem at hand.

---

2 A logiC-based ONtology inTEgratioN Tool using MAPpings.

Table 1. Example Ontologies

	Ontology $\mathcal{O}_1$
α_1	Juvenile_Arthritis $\sqsubseteq$ Systemic_Disease $\sqcap$ Rheumatoid_Arthritis
α_2	Multi_Joint_Disease $\sqsubseteq$ Disease $\sqcap \geqslant 5$ affects.Joint
α_3	Rheumatoid_Arthritis $\sqsubseteq$ Disease $\sqcap\ \exists$has_Factor.$\top$
α_4	Poly_Juvenile_Arthritis $\sqsubseteq$ Juvenile_Arthritis $\sqcap$ Multi_Joint_Disease
α_5	Oly_Juvenile_Arthritis $\sqsubseteq$ Juvenile_Arthritis $\sqcap\ \neg$Poly_Juvenile_Arthritis
α_6	Oly_Juvenile_Arthritis $\sqsubseteq\ \forall$has_Factor.Negative_Factor
α_7	Negative_Factor $\sqcap$ Positive_Factor $\sqsubseteq \bot$
	Ontology $\mathcal{O}_2$
β_1	Juv_Rheum_Arthritis $\sqsubseteq$ Rheum_Arthritis $\sqcap$ Juv_Disease
β_2	Rheum_Arthritis $\sqsubseteq$ Disease $\sqcap\ \exists$has_Rheum_Factor.$\top$
β_3	Poly_Juv_Rheum_Arthritis $\sqsubseteq$ Juv_Rheum_Arthritis $\sqcap = 3$ affects.Joint
β_4	Poly_Juv_Rheum_Arthritis $\sqsubseteq\ \forall$has_Rheum_Factor.Positive_Rheum_Factor
β_5	Negative_Rheum_Factor $\sqcap$ Positive_Rheum_Factor $\sqsubseteq \bot$

Table 2. Example Mappings

Intended Mappings
$\langle\mu_1$, 1:Joint, 2:Joint, 1.0, $(\equiv)\rangle$;
$\langle\mu_2$, 1:Disease, 2:Disease, 1, $(\equiv)\rangle$;
$\langle\mu_3$, 1:Disease, 2:Disease, 1.0, $(\equiv)\rangle$;
$\langle\mu_4$, 1:affects, 2:affects, 0.87, $(\equiv)\rangle$;
$\langle\mu_5$, 1:Rheumatoid_Arthritis, 2:Rheum_Arthritis, 1.0, $(\equiv)\rangle$;
$\langle\mu_6$, 2:Juv_Disease, 1:Disease, 0.63, $(\sqsubseteq)\rangle$;
$\langle\mu_7$, 1:Poly_Juvenive_Arthritis, 2:Poly_Juv_Rheum_Arthritis, 0.7, $(\equiv)\rangle$;
$\langle\mu_8$, 1:has_Factor, 2:has_Rheum_Factor, 0.7, $(\equiv)\rangle$;
$\langle\mu_9$, 1:Positive_Factor, 2:Positive_Rheum_Factor, 0.75, $(\equiv)\rangle$;
$\langle\mu_{10}$, 1:Negative_Factor, 2:Negative_Rheum_Factor, 0.75, $(\equiv)\rangle$;
$\langle\mu_{11}$, 2:Juv_Rheum_Arthritis, 1:Rheumatoid_Arthritis, 0.5, $(\sqsubseteq)\rangle$;
Erroneous Mappings
$\langle\mu_{12}$, 1:Rheumatoid_Arthritis, 2:Juv_Rheum_Arthritis, 0.5, $(\sqsubseteq)\rangle$;
$\langle\mu_{13}$, 1:Disease, 2:Juv_Disease, 0.63, $(\sqsubseteq)\rangle$;
$\langle\mu_{14}$, 1:Positive_Factor, 2:Negative_Rheum_Factor, 0.63, $(\equiv)\rangle$;
$\langle\mu_{15}$, 2:Positive_Factor, 1:Negative_Rheum_Factor, 0.63, $(\equiv)\rangle$;

As shown in Table 2, some of the mappings are clearly erroneous and, under any reasonable semantics for the mappings, will lead to unintended logical consequences. Indeed, the mappings μ_{14} and μ_{15} entail the logical equivalence of two concepts that are disjoint in both $\mathcal{O}_1$ and $\mathcal{O}_2$ (see also intended mappings μ_9 and μ_{10}); also, the mapping μ_{12} will lead to unintended subsumptions since not all forms of rheumatoid arthritis affect only children. Furthermore, even if only intended mappings had been generated, errors would still have occurred due to conflicting descriptions of some of the entities that occur in both ontologies. For example, the descriptions of Polyarticular Juvenile Rheumatoid Arthritis (axioms α_4, α_2 in $\mathcal{O}_1$ and β_3 in $\mathcal{O}_2$) are contradictory: in $\mathcal{O}_1$ it is described as a disease that affects at least 5 different joints, whereas in $\mathcal{O}_2$ it is said to affect exactly 3 joints. The intended mapping μ_7 will reveal this contradiction.

This example suggests some requirements for a tool supporting ontology integration using mappings. A suitable tool should provide support for: *(i)* generating sets of mappings, either manually or by selecting one or more mapping algorithms and setting their parameters; *(ii)* editing sets of mappings and filtering them according to different criteria; *(iii)* reasoning with the ontologies to be integrated together with the relevant mappings; *(iv)* comparing the entailments holding before and after the integration and detecting possible unintended entailments; *(v)* suggesting possible ways to repair the identified errors.

3 Proposed Method, Algorithms and Tool Support

In this paper, we present a novel approach for integrating ontologies using mappings that addresses the requirements from Section 2. To this end, a formal representation of ontology mappings should be adopted. This is crucial, for example, to reason unambiguously with the mappings together with the ontologies to be integrated, and thus to address requirements *(iii)*–*(v)* in Section 2.

A number of specialised semantics for ontology mappings have been proposed in the literature (e.g. [6, 7]). In this paper, however, we have chosen to represent ontology mappings as OWL 2 axioms [4]. Such a representation seems semantically coherent, and allows us to reuse the extensive range of OWL algorithmic techniques and infrastructure that is currently available. The general approach

Table 3. Ontology Integration Method

Input: $\mathcal{O}_1, \mathcal{O}_2$: ontologies with $\mathsf{Sig}(\mathcal{O}_1) = \Sigma_1, \mathsf{Sig}(\mathcal{O}_2) = \Sigma_2$ and $\Sigma_1 \cap \Sigma_2 = \{\top, \bot\}$
Output: $\mathcal{O}'_1, \mathcal{O}'_2$: modified ontologies
$\mathcal{M}'$: Mappings between Σ_1 and Σ_2

1: $(\star)$ Select mapping algorithm $\mathsf{map}(\mathcal{O}_1, \mathcal{O}_2)$
2: $(\star)$ Select $\mathcal{M} \subseteq \mathsf{map}(\mathcal{O}_1, \mathcal{O}_2)$
3: $(\star)$ **if** $\mathcal{O}_1, \mathcal{O}_2, \mathcal{M}$ satisfactory, **then return** $\mathcal{O}'_1 := \mathcal{O}_1, \mathcal{O}'_2 := \mathcal{O}_2, \mathcal{M}' := \mathcal{M}$
4: $\mathcal{U} := \mathcal{O}_1 \cup \mathcal{O}_2 \cup \mathcal{M}$
5: $(\star)$ Select approximation functions $\mathsf{diff}^{\approx}$ and $\mathsf{mdiff}^{\approx}$
6: Compute $\Lambda = \mathsf{diff}^{\approx}_{\Sigma_1}(\mathcal{O}_1, \mathcal{U}) \cup \mathsf{diff}^{\approx}_{\Sigma_2}(\mathcal{O}_2, \mathcal{U}) \cup \mathsf{mdiff}^{\approx}_{\Sigma_1, \Sigma_2}(\mathcal{M}, \mathcal{U})$
7: $(\star)$ Select $\Im^+, \Im^- \subseteq \Lambda$
8: $(\star)$ Select $\mathcal{O}^- \subseteq \mathcal{U}$
9: $\mathbb{P} :=$ all minimal plans for $\mathcal{U}$ given $\Im^+$, $\Im^-$, and $\mathcal{O}^-$
10: $(\star)$ **if** no satisfactory plan in $\mathbb{P}$, **then** come back to either Step 2 or Step 7
11: $(\star)$ Select $\mathcal{P} \in \mathbb{P}$
12: **return** $\mathcal{O}'_1 := \mathcal{O}_1 \setminus \mathcal{P}, \mathcal{O}'_2 := \mathcal{O}_2 \setminus \mathcal{P}, \mathcal{M}' := \mathcal{M} \setminus \mathcal{P}$

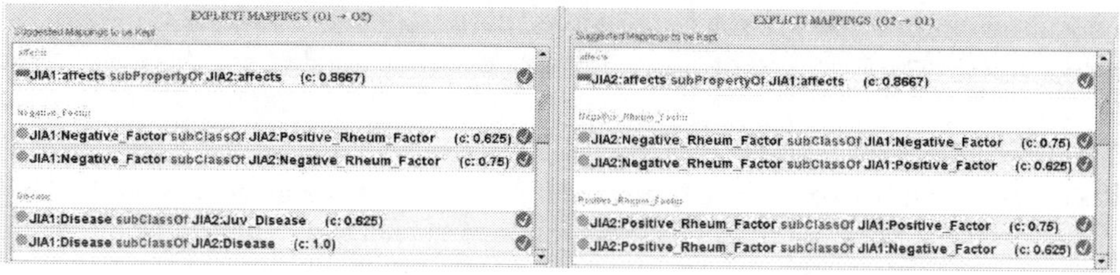

Fig. 1. GUI for Visualising Explicit Mappings in ContentMap

we present here, however, could be adapted to alternative formal representations of mappings (see Section 5 for further discussion).

We assume from now on that a set of mappings is represented as an OWL 2 ontology $\mathcal{M}$, where mappings $\langle id, e_1, e_2, n, \rho \rangle$ are given as axioms of the form SubClassOf(e_1 e_2), EquivalentClasses(e_1 e_2), or DisjointClasses(e_1 e_2), for ρ of the form ($\sqsubseteq$), ($\equiv$), or ($\perp$) respectively, with id (the mapping id) and n (the confidence value) added as axiom annotations [4]. We denote with $\mathsf{conf}(\alpha)$ the confidence value with which an axiom α is annotated. Representing mappings in this way gives them a standard "crisp" semantics, with ids and confidence values being represented only as annotations and thus having no effect on entailments.

Our general approach is given in Table 3. It is an interactive process where some steps involve computations that tools should perform automatically, while others (marked with ($\star$)) may require manual intervention. It consists of four main parts: *i)* Computation of mappings (Steps 1–3); *ii)* Computation of new entailments (Steps 4–6); *iii)* Detection of errors (Step 7); *iv*) Repair of the identified errors (Steps 8–12). We next present our method and describe its realisation in ContentMap—a Protégé 4 plugin available for download (see [5]).

3.1 Computation of the Mappings

General Description. Ontology mappings can be computed using one or more mapping tools (see Step 1), and can subsequently be refined (Step 2) in various ways (e.g., via manual selection or filtering according to a given threshold). After obtaining and possibly refining the mappings, a user may decide that the integration process is complete (Step 3), in which case it remains a purely syntactic process. They may, however, want to analyse the semantic consequences of the integration, in which case a reasoner is required.

Tool Support. ContentMap provides for the input ontologies to be loaded, and for one or more mapping tools to be selected. Different weights can also be assigned to each of the mapping tools. It is also possible to load pre-computed mappings in the form of an OWL 2 ontology.

ContentMap provides a GUI for visualising the mappings (Fig. 1) and showing those that are to be accepted (marked with ✓) and those that are to be rejected (marked with ⊗). Users can automatically filter the mappings by setting a confidence threshold τ: mappings with a confidence value lower than τ are automatically marked for rejection while those with a confidence value greater than τ are marked for acceptance. These selections can then be refined at will.

3.2 Computation of New Entailments

General Description. To help users understand the semantic consequences of the integration, they should be informed about new entailments that hold in the merged ontology $\mathcal{U}$, but not in $\mathcal{O}_1$, $\mathcal{O}_2$ and $\mathcal{M}$ alone. To this end, we use the notion of *deductive difference* [8, 9]. Intuitively, the deductive difference between $\mathcal{O}$ and $\mathcal{O}'$ w.r.t a signature Σ is the set of entailments constructed over Σ that do not hold in $\mathcal{O}$, but do hold in $\mathcal{O}'$.

Definition 1. *Let $\mathcal{DL}$ be a DL and Σ be a $\mathcal{DL}$-signature. The Σ-deductive difference between $\mathcal{DL}$-ontologies $\mathcal{O}$ and $\mathcal{O}'$ is given as follows:*

$$\mathsf{diff}_{\Sigma}(\mathcal{O},\mathcal{O}') = \{\alpha \mid \alpha \text{ a } \mathcal{DL}\text{-axiom}, \mathcal{O} \not\models \alpha, \mathcal{O}' \models \alpha \text{ and } \mathsf{Sig}(\alpha) \subseteq \Sigma\} \quad (1)$$

In our integration scenario, we want to inform the user about new entailments that hold in $\mathcal{O}_1$ or $\mathcal{O}_2$ as a result of the integration, i.e., $\mathsf{diff}_{\Sigma_1}(\mathcal{O}_1,\mathcal{U})$ and $\mathsf{diff}_{\Sigma_2}(\mathcal{O}_2,\mathcal{U})$. In our example from Tables 1 and 2, where the mappings are given as an OWL 2 ontology as described above, the entailments ($\mathsf{Positive\_Factor} \sqsubseteq \bot$) and ($\mathsf{Rheum\_Arthritis} \sqsubseteq \mathsf{Juv\_Disease}$) belong to $\mathsf{diff}_{\Sigma_1}(\mathcal{O}_1,\mathcal{U})$ and $\mathsf{diff}_{\Sigma_2}(\mathcal{O}_2,\mathcal{U})$ respectively. In addition, we want to inform the user about new entailments that contain symbols from both $\mathcal{O}_1$ and $\mathcal{O}_2$. These entailments can be seen as inferred mappings provided that the notion of a mapping is generalised to be an arbitrary DL axiom using terms from two signatures:

Definition 2. *Let $\mathcal{DL}$ be a DL and let Σ_1, Σ_2 be two $\mathcal{DL}$-signatures s.t. $\Sigma_1 \cap \Sigma_2 = \{\top,\bot\}$. Let $\Sigma = \Sigma_1 \cup \Sigma_2$. A $\mathcal{DL}$-mapping between Σ_1 and Σ_2 is a $\mathcal{DL}$-axiom α over Σ such that $(\mathsf{Sig}(\alpha) \cap \Sigma_i) \setminus \{\top,\bot\} \neq \emptyset$ for each $i \in \{1,2\}$. The axiom α is annotated with an id and with a confidence value $0 < \mathsf{conf}(\alpha) \leq 1$.*

Definition 1 can now be extended to take into account the new mappings, i.e.:

$$\mathsf{mdiff}_{\Sigma_1,\Sigma_2}(\mathcal{O},\mathcal{O}') = \{\alpha \in \mathsf{diff}_{\Sigma}(\mathcal{O},\mathcal{O}') \mid \alpha\ \mathcal{DL}\text{-mapping between } \Sigma_1,\ \Sigma_2\} \quad (2)$$

In our scenario, inferred mappings are captured by $\mathsf{mdiff}_{\Sigma_1,\Sigma_2}(\mathcal{M},\mathcal{U})$. For example, the mapping ($\mathsf{1{:}Juvenile\_Arthritis} \sqsubseteq \mathsf{2{:}Rheum\_Arthritis}$) would belong to this difference, because it is entailed by $\mathcal{U}$ but not by $\mathcal{M}$ alone.

The notion of deductive difference, however, has several drawbacks in practice. First, there is no algorithm for computing $\mathsf{diff}(\mathcal{O},\mathcal{O}')$ in expressive DLs such as $\mathcal{SROIQ}$ (OWL 2) and $\mathcal{SHOIQ}$ (OWL DL) [9]. For these languages, checking whether $\mathsf{diff}(\mathcal{O},\mathcal{O}') = \emptyset$ is undecidable. Currently, algorithms only exist for (fragments of) the OWL 2 EL and QL profiles [10, 9, 8]. Second, the number of entailments in the difference can be huge (even infinite), and so likely to overwhelm users. These drawbacks motivate the need for *approximations*—subsets of the difference—that the user can select in Step 5 from Table 3.

Definition 3. *A function $\mathsf{diff}^{\approx}_{\Sigma}(\mathcal{O},\mathcal{O}')$ (resp. $\mathsf{mdiff}^{\approx}_{\Sigma_1,\Sigma_2}(\mathcal{O},\mathcal{O}')$) is an approximation for $\mathsf{diff}_{\Sigma}(\mathcal{O},\mathcal{O}')$ ($\mathsf{mdiff}_{\Sigma_1,\Sigma_2}(\mathcal{O},\mathcal{O}')$) if for any two $\mathcal{DL}$-ontologies $\mathcal{O},\mathcal{O}'$, $\mathsf{diff}^{\approx}_{\Sigma}(\mathcal{O},\mathcal{O}') \subseteq \mathsf{diff}_{\Sigma}(\mathcal{O},\mathcal{O}')$ ($\mathsf{mdiff}^{\approx}_{\Sigma_1,\Sigma_2}(\mathcal{O},\mathcal{O}') \subseteq \mathsf{mdiff}_{\Sigma_1,\Sigma_2}(\mathcal{O},\mathcal{O}')$).*

Tool Support. A useful approximation should be easy to compute, yet still provide meaningful information to the user. One possibility is to define an approximation by considering only entailments of a certain form.

The GUI implemented in ContentMap allows users to customise approximations by selecting among the following simple kinds of entailment, where A, B are atomic concepts (including $\top, \bot$) and R, S atomic roles or inverses of atomic roles: *(i)* $A \sqsubseteq B$, *(ii)* $A \sqsubseteq \neg B$, *(iii)* $A \sqsubseteq \exists R.B$, *(iv)* $A \sqsubseteq \forall R.B$, and *v)* $R \sqsubseteq S$. The smallest implemented approximation considers only axioms of the form *(i)*, while the largest one considers all types *(i)*—*(v)*.

All these approximations can be algorithmically computed, and ContentMap uses a number of optimisations for computing them. Clearly, the larger the class of entailments presented to the user, the more errors could be detected. The corresponding differences, however, are harder to compute, harder to present to the user, and may be harder for the user to understand.

3.3 Detection of Errors

General Description. Some entailments in the relevant (approximate) differences may be intended, while others may reveal potential errors in the merged ontology $\mathcal{U}$. Step 7 therefore involves selecting entailments that: *(i)* are intended and should be entailed in $\mathcal{U}$ (written $\Im^+$ in Table 3), and *(ii)* are unintended and should not be entailed by $\mathcal{U}$ (written $\Im^-$).

In our example, the entailments (Positive_Factor $\sqsubseteq \bot$) and (Rheum_Arthritis $\sqsubseteq$ Juv_Disease) in $\mathsf{diff}_{\Sigma_1}(\mathcal{O}_1, \mathcal{U})$ and $\mathsf{diff}_{\Sigma_2}(\mathcal{O}_2, \mathcal{U})$ respectively are unintented and should belong to $\Im^-$. In contrast, the entailed mapping (1:Rheum_Arthritis $\sqsubseteq$ 2:Disease) is intended and should be included in $\Im^+$.

Tool Support. The development of techniques to help users understand the relevant differences and subsequently select the sets of intended and unintended entailments is especially challenging.

First, the tool should explain why the new entailments that hold in $\mathcal{U}$ do not hold in $\mathcal{O}_1$, $\mathcal{O}_2$ and $\mathcal{M}$ alone. The notion of a justification has proved very useful in understanding why an ontology entails a certain axiom [11, 12].

Definition 4. *Let $\mathcal{O} \models \alpha$. A justification for α in $\mathcal{O}$ is an ontology $\mathcal{O}' \subseteq \mathcal{O}$ satisfying the following properties:* (i) *$\mathcal{O}' \models \alpha$, and* (ii) *there is no $\mathcal{O}'' \subset \mathcal{O}'$ s.t. $\mathcal{O}'' \models \alpha$. We denote by* Just$(\alpha, \mathcal{O})$ *the set of all justifications for α in $\mathcal{O}$.*

In order to explain a given entailment, ContentMap presents all its justifications. Computing justifications is expensive, so ContentMap uses the optimisations proposed in [13, 14] which have proved effective in practice.

Second, the potentially large number of relevant entailments may overwhelm the user. Hence, on the one hand, the tool should try to present these entailments in a way that makes them easier to understand and, on the other hand, it should provide suggestions about which entailments to include in $\Im^+$ and $\Im^-$.

ContentMap exploits the dependencies between entailments to organise the way in which they are presented. Intuitively an entailment β depends on α in $\mathcal{O}$

Algorithm 1. Heuristic suggestions for $\Im^+$, $\Im^-$

Input: $\Lambda = \mathsf{diff}^{\approx}_{\Sigma_1}(\mathcal{O}_1,\mathcal{U}) \cup \mathsf{diff}^{\approx}_{\Sigma_2}(\mathcal{O}_2,\mathcal{U}) \cup \mathsf{mdiff}^{\approx}_{\Sigma_1,\Sigma_2}(\mathcal{M},\mathcal{U})$ as in Table 3
τ_1, τ_2: real values with $0 < \tau_1 \leq \tau_2 \leq 1$
$\rhd$: dependency relation between axioms in Λ w.r.t. $\mathcal{U}$
Output: $\Im^+, \Im^-$: entailments suggested to be hold or not

```
Ɔ+, Ɔ- := ∅
for each α ∈ Λ do
    if α of the form A ⊑ ⊥ for A an atomic concept then
        Ɔ- := Ɔ- ∪ {α}
        for each β ∈ Λ such that α ▷* β do Ɔ- := Ɔ- ∪ {β}
    end if
    conf(α) := confidence of α as in Definition 7
    if conf(α) ≤ τ1 then Ɔ- := Ɔ- ∪ {α}
    if conf(α) ≥ τ2 and α ∉ Ɔ- then Ɔ+ := Ɔ+ ∪ {α}
end for
return Ɔ+, Ɔ-
```

if, whenever α is invalidated by removing a set of axioms from $\mathcal{O}$, then β is also invalidated. This implies that if $\alpha \in \Im^-$, then β should also be included in $\Im^-$ (and never in $\Im^+$). This idea can be formalised in our setting as follows:

Definition 5. *Let $\mathcal{O} \models \alpha, \beta$. The axiom β depends on α w.r.t. $\mathcal{O}$, written $\alpha \rhd \beta$, iff for each $\mathcal{J}_\beta \in \mathsf{Just}(\beta, \mathcal{O})$ there is $\mathcal{J}_\alpha \in \mathsf{Just}(\alpha, \mathcal{O})$ s.t. $\mathcal{J}_\alpha \subseteq \mathcal{J}_\beta$.*

The relation $\rhd$ is consistent with our intuitions, as shown next:

Proposition 6. *Consider Table 3 and Definition 5. Let $\mathcal{U} \models \alpha, \beta$, $\mathcal{U}' \subset \mathcal{U}$ and $\alpha \rhd \beta$. Then: 1) $\mathcal{U}' \not\models \alpha$ implies $\mathcal{U}' \not\models \beta$, and 2) $\mathcal{U}' \models \beta$ implies $\mathcal{U}' \models \alpha$.*

In order to suggest to the user which entailments to include in $\Im^+$ and $\Im^-$, ContentMap exploits both the dependencies between entailments in the differences and the confidence values for each explicit mapping. We use the confidence values in the mappings to compute confidence values in the entailments from each of the obtained differences: we consider our confidence in an entailment to be equal to the maximum confidence we have in one of its justifications, and our confidence in a justification to be the product of our confidences in each of the axioms it contains. We formalise this in the following definition, in which we extend conf() to arbitrary (explicit) axioms and justifications. We assume that we have complete confidence in an axiom (i.e., $\mathsf{conf}(\alpha) = 1$) if it is not a mapping.

Definition 7 (Confidence in an Entailment). *Let $\mathcal{O}$ be an ontology, α an axiom in $\mathcal{O}$ that is not annotated with a confidence value, $\mathcal{J}$ a justification, and β an entailment s.t. $\mathcal{O} \models \beta$. We define $\mathsf{conf}(\alpha) = 1$, and $\mathsf{conf}(\mathcal{J})$ and $\mathsf{conf}(\beta)$ as follows:*

$$\mathsf{conf}(\mathcal{J}) = \prod_{\gamma \in \mathcal{J}} \mathsf{conf}(\gamma) \quad \textit{and} \quad \mathsf{conf}(\beta) = \max\Big(\bigcup_{\mathcal{J} \in \mathsf{Just}(\beta,\mathcal{O})} \mathsf{conf}(\mathcal{J})\Big) \qquad (3)$$

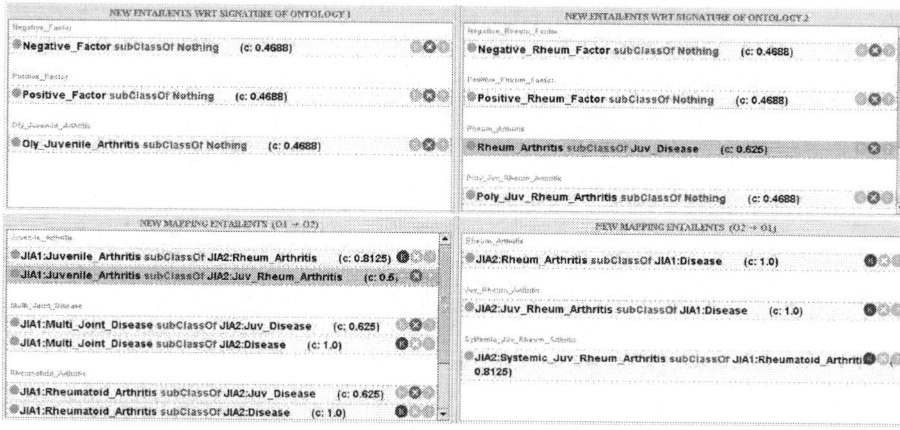

Fig. 2. GUI for Visualising New Entailments in ContentMap

The following proposition shows that the confidence values in Definition 7 are well-behaved w.r.t. Definition 5: if β depends on α then the confidence in β is smaller than the confidence in α.

Proposition 8. *If* $\alpha \rhd^* \beta$, *then* $\mathsf{conf}(\beta) \leq \mathsf{conf}(\alpha)$

In order to compute initial suggestions for $\Im^-$ and $\Im^+$, ContentMap implements the heuristics in Algorithm 1, which are based on Definitions 5, 7 and Propositions 6, 8. The algorithm accepts as input the entailments in the relevant differences, their dependencies, and two confidence thresholds τ_1 and τ_2. For each input entailment α, Algorithm 1 uses $\rhd$, τ_1 and τ_2 to either include it in $\Im^+$ (i.e. α is intended), in $\Im^-$ (i.e. α is unintended) or in neither of them. An entailment α is included in $\Im^-$ if it reveals a contradiction (Line 3), depends on a contradiction (Line 5), or if its confidence according to Definition 7 is lower than τ_1 (Line 8). In contrast, α is included in $\Im^+$ if it is not contained in $\Im^-$ and its confidence is higher than the threshold τ_2 (Line 9).

Figures 2 and 3 show the ContentMap GUI for selecting $\Im^+$ and $\Im^-$. Figure 2 shows the new entailments in each of the computed differences. The left-hand side of Figure 3(a) displays the dependency relation as a hierarchy, which can be expanded and contracted in the usual way; on the right-hand side of Figure 3(a), the user can select an entailment to be in $\Im^+$ or $\Im^-$ and show its justifications (Figure 3(b)). When displaying justifications, the GUI indicates whether the axioms in these justifications come from $\mathcal{O}_1$, $\mathcal{O}_2$, or the mappings $\mathcal{M}$ (marked with '1', '2' and 'M' respectively). The entailments in Figures 2 and 3(a) already marked as intended/unintended are those suggested by Algorithm 1.

3.4 Repair

General Description. If the user has selected one or more unintended entailments (i.e., $\Im^- \neq \emptyset$), then $\mathcal{U}$ clearly contains errors. These errors can always

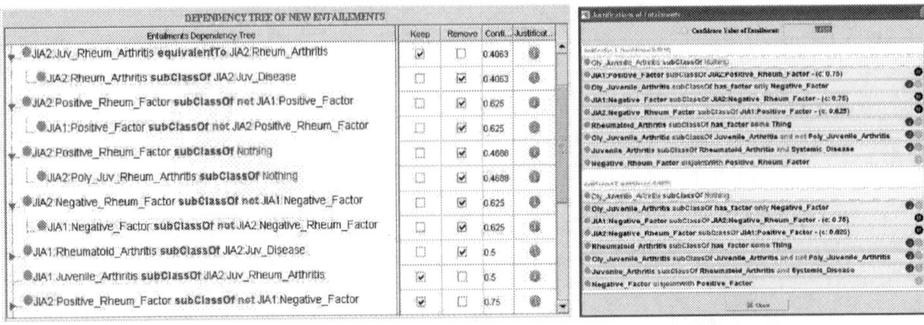

(a) Dependency Relation (b) Justifications

Fig. 3. GUI for Visualising the Dependency Relation and the Justifications

be repaired by removing axioms from $\mathcal{U}$—in the limit, removing *all* the axioms from $\mathcal{U}$ will eliminate all entailments. However, any removal of axioms should also respect $\Im^+$, the user's selection of intended entailments, i.e., the entailments in $\Im^+$ should still hold after any removal of axioms.

As mentioned in Section 1, errors could be due to erroneous mappings, to inherently conflicting information in the two ontologies, or to some combination of both. Repairing such errors might, therefore, require the removal of axioms from $\mathcal{M}$, $\mathcal{O}_1$ and $\mathcal{O}_2$. The user may, however, have a (principled or pragmatic) preference regarding the source of axioms to be removed; e.g., $\mathcal{M}$ might be considered the most likely source of errors, or it may be impossible to change (one or both of) $\mathcal{O}_1$ and $\mathcal{O}_2$. We therefore define the notion of a *repair plan* (or *plan* for short) for an ontology $\mathcal{O}$ w.r.t. the (un-) intended entailments $\Im^-$ and $\Im^+$, and a subset $\mathcal{O}^-$ of $\mathcal{O}$, where the axioms in $\mathcal{O}^-$ are those that are allowed to be removed from $\mathcal{O}$. Such a plan is simply a subset of the axioms in $\mathcal{O}^-$ whose removal from $\mathcal{O}$ both eliminates the entailments in $\Im^-$ and preserves those in $\Im^+$. In general, there may be zero or more such plans.

Definition 9. *Let $\mathcal{O}$, $\Im^+$, $\Im^-$, and $\mathcal{O}^-$ be finite sets of axioms such that $\mathcal{O}^- \subseteq \mathcal{O}$, $\mathcal{O} \models \Im^-$, $\mathcal{O} \models \Im^+$, and $\Im^+ \cap \Im^- = \emptyset$.*

A repair plan for $\mathcal{O}$ given $\mathcal{O}^-$, $\Im^+$ and $\Im^-$ is a set $\mathcal{P} \subseteq \mathcal{O}^-$ such that: 1) *$(\mathcal{O} \setminus \mathcal{P}) \models \alpha$ for each $\alpha \in \Im^+$, and* 2) *$(\mathcal{O} \setminus \mathcal{P}) \not\models \beta$ for each $\beta \in \Im^-$. A plan $\mathcal{P}$ is minimal if there is no $\mathcal{P}_1$ such that $\mathcal{P}_1 \subset \mathcal{P}$.*

In our method, $\Im^-$, $\Im^+$ and $\mathcal{O}^-$ are selected by the user in Steps 7 and 8 from Table 3. In our example, if Peter imports $\mathcal{O}_1$ and $\mathcal{O}_2$ from the Web, and if he is not willing to copy and modify them (e.g., if he wants to always import the latest versions), then plans should only remove axioms from the mappings (i.e., $\mathcal{O}^- = \mathcal{M}$). However, the descriptions of Polyarticular Juvenile Rheumatoid Arthritis in $\mathcal{O}_1$ and $\mathcal{O}_2$ are inherently in contradiction and hence, if $\mathcal{O}^- = \mathcal{M}$, the corresponding error cannot be fixed without deleting reasonable mappings. In this case a more sensible choice would be to repair one or both of $\mathcal{O}_1$ and $\mathcal{O}_2$.

Algorithm 2. Computing All Plans Using Justifications

Procedure all_Plans($\mathcal{O}, \mathcal{O}^-, \Im^+, \Im^-$)
Input: $\mathcal{O}$, $\mathcal{O}^-$, $\Im^+$, $\Im^-$ as in Definition 9, $\mathsf{Just}(\gamma, \mathcal{O})$ for each $\gamma \in \Im^+ \cup \Im^-$
Output: P: set of all plans
1: $\mathbf{P} := \emptyset$
2: **for each** $\mathcal{P} \subseteq \mathcal{O}^-$ **do**
3: validPlan := true
4: **for each** $\alpha \in \Im^+$ **do**
5: foundJust := false
6: **for each** $\mathcal{J}_\alpha \in \mathsf{Just}(\alpha, \mathcal{O})$ **do**
7: **if** $\mathcal{J}_\alpha \subseteq \mathcal{O} \setminus \mathcal{P}$ **then** foundJust := true
8: **end for**
9: **if** foundJust = false **then** validPlan := false
10: **end for**
11: **for each** $\beta \in \Im^-$ **and** $\mathcal{J}_\beta \in \mathsf{Just}(\beta, \mathcal{O})$ **do**
12: **if** $\mathcal{J}_\beta \subseteq \mathcal{O} \setminus \mathcal{P}$ **then** validPlan := false
13: **end for**
14: **if** validPlan = true **then** $\mathbf{P} := \mathbf{P} \cup \{\mathcal{P}\}$
15: **end for**
16: **return P**

Tool Support. Definition 9 suggests a simple procedure for computing all plans: for each possible $\mathcal{P} \subseteq \mathcal{O}^-$, use a reasoner to check if $\mathcal{P}$ is a valid plan, i.e., satisfies Conditions 1 and 2 from Definition 9. ContentMap, however, implements an algorithm (see Algorithm 2) that avoids potentially expensive entailment checks by reusing the justifications already computed when obtaining the dependency relation from Definition 5. The algorithm is based on the following principles: *(i)* in order for $\alpha \in \Im^+$ to hold after the execution of a plan $\mathcal{P}$, the ontology $\mathcal{O} \setminus \mathcal{P}$ must contain at least one justification for α in $\mathcal{O}$ (see Lines (6-8)); *(ii)* in order for $\beta \in \Im^-$ not to hold after the execution of $\mathcal{P}$, it is sufficient to show that no justification for β in $\mathcal{O}$ is in $\mathcal{O} \setminus \mathcal{P}$ (Lines 11-13). The set of all minimal plans can be straightforwardly computed once all the plans have been obtained.

Proposition 10. *Algorithm 2 computes all the plans for $\mathcal{O}$ given $\mathcal{O}^-$, $\Im^+$, $\Im^-$.*

Note that conflicting choices in $\Im^+$ and $\Im^-$ may make it impossible to find any plans. Some of these conflicts can be detected using the dependency relation $\rhd$, as shown in the following proposition:

Proposition 11. *Let $\mathcal{O}$, $\Im^+$, $\Im^-$ and $\mathcal{O}^-$ be as in Definition 9 and let $\mathcal{O} \models \alpha, \beta$. If $\alpha \rhd^* \beta$ w.r.t. $\mathcal{O}$, $\alpha \in \Im^-$ and $\beta \in \Im^+$, then no plan exists.*

There may also be particular selections of $\mathcal{O}^-$, $\Im^+$ and $\Im^-$ for which the number of minimal plans is very large. In this case, selecting the most suitable minimal plan becomes difficult for users.

In order to assist users in the selection of a minimal plan (Steps 10-11), ContentMap implements a number of heuristics based on the confidence of the mappings in the plans and the total number of axioms in the plan, among others.

4 Evaluation

In our experiments, we used a suite of four ontologies adapted from the 2004 EON Ontology Alignment Contest[3] which describe the domain of bibliographic references. These ontologies have been developed independently by INRIA ($\mathcal{O}_{\mathsf{INR}}$), MIT ($\mathcal{O}_{\mathsf{MIT}}$), UMBC ($\mathcal{O}_{\mathsf{UMBC}}$) and AIFB Karlsruhe ($\mathcal{O}_{\mathsf{AIFB}}$) respectively. Their sizes vary from 58 classes, 46 object properties, 26 data properties and 235 axioms in $\mathcal{O}_{\mathsf{AIFB}}$ to 18 classes, 12 object properties, 19 data properties and 96 axioms in $\mathcal{O}_{\mathsf{UMBC}}$. Even if small, all these ontologies are fairly expressive ($\mathcal{O}_{\mathsf{INR}}$ and $\mathcal{O}_{\mathsf{MIT}}$ are expressible in the DL $\mathcal{ALCHQ}(\mathcal{D})$, $\mathcal{O}_{\mathsf{UMBC}}$ in $\mathcal{ALCIN}(\mathcal{D})$ and $\mathcal{O}_{\mathsf{AIFB}}$ in $\mathcal{ALCI}(\mathcal{D})$ respectively). Their average classification time is less than a second when using Pellet 1.5.

In the 2004 EON contest, $\mathcal{O}_{\mathsf{INR}}$ was used as reference ontology, and the other ontologies were independently aligned with it using each of the competing tools. For evaluation, a manually produced gold standard containing the agreed-upon, correct mappings was provided for each pair of aligned ontologies. The gold standards for $\mathcal{O}_{\mathsf{MIT}}$, $\mathcal{O}_{\mathsf{UMBC}}$ and $\mathcal{O}_{\mathsf{AIFB}}$ contain 119, 83 and 98 mappings respectively. All the test ontologies and gold standards are available at our web page [5].

The experiments are organised in two parts, which we specify next, and were performed on a laptop with a 1.82 GHz processor and 3GB of RAM.

Repair of Ontologies Using Gold Standard. First, we have evaluated the semantic consequences of integrating each of $\mathcal{O}_{\mathsf{MIT}}$, $\mathcal{O}_{\mathsf{UMBC}}$ and $\mathcal{O}_{\mathsf{AIFB}}$ with $\mathcal{O}_{\mathsf{INR}}$ using the corresponding gold standard to detect inherent disagreements between them. In each case we have applied our method from Table 3, with the mappings in Step 2 being the corresponding gold standard. We have used both the smallest and the largest approximations of the deductive differences implemented in ContentMap (see Section 3.2), and obtained the following results.

1. Alignment $\mathcal{O}_{\mathsf{MIT}}$, $\mathcal{O}_{\mathsf{INR}}$: for the smallest (resp. the largest) approximation, there were 3 (resp. 33) new entailments in $\mathcal{O}_{\mathsf{MIT}}$, 13 (resp. 85) in $\mathcal{O}_{\mathsf{INR}}$ and 35 (resp. 189) new mappings.
2. Alignment $\mathcal{O}_{\mathsf{UMBC}}$, $\mathcal{O}_{\mathsf{INR}}$: for the smallest (largest) approximation, there were 2 (7) new entailments in $\mathcal{O}_{\mathsf{UMBC}}$, 10 (300) in $\mathcal{O}_{\mathsf{INR}}$ and 28 (176) new mappings.
3. Alignment $\mathcal{O}_{\mathsf{AIFB}}$, $\mathcal{O}_{\mathsf{INR}}$:for the smallest (largest) approximation, there were 4 (37) new entailments in $\mathcal{O}_{\mathsf{AIFB}}$, 2 (19) in $\mathcal{O}_{\mathsf{INR}}$ and 46 (140) new mappings.

Note that the number of new entailments (and hence the amount of information that ContentMap shows to users) largely depends on the selected approximation.

In all cases we found a significant number of obviously unintended inferences due to inherent disagreements between the ontologies being integrated. For example, when integrating $\mathcal{O}_{\mathsf{AIFB}}$ and $\mathcal{O}_{\mathsf{INR}}$, our tool detected a total of 34 newly unsatisfiable concepts in the two ontologies. The large number of (rather complex) justifications for some of these entailments would make manual repair extremely challenging. When computing the corresponding plans, we found that

[3] `http://oaei.ontologymatching.org/2004/Contest/`

Table 4. Synthetic Experiments

Input: $\mathcal{O}_1, \mathcal{O}_2$: ontologies with $\mathsf{Sig}(\mathcal{O}_1) = \Sigma_1, \mathsf{Sig}(\mathcal{O}_2) = \Sigma_2$ and $\Sigma_1 \cap \Sigma_2 = \{\top, \bot\}$
$\mathcal{M}$: automatically-generated mappings between Σ_1 and Σ_2, $\mathcal{M}_{\mathcal{G}}$: gold standard

Filter $\mathcal{M}$ given a confidence threshold τ
Compute recall and precision of $\mathcal{M}$ with respect to $\mathcal{M}_{\mathcal{G}}$
Compute $\mathsf{diff}^{\approx}_{\Sigma_1}(\mathcal{O}_1, \mathcal{U})$, $\mathsf{diff}^{\approx}_{\Sigma_2}(\mathcal{O}_2, \mathcal{U})$ and $\mathsf{mdiff}^{\approx}_{\Sigma_1,\Sigma_2}(\mathcal{M}, \mathcal{U})$ for $\mathcal{U} := \mathcal{O}_1 \cup \mathcal{O}_2 \cup \mathcal{M}$
Compute all minimal plans for $\mathcal{U}$ given $\mathcal{O}^- = \mathcal{M}$, $\Im^+ = \emptyset$, and $\Im^- :=$ new concept unsatisfiability entailments.
Compute $\mathcal{O}_1' := \mathcal{O}_1 \setminus \mathcal{P}, \mathcal{O}_2' := \mathcal{O}_2 \setminus \mathcal{P}, \mathcal{M}' := \mathcal{M} \setminus \mathcal{P}$, for $\mathcal{P}$ the best plan.
Compute recall and precision of $\mathcal{M}'$ with respect to $\mathcal{M}_{\mathcal{G}}$
Compute $\Lambda = \mathsf{diff}^{\approx}_{\Sigma_1}(\mathcal{O}_1', \mathcal{U}') \cup \mathsf{diff}^{\approx}_{\Sigma_2}(\mathcal{O}_2', \mathcal{U}') \cup \mathsf{mdiff}^{\approx}_{\Sigma_1,\Sigma_2}(\mathcal{M}', \mathcal{U}')$ for $\mathcal{U}' := \mathcal{O}_1' \cup \mathcal{O}_2' \cup \mathcal{M}'$; and Compute dependency relation $\rhd$ over Λ
Compute all minimal plans for $\mathcal{U}'$ given $\mathcal{O}^- = \mathcal{M}'$, $\Im^+ = \emptyset$ and $\Im^- :=$ as obtained in **Algorithm 1** for a pre-fixed confidence threshold τ_1.
Compute $\mathcal{O}_1'' := \mathcal{O}_1' \setminus \mathcal{P}, \mathcal{O}_2'' := \mathcal{O}_2' \setminus \mathcal{P}, \mathcal{M}'' := \mathcal{M}' \setminus \mathcal{P}$, for $\mathcal{P}$ the best plan.
Compute recall and precision of $\mathcal{M}''$ with respect to $\mathcal{M}_{\mathcal{G}}$

the problem originated in the incompatible ranges of two datatype properties; hence, the modification of two axioms was sufficient to eliminate all the errors. In most cases, a significant number of obviously unintended new subsumptions were also detected. For example, merging $\mathcal{O}_{\mathsf{MIT}}$ and $\mathcal{O}_{\mathsf{INR}}$ resulted in the new subsumptions $\mathsf{TechnicalReport} \sqsubseteq \mathsf{Date}$ and $\mathsf{TechnicalReport} \sqsubseteq \exists \mathsf{date.Reference}$. Again, the obtained plans revealed the origin of the problems and facilitated their repair.

Synthetic Repair of Automatically Generated Mappings. Having repaired the test ontologies using the gold standard mappings, we used ContentMap to automatically detect and repair errors resulting from the generation of new mappings using the mapping tools OLA , AROMA[4] and CIDER[5]. Our goal was twofold: first, to show that the algorithms in ContentMap are practical; second, to show that ContentMap can be used to automatically detect and repair errors, as well as to improve the quality of automatically generated mappings.

For each pair of test ontologies, and for various sets of automatically generated mappings, we performed the synthetic experiments described in Table 4, which closely follow our proposed method from Table 3. In contrast to Table 3, however, the repair of errors is performed in two stages: first, the obvious errors (i.e. unsatisfiable concepts); then, those subsumptions that ContentMap heuristically found unintended. For these experiments we have used the smallest approximation of the deductive difference available in ContentMap.

Due to space limitations, here we only summarise the main conclusions; the complete results are, however, available in our technical report [5]. First, from a computational point of view, the main bottleneck is the computation of all the justifications for the entailments of interest. Once the justifications have been computed, the time needed for computing the plans is relatively short. Hence,

[4] AROMA: http://www.inrialpes.fr/exmo/people/jdavid/
[5] CIDER: http://sid.cps.unizar.es/SEMANTICWEB/ALIGNMENT/

it is important to investigate optimisations for computing all justifications; first steps in this direction have been taken in [13, 14].

Second, the number of entailments in the computed differences is similar to the results obtained using the gold standard with the same approximation, and the number of obtained minimal plans varies from 1-50. Furthermore, the use of the dependencies given in Definition 5 significantly reduces the amount of information that the user would need to consider.

Third, the use of automatically generated mappings did result in the occurrence of a significant number of unintended entailments. For example, when aligning $\mathcal{O}_{\mathsf{AIFB}}$ and $\mathcal{O}_{\mathsf{INR}}$ using CIDER with confidence threshold $\tau = 0.1$, we found 55 new unsatisfiable concepts. After fixing these errors, ContentMap found 34 unintended subsumption relationships using Algorithm 1 with a threshold $\tau_1 = 0.3$. Moreover, we checked manually selected examples and found out that the heuristically detected unintended subsumptions were indeed errors; furthermore, these errors were mainly caused by incorrect mappings.

Fourth, the application of the plans resulted in the automatic correction of the identified errors, which resulted in an improvement in the precision of the automatically generated mappings. The improvement in precision occurred in all cases and varied from 1%-5%, achieving a good balance with respect to the recall, which remained unchanged in most cases, although a decrease from 1%-3% was observed in a few isolated cases.

5 Related Work and Conclusion

In the last few years, the problem of automatically generating mappings between ontologies has been extensively investigated. An comprehensive and up-to-date source of information about the topic (including papers, tools, ontologies for evaluation, etc.) can be found at `http://www.ontologymatching.org/`

The problem of reasoning with the generated mappings has also received significant attention. Different semantics have been proposed for assigning formal meaning to ontology mappings (e.g. [6, 7]). As mentioned in Section 3, our general approach is applicable regardless of the semantics adopted for the mappings.

Recently, research has been conducted on the debugging and revision of mappings [1, 2]. This research has so far been focused on mappings represented using Distributed Description Logics (DDL) [7]. Our work is also related to the existing approaches for debugging and repairing inconsistencies in OWL ontologies (e.g. [15, 12, 11, 13]). In both of these lines of research, the detected and repaired errors are limited to unsatisfiable concepts and inconsistent ontologies.

When compared to this existing work, we believe that our approach presents a number of improvements. First,the entailments to be repaired are not restricted to obvious inconsistencies, but can include any unintended entailment. Second, users can customise the kinds of entailments to be taken into account when comparing the integrated ontology to the individual ones in order to detect errors (i.e., select the approximation of the deductive difference). Third, users can select not only which entailments should be invalidated, but also which ones should necessarily hold upon completion of the repair process. To this end, we provide a number of novel techniques for helping the user to select which entailments are (un)intended, such as the computation of the dependencies between entailments (the relation $\rhd$), confidence in entailments according to the confidence in the mappings, etc. Fourth, we provide efficient algorithms for computing *all* the repair plans and to help the user select the most suitable plan from amongst those computed. Fifth, when compared to existing work on debugging and repair in OWL, we provide a clear distinction between the ontologies being integrated and the mappings, and users can customise the ontologies from which the plans are allowed to delete axioms. Finally, we provide a fully-fledged editor and reasoning infrastructure integrated with Protégé 4.

References

[1] Meilicke, C., Stuckenschmidt, H., Tamilin, A.: Reasoning Support for Mapping Revision. Journal of Logic and Computation (2008)
[2] Meilicke, C., Stuckenschmidt, H., Tamilin, A.: Repairing ontology mappings. In: Proc. of AAAI, pp. 1408–1413 (2007)
[3] Horrocks, I., Patel-Schneider, P.F., van Harmelen, F.: From $\mathcal{SHIQ}$ and RDF to OWL: the making of a web ontology language. J. Web Sem. 1(1), 7–26 (2003)
[4] Cuenca Grau, B., Horrocks, I., Motik, B., Parsia, B., Patel-Schneider, P., Sattler, U.: OWL 2: The next step for OWL. J. Web Semantics 6(4), 309–322 (2008)
[5] Jimenez-Ruiz, E., Cuenca Grau, B., Horrocks, I., Berlanga, R.: Ontology integration using mappings: Towards getting the right logical consequences. Technical report The ContentMap tool is also available at this location (2008), `http://krono.act.uji.es/people/Ernesto/contentmap`
[6] Euzenat, J.: Semantic precision and recall for ontology alignment evaluation. In: Proc. of IJCAI, pp. 348–353 (2007)
[7] Borgida, A., Serafini, L.: Distributed description logics: Assimilating information from peer sources. J. Data Semantics 1, 153–184 (2003)
[8] Kontchakov, R., Wolter, F., Zakharyaschev, M.: Can you tell the difference between DL-Lite ontologies? In: Proc. of KR (2008)
[9] Konev, B., Walther, D., Wolter, F.: The logical difference problem for description logic terminologies. In: Armando, A., Baumgartner, P., Dowek, G. (eds.) IJCAR 2008. LNCS, vol. 5195, pp. 259–274. Springer, Heidelberg (2008)
[10] Motik, B., Cuenca Grau, B., Horrocks, I., Wu, Z., Fokoue, A., Lutz, C.: OWL 2 Web Ontology Language: Profiles. W3C Working Draft (2008)
[11] Kalyanpur, A., Parsia, B., Sirin, E., Hendler, J.A.: Debugging unsatisfiable classes in OWL ontologies. J. Web Semantics 3(4), 268–293 (2005)
[12] Schlobach, S., Huang, Z., Cornet, R., van Harmelen, F.: Debugging incoherent terminologies. J. Autom. Reasoning 39(3), 317–349 (2007)
[13] Kalyanpur, A., Parsia, B., Horridge, M., Sirin, E.: Finding all justifications of OWL DL entailments. In: Aberer, K., Choi, K.-S., Noy, N., Allemang, D., Lee, K.-I., Nixon, L., Golbeck, J., Mika, P., Maynard, D., Mizoguchi, R., Schreiber, G., Cudré-Mauroux, P. (eds.) ASWC 2007 and ISWC 2007. LNCS, vol. 4825, pp. 267–280. Springer, Heidelberg (2007)
[14] Suntisrivaraporn, B., Qi, G., Ji, Q., Haase, P.: A modularization-based approach to finding all justifications for OWL DL entailments. In: Proc. of ASWC (2008)
[15] Kalyanpur, A., Parsia, B., Sirin, E., Grau, B.C.: Repairing unsatisfiable concepts in OWL ontologies. In: Sure, Y., Domingue, J. (eds.) ESWC 2006. LNCS, vol. 4011, pp. 170–184. Springer, Heidelberg (2006)

Using Partial Reference Alignments to Align Ontologies

Patrick Lambrix and Qiang Liu

Department of Computer and Information Science
Linköpings universitet
581 83 Linköping, Sweden

Abstract. In different areas ontologies have been developed and many of these ontologies contain overlapping information. Often we would therefore want to be able to use multiple ontologies. To obtain good results, we need to find the relationships between terms in the different ontologies, i.e. we need to align them.

Currently, there already exist a number of ontology alignment systems. In these systems an alignment is computed from scratch. However, recently, some situations have occurred where a partial reference alignment is available, i.e. some of the correct mappings between terms are given or have been obtained. In this paper we investigate whether and how a partial reference alignment can be used in ontology alignment. We use partial reference alignments to partition ontologies, to compute similarities between terms and to filter mapping suggestions. We test the approaches on previously developed gold standards and discuss the results.

1 Introduction

In recent years many ontologies have been developed. The benefits of using ontologies include reuse, sharing and portability of knowledge across platforms, and improved documentation, maintenance, and reliability (e.g. [5]). Ontologies lead to a better understanding of a field and to more effective and efficient handling of information in that field. Many of the currently developed ontologies contain overlapping information. For instance, Open Biomedical Ontologies (OBO, http://www.obofoundry.org/) lists 26 different anatomy ontologies (October 2008). Often we would want to be able to use multiple ontologies. For instance, companies may want to use community standard ontologies and use them together with company-specific ontologies. Applications may need to use ontologies from different areas or from different views on one area. Ontology builders may want to use already existing ontologies as the basis for the creation of new ontologies by extending the existing ontologies or by combining knowledge from different smaller ontologies. In each of these cases it is important to know the relationships between the terms in the different ontologies. Further, the data in different data sources in the same domain may have been annotated with different but similar ontologies. Knowledge of the inter-ontology relationships would in this case lead to improvements in search, integration and analysis of data. It has been realized that this is a major issue and much research has recently been done on ontology alignment, i.e. finding mappings between terms in different ontologies (e.g. [4]).

L. Aroyo et al. (Eds.): ESWC 2009, LNCS 5554, pp. 188–202, 2009.

In the current alignment systems an alignment is computed from scratch. However, recently, some situations have occurred where a partial reference alignment is available, i.e. some of the correct mappings between terms are given or have been obtained. One example is the development of Bioportal (bioportal.bioontology.org, [13]) where mappings between different ontologies in the biomedical domain have been collected. Bioportal also supports collaborative ontology alignment (one of the challenges for ontology alignment described in [15]) where experts can focus on their piece of expertise. In this case for some parts of the ontologies mappings will be available while they are still lacking for other parts of the ontologies. Another situation is an iterative ontology alignment methodology where people and ontology alignment systems interact to iteratively align and improve the quality of the mappings. In systems such as SAMBO [7], users can input their own mappings as well as accept or reject mapping suggestions generated by the system. Both the mappings given by the users and the rejection and acceptance of system-generated mappings by users influence future iterations of ontology alignment. Finally, the Ontology Alignment Evaluation Initiative (OAEI, http://oaei.ontologymatching.org/) introduced in 2008 a new task in the Anatomy track [10] where a partial reference alignment was given and participating systems could use the partial reference alignment to improve the quality of their mapping suggestions.

In this paper we investigate whether and how a partial reference alignment (PRA) can be used in ontology alignment. We use PRAs in the different steps of ontology alignment. We use PRAs in a preprocessing step to partition the ontologies into mappable parts that are likely to contain correct mappings and therefore not every term in the first ontology needs to be compared to every term in the second ontology. We also use PRAs in the computation of similarities between terms. Further, we use PRAs to filter mapping suggestions. We test the approaches on previously developed gold standards (from [7] and the Anatomy track of OAEI) and discuss the results.

The remainder of the paper is organized as follows. In section 2 we give background information. We describe a framework for ontology alignment as well as the systems, the gold standards and the evaluation measures that we have used in our experiments. The experiments are described in section 3. We investigate the use of PRAs in the different components of the ontology alignment framework. We also investigate the influence of the size of the PRA on the quality of the alignments. The results of the experiments are shown and discussed in section 4. The paper concludes in section 5.

2 Background

2.1 Ontology Alignment

A large number of ontology alignment systems have been developed (see, for instance, review papers [7,14,12,6], the book [4] on ontology matching, and the ontology matching web site at http://www.ontologymatching.org/). Many ontology alignment systems are based on the computation of similarity values between terms in different ontologies and can be described as instantiations of the general framework in figure 1. The framework consists of two parts. The first part (*I* in figure 1) computes mapping suggestions. The second part (*II*) interacts with the user to decide on the final alignment.

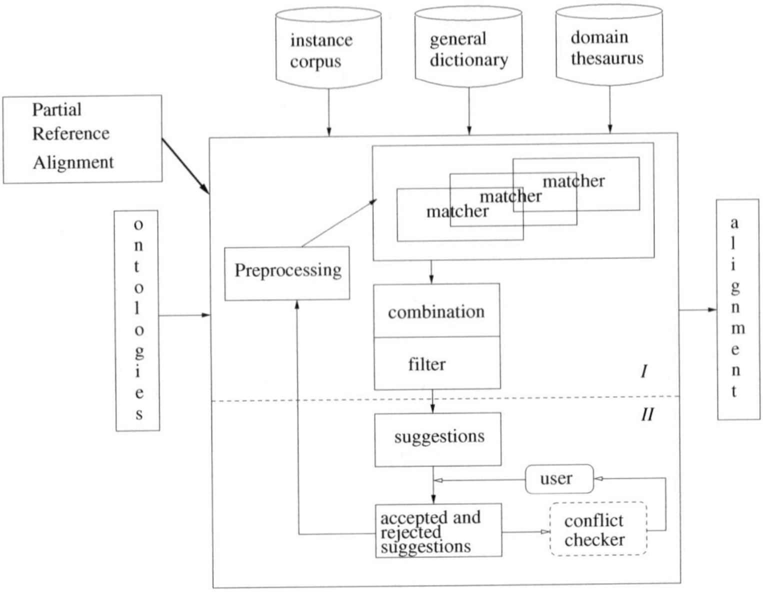

Fig. 1. Alignment framework (modified from [7])

An alignment algorithm receives as input two source ontologies. A preprocessing step can be used to modify the original ontologies or to partition the ontologies into mappable parts. The algorithm can include several matchers that calculate similarities between the terms from the different source ontologies or mappable parts of the ontologies. They can implement strategies based on linguistic matching, structure-based strategies, constraint-based approaches, instance-based strategies, strategies that use auxiliary information or a combination of these. Each matcher utilizes knowledge from one or multiple sources. Mapping suggestions are then determined by combining and filtering the results generated by one or more matchers. By using different matchers and combining and filtering the results in different ways we obtain different alignment strategies. The suggestions are then presented to the user who accepts or rejects them. The acceptance and rejection of suggestions may influence further suggestions. Further, a conflict checker is used to avoid conflicts introduced by the mapping suggestions. The output of the alignment algorithm is a set of mappings between terms from the source ontologies. In this paper we add a partial reference alignment as input and study how this can be used in the different components of this framework.

2.2 SAMBO and SAMBOdtf

SAMBO and SAMBOdtf are ontology alignment systems based on the framework described in the previous section. They both currently contain five basic matchers [7]: two terminological matchers (a basic matcher and an extension using WordNet; extension described below), a structure-based matcher (which uses the is-a and part-of hierarchies of the source ontologies), a matcher based on domain knowledge (described below), and a learning matcher (which uses literature that is related to the concepts in the ontologies to define a similarity value between the concepts). In addition to these techniques we have also experimented with other matchers [9,17].

In our evaluations we use the versions of SAMBO and SAMBOdtf as used in OAEI 2008 [8]. These systems performed best and second best, respectively, in the Anatomy track of OAEI 2008. These versions contain the matchers *TermWN* and *UMLSKSearch*. The matcher *TermWN* contains matching algorithms based on the textual descriptions (names and synonyms) of concepts and relations. In the current implementation, the matcher includes two approximate string matching algorithms (n-gram and edit distance), and a linguistic algorithm that also uses WordNet (http://wordnet.princeton.edu/) to find synonyms and is-a relations. Our matcher *UMLSKSearch* uses the Metathesaurus in the Unified Medical Language System (UMLS, http://www.nlm.nih.gov/research/-umls/). The similarity of two terms in the source ontologies is determined by their relationship in UMLS. In our experiments we used the UMLS Knowledge Source Server to query the UMLS Metathesaurus with source ontology terms. The querying is based on searching the normalized string index and normalized word index provided by the UMLS Knowledge Source Server. We used version 2008AA of UMLS. As a result we obtain concepts that have the source ontology term as their synonym. We assign a similarity value of 0.99 if the source ontology terms are synonyms of the same concept and 0 otherwise.

The combination algorithm used for the OAEI versions of SAMBO and SAMBOdtf is a maximum-based algorithm. The similarity value for a pair of terms is the maximum of the values obtained from TermWN and UMLSKSearch for this pair of terms.

The filtering method in SAMBO is single threshold filtering. Pairs of concepts with a similarity value higher than or equal to a given threshold value are returned as mapping suggestions to the user. SAMBOdtf implements the double threshold filtering method developed in [1]. The double threshold filtering approach uses the structure of the ontologies. It is based on the observation that (for the different approaches in the evaluation in [7]) for single threshold filtering the precision of the results is decreasing and the recall is increasing when the threshold is decreasing. Therefore, we propose to use two thresholds. Pairs with similarity value equal or higher than the upper threshold are retained as suggestions. The intuition is that this gives suggestions with a high precision. Further, pairs with similarity values between the lower and the upper threshold are filtered using structural information and the rest is discarded. We require that the pairs with similarity values between the two thresholds are 'reasonable' from a structural point of view.[1] The intuition here is that the recall is augmented by adding new suggestions, while at the same time the precision stays high because only structurally reasonable suggestions are added. The double threshold filtering approach contains the following three steps. (i) Find a consistent suggestion group from the pairs with similarity value higher or equal than the upper threshold. We say that a set of suggestions is a consistent suggestion group if each concept occurs at most once as first argument in a pair, at most once as second argument in a pair and for each pair of suggestions <A,A' > and <B,B' > where A and B are concepts in the first ontology and A' and B'

[1] In our implementation we have focused on the is-a relation.

are concepts in the second ontology: A ⊂ B iff A' ⊂ B'. (ii) Use the consistent suggestion group to partition the original ontologies. (iii) Filter the pairs with similarity values between the lower and upper thresholds using the partitions. Only pairs of which the elements belong to corresponding pieces in the partitions are retained as suggestions. For details we refer to [1].

In contrast to the original versions of SAMBO and SAMBOdtf where a term in one ontology can be suggested to be mapped to different terms in the other ontology, in the OAEI versions this is not the case. We retain only suggestions where the similarity between the terms in the suggestion is higher than or equal to the similarity of these terms to any other term according to the suggestion list. In the case there are different possibilities, one is randomly chosen. (In the implementation the first in the list is chosen.)

2.3 Gold Standard Alignments

In our tests we use alignments that have been used in previous evaluations ([7] and OAEI). Essentially, for our purpose all used ontologies can be seen as taxonomies (concepts and is-a and part-of relations). In [7] a number of smaller test cases were introduced to evaluate different alignment strategies. For the first two cases we use a part of a Gene Ontology ontology (GO) [2] together with a part of Signal Ontology (SigO) [16]. The first case, *B* (behavior), contains 57 terms from GO and 10 terms from SigO. Its reference alignment (RA) contains 4 mappings. The second case, *ID* (immune defense), contains 73 terms from GO and 17 terms from SigO. Its RA contains 8 mappings. The other cases are taken from the anatomy category of Medical Subject Headings (MeSH, [11]) and the Adult Mouse Anatomy (MA, available from OBO): *nose* (15 terms from MeSH, 18 terms from MA, and 7 mappings in the RA), *ear* (39 terms from MeSH, 77 terms from MA, and 27 mappings in the RA), and *eye* (45 terms from MeSH, 112 terms from MA, and 27 mappings in the RA).

A larger test case that we use is the case of the Anatomy track of OAEI. OAEI is a yearly initiative that was started in 2004. The goals are, among others, to assess the strengths and weaknesses of alignment systems, to compare different techniques and to improve evaluation techniques. This is to be achieved through controlled experimental evaluation. For this purpose OAEI publishes different cases of ontology alignment problems, some of which are open (RA is known beforehand), but most are blind (RA is not known - participants send their alignment suggestions to organizers who evaluate the performance). In the *Anatomy* case (version 2008) participants were required to align the Adult Mouse Anatomy (2744 concepts) and the NCI Thesaurus - anatomy (3304 concepts). The anatomy case is a blind case. The RA contains 1523 equivalence mappings of which 934 are deemed trivial (i.e. they can be found by a relatively basic string-based matcher).

2.4 Evaluation Measures

The results of our experiments are given in terms of the quality of the mapping suggestions. We use precision, recall, recall$_{PRA}$ and f-measure. *Precision* measures how many of the mapping suggestions were correct. It is defined as the number of correct suggestions divided by the number of suggestions. *Recall* measures how many of the correct mappings are found by the alignment algorithm. It is defined as the number of correct suggestions divided by the number of correct mappings. We also introduce the measure $recall_{PRA}$ which measures how many of the correct mappings that are not in a PRA are found by the alignment algorithm. It is defined as the number of correct mapping suggestions not in the PRA divided by the number of correct mappings not in the PRA. *F-measure* is the weighted harmonic mean of precision and recall. In our tests precision and recall are weighted evenly.

3 Experiments

In this section we define a number of experiments to test the usefulness of using a PRA for ontology alignment. We consider the use of PRAs in the different components of the alignment framework in figure 1. As base systems we use SAMBO and SAMBOdtf as described in section 2.2 and modify their components. We describe the experiments and provide an overview of the resulting strategies in table 1.

An immediate observation regarding a PRA is that the mappings in the PRA are deemed to be correct and therefore should be included in the final result. For the same reason, if there are different suggestions for mapping a term, then a suggestion contained in the PRA is preferred. Therefore, in all alignment strategies using a PRA, we add the mappings in the PRA to the list of suggestions with a special status. These mappings cannot be removed in any filtering step. This technique we call *filter with PRA*.

Table 1. Alignment strategies

	preprocessing	matchers	combination	filter
SAMBO	none	TermWN + UMLSKSearch	maximum	single threshold
SAMBOdtf	none	TermWN + UMLSKSearch	maximum	double threshold
mgPRA	partitioning	TermWN + UMLSKSearch	maximum	single threshold filter with PRA
mgfPRA	fixing and partitioning	TermWN + UMLSKSearch	maximum	single threshold filter with PRA
pmPRA	none	TermWN + UMLSKSearch pattern-based augmentation	maximum	single threshold filter with PRA
fPRA	none	TermWN + UMLSKSearch	maximum	single threshold filter with PRA
dtfPRA	none	TermWN + UMLSKSearch	maximum	double threshold with PRA filter with PRA
pfPRA	none	TermWN + UMLSKSearch	maximum	filter based on EM and PRA filter with PRA

3.1 Use of PRA in the Preprocessing Step

A first question we want to investigate is whether we can use a PRA in the preprocessing phase. Most systems compute a similarity value between all terms from the first ontology and all terms from the second ontology. Some work has been done on partitioning the ontologies to find mappable parts of the ontologies (e.g. [3]). The motivation for that work is scalability. When the size of the ontologies grows, some techniques may become infeasible. In our work we investigate whether we can use a PRA to partition the ontologies into mappable parts and test whether, in addition to the fact that we do not have to compute similarity values between all terms from the first ontology and all terms from the second ontology, this also leads to a better quality of the mapping suggestions.

In the first approach we partition the ontologies into mappable parts using the partitioning step of the double threshold filtering described in section 2.2 and [1]. A part of the PRA satisfying the consistent group property is used as a consistent group. The resulting alignment strategy (this preprocessing, the SAMBO matchers, combination and filters, and filter with PRA) we call *mgPRA* (mappable groups with PRA).

According to our experience in aligning ontologies we know that the structure of the source ontologies is not always perfect. For instance, given the two ontologies and the PRA in the Anatomy task of OAEI 2008, it can be deduced that many is-a relations are missing in at least one of the source ontologies. Based on this observation we experiment with a second approach where we add to the source ontologies the missing is-a relationships that can be deduced from the source ontologies and the PRA.[2] After this 'fixing' of the source ontologies the PRA will satisfy the consistent group property. The resulting strategy is called *mgfPRA* (mappable groups and fixing with PRA).

3.2 Use of PRA in a Matcher

One way to create a matcher based on a PRA, is to use underlying properties of the mappings in the PRA. We have previously observed that sometimes for two given source ontologies, common patterns can be found between the correct mappings. For instance, in the PRA of the OAEI 2008 Anatomy we find the mappings <lumbar vertebra 5, l5 vertebra> and <thoracic vertebra 11, t11 vertebra> which share a similar linguistic pattern. Also the mappings <forebrain, fore brain>, <gallbladder, gall bladder>, and the mappings <stomach body, body stomach> and <stomach fundus, fundus stomach> share similar linguistic patterns, respectively. When using different linguistic matchers the similarity values according to these matchers for the mappings sharing similar patterns are therefore very similar as well.

Based on this observation we developed a matcher that augments previously generated similarity values for term pairs when these term pairs display a similar (linguistic) pattern as mappings in the PRA. For the experiment we used TermWN and UMLSKSearch to compute the original similarity values. Further, for each term pair we compute a vector with, in this case, three similarity values, based on the different components of TermWN (n-gram, edit distance and a linguistic algorithm that uses WordNet; as described in section 2.2). Based on this similarity vector we compute the Euclidean distance of a term pair to the mappings in the PRA and count how many of the PRA mappings are within a predefined radius. (In the experiment we used 0.1 for the radius.) Based on the value of this count, the matcher may augment the original

[2] This is in line with the challenge for ontology alignment on discovering missing background knowledge as described in [15].

similarity value for the term pair. (In the experiment we used (count * 0.06) as augmentation value and augmented only pairs with a similarity value lower than 0.9. We also set the limit for the augmented similarity value to 0.9.) The resulting alignment strategy is called *pmPRA* (pattern matcher with PRA).

3.3 Use of PRA in the Filter Step

Another question that we want to investigate is whether we can use a PRA for filtering the list of mapping suggestions.

As mentioned before, all strategies using PRAs implement the filter with PRA approach. In a first experiment we only add this filter approach to SAMBO. The resulting alignment strategy is called *fPRA*.

In addition to filter with PRA, the second strategy also uses a variant of the double threshold filtering approach of SAMBOdtf. While in the original double threshold filtering approach, a consistent suggestion group is computed based on mapping suggestions with a high similarity value, in this approach a part of the PRA satisfying the consistent group property is used as a consistent group. The resulting alignment strategy is called *dtfPRA*.

The last strategy we use is called *pfPRA* (pattern filter with PRA). This strategy is based on the observation described above that some correct mappings share similar patterns. Similar to pmPRA, we assign a vector with similarity values based on the different components of TermWN to each term pair in the suggestion list and the PRA. We compute clusters based on these similarity vectors of the term pairs in the suggestion list using the expectation-maximization algorithm. (For the small cases we set the number of clusters. For *Anatomy* we did not use a predefined number of clusters.) Then, each mapping in the PRA is assigned to the cluster for which the distance between the cluster center and the PRA mapping is the smallest. Further, we compute for each cluster the average distance between a PRA mapping in the cluster and its cluster center. The filter strategy retains the suggestions in a cluster with a distance to the cluster center that is smaller or equal to the computed average for that cluster and discards the others.

3.4 Influence of the Size of the PRA

We also want to investigate the influence of the size of the PRA. For this purpose we compare the results of the approaches for the *Anatomy* case with two PRAs: PRA-F and its subset PRA-H. For PRA-F we take the PRA as provided by the OAEI 2008 Anatomy task. PRA-H contains half of the trivial and half of the non-trivial mappings from PRA-F.

4 Results and Discussion

Table 2 shows the number of suggestions generated by the algorithms in the experiments. For SAMBO and SAMBOdtf we also added the number of mappings in the PRA that were found by those algorithms. We note that for all thresholds for *B*, *nose* and *ear*, for thresholds 0.4 and 0.6 for *ID* and for threshold 0.4 for *eye* SAMBO actually

Table 2. Number of mapping suggestions. For SAMBO and SAMBOdtf also the number of PRA mappings found.

Case	*Th*	SAMBO	SAMBOdtf	mgPRA	mgfPRA	pmPRA	fPRA	dtfPRA	pfPRA
B	0.4	6/2	6/2	6	2	6	6	4	3
	0.6	5/2	5/2	5	2	5	5	4	3
	0.8	4/2	-	4	2	4	4	-	3
ID	0.4	12/4	11/3	12	12	12	12	11	12
	0.6	8/4	7/3	5	5	8	8	8	8
	0.8	7/3	-	4	5	8	8	-	8
nose	0.4	7/4	7/4	7	4	7	7	7	6
	0.6	7/4	7/4	7	4	7	7	7	6
	0.8	7/4	-	7	4	7	7	-	6
ear	0.4	30/14	29/14	28	18	4	30	30	25
	0.6	29/14	29/14	27	18	3	29	29	25
	0.8	26/14	-	24	18	1	26	-	24
eye	0.4	31/13	30/13	30	13	6	31	31	23
	0.6	36/12	26/12	26	13	3	27	27	22
	0.8	24/11	-	25	13	2	26	-	22
Anatomy	0.4	1575/943	1527/940	1690	1663	1625	1601	1552	1251
	0.6	1466/942	1438/940	1488	1444	1598	1498	1474	1221
	0.8	1297/933	-	1308	1271	1528	1342	-	1139

finds all mappings in the PRA. This means that in these cases adding the PRA to the solutions in itself does not improve recall. For *Anatomy* SAMBO does not find between 45 (threshold 0.4) and 55 (threshold 0.8) of the 988 mappings in the PRA.

The results in terms of precision, recall, f-measure and recall$_{PRA}$ are given in tables[3] 3 to 7. The results in the tables are truncated values. Note also that as SAMBO and SAMBOdtf do not use a PRA, we have set recall = recall$_{PRA}$ for SAMBO and SAMBOdtf. *Th* is the threshold for the filtering for the single threshold approaches. For SAMBOdtf and dtfPRA the upper threshold is always 0.8 while *Th* is the lower threshold. There are no results for SAMBOdtf and dtfPRA for upper threshold 0.8 as this would be the same as using SAMBO and fPRA with single threshold 0.8, respectively.

4.1 PRA in Preprocessing

The results of the experiments for using PRAs in the preprocessing step are given in table 3. The intuition behind mgPRA and mgfPRA is to partition the ontologies into mappable parts. Therefore, we can only generate mapping suggestions that are reasonable from a structural point of view. This suggests that, comparing to the base systems, the precision may become higher as suggestions that do not conform to the structure of the source ontologies cannot be made. As we add the PRA to the result, the recall may be increased as some of the PRA mappings may not be found by the base systems.

[3] As *Anatomy* is a blind case at OAEI we do not have the RA available. Therefore, we have sent the mapping suggestions for *Anatomy* to the organizers of the OAEI 2008 Anatomy track who have returned the values for the different evaluation measures.

Table 3. Using the PRA in the preprocessing phase (precision/recall/f-measure/recall$_{PRA}$)

Case	RA	PRA	*Th*	SAMBO	mgPRA	mgfPRA
B	4	2	0.4	0.66/1.00/0.80/1.00	0.66/1.00/0.80/1.00	1.00/0.50/0.66/0.00
			0.6	0.80/1.00/0.88/1.00	0.80/1.00/0.88/1.00	1.00/0.50/0.66/0.00
			0.8	1.00/1.00/1.00/1.00	1.00/1.00/1.00/1.00	1.00/0.50/0.66/0.00
ID	8	4	0.4	0.50/0.75/0.60/0.75	0.41/0.62/0.50/0.25	0.41/0.62/0.50/0.25
			0.6	0.75/0.75/0.75/0.75	1.00/0.62/0.76/0.25	1.00/0.62/0.76/0.25
			0.8	0.71/0.62/0.66/0.62	1.00/0.62/0.76/0.25	1.00/0.62/0.76/0.25
nose	7	4	0.4	1.00/1.00/1.00/1.00	1.00/1.00/1.00/1.00	1.00/0.57/0.72/0.00
			0.6	1.00/1.00/1.00/1.00	1.00/1.00/1.00/1.00	1.00/0.57/0.72/0.00
			0.8	1.00/1.00/1.00/1.00	1.00/1.00/1.00/1.00	1.00/0.57/0.72/0.00
ear	27	14	0.4	0.86/0.96/0.91/0.96	0.85/0.88/0.87/0.76	1.00/0.66/0.80/0.30
			0.6	0.89/0.96/0.92/0.96	0.88/0.88/0.88/0.76	1.00/0.66/0.80/0.30
			0.8	0.96/0.92/0.94/0.92	1.00/0.88/0.94/0.76	1.00/0.66/0.80/0.30
eye	27	13	0.4	0.80/0.92/0.86/0.92	0.80/0.88/0.84/0.78	1.00/0.48/0.65/0.00
			0.6	0.92/0.88/0.90/0.88	0.92/0.88/0.90/0.78	1.00/0.48/0.65/0.00
			0.8	0.91/0.81/0.86/0.81	0.92/0.85/0.88/0.71	1.00/0.48/0.65/0.00
Anatomy	1523	988	0.4	0.82/0.85/0.83/0.85	0.78/0.87/0.82/0.64	0.78/0.85/0.81/0.58
			0.6	0.88/0.84/0.86/0.84	0.88/0.86/0.87/0.61	0.88/0.84/0.86/0.55
			0.8	0.94/0.80/0.87/0.80	0.96/0.82/0.89/0.50	0.96/0.80/0.88/0.45

However, the similarity values between the terms do not change and it is therefore not likely that new mappings are found. (The only way to find new mappings compared to the base system is when a mapping suggestion with high similarity in the base system cannot be suggested by using mgPRA and mgfPRA because the terms were in incompatible parts of the ontologies. In that case other suggestions involving these terms may be generated.) The results give some support to these intuitions. For threshold 0.8 the precision of mgPRA and mgfPRA is always equal to or higher than the precision for SAMBO. This is also almost always the case for threshold 0.6. For threshold 0.4 there is no conclusive result. We also notice that, except for threshold 0.4 in *ID* and *Anatomy*, mgfPRA always gives better precision than SAMBO.

As expected, the recall for mgPRA and mgfPRA is equal to or less than the recall for SAMBO in most cases. For the large test case the recall is always higher for mgPRA and equal for mgfPRA. In the cases there is a loss of recall, this is due to a different modeling in the source ontologies. When the PRA satisfies the consistent group property, mgPRA and mgfPRA give the same results. When this is not the case, mgfPRA 'fixes' the source ontologies by adding missing is-a relationships. In most cases this has led to an improvement in precision, in the other cases the precision stayed the same. However, contrary to the intuition in many cases fixing the source ontologies has led to a decrease in recall. This is due to the use of is-a in the source ontologies. For instance, the hierarchical relation in MeSH covers both is-a and part-of. Therefore, not all of these relations should be treated as is-a. However, as mgfPRA cannot distinguish between these, it may fix the source ontology in the wrong way, by adding a hierarchical link (which mgfPRA interprets as is-a, but which should have been interpreted as part-of). For instance, in the *nose* case, having <nose, nose> in the PRA would lead to introducing is-a relations in MA between nose and its parts. Therefore, fixing the ontologies may lead to worse results. As for all alignment strategies using structural information, the quality of the underlying ontologies, the completeness of the structure and the correct use of the structural relations, has an important influence on the quality of the results.

4.2 PRA in a Matcher

The results of the experiment using PRAs in a matcher are found in table 4. The intuition of the matcher was to augment the similarity values of suggestions that had a similar linguistic pattern as mappings in the PRA. For *B*, *nose* and *ear* the results for SAMBO and pmPRA are the same. The augmentation did not have any influence. For *B* and *nose* the recall was already 1, so that no influence was expected. For *ear* the found correct mappings had already a high similarity value and the missed correct mappings (two for threshold 0.8 and one for thresholds 0.4 and 0.6) did not have a similar linguistic pattern as the mappings in the PRA. For *ID* the recall and the precision were equal or became higher (because of the addition of the PRA to the results). For *eye* the recall improved or was the same. For *Anatomy* the precision and, for low thresholds, also the recall decreased. The recall increased for high thresholds. To investigate the increase of recall for *Anatomy*, we compared the results of using PRA-F with PRA-H. In this case augmenting allowed to find <lateral cuneiform, external cuneiform bone foot> and <brain arachnoid matter, cerebral arachnoid membrane>. For each of these there were 7 mappings in PRA-H with similar linguistic patterns.

Table 4. Using the PRA in a matcher (precision/recall/f-measure/recall$_{PRA}$)

Case	RA	PRA	*Th*	SAMBO	pmPRA
B	4	2	0.4	0.66/1.00/0.80/1.00	0.66/1.00/0.80/1.00
			0.6	0.80/1.00/0.88/1.00	0.80/1.00/0.88/1.00
			0.8	1.00/1.00/1.00/1.00	1.00/1.00/1.00/1.00
ID	8	4	0.4	0.50/0.75/0.60/0.75	0.50/0.75/0.60/0.50
			0.6	0.75/0.75/0.75/0.75	0.75/0.75/0.75/0.50
			0.8	0.71/0.62/0.66/0.62	0.75/0.75/0.75/0.50
nose	7	4	0.4	1.00/1.00/1.00/1.00	1.00/1.00/1.00/1.00
			0.6	1.00/1.00/1.00/1.00	1.00/1.00/1.00/1.00
			0.8	1.00/1.00/1.00/1.00	1.00/1.00/1.00/1.00
ear	27	14	0.4	0.86/0.96/0.91/0.96	0.86/0.96/0.91/0.92
			0.6	0.89/0.96/0.92/0.96	0.89/0.96/0.92/0.92
			0.8	0.96/0.92/0.94/0.92	0.96/0.92/0.94/0.84
eye	27	13	0.4	0.80/0.92/0.86/0.92	0.80/0.92/0.86/0.85
			0.6	0.92/0.88/0.90/0.88	0.89/0.92/0.90/0.85
			0.8	0.91/0.81/0.86/0.81	0.92/0.88/0.90/0.78
Anatomy	1523	988	0.4	0.82/0.85/0.83/0.85	0.78/0.83/0.81/0.54
			0.6	0.88/0.84/0.86/0.84	0.79/0.83/0.81/0.54
			0.8	0.94/0.80/0.87/0.80	0.83/0.83/0.83/0.52

Table 5. Using the PRA during the filter phase - 1 (precision/recall/f-measure/recall$_{PRA}$)

Case	RA	PRA	*Th*	SAMBO	fPRA	pfPRA
B	4	2	0.4	0.66/1.00/0.80/1.00	0.66/1.00/0.80/1.00	1.00/0.75/0.85/0.50
			0.6	0.80/1.00/0.88/1.00	0.80/1.00/0.88/1.00	1.00/0.75/0.85/0.50
			0.8	1.00/1.00/1.00/1.00	1.00/1.00/1.00/1.00	1.00/0.75/0.85/0.50
ID	8	4	0.4	0.50/0.75/0.60/0.75	0.50/0.75/0.60/0.50	0.50/0.75/0.60/0.50
			0.6	0.75/0.75/0.75/0.75	0.75/0.75/0.75/0.50	0.75/0.75/0.75/0.50
			0.8	0.71/0.62/0.66/0.62	0.75/0.75/0.75/0.50	0.75/0.75/0.75/0.50
nose	7	4	0.4	1.00/1.00/1.00/1.00	1.00/1.00/1.00/1.00	1.00/0.85/0.92/0.66
			0.6	1.00/1.00/1.00/1.00	1.00/1.00/1.00/1.00	1.00/0.85/0.92/0.66
			0.8	1.00/1.00/1.00/1.00	1.00/1.00/1.00/1.00	1.00/0.85/0.92/0.66
ear	27	14	0.4	0.86/0.96/0.91/0.96	0.86/0.96/0.91/0.92	1.00/0.92/0.96/0.84
			0.6	0.89/0.96/0.92/0.96	0.89/0.96/0.92/0.92	1.00/0.92/0.96/0.84
			0.8	0.96/0.92/0.94/0.92	0.96/0.92/0.94/0.84	1.00/0.88/0.94/0.76
eye	27	13	0.4	0.80/0.92/0.86/0.92	0.80/0.92/0.86/0.85	0.95/0.81/0.88/0.64
			0.6	0.92/0.88/0.90/0.88	0.92/0.92/0.92/0.85	1.00/0.81/0.89/0.64
			0.8	0.91/0.81/0.86/0.81	0.92/0.88/0.90/0.78	1.00/0.81/0.89/0.64
Anatomy	1523	988	0.4	0.82/0.85/0.83/0.85	0.83/0.88/0.86/0.66	0.91/0.74/0.82/0.28
			0.6	0.88/0.84/0.86/0.84	0.89/0.87/0.88/0.64	0.93/0.74/0.82/0.27
			0.8	0.94/0.80/0.87/0.80	0.95/0.84/0.89/0.54	0.97/0.72/0.83/0.22

Table 6. Using the PRA during the filter phase - 2 (precision/recall/f-measure/recall$_{PRA}$)

Case	RA	PRA	*Th*	SAMBOdtf	dtfPRA
B	4	2	0.4	0.66/1.00/0.80/1.00	1.00/1.00/1.00/1.00
			0.6	0.80/1.00/0.88/1.00	1.00/1.00/1.00/1.00
ID	8	4	0.4	0.45/0.62/0.52/0.62	0.54/0.75/0.63/0.50
			0.6	0.71/0.62/0.66/0.62	0.75/0.75/0.75/0.50
nose	7	4	0.4	1.00/1.00/1.00/1.00	1.00/1.00/1.00/1.00
			0.6	1.00/1.00/1.00/1.00	1.00/1.00/1.00/1.00
ear	27	14	0.4	0.89/0.96/0.92/0.96	0.86/0.96/0.91/0.92
			0.6	0.89/0.96/0.92/0.96	0.89/0.96/0.92/0.92
eye	27	13	0.4	0.83/0.92/0.87/0.92	0.80/0.92/0.86/0.85
			0.6	0.92/0.88/0.90/0.88	0.92/0.92/0.92/0.85
Anatomy	1523	988	0.4	0.84/0.84/0.84/0.84	0.86/0.87/0.87/0.65
			0.6	0.89/0.84/0.86/0.84	0.90/0.87/0.88/0.64

4.3 PRA in a Filter

The results for the experiments regarding the use of PRAs in the filter step are given in tables 5 and 6. In the filter phase mapping suggestions are removed. Both correct and wrong suggestions can be removed and therefore both precision and recall can change compared to the base system. All strategies implement filter with PRA. Therefore, as we add the PRA to the result, the recall may be increased as some of the PRA mappings may not be found by the base systems. The precision and recall may increase when wrong suggestions are removed because there are mappings in the PRA involving the same terms. This intuition is supported by the test cases. For all cases fPRA has a higher or equal precision and recall than SAMBO.

In dtfPRA we use, in addition to filter with PRA, also the structure of the source ontologies to filter the suggestion list by filtering out the suggestions which are not reasonable with respect to the structure of the ontologies and the given PRA. The intuition is that wrong suggestions may be removed and that it is better to use a PRA than a computed suggestion group. For *B*, *ID* and *Anatomy*[4] dtfPRA has a higher precision and recall than SAMBOdtf. For *ear* and *eye* the recall is equal or higher for dtfPRA than for SAMBOdtf. For lower threshold 0.4 the precision is lower for dtfPRA than for SAMBOdtf, while it is higher or equal for lower threshold 0.6. The lower precision for *ear* at threshold 0.4 comes from a suggestion <inner ear epithelium, inner hair cell> which was filtered out by SAMBOdtf, but not by dtfPRA. One reason could be that in the SAMBOdtf case the consistent group consisted of 17 mapping suggestions while the consistent part of the PRA only consisted of 9 mappings (as the whole PRA did not satisfy the consistent group property). The partitioning for SAMBOdtf could therefore result in smaller mappable parts. We also compared dtfPRA to SAMBOdtf with just adding the PRA to the results. In most cases we have the same precision and recall. For *ID* threshold 0.4 precision for dtfPRA is slightly better, while for *ear* threshold 0.4 it is slighlty worse. For *Anatomy* just adding the PRA to SAMBOdtf actually gives a slightly higher recall than dtfPRA (3 more correct mappings are found), but a lower precision.

The third filter strategy removes suggestions that do not have similar linguistic patterns than the mappings in the PRA. We expect therefore that some correct suggestions obtained through UMLS will be removed and therefore the recall may go down. This is indeed the case in our experiments. The precision for pfPRA is, however, always higher or equal to the precision for SAMBO. This is because the suggestions that had a linguistically similar pattern as mappings in the PRA were usually correct.

4.4 Influence of the Size of the PRA

The results for the experiment regarding the influence of the size of the PRA are shown in table 7. Intuitively, the more correct mappings we have, the higher the recall should be. This is supported by all test cases. As shown in the last column of table 7, several mappings in PRA-F are not found by the algorithms using PRA-H. However, using PRA-H seems to generate more mappings from the unknown part of the RA. On the other hand, it also generates more wrong suggestions. One explanation could be that for most strategies the larger the PRA, the more constraints the suggestions need to satisfy and thus the fewer suggestions (correct and wrong) are generated. For the preprocessing strategies mgPRA and mgfPRA the precision is lower for the larger PRA when we use a low threshold. However, the precision is better for the larger PRA when the threshold is high. The matcher strategy pmPRA shows a similar tendency. For the three filtering

[4] According to the results of OAEI 2008 Anatomy task [10], with respect to the unknown part of the RA, dtfPRA's precision increased with 0.040, its recall with 0.008 and its f-value with 0.025. dtfPRA was the system with the highest increase in f-value and was the only system that used the PRA to increase both precision and recall. In [10] dtfPRA is called 'SAMBOdtf for task 4'.

Table 7. *Anatomy* (1523 correct mappings in the RA) with PRA-F (988 mappings) and PRA-H (494 mappings) - (precision/recall/f-measure). new-X represents the number of correct mappings not in PRA-F found by using PRA-X. NF is the number of mappings in PRA-F not found by the algorithms using PRA-H.

Strategy	*Th*	PRA-F	new-F	PRA-H	new-H	NF
mgPRA	0.4	0.78/0.87/0.82	345	0.80/0.85/0.82	351	44
	0.6	0.88/0.86/0.87	327	0.88/0.83/0.85	337	46
	0.8	0.96/0.82/0.89	281	0.95/0.80/0.86	281	50
mgfPRA	0.4	0.78/0.85/0.81	313	0.79/0.81/0.80	336	85
	0.6	0.88/0.84/0.86	295	0.87/0.80/0.83	321	87
	0.8	0.96/0.80/0.88	243	0.95/0.76/0.84	268	89
pmPRA	0.4	0.78/0.83/0.81	290	0.77/0.83/0.80	313	26
	0.6	0.79/0.83/0.81	290	0.79/0.83/0.81	312	26
	0.8	0.83/0.83/0.83	282	0.84/0.82/0.83	294	28
fPRA	0.4	0.83/0.88/0.86	356	0.83/0.86/0.84	357	25
	0.6	0.89/0.87/0.88	347	0.88/0.86/0.87	348	26
	0.8	0.95/0.84/0.89	293	0.95/0.82/0.88	294	30
pfPRA	0.4	0.91/0.74/0.82	152	0.90/0.74/0.81	179	32
	0.6	0.93/0.74/0.82	148	0.92/0.74/0.82	175	33
	0.8	0.97/0.72/0.83	118	0.96/0.71/0.82	136	34
dtfPRA	0.4	0.86/0.87/0.87	350	0.84/0.86/0.85	355	26
	0.6	0.90/0.87/0.88	344	0.89/0.86/0.87	348	26

strategies the precision for the larger PRA is always better or equal than the precision of the smaller PRA.

5 Conclusion

In this paper we have investigated whether and how a PRA can be used in ontology alignment by experimenting with using a PRA in the different components of ontology alignment systems. The use of PRA in preprocessing and filtering reduces the number of suggestions and in most cases leads to an improvement in precision. In some cases also the recall improved. Filter with PRA should always be used. For approaches using structural information the quality of the structure in the underlying ontologies has a large impact. The matcher using linguistic patterns in the PRA mappings can be used for finding new suggestions. The differences between the results for the algorithms that use a PRA and the base systems are relatively small. However, considering the nature of the test cases and the fact that SAMBO and SAMBOdtf perform already well on their own, even small improvements are valuable. Also, for the large test case, due to the choice of the PRA all newly found mappings are non-trivial. There are different directions for future work. The algorithms should be tested on other ontologies and with different base algorithms. Combinations and interactions of the methods should be investigated. It would also be interesting to look at other kinds of patterns in alignment data. The approach should also be integrated in an iterative ontology alignment framework.

Acknowledgements. We are indebted to Christian Meilicke for running our algorithms on the Anatomy track test case of the OAEI. We also acknowledge the financial support of the Swedish Research Council (Vetenskapsrådet).

References

1. Chen, B., Tan, H., Lambrix, P.: Structure-based filtering for ontology alignment. In: Proc. IEEE WETICE Workshop on Semantic Technologies in Collaborative Applications, pp. 364–369 (2006)
2. The Gene Ontology Consortium. Gene ontology: tool for the unification of biology. Nature Genetics 25(1), 25–29 (2000)
3. Ehrig, M., Staab, S.: QOM - quick ontology mapping. In: McIlraith, S.A., Plexousakis, D., van Harmelen, F. (eds.) ISWC 2004. LNCS, vol. 3298, pp. 683–697. Springer, Heidelberg (2004)
4. Euzenat, J., Shvaiko, P.: Ontology Matching. Springer, Heidelberg (2007)
5. Jasper, R., Uschold, M.: A framework for understanding and classifying ontology applications. In: Proc. IJCAI Workshop on Ontologies and Problem-Solving Methods: Lessons Learned and Future Trends (1999)
6. Kalfoglou, Y., Schorlemmer, M.: Ontology mapping: the state of the art. The Knowledge Engineering Review 18(1), 1–31 (2003)
7. Lambrix, P., Tan, H.: SAMBO - a system for aligning and merging biomedical ontologies. Journal of Web Semantics 4(3), 196–206 (2006)
8. Lambrix, P., Tan, H., Liu, Q.: SAMBO and SAMBOdtf results for the ontology alignment evaluation initiative 2008. In: Proc. 3rd International Workshop on Ontology Matching (2008)
9. Lambrix, P., Tan, H., Xu, W.: Literature-based alignment of ontologies. In: Proc. 3rd International Workshop on Ontology Matching (2008)
10. Meilicke, C., Stuckenschmidt, H.: Anatomy track at the Ontology Alignment Evaluation Initiative (2008), http://oaei.ontologymatching.org/2008/
11. MeSH. Medical subject headings, http://www.nlm.nih.gov/mesh/
12. Noy, N.F.: Semantic integration: A survey of ontology-based approaches. Sigmod Record 33(4), 65–70 (2004)
13. Noy, N.F., Griffith, N., Musen, M.: Collecting community-based mappings in an ontology repository. In: Proc. 7th International Semantic Web Conference, pp. 371–386 (2008)
14. Shvaiko, P., Euzenat, J.: A survey of schema-based matching approaches. Journal on Data Semantics IV, 146–171 (2005)
15. Shvaiko, P., Euzenat, J.: Ten challenges for ontology matching. In: Proc. 7th International Conference on Ontologies, Databases, and Applications of Semantics (2008)
16. Takai-Igarashi, T., Nadaoka, Y., Kaminuma, T.: A database for cell signaling networks. Journal of Computational Biology 5(4), 747–754 (1998)
17. Tan, H., Jakonienė, V., Lambrix, P., Aberg, J., Shahmehri, N.: Alignment of biomedical ontologies using life science literature. In: Bremer, E.G., Hakenberg, J., Han, E.-H(S.), Berrar, D., Dubitzky, W. (eds.) KDLL 2006. LNCS (LNBI), vol. 3886, pp. 1–17. Springer, Heidelberg (2006)

Semantic Matching Using the UMLS

Jetendr Shamdasani*, Tamás Hauer, Peter Bloodsworth, Andrew Branson, Mohammed Odeh, and Richard McClatchey

CCCS Research Centre, CEMS Faculty, University of the West of England
Coldharbour Lane, Frenchay, Bristol BS16 1QY, UK
firstname.lastname@cern.ch

Abstract. Traditional ontology alignment techniques enable equivalence relationships to be established between concepts in two ontologies with some confidence value. With semantic matching, however, it is possible to identify not only equivalence ($\equiv$) relationships between concepts, but less general ($\sqsubseteq$) and more general relationships ($\sqsupseteq$). This is beneficial since more expressive relationships can be discovered between ontologies thus helping us to resolve heterogeneity between differing semantic representations at a finer level of granularity. This work concerns the application of semantic matching to the medical domain. We have extended the SMatch algorithm to function in the medical domain with the use of the UMLS metathesaurus as the background resource, hence removing its previous reliance on WordNet, which does not cover the medical domain in a satisfactory manner. We describe the steps required to extend the SMatch algorithm to the medical domain for use with UMLS. We test the accuracy of our approach on subsets of the FMA[1] and MeSH[2] ontologies, with both precision and recall showing the accuracy and coverage of different versions of our algorithm on each dataset.

1 Introduction

Semantic Matching [1] is the process of discovering set theoretic based relationships between differing concepts within two or more ontologies. It is richer than identifying equivalent concept pairs since finer relationships between concepts may be discovered, in particular *less general* ($\sqsubseteq$) and *more general* ($\sqsupseteq$). The framework is well understood, yet to our knowledge it has not yet been exploited in medical applications. Ontologies are commonly used in the medical domain for creating widely agreed terminologies, taxonomies, for annotating medical data from publications to patient records and in reformulating clinicians' queries. The value of ontology matching lies in the fact that there are numerous, largely independent terminologies, taxonomies and ontologies covering the various subdomains of medicine. In order for interoperability between these semantic models to be achieved they need to be matched. Alignment takes

* Corresponding author.
[1] http://sig.biostr.washington.edu/projects/fm/AboutFM.html
[2] http://www.nlm.nih.gov/mesh/

L. Aroyo et al. (Eds.): ESWC 2009, LNCS 5554, pp. 203–217, 2009.

this a step further and allows us to describe the overlap between subdomains. As an example if we consider the case of two clinicians trying to gain access to a clinical system, each clinician has their own ontology of a subset of a domain and the clinical system has a more general view of the clinical world. For these clinicians to be able to access and share data they need to speak the same language. An alignment between these varying models is beneficial since it facilitates the sharing of information.

In this paper we present a semantic matching framework for use in the medical domain. The current work is an extension of the SMatch approach [2] and has been motivated by the requirements of the E.C. funded Health-e-Child project[3]. The SMatch framework is based on two cornerstones: firstly, it uses a predicate logic framework to unify node-wise string matching with structural matching (thus exploiting their interplay.) and secondly, it uses a background thesaurus as an anchor for tokenization, sense filtering and knowledge seeding. The direct application of the algorithm to medical ontologies is somewhat weak because of its reliance on the WordNet thesaurus [3]; this was shown in our previous work [4]. Previously we only used the UMLS Concept hierarchy for disambiguation. In this work we present results using different features of the UMLS for hierarchical disambiguation with our structural filtering implementation which differs from the original SMatch approach. We have formalised the original steps for semantic matching and created a framework built entirely on the logic behind the original algorithm.

We have chosen the UMLS metathesaurus [5] as our background resource because of its wide coverage of clinical terms and its disambiguation of these terms. We have built a prototype matcher that has been evaluated on test ontologies with some initial results. We have formalised the steps to derive a semantic match between medical ontologies; these are shown with a worked example. Section 2 presents previous work conducted in the area of semantic matching and ontology alignment. Section 3 provides a brief introduction to the UMLS. Section 4 explains the idea behind semantic matching. Section 5 describes the modifications to the original SMatch algorithm for use with the UMLS. Section 6 presents our evaluation methodology for our algorithm. Section 7 presents the results with different versions of our algorithm. Finally, conclusions and directions for further work are presented in section 8.

2 Previous Work

Previous work within the area of ontology alignment has mostly focused on discovering equivalence relationships between different concepts and roles in an ontology. Their output is normally in the form of an equivalence relationship with a confidence value between concepts within the range 0 to 1. There are many different approaches within the literature which rely on different features of an ontology (its schema, instances etc..). A diverse number of approaches have been attempted in the ontology alignment field [1] and significant progress has been

[3] `http://www.health-e-child.org` (IST 2004-027749)

made. Much of this progress has been in the form of using differing background knowledge sources. Background knowledge has been used in the form of free text [6], previous alignments [7], search engines [8] and other ontologies. In our case *Semantic Matching* uses other ontologies as background resources. There have been approaches using single or multiple resources. A standard semantic matching approach is composed of an anchoring step, deriving semantics by using background knowledge and a set of reasoning rules. Semantic matching differs from other approaches because the goal is not to look for equivalence but rather to discover other relationships.

Aleskovsi et al [9] use a single background resource to match two flat lists of medical terms using a single background ontology. Their process consists of an anchoring step where they firstly tokenize the string to be matched and then "anchor" these to concepts in the background ontology. This is done for all strings in both lists. Once this has been completed they are able to derive a semantic match between two terms using the background resource. They use a set of predefined reasoning rules to achieve this, for example, using transitive reasoning. They have shown the application of these rules in their follow-up work [10] where they matched the anatomy parts of the CRISP to MeSH and used the FMA as a background resource.

Sabou et al [11] have extended the approach by Alekovsi et al. to use multiple background ontologies using the Swoogle ontology search engine. They firstly search for the concepts they are going to match then search for ontologies which contain both of these terms. If a suitable resource cannot be found, then separate ontologies are searched for and are then matched against each other, this process is repeated until a match is found. They semi-formalise the steps involved in their framework. Their search approach has been improved in a later work [12] by using the senses available from WordNet, therefore allowing a better ontology selection process for their background knowledge. SMatch is another semantic matching approach. Instead of defining inference rules, Giunchiglia et al. cast the ontology alignment problem as a propositional satisfiability problem. Here labels of nodes in trees are tokenized and the lexical structure in addition to the tree structure are represented by logical formulae built from WordNet senses that are associated with the tokens. These formulae are fed into a standard SAT solver to check the validity of the matches based on predefined reasoning rules. One strength of this approach is that there does not necessarily have to be a complete lexical overlap between the ontologies to be matched and WordNet. SMatch and its modification to use the UMLS is explained further in section 5 with a worked example.

3 The UMLS

The UMLS is a thesaurus of biomedical knowledge. Its purpose is to integrate conceptual information from differing sources of biomedical terminology. These can be complex ontologies or simple lists, being known as source vocabularies in the schema. The UMLS contains relationships between these concepts organised into its own model and consequently it is a conceptual resource about the biomedical domain. A concept in the UMLS is identified by a CUI (Concept Unique Identifier) which is a unique value identifying a Concept. Every Concept has a set of Terms (or Lexical Groupings) identified by an LUI (Lexical Unique Identifier) and a set of Strings identified by a SUI (String Unique Identifier). Every concept has an Atom identified by an AUI (Atomic Unique Identifier) which are the original definitions of a concept from its source vocabulary. Figure 1 shows how these relate to each other via a UML diagram.

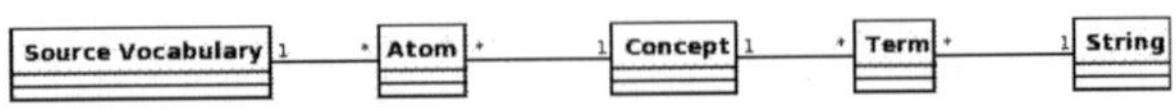

Fig. 1. A UML based representation of the UMLS

Every Concept in the UMLS is related to another concept in the UMLS hierarchy via Broader Than (RB), Narrower Than (RN), Parent (PAR), Child (CHD) and Sibling (SIB) relationships, this information being contained within the MRREL table of the UMLS. As well as relationships between concepts the UMLS also contains hierarchical information between Atoms in their original source vocabularies. Atoms are related to one another in simple taxonomic hierarchies with varying semantics for example "isa" or "part_of", this information is contained within the MRHIER table. Also for disambiguation purposes there is the MRCOC table which contains co-occurrences relationships between UMLS Concepts in text. Section 5 describes how the UMLS can be applied to semantic matching. The next section explains the idea behind semantic matching.

4 Semantic Matching Explained

Semantic Matching is the process of discovering $\equiv$, $\sqsubseteq$, $\sqsupseteq$ relationships between concepts in two ontologies. We have chosen to use the propositional reasoning based approach of the SMatch algorithm. This section will explain what semantic matching is. At the highest level, the SMatch algorithm assigns a logical formula to each node of a tree, such that the meaning of the node that is inferred from both the lexical and the tree structures which are encoded by the predicate calculus. These formulae are in turn input into reasoning rules to derive a match.

Fig. 2. String Matching Example

If we consider the matching of two strings in figure 2, these strings represent two concepts which are the concepts of "Examination" and "Kidney Examination" respectively. The expected result for this case would be that "Kidney Examination" is a $\sqsubseteq$ of "Examination". SMatch is able to discover this relationship due to its logic based representation of strings which are converted into atomic formulae. The atomic formula for the node "Examination" would be the name of the node i.e. *examination* and the atomic formula for the node "Kidney Examination" would be the name of the node as well, however the white space character is converted to $\sqcap$ which means the meaning of the concept is the conjunction of the concepts of "Kidney"and "Examination" i.e. $kidney \sqcap examination$. The shaded region in figure 3(a) shows the meaning of this concept and figure 3(b) shows the match result.

Figure 4 shows an example of matching two trees taking their structures into account, Tree 1 (on the left) has no structure with only one node with the label "Examination" whereas Tree 2 does have structure. If we were matching the nodes "Examination" and "Midbrain" the correct result would be that "Examination" $\sqsupseteq$ "Midbrain" due to the fact that the meaning of the node "Midbrain" is influenced by its parent nodes. To achieve this SMatch takes two factors into account, firstly from our background knowledge we know that "Examination" is equivalent to "Anatomical Examination"; this is the background theory that is implicit and is used in the reasoning process. Also the node with structure ("Midbrain") is given a *context*, the manner in which this is achieved is that the logical formula of "Midbrain" is altered to take into account its relationship with its parents. This constraint is applied by an intersection with its logical formula to its path to the root, the new formula for this node is $midbrain \sqcap brain \sqcap anatomical\ examination$. The shaded area in figure 5 shows the meaning of this concept visually.

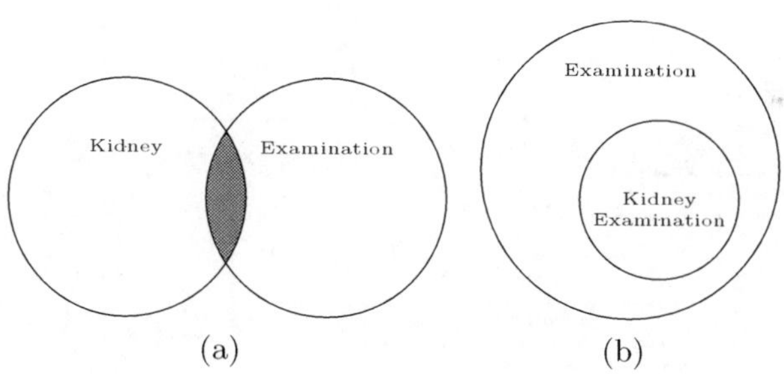

Fig. 3. Venn diagrams showing the meaning of the concept "Kidney Examination" (a) and the result of the matching process (b)

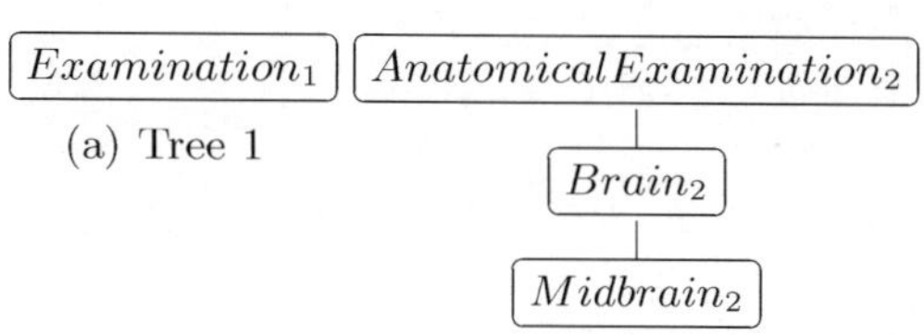

Fig. 4. Two example trees showing how SMatch takes structure into account

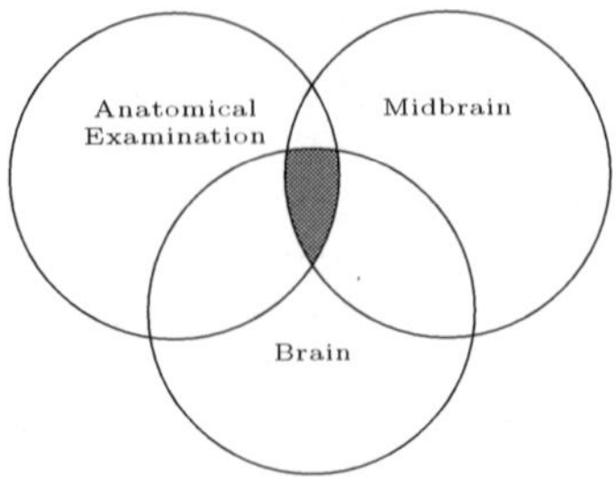

Fig. 5. The meaning of the concept Midbrain after constraining its mean via its relationship to its parents

5 The Modified SMatch Algorithm

Our modified algorithm takes two trees as input and as an output it produces a matrix of relationships between them. Figure 6 shows our input trees for this purpose. The algorithm works on the same 4-step premise as the original algorithm. At the highest level in the first step the label of a node in a tree is converted into a logical formula using specific rules with some filtering performed and concepts being attached via a background resource. Step 2 consists of creating a node formula and applying context in the form of filtering out irrelevant concepts. The third step involves the matching of atomic formulae using a background resource and the fourth and final step is the reasoning process. These will now be discussed in full.

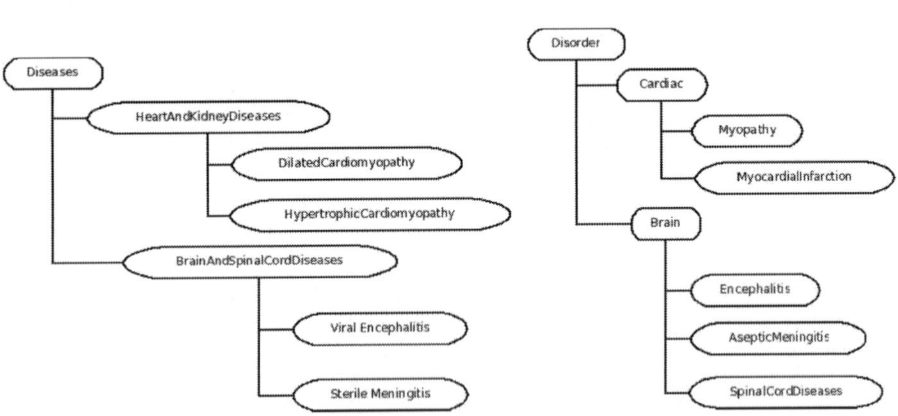

(a) Tree 1 (b) Tree 2

Fig. 6. The two input trees for our matching process

5.1 Step 1 – String to Formula Conversion

This step, as stated previously, works exclusively on the label of a node. Firstly the label of a node is normalized then tokenized. For example we consider the node with the label "HeartAndKidneyDiseases" from Tree 1; its tokens would be $\{heart, and, kidney, disease\}$. Once this has been completed tokens are converted to *atomic formulae* and a look up in the UMLS is performed to attach Concepts to these. Words which denote preposition, conjunctions etc... are ignored i.e. are not annotated with concepts from the UMLS, however they are kept for logical formula conversion. For example if we query the UMLS for the term *heart* we find that heart has a total of 5 concepts from the UMLS. Hence the formula of *heart* has a total of 5 concepts attached to it. The notation for this is $heart\{cui\#5\}$. For the node "HeartAndKidneyDiseases" we would have three atomic formulae for this node which would be $heart\{cui\#5\}$, $kidney\{cui\#1\}$ and $disease\{cui\#5\}$. As a final step these atomic formulae are joined with boolean connectives ($\sqcup$, $\sqcap$ and $\neg$). The rules for this are that tokens denoting disjunction (e.g. or) are converted to logical disjunction, tokens denoting conjunctive terms (e.g. for) are converted to logical conjunction and terms denoting negation (e.g. not) are converted to negation. Hence the final formula for the node "HeartAndKidneyDiseases" would be $((heart \sqcup renal) \sqcap disease)$. Notice how the term "and", despite being a connective word in the English language has been converted to disjunction, this captures the meaning of the concept. A change to this step is that before the tokenization step we check to see if a concept exists "as is" within the UMLS i.e. there is an existing match in the background resource for the entire label. For example of we consider the node labelled "MyocardialInfarction" from Tree 2, its atomic formula would be $myocardial\ infarction$ with its corresponding Concepts attached from the UMLS. This would be the single atomic formula for the node.

5.2 Step 2 – Context Creation and Filtering

The purpose of this step is to capture the *meaning* of a concept according to its relationships with its parents. The manner in which this is achieved is that the conjunction of the current node to its parent is taken, since in essence the meaning of this node is the intersection with all of its parents. For example for the node named "HyperthrophicCardiomyopathy" in Tree 2 the conjunction of the formula of this node is taken to its parent node would be $((disease) \sqcap ((heart \sqcup kidney) diesease) \sqcap (hypertrophic\ cardiomyopathy))$. This is the *context* of the node with its UMLS concepts attached to each atomic formula. Once the contextual formula has been created we can then perform a filtering of the Concepts, a necessary step in removing unrelated Concepts attached to atomic formulae and hence improving the matching process.

We filter the Concepts based on information we have available from the UMLS. The UMLS itself has three tables for disambiguation: the MRREL (Concept relationships), MRHIER (Atom relationships) and MRCOC (Co-Occurrence relationships). Our concept filtering algorithm is based on the *tree context* of the node, hence its path to the root and all of its children (not be confused with its context which is its path to root). For each concept, c, for an atomic formula a we get all the concepts attached to a in their tree context and we give a score to c if it is related to another concept in its tree context from either the

Concept, Atom or Co-Occurrences relationships to *c*. We currently retain all related concepts i.e. with a score greater than zero and disregard all unrelated concepts. Only one table at a time is used i.e. they are not used in conjunction with each other. This step is beneficial since it is able to drop concepts attached to an atomic formula which have no relationships in its current tree context. It should be noted that if an atomic formula only has one concept we retain that.

5.3 Step 3 – Atomic Formula Matching with the UMLS

Once the above steps are completed, the *concepts*, related to atomic formulae, in the two trees, are matched against each other using the UMLS hierarchy. As stated previously there is more than one source of hierarchical information in the UMLS; these are the MRREL and MRHIER tables. MRREL describes the relationships between concepts, whereas and the MRHIER describes the relationships between atoms.

If we wish to use the Atom hierarchy, the mapping rules are as follows:

- $\equiv$ **rule** - If a Concept from A contains a Concept from B then a $\equiv$ relationship is returned.
- $\sqsubseteq$ **rule** - If an Atom from a Concept in A is a subclass of an Atom from a Concept B in a single source vocabulary then a $\sqsubseteq$ relationship is returned.
- $\sqsupseteq$ **rule** - If an Atom from a Concept in A is a superclass of an Atom from a Concept in B in a single source vocabulary then a $\sqsupseteq$ relationship is returned.

where A and B are both atomic formulae. If we use the Concept hierarchy of the UMLS, the mapping rules are:

- $=$ **rule** - If a Concept from A contains a Concept from B then a $\equiv$ relationship is returned.
- $\sqsubseteq$ **rule** - If a Concept from A is a subclass of a Concept from B i.e. it is related via a RN or CHD relationship then a $\sqsubseteq$ relationship is returned.
- $\sqsupseteq$ **rule** - If a Concept from A is a superclass of a Concept from B i.e. it is related via a PAR or RB relationship then a $\sqsupseteq$ relationship is returned.

where A and B are both atomic formulae. As can be seen there is no $\perp$ test because unlike WordNet the UMLS does not explicitly state antonymy between concepts. We select the first relationship discovered with equivalence taking the highest level of precedence.

Tables 1 and 2 show a subset of results of matching atomic formulae from the two trees. In addition to showing equivalence and subsumption relationships there is also a "NF" relationship, this means that a relationship has not been found between atomic formulae using the UMLS. It is of interest to note that some of the relationships discovered using the differing hierarchies have presented different results e.g. *heart* is $\sqsupseteq$ *brain* in table 2, whereas this relationship does not exist in table 1. Using the atom hierarchy fewer relationships are discovered between atomic formulae, whereas using the concept hierarchy many more relationships are discovered. Whether or not these are correct is, of course another matter altogether.

Table 1. This is the output of step three of the algorithm where the labels are matched using the UMLS **Atom** hierarchy

$Tree_2$ / $Tree_1$	*disorder*	*cardiac*	*myopathy*	*myocardial infarction*	*brain*
disease	$\equiv$	NF	$\sqsupseteq$	$\sqsupseteq$	$\sqsupseteq$
heart	$\sqsubseteq$	$\equiv$	NF	NF	NF
kidney	$\sqsubseteq$	NF	NF	NF	NF
dilated cardiomyopathy	$\sqsubseteq$	$\sqsubseteq$	NF	NF	NF
hypertrophic cardiomyopathy	$\sqsubseteq$	$\sqsubseteq$	NF	NF	NF

Table 2. This is the output of step three of the algorithm where the labels are matched using the UMLS **Concept** hierarchy

$Tree_2$ / $Tree_1$	*disorder*	*cardiac*	*myopathy*	*myocardial infarction*	*brain*
disease	$\equiv$	$\sqsupseteq$	$\sqsupseteq$	$\sqsupseteq$	$\sqsupseteq$
heart	$\sqsubseteq$	$\equiv$	$\sqsupseteq$	$\sqsupseteq$	$\sqsupseteq$
kidney	$\sqsubseteq$	NF	NF	NF	NF
dilated cardiomyopathy	$\sqsubseteq$	$\sqsubseteq$	$\sqsubseteq$	$\sqsubseteq$	$\sqsubseteq$
hypertrophic cardiomyopathy	$\sqsubseteq$	$\sqsubseteq$	$\sqsubseteq$	$\sqsubseteq$	$\sqsubseteq$

5.4 Step 4 – Reasoning

The purpose of this step is to prove a relationship between two nodes in the trees. To achieve this equation 1 is used. If we were trying to prove a less general relationship ($\sqsubseteq$) between two nodes we would try and prove a $\rightarrow$ relationship which is its propositional equivalent and $\leftarrow$ for a more general ($\sqsupseteq$). For equivalence we would prove a $\leftrightarrow$ relationship. Before this is done however, *axioms* (eq 1) or the implicit background theory from the end of the previous step, have to be created. The relationships between atomic formulae discovered from the UMLS are converted to their propositional equivalents. Axioms are only created for the matching task at hand. For example if we were matching the node "HeartAndKidney-Diseases" from Tree 1 in figure 6 to the node "MyocardialInfarction" in Tree 2 the axioms would be $((disease \leftrightarrow disorder) \wedge (heart \leftrightarrow cardiac) \wedge (disease \leftarrow myocardial\ infarction))$ this being derived from table 1. Once this is completed the contexts from the end of the second step for each tree are fed into equation 1, *rel* is relationship we are trying to prove ($=$,$\sqsubseteq$,$\sqsupseteq$). Equation 3 shows a more general relationship between the two nodes. To prove this relationship we have to show that its negation is not satisfiable (Equation 2). Equation 4 shows the full propositional formula that is fed into a SAT solver before it is converted into its conjunctive normal form. We can see from equation 4 that this is not satisfiable and therefore the more general relationship between these two nodes holds. It is interesting to note that this information was not taken from the UMLS (table 1) but that this relationship was inferred. This step stays the same regardless of which features of the UMLS we use for disambiguation.

$$axioms \rightarrow rel(context_A, context_B) \tag{1}$$

$$axioms \wedge \neg rel(context_A, context_B) \tag{2}$$

$$\begin{aligned} &((disease \leftrightarrow disorder) \wedge (heart \leftrightarrow \\ &cardiac) \wedge (disease \leftarrow myocardial\ infarction)) \\ &\rightarrow ((disease) \wedge ((heart \vee kidney) \wedge disease)) \\ &\leftarrow ((disorder) \wedge (cardiac) \wedge \\ &(myocardial\ infarction)) \end{aligned} \tag{3}$$

$$\begin{aligned} &((disease \leftrightarrow disorder) \wedge (heart \leftrightarrow \\ &cardiac) \wedge (disease \leftarrow myocardial\ infarction)) \wedge \\ &\neg(((disease) \wedge ((heart \vee kidney) \wedge \\ &disease)) \leftarrow ((disorder) \wedge (cardiac) \wedge \\ &(myocardial\ infarction))) \end{aligned} \tag{4}$$

Table 3. This is the final output of the algorithm when the Atom hierarchies are used for both filtering and matching of the concepts

$Tree_1$ \ $Tree_2$	Disorder	Cardiac	Myopathy	Myocardial Infarction	Brain	Encephalitis	Aseptic Meningitis	Spinal Cord Diseases
Diseases	$\equiv$	$\sqsupseteq$	$\sqsupseteq$	$\sqsupseteq$	$\sqsupseteq$	$\sqsupseteq$	$\sqsupseteq$	$\sqsupseteq$
HeartAndKidneyDiseases	$\sqsubseteq$	$\sqsupseteq$	$\sqsupseteq$	$\sqsupseteq$	NF	NF	NF	NF
DilatedCardiomyopathy	$\sqsubseteq$	$\sqsubseteq$	NF	NF	NF	NF	NF	NF
HypertrophicCardiomyopathy	$\sqsubseteq$	$\sqsubseteq$	NF	NF	NF	NF	NF	NF
BrainAndSpinalCordDiseases	$\sqsubseteq$	NF	NF	NF	NF	NF	NF	NF
ViralEncephalitis	$\sqsubseteq$	NF	NF	NF	$\sqsubseteq$	$\sqsubseteq$	NF	NF
SterileMeningitis	$\sqsubseteq$	NF	NF	NF	NF	NF	NF	NF

Table 4. This is the output of the algorithm with filtering using the Atom Hierarchy of the UMLS and the Concept Matching of the third step

$Tree_1$ \ $Tree_2$	Disorder	Cardiac	Myopathy	Myocardial Infarction	Brain	Encephalitis	Aseptic Meningitis	Spinal Cord Diseases
Diseases	$\equiv$	$\sqsupseteq$	$\sqsupseteq$	$\sqsupseteq$	$\sqsupseteq$	$\sqsupseteq$	$\sqsupseteq$	$\sqsupseteq$
HeartAndKidneyDiseases	$\sqsubseteq$	$\sqsupseteq$	$\sqsupseteq$	$\sqsupseteq$	$\sqsupseteq$	$\sqsupseteq$	$\sqsupseteq$	$\sqsupseteq$
DilatedCardiomyopathy	$\sqsubseteq$	$\sqsubseteq$	$\sqsubseteq$	$\sqsubseteq$	$\sqsubseteq$	NF	NF	NF
HypertrophicCardiomyopathy	$\sqsubseteq$	$\sqsubseteq$	$\sqsubseteq$	$\sqsubseteq$	$\sqsubseteq$	NF	NF	NF
BrainAndSpinalCordDiseases	$\sqsubseteq$	$\sqsubseteq$	$\sqsubseteq$	NF	$\sqsubseteq$	NF	NF	NF
ViralEncephalitis	$\sqsubseteq$	$\sqsubseteq$	$\sqsubseteq$	NF	$\sqsubseteq$	$\sqsubseteq$	NF	NF
SterileMeningitis	$\sqsubseteq$	$\sqsubseteq$	$\sqsubseteq$	NF	$\sqsubseteq$	NF	NF	NF

Tables 3 and 4 show the results of matching the two trees in figure 6, using the Atom and Concept hierarchies for matching atomic formulae. As can be seen the results of using either the concept hierarchy or the atom hierarchy in the third step are different, hence the axioms, or information from a background resource can have a great influence on the final result. Our evaluation focuses on using differing features of the UMLS for the filtering in the second step and the matching of atomic formulae in the third step. This will be discussed further in the next section.

6 Evaluation Methodology

Our evaluation methodology focuses on the accuracy of the differing hierarchical matching and filtering schemes used by our algorithm. We have selected subsets of the FMA and MeSH ontologies for our matching process and have created a reference alignment for our matching task. The subset of the FMA we have chosen is rooted at the concept of Brain, and follows the "regional_part_of" relationship to its leaf nodes, this contains 476 concepts in total. The subset of MeSH we have chosen is rooted at the concept of brain as well and traverses down to its leaf nodes, this contains 181 concepts in total. Both of these subsets cover the same domain, as they describe anatomical relationships with the organ "Brain" however, their level of coverage, naming of terms and models are different. All the matches are from the FMA to MeSH, i.e the FMA is the source ontology and MeSH is the target. We have used the 2008AA version of the UMLS in all of our experiments. The SMatch algorithm has been reimplemented from scratch with our modifications incorporated and we have used the SAT4J[4] library as our SAT solver.

For the evaluation of our algorithm we have chosen to use different sources for the filtering of concepts (Concepts, Atoms or Co-Occurrences, none) and for the hierarchical matching as well (Concepts or Atoms). We created our reference alignment with the aid of a medical expert, our process for creating our standard has been to select 20 random concepts from our FMA subset and 40 random concepts from our MeSH subset and manually derive a match in the semantic matching range.

7 Results

The following versions of the algorithm were run:

1. **Concept_NoFilter:** This uses the concept hierarchy for the matching of atomic formulae with no filtering of concepts.
2. **Atom_NoFilter:** This uses the atom hierarchy for the matching of atomic formulae with no filtering of concepts.
3. **Concept_ConceptFilter:** This uses the concept hierarchy of the UMLS for atomic formulae matching and the concept hierarchy for the filtering of concepts.

[4] http://www.sat4j.org

4. **Atom_AtomFilter:** This uses the Atom hierarchy of the UMLS for matching atomic formulae and the Atom hierarchy for filtering.
5. **Concept_AtomFilter:** This uses the Concept hierarchy for the matching of atomic formulae and the Atom hierarchy for the filtering of concepts.
6. **Atom_ConceptFilter:** This uses the Atom hierarchy of the UMLS for the matching of atomic formulae and the Concept hierarchy for the filtering of Concepts.
7. **Concept_COCFilter:** This uses the Concept hierarchy of the UMLS for the matching of atomic formulae and the Co-Occurrence relationships for filtering.
8. **Atom_COCFilter:** This uses the Atom hierarchy of the UMLS for the matching of atomic formulae and the Co-Occurrence relationships for filtering.

Table 5. The number of matches found for each version of our algorithm

Test	No. Equivalences	No. Less General	No. More General	No. Not Found
Concept_NoFilter	65	4069	937	81085
Atom_NoFilter	65	2581	726	82784
Concept_ConceptFilter	65	4069	937	81085
Atom_AtomFilter	65	2587	726	82778
Concept_AtomFilter	65	4069	937	81085
Atom_ConceptFilter	65	2587	726	82778
Concept_COCFilter	65	4069	935	81087
Atom_COCFilter	65	2581	726	82784

Table 5 shows the total number of matches found for each matching task. Overall using the Concept hierarchy for the matching of atomic formulae in the third step resulted in many more Less General and More General relationships being discovered than using the Atom hierarchy; this is because the Concept hierarchy is more expressive than the Atom hierarchy of the UMLS. The interesting thing to note is that the filtering using the Co-Occurrence relationships within the UMLS resulted in a slightly reduced number of alignments being classified as a less general relationship in the Atom hierarchy tests. Also these results are equivalent to no filtering of Concepts attached to atomic formulae. When using the UMLS Concept hierarchy for the matching of atomic formulae there seems to be no significant difference in the number of results regardless of filtering scheme employed. Our precision and recall values are identical for each version of the algorithm (both at 1), we think this is a fault with our reference alignment where there were many "NF" relationships discovered during the manual matching task by our clinical expert. Also the reference alignment did not contain any of the differing results discovered (for example the extra less general relationships discovered by the Concept matching of atomic formulae). However, with this preliminary evaluation the values are still significantly high for both precision and recall which is promising once a larger scale evaluation has been conducted.

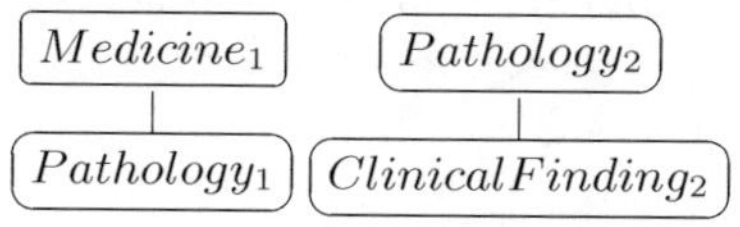

(a) Tree 1 (b) Tree 2

Fig. 7. Two trees to show where filtering can dramatically affect the results

The filtering however is useful although this is just a special case, if we consider the example in figure 7. If there was no filtering scheme involved at the hierarchical level we would have the following relationships discovered:

- $Medicine_1 \sqsupseteq Pathology_2$
- $Medicine_1 \sqsupseteq ClinicalFinding_2$
- $Pathology_1 \equiv Pathology_2$
- $Pathology_1 \sqsupseteq ClinicalFinding_2$

However, if we use Atom filtering we would have the following relationships discovered:

- $Medicine_1 NF Pathology_2$
- $Medicine_1 NF ClinicalFinding_2$
- $Pathology_1 NF Pathology_2$
- $Pathology_1 NF ClinicalFinding_2$

This is because when we query the UMLS for the terms in Tree 1 which are "Medicine" we get the following Concepts returned:

- *C0013227:Pharmaceutical Preparations*
- *C0025118:Medicine*

For "Pathology" we have the following:

- *C0677042:Pathology processes*
- *C0334866:Medical pathologist*
- *C0205469:Pathological aspects*
- *C0919386:Pathology procedure*
- *C0030664:Pathology*

The filtering scheme retains the concept of *0025118:Medicine* and drops *C0013227:Pharmaceutical Preparations* since this concept is not related to any of the concepts of Pathology. For the node labelled "Pathology" the concept of *C0030664:Medical pathologist* is kept since it is the only concept related to any of the concepts of "Medicine". The meaning of this tree has now been constrained and given context since each node only has these concepts attached.

An identical process occurs between the nodes in tree 2 where "ClinicalFinding", only has one concept attached so it is kept which is *C0037088:Signs and Symptoms* and the node "Pathology" has the same concepts attached as above, however, now the concept that is retained during the filtering process is *C0677042:Pathology processes*. Therefore the atomic formula for the node "Pathology" in tree 2 now has a different concept attached when compared to the node "Pathology" in tree 1. According to our rules in step 3 there is now no relationship discovered, hence the axioms are empty and no relationship can be deduced during the reasoning process. This is an ideal case where the filtering of concepts is useful, however when matching our subsets of FMA to MeSH they both cover the same domain hence the filtering (although dropping irrelevant concepts) has not been effective in the number of results discovered since identical concepts were dropped in both trees. In our future work we will match two highly unrelated trees to gauge the effectiveness of our filtering scheme.

8 Conclusion

In this work we have shown how to apply the UMLS to a semantic matching task by modifying the SMatch algorithm to its use. For this we have detailed the steps involved and how the UMLS can be used. We have found by the number of results shown that the filtering has a minimal effect on the number of results discovered, however, in some situations this is not the case. Overall the accuracy is our method is high, this is evidenced by the high precision values of our algorithm and recall is high as well (both are perfect at 1), which is to be expected on such a small reference alignment. Currently our results are preliminary, our future work will consist of a further evaluation of our approach on a much larger dataset with a complete reference alignment. This however is difficult since there are no commonly agreed upon reference alignments or datasets for semantic matching in the medical domain. We intend to submit our matching system to the OAEI [13] medical track to aid our evaluation. We will also test our algorithm on different datasets with varying degrees of overlap.

The significance of this work is that it is possible to conduct semantic matching in the medical domain, we have shown this with the use of one background resource which is the UMLS. This is beneficial because the expressivity of semantic matching is able to increase the number of alignments discovered between elements, hence increasing the number of potential matches between concepts. We will later investigate how our approach can apply to differing sources of medical knowledge and outline the steps to achieve this.

References

1. Euzenat, J., Shvaiko, P.: Ontology Matching. Springer, Heidelberg (2007)
2. Giunchiglia, F., Yatskevich, M., Shvaiko, P.: Semantic Matching: Algorithms and Implementation. Journal of Data Semantics, 1–38 (2007)
3. Fellbaum, C.: WordNet: An Electronic Lexical Database. MIT Press, Cambridge (1998)
4. Shamdasani, J., Bloodsworth, P., McClatchey, R.: Semantic Matching for the Medical Domain. In: Gray, A., Jeffery, K., Shao, J. (eds.) BNCOD 2008. LNCS, vol. 5071, pp. 198–202. Springer, Heidelberg (2008)

5. Bodenreider, O.: The Unified Medical Language System (UMLS): Integrating Biomedical Terminology. Nucleic Acids Research 32 (2004)
6. Tan, H., Jakoniene, V., Lambrix, P., Aberg, J., Shahmehri, N.: Alignment of Biomedical Ontologies Using Life Science Literature. In: Bremer, E.G., Hakenberg, J., Han, E.-H.(S.), Berrar, D., Dubitzky, W. (eds.) KDLL 2006. LNCS (LNBI), vol. 3886, pp. 1–17. Springer, Heidelberg (2006)
7. Do, H., Rahm, E.: COMA a system for flexible combination of schema matching approaches. In: 28th International Conference of Very Large Databases (2002)
8. Risto, G., Aleksovski, Z., ten Kate, W., van Harmelen, F.: Using Google Distance to Weight Approximate Ontology Matches. In: BNACI 2007 (2007)
9. Aleksovski, Z., Klein, M., ten Kate, W., van Harmelen, F.: Matching Unstructured Vocabularies using a Background Ontology. In: Staab, S., Svátek, V. (eds.) EKAW 2006. LNCS, vol. 4248, pp. 182–197. Springer, Heidelberg (2006)
10. Aleksovski, Z., ten Kate, W., van Harmelen, F.: Exploiting the structure of background knowledge used in ontology matching. In: Ontology Matching Workshop (ISWC 2006) (2006)
11. Sabou, M., d'Aquin, M., Motta, E.: Exploring the Semantic Web as Background Knowledge for Ontology Matching. International Journal of Data Semantics (2008)
12. Garcia, J., Lopez, V., d'Aquin, M., Sabou, M., Motta, E., Mena, E.: Solving Semantic Ambiguity to Improve Semantic Web based Ontology Matching. In: International Workshop on Ontology Matching (OM 2007) (2007)
13. Caracciolo, C., Euzenat, J., Hollink, L., Ichise, R., Isaac, A., Malais, V., Meilicke, C., Pane, J., Shvaiko, P., Stuckenschmidt, H., Svab-Zamazal, O., Svatek, V.: Results of the Ontology Alignment Evaluation Initiative 2008. In: Ontology Matching Workshop (ISWC 2008) (2008)

Embedding Knowledge Patterns into OWL

Luigi Iannone, Alan Rector, and Robert Stevens

School of Computer Science,
University of Manchester,
Manchester,
M13 9PL UK
{iannone,rector,robert.stevens}@cs.manchester.ac.uk

Abstract. We describe the design and use of the Ontology Pre-Processor Language (OPPL) as a means of embedding the use of Knowledge Patterns in OWL ontologies. We illustrate the specification of patterns in OPPL and discuss the advantages of its adoption by Ontology Engineers with respect to ontology generation, transformation, and maintainability. The consequence of the declarative specification of patterns will be their unambiguous description inside an ontology in OWL. Thus, OPPL enables an ontology engineer to work at the level of the pattern, rather than of the raw OWL axioms. Moreover, patterns can be analysed rigorously, so that the repercussions of their reuse can be better understood by ontology engineers and tools implementers. Thus the delivery of patterns with OPPL can provide a means of addressing the opacity and sustainability of OWL ontologies.

1 Introduction

The current focus of discussion about ontology engineering is on topics such as: the *sustainability* of the knowledge maintenance process; the *opacity* of the design of the underlying reusable knowledge models; and the lack of tool support in these and other areas. To cite one recent work:

> Today, one of the most challenging and neglected areas of ontology design is reusability. The possible reasons include at least: size and complexity of the major reusable ontologies, opacity of design rationales in most ontologies, lack of criteria in the way existing knowledge resources (e.g. thesauri, database schemata, lexica) can be reengineered, and brittleness of tools that should assist ontology designers. [1]

The evidence that knowledge models expressed in OWL can be, and indeed are, opaque to their users comes (indirectly) from the interest sparked by work on entailment justifications. Some models are so hard to interpret that researchers have recently focused on automatically generating explanations that are better suited for users' interpretation and understanding [2]. Whereas justifications seek to make the inferences from the raw OWL more comprehensible, patterns seek to raise the level of abstraction at which the ontology is formulated. This is

L. Aroyo et al. (Eds.): ESWC 2009, LNCS 5554, pp. 218–232, 2009.

achieved by encapsulating a particular *pattern* of axioms that captures a particular modelling issue and enables the desired inferences to be made. Understanding may still be required, but the solution is ready-made and its reuse provides consistency of style.

Knowledge Patterns, introduced in [3], have been succinctly described, in [4], as:

> [...] namely representations which capture recurring structure within and across ontologies

This notion, coupled with the concept from Software Engineering of Design Patterns, led to the introduction of Ontology Design Patterns (ODPs) [5] that are collections of best practices in modeling knowledge. ODP have been further categorised into *Logic* and *Content* patterns [1]. Current work in ODPs aims to collect reusable fragments of patterns of knowledge representation that can be accessed, documented, validated, and, finally, deployed by knowledge engineers. Portals[1] have recently been launched to host centralised collections of ODPs to facilitate their uptake. Although this could be beneficial for the creation of patterns validated by the community, we argue that the aspects more strictly related to the pragmatics of the reuse are still largely unaddressed. We observe that, apart from the documentation, there is no way, to the best of our knowledge, to specify that an ontology uses a pattern or the way a pattern should be used. This means that if ontology engineers use a pattern in an ontology they might have to document the details of its application or, in case of reuse, the ontology's maintainers will have to examine the knowledge model to understand the pattern instantiation. We argue that what is needed is a language and a framework that allows the explicit definition of patterns and the ability to refer to them in OWL ontologies. Such a language would have the following advantages:

- Declarative specification of patterns provides the possibility to describe unambiguously what it means, in OWL, to use a pattern inside an ontology and therefore
  - to produce pattern instantiations automatically;
  - to add pattern support to ontology engineering tools.
- Patterns can be analysed formally, and the repercussions of their reuse can be better understood by ontology engineers.
- Specifying those patterns that are to be reused in an ontology represents a viable means (besides natural language documentation) of expressing the ways such ontology should evolve according to the intentions of its original modelers.

Note that the word *pattern* is more often used as a synonym for Ontology Design Pattern than in its general accepted meaning of *knowledge pattern*. From what we said above, an ODP is a *knowledge pattern*, but the converse does not necessarily hold in general. The main contribution of this work is to propose a

[1] E.g.: `http://ontologydesignpatterns.org`

language to specify knowledge patterns in OWL, therefore, in the following, the occurrence of the term *pattern* will not refer to an ODP, unless it is explicitly indicated.

The remainder of this paper is organised as follows:

- we specify our proposal for such a language;
- we propose a framework for embedding it into OWL ontologies;
- we present some example of patterns coming from various ontologies and try to give an idea of the extent of the impact of pattern introduction;
- we discuss briefly some motivating applications;
- finally, we discuss some promising research directions that the introduction of this or a similar language could enable.

2 OPPL for Patterns

In [6] we introduced the new version of the Ontology Pre-Processing Language initially proposed in [7] and applied in [8] as a declarative manipulation language for ontologies. In [8] OPPL was initially motivated by the need of ontology developers to transform one ontology to an axiomatically richer form. Here the aim is to design a language to encapsulate recurring knowledge structures (set of parametrized operations on OWL axioms) expressed in OWL. The OWL axiom is the basic unit upon which OPPL acts. Each OPPL construct and its effects can be simply expressed in terms of *additions* and *removals* of axioms. This means that:

a. The semantics underneath patterns in OPPL are OWL-DL semantics, which is both familiar to both OWL users and well-understood by implementers and language designers;
b. The effects of using a pattern can be immediately computed/visualized without needing any intermediate translation.

The grammar for the whole of OPPL can be found at `http://www.cs.man.ac.uk/~iannonel/oppl/documentation.html`. Here, we limit ourselves to illustrating, by means of examples, the most relevant features of the language when used for pattern specification. (Examples can be found in Figures 1 and 2).

2.1 Language Overview

The pattern sub-language of OPPL has two main components, namely:

1. Variables;
2. Actions.

Variables, just as in full OPPL, can have the following types:

1. CLASS;
2. OBJECTPROPERTY;

3. DATAPROPERTY;
4. INDIVIDUAL;
5. CONSTANT.

they cover all the logical entity types in an OWL ontology[2].

Actions can either ADD or REMOVE an axiom that, in its turn can be any axiom in OWL-DL built upon a combination of variables and entities from an ontology. We will introduce the OPPL syntax with an example pattern using the widely known Wine Ontology from W3C OWL Guide[3]. Suppose a knowledge engineer wants to provide a pattern to build new sub-classes of `wine:Wine`, so that the ontology could be enriched with wine classes that do not appear in the initial model. He could specify the OPPL Pattern shown in Figure 1[4].

```
?region:CLASS,                                    Variables
?winery:CLASS[subClassOf Winery],
?grape:CLASS,
?body:INDIVIDUAL,
?color:INDIVIDUAL,
?flavor:INDIVIDUAL,
?sugar:INDIVIDUAL
BEGIN                                               Actions
ADD $thisClass subClassOf Wine,
ADD $thisClass subClassOf  locatedIn some ?region,
ADD $thisClass subClassOf hasMaker some ?winery,
ADD $thisClass subClassOf madeFromGrape some ?grape,
ADD $thisClass subClassOf hasBody value ?body,
ADD $thisClass subClassOf hasColor value ?color,
ADD $thisClass subClassOf hasFlavor value ?flavor,
ADD $thisClass subClassOf hasSugar value ?sugar
END;
Wine with located in ?region made by ?winery from ?grape grape   Rendering
with ?body body, ?color, color , ?flavor flavor, and ?sugar sugar
```

Fig. 1. Hypothetical pattern for the creation of a kind of Wine in the W3C Wine ontology

[2] With entity here we denote: named classes, object and data properties, individuals, and constants. Note that annotations are not considered as they are currently not part of the logical structure of OWL and their specification is changing rapidly in the revision from OWL 1.0 to OWL 2.0.

[3] `http://www.w3.org/TR/2003/PR-owl-guide-20031209/wine`

[4] Notice that OPPL relies on Manchester OWL Syntax [9], from which its own syntax has been derived by adding variable support. Hence, all the axioms in the example ontologies in this paper will also be reported using their Manchester OWL Syntax rendering, with the exception of Figure 5 that uses RDF/XML-ABBREV as axiom annotations are not yet supported in Manchester Syntax.

The pattern above defines the new wine in terms of 7 variables that correspond to the features specified, in the starting ontology, by the class `wine:Wine` whose description we report below for completeness:

```
Wine subClassOf PotableLiquid
Wine subClassOf locatedIn some Region
Wine subClassOf hasMaker only Winery
Wine subClassOf madeFromGrape min 1 Thing
Wine subClassOf hasBody exactly 1 Thing
Wine subClassOf hasColor exactly 1 Thing
Wine subClassOf hasFlavor exactly 1 Thing
Wine subClassOf hasMaker exactly 1 Thing
Wine subClassOf hasSugar exactly 1 Thing
```

After the variable definition block, the OPPL pattern specifies the operations on the ontology that its use (henceforth *instantiation*) requires to be executed. In our case they consist of adding 8 axioms. This pattern is peculiar because it is supposed to be applied to a class description. It means that the ontology engineer, in order to use this pattern, will have to pick a class already present in the ontology to which the pattern will be applied. The way to refer to such a class inside a pattern is to use the reserved variable name `$thisClass`. An equivalent pattern that is not supposed to be applied to a class can be obtained by adding another variable to the one in Figure 1, and editing the pattern as reported in Figure 2.

The block after the actions is called *rendering*. Its purpose is merely to provide tools with more user-friendly presentations for patterns. Its content is, in fact, free text in which variable names could be inserted. In a pattern instantiation, such variable names will be replaced with the values assigned to them.

Now suppose our knowledge engineer decides to instantiate the pattern in Figure 2 assigning to its variable the following values:

```
?wine → PrimitivoManduriaDolceNaturale
?region → PugliaRegion
?winery → PuglieseWinery
?grape → Primitivo
?body → Full
?color → Red
?flavor → Strong
?sugar → Sweet
```

The effects of such an instantiation will be the addition of the following axioms:

```
ADD PrimitivoManduriaDolceNaturale subClassOf Wine
ADD PrimitivoManduriaDolceNaturale subClassOf locatedIn some PugliaRegion
ADD PrimitivoManduriaDolceNaturale subClassOf hasMaker some PuglieseWinery
ADD PrimitivoManduriaDolceNaturale subClassOf madeFromGrape some Primitivo
ADD PrimitivoManduriaDolceNaturale subClassOf hasBody value Full
```

```
?wine:CLASS,
?region:CLASS,
?winery:CLASS[subClassOf Winery],
?grape:CLASS,
?body:INDIVIDUAL,
?color:INDIVIDUAL,
?flavor:INDIVIDUAL,
?sugar:INDIVIDUAL
BEGIN
ADD ?wine subClassOf Wine,
ADD ?wine subClassOf  locatedIn some ?region,
ADD ?wine subClassOf hasMaker some ?winery,
ADD ?wine subClassOf madeFromGrape some ?grape,
ADD ?wine subClassOf hasBody value ?body,
ADD ?wine subClassOf hasColor value ?color,
ADD ?wine subClassOf hasFlavor value ?flavor,
ADD ?wine subClassOf hasSugar value ?sugar
END;
?wine Wine with located in ?region made by ?winery from ?grape grape
with ?body body, ?color color , ?flavor flavor, and ?sugar sugar
```

Fig. 2. Pattern for the creation of a kind of Wine in the W3C Wine ontology not applicable to a class

```
ADD PrimitivoManduriaDolceNaturale subClassOf hasColor value Red
ADD PrimitivoManduriaDolceNaturale subClassOf value Strong
ADD PrimitivoManduriaDolceNaturale subClassOf value Sweet
```

And its rendering will be:
`PrimitivoManduriaDolceNaturale` Wine located in `PugliaRegion` made by `PuglieseWinery` from `Primitivo` grape with `Full` body, `Red` color, `Strong` flavor, and `Sweet` sugar.

An OPPL pattern may reference other OPPL patterns. To illustrate this consider another example ontology, this time about food. For the sake of simplicity let us reduce it to the following axioms:

```
Food subClassOf Thing
Meat subClassOf Food
Egg subClassOf Food
Vegetables subClassOf Food
Objectproperty contains
Course subClassOf contains some Food
Menu subClassOf contains some Course
```

Let us suppose that we want to specify sub classes of `Course` and `Menu` on the basis of what they do or do not contain. Let us consider the following pattern (called `freeFromCourse` and shown in Figure 3).

This pattern uses the `RETURN` clause. This clause defines a particular subset of patterns that, when instantiated, not only executes their actions, but

```
?x:CLASS,
?forbiddenContent:CLASS=createUnion(?x.VALUES)
BEGIN
ADD $thisClass equivalentTo Food and contains only (not ?forbiddenContent)
END;
A ?x free stuff;
RETURN $thisClass
```

Fig. 3. Our `FreeFromPattern`

also *returns* a value. The instantiation of the pattern above, when applied to a given class, makes it equivalent to the class of all the individuals whose fillers, for the property `contains`, are all in the complement of the class variable `?forbiddenContent`. This variable is 'generated'–this means that its values will not be provided by the users in the instantiation, but will come from other variables/entities in the pattern. In our case, this variable is the disjunction of all the values assumed by the `CLASS` variable `?x`. Now suppose we instantiate this pattern, assigning `Meat` and `Egg` to `?x`, and apply it to `VegetarianCourse`. The ontology, after the instantiation will include the following axioms:

`VegetarianCourse equivalentTo contains only (not (Egg or Meat))`

Although kept simple for the sake of clarity, this example shows the potential of patterns and the advantages of having a language for their expression. The only thing the knowledge engineer had to specify was a single pattern. After that, enriching the ontology with all the plausible combinations of food content one may need does not require an understanding of all of the underlying model. All that needs to be done is the assignment of the desired values to the (in this case) unique input variable.

The return clause, however, makes the above pattern suitable to be referenced. This means that other patterns can refer to it in their actions. For instance, let us say we want to specify a pattern for creating `Menu` sub-classes that do not contain a certain food. We can build that pattern upon the pattern already created, using the syntax in Figure 4.

```
?x:CLASS[subClassOf Food]
BEGIN
ADD $thisClass subClassOf Menu,
ADD $thisClass subClassOf contains Course and only ($FreeFromPattern(?x))
END;
A ?x - free Menu
```

Fig. 4. Pattern on Menu reusing the pattern in Figure 3, FreeFromPattern

Let us suppose we want to create a `VegetarianMenu` class and instantiate it using again `Egg` and `Meat` as values for our input variable `?x`.

The following axioms would be added to the ontology on instantiation:

```
VegetarianMenu subClassOf Menu
VegetarianMenu subClassOf contains Course and only
  (contains only not (Egg or Meat))
```

In other words, an inner pattern instantiation retrieves all the class equivalence axioms that have the return variable as member and in place of the reference in the outer pattern[5] it puts the conjunction of all the equivalent classes that appear in such axioms[6].

2.2 Embedding Patterns into OWL Ontologies

This section briefly describes our framework that allows the embedding patterns, specified using OPPL, and their instantiations into ontologies. We will illustrate it using screenshots from a tool we developed as a plug-in for Protégé 4 (`http://www.cs.man.ac.uk/~iannonel/oppl/patterns`). This tool allows the creation, editing and instantiation of such patterns. This therefore facilitates the maintenance of an ontology when using Protégé 4. Trivial as it may appear in the examples above, we provided a way to encapsulate the semantics of being *something which does not contain* `?x`, in the pattern described in Figure 3. Should the way of modeling that change in the ontology, knowledge engineers would need to maintain only the pattern and the framework should update the referencing entities accordingly.

There are three main pieces of information that need to be stored into an ontology in order to enable the use of OPPL patterns, they are:

1. Pattern definitions, i.e.: the OPPL specifications, rendering, and, optionally, return clauses described in the previous section;
2. Pattern instantiations, i.e: the set of values assigned to each variable when using a pattern;
3. Pattern generated effects, i.e.: the reference to the creating pattern for each axiom generated by a pattern instantiation.

All three pieces of information are stored in different annotations. Pattern definitions and pattern instantiations are serialized into ontology annotations, unless the pattern to be instantiated uses the reserved `$thisClass` variable (pattern applicable to classes). If `$thisClass` is used, the pattern instantiation refers to a particular class in the ontology—therefore, it is stored as an annotation for that class.

Finally, every axiom added by a pattern instantiation is annotated with enough information to reference what instantiation caused the assertion of such an axiom. An example of the resulting OWL serialization of the pattern in Figure 3, its instantiation and the corresponding annotated axiom is reported in Figure 5.

[5] Currently, in our language only class variables can be returned.

[6] In our case, there is only one axiom and one equivalent class in the pattern in Figure 3.

```
[...]
    <owl:Ontology rdf:about="">
        <patterns:Free rdf:datatype="&xsd;string"
            >?x:CLASS,?forbiddenContent:CLASS=createUnion(?x.VALUES)
            BEGIN
             ADD $thisClass equivalentTo contains only (not ?forbiddenContent )
 END;
 A ?x free stuff ;
 RETURN $thisClass
  </patterns:Free>
    </owl:Ontology>

[...]
    <!-- http://www.coode.org/patterns/examples/food#VegetarianCourse -->

    <owl:Class rdf:about="#VegetarianCourse">
        <rdfs:subClassOf rdf:resource="#Course"/>
        <patterns:VegetarianCourseFreePatternInstantiation rdf:datatype="&xsd;string">
         $Free(Meat)
        </patterns:VegetarianCourseFreePatternInstantiation>
    </owl:Class>
    <rdf:Description>
        <rdf:type rdf:resource="&owl;Axiom"/>
        <rdf:subject rdf:resource="#VegetarianCourse"/>
        <rdf:predicate rdf:resource="&owl;equivalentClass"/>
        <rdf:object>
            <owl:Restriction>
                <owl:onProperty rdf:resource="#contains"/>
                <owl:allValuesFrom>
                    <owl:Class>
                        <owl:complementOf rdf:resource="#Meat"/>
                    </owl:Class>
                </owl:allValuesFrom>
            </owl:Restriction>
        </rdf:object>
        <patterns:createdBy rdf:datatype="&xsd;string">
        http://www.co-ode.org/patterns#VegetarianCourseFreePatternInstantiation
        </patterns:createdBy>
    </rdf:Description>
```

Fig. 5. Pattern specification, instantiation and storage effects in OWL excerpts

Methodologies have recently been proposed to create and reuse knowledge patterns. In particular, in [1] (see Section 3), a *common set of operations* for creating and using Ontology Content Patterns (OCPs) has been described. An ontology content pattern, is, informally, a knowledge pattern that, although domain specific, solves recurrent common modeling problems. In the same paper, the authors illustrate the creation and the usage of patterns by means of two examples extracted from the Dolce Ultra Light ontology (DUL[7]): namely the *Information Realization* and the *Time Indexed Person Role* content patterns. The proposed extraction process produces entries for an online catalogue to be browsed by knowledge engineers. The typical entry of such a catalogue looks like a form with fields specifying:

[7] http://www.loa-cnr.it/ontologies/DUL.owl

- Name;
- Intent;
- Extracted from;
- Examples;
- Diagram;
- Elements.

There seems, however, not to be a concrete representation of the captured pattern so that it could be reused *off the shelf*. We claim that besides the information already provided by catalogues of patterns, encoding the patterns in OPPL could further enhance their usability. Currently the instantiation technique suggested by the authors relies on cloning (partially, or completely) entities from the pattern origin ontology and constructing the resulting axioms using SPARQL [8]. The authors acknowledge that the problem of cloning has not been fully addressed so far and acknowledge it may happen that the ontology engineers have to manually edit SPARQL results. Using OPPL allows the expression of patterns in the same ontological language and therefore avoids the need for custom SPARQL adjustments as the patterns creators can specify which axioms will correspond to a pattern instantiation.

Let us consider, now, the first OCP, *Information Realization*. In a nutshell, the DUL ontology says that:

> [Information Realization] represents the relations between information objects like poems, songs, formulas, etc., and their physical realizations like printed books, registered tracks, physical les, etc.

In Figure 6 we report the OPPL encoding of this pattern. Notice that this pattern when instantiated in an ontology that imports the DUL will assert the opportune sub-class relationships between the variables and the classes/properties in the DUL ontology (i.e: `InformationRealization`, `InformationObject`, and `realizationProperty`) and then the required axiom that relates the information object to its realization. The advantage of this is that it can be directly stored in the ontology, and applied, just assigning values to the three variables. Once validated by the ontology engineer, this pattern can be routinely instantiated to produce as many `InformationRealization` sub-classes as required—all its users have to know is what values they should assign to the variables.

Likewise, we can consider the other example in the paper referenced above: *Time Indexed Person Role* described as:

> a CP that represents time indexing for the relation between persons and roles they play

Looking at the ontology, a *Time Indexed Person Role* is a sub-class of `Situation` whose settings include a sub-class of `Person`, of `Role`, and of `TimeInterval`. A generic pattern for *Time Indexed Person Role* could be the one shown in Figure 7. Notice that there is no particular reason to interpret the

[8] `http://www.w3.org/TR/rdf-sparql-query/`

```
?informationObject:CLASS,
?informationRealization:CLASS,
?realizationProperty:OBJECTPROPERTY
BEGIN
ADD ?informationRealization subClassOf InformationRealization,
ADD ?informationObject subClassOf InformationObject,
ADD ?realizationProperty subPropertyOf realizes,
ADD ?informationRealization  subClassOf PhysicalObject
     and ?realizationProperty some ?InformationObject
END;
Information Realization Pattern:
?informationRealization ?realizationProperty ?informationObject
```

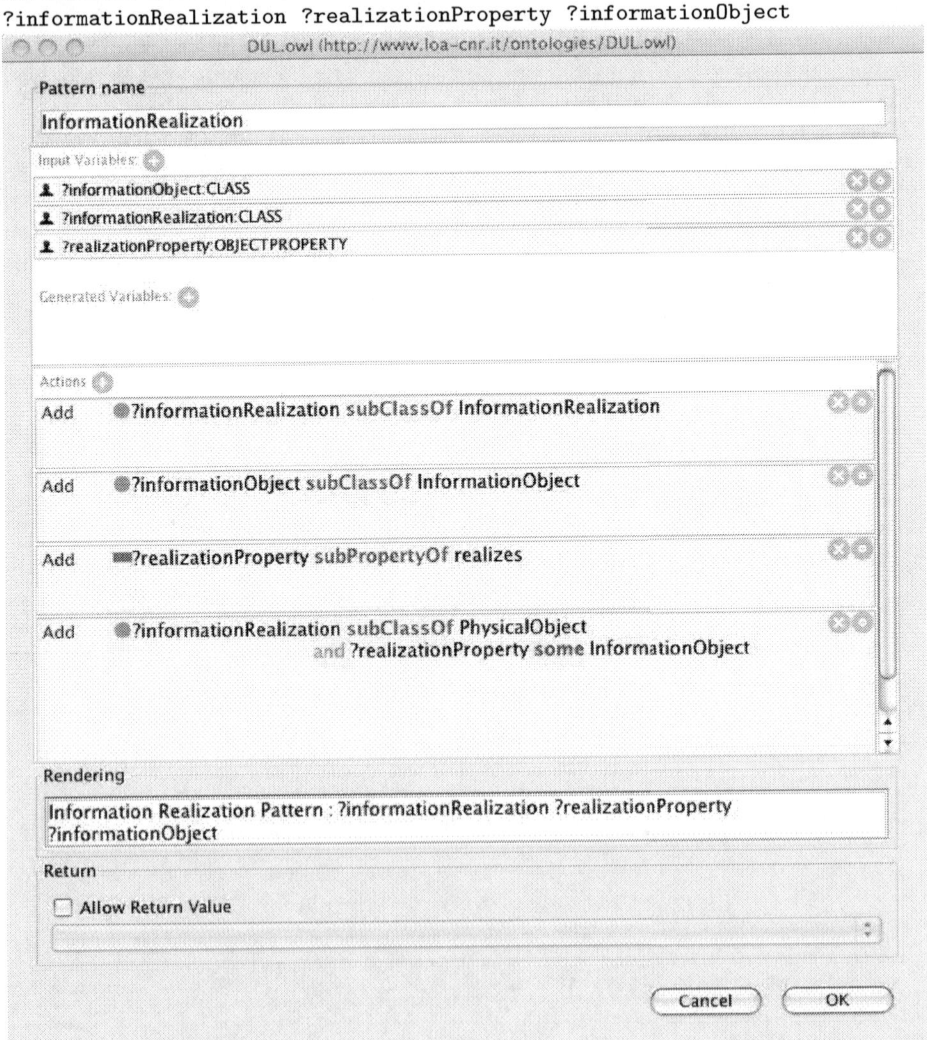

Fig. 6. Information Object Pattern in OPPL

latter pattern as a class, the purpose was only to show an alternative to the plain, class independent implementation of the previous *Information Realization* example. There is, however, another important difference between the two exemplar patterns: their level of generality. In the former both the classes and the property involved have been replaced by variables. In the latter a fixed reference to the property `isSettingFor` appears. The introduction of a `OBJECTPROPERTY` variable that replaces such property would produce a more general pattern. Conversely, the elimination of a variable and its replacement with an entity would result in a more specific version of the *Information Realization Pattern* described above. These two operations correspond to implementing in OPPL what, in [1], is meant by pattern *generalization* and *specialization*.

3 Related Work

The literature about knowledge patterns, even when restricted to the work specifically aimed at Semantic Web formalisms, is vast. Therefore, the survey in this Section is not intended to be exhaustive. Its aim is rather to identify where our contribution stands with respect to the state-of-the-art. Clark et *al.* were the first, to the best of our knowledge, to use the label *knowledge patterns*, defining them as:

> first order theories whose axioms are not part of the target knowledge-base, but can be incorporate via renaming of their non logical symbols (see Section 2 in [3]).

Explicit functions called *morphisms* were required to use patterns in knowledge bases. In the following years, researchers seemed to focus on pattern categorisation (see, for instance, [10] and its references) aiming at creating organised catalogues that engineers could browse and adopt in their knowledge bases. Recently, patterns have also been employed in semi-automatic knowledge acquisition tasks. In [11], they are combined with Case-Based Reasoning in order to guide the extraction of terminological knowledge from text documents. However, as far as we are aware, in the last decade, we can cite only two works that investigate the problem we discussed in this paper: pattern representation languages. Staab *et al.*, in [12], propose a framework based on RDF for creating catalogues of patterns that are (partially - according to the authors themselves) *executable/portable* in any implementation language. Vrandecic, in [13], proposes the creation of scripts (called *macros* in the paper) that, without changing the semantics of the underlying knowledge representation language – OWL-DL in this case – capture sets of axioms and can be reused. Our approach is less ambitious than Staab *et al.*'s as our pattern specification language is targeted to cover OWL-DL ontologies only. Therefore, in Staab *et al.*'s framework, ours could be the component responsible for translating a semantic pattern into OWL-DL.

With respect to Vrandecic's work, by contrast, our framework seems a step forward, providing a concrete language for encoding patterns, although we differ in the interpretation of in what way a macro (pattern) should be dealt. Vrandecic hypothesizes that patterns should be encoded into the ontology, then interpreted

```
?person:CLASS,
?role:CLASS,
?timeInterval:CLASS
BEGIN
ADD $thisClass subClassOf Situation,
ADD $thisClass subClassOf isSettingFor some ?person,
ADD $thisClass subClassOf isSettingFor some ?role,
ADD $thisClass subClassOf isSettingFor some ?timeInterval,
END;
Situation where ?person play the role ?role during the time interval ?timeInterval
```

Fig. 7. Time Indexed Person Role Pattern

and resolved using external tools (maybe rule or template based) every time an ontology is opened to be sent to any reasoning service. On the contrary, we believe that patterns should be instantiated and references resolved as soon as they are used in the ontology.

As we explained above, when a pattern is instantiated, our framework will carry out the resulting actions on the ontology, and there will be no need of interpreters after that point.

4 Conclusions and Future Work

In this paper we used OPPL as a means of encoding patterns. OPPL patterns are a set of operations to be executed on the ontology to which they are applied. Such operations are specified in terms of the axioms to be added/removed with variables that have to be instantiated, assigning values to each. Having a language and a concrete framework to encode patterns into ontologies has the added value of enabling an ontology engineer to specify both what could be extended in an ontology and how this could be accomplished. Patterns, besides representing abstractions over recurrent modeling problems, could also function as ontology extension points. Knowledge engineers could systematically create patterns that indicate the way they foresee further axiomatizations in the ontology. this would implement a sort if *interface* - in its software engineering accepted meaning - for ontology expansion and at the same time hiding modeling details.

Furthermore, a declarative language for encoding patterns makes it possible to explore its expressivity limitations as well as the effects that certain kinds of patterns could have when instantiated within an ontology. Promising research results on the evaluation of the impact of ontology modifications [14,15] could be reused and applied to assess the consequences of the instantiation of a pattern. Such analysis could be automated and embedded into knowledge engineering tools that will provide more information, at a higher level of abstraction, to their users interested in maintaining an ontology exposing patterns.

Introducing a pattern language opens another interesting direction: pattern matching and induction. Matching is meant here as deciding whether either a given ontology or its part is compliant (*matches*) with a given pattern. Pattern induction, in contrast, is intended as detecting the regularities in an ontology, seeking recurring patterns. The application of Machine Learning techniques to study the complexity of the induction for this language is certainly worth investigating.

Last, but not least, its adoption as a language for patterns will contribute in steering the further developments of the OPPL language itself. OPPL was born as a scripting language to automate bulk modifications to ontologies. The language creators main concern was to maintain its decidability keeping a very parsimonious attitude with respect to the set of possible language constructs. It is likely that patterns will push in the opposite direction, demanding more and more expressivity. Reaching a useful trade-off will be an interesting research topic in itself. We therefore see this new version of OPPL, with its ability to declare knowledge patterns and instantiate them within an ontology, as the starting point for supporting the sustainable development of richly axiomatised ontologies.

References

1. Presutti, V., Gangemi, A.: Content ontology design patterns as practical building blocks for web ontologies. In: Li, Q., Spaccapietra, S., Yu, E., Olivé, A. (eds.) ER 2008. LNCS, vol. 5231, pp. 128–141. Springer, Heidelberg (2008)

2. Horridge, M., Parsia, B., Sattler, U.: Laconic and precise justifications in owl. In: Sheth, A.P., Staab, S., Dean, M., Paolucci, M., Maynard, D., Finin, T.W., Thirunarayan, K. (eds.) International Semantic Web Conference. LNCS, vol. 5318, pp. 323–338. Springer, Heidelberg (2008)
3. Clark, P., Thompson, J., Porter, B.W.: Knowledge patterns. In: KR, pp. 591–600 (2000)
4. Clark, P.: Knowledge patterns. In: [16], pp. 1–3
5. Gangemi, A.: Ontology design patterns for semantic web content. In: Gil, Y., Motta, E., Benjamins, V.R., Musen, M.A. (eds.) ISWC 2005. LNCS, vol. 3729, pp. 262–276. Springer, Heidelberg (2005)
6. Iannone, L., Egaña, M., Rector, A., Stevens, R.: Augmenting the Expressivity of the Ontology Pre-Processor Language (2008), http://www.webont.org/owled/2008/papers/owled2008eu_submission_16.pdf
7. Egaña, M., Antezana, E., Stevens, R.: Transforming the Axiomisation of Ontologies: The Ontology Pre-Processor Language. In: Proceedigns of OWLED 2008 DC OWL: Experiences and Directions, Washington, DC (2008)
8. Egaña, M., Rector, A.L., Stevens, R., Antezana, E.: Applying ontology design patterns in bio-ontologies. In: [16], pp. 7–16
9. Horridge, M., Drummond, N., Godwin, J., Rector, A., Stevens, R., Wang, H.: The Manchester OWL Syntax. In: Proceedigns of OWLED 2006 OWL: Experiences and Directions, Athens GA, USA (2006)
10. Blomqvist, E., Sandkuhl, K.: Patterns in ontology engineering: Classification of ontology patterns. In: Chen, C.S., Filipe, J., Seruca, I., Cordeiro, J. (eds.) ICEIS (3), pp. 413–416 (2005)
11. Blomqvist, E.: Ontocase - a pattern-based ontology construction approach. In: Meersman, R., Tari, Z. (eds.) OTM 2007, Part I. LNCS, vol. 4803, pp. 971–988. Springer, Heidelberg (2007)
12. Staab, S., Erdmann, M., Maedche, A.: Engineering ontologies using semantic patterns. In: IJCAI Workshop on E-business & The Intelligent Web (2001)
13. Vrandecic, D.: Explicit knowledge engineering patterns with macros. In: Proceedings of the Ontology Patterns for the Semantic Web Workshop at the ISWC 2005, Galway, Ireland (November 2005)
14. Grau, B.C., Horrocks, I., Kutz, O., Sattler, U.: Will my ontologies fit together? In: Parsia, B., Sattler, U., Toman, D. (eds.) Description Logics. CEUR Workshop Proceedings, vol. 189, CEUR-WS.org (2006)
15. Grau, B.C., Horrocks, I., Kazakov, Y., Sattler, U.: Ontology reuse: Better safe than sorry. In: Calvanese, D., Franconi, E., Haarslev, V., Lembo, D., Motik, B., Turhan, A.Y., Tessaris, S. (eds.) Description Logics. CEUR Workshop Proceedings, vol. 250, CEUR-WS.org (2007)
16. Gangemi, A., Euzenat, J. (eds.): EKAW 2008. LNCS, vol. 5268. Springer, Heidelberg (2008)

A Core Ontology of Knowledge Acquisition

José Iria

Department of Computer Science, The University of Sheffield, UK
jiria@dcs.shef.ac.uk

Abstract. Semantic descriptions of knowledge acquisition (KA) tools and resources enable machine reasoning about KA systems and can be used to automate the discovery and composition of KA services, thereby increasing interoperability among systems and reducing system design and maintenance costs. Whilst there are a few general-purpose ontologies available that could be combined for describing knowledge acquisition, albeit at an inadequate abstraction level, there is as yet no KA ontology based on Semantic Web technologies available. In this paper, we present OAK, a well-founded, modular, extensible and multimedia-aware ontology of knowledge acquisition which extends existing foundational and core Semantic Web ontologies. We start by using a KA tool development scenario to illustrate the complexity of the problem, and identify a number of requirements for OAK. After we present the ontology in detail, we evaluate it with respect to the identified requirements.

1 Introduction

The goal of Knowledge Acquisition (KA) is to develop methods and tools that make the arduous task of capturing and validating an expert's knowledge as efficient and effective as possible. Of special relevance to the fulfillment of the Semantic Web vision is automating KA from text and image resources. Automated KA systems take as input multimedia documents originally intended for human consumption only and provide as output knowledge that machines can reason about.

Despite the growing need for KA and recent advances in the field, the development of KA systems remains a complex and costly affair. Previous research on KA for the Semantic Web has followed three main directions: 1) increasing the accuracy of the extracted knowledge [9], 2) scaling up KA systems to address the knowledge acquisition bottleneck in populating large knowledge bases [12], and 3) providing semantically rich and well-founded means of representing multimedia data and annotations thereof [1]. Unfortunately, none of these research directions addresses the core problem of reducing the complexity in developing and maintaining these systems.

Research on Service-Oriented Architectures (SOA) addresses this problem. Frequently realized through (semantic) Web services concepts and technologies, SOA encourage following modularization principles for dealing with the complexity of software systems. Ontological descriptions of services serve as a basis

L. Aroyo et al. (Eds.): ESWC 2009, LNCS 5554, pp. 233–247, 2009.

for interoperability between systems at various levels of abstraction, and this enables important functionalities such as (automatic) semantic service discovery and composition. In this context, ontologies like OWL-S [8] and the Core Software Ontology [11] provide generic top-level constructs to formally describe software systems. However, to the best of our knowledge, there is currently no specialized ontology available for semantically describing KA tasks and systems.

The advantages of an ontology of KA are threefold. Firstly, it provides an agreed-upon means of referring to and describing KA tasks and KA systems, their inputs and outputs and their internal constituents (components, subsystems, auxiliary resources, and so on). Secondly, making such descriptions available greatly increases the interoperability between KA systems, by enabling both humans and machines to search for and reason upon them. Lastly, it cuts down system development and maintenance costs, and promotes optimal decisions both at design time and at runtime, because systems, subsystems and components of a KA system can more easily be found and put together.

Our contribution in this paper is thus to provide a core Ontology of Acquisition of Knowledge (OAK). In the design of OAK we have reused and extended existing foundational and core Semantic Web ontologies. Concretely, we have used DOLCE [5], the ontology of Descriptions & Situations, the Ontology of Plans, and the Ontology of Information Objects; and the core ontologies of multimedia (COMM) [1] and software (CSO) [11]. The relationship between OAK and these other ontologies is depicted in Figure 1.

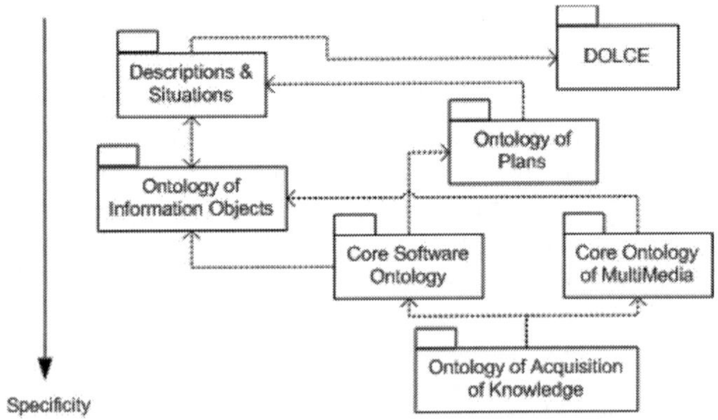

Fig. 1. Positioning the Ontology of Acquisition of Knowledge (OAK) in the framework of existing foundational and core ontologies. OAK extends the Core Software Ontology (CSO) and the Core Ontology of MultiMedia (COMM).

The rest of the paper is structured as follows. In the next section, we motivate the need for semantically describing KA systems. We center our discussion on a simple KA task to populate a knowledge base. In section 3, we review related work. In section 4, we identify a number of requirements that an ontology of knowledge acquisition should satisfy. We then present OAK using a collection of

UML diagrams and describe it in detail. In section 6, we show how the simple KA task is encoded with the proposed ontology. Finally, we present our conclusions and outline future work.

2 Developing a Simple Knowledge Acquisition System

Let us imagine that a team of developers is working on a new KA system. The system's goal is to extract facts about sports celebrities from online multimedia documents. Available to the developers is a set of sample facts extracted from a few hundred documents that can be used as training examples in a machine learning-based approach to the problem. The system should make use of both text and image features, as both are potentially valuable – for example, the presence of images about sports increase the likelihood that the surrounding text contains the facts of interest (expressed in natural language).

Developing a new KA system involves a number of decisions regarding subsystems and components to use. Even the development of a simple system that works in the above scenario requires addressing at least the following concerns:

Decomposition: Both text and images need to be decomposed into finer-grained media segments, which constitute the building blocks for deriving Information Extraction (IE) patterns and models, be it manually or via induction algorithms. Text is typically decomposed into sentences, noun phrases and tokens. Images are typically segmented into regions of interest.

Tagging/Annotation: Media segments need to be tagged or annotated, that is, extra information, obtained through some form of analysis and/or use of external resources, is attached to the segment, with the purpose of enriching IE patterns and models. Typical text tagging tools are part-of-speech and orthography taggers, which attach tags at the token level. Typical image analysis tools are color and texture analyzers, and edge detectors.

Data Modeling: A predictive model, capable of extracting facts from the decomposed and tagged input media, needs to be constructed, either manually or through automatic induction methods. Manually built models consist mainly of text patterns, carefully created, tested and maintained by domain and linguistic experts. Induced models are created by machine learning-based systems, which tap into a wealth of machine learning literature for a choice of algorithms and meta-algorithms, and are able to work on both text and images.

Semantic Annotation: Models need to be applied over unseen media to extract novel facts, typically expressed in the form of semantic annotations. The application of predictive models may sometimes not directly yield the facts of interest, but some form of validation, consistency checking, and merging of intermediate facts is required as a last step. This is typically achieved through reasoning on the output of the models.

2.1 Problems with Existing Solutions

State-of-the art middleware for the development of IE and KA systems do not offer facilities for semantically describing the subsystems and components that make a KA system, nor is there a specialized ontology of KA available that can be used to do that (see Section 3). This means that the developers in our example will find it hard to integrate the required tools and components obtained from several parties. Furthermore, they will not be able to publish the semantic descriptions of their system, to ease the effort by third parties in finding and using them in the future. Given the sheer number and diversity of subsystems involved, it would be desirable to automate service discovery and composition, not only to speed up development, but also to improve systemic qualities such as robustness to failure of one of the subsystems.

Our developers will use the XML plugin descriptor files or equivalent mechanisms (e.g., Java language annotations) which are offered by the current middleware solutions. Unfortunately, these descriptors present several problems:

Scope: The descriptors tend to mimic the native programming language, thus describing the components in terms of their inputs and outputs, and design time and runtime parameters. There is seldomly a way of describing aspects of the KA subsystems and components that fall outside the scope of those constructs. For instance, it would prove difficult to represent that the output of the component is a relationship between a given decomposition of text and images and a statement in the knowledge base.

Semantics: The descriptors lack formal semantics. While appropriate for describing the components within the context of their respective middleware, it is not possible to guarantee that the descriptors generated by different agents will be mutually understood – potentially, the same descriptor syntax can be used to mean a number of semantically distinct things.

Web Interoperability: The descriptors are not interoperable with existing (semantic) web standards. For example, there is no formal way of referring to a concept in a domain-specific ontology. This limits the interoperability with other web systems.

3 Related Work

In recent years, new directions in software engineering have started to be explored to combat the increasing complexity and rapid rate of change in modern systems development. Among these new paradigms is Semantic Web Enabled Software Engineering (SWESE), which tries to apply Semantic Web technologies (such as ontologies and reasoners) in mainstream software engineering [7]. SWESE hopes to provide stronger logical foundations and precise semantics for software models and other development artifacts. The work presented in this paper can be viewed as a contribution to this goal, with applications in the domain of knowledge acquisition systems engineering.

There are currently several state-of-the-art software frameworks available that are suitable for developing tools for knowledge acquisition from text and/or multimedia documents. The Unstructured Information Management Architecture (UIMA) [4] is an open-source platform for integrating components that analyze unstructured sources such as multimedia documents. UIMA-based systems define type systems (i.e., ontologies with extremely limited semantic commitments) to specify the kinds of information that they manipulate [6]. UIMA type systems include no more than a single-inheritance type/subtype hierarchy, thus to do substantive reasoning over the results of UIMA-based extraction, one needs to convert results into a more expressive representation. GATE [2] is a framework and graphical development environment for NLP tools and applications that has been extended over the years to include support for ontologies, multimedia data, and machine learning. Unlike UIMA, GATE does not provide a specialized data layer, using instead native data structures (the current version of GATE is implemented in Java). It features a plugin-based architecture in which components are described using so-called CREOLE (XML) descriptor files, which suffer from the problems identified in the previous section. In our survey of the state-of-the-art, we found that several other frameworks, e.g. [15], [16], presented identical limitations. In face of this, we believe that the formalization of the knowledge acquisition domain presented here is an important first step in the direction of interoperability between systems built on top of these middleware platforms.

Early research approaches to using ontologies to model the knowledge acquisition problem, e.g. [13], would focus on the interaction between a KA user interface and the domain experts. The work in this paper belongs to the class of more recent approaches that attempt to automate KA from documents written by the experts.

Our work is targeted at the fields of Knowledge Acquisition and Ontology-based Information Extraction [14]. The design of OAK was greatly inspired by working with a number of systems in these fields and by designing our own systems. For example, the system in [3] takes the approach that a large class of data sources on the web can be viewed as semantically related in the context of a given knowledge acquisition task, and design a general strategy for learning classifiers from those sources. This and other systems, such as those described in [10], [9] and [12], which at aim populating small ontologies from (sometimes very large) repositories of text or multimedia data, are the kinds of systems that would benefit from our work. None of these systems have been explicitly semantically described, and therefore machine reasoning about their functionality is not possible – and it is not uncommon for even humans to have difficulty in replicating the systems from their natural language descriptions found in the literature.

Other, more general, ontologies have been developed which could be used together to describe KA tasks and systems, albeit at an inadequate abstraction level. The Core Ontology of MultiMedia [1] is a well-founded ontology and API that provides constructs to semantically describe multimedia assets available on the Web. The Core Ontology of Software [11] formalizes the most fundamental concepts which are required to model both software components and Web services, including concepts such as software, data, users, access rights or interfaces. Roughly speaking, COMM covers the data part of our ontology, while CSO covers the processing part. We have extended both ontologies and provided more specialized semantic constructs for the domain of automated KA from multimedia data.

4 Requirements for the Knowledge Acquisition Ontology

Here we compile a list of requirements that an ontology of knowledge acquisition should satisfy, given our discussion in the previous sections:

Multimedia: The proposed ontology should describe KA systems that work with multimedia data. In particular, it should support text and images, and be designed in such a way that it can be extended to other types of media as well.

Semantic Interoperability: An ontology of KA, much like any other ontology, must ensure that the intended meaning of the captured semantics can be shared among different systems. Reasoning processes about concepts and relations in different environments can only be guaranteed to yield identical results if the semantics is sufficiently explicitly described. Only when the captured semantics can be shared among multiple systems and applications are the KA system descriptions truly re-usable.

Syntactic Interoperability: The semantics of the KA system descriptions are only shareable among different systems if there is some agreed-upon syntax in which to convey it. In our case, the descriptions should be expressed in a semantic web language, such as OWL or RDF/XML, because our goal is to take advantage of other Semantic Web technologies such as semantic service discovery and composition.

Separation of Concerns: Domain knowledge should be kept separate from knowledge about the KA system. Moreover, as indicated in section 2, the development of a KA system should address at least the concerns of decomposition, tagging, data modeling, and semantic annotation. These should also be kept separate, if possible, in the design of the ontology, in form of patterns.

Modularity: Modularity is a key engineering principle which arises whenever dealing with large systems, be it software systems or ontologies. Since the ontology of knowledge acquisition can become very large, its design should be made modular from its inception.

Extensibility: Just like we have extended COMM and CSO in designing our KA ontology, we also expect OAK to be eventually extended and adapted to specific domains and applications. As ontology development methodologies show, ontologies are inherently incomplete, and for that reason extensibility is a key and pervasive requirement in ontology engineering.

5 The OAK Ontology

In this section we present OAK, a core ontology of knowledge acquisition from multimedia data. The ontology is organized into several patterns, described below with the aid of UML diagrams.

5.1 Foundational and Core Ontologies Used

The Descriptive Ontology for Linguistic and Cognitive Engineering (DOLCE) [5] is used as a modeling basis. DOLCE is a well-founded ontology which models three ontological patterns which we require: the Descriptions & Situations (D&S) ontology, the Ontology of Information Objects (OIO) and the Ontology of Plans (OoP). The first pattern is used to formalize contextual knowledge. The second pattern implements a semiotics model of communication theory. The third pattern characterizes planning concepts.

As previously mentioned, we extend the COMM and CSO core ontologies. Please consult Section 3 for the references to the publications which describe them, a reading of which may prove necessary to understand some of the concepts discussed in this section.[1]

5.2 KA Task and KA System

We start by characterizing core KA concepts by answering the questions: What is a knowledge acquisition task? What is a knowledge acquisition system? What is a knowledge base (KB) statement? We define these concepts in terms of the Descriptions & Situations pattern, which contextualizes an interplay of COMM and CSO concepts. This is depicted in Figure 2.

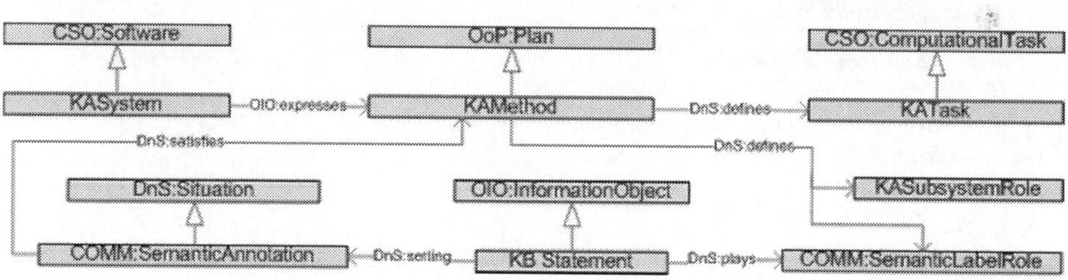

Fig. 2. Ontology pattern for describing a knowledge acquisition system

A `KASystem` is a `CSO:Software` that expresses a `OoP:Plan`, namely that of a `KAMethod`. A `KATask` is a `CSO:ComputationalTask` defined by the method. A statement in a knowledge base, `KBStatement`, is a `DnS:Role`,

[1] Throughout the paper, concepts and associations are written in `sans serif` and are labelled in a namespace-like manner. Namespace prefixes indicate the ontology where those concepts and associations are defined. If no namespace is used, they are assumed to be defined in OAK.

namely the COMM:SemanticLabelRole, played by an OIO:InformationObject. In other words, the statement is an information object expressing a fact that provides a semantic label to some media segment – the segment which contributed to that fact being extracted or derived. A KAMethod is a DnS:Description of a situation. In the KA domain, the situation it describes is a COMM:SemanticAnnotation. This provides a DnS:setting where the statements and the CSO:ComputationalActivity sequenced by the KATask (not shown in the figure) exist.

Additionally, the KAMethod defines the KASubsystemRole, the meaning of which will become clear in the next section.

5.3 KA Subsystem

Knowledge acquisition systems are complex systems composed of many subsystems. Our ontology should capture these as well. This is depicted in Figure 3, for natural language processing (NLP) tools, arguably the most commonly used type of KA subsystem.

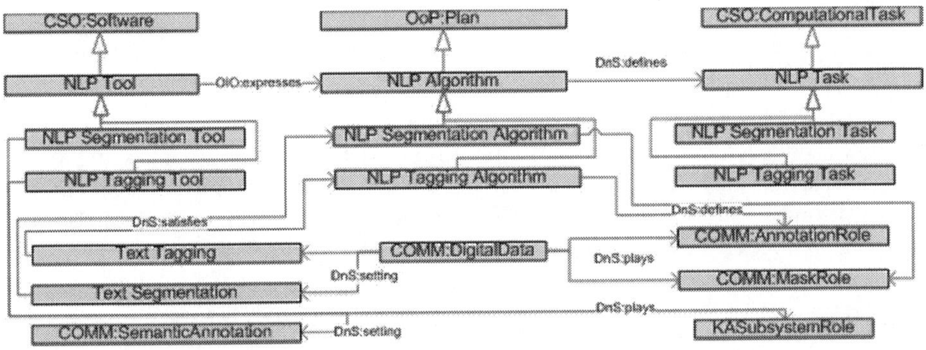

Fig. 3. Ontology pattern for describing one of the possible types of KA subsystems: NLP tools

A NLPTool is a software that OIO:expresses a NLPAlgorithm which DnS:defines an NLPTask. Following COMM patterns of decomposition and annotation, NLP tools can be specialized into segmentation and tagging tools. Examples of the former include the sentence splitter, the chunker and the tokenizer, while examples of the latter include tools such as the part-of-speech tagger or the orthography tagger. Both NLPTaggingTools and NLPSegmentationTools play the KASubsystemRole in the setting of a COMM:SemanticAnnotation (defined above). A NLPSegmentationTool expresses a NLPSegmentationAlgorithm that defines a NLPSegmentationTask and the COMM:MaskRole, which is played by some COMM:DigitalData in the setting of a TextSegmentation situation satisfied by the algorithm. Conversely, a NLPTaggingTool expresses a NLPTaggingAlgorithm which defines a NLPTaggingTask and the

COMM:AnnotationRole, played by some COMM:DigitalData in the setting of a TextTagging situation. The TextSegmentation and TextTagging situations are specializations of the COMM:SegmentDecomposition and COMM:Annotation situations, respectively (see Figure 4).

Many other tools may form part of a KA system. Of particular relevance to the simple KA system described in Section 2 are image analysis tools, which, together with NLP tools, enable handling multimedia documents. Due to space constraints, we do not show figures about these tools, but it should be mentioned that their characterization follows a similar pattern. Thus, a IATool is a software that OIO:expresses a IAAlgorithm which DnS:defines an IATask. We define two kinds of image analysis tools, the IADecompositionTool, e.g., a region of interest classifier, and the IAAnnotationTool, e.g., an edge detector . Like NLP tools, both play a KASubsystemRole in the setting of a COMM:SemanticAnnotation.

5.4 Decomposition

The subsystems of a KA system are typically regarded as blackboxes, and this lack of detail makes it impossible for machines to reason about them. Hence, we need semantic constructs to declaratively specify the data resources and internal processes used by these subsystems. Here, we specialize a number of generic COMM constructs to the KA domain. This is depicted in Figure 4.

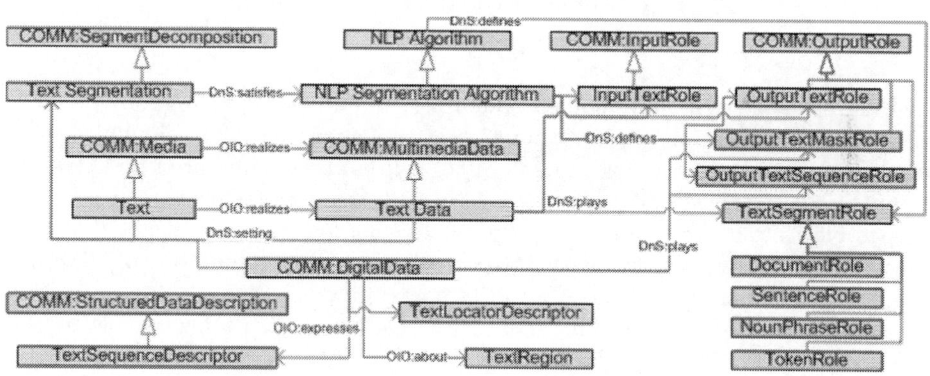

Fig. 4. Specializing the COMM decomposition pattern to the KA domain

Text is a type of COMM:Media that realizes TextData (a specialization of COMM:MultimediaData). A NLPAlgorithm defines a TextSegmentRole which TextData can play in the setting of some situation. In the case of the TextSegmentation situation depicted, TextData also plays the InputTextRole and OutputTextRole roles defined by the NLPSegmentationAlgorithm. In other words, the segmentation algorithm can, for example, take a document and split it into sentences (sentence splitter), or it can take a sentence and split it into tokens (sentence-level tokenizer), or it can take a document and split it into tokens (document-level tokenizer), and so on. Two other types of outputs are defined by a segmentation algorithm: the `OutputTextMaskRole` and the `OutputTextSequenceRole`. The former is played by a `COMM:DigitalData` object that expresses a `TextLocatorDescriptor` and is about some `TextRegion`, e.g. a descriptor about a sentence that provides a means of locating it in the document. The latter is played by a `COMM:DigitalData` object that expresses a `TextSequenceDescriptor`, e.g. a descriptor about the order in which tokens appear in the document[2].

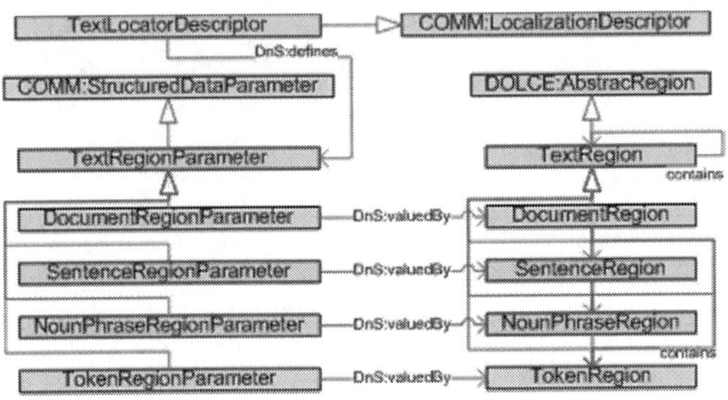

Fig. 5. Specializing COMM locators for text data

Figure 5 shows the `TextLocatorDescriptor` and the `TextRegion`. The former is a specialization of `COMM:LocalizationDescriptor` for text data. This descriptor enables specifying regions in the `TextData` manipulated by the algorithm. It defines a number of types of `TextRegionParameters` valued by their respective `TextRegions`. This is by no means an exhaustive list of possible subclasses, but merely illustrative of a pattern that suits a number of different concrete implementations and can easily be extended.

For the decomposition of image data, the constructs introduced in [1] by the authors of COMM are sufficient for the purposes of OAK, and thus do not need to be extended.

5.5 Annotation

Addressing the annotation concern requires a very similar pattern to the decomposition pattern. Hence, we just outline the differences with respect to the semantic constructs depicted in Figure 4.

[2] In practice, this descriptor is often expressed implicitly in the native programming language. For example, returning a `List` of `Token` objects in Java implicitly encodes the sequence information, since lists are ordered collections.

In a `TextTagging` situation satisfying a `NLPTaggingAlgorithm`, the latter can take as input and return as output `TextData` that play different `TextSegmentRoles`. For instance, a part-of-speech tagger tags tokens, while a sentiment classifier may tag sentences. A tagging algorithm defines a `InputTextRole` and a `OutputTextTagRole`. The latter is played by a structured `COMM:DigitalData` object that `OIO:expresses` a `TextTagDescriptor` (a subclass of `COMM:StructuredDataDescription`). The `COMM:DigitalData` object mentioned is `OIO:about` a `DOLCE:Particular`.

As with the decomposition pattern, we found that there was no need to extend the constructs introduced in [1] for the annotation of image data, as they are sufficient for the purposes of OAK. Moreover, for the same reasons we also directly adopt the semantic annotation pattern of COMM, as we find it requires no specialization in order to address the semantic annotation concern mentioned in Section 2.

5.6 Data Modeling

Another identified concern was that of managing models of the data. An ontology pattern that addresses this concern is depicted in Figure 6.

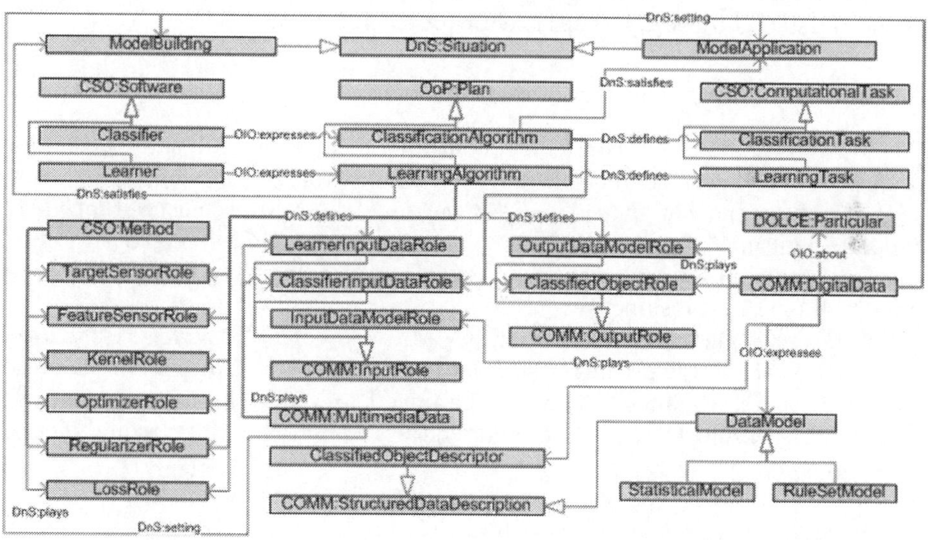

Fig. 6. Ontology pattern for describing building and applying data models

There are two situations to describe in data modeling: `ModelBuilding` and `ModelApplication`. A `Learner` is a `CSO:Software` that expresses a `LearningAlgorithm`, which defines a `LearningTask` and satisfies the `ModelBuilding` situation. Conversely, a `Classifier` is a `CSO:Software` that expresses a `ClassificationAlgorithm`, which defines a `ClassificationTask` and satisfies the `ModelApplication` situation.

For the purposes of KA from multimedia documents, a `LearningAlgorithm` takes `COMM:MultimediaData` as input and outputs a `DataModel` (expressed by some `COMM:DigitalData`). Data models are often automatically induced via machine learning methods, resulting in `StatisticalModels`, or often manually crafted by domain experts in the form of a set of rules, resulting in `RuleSetModels`. Learning algorithms tend to be highly configurable, and we tried to capture that by defining a number of roles for well-known basic functions in the machine learning literature (here semantically described by a `CSO:Method`). The `KernelRole` is played by a similarity metric or kernel function, e.g. cosine similarity, radial kernel. The `OptimizerRole` is played by an optimization algorithm, e.g. quasi-Newton method. The `RegularizerRole` is played by a regularization function used to keep model complexity low and prevent over-fitting. The `LossRole` is played by a loss function that defines the penalty of miss-prediction, e.g. hinge loss. The `FeatureSensorRole` is played by a function that extracts features from the data, while the `TargetSensorRole` is played by a function that determines the class which the data belongs to by consulting some oracle, e.g. manually annotated data.

A `ClassificationAlgorithm` takes `COMM:MultimediaData` and a `DataModel` as input, and outputs `COMM:DigitalData` that expresses a structured description of the classified object, a `ClassifiedObjectDescriptor`.

5.7 Fulfillment of the Requirements

We now discuss how the requirements outlined in Section 4 are satisfied by our proposed modeling of the knowledge acquisition ontology.

By carefully choosing to model OAK on top of COMM, we ensure that OAK supports multimedia data. We have shown how to specialize the core COMM constructs to support text. OAK can easily be extended to accommodate more types of media in the same way.

All the ontologies chosen – DOLCE, COMM and CSO – provide a rich axiomatization of each pattern using first order logic. Moreover, through the semantic annotation pattern, our ontology can be linked to any Web-based domain ontology. These fulfill the semantic and syntactic interoperability requirements, respectively.

The use of ontology patterns ensures a clear separation of concerns. The KA system and subsystem patterns define the core concepts. The decomposition pattern, the annotation pattern and the data modeling pattern each address the homonymous concerns mentioned in Section 2.

The modularity requirement is satisfied, since these patterns form modules in the core of the architecture of the KA ontology. The extensibility requirement is fulfilled in several ways. First, OAK allows accommodating further media types as mentioned above. Second, it is also straightforward to define new types of KA subsystems. Third, it is equally easy to define new types of segment roles and regions. Finally, new types of learning and classification algorithms can also be added without much effort. The modularity and extensibility of OAK is in great part due to our patterns being grounded in the D&S pattern, which enables adding further contextual knowledge in such a way that will not change the patterns (mainly by defining new roles or parameters), so that legacy descriptions remain valid.

6 Describing the Simple KA System with OAK

Let us revisit the scenario described in Section 2. The developers have decided to throw away any documents that do not contain sports pictures as a prior filtering step. On the remaining documents, a sentence splitter and a named entity recognizer are run to split each document into sentences and tag occurrences of people and location names. Each sentence is then passed to a classifier which generates statements of the type *born_in(person, location)* as output. Figure 7 shows how to semantically describe such a simple KA system.

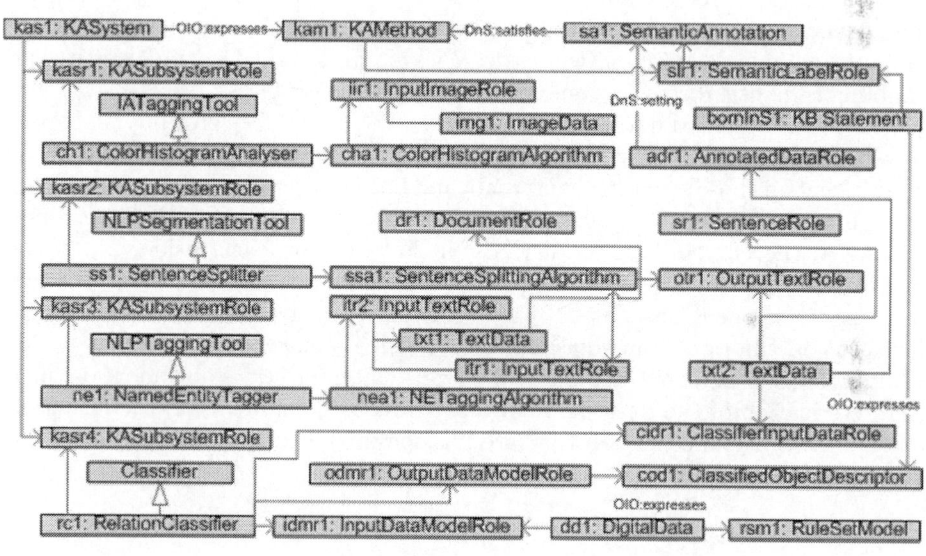

Fig. 7. Semantic description of a simple KA system that populates the KB with *born_in* relations. All the `DnS:plays`, `DnS:defines` and some `DnS:setting` and `OIO:expresses` labels were omitted for clarity. Due to space limitations, the outputs of the color histogram analyser and the named entity tagger were also omitted.

The system defines four subsystem roles, played by a color histogram analyser (an image analysis tagging tool), a sentence splitter (a nlp segmentation tool), a named entity tagger (a nlp tagging tool) and a relation classifier (a classifier). The image analyser takes the image data from the multimedia document and outputs a color histogram, which is used to decide whether the document should be discarded. The input to both the sentence splitter and the named entity tagger is the text from the whole document (hence playing a `DocumentRole`), whereas the output of the sentence splitter is the text from a sentence (hence playing a `SentenceRole`). The sentence text also plays the `ClassifierInputDataRole` defined by the relation classifier. The latter also takes a set of manually built text extraction rules as input (via the `InputDataModelRole`). The generated KB statements express the `ClassifiedObjectDescriptor` output by the classifier. Finally, the sentence text plays the `COMM:AnnotatedDataRole` and the KB statement the `COMM:SemanticLabelRole` in this `COMM:SemanticAnnotation` setting, which satisfies the KA method expressed by the system.

7 Conclusion and Future Work

In this paper we presented OAK, a well-founded, modular, extensible and multimedia-aware ontology of knowledge acquisition which extends existing foundational and core Semantic Web ontologies. The ontology addresses concrete problems with current state-of-the-art middleware for KA systems development. The advantages of its use include enhanced interoperability among systems, enabling machine reasoning on what the KA system does and how it does it, and enabling semantic discovery and composition of KA services.

The work presented in this paper just scratches the surface on what is possible to represent about KA systems. Nevertheless, as future work, we intend to keep on focusing on those semantic constructs and ontology patterns which are generic enough to constitute the core of KA. One particular pattern that we found lacking in the current version of OAK is that of knowledge fusion, i.e., to be able to describe the situation in which a KB statement is not directly related to a media segment (because it was extracted from it), but rather receives a indirect contribution from the segment towards its existence and validity (e.g., via merging of several extracted facts). We also intend to work on a more detailed characterization of the data modelling pattern, by including constructs such as dataset, instance and pattern that are missing in the current version.

Acknowledgments

This work was funded by the X-Media project (www.x-media-project.org) sponsored by the European Commission as part of the Information Society Technologies (IST) programme under EC grant number IST-FP6-026978.

References

1. Arndt, R., Troncy, R., Staab, S., Hardman, L., Vacura, M.: COMM: Designing a Well-Founded Multimedia Ontology for the Web. In: Aberer, K., Choi, K.-S., Noy, N., Allemang, D., Lee, K.-I., Nixon, L., Golbeck, J., Mika, P., Maynard, D., Mizoguchi, R., Schreiber, G., Cudré-Mauroux, P. (eds.) ASWC 2007 and ISWC 2007. LNCS, vol. 4825, pp. 30–43. Springer, Heidelberg (2007)

2. Bontcheva, K., Tablan, V., Maynard, D., Cunningham, H.: Evolving GATE to Meet New Challenges in Language Engineering. Natural Language Engineering 10(3/4), 349–373 (2004)
3. Caragea, D., Bao, J., Honavar, V.: A General Strategy for Knowledge Acquisition from Semantically Heterogeneous Data Sources. In: AAAI 2006 Fall Symposium on Semantic Web for Collaborative Knowledge Acquisition (2006)
4. Ferrucci, D., Lally, A.: UIMA: an architectural approach to unstructured information processing in the corporate research environment. Natural Language Engineering 10(3/4), 327–348 (2004)
5. Gangemi, A., Borgo, S., Catenacci, C., Lehmann, J.: Task Taxonomies for Knowledge Content. Metokis Deliverable D 2007 (September 2004)
6. Götz, T., Suhre, O.: Design and implementation of the UIMA Common Analysis System. IBM Systems Journal 43(3), 476–489 (2004)
7. Happel, H.-J., Seedorf, S.: Applications of Ontologies in Software Engineering. In: Second International Workshop on Semantic Web Enabled Software Engineering (SWESE 2006), held at the Fifth International Semantic Web Conference (November 2006)
8. Martin, D., Burstein, M., Hobbs, J., Lassila, O., McDermott, D., McIlraith, S., Narayanan, S., Paolucci, M., Parsia, B., Payne, T., Sirin, E., Srinivasan, N., Sycara, K.: OWL-S: Semantic Markup for Web Services (November 2004), `http://www.w3.org/Submission/OWL-S/`
9. Maynard, D.: Benchmarking ontology-based annotation tools for the Semantic Web. In: AHM 2005 Workshop on Text Mining, e-Research and Grid-enabled Language Technology (2005)
10. Maynard, D., Yankova, M., Kourakis, A., Kokossis, A.: Ontology-based information extraction for market monitoring and technology watch. In: ESWC 2005 Workshop on End User Aspects of the Semantic Web (May 2005)
11. Oberle, D., Lamparter, S., Grimm, S., Vrandecic, D., Staab, S., Gangemi, A.: Towards Ontologies for Formalizing Modularization and Communication in Large Software Systems. Journal of Applied Ontology 1(2), 163–202 (2006)
12. Popov, B., Kiryakov, A., Ognyanoff, D., Manov, D., Kirilov, A.: KIM - A Semantic Platform for Information Extraction and Retrieval. Journal of Natural Language Engineering 10(3/4), 375–392 (2004)
13. Puerta, A.R., Neches, R., Eriksson, H., Szekely, P., Luo, P., Musen, M.A.: Toward Ontology-Based Frameworks for Knowledge-Acquisition Tools. In: Eighth Knowledge Acquisition Workshop for Knowledge-Based Systems (1994)
14. Yildiz, B., Miksch, S.: Motivating Ontology-Driven Information Extraction. In: International Conference on Semantic Web and Digital Libraries (2007)
15. `http://t-rex.sourceforge.net`
16. `http://alias-i.com/lingpipe`

ONTOCOM Revisited: Towards Accurate Cost Predictions for Ontology Development Projects

Elena Simperl, Igor O. Popov, and Tobias Bürger

Semantic Technology Institute (STI) Innsbruck, University of Innsbruck,
6020 Innsbruck, Austria
{elena.simperl,igor.popov,tobias.buerger}@sti2.at

Abstract. Reliable methods to assess the costs and benefits of ontologies are an important instrument to demonstrate the tangible business value of semantic technologies within enterprises, as an argument to encourage their wide-scale adoption. The economic aspects of ontologies have been investigated in previous work of ours. With ONTOCOM we proposed a cost estimation model for ontologies and ontology development projects. This paper revisits this model and presents its latest achievements. We report on a comprehensive calibration of ONTOCOM based on a considerably larger data set of 148 ontology development projects. The calibration used a combination of statistical methods, ranging from preliminary data analysis to regression and Bayes analysis, and resulted a significant improvement of the prediction quality of up to 50%. In addition, the availability of a representative data set allowed us to identify meaningful directions for customizing the generic cost model along particular types of ontologies, and ontology-like structures as those specific to the emerging Web 3.0. Last but not least, we developed a software tool that allows ontology development project managers to easily use and adapt and to systematically calibrate the model, thus facilitating its adoption in real-world projects.

1 Introduction

Ontology-based technologies use machine-understandable formalizations of expert knowledge to provide intelligent and economically feasible solutions to many prevailing problems in today's information and knowledge management systems. In order to encourage the wide uptake of ontology-based technologies at corporate level, instruments to assess their economic benefits and to predict the total costs of their development and deployment are a must.

CIOs are starting to acknowledge the technical value of these technologies for enterprises. In the last years, early adopters have been increasingly using them in various application settings ranging from content management to social networks and enterprise application integration. Core technological building blocks and development platforms are meanwhile available from established vendors. Despite this promising position, it is still difficult to argue in favor of ontology-based technologies in front of CFOs because of the lack of convincing measurable business benefits, and of proven-and-tested methods to determine and quantify these.

The economic aspects of ontologies have been investigated in previous work of ours. With ONTOCOM we presented the first cost estimation model for ontologies and

L. Aroyo et al. (Eds.): ESWC 2009, LNCS 5554, pp. 248–262, 2009.

ontology development projects [19]. Essentially, estimating the efforts required to build an ontology can be undertaken in similar ways as in every other engineering discipline dealing with the question of costs and benefits. It requires the identification and study of those aspects of the engineered artifact, in our case, the ontology, and the engineering process thereof, that are likely to have an impact on the total development costs. Moreover, it uses specific methods and techniques to make effort predictions based on these so-called "cost drivers". Given these baseline commonalities, it is understandable that ONTOCOM largely benefits from the experiences in cost estimation made in the related area of software engineering. Through expert interviews, we identified the most relevant cost drivers for a wide class of ontology development projects. In a second step we acquired relevant data from 36 ontology-related projects and calibrated a parametric cost prediction equation. This resulted in a first release of the ONTOCOM model, which was used as a baseline for further studies. This model was presented in [19].

In this paper we revisit the model and presents its latest achievements. We report on a second, more comprehensive calibration of ONTOCOM based on a considerably larger data set of 148 ontology development projects. The calibration used a combination of statistical methods, ranging from preliminary data analysis to regression and Bayes analysis, and resulted in a significant improvement of the prediction quality of up to 50%. In addition, the availability of a critical mass of data allowed us to identify meaningful directions for customizing the generic model along particular types of ontologies, and ontology-like structures as those specific to the emerging Web 3.0. Last but not least, we developed a software tool that allows ontology development project managers to easily use and adapt, and to systematically calibrate the model, thus facilitating its adoption in real-world projects.

The remainder of this paper is structured as follows: in Section 2 we briefly introduce the ONTOCOM model, its assumptions, main cost drivers, and prediction algorithm. Section 3 gives full particulars on the newest calibration of the model and analyzes the implication of its results. The software tool for using, adapting and calibrating ONTOCOM is presented in Section 4, whilst related work is discussed in Section 5. The paper is concluded with a summary of the most important findings and planned future work (cf. Section 6).

2 ONTOCOM in a Nutshell

ONTOCOM is a generic cost model for ontology development. The model is generic in that it assumes a sequential ontology life cycle, according to which, an ontology development project starts with the analysis of the domain and of the requirements to be fulfilled by the ontology; once an initial set of requirements is specified, the ontology is conceptualized, implemented and evaluated. The outcomes of the evaluation can lead to further iterations of the process either at the domain analysis, the conceptualization or the implementation phase. This simplified ontology development process is depicted in Figure 1. Alternative engineering strategies such as rapid prototyping or agile methods, which are based on different life cycles, are not yet considered.[1] This limitation

[1] We refer to [7] for a discussion on the relation between this process model and the IEEE standards [8].

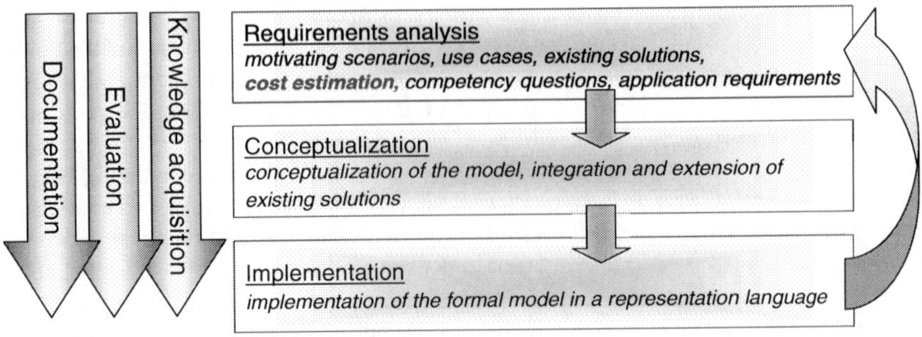

Fig. 1. Typical Ontology Development Process

was addressed in previous work of ours; the interested reader could refer to [18], which describes how the generic model should be customized to suit collaborative ontology engineering scenarios. The cost estimation model is realized as follows. First, a work breakdown structure for ontology development processes is defined. This is required in order to reduce the complexity of project budgetary planning and controlling operations down to more manageable units [1]. The global costs can then be derived using various prediction methods. Examples thereof include the expert judgement method, the analogy method or the parametric/algorithmic method [19]. ONTOCOM applies primarily the latter. The effort estimates are calculated using a parameterized mathematical formula. The parameters of this formula are given start values in predefined intervals, and are subsequently calibrated on the basis of previous project data. This empirical information complemented by expert opinions is used to evaluate and revise the predictions of the initial *a-priori model*, thus creating a validated *a-posteriori model*.

$$PM = A * Size^{\alpha} * \prod CD_i \quad (1)$$

According to the parametric method, the total development effort is associated with cost drivers specific for the ontology development process and its main steps and activities. Experiences in related engineering areas [1,9,20] let us assume that the most significant cost driver is the *size of the ontology*. This is reflected in Equation 1 in the parameter $Size$ that corresponds to the size of the ontology, that is, the number of kilo primitives which are expected to result from the conceptualization phase (including fragments built by reuse or other knowledge acquisition methods). The possibility of a nonlinear behavior of the model with respect to the size of the ontology is covered by the parameter α. The constant A represents a baseline multiplicative calibration constant in person months and stands for the costs which occur "if everything is normal" for developing an ontology with the size of 1 (1000 ontological primitives). The remaining *cost drivers* CD_i have a rating level (from $VeryLow$ to $VeryHigh$) that expresses their impact on the development effort. For the purpose of a quantitative analysis each rating level of each cost driver is associated to a weight referred to as *effort multiplier* EM_i. The *productivity range* PR_i of a cost driver stands for the ratio between the highest and the lowest effort multiplier of a cost driver $PR_i = \frac{max(EM_i)}{min(EM_i)}$) and is an indicator for the relative importance of the respective cost driver for the overall cost estimation [1].

The ONTOCOM cost drivers can be divided into three categories [19]:

Product-related cost drivers that account for the impact of the characteristics of the ontology on the overall costs. Cost drivers that were identified for the task of ontology development are, for instance, the domain analysis complexity, the conceptualization complexity, or the documentation needs.

Personnel-related cost drivers that emphasize the role of team experience, ability and continuity with respect to the effort required by an ontology development project. In this category, we mention the ontologist/domain expert capability, the level of expertise with respect to languages or tools, and the personnel turnover.

Project-related cost drivers that are related to the characteristics of the overall ontology development project and their impact on the total costs. Relevant cost drivers include the level of support and automation provided by ontology engineering software, and the additional efforts related to operating the project in a decentralized environment.

The ONTOCOM cost drivers were defined after extensively surveying ontology engineering literature from the last decade and conducting expert interviews, and from empirical findings of numerous case studies in the field [16]. For each cost driver we specified in detail the decision criteria which have to be taken into account by the user of the model in order for her to determine the concrete rating level of the driver in a particular project setting. For example, for the cost driver CCPLX, accounting for the costs produced by a particularly complex conceptualization, we predefined the meaning of the rating levels as shown in Table 1. The decision criteria associated with a cost driver are typically more complex than for the CCPLX example, and might be even divided into further subcategories, whose impact is aggregated to a final rating value by means of normalized weights. The first release of the ONTOCOM model was evaluated on a data set of 36 ontologies. Using regression and Bayes analysis as described in [19] we calibrated

Table 1. The Conceptualization Complexity Cost Driver **CCPLX**

Rating Level	Description
Very Low	The conceptual model is a concept list.
Low	The conceptual model is a taxonomy. A high number of patterns supporting the creation of the taxonomy are available. No special modeling constraints are imposed through application requirements.
Nominal	The model contains a taxonomical structure and domain properties. Again, modeling patterns are available, while the application setting imposes some simple constraints which produce additional modeling overload.
High	The model contains in addition to the previous case axioms. the engineering team can not resort to a feasible number of modeling patterns to ease the conceptualization task. In the same time, the number of application-driven constrains increases.
Very High	The conceptual model is an axiomatized ontology containing both schema and instance data. There are few to no modeling patterns to support the conceptualization task, while the number of application-driven constrains is significantly higher.

the model and obtained an initial prediction quality of 75% of the data points within a range of ±75% of the estimates. For the corresponding ±30% range the model covered 32% of the real-world data. These results, thus promising, required further optimization, both with respect to the size of the data set, and the statistical methods used. The next section explains how the second calibration of ONTOCOM overcame these limitations.

3 Towards Accurate Cost Predictions

The second calibration of ONTOCOM started with a comprehensive survey, in which participants in ontology development projects were interviewed with respect to aspects related to the different cost drivers of the model. The survey was supported by an online questionnaire. The questionnaire consisted of 42 questions divided into four parts. The first part referred to general aspects of the ontology development project at hand, including the size of the resulting ontology, its purpose or the effort needed to build it. The second part consisted of questions related to particular ontology development phases, such as domain analysis, conceptualization, implementation, documentation and evaluation. The third part of the questionnaire was related to characteristics of the development team which was engaged in the project. Finally, the fourth part contained questions about the software used to support and automatize process. A detailed presentation of the structure of the ONTOCOM questionnaire can be found in [17]. This questionnaire was used, to a large extent, in the initial calibration of the model in 2006 as well. Minor modifications were made based on feedback received in the first round.

The survey contains data from 148 ontology development projects. Experiences in effort estimation in related engineering disciplines, notably software engineering, let us assume that this data set is large enough to form the basis for a reliable prediction model of the size of ONTOCOM [5]. Approximately 50% of the data was collected during face-to-face or telephone interviews, the rest via a self-administered online questionnaire.[2] Approximately 95% of the ontologies collected were built in Europe, whilst nearly 35% originated from industry parties. The size of the ontologies in the data set varied from 60 to 11 million entities. Most ontologies were implemented in OWL DL (30%), followed by WSML DL and WSML Flight (around 10% each) and RDF(S) (9%). The duration of the projects varied from 0.02 to 156 person months.

We then calibrated the model on this enlarged data set using the same statistical methods as in the previous ONTOCOM release. This led to a comparatively minor improvement in the prediction accuracy. The previous calibration of the model, on 36 data points, resulted in an accuracy of 31.82% within a ±30% error margin. The new calibration, based on a three-times larger data set, led to an accuracy improvement of around 2% in the same margin. The reason for this minimal change could be traced back to the issue of unbalanced data sets [2]. To solve this issue, and to enable the model to reach an accuracy level which allows it to be used in real-world enterprise settings, we needed to further analyze the nature of the data collected, as described for instance in [10,11,14,13].

Consequently, we conducted a preliminary data analysis on the enlarged data set following the framework of Lin & Mintram [13]. We were particularly interested in the first

[2] The questionnaire is available at `http://ontocom.sti-innsbruck.at`

six steps of the framework, which dealt with the identification and removal of outliers from data sets and with the determination of the cost drivers with the greatest impact on the estimated effort. The preliminary data analysis resulted in a considerable improvement of our predictions. First, outliers in the numerical variables were removed and the remaining data was transformed to a normal distribution. The same was applied to ordinal variables and the entire process was repeated until all outliers were eliminated. Once this was achieved, we performed a correlation and regression analysis, followed by an analysis of variance (ANOVA) on the ordinal variables. Finally the data set was calibrated (using multivariate regression and Bayes analysis) in order to combine the empirical data with the expert data.

The ONTOCOM model contains 2 numerical variables and 11 ordinal ones. The numerical variables are *Person Months (PM)*, which is a response variable, and *Size* which is an exploratory variable.[3] For the calibration of the model we considered only those cost drivers for which expert data was available. These are listed in Table 2. The rating values collected for each cost driver for each data point (ranging from Very Low to Very High) were transformed to nominal numbers (ranging from 1 to 5).

Table 2. Variables Used in the Preliminary Data Analysis

Abbreviation	Cost Driver	Type
PM	Person months	Numerical
SIZE	Size of the ontology	Numerical
DCPLX	Domain Complexity	Ordinal
CCPLX	Conceptualization Complexity	Ordinal
ICPLX	Implementation Complexity	Ordinal
REUSE	Required Reusability	Ordinal
DOCU	Documentation Needs	Ordinal
OE	Ontology Evaluation	Ordinal
OCAP/DECAP	Ontologist/Domain Expert Capability	Ordinal
OEXP/DEEXP	Ontologist/Domain Expert Experience	Ordinal
PCON	Personnel Continuity	Ordinal
LEXP/TEXP	Language/Tool Experience	Ordinal
SITE	Multisite Development	Ordinal

In the following we briefly describe each step of the framework, and discuss the results of the subsequent calibration.

3.1 Step 1: Outlier Detection in Numerical Variables

Table 3 summarizes the analysis of the numerical variables *PM* and *Size*. For each variable the name of the variable, the number of observations, the range of values and the mean and standard deviation are shown. The *PM* variable ranges from 0.02 to 156

[3] Response variables are those that are observed to change in response to the explanatory variables. In our case the effort variable (PM) is the response variable; all other are explanatory variables.

Table 3. ONTOCOM Numerical Variables

Var. name	Obs.	Minimum	Maximum	Mean	Std. Dev.
PM	148	0.02	156	10.22	25.60
Size	148	0.006	11000	76.14	904.09

person-months. The average size of *PM* is 10.22 person-months with a standard deviation of 25.60. The *Size* variable ranges from 0.06 to 11000 kilo entities, with an average size of 76.14 kilo entities with a standard deviation of 904.09.

The outlier detection was performed using box plots, which is a graphical way of depicting groups of numerical data. Our initial run identified 5 data points as outliers which were removed from the analysis.

3.2 Step 2: Data Transformation

In this step we transformed the data so that it approximates a normal distribution. As both numerical variables were not normally distributed, we created new variables, named *LnPM* and *LnSize*, by applying the natural logarithm, and tested their distribution using histograms (cf. Figures 2 and 3, respectively). The new variables *LnPM* and *LnSize* were used in the subsequent steps of the preliminary data analysis.

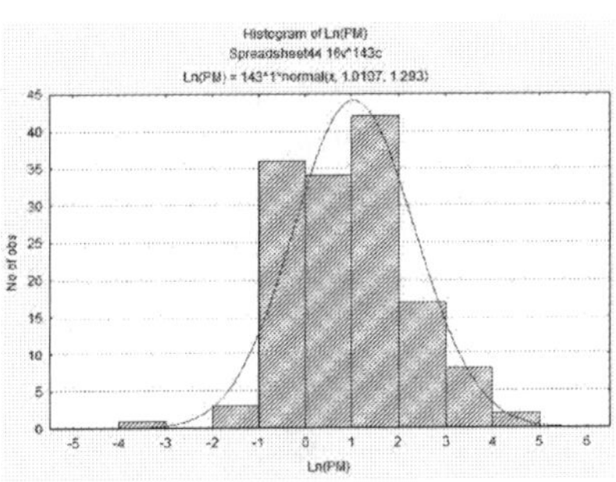

Fig. 2. Histogram of the Variable *LnPM*

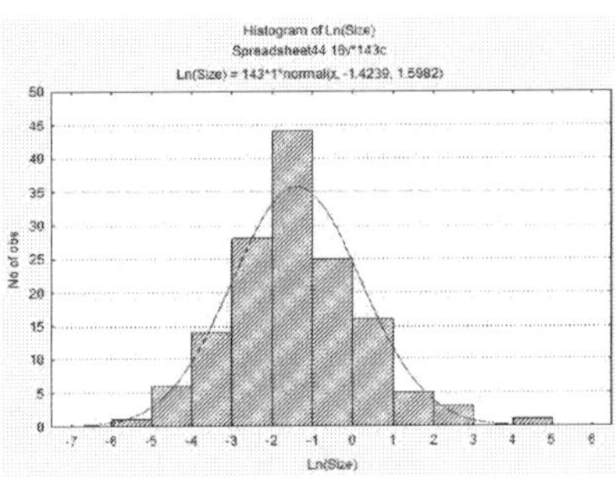

Fig. 3. Histogram of Variable *LnSize*

3.3 Step 3: Outlier Detection in Ordinal Variables

Ordinal variables correspond to the three categories of cost drivers in the ONTOCOM model introduced in Section 2. The outlier analysis was performed again using box plot diagrams and led to the removal of 44 data points (in two iterations). The box plot in Figure 4 depicts the remaining data set without outliers in the ordinal variables. We also remeasured the normal distribution of this data set through Quantile-Quantile (Q-Q) normality plots for the numerical variables *LnPM* and *LnSize* (Figures 5 and 6, respectively). Steps 4 to 6 of the preliminary data analysis and the subsequent calibration were performed on the remaining data set of 99 ontology development projects.

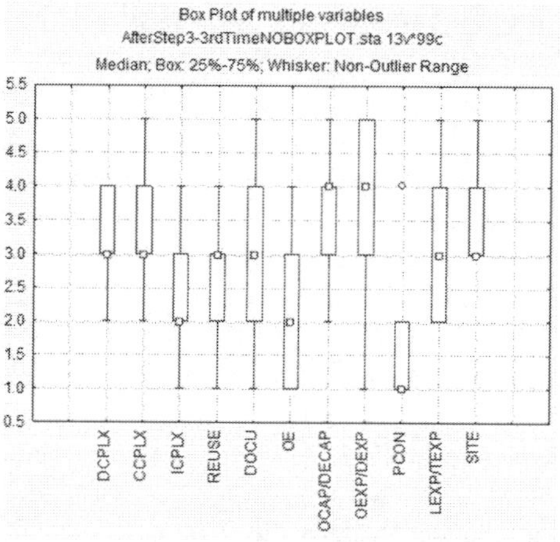

Fig. 4. Boxplot for Outlier Detection of Ordinal Variables

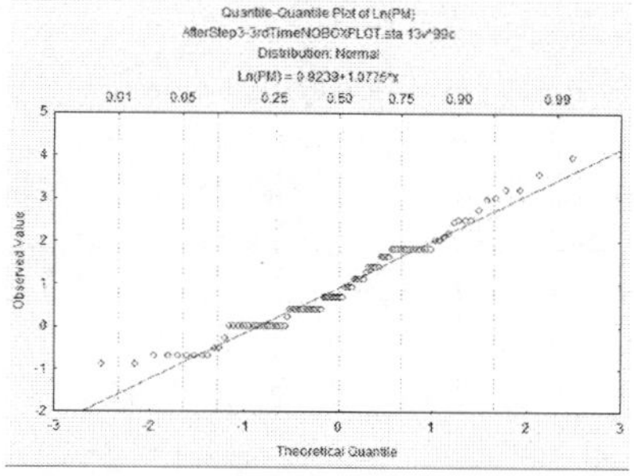

Fig. 5. Normal Q-Q plot for *LnPM*

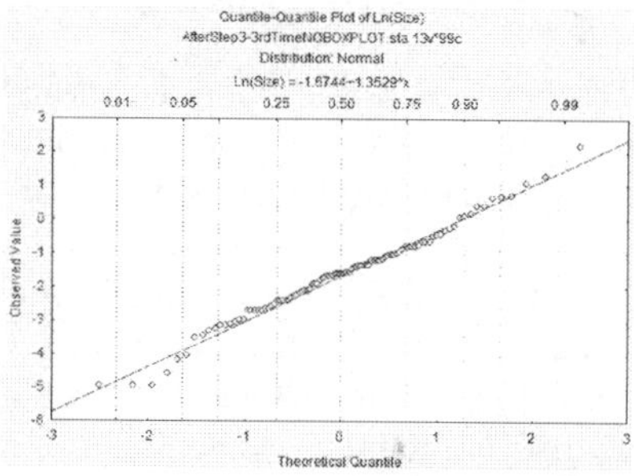

Fig. 6. Normal Q-Q plot for *LnSize*

3.4 Steps 4 and 5: Correlation Analysis and Regression Analysis

Correlation analysis involves calculating a correlation coefficient which measures the relationship between the response and the explanatory variable. The coefficient has a value between $+1$ and -1. If the value of the coefficient if greater than 0 then the variables are positively correlated, in other words, a rise of the explanatory variable will result in a rise of the response variable, whilst a decrease in the explanatory variable will result in a decrease of the response variable. The value is an indicator on how strong this relationship is, with 1 staying for a pair of perfectly positively correlated variables. Conversely, if the value is below 0 then an increase in the explanatory variable will result in a decrease of the response variable and viceversa, with -1 standing for variables that are perfectly negatively correlated. 0 means that the variables are not correlated. Two commonly used measures for correlation analysis are the Spearmans's rank coefficient (for ordinal variables) and Pearson's rank coefficient (for interval or ration type data). Both coefficients are relevant to our data set. The threshold values are set to $Pr > |0.65|$ and $Pr < |0.1|$, for high and low correlation, respectively. The correlation analysis indicated no correlation in the numerical or ordinal variables. A subsequent regression analysis confirmed that the correlation between *LnSize* and *LnPM* is in the allowed intervals.

3.5 Step 6: Stepwise ANOVA Analysis

In the next step we performed a stepwise ANOVA (Analysis of Variance) analysis to compute the *best* 6*-variable* ONTOCOM *model*. ANOVA is a statistical method commonly used to analyze data sets with ordinal type variables [2,10,14,13]. The data set variables are referred to as "factors" and the scale values of the variables as "factor levels". ANOVA tests whether there are significant differences between factor level means. It is assumed that the variance of the level means will be large if there are differences between the groups, and small if there are none. The variance of the means between groups is divided by the variance within the groups to calculate the F-statistic measure(Equation 2). The result will be close to 1 if the group means are not significantly different from one another.

$$F = \frac{Variance \quad between}{Variance \quad within} \tag{2}$$

The *best* 6*-variable model* is defined iteratively. In each iteration one identifies the next explanatory variable that has the most impact on the response variable. This is done by examining the adjusted coefficient of determination (i.e., the R^2 value). The variable with the highest adjusted R^2 is the variable that best explains the variation in the response variable and is thus kept in the model [13]. In each iteration the adjusted R^2 values are recalculated in order to account for the variables previously selected.

Applying this procedure to ONTOCOM and our data set revealed that the six variables that best explain the behavior of the effort are (in this order): Domain Complexity (DCPLX),Ontology Evaluation (OE), Language\Tool Experience (LEXP\TEXP), Ontologist\Domain Expert Capability (OCAP\DECAP), Documentation needs (DOCU), and Personnel Continuity (PCON). Our analysis based on the adjusted R^2 values also pointed out that any additional variables added to the model would not have a significant impact on the response variable.

3.6 Calibration Results and Discussion

The preliminary data analysis resulted in a set of 99 data points and identified the six variables which best explain the variation in the response variable. We performed a calibration on this best 6-variable model to combine the expert opinion with the data collected. Our results showed a substantial improvement of the accuracy of predicting results within a $\pm 30\%$ margin from 33.78% to 45.82% and within a $\pm 75\%$ margin from 66.89% to 79.80. We then performed our calibration with all eleven variables to examine the impact of the other variables. Our result was a slight increase in prediction accuracy of 46.46% within the $\pm 30\%$ error margin, whilst for the $\pm 75\%$ margin the differences were even less significant.

The analysis of the behavior of the cost drivers is consistent with our previous findings in [19], based on a 36 data point set, particularly for two of the three most dominant cost drivers, Domain Complexity (DCPLX) and Ontology Evaluation (OE). Additionally, we examined the behavior of productivity, i.e., the ratio between the *Size* and *PM* variables, and prediction accuracy for the larger margin error of ±75%. The box plot of the productivity factor showed a stable set with only 7 outliers from the full data set. The accuracy for predicting results within ±75% increased from 66.67% to 79.80% after the preliminary data analysis. This supports the fact that ontology development projects rarely exhibit any erratic behavior in their costs, in other words, there are only rare cases when the actual costs might exceed the anticipated costs by several times. Exponential behavior of the effort w.r.t. *Size* confirmed our previous findings that ontology development projects exhibit economies of scale. This suggests that building larger ontologies is economically feasible.

The dominance of the cost drivers DCPLX and OE indicates that any facilitation in these tasks may result in major efficiency gains. While tools such as wikis may be helpful especially in collaborative settings, they are still rarely used for the purpose of domain analysis. More generally, the results of the interviews indicated a low tool support for this task. This situation could be improved by applying methods such as automated document analysis, or ontology learning approaches, to support the analysis of the domain and most notably the assessment of the information sources available. A similar conclusion can be drawn for ontology evaluation. Despite of the availability of automated approaches such as unit tests, consistency checks or taxonomy cleaning techniques, ontology evaluation still lacks tools which are easy to use and comprehensible for most users. The dominance of personnel-related cost drivers like PCON, LEXP/TEXP, or OCAP/DECAP suggests that more training programs in the area of ontology engineering, better collaboration support, and an improved, more fine granular documentation of decisions taken during ontology engineering may also result in efficiency gains. In this context, interviews also revealed that ontology engineering methodologies are rarely used. A more structured process-level support within development environments may again lead to an improvement.

It should be noted that the data in our set was collected from ontology development projects from various domains, and that the ontologies were built for different purposes and used in different environments. Experiences with cost estimation models suggest that a customized model, optimized for particular project or enterprise settings, would yield even better prediction results [5]. This potential mismatch between the generality of the model and its suitability for specific environments can be (partially) compensated by a large, representative data set used to calibrate the model. The data set in our current calibration enabled us to develop a second release of the ONTOCOM with a significantly improved prediction quality. This can be traced back to the preliminary data analysis performed, but also to the size of the data set, which is sufficiently high for the number of variables in the model. The results we obtained on the calibration on 148 data points suggest, nevertheless, that the predictions of ONTOCOM as a generic model should improve with the availability of additional data. Data analysis on an extended set might show slight differences with respect to the dominant behavior of certain variables, however, our experience shows us that the relevancy of the cost factors is very much consistent. The quality of the ONTOCOM predictions could be further optimized by establishing variants of the model tailored to reflect particularities of certain classes of ontological structures. Following this line of reasoning, we have developed ONTOCOM -Lite, a cost model for lightweight ontologies, in particular folksonomies and taxonomies. Taxonomies are widely used in various commercial environments and the process of developing taxonomies is mature (such well defined development processes are a prerequisite for adapting our model, notably to modify the work breakdown structure and to identify new cost drivers, cf. Section 2). Folksonomies, on the other side, are semi-structured, collaboratively emerging knowledge structures which are increasingly used for information and knowledge management purposes. Despite their popularity among users, the total costs of tagging are an interesting figure to argue in favor of this new technology within an enterprise.

4 ONTOCOM Tooling

We developed a software tool in order to allow project managers to automatically perform the calibration process on an arbitrary data set, and to use the model to make

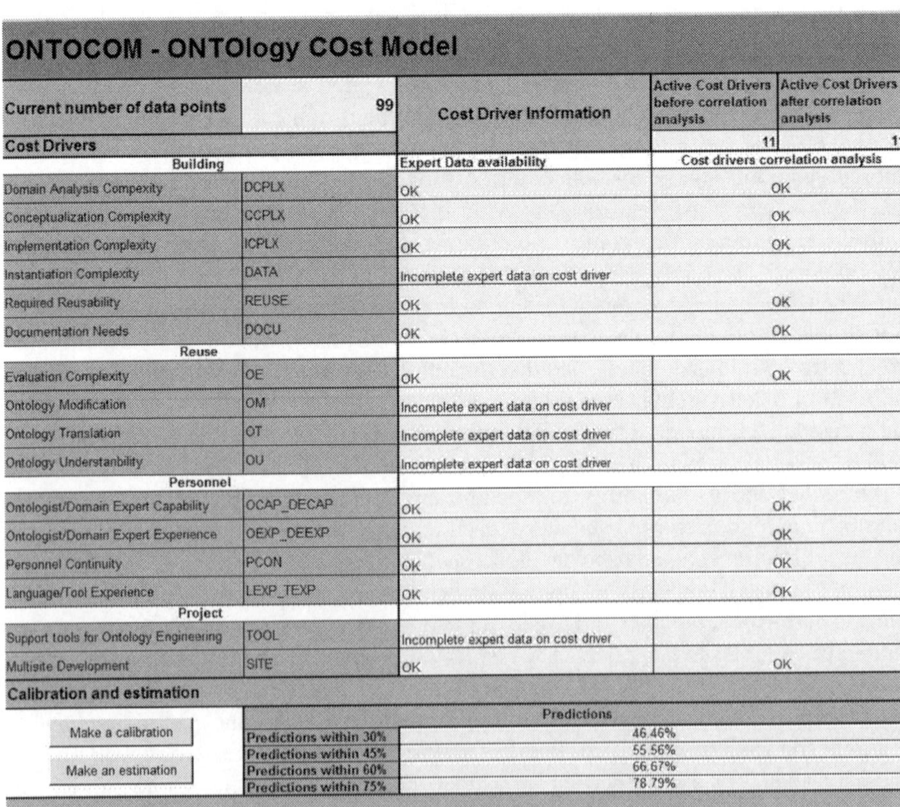

Fig. 7. The ONTOCOM Main User Interface

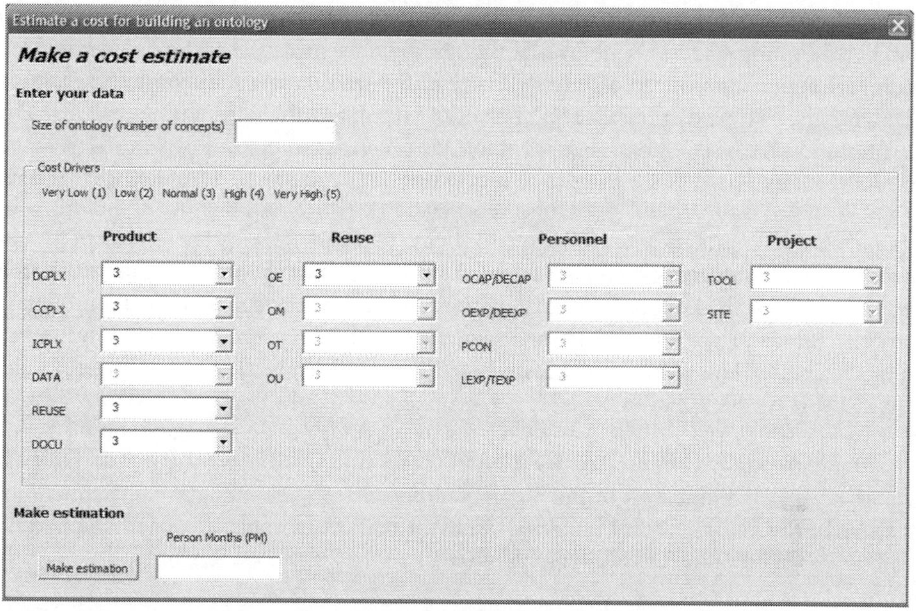

Fig. 8. The ONTOCOM cost predictions UI

estimations for planned ontology development projects.[4] The tool provides an easy-to-use user interface for running calibrations and predictions. It provides support for the calibration of the ONTOCOM model using either existing, self-owned data, or a combination of both. Furthermore, users are able to customize the model, selecting the cost drivers to be taken into account for a calibration. The ONTOCOM tool is based on Microsoft Excel. It consists of 3 spreadsheets reflecting the user interface, and additional spreadsheets that are automatically created in the course of the calibration of the model. The first spreadsheet, depicted in Figure 7, can be used to perform a calibration or to calculate predictions. Furthermore, it contains information about the calibration results, including information on the number of data points used in the calibration, the use of cost drivers, and on the accuracy of the predictions for different thresholds.

The second spreadsheet shows the expert data available for each cost driver and, in the subsequent columns, the updated productivity range and effort multipliers based on the calibration of the model. The user is allowed to change the expert values to better reflect her own opinion. The calibrated values of the cost drivers are used for predictions. On the third spreadsheet the user has the option to add or remove data points to the calibration data set.

When the user starts a new calibration the tool creates additional spreadsheets, reflecting the individual calibration steps performed. In the fourth spreadsheet (Empirical Data (Mapping)) categorical values of the empirical data are mapped to their corresponding values from the expert data. If the latter is not available for a cost driver, the cost driver will not be used in subsequent calibration steps. In the user interface

[4] The tool is available online at `http://ontocom.sti-innsbruck.at/tools.htm`

spreadsheet (cf. Figure 7) the user is then notified which cost drivers were not used in the calibration due to this unavailability. In the next spreadsheet the data is mapped to logarithmic values, a procedure that is required when using multivariate regression analyis. The two next spreadsheets present the results of the correlation analysis. The "Correlation analysis" spreadsheet contains information about the correlations between explanatory and predictor variables and between the explanatory variables themselves. The "Post-correlation data" contains the selected (logarithmic) data from the previous spreadsheet based on the result of the correlation analysis. The next spreadsheet is dedicated to the "Statistical Analysis". It shows the results of the multivariate regression and Bayes analysis. For each cost driver, the result is an exponent which is used on the original values to get a new calibrated value for each of its 5 rating levels. In the last spreadsheet the data points are tested against the predictions which are derived by the model using the new calibrated values. The prediction results are evaluated for different thresholds, which are presented as an accuracy percentage in the interface spreadsheets.

After calibration a user can make predictions for future projects by using an interface where he selects the values of the parameters for his project, including an estimate of the size of the ontology (cf. Figure 8). Using the calibrated values the user is presented with an estimate of the costs in person months.

5 Related Work

Cost estimation has a long-standing tradition in more mature engineering disciplines such as software engineering or industrial production [1,9,21]. Although the importance of cost issues is well-acknowledged in the semantic technologies community, as to the best of our knowledge, no cost estimation model for ontology development besides ONTOCOM has been proposed so far. Analogue models for the development of knowledge-based systems (e.g., [6]) implicitly assume the availability of the underlying conceptual structures. [15] provides a qualitative analysis of the costs and benefits of ontology usage in application systems, but does not offer any model to estimate the efforts. [4] presents empirical results for quantifying ontology reuse. [3] proposes a model for assessing the benefits of using ontologies in particular settings, but the model proposed in not quantitative, and the analysis of cost and benefit information is not straightforward. [23] proposes a cost and benefit analysis model for personal knowledge management which, most notably, takes externalization of knowledge in the form of formal statements into account. They apply their model to semantic wikis, but do not provide any empirical evaluation. [12] adjusts a cost estimation model for Web applications with respect to the usage of ontologies. The resulting cost drivers are, however, not adapted to the requirements of ontology development and no evaluation is provided. Nevertheless, the question of cost estimation for ontology-based software application is relevant, and a possible direction of research for our ontology engineering economics work. Wolff et al. address the efforts related to the engineering of Semantic Web Services in [24]. They argue that Semantic Web Services projects require less time compared to their traditional Web services equivalents, but do not treat ontologies, or any other semantic technology, as a separate part of their observations. Other researchers recently started to investigate economic issues pertaining to the development of ontology-based applications: In [22] gOnntt, a planning tool for ontology projects using GANTT charts is introduced in which activities related to the life cycle of networked ontologies are mapped to existing life cycle models. These activities should be augmented with effort-related information in one of the next releases of the tool. ONTOCOM could be used to calculate the required effort information.

6 Conclusions

The adoption of ontology-based technologies in commercial settings depends on the availability of appropriate methodologies and software assisting the ontology engineering process, and on methods for an effective cost benefit management. In earlier work, we proposed a parametric cost estimation model for ontologies which we recalibrated to improve the accuracy of effort predictions. By applying a careful preliminary data analysis we could improve the accuracy of our model in a $\pm 30\%$ margin by 35% compared to a first calibration reported in 2006. The calibration confirmed earlier observations regarding the impact of particular cost drivers and furthermore showed that prediction accuracy improves with the availability of more data points. The cost drivers that had the most impact on the results are Domain Complexity (DCPLX), Ontology Evaluation (OE), Language$\backslash$Tool Experience (LEXP$\backslash$TEXP), Ontologist$\backslash$Domain Expert Capability (OCAP$\backslash$DECAP), Documentation needs (DOCU), and Personnel Continuity (PCON). We also introduced an Excel-based software tool which offers on-the-fly cost estimation for ontology projects and automatic calibration support. Future tool development plans include integrating cost estimation into ontology development and management environments such as gOnntt in which the user can assess costs in the same platform in which the overall ontology development process is managed.

In the context of the EU IP project ACTIVE we will design ONTOCOM Lite, an adaptation of the generic model for lightweight knowledge structures such as folksonomies and taxonomies. Given the increased number of useful ontologies available on the Web, ontology reuse slowly becomes feasible. A cost model to estimate the efforts related to this engineering strategy and to compare it to alternatives such as development from scratch is a natural extension of the current ONTOCOM model.[5] Last, but not least, we will use ONTOCOM to provide information about the total development costs of software applications using ontologies and related technologies. It remains to be investigated whether this will result into an extension of established software engineering cost models such as COCOMO [1] or whether a new model will be more suitable.

Acknowledgements. The research leading to this paper was partially supported by the European Commission under the contract FP7-215040 "ACTIVE".

References

1. Boehm, B.W.: Software Engineering Economics. Prentice-Hall, Englewood Cliffs (1981)
2. Briand, L.C., Basili, V.R., Thomas, W.M.: A pattern recognition approach for software engineering data analysis. IEEE Transactions on Software Engineering 11(18), 931–942 (1992)

[5] The current model contains cost drivers specific for ontology reuse. However, they are not used for calibration purposes as no empirical data is available.

3. Bürger, T., Simperl, E.: Measuring the benefits of ontologies. In: Proceedings of Ontology Content and Evaluation in Enterprise (OntoContent 2008), In conjunction with OnTheMove Federated Conferences (OTM 2008), Monterrey, Mexico, November 9-14 (2008)
4. Cohen, P.R., Chaudhri, V.K., Pease, A., Schrag, R.: Does prior knowledge facilitate the development of knowledge-based systems? In: AAAI/IAAI, pp. 221–226 (1999)
5. Devnani-Chulani, S.: Bayesian analysis of software cost and quality models. PhD thesis, University of Southern California (1999)
6. Felfernig, A.: Effort estimation for knowledge-based configuration systems. In: Proc. of the 16th Int. Conf. of Software Engineering and Knowledge Engineering, SEKE 2004 (2004)
7. Gomez-Perez, A., Fernandez-Lopez, M., Corcho, O.: Ontological Engineering – with examples form the areas of Knowledge Management, e-Commerce and the Semantic Web. Springer, Heidelberg (2004)
8. IEEE Computer Society. IEEE Standard for Developing Software Life Cycle Processes. IEEE Std 1074-1995 (1996)
9. Kemerer, C.F.: An Empirical Validation of Software Cost Estimation Models. Communications of the ACM 30(5) (1987)
10. Kitchenham, B.: A procedure for analyzing unbalanced datasets. IEEE Transactions on Software Engineering 4(24), 278–301 (1998)
11. Kitchenham, B., Pfleeger, S.L., Pickard, L., Jones, P., Hoaglin, D.C., Emam, K.E., Rosenberg, J.: Preliminary guidelines for empirical research in software engineering. IEEE Transactions on Software Engineering 8(28), 721–734 (2002)
12. Korotkiy, M.: On the effect of ontologies on web application development effort. In: Proc. of the Knowledge Engineering and Software Engineering Workshop (2005)
13. Lin, Q., Mintram, R.: Preliminary data analysis methods in software estimation. Software Quality Journal 13(1), 91–115 (2005)
14. Maxwell, K.: Applied Statistics for Software Managers. Pearson Education, UpperSaddle River (2002)
15. Menzies, T.: Cost benefits of ontologies. Intelligence 10(3), 26–32 (1999)
16. Paslaru Bontas, E., Tempich, C.: How Much Does It Cost? Applying ONTOCOM to DILIGENT. Technical Report TR-B-05-20, Free University of Berlin (October 2005)
17. Paslaru Bontas, E., Tempich, C.: Ontology engineering: A reality check. In: 5th Int. Conf. on Ontologies, DataBases, and Applications of Semantics (ODBASE 2006) (2006)
18. Paslaru Bontas Simperl, E., Tempich, C., Mochol, M.: Cost estimation for ontology development: Applying the ontocom model. In: Technologies for Business Information Systems, pp. 327–339 (2007)
19. Paslaru Bontas Simperl, E., Tempich, C., Sure, Y.: Ontocom: A cost estimation model for ontology engineering. In: Cruz, I., Decker, S., Allemang, D., Preist, C., Schwabe, D., Mika, P., Uschold, M., Aroyo, L.M. (eds.) ISWC 2006. LNCS, vol. 4273, pp. 625–639. Springer, Heidelberg (2006)
20. Putnam, L.H., Myers, W.: Five Core Metrics: Intelligence behind Successful Software Management. Dorset House Publishing Co., Inc. (2003)
21. Stewart, R.D., Wyskida, R.M., Johannes, J.D.: Cost Estimator's Reference Manual. Wiley, Chichester (1995)
22. Suárez-Figueroa, M.C., Gomez-Perez, A.: Building ontology networks: How to obtain a particular ontology network life cycle? In: Proceedings of the third International Conference on Semantic Systems (I-Semantics) (2008)
23. Völkel, M., Abecker, A.: Cost-benefit analysis for the design of personal knowledge management systems. In: Proceedings of 10th International Conference on Enterprise Information Systems (ICEIS 2008), pp. 95–105 (2008)
24. Wolff, F., Oberle, D., Lamparter, S., Staab, S.: Economic reflections on managing web service using semantics. In: EMISA, pp. 194–207 (2005)

Ranking Approximate Answers to Semantic Web Queries

Carlos A. Hurtado[1], Alexandra Poulovassilis[2], and Peter T. Wood[2]

[1] Faculty of Engineering and Sciences
Universidad Adolfo Ibáñez, Chile
carlos.hurtado@uai.cl

[2] School of Computer Science and Information Systems
Birkbeck, University of London, UK
{ap,ptw}@dcs.bbk.ac.uk

Abstract. We consider the problem of a user querying semistructured data such as RDF without knowing its structure. In these circumstances, it is helpful if the querying system can perform an approximate matching of the user's query to the data and can rank the answers in terms of how closely they match the original query. Our approximate matching framework allows us to incorporate standard notions of approximation such as edit distance as well as certain RDFS inference rules, thereby capturing semantic as well as syntactic approximations. The query language we adopt comprises conjunctions of regular path queries, thus including extensions proposed for SPARQL to allow for querying paths using regular expressions. We provide an incremental query evaluation algorithm which runs in polynomial time and returns answers to the user in ranked order.

1 Introduction

The volume of semistructured data available to users on the web continues to grow significantly. In particular, there has recently been a substantial increase in the amount of RDF data being generated and made available, often in the form of *linked data* [1]. Given the volume and heterogeneity of this data, users are frequently unaware of the structure of the data. This has given rise to linked data browsers such as Tabulator [1]. However, users may also wish to query such large repositories of linked data and so there is a need to assist them by providing querying systems which do not require that users' queries necessarily match exactly the data structures being queried.

In this paper we consider general semistructured data modelled as a graph structure, with RDF linked data being a particular application of this model. Our data model is that of a directed graph $G = (V, E)$, where each node in V is labelled with a constant and each edge e is labelled with a symbol $l(e)$ drawn from a finite alphabet Σ. Such a model does not allow for the representation of RDF blank nodes, e.g., but these are specifically discouraged for linked data [2].

We are interested in developing efficient algorithms which allow for the approximate matching of users' queries on such data, with the answers to queries

L. Aroyo et al. (Eds.): ESWC 2009, LNCS 5554, pp. 263–277, 2009.

being returned to users in ranked order. In order to achieve this, we restrict our query language to that of conjunctive regular path queries [3]. A conjunctive regular path query Q consisting of n query triples (or conjuncts) is the form

$$(Z_1, \ldots, Z_m) \leftarrow (X_1, R_1, Y_1), \ldots, (X_n, R_n, Y_n)$$

in which each X_i and Y_i, $1 \leq i \leq n$, is a variable or constant, each Z_i, $1 \leq i \leq m$, is a variable appearing in the body of Q, and each R_i, $1 \leq i \leq n$, is a regular expression over Σ.

The idea of using regular expressions to specify paths through graphs has been around for a long time (e.g. [4]), having been taken up by the semistructured data community (e.g. [3]), and more recently by the RDF community (e.g. [5,6]) in proposed extensions to SPARQL [7].

The exact answer to a conjunctive regular path query Q on a graph G is specified as follows. We first find, for each $1 \leq i \leq n$, a relation r_i over the scheme (X_i, Y_i) such that tuple $t \in r_i$ iff there is a path from $t[X_i]$ to $t[Y_i]$ in G that satisfies R_i, that is, whose concatenation of edge labels is in $L(R_i)$. We then form the natural join of relations $r_1, \ldots, r_n$ and project over Z_1 to Z_m.

Example 1. As an example, consider an RDF graph G involving information about a transport network (Fig. 1). Some nodes represent cities (denoted by URIs) and others represent names of cities. Edges state that we can travel from one city to another city by train, bus, or by airplane.

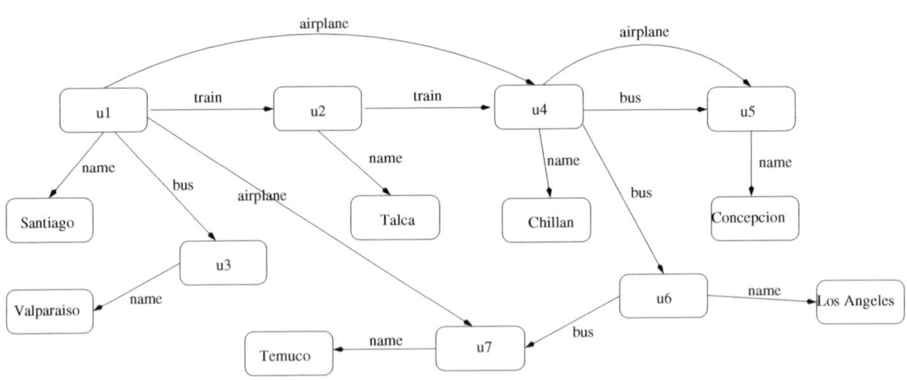

Fig. 1. An RDF graph of a basic transport network

Suppose that we want to find the cities from which we can travel to city u5 using only airplanes as well as to city u6 using only trains or busses. This can be expressed by the following conjunctive regular path query[1] Q:

$$?X \leftarrow (?X, (airplane)^+, u5), (?X, (train|bus)^+, u6)$$

[1] In the concrete syntax of queries, we precede variable names with '?' as in SPARQL.

When Q is evaluated on G, the set of bindings for $?X$ generated by the first conjunct is $\{u1, u4\}$, while that for the second conjunct is $\{u1, u2, u4\}$. Hence the answer is $\{u1, u4\}$. □

As stated above, we also want to allow for approximate matching of queries on graphs. Such approximate (or similarity) matching can take many forms and has been the subject of much study [8,9,10,11,12,13]. In general, we are interested in using *weighted regular transducers* to model the kinds of approximations allowed since single conjunct queries with a transducer applied can be evaluated *incrementally* in polynomial time [9]. Incremental evaluation allows for answers to be returned incrementally to a user in ranked order, without the user having to wait for entire query evaluation to end. In this paper, we extend these results to show that *multiple* conjunct queries can also be evaluated incrementally in polynomial time, using ideas from [14].

A weighted regular transducer is basically a finite state automaton in which the transitions are labelled with triples rather than single symbols. A transition from state s to state t labelled with triple (a, i, b) means that if the transducer is in state s, it can move to state t on input a with cost i while outputting b. In our context, such a transition would be interpreted as specifying that symbol a in a query can match the label b of a graph edge with cost i.

Weighted regular transducers provide for greater generality than we need for most of our examples. They also make the agorithms more difficult to explain. Hence for most of the paper we restrict ourselves to using approximate regular expression matching [15], which can easily be specified using weighted regular transducers [9]. The edit operations which we allow in approximate matching are insertions, deletions and substitutions of symbols, along with insertions of inverted symbols (corresponding to edge reversals) and transpositions of adjacent symbols, each with an assumed cost of 1.

Example 2. Consider again the application in Example 1 and assume that a user who has little knowledge of the structure of the data formulates the following query to retrieve all the cities reachable from Santiago by non-stop flights:

$$?X \leftarrow (Santiago, airplane, ?X)$$

The query does not return any answers because it does not match the structure of the graph. However, allowing edit operations such as insertions of symbols and inverted symbols (indicated by using '−' as a superscript to the symbol and corresponding to matching an edge in the reverse direction), each at an assumed cost of 1, the regular expression *airplane* can be successively relaxed to the regular expression $((name)^- \cdot airplane \cdot name)$, which captures as answers the city names of Temuco and Chillan. Since the cost of each edit operation is set to 1, these answers can be regarded as having a distance of 2 from the original query (two insertions).

Suppose now that, after acquiring some knowledge of the structure of the data, the user formulates the following query to find cities reached from Santiago by train:

$$?X \leftarrow (Santiago, ((name)^- \cdot (train)^+ \cdot name), ?X)$$

If the user is prepared to consider replacing a train journey by a bus journey, but at a cost of 1 per replacement, this can be specified as the only edit operation allowed on the user's query. The same effect can be achieved by a transducer with a single state with two self-edges to/from it: one labelled with $(train, 0, train)$, i.e. $train$ can be matched by $train$ with cost 0, and one labelled with $(train, 1, bus)$, i.e. $train$ can be matched by bus with cost 1. Now, the results of the query will be returned to the user ranked by the number of bus journeys: Talca and Chillan will have rank 0, Valparaiso, Concepcion and Los Angeles will have rank 1, and Temuco will have rank 2. □

We do not expect that a user would necessarily want or be able to design a weighted regular transducer in order to capture their requirements in terms of approximate matching. Instead, we would expect that a query interface would provide a user with readily understandable options from which to select their requirements. Our concern in this paper is in showing that weighted regular transducers and conjunctive regular path queries provide a good framework within which to investigate approximate querying of graph-structured data.

In the next section, we consider the special case of computing approximate answers for queries consisting only of a single conjunct. We show that approximate answers can be computed in time which is polynomial in the size of the query and the input graph. We also show that these answers can be computed incrementally and returned to the user in ranked order. Section 3 generalises to the case of multi-conjunct queries and shows that incremental computation can still be achieved in polynomial time as long as the queries are acyclic and have a fixed number of head variables. Related work is covered in Sect. 4 which also shows how our approach generalises previous work in a number of areas. Conclusions and future work are described in Sect. 5.

2 Single-Conjunct Queries

Recall that our data model is that of a directed graph $G = (V, E)$, where each node in V is labelled with a constant and each edge e is labelled with a symbol $l(e)$ drawn from a finite alphabet Σ. In queries, we will allow edges to be traversed both from their source to their target node and in reverse, from their target to their source node. So we now introduce the notion of the *inverse* of an edge label l, denoted by l^-, which is used to specify a reverse traversal of an edge. Let $\Sigma^- = \{l^- \mid l \in \Sigma\}$. If $l \in \Sigma \cup \Sigma^-$, we use l^- to mean the *inverse* of l, that is, if l is a for some $a \in \Sigma$, then l^- is a^-, while if l is a^- for some $a \in \Sigma$, then l^- is a.

A *single-conjunct query* consists of an expression of the form:

$$Z_1, Z_2 \leftarrow (X, R, Y)$$

where X and Y are constants or variables, R is a regular expression, and each of Z_1 and Z_2 is one of X or Y. A *regular expression* R over Σ is defined as follows:

$$R := \epsilon \mid a \mid a^- \mid \_ \mid (R1 \cdot R2) \mid (R1 | R2) \mid R^* \mid R^+$$

where ϵ is the empty string, a is any symbol in Σ, a^- is the inverse of a, "_" denotes the disjunction of all constants in Σ, and the operators have their usual meaning.

2.1 Exact Semantics

Because of the presence of inverse operators in our queries, we need to introduce the notion of a *semipath* in G in order to define the semantics of a query [3]. A *semipath* p in $G = (V, E)$ from $x \in V$ to $y \in V$ is a sequence of the form $(v_1, l_1, v_2, l_2, v_3, \ldots, v_n, l_n, v_{n+1})$, where $n \geq 0$, $v_1 = x$, $v_{n+1} = y$ and for each v_i, l_i, v_{i+1} either $v_i \xrightarrow{l_i} v_{i+1} \in E$ or $v_{i+1} \xrightarrow{l_i^-} v_i \in E$. The semipath p *conforms* to regular expression R if $l_1 \cdots l_n \in L(R)$, the language denoted by R.

Given a single-conjunct query Q and graph G, let θ be a matching from variables and constants of Q to nodes of G that maps each constant to itself. We say that the tuple $\theta(Z_1, Z_2)$ *satisfies* Q on G if there is a semipath from $\theta(X)$ to $\theta(Y)$ which conforms to R. The *exact answer* of Q on G, denoted $Q(G)$, is the set of tuples which satisfy Q on G.

2.2 Approximate Semantics

We also allow for approximate answers to queries using approximate regular expression matching. As stated in the Introduction, we can in fact use a weighted regular transducer to specify the approximations, but it is easier to explain using only approximate regular expression matching. The edit operations that we allow for approximate matching are insertions, deletions, inversions, substitutions, and transpositions of adjacent symbols. The user can specify which, if any, of these edit operations should be undertaken by the system when answering a particular query. A different cost can be associated with applying each operation.

For strings, the edit distance from string w to string v is the minimum cost of any sequence of edit operations which transforms w to v. In our case, we treat the labels on edges as atomic and it is sequences of such labels which are transformed using edit operations. The edit distance from a semipath w to a semipath v is the minimum cost of any sequence of edit operations which transforms the sequence of edge labels of w to the sequence of edge labels of v; for the sake of simplicity, in this paper we assume that all edit operations have a cost of 1. The distance of a semipath w to a regular expression R, denoted *dist*(w, R), is the minimum edit distance from w to any semipath that conforms to R.

Given a matching θ from variables and constants of a query Q to nodes in a graph G, we have that the tuple $\theta(Z_1, Z_2)$ has distance *dist*(p, R) to Q, where p is a semipath from $\theta(X)$ to $\theta(Y)$ which has the minimum edit distance to R of any semipath from $\theta(X)$ to $\theta(Y)$ in G. Note that if p conforms to R, then $\theta(Z_1, Z_2)$ has distance zero to Q.

Given a graph $G = (V, E)$ and single-conjunct query Q, the *approximate top-k answer* of Q on G is a list containing the k tuples $\theta(Z_1, Z_2)$ with minimum distance to Q, ranked in order of increasing distance to Q. The *approximate answer* of Q on G contains all the tuples at any distance to the query (ranked by distance). Notice that for $k = |E|^2$, the approximate answer is equal to the approximate top-k answer.

2.3 Approximate Automaton

In this section we will introduce the notion of the approximate automaton of a regular expression R: the *approximate automaton of* R *at distance* d, where d is an integer, accepts all strings at distance at most d from R.

For any regular expression R we can construct an NFA M_R to recognise $L(R)$ using Thompson's construction. This ensures that M_R has a single initial state, denoted s_0, a single final state, denoted s_f, and $O(|R|)$ states. Then, the approximate automaton at distance d is obtained following a standard construction used in approximate string matching [15].

We make d copies of M_R, each copy denoted M_R^j, $0 \leq j \leq d$, whose states are those of M_R with superscript j, representing distance j from the original automaton M_R. The only initial state in the approximate automaton M is s_0^0, in other words the initial state of M_R at distance 0. The final state of each M_R^j remains a final state in the approximate automaton.

Subautomaton M_R^j is connected to subautomaton M_R^{j+1} by $O(|R|)$ transitions for deletions, $O(|\Sigma| \cdot |R|)$ transitions for insertions, $O(|\Sigma| \cdot |R|)$ transitions for substitutions, $O(|R|)$ transitions for inversions, and $O(|R|^2)$ transitions for transpositions.

Figure 2 shows the skeleton of the construction for insertions and deletions. Transitions for deletions are labelled with ϵ, and transitions for insertions are labelled with Σ, Σ^- (because we may insert a label or an inverse label). Figure 3 shows the additional states and transitions needed for transpositions, inversions, and substitutions.

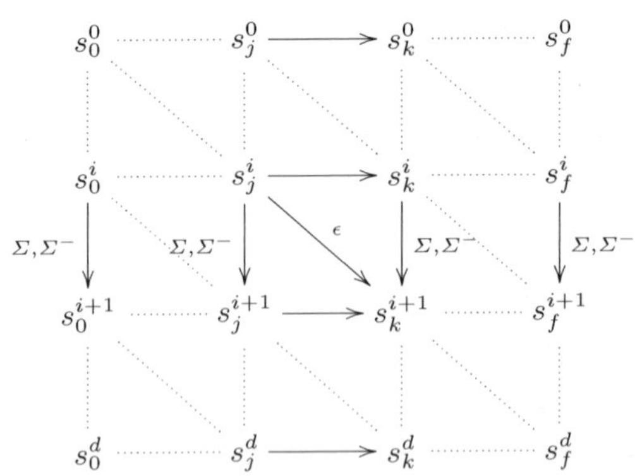

Fig. 2. Approximate automaton at level d (insertions and deletions only)

The following lemma shows two basic properties of the approximate automaton.

Lemma 1. *Let* $G = (V, E)$ *be a database,* R *be a regular expression, and* M *be the approximate automaton of* R *at distance* d. *(i)* M *has size* $O(d \cdot (|R| + |\Sigma| \cdot |R| + |R|^2))$. *(ii) If* $d = |R| + |E|$, *then every semipath in* G *conforms to* M.

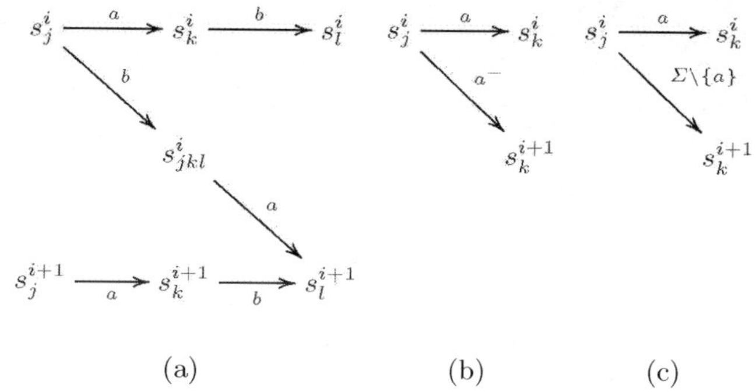

Fig. 3. Transitions for (a) transpositions, (b) inversions and (c) substitutions

Proof. (i) This follows directly from the definition of the approximate automaton. (ii) Any semipath w in G can be accepted by the approximate automaton through transitions labelled with ϵ that delete all the symbols that appear in R followed by transitions that add all the labels of w. □

2.4 Evaluation of Single-Conjunct Queries

Consider the single-conjunct query Q:

$$Z_1, Z_2 \leftarrow (X, R, Y)$$

In order to evaluate Q on graph G, we start by constructing the approximate automaton M of R at distance $d = |R| + |E|$. We next form the product automaton $H = M \times G$, viewing each node in G as both an initial and a final state. H contains as nodes all the pairs (x, y) such that x is a node of M and y is a node of G, and edges $((x_1, y_1), (x_2, y_2))$ labelled l such that there is an edge (x_1, x_2) labelled l in M and there is an edge (y_1, y_2) labelled l in G.

We associate a cost with each transition in the product automaton as follows. Any transition between a pair of states such that both derive from states of the approximate automaton M with the same superscript has cost 0, while all other transitions have cost 1.

Assume for the moment that X is a node n of G. We then perform a uniform cost traversal of H starting from vertex (s_0^0, n). We keep a list of visited nodes, so that no node is visited twice. The size of the search tree is in $O(|V|)$, where V is the set of vertices of H. Whenever we reach a vertex (s_f^i, m) in H we output m. The distance of m to the query Q is given by the total cost of the path from (s_0^0, n) to the vertex (s_f^i, m) in the traversal tree.

When both Z_1 and Z_2 are variables and assuming that Z_1 equals X, Q can be evaluated by answering the query $n, Z_2 \leftarrow (n, R, Y)$ for each node n in G.

Lemma 2. *Let $G = (V, E)$ be a graph and Q be a single-conjunct query using regular expression R. Assuming $|E| > |R|$, the approximate answer of Q on G can be computed in time $O(|V| \cdot |E|^2 \cdot |R|^2)$.*

Proof. Assuming $|E| > |R|$, the size of the approximation automaton M of R at distance $|R| + |E|$ is $O(|E| \cdot |R|^2)$. The size of $H = M \times G$ is $O(|E|^2 \cdot |R|^2)$ (we can discard disconnected nodes from H). So computing the approximate answer takes $O(|V| \cdot |E|^2 \cdot |R|^2)$, that is $|V|$ traversals of H, each one at cost equal to the size of H. □

Example 3. Consider the graph depicted in Fig. 1 and a slight modification of the query from Example 2:

$$?X \leftarrow (Santiago, (airplane)^+, ?X)$$

Figure 4 shows the approximate automaton of R at distance $d = 2$, where we use a for *airplane* and n for *name* and, for simplicity, we consider only insertions of n or n^- and inversion of a.

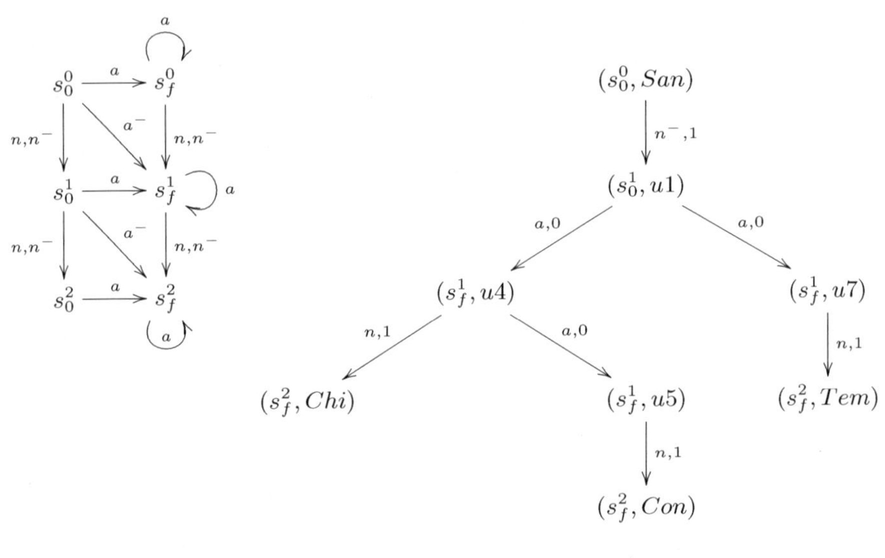

Fig. 4. The approximate automaton ($d = 2$) for $R = a^+$ and a traversal tree

Figure 4 also shows the traversal tree starting from (s_0^0, San), where *San*, *Chi*, *Con* and *Tem* are abbreviations for *Santiago*, *Chillan*, *Concepcion* and *Temuco*, respectively. Here we assume that all operations have cost 1.

As we saw previously, there are no answers at rank 0. The answers at rank 1 (indicated by being associated with state s_f^1) are $u4$, $u5$ and $u7$, while those at rank 2 are *Chi*, *Con* and *Tem*. □

2.5 Incremental Evaluation

We may compute edges of the graph $H = M \times G$ above incrementally, avoiding the precomputation and materialization of the entire graph H. Recall that a vertex in H is a pair (s_i^j, n), where s_i^j is a state of automaton M and n is a node of G. Each edge of H is labelled with a symbol and a weight.

The on-demand computation of edges of H is performed by calling a function *Succ* with a node (s_i^j, n) of H. The function returns a set of transitions $\overset{e,w}{\rightarrow} (s_k^l, m)$, such that there is an edge in H from (s_i^j, n) to (s_k^l, m) with label e and weight w. As an example, in Fig. 4 $Succ(s_f^1, u4)$ returns the following transitions: $\overset{a,0}{\rightarrow} (s_f^1, u5)$ (without relaxation—cost 0) and $\overset{n,1}{\rightarrow} (s_f^2, Chi)$ (insertion of label n at cost 1).

A procedure for computing *Succ* is shown overleaf. Here, M_R is the original NFA constructed to recognise $L(R)$. The function $\texttt{nextStates}(M_R, s, c)$ returns the set of states in M_R that can be reached from state s on reading input c. We assume that this function can also take as an argument a state s^i in the approximate automaton at distance i from M_R. If $\texttt{nextStates}(M_R, s, c)$ returns $\{s_1, s_2, \ldots, s_k\}$, then $\texttt{nextStates}(M_R, s^i, c)$ returns $\{s_1^i, s_2^i, \ldots, s_k^i\}$.

Incremental evaluation proceeds by performing the initialisations shown below, where d is used to record distances of nodes, $\texttt{visited}_R$ stores triples (v, n, s) representing the fact that node n was visited in state s starting from node v, and Q_R is a priority queue ordered by distance. The procedure *getNext* (see overleaf) returns the next tuple for a conjunct in ranked order.

Incremental evaluation for a single-conjunct query

---

begin
    construct NFA M_R for R, with initial state s_0 and final state s_f
    $\texttt{visited}_R \leftarrow \emptyset$
    $d \leftarrow 0$
    foreach *node n in G* **do** *enqueue*$(Q_R, (n, n, s_0, d))$
    while $(v, n, d) \leftarrow$ *getNext*$(X, R, Y) \neq$ *null* **do** output (v, n, d)
end

In particular, when called with conjunct (X, R, Y), the procedure *getNext* begins by dequeuing a tuple (v, n, s, d) from Q_R, where v and n are nodes in G, s is a state in the NFA M_R, and d is the distance associated with visiting n in state s having started from v. If s is the final state then the triple (v, n, d) is returned. Otherwise we continue to traverse G from n.

Example 4. If we apply our incremental evaluation algorithm to Example 3, we only need to enqueue $(San, San, s_0, 0)$ initially because X is the constant San. This tuple is dequeued by *getNext* and $Succ(s_0, San)$ is called. Since the only edge adjacent to San in G is one labelled *name* with target San and source $u1$, and there is no transition labelled with n^- from s_0 in M_R, *Succ* adds $\overset{n^-,1}{\rightarrow} (s_0^1, u1)$ to W and returns. This results in $(San, u1, s_0^1, 1)$ being enqueued.

The while loop now repeats and $(San, u1, s_0^1, 1)$ is dequeued. $Succ(s_0^1, u1)$ is called and because there is a transition labelled *airplane* from s_0^1 to s_f^1 in M_R and edges in G from $u1$ to $u4$ and $u7$ labelled *airplane*, $\overset{a,0}{\rightarrow} (s_f^1, u4)$ and $\overset{a,0}{\rightarrow} (s_f^1, u7)$ are

Procedure `Succ`(s^i, n)

Input: state s^i and graph node n
Output: set of transitions which are successors of (s^i, n)

```
W ← ∅
for (n, a, m) ∈ G do
    for p^i ∈ nextStates(M_R, s^i, a) do
        add -a,0-> (p^i, m) to W ;                          /* normal traversal */
for (m, a, n) ∈ G do
    for p^i ∈ nextStates(M_R, s^i, a^-) do
        add -a^-,0-> (p^i, m) to W ;                        /* reverse traversal */
for (n, a, m) ∈ G such that nextStates(M_R, s^i, a) = ∅ do
    add -a,1-> (s^{i+1}, m) to W ;                          /* insertion of a */
for (m, a, n) ∈ G such that nextStates(M_R, s^i, a^-) = ∅ do
    add -a^-,1-> (s^{i+1}, m) to W ;                        /* insertion of a^- */
for p^i ∈ nextStates(M_R, s^i, b) for each b ∈ Σ do
    add -ε,1-> (p^{i+1}, n) to W ;                          /* deletion of b */
for (n, a, m) ∈ G and (m, b, u) ∈ G do
    for p^i ∈ nextStates(M_R, s^i, a) and q^i ∈ nextStates(M_R, p^i, b) do
        add -ba,1-> (q^{i+1}, u) to W ;                     /* swap of a and b */
return W
```

Procedure `getNext`(X, R, Y)

Input: node in query evaluation tree corresponding to conjunct (X, R, Y)
Output: triple (v, n, d), where v and n are instantiations of X and Y

```
while nonempty(Q_R) do
    (v, n, s, d) ← dequeue(Q_R)
    add (v, n, s) to visited_R
    if s = s^i_f for some i then return (v, n, d)
    foreach -c,w-> (s', m) ∈ Succ(s, n) s.t. (v, m, s') ∉ visited_R do
        enqueue(Q_R, (v, m, s', d + w))
return null
```

added to W. Although the *Succ* procedure shown would also consider traversals to $u2$ and $u3$ by inserting *bus* and *train* respectively, recall that we are assuming for this example that only *name* and $(name)^-$ can be added. After *Succ* returns, $(San, u4, s_f^1, 1)$ and $(San, u7, s_f^1, 1)$ are enqueued. These tuples are at the front of the queue and when dequeued result in $(San, u4, 1)$ and $(San, u7, 1)$ being returned because s_f^1 is a final state. The algorithm continues in this way until the entire tree shown in Fig. 4 has been traversed. □

3 Multi-conjunct Queries

Assume that we have a conjunctive query Q consisting of n conjuncts:

$$(Z_1, \ldots, Z_m) \leftarrow (X_1, R_1, Y_1), \ldots, (X_n, R_n, Y_n)$$

in which each X_i and Y_i, $1 \leq i \leq n$, is a variable or constant, and each Z_i is a variable appearing in the body of Q. Given a matching θ from variables and constants of Q to nodes in a graph G, we have that the tuple $\theta(Z_1, \ldots, Z_m)$ has distance $dist(p_1, R_1) + \cdots + dist(p_n, R_n)$ to Q, where each p_i is the semipath from $\theta(X_i)$ to $\theta(Y_i)$ which has the minimum edit distance to R_i of any semipath from $\theta(X_i)$ to $\theta(Y_i)$ in G. The *approximate top-k answer* of Q on G is the list of k tuples of the form $\theta(Z_1, \ldots, Z_m)$ with minimum distance to Q, ranked in order of increasing distance to Q.

We also require that the conjuncts of a query Q are *acyclic*, which is needed to ensure polynomial-time evaluation[2] and means that there is a join tree T whose nodes are the conjuncts of Q such that, for every variable in Q, the subgraph of T induced by the conjuncts containing the variable is connected.

Our query evaluation algorithm is based on one of the rank-join algorithms developed by Ilyas et al. [14], specifically their hash ripple join. That algorithm takes as input m relations, each of whose sets of tuples has already been ranked according to some citerion, a join condition, a monotone combining ranking function and the number k of ranked results to be output. Ranking in their case is by decreasing value of some score, whereas in our case it is by increasing distance from Q.

In our case, given a graph $G = (V, E)$, for each conjunct (X_i, R_i, Y_i) we can use the algorithm described in Sect. 2.5 to compute a relation r_i over scheme (X_i, Y_i, D_i) where, for tuple $t \in r_i$, $t[D_i]$ represents the minimum distance from R_i of any path from $t[X_i]$ to $t[Y_i]$ in G. Joining the results then uses the hash ripple join of [14].

Lemma 3. *Given a graph G and a multi-conjunct query Q with a fixed number of head variables m, the approximate answer of Q on G can be computed in time polynomial in the size of Q and G.*

Proof. Each conjunct in Q can be evaluated in polynomial time, as shown in Lemma 2. The result follows from the acyclicity condition and the fixed number of head variables [16]. □

Because Q is acyclic, we can construct a query evaluation tree E for Q, which consists of nodes denoting join operators and nodes representing conjuncts of Q. We initialise all the algorithm structures by calling the procedure *open*, shown below, with the root of E. The part of procedure *open* dealing with nodes denoting joins is adapted from [14]: hash tables are built for each child node (operands LN and RN) and *open* is called recursively for each; the $firstTuple$ property of

[2] It is well known that the problem of deciding if the natural join of n relations is nonempty is NP-complete.

each is set to *true* (this is used in the *getNext* procedure—see below). The part of *open* dealing with nodes denoting conjuncts (X, R, Y) performs the same initialisations as in Sect. 2.5. For notational simplicity, we assume that each regular expression in a conjunctive query Q is distinct. Hence, for conjunct (X, R, Y) we can use M_R to denote its NFA, Q_R its priority queue, etc.

Procedure `open(`N`)`

Input: node N in the query evaluation tree
allocate a priority queue Q_N associated with node N
if N *is* $LN \bowtie RN$ **then**
  build hash tables for LN and RN
  $open(LN)$; $LN.firstTuple \leftarrow true$
  $open(RN)$; $RN.firstTuple \leftarrow true$
  $T_{LN \bowtie RN} \leftarrow 0$; `/*` T `is the threshold */`
else `/*` N `is a conjunct` (X, R, Y) `*/`
  construct NFA M_R for R, with initial state s_i and final state s_f
  `visited`$_R \leftarrow \emptyset$
  $d \leftarrow 0$
  foreach *node* n *in* G **do** $enqueue(Q_R, (n, n, s_i, d))$

Incremental evaluation of query Q proceeds by calling the procedure *getNext* with the root of the join tree E. There are two versions of *getNext*, one for when the evaluation tree node represents a join, shown overleaf and adapted from [14], and one for when it represents a conjunct, which was discussed previously in Sect. 2.5.

Procedure `getNext(`$LN \bowtie RN$`)`

Input: node N (corresponding to a join) in the query evaluation tree
while $empty(Q_{LN \bowtie RN})$ **or** $head(Q_{LN \bowtie RN})[D] > T_{LN \bowtie RN}$ **do**
  determine next input I // either LN or RN
  $t \leftarrow getNext(I)$
  if $I.firstTuple$ **then**
    $I_{top} \leftarrow t[D]$
    $I.firstTuple \leftarrow false$
  $I_{bottom} \leftarrow t[D]$
  $T_{LN \bowtie RN} \leftarrow \min(LN_{top} + RN_{bottom}, LN_{bottom} + RN_{top})$
  insert t in hash table for I
  probe other hash table with t
  foreach *valid join combination* u *of* t *with* s, *say* **do**
    $u[D] \leftarrow t[D] + s[D]$
    $enqueue(Q_{LN \bowtie RN}, u)$
return $dequeue(Q_{L \bowtie R})$

The *getNext* procedure for a join begins by choosing from which operand of the join to retrieve tuples. We do not consider here the various heuristics that might be used to decide this. For a tuple t, we denote the distance value of t by $t[D]$. For input I (either LN or RN), I_{top} represents the distance value of the first tuple retrieved from I (i.e., the smallest distance in I), while I_{bottom} represents the distance of the most recently retrieved tuple from I. Although I_{top} will always be 0 when I is a conjunct, this will not necessary be the case when I is a join. The threshold value T represents the smaller of the two distances given by $LN_{top} + RN_{bottom}$ and $LN_{bottom} + RN_{top}$. These two values give the possible distances arising from joining the first tuple of LN with the most recent tuple from RN, or vice versa, either of which remains possible until the end of the operation. The smaller of these two distances gives the smallest possible distance for a join tuple that has yet to be computed; in other words, no tuple that might result from the join of tuples yet to be retrieved with tuples already retrieved or yet to be retrieved can have a distance less than T. It is therefore safe to output a join tuple whose distance is equal to T.

4 Related Work and Generalisations

Jagadish et al. [11] develop a framework for similarity-based queries, but it turns out to be too powerful to permit efficient query evaluation. For this reason, Grahne and Thomo [9] restrict themselves to using weighted regular transducers for performing transformations to regular path queries in order for them to match semi-structured data approximately. Their regular path queries consist only of a single conjunct and do not include inversions.

Similarity-based querying is also the focus of iSPARQL [13], where resources are compared using similarity measures. One similarity measure used is that of edit distance between strings. However, similarity is measured with respect to the resources themselves rather than the paths connecting resources; thus it is complementary to our approach.

SPARQLeR extends SPARQL with regular path expressions designed for querying semantic associations [5]. Only exact answers are considered. nSPARQL adds *nested* regular expressions to SPARQL and shows that these are necessary in order to answer queries using the semantics of the RDFS vocabulary by directly traversing the RDF graph [6]. It would be interesting to see whether our approach of augmenting ordinary regular expressions with a regular transducer could similarly achieve this.

Kanza and Sagiv consider querying semistructured data using flexible matchings which allow paths whose edge labels simply contain those appearing in the query to be matched [12]. Such semantics can be captured by our approach by allowing transpositions and insertions as the only edit operations.

In cooperative query answering, overconstrained queries are automatically relaxed [8,10]. Dolog et al. [8] use conditional rewriting rules on query patterns to perform both query refinement by including user preferences as well as query relaxation. Our edit operations and, more generally, regular transducers also effectively introduce a form of query rewriting, so it would be interesting to compare the relative expressive power of the two approaches. Our previous work in [10] was concerned with relaxing queries on RDF by considering query generalisation with respect to an RDFS vocabulary (see that paper for references to other work adopting a similar approach). The substitution edit operation of our work here subsumes the property relaxation semantics used in [10].

As suggested by Example 2 where a user was prepared to consider travelling by bus rather than by train, our approach in this paper has some overlap with work which allows users to specify preferences in queries (e.g. [17]).

We can also apply our work to keyword search in relational databases. Given nodes v_1 to v_n in G and assuming that only insertion edit operations are allowed in approximate matchings, the query

$$X \leftarrow (X, \epsilon, v_1), \ldots, (X, \epsilon, v_n)$$

returns as its first result a *median* of G, that is, a node with minimum total distance to v_1 to v_n. A median is effectively what the BANKS algorithm finds [18], where v_1 to v_n represent keywords in a relational database.

5 Conclusions

In this paper, we have developed a framework within which we can model various aspects of approximate answers to queries on semistructured resources including RDFS. We have presented incremental algorithms which allow answers to be returned to a user in ranked order in polynomial time.

There are several improvements and generalisations which we are investigating. One obvious one is a mechanism to allow users and/or interface designers to specify their requirements in terms of approximation. Another is the ability to have the paths themselves returned to users, rather than simply nodes.

We are currently working on an implementation of the algorithms presented above. This will allow us to determine both the utility and the practical efficiency of the various operations for approximate matching.

Acknowledgements. This work was supported by the Royal Society under their International Joint Projects Grant Programme. In addition, Carlos Hurtado was partially funded by Fondecyt project number 1080672.

References

1. Berners-Lee, T., Chen, Y., Chilton, L., Connolly, D., Dhanaraj, R., Hollenbach, J., Lerer, A., Sheets, D.: Tabulator: Exploring and analyzing linked data on the semantic web. In: Proc. 3rd Int. Semantic Web User Interaction Workshop (2006)
2. Heath, T., Hausenblas, M., Bizer, C., Cyganiak, R.: How to publish linked data on the web (tutorial). In: Proc. 7th Int. Semantic Web Conf. (2008)
3. Calvanese, D., Giacomo, G.D., Lenzerini, M., Vardi, M.Y.: Containment of conjunctive regular path queries with inverse. In: Proc. Seventh Int. Conf. on Principles of Knowledge Representation and Reasoning, pp. 176–185 (2000)
4. Cruz, I.F., Mendelzon, A.O., Wood, P.T.: A graphical query language supporting recursion. In: Proc. ACM SIGMOD Conf., pp. 323–330 (1987)

5. Kochut, K., Janik, M.: SPARQLeR: Extended SPARQL for semantic association discovery. In: Franconi, E., Kifer, M., May, W. (eds.) ESWC 2007. LNCS, vol. 4519, pp. 145–159. Springer, Heidelberg (2007)
6. Pérez, J., Arenas, M., Gutierrez, C.: nSPARQL: A navigational language for RDF. In: Sheth, A., et al. (eds.) ISWC 2008. LNCS, vol. 5318, pp. 66–81. Springer, Heidelberg (2008)
7. Prud'hommeaux, E., Seaborne, A. (eds.): SPARQL Query Language for RDF. W3C Candidate Recommendation (April 6, 2006)
8. Dolog, P., Stuckenschmidt, H., Wache, H.: Robust query processing for personalized information access on the semantic web. In: Larsen, H.L., Pasi, G., Ortiz-Arroyo, D., Andreasen, T., Christiansen, H. (eds.) FQAS 2006. LNCS, vol. 4027, pp. 343–355. Springer, Heidelberg (2006)
9. Grahne, G., Thomo, A.: Approximate reasoning in semi-structured databases. In: Proc. 8th Int. Workshop on Knowledge Representation meets Databases (2001)
10. Hurtado, C.A., Poulovassilis, A., Wood, P.T.: Query relaxation in RDF. Journal on Data Semantics X, 31–61 (2008)
11. Jagadish, H.V., Mendelzon, A.O., Milo, T.: Similarity-based queries. In: Proc. Fourteenth ACM Symp. on Principles of Databases Systems, pp. 36–45 (1995)
12. Kanza, Y., Sagiv, Y.: Flexible queries over semistructured data. In: Proc. Twentieth ACM Symp. on Principles of Databases Systems, pp. 40–51 (2001)
13. Kiefer, C., Bernstein, A., Stocker, M.: The fundamentals of iSPARQL: A virtual triple approach for similarity-based semantic web tasks. In: Aberer, K., Choi, K.-S., Noy, N., Allemang, D., Lee, K.-I., Nixon, L., Golbeck, J., Mika, P., Maynard, D., Mizoguchi, R., Schreiber, G., Cudré-Mauroux, P. (eds.) ASWC 2007 and ISWC 2007. LNCS, vol. 4825, pp. 295–309. Springer, Heidelberg (2007)
14. Ilyas, I.F., Aref, W.G., Elmagarmid, A.K.: Supporting top-k join queries in relational databases. The VLDB Journal 13, 207–221 (2004)
15. Wu, S., Manber, U.: Fast text searching allowing errors. Commun. ACM 35(10), 83–91 (1992)
16. Gottlob, G., Leone, N., Scarcello, F.: The complexity of acyclic conjunctive queries. J. ACM 43(3), 431–498 (2001)
17. Siberski, W., Pan, J.Z., Thaden, U.: Querying the semantic web with preferences. In: Cruz, I., Decker, S., Allemang, D., Preist, C., Schwabe, D., Mika, P., Uschold, M., Aroyo, L.M. (eds.) ISWC 2006. LNCS, vol. 4273, pp. 612–624. Springer, Heidelberg (2006)
18. Bhalotia, G., Hulgeri, A., Nakhe, C., Chakrabarti, S., Sudarshan, S.: Keyword searching and browsing in databases using BANKS. In: Proc. 18th Int. Conf. on Data Engineering, pp. 431–440 (2002)

Tempus Fugit*

Towards an Ontology Update Language

Uta Lösch, Sebastian Rudolph, Denny Vrandečić, and Rudi Studer

Institut AIFB – Universität Karlsruhe (TH)
Karlsruhe, Germany
{uhe,sru,dvr,rst}@aifb.uni-karlsruhe.de

Abstract. Ontologies are used to formally describe domains of interest. As domains change over time, the ontologies have to be updated accordingly. We advocate the introduction of an Ontology Update Language that captures frequent domain changes and hence facilitates regular updates to be made in ontologies. We thoroughly discuss the general design choices for defining such a language and a corresponding update framework. Moreover, we propose a concrete language proposal based on SPARQL Update and provide a reference implementation of the framework.

1 Introduction and Motivation

Ontologies (see [12] for a comprehensive overview) are formal specifications of the knowledge about a domain of interest. They constitute one of the central concepts in the field of Semantic Web technologies and facilitate information integration and exchange as well as semantic search. Usually, their expressive power exceeds that of traditional databases and allows to infer new information that is not explicitly present in the specification but a logical consequence of it (the so-called *implicit knowledge*).

Currently, the most commonly used ontology languages are RDF and OWL, both standardized by the World Wide Web Consortium (W3C). RDF ([7], in particular the RDF Schema extension) constitutes a so-called "leightweight" ontology language, providing basic modeling features for assertional (instance-related) and terminological knowledge handling classes, binary relations (so-called properties), hierarchies of classes (also referred to as taxonomies) and properties as well as domain and range specifications for properties. OWL (the Web Ontology Language, [8]) constitutes the other well-established knowledge representation formalism for the Semantic Web, based on Description Logics [1], a family of decidable yet expressive logics for which highly optimized inferencing implementations exist. Syntactically, OWL can be represented in RDF(S) where certain vocabulary elements impose stronger restrictions on the semantics.

* Supported by the Deutsche Forschungsgemeinschaft (DFG) under the ReaSem project and the Graduate School Information Management and Market Engineering and by the European Commission through the IST project ACTIVE (ICT-FP7-215040).

L. Aroyo et al. (Eds.): ESWC 2009, LNCS 5554, pp. 278–292, 2009.

Following the uptake of ontology-based technologies by industry, the comprehensive modeling of complex domains will become necessary. We expect future ontologies to rarely be developed by a single person starting from scratch. Rather collaborative design and development of ontologies and continuous refinement will become the usual scenarios for which elaborate methodologies and appropriate tool support will become crucial (see for example [15,14]).

In this spirit, *ontology change* is one of the fundamental issues to be addressed. Obviously, there are diverse reasons for changing an ontology. We argue that it is helpful to conceptually distinguish two cases by answering the question:

Is the ontology changed due to an according change in the described domain?

Note that there are many cases where the answer to that question would be no: an ontology change might be the consequence of the discovery of modeling errors (*ontology repair*) or the acquisition of new additional domain knowledge (*ontology amendment*). Obviously, this type of ontology changes reflects a change in the way the modeler conceives or formalizes the domain of interest. As an example, consider the case that new findings in genetics might imply that the taxonomy of living beings has to be corrected in order to properly reflect the current knowledge of the real situation. A lot of research has been devoted to this kind of ontology change (employing techniques from belief revision, knowledge acquisition, ontology learning and ontology evolution to name just a few). We propose the term *ontology refinement* to subsume all those activities.

As opposed to those, we will be concerned with the task that we refer to as *(temporal) ontology update*: changes to the ontology might become necessary as the underlying domain changes over time. As time passes, the state of affairs in the domain like a person's employer or her academic title may change. However, such changes are not restricted to assertional knowledge, but may also concern the schema. As an example consider EU membership: it is expected that additional countries become member of the EU, thereby changing the terminological definition of the class of EU citizens.

In many cases it will be possible to come up with *change patterns* which describe typical ways in which a domain may evolve. For example, a domain individual recorded as underage may turn into an adult, while the opposite is impossible. While most of these patterns will probably deal with the update of assertional knowledge, they may also occur on the schema level (as an example consider a country becoming member of the EU).[1]

The observation that such change patterns exist leads to the idea to formally specify typical ways in which an ontology may be updated over time. As an example, in a biological domain, an individual might cease to be member of the class `Caterpillar` and become member of the class `Butterfly` instead. These update specifications may concern schema knowledge as well as fact knowledge. On a more general level, update specifications allow to encode process knowledge and associate it with the ontology, such that it can be used for updates.

---

[1] While here we focus on change patterns that reflect temporal and domain-specific changes, we are aware that such patterns may be identified beyond this use case (our approach may thus be applicable in other change scenarios, too).

The specifications can be seen as operational descriptions how to update an ontology as a consequence to information entered into the system. However, in contrast to common Ontology Evolution approaches [13], we propose to base those updates on domain specific knowledge about temporal changes. Using the above mentioned information, a natural reaction to asserting that an individual `willie` is now a class member of `Butterfly` would be to also retract the information that `willie` is an instance of `Caterpillar` [3].

This way, an ontology change requested by a person responsible for ontology maintenance can be supplemented by additional changes. This allows to prevent modeling flaws that might occur due to only partially entered information. Generally, an ontology update specification allows constraining ontology changes to clearly defined, foreseen cases in a domain-specific way.

Then, frequent maintenance or update tasks can be transferred from knowledge engineers (roughly: the "ontology administrator") to knowledge workers (possibly formally less skilled users in charge of monitoring changes in the domain of interest and transferring them into the ontology) while minimizing the risk of introducing errors. Like the actual ontology, the according update specification has to be created by a knowledge engineer who is also in charge of ontology refinement activities as well as addressing unforeseen changes not anticipated by the update specification that might become necessary.

In this paper, we elaborate on our idea of a framework and a language for ontology updates. Section 2 explains our design choices when working out the overall and more detailed aspects of an ontology update language. In Section 3 we briefly sketch the architecture of a system where ontology update specifications are employed. Section 4 provides the formal specification of our proposed Ontology Update Language and Section 6 describes the details of our reference implementation of an according ontology update system. In Section 5 the benefits of our approach will be illustrated by means of a small example before we conclude and provide directions for future research in Section 7.

2 Design Choices

In this section, we review the major choices and questions to be addressed when designing a framework for ontology update management.

2.1 Ontology-Inherent Temporal Knowledge vs. External Specification

One approach to capture temporal changes in a domain is to use logic formalisms that allow for their description inside ontological specifications. There is a plethora of formalisms and approaches such as temporal logics or situation calculi that provide a logic-inherent way of describing temporal and dynamic phenomena in the domain. Clearly, these approaches have advantages whenever the intended use of the ontology includes reasoning over domain changes (maybe even planning). However, besides the more complicated formalisms, a usual drawback of such formalisms is the high reasoning complexity.

Note that our goal is much more moderate: from the above described, it becomes clear that we aim at designing an operational formalism that – given a change request – deterministically comes up with an updated ontology in a timely manner. Moreover, we would like to stick to the usual approach that an ontology encodes a static description of the domain, hence every state of the actual ontology should be considered a kind of "snapshot".

Therefore, we adopt an approach keeping the actual ontology and the according update specification distinct, which also enables to use off-the-shelf reasoners for dealing with the ontology part.

2.2 Unguided Belief Revision vs. Guided Update

Although most of the work done in the area of belief revision has dealt with scenarios of ontology refinement (see [6] for an RDFS-based formal framework and [10] for a survey), some proposals have been made to also address the update scenario [5]. Notwithstanding, belief revision approaches try to resolve inconsistencies that were introduced by an update. However, many changes in the ontology will not lead to an inconsistent state, thus no additional changes are performed. Therefore, the applicability of belief revision is restricted to formalisms expressive enough to cause inconsistencies. While this is certainly the case for OWL, causing "meaningful" inconsistencies in RDF(S) is virtually impossible.[2] To a certain extent, this fallback can be mitigated by adding additional constraints beyond RDF(S) on top of an RDF(S) knowledge base[3] [6].

As another downside of belief revision, note that the strategies to restore consistency do not take domain specifics into account. To illustrate that, consider the following example: let a knowledge base contain the disjointness of the classes `Adult` and `Child` and the fact that `Peter` belongs to the class `Child`. If we now add that `Peter` is also an instance of the class `Adult`, the knowledge base becomes inconsistent and (if configured appropriately) a belief revision approach would retract `Child(Peter)`, as newly added facts override those already present. While this is the desired behavior, re-adding `Child(Peter)` to the new knowledge base would lead to the deletion of `Adult(Peter)` irrespective of the actual irreversibility of this development in the described domain. Clearly, a more appropriate "reality-aware" reaction of an update mechanism would be to reject the second change request.

As opposed to belief revision, our approach aims at preventing inconsistent ontology states that might arise from incomplete change requests. To this end, change requests are completed by further ontology changes based on specified knowledge about how a domain may develop. This way, consistency can be preserved; as a worst case, the change request can be denied.

[2] More precisely: in RDF(S) inconsistencies can only be provoked by so-called XML-clashes, which is more a datatype-related unintentional peculiarity than a design feature.

[3] Note that we use the terms ontology and knowledge base interchangeably throughout the paper.

Note however, that those two approaches are not mutually exclusive but could be combined: in case of a change request not matching any of the anticipated update patterns, applying a belief revision "fallback solution" might sometimes be preferable to simply rejecting the request.

2.3 Syntactic vs. Semantic Preconditions

In most cases, the best way to react to a change request will depend on the current state of the ontology. Hence it is crucial to provide the opportunity to formulate respective preconditions for triggering changes. There are essentially two distinct kinds of checks that can be done against an ontology: *semantic* and *syntactic*[4] ones.

If some changes should be made depending on the validity of some statement in the ontology, we have to employ reasoning in order to decide whether the statement is logically entailed by the given information. As a special case of this, one could diagnose whether an intended change would turn the ontology inconsistent and reject the requested change on these grounds. Semantic checks provide the more thorough way of testing the knowledge contained in an ontology, however the reasoning to be employed may be expensive with respect to memory and runtime.

The alternative would be to just syntactically determine whether certain axioms are literally contained in a knowledge base. This would be less expensive than the semantic approach. Yet usually, there are many possible ways to syntactically express one piece of semantic information making a naïve syntactic "pattern matching" approach problematic at best. One way to mitigate that problem while still avoiding to engage in heavy-weight reasoning would be to syntactically normalize the ontology and the change request. That is, the ontology is transformed into a semantically equivalent, but syntactically more constrained form, facilitating to identify and manipulate pieces of semantic information by purely syntactic analyses.

Since both kinds of preconditions are useful under different circumstances, we argue that an ontology update formalism should offer both options leaving to the knowledge engineer to decide which one should be used in a specific case.

2.4 Change Feedback

It has to be expected that the changes mediated by an ontology update specification might not be directly obvious for the knowledge worker. However, it is clearly crucial to ensure that the system's behavior is as transparent as possible to the knowledge worker. For this reason, feedback about the automated reactions to a change request should be an essential part in any practically employable ontology update framework.

[4] In our actual proposal, we refrain from taking a purely syntactic approach, but rather the structural level of the RDF graph that is used by SPARQL.

In order to provide informative feedback, the ontology engineer has to provide template-like explanation snippets commenting on the nature of the change patterns contained in the update specification and the changes triggered by them. At runtime, those templates instantiated with the actually changed ontology elements can be presented to the knowledge worker in order to explain what actually happened to the ontology.

3 System Architecture

In this section, we propose an abstract architecture for an ontology update framework taking into account the design choices made in the previous section. A concrete instantiation of this architecture is described in the subsequent sections. The suggested architecture is sketched in Fig. 1. Therein, the usual unguided interaction mode of committing changes directly to the ontology is complemented by an additional update management component as editing interface for the knowledge worker. Still, the knowledge engineer will be able to directly change both the ontology and the ontology update specification.

The typical work flow of an ontology update step is carried out as follows: Initially, the knowledge worker issues a change request by providing a piece of knowledge to be added to or deleted from the ontology.

A change request will only be acted upon if the accompanying Ontology Update Specification accounts for it. The Ontology Update Manager will scan the Update Specification for an update rule whose applicability conditions are satisfied by the uttered change request and the ontology. Those applicability conditions might contain syntactic as well as semantic checks. If there are several applicable update rules, only one of them is applied.

If an applicable update rule has been determined, the change request can be acted on accordingly by denying or accepting it but possibly also by carrying out more ontology changes than explicitly requested.

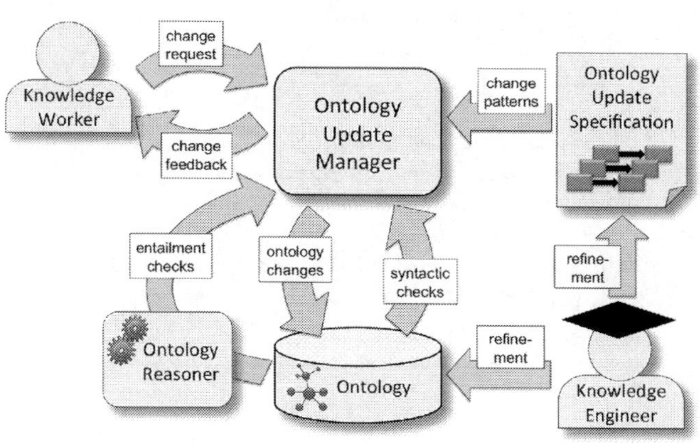

Fig. 1. Ontology Update Architecture

Finally, a feedback message describing the activated change pattern and containing the actually performed changes is generated and sent back to the knowledge worker.

In case no applicable update specification was found, the change request is denied. It is however logged such that the ontology engineer can take care of it later and also refine the Ontology Update Specifications if needed.

The proposed framework is inspired by database triggers as e.g. described in the SQL standard [4]. Database triggers are stored procedures that are activated by changes that are submitted to the database. They may be defined for insertions, updates and deletions of instance data in the database. Our approach provides the same kind of functionality for instance data in ontologies while additionally allowing for defining update specifications for schema changes, which is usually not possible with database triggers.

4 A Language Proposal

In this section, we instantiate our previous general considerations by providing an ontology update framework based on RDF(S) and SPARQL, as well as SPARQL Update. This framework consists of the syntax of the Ontology Update Language (OUL, specified in Fig. 2) together with the precise description how ontology change requests are to be handled by the ontology management component (cf. Alg. 1).

```
CREATE CHANGEHANDLER <name>
FOR <changerequest>
AS
  [ IF <precondition>
  THEN ] <actions>
<changerequest>  ::== add    [unique] (<SPARQL>)
                   |   delete [unique] (<SPARQL>)
<precondition>   ::== contains(<SPARQL>)
                   | entails(<SPARQL>)
                   | entailsChanged(<SPARQL>)
                   | (<precondition>)
                   | <precondition> and <precondition>
                   | <precondition> or <precondition>
<actions>        ::== [<action>]|<action><actions>
<action>         ::== <SPARQL update>
                   | for( <precondition> ) <actions> end;
                   | feedback(<text>)
                   | applyRequest
<SPARQL>         ::== where clause of a SPARQL query
<SPARQL update>  ::== a modify action (in SPARQL Update)
<text>           ::== string (may contain SPARQL variables)
```

Fig. 2. Ontology Update Language syntax specification in BNF

Algorithm 1. Processing of Change Request

Input: ontology $\mathcal{O}$ consisting of axioms (RDF triples – Note that in RDFS every axiom is represented by exactly one triple),
ontology update specification $\mathcal{US}$ treated as list of changehandlers,
change request $op(Ax)$ where $op \in \{\texttt{add}, \texttt{del}\}$ and Ax is a set of axioms resp. triples.
Data: *candidate* changehandler that is checked for applicability
toExecute container to store the activated changehandler
updateList list of SPARQL Updates to be carried out, initially empty
Result: Updated ontology $\mathcal{O}$

```
1  //find an appropriate changehandler
2  while toExecute.isEmpty and not US.endOfDocument do
3      candidate ← US.nextChangeHandler
4      matches ← SPARQLmatch(candidate.changerequest, op(Ax))
5      if not matches.isEmpty then
6          if matches.count == 1 or not candidate.changerequest.unique then
7              instPrecondition ←
               Substitute(candidate.precondition, matches.first)
8              if not evaluate(instPrecondition, O).isEmpty then
9                  toExecute.add(candidate)
10             end
11         end
12     end
13 end
14 //execute actions, if applicable
15 if not toExecute.isEmpty then
16     todo ← Substitute(toExecute.first.actions, matches.first)
17     cumulateActions(O, todo, updateList)
18     foreach update ∈ updateList do
19         apply update to O
20     end
21 end
22 return O
```

Every update rule (also called changehandler) has an identifier. It carries a change request pattern, expressing for which change request it can be applied and some preconditions that define whether a change request can be handled by this rule depending on the current ontology state. If several matching changehandlers exist, the first one occurring in the update specification will be applied.[5]

A change request pattern is defined by the `WHERE`-clause of a SPARQL select query that is to be evaluated on the change request. That means that the statements submitted for a change are interpreted as an RDF graph and it is checked whether the change request pattern executed as SPARQL query yields any (or, if the `unique` option is set: exactly one) result on this change graph (lines 4–6

[5] We are aware that we thereby deviate from pure declarativity. However, for a first proposal, this kind of implicit priority declaration seems both intuitive and computationally feasible.

in the Algorithm) The result of this queries are bindings of all variables that are present in the `WHERE`-clause. Those bound variables can be reused later in the preconditions and actions part.

If a match is found, the precondition of the changehandler is evaluated. This determines whether the changehandler is applicable or whether another matching one has to be found. There are three basic types of preconditions: a syntactic check (that simply verifies whether certain triples are contained in the ontology) is performed via `contains`. Semantic entailment checks may be performed on the ontology in its current, unaltered state (`entails`) or on the "hypothetical" ontology that would result from carrying out the changes as requested (`entailsChanged`). As explained above, it is desirable to provide syntactic and semantic checks on the ontology, as syntactic checks are less expensive but also less accurate than semantic checks. Syntactically, basic preconditions are also the `WHERE`-part of a SPARQL query.

Basic preconditions can be combined by `and` and `or`. As it is reasonable to allow for (yet unbound) variables to occur in several basic preconditions, `and` and `or` are realized as join and union on the result sets of the basic preconditions. In the end, a precondition is considered to be successful (line 8) if its result set contains at least one entry. Before evaluation of the precondition, all variables that occurred before in the change request pattern are substituted by their binding (line 7, in the case of several existing bindings, the first one is chosen).

If the precondition is evaluated successfully, the actions specified in the changehandler's body are applied to the ontology.[6]

As for the evaluation of preconditions, all variables occurring in the action part of the changehandler that were bound before (i.e. that were present in either the change request or in the precondition) are replaced by their binding before the action part is applied (line 16). If several possible bindings were found for the change request or the precondition, the first binding is chosen.

Elementary actions that can be carried out are knowledge base changes (expressed as SPARQL Update operations), the `applyRequest` action (carrying out the originally uttered change request), and feedback messages. Moreover, elementary actions can be nested into loops which iterate over the result set of a precondition. While executing the action part of a changehandler, the activated ontology changes are not directly applied but first assembled in a list (line 17) and applied thereafter. This way all the loop preconditions are evaluated against the original ontology (or, in the case of `entailsChanged`, against the ontology that has been altered in the initially proposed way), thereby preserving a declarative flavor. The application of the changes is done in an atomic manner after assembling is complete.

As it would not be easy to ensure termination or avoid high computational cost if the *actions* part of a change request was allowed to trigger other change requests, no other changehandlers are triggered during the execution of a change-

[6] As it is possible to specify a changehandler which matches any request, it is up to the knowledge worker what should be done with change requests for which no matching changehandler is found.

handler. While this ensures termination, it makes the ontology engineer responsible for "manually" handling all additional changes that might become necessary due to the changes during the execution of the changehandler.

As a preliminary solution, the association of changehandlers with an ontology works similar to the association of DTDs with a XML document [2]. They can either be defined inline in the document specifying the data or they can be defined in external files. In either case, the changehandler is defined in a comment (such that the rdf file can also be parsed by ontology management systems that do not support OUL). All comments that have an extra '-' at the beginning are parsed as changehandler definitions. This begin of the comment may be followed by a file name enclosed in quotation marks, which defines an external changehandler, or by the definition of a changehandler enclosed in square brackets, defining the changehandler inline.

5 Examples

In this section, we provide a small example aimed at both advocating the potential usefulness of our proposed update framework and demonstrating the concrete work flow. We start with the knowledge base from Fig. 3. This RDFS specification, where we use Turtle syntax for better readability, contains the knowledge of the current status of our domain. Furthermore suppose this ontology is accompanied by the ontology update specification shown in Fig. 4. The first changehandler therein deals with the case that somebody leaves his/her current affiliation. In that case, the deletion of the affiliation information has to trigger further changes: the person will not continue to lead projects at the institution he/she leaves nor to supervise persons. Now suppose the following change request, indicating that Philipp is leaving the AIFB institute, is entered into the system:

```
delete data {philipp hasAffiliation aifb .}
```

The system will now check whether this change request matches the first changehandler's change pattern `del { ?x hasAffiliation ?y }`. This is the

```
philipp     rdf:type            PhDStudent .
philipp     hasAffiliation      aifb .
philipp     leads               xmedia .
philipp     worksOn             multipla .
philipp     supervises          thanh .
thanh       rdf:type            PhDStudent .
thanh       hasAffiliation      aifb .
thanh       worksOn             xmedia .
xmedia      rdf:type            Project .
xmedia      assocInstitution    aifb .
multipla    rdf:type            Project .
multipla    assocInstitution    aifb .
```

Fig. 3. Example knowledge base

```
CREATE CHANGEHANDLER leavesInstitution
  FOR del { ?x  hasAffiliation  ?y }
AS applyRequest;
  feedback("?x is no longer affiliated to ?y");
  for(contains(?x ?wol ?z . ?z rdf:type Project .
               ?z  assocInstitution ?y .
               FILTER(?wol=worksOn || ?wol=leads)))
    delete data {?x ?wol ?z};
    feedback("Thus, ?x does not lead/work on project ?z anymore."); end;
  for(contains(?x  supervises ?z .  ?z  hasAffiliation ?y))
    delete data {?x supervises ?z};
    feedback("Thus, ?x does not supervise ?z anymore"); end;

CREATE CHANGEHANDLER authorsPhD
  FOR add { ?x  swrc:authorOf  ?y }
AS IF entailschanged( ?y  rdf:type swrc:PhDThesis . )
  THEN applyRequest;
      delete data { ?x  rdf:type  swrc:PhDStudent};
      feedback("Change accepted. ?x authored a PhDThesis,
                 so he is no PhD student anymore.");
```

Fig. 4. Example ontology update specification

case, as the corresponding SPARQL query yields a result which binds `philipp` to `?x` and `aifb` to `?y`. The considered changehandler is now executed as it does not contain further preconditions for activation. So, the specified actions will be carried out: `applyRequest` means that the initial change request is granted and added to the list of updates to be executed. After that the following message is displayed:

`philipp is no longer affiliated to aifb.`

Next, we consider the graph pattern in the first loop. Note that it contains variables that have already been bound by the change pattern matching. Before evaluating the loop action, those variables are substituted by their bindings, in our case resulting in the following rewritten loop action:

```
for(contains(philipp ?wol ?z . ?z rdf:type Project .
             ?z  assocInstitution aifb .
             FILTER(?wol=worksOn || ?wol=leads)))
  delete data {philipp ?wol ?z};
  feedback("philipp does not lead/work on project ?z anymore"); end;
```

Now, the conditional part of the rewritten loop action is matched against the database, yielding the following two variable bindings: `?wol`↦`leads`, `?z` ↦`xmedia` and `?wol`↦`worksOn`, `?z` ↦`multipla`. Next, for each of these two bindings, the subsequent actions are executed: therefore, the triples `philipp leads xmedia.` and `philipp worksOn multipla.` are scheduled for deletion and the following two messages are prompted to the user:

```
Thus, philipp does not lead/work on project xmedia anymore.
Thus, philipp does not lead/work on project multipla anymore.
```

In analogy to that, by executing the second loop of the activated changehandler, `philipp supervises thanh.` is scheduled for deletion and the message

```
Thus, philipp does not supervise thanh anymore.
```

prompted to the user. Finally, all the scheduled changes are carried out.

Finally, we will present some standard situations or decisions which might occur in an ontology update setting and how they can be realized by means of the Ontology Update Language as presented in this paper.

Restricting the Size of the Change. Clearly, it is not always desirable to permit change requests of arbitrary size. In principle, an entire ontology could be added in one step, which would be a situation hard to handle with update rules. One solution to this is to restrict the size of the change a priori. As an extreme case of this, only one RDF triple per change might be allowed. While this constraint can be imposed by external means, our formalism is flexible enough to handle it. In order to allow only add changes consisting of one triple, the following changehandler would have to be inserted at the beginning of an ontology update specification:

```
CREATE CHANGEHANDLER tooMuchForOneBite
   FOR add ( { ?a ?b ?c . ?d ?e ?f .
        !(sameTERM(?a,?d) && sameTERM(?b,?e) && sameTERM(?c,?f))} )
AS feedback("Request denied. Only one triple per change!");
```

In words, the change request pattern checks whether there are two distinct triples contained in the change request. If so, the changehandler is activated without doing any changes (thereby effectively rejecting the request). Note that the implemented selection strategy also prevents any subsequent changehandler in the document from being activated.

Handling of Change Requests. Of course, change requests might occur which do not activate any of the specified changehandlers. In the presented implementation, the request will be tacitly denied in this case. It is however possible to create changehandlers that match any `add` (resp. `delete`) request. Placed at the bottom of an ontology update specification, those can be used to provide feedback whenever no other changehandler was activated:

```
CREATE CHANGEHANDLER noMatchRestrictive FOR add ( { ?a ?b ?c . } )
AS feedback("Request denied. No matching change rule found!");
```

This is just an explication of the default restrictive strategy: every unforeseen request will be denied. In the same way, it is of course possible to realize a permissive strategy by stating

```
CREATE CHANGEHANDLER noMatchPermissive FOR add ( { ?a ?b ?c . } )
AS applyrequest;
   feedback("Request accepted. No matching change rule found.");
```

instead. This way, all requests not matching any of the preceding changehandlers will be complied with.

6 Implementation

We provide an implementation of our architecture and our language proposal at `http://www.aifb.uni-karlsruhe.de/WBS/uhe/OUL/`. The implementation uses Jena as underlying framework for ontology management. This framework was chosen, as Jena provides an implementation of SPARQL update [11].

SPARQL update is an extension of the ontology query language SPARQL [9]. While SPARQL's purpose is to find data in a RDF graph, SPARQL update provides functionality for updating and managing RDF graphs using a SPARQL-like syntax. Update operations allow changing existing graphs by adding data (`insert`), deleting data (`delete`) or both (`modify`). Besides directly inserting/deleting a set of triples, these operations allow to change an ontology based on a pattern. A pattern is a description of a RDF graph, which may contain variables that can be used to describe elements of the graph and can then be determined as result. These results are used in SPARQL update queries to determine which triples should be added resp. deleted from the ontology.

Our implementation provides a wrapper for Jena's SPARQL update endpoint, which implements the ontology update management as we proposed it. SPARQL update requests can be submitted as in the original implementation, but instead of directly executing them, the graph of changes that will have to be applied is constructed and a suitable change handler is searched for as explained in Section 4 and the respective actions are performed. By default, if no change handler is found, the ontology remains unchanged. This approach makes the update management as transparent as possible for the user. The only difference with respect to using the original implementation is using another endpoint and getting all the described advantages. This allows to open SPARQL endpoints for writing access more liberally.

7 Conclusions and Future Work

In this paper, we have addressed the task of updating an ontology due to changes in the described domain. We have argued for a formalism that allows for specifying the domain-dependent ways in which a specific ontology may evolve over time. We thoroughly discussed the crucial design decisions to be made for an ontology update framework that would automatically align change requests with change patterns and thereby allow to delegate simple ontology maintenance tasks to users not necessarily possessing the expertise of a knowledge engineer. We presented and implemented a proposal for such a framework.

Being aware that our proposal constitutes just a first step towards a much needed functionality for semantic technologies, we identify several directions for future research:

Extending the implementation to OWL. Currently, our implementation works with RDF(S) and SPARQL. Extending it to OWL would require to extend

SPARQL accordingly, and to allow Algorithm 1 to use multiple-triple axioms as they often occur in OWL DL knowledge bases.

Interactivity. It might be the case that a change request cannot be unequivocally assigned to one single change pattern. As an example, the change request to retract the fact that an individual is a PhD student might be due to the fact that he got his PhD degree or he dropped out; either of the variants requiring different further ontology changes. While this alternative could be easily accommodated by extending the update language, the ultimate choice which change pattern to apply in a concrete case should be left to the person uttering the change request. Therefore, also additional interactive control loops would have to be introduced into the framework.

Combination with belief revision. Although we have argued that the rationale of belief revision does not fit well with our purpose, there are certainly cases where a combination of both is beneficial. Belief revision could be used as a fall-back strategy if a change request would lead to an inconsistent ontology and is not tackled by any of the update specification's change patterns. Instead of simply rejecting the change, belief revision techniques could be more appropriate. In general, belief revision and other coping strategies could be incorporated into the proposed formalism in a plugin-manner as additional actions next to adding and deleting axioms.

Higher order constructs. Our proposal of an ontology update has a rather operational flavor. While this arguably facilitates the employment and allows for an efficient and straightforward implementation, a more declarative way of describing the possible domain changes would be more in the spirit of the current ontology languages. Moreover a specification in OWL would abide by the rationale to reuse formalisms (just as the XML syntax is also used for XML Schema). Hence it seems sensible to introduce a more abstract description layer for complex changes, preferably in OWL. The underlying model for such a declarative framework could be inspired by the usual ways of describing discrete dynamic systems such as finite automata or petri nets. A simple example would be to relate the two classes Child and Adult with each other with a property allowing the transformation of instances of the one class to an instance of the other, e.g `Child disjointTransformationTo Adult`. Note that in OWL2 such a property is legal due to punning.

Learning Change Patterns. Clearly, the success of the proposed framework depends on the quality of the update specification. While in certain domains the development of such a specification might be straight forward (possibly because there are already informal documents describing the standard processes and work flows) there might be scenarios where this is not the case. Under those circumstances, frequent change patterns could be extracted from ontology change logs by some machine learning techniques. Those findings could then be presented to the knowledge engineer as suggestions for ontology update rules to be incorporated into the specification.

An elaborate ontology update mechanism as presented here allows ontologies to be updated in a more predictable and quality preserving way. Administrators of ontology based system may choose to allow a wider audience to edit their ontologies in a controlled manner, thus extending the collaborative aspect of ontology maintenance.

References

1. Baader, F., et al. (eds.): The Description Logic Handbook: Theory, Implementation and Applications. Cambridge University Press, Cambridge (2007)
2. Bray, T., et al.: Extensible markup language (xml) 1.0, 5th edn. W3C Recommendation, November 26 (2008), http://www.w3.org/TR/REC-xml/
3. Carle, E.: The Very Hungry Caterpillar. Philomel Books (1969)
4. Date, C.J., Darwen, H.: ISO/IEC 9075-2:2008 (SQL - Part 2: Foundations), The SQL Standard, 3rd edn. Addison-Wesley Publishing Company, Reading (1993)
5. Katsuno, H., Mendelzon, A.O.: On the difference between updating a knowledge base and revising it. In: Gärdenfors, P. (ed.) Belief Revision, pp. 183–203. Cambridge University Press, Cambridge (1992)
6. Konstantinidis, G., et al.: A formal approach for RDF/S ontology evolution. In: Ghallab, M., et al. (eds.) Proc. ECAI 2008, pp. 70–74. IOS Press, Amsterdam (2008)
7. Manola, F., Milner, E.: RDF Primer. W3C Recommendation (February 10, 2004), http://www.w3.org/TR/REC-rdf-syntax/
8. McGuinness, D., Harmelen, F.v.: OWL Web Ontology Language Overview. W3C Recommendation (February 10, 2004), http://www.w3.org/TR/owl-features/
9. Prud'hommeaux, E., Seaborne, A.: SPARQL query language for RDF. W3C Recommendation (January 15, 2008), http://www.w3.org/TR/rdf-sparql-query/
10. Qi, G., Yang, F.: A survey of revision approaches in description logics. In: Calvanese, D., Lausen, G. (eds.) Proc. RR 2008. LNCS, vol. 5341, pp. 74–88. Springer, Heidelberg (2008)
11. Seaborne, A., et al.: SPARQL Update. W3C Member Submission (July 15, 2008), http://www.w3.org/Submission/2008/SUBM-SPARQL-Update-20080715/
12. Staab, S., Studer, R. (eds.): Handbook on Ontologies. International Handbooks on Information Systems. Springer, Heidelberg (2004)
13. Stojanovic, L.: Methods and Tools for Ontology Evolution. PhD thesis, Universität Karlsruhe, TH (2004)
14. Tempich, C., et al.: Argumentation-based ontology engineering. IEEE Intelligent Systems 22(6), 52–59 (2007)
15. Tudorache, T., et al.: Supporting collaborative ontology development in protégé. In: International Semantic Web Conference, pp. 17–32 (2008)

Representing, Querying and Transforming Social Networks with RDF/SPARQL*

Mauro San Martín[1,2] and Claudio Gutierrez[2]

[1] Departamento de Matemáticas, Universidad de La Serena
[2] Computer Science Department, Universidad de Chile
{msanmart,cgutierr}@dcc.uchile.cl

Abstract. As *social networks* are becoming ubiquitous on the Web, the Semantic Web goals indicate that it is critical to have a standard model allowing exchange, interoperability, transformation, and querying of social network data.

In this paper we show that RDF/SPARQL meet this *desiderata*. Building on developments of *social network analysis*, *graph databases* and *Semantic Web*, we present a social networks data model based on RDF, and a query and transformation language based on SPARQL meeting the above requirements. We study its expressive power and complexity showing that it behaves well, and present an illustrative prototype.

1 Introduction

The recording of social interactions over the Web, and the tagging and annotation of data are creating networks of data with the structure of what is classically known as *social networks* [1,2,3]. Examples include Wikipedia, del.icio.us, YouTube, DBLP, Flickr, Facebook, and the blogosphere. In all these services, each time a user sends a message, chooses a tag or annotates a resource, one or multiple relations are recorded. Researchers, developers, and users have realized that there is useful information in these underlying networks [1], consequently some sort of structural analysis is performed over them [3,4]. Many of the techniques used come from the established area of social networks analysis (SNA), which Breiger [5] defines as the disciplined inquiry into the patterning of relations among social actors of different kinds and at different levels (see also [6,7,8]).

Nevertheless, the processing and managing of this huge amount of data is still an unsolved problem for the common user and developer. The problem is not only the format (today ranging from spreadsheets to files in proprietary formats), but also the means of querying, mixing and transforming such data with standard models. Furthermore, most data formats currently in use do not have explicit semantics nor support for provenance [1]. Advances have been done in the field of Social Network Analysis, e.g. Pajek [9], Ucinet [10], and *network* and *sna* packages of R [11,12]. These tools are focused in implementing techniques and algorithms for data previously prepared and formated to fit their

* This research was supported by Fondecyt 1070348, RDF Databases.

L. Aroyo et al. (Eds.): ESWC 2009, LNCS 5554, pp. 293–307, 2009.

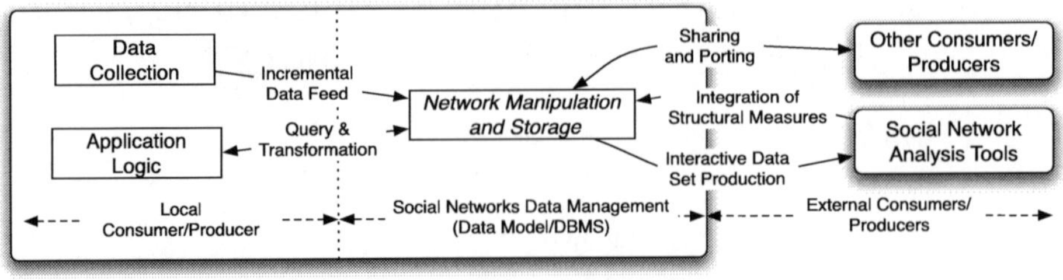

Fig. 1. *Data Management Needs for Social Networks.* Each social application is a consumer/producer of social networks, producing and/or collecting network data, and consuming data produced by other applications. The need for a standard representation and query language follows.

proprietary formats [10]. Another group of developments in these lines, proposals using RDF for covering different applications of the social Web, currently are focused only on the representational aspects of the problem (FOAF, XFN, RELATIONSHIP, etc.) This amounts to the definition of custom vocabularies and ontologies tailored for specific applications. The question remains how they may work together, and at what level of abstraction [13].

Thus a basic *desiderata* for managing social network data should include at least the following: 1) to be able to represent and store any kind of social network, including provenance information; 2) to be able to share and mix social networks (or parts of them) among users and applications; 3) to have a set of standard operations over these social networks, to query, transform and update them (a query language). We show in this paper that it is possible to fulfill such desiderata based on the experiences of social networks and database communities, and that of RDF/SPARQL.

The strategy to reach this goal is as follows. In the first place, we use the fact that all social networks of data have a stable common underlying model given by a characteristic practice of the SNA community, which is based in well developed methodologies and techniques (see Freeman [14] and [8, ch.1]). From here, and based on the Semantic Web practice, we abstract a model of data management for social networks based in consumers and producers of networks. It provides data management services to each consumer/producer: query, transform and update their own data; and between pair of consumers/producers: porting and sharing of data (see Fig. 1). For instance, SNA can be viewed as a consumer of social networks to be analyzed; annotators and users are producing and updating social networks; developers of applications are consuming and producing networks, etc. Finally, using the findings of database modeling [15], a graph structure emerges as the natural choice for such data model. In particular, RDF offers a standard model and SPARQL can be used as an off-the-shelf language. SPARQL fulfills almost all query and transformation requirements for social networks except for aggregation. Note also that relational model by itself does not intrinsically support a graph data structure, and consequently it is not straightforward to define relevant social network queries in SQL.

Contributions. In this paper we show that RDF/SPARQL are excellent data model/query language candidates for data processing social networks on the Web and cope with their basic requirements. First, we gather and study SN requirements of representation and storage, manipulation (querying, transforming and updating), and sharing/porting. Based on them, we define a conceptual graph data model for social networks. Second, we show that RDF/SPARQL fulfill these requirements and model, by mapping the conceptual data model data structure to RDF, and the conceptual query language to a composition of SPARQL and basic SQL queries (essentially used to provide aggregation which is currently missing in SPARQL). Third, we study the expressive power and the complexity of the query language showing that the model scales in the database sense. Finally, we provide a formal and practical bridge between Semantic Web framework and Social networks applications (which are key consumers/producers of social network data). We sketch a prototype implementation of a DBMS for social networks over RDF/SPARQL.

Related Work. There are three main areas related to our research:

1. Networks and Social Network Analysis. The closest development to our work are SNA software tools. One of the most popular and complete is Pajek [9], which offers an extensive set of analysis algorithms and visualization options. It has basic elements of data manipulation, for instance to filter out nodes and edges. The *R* statistical software also has packages to deal with network data [11]. In all these SNA tools, data storage is based on text files in various formats, which limits the data managing options available to the user, and in the long term sensibly rises the costs of network data curation. Low level storage restrictions are exposed to the user complicating unnecessarily the data manipulation. None of these tools provide the required data management services. There are many custom made applications that solve specific problems, for instance Klink et al. [16] use explicit social network data to improve DBLP navigation.

2. Network Data Models and Databases. Although there is an increasing awareness of the need of data management support in several fields which deal with networks[17,18,19], to the best of our knowledge there are no works discussing systematically data models for social networks. Jensen and Neville [20] propose Proximity, a data mining tool that supports statistical models over network data, which in turn uses MonetDB, a database management system (DBMS) able to store network data at low level, but lacking the appropriate level of abstraction for SNA use. Proximity uses a graph query language called QGraph [21]. Tsvetovat et al. [22] propose a specific application based on the *relational data model*, and also propose DyNetML [23] , an XML based format, to store rich social network data. No query or transformation issues are addressed in these works.

3. Semantic Web. In this area representational aspects present in our work have been addressed for specific applications with several standards, like for instance FOAF. How to integrate these standards and at which abstraction level remains an open question [13]. Mika [1, ch. 5] discusses several possible representation models, tools and standards from database, SNA, and Semantic Web communities, and reaches the same conclusion as us: current tools and data formats do not solve data representation and aggregation requirements. However, he proposes a solution based on ontologies and automated reasoning, whereas we propose a data model and the corresponding querying and transformation language. We show that having explicit metadata and a query language suffices for most data management requirements. Erétéo et al. [4] present a framework to provide "semantic aware social network analysis". Jung and Euzenat [2] propose a three layer model to represent and extract information from social networks. In terms of data management as previously defined, none of these models cope with the requirements. Producer/consumer has been addressed in other works, for instance, Polleres et al. propose XSPARQL [24] as a standard query language for lifting and lowering between XML and RDF data sources. Finally there are many works that extract, model, manipulate, and analyze social networks in the Semantic Web for specific applications, see for example: Aleman-Meza et al. [25], Kinsella et al. [26], and Mika [27].

The paper is organized as follows: In Section 2 the data structure of the data model is presented; Section 3 studies the corresponding query language, and discusses implementation issues; Section 4 presents conclusions.

2 Representing Social Networks over RDF

Social Networks data management requirements have come to the foreground with the advent of the "Social Web" [13], particularly due to the needs of interoperability of SN data (weak or inexistent in applications like Flickr, Delicious, etc.) Some applications like DBLP and FOAF have data available in several formats (particularly XML and RDF). Nevertheless, they have different vocabularies and model design, making interoperability non automatic. The core of these applications are their social network characteristic. Thus a natural source to look for requirements is the work that the SN community has done.

Example 1 (The KHTM Network). To illustrate the ideas we will use the *Krackhardt's High-tech Managers* (KHTM) network, a well known data set in the field of SNA. We use the version provided by Wasserman and Faust [7] of the data gathered by Krackhardt's in 1987 in a small manufacturing organization on the west coast of the U.S. The KHTM data set is a one mode network, with three kinds of relations (friendship, advice, reports_to) collected among the twenty one managers of the organization. For each manager are also recorded four attributes: age, tenure, level in corporate hierarchy, and the department in which the manager works (see Fig. 2). A total of 624 ties were recorded.

Requirements for a Social Networks Data Model. An appropriate departure point for SNA model requirements is Freeman's *maximal structure experiment* [8, ch.1]: one or more kinds of relations; one or more types or levels of social units; structures that change through time; sets of social units that grow or shrink; attributes of social units, and attributes that change. Two elements should be added to this model: attributes of relations, and n-ary relations, i.e. relations linking more than two actors [28].

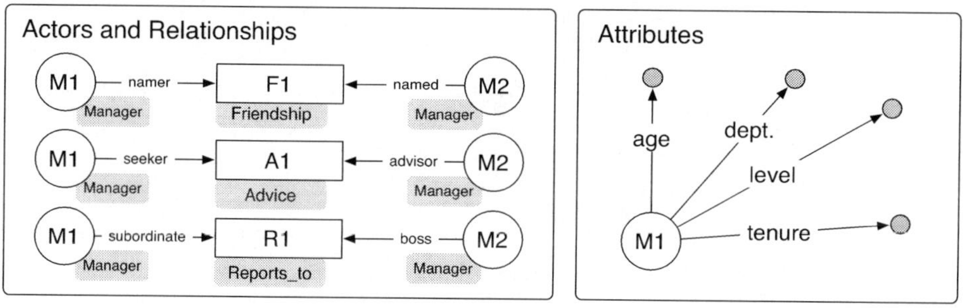

Fig. 2. *Schema of the KHTM data set.* Left hand side shows the three types of binary relations (rectangular nodes) between pairs of distinct managers (round nodes) in the original network data set. Note that relations are also represented as nodes, and arcs' labels represent the role that an actor perform in the relation. On the right hand side, attributes are represented as part of the graph.

In addition, the SN data model must fulfill the requirements that arise from current practices in the SN community (see [7], [6] and [29]), for instance, it must provide data management support to extract input data sets for popular SNA tools, e.g. Pajek [9], as a recurrent stage in the workflow depicted in the Fig. 1.

The implemented model must also address the problems that arise in the long term scientific data curation, e.g. provenance, reuse, and archiving [19]. Portability is another crucial requirement, particularly for the Semantic Web, from user profiles to more recently reuse of whole networks (e.g., in eScience). This is achieved via a standard underlying data structure and data manipulation language. Time is another important parameter to be included as desired. It can be added as metadata to actors or relations, although there are subtleties that have to be considered [30].

The requirements discussed above demand representations more elaborate than the simple graphs (or matrices) of the SNA classical modeling. This challenge is not exclusive of the domain of SN and has been explored extensively in other contexts [31,21,15]. From this background and trends in information exchange indicating that all the information should be in the same data structure and that it should support the addition of arbitrary metadata (e.g. provenance), basic requirements for such a model emerge. First, attributes values should be part of the graph, and second, relations should be represented as nodes instead of edges or arcs, allowing the seamlessly representation of n-ary relations and attributes on relations over the same data structure (see Fig. 2 and 3). Formally a *social network* can be represented by a special type of directed labeled graph.

Definition 1 (Social Network Data model). *A* social network *is a triple* $S = (V, E, L)$ *where* (V, E) *is a directed graph and* L *a set of labels (and labeling functions), specified as follows:*

- *The set of nodes* $V = A \cup R \cup C$ *is a disjoint union of the set* A *of actors,* R *of relations and* C *of attributes.*

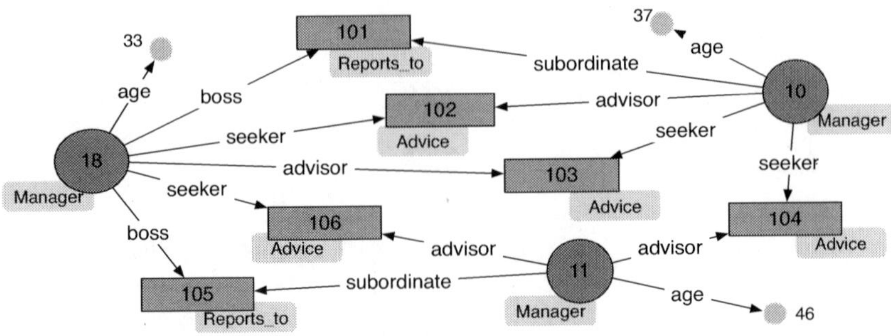

Fig. 3. *Subnetwork of KHTM data set: A view of a subnetwork of all relations among members of Department 3.* Actors are round nodes (black outline), and relations are square nodes. Attributes are small round nodes (grey outline). Labels indicate participation roles and meanings (attributes). To keep the diagram clean, only age attribute is shown and family belonging is depicted as a second label with grey background in each node (instead of the corresponding node and arc). Note that this is already an RDF graph. Arrows go from subject to objects and are labeled with predicates.

Each of this set is partitioned in families (or types) as follows: Actor set $A = \bigcup_i A_i$*, where* A_i *defines a family of actors. Relation set* $R = \bigcup_i R_i$*, where* R_i *defines a family of relations. Attribute set* C *is usually as well partitioned into different data types.*

- *The set of edges* $E = E_{AR} \cup E_{AC} \cup E_{RC}$ *is a disjoint union of the following types of edges:*
  - E_{AR} *is a multiset of elements of* $A \times R$*, the* participation edges *(i.e. we allow multiple edges between an actor and a relation.)*
  - $E_{AC} \subseteq A \times C$ *and* $E_{RC} \subseteq R \times C$ *are the set of* meaning edges.
- *The set* L *of labels is the union of sets of labels for the different types of edges:* L_P *(participation) for labels on* E_{AR}*, and* L_M *(meanings) for* E_{AC} *and* E_{RC}*, with their corresponding labeling functions.*

This model can be naturally viewed as a set of triples in the style of RDF. The triples corresponding to the view in Fig. 3 are shown in the Fig. 4. The formal specification is as follows.

Definition 2 (Social Network Triple Representation). *Consider the vocabulary* $\Sigma = A \cup R \cup C \cup \{isA, isR\} \cup L_P \cup L_M$*. Then define the following:*

1. *List of Nodes and Family Belonging triple set:*
   $N_A \subseteq A \times \{isA\} \times \{A_1, \ldots, A_{m_A}\}$,
   $N_R \subseteq R \times \{isR\} \times \{R_1, \ldots, R_{m_R}\}$.
2. *Participation triple set:* $P \subseteq A \times L_P \times R$.
3. *Meanings (attributes) triple set:* $M \subseteq (A \cup R) \times L_M \times C$.

N_A
(10, isA, Manager)
(11, isA, Manager)
(18, isA, Manager)

N_R
(102, isR, Advice)
(103, isR, Advice)
(104, isR, Advice)
(106, isR, Advice)
(101, isR, Reports_to)
(105, isR, Reports_to)

P
(10, seeker, 103)
(10, seeker, 104)
(10, advisor, 102)
(10, subordinate, 101)
(11, advisor, 104)
(11, advisor, 106)
(11, subordinate, 105)
(18, advisor, 103)
(18, seeker, 102)
(18, seeker, 106)
(18, boss, 101)
(18, boss, 105)

M
(10, age, 37)
(11, age, 46)
(18, age, 33)

Fig. 4. *KHTM - Department 3 subnetwork (Fig. 3) as triples.* The separation in four groups (tables) of triples represents the social structure of the RDF graph. Tables P and M represent the node-edge-node triples shown in Fig. 3. Tables N_A and N_R represent the typing of each node (actors and relations respectively).

From the definitions above it is not difficult to show:

Lemma 1. *A social network can be represented by four sets of triples* (N_A, N_R, P, M).

In what follows, we will associate social networks with their triple representation. Regarding the RDF implementation, the vocabulary, that is, actors and relations and the sets L_P and L_M, are implemented as URIs, and attributes values as literals. The predicates isA and isR can be either implemented as predicates or replaced by the RDF keyword `rdf:type`. Then the set of triples is precisely the union of the sets N_A, N_R, P and M.

3 Transforming Social Networks with SPARQL

Query and Transformation Requirements. Querying and transforming social networks turns out to be a non-trivial task due to the intrinsic complexity of the networked data [15]. The good news is that the requirements are uniform, as shown by use cases collected from bibliographic sources [7,6,9], the papers from last three years of the Social Networks Journal and current software tools for SNA, particularly Pajek and its reference book *Exploratory Social Network Analysis with Pajek* [9] (see Table 1).

The range of queries that SN practitioners need are diverse, and can be roughly analyzed along two axis. First, queries that return values or measures of the whole input network or of some subnetwork (e.g. centrality, diameter, etc.) They fall into structural analysis and can be more properly treated by analysis-tools for final users. The second axis regards queries and transformations which output networks and constitutes properly the data management part of processing SN. (cf. Table 1). We focused on this latter group, which is the one currently missing from the applications and corresponds properly to data management needs discussed above.

Table 1. Archetypical social network data management operations from the book *Exploratory Social Network Analysis with Pajek* [9], which also recurrently appear in literature (see [7,6,9])

Chapter Title	Use Case	Query Description (see Definition 3)
Looking for Social Structure	Directed to undirected Binary Relations (*arcs to edges* in Pajek)	In P_I both participation roles have different participation roles, in P_O both roles are the same.
	Remove relations (*remove edges* in Pajek)	Some relation nodes matched by P_I are not replicated in P_O
Attributes and Relations	Extract a subnetwork based on attributes	P_I include triples from M with fixed literals, or with filters on the corresponding variables.
	Group actors based on attributes (*shrink network* in Pajek)	P_I matches a subnetwork, including at least one actor attribute, P_O produces only one actor per value of the attribute. Aggregate functions may be required to count the size of the groups or summarize other attributes.
	Selective grouping of actors based on attributes (*contextual view* in Pajek)	A variant of the previous, here some portion of the subnetwork matched is not grouped
Cohesive Subgroups	Extract the subnetwork induced by cliques of size n	P_I is the desired clique for a fixed n, $P_O = P_I$
Sentiments and Friendship	Extract subnetwork by time	Time is represented by an attribute. M triples in P_I have fixed literals or filters on the corresponding variables.
Affiliations	Two-mode network to one-mode network	P_I represents a transitive relation, P_O eliminates the actor in the middle. It may need a function to generate ids for the new relation.
Center and Periphery	Group multiple binary relations (*remove multiple lines* in Pajek)	P_I selects all the required binary relations. P_O groups and typically counts them by pair of actors.
Brokers and Bridges	Extract egonetwork of an actor	P_I selects all relations and neighbors, and the relations involving those. $P_O = P_I$. Requires OPTIONAL or UNION. Each radius requires different set of patterns.
	Remove relations between groups (*remove lines between clusters* in Pajek)	Requires that the groups are represented as attributes first. P_I selects the network except the relations between actors of different groups.
Diffusion	Selective counting of neighbors	P_I selects the neighbors given certain condition. P_O requires aggregate functions.
	Operations between attribs. (*divide vectors* in Pajek)	Requires functions over literals in P_O.
	Change relation direction based on attributes	P_I selects the required network, P_O swaps labels of participation roles.
Prestige	Discretize an attribute	Requires functions over literals in P_O.
Ranking	Find triads by type	P_I is a three actor pattern, a triad. In each triad some relations are allowed and some are forbidden. Requires negation.
Genealogies and Citations	Loop removal	P_I selects all relations except those with arity equals to 1.

Query Language Definition. The ideal design is to have a set of simple primitives from where to compose by simple operations all needed queries. In our case, pattern matching fulfills this need. Informally speaking, the language consists of three parts: an *input pattern* P_I, which identifies (matches) all pieces of information needed and extracts the corresponding values; a *transformation* T of values, which updates and aggregates the values obtained in the previous step; and finally, an *output pattern* P_O, which defines the form of the data that will constitute the social network to be outputted.

Definition 3. *Let $\Sigma^X = \Sigma \cup X$ be the extension of the alphabet Σ with the set of variables X (disjoint of Σ).*

- *A* basic pattern *P is a social network over the vocabulary Σ^X.*
- *A* composite pattern *is defined by the following grammar:*

$$P ::= P \,\mathrm{AND}\, P \mid P \,\mathrm{OR}\, P \mid P \,\mathrm{MINUS}\, P \mid P(E),$$

where E is a Boolean expression of atoms of the form $(x \,\mathrm{op}\, y)$ or $(x \,\mathrm{op}\, c)$, where $x, y \in X$, c is a constant, and op *is in the set $\{=, \leq, \geq\}$.*

A query is a triple of the form $((P_I, X_I), T, (P_O, X_O))$, where

1. *(P_I, X_I) is a pattern over Σ^{X_I} (the* input *pattern). (Note that we write explicitly in the query the set of variables X_I occurring in the pattern.)*
2. *T is a* basic *SQL query from a table with attributes X_I which outputs a table with attributes X_O. A 'basic SQL' means here single table, no nesting; the selected attributes Aj could be functions of the attributes of X_I.*
3. *(P_O, X_O) is a basic pattern over Σ^{X_O} (the* output *pattern).*

Query Language Semantics. We will give an operational semantics described by the following query evaluation procedure using SPARQL and SQL. P_I and P_O will be SPARQL patterns, and P_O a basic graph pattern that may also contains filters. The procedure is composed by three stages:

Stage 1: Extraction. This corresponds to a SPARQL query that captures the values of the input pattern. The output is a solution sequence (a table).

```
SELECT X_I  FROM G  WHERE P_I
```

The result of this query is a table $R(X_I)$ containing all the bindings of the variables in X_I.

Stage 2: Transformation. The table obtained in the previous step is transformed by an SQL query (representing T) to output another table, whose attributes correspond to the values $f_i(X_I)$ for certain functions f_j.

```
SELECT f1(X_I) AS att1,...,fk(X_I) AS attk
FROM R(X_I) WHERE C_1
GROUP BY <...>  HAVING C_2
```

where `C_1` and `C_2` are conditions. From this table the final graph will be constructed in the next step. Note that this transformation is currently not expressible in SPARQL (even if there are no aggregate functions).

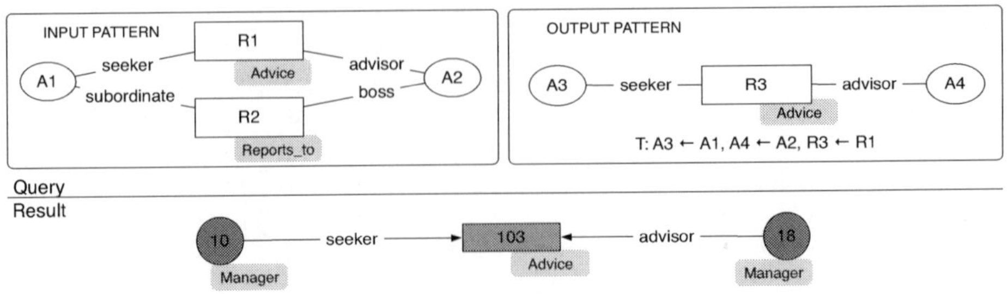

Fig. 5. *Query and Result: "Find all managers that seek advice with their bosses".* The query is composed by the input and output patterns, and the transformation T. The result network when applied to network in Fig. 3 is depicted below.

STAGE 3: CONSTRUCTION. Here the CONSTRUCT query form of SPARQL is used. The idea is to create an RDF graph which is the union of the instantiations of the output pattern P_O for each tuple of values of the table produced in the previous step.

Note that we cannot input a table to a SPARQL query. To handle this problem, we need to construct an auxiliary RDF graph H. Assuming $X_O = \{X_O^1, \ldots, X_O^k\}$, the specification of the final query is:

```
CONSTRUCT P_O  FROM H
WHERE {(att1, ?Tuple, ?X_O^1). ... (attk, ?Tuple, ?X_O^k)}
```

where the graph H is defined as follows: For each tuple $t_j = (v_{1j}, \ldots, v_{kj})$ in the output table in the previous step, the following set of triples T_j will be produced, $T_j = \bigcup_{i=1}^{k} \{(att_i, tuple_j, v_{ij})\}$. Then $H = \bigcup_j T_j$.

Example 2 (Simple Case). Figure 5 graphically represents the query: "extract the subnetwork of relations seeker-advisor, where the actor A, seeking advise with B, have also the relation subordinate-boss with B." In this query $X_I = \{A1, A2, R1, R2\}$, $P_I = \{(A1, seeker, R1), (A2, advisor, R1), (A1, subordinate, R2), (A2, boss, R2)\}$, $T = \{A3 \leftarrow A1,\ A4 \leftarrow A2,\ R3 \leftarrow R1\}$, $X_O = \{A3, R3, A4\}$, and $P_O = \{(A3, seeker, R3), (A4, advisor, R3)\}$. In this and the following examples we assume, for the sake of clarity, that the types of the actors and relations are implicit in the pattern (grey labels in Fig. 5), i.e., all the triples of the form (a, isA, X) and (r, isR, Y) –for a and r in the pattern– are implicit in the pattern and are carried as needed to the output following the transformation. The SPARQL query is:

```
Construct {?A1 KHTM:seeker ?R1. ?A2 KHTM:advisor ?R1}
From G
Where     {?A1 KHTM:seeker ?R1. ?A2 KHTM:advisor ?R1.
           ?A1 KHTM:subordinate ?R2. ?A2 KHTM:boss ?R2 }
```

If G is the network data set in Fig. 3, the result is the following set of triples (namespace KHTM omitted): $\{(10, seeker, 103), (18, advisor, 103)\}$ (See Fig. 5).

However, in the general case T is not a simple renaming mapping, the variables X_O are functions of X_I, and T may include aggregate functions. The following example covers this.

Example 3 (Filtering and Counting). "Extract the subnetwork of relations seeker-advisor for those advisors which have at least 2 advisees and add the number of advisees as an attribute to the advisors". G is again the data in Fig. 3.

Stage 1. First SPARQL query (SELECT)

```
SELECT ?A1, ?R1, ?A2
FROM G
WHERE {?A1 KHTM:seeker ?R1.
       ?A2 KHTM:advisor ?R1}
```

Result: table R

A1	R1	A2
10	104	11
18	106	11
10	103	18
18	102	10

Stage 2. SQL query

```
SELECT  A1 as att1, R1 as att2,
        A2 as att3, COUNT(*) as att4
FROM R
GROUP BY A2  HAVING COUNT(*) >= 2
```

Result: table S

att1	att2	att3	att4
10	104	11	2
18	106	11	2

Stage 3. Second SPARQL query (CONSTRUCT)

```
CONSTRUCT {?A1 KHTM:seeker ?R1. ?A2 KHTM:advisor ?R1.
           ?A2 KHTM:numAdv ?N}
FROM H
WHERE  {att1 ?T ?A1. att2 ?T ?R1. att3 ?T ?A2. att4 ?T ?N }
```

Where H is the set of triples { (att1, t1, 10), (att2, t1, 104), (att3, t1, 11), (att4, t1, 2), (att1, t2, 18), (att2, t2, 106), (att3, t2, 11), (att4, t2, 2) }, which is simply a coding of Table S as a set of triples.

Example 4 (Queries expressible in the language). Table 1 describes typical data management queries in SNA. All of them are expressible in the language (indications given in the table). Note that some of them require negation of patterns, which can be expressed in SPARQL, despite not being explicit in its syntax [32].

3.1 Expressive Power and Complexity of the Language

Is the language rich enough for performing the usual queries in the area? Does the language have an efficient evaluation? These are two key questions we must answer to prove that the language works in general. We will see that while the latter can be answered in the positive, the answer for the first is a yes/no. Using the result in Angles [32] which shows that SPARQL has the same expressive power as Relational Algebra, it is not difficult to prove the following result:

Theorem 1 (Expressive Power). *The expressive power of the query language defined above in this section (see Definition 3) contains Relational Algebra and is contained in Relational Algebra plus aggregation.*

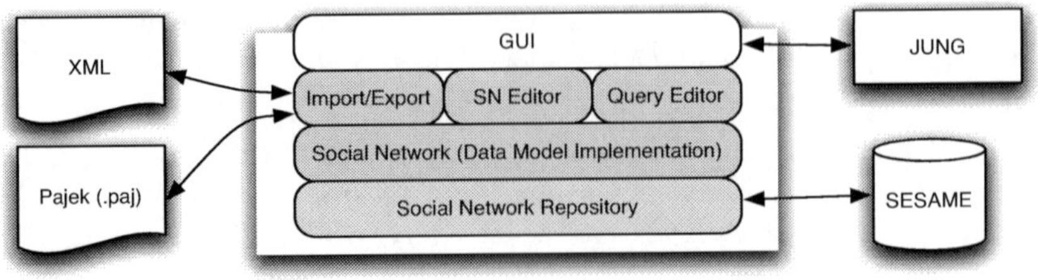

Fig. 6. *Prototype Architecture.* The figure shows the packages that form the SNDB.

This result shows that the language is expressive enough to make all sensible transformations of networks (see Table 1).

On the negative aspects, it shows that "global" queries cannot be expressed in it. These are queries found in SNA practice ([7,6,9]) corresponding to analytical measures of the network as a *whole* (i.e. they are attributes of the entire network): density, balance theory (based in paths and cycles), betweenness centrality (depends on paths), exposure/ threshold, critical mass (diffusion), etc. Technically they are not expressible in relational algebra, see e.g. [33]. There is work both on trying to incorporate some features for retrieving paths in SPARQL [34,35], and also on retrieving paths in the spirit of SNA [4].

Complexity. The cost in time of the evaluation of a query of the language, based on the translation to relational algebra we provided, is as follows:

1. First part: relational algebra. Essentially has the same cost as evaluating relational algebra. Note that the size of the output table $R_I(X_I)$ is, in the worst case, $O(|SN|^{X_I})$, where X_I is the number of variables of the input pattern of the query.
2. Second part, the SQL transformation T. The complexity of evaluating this transformation is linear in the number of rows of $R_I(X_I)$ once it is ordered by the `GROUP-BY` condition.
3. Third part: construction of the output network. The cost is asymptotically linear in the size of (P_O, X_O) times the size of the output of the table in (2).

3.2 A Prototype – Implementation Remarks

To illustrate the model, we built a proof of concept prototype of DBMS for the social network data model: SNDB. This prototype has two versions, one maps the conceptual model to the relational model, using as data back-end an RDBMS; and the other maps the model to RDF/SPARQL. The latter is described in this section.

The central module is *Social Network* (see Fig. 6) which implements the model and the query language, and handles main memory data structures needed to interact with upper layers. *SocialNetwork Repository* module works as an interface between *SocialNetwork* and the data services provider, in this case Sesame. Here model structures are translated to RDF, and model operations and queries are translated to Sesame update methods and SPARQL queries. *Network Import*

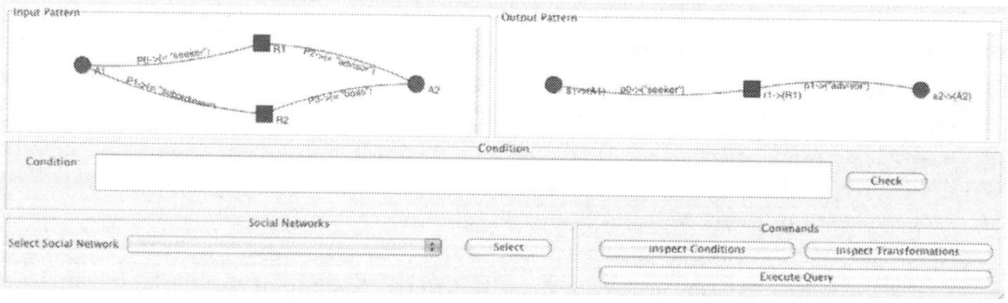

Fig. 7. *Protoype Query Editor.* Query editor snapshot showing query in *Example 2.*

and Export is composed by several modules, each one translating to and from an external format. Currently there are two modules implemented: XML, which handles the serialization and reading of social networks and queries to and from XML files (under a predefined schema); and Pajek, which imports and exports networks from and to Pajek files. The *Social Networks Editor* is a basic graphic social network editor which provides an interface to insert, delete and update individual social network components. Finally, the *Query Editor* (see Fig. 7) implements a graphical query editor, where it is also possible to execute the query over a given social network, and inspect the result.

The SNDB prototype was implemented using Java 1.5. It uses JUNG 1.7.6 network library to implement the GUI (analysis features are not used). Sesame 2.2.2 is used to implement the triple store. The choice of using SPARQL as the query language limits the queries that can be executed in the triple store to those that do not require aggregated functions. Figure 7 shows the query editor with the query in the *Example 2.*

The prototype has shown the advantages of having an abstract SNA model intuitive language, and a friendly query editor and processor. The problems we faced deal with the current lack of aggregation in SPARQL, and scalability, both at the visualization and response time level for big RDF graphs. There is plenty of optimizations to be done. For example, preliminary trials over a data set containing part of current DBLP (aprox. 100.000 records) in an average PC shows great flexibility for querying, however response times for complex queries are far from being satisfactory..

Overall, the prototype pinpoint several advantages of the approach: to interactively build networks and queries, to import and export data sets, and to execute queries and inspect results. Certainly it is mandatory to address the issues of optimization and aggregation in SPARQL.

4 Conclusions

We presented a data model and a query language for social networks data over RDF/SPARQL. This model provides a common data structure which support data interoperation, and the query language support the separation between data management tasks and application logic.

This query language formalizes requirements; hence, it may work as an implementation guide. These requirements where collected from SNA community practice, including published research, SNA software tools, and well known bibliographic sources, and are coincident with the main ideas suggested by other authors in the context of Semantic Web [1]. Finally we show that the RDF/SPARQL technology presents excellent features (except for the missing one of aggregation in SPARQL) for processing social networks.

Given that the implementation of the model may rest over the RDF/SPARQL stack, it can be almost seamlessly integrated with existing and new vocabularies, ontologies and their applications, leveraging the efforts to handle the enormous amount of heterogeneous social network data becoming available. Having a transformation and query language that has the flexibility and expressive power of relational algebra, clearly the main efforts should be directed to optimizations for big data sets.

References

1. Mika, P.: Social Networks and the Semantic Web. Semantic Web And Beyond Computing for Human Experience, vol. 5. Springer, Heidelberg (2007)
2. Jung, J.J., Euzenat, J.: Towards semantic social networks. In: Franconi, E., Kifer, M., May, W. (eds.) ESWC 2007. LNCS, vol. 4519, pp. 267–280. Springer, Heidelberg (2007)
3. Finin, T., Ding, L., Zou, L.: Social networking on the semantic web. Learning Organization Journal Ubiquitous Business Intelligence (2005)
4. Erétéo, G., et al.: A state of the art on social network analysis and its applications on a semantic web. In: SDoW 2008 at ISWC 2008 (2008)
5. Breiger, R.L.: The analysis of social networks. In: Hardy, M., Bryman, A. (eds.) Handbook of Data Analysis, pp. 505–526. Sage Publications, Thousand Oaks (2004)
6. Scott, J.: Social Network Analysis, 2nd edn. SAGE Publications, Thousand Oaks (2000)
7. Wasserman, S., Faust, K.: Social Network Analysis: Methods and Applications. Structural Analysis in the Social Sciences. Cambridge University Press, Cambridge (1994)
8. Freeman, L., Romney, A.K., White, D.R. (eds.): Research Methods in Social Network Analysis. Transaction Publishers (1992)
9. de Nooy, W., Mrvar, A., Batagelj, V.: Exploratory Social Network Analysis with Pajek. Cambridge University Press, Cambridge (2005)
10. Huisman, M., van Duijn, M.: Software for Social Network Analysis. In: [29], pp. 270–316
11. Handcock, M.S., Hunter, D.R., Butts, C.T., Goodreau, S.M., Morris, M.: Statnet: Software tools for the representation, visualization, analysis and simulation of network data. Journal of Statistical Software 24(1), 1–11 (2008)
12. Butts, C.T.: Network: A package for managing relational data in R. Journal of Statistical Software 24(2), 1–36 (2008)
13. Halpin, H.: Beyond walled gardens: Open standards for the social web. In: SDoW 2008 at ISWC 2008 (2008)
14. Freeman, L.: The Development of Social Network Analysis. Empirical Press (2004)

15. Angles, R., Gutierrez, C.: Survey of graph database models. ACM Computing Surveys (CSUR) 40(1), 1–39 (2008)
16. Klink, S., Reuther, P., Weber, A., Walter, B., Ley, M.: Analysing social networks within bibliographical data. In: Bressan, S., Küng, J., Wagner, R. (eds.) DEXA 2006. LNCS, vol. 4080, pp. 234–243. Springer, Heidelberg (2006)
17. Carley, K.M.: Linking capabilities to needs. In: [36], pp. 363–370
18. Jagadish, H.V., Olken, F.: Database management for life science research: Summary report of the workshop on data management for molecular and cell biology. OMICS 7(1), 131–137 (2003)
19. Gray, J., Liu, D., Nieto-Santisteban, M., Szalay, A., DeWitt, D., Heber, G.: Scientific data management in the coming decade. SIGMOD Record 34(4), 34–41 (2005)
20. Jensen, D., Neville, J.: Data mining in social networks. In: [36], pp. 289–302
21. Blau, H., Immerman, N., Jensen, D.: A visual language for querying and updating graphs. CS Technical Report 2002-037, University of Massachusetts (2002)
22. Tsvetovat, M., Diesner, J., Carley, K.M.: Netintel: A database for manipulation of rich social network data. CMU-ISRI-04-135 (March 2005)
23. Tsvetovat, M., Reminga, J., Carley, K.M.: Dynetml: Interchange format for rich social network data. CMU-ISRI-04-105 (2004)
24. Polleres, A., Krennwallner, T., Lopes, N., Kopecký, J., Drecker, S.: XSPARQL language specification (2009), `http://xsparql.deri.org/spec/`
25. Aleman-Meza, B., Nagarajan, M., Ramakrishnan, C., Ding, L., Kolari, P., Sheth, A.P., Arpinar, I., Joshi, A., Finin, T.: Semantic analytics on social networks: Experiences in addressing the problem of conflict of interest detection. In: WWW 2006, pp. 407–416 (2006)
26. Kinsella, S., Harth, A., Troussov, A., Sogrin, M., Judge, J., Hayes, C., Breslin, J.G.: Navigating and annotating semantically-enabled networks of people and associated objects. In: ASNA 2007 (2007)
27. Mika, P.: Social networks and the semantic web. In: Web Intelligence, pp. 285–291. IEEE Computer Society, Los Alamitos (2004)
28. Bonacich, P., Cody Holdren, A., Johnston, M.: Hyper-edges and multidimensional centrality. Social Networks 26, 189–203 (2004)
29. Carrington, P.J., Scott, J., Wasserman, S. (eds.): Models and Methods in Social Network Analysis. Cambridge (2005)
30. Gutierrez, C., Hurtado, C.A., Vaisman, A.: Introducing time into RDF. IEEE Transactions on Knowledge and Data Engineering 19(2), 207–218 (2007)
31. Guting, R.: Graphdb: modeling and querying graphs in databases. In: 20th VLDB Conference, pp. 297–308 (1994)
32. Angles, R., Gutierrez, C.: The expressive power of SPARQL. In: Sheth, A., et al. (eds.) ISWC 2008. LNCS, vol. 5318, pp. 114–129. Springer, Heidelberg (2008)
33. Courcelle, B.: Graph Rewriting: an Algebraic and Logic Approach. In: Handbook of Theoretical Computer Science. Formal Models and Sematics, vol. B, pp. 193–242. Elsevier and MIT Press (1990)
34. Pérez, J., Arenas, M., Gutierrez, C.: nSPARQL: A navigational language for RDF. In: ISWC 2008, pp. 66–81 (2008)
35. Alkhateeb, F., Baget, J., Euzenat, J.: Constrained regular expressions in SPARQL. In: SWWS 2008, pp. 91–99 (2008)
36. Breiger, R., Carley, K., Pattison, P. (eds.): Dynamic Social Network Modeling and Analysis: Workshop Summary and Papers. The National Academies Press, Washington (2003)

Applied Temporal RDF: Efficient Temporal Querying of RDF Data with SPARQL

Jonas Tappolet* and Abraham Bernstein

University of Zurich, Switzerland
lastname@ifi.uzh.ch

Abstract. Many applications operate on time-sensitive data. Some of these data are only valid for certain intervals (e.g., job-assignments, versions of software code), others describe temporal events that happened at certain points in time (e.g., a person's birthday). Until recently, the only way to incorporate time into Semantic Web models was as a data type property. Temporal RDF, however, considers time as an additional dimension in data preserving the semantics of time.

In this paper we present a syntax and storage format based on named graphs to express temporal RDF. Given the restriction to preexisting RDF-syntax, our approach can perform any temporal query using standard SPARQL syntax only. For convenience, we introduce a shorthand format called τ-SPARQL for temporal queries and show how τ-SPARQL queries can be translated to standard SPARQL. Additionally, we show that, depending on the underlying data's nature, the temporal RDF approach vastly reduces the number of triples by eliminating redundancies resulting in an increased performance for processing and querying. Last but not least, we introduce a new indexing approach method that can significantly reduce the time needed to execute time point queries (e.g., what happened on January 1st).

1 Introduction

Time, intervals, and versioning are central aspects of many applications. People in organizations, for example, hold many different positions over time, software, comes in many revisions, or services have temporal constraints (such as guaranteed executions times). As a consequence, there exist many data representations of time. Today's approaches, however, code temporal information as additional data inside the data model. Therefore, temporal information is implicit in the data and only difficult to access by programs. Indeed, the semantics of the temporal information needs to be completely coded in the client programs accessing the data. The Semantic Web, however, claims the accessibility of semantics by machines / agents and, therefore, an explicit time representation should support and preserve this central pillar of the Semantic Web.

* Partial support provided by Swiss National Science Foundation award number 200021-112330.

L. Aroyo et al. (Eds.): ESWC 2009, LNCS 5554, pp. 308–322, 2009.

In this paper we propose to use time as an additional semantic dimension of data. Therefore, it needs to be regarded as an element of the meta model instead of being just part of the data model. By applying the foundations established by Gutierrez et al.[1,2], our approach proposes to implement, first, an RDF [3] compatible syntax for temporal data. Second, we provide strategies for efficiently storing temporal RDF data (including an interval based index structure). Last but not least, we introduce a temporal extension to SPARQL [4] called τ-SPARQL and show how τ-SPARQL queries can be automatically translated to standard SPARQL operating on our storage scheme leveraging existing RDF/SPARQL infrastructure.

The paper first discusses the relevant related work. Then, it covers the proposed syntax for temporal RDF and τ-SPARQL. After presenting the mapping of τ-SPARQL to SPARQL it introduces a specialized temporal index structure for temporal RDF, which significantly improves retrieval performance of some types of temporal queries. The paper closes with an empirical evaluation of the introduced techniques and a discussion of the limitations and future work.

2 Related Approaches

The introduction of time into Semantic Web Data structures is not a novel challenge. In this section we will discuss some of these propositions and highlight their strengths and weaknesses. Specifically, we will discuss temporal extensions to OWL and Description Logic, methods for encoding time in the data model, and maintenance of graph versions.

2.1 Temporal Extensions

Different approaches cope with the temporal extension of Semantic Web or Semantic Web related technologies. For example, Artale and Franconi [5] propose to extend Description Logics with temporal features, while Welty and Fikes [6], Kim et al. [7] and the tOWL [8] project aim at introducing temporal entities into OWL. O'Connor et al.[9] investigated the querying of OWL data from the medical domain. In their approach, unlike ours, they use SWRL rules to query the KB. Our goal is to embed time into RDF, i.e., lower Semantic Web layers. We, therefore, base our approach on the foundations introduced by Gutierrez et. al. [1,2]. Gutierrez and colleagues also present a formal sketch for querying temporally enhanced RDF-graphs but did not propose a syntax to be used in, e.g., SPARQL. Even though temporal query languages have been extensively researched in the database community and lead to the TSQL2 query language [10], we were not able to find any analogous extension of the RDF query language SPARQL.

2.2 Time Encoded in the Data Model

A commonly used alternative to encoding time as a semantic element in the representation is to express temporal restrictions and validities, time is modeled as a part of the user's data model. Usually, a datatype property is defined with a range to one of xsd's date datatypes [11]. Pursuing this approach, two different problems arise:

First, we *loose the semantics of time.* Only a human understands that, e.g., the `hasBirthday` property denotes the start of the validity of an instance of the `Human` concept and that the completely different property `manufacturingDate` of the concept `Car` bears semantically comparable information.

Second, *temporal properties can only be attached to concepts or instances.* Whenever a relation needs to be annotated with temporal validity information, work-around solutions such as relationship-reification need to be introduced. As an example, consider the property `isMemberOf`, which is valid in a certain time interval. One work-around would be the reification of the property using a "Membership" blank-node with the properties `start`, `end`, `memberOf`, and `member`. This is cumbersome and leads to an increased query complexity requiring the inclusion of the blank-node in the query's graph pattern.

2.3 Graph Versions and Version Management Systems

Another approach to model time is to maintain multiple versions or temporal snapshots of the graph. Indeed version management systems are very popular to track the evolution of files (such as program source code) and their content. *SemVersion* [12] is such a version management system with a focus on ontology evolution management based on named graphs. However, *SemVersion* uses its own data model that contains the user's ontology and its evolution. Our proposition is not to introduce a new data model, but, storing the elements of the users' data model inside different named graphs according to their temporal context.

The advantage of versioning graphs is that the data model (usually) stays untouched. The major disadvantages are that they (1) limit the temporal semantics to certain time-points (i.e., snap-shots at given points in time) making the use of even slightly more involved temporal semantics, such as interval queries, impossible, (2) do not expose time explicitly but implicitly through snapshots making temporal reasoning complicated, and (3) will force users to access many snapshots if an interval of interest spans many versions forcing them to incur a potentially huge overhead due to redundant information in multiple versions. Our proposed solution will avoid all of these disadvantages.

3 A Temporal Syntax for RDF

In this section we introduce our temporal representations of time. Specifically, we introduce two different representations. The first is for internal use only and facilitates the addressing and indexing of temporal entities (e.g. for a triple store). The second provides the basis for a machine processable exchange format considering the semantics of time.

3.1 Internal Representation

As in most of the approaches dealing with temporal entities, we model time as a 1-dimensional discrete value (i.e. no branching time). More formally, each valid time point is defined as $\tau \in \mathbb{N}_0$. A time interval consists of two time points s (for start) and e (for end) such that $s, e \in \tau | s \leq e$. By allowing s to be equals e, we are able to express time points as intervals. Consequently, we will henceforth only use the interval notation. Note that this integer representation is an abstracted, simplified form of time which looses the exact semantics of time (e.g., whether it is a year number or version number). For most *internal* representations this is sufficient, since we only need to operate on the relative relations of values, i.e., $>$, $=$ and $<$. Furthermore, we allow half-bounded and unbounded intervals. We employ three special intervals that denote time periods which are open in at least one direction: $[0, e]$, $[s, \infty]$, and $[0, \infty]$. Since we restricted the range of valid time points to $\mathbb{N}_0$, 0 denotes the lowest possible value (i.e., the beginning of time). There are no time points before 0. The maximum time value, different from what ∞ might imply, does not lie ahead. It is allocated with the actual time during execution.

3.2 Exposed Representation

As a machine and human legible exchange format an exposed, semantic representation / interpretation is necessary, which may consist of the well-known xsd datatypes. In this approach we use the OWL-Time [13] ontology that defines `Timepoints`, `Intervals` and different date formats. However, we extended OWL-Time with a new date format type called `IntegerTime` to express non-calendaric time representations such as version numbers.

Analogous to the internal and conceptual schema of a DBMS the mapping of the internal to the exposed representation and vice-versa is a storage system dependent strategic decision that impacts the storage and retrieval performance. To map the above mentioned borderline values 0 and ∞ to semantically more expressive external representations, we use the literal values EVER and NOW. Hence, the internal $[0, \infty]$ interval is mapped to its semantic counterpart [EVER,NOW] which refers to a time interval whose elements have always been valid. Any data set without temporal information will be considered to be valid in this "eternal" interval.

3.3 Storage Format and Syntax

As mentioned before, we use named graphs [14] to enable the temporal data representation. We refer to a set of temporally related named graphs as a temporal graph. Each time interval is represented by exactly one named graph, where all triples belonging to this graph share the same validity period. To add an element to a temporal graph, the algorithm shown in Listing 1 is executed.

```
# retrieve validity interval
[start, end] = tripleToAdd.validity
if tripleToAdd containsOneOrMore blankNode
        # generate URI for blank-node
        tripleToAdd = convertBlankToURI(tripleToAdd)
endif
# if no named graph for the interval exists then make one
if !namedGraph[start, end].exists
    create new namedGraph[start, end]
    # add the new named graph to directory in default graph
    namedGraph.defaultGraph += namedGraph[start,end].temporalInfo
endif
namedGraph[start,end] +=  tripleToAdd
```

Listing 1. Pseudo Code for an insert operation into a temporal graph

As the listing shows, we first ascertain the triple's temporal validity. Note that the extraction of the validity is dependant on the way time is represented in a data set. Then, we check, if it contains a blank node. Blank nodes are especially problematic as their validity is restricted to the node's parent graph. In our case, however, where the collection of the interval-representing named graphs logically form the overall graph, it might make sense to have blank-nodes that span multiple intervals and, thus, multiple named graphs. This requirement is akin to the implementation employed by the Intellidimension RDF Gateway's quad approach [14], which deviates from the *W3C* recommendation in that respect. To adhere to the W3C recommendation we simply assign a URI to any blank node circumventing the blank node's spirit but fulfilling our premise of using off-the-shelf tools – in our opinion an acceptable compromise between adhering to the spirit of RDF and practicality. Next, we establish, if a named graph representing the interval already exists. If not, then we generate a new named graph for this interval and add information about it to the named graph directory in the default graph. Finally, we add the triple to the appropriate named graph. Note that we do not allow the same triple to appear in multiple temporal contexts. Therefore, a temporal RDF framework needs to take care that a triple only exists once, but maybe, due to an update of the temporal information travels from one named graph to another.

As apparent from this procedure we use the default graph as a directory or container for the semantic information about the intervals. Since every interval is uniquely identified by the URI of its named graph, vital as well as additional information can be provided in the default graph. Vital information includes the start and end time of an interval in any time format allowed by OWL-Time. Additional information includes relations between intervals such as `after`, `before`, or `overlaps`. Since the additional information about intervals can be inferred from the start and end time of an interval, a temporal reasoning system could entail these relations.

With this approach, we can benefit from well-established, scalable support of named graphs in storage systems without having to implement any adaption to temporal RDF. As mentioned above, temporal validity information could also be provided using reification. However, to express the same amount of information, reification would consume 15 times more triples than named graphs seriously questioning the scalability both in terms of storage space and, more importantly, in terms of retrieval run-time.

4 The τ-SPARQL Query Language

Having established a storage format in Section 3.3, in this section we focus on the retrieval of these triples. It introduces τ-SPARQL, an extended syntax for SPARQL to express temporal queries. We introduce τ-SPARQL by discussing its two major usage formats: time point Queries and temporal queries.

4.1 Time Point Queries

Time point queries aim at retrieving information valid at a specified point in time. This is of special importance whenever a snapshot in the past needs to be retrieved from a dataset. Similar to the `FROM` statement in SPARQL to select a dataset (or graph), we define a `FROM SNAPSHOT` τ expression to select a specific point τ in time. The `FROM SNAPSHOT` τ expression signals the query engine to evaluate the query's graph pattern only on graphs-elements valid at the time point τ, where τ has to be a literal time value (the literal type is depending on the underlying time format). The query in Listing 2, for example, retrieves all `foaf:Persons` that were valid (i.e. alive) in 1995.

```
SELECT ?person FROM SNAPSHOT 1995 WHERE{
        ?person a foaf:Person .
}
```

Listing 2. τ-SPARQL: Time point query

4.2 Temporal Queries

Complementing time point queries that query the graph valid at a given point in time, temporal queries allow the usage of wild card intervals and time points. These wild cards can be used to bind a variable to the validity period of a triple or to express temporal relationships between between intervals. τ-SPARQL allows one form of temporal wildcards, `[?s,?e]`, which binds the literal start and end values. Note that whilst τ-SPARQL explicitly binds the two variables `?s` and `?e` that can be used elsewhere in the query, a third implicit variable `?g` is internally bound to the URI of the named graph that represents the interval defined by *[?s,?e]*. Hence, the expression used in the interval context *[?s,?e]* is bound to the named graph URI while the partial variables *?s* and *?e* are standard SPARQL variables which are bound to the literal start and end value of the interval's validity. Therefore, `?s` and `?e` can occur in any part of the query specified by the SPARQL syntax, e.g. in `FILTER` or `ORDER BY` expressions. In Listing 3, an example query is shown, which retrieves a person's URI and the start of the validity period (i.e. the person's birthday). Since a temporal annotation is based on triple level, the interval variables (or specifications) could be defined before or after each subject, predicate, or object. For the sake of consistency, we define that a temporally annotated triple pattern (i.e., quad pattern) must always start with the temporal variable part.

```
SELECT ?s ?person WHERE{
        [?s,?e] ?person a foaf:Person .
}
```

Listing 3. τ-SPARQL: Selection of validity period

To query for relations between intervals, two interval wildcards can be connected with the properties defined by the OWL-Time ontology. These properties are the well-know Allen Interval Relations [15]. Listing 4 shows a query to retrieve all the `foaf:Persons` that could have known Albert Einstein because their lifespan overlaps with Einstein's. In literature, this kind of patterns are referred to as *temporal join* [10].

```
<PREFIX time: http://www.w3.org/2006/time#>
<PREFIX foaf: http://xmlns.com/foaf/0.1/#>
SELECT  ?s2 ?e2 ?person WHERE{
        [?s1,?e1] ?einstein foaf:name "Albert Einstein" .
        [?s2,?e2] time:intervalOverlaps [?s1,?e1] .
        [?s2,?e2] ?person a foaf:Person .
}
```

Listing 4. τ-SPARQL: Interval relations

5 Mapping τ-SPARQL to Standard SPARQL

In this section we show, that the τ-SPARQL syntax extension can be rewritten without any loss of expressivity to standard SPARQL. Therefore, τ-SPARQL offers an intuitive syntax for the composition of queries (for human consumption), while the rewritten form can be evaluated as standard SPARQL using any query engine supporting named graphs using their built-in standard query optimization and storage formats scalability.

5.1 Mapping Time Point Queries

Rewriting a time point query requires three steps: First, the respective intervals for a time point needs to be determined. Second, the elements belonging to these intervals need to be combined to a new, virtual data set. Finally, the graph pattern needs to be matched against this new data set.

As described above, the information about the validity of an interval is encoded in the default graph of the named graph set. To extract the intervals valid at a given time point, the partial query shown in Listing 5 has to be added to the initial query in the `WHERE` clause.

```
?g time:hasBeginning ?start .
?start time:[inInteger|inXsdDateTime|...] ?s .
FILTER(?s <= <TP> || ?start = NOW) .
?g time:hasEnd ?end .
?end time:[inInteger|inXsdDateTime|...] ?e .
FILTER(?e >= <TP> || ?end = EVER) .
```

Listing 5. Partial query to determine valid intervals at a given time point

The pattern in Listing 5 will bind variable ?g to all the named graphs containing triples valid in time point <TP>. The expression [inInteger | inXsdDateTime | ...] is evaluated according to the literal type of <TP>. If it is a xsd:Integer, then the time:inInteger property is selected. Analogously, a xsd:dateTime will be represented by the inXsdDateTime property. Note that both the τ-SPARQL query language and the representation format are open to additional time formats as long as the rewriting algorithm is aware of the acceptable correct formats and their respective mappings.

In a last step of the rewriting process, the initial graph pattern of the query needs to be restricted to the intervals bound in ?g. SPARQL offers this functionality with the GRAPH keyword. This is done by extracting the pattern inside the WHERE clause of the query and inserting it enclosed by GRAPH ?g {<pattern>}.

5.2 Mapping Temporal Queries

To map the queries that select the validity of a triple pattern and/or relations between validities, the rewriting process is slightly more complex. We define that all classic SPARQL variables ?v belong to a set C. The intervals [?s,?e] to belong to the set of intervals I, and ?s and ?e to belong to the set of point variables P. For each $i \in I$ there exists exactly one URI g belonging to the set of URIs of named graphs G and a pair of time points ?s and ?e, which mark its start and end. Any pair of time points ?s and ?e can be mapped to an interval i, which in turn can be mapped to a URI from G. Consequently, a quad pattern in the form [?s,?e] <triple pattern> can be rewritten as GRAPH ?x{<triple pattern>} where ?x is the URI $\in G$ that corresponds to the interval defined by [?s,?e]. Note, that a new variable "?x" needs to be chosen for every rewrite. Please note that temporal variables are not allowed in the predicate part.

```
<PREFIX time: http://www.w3.org/2006/time#>
<PREFIX foaf: http://xmlns.com/foaf/0.1/#>
SELECT  ?s2 ?e2 ?person WHERE{
        GRAPH g1 { ?einstein foaf:name "Albert Einstein" .}
        ?g2 time:intervalOverlaps ?g1 .
        GRAPH g2 { ?person a foaf:Person }.
        ?g2 time:hasBeginning ?start .
        ?start time:inInteger ?s2 .
        ?g2 time:hasEnd ?end .
        ?end time:inInteger ?e2 .
}
```

Listing 6. Rewritten τ-SPARQL query

It is important that whenever a temporal variable is used outside the interval context (i.e., outside a `GRAPH ?x{<triple pattern>}`) then it needs to be treated like a time point statement. Hence, the partial query shown in Listing 5 needs to be added to the rewritten τ-SPARQL query.

As an example, Listing 6 shows the rewritten standard SPARQL syntax of the query in Listing 4.

6 An Index Structure for Time Intervals

In the previous section we demonstrated how to rewrite τ-SPARQL time point queries to standard SPARQL. However, the performance of time point queries can be raised by providing a dedicated index structure. Temporal index structures have been discussed extensively in the database community. In a recent publication, [16] proposes the tGRIN index structure for temporal RDF w.r.t. the temporal as well as the structural neighborhood of triples. Our requirement, however, is the retrieval of intervals valid in a certain time instant. Therefore, we propose such an index structure that supports the efficient retrieval of intervals valid in certain time instants. This is not a base index of triples, it is rather a meta-index containing named graphs which, by themselves, have indices of their triples.

Specifically, we use a tree-based index structure which stores the validity start and end values in an ordered manner. To achieve a fast retrieval of all valid intervals at a specified time point we would need to store each interval multiple times into this index: in each time point of its validity. E.g., an interval [0,100] would appear 100 times in the index. This is an unpractical approach as it requires a large amount of space. Similar to the mpeg [17] compression algorithm we apply the concept of key frames (more precisely: I frames) to our interval index structure. Similar to a key frame, a key index element stores all the intervals valid at this time point. Between the key indices only the deltas are stored. We refer to this index structure as keyTree index. This index structure is very similar to the Time Index proposed by [18]. The Time index maintains incremental buckets which refer to the deltas between two index elements. However, unlike our keyTree, the Time Index does not allow to select a key index distance, it rather selects the key indexing points whenever an interval starts or ends. This can lead, in datasets with very dense intervals, to oversized indices. Our index offers the parameter *key index distance* which permits a storage system to adapt to different temporal kinds of datasets, e.g., whether the time intervals distribution is dense or sparse. The retrieval strategy of the keyTree index can be summarized as follows:

- if the asked time instant coincides with a key index, return all elements in the key index
- if the asked time instant does not coincide with a key index, go back to the immediately preceding key index. From there, walk forward and apply all the deltas until the asked index element is reached and return the set with the applied deltas.

We will show an evaluation of this index structure in the next section with a comparison to an alternative index.

7 Evaluation

For the evaluation of our approach we are using two different data sources. We used the EvoOnt data [19,20] that consists of software source code information. Specifically, we used information about the source code of the *org.eclipse.compare* plugin of the Eclipse Project. 306 releases created in the period between 2001 and 2007 were available. The second data source is a data set describing Swiss parliamentarians (with information about year of birth and death as well as start and end of service periods) between 1848 and 2008. The latter data has to be seen, for this approach, as a worst case, as almost every triple is defined in its own time interval. Table 1 lists the characteristics of the data sources.

Table 1. Datasets

Data Source	Time Model	Data Period	# Triples	# Intervals
EvoOnt	Versioned graphs	2001 - 2007	22 millions	2505
Parliamentarians	Encoded in data model	1848 - 2008	50'000	38'182

All evaluations in this section were computed on in-memory graphs using the ng4j [21] named graph API. To ensure that we compare different approaches rather than the quality of the implementation of the RDF APIs, we also load the comparison data sets into one single named graph and run queries against the ng4j implementation (which is based on Jena [22]). All the evaluations were executed on an Intel Core2 Duo 1.6 GHz system with 4 GB of RAM running Windows Vista.

7.1 Dataset Conversion

Our first evaluation covers the conversion from the respective time representation format into temporal RDF according to the method presented in this work. First, we converted the EvoOnt data set into a temporal graph. Figure 1(a) shows the vast reduction of triples in the temporal graph approach due to the reduction of redundancies. Having a graph that contains the information about all the 306 versions, the number of triples reduce from 22 millions in the versioned graph approach to 385'000 in the temporal graph. The reason for this huge reduction is depicted in Figure 1(b). The fraction of triples changing between versions of the graph (i.e. between the software versions) is usually very small (far below 5%). As a consequence, 95% of the graph (i.e., the unchanged triples) are stored redundantly.

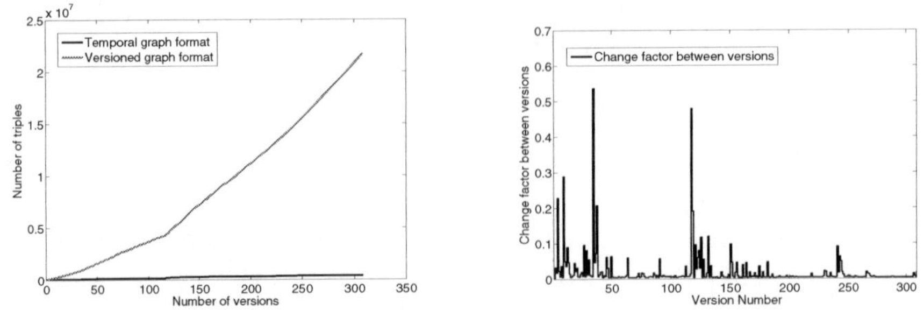

(a) Growth of number of triples. Gray line: versioned graph, black line: temporal graph.

(b) Change factor between the different versions in the dataset.

Fig. 1. Conversion from graph versions to a temporal graph

7.2 keyTree Index

To evaluate the efficiency of our keyTree index structure introduced in Section 6, we compared our index structure to an in-memory ordered list (Java's TreeList implementation) that contains the intervals' start and end times. Additionally, we added three special sets of intervals containing the borderline values [EVER,x], [x,NOW] and [EVER,NOW] to this baseline implementation.

We evaluated two different scenarios on our datasets. The first is the build up time for the indices with different values for the keyTree index' parameter *key index distance* (i.e. the number of time instants between two key indices). Secondly, we measured the retrieval time for the set of valid intervals in a certain time instant. The following figures show the build and retrieval times for the Parliamentarians (Figure 2) and EvoOnt (Figure 3) datasets. All values were calculated by running each operation 100 times and plotting the average time of these runs.

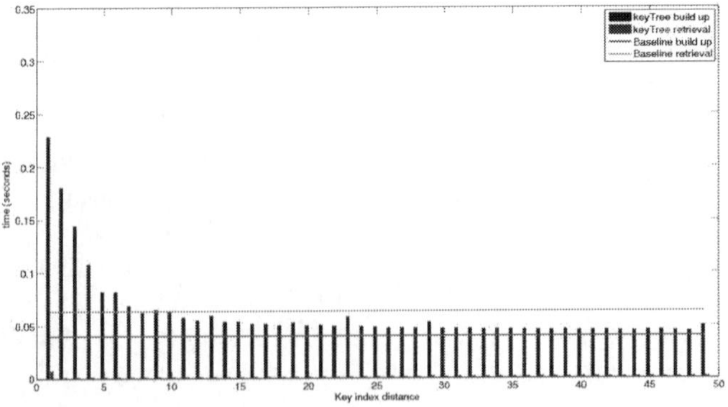

Fig. 2. keyTree index build up and retrieval of the Parliamentarians dataset

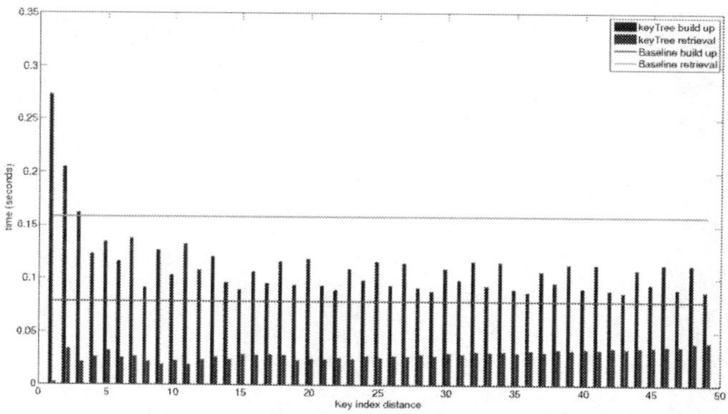

Fig. 3. keyTree index build up and retrieval of the EvoOnt dataset

The figures show, that the keyTree index structure never outperforms the baseline implementation in terms of build time. However, when amortizing the build time over multiple retrievals the use of the more sophisticated keyTree seems to pay off. As a conclusion, the keyTree index is at least three times faster than a "standard" one. The reason for the decreased performance of the baseline is that for each retrieval request the whole index needs to be traversed to check whether there are elements that are valid during the asked time point.

7.3 Timepoint Queries

Next, we run time point queries using both, the Parliamentarians and EvoOnt data sets. Since a time point in the EvoOnt dataset is represented by a snapshot graph, we equate the time to load the graph into memory with the querying time. We also compare the querying time for the temporal graph with the τ-SPARQL triple pattern approach to determine the valid intervals. The evaluated query consisted of retrieving all the triples in a defined time instant (Parliamentarians dataset: 1995, EvoOnt: 5). Additionally we queried the datasets using our keyTree index. Figure 4 shows the execution times of the queries in the different setups and datasets.

Again, the versioned graph approach does not allow to run time point queries, instead, we listed the time needed to load the graph into memory which is the equivalent task needed to select the valid triples in a specific time instant. The results shown in Figure 4 reflect the different nature of the datasets. The Parliamentarian data consists of many different time intervals with relatively few triples in each interval whereas the EvoOnt dataset has less intervals but many more triples per interval. On this note, the keyTree index structure can greatly reduce the query execution time in the EvoOnt dataset because the query engine encounters a vastly reduced number of triples to run the pattern against. The long query execution time for the non-temporal query on the Parliamentarians dataset is mostly due to the complex query pattern using UNIONs to retrieve all elements valid at the desired time point.

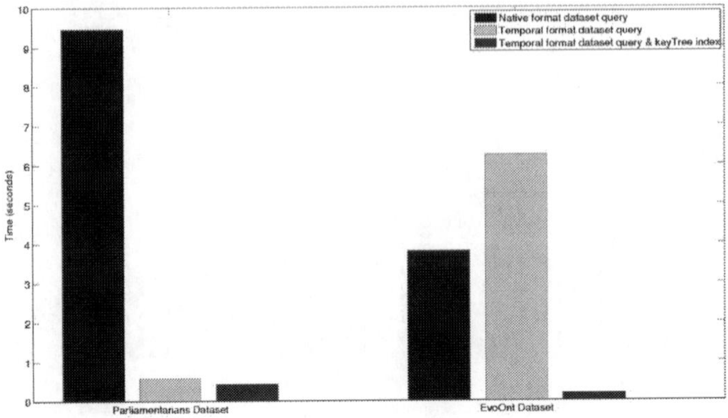

Fig. 4. Execution times of time point queries using different datasets and query strategies

7.4 Temporal Queries

In a final set of evaluations we cover temporal queries. Since such queries (e.g. after, overlaps) are not possible to run against a dataset that uses versioned graphs, we only used the Parliamentarians dataset. The query we evaluated was to retrieve all the parliamentarians complete service periods while "Kurt Furgler" was alive (1924-2008). The τ-SPARQL query uses the `time:intervalDuring` relation, while the classical SPARQL comparison query checks every service period whether it is enclosed by the life time of "Kurt Furgler". Figure 5 shows the execution times of the two queries. Again, it can be seen, that the temporal approach is performing (almost 50%) better because of the reduced complexity of the query pattern. However, the execution time increases whenever all the interval relations are entailed and written into the default graph. For the query, we only entailed the interval relations asked by the query and omitted additional ones. But, as mentioned above, this can be absorbed by using a temporal reasoner that provides the interval relations without materializing them into triples.

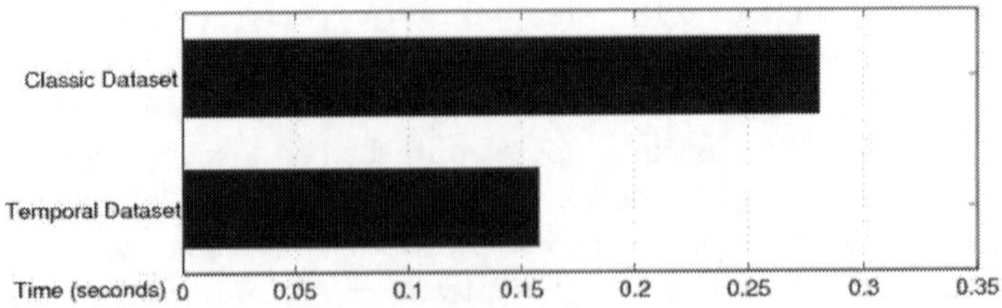

Fig. 5. Execution times of temporal queries using temporal and non-temporal query strategies

8 Conclusion, Limitations and Future Work

We presented an applied temporal RDF using named graphs. In detail, we presented a syntax as storage format which allows the annotation of RDF triples with temporal validity intervals. This approach is as minimally invasive as possible to preexisting storage and querying systems. Furthermore, we introduced T-SPARQL, an extension to express temporal queries. We showed, that τ-SPARQL can be directly mapped to standard SPARQL. Additionally we presented a specific index structure for temporal intervals which improves the retrieval time of time point queries. Finally we evaluated our approach comparing our approach to classical graph versioning and time representation and could show that our temporal RDF and τ-SPARQL approach enables queries that are either impossible (versioned graph) or only achievable with very complex query patterns (time in data model) resulting in increased performance in terms of answering time of queries. We showed cases where our approach outperforms existing ones and circumstances where our approach does not completely outperform existing ones. However, we showed, that a combination with our keyIndex structure can overcome this lack of performance. Moreover, when using temporal RDF, the total number of triples can be largely reduced by removing redundancies which releases resources in each of the storage, reasoning and querying systems. This has a positive effect on disk space usage and response time.

As a limitation we have to mention the lacking availability of temporal reasoners. When materializing the interval relations such as *overlaps*, *meets*, or *before*, the number of temporal relations grows exponentially which makes the materialization in triple form unpractical. Furthermore, our approach performs best on data that has multiple data elements valid within the same interval. If each triple's validity is distinct then one named graph needs to be defined for each triple, which can lead to a decreased performance. However, we believe that many application scenarios exist, where such an extreme interval mapping is not the case.

Our future work will have to concentrate at addressing the current limitations. One possibility is to explore the the entailment of new temporal relations based on our keyTree index. Additionally, we plan to further evaluate our approach with more complex query patterns from different domains of application.

References

1. Gutierrez, C., Hurtado, C.A., Vaisman, A.: Introducing time into RDF. IEEE Transactions on Knowledge and Data Engineering 19(2) (2007)
2. Gutierrez, C., Hurtado, C., Vaisman, A.: Temporal RDF. In: Gómez-Pérez, A., Euzenat, J. (eds.) ESWC 2005. LNCS, vol. 3532, pp. 93–107. Springer, Heidelberg (2005)
3. Klyne, G., Carroll, J.J.: Resource description framework (RDF): Concepts and abstract syntax. Technical report, W3C Recommendation (2004)
4. Prud'hommeaux, E., Seaborne, A.: SPARQL query language for RDF. Technical report, W3C Recommendation (2008)

5. Artale, A., Franconi, E.: Temporal description logics. In: Handbook of Time and Temporal Reasoning in Artificial Intelligence. Elsevier, Amsterdam (2005)
6. Welty, C.A., Fikes, R.: A reusable ontology for fluents in OWL. In: Proc. of the 4th Int'l. Conference On Formal Ontology In Information Systems (2006)
7. Kim, S.K., Song, M.Y., Kim, C., Jang, H.C.: Temporal ontology language for representing and reasoning interval-based temporal knowledge. In: Proc. of the 3rd Asian Semantic Web Conference (2008)
8. Milea, V., Frasincar, F., Kaymak, U.: Knowledge engineering in a temporal semantic web context. In: The Eighth Int'l. Conference on Web Engineering (2008)
9. O'Connor, M., Shankar, R., Parrish, D., Das, A.: Knowledge-level querying of temporal patterns in clinical research systems. In: Proc. of the 12th World Congress on Health (Medical) Informatics (2007)
10. Snodgrass, R.T. (ed.): The TSQL2 Temporal Query Language. Kluwer, Dordrecht (1995)
11. Campbell, C.E., Eisenberg, A., Melton, J.: Xml schema. Technical report, W3C Recommendation
12. Völkel, M., Enguix, C.F., Kruk, S.R., Zhdanova, A.V., Stevens, R., Sure, Y.: Semversion - versioning RDF and ontologies. Technical report, Institute AIFB, University of Karlsruhe (2005)
13. Hobbs, J.R., Pan, F.: Time ontology in OWL. Technical report, W3C Working Draft (2005)
14. Carroll, J.J., Bizer, C., Hayes, P., Stickler, P.: Named graphs. Journal of Web Semantics 3(3) (2005)
15. Allen, J.F.: Maintaining knowledge about temporal intervals. Communications of the ACM 26(11) (1983)
16. Pugliese, A., Udrea, O., Subrahmanian, V.S.: Scaling RDF with time. In: Proc. of the 17th Int'l. World Wide Web Conference (2008)
17. Pan, D.: A tutorial on mpeg/audio compression. IEEE MultiMedia 2(2) (1995)
18. Elmasri, R., Wuu, G., Kim, Y.: The time index: An access structure for temporal data. In: Proc. of the 16th Int'l. Conference on Very Large Data Bases (1990)
19. Kiefer, C., Bernstein, A., Tappolet, J.: Analyzing software with iSPARQL. In: Proc. of the 3rd Int'l. Workshop on Semantic Web Enabled Software Engineering. Springer, Heidelberg (2007)
20. Kiefer, C., Bernstein, A., Tappolet, J.: Mining Software Repositories with iSPARQL and a Software Evolution Ontology. In: Proc. of the 2007 Int'l. Workshop on Mining Software Repositories. IEEE Computer Society, Los Alamitos (2007)
21. Bizer, C., Cyganiak, R., Watkins, R.: NG4J - Named Graphs API for Jena (Poster). In: Proc. of the 2nd European Semantic Web Conference (2005)
22. Carroll, J.J., Dickinson, I., Dollin, C., Seaborne, A., Wilkinson, K., Reynolds, D.: Jena: Implementing the semantic web recommendations (2004)
23. Emerson, E.A.: Temporal and modal logic. In: Handbook of Theoretical Computer Science. Elsevier, Amsterdam (1990)

ReduCE: A Reduced Coulomb Energy Network Method for Approximate Classification

Nicola Fanizzi, Claudia d'Amato, and Floriana Esposito

Dipartimento di Informatica, Università degli studi di Bari
Campus Universitario, Via Orabona 4, 70125 Bari, Italy
{fanizzi,claudia.damato,esposito}@di.uniba.it

Abstract. In order to overcome the limitations of purely deductive approaches to the tasks of classification and retrieval from ontologies, inductive (instance-based) methods have been proposed as efficient and noise-tolerant alternative. In this paper we propose an original method based on non-parametric learning: the Reduced Coulomb Energy (RCE) Network. The method requires a limited training effort but it turns out to be very effective during the classification phase. Casting retrieval as the problem of assessing the class-membership of individuals w.r.t. the query concepts, we propose an extension of a classification algorithm using RCE networks based on an *entropic* similarity measure for OWL. Experimentally we show that the performance of the resulting inductive classifier is comparable with the one of a standard reasoner and often more efficient than with other inductive approaches. Moreover, we show that new knowledge (not logically derivable) is induced and the likelihood of the answers may be provided.

1 Introduction

The inherent incompleteness and accidental inconsistency of knowledge bases (KBs) in the Semantic Web (SW) requires approximate query answering methods for retrieving resources. This activity is generally performed by means of logical approaches that try to cope with the problems mentioned above. This has given rise to alternative methods [15, 10, 9, 13, 11, 14].

Inductive methods applied to specific tasks are known to be often more efficient and noise-tolerant. Lately, first steps have been taken to apply classic machine learning techniques for building inductive classifiers to the complex representation and semantics adopted in the context of the SW [7], especially through non-parametric methods [4, 8]. Instance-based inductive methods may help a knowledge engineer populate ontologies [2]. Some methods are also able to complete ontologies with probabilistic assertions, allowing for sophisticate approaches to dealing with uncertainty [12].

In this paper we propose a novel method for inducing a classifier that can have several applications. Mainly it may naturally be employed as an alternative method for performing *concept retrieval* [1]. Even more so, like its predecessors mentioned above, it is able to determine a likelihood measure of the

L. Aroyo et al. (Eds.): ESWC 2009, LNCS 5554, pp. 323–337, 2009.

induced class-membership assertions which is important for *approximate query answering*. Some assertions could not be logically derived, but may be highly probable according to the classifier; this may help to cope with the uncertainty caused by the inherent incompleteness of the KBs even in absence of an explicit probabilistic model.

Specifically, we propose to answer queries adopting an *instance-based* classifier induced by a non-parametric learning method, the REDUCED COULOMB ENERGY (RCE) network [5], that has been extended to be applied to the standard representations of the SW via a semantic similarity measure for individual resources. As with other similarity-based methods, a retrieval procedure may seek for individuals belonging to query concepts, exploiting the analogy with other training instances, namely the classification of the nearest ones (w.r.t. the similarity measure).

Differently from lazy-learning approaches experimented in the past [4], that do not require training, yet more similarly to the methods based on *kernel machines* [3, 8], the new method is organized in two phases: 1) in the *training* phase a simple network based on prototypical individuals (parametrized for each prototype) is trained to detect the membership of further individuals w.r.t. some query concept; 2) this network is then exploited by the classifier, during the *classification* phase, to make a decision on the class-membership of an individual w.r.t. the query concept on the grounds of a likelihood estimate. The network parameters correspond to the radii of hyperspheres centred at the training individuals in the context of the metric space determined by some measure. The classification of an individual is induced observing the balls it belongs to and on its distance from their centers (training prototypes). The efficiency of the method derives from limiting the expensive model construction to a small set of training individuals, while the resulting RCE network can be exploited in the next phase to efficiently classify a large number of individuals.

A similarity measure derived from a family of semantic pseudo-metrics [4] was exploited. This language-independent measure assesses the similarity of two individuals by comparing them on the grounds of their behavior w.r.t. a committee of features (concepts), namely (a subset of) those defined in the KB which can be thought as a context; alternatively, it may be generated to this purpose through randomized search algorithms aiming at optimal committees to be run in advance [6]. Besides, since different features may exhibit a different relevance w.r.t. the similarity value to be determined, each feature is weighted on the grounds of the *quantity of information* it conveys. This weight is determined by an *entropic* measure. The rationale is that the more general a feature (or its negation) the less likely is should be used for distinguishing the individuals.

The resulting system REDUCE allowed for an intensive experimentation of the method on performing approximate query answering with a number ontologies from public repositories. Like in previous works [4, 8], the predictions were compared to assertions that were logically derived by a deductive reasoner. The experiments showed that the classification results are comparable (although slightly less complete) and also that the classifier is able to induce new knowledge that is not logically derivable.

The paper is organized as follows. The basics of the instance-based approach applied to the standard representations are recalled in Sect. 2. The next Sect. 3 presents the semantic similarity measures adopted in the retrieval procedure. Sect. 4 reports the outcomes of the experiments performed with the implementation of the procedure. Possible developments are finally examined in Sect. 5.

2 RCE Network Training and Classification

In the following sections, we assume that concept descriptions are defined in terms of a generic language based on OWL-DL that may be mapped to *Description Logics* (DLs) with the standard model-theoretic semantics (see the handbook [1] for a thorough reference).

A *knowledge base* $\mathcal{K} = \langle \mathcal{T}, \mathcal{A} \rangle$ contains a *TBox* $\mathcal{T}$ and an *ABox* $\mathcal{A}$. $\mathcal{T}$ is a set of axioms that define concepts (and relationships). $\mathcal{A}$ contains factual assertions concerning the individuals (resources). The set of the individuals occurring in $\mathcal{A}$ is denoted with $\mathsf{Ind}(\mathcal{A})$. The *unique names assumption* may be made on the such individuals, that are represented in OWL by their URIs.

As regards the required inference services, like all other instance-based methods, the inductive procedure requires performing *instance-checking* [1], which roughly amounts to determining whether an individual, say a, belongs to a concept extension, i.e. whether $C(a)$ holds for a certain concept C. This is denoted by $\mathcal{K} \models C(a)$. This service is generally provided by a reasoner. Note that because of the open-world semantics, it may be unable to give a positive or negative answer to a class-membership query.

2.1 Basics

An approximate classification method is proposed for individuals in the context of semantic knowledge bases. The method requires training an RCE network, which can be successively exploited for an inductive classification procedure. Moreover, a likelihood measure of the decisions made by the inductive procedure can be provided.

In *instance-based* learning [5, 16], the basic mechanism determines the membership of a query instance to some concept in accordance with the membership of the most similar instance(s) with respect to a similarity measure.

This may be formalized as the induction of an estimate for a discrete-valued target hypothesis function $h : \mathit{TrSet} \to V$ from a space of instances TrSet to a set of values $V = \{v_1, \ldots, v_s\}$, standing for alternative classes to be predicted. In the specific case of interest [4], the adopted value set is $V = \{+1, -1, 0\}$, where the three values denote, respectively, membership, non-membership, and uncertainty. The values of h for the training instances are determined as follows:

$$\forall x_i \in \mathit{TrSet}: \qquad h_Q(x_i) = \begin{cases} +1 & \mathcal{K} \models Q(x_i) \\ -1 & \mathcal{K} \models \neg Q(x_i) \\ 0 & \mathit{otherwise} \end{cases}$$

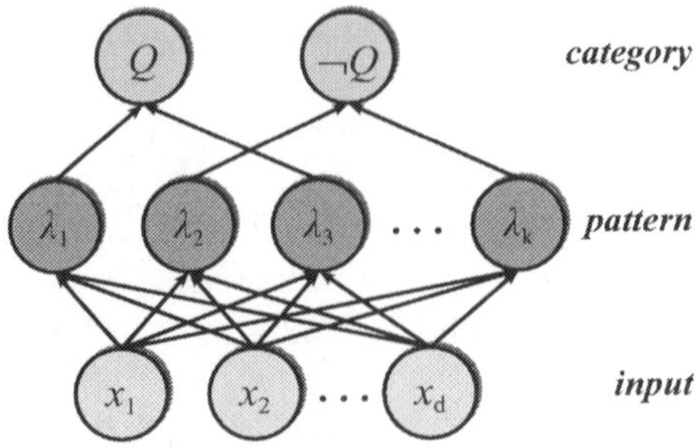

Fig. 1. The structure of a Reduced Coulomb Energy network

Note that normally $|\mathit{TrSet}|$ is much less than $|\mathsf{Ind}(\mathcal{A})|$ i.e. only a limited number of training instances (exemplars) is needed, especially if they can be carefully selected among the prototypical ones for the various regions of the search space.

Let x_q be the query instance whose class-membership is to be determined. Given a dissimilarity measure d, in the *k-Nearest Neighbor* method (k-NN) [4] the estimate of the hypothesis function for the query individual may be determined by: $\hat{h}_Q(x_q) = \mathrm{argmax}_{v \in V} \sum_{i=1}^{k} \delta(v, h_Q(x_i))/d(x_i, x_q)^2$ where δ returns 1 in case of matching arguments and 0 otherwise.

2.2 Training

The RCE method[1] adopts a similar non-parametric approach. The construction of the RCE model can be implemented as training a (sort of neural) network whose structure, for the binary problem of interest, is sketched in Fig. 1.

The input layer receives the information from the individuals. The middle layer of patterns represents the features that are constructed for the classification; each node in this layer is endowed with a parameter λ_j representing the radius of the hypersphere centered in the j-th prototype individual of the training set. During the training phase, each coefficient λ_j is adjusted so that the spherical region is as large as possible, provided that no training individual of a different class is contained. Each pattern node is connected to one of the output nodes representing the predicted class. In our case we have only two categories representing membership and non-membership w.r.t. the query concept Q.

The training algorithm is sketched in Fig. 2. Suppose the training instances (a.k.a. examples) are labeled with their correct classification $\langle x_i, h_Q(x_i)\rangle$, where $h_Q(x_i) \in V$, as seen before. In this phase, each parameter λ_j, that represents the radius of a hypothetical hypersphere centered at the input example, is adjusted to be as large as possible (they are initialized with a maximum radius), provided that the resulting region does not enclose counterexamples. As new individuals are processed, each such radius λ_j is decreased accordingly (and can never increase). In this way, each pattern unit can enclose several prototypes, but only those having the same category label.

[1] Borrowing its name from electrostatics, where energy is associated with charged particles [5].

input
   $TrSet = \{\langle x_i, h_Q(x_i)\rangle\}$: set of training examples
output
   w_{jk}, λ_j, a_{cj}: RCE Network weights

1. **begin**
2. initialize $\epsilon \leftarrow$ small parameter; $\lambda_{\max} \leftarrow$ max radius
3. **for** $j \leftarrow 1$ **to** $|TrSet|$ **do**
   (a) train weight: $w_{jk} \leftarrow x_k$
   (b) find nearest counterexample: $\bar{x} \leftarrow \arg\min_{x \in \mathcal{C}_j} d(x, x_j)$
   where $\mathcal{C}_j = \{x \in TrSet \mid h_Q(x_j) \neq h_Q(x)\}$
   (c) set radius: $\lambda_j \leftarrow \min[\max(d(\bar{x}, x_j), \epsilon), \lambda_{\max}]$
   (d) **if** $(h_Q(x_j) = +1)$ **then** $a_{Qj} \leftarrow 1$ **else** $a_{\neg Qj} \leftarrow 1$
4. **end**

Fig. 2. RCE training algorithm

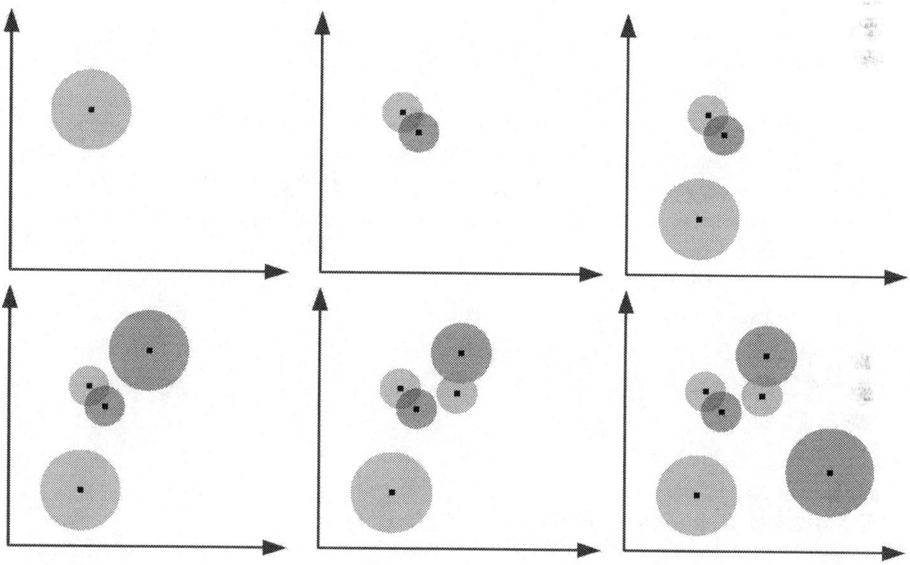

Fig. 3. Evolution of the model built by a RCE network: the centers of the hyperspheres represent the prototypical individuals

Fig. 3 shows the evolution of the model (two-colored decision regions for binary problems) that becomes more and more complex as long as training instances are processed. Different colors represent the different membership predictions. Uncertain regions present overlapping colors.

It is easy to see that the complexity of the method, when no particular optimization is made, is $O(n^2)$ in the number of training examples ($n = |TrSet|$) as the weights are to be adjusted for all of them and each modification of the radius λ_j requires a search for the counterexample with minimal distance among the other training instances.

The method can be used both in *batch* and *on-line* mode. The latter incremental model is particularly appealing when the application may require performing intensive queries on the same query concept and new instances are likely to be made available along the time.

There are several subtleties that may be considered. For instance, if the radius of a pattern unit becomes too small (i.e., less than some threshold $\lambda_{\min}$), then it indicates highly overlapping different categories. In that case, the pattern unit is called a *probabilistic* unit, and so marked.

The method is related to other non-parametric methods, *k-NN estimation* and *Parzen windows* [5]. In particular the Parzen windows method uses fixed window sizes that could lead to some difficulties: in some regions a small window width is appropriate while elsewhere a large one would be better. The k-NN method uses variable window sizes increasing the size until enough samples are enclosed. This may lead to unnaturally large windows when sparsely populated regions are targeted. In the RCE method, the window sizes are adjusted until points of a different category are encountered.

2.3 Classification

The classification of a query individual x_q using the trained RCE network is quite simple in principle. As shown in the basic (*vanilla*) form of the classification algorithm depicted in Fig. 4, the set $N(x_q) \subseteq$ *TrSet* of the nearest training instances is built on the grounds of the hyperspheres (determined by the λ_j weights) the query instance belongs to. Each hypersphere has a related classification determined by the prototype at its center (see Fig. 5). If all prototypes agree on the classification this value is returned as the induced estimate, otherwise the query individual is deemed as ambiguous w.r.t. Q, which represents the default case.

input
 x_q: query individual
 TrSet: set of training examples
 λ_j: parameters of the trained RCE network
output
 $\hat{h}_Q(x_q)$: estimated classification

1. **begin**
2. initialize $k \leftarrow 0$; $N(x_q) \leftarrow \emptyset$
3. **for** $j \leftarrow 1$ to $|TrSet|$ **do**
   – **if** $d(x_q, x_j) < \lambda_j$
     then $N(x_q) \leftarrow N(x_q) \cup \{x_j\}$
4. **if** $(\forall x, x' \in N(x_q) : h_Q(x) = h_Q(x'))$ all share the same class
   then return $h_Q(x)$, shared class of all $x \in N_{set}$
   else return 0 // *uncertain* case
5. **end**

Fig. 4. Vanilla RCE classification

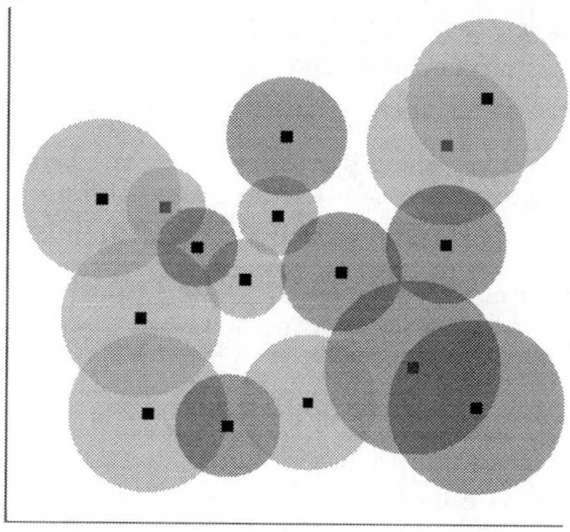

Fig. 5. A representation of the model built by an RCE network used for classification: regions with different colors represent different classifications for instances therein. The areas with overlapping colors or outside of the scope of the hyperspheres represent regions of uncertain classification (see the proposed enhanchment of the procedure in this case).

It is easy to see that this inductive classification procedure is linear in the number of training instances and it could be optimized by employing specific data structures, such as *kD-tress* or *ball-trees* [16], which would allow a faster search of the including hyperspheres and their centers.

In case of uncertainty, this procedure may be enhanced in the spirit of k-NN classification recalled above. If a query individual is located in more than one hypersphere, instead of a catch-all decision like the one made at step 4. of the algorithm, requiring all involved prototypes to agree on their classification in a sort of *voting* procedure, each vote may be weighted by the similarity of the query individual w.r.t. the hypersphere center in terms of a similarity measure[2] s, and the classification should be defined by the class whose prototypes are globally the closest, considering the difference between the closeness values of the query individual from the centers classified by either class. Indeed, one may also consider the *signed* vote, where the sign is determined by the classification of each selected training prototype, and sum up these votes determining the classification with a sign function.

Formally, suppose the nearest prototype set $N(x_q)$ has been determined. The decision function is defined:

$$g(x_q) = \sum_{x_j \in N(x_q)} h_Q(x_j) \cdot s(x_j, x_q)$$

Then step 4. in the procedure becomes:

4. **if** $(|g(x_q)| > \theta)$ **then return** $\mathrm{sgn}(g(x_q))$ **else return** 0

[2] It may be derived from a distance or dissimilarity measure by means of simple transformations, as shown later.

where $\theta \in]0, 1]$ is a tolerance threshold for the uncertain classification (*uncertainty threshold*). We may foresee that higher values of this threshold make the classifier more *skeptical* in uncertain cases, while lower values make it more credulous in suggesting ± 1.

2.4 Likelihood of the Inductive Classification

The analogical inference made by the procedure shown above is not guaranteed to be deductively valid. Indeed, inductive inference naturally yields a certain degree of uncertainty.

In order to measure the likelihood of the decision made by the inductive procedure, one may resort to an approach that is similar to the one applied with the kNN procedure [4]. It is convenient to decompose the decision function $g(x)$ in three components corresponding to the values $v \in V : g_v(x)$ and use those weighted votes. Specifically, given the nearest training individuals in $N(x_q) = \{x_1, \ldots, x_k\}$, the values of the decision function should be normalized as follows, producing a likelihood measure:

$$\ell(\hat{h}(x_q) = v \mid N(x_q)) = \frac{g_v(x_q)}{\sum_{u \in V} g_u(x_q)}$$

which can be written also:

$$\ell(\hat{h}(x_q) = v \mid N(x_q)) = \frac{\sum_{j=1}^{k} \delta(v, h_Q(x_j)) \cdot s(x_q, x_j)}{\sum_{u \in V} \sum_{h=1}^{k} \delta(u, h_Q(x_h)) \cdot s(x_q, x_h)}$$

For instance, the likelihood of the assertion $Q(x_q)$ corresponds to the case when $v = +1$ (i.e. to $g_1(x_q)$). This could be used in case the application requires that the hits be ranked along with their likelihood values.

3 Similarity Measures

The method described in the previous section relies on a notion of similarity which should be measured by means of specific metrics which are to be sensible to the similarity/difference between the individuals. to this purpose various definitions have been proposed [6, 4, 8].

Given a dissimilarity measure d with values in $[0, 1]$ belonging to the family of pseudo-metrics defined in [4], the easiest way to derive a similarity measure would be: $\forall a, b$: $s(a, b) = 1 - d(a, b)$ or $s(a, b) = 1/d(a, b)$. The latter case needs some correction to avoid the undefined cases.

A more direct way follows the same ideas that inspired the mentioned metrics. Indeed, these measures are based on the idea of comparing the semantics of the input individuals along a number of dimensions represented by a committee of concept descriptions. Indeed, on a semantic level, similar individuals should behave similarly with respect to the same concepts.

More formally, the individuals are compared on the grounds of their semantics w.r.t. a collection of concept descriptions, say $\mathsf{F} = \{F_1, F_2, \ldots, F_m\}$, which stands as a group of discriminating *features* expressed in the language taken into account. In its simple formulation, a family of similarity functions for individuals inspired to Minkowski's norms can be defined as follows [8]:

Definition 3.1 (family of similarity measures). *Let $\mathcal{K} = \langle \mathcal{T}, \mathcal{A} \rangle$ be a knowledge base. Given a set of concept descriptions $\mathsf{F} = \{F_i\}_{i=1}^m$ and a normalized vector of weights $\boldsymbol{w} = (w_1, \ldots, w_m)^t$, a family of similarity functions*

$$s_p^{\mathsf{F}} : \mathsf{Ind}(\mathcal{A}) \times \mathsf{Ind}(\mathcal{A}) \to [0, 1]$$

is defined as follows:
$\forall a, b \in \mathsf{Ind}(\mathcal{A})$

$$s_p^{\mathsf{F}}(a, b) = \frac{1}{m} \left[\sum_{i=1}^{m} w_i \mid \sigma_i(a, b) \mid^p \right]^{1/p}$$

where $p > 0$ and $\forall i \in \{1, \ldots, m\}$ the similarity function σ_i is defined by:
$\forall a, b \in \mathsf{Ind}(\mathcal{A})$

$$\sigma_i(a, b) = \begin{cases} 0 & \textit{if } [\mathcal{K} \models F_i(a) \textit{ and } \mathcal{K} \models \neg F_i(b)] \textit{ or } [\mathcal{K} \models \neg F_i(a) \textit{ and } \mathcal{K} \models F_i(b)] \\ 1 & \textit{if } [\mathcal{K} \models F_i(a) \textit{ and } \mathcal{K} \models F_i(b)] \textit{ or } [\mathcal{K} \models \neg F_i(a) \textit{ and } \mathcal{K} \models \neg F_i(b)] \\ \frac{1}{2} & \textit{otherwise} \end{cases}$$

The rationale of the measure is summing the partial similarity w.r.t. the single concepts. Functions σ_i assign the weighted maximal similarity to the case when the individuals exhibit the same behavior w.r.t. the given feature F_i, and null similarity when they belong to disjoint features. An intermediate value is assigned to the case when reasoning cannot ascertain one of such required class-memberships.

As regards the vector of weights $\boldsymbol{w}$ employed in the family of measures, they should reflect the impact of the single feature concept w.r.t. the overall similarity. As mentioned, this can be determined by the quantity of information conveyed by a feature, which can be measured as its entropy. Namely, the extension of a feature F_i w.r.t. the whole domain of objects may be probabilistically quantified as $P_i^+ = |F_i^{\mathcal{I}}|/|\Delta^{\mathcal{I}}|$ (w.r.t. the canonical interpretation $\mathcal{I}$ whose domain is made up by the very individual names occurring in the ABox [1]). This can be roughly approximated with: $P_i^+ = |\mathsf{retrieval}(F_i)|/|\mathsf{Ind}(\mathcal{A})|$. Hence, considering also the probability P_i^- related to its negation and that related to the unclassified individuals (w.r.t. F_i), $P_i^U = 1 - (P_i^+ + P_i^-)$, one may determine an entropic measure for the discernibility yielded by the feature:

$$h_i = -\left[P_i \log(P_i) + P_i^- \log(P_i^-) + P_i^U \log(P_i^U)\right]$$

These measures may be normalized for providing a good set of weights for distance or similarity measures $w_i = h_i/||\boldsymbol{h}||$.

4 Experimentation

The REDUCE system implements the training and classification procedures explained in the previous sections, borrowing the simplest metrics of the family (i.e., with $p = 1$), for the sake of efficiency.

Table 1. Facts concerning the ontologies employed in the experiments

ontology	DL language	#concepts	#object prop.	#data prop.	#individuals
SWM	$\mathcal{ALCOF}(\mathcal{D})$	19	9	1	115
BioPAX	$\mathcal{ALCHF}(\mathcal{D})$	28	19	30	323
LUBM	$\mathcal{ALR}^{+}\mathcal{HI}(\mathcal{D})$	43	7	25	555
NTN	$\mathcal{SHIF}(\mathcal{D})$	47	27	8	676
SWSD	$\mathcal{ALCH}$	258	25	0	732
Financial	$\mathcal{ALCIF}$	60	17	0	1000

Its performance has been tested in a number of classification problems that had been previously approached with other inductive methods [4, 8]. This allows the comparison of the new system to other inductive methods.

4.1 Experimental Setting

A number of ontologies from different domains represented in OWL have been selected, namely: Surface-Water-Model (SWM), NewTestamentNames (NTN) from the Protégé library[3], the Semantic Web Service Discovery dataset[4] (SWSD), an ontology generated by the Lehigh University Benchmark[5] (LUBM), the BioPax glycolysis ontology[6] (BioPax) and the Financial ontology[7]. Table 1 summarizes details concerning these ontologies.

For each ontology, 100 satisfiable query concepts were randomly generated by composition (conjunction and/or disjunction) of (2 through 8) primitive and defined concepts in each knowledge base. The query concepts were also guaranteed to have instances in the ABox. Query concepts were constructed so to be endowed with at least a 50% of the ABox individuals classified positive and negative instances. The performance of the inductive method was evaluated by comparing its responses to those returned by a standard reasoner[8] as a baseline.

Experimentally, it was observed that large training sets make the similarity measures (and consequently the inductive procedure) very accurate. The simplest similarity measure (s_1^{F}) was employed from the *entropic* family, using all the named concepts in the knowledge base for determining the committee of features F with no further optimization.

In order to lower the variance due to the composition of the specific training/test sets during the various runs, for each choice of query concepts, the experiments were replicated (250 folds) and the rates averaged according to the standard *.632+ bootstrap* procedure [5], which is based on sampling with replacement,

[3] `http://protege.stanford.edu/plugins/owl/owl-library`

[4] `https://www.uni-koblenz.de/FB4/Institutes/IFI/AGStaab/Projects/xmedia/dl-tree.htm`

[5] `http://swat.cse.lehigh.edu/projects/lubm`

[6] `http://www.biopax.org/Downloads/Level1v1.4/biopax-example-ecocyc-glycolysis.owl`

[7] `http://www.cs.put.poznan.pl/alawrynowicz/financial.owl`

[8] Pellet v. 2.0.0rc3 was employed.

hence producing tests sets of, approximately, one third of the individuals occurring in the ABoxes.

4.2 Results

In order to be able to compare the outcomes with those of previous experiments, we adopted the same evaluation indices which are briefly recalled here:

- *match rate*: rate of individuals that were classified with the same value ($v \in V$) by both the inductive and the deductive classifier;
- *omission error rate*: rate of individuals for which the inductive method could not determine whether they were relevant to the query or not while they were actually relevant according to the reasoner (0 vs. ± 1);
- *commission error rate*: rate of individuals that were classified as belonging to the query concept, while they belong to its negation or vice-versa ($+1$ vs. -1 or -1 vs. $+1$);
- *induction rate*: rate of individuals inductively classified as belonging to the query concept or to its negation, while either case is not logically derivable from the knowledge base (± 1 vs. 0);

We report in the following the outcomes of three different sessions with different choices of the parameters.

First Session. Table 2 reports the outcomes in terms of the indices in an experimentation where the uncertainty threshold was set to 0.3 and $\epsilon = .01$, which corresponds to a propensity for credulous classification.

Preliminarily, it is important to note that, in each experiment, the commission error was quite low. This means that the inductive search procedure is quite accurate, namely it did not make critical mistakes attributing an individual to a concept that is disjoint with the right one. The most difficult ontologies in this respect are those that contain many disjointness axioms, thus making more unlikely to have individuals with an unknown classification. Also the omission error rate was generally quite low, yet more frequent than the previous type of error.

It is noticeable also the induction rate which is generally quite high since a low uncertainty threshold makes the inductive procedure more prone to give a positive/negative classification in case a query individual is located in more than one hypersphere labeled with different classes.

From the instance retrieval point of view, the cases of induction are interesting because they suggest new assertions which cannot be logically derived by using a deductive reasoner yet they might be used to complete a knowledge base [2], e.g. after being validated by an ontology engineer. For each new candidate assertion, the estimated likelihood measure may be employed to assess its probability and hence decide on its inclusion in the KB.

Second Session. In another session the same experiment was repeated with a different thresholds, namely $\theta = .7$ and $\epsilon = .01$, which made the inductive classification much more *cautious*. The outcomes are reported in Table 3.

Table 2. Results of the first session with uncertainty threshold $\theta = .3$ and minimum ball radius $\epsilon = .1$: average values ± average standard deviations per query

ontology	match rate	commission rate	omission rate	induction rate
SWM	83.99±01.06	00.00±00.00	04.80±00.47	11.21±00.75
BIOPAX	85.43±00.43	03.49±00.23	05.32±00.02	05.76±00.25
LUBM	89.77±00.26	00.00±00.00	06.68±00.21	03.55±00.06
NTN	86.71±00.32	00.08±00.00	05.48±00.21	07.73±00.33
SWSD	98.12±00.05	00.00±00.00	01.30±00.05	00.58±00.00
FINANCIAL	90.26±00.09	04.16±00.05	02.57±00.01	03.01±00.05

Table 3. Results of the second session with uncertainty threshold $\theta = .7$ and minimum ball radius $\epsilon = .01$: average values ± average standard deviations per query

ontology	match rate	commission rate	omission rate	induction rate
SWM	93.52±00.58	00.00±00.00	06.19±00.59	00.29±00.05
BIOPAX	81.42±04.83	00.80±00.18	13.00±04.86	04.78±00.35
LUBM	91.59±00.24	00.00±00.00	07.80±00.23	00.62±00.02
NTN	83.78±01.51	00.00±00.00	14.23±02.31	01.99±00.83
SWSD	98.29±00.05	00.00±00.00	01.71±00.05	00.00±00.00
FINANCIAL	82.65±00.70	01.56±00.10	13.72±00.97	02.08±00.27

In this session, it is again worthwhile to note that commission errors seldom occurred quite rarely during the various runs, yet the omission error rate is higher than in the previous session (although generally limited and less than 14%). This is due to the procedure becoming more cautious because of the choice of the thresholds: the evidence for a positive or negative classification was not enough to decide on uncertain cases that, with a lower threshold, were solved by the voting procedure. On the other hand, for the same reasons, the induction rate decreased as well as the match rate (with some exceptions). Another difference is given by the increased variance in the results for all indices.

Third Session. In yet another session the experiment was repeated with thresholds $\theta = .5$ and $\epsilon = .01$, as a tradeoff between the cautious and the credulous modes. The outcomes are reported in Table 4.

In this case the match and induction rates are better than in the previous sessions. while commission errors are again nearly absent, omission errors decrease w.r.t. the previous session. it is also worthwhile to note that the standard deviations again decrease w.r.t. the outcomes of the previous session.

Summarizing, on a comparison of these tables of outcomes to those reported in previous works on inductive classification with other methods [4, 8], an increase of the performance due to the accuracy of the new method.

Besides, the advantage of this new method is that more parameters are available to be tweaked to get better results depending on the ontology at hand.

Table 4. Results of the third session with uncertainty threshold $\theta = .5$ and minimum ball radius $\epsilon = .01$: average values ± average standard deviations per query

ontology	match rate	commission rate	omission rate	induction rate
SWM	94.24±00.83	00.00±00.00	05.26±00.86	00.51±00.24
BioPax	85.11±00.95	01.36±00.29	08.21±00.90	05.31±00.44
LUBM	97.49±00.74	00.00±00.00	02.47±00.73	00.04±00.02
NTN	86.85±00.24	00.00±00.00	06.57±00.74	06.58±00.63
SWSD	98.29±00.05	00.00±00.00	01.71±00.05	00.00±00.00
Financial	87.98±01.84	03.18±00.71	06.12±02.72	02.72±00.32

Of course, these parameters may be also the subject of a preliminary learning session (e.g. through cross-validation).

Also the elapsed time (not reported here) was comparable, even though a more complex training phase was performed before classification, also similarly to the kernel-based methods [8].

5 Concluding Remarks and Possible Extensions

This paper explored the application of a novel similarity-based classification applied to knowledge bases represented in OWL. The similarity measure employed derives from the extended family of semantic pseudo-metrics based on feature committees [4]: weights are based on the amount of information conveyed by each feature, on the grounds of an estimate of its entropy. The measures were integrated in a similarity-based classification procedure that builds models of the search-space based on prototypical individuals. The resulting system was exploited for the task of approximate instance retrieval which can be efficient and effective even in the presence of incomplete (or noisy) information.

An extensive evaluation performed on various ontologies showed empirically the effectiveness of the method and also that, while the performance depends on the number (and distribution) of the available training instances, even working with limited training sets guarantees a good outcomes in terms of accuracy. Besides, the procedure appears also robust to noise since it seldom made critical mistakes.

The utility of alternative methods for individual classification are manifold. On one hand they can be more robust to noise than purely logical methods, and thus they can exploited in case the knowledge base contains some incoherence which would hinder deriving correct conclusions. On the other hand, instance-based methods are also very efficient compared to the complexity of purely logical approaches. One of the possible applications of inductive methods may regards the task of ontology population [2] known as particularly burdnesome for the knowledge engineer and experts. In particular, the presented method may even be exploited for completing ontologies with probabilistic assertions, allowing further more sophisticate approaches to dealing with uncertainty [12].

Extensions of the similarity measures can be foreseen. Since they essentially depend on the choice of the features for the committee, two enhancements may be investigated: 1) constructing a limited number of discriminating feature concepts [6]; 2) investigating different forms of feature weighting, e.g. based on the notion of *variance*. Such objectives can be accomplished by means of machine learning techniques, especially when ontologies with large sets of individuals are available. Namely, part of the entire data can be saved in order to learn optimal feature sets or a good choice for the weight vectors, in advance with respect to their usage.

As mentioned before, largely populated ontologies may also be exploited in a preliminary cross-validation phase in order to provide optimal values for the parameters of the algorithm also depending on the preferred classification mode (cautious or credulous). The method may also be optimized by reducing the set of instances used for the classification limiting them to the prototypical exemplars and exploiting *ad hoc* data structures for maintaining them so to facilitate their retrieval [16].

References

[1] Baader, F., Calvanese, D., McGuinness, D., Nardi, D., Patel-Schneider, P. (eds.): The Description Logic Handbook. Cambridge University Press, Cambridge (2003)

[2] Baader, F., Ganter, B., Sertkaya, B., Sattler, U.: Completing description logic knowledge bases using formal concept analysis. In: Veloso, M. (ed.) Proceedings of the 20th International Joint Conference on Artificial Intelligence, Hyderabad, India, pp. 230–235 (2007)

[3] Bloehdorn, S., Sure, Y.: Kernel methods for mining instance data in ontologies. In: Aberer, K., Choi, K.-S., Noy, N., Allemang, D., Lee, K.-I., Nixon, L., Golbeck, J., Mika, P., Maynard, D., Mizoguchi, R., Schreiber, G., Cudré-Mauroux, P. (eds.) ASWC 2007 and ISWC 2007. LNCS, vol. 4825, pp. 58–71. Springer, Heidelberg (2007)

[4] d'Amato, C., Fanizzi, N., Esposito, F.: Query answering and ontology population: An inductive approach. In: Bechhofer, S., Hauswirth, M., Hoffmann, J., Koubarakis, M. (eds.) ESWC 2008. LNCS, vol. 5021, pp. 288–302. Springer, Heidelberg (2008)

[5] Duda, R.O., Hart, P.E., Stork, D.G.: Pattern Classification, 2nd edn. Wiley, Chichester (2001)

[6] Fanizzi, N., d'Amato, C., Esposito, F.: Induction of optimal semi-distances for individuals based on feature sets. In: Calvanese, D., et al. (eds.) Working Notes of the 20th International Description Logics Workshop, DL 2007, Bressanone, Italy. CEUR Workshop Proceedings, vol. 250 (2007)

[7] Fanizzi, N., d'Amato, C., Esposito, F.: DL-Foil: Concept learning in description logics. In: Železný, F., Lavrač, N. (eds.) ILP 2008. LNCS(LNAI), vol. 5194, pp. 107–121. Springer, Heidelberg (2008)

[8] Fanizzi, N., d'Amato, C., Esposito, F.: Statistical learning for inductive query answering on OWL ontologies. In: Sheth, A., et al. (eds.) ISWC 2008. LNCS, vol. 5318, pp. 195–212. Springer, Heidelberg (2008)

[9] Haase, P., van Harmelen, F., Huang, Z., Stuckenschmidt, H., Sure, Y.: A framework for handling inconsistency in changing ontologies. In: Gil, Y., Motta, E., Benjamins, V.R., Musen, M.A. (eds.) ISWC 2005. LNCS, vol. 3729, pp. 353–367. Springer, Heidelberg (2005)

[10] Hitzler, P., Vrandečić, D.: Resolution-based approximate reasoning for OWL DL. In: Gil, Y., Motta, E., Benjamins, V.R., Musen, M.A. (eds.) ISWC 2005. LNCS, vol. 3729, pp. 383–397. Springer, Heidelberg (2005)
[11] Huang, Z., van Harmelen, F.: Using semantic distances for reasoning with inconsistent ontologies. In: Sheth, A., et al. (eds.) Proceedings of the 7th International Semantic Web Conference, ISWC 2008. LNCS, vol. 5318, pp. 178–194. Springer, Heidelberg (2008)
[12] Lukasiewicz, T.: Expressive probabilistic description logics. Artificial Intelligence 172(6-7), 852–883 (2008)
[13] Möller, R., Haarslev, V., Wessel, M.: On the scalability of description logic instance retrieval. In: Parsia, B., Sattler, U., Toman, D. (eds.) Proceedings of the 2006 International Workshop on Description Logics, DL 2006. CEUR Workshop Proceedings, vol. 189. CEUR (2006)
[14] Tserendorj, T., Rudolph, S., Krötzsch, M., Hitzler, P.: Approximate owl-reasoning with screech. In: Calvanese, D., Lausen, G. (eds.) RR 2008. LNCS, vol. 5341, pp. 165–180. Springer, Heidelberg (2008)
[15] Wache, H., Groot, P., Stuckenschmidt, H.: Scalable instance retrieval for the semantic web by approximation. In: Dean, M., Guo, Y., Jun, W., Kaschek, R., Krishnaswamy, S., Pan, Z., Sheng, Q.Z. (eds.) WISE 2005 Workshops. LNCS, vol. 3807, pp. 245–254. Springer, Heidelberg (2005)
[16] Witten, I.H., Frank, E.: Data Mining, 2nd edn. Morgan Kaufmann, San Francisco (2005)

Hybrid Reasoning with Forest Logic Programs*

Cristina Feier and Stijn Heymans

Knowledge-Based Systems Group, Institute of Information Systems
Vienna University of Technology
Favoritenstrasse 9-11, A-1040 Vienna, Austria
{feier,heymans}@kr.tuwien.ac.at

Abstract. Open Answer Set Programming (OASP) is an attractive framework for integrating ontologies and rules. Although several decidable fragments of OASP have been identified, few reasoning procedures exist. In this paper, we provide a sound, complete, and terminating algorithm for satisfiability checking w.r.t. forest logic programs, a fragment where rules have a tree shape and allow for inequality atoms and constants. We further introduce f-hybrid knowledge bases, a hybrid framework where $\mathcal{SHOQ}$ knowledge bases and forest logic programs co-exist, and we show that reasoning with such knowledge bases can be reduced to reasoning with forest logic programs only. We note that f-hybrid knowledge bases do not require the usual (weakly) DL-safety of the rule component, providing thus a genuine alternative approach to hybrid reasoning.

1 Introduction

Integrating Description Logics (DLs) with rules for the Semantic Web has received considerable attention with approaches such as *Description Logic Programs* [7], *DL-safe rules* [17], *$\mathcal{DL}$+log* [19], *dl-programs* [3], *Description Logic Rules* [13], and Open Answer Set Programming (OASP) [11]. OASP combines attractive features from the DL and the Logic Programming (LP) world: an open domain semantics from the DL side allows for stating generic knowledge, without the need to mention actual constants, and a rule-based syntax from the LP side supports nonmonotonic reasoning via *negation as failure*.

Several decidable fragments of OASP were identified by syntactically restricting the shape of logic programs, while carefully safe-guarding enough expressiveness for integrating rule- and ontology-based knowledge. Notable fragments are *Conceptual Logic Programs (CoLPs)* [9] that are able to simulate reasoning in the DL $\mathcal{SHIQ}$ and *Forest Logic Programs (FoLPs)* [10] that are expressive enough to deal with $\mathcal{SHOQ}$. Note that both $\mathcal{SHOQ}$ and $\mathcal{SHIQ}$ are close family of $\mathcal{SHOIN}(\mathbf{D})$, the DL underlying the Web Ontology language OWL-DL [21]. A serious shortcoming of these decidable fragments is their lack of effective reasoning procedures. In [5], we took a first step in mending this by providing a

* This work is partially supported by the Austrian Science Fund (FWF) under the projects P20305 and P20840, and by the European Commission under the project OntoRule (IST-2009-231875).

L. Aroyo et al. (Eds.): ESWC 2009, LNCS 5554, pp. 338–352, 2009.

sound and complete algorithm for *simple CoLPs*. Simple CoLPs restrict CoLPs by disallowing the use of inverse predicates, inequality and restricting the literal cyclity, but that are still expressive enough to simulate the DL $\mathcal{ALCH}$.

In this paper, we extend the algorithm of [5] to *Forest Logic Programs*, an extension of simple CoLPs with constants and inequality and no cyclicity restriction. FoLPs are able to simulate the DL $\mathcal{SHOQ}$. They serve as well as an underlying integration vehicle for ontologies and rules. In order to illustrate this, we define *f-hybrid knowledge bases (fKBs)*, consisting of a $\mathcal{SHOQ}$ knowledge base and a rule component that is a FoLP, with a nonmonotonic semantics similar to the semantics of *g-hybrid knowledge bases* [8], *r-hybrid knowledge bases* [20], and $\mathcal{DL}$+*log* [19]. Our approach differs in two points with current other proposals: (1) In contrast with Description Logic Programs, DL-safe rules, and Description Logic Rules, fKBs have, in line with traditional logic programming, a minimal model semantics for the rule component, thus allowing for nonmonotonic reasoning. (2) To ensure effective reasoning, we do not rely on a (weakly) DL-safeness condition such as [17,19,20], which restricts the interaction of the rule component with the DL component. Instead, we rely on a translation to FoLPs.

The major contributions of the paper can be summarized as follows:

- We define in Section 4 a nondeterministic algorithm for deciding satisfiability w.r.t. FoLPs, inspired by tableaux-based methods from DLs. We show that this algorithm is terminating, sound, complete, and runs in 2-NEXPTIME. The algorithm is non-trivial from two perspectives: both the minimal model semantics of OASP, compared to the model semantics of DLs, as well as the open domain assumption, compared to the closed domain assumption of ASP [6], pose specific challenges.
- We show in Section 5 that FoLPs are expressive enough to simulate the DL $\mathcal{SHOQ}$ with *fKBs*, an alternative characterization for hybrid representation and (nonmonotonic) reasoning of knowledge, that supports a tight integration of ontologies and rules.

Extended examples, explanations, and detailed proofs, can be found in [4].

2 Preliminaries

We recall the open answer set semantics [11]. *Constants* $a, b, c, \ldots$, *variables* $x, y, \ldots$, *terms* $s, t, \ldots$, and *atoms* $p(t_1, \ldots, t_n)$ are as usual. We allow for *equality atoms* $s = t$. A *literal* is an atom L or a negated atom $not\ L$. An *inequality literal* $not\ (s = t)$ will often be denoted with $s \neq t$. An atom (literal) that is not an equality atom (inequality literal) will be called a *regular atom (literal)*. For a set α of literals or (possibly negated) predicates, $\alpha^+ = \{l \mid l \in \alpha\}$ and $\alpha^- = \{l \mid not\ l \in \alpha\}$. For a set X of atoms, $not\ X = \{not\ l \mid l \in X\}$. For a set of (possibly negated) predicates α, $\alpha(x) = \{a(x) \mid a \in \alpha\}$ and $\alpha(x, y) = \{a(x, y) \mid a \in \alpha\}$.

A *program* is a countable set of rules $\alpha \leftarrow \beta$, where α is a finite set of regular literals and β is a finite set of literals. The set α is the *head* and represents a disjunction, while β is the *body* and represents a conjunction. If $\alpha = \emptyset$, the rule is called a *constraint*. *Free rules* are rules $q(t_1, \ldots, t_n) \vee not\ q(t_1, \ldots, t_n) \leftarrow$ for terms $t_1, \ldots, t_n$; they enable a choice for the inclusion of atoms. We call a predicate q *free* if there is a $q(x_1, \ldots, x_n) \vee not\ q(x_1, \ldots, x_n) \leftarrow$, with variables $x_1, \ldots, x_n$. Atoms, literals, rules, and programs that do not contain variables are *ground*. For a rule or a program X, let $cts(X)$ be the constants in X, $vars(X)$ its variables, and $preds(X)$ its predicates with $upreds(X)$ the unary and $bpreds(X)$ the binary predicates. A *universe* U for P is a non-empty countable superset of the constants in P: $cts(P) \subseteq U$. We call P_U the ground program obtained from P by substituting every variable in P by every element in U. Let $\mathcal{B}_P$ ($\mathcal{L}_P$) be the set of regular atoms (literals) that can be formed from a ground program P.

An *interpretation* I of a ground P is a subset of $\mathcal{B}_P$. We write $I \models p(t_1, \ldots, t_n)$ if $p(t_1, \ldots, t_n) \in I$ and $I \models not\ p(t_1, \ldots, t_n)$ if $I \not\models p(t_1, \ldots, t_n)$. Also, for ground terms s, t, we write $I \models s = t$ if $s = t$ and $I \models not\ s = t$ or $I \models s \neq t$ if $s \neq t$. For a set of ground literals X, $I \models X$ if $I \models l$ for every $l \in X$. A ground rule $r : \alpha \leftarrow \beta$ is *satisfied* w.r.t. I, denoted $I \models r$, if $I \models l$ for some $l \in \alpha$ whenever $I \models \beta$. A ground constraint $\leftarrow \beta$ is satisfied w.r.t. I if $I \not\models \beta$.

For a ground program P without *not*, an interpretation I of P is a *model* of P if I satisfies every rule in P; it is an *answer set* of P if it is a subset minimal model of P. For ground programs P containing *not*, the *GL-reduct* [6] w.r.t. I is defined as P^I, where P^I contains $\alpha^+ \leftarrow \beta^+$ for $\alpha \leftarrow \beta$ in P, $I \models not\ \beta^-$ and $I \models \alpha^-$. I is an *answer set* of a ground P if I is an answer set of P^I.

Below, a program is assumed to be a finite set of rules; infinite programs only appear as byproducts of grounding with an infinite universe. An *open interpretation* of a program P is a pair (U, M) where U is a universe for P and M is an interpretation of P_U. An *open answer set* of P is an open interpretation (U, M) of P with M an answer set of P_U. An n-ary predicate p in P is *satisfiable* if there is an open answer set (U, M) of P s. t. $p(x_1, \ldots, x_n) \in M$.

We introduce notation for trees which extend those in [23]. Let $\cdot$ be a concatenation operator between sequences of constants or natural numbers. A *tree* T with root c (T_c), where c is a specially designated constant, has as nodes sequences of the form $c \cdot s$, where s is a (possibly empty) sequence of positive integers formed with the concatenation operator; for $x \cdot d \in T$, $d \in \mathbb{N}^*$, we have that $x \in T$. For $x \in T$, we call $succ_T(x) = \{x \cdot n \in T \mid n \in \mathbb{N}^*\}$[1], *successors* of x. The set $A_T = \{(x, y) \mid x, y \in T, \exists n \in \mathbb{N}^* : y = x \cdot n\}$ is the set of arcs of a tree T. For $x, y \in T$, we say that $x <_T y$ iff x is a prefix of y and $x \neq y$.

A *forest* F is a set of trees $\{T_c \mid c \in C\}$, where C is a set of distinguished constants. $N_F = \cup_{T \in F} T$ and $A_F = \cup_{T \in F} A_T$ are the set of nodes and the set of arcs of a forest F. For $x \in F$, its successors in F are $succ_F(x) = succ_T(x)$ for $x \in T$. Let $<_F$ be a strict partial order relationship on the set of nodes N_F of a forest F where $x <_F y$ iff $x <_T y$ for some tree T in F. An extended forest EF is a tuple (F, ES) where $F = \{T_c \mid c \in C\}$ is a forest and $ES \subseteq N_F \times C$. ES extends the successor relation: $succ_{EF}(x) = \{y \mid y \in succ_F(x)$ or $(x, y) \in ES\}$. A path in EF is a sequence of nodes linked by the successor relation. We denote by $N_{EF} = N_F$ the nodes of EF and by $A_{EF} = A_F \cup ES$ its arcs.

[1] $\mathbb{N}^*$ is the set of positive integers.

Finally, for a directed graph G, $paths_G$ is the set of pairs of nodes for which there exists a path in G from the first node in the pair to the second one.

3 Forest Logic Programs

Forest Logic Programs (FoLPs) were introduced in [10] as a syntactical fragment of OASP. FoLPs are a generalization of Conceptual Logic Programs [9] which are logic programs with tree-shaped rules for which satisfiability checking under the open answer set semantics is decidable. FoLPs impose the same structure for rules as CoLPs, but also allow for constants.

Definition 1. *A* forest logic program (FoLP) *is a program with only unary and binary predicates, and such that a rule is either a* free rule $a(s) \vee not\ a(s) \leftarrow$ *or* $f(s,t) \vee not\ f(s,t) \leftarrow$ *, where* s *and* t *are terms such that if* s *and* t *are both variables, they are different* [2], *a unary rule*

$$r : a(s) \leftarrow \beta(s), (\gamma_m(s, t_m), \delta_m(t_m))_{1 \leq m \leq k}, \psi \tag{1}$$

where s *and* t_m, $1 \leq m \leq k$, *are terms (again, if both* s *and* t_m *are variables, they are different; similarly for* t_i *and* t_j*), where*

1. $\psi \subseteq \bigcup_{1 \leq i \neq j \leq k} \{t_i \neq t_j\}$ *and* $\{\neq\} \cap \gamma_m = \emptyset$ *for* $1 \leq m \leq k$,
2. $\forall t_i \in vars(r) : \gamma_i^+ \neq \emptyset$, *i.e., for variables* t_i *there is a positive atom that connects* s *and* t_i,

or a binary rule

$$f(s,t) \leftarrow \beta(s), \gamma(s,t), \delta(t) \tag{2}$$

with $\{\neq\} \cap \gamma = \emptyset$ *and* $\gamma^+ \neq \emptyset$ *if* t *is a variable (*s *and* t *are different if both are variables), or a constraint* $\leftarrow a(s)$ *or* $\leftarrow f(s,t)$ *where* s *and* t *are different if both are variables).*

Constraints can be left out without losing expressiveness: $\leftarrow body$ can be replaced by a rule of the form $co(x) \leftarrow not\ co(x), body$, for a new predicate co. As their name suggests FoLPs have the *forest model property*:

Definition 2. *Let* P *be a program. A predicate* $p \in upreds(P)$ *is* forest satisfiable *w.r.t.* P *if there is an open answer set* (U, M) *of* P *and there is an extended forest* $EF \equiv (\{T_\varepsilon\} \cup \{T_a \mid a \in cts(P)\}, ES)$*, where* ε *is a constant, possibly one of the constants appearing in* P[3]*, and a labeling function* $\mathcal{L} : \{T_\varepsilon\} \cup \{T_a \mid a \in cts(P)\} \cup A_{EF} \rightarrow 2^{preds(P)}$ *s. t.*

- $U = N_{EF}$, *and*
- $p \in \mathcal{L}(\varepsilon)$,

[2] A rule $f(X,X) \vee not\ f(X,X) \leftarrow$ is not allowed.

[3] Note that in this case $T_\varepsilon \in \{T_a \mid a \in cts(P)\}$. Thus, the extended forest contains for every constant from P a tree which has as root that specific constant and possibly, but not necessarily, an extra tree with unidentified root node.

– $z \cdot i \in T \in EF$, $i > 0$, *iff there is some* $f(z, z \cdot i) \in M$, $z \in T_x$, *and*
– *for* $y \in T \in EF$, $q \in upreds(P)$, $f \in bpreds(P)$, *we have that*
  - $q(y) \in M$ *iff* $q \in \mathcal{L}(y)$, *and*
  - $f(y, u) \in M$ *iff* $(u = y \cdot i \vee u \in cts(P)) \wedge f \in \mathcal{L}(y, u)$.

We call such a (U, M) *a* forest model *and a program* P *has the* forest model property *if the following property holds: if* $p \in upreds(P)$ *is satisfiable w.r.t.* P *then* p *is forest satisfiable w.r.t.* P.

Proposition 1. *FoLPs have the forest model property.*

In [5], we introduced the class of simple Conceptual Logic Programs. It is easy to see that every simple CoLP is an FoLP. As satisfiability checking w.r.t. simple Conceptual Logic Programs is EXPTIME-hard, the following property follows:

Proposition 2. *Satisfiability checking w.r.t. FoLPs is* EXPTIME*-hard.*

4 An Algorithm for Forest Logic Programs

In this section, we define a sound, complete, and terminating algorithm for satisfiability checking w.r.t. FoLPs. In [10] it has been shown that several restrictions of FoLPs which have the finite model property are decidable, but there was no result so far regarding the whole fragment. Thus, the algorithm described in this section also establishes a decidability result for FoLPs. The algorithm also relies on the finite model property of FoLPs the proof of which is one of the main steps of the completeness proof.

For every non-free predicate q and a FoLP P, let P_q be the rules of P that have q as a head predicate. For a predicate p, $\pm p$ denotes p or *not* p, whereby multiple occurrences of $\pm p$ in the same context will refer to the same symbol (either p or *not* p). The negation of $\pm p$ (in a given context) is $\mp p$, that is, $\mp p =$ *not* p if $\pm p = p$ and $\mp p = p$ if $\pm p =$ *not* p.

The basic data structure for our algorithm is a *completion structure*.

Definition 3. *A completion structure for a FoLP* P *is a tuple* $\langle EF, G, \text{CT}, \text{ST}, bl\rangle$ *where* $EF = (F, ES)$ *is an extended forest which together with the labeling functions* CT *and* ST *and the set* bl *of blocking nodes is used to represent/construct a tentative forest model.* $G = \langle N, A\rangle$ *is a directed graph with nodes* $N \subseteq \mathcal{B}_{P_{N_{EF}}}$ *and edges* $A \subseteq N \times N$ *which is used to keep track of dependencies between elements of the constructed model (the positive atom dependency graph of* $P_{N_{EF}}$*):* N *represents the tentative model, while* N_{EF} *represents the tentative universe.*

The content *function* $\text{CT} : N_{EF} \cup A_{EF} \to 2^{preds(P) \cup not\ (preds(P))}$ *maps a node to a set of (possibly negated) unary predicates and an arc to a set of (possibly negated) binary predicates such that* $\text{CT}(x) \in 2^{upreds(P) \cup not(upreds(P))}$ *if* $x \in N_{EF}$*, and* $\text{CT}(x) \in 2^{bpreds(P) \cup not(bpreds(P))}$ *if* $x \in A_{EF}$*. Every presence of a non-negated predicate symbol* p *in the content of some node/arc* x *of* EF *indicates that* $p(x)$ *is part of the tentative model represented by* EF*.*

The status *function* $\text{ST} : \{(x, q) \mid q \in \text{CT}(x), x \in N_{EF} \cup A_{EF}\} \cup \{(x, q) \mid not\ q \in \text{CT}(x), x \in A_{EF}\} \cup \{(x, not\ q, r) \mid not\ q \in \text{CT}(x), x \in N_{EF}, r \in P_q\} \rightarrow \{exp, unexp\}$ *attaches to every (possibly negated) predicate which appears in the content of a node/arc x of EF a status value which indicates whether the predicate has already been expanded in that node/edge. As it will be indicated later, the completion structure is evolved such that the presence/absence of any predicate symbol in the content of some node/arc is justified, so it is necessary to keep track which predicate symbols have already been justified in every node/arc of EF. As negative unary predicates have to be justified by showing that no rule which defines them can be applied and this can be done in several steps, the function takes as an argument also a rule for such negated predicate symbols.*

An initial completion structure *for checking satisfiability of a unary predicate p w.r.t. a FoLP P is a completion structure* $\langle EF, G, \text{CT}, \text{ST}, bl\rangle$ *with* $EF = (F, ES)$, $F = \{T_\varepsilon\} \cup \{T_a \mid a \in cts(P)\}$, *where ε is a constant, possibly in $cts(P)$,* $T_x = \{x\}$, *for* $x \in \{\varepsilon\} \cup cts(P)$, $ES = \emptyset$, $G = \langle V, E\rangle$, $V = \{p(\varepsilon)\}$, $E = \emptyset$, *and* $\text{CT}(\varepsilon) = \{p\}$, $\text{ST}(\varepsilon, p) = unexp$, $bl = \emptyset$.

The forest is initialized with single-node trees with roots constants from P and, possibly, a new single-node tree with an anonymous root. The anonymous root, in case it exists, contains p, the predicate to satisfy. Otherwise the root of one of the other trees contains p. G is a graph with a single vertex $p(\varepsilon)$.

Next, we will show how to evolve an initial completion structure for checking the satisfiability of a unary predicate p w.r.t. a FoLP P by means of *expansion rules* to a complete clash-free structure that corresponds to a finite representation of an open answer set in case p is satisfiable w.r.t. P. *Applicability rules* state the necessary conditions to apply these expansion rules.

4.1 Expansion Rules

The expansion rules update the completion structure whenever justifying a literal l in the current model leads to the presence of new literal $\pm p(z)$ in the model. Thus, $\pm p$ is inserted in the content of z if it is not already there and marked as unexpanded (w.r.t. all rules which define p in case it is a negative instance and z is a node in EF). Also, in case $\pm p(z)$ is an atom, it is ensured that it is a node in G; moreover, if l is also an atom, a new arc from l to $\pm p(z)$ is created to indicate the dependency between l and $\pm p(z)$ in the model. Formally:

- if $\pm p \notin \text{CT}(z)$, then $\text{CT}(z) = \text{CT}(z) \cup \{\pm p\}$ and $\text{ST}(z, \pm p) = unexp$ in case $\pm p = p$ or $b \in bpreds(()P)$ and $\text{ST}(z, \pm p, r) = unexp$, for all $r \in P_p$ in case $\pm p = not\ p$ and $p \in upreds(()P)$,
- if $\pm p = p$ and $\pm p(z) \notin V$, then $V = V \cup \{\pm p(x)\}$,
- if $l \in \mathcal{B}_{P_{N_{EF}}}$ and $\pm p = p$, then $E = E \cup \{(l, \pm p(z))\}$.

As a shorthand, we denote this sequence of operations as $update(l, \pm p, z)$; more general, $update(l, \beta, z)$ for a set of (possibly negated) predicates β, denotes $\forall \pm a \in \beta, update(l, \pm a, z)$.

In the following, for a completion structure $\langle EF, G, \text{CT}, \text{ST}, bl\rangle$, let $x \in N_{EF}$ and $(x, y) \in A_{EF}$ be the node, resp. arc, under consideration.

(i) Expand unary positive. For a unary positive (non-free) $p \in \text{CT}(x)$ s. t. $\text{ST}(x, p) = \mathit{unexp}$,

- nondeterministically choose a rule $r \in P_p$ of the form (1) s. t. s unifies with x. The rule will motivate $p(x)$ in the tentative open answer set.
- for the β in the body of r, $\mathit{update}(p(x), \beta, x)$,
- pick up (or define when needed) k successors for x, $(y_m)_{1 \leq m \leq k}$, s. t.:
  - for every $1 \leq i, j \leq k$ s. t. $t_i \neq t_j \in \psi$: $y_i \neq y_j$;
  - for every $1 \leq m \leq k$: $y_m \in \mathit{succ}_{EF}(x)$ or y_m is a new EF-successor of x:
    * $y_m = x \cdot n$, $n \in \mathbb{N}^*$ and $x \cdot n \notin \mathit{succ}_{EF}(x)$. Update T_c, where $x \in T_c$: $T_c = T_c \cup \{y_m\}$, or
    * $y_m = a$, where $a \in \mathit{cts}(P)$ and $a \notin \mathit{succ}_{EF}(x)$; $ES = ES \cup (x, a)$.
- for every successor y_m of x, $1 \leq m \leq k$: $\mathit{update}(p(x), \gamma_m, (x, y_m))$ and $\mathit{update}(p(x), \delta_m, y_m)$.
- set $\text{ST}(x, p) = \mathit{exp}$.

(ii) Expand unary negative. Justifying a negative unary predicate $\mathit{not}\ p \in \text{CT}(x)$ (the absence of $p(x)$ in the constructed model) means refuting the body of every ground rule which defines $p(x)$. The body of such a rule is refuted either

- locally: a literal from $\beta(x)$ has to be refuted which amounts to a certain $\pm q \in \beta$ not appearing in $\text{CT}(x)$, or
- depending on the particular grounding of the rule, in an outgoing arc of x, or in a successor of x: thus, some $\pm f \in \delta_m$ should not appear in $\text{CT}(x, y_j)$ or some $\pm q \in \gamma_m$ should not appear in $\text{CT}(y_j)$, where $y_j \in \mathit{succ}_F(x)$, or
- if there is no valid assignment of successors of x to successors of s in the rule which fulfill the inequalities in the rule.

Formally, for a unary negative (non-free) $\mathit{not}\ p \in \text{CT}(x)$ and a rule $r \in P_p$ of the form (1) s. t. x unifies with s (s is the term from the head of the rule) and $\text{ST}(x, \mathit{not}\ p, r) = \mathit{unexp}$ do one of:

- choose a $\pm q \in \beta$, $\mathit{update}(\mathit{not}\ p(x), \mp q, x)$, and set $\text{ST}(x, \mathit{not}\ p, r) = \mathit{exp}$, or
- if for all $p \in \mathit{upreds}(P)$, $p \in \text{CT}(x)$ or $\mathit{not}\ p \in \text{CT}(x)$, and for all $p \in \text{CT}(x)$, $\text{ST}(p, x) = \mathit{exp}$, then for all $y_{i_1}, \ldots, y_{i_k}$ s. t. $(1 \leq i_j \leq n)_{1 \leq j \leq k}$, where $\mathit{succ}_{EF}(x) = \{y_1, \ldots y_n\}$: if for all $1 \leq j, l \leq k$, $t_j \neq t_l \in \psi \Rightarrow y_{i_j} \neq y_{i_l}$, then do one of the following:
  - for some m, $1 \leq m \leq k$, pick up a binary (possibly negated) predicate symbol $\pm f$ from δ_m and $\mathit{update}(\mathit{not}\ p(x), \mp f, (x, y_{i_m}))$, or
  - for some m, $1 \leq m \leq k$, pick up a unary negated predicate symbol $\mathit{not}\ q$ from γ_m and $\mathit{update}(\mathit{not}\ p(x), q, y_{i_m})$.

  Set $\text{ST}(x, \mathit{not}\ p, r) = \mathit{exp}$.

One can see that once the body of a ground version of a unary rule $r \in P_p$, for which the head term s is substituted with the current node x, is locally refuted, the bodies of all ground versions of this rule, for which s is substituted with x are locally refuted, too. For the other two refutation cases all possible groundings of a rule have to be considered and this is not possible until all the successors of x are known. This is the case when all positive predicates in the content of the current node have been expanded and no positive predicate will be further inserted in $\text{CT}(x)$:then, an iteration over all possible groundings of the rule r is triggered. For every grounding if the inequality constraints are violated the body of that particular instantiation of the rule is already refuted; otherwise, one of the body literals from the non-local part of the rule (γs or δs) has to be refuted.

(iii) Choose a unary predicate. If for all $a \in \text{CT}(x)$, $\text{ST}(x, a) = exp$ or a is free, and for all $(x, y) \in A_{EF}$, and for all $\pm f \in \text{CT}(x, y)$, $\text{ST}((x, y), \pm f) = exp$ or f is free, and there is a $p \in upreds(P)$ s. t. $p \notin \text{CT}(x)$ and $not\ p \notin \text{CT}(x)$, then add p to $\text{CT}(x)$ with $\text{ST}(x, p) = unexp$ or add $not\ p$ to $\text{CT}(x)$ with $\text{ST}(x, not\ p, r) = unexp$, for every rule $r \in P_p$.

In other words, if there is a node x for which all positive predicates in its content and all predicates in the contents of its outgoing arcs are free or have already been expanded and if there are still unary predicates which do not appear in $\text{CT}(x)$, one has to pick such a unary predicate symbol p and inject either p or $not\ p$ in $\text{CT}(x)$. This is needed for consistency: it does not suffice to find a justification for the predicate to satisfy, but one also has to show that this justification is part of an actual open answer set, which is done by effectively constructing it. We do not impose that all negative predicate symbols are expanded as that would constrain all the ensuing literals to be locally refuted.

Similarly to rules (i), (ii), and (iii) we define the expansion rules for binary predicates: (iv) *Expand binary positive*, (v) *Expand binary negative*, and (vi) *Choose binary*.

4.2 Applicability Rules

The applicability rules restrict the use of the expansion rules.

(vii) Saturation. A node $x \in N_{EF}$ is *saturated* if for all $p \in upreds(P)$, $p \in \text{CT}(x)$ or $not\ p \in \text{CT}(x)$, and no $\pm q \in \text{CT}(x)$ can be expanded with rules (i-iii), and for all $(x, y) \in A_{EF}$ and $p \in bpreds(P)$, $p \in \text{CT}(x, y)$ or $not\ p \in \text{CT}(x, y)$, and no $\pm f \in \text{CT}(x, y)$ can be expanded with (iv-vi). No expansions should be performed on a node from N_{EF} which does not belong to $cts(P)$ until its predecessor is saturated (constants can have more than one predecessor in the completion, including themselves).

(viii) Blocking. A node $x \in N_{EF}$ is *blocked* if there is an ancestor y of x in F, $y <_F x$, $y \notin cts(P)$, s.t. $\text{CT}(x) \subseteq \text{CT}(y)$ and the set $paths_G(x, y) = \{(p, q) \mid (p(x), q(y)) \in paths_G\}$ is empty. We call (y, x) a *blocking pair* and update bl: $bl = bl \cup \{(y, x)\}$. No expansions can be performed on a blocked node. Intuitively, if there is an ancestor y of x which is not a constant, whose content includes the content of x, and there are no paths in G from a positive literal $p(x)$ to another positive literal $q(y)$ one could reuse the justification for x when dealing with y.

(ix) Redundancy. A node $x \in N_{EF}$ is *redundant* if it is saturated and there are k ancestors of x in F, $(y_i)_{1 \le i \le k}$, where $k = 2^p(2^{p^2} - 1)$, and $p = |upreds(P)|$, s. t. $\text{CT}(x) = \text{CT}(y_i)$ and (y_i, x) is not a blocking pair for every $1 \le i \le n$. In other words, a node is redundant if there are other k nodes on the same branch with the current node which all have content equal to the content of the current node. The presence of a redundant node stops the expansion process. In the completeness proof we show that any forest model of a FoLP P which satisfies p can be reduced to another forest model which satisfies p and has at most $k+1$ nodes with equal content on any branch of a tree from the forest model, and furthermore the $(k+1)st$ node, in case it exists, is blocked. One can thus search for forest models only of the latter type. This rule exploits that result: we discard models which are not in this shrunk search space.

4.3 Termination, Soundness, and Completeness

A completion structure is *contradictory* if for some $x \in N_{EF}$ and $a \in upreds(P)$, $\{a, not\ a\} \subseteq \text{CT}(x)$ or for some $(x, y) \in A_{EF}$ and $f \in bpreds(P)$, $\{f, not\ f\} \subseteq \text{CT}(x, y)$. A *complete completion structure* for a FoLP P and a $p \in upreds(P)$, is a completion structure that results from applying the expansion rules to the initial completion structure for p and P, taking into account the applicability rules, s. t. no expansion rules can be further applied. Also, a complete completion structure $CS = \langle EF, G, \text{CT}, \text{ST}, bl \rangle$ is *clash-free* if: (1) CS is not contradictory (2) EF does not contain redundant nodes (3) G does not contain cycles.

We show that an initial completion structure for a unary predicate p and a FoLP P can always be expanded to a complete completion structure (*termination*), that, if p is satisfiable w.r.t. P, there is a clash-free complete completion structure (*soundness*), and, finally, that, if there is a clash-free complete completion structure, p is satisfiable w.r.t. P (*completeness*).

Proposition 3 (termination). *Let P be FoLP and $p \in upreds(P)$. Then, one can construct a finite complete completion structure by a finite number of applications of the expansion rules to the initial completion structure for p w.r.t. P, taking into account the applicability rules.*

Proof Sketch. One cannot introduce infinitely many successors to a node, as the arity of the trees in the completion structure is bound by the number of unary predicates in P and the number of variables in unary rules. Also, every infinite path in the extended forest eventually contains $|k+1|$ saturated nodes with equal content, where k is as in the redundancy rule, and thus either a blocked or a redundant node, which is not further expanded. □

Proposition 4 (soundness). *Let P be a FoLP and $p \in upreds(P)$. If there is a clash-free complete completion structure for p w.r.t. P, then p is satisfiable w.r.t. P.*

Proposition 5 (completeness). *Let P be a FoLP and $p \in upreds(P)$. If p is satisfiable w.r.t. P, then there exists a clash-free complete completion structure for p w.r.t. P.*

4.4 Complexity Results

Let $CS = \langle EF, G, \text{CT}, \text{ST}, bl\rangle$ be a complete completion structure. Every path of a tree in EF contains at most $k+1$ nodes with equal content (as suggested by the applicability rules (viii) and (ix)), where k is as defined in the redundancy rule; thus, there are at most $(k+1)2^n$ nodes on every such path, where $n = |upreds(P)|$. The branching of every tree is bound by a constant q which is a linear function of the number of variables in unary rules from P. Thus, there are at most $q^{(k+1)2^n}$ nodes in a tree, and at most $(c+1)q^{(k+1)2^n}$ nodes in EF, where c is the number of constants present in the program at hand. So the amount of nodes in CS is double exponential in the size of P, and the algorithm runs in 2-NEXPTIME.

Note that such a high complexity is expected when dealing with tableaux-like algorithms. For example in Description Logics, although satisfiability checking in $\mathcal{SHIQ}$ is EXPTIME-complete, practical algorithms run in 2-NEXPTIME [22].

5 F-Hybrid Knowledge Bases

In this section, we introduce *f-hybrid* knowledge bases (fKBs), a formalism that combines KBs expressed in the Description Logic $\mathcal{SHOQ}$ with FoLPs. We assume the reader is familiar with $\mathcal{SHOQ}$; details can be found in [4].

Note that we assume the *unique name assumption* by imposing that $o^\mathcal{I} = o$ for individuals $o \in \mathbf{I}$. Individuals are thus assumed to be part of any domain $\Delta^\mathcal{I}$. OWL does not have the unique name assumption [21], and thus different individuals can point to the same resource. However, the open answer set semantics gives a Herbrand interpretation to constants and for consistency we assume that also DL nominals are interpreted this way.

Example 1. Consider the $\mathcal{SHOQ}$ KB Σ with $Father \sqsubseteq \exists child.Human \sqcap \neg Female$ and $\{john\} \sqsubseteq (\leq 2child.Human)$ as axioms. Intuitively, the first terminological axiom says that fathers have a human child and are not female. The second axiom says that *john* has not more than 2 human children.

Definition 4. *An* f-hybrid knowledge base (fKB) *is a pair* $\langle \Sigma, P\rangle$ *where* Σ *is a* $\mathcal{SHOQ}$ *KB and* P *is a FoLP.*

The predicate symbols in P can be concept or role names from Σ, in which case the corresponding atoms/literals are called *DL atoms/literals*, but w.l.o.g. we assume they do not coincide with complex concept or role descriptions. Note that we do not impose Datalog safeness or *(weakly) DL safeness* [18,20,19] for the rule component. Intuitively, the restricted shape of FoLPs suffices to guarantee decidability; FoLPs are in general neither Datalog safe nor weakly DL-safe; we discuss the relation with weakly DL-safeness in detail in Section 6.

Example 2. An fKB $\langle \Sigma, P\rangle$, with Σ as in Ex. 1 and P the FoLP $unhappy(X) \leftarrow not\ Father(X)$ indicates that persons that are not fathers are unhappy, where $Father(X)$ is a DL literal.

Similarly as in [8], we define, given a DL interpretation $\mathcal{I} = (\Delta^{\mathcal{I}}, \cdot^{\mathcal{I}})$ and a ground program P, the *projection* $\Pi(P, \mathcal{I})$ of P with respect to $\mathcal{I}$, as follows: for every rule r in P,

(a) if there exists a DL literal in the head of the form $A(t_1, \ldots, t_n)$ with $(t_1, \ldots, t_n) \in A^{\mathcal{I}}$, or *not* $A(t_1, \ldots, t_n)$ with $(t_1, \ldots, t_n) \notin A^{\mathcal{I}}$, then delete r,

(b) if there exists a DL literal in the body of the form $A(t_1, \ldots, t_n)$ with $(t_1, \ldots, t_n) \notin A^{\mathcal{I}}$, or *not* $A(t_1, \ldots, t_n)$ with $(t_1, \ldots, t_n) \in A^{\mathcal{I}}$, then delete r, (c) otherwise, delete all DL literals from r. Intuitively, the projection "evaluates" the program with respect to $\mathcal{I}$ by removing (evaluating) rules and DL literals consistently with $\mathcal{I}$; conceptually this is similar to the reduct, which removes rules and negative literals consistently with an interpretation of the program.

Definition 5. *Let $\langle \Sigma, P \rangle$ be an fKB. An* interpretation *of $\langle \Sigma, P \rangle$ is a tuple $(U, \mathcal{I}, M)$ s. t. U is a universe for P, $\mathcal{I} = (U, \cdot^{\mathcal{I}})$ is an interpretation of Σ, and M is an interpretation of $\Pi(P_U, \mathcal{I})$. Then, $(U, \mathcal{I}, M)$ is a* model *of an fKB $\langle \Sigma, P \rangle$ if $\mathcal{I}$ is a model of Σ and M is an answer set of $\Pi(P_U, \mathcal{I})$.*

The semantics of a fKB $\langle \Sigma, P \rangle$ is s. t. if $\Sigma = \emptyset$, a model of $\langle \Sigma, P \rangle$ corresponds to an open answer set of P, and if $P = \emptyset$, a model of $\langle \Sigma, P \rangle$ corresponds to a DL model of Σ. In this way, the semantics of fKBs is nicely layered on top of both the DL semantics and the open answer set semantics.

Example 3. For the fKB $\langle \Sigma, P \rangle$ in Example 2, take a universe $U = \{john, x\}$ and $\cdot^{\mathcal{I}}$ s. t. $Father^{\mathcal{I}} = \{x\}$, $child^{\mathcal{I}} = \{(x, john)\}$, $Female^{\mathcal{I}} = \emptyset$, $Human^{\mathcal{I}} = U$, and $john^{\mathcal{I}} = john$. It is easy to see that $\mathcal{I} = (U, \cdot^{\mathcal{I}})$ is indeed a model of Σ. With $M = \{unhappy(john)\}$, one can check that $(U, \mathcal{I}, M)$ is a model of $\langle \Sigma, P \rangle$.

For p a concept expression from Σ or a predicate from P, we say that p is *satisfiable* w.r.t. (Σ, P) if there is a model $(U, \mathcal{I}, M)$ s. t. $p^{\mathcal{I}} \neq \emptyset$ or $p(x_1, \ldots, x_n) \in M$ for some $x_1, \ldots, x_n$ in U, resp.

We can reduce satisfiability checking w.r.t. fKBs to satisfiability checking of FoLPs only. Roughly, for each concept expression one introduces a new predicate together with rules that define the semantics of the corresponding DL construct. Constraints then encode the axioms, and the first-order interpretation of DL concept expressions is simulated using free rules.

For the formal translation, we define the *closure* $clos(\Sigma)$ of a $\mathcal{SHOQ}$ KB Σ as the smallest set satisfying the following conditions:

- for each $C \sqsubseteq D$ an axiom in Σ (role or terminological), $\{C, D\} \subseteq clos(\Sigma)$,
- for each $\mathsf{Trans}(R)$ in Σ, $\{R\} \subseteq clos(\Sigma)$,
- for every D in $clos(\Sigma)$, we have
  - if $D = \neg D_1$, then $\{D_1\} \subseteq clos(\Sigma)$,
  - if $D = D_1 \sqcup D_2$, then $\{D_1, D_2\} \subseteq clos(\Sigma)$,
  - if $D = D_1 \sqcap D_2$, then $\{D_1, D_2\} \subseteq clos(\Sigma)$,
  - if $D = \exists R.D_1$, then $\{R, D_1\} \cup \{\exists S.D_1 \mid S \sqsubseteq^{*} R, S \neq R, \mathsf{Trans}(S) \in \Sigma\} \subseteq clos(\Sigma)$,
  - if $D = \forall R.D_1$, then $\{\exists R.\neg D_1\} \subseteq clos(\Sigma)$,
  - if $D = (\leq n\ Q.D_1)$, then $\{(\geq n+1\ Q.D_1)\} \subseteq clos(\Sigma)$,
  - if $D = (\geq n\ Q.D_1)$, then $\{Q, D_1\} \subseteq clos(\Sigma)$.

Concerning the addition of the extra $\exists S.D_1$ for $\exists R.D_1$ in the closure, note that $x \in (\exists R.D_1)^{\mathcal{I}}$ holds if there is some $(x, y) \in R^{\mathcal{I}}$ with $y \in D_1^{\mathcal{I}}$, i.e., there is some $S \sqsubseteq^* R$ with S transitive (or $S = R$) s. t. $(x, u_0) \in S^{\mathcal{I}}, \ldots, (u_n, y) \in S^{\mathcal{I}}$ with $y \in D_1^{\mathcal{I}}$. This amounts to $x \in (\exists S.D_1)^{\mathcal{I}}$. Thus, in the open answer set setting, we have that $\exists R.D_1(x)$ is in the open answer set if $R(x, y)$ and $D_1(y)$ hold or $\exists S.D_1(x)$ holds for some transitive subrole S of R. The predicate $\exists S.D_1$ will be defined by adding recursive rules, hence the inclusion of such predicates in the closure. Furthermore, for a $(\leq n\ Q.D_1)$ in the closure, we add $\{(\geq n+1\ Q.D_1)\}$, since we will base our definition of the former predicate on the DL equivalence $(\leq n\ Q.D_1) \equiv \neg(\geq n+1\ Q.D_1)$.

We define the FoLP $\Phi(\Sigma)$, obtained from the $\mathcal{SHOQ}$ KB Σ, as follows:

- For each terminological axiom $C \sqsubseteq D \in \Sigma$, add $\leftarrow C(X), not\ D(X)$.
- For each role axiom $R \sqsubseteq S \in \Sigma$, add $\leftarrow R(X, Y), not\ S(X, Y)$.
- Next, we distinguish between the types of concept expressions that appear in $clos(\Sigma)$. For each $D \in clos(\Sigma)$:
  - if D is a concept name, add $D(X) \vee not\ D(X) \leftarrow$,
  - if D is a role name, add $D(X, Y) \vee not\ D(X, Y) \leftarrow$,
  - if $D = \{o\}$, add $D(o) \leftarrow$,
  - if $D = \neg E$, add $D(X) \leftarrow not\ E(X)$,
  - if $D = E \sqcap F$, add $D(X) \leftarrow E(X), F(X)$,
  - if $D = E \sqcup F$, add $D(X) \leftarrow E(X)$ and $D(X) \leftarrow F(X)$,
  - if $D = \exists Q.E$, add $D(X) \leftarrow Q(X, Y), E(Y)$ (5), and for all $S \sqsubseteq^* Q$, $S \neq R$, with $\mathsf{Trans}(S) \in \Sigma$, add rules $D(X) \leftarrow (\exists S.E)(X)$ (6). If $\mathsf{Trans}(Q) \in \Sigma$, we further add the rule $D(X) \leftarrow Q(X, Y), D(Y)$ (7),
  - if $D = \forall R.E$, add $D(X) \leftarrow not\ (\exists R.\neg E)(X)$
  - if $D = (\leq n\ Q.E)$, add $D(X) \leftarrow not\ (\geq n+1\ Q.E)(X)$,
  - if $D = (\geq n\ Q.E)$, add $D(X) \leftarrow Q(X, Y_1), \ldots, Q(X, Y_n), E(Y_1), \ldots, E(Y_n), (Y_i \neq Y_j)_{1 \leq i \neq j \leq n}$

Rule (5) is what one would intuitively expect for the exists restriction. However, in case Q is transitive this rule is not enough. Indeed, if $Q(x, y)$, $Q(y, z)$, $E(z)$ are in an open answer set, one expects $(\exists Q.E)(x)$ to be in it as well if Q is transitive. However, we have no rules enforcing $Q(x, z)$ to be in the open answer set without violating the FoLP restrictions. We can solve this by adding to (5) rule (6) s. t. the chain $Q(x, y)$, $Q(y, z)$, with $E(z)$ in the open answer set correctly deduces $D(x)$. It may still be that there are transitive subroles of Q that need the same recursive treatment as above. To this end, we introduce rule (7). We do not need such a trick with the number restrictions since the roles Q in a number restriction are required to be simple, i.e., without transitive subroles.

Proposition 6. *Let $\langle \Sigma, P \rangle$ be a $\mathcal{SHOQ}$ KB. Then, $\Phi(\Sigma) \cup P$ is a FoLP, and has a size that is polynomial in the size of Σ.*

Proposition 7. *A predicate p is satisfiable w.r.t. a fKB $\langle \Sigma, P \rangle$ iff p is satisfiable w.r.t. $\Phi(\Sigma) \cup P$.*

Proposition 7 also holds for satisfiability checking of concept expressions C: introduce a rule $p(X) \leftarrow C(X)$ in P and check satisfiability of p.

Using the translation from fKBs to FoLPs in Proposition 7 and the polynomiality of this translation (Proposition 6), together with the complexity of the algorithm for satisfiability checking w.r.t. FoLPs, we have the following result:

Proposition 8. *Satisfiability checking w.r.t. fKBs is in* 2-NEXPTIME.

As satisfiability checking of $\mathcal{ALC}$ concepts w.r.t. an $\mathcal{ALC}$ TBox (note that $\mathcal{ALC}$ is a fragment of $\mathcal{SHOQ}$) is EXPTIME-complete ([1, Chapter 3]), we have that satisfiability checking w.r.t. fKBs is EXPTIME-hard.

Proposition 9. *Satisfiability checking w.r.t. fKBs is* EXPTIME*-hard.*

6 Discussion and Outlook

We compare fKBs to the r-hybrid KBs from [20], which extends $\mathcal{DL}$+*log* from [19] with inequalities and negated DL atoms.

In [20], a r-hybrid KB consists of a DL knowledge base and a disjunctive Datalog program where each rule is *weakly DL-safe*: (a) every variable in the rule appears in a positive atom in the body of the rule (*Datalog safeness*), and (b) every variable either occurs in a positive non-DL atom in the body of the rule, or it only occurs in positive DL atoms in the body of the rule.

The semantics of r-hybrid and fKBs correspond to a large extent. A main difference is that fKBs do not make the *standard names assumption*, in which basically the domain of every interpretation is the same infinitely countable set of constants. Some more key differences to note:

- We do not require Datalog safeness neither do we require weakly DL-safeness. Indeed, fKBs may have a rule component that is not weakly DL-safe. Take the fKB $\langle \Sigma, P \rangle$ from Example 2 with P. The atom $Father(X)$ is a DL-atom s. t. the rule is neither Datalog safe nor weakly DL-safe. Modifying the program to $unhappy(X) \leftarrow Human(X), not\ Father(X)$ leads to a Datalog safe program, however, it is still not weakly DL-safe as X is not appearing only in positive DL-atoms. On the other hand, both the above rules are FoLPs and thus constitute a valid component of an fKB.
- For r-hybrid KBs, due to the safeness conditions, it suffices for satisfiability checking to ground the rule component with the constants appearing explicitly in the KB.[4] One does not have such a property for fKBs. Consider the fKB $\langle \Sigma, P \rangle$ with $\Sigma = \emptyset$ and the program P with rules $a(X) \leftarrow not\ b(X)$ and $b(0) \leftarrow$. This program is a FoLP, but it is not Datalog safe nor is it weakly DL-safe. Grounding only with the constants in the program yields that a is not satisfiable; however, grounding with, e.g., $\{0, x\}$, makes a satisfiable.

[4] [20,19] considers checking satisfiability of KBs rather than satisfiability of predicates. However, the former can easily be reduced to the latter.

– Decidability for satisfiability checking of r-hybrid KBs is guaranteed if decidability of the conjunctive query containment problem is guaranteed for the DL at hand. In contrast, we relied on a translation of DLs to FoLPs for establishing decidability, and not all DLs can be translated this way; we illustrated the translation for $\mathcal{SHOQ}$.

Hybrid MKNF KBs [16,15] consist of a DL component and a component of so-called *MKNF* rules. Such MKNF rules allow for modal operators **K** and **not** in front of atoms; for a detailed definition, we refer the reader to [16]. Hybrid MKNF KBs generalize approaches to integrating ontologies and rules such as CARIN [14], $\mathcal{A}$-log [2], DL-safe rules [17], and the Semantic Web Rule Language (SWRL) [12], as well as r-hybrid KBs [20]. In particular, one can write the latter as equisatisfiable hybrid MKNF KBs [16, Theorem 4.8].

In contrast with fKBs, hybrid MKNF KBs, like r-hybrid KBs, make the standard names assumption. Every variable in a rule has to appear in a non-DL atom of the body of the rule. As with r-hybrid KBs, our fKBs do not have such a restriction of the interaction between the structural DL component and the rule component, but rely instead on the existence of an integrating framework (FoLPs under an open answer set semantics) that has reasoning support; reasoning support that we provided in this paper.

A formalism related to FoLPs is $\mathbb{FDNC}$ [24]. $\mathbb{FDNC}$ is an extension of ASP with function symbols where rules are syntactically restricted in order to maintain decidability. While the syntactical restriction is similar to the one imposed on FoLP rules, predicates having arity maximum two, and the terms in a binary literal can be seen as arcs in a forest (imposing the Forest Model Property), the direction of deduction is different: while for FoLPs, all binary literals in a rule body have an identical first term which is also the term which appears in the head, for $\mathbb{FDNC}$ (with the exception of one rule type) the second term is the one which also appears in the head. $\mathbb{FDNC}$ rules are required to be safe unlike FoLP ones. The complexity for standard reasoning tasks for $\mathbb{FDNC}$ is EXPTIME-complete and worst-case optimal algorithms are provided.

For the future, we intend to look into an extension of f-hybrid knowledge bases and its reasoning algorithm, from $\mathcal{SHOQ}$ towards $\mathcal{SROIQ}(\mathbf{D})$ the DL underlying OWL-DL in OWL 2[5]. A prototype implementation of the algorithm is planned, and will feed the need for optimization strategies.

References

1. Baader, F., Calvanese, D., McGuinness, D.L., Nardi, D., Patel-Schneider, P.F. (eds.): The DL Handbook: Theory, Implementation, and Applications (2003)
2. Donini, F., Lenzerini, M., Nardi, D., Schaerf, A.: AL-log: Integrating Datalog and Description Logics. J. of Intelligent and Cooperative Information Systems 10, 227–252 (1998)
3. Eiter, T., Ianni, G., Lukasiewicz, T., Schindlauer, R., Tompits, H.: Combining answer set programming with description logics for the semantic web. Artificial Intelligence 172(12-13), 1495–1539 (2008)

[5] `http://www.w3.org/2007/OWL`

4. Feier, C., Heymans, S.: Hybrid reasoning with forest logic programs. Technical Report Infsys Research Report 184-08-14, KBS Group, Vienna University of Technology, Austria (December 2008), www.kr.tuwien.ac.at/staff/heymans/priv/projects/fwf-doasp/eswc2009-tr.pdf
5. Feier, C., Heymans, S.: A sound and complete algorithm for simple conceptual logic programs. In: Proc. of ALPSWS 2008 (2008)
6. Gelfond, M., Lifschitz, V.: The Stable Model Semantics for Logic Programming. In: Proc. of ICLP 1988, pp. 1070–1080 (1988)
7. Grosof, B.N., Horrocks, I., Volz, R., Decker, S.: Description logic programs: combining logic programs with description logic. In: Proc. of the World Wide Web Conference (WWW), pp. 48–57 (2003)
8. Heymans, S., de Bruijn, J., Predoiu, L., Feier, C., Van Nieuwenborgh, D.: Guarded hybrid knowledge bases. TPLP 8(3), 411–429 (2008)
9. Heymans, S., Van Nieuwenborgh, D., Vermeir, D.: Conceptual logic programs. Annals of Mathematics and Artificial Intelligence (Special Issue on Answer Set Programming) 47(1–2), 103–137 (2006)
10. Heymans, S., Van Nieuwenborgh, D., Vermeir, D.: Open answer set programming for the semantic web. J. of Applied Logic 5(1), 144–169 (2007)
11. Heymans, S., Van Nieuwenborgh, D., Vermeir, D.: Open answer set programming with guarded programs. ACM Trans. on Comp. Logic 9(4) (October 2008)
12. Horrocks, I., Patel-Schneider, P.F.: A proposal for an OWL rules language. In: Proc. of the World Wide Web Conference (WWW), pp. 723–731. ACM, New York (2004)
13. Krötzsch, M., Rudolph, S., Hitzler, P.: Description logic rules. In: Proc. 18th European Conf. on Artificial Intelligence (ECAI 2008), pp. 80–84 (2008)
14. Levy, A.Y., Rousset, M.: CARIN: A Representation Language Combining Horn Rules and Description Logics. In: Proc. of ECAI 1996, pp. 323–327 (1996)
15. Motik, B., Horrocks, I., Rosati, R., Sattler, U.: Can OWL and logic programming live together happily ever after? In: Cruz, I., Decker, S., Allemang, D., Preist, C., Schwabe, D., Mika, P., Uschold, M., Aroyo, L.M. (eds.) ISWC 2006. LNCS, vol. 4273, pp. 501–514. Springer, Heidelberg (2006)
16. Motik, B., Rosati, R.: Closing Semantic Web Ontologies. Technical report (2006)
17. Motik, B., Sattler, U., Studer, R.: Query answering for OWL-DL with rules. Journal of Web Semantics 3(1), 41–60 (2005)
18. Rosati, R.: On the decidability and complexity of integrating ontologies and rules. Web Semantics 3(1), 41–60 (2005)
19. Rosati, R.: DL+log: Tight integration of description logics and disjunctive datalog. In: Proc. of the Int. Conf. on Principles of KR and Reas (KR), pp. 68–78 (2006)
20. Rosat, R.: On combining description logic ontologies and nonrecursive datalog rules. In: Proc. of the 2nd Int. Conf. on Web Reasoning and Rule Systems (2008)
21. Smith, M., Welty, C., McGuinness, D.: OWL Web Ontology Language Guide (2004), http://www.w3.org/TR/owl-guide/
22. Tobies, S.: Complexity Results and Practical Algorithms for Logics in Knowledge Representation. PhD thesis, RWTH-Aachen (2001)
23. Vardi, M.Y.: Reasoning about the past with two-way automata. In: Larsen, K.G., Skyum, S., Winskel, G. (eds.) ICALP 1998. LNCS, vol. 1443, p. 628. Springer, Heidelberg (1998)
24. Šimkus, M., Eiter, T.: $\mathbb{FDNC}$: Decidable non-monotonic disjunctive logic programs with function symbols. In: Dershowitz, N., Voronkov, A. (eds.) LPAR 2007. LNCS, vol. 4790, pp. 514–530. Springer, Heidelberg (2007)

SIM-DL$_A$: A Novel Semantic Similarity Measure for Description Logics Reducing Inter-concept to Inter-instance Similarity

Krzysztof Janowicz and Marc Wilkes

Institute for Geoinformatics, University of Muenster, Germany
{janowicz,marc.wilkes}@uni-muenster.de

Abstract. While semantic similarity plays a crucial role for human categorization and reasoning, computational similarity measures have also been applied to fields such as semantics-based information retrieval or ontology engineering. Several measures have been developed to compare concepts specified in various description logics. In most cases, these measures are either structural or require a populated ontology. Structural measures fail with an increasing expressivity of the used description logic, while several ontologies, e.g., geographic feature type ontologies, are not populated at all. In this paper, we present an approach to reduce inter-concept to inter-instance similarity and thereby avoid the canonization problem of structural measures. The novel approach, called SIM-DL$_A$, reuses existing similarity functions such as co-occurrence or network measures from our previous SIM-DL measure. The required instances for comparison are derived from the completion tree of a slightly modified DL-tableau algorithm as used for satisfiability checking. Instead of trying to find one (clash-free) model, the new algorithm generates a set of proxy individuals used for comparison. The paper presents the algorithm, alignment matrix, and similarity functions as well as a detailed example.

1 Introduction and Motivation

Semantic similarity measurement plays an important role in information retrieval on the semantic Web [1]. It supports users in searching and browsing through structured data. To improve human computer interaction, i.e., to be successfully integrated into semantics-aware user interfaces, such measures need to fulfill two criteria. First, they need to be able to handle the expressivity of the used description logics. Second, the similarity rankings produced by these measures have to correlate with human similarity rankings for the same set of compared concepts or individuals. In our previous work [2,3], we developed a structural similarity measure (SIM-DL[1]) for information retrieval on the semantic geospatial Web which fulfills these requirements [4]. It defines similarity as measure of conceptual overlap. As SIM-DL compares concept definitions, it

[1] The SIM-DL similarity server and Protégé plug-in are free and open-source software and can be downloaded at http://sim-dl.sourceforge.net/downloads/.

L. Aroyo et al. (Eds.): ESWC 2009, LNCS 5554, pp. 353–367, 2009.

does not require a populated ontology. This is especially important for various applications in geoinformation science (GIScience) such as Web gazetteers. In addition to simple place name queries, gazetteers also support type queries (such as for *archaeological sites in Crete*). In the past, thesauri have been used for the definition and retrieval of these types. Advanced gazetteer interfaces support semantics-based retrieval and rely on geographic feature type ontologies [3]. For various reasons not discussed here in detail, the millions of geographic places defined over the last decades cannot be directly mapped to instances within such ontologies. While a semi-automatic mapping is part of the long term agenda towards a semantic geospatial Web, many geographic feature type ontologies remain unpopulated until now. As SIM-DL is a structural measure, its major shortcoming is the need for canonization, i.e., concept definitions have to be rewritten to a common form to eliminate syntactic influence. Hence, SIM-DL fails to handle expressive knowledge representation languages such as OWL. In this paper, we propose a novel measure called SIM-DL_A. It overcomes these difficulties by reducing inter-concept to inter-instance similarity and reuses the similarity functions defined for SIM-DL. Our approach is based on a modified tableau algorithm which generates a set of proxy individuals for comparison.

2 Semantic Similarity Measurement

The theory of similarity has its roots in philosophy and psychology and was established to determine why and how entities are grouped into categories, and why some categories are comparable to each other while others are not [5,6]. Nowadays, similarity measures play an important role for ontology engineering, alignment and matching [7,8,9,10], as well as for information retrieval [3,11]. The main challenge for *semantic* similarity measurement is the comparison of meanings as opposed to a purely structural (syntactical) comparison. Depending on the representation language, concepts are specified as collections of features [12], regions in a multidimensional space [13], or formal restrictions specified on sets using description logics [14,15,16,2]. Besides representation, context is a major challenge for similarity. In most cases, meaningful measures cannot be defined without specifying a context in which similarity is measured [6,17,18,19,20].

2.1 Semantics of Similarity

As argued by Goodman [21], there is no global and application independent law on how similarity is measured. Strictly speaking, there is even no single definition of what similarity measures [6]. This makes the selection of an appropriate measure for a particular application area and also the comparison of existing similarity measures difficult. In the past, we have examined several measures for different applications and found generic patterns which jointly form a framework explaining how (inter-concept) similarity is measured [3]. It consists of the following seven steps. Their concrete realization depends on the similarity measure and the used representation language. While some of these steps are important for a particular measure, they may play a marginal role for another.

1. Definition of application area and intended audience
2. Selection of search (query) and target concepts
3. Transformation of concepts to canonical form
4. Definition of an alignment matrix for concept descriptors
5. Application of constructor specific similarity functions
6. Determination of standardized overall similarity
7. Interpretation of the resulting similarity value(s)

Every similarity measure should define in which way it implements the proposed steps and thereby specifies the semantics of similarity (values) as well as its properties, i.e., whether it is reflexive, symmetric, transitive, strict, minimal, etc.; see [22,23,5] for a detailed discussion. It is interesting to note that researchers from both cognitive science and artificial intelligence claim that most of these properties contradict with the nature of (human) similarity judgments. Finally, the framework allows for a better separation between the process of measuring similarity (i.e., what is measured) and the used similarity functions (i.e., how it is measured). In the following, a brief description of the steps is given.

Application Area and Intended Audience. While similarity measurement is not restricted to solve a particular task, most similarity functions have been developed for a specific purpose. Therefore, the question which functions should be selected depends on the application area. Theories developed for information retrieval and in cognitive science tend to use asymmetric similarity functions to explaining human similarity reasoning [5]. In contrast, measures developed for ontology alignment prefer symmetric functions (as there is no specific role between compared ontologies). Additionally, in case of disjunction one can choose between computing maximum, minimum [24], or average similarity [3]. Finally, human similarity judgments are influenced by age, language, and cultural background which may play a crucial role in human computer interaction [6,18].

Search and Target Concepts. Before similarity can be measured the compared concepts have to be selected. Depending on the application area, the search (or query) concept C_s can be either part of the examined ontology or phrased using a shared vocabulary [2,3]. The target concepts $C_{t_1}, ..., C_{t_i}$ form the so-called context of discourse $\mathcal{C}_d$ [18] and are selected by hand or automatically determined by specifying a context concept C_c. In the latter case, the target concepts are all concepts subsumed by C_c ($\mathcal{C}_d = \{C_t \mid C_t \sqsubseteq C_c\}$). The distinction between search and target concept is especially important for asymmetric similarity. The selection of a context concept does not only determine which concepts are compared, it also affects the measured similarity (see section 3.4). Based on search, target, and context concept similarity queries may look like the following ones:

– How similar is *Canal* (C_s) to *River* (C_t)?
– Which kind of *Waterbody* (C_c) is most similar to *Canal* (C_s)?

In the first case, *Canal* would be compared to *River*, while it is compared to all subconcepts of *Waterbody* (e.g., *River*, *Lake*, *Reservoir*) in the second case.

Canonical Form. Semantic similarity should depend on what is said about concepts, not how it is said. If two concept descriptions (specified in a given language) denote the same facts using different language elements, they need to be rewritten to a common form to eliminate unintended syntactic influences. This step mainly depends on the underlying representation language and is most important for structural similarity measures. A simple examples is:

Condition $\forall R.C \sqcap \forall R.C'$ **Rewrite** $\forall R.C \sqcap \forall R.C'$ to $\forall R.(C \sqcap C')$

The complexity of such rewriting rules increases with the expressivity of the used language and is a major obstacle in defining structural measures for concepts phrased in high expressive description logics.

Alignment Matrix. While the first steps of the framework determine which concepts are selected for comparison, the alignment matrix specifies which and how concept descriptors (e.g., dimensions, features, super/subconcepts) are compared. The term alignment is chosen here following research from psychology which investigates how structure and correspondence influences similarity judgments [25,6]. For instance, $\exists overlaps.C$ can be compared to $\exists inside.C$ because the involved roles are part of the same (topological) conceptual neighborhood [26]. In contrast, both cannot be compared to $\exists likes.C$. Likewise, the green wheel of a car cannot be aligned for comparison to the green hood of a truck to increase their similarity [5]. The term matrix points out that the selection of comparable tuples of descriptors requires a matrix $C_{Ds} \times C_{Dt}$ (where C_{Ds} and C_{Dt} are the sets of descriptors forming C_s, and C_t, respectively).

Similarity Functions. After selecting the compared concepts and aligning their descriptors, similarity is measured for each selected tuple. Depending on the used representation language different similarity functions have to be applied. For instance, in case of the Matching Distance Similarity Measure (MDSM) [12], features are distinguished into different types during the alignment process: parts, attributes, and functions. Although a context weighting is computed for each of these types, the same similarity function is applied to all of them. SIM-DL distinguishes between several similarity functions for roles and their fillers, e.g., functions for conceptual neighborhoods, role hierarchies, or co-occurrence of primitives. While we focus on inter-concept similarity here, some similarity functions also take individuals into account [15,24,20]. In most cases, each similarity function takes care of standardization (to values between 0 and 1) itself.

Overall Similarity. Next, the single similarity values derived from applying the similarity functions to all selected tuples (of the compared concepts) are combined to an overall similarity. In most theories this step is implemented as weighted summation function. For MDSM, the overall similarity is the weighted sum of the similarities determined for functions, parts, and attributes. The weights ω_f, ω_p, and ω_a indicate the relative importance of each feature type using either a commonality or variability model [12]. At the same time, the weights act as standardization factors ($\sum \omega = 1$). For conceptual space-based approaches, the overall similarity is given by the normalized, i.e., z-transformed sum of compared values [13]. In case of SIM-DL, the overall similarity is simply the sum of the similarities computed for the compared tuples.

Interpretation. Finally, a single similarity value is difficult to interpret. For instance, it does not answer the question whether there are more or less similar target concepts in the examined ontology, i.e., the distribution of similarities within the ontology is unknown. Moreover, an isolated comparison puts too much stress on the specific similarity value. It is difficult to argue that and why the result is (cognitively) plausible without other reference values. Therefore, SIM-DL focuses on similarity rankings visualized as lists of descending similarities, font-size scaling as known from tag clouds, or the clustering of the results.

3 The Novel SIM-DL_A-Measure

In this section, the novel SIM-DL_A-Measure is introduced. It reduces inter-concept to inter-instance similarity to avoid the problem of canonization known from purely structural measures. As SIM-DL_A should not require a populated ontology, the instances have to be generated before similarity is measured. This is achieved by creating a representative set of instances for both the search and target concept(s) using a slightly modified tableau algorithm as known from satisfiability checking. The algorithm produces additional models, if necessary, in order to account for all possible interpretations of a concept. To be tailored to the user's needs and application area (see step one of the framework), SIM-DL_A supports symmetric and asymmetric versions of maximum and minimum similarity. For reasons of simplification, i.e., to avoid substraction operations [27] on DL concepts, the context concept is restricted to primitive concepts. In this paper, we discuss the modified tableau algorithm for the description logic $\mathcal{SHI}$ and focus on step three to six of the similarity framework. Hence, we start with the alignment matrix and introduce the applied similarity functions afterwards. A detailed example is given in section 4.

3.1 Syntax and Semantics of $\mathcal{SHI}$

Primitive concepts and roles can be combined using constructors to build complex concepts. The language $\mathcal{SHI}$ provides the constructors *intersection*, *union existential quantification*, *value restriction*, and *complex concept negation* (table 1). In addition, roles can be defined to be *transitive*, *inverse* or form a *role hierarchy.* In the following, the letters C, D represent concepts and R, S roles. N_C denotes the set of concept names, accordingly N_R represents the set of role names. N_R^+ is the subset of N_R which contains only transitive roles. Finally, the letters x, y, and z represent individuals. The formal semantics of a $\mathcal{SHI}$-concept is given by its interpretation $\mathcal{I}$. $\mathcal{I}$ consists of a non-empty set $\triangle^{\mathcal{I}}$, which is called the domain of interpretation, and an interpretation function which maps a concept name C to a set of individuals ($C^{\mathcal{I}} \subseteq \triangle^{\mathcal{I}}$) and a role R to a set of pairs of individuals ($R^{\mathcal{I}} \subseteq \Delta^{\mathcal{I}} \times \Delta^{\mathcal{I}}$). An interpretation that satisfies the axioms of a TBox $\mathcal{T}$ is called a *model* of $\mathcal{T}$. Accordingly, in the presence of an ABox $\mathcal{A}$, an interpretation is called a model of $\mathcal{A}$ if it satisfies $(x^{\mathcal{I}}, y^{\mathcal{I}}) \in R^{\mathcal{I}}$ for all role assertions $R(x, y) \in \mathcal{A}$, and $x^{\mathcal{I}} \in C^{\mathcal{I}}$ for all concept assertions $C(x) \in \mathcal{A}$.

Table 1. Syntax and semantics of $\mathcal{SHI}$

Name	Syntax	Semantics
Atomic concept	A	$A^{\mathcal{I}} \subseteq \Delta^{\mathcal{I}}$
Atomic role	R	$R^{\mathcal{I}} \subseteq \Delta^{\mathcal{I}} \times \Delta^{\mathcal{I}}$
Concept negation	$\neg C$	$\Delta^{\mathcal{I}} \backslash C^{\mathcal{I}}$
Concept intersection	$C \sqcap D$	$C^{\mathcal{I}} \cap D^{\mathcal{I}}$
Concept union	$C \sqcup D$	$C^{\mathcal{I}} \cup D^{\mathcal{I}}$
Existential quantification	$\exists R.C$	$\{x \in \Delta^{\mathcal{I}} \mid \exists y.(x, y) \in R^{\mathcal{I}} \wedge y \in C^{\mathcal{I}}\}$
Value restriction	$\forall R.C$	$\{x \in \Delta^{\mathcal{I}} \mid \forall y.(x, y) \in R^{\mathcal{I}} \rightarrow y \in C^{\mathcal{I}}\}$
Inverse role	R^-	$\{(x, y) \in \Delta^{\mathcal{I}} \times \Delta^{\mathcal{I}} \mid (y, x) \in R^{\mathcal{I}}\}$
Transitive roles	$R \in N_R^+$	$\{(x, z) \mid (x, y) \in R^{\mathcal{I}} \wedge (y, z) \in R^{\mathcal{I}}\}$
Role inclusion	$R \sqsubseteq S$	$R^{\mathcal{I}} \subseteq S^{\mathcal{I}}$

3.2 Similarity Tableau for $\mathcal{SHI}$

The tableau algorithm introduced in this section is part of the fourth step of the similarity framework – the definition of an alignment matrix. Before the matrix can be constructed, the third step requires the concepts to be rewritten to a canonical form. Sim-DL$_A$ is designed to avoid a complex canonization, i.e., the concepts are simply rewritten to *negation normal form (NNF)*. Additionally, concept inclusion is rewritten to equivalence using the *primitiveness* of concepts [28]. For example, $C \sqsubseteq A$ can be rewritten as $C \equiv A \sqcup C'$.

Tableau algorithms prove the satisfiability of a concept C by trying to construct a model of C. More precisely, as soon as the algorithm constructed *one* model of C the satisfiability is affirmed and the algorithm terminates. In contrast, the expansion rules presented in this section do not aim at constructing only one model of C, but rather a number of models which act as proxies for the extension of C. These models are called *proxy models* here. In the following, the tableau expansion rules are presented. Since these rules differ only slightly from those used for checking satisfiability, readers interested in more details are referred to the work of Horrocks et al. [29,30].

The application of the expansion rules yields a *completion tree* **T**. For a concept C, this is a tree where each node is labeled with a set $\mathcal{L}(x)$ of concept expressions occurring in C. An edge $\langle x, y\rangle$ is either labeled with $\sqcup$ or $\forall$ (and indicates another potential proxy model starting at the y-node), or with $\mathcal{L}(\langle x, y\rangle) = R$, where R is a role occurring in C [29]. For a node x, $\mathcal{L}(x)$ contains a *clash* if, for some concept C, $\{C, \neg C\} \subseteq \mathcal{L}(x)$. Those branches of the tree **T** which do not contain a clash are proxy models and enter the similarity measurement process. A node y is called a R-neighbor of x if either y is a *successor* of x (i.e., x and y are connected by an edge $\langle x, y\rangle$) and $\mathcal{L}(\langle x, y\rangle) = S$ or y is a *predecessor* of x (x and y are connected by an edge $\langle y, x\rangle$) and $\mathcal{L}(\langle y, y\rangle) = Inv(S)^2$ for some S within the transitive closure of R ($S \sqsubseteq^* R$). A node x is *blocked* if one of the *ancestors* is blocked or $\mathcal{L}(x) = \mathcal{L}(y)$, where *ancestor* is the transitive closure of *predecessor*.

In contrast to the original tableau algorithm [30], two expansion rules are modified. First, the $\sqcup$-rule is modified in order to generate a third possible new model allowing an individual for being an instance of both concepts participating in the union. Second, the $\forall$-rule is modified. In the absence of $\exists$-expressions, the modified $\forall$-rule generates two possible models. One of these models will be unchanged (as in standard tableau algorithms), the other will be enriched by an existential quantification. When a value restriction was explicitly specified for a given concept, this should impact similarity and hence has to be reflected by a proxy model. The following expansions rules are applied:

The $\sqcap$-rule:
Condition: $C_1 \sqcap C_2 \in \mathcal{L}(x)$, x is not indirectly blocked, and $\{C_1, C_2\} \notin \mathcal{L}(x)$.
Action: $\mathcal{L}(x) := \mathcal{L}(x) \cup \{C_1, C_2\}$.

The $\sqcup$-rule:
Condition: $C_1 \sqcup C_2 \in \mathcal{L}(x)$ and x is not indirectly blocked.
Action: Create three $\sqcup$-successors w, y, z of x with:
$\mathcal{L}(w) := (\mathcal{L}(x)\backslash\{C_1 \sqcup C_2\}) \cup \{C_1\}$
$\mathcal{L}(y) := (\mathcal{L}(x)\backslash\{C_1 \sqcup C_2\}) \cup \{C_2\}$
$\mathcal{L}(z) := (\mathcal{L}(x)\backslash\{C_1 \sqcup C_2\}) \cup \{C_1, C_2\}$

The $\exists$-rule:
Condition: $(\exists R.C) \in \mathcal{L}(x)$, x is not blocked, and x has no R-neighbor y with $C \in \mathcal{L}(y)$.
Action: Create a new node y with $\mathcal{L}(y) = \{C\}$ and the edge $\mathcal{L}(\langle x, y\rangle) = R$.

The $\forall$-rule:
Condition: $(\forall R.C) \in \mathcal{L}(x)$, x is not indirectly blocked.
Action:
If there is an R-neighbor y of x and $C \notin \mathcal{L}(y)$:
$\mathcal{L}(y) := \mathcal{L}(y) \cup \{C\}$.
If there is no R-neighbor y of x, create two $\forall$-successors y, z of x with:
$\mathcal{L}(y) := \mathcal{L}(x)$ (y will then be blocked)
$\mathcal{L}(z) := \mathcal{L}(x) \cup \{\exists R.C\}$.

The $\forall_+$-rule:
Condition: $(\forall R.C) \in \mathcal{L}(x)$, x is not indirectly blocked, there is some S which is transitive and $S \sqsubseteq R$, and there is an S-neighbor y of x with $\forall S.C \notin \mathcal{L}(y)$.
Action: $\mathcal{L}(y) := \mathcal{L}(y) \cup \{\forall S.C\}$.

3.3 Alignment Matrices

For a search concept C_s and a target concept C_t, the completion tree obtained from applying the modified tableau algorithm serves as starting point to derive

[2] $Inv(S) := S^-$, if S is a role name. $Inv(S) := R$, if $S = R^-$ for a role name R [30].

the alignment matrices (the fourth step of the framework). We distinguish two levels of matrices, the model level and assertion level matrices.

Model Level Matrix – Selecting Models for Comparison. For C_s and C_t, each completion tree contains a set of proxy models. These proxy models define the matrix M_M (where $_M$ indicates that models are compared). More precisely, the number of columns is determined by the number of models created for the search concept. Accordingly, each model created for the target concept corresponds to one row in the matrix. For each field in M_M, the similarity between the two models is computed and entered into the matrix. Which of these tuples are selected for comparison depends on the application and user. For example, if an application requires maximum similarity, the tuple with the highest value is returned. For minimum similarity, the tuple with the lowest similarity is selected.

Assertion Level Matrix – Computing Similarity between Models. While the first matrix is concerned with selecting appropriate models for comparison, the following matrix M_A (where $_A$ indicates that assertions are compared) addresses the comparison of two actual instances. Therefore, each primitive concept that is instantiated by a model's root individual is added as a column/row to the matrix. Furthermore, each outgoing edge (representing a role assertion) is added as well. Now, the similarity between two primitive concepts can be computed. The similarity between role assertions consists of two steps. First, the similarity between the roles is computed. The role similarity determines which role assertions are aligned, i.e., filler similarity is only computed when appropriate. If the filler similarity needs to be determined, another assertion level alignment matrix for the two filler individuals is created. Once the matrix is filled, those pairs of assertions with the highest similarity values are selected for comparison. It is important to note that each column and each row is only selected once, avoiding a too heavy influence of single assertions. The similarity values are summed up and standardized. The standardization factor σ (see equation (4)) depends on whether the symmetric or asymmetric measure is chosen [2,3]. In the latter case, σ is the number of assertions of the search individual (i.e., the number of columns). In the symmetric case, the maximum number of either columns or rows is used.

With respect to step one of the framework, the distinction between maximum and minimum similarity is reflected in the model level matrix. It determines which of the proxy models are compared. Symmetry and asymmetry are reflected in the assertion level matrix which selects assertions for comparison. This way, and in contrast to measures such as MDSM [12], SIM-DL$_{\mathrm{A}}$ can be tailored to a specific application area without altering the similarity functions themselves.

3.4 Measuring Inter-instance Similarity

In this section, we introduce the concrete similarity functions used to determine the tuples selected for comparison. Each assertion is either of the form $A(x)$ or $r(x, y)$, where A is a primitive concept instantiated by x and $r(x, y)$ is an ordered pair of individuals in $R^{\mathcal{I}}$. Following the introduced framework, three similarity functions are necessary: one for primitives, roles, and role-filler-pairs.

Primitive Concepts. Primitives (base symbols) occur only on the right-hand side of axioms, i.e., they are not definable. To measure similarity between primitives (sim_p; see equation (1)), an adapted version of the Jaccard Similarity Coefficient is used. It measures the degree of overlap between two sets S_1 and S_2 as ratio of the cardinality of shared members from $S_1 \wedge S_2$ to the cardinality retrieved from $S_1 \vee S_2$. MDSM [12] uses an asymmetric version of Jaccard's coefficient, while in case of SIM-DL [3] it is adapted to compute the context-aware co-occurrence of primitives within the definitions of other (non-primitive) concepts. Two primitives are the more similar, the more complex concepts are defined using both (and not only one) of them. If $sim_p(A, B) = 1$, both primitives always co-occur in complex concepts and cannot be distinguished. Hence, to measure the similarity between assertions of the type $A(x)$ and $B(y)$, the similarity between A and B is measured. As similarity also depends on the context of discourse [18], we only consider those concepts C_i which are subconcepts of C_c (see step two of the similarity framework).

$$sim(A(x), B(y)) = sim_p(A, B) = \frac{\mid \{C \mid (C \sqsubseteq C_c) \wedge (C \sqsubset A) \wedge (C \sqsubset B)\} \mid}{\mid \{C \mid (C \sqsubseteq C_c) \wedge ((C \sqsubset A) \vee (C \sqsubset B))\} \mid} \quad (1)$$

Role Hierarchy. $\mathcal{SHI}$ supports role hierarchies, i.e., role inclusion, but does not support intersection or composition. Same as argued for primitives, there are no role definitions which can be compared for similarity. Because of the missing intersection constructor we cannot apply Jaccard's coefficient here. Instead, a network-based approach [31] is taken to compute the similarity of roles (R and S) within a hierarchy. Similarity (sim_r; see equation (2)) is defined as ratio between the shortest path from R to S and the maximum path within the graph representation of the role hierarchy; where the universal role U ($U \equiv \Delta^{\mathcal{I}} \times \Delta^{\mathcal{I}}$) forms the graph's root. Compared to sim_p, similarity between roles is defined without reference to the context. This would require to take only such roles into account which are used within quantifications or restrictions of concepts within the context. The standardization in equation (2) is depth-dependent to indicate that the distance from node to node decreases with increasing depth level of R and S within the hierarchy. In other words, the weights of the edges used to determine the path between R and S decrease with increasing depth of the graph. If a path between two roles crosses U, similarity is 0. The $lcs(R, S)$ is the least common subsumer, in this case the first common super role of R and S.

$$sim_r(R, S) = \frac{depth(lcs(R, S))}{depth(lcs(R, S)) + edge\_distance(R, S)} \quad (2)$$

SIM-DL uses a second function to compare roles within a conceptual neighborhood [2]. This is necessary for topological and temporal relations as used in GIScience[3]. One could also define network-based similarity functions for other roles such as *part-of*. These functions can be integrated into SIM-DL$_A$.

[3] See [32] on the problems of integrating topological relations and reasoning in DL.

Roles and Fillers. The similarity between assertions of the type $r(x, y)$ and $s(w, z)$, is the similarity of the involved roles R and S times the overall similarity of the assertions about y and z; where x and w are the individuals to be compared (see equation (3)).

$$sim(r(x,y), s(w,z)) = sim_r(R,S) * sim_o(y,z) \tag{3}$$

Some similarity measures define role-filler similarity as weighted average of the role and filler similarities, but the multiplicative approach has turned out to be cognitively plausible [4] and allows for simple approximation and optimization techniques not discussed here in detail.

3.5 Overall Similarity

As sim_p and sim_r deliver similarity values between 0 and 1, the overall instance similarity (sim_o; see equation (4)) in SIM-DL$_A$ is simply the standardized sum of the similarities computed for all assertion tuples selected during the alignment; see section 3.3. The similarity of compared concepts is either the similarity of the most or least similar proxy individuals (computed using sim_o).

$$sim_o(x,y) = \frac{\sum M_{A_{ij}}}{\sigma}; \text{ where } M_{A_{ij}} \text{ is a selected tuple from the matrix } M_A. \tag{4}$$

4 Exemplary Application of SIM-DL$_A$

The previous section presented the novel SIM-DL$_A$ measure. In order to illustrate the steps involved in a comparison process between a search concept C_s and a target concept C_t, this section provides a detailed example. A, B, D, and E are primitive concepts, R and S are roles with $R \sqsubset S$.

$C_s \equiv A \sqcap (B \sqcup D) \sqcap \exists R.E$ $\qquad$ $C_t \equiv A \sqcap (B \sqcup \neg A) \sqcap \forall S.E \sqcap \forall S.D$

For C_s, the expansion rules are applied as follows, the expansion tree $\mathbf{T}_s$ and resulting models are shown in figure 1.

1. Unfold C_s and convert it to NNF.	$A \sqcap (B \sqcup D) \sqcap \exists R.E$
2. Initialize $\mathbf{T}_s$ containing a node s_0.	$\mathcal{L}(s_0) = \{A \sqcap (B \sqcup D) \sqcap \exists R.E\}$
3. Apply the $\sqcap$-rule to $A \sqcap (B \sqcup D) \sqcap \exists R.E \in \mathcal{L}(s_0)$.	$\mathcal{L}(s_0) := \mathcal{L}(s_0) \cup \{A, B \sqcup D, \exists R.E\}$
4. Apply the $\sqcup$-rule to $B \sqcup D \in \mathcal{L}(s_0)$. Create three $\sqcup$-successors s_1, s_2, s_3 of s_0.	$\mathcal{L}(s_1) := \mathcal{L}(s_0) \backslash \{B \sqcup D\} \cup \{B\}$ $\mathcal{L}(s_2) := \mathcal{L}(s_0) \backslash \{B \sqcup D\} \cup \{D\}$ $\mathcal{L}(s_3) := \mathcal{L}(s_0) \backslash \{B \sqcup D\} \cup \{B, D\}$
5. Apply the $\exists$-rule to $\exists R.E \in \mathcal{L}(s_1)$. Create a node s_4 and an edge $\langle s_1, s_4 \rangle$.	$\mathcal{L}(s_4) = \{E\}$ $\mathcal{L}(\langle s_1, s_4 \rangle) = \{R\}$
6. Apply the $\exists$-rule to $\exists R.E \in \mathcal{L}(s_2)$. Create a node s_5 and an edge $\langle s_2, s_5 \rangle$.	$\mathcal{L}(s_5) = \{E\}$ $\mathcal{L}(\langle s_2, s_5 \rangle) = \{R\}$
7. Apply the $\exists$-rule to $\exists R.E \in \mathcal{L}(s_3)$. Create a node s_6 and an edge $\langle s_3, s_6 \rangle$.	$\mathcal{L}(s_6) = \{E\}$ $\mathcal{L}(\langle s_3, s_6 \rangle) = \{R\}$

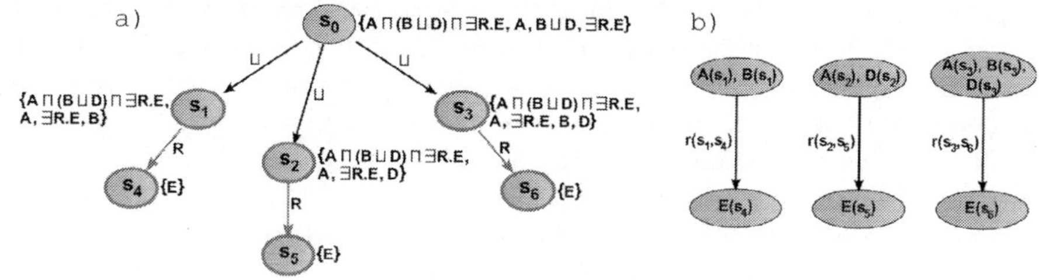

Fig. 1. a) Expansion tree for C_s. b) Models for C_s.

For C_t, the expansion tree $\mathbf{T}_t$ is slightly more complex. It results from applying the expansion rules as listed below. The tree as well as the models are shown in figure 2.

1. Unfold C_t and convert it to NNF.	$A \sqcap (B \sqcup \neg A) \sqcap \forall S.E \sqcap \forall S.D$
2. Initialize $\mathbf{T}_t$ containing a node t_0.	$\mathcal{L}(t_0) = \{A \sqcap (B \sqcup \neg A) \sqcap \forall S.E \sqcap \forall S.D\}$
3. Apply the $\sqcap$-rule to $A \sqcap (B \sqcup \neg A) \sqcap \forall S.E \sqcap \forall S.D \in \mathcal{L}(t_0)$.	$\mathcal{L}(t_0) := \mathcal{L}(t_0) \cup \{A, B \sqcup \neg A, \forall S.E, \forall S.D\}$
4. Apply the $\sqcup$-rule to $B \sqcup \neg A \in \mathcal{L}(t_0)$. Create three $\sqcup$-successors t_1, t_2, t_3 of t_0.	$\mathcal{L}(t_1) := \mathcal{L}(t_0) \setminus \{B \sqcup \neg A\} \cup \{B\}$ $\mathcal{L}(t_2) := \mathcal{L}(t_0) \setminus \{B \sqcup \neg A\} \cup \{\neg A\}$ (clash) $\mathcal{L}(t_3) := \mathcal{L}(t_0) \setminus \{B \sqcup \neg A\} \cup \{B, \neg A\}$ (clash)
5. Apply the $\forall$-rule to $\forall S.E \in \mathcal{L}(t_1)$. Create two $\forall$-successors t_4, t_5 of t_1.	$\mathcal{L}(t_4) := \mathcal{L}(t_1)$ (t_4 is blocked by t_1) $\mathcal{L}(t_5) := \mathcal{L}(t_1) \cup \{\exists S.E\}$
6. Apply the $\exists$-rule to $\exists S.E \in \mathcal{L}(t_5)$. Create a node t_6 and an edge $\langle t_5, t_6 \rangle$.	$\mathcal{L}(t_6) = \{E\}$ $\mathcal{L}(\langle t_5, t_6 \rangle) = \{S\}$
7. Apply the $\forall$-rule to $\forall S.D \in \mathcal{L}(t_5)$. There is a S-neighbor t_6 of t_5, but $D \notin \mathcal{L}(t_6)$.	$\mathcal{L}(t_6) := \mathcal{L}(t_6) \cup \{D\}$

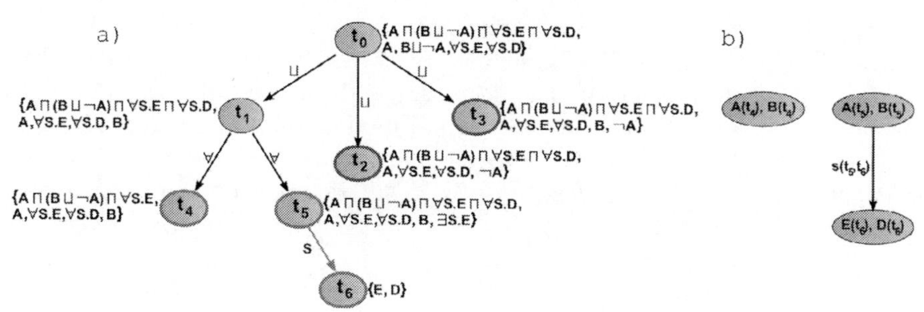

Fig. 2. a) Expansion tree for C_t. b) Models for C_t.

The application of the tableau algorithm generates three proxy models for C_s and two for C_t. As shown in figure 3a), the similarity between each possible tuple of models is computed. In the model level matrix, a model is represented by its root node. Given that the user queries for the maximum similarity between C_s and C_t, $M_{M_{12}}$ is selected and yields the similarity value 0.83. The assertion level matrix is shown in figure 3b). As described in section 3.3, each concept and role assertion is added as a column (or row, respectively) in the alignment matrix. For simplification we assume that $C_c = \top$ and $sim(A, B) = 0.5$. Whereas the similarity between primitive concepts can directly be computed using function (1), those fields in the alignment matrix representing tuples of role assertions are computed using similarity function (3), hence demanding for two calculations, $sim_r(R, S)$ and $sim_o(s_4, t_6)$. The former function calculates the similarity between two roles. R is a direct subrole of S and for simplification we assume that $sim(R, S)$ yields 0.5. As the role similarity is above 0 and there are no other role assertion tuples, the filler similarity is computed. To do so, another alignment matrix for s_4 and t_6 is created, resulting in $sim_o(s_4, t_6) = 1$ (see figure 3c).

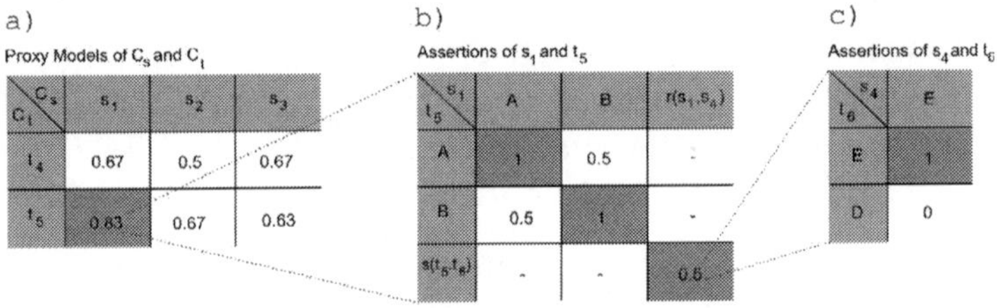

Fig. 3. The alignment matrices for comparing C_s and C_t. Given that asymmetric maximum similarity is chosen, the green fields show the tuples selected for comparison.

5 Conclusions and Further Work

In this paper, we introduced the novel Sim-DL$_A$ measure which reduces inter-concept to inter-instance similarity by generating proxy models using a modified tableau algorithm for $\mathcal{SHI}$. The algorithm can still be used for satisfiability checking, although its performance decreases with an extensive use of disjunctions and value restrictions, since additional models have to be created. Consequently, it should be implemented as optional *completion strategy* as known from the Pellet reasoner. The complexity of Sim-DL$_A$ depends on the used DL (i.e., the tableau rules) and the complexity of computing the similarity for the matrices. Nevertheless, structural measures would require to define and implement a complete set of canonization rules. Following the introduced similarity framework, Sim-DL$_A$ clearly separates similarity modes such as maximum/minimum similarity and symmetry/asymmetry, which are used to tailor the measure to the user's needs, from the concrete similarity functions. This allows to replace particular functions without changing the measurement process.

Our next steps involve the extension of the presented approach to more expressive description logics such as $\mathcal{SHIQ}$, i.e., to account for qualified number restrictions. We also work on the integration of Sim-DL$_A$ into the SIM-DL server and Protégé plug-in. In this paper we focused on the theory. While the new measure relies on the same similarity functions, the alignment process differs. Hence, we have to redo our human participants tests [4] to verify that the Sim-DL$_A$ similarity rankings correlate with human rankings. We also need to investigate how the creation of proxy individuals impacts similarity. Like the work of Araújo and Pinto [16] our measure could also be extended to compare ontologies. Finally, one could try to express similarity as the ratio of clashing versus clash-free models of compared concepts. The clashes can then be presented (in natural language) to the end-user to explain the resulting similarity values.

Acknowledgments

The comments from four anonymous reviewers as well as from Simon Scheider provided useful suggestions to improve the content and clarity of the paper. This work is funded by the *SimCat* project granted by the German Research Foundation (DFG Ra1062/2-1), as well as the International Research Training Group on *Semantic Integration of Geospatial Information* (DFG GRK 1498).

References

1. Rissland, E.L.: AI and similarity. IEEE Intelligent Systems 21(3), 39–49 (2006)
2. Janowicz, K.: Sim-DL: Towards a semantic similarity measurement theory for the description logic $\mathcal{ALCNR}$ in geographic information retrieval. In: Meersman, R., Tari, Z., Herrero, P. (eds.) OTM 2006 Workshops. LNCS, vol. 4278, pp. 1681–1692. Springer, Heidelberg (2006)
3. Janowicz, K., Keßler, C., Schwarz, M., Wilkes, M., Panov, I., Espeter, M., Baeumer, B.: Algorithm, implementation and application of the SIM-DL similarity server. In: Fonseca, F., Rodríguez, M.A., Levashkin, S. (eds.) GeoS 2007. LNCS, vol. 4853, pp. 128–145. Springer, Heidelberg (2007)
4. Janowicz, K., Keßler, C., Panov, I., Wilkes, M., Espeter, M., Schwarz, M.: A study on the cognitive plausibility of SIM-DL similarity rankings for geographic feature types. In: Bernard, L., Friis-Christensen, A., Pundt, H. (eds.) 11th AGILE International Conference on Geographic Information Science, Girona, Spain. Lecture Notes in Geoinformation and Cartography, pp. 115–133. Springer, Heidelberg (2008)
5. Goldstone, R.L., Son, J.: Similarity. In: Holyoak, K., Morrison, R. (eds.) Cambridge Handbook of Thinking and Reasoning, pp. 13–36. Cambridge University Press, Cambridge (2005)
6. Medin, D., Goldstone, R., Gentner, D.: Respects for similarity. Psychological Review 100(2), 254–278 (1993)
7. Cruz, I., Sunna, W.: Structural alignment methods with applications to geospatial ontologies. Transactions in GIS 12(6), 683–711 (2008)
8. Euzenat, J., Valtchev, P.: Similarity-based ontology alignment in OWL-lite. In: de Mántaras, R.L., Saitta, L. (eds.) Proceedings of the 16th European Conference on Artificial Intelligence (ECAI 2004), pp. 333–337. IOS Press, Amsterdam (2004)

9. Shvaiko, P., Euzenat, J.: Ten challenges for ontology matching. In: Meersman, R., Tari, Z. (eds.) On the Move to Meaningful Internet Systems: OTM 2008. LNCS, vol. 5332, pp. 1164–1182. Springer, Heidelberg (2008)
10. Falconer, S., Noy, N., Storey, M.A.: Ontology mapping - a user survey. In: Shvaiko, P., Euzenat, J., Giunchiglia, F., He, B. (eds.) Proceedings of the Workshop on Ontology Matching (OM 2007) at ISWC/ASWC 2007, Busan, South Korea (2007)
11. Ricklefs, M., Blomqvist, E.: Ontology-based relevance assessment: An evaluation of different semantic similarity measures. In: Meersman, R., Tari, Z. (eds.) On the Move to Meaningful Internet Systems: OTM 2008. OTM Conferences (2). LNCS, vol. 5332, pp. 1235–1252. Springer, Heidelberg (2008)
12. Rodríguez, A., Egenhofer, M.: Comparing geospatial entity classes: an asymmetric and context-dependent similarity measure. International Journal of Geographical Information Science 18(3), 229–256 (2004)
13. Raubal, M.: Formalizing conceptual spaces. In: Varzi, A., Vieu, L. (eds.) Formal Ontology in Information Systems, Proceedings of the Third International Conference (FOIS 2004), vol. 114, pp. 153–164. IOS Press, Torino (2004)
14. Borgida, A., Walsh, T., Hirsh, H.: Towards measuring similarity in description logics. In: International Workshop on Description Logics (DL 2005), July 2005. CEUR Workshop Proceedings, vol. 147. CEUR, Edinburgh (2005)
15. d'Amato, C., Fanizzi, N., Esposito, F.: A semantic similarity measure for expressive description logics. In: CILC 2005, Convegno Italiano di Logica Computazionale, Rome, Italy (2005)
16. Araújo, R., Pinto, H.S.: Semilarity: Towards a model-driven approach to similarity. In: International Workshop on Description Logics (DL 2007), vol. 20, pp. 155–162. Bolzano University Press, Bozen-Bolzano (2007)
17. Albertoni, R., Martino, M.D.: Semantic similarity of ontology instances tailored on the application context. In: Meersman, R., Tari, Z. (eds.) OTM 2006. LNCS, vol. 4275, pp. 1020–1038. Springer, Heidelberg (2006)
18. Janowicz, K.: Kinds of contexts and their impact on semantic similarity measurement. In: Proceedings of the 6th IEEE International Conference on Pervasive Computing and Communications; 5th Workshop on Context Modeling and Reasoning (CoMoRea 2008), Hong Kong, pp. 441–446. IEEE Computer Society, Los Alamitos (2008)
19. Keßler, C.: Similarity measurement in context. In: Kokinov, B., Richardson, D.C., Roth-Berghofer, T.R., Vieu, L. (eds.) CONTEXT 2007. LNCS, vol. 4635, pp. 277–290. Springer, Heidelberg (2007)
20. d'Amato, C., Fanizzi, N., Esposito, F.: Query answering and ontology population: An inductive approach. In: Bechhofer, S., Hauswirth, M., Hoffmann, J., Koubarakis, M. (eds.) ESWC 2008. LNCS, vol. 5021, pp. 288–302. Springer, Heidelberg (2008)
21. Goodman, N.: Seven strictures on similarity. In: Problems and projects, pp. 437–447. Bobbs-Merrill, New York (1972)
22. Ashby, F.G., Perrin, N.A.: Toward a unified theory of similarity and recognition. Psychological Review 95, 124–150 (1988)
23. Cross, V., Sudkamp, T.: Similarity and Computability in Fuzzy Set Theory: Assessments and Applications. Studies in Fuzziness and Soft Computing, vol. 93. Physica-Verlag (2002)
24. d'Amato, C., Fanizzi, N., Esposito, F.: A dissimilarity measure for $\mathcal{ALC}$ concept descriptions. In: SAC 2006: Proceedings of the 2006 ACM symposium on Applied computing, pp. 1695–1699. ACM Press, New York (2006)

25. Markman, A.B.: Structural alignment, similarity, and the internal structure of category representations. In: Similarity and Categorization, pp. 109–130. Oxford University Press, Oxford (2001)
26. Egenhofer, M., Al-Taha, K.: Reasoning about gradual changes of topological relationships. In: Frank, A.U., Formentini, U., Campari, I. (eds.) GIS 1992. LNCS, vol. 639, pp. 196–219. Springer, Heidelberg (1992)
27. Teege, G.: Making the difference: A subtraction operation for description logics. In: Doyle, J., Sandewall, E., Torasso, P. (eds.) 4th International Conference on Principles of Knowledge Representation and Reasoning (KR 1994), Bonn, Germany, pp. 540–550. Morgan Kaufmann, San Francisco (1994)
28. Horrocks, I.: Implementation and Optimization Techniques. In: The Description Logic Handbook: Theory, Implementation and Applications, pp. 306–346. Cambridge University Press, Cambridge (2003)
29. Horrocks, I.: Optimising Tableaux Decision Procedures for Description Logics. PhD thesis, University of Manchester (1997)
30. Horrocks, I., Sattler, U., Tobies, S.: A description logic with transitive and converse roles, role hierarchies and qualifying number restrictions. LTCS-Report LTCS-99-08, LuFG Theoretical Computer Science, RWTH Aachen (1999)
31. Rada, R., Mili, H., Bicknell, E., Blettner, M.: Development and application of a metric on semantic nets. IEEE Transactions on Systems, Man and Cybernetics 19, 17–30 (1989)
32. Lutz, C., Möller, R.: Defined topological relations in description logics. In: Rousset, M.C., Brachman, R., Donini, F., Franconi, E., Horrocks, I., Levy, A. (eds.) Proceedings of the International Workshop on Description Logics, Gif sur Yvette, Paris, France, Université Paris-Sud, Centre d'Orsay, September 1997, pp. 15–19 (1997)

Decidability of $\mathcal{SHI}$ with Transitive Closure of Roles

Chan Le Duc

INRIA Grenoble Rhône-Alpes - LIG
Chan.Leduc@inrialpes.fr

Abstract. This paper investigates a Description Logic, namely $\mathcal{SHI}_+$, which extends $\mathcal{SHI}$ by adding transitive closure of roles. The resulting logic $\mathcal{SHI}_+$ allows transitive closure of roles to occur not only in concept inclusion axioms but also in role inclusion axioms. We show that $\mathcal{SHI}_+$ is decidable by devising a terminating, sound and complete algorithm for deciding satisfiability of concepts in $\mathcal{SHI}_+$ with respect to a set of concept and role inclusion axioms.

Keywords: OWL, Description Logics, Tableaux, Decidability.

1 Introduction

The ontology language OWL-DL [10] is widely used for formalizing resources on Semantic Web. This language is mainly based on the description logic $\mathcal{SHOIN}$ which is known to be decidable [12]. Although $\mathcal{SHOIN}$ is expressive and provides *transitive roles* to model transitivity of relations, we can find several applications in which *the transitive closure of roles*, that is more expressive than transitive roles, is necessary. An example in [11] describes two categories of devices as follows: (1) Devices have as their direct part a battery: $\mathsf{Device} \sqcap \exists \mathsf{hasPart.Battery}$, (2) Devices have at some level of decomposition a battery: $\mathsf{Device} \sqcap \exists \mathsf{hasPart}^+.\mathsf{Battery}$. However, if we now define $\mathsf{hasPart}$ a *transitive role*, the concept $\mathsf{Device} \sqcap \exists \mathsf{hasPart.Battery}$ does not represent the devices as described above since it does not allow one to distinguish these two categories of devices.

We now consider a more specific example in which we need the transitive closure of roles to occur in role inclusion axioms.

Example 1. Let us consider a LOTOS description [9] (producer/consumer):

$$\mathsf{P}[\mathsf{put}, \mathsf{get}] := \mathsf{put}; (\ (\mathsf{get}; \mathbf{stop})\ |||\ \mathsf{P}[\mathsf{put}, \mathsf{get}]\ )$$

Signs ";" and "|||" represent sequential and parallel operators respectively. An execution trace of P can be generated by executing an action put, and then the parallel block (get; **stop**) ||| P[put, get]. This block can be started by executing either action get or a recursive call P. The special action **stop** will be performed if no action of the description can be executed, which is never the case in this example because, according to the semantics of parallel operator, following a get a recursive call P can be invoked. This means that execution traces of P

L. Aroyo et al. (Eds.): ESWC 2009, LNCS 5554, pp. 368–383, 2009.

are infinite and their prefixes verify the following property: if n_1 and n_2 are the number of put and get in a given prefix, $n_1 \geq n_2$ always holds i.e. the intersection of the execution traces with the language $\mathsf{put}^*.\mathsf{get}^*$ is the irregular language $\{\mathsf{put}^{n_1}.\mathsf{get}^{n_2} \mid n_1 \geq n_2\}$.

If one uses a DL to describe these execution traces then roles put, get and disjoint concepts **put**, **put′**, **get** would be needed. A role next subsuming put, get would be used to impose general properties on both of them. In addition, an artificial concept $\mathsf{g_{start}}$ and role $\overrightarrow{\mathsf{g_{start}}}$ subsumed by next can be added for the sake of axiom construction.

An expected model of concept $\mathsf{g_{start}}$ w.r.t. a set of DL axioms is depicted in Fig. 1.

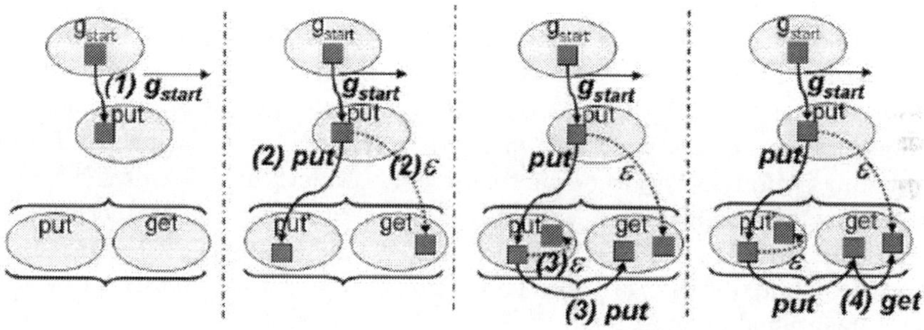

Fig. 1. A model of $\mathsf{g_{start}}$ to represent an execution trace

The essential point is that each execution of the parallel block leads to two executions corresponding to blocks (get; **stop**) and P[put, get]. Since only one action can be executed at one moment, the non-consumed execution must be memorized and will be performed in the future. This characteristic of execution traces is expressed by a role inclusion axiom $\varepsilon \sqsubseteq \mathsf{next}^+$: relation ε memorizes the non-consumed execution which is initialized by the execution of the parallel block. A sequence of next, i.e. next^+, starts from this point of execution and joints to the memorized non-consumed execution at *some moment in the future*. For instance, the following axioms:

(1) $\varepsilon \sqsubseteq \mathsf{next}^+$; $\mathsf{g_{start}} \sqsubseteq \exists\overrightarrow{\mathsf{g_{start}}}.\mathbf{put}$; $\mathsf{next}, \mathsf{next}^-, \varepsilon, \varepsilon^-$: functional
(2) $\mathbf{put} \sqsubseteq (\exists \mathsf{put}.\mathbf{put'} \sqcap \exists\varepsilon.\mathbf{get}) \sqcup (\exists\mathsf{put}.\mathbf{get} \sqcap \exists\varepsilon.\mathbf{put'})$;
(3) $\mathbf{put'} \sqsubseteq (\exists \mathsf{put}.\mathbf{put} \sqcap \exists\varepsilon.\mathbf{get}) \sqcup (\exists\mathsf{put}.\mathbf{get} \sqcap \exists\varepsilon.\mathbf{put})$;
(4) ...

describe a trace as a model of $\mathsf{g_{start}}$. Since individuals are distinct it holds that the number of instances put from **put** to **put′** equals to the number of instances ε from **put** to **get**. Note that if we replace the term $\exists\varepsilon.\mathbf{get}$ in (2) by $\exists\mathsf{next}^+.\mathbf{get}$ then this property no longer holds.

Such examples motivate the study of the logic $\mathcal{SHI}_+$ which is obtained by allowing the transitive closure of roles to occur in both concept and role inclusion axioms in $\mathcal{SHI}$. To the best of our knowledge, the decidability of $\mathcal{SHI}_+$

is unknown. In the literature, many decidability results in Description Logics (DL) can be obtained from their counterparts in modal logics (see [4], [5], [6]). However, these counterparts do not take into account expressive role inclusion axioms. In particular, [5] has shown decidability of a very expressive DL, so-called $\mathcal{CATS}$, including $\mathcal{SHIQ}$ with the transitive closure of roles but not allowing it to occur in role hierarchies.

In addition, tableaux-based algorithms for expressive DLs like $\mathcal{SHIQ}$ [8] and $\mathcal{SHOIQ}$ [7] result in efficient implementations. This kind of algorithms relies on two structures, so-called *tableau* and *completion graph*. Roughly speaking, a tableau for a concept represents a model for the concept and it is possibly infinite. A tableau translates satisfiability of all given concept and role inclusion axioms into *local* satisfiability of semantic constraints imposed on each individual of the tableau. This characteristic of tableaux will be called *local satisfiability property*. In turn, a completion graph for a concept is a *finite* representation from which a tableau can be built. To check satisfiability of a concept, tableaux-based algorithms try to build a completion graph whose finiteness is ensured by a technique, so-called *blocking technique*. It provides a termination condition by guaranteeing soundness and completeness. The underlying idea of the blocking mechanism is to detect "loops" which are repeated pieces of a completion graph. The algorithm in [2] for satisfiability in $\mathcal{ALC}_{reg}$ (including the transitive closure of roles and other role operators) introduced a method to deal with loops which can hide unsatisfiable nodes.

The contribution of the present paper is to propose an algorithm for concept satisfiability in $\mathcal{SHI}_+$. Reasoning in $\mathcal{SHI}_+$ is harder than in $\mathcal{SHI}$ because of the two following reasons: (i) adding to a logic transitive closure brings this logic beyond first order logic. This is so because [1] has shown that the transitive closure of roles cannot be expressed in first-order predicate logic, (ii) the presence of the transitive closure of roles in role inclusion axioms, such as $\varepsilon \sqsubseteq a^+$, leads to the loss of the *tree model property*.

2 The Logic $\mathcal{SHI}_+$

In this section, we present the syntax and semantics of the logic $\mathcal{SHI}_+$. This includes the definitions of inference problems and how they are interrelated. The definitions reuse notation introduced in [8].

Definition 1. *Let* $\mathbf{C}$ *and* $\mathbf{R}$ *be sets of* concept *and* role names, *respectively.*
∗ *The set of* $\mathcal{SHI}_+$*-roles is* $\mathbf{R}' \cup \{R^- \mid R \in \mathbf{R}'\}$ *with* $\mathbf{R}' = \mathbf{R} \cup \{P^+ \mid P \in \mathbf{R}\}$. *A* role inclusion axiom *is of the form* $R \sqsubseteq S$ *for two* $\mathcal{SHI}_+$*-roles* R *and* S. *A* role hierarchy $\mathcal{R}$ *is a finite set of role inclusion axioms. A role* $R \in \mathbf{R}$ *is called* transitive *if* $R^+ \sqsubseteq R \in \mathcal{R}$.
∗ *Function* Inv *returns the inverse of a role as follows:*

$$\mathsf{Inv}(R) := \begin{cases} R^- & \text{if } R \in \mathbf{R} \cup \{P^+ \mid P \in \mathbf{R}\} \\ S & \text{if } R = S^- \text{ where } S \in \mathbf{R} \cup \{P^+ \mid P \in \mathbf{R}\} \end{cases}$$

∗ *A relation* $\underline{\circledast}$ *is defined as the transitive-reflexive closure of* $\sqsubseteq$ *on* $\mathcal{R} \cup \{\mathsf{Inv}(R) \sqsubseteq \mathsf{Inv}(S) \mid R \sqsubseteq S \in \mathcal{R}\} \cup \{P \sqsubseteq P^+ \mid P \in \mathbf{R}\}$. *We denote* $S \equiv R$ *iff* $R \underline{\circledast} S$ *and* $S \underline{\circledast} R$.

∗ *Function* Trans *is defined as follows:*

$$\mathsf{Trans}(R) := \begin{cases} true & \text{if there is some } S \text{ such that } S \equiv R \text{ and} \\ & \{S,\ \mathsf{Inv}(S)\} \cap \{\{R \mid R^+ \sqsubseteq R \in \mathcal{R}\} \cup \{P^+ \mid P \in \mathbf{R}\}\} \neq \emptyset \\ false, & \text{otherwise} \end{cases}$$

∗ *The set of* $\mathcal{SHI}_+$*-concepts is inductively defined as the smallest set containing all* C *in* $\mathbf{C}$, $\top$, $C \sqcap D$, $C \sqcup D$, $\neg C$, $\exists R.C$ *and* $\forall R.C$ *where* C *and* D *are* $\mathcal{SHI}_+$*-concepts and* R *is an* $\mathcal{SHI}_+$*-role. We denote* $\bot$ *for* $\neg\top$.

∗ *An interpretation* $\mathcal{I} = (\Delta^{\mathcal{I}}, \cdot^{\mathcal{I}})$ *consists of a non empty set* $\Delta^{\mathcal{I}}$ (domain) *and a function* $\cdot^{\mathcal{I}}$ *which maps each role* R *to a subset of* $\Delta^{\mathcal{I}} \times \Delta^{\mathcal{I}}$ *such that, for* $R, P \in \mathbf{R}'$,

$$R^{-\mathcal{I}} = \{\langle x, y\rangle \in (\Delta^{\mathcal{I}})^2 \mid \langle y, x\rangle \in R^{\mathcal{I}}\},$$
$$P^{+\mathcal{I}} = \bigcup_{n>0} (P^n)^{\mathcal{I}} \text{ with } (P^1)^{\mathcal{I}} = P^{\mathcal{I}},\ (P^n)^{\mathcal{I}} = (P^{n-1})^{\mathcal{I}} \circ P^{\mathcal{I}} \text{ and}$$
$$(P^{n-1})^{\mathcal{I}} \circ P^{\mathcal{I}} = \{\langle x, y\rangle \in (\Delta^{\mathcal{I}})^2 \mid \exists z \in \Delta^{\mathcal{I}}, (\langle x, z\rangle \in (P^{n-1})^{\mathcal{I}}, \langle z, y\rangle \in P^{\mathcal{I}})\}$$

∗ *The function* $\cdot^{\mathcal{I}}$ *maps also each concept to a subset of* $\Delta^{\mathcal{I}}$ *such that* $\top^{\mathcal{I}} = \Delta^{\mathcal{I}}$, $(C \sqcap D)^{\mathcal{I}} = C^{\mathcal{I}} \cap D^{\mathcal{I}}$, $(C \sqcup D)^{\mathcal{I}} = C^{\mathcal{I}} \cup D^{\mathcal{I}}$, $(\neg C)^{\mathcal{I}} = \Delta^{\mathcal{I}} \backslash C^{\mathcal{I}}$,
$(\exists R.C)^{\mathcal{I}} = \{x \in \Delta^{\mathcal{I}} \mid \exists y \in \Delta^{\mathcal{I}}, (\langle x, y\rangle \in R^{\mathcal{I}} \wedge y \in C^{\mathcal{I}})\}$,
$(\forall R.C)^{\mathcal{I}} = \{x \in \Delta^{\mathcal{I}} \mid \forall y \in \Delta^{\mathcal{I}}, (\langle x, y\rangle \in R^{\mathcal{I}} \Rightarrow y \in C^{\mathcal{I}})\}$

∗ *An interpretation* $\mathcal{I}$ *satisfies a role hierarchy* $\mathcal{R}$ *iff* $R^{\mathcal{I}} \subseteq S^{\mathcal{I}}$ *for each* $R \sqsubseteq S \in \mathcal{R}$. *Such an interpretation is called a* model *of* $\mathcal{R}$, *denoted by* $\mathcal{I} \models \mathcal{R}$.

∗ $C \sqsubseteq D$ *is called a general concept inclusion (GCI) where* C, D *are* $\mathcal{SHI}_+$*-concepts (possibly complex), and a finite set of GCIs is called a terminology* $\mathcal{T}$. *An interpretation* $\mathcal{I}$ *satisfies a GCI* $C \sqsubseteq D$ *if* $C^{\mathcal{I}} \subseteq D^{\mathcal{I}}$ *and* $\mathcal{I}$ *satisfies a terminology* $\mathcal{T}$ *if* $\mathcal{I}$ *satisfies each GCI in* $\mathcal{T}$. *Such an interpretation is called a* model *of* $\mathcal{T}$, *denoted by* $\mathcal{I} \models \mathcal{T}$.

∗ *A concept* C *is called* satisfiable *w.r.t. a role hierarchy* $\mathcal{R}$ *and a terminology* $\mathcal{T}$ *iff there is some interpretation* $\mathcal{I}$ *such that* $\mathcal{I} \models \mathcal{R}$, $\mathcal{I} \models \mathcal{T}$ *and* $C^{\mathcal{I}} \neq \emptyset$. *Such an interpretation is called a* model *of* C *w.r.t.* $\mathcal{R}$ *and* $\mathcal{T}$. *A pair* $(\mathcal{T}, \mathcal{R})$ *is called an* $\mathcal{SHI}_+$ ontology *and said to be consistent if there is a model of* $(\mathcal{T}, \mathcal{R})$.

∗ *A concept* D subsumes *a concept* C *w.r.t.* $\mathcal{R}$ *and* $\mathcal{T}$, *denoted by* $C \sqsubseteq D$, *if* $C^{\mathcal{I}} \subseteq D^{\mathcal{I}}$ *holds in each model* $\mathcal{I}$ *of* $(\mathcal{T}, \mathcal{R})$.

Since negation is allowed in the logic $\mathcal{SHI}_+$, unsatisfiability and subsumption w.r.t. $(\mathcal{T}, \mathcal{R})$ can be reduced each other: $C \sqsubseteq D$ iff $C \sqcap \neg D$ is unsatisfiable. In addition, we can reduce ontology consistency to concept satisfiability w.r.t. an ontology: $(\mathcal{T}, \mathcal{R})$ is consistent if $A \sqcup \neg A$ is satisfiable w.r.t. $(\mathcal{T}, \mathcal{R})$ for some concept name A.

For the ease of construction, we assume all concepts to be in *negation normal form* (NNF) i.e. negation occurs only in front of concept names. Any $\mathcal{SHI}_+$-concept can be transformed to an equivalent one in NNF by using DeMorgan's laws and some equivalences as presented in [8]. For a concept C, we denote the nnf of C by $\mathsf{nnf}(C)$ and the nnf of $\neg C$ by $\dot{\neg} C$.

Let D be an $\mathcal{SHI}_+$-concept in NNF w.r.t. an ontology $(\mathcal{T}, \mathcal{R})$. We define $\mathsf{sub}(D)$ to be the smallest set that contains all sub-concepts of D including D. For an ontology $(\mathcal{T}, \mathcal{R})$, we define the set of all sub-concepts $\mathsf{sub}(\mathcal{T}, \mathcal{R})$ as follows:

$$\mathsf{sub}(\mathcal{T}, \mathcal{R}) := \bigcup_{C \sqsubseteq D \in \mathcal{T}} \mathsf{sub}(\mathsf{nnf}(\neg C \sqcup D), \mathcal{R})$$

$$\begin{aligned}\mathsf{sub}(E, \mathcal{R}) := {} & \mathsf{sub}(E) \cup \{\dot{\neg} C \mid \neg C \in \mathsf{sub}(E)\} \ \cup \\ & \{\forall S.C \mid (\forall R.C \in \mathsf{sub}(E), S \,\underline{\circledast}\, R) \vee (\dot{\neg} \forall R.C \in \mathsf{sub}(E), S \,\underline{\circledast}\, R) \\ & \quad \text{and } S \text{ occurs in } \mathcal{T} \text{ or } \mathcal{R}\}\end{aligned}$$

For the sake of simplicity, for each concept D w.r.t. $(\mathcal{T}, \mathcal{R})$ we denote $\mathsf{sub}(\mathcal{T}, \mathcal{R}, D)$ for $\mathsf{sub}(\mathcal{T}, \mathcal{R}) \cup \mathsf{sub}(D)$, and $\mathbf{R}_{(\mathcal{T}, \mathcal{R}, D)}$ for the set of roles occurring in $\mathcal{T}, \mathcal{R}, D$ with their inverse and transitive closure. If it is clear from the context we use $\mathbf{R}$ instead of $\mathbf{R}_{(\mathcal{T}, \mathcal{R}, D)}$.

3 Deciding Satisfiability in $\mathcal{SHI}_+$

In this section, we establish decidability of $\mathcal{SHI}_+$ by devising an algorithm for checking satisfiability of $\mathcal{SHI}_+$ concepts w.r.t. a terminology and role hierarchy.

In our approach, we first extend the tableau definition presented in [8] by adding a *global* property for expressing the transitive closure of roles. This causes the tableaux to lose the local satisfiability property. Next, we define a sub-structure of graphs, called *neighborhood*, which consists of a node together with its neighbors. Such a neighborhood captures all semantic constraints imposed by the logic constructors of $\mathcal{SHI}$. A tree-like structure obtained by "tiling" neighborhoods together allows us to represent in some way a model for a concept in $\mathcal{SHI}_+$. In fact, we embed in this tree-like structure another structure, called *cyclic path*, to express the transitive closure of roles. Since all expansion rules for $\mathcal{SHI}$ can be translated into building neighborhoods, the algorithm presented in this paper focuses on defining cyclic paths over such a tree-like structure. By this way, the non-determinism resulting from satisfying the transitive closure of roles can be translated into the search in a space of all possible tree-like structures obtained from tiling neighborhoods.

3.1 $\mathcal{SHI}_+$-Tableau

Tableau structure is introduced to describe a model of a concept w.r.t. a terminology and role hierarchy. Properties in the tableau definition express semantic constraints resulting directly from the logic constructors in $\mathcal{SHI}_+$.

Considering the tableau definition for $\mathcal{SHIQ}$ presented in [8], the following definition adopts an additional property, namely P9, that imposes a global constraint on a set of individuals of a tableau.

Definition 2. *Let $(\mathcal{T}, \mathcal{R})$ be an $\mathcal{SHI}_+$ ontology. A tableau T for a concept D w.r.t $(\mathcal{T}, \mathcal{R})$ is defined to be a triplet $(\mathbf{S}, \mathcal{L}, \mathcal{E})$ such that $\mathbf{S}$ is a set of individuals,*

$\mathcal{L}$: $\mathbf{S} \rightarrow 2^{\mathsf{sub}(\mathcal{T},\mathcal{R},D)}$ and $\mathcal{E}$: $\mathbf{R} \rightarrow 2^{\mathbf{S}\times\mathbf{S}}$, and there is some individual $s \in \mathbf{S}$ such that $D \in \mathcal{L}(s)$. For all $s \in \mathbf{S}$, $C, C_1, C_2 \in \mathsf{sub}(\mathcal{T},\mathcal{R},D)$, and $R, S, P^+ \in \mathbf{R}$, T satisfies the following properties:

- P1 *if* $C_1 \sqsubseteq C_2 \in \mathcal{T}$ *and* $s \in \mathbf{S}$ *then* $\mathsf{nnf}(\neg C_1 \sqcup C_2) \in \mathcal{L}(s)$
- P2 *if* $C \in \mathcal{L}(s)$, *then* $\neg C \notin \mathcal{L}(s)$
- P3 *if* $C_1 \sqcap C_2 \in \mathcal{L}(s)$, *then* $C_1 \in \mathcal{L}(s)$ *and* $C_2 \in \mathcal{L}(s)$
- P4 *if* $C_1 \sqcup C_2 \in \mathcal{L}(s)$, *then* $C_1 \in \mathcal{L}(s)$ *or* $C_2 \in \mathcal{L}(s)$
- P5 *if* $\forall S.C \in \mathcal{L}(s)$ *and* $\langle s,t\rangle \in \mathcal{E}(S)$, *then* $C \in \mathcal{L}(t)$
- P6 *if* $\exists S.C \in \mathcal{L}(s)$, *there is* $t \in \mathbf{S}$ *such that* $\langle s,t\rangle \in \mathcal{E}(S)$ *and* $C \in \mathcal{L}(t)$
- P7 *if* $\forall S.C \in \mathcal{L}(s)$ *and* $\langle s,t\rangle \in \mathcal{E}(R)$ *for* $R \underline{\boxast} S$ *and* $\mathsf{Trans}(R)$, *then* $\forall R.C \in \mathcal{L}(t)$
- P8 $\langle s,t\rangle \in \mathcal{E}(R)$ *iff* $\langle t,s\rangle \in \mathcal{E}(\mathsf{Inv}(R))$
- P9 *if* $\langle s,t\rangle \in \mathcal{E}(P^+)$ *then either* $\langle s,t\rangle \in \mathcal{E}(P)$, *or* *there exist* $t_1, \cdots, t_n \in \mathbf{S}$ *such that* $\langle s,t_1\rangle, \cdots, \langle t_n,t\rangle \in \mathcal{E}(P)$
- P10 *if* $\langle s,t\rangle \in \mathcal{E}(R)$, $R \underline{\boxast} S$ *then* $\langle s,t\rangle \in \mathcal{E}(S)$

Note that when only *transitive roles* are added without transitive closure, the local satisfiability property of tableaux can be preserved by the property P7 which was already introduced in [8]. The property P9 in Def. 2 expresses explicitly a cycle for each transitive closure occurring in the label of an edge $\langle s,t\rangle$. A tableau for a concept represents exactly a model for the concept, that is affirmed by the following lemma.

Lemma 1. *An $\mathcal{SHI}_+$-concept D is satisfiable w.r.t. $(\mathcal{T},\mathcal{R})$ iff D has a tableau.*

Lemma 1 is obvious since the properties of tableaux represent directly semantic constraints imposed by the logic constructors in $\mathcal{SHI}_+$.

3.2 Neighborhood

Tableau-based algorithms, as presented in [8], use expansion rules to build a completion tree. Applying expansion rules makes all nodes of a completion tree satisfy semantic constraints imposed by concept definitions in the label associated with each node. This means that *local* satisfiability in such completion trees is sufficient to ensure *global* satisfiability. The notion of *neighborhood* introduced in Def. 3 expresses exactly the expansion rules for $\mathcal{SHI}$, consequently, guarantees local satisfiability. Therefore, a completion tree built by a tableau-based algorithm can be considered as set of neighborhoods which are tiled together. In other terms, building a completion tree by applying expansion rules is equivalent to the search of a tiling of neighborhoods.

Definition 3 (Neighborhood). *Let D be an $\mathcal{SHI}_+$ concept w.r.t. $(\mathcal{T},\mathcal{R})$. Let $\mathbf{R}$ be the set of roles occurring in D and $\mathcal{T},\mathcal{R}$ together with their inverse and transitive closure. A* neighborhood, *denoted (v_B, N_B, l), for D w.r.t. $(\mathcal{T},\mathcal{R})$ is formed from a* core *node v_B, a set of* neighbor *nodes N_B, edges $\langle v_B, v\rangle$ with $v \in N_B$ and a labelling function l such that $l(u) \in 2^{\mathsf{sub}(\mathcal{T},\mathcal{R},D)}$ with $u \in \{v_B\}\cup N_B$ and $l(\langle v_B, v\rangle) \in 2^{\mathbf{R}}$ with $v \in N_B$.*

1. *A node v is* valid *w.r.t. D and $(\mathcal{T}, \mathcal{R})$ iff*
   (a) *tbox-rule: If $C \sqsubseteq D \in \mathcal{T}$ then $\mathsf{nnf}(\neg C \sqcup D) \in l(v)$, and*
   (b) *clash-rule: $\{A, \neg A\} \not\subseteq l(v)$ with any concept name A, and*
   (c) *$\sqcap$-rule: If $C_1 \sqcap C_2 \in l(v)$ then $\{C_1, C_2\} \subseteq l(v)$, and*
   (d) *$\sqcup$-rule: If $C_1 \sqcup C_2 \in l(v)$ then $\{C_1, C_2\} \cap l(v) \neq \emptyset$.*
2. *A neighborhood $B = (v_B, N_B, l)$ is* valid *iff all nodes $\{v_B\} \cup N_B$ are valid and the following conditions are satisfied:*
   (a) *$\exists$-rule: If $\exists R.C \in l(v_B)$ then there is a neighbor $v \in N_B$ such that $C \in l_B(v)$ and $R \in l(\langle v_B, v\rangle)$;*
   (b) *rbox-rule: For each $v \in N_B$, if $R \in l(\langle v_B, v\rangle)$ and $R \underline{\circledast} S$ then $S \in l(\langle v_B, v\rangle)$;*
   (c) *$\forall$-rule: For each $v \in N_B$, if $R \in l(\langle v_B, v\rangle)$ (resp. $R \in \mathsf{Inv}(l(\langle v_B, v\rangle)))$ and $\forall R.C \in l(v_B)$ (resp. $\forall R.C \in l(v)$) then $C \in l(v)$ (resp. $C \in l(v_B)$);*
   (d) *$\forall^+$-rule: For each $v \in N_B$, if $R \in l(\langle v_B, v\rangle)$ (resp. $R \in \mathsf{Inv}(l_B(\langle v_B, v\rangle)))$, $S \underline{\circledast} R$, $\mathsf{Trans}(S)$, $\forall R.D \in l(v_B)$ (resp. $\forall \mathsf{Inv}(R).D \in l(v_B)$) then $\forall S.D \in l_B(v)$ (resp. $\forall \mathsf{Inv}(S).D \in l(v_B)$);*
   (e) *There are at most two nodes $v, v' \in N_B$ such that $l(v) = l(v')$ and $l(\langle v_B, v\rangle) = l(\langle v_B, v'\rangle)$.*

We denote $\mathbf{B}_{(\mathcal{T},\mathcal{R},D)}$ for a set of all valid neighborhoods for D w.r.t. $(\mathcal{T}, \mathcal{R})$.

The condition 2e in Def. 3 ensures that any neighborhood has a finite number of neighbors. Moreover, we could replace "at most two nodes" by "at most one node" since $\mathcal{SHI}_+$ does not allow for qualifying number restrictions but the former can simplify some constructions later on (Def. 5 and Rem. 1).

A valid neighborhood as presented in Def. 3 captures all necessary information related to an individual of a tableau. When a global property like P9 is added to tableaux, it imposes just a relationship on a set of individuals of the tableau but each individual is still characterized by a neighborhood which is built from its neighbors and itself. For this reason, neighborhoods can be still used to tile a completion tree for $\mathcal{SHI}_+$ without taking care of expansion rules for $\mathcal{SHI}$. In other terms, the neighbourhood notion expresses the local satisfiability in a sufficient way for being used in a global context.

Lemma 2. *Let D be an $\mathcal{SHI}_+$ concept w.r.t. $(\mathcal{T}, \mathcal{R})$. Let $(v_B, N_B, l), (v_{B'}, N_{B'}, l)$ be two valid neighborhoods with $l(v_B) = l(v_{B'})$. If there is $v \in N_B$ such that there does not exist any $v' \in N_{B'}$ satisfying $l(v') = l(v)$ and $l(\langle v_B, v\rangle) = l(\langle v_{B'}, v'\rangle)$ then the neighborhood $(v_{B'}, N_{B'} \cup \{u\}, l)$ is valid where $l(u) = l(v)$ and $l(\langle v_B, u\rangle) = l(\langle v_B, v\rangle)$.*

This lemma holds due to the facts that (i) a valid neighbor in a valid neighborhood B is also a valid neighbor in another valid neighborhood B' if the labels of two core nodes of B and B' are identical, (ii) since $\mathcal{SHI}_+$ does not allow for qualifying number restrictions hence Def. 3 has no restriction on the number of neighbors of a core node.

3.3 Completion Tree with Cyclic Paths

As discussed in [3], the blocking technique fails in treating DLs with the transitive closure of roles. It works correctly only if satisfiability of a node in completion tree can be decided from its neighbors and itself i.e. *local* satisfiability must be sufficient for such completion graphs. However, the presence of the transitive closure of roles makes satisfiability of a node depend on further nodes which can be arbitrarily far from it.

More precisely, satisfying the transitive closure P^+ in an edge $\langle x, y\rangle$ (i.e. $P^+ \in L(\langle x, y\rangle)$) is related to a set of nodes on a path rather than a node with its neighbors i.e. it imposes a semantic constraint on a set of nodes $x, x_1, \cdots, x_n, y$ such that they are connected together by P-edges. In general, satisfying the transitive closure is quite non-deterministic since the semantic constraint can lead to apply to an *arbitrary* number of nodes. In addition, the presence of the transitive closure of roles in a role hierarchy makes this difficulty worse. For instance, if $P \sqsubseteq Q^+, Q \sqsubseteq S^+$ are axioms in a role hierarchy then each Q-edge generated for satisfying Q^+ may lead to generate an arbitrary number of S-edges for satisfying S^+.

The most common way for dealing with a new logic constructor is to add a new expansion rule for satisfying the semantic constraint imposed by the new constructor. Such an expansion rule for the transitive closure of roles must:

1. find or create a set of P-edges forming a path for each occurrence of P^+ in the label of edges,
2. deal with non-deterministic behaviours of the expansion rule resulting from the semantics of the transitive closure of roles,
3. enable to control the expansion of completion trees by a new blocking technique which has to take into account the fact that satisfying the transitive closure of a role may add an arbitrary number of new transitive closures to be satisfied.

To avoid these difficulties, our approach does not aim to directly extend the construction of completion trees by using a new expansion rule, but to translate this construction into selecting a "good" normalization tree, namely *completion tree with cyclic paths*, from a finite set of completion trees without taking into account the semantic constraint imposed by the transitive closure of roles. The process of selecting a "good" normalization tree is guided by finding in this tree a *cyclic path* for each occurrence of the transitive closure of a role.

Summing up, a completion tree with cyclic paths will be built in two stages. The first one consists of tiling valid neighborhoods together such that two neighborhoods are tiled if they have *compatible* neighbors. This stage yields a *normalization tree* as described in Def. 4. The second stage given in Def. 5 deals with the transitive closure of roles by defining cyclic paths over normalization trees.

Definition 4 (Normalization tree). *Let D be an $\mathcal{SHI}_+$ concept w.r.t. $(\mathcal{T}, \mathcal{R})$. Let $\mathbf{B}_{(\mathcal{T},\mathcal{R},D)}$ be the set of all valid neighborhoods for D w.r.t. $(\mathcal{T}, \mathcal{R})$.* A normalization tree $\mathbf{T} = (V, E, L)$ *for D w.r.t. $(\mathcal{T}, \mathcal{R})$ is built from $\mathbf{B}_{(\mathcal{T},\mathcal{R},D)}$ as follows:*

- *A root node x_0 of* $\mathbf{T}$ *is built from a valid neighborhood* $(v_0, N_0, l) \in \mathbf{B}_{(\mathcal{T},\mathcal{R},D)}$ *such that* $L(x_0) = l(v_0)$ *with* $D \in l(v_0)$. *Additionally, for each* $v \in N_0$ *a successor* x *of* x_0 *is added to* V *with* $L(x) = l(v)$ *and* $L(\langle x_0, x\rangle) = l(\langle v_0, v\rangle)$,
- *For each node* $x \in V$ *with its predecessor* x',
  - *If there are ancestors* y, y' *of* x' *such that* y' *is the predecessor of* y *and* $L(y) = L(x), L(y') = L(x'), L(\langle y', y\rangle) = L(\langle x', x\rangle)$, *then* x *is* blocked *by* y. *In this case,* x *is a leaf node;*
  - *Otherwise, we find a neighborhood* $B_x = (v_{B_x}, N_{B_x}, l)$ *from* $\mathbf{B}_{(\mathcal{T},\mathcal{R},D)}$ *such that* $l(v_{B_x}) = L(x)$, $l(v) = L(x')$, $\mathsf{Inv}(\ l(\langle v_{B_x}, v\rangle)\ ) = L(\langle x', x\rangle)$ *for some* $v \in N_{B_x}$, *and build a successor* y *for each* $u \in N_{B_x} \setminus \{v\}$ *such that* $L(y) = l(u)$ *and* $L(\langle x, y\rangle) = l(\langle v_{B_x}, u\rangle)$.

We say a node x *is a* R*-successor of* $x' \in V$ *if* $R \in L(\langle x', x\rangle)$. *A node* x *is called a* R*-neighbor of* x' *if* x *is a* R*-successor of* x' *or* x' *is a* $\mathsf{Inv}(R)$*-successor of* x.

Note that the construction of normalization trees uses the blocking technique for termination condition. The following definition embeds cyclic paths into normalization trees for satisfying the semantic constraint imposed by the transitive closure of roles.

Definition 5 (Completion tree with cyclic paths). *Let* D *be an* $\mathcal{SHI}_+$ *concept w.r.t.* $(\mathcal{T}, \mathcal{R})$. *Let* $\mathbf{T} = (V, E, L)$ *be a normalization tree for* D *w.r.t.* $(\mathcal{T}, \mathcal{R})$.

- *A path, denoted* $\varphi = \langle x_0, \cdots, x_k, \cdots, x_{n+1}\rangle$, *is formed from nodes* $x_i \in V$ *if*
  - x_i *is not blocked for all* $i \in \{0, \cdots, n+1\}$;
  - x_{i+1} *is a successor of* x_i *or blocks a successor of* x_i *for all* $k \leq i \leq n$;
  - x_{i-1} *is a successor of* x_i *or blocks a successor of* x_i *for all* $1 \leq i \leq k$.
- *A path* $\langle x_0, \cdots, x_{n+1}\rangle$ *is called* non-duplicated *if there do not exist* $i, j \in \{0, \cdots, n\}$ *with* $j > i$ *such that* $L'(\langle x_i, x_{i+1}\rangle) = L'(\langle x_j, x_{j+1}\rangle)$ *and* $L(x_i) = L(x_j), L(x_{i+1}) = L(x_{j+1})$;
- *A path* $\langle x_0, \cdots, x_k, \cdots, x_{n+1}\rangle$ *is* cyclic *if* $L(\langle x_1, x_0\rangle) = \mathsf{Inv}(\ L(\langle x_n, x_{n+1}\rangle)\ )$ *and* $L(x_0) = L(x_n), L(x_1) = L(x_{n+1})$.

$\mathbf{T} = (V, E, L)$ *is called a* completion tree with cyclic paths *if for each* $\langle u, v\rangle \in E$ *such that* $Q^+ \in L(\langle u, v\rangle)$ *and* $Q \notin L(\langle u, v\rangle)$ *with* $Q \in \mathbf{R} \cup \{\mathsf{Inv}(P) \mid P \in \mathbf{R}\}$ *there exists a cyclic non-duplicated path* $\varphi = \langle x_0, \cdots, x_k, \cdots, x_{n+1}\rangle$ *which satisfies:*

- $u = x_k$, *and* $x_{k-1} = v$ *if* v *is not blocked or* $x_{k-1} = z$ *if* z *blocks* v;
- $\mathsf{Inv}(Q) \in L'(\langle x_{k-i}, x_{k-i-1}\rangle)$ *for all* $1 \leq i \leq k-1$;
- $Q \in L'(\langle x_i, x_{i+1}\rangle)$ *for all* $k \leq i \leq n$.

where $L'(\langle x, y\rangle) = L(\langle x, y\rangle)$ *if* y *is a successor of* x, $L'(\langle x, y\rangle) = \mathsf{Inv}(\ L(\langle y, x\rangle)\ )$ *if* x *is a successor of* y, *and* $L'(\langle x, y\rangle) = L(\langle x, z\rangle)$ *if* y *blocks a successor* z *of* x.
In this case, φ *is called a cyclic* Q*-path and denoted by* $\varphi_{Q\langle u,v\rangle}$.

By its name, we mean that each cyclic path $\langle x_0, \cdots, x_{n+1}\rangle$ becomes a cycle if x_1 and x_0 are respectively pasted to x_{n+1} and x_n.

Remark 1. From Def. 5, each node x_i of a cyclic non-duplicated path $\varphi_{Q\langle u,v\rangle} = \langle x_0, \cdots, x_k, \cdots, x_{n+1}\rangle$ with $x_k = u$ can be reached through a path which only goes *down* either from u to a blocked node, or from a blocking to blocked node. This ensures that for any occurrence of $\langle u, v\rangle$ with $Q^+ \in L(\langle u, v\rangle)$ in a possibly infinite path crossing blocked, blocking nodes, we can define a cyclic Q-path in a tableau for satisfying this occurrence. This property guarantees the soundness of Alg. 1. For the completeness, when building a cyclic Q-path as down-going one from a Q-path in a tableau, we need to add a successor which is identical to the predecessor of a node. This is allowed by Def. 3 which accepts a neighborhood with two identical neighbors.

A completion tree with cyclic paths encapsulates the following notions: neighborhood, blocking condition and cyclic path. The first one captures the semantics of all logic constructors except for the transitive closure of roles. The second one which was introduced in [8] is crucial for obtaining a finite representation of a possibly infinite model. The third one represents the transitive closure of roles.

Lemma 3 (Soundness and completeness). *Let D be an $\mathcal{SHI}_+$-concept. Let $\mathcal{T}$ and $\mathcal{R}$ be a terminology and role hierarchy.*

1. *If there exists a completion tree with cyclic paths for D w.r.t. $(\mathcal{T}, \mathcal{R})$ then D has a tableau.*
2. *If D has a tableau w.r.t. $(\mathcal{T}, \mathcal{R})$ then there exists a completion tree with cyclic paths.*

To prove the soundness (1.), we define a "quasi-tableau" for D from a set of Paths in a given completion tree with cyclic paths. This technique is already used in [8]. Next, in order to introduce cycles to the quasi-tableau from cyclic paths of the completion tree, we use a function which performs a contraction of the set of paths by unifying the ending points of cyclic paths Paths. Since qualifying number restrictions are not allowed in $\mathcal{SHI}_+$, such a contraction function preserves all properties of the quasi-tableau. For the completeness (2.), we define directly neighborhoods from individuals of a given tableau and build a normalization tree from them. Next, cyclic paths are embedded into the normalization tree by devising non-duplicated paths from finite cycles for the transitive closure of roles in the tableau. A more complete proof of Lem. 3 can be found in Appendix.

From the construction of completion trees with cyclic paths according to Def. 5 and Lem. 3 we can devise immediately Alg. 1 for concept satisfiability in $\mathcal{SHI}_+$.

Lemma 4 (Termination). *Algorithm 1 terminates.*

Termination of Alg. 1 is a consequence of the following facts: (i) the number of neighborhoods is bounded, (ii) the size of normalization trees which are tiled from neighbourhoods is bounded.

Alg. 1 is highly complex since it is not a goal-directed decision procedure at all. Such an exhaustive behaviour is very different from that of tableaux-based algorithms in which the construction of a completion tree is inherited from step to step. In Alg. 1, when a normalization tree cannot satisfy an occurrence of the

Input : Concept D, terminology $\mathcal{T}$ and role hierarchy $\mathcal{R}$
Output: IsSatisfiable(D)

1 **foreach** *Normalization tree* $\mathbf{T} = (V, E, L)$ **do**
2   **if** *For each* $\langle x,y \rangle \in E$ *with* $Q^+ \in L(\langle x,y \rangle), Q \notin L(\langle x,y \rangle)$, $\mathbf{T}$ *has a* $\varphi_{Q\langle x,y \rangle}$ **then**
3   `return true;`
4 `return false;`

Algorithm 1. Deciding concept satisfiability in $\mathcal{SHI}_+$

transitive closure of a role (after satisfying others), a new normalization tree will be picked and embedding cyclic paths into this one has to restart. The following theorem is a direct consequence of Lem. 3 and 4.

Theorem 1. *Algorithm 1 decides the satisfiability of $\mathcal{SHI}_+$-concepts w.r.t. a terminology and role hierarchy.*

4 Conclusion

We have proved decidability of the concept satisfiability in $\mathcal{SHI}_+$ by providing a sound and complete algorithm. The establishment of the algorithm relies on the neighborhood notion which is an abstraction of the local satisfiability property of tableaux. This abstraction enables us to encapsulate all semantic constraints imposed by the logic constructors in $\mathcal{SHI}$, and thus to deal with the transitive closure of roles independently from the other constructors.

This work is a first step toward an empirical algorithm whose behaviour is more goal-directed i.e. the construction of a completion tree would be refined along with satisfying the transitive closure of roles, e.g., the non-impacted parts of the tree when rebuilding it would be reused.

Acknowledgements. This work has been partially supported by the European integrated project NeOn (IST-2005-027595). Thanks to Sophie Coudert and Ludovic Apvrille for the motivating example, to Jérôme Euzenat, Marie-Christine Rousset and the anonymous reviewers for their comments.

References

1. Aho, A.V., Ullman, J.D.: Universality of data retrieval languages. In: Proceedings of the 6th of ACM on Principles of Programming Language (1979)
2. Baader, F.: Augmenting concept languages by transitive closure of roles: An alternative to terminological cycles. In: Proceedings of the Twelfth International Joint Conference on Artificial Intelligence (1991)
3. Baader, F., Sattler, U.: Tableau algorithms for description logics. In: Dyckhoff, R. (ed.) TABLEAUX 2000. LNCS(LNAI), vol. 1847, p. 118. Springer, Heidelberg (2000)

4. de Giacomo, G., Lenzerini, M.: Boosting the correspondence between description logics and propositional dynamic logics. In: Proceedings of the 12th National conference on Artificial Intelligence, pp. 205–212 (1994)
5. de Giacomo, G., Lenzerini, M.: What's in an aggregate: Foundations for description logics with tuples and sets. In: Proceedings of the Fourteenth International Joint Conference On Intelligence Artificial 1995 (IJCAI 1995) (1995)
6. de Giacomo, G., Massacci, F.: Combining deduction and model checking into tableaux and algorithms for converse-PDL. Information and Computation (1) (1998)
7. Horrocks, I., Sattler, U.: A tableau decision procedure for $\mathcal{SHOIQ}$. Journal Of Automated Reasoning 39(3), 249–276 (2007)
8. Horrocks, I., Sattler, U., Tobies, S.: Practical reasoning for expressive description logics. In: Ganzinger, H., McAllester, D., Voronkov, A. (eds.) LPAR 1999. LNCS, vol. 1705. Springer, Heidelberg (1999)
9. ISO. Standard ISO 8807: LOTOS, a formal description technique based on temporal ordering of observational behaviour (1988)
10. Patel-Schneider, P., Hayes, P., Horrocks, I.: OWL web ontology language semantics and abstract syntax. W3C Recommendation (2004)
11. Sattler, U.: A concept language extended with different kinds of transitive roles. In: Proceedings of the 20th German Annual Conf. on Artificial Intelligence (KI 2001), vol. 1137, pp. 199–204. Springer, Heidelberg (2001)
12. Tobies, S.: The complexity of reasoning with cardinality restrictions and nominals in expressive description logics. Journal of Artificial Intelligence Research 12, 199–217 (2000)

Appendix

Proof of soundness (Lem. 3). Let $\mathbf{T} = (V, E, L)$ be a completion tree with cyclic paths for D w.r.t. $\mathcal{T}$ and $\mathcal{R}$. A path is of the form $p = [(x_0, x'_0), \cdots, (x_n, x'_n)]$ where $x_i, x'_i \in V$. For such a path we define $\mathsf{Tail}(p) = x_n$, $\mathsf{Tail}'(p) = x'_n$. We denote $[p|(x_{n+1}, x'_{n+1})]$ for the path $[(x_0, x'_0), \cdots, (x_n, x'_n), (x_{n+1}, x'_{n+1})]$. The set $\mathsf{Paths}(\mathbf{T})$ is defined as follows:

1. For the root v_0 we define $[(v_0, v_0)] \in \mathsf{Paths}(\mathbf{T})$, and
2. For a path $p \in \mathsf{Paths}(\mathbf{T})$ and a node $v' \in V$,
   (a) If $\langle \mathsf{Tail}(p), v' \rangle \in E$, v' is not a blocked node then $[p|(v', v')] \in \mathsf{Paths}(\mathbf{T})$,
   (b) If $\langle \mathsf{Tail}(p), v' \rangle \in E$ and v' is blocked by z then $[p|(z, v')] \in \mathsf{Paths}(\mathbf{T})$.

From the definition of Paths, for $p \in \mathsf{Paths}(\mathbf{T})$ and $p = [q|(v, v')]$ we have $\mathcal{L}(v) = \mathcal{L}(v')$, and $v \neq v'$ iff v' is blocked by v.

We define a quasi-tableau $T' = (\mathbf{S}', \mathcal{L}', \mathcal{E}')$ as follows:

$$\mathbf{S}' = \mathsf{Paths}(\mathbf{T})$$
$$\mathcal{L}'(p) = \mathcal{L}'(\mathsf{Tail}(p))$$
$$\mathcal{E}'(R) = \{\langle p, q\rangle \in \mathsf{Paths}^2(\mathbf{T}) \mid$$
$$1.\ q = [p|(v, v')],\ R \in L(\langle \mathsf{Tail}(p), v'\rangle), \text{ or}$$
$$2.\ p = [q|(v, v')] \text{ and } \mathsf{Inv}(R) \in L(\langle \mathsf{Tail}(q), v'\rangle)\ \}$$

From $T' = (\mathbf{S}', \mathcal{L}', \mathcal{E}')$, we denote $\mathcal{L}'(\langle s,t\rangle) = \{R \mid \langle s,t\rangle \in \mathcal{E}(R)\}$. We now build a tableau $T = (\mathbf{S}, \mathcal{L}, \mathcal{E})$ from this quasi-tableau $T' = (\mathbf{S}', \mathcal{L}', \mathcal{E}')$ with help of function $\pi : \mathbf{S}' \rightarrow \mathbf{S}$ that is inductively defined on the length of paths from the root v_0. For $p_0, \cdots, p_{n+1} \in \mathbf{S}'$ such that there is a cyclic path $\langle x_0, \cdots, x_{n+1}\rangle$ with $\mathsf{Tail}(p_i) = x_i$, the function π merges p_{n+1} with p_1, and p_n with p_0.

1. $[(v_0, v_0)] \in \mathbf{S}'$ and $\pi([(v_0, v_0)]) = [(v_0, v_0)]$ where v_0 is the root of $\mathbf{T}$. Assume that $\pi(p)$ are define for all $p \in \mathbf{S}'$ such that $|p| \leq K$ with $|[(v_0, v_0)]| = 0$ and $|p| = |q| + 1$, $p = [q|(v, v')]$.
2. For each $p \in \mathbf{S}'$ with $\pi(p) = q$, $|p| = K$ and for each $p' = [p|(v, v')] \in \mathbf{S}'$,
   (a) If there is a cyclic Q-path $\langle x_0, \cdots, x_k, \cdots, x_{n+1}\rangle$ where $p_0, \cdots, p_k, \cdots, p_{n+1} \in \mathbf{S}'$, $\mathsf{Tail}(p_k) = x_k$ and $p_{i-1} = [p_i|(x_i, x_i')]$ for all $1 \leq i \leq k$, $p_{j+1} = [p_j|(x_j, x_j')]$ for all $k \leq j \leq n$, such that $p = p_1, p' = p_0$ or $p = p_n, p' = p_{n+1}$ then
      – If $p = p_1, p' = p_0$ we define $\pi(p_{n+1}) = \pi(p_1) = q$, $\pi(p') = \pi(p_n)$ and $\mathcal{L}(\langle q, \pi(p')\rangle) = \mathcal{L}'(\langle p, p_0\rangle)$, $\mathcal{L}(\langle \pi(p_n), \pi(p_{n+1})\rangle) = \mathcal{L}'(\langle p_n, p_{n+1}\rangle)$,
      – If $p = p_n, p' = p_{n+1}$ we define $\pi(p_0) = \pi(p_n) = q$, $\pi(p') = \pi(p_1)$ and $\mathcal{L}(\langle q, \pi(p_{n+1})\rangle) = \mathcal{L}'(\langle p, p_{n+1}\rangle)$, $\mathcal{L}(\langle \pi(p_1), \pi(p_0)\rangle) = \mathcal{L}'(\langle p_1, p_0\rangle)$.
   (b) Otherwise, we define $q' \in \mathbf{S}$ such that $\pi(p') = q'$, $\mathcal{L}(q') = \mathcal{L}'(p')$, $\mathcal{L}(\langle q, q'\rangle) = \mathcal{L}'(\langle p, p'\rangle)$.

We can define $\mathcal{E}$ from $\mathcal{L}$ as follows: $\mathcal{E}(R) = \{\langle s,t\rangle \in \mathbf{S} \times \mathbf{S} \mid R \in \mathcal{L}(\langle s,t\rangle)\}$ for all $R \in \mathbf{R}$.

The function π is well defined since (i) $\mathcal{L}'(p_0) = \mathcal{L}'(p_n) = \mathcal{L}(\pi(p_0)) = \mathcal{L}(\pi(p_n))$, $\mathcal{L}'(p_1) = \mathcal{L}'(p_{n+1}) = \mathcal{L}(\pi(p_1)) = \mathcal{L}(\pi(p_{n+1}))$ by the definition of cyclic Q-paths, and (ii) $\mathcal{L}(\langle \pi(p_1), \pi(p_0)\rangle) = \mathcal{L}'(\langle p_1, p_0\rangle)$ with $p_0 = [p_1|(x, x')]$ and $\mathcal{L}(\langle \pi(p_n), \pi(p_{n+1})\rangle) = \mathcal{L}'(\langle p_n, p_{n+1}\rangle)$ with $p_{n+1} = [p_n|(y, y')]$ i.e. the construction of π never changes the label of nodes and edges defined previously. However, it may merge some nodes.

Claim. 4. For all $p, q \in \mathbf{S}'$ with $q = [p|(x, x')]$ it holds that $\mathcal{L}(\pi(p)) = \mathcal{L}'(p), \mathcal{L}(\pi(q)) = \mathcal{L}'(q)$ and $\mathcal{L}(\langle \pi(p), \pi(q)\rangle) = \mathcal{L}'(\langle p, q\rangle)$.

Claim 4. asserts that π preserves all label of nodes and edges.

Proof of Claim 4. We prove the claim by induction on the length of $p \in \mathbf{S}'$. If $|p| = 0$ then Claim 4. is verified. Assume that Claim 4. is verified for all p with $|p| \leq m$. Let $q = [p|(x, x')]$ with $\langle \mathsf{Tail}(p), x\rangle \in E$. By the induction hypothesis, we have $\mathcal{L}(\pi(p)) = \mathcal{L}'(p)$.

According to the definition of π, we consider the two following cases:

– There does not exist any cyclic Q-path $\varphi = \langle x_0, \cdots, x_k, \cdots, x_{n+1}\rangle$ such that $p = p_1, q = p_0$ with $\mathsf{Tail}(p_1) = x_1, \mathsf{Tail}(q) = x_0$ or $p = p_n, q = p_{n+1}$ with $\mathsf{Tail}(p_n) = x_n, \mathsf{Tail}(p_{n+1}) = x_{n+1}$. From the definition of π, we have $\mathcal{L}(\pi(q)) = \mathcal{L}'(q)$ and $\mathcal{L}(\langle \pi(p), \pi(q)\rangle) = \mathcal{L}'(\langle p, q\rangle)$.
– Otherwise, assume that $p = p_1, q = p_0$ with $\mathsf{Tail}(p_1) = x_1, \mathsf{Tail}(q) = x_0$ (if $p = p_n, q = p_{n+1}$ with $\mathsf{Tail}(p_n) = x_n, \mathsf{Tail}(p_{n+1}) = x_{n+1}$ it will be similarly treated). By the definition of π, we have $\pi(q) = \pi(p_n)$, $\mathcal{L}(\langle \pi(p), \pi(q)\rangle) = \mathcal{L}'(\langle p, q\rangle)$, and thus, $\mathcal{L}(\pi(q)) = \mathcal{L}'(\pi(p_n))$ and $\mathcal{L}(\langle \pi(p), \pi(q)\rangle) = \mathcal{L}'(\langle p, q\rangle)$. ■

Claim. 5

1. For all cyclic Q-path $\varphi = \langle x_0, \cdots, x_k, \cdots, x_{n+1}\rangle$, *there are* $p_i \in \mathbf{S}'$ *with* $\mathsf{Tail}(p_i) = x_i$ *for all* $i \in \{0, \cdots, n+1\}$ *such that* $\mathcal{L}(\langle \pi(p_i), \pi(p_{i+1})\rangle) = \mathcal{L}'(\langle p_i, p_{i+1}\rangle)$ *for all* $i \in \{0, \cdots, n\}$, *and* $\pi(p_0) = \pi(p_n), \pi(p_{n+1}) = \pi(p_1)$.

2. For all $\pi(p), \pi(q)$ *with* $p, q \in \mathbf{S}'$ *such that* $Q^+ \in \mathcal{L}(\langle \pi(p), \pi(q)\rangle)$ *and* $Q \notin \mathcal{L}(\langle \pi(p), \pi(q)\rangle)$, *there are* $p_0, \cdots, p_k, \cdots, p_{n+1} \in \mathbf{S}'$ *with* $\pi(p_0) = \pi(p_n), \pi(p_{n+1}) = \pi(p_1)$ *such that* $p_k = p, p_{k-1} = q$ *and* $\mathsf{Inv}(Q) \in \mathcal{L}(\langle \pi(p_i), \pi(p_{i-1})\rangle)$ *for all* $i \in \{k-1, \cdots, 1\}$, $Q \in \mathcal{L}(\langle \pi(p_i), \pi(p_{i+1})\rangle)$ *for all* $j \in \{k, \cdots, n\}$.

Item (1.) asserts that π transforms all cyclic paths in $\mathbf{S}'$ into cycles in $\pi(\mathbf{S}')$. Item (2.) asserts that each occurrence Q^+ in the label of an edge has a cycle formed of Q-edges.
Proof of Claim 5.

1. This is directly implied from the definition of π and Claim 4.
2. Let $\pi(p), \pi(q) \in \pi(\mathbf{S})$ such that $Q^+ \in \mathcal{L}(\langle \pi(p), \pi(q)\rangle)$ and $Q \notin \mathcal{L}(\langle \pi(p), \pi(q)\rangle)$. From Claim 4. we have $Q^+ \in \mathcal{L}(\langle p, q\rangle)$ and $Q \notin \mathcal{L}(\langle p, q\rangle)$, and thus, $Q^+ \in L(\langle \mathsf{Tail}(p), \mathsf{Tail}(q)\rangle)$ and $Q \notin L(\langle \mathsf{Tail}(p), \mathsf{Tail}(q)\rangle)$.
   By the definition of cyclic Q-paths, there is a cyclic Q-path $\langle x_0, \cdots, x_k, \cdots, x_{n+1}\rangle$ such that $\mathsf{Tail}(p) = x_k, \mathsf{Tail}(q) = x_{k-1}$. In addition, by the definition of cyclic paths in Def. 5 we have x_{k-1}, x_{k+1} are successors of x_k or blocking nodes of a successor of of x_k. Thus, from the definition of the quasi-tableau we have $p_0, \cdots, p_{n+1} \in \mathbf{S}'$ with $p = p_k, q = p_{k-1}$, $\mathsf{Tail}(p_i) = x_i$ for all $i \in \{0, \cdots, n+1\}$ and $\mathsf{Inv}(Q) \in \mathcal{L}'(\langle p_i, p_{i-1}\rangle)$ for all $i \in \{1, \cdots, k-1\}$, and $Q \in \mathcal{L}'(\langle p_j, p_{j+1}\rangle)$ for all $j \in \{k+1, \cdots, n\}$.
   From the item .1 it holds that $\pi(x_0) = \pi(x_n), \pi(x_{n+1}) = \pi(x_1)$. Moreover, by Claim 4. we obtain $\mathsf{Inv}(Q) \in \mathcal{L}(\langle \pi(p_i), \pi(p_{i-1})\rangle)$ for all $i \in \{1, \cdots, k-1\}$, $Q \in \mathcal{L}(\langle \pi(p_i), \pi(p_{i+1})\rangle)$ for all $i \in \{k, \cdots, n\}$. ■

To show that T is a tableau, we have to prove that T satisfies all the properties from Def. 2. The property P9 is a direct consequence of the claims. For instance, we check the property P6.

Assume $\exists R.C \in \mathcal{L}(\pi(p))$ and $v = \mathsf{Tail}(p)$. We show that there exists $\pi(q) \in \mathbf{S}$ such that $C \in \mathcal{L}(\pi(q))$ and $\langle \pi(p), \pi(q)\rangle \in \mathcal{E}(R)$. According to Claim 4. it suffices to show that there exists $q \in \mathbf{S}'$ such that $C \in \mathcal{L}'(q)$ and $\langle p, q\rangle \in \mathcal{E}(R)$. By the construction of neighborhoods and Def. 3 there exists some neighbor $v' \in N$ in the corresponding neighborhood (v, N, l) such that $C \in l(v')$ and $R \in l(\langle v, v'\rangle)$. The definition of $\mathcal{E}'$ implies $R \in L(\langle v, v'\rangle) = l(\langle v, v'\rangle)$ or $\mathsf{Inv}(R) \in L(\langle v', v\rangle) = l(\langle v, v'\rangle)$ i.e. v' can be a R-successor of v or v is a R-successor of v' in the completion tree. We consider the following cases.

(i) Assume that $R \in L(\langle v, v'\rangle)$. If v' is not blocked, this is trivial since there is $q = [p|(v', v')]$, $\langle p, q\rangle \in \mathcal{E}(R)$ and $C \in \mathcal{L}(q)$. Assume that $q = [p|(z, v')] \in \mathbf{S}$ and v' is blocked by z. Since v' is a R-successor of v, by the definition of $\mathcal{E}$, we have $\langle p, q\rangle \in \mathcal{E}(R)$. Furthermore, we have $C \in \mathcal{L}(q) = L(z) = L(v')$.

(ii) Assume $\mathsf{Inv}(R) \in L(\langle v', v\rangle)$. If $p = [q|(v, v)] \in \mathbf{S}$ i.e. v is not blocked, this is trivial since $\langle q, p\rangle \in \mathcal{E}(R)$ and $C \in \mathcal{L}(q) = L(v')$. Note that v is never blocked since $v = \mathsf{Tail}(p)$. Therefore, assume that $p = [q|(v,x)] \in \mathbf{S}$ such that v blocks x. By the blocking condition we have $L(v') = L(\mathsf{Tail}(q))$ and $L(\langle \mathsf{Tail}(q), x\rangle) = L(\langle v', v\rangle)$. This implies that $\mathsf{Inv}(R) \in L(\langle \mathsf{Tail}(q), x\rangle)$. By the definition of $\mathcal{E}$, we have $\langle q, p\rangle \in \mathcal{E}(R)$ and $C \in \mathcal{L}(q) = L(v')$.

Proof of completeness(Lem. 3). Assume that $\mathcal{SHI}_+$-concept D has a tableau $T = (\mathbf{S}, \mathcal{L}, \mathcal{E})$. A normalization tree $\mathbf{T} = (V, E, L)$ can be inductively built from T together with a function π from V to $\mathbf{S}$. This construction is quite intuitive since we can define a valid neighborhood from each individual $s \in \mathbf{S}$ as follows:

- We define $l(v) := \mathcal{L}(s)$. v is valid since any node whose label is included in the label of a node in the tableau T is always valid.
- Let $S'(s) \subseteq \mathbf{S}$ such that $s' \in S'(s)$ iff $\mathcal{L}(\langle s, s'\rangle) \neq \emptyset$ where $\mathcal{L}(\langle s, s'\rangle) := \{R \mid \langle s, s'\rangle \in \mathcal{E}(R) \text{ for some } R \in \mathbf{R}_{(\mathcal{T},\mathcal{R},D)}\}$.
  Let $S(s) \subseteq S'(s)$ such that for each $\mathscr{C} \in 2^{\mathsf{sub}(\mathcal{T},\mathcal{R},D)}$ and $\mathscr{R} \in 2^{\mathbf{R}}$ if there is a $t \in S'(s)$ with $\mathcal{L}(t) = \mathscr{C}$ and $\mathcal{L}(\langle s, t\rangle) = \mathscr{R}$ then there is a unique node $t' \in S(s)$ with $\mathcal{L}(t') = \mathcal{L}(t)$ and $\mathcal{L}(\langle s, t\rangle) = \mathcal{L}(\langle s, t'\rangle)$. This implies that $S(s)$ is finite.
- For each $t \in S(s)$ we define a node $u \in N_0$ such that $l(u) = \mathcal{L}(t)$ and $l(\langle v, u\rangle) = \mathcal{L}(\langle s, s'\rangle)$. From the construction, (v, N_0, l) is valid.

A normalization tree $\mathbf{T} = (V, E, L)$ will be obtained by tiling neighborhoods built from connected individuals started at $s_0 \in \mathbf{S}$ with $D \in \mathcal{L}(s_0)$ and a function π from V to $\mathbf{S}$. Note that if u, v are neighbors in $\mathbf{T}$ then $\pi(u), \pi(v)$ are also neighbors in T. The blocking condition ensures this construction terminates.

We now build cyclic paths for the transitive closure of roles. By the construction of $\mathbf{T}$, for each $x, y \in V$ such that $Q^+ \in L(\langle x, y\rangle)$, $Q \notin L(\langle x, y\rangle)$ with $Q \in \mathbf{R} \cup \{\mathsf{Inv}(P) \mid P \in \mathbf{R}\}$ we have $\langle \pi(x), \pi(y)\rangle \in \mathcal{E}(Q^+)$.

According to T9 there are $t_1, \cdots, t_n \in \mathbf{S}$ such that $\langle s, t_1\rangle, \cdots, \langle t_n, t\rangle \in \mathcal{E}(Q)$ and $\pi(x) = s, \pi(y) = t$. From this set of edges, we can pick $\langle s, t_1\rangle, \cdots, \langle t_k, t_{k+1}\rangle$ and $\langle t_l, t_{l+1}\rangle, \cdots, \langle t_n, t\rangle \in \mathcal{E}(Q)$ such that

1. $\mathcal{L}(t_k) = \mathcal{L}(t_l), \mathcal{L}(t_{k+1}) = \mathcal{L}(t_{l+1}), \mathcal{L}(\langle t_k, t_{k+1}\rangle) = \mathcal{L}(\langle t_l, t_{l+1}\rangle)$,
2. there are not $i, j \in \{0, \cdots, k, l+1, \cdots, n\}, i < j$ with $t_0 = s, t_{n+1} = t$ such that $\mathcal{L}(t_i) = \mathcal{L}(t_j), \mathcal{L}(t_{i+1}) = \mathcal{L}(t_{j+1}), \mathcal{L}(\langle t_i, t_{i+1}\rangle) = \mathcal{L}(\langle t_j, t_{j+1}\rangle)$.

We now build a cyclic non-duplicated Q-path from $\{s, t_1, \cdots, t_k, t_{k+1}, t_l, t_{l+1}, \cdots, t_n, t\}$. Since x is not blocked (x has a successor y), by the construction of $\mathbf{T}$ with $\pi(x) = s$, $\langle s, t_1\rangle \in \mathcal{E}(Q)$, $L(x) = \mathcal{L}(s)$, there exists a neighbor w of x such that $L(w) = \mathcal{L}(t_1)$ and $L'(\langle x, w\rangle) = \mathcal{L}(\langle s, t_1\rangle)$ where $L'(\langle x, w\rangle) = L(\langle x, w\rangle)$. If w is a not successor of x, by Lem.2, we can add a new successor w' of x (and the subtree whose root is w' can be built as above) such that $L(w) = \mathcal{L}(t_1)$ and $L(\langle x, w\rangle) = \mathcal{L}(\langle s, t_1\rangle)$. Thus, we can assume that w is a Q-successor of x since $\langle s, t_1\rangle \in \mathcal{E}(Q)$, and define $x_1 = w$ and $\pi(w) = t_1$.

Assume that there is $x_k \in V$ (x_k is not blocked by construction) with $\pi(x_k) = t_k$ such that $L(x_k) = \mathcal{L}(t_k), L(x_{k-1}) = \mathcal{L}(t_{k-1})$, $L(\langle x_{k-1}, x_k\rangle) = \mathcal{L}(\langle t_{k-1}, t_k\rangle)$. We consider the following cases:

1. x_k has a successor w' such that $L(w') = \mathcal{L}(t_{k+1})$ and $L'(\langle x_k, w'\rangle) = \mathcal{L}(\langle t_k, t_{k+1}\rangle)$. If w' is not blocked then define $x_{k+1} = w'$. If w' is blocked by z then define $x_{k+1} = z$. Since $\langle t_k, t_{k+1}\rangle \in \mathcal{E}(Q)$ hence w' is a Q-successor of x_k.
2. x_k has no successor w' such that $L(w') = \mathcal{L}(t_{k+1})$ and $L'(\langle x_k, w'\rangle) = \mathcal{L}(\langle t_k, t_{k+1}\rangle)$. Since $L(x_k) = \mathcal{L}(t_k)$, by Lem.2, we can add a successor w' of x_k such that $L(w') = \mathcal{L}(t_{k+1})$ and $L'(\langle x_k, w'\rangle) = \mathcal{L}(\langle t_k, t_{k+1}\rangle)$. If w' is not blocked then define $x_{k+1} = w'$. If w' is blocked by z then define $x_{k+1} = z$. In addition, we define $\pi(w') = t_{k+1}$. Since $\langle t_k, t_{k+1}\rangle \in \mathcal{E}(Q)$ hence w' is a Q-successor of x_k.

Consequently, we obtain $x_1, \cdots, x_{k+1}$ such that $L(x_i) = \mathcal{L}(t_i)$ for all $i \in \{1, \cdots, k+1\}$ and $Q \in L(\langle x_i, x_{i+1}) = \mathcal{L}(\langle t_i, t_{i+1})$, for all $i \in \{1, \cdots, k\}$ with $x_0 = x$ and $t_0 = t$.

By the same way, we can obtain $x_l, \cdots, x_n$ such that $L(x_i) = \mathcal{L}(t_i)$ and $\mathsf{Inv}(Q) \in L(\langle x_i, x_{i+1}) = \mathcal{L}(\langle t_i, t_{i+1})$ for all $i \in \{l, \cdots, n\}$ where $t_{n+1} = t$ and $x_{n+1} = y$ if y is not blocked or $x_{n+1} = z$ if z blocks y.

According to Def. 5, $\langle x_{l+1}, x_l, \cdots, x_{n+1}, x, x_1, \cdots, x_{k+1}\rangle$ form a cyclic non-duplicated Q-path. Thus, $\mathbf{T}$ is a completion tree with cyclic paths for D. □

FO(ID) as an Extension of DL with Rules

Joost Vennekens* and Marc Denecker

Dept. of Computer Science, K.U. Leuven, Belgium
{joost.vennekens,marc.denecker}@cs.kuleuven.be

Abstract. There are many interesting Knowledge Representation questions surrounding rule languages for the Semantic Web. The most basic one is of course: which kind of rules should be used and how do they integrate with existing Description Logics? Similar questions have already been addressed in the field of Logic Programming, where one particular answer has been provided by the language of FO(ID). FO(ID) is an extension of first-order logic with a rule-based representation for inductive definitions. By offering a general integration of first-order logic and Logic Programs, it also induces a particular way of extending Description Logics with rules. The goal of this paper is to investigate this integration and discover whether there are interesting extensions of DL with rules that can be arrived at by imposing appropriate restrictions on the highly expressive FO(ID).

1 Introduction

Over the past decades, Description Logics (DL) have emerged as an important Knowledge Representation (KR) technology. More recently, they have also had a significant impact on industry, most notably with the adoption of OWL as a W3C standard. In current research, we find a trend to investigate extensions of OWL with *rules* (e.g. [11]), and, in fact, the hierarchical Semantic Web architecture already prescribes a *rule layer* on top of the *ontology layer* formed by OWL. There are a number of interesting KR questions surrounding this topic:

- *Which kind of rules are to be used?* There are numerous kinds of rules known in the literature (inference rules, rewrite rules, ...), with subtle differences between them.
- *What precisely do this kind of rules mean?* It should be possible to explain exactly the information content of such a rule and, obviously, this explanation should be consistent with the formal semantics of the rules.
- *How do the rules complement DL?* We should be able to clearly indicate how the rules extend the class of knowledge that can be represented by the logic.

In this paper, we will present an answer to these questions based on the language of FO(ID), a general integration of classical first-order logic (FO) and Logic Programming (LP). Conceptually, FO(ID) is an extension of FO with

* Joost Vennekens is a postdoctoral researcher of the FWO.

L. Aroyo et al. (Eds.): ESWC 2009, LNCS 5554, pp. 384–398, 2009.

inductive definitions. To explain, let us consider the following example of an inductive definition, taken from the wikipedia page on the topic[1]:

> *The prime numbers can be defined as consisting of:*
> - *2, the smallest prime;*
> - *each positive integer which is not evenly divisible by any of the primes smaller than itself.*

FO(ID) offers a formal syntax for representing such an inductive definition as a set of *definitional rules*:

$$\left\{\begin{array}{r} Prime(2) \leftarrow \\ \forall x\ Prime(x) \leftarrow x > 0 \wedge \neg \exists y\ y < x \wedge Prime(y) \wedge Divisible(x, y) \end{array}\right\}$$

In natural language, an inductive definition consists of a set of cases in which the defined relation, *Prime* in this case, holds. Each of these cases corresponds to a definitional rule in the formal syntax of FO(ID). We remark that the '$\leftarrow$'-symbol in the above expression therefore conveys more meaning than the normal material implication of FO: it not only states that certain numbers are prime, but also that no number can be prime *unless* its "primeness" can be constructively derived by applying the two rules. This additional meaning of such a rule is formalized in FO(ID) using the (parameterized) *well-founded model* construction from Logic Programming. It was argued in [5] that this formal construction correctly captures the common-sense meaning of such a definition, as it is understood throughout mathematics.

The "job" of an inductive definition is, obviously, to define certain relation(s) (e.g., $Prime/1$) in terms of some other relation(s) (e.g., $</2$ and $Divisible/2$). In this sense, it has the same semantic status as an FO formula: it expresses a relation between a number of predicates, which may or may not be satisfied by a given interpretation for these predicates. Based on this observation, FO(ID) now integrates inductive definitions into FO: a formula of FO(ID) is either a regular FO formula or an inductive definition. In this way, FO is extended with a particular form of rules, namely, definitional rules, whose meaning is that each such rule represent a "case" of a particular inductive definition. This enlarges the expressive power of the logic: it is well-known that inductive definitions, such as that of transitive closure, cannot be expressed in FO.

The integration of FO and LP offered by FO(ID) has a number of properties that are quite appealing from the point-of-view of extending DL with rules:

- As opposed to the many hybrid approaches that exist, FO(ID) offers a strong semantic integration of FO and LP.
- In this integration, both components have a clear knowledge representation "task": the LP component is to be used to *define* concepts, whereas the FO component can be used to assert additional properties of both the defined concepts and concepts for which no definition is provided.

[1] `http://wikipedia.org/wiki/Inductive_definition`

- FO(ID) is particularly natural from a DL perspective: the ability to properly define concepts is already considered to be an important and characteristic feature of DL[2], and FO(ID) essentially just extends it to definitions that cannot be represented by a normal FO equivalence.
- FO(ID) does not impose any arbitrary restrictions on expressivity: it contains full FO and allows general FO formulas in the bodies of its definitional rules.

These properties suggest that FO(ID) is well-suited to serve as a kind of upperbound, from which we can derive meaningful extensions of DL with rules, by imposing suitable restrictions on it. We believe that, in general, this is better than the opposite approach of gradually extending a small tractable language into a more expressive one, since it allows the trade-off between expressivity and complexity to be made more consciously and informedly, which reduces the risk of creating an *ad hoc* language, whose boundaries are decided more by coincidence than design. In the rest of this paper, we will try to make these claims more concrete, by defining two fragments of FO(ID), that seem to be interesting DL rule languages.

2 Preliminaries: FO(ID)

This section summarizes the definition of FO(ID) as found in [5]. Syntactically, a definition in FO(ID) is a set of *definitional rules*, which are of the form:

$$\forall \boldsymbol{x}\ P(\boldsymbol{x}) \leftarrow \varphi(\boldsymbol{x}).$$

Here, $\varphi(\boldsymbol{x})$ is an FO formula whose free variables are $\boldsymbol{x}$, and '$\leftarrow$' is a new symbol, the *definitional implication*, which is to be distinguished from the material implication of FO, which we denote by '$\Leftarrow$'. We refer to $P(\boldsymbol{x})$ as the *head* of the rule r, denoted $head(r)$, and to the formula φ as its *body*, denoted $body(r)$.

An FO(ID) *formula* is any expression that can be formed by combining atoms and definitions, using the standard FO connectives and quantifiers. The meaning of the FO formulas and boolean connectives is standard; we therefore only need to define the semantics of a definition in order to define that of FO(ID).

Let us first introduce some semantical concepts. An *interpretation* $\mathcal{S}$ for a vocabulary Σ consists of a non-empty domain D, a mapping from each function symbol f/n to an n-ary function on D, and a mapping from each predicate symbol P/n to a relation $R \subseteq D^n$. A *three-valued* interpretation ν is the same as a two-valued one, except that it maps each predicate symbol P/n to a function P^ν from D^n to the set of truth values $\{\mathbf{t}, \mathbf{f}, \mathbf{u}\}$. Such a ν assigns a truth value to each logical atom $P(\boldsymbol{c})$, namely $P^\nu(c_1^\nu, \dots, c_n^\nu)$. This assignment can be extended to an assignment $\nu(\varphi)$ of a truth value to each formula φ, using the standard Kleene truth tables for the logical connectives:

[2] For instance, the DL handbook [1] has the following to say about DL's '$\equiv$'-connective: "This form of definition is much stronger than the ones used in other kinds of representations of knowledge, which typically impose only necessary conditions; the strength of this kind of declaration is usually considered a characteristic feature of DL knowledge bases."

φ, ψ	t, t	t, f	t, u	u, f	u, u	f, f
$\varphi \vee \psi$	t	t	t	u	u	f

φ	t	u	f
$\neg\varphi$	f	u	t

and so on. The three truth values can be partially ordered according to *precision*: $\mathbf{u} \leq_p \mathbf{t}$ and $\mathbf{u} \leq_p \mathbf{f}$. This order induces a precision order $\leq_p$ on interpretations: $\nu \leq_p \nu'$ if for each predicate P/n and tuple $\boldsymbol{d} \in D^n$, $P^\nu(\boldsymbol{d}) \leq_p P^{\nu'}(\boldsymbol{d})$.

For a predicate P/n and a tuple $\boldsymbol{d} \in D^n$, we denote by $\nu[P(\boldsymbol{d})/\mathbf{v}]$ the three-valued interpretation ν' that coincides with ν on all symbols apart from P/n, and for which $P^{\nu'}$ maps $\boldsymbol{d}$ to $\mathbf{v}$ and all other tuples $\boldsymbol{d}'$ to $P^\nu(\boldsymbol{d}')$. We also extend this notation to sets $\{(P_1(\boldsymbol{d}_1), \ldots, P_n(\boldsymbol{d}_n)\}$ of such pairs.

Our goal is to define when a (two-valued) interpretation $\mathcal{S}$ is a model of a definition Δ. We call the predicates that appear in the head of a rule of Δ its *defined predicates* and we denote the set of all these by $Def(\Delta)$; all other symbols are called *open* and the set of open symbols is written $Op(\Delta)$. The purpose of Δ is now to define the predicates $Def(\Delta)$ in terms of the symbols $Op(\Delta)$, i.e., we should assume the interpretation of $Op(\Delta)$ as given and try to construct a corresponding interpretation for $Def(\Delta)$. Let O be the restriction $\mathcal{S}|_{Op(\Delta)}$ of $\mathcal{S}$ to the open symbols. We are now going to construct a sequence of three-valued interpretations $(\nu^O_\alpha)_{0\leq\alpha}$, each of which extends O; we will use the limit of this sequence to interpret $Def(\Delta)$.

- ν^O_0 assigns $O(P(\boldsymbol{d})) \in \{\mathbf{t}, \mathbf{f}\}$ to $P(\boldsymbol{d})$ if $P \in Op(\Delta)$ and $\mathbf{u}$ if $P \in Def(\Delta)$;
- ν^O_{i+1} is related to ν^O_i in one of two ways:
  - Either $\nu^O_{i+1} = \nu^O_i[P(\boldsymbol{d})/\mathbf{t}]$, such that Δ contains a rule $\forall \boldsymbol{x}\ P(\boldsymbol{x}) \leftarrow \varphi(\boldsymbol{x})$ with $\nu^O_i(\varphi[\boldsymbol{d}]) = \mathbf{t}$
  - Or $\nu^O_{i+1} = \nu^O_i[U/\mathbf{f}]$, where U is any *unfounded set*, meaning that it consists of pairs of predicates P/n and tuples $\boldsymbol{d} \in D^n$ for which $P^{\nu^O_i}(\boldsymbol{d}) = \mathbf{u}$, and for each rule $\forall \boldsymbol{x}\ P(\boldsymbol{x}) \leftarrow \varphi(\boldsymbol{x})$, we have that $\nu^O_{i+1}(\varphi[\boldsymbol{d}]) = \mathbf{f}$.
- For each limit ordinal λ, ν^O_λ is the least upper bound w.r.t. $\leq_p$ of all ν^O_δ for which $\delta < \lambda$.

We call such a sequence a *well-founded induction* of Δ in O. Each such sequence eventually reaches a limit ν^O_β. It was shown in [5] that all sequences reach the same limit. It is now this ν^O_β that tell us how to interpret the defined predicates. To be more precise, we define that:

$$\mathcal{S} \models \Delta \text{ iff } \mathcal{S}|_{Def(\Delta)} = \nu^O_\beta|_{Def(\Delta)}, \text{ with } O = \mathcal{S}|_{Op(\Delta)}.$$

This tells us when a definition Δ is satisfied in a structure $\mathcal{S}$. The semantics of full FO(ID), in which definitions and atoms can be combined using the standard FO connectives, is now simply that which is obtained by adding the usual recursive cases for the connectives (e.g., $\mathcal{S} \models \varphi \wedge \psi$ if $\mathcal{S} \models \varphi$ and $\mathcal{S} \models \psi$, and so). Note that if there is some predicate P/n and $\boldsymbol{d} \in D^n$ such that ν_β still assigns $\mathbf{u}$ to $P(\boldsymbol{d})$, then the definition has no models extending O. Intuitively, this means that, for this particular interpretation of its open symbols, Δ does not manage to unambiguously define the predicates $Def(\Delta)$, due to some non well-founded use of negation.

In the rest of this paper, we will only consider relational vocabularies, that is, there will be no function symbols of arity > 0.

3 From FO(ID) to DL(ID)

As a language that extends full FO, FO(ID) is too expressive to serve as a DL itself. In this section, we discuss the problem of restricting FO(ID) to a smaller, more DL-like language. In doing this, it is important to be clear about the goals we are trying to achieve. DLs themselves are fragments of FO that are interesting for essentially two reasons:

- They are decidable;
- They are tailored towards a concept-centric modeling style, which they support by means of an intuitive syntax sugar that hides away many of the complexities of FO.

The first of these properties is only relevant for computation, whereas the second concerns knowledge representation. To a certain extent, they are therefore orthogonal concerns.

Most of our attention in this paper will go to the knowledge representation issue, simply because we feel that it *comes first*. If a logic is to be used at all, for whatever purpose, then theories will have to be written in it. If this is difficult, for instance because the language is too complex or because the meaning of its statements is unclear, then it is unlikely the language will enjoy enduring success, no matter how efficient certain inference task for it might be. Therefore, a primary goal of designing a rule language for DL should be to ensure that it provides a clean and coherent integration of rules into DL, which is easy to understand for those with a working knowledge of DL.

This does not mean that decidability is unimportant, but in the end it is of secondary nature, since also undecidable languages can have interesting uses. Indeed, decidability is only relevant if the goal is to perform deductive inference (i.e., to decide whether, for formulas φ and ψ, it is the case that $\varphi \models \psi$, in the sense that each model of φ is also a model of ψ) in an *unknown* or *infinite* domain. Formally, we know that:

- Given an FO theory T and an FO formula φ, it is undecidable whether for all structures S, it holds that $S \models T$ implies $S \models \varphi$;
- However, given *a finite set* D, an FO theory T and an FO formula φ, it is decidable (and in co-NP) whether for all structures S *such that the domain of* S *is* D, it holds that $S \models T$ implies $S \models \varphi$;

When we go from FO to FO(ID), both of these theorems still hold. This is important, because the second situation arizes in numerous interesting applications. For instance, when reasoning in the context of a particular database, we only want to consider the domain D consisting of all the objects in the database. Similarly, if the goal is to query some data from Semantic Web sites, we should only consider the domain D of objects about which we are able to retrieve information. A second example is that for many combinatorial problems, the domain is given as part of the specific problem instance that is to be solved; e.g., in graph colouring, D would be the set of nodes, which is fixed once we get a specific graph to colour. Moreover, even when the domain is not known, there might still be approximate algorithms, that are able to derive useful information without guarantees of completeness.

We will therefore now first present a fragment of FO(ID) that allows a DL-like syntactic sugar and offers a DL-like modeling methodology. Only afterwards will we (more briefly) turn to the topic of inference.

4 A Fragment of FO(ID): $\mathcal{ALCI}$(ID)

The goal of this section is to present a fragment of FO(ID), that supports a DL-like modeling style. This will be an extension of the Description Logic $\mathcal{ALCI}$. More expressive logics, such as $\mathcal{SHOIN}$(D) which underlies OWL-DL, can be extended in the same way; we restrict to $\mathcal{ALCI}$ merely for simplicity.

Let us first briefly recall $\mathcal{ALCI}$. We start from a set of unary predicates $\{A_1, A_2, \ldots\}$ called (atomic) *concepts* and a set of binary predicates $\{B_1, B_2, \ldots\}$ called (atomic) *roles*. From the atomic concepts, we build more complex ones using the connectives $\sqcup$, $\sqcap$, $\neg$, $\exists$ and $\forall$; their meaning can be inductively defined by the mapping to FO given in Fig. 1. A role is either an atomic role B or its inverse B^-, as in Fig. 2. A *TBox* then consists of a set of inclusions and equivalences as in Fig. 3.

C	$\langle C\rangle(x)$	
A	$A(x)$	with A an atomic concept
$C_1 \sqcap C_2$	$\langle C_1\rangle(x) \wedge \langle C_2\rangle(x)$	with C_1, C_2 concepts
$C_1 \sqcup C_2$	$\langle C_1\rangle(x) \vee \langle C_2\rangle(x)$	with C_1, C_2 concepts
$\neg C$	$\neg\langle C\rangle(x)$	with C a concept
$\forall R.C$	$\forall y\ \langle R\rangle(x,y) \Rightarrow \langle C\rangle(y)$	with R a role and C a concept
$\exists R.C$	$\exists y\ \langle R\rangle(x,y) \wedge \langle C\rangle(y)$	with R a role and C a concept

Fig. 1. A concept C of $\mathcal{ALCI}$ represents a unary formula $\langle C\rangle(x)$ of FO

R	$\langle R\rangle(x,y)$	
B	$B(x,y)$	with B an atomic role
B^-	$B(y,x)$	with B an atomic role

Fig. 2. A role R of $\mathcal{ALCI}$ represents a binary formula $\langle R\rangle(x,y)$ of FO

φ	$\langle\varphi\rangle$	
$C_1 \sqsubseteq C_2$	$\forall x\ \langle C_1\rangle(x) \Rightarrow \langle C_2\rangle(x)$	with C_1, C_2 concepts
$C_1 \equiv C_2$	$\forall x\ \langle C_1\rangle(x) \Leftrightarrow \langle C_2\rangle(x)$	with C_1, C_2 concepts

Fig. 3. A statement φ in a TBox correspond to an FO sentence $\langle\varphi\rangle$

We now form the language $\mathcal{ALCI}$(ID) by extending $\mathcal{ALCI}$. First, we add the connectives in Fig. 1 for constructing more complex roles: The first three are the obvious analogues of the connectives for concepts, the fourth takes the join of two roles, and the last one takes the Cartesian product of two concepts. Second, and more importantly, we add two new connectives for representing inductive definitions. The first is the symbol '$\doteq$', which is conceptually the same as '$\equiv$', with the difference that it also works for inductive definitions. Formally, it is defined as an abbreviation for an FO(ID) definition containing a single rule, as shown in the first two entries of Fig. 5.

R	$\langle R\rangle(x,y)$	
$R_1 \sqcap R_2$	$\langle R_1\rangle(x,y) \wedge \langle R_2\rangle(x,y)$	with R_1, R_2 roles
$R_1 \sqcup R_2$	$\langle R_1\rangle(x,y) \vee \langle R_2\rangle(x,y)$	with R_1, R_2 roles
$\neg R$	$\neg\langle R\rangle(x,y)$	with R a role
$R_1.R_2$	$\exists z\ \langle R_1\rangle(x,z) \wedge \langle R_2\rangle(z,y)$	with R_1, R_2 roles
$C_1 \times C_2$	$C_1(x) \wedge C_2(y)$	with C_1, C_2 concepts

Fig. 4. Additional ways of constructing roles

For instance, we can define the concept of an uncle as the brother of a parent:

$$Uncle \doteq Brother.Parent$$

This abbreviates the FO(ID) definition:

$$\{\forall x, y\ Uncle(x,y) \leftarrow \exists z\ Brother(x,z) \wedge Parent(z,y)\}$$

Such a non-inductive definition (i.e., the role/concept on the left-hand side does not appear in the right-hand side) is equivalent to a regular FO equivalence:

$$\forall x, y\ Uncle(x,y) \Leftrightarrow \exists z\ Brother(x,z) \wedge Parent(z,y).$$

However, '$\doteq$' can also correctly represent inductive definitions. For instance, we can define the role *Ancestor* as the transitive closure of *Parent*, by saying that an *Ancestor* is either a *Parent* or the *Parent* of an *Ancestor*:

$$Ancestor \doteq Parent \sqcup Parent.Ancestor$$

Because of the translation to FO(ID), this too has the correct semantics. Therefore, '$\doteq$' can be used instead of a transitive closure construct such as $\cdot^+$ or the reflexive-transitive closure $\cdot^*$ of e.g. [12]; it can also replace non-nested uses of the explicit least fixpoint operator μ.

In addition to $\doteq$, we also add the new connective $\leftarrow$ to represent a definitional rule and allow sets of such rules as statements in our language; this is shown in Fig. 5. For instance, extending the notion of an uncle to also include uncles-by-marriage, we could then define this concept by saying that an uncle is

φ	$\langle\varphi\rangle$	
$R_1 \doteq R_2$	$\{\forall x, y\ \langle R_1\rangle(x,y) \leftarrow \langle R_2\rangle(x,y)\}$	with R_1, R_2 roles
$C_1 \doteq C_2$	$\{\forall x\ \langle C_1\rangle(x) \leftarrow \langle C_2\rangle(x)\}$	with C_1, C_2 roles
$\{\varphi_1, \ldots, \varphi_n\}$	$\{\langle\varphi_1\rangle, \ldots, \langle\varphi_n\rangle\}$	with $\varphi_1, \ldots, \varphi_n$ definitional rules (Fig.6)

Fig. 5. A definition φ in $\mathcal{ALCI}$(ID) corresponds to a definition $\langle\varphi\rangle$ in FO(ID)

φ	$\langle\varphi\rangle$	
$R_1 \leftarrow R_2$	$\forall x, y\ \langle R_1\rangle(x,y) \leftarrow \langle R_1\rangle(x,y)$	with R_1, R_2 roles
$C_1 \leftarrow C_2$	$\forall x\ \langle C_1\rangle(x) \leftarrow \langle C_2\rangle(x)$	with C_1, C_2 concepts

Fig. 6. A definitional rule φ in $\mathcal{ALCI}$(ID) corresponds to a definitional rule $\langle\varphi\rangle$ in FO(ID)

– Either a brother of a parent;
– Or a husband of an aunt.

We can write this down in $\mathcal{ALCI}$(ID) as:

$$\left\{\begin{array}{l} Uncle \leftarrow Brother.Parent \\ Uncle \leftarrow Husband.Aunt \end{array}\right\}$$

This abbreviates the following definition in FO(ID):

$$\left\{\begin{array}{l} \forall x, y\ Uncle(x,y) \leftarrow \exists z\ Brother(x,z) \wedge Parent(z,y) \\ \forall x, y\ Uncle(x,y) \leftarrow \exists z\ Husband(x,z) \wedge Aunt(z,y) \end{array}\right\}$$

In general, if a definition contains multiple rules with the same predicate in the head, these can always be replaced by a single rule whose body is the disjunction of the bodies of the original rules. Therefore, we could rephrase the above definition of *Uncle* as

$$Uncle \doteq Brother.Parent \sqcup Husband.Aunt$$

However, this loses the natural case-based structure of the definition, i.e., it makes it more difficult to see that there are actually two separate sufficient conditions, which together also form a necessary condition. Moreover, the rule-based format is also more elaboration tolerant, since rules can more easily be added or removed. For instance, a bank might define the class of persons eligible for a loan as consisting of people with a large income, people who own a house and people with a good credit history; each time the bank now tightens or relaxes its policy, certain rules would have to be removed or added to this definition.

The rule-based representation is also more general than '$\doteq$', since it allows definitions by *simultaneous* induction as well. For instance, given a two-player game whose move tree is described by the role *Successor*, we can define the nodes in which I move and the nodes in which my opponent moves by the following simultaneous induction (assuming I start):

$$\left\{\begin{array}{l} HisMove \leftarrow \exists Successor.MyMove \\ MyMove \leftarrow \exists Successor.HisMove \\ MyMove \leftarrow \forall Successor.\bot \end{array}\right\}$$

Here, $\bot$ denotes the empty concept.

5 An Example

In this section, we will illustrate the language $\mathcal{ALCI}$(ID) by means of a simple game. The player is presented a grid of coloured balls. He makes a move by selecting one of these balls. The effect of this is that the entire colour-group to which the ball belongs disappears; the remaining balls then fall down, yielding the next position of the game, as depicted in Fig. 7. The goal of the game is to remove all balls from the grid in such a way as to score as many points as possible. What we will do here is construct an $\mathcal{ALCI}$(ID) theory that describes the effect of a single move of this game.

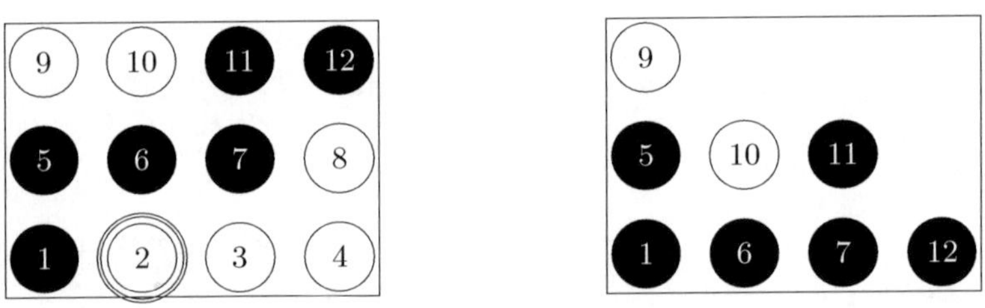

Fig. 7. The result of selecting ball 2

To make things more concrete, let us fix a representation for a state of the game. We represent the grid by roles Up and $Left$, both with the obvious meaning. The starting grid in Fig. 7, for instance, would correspond to the following interpretations:

$$\begin{aligned} Up =& \{(5,1),(9,5),(6,2),(10,6),(7,3), \\ & (11,7),(8,4),(12,8)\}; \\ Left =& \{(1,2),(2,3),(3,4),(5,6),(6,7), \\ & (7,8),(9,10),(10,11),(11,12)\}. \end{aligned}$$

We represent the player's move by a concept $Chosen$; the move made in Fig. 7 would correspond to $Chosen = \{2\}$. Our goal is now to define roles Up' and $Left'$, representing the next state of the game.

We first define some useful auxiliary relations. We begin by defining $Above$ as the transitive closure of Up. One ball is above another if it is either directly on top of it, or on top of a ball that is already above it:

$$Above \doteq Up \sqcup Up.Above$$

A ball is next to another ball if it is either to the left, to the right, underneath, or on top of it:

$$NextTo \doteq Left \sqcup Right \sqcup Up \sqcup Down$$

Of course, *Right* and *Down* are the inverses of, respectively $Left$ and Up:

$$Right \doteq Left^{-} \qquad Down \doteq Up^{-}$$

We remark that of these four definitions, *Above* is defined inductively, while the others are not. However, in $\mathcal{ALCI}$(ID) this difference hardly matters: all four definitions have the same "look and feel".

The balls that disappear after the move are the chosen ball itself and all balls belonging to the same colour-group:

$$Disappears \doteq Chosen \sqcup \exists InColourGroup.Chosen$$

Being in the same colour group means being connected through a sequence of balls of the same colour:

$$\left\{ \begin{array}{l} InColourGroup \leftarrow SameColour \sqcap NextTo \\ InColourGroup \leftarrow InColourGroup.InColourGroup \end{array} \right\}$$

Here, we use the $\leftarrow$ connective to separate the base case and the inductive step of this definition; the same could be done with our earlier definition of *Above*.

Two balls have the same colour if the colour of one is also that of the other:

$$SameColour \doteq HasColour.HasColour^{-}$$

Having now defined which balls disappear, we define the remaining balls as the complement thereof:

$$Remains \doteq \neg Disappears$$

We now define the relation *Above'*, i.e., the "above"-relation as it will be in the next state. This will hold for any two remaining balls that were originally above each other:

$$Above' \doteq Above \sqcap (Remains \times Remains)$$

We can now define the relation Up' as the intransitive relation of which *Above'* is the transitive closure:

$$Up' \doteq Above' \sqcap \neg(Above'.Above')$$

We define an auxiliary concept *OnGround'* as consisting of those balls that will form the bottom row in the new situation:

$$OnGround' \doteq Remains \sqcap \neg\exists Above^{-}.Remains$$

All that remains now is to define $Left'$. Let us first define when a ball x is in the column to the left of some other ball y. For those y that are on a height not greater than that of x, this is the case if x is either directly to the left of y, or below a ball directly to the left of y, or above a ball directly to the left of y. To cover that situation that y is heigher than x's, we also add the recursive case that y is above a ball for which we already know that x is in its left column.

$$\left\{\begin{array}{l} InLeftColumn \leftarrow Left \\ InLeftColumn \leftarrow Above.Left \\ InLeftColumn \leftarrow Above^{-}.Left \\ InLeftColumn \leftarrow LeftColumn.Above' \end{array}\right\}$$

We now define $Left'$ as consisting of all pairs of balls such that (1) one ball is in the column to the left of the other, and (2) both are on the same height:

$$Left' \doteq \underbrace{InLeftColumn}_{(1)} \sqcap \underbrace{((OnGround' \times OnGround') \sqcup Up'.(Left'.Up'^{-}))}_{(2)}$$

Note that this too is an inductive definition, which proceeds along the rows of the grid: the base case is the bottom row, whereas the inductive case defines that x is to the left of y in the following situation:

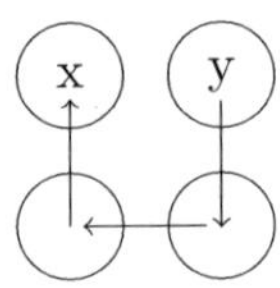

This concludes our representation of the game. As this example shows, $\mathcal{ALCI}$(ID) can represent many different kinds of definitions (case based definitions, non-inductive definitions, monotone inductive definitions, non-monotone inductive definitions) in a uniform way in, leading to a clean and coherent language.

6 Inference in $\mathcal{ALCI}$(ID)

We lack space for a proof, but it is not hard to show that $\mathcal{ALCI}$(ID) is undecidable. However, this does not mean that is cannot be used in practice.

6.1 Model Expansion

[13] considered the inference task of *model expansion* for FO(ID): given an interpretation $\mathcal{S}$ for some subset Σ_0 of the alphabet Σ of a theory T, extend $\mathcal{S}$ with an interpretation for the remaining symbols $\Sigma \setminus \Sigma_0$, such that the resulting interpretation is a model of T. Here, $\mathcal{S}$ is required to have a finite domain. Note that, in general, a model expansion problem may have many different solutions, because even though $\mathcal{S}$ has to fix the domain of discourse, it does not fix the interpretation of $\Sigma \setminus \Sigma_0$ in this domain, but only restrains it to satisfy T. This form of inference is decidable for full FO(ID) and in fact captures the complexity class NP.

To illustrate the usefulness of this inference task, let us consider again the example of the previous section. Here, we defined the next state of a game ($Left'$ and Up') in terms of its old state ($Left$ and Up) and a given move ($Chosen$). We can therefore compute a new state of the game by performing model expansion on the structure $\mathcal{S}$ for the alphabet $\Sigma_0 = \{Left, Up, Chosen\}$. Because our representation of the game defines all of its predicates except the ones in Σ_0, this computation can actually be done in polynomial times.

The IDP-system[3] implements the task of model generation of FO(ID). A program transforming $\mathcal{ALCI}(ID)$ syntax to input for this system is available.[4] Together, these two programs are able to, for instance, use our formalizaton of Section 5 to compute state transitions for this game.

6.2 Guarded $\mathcal{ALCI}$(ID)

As discussed, there are interesting applications, that only require reasoning in a fixed, finite domain and for which the undecidability of FO(ID) is therefore not a problem. However, because there are of course also other cases, this section will develop a decidable fragment of FO(ID).

This fragment is based on the *guarded fragment* of FO. We recall that an FO formula ψ is *guarded* if every one of its quantified subformulas is either of the form $\exists \boldsymbol{x}\ G(\boldsymbol{x}, \boldsymbol{y}) \wedge \varphi(\boldsymbol{x}, \boldsymbol{y})$ or $\forall \boldsymbol{x}\ G(\boldsymbol{x}, \boldsymbol{y}) \Rightarrow \varphi(\boldsymbol{x}, \boldsymbol{y})$, where $G(\boldsymbol{x}, \boldsymbol{y})$ is an atom, called the *guard*, such that the free variables $free(\varphi(\boldsymbol{x}, \boldsymbol{y})) \subseteq free(G(\boldsymbol{x}, \boldsymbol{y}))$.

We now define a similar guarded fragment of FO(ID). This fragment allows only theories consisting of precisely one definition and precisely one FO formula.

Definition 1. *Let T be an FO(ID) theory, consisting of precisely one definition Δ and one FO formula φ. T is* guarded *if*

- *φ is a guarded formula;*
- *for each rule $\forall \boldsymbol{x}\ P(\boldsymbol{x}) \leftarrow \psi$, it holds that ψ is a guarded formula and $\boldsymbol{x} \subseteq free(\psi)$;*
- *none of the guards is defined by Δ.*

Exploiting a theorem from [8], we can prove the following result.

Theorem 1. *The guarded fragment of FO(ID) is decidable. More precisely, deductive reasoning for FO(ID) is 2*EXPTIME*-complete.*

Proof. Omitted for space.

This of course raises the question of how this guarded fragment of FO(ID) relates to $\mathcal{ALCI}$(ID). It is easy to see that neither is subsumed by the other. For instance, $P \doteq P.P$ translates to

$$\{\forall x, y\ P(x,y) \leftarrow \exists z\ P(x,z) \wedge P(z,y).\},$$

[3] http://www.cs.kuleuven.be/~dtai/krr/software/idp.html

[4] http://www.cs.kuleuven.be/~joost/alc_id.tar.gz

which is not guarded. Conversely, the guarded fragment is not subsumed by $\mathcal{ALCI}$(ID) either, since, for instance, the latter only allows unary and binary predicates, while the former allows arbitrary arities.

The two fragments do, however, have a significant intersection. Let us consider a theory T in $\mathcal{ALCI}$(ID) and examine which additional conditions can ensure that its translation is guarded.

The guarded fragment of FO(ID) only allows theories to contain a *single* definition, whereas an $\mathcal{ALCI}$(ID) theory can have multiple definitions. In [17], we examined when a set of definitions $(\Delta_i)_{1 \leq i \leq n}$ can be merged into a single equivalent definition $\cup_i \Delta_i$. We showed that the following condition is sufficient.

Condition 1. *There exists a preorder $\leq$ on the predicates of the theory, such that for all Δ_i and $P \in Def(\Delta_i)$:*

- *if $Q \in Def(\Delta_i)$ then $P \geq Q$;*
- *if Q appears in Δ_i but does not belong to $Def(\Delta_i)$, then $P > Q$ (i.e., $P \geq Q$ and $P \not\leq Q$).*

By imposing this condition, we can merge all definitions into a single definition, thus already satisfying that particular requirement of our guarded fragment. Looking at the translation of $\mathcal{ALCI}$(ID) into FO(ID), it is quite clear that the only other cause for falling outside of the guarded fragment is the fact that defined predicates might be used as guards.

Condition 2. *For every construct $\exists R.C$, $\forall R.C$ or $R.S$ that appears in T, it must be the case that R is not defined, i.e., T does not contain any formulas of the form $R \doteq \varphi$ or any definitions in which a rule $R \leftarrow \varphi$ appears.*

We will say that an $\mathcal{ALCI}$(ID) theory is *guarded* if it satisfies both Condition 1 and Condition 2.

Theorem 2. *Let T be a guarded $\mathcal{ALCI}$(ID) theory and let T' be the result of merging all definitions in $\langle T \rangle$ into a single definition. Then T' is equivalent to $\langle T \rangle$ itself and belongs the guarded fragment of FO(ID).*

We remark that our Condition 2 has a somewhat similar flavour to the decidable fragment of SWRL that is formed by *(strongly) safe* rules [15]: a SWRL-rule (i.e., Horn clause) is safe if every one of its variables appears in an atom whose predicate is not used anywhere in the TBox (i.e., it may only appear in other rules or in the ABox). This too requires that variables are "guarded" by atoms about which there is not "too much" information present in the theory.

7 Related Work

The most widely known DL rule-language is currently that of SWRL [11], which extends OWL with Horn clauses. Since its rules only state sufficient conditions, and cannot express necessary ones, they cannot be used to define concepts. Because definitions are an important feature of DL (see footnote on p.386), this is a curious lack. $\mathcal{ALCI}$(ID), by contrast, also allows the definitional rule construct $\leftarrow$, in addition to $\sqsupseteq$. Moreover, it also allows more that just Horn clauses, since the body of a rule may contain negation and quantifiers.

In [6], a combination of DL and LP under the well-founded semantics is investigated, while [7] presents an integration of DL and ASP. Unlike FO(ID)'s semantic integration of LP and DL, these two approaches foster a strong separation between the two components: essentially, they allow a logic program to pose queries to a description logic theory, with the latter acting more-or-less as a black box towards the former. In contrast, FO(ID) provides a full semantic integration in which both both logic programs and description logic axioms are first-class citizens.

Description logic programs [9] and $\mathcal{DL}$+log [16] are two strong integrations, which both consider only a quite restricted language, with Grosof et al. focusing on the intersection, rather than the union, of DL and LP, and Rosati only considering Datalog. By contrast, FO(ID) allows full first-order logic in the body of rules, using the machinery of the well-founded semantics to properly handle negation.

[3], [14] and [10] both provide strong semantic integrations for a more general kind of rules, with [10] also defining a guarded fragment that, like ours, can be mapped into guarded fixpoint logic. The main feature that distinguishes our approach from these is of a KR nature. We offer rules that have a clear and intuitive interpretation as *definitional rules*. On the one hand, this means that it easy to explain the total information content of such a rule: it represents a "case" of an (inductive) definition. On the other hand, we also have clear guidelines as to what they should be used for: the definitional rules should be used to define concepts, and other statements (regular inclusions) can serve to state additional properties, such conditions that are only sufficient. The KR picture for the other approaches is less clear.

8 Conclusions and Future Work

In this paper, we have investigated the extension of DL with rules that is induced by FO(ID). We first argued that this is quite natural for the following reasons:

- There is an appealing match between the intuitive notion of a "case" in an inductive definition and the formal construct of a definitional rule in FO(ID);
- Since non-inductive definitions are already a key feature of DL, it makes sense to exploit this match by adding definitional rules to DL, in order to extend the class of definitions that can be represented.

Motivated by these arguments, we have defined a fragment $\mathcal{ALCI}$(ID) of FO(ID), which offers a DL-like syntax for representing (inductive) definitions. For FO(ID), and therefore also for $\mathcal{ALCI}$(ID), there exist useful inference tasks, such as model expansion, that can be performed efficiently (see Section 6. This is despite the fact that the language as a whole is undecidable for deductive inference. We have defined a decidable guarded fragment of $\mathcal{ALCI}$(ID). A more comprehensive analysis of this and other decidable fragments of FO(ID) is left for future work. The goal of the current paper is not to present a single DL rule-language that is suitable for all purposes, but rather to point towards FO(ID) in general as an interesting foundation from which such languages can be derived.

References

1. Baader, F., Calvanese, D., McGuinness, D., Nardi, D., Patel-Schneider, P. (eds.): The Description Logic Handbook. Theory, Implementation and Applications. Cambridge University Press, Cambridge (2002)
2. Brachman, R.J., Levesque, H.J.: Competence in Knowledge Representation. In: AAAI, pp. 189–192 (1982)
3. de Bruijn, J., Pearce, D., Polleres, A., Valverde, A.: Quantified equilibrium logic and hybrid rules. In: Marchiori, M., Pan, J.Z., Marie, C.d.S. (eds.) RR 2007. LNCS, vol. 4524, pp. 58–72. Springer, Heidelberg (2007)
4. Denecker, M., Ternovska, E.: A logic of non-monotone inductive definitions. Transactions On Computational Logic (TOCL) (2008)
5. Denecker, M., Vennekens, J.: Well-founded semantics and the algebraic theory of non-monotone inductive definitions. In: Baral, C., Brewka, G., Schlipf, J. (eds.) LPNMR 2007. LNCS, vol. 4483, pp. 84–96. Springer, Heidelberg (2007)
6. Drabent, W., Henriksson, J., Maluszynski, J.: HD-rules: a hybrid system interfacing Prolog with DL-reasoners. In: 2nd Int'l. Workshop on Applications of Logic Programming to the Web, Semantic Web and Semantic Web Services (2007)
7. Eiter, T., Lukasiewicz, T., Schindlauer, R., Tompits, H.: Combining answer set programming with description logics for the semantic web. In: KR (2004)
8. Grädel, E., Walukiewicz, I.: Guarded fixed point logic. In: LICS, pp. 45–55 (1999)
9. Grosof, B., Horrocks, I., Volz, R., Decker, S.: Description logic programs: combining logic programs with description logic. In: WWW (2003)
10. Heymans, S., Van Nieuwenborgh, D., Vermeir, D.: Open Answer Set Programming with guarded programs. ACM TOCL 9(4) (2008)
11. Horrocks, I., Patel-Schneider, P.F., Boley, H., Tabet, S., Grosof, B., Dean, M.: SWRL: A semantics web rule language combining OWL and RuleML. W3C Submission (2004), `http://www.w3.org/Submission/2004/SUBM-SWRL-20040521/`
12. De Giacomo, G., Lenzerini, M.: Tbox and ABox reasoning in expressive description logics. In: KR, pp. 316–327. Morgan Kaufmann, San Francisco (1996)
13. Mitchell, D., Ternovska, E.: A framework for representing and solving NP search problems. In: AAAI, pp. 430–435. AAAI Press/MIT Press (2005)
14. Motik, B., Rosati, R.: A faithful integration of description logics with logic programming. In: IJCAI (2007)
15. Motik, B., Sattler, U., Studer, R.: Query answering for OWL-DL with rules. In: McIlraith, S.A., Plexousakis, D., van Harmelen, F. (eds.) ISWC 2004. LNCS, vol. 3298, pp. 549–563. Springer, Heidelberg (2004)
16. Rosati, R.: DL+log: Tight integration of description logics and disjunctive datalog. In: KR, pp. 68–78 (2006)
17. Vennekens, J., Denecker, M.: An algebraic account of modularity in ID-logic. In: Baral, C., Greco, G., Leone, N., Terracina, G. (eds.) LPNMR 2005. LNCS, vol. 3662, pp. 291–303. Springer, Heidelberg (2005)

A Tableau Algorithm for Handling Inconsistency in OWL*

Xiaowang Zhang[1,2], Guohui Xiao[1], and Zuoquan Lin[1]

[1] School of Mathematical Sciences, Peking University, Beijing, China
[2] School of Mathematical Sciences, Anhui University, Hefei, China
{zxw,xgh,lzq}@is.pku.edu.cn

Abstract. In Semantic Web, the knowledge sources usually contain inconsistency because they are constantly changing and from different view points. As is well known, as based on the description logic, OWL is lack of the ability of tolerating inconsistent or incomplete data. Recently, the research in handling inconsistency in OWL becomes more and more important. In this paper, we present a paraconsistent OWL called quasi-classical OWL to handle inconsistency with holding important inference rules such as modus tollens, modus ponens, and disjunctive syllogism. We propose a terminable, sound and complete tableau algorithm to implement paraconsistent reasoning in quasi-classical OWL. In comparison with other approaches to handle inconsistency in OWL, our approach enhances the ability of reasoning by integrating paraconsistent reasoning with important classical inference rules.

1 Introduction

In recent years, the problem of inconsistency handling in OWL is attracting a lot of attention in logics and Semantic Web. Many reasons cause the occurrence of inconsistency such as modeling errors, migration from other formalisms, merging ontologies, and ontology evolution [1]. In practical reasoning, it is common to have "too much" information about some situation. In other words, it is common for there to be classically inconsistent information in practical reasoning ontologies [2]. According the fact *ex contradictione quodlibet* in classical logic, if ontologies contain inconsistencies then the classical entailment in logics is explosive. That is to say, any formula is a logical consequence of a contradiction. Therefore, conclusions drawn from an inconsistent knowledge base may be completely meaningless [3]. This is particularly important if the full power of logic-based approaches like the Web Ontology Language (short OWL) [4] shall be employed, as classical logic breaks down in the presence of inconsistent knowledge. Not surprisingly, the study of handling inconsistency in OWL becomes more and more important.

There are serval approaches to handling inconsistency in OWL, which can be generally divided into two fundamentally different approaches. The first is based on the assumption that inconsistencies indicate erroneous data which are to be repaired in order to obtain a consistent knowledge base, e.g. by selecting consistent subsets for the

* We acknowledge support by Major Program of National Natural Science Foundation of China (NSFC) (No. 60496322).

L. Aroyo et al. (Eds.): ESWC 2009, LNCS 5554, pp. 399–413, 2009.

reasoning process [3,5,6]. Another approach, called paraconsistent approach, is not to simply avoid the inconsistency but to apply a non-standard reasoning method to obtain meaningful answers [7,8,9,10,11]. In the paraconsistent approach, inconsistencies are treated as a natural phenomenon in realistic data and are tolerated in reasoning. Compared with the former, the latter acknowledges and distinguishes the different epistemic statuses between "the assertion is true" and "the assertion is true with conflict". So far, the main idea of paraconsistent approach of handling inconsistency in OWL is borrowing Belnaps four-valued semantics [12] for OWL. The most prominent of them are based on the use of additional truth values standing for $underdefined$ (i.e. neither true nor false) and $overdefined$ (or $contradictory$, i.e, both true and false). Four-valued semantics proves useful for measuring inconsistency of ontologies [13], which can provide context information for facilitating inconsistency handling. To a certain extent four-valued semantics handles inconsistency, however the capability of reasoning is weaker than classical OWL because four-valued OWL doesn't hold three basic inference rules such as modus ponens, disjunctive syllogism, modus tollens or intuitive equivalences. In [8,9], a total negation is introduced to strengthen the capability of paraconsistent reasoning in four-valued OWL by applying resolution principles. However, four-valued OWL with the total negation don't hold intuitive equivalences. These shortcomings are inherent limitations of four-valued semantics in reasoning.

Naturally, we expect that there is a paraconsistent logic which has the same capability of reasoning in classical logics to handle inconsistency in OWL. However, it is still an open problem now. Motivation of this paper is to find a paraconsistent approach to OWL with three basic inference rules and intuitive equivalences.

OWL is based on description logics and the description logic $\mathcal{ALCNQ}$ is considered to be the most foundational one and comprises a large fragment of OWL. In this paper, based on [14], we study a description logic $\mathcal{ALCNQ}$ version of paraconsistent logic, called quasi-classical (or QC) $\mathcal{ALCNQ}$, which is an extension of quasi-classical logic [15,16] and a paraconsistent version of $\mathcal{ALCNQ}$. We contribute to the inconsistency handling for $\mathcal{ALCNQ}$ in terms of the QC semantics in the following aspects.

- A new semantics called QC semantics including two semantics "QC weak semantics" and "QC strong semantics" is introduced to description logic $\mathcal{ALCNQ}$ in this paper. QC weak semantics inherits the characters from four-valued semantics in order to reason paraconsistently, while QC strong semantics is introduced to strengthen the capability of reasoning in ontologies. Compared to QC weak semantics, QC strong semantics refines the disjunction of concepts in order to hold three basic inference rules. Moreover, concept subsumption redefined in QC strong semantics is different from four-valued description logic in order to hold intuitive equivalences, i.e., $C \sqsubseteq D$ iff $\neg C \sqcup D(a)$ for any individual a occurring in ontologies.
- A QC entailment (written by "$\models_Q$") between an ontology and an axiom is presented by applying QC semantics in this paper. "$\models_Q$" is a nontrivial entailment, i.e., $\{C(a), \neg C(a)\} \not\models_Q \alpha$ for any axiom α. Compared with four-valued description logic, QC entailment holds three basic inference rules such as modus ponens: $\{C(a), C \sqsubseteq D\} \models_Q D(a)$, modus tollens: $\{\neg D(a), C \sqsubseteq D\} \models_Q \neg C(a)$ and disjunctive syllogism: $\{\neg C(a), C \sqcup D\} \models_Q D(a)$. So as a paraconsistent description

logic, QC description logic based on QC entailment has the approximate ability of classical description logic in reasoning.

- A complement of concept is defined to QC semantics in order to reverse both the information of being true and of being false. The complement of concept plays the same role as negation in classical description logics.
- A tableau algorithm for QC $\mathcal{ALCNQ}$ called QC tableau algorithm based on the complement of concept is proposed to implement paraconsistent reasoning in ontologies. QC transformation rules are developed from transformation rules in classical description logics by modifying them moderately. Furthermore, we prove that QC tableau algorithm for QC $\mathcal{ALCNQ}$ is terminable, sound and complete. Finally, we show that the complexity of QC tableau algorithm for QC $\mathcal{ALCNQ}$ is not higher than the complexity of tableau algorithms for $\mathcal{ALCNQ}$.

The paper is structured as follows. In Section 2, description logic $\mathcal{ALCNQ}$ as basic knowledge is briefly reviewed. In Section 3, we introduce QC semantics including QC weak semantics and QC strong semantics for description logic $\mathcal{ALCNQ}$ and define the QC entailment relationship. In Section 4, a terminable, sound and complete QC tableau algorithm with blocking technique to implement the querying. In Section 5, some related works on comparing the QC description logic to four-valued description logic are discussed. In Section 6, we conclude the main contributions in this paper and discuss the future work. Due to paper space limitations, proofs are absent, and detailed proofs are available in an Online Appendix.[1]

2 Preliminaries

In this section, we briefly review notation and terminology of the description logic $\mathcal{ALCNQ}$, but we basically assume that the reader is familiar with description logics. For comprehensive background reading, please refer to [17].

We assume that we are given a set of atomic concepts (or concept names), denoted by N_C, a set of roles (or role names), denoted by N_R and a set of individuals, denoted by N_I. With the symbols $\top$ and $\bot$, we furthermore denote the top concept and the bottom concept, respectively.

Complex concepts in $\mathcal{ALCNQ}$ can be formed from these inductively as follows.

- $\top$, $\bot$, and each atomic concept are concepts;
- If C, D are concepts, then $C \sqcup D$, $C \sqcap D$, and $\neg C$ are concepts;
- If C is a concept and R is a role, then $\forall R.C$, $\exists R.C$, $\geq nR.C$ and $\leq nR.C$ with n a non-negative integer are concepts.

Let C, D be concepts, a, b individuals and R a role. In description logic $\mathcal{ALCNQ}$, an ontology $\mathcal{O}$ is a pair $\langle \mathcal{T}, \mathcal{A} \rangle$, where $\mathcal{T}$ is called the $TBox$ (or $terminology$) of the ontology and $\mathcal{A}$ is called the $ABox$ of the ontology. $Assertions$ are of the form $C(a)$ or $R(a, b)$. $Inclusion\ axioms$ are of the form $C \sqsubseteq D$ which is called a $general\ concept\ inclusion$ (or GCI) axiom. Informally, an axiom $C(a)$ means that the individual a is an instance of concept C, and an axiom $R(a, b)$ means that individual a is related with individual b via the property R. The inclusion axiom $C \sqsubseteq D$ means that each individual of C is an individual of D.

[1] `http://www.is.pku.edu.cn/~zxw/QCALCNQAppendix.pdf`

Table 1. Syntax and semantics of concept constructors and axioms in $\mathcal{ALCNQ}$

Constructor Name	Syntax	Semantics
atomic concept A	A	$A^{\mathcal{I}} \subseteq \Delta^{\mathcal{I}}$
abstract role R	R	$R^{\mathcal{I}} \subseteq \Delta^{\mathcal{I}} \times \Delta^{\mathcal{I}}$
individuals I	o	$o^{\mathcal{I}} \in \Delta^{\mathcal{I}}$
inverse role	R^-	$(R^-)^I \subseteq \Delta^I \times \Delta^I$
top concept	$\top$	Δ^I
bottom concept	$\bot$	$\emptyset$
conjunction	$C_1 \sqcap C_2$	$C_1^{\mathcal{I}} \cap C_2^{\mathcal{I}}$
disjunction	$C_1 \sqcup C_2$	$C_1^{\mathcal{I}} \cup C_2^{\mathcal{I}}$
negation	$\neg C$	$\Delta^{\mathcal{I}} \setminus C^I$
exists restriction	$\exists R.C$	$\{x \mid \exists y, (x,y) \in R^{\mathcal{I}} \text{ and } y \in C^{\mathcal{I}}\}$
value restriction	$\forall R.C$	$\{x \mid \forall y, (x,y) \in R^{\mathcal{I}} \text{ implies } y \in C^{\mathcal{I}}\}$
number atleast restriction	$\geq nR$	$\{x \mid \sharp(\{y.(x,y) \in R^I\}) \geq n\}$
number atmost restriction	$\leq nR$	$\{x \mid \sharp(\{y.(x,y) \in R^I\}) \leq n\}$
qualifying atleast restriction	$\geq nR.C$	$\{x \mid \sharp(\{y.(x,y) \in R^I\} \text{ and } y \in C^{\mathcal{I}}) \geq n\}$
qualifying atmost restriction	$\leq nR.C$	$\{x \mid \sharp(\{y.(x,y) \in R^I\} \text{ and } y \in C^{\mathcal{I}}) \leq n\}$
Axiom Name	**Syntax**	**Semantics**
concept assertion	$C(a)$	$a^{\mathcal{I}} \in C^{\mathcal{I}}$
role assertion	$R(a,b)$	$(a^{\mathcal{I}}, b^{\mathcal{I}}) \in R^{\mathcal{I}}$
concept inclusion	$C_1 \sqsubseteq C_2$	$C_1^{\mathcal{I}} \subseteq C_2^{\mathcal{I}}$
individual equality	$a = b$	$a^{\mathcal{I}} = b^{\mathcal{I}}$
individual inequality	$a \neq b$	$a^{\mathcal{I}} \neq b^{\mathcal{I}}$

Remark 1. In Table 1, number restrictions can be taken as qualifying restrictions, i.e., $\leq nR \equiv \leq nR.\top$ and $\geq nR \equiv \geq nR.\top$, where $\equiv$ is the logically equivalent relationship.

The formal definition of the (model-theoretic) semantics of $\mathcal{ALCNQ}$ is given by means of interpretations $\mathcal{I} = (\Delta^{\mathcal{I}}, \cdot^{\mathcal{I}})$ consisting of a non-empty domain $\Delta^{\mathcal{I}}$ and a mapping $\cdot^{\mathcal{I}}$ satisfying the conditions in Table 1 where $\sharp M$ denotes the cardinality of a set M, interpreting concepts as subsets of the domain and roles as binary relations on the domain. An interpretation $\mathcal{I}$ satisfies $C \sqsubseteq D$ (written $\mathcal{I} \models C \sqsubseteq D$) iff $C^{\mathcal{I}} \subseteq D^{\mathcal{I}}$ and it satisfies a $TBox$ $\mathcal{T}$ if it satisfies every assertion in $\mathcal{T}$. Such an interpretation is called a *model* of $\mathcal{T}$ (written $\mathcal{I} \models \mathcal{T}$). A concept C is called $satisfiable$ w.r.t. a terminology $\mathcal{T}$ iff there is a model $\mathcal{I}$ of $\mathcal{T}$ with $C^{\mathcal{I}} \neq \emptyset$. A concept D $subsumes$ a concept C w.r.t $\mathcal{T}$ iff $C^{\mathcal{I}} \subseteq D^{\mathcal{I}}$ holds for each model $\mathcal{I}$ of $\mathcal{T}$. For an interpretation $\mathcal{I}$, an element $a^{\mathcal{I}} \in \Delta^{\mathcal{I}}$ is called an $instance$ of a concept C iff $a^{\mathcal{I}} \in C^{\mathcal{I}}$. An interpretation $\mathcal{I}$ $satisfies$ $C(a)$ iff $a^{\mathcal{I}} \in C^{\mathcal{I}}$, it satisfies $R(a,b)$ iff $(a^{\mathcal{I}}, b^{\mathcal{I}}) \in R^{\mathcal{I}}$ and it $satisfies$ an ABox $\mathcal{A}$ (written $\mathcal{I} \models \mathcal{A}$) if it satisfies every assertion in $\mathcal{A}$. An interpretation $\mathcal{I}$ $satisfies$ an $\mathcal{ALCNQ}$ ontology (i.e. is a model of the ontology) (written $\mathcal{I} \models \mathcal{O}$) iff it satisfies each axiom in both the ABox and the TBox.

An ontology is called $satisfiable$ ($unsatisfiable$) iff there exists (does not exist) such a model. An ABox is $consistent$ iff it has a model; and $inconsistent$ otherwise. Using the definition of satisfiability, an assertion α is said to be a $logical\ consequence$

of an ontology $\mathcal{O}$ (written $\mathcal{O} \models \alpha$) iff α is satisfied by every interpretation that satisfies $\mathcal{O}$. In $\mathcal{ALCNQ}$, reasoning tasks, i.e. the derivation of logical consequences, can be reduced to check satisfiability of ontologies. Tableau algorithms can implement checking satisfiability of ABox. Ian Horrocks et al [18] have developed an algorithm for combined TBox and ABox reasoning in description logic based on the following lemma.

Lemma 1. *Let C, D be concepts, $\mathcal{A}$ an ABox, $\mathcal{T}$ a TBox. We define*

$$C_{\mathcal{T}} := \sqcap_{C_i \sqsubseteq D_i \in \mathcal{T}} \neg C_i \sqcup D_i.$$

Then the following properties hold.

1. *C is satisfiable w.r.t. $\mathcal{T}$ iff $C \sqcap C_{\mathcal{T}}$ is satisfiable.*
2. *D subsumes C w.r.t. $\mathcal{T}$ iff $C \sqcap \neg D \sqcap C_{\mathcal{T}}$ is unsatisfiable.*
3. *$\mathcal{A}$ is consistent w.r.t. $\mathcal{T}$ iff $\mathcal{A} \cup \{a : C_{\mathcal{T}} \mid a \text{ occurs in } \mathcal{A}\}$ is consistent.*

3 Quasi-Classical Description Logic $\mathcal{ALCNQ}$

Quasi-classical description logic is a development of quasi-classical logic as defined in [15]. We assume that the language of QC description logic $\mathcal{ALCNQ}$ is $\mathcal{L}$.

3.1 Basic Definitions

Let A be an atomic concept and R be a role. A and $\neg A$ are *concept literals*. A concept C is in NNF, i.e., negation ($\neg$) occurs only in front of concept names. A *role literal* has the form $\forall R.C$, $\exists R.C$, $\geq nR$, $\leq nR$, $\geq nR.C$ or $\leq nR.C$ with C a concept in NNF. A *literal* is either a concept literal or a role literal, written by L. A *clause* is the disjunction of finite literals. Let $L_1 \sqcup \cdots \sqcup L_n$ be a clause, then $Lit(L_1 \sqcup \cdots \sqcup L_n)$ is the set of literals $\{L_1, \ldots, L_n\}$ that are in the clause. A clause is *empty clause*, denoted by $\Diamond$, if it has no literals. We define $\sim$ be a *complementation operation* such that $\sim A$ is $\neg A$ and $\sim (\neg A)$ is A. The $\sim$ operator is used to make some definitions clearer.

Definition 1. *Let $L_1 \sqcup \cdots \sqcup L_n$ be a clause that includes a literal disjunct L_i. The focus of $L_1 \sqcup \cdots \sqcup L_n$ by L_i, denoted $\otimes(L_1 \sqcup \cdots \sqcup L_n, L_i)$, is defined as the clause obtained by removing L_i from $Lit(L_1 \sqcup \cdots \sqcup L_n)$. In the case of a clause with just one disjunct, we assume $\otimes(L, L) = \bot$.*

Example 1. *Given a clause $L_1 \sqcup L_2 \sqcup L_3$, $\otimes(L_1 \sqcup L_2 \sqcup L_3, L_2) = L_1 \sqcup L_3$.*

In the following, we define strong interpretation and weak interpretation over domain $\Delta^{\mathcal{I}}$ by assigning to each concept C a pair $\langle +C, -C \rangle$ of (not necessarily disjoint) subsets of $C^{\mathcal{I}}$. Intuitively, $+C$ is the set of elements known to belong to the extension of C, while $-C$ is the set of elements known to be not contained in the extension of C. $+C$ and $-C$ are not necessarily disjoint and mutual complemental with respect to the domain. The *complemental set* of a set S w.r.t. an interpretation $\mathcal{I}$, denoted by $\overline{S}$, if $\overline{S} = \Delta^{\mathcal{I}} \setminus S$.

Definition 2. *Let $\mathcal{I}$ be a pair $\mathcal{I} = (\Delta^{\mathcal{I}}, \cdot^{\mathcal{I}})$ with $\Delta^{\mathcal{I}}$ as domain, where $\cdot^{\mathcal{I}}$ is a function assigning elements of $\Delta^{\mathcal{I}}$ to individuals, and subsets of $\Delta^{\mathcal{I}} \times \Delta^{\mathcal{I}}$ to concepts. $\mathcal{I}$ is a weak interpretation in QC $\mathcal{ALCNQ}$ if the conditions in Table 2 are satisfied.*

Table 2. Weak Semantics of QC $\mathcal{ALCNQ}$

Constructor Syntax	Weak Semantics
A	$A^{\mathcal{I}} = \langle +A, -A \rangle$, where $+A, -A \subseteq \Delta^{\mathcal{I}}$
R	$R^{\mathcal{I}} = \langle +R, -R \rangle$, where $+R, -R \subseteq \Delta^{\mathcal{I}} \times \Delta^{\mathcal{I}}$
o	$o^{\mathcal{I}} \in \Delta^{I}$
$\top$	$\langle \Delta^{I}, \emptyset \rangle$
$\bot$	$\langle \emptyset, \Delta^{I} \rangle$
$C_1 \sqcap C_2$	$\langle +C_1 \cap +C_2, -C_1 \cup -C_2 \rangle$,if $C_i^{\mathcal{I}} = \langle +C_i, -C_i \rangle$ for $i = 1, 2$
$C_1 \sqcup C_2$	$\langle +C_1 \cup +C_2, -C_1 \cap -C_2 \rangle$,if $C_i^{\mathcal{I}} = \langle +C_i, -C_i \rangle$ for $i = 1, 2$
$\neg C$	$(\neg C)^{\mathcal{I}} = \langle -C, +C \rangle$, if $C^{\mathcal{I}} = \langle +C, -C \rangle$
$\exists R.C$	$\langle \{x \mid \exists y, (x, y) \in +R$ and $y \in +C\}, \{x \mid \forall y, (x, y) \in +R$ implies $y \in -C\} \rangle$, if $C^{\mathcal{I}} = \langle +C, -C \rangle$ and $R^{\mathcal{I}} = \langle +R, -R \rangle$
$\forall R.C$	$\langle \{x \mid \forall y, (x, y) \in +R$ implies $y \in +C\}, \{x \mid \exists y, (x, y) \in +R$ and $y \in -C\} \rangle$, if $C^{\mathcal{I}} = \langle +C, -C \rangle$ and $R^{\mathcal{I}} = \langle +R, -R \rangle$
$\geq nR$	$\langle \{x \mid \sharp(\{y.(x, y) \in +R\}) \geq n\}, \{x \mid \sharp(\{y.(x, y) \in +R\}) < n\} \rangle$, if $R^{\mathcal{I}} = \langle +R, -R \rangle$
$\leq nR$	$\langle \{x \mid \sharp(\{y.(x, y) \in +R\}) \leq n\}, \{x \mid \sharp(\{y.(x, y) \in +R\}) > n\} \rangle$, if $R^{\mathcal{I}} = \langle +R, -R \rangle$
$\geq nR.C$	$\langle \{x \mid \sharp(\{y.(x, y) \in +R\}$ and $y \in +C) \geq n\}$, $\{x \mid \sharp(\{y.(x, y) \in +R\}$ and $y \notin -C) < n\} \rangle$ if $C^{\mathcal{I}} = \langle +C, -C \rangle$ and $R^{\mathcal{I}} = \langle +R, -R \rangle$
$\leq nR.C$	$\langle \{x \mid \sharp(\{y.(x, y) \in +R\}$ and $y \notin -C) \leq n\}$, $\{x \mid \sharp(\{y.(x, y) \in +R\}$ and $y \in +C) > n\} \rangle$ if $C^{\mathcal{I}} = \langle +C, -C \rangle$ and $R^{\mathcal{I}} = \langle +R, -R \rangle$

Definition 3. *Let $\models_w$ be a satisfiability relation called weak satisfaction. For a weak interpretation $\mathcal{I}$, we define $\models_w$ as follows:*

- $\mathcal{I} \models_w C(a)$ *iff* $a^{\mathcal{I}} \in +C, C^{\mathcal{I}} = \langle +C, -C \rangle$;
- $\mathcal{I} \models_w R(a, b)$ *iff* $(a^{\mathcal{I}}, b^{\mathcal{I}}) \in +R, R^{\mathcal{I}} = \langle +R, -R \rangle$;
- $\mathcal{I} \models_w C_1 \sqsubseteq C_2$ *iff* $+C_1 \subseteq +C_2$, *for* $i = 1, 2$, $C_i^{\mathcal{I}} = \langle +C_i, -C_i \rangle$;
- $\mathcal{I} \models_w a = b$ *iff* $a^{\mathcal{I}} = b^{\mathcal{I}}$;
- $\mathcal{I} \models_w a \neq b$ *iff* $a^{\mathcal{I}} \neq b^{\mathcal{I}}$;

where C, C_1, C_2 are concepts, R is a role and a is an individual.

A weak interpretation $\mathcal{I}$ *satisfies* a terminology $\mathcal{T}$ iff $\mathcal{I} \models_w C \sqsubseteq D$ for each GCI $C \sqsubseteq D$ in $\mathcal{T}$. Such a weak interpretation is called a *QC weak model* of $\mathcal{T}$ (written $\mathcal{I} \models_w \mathcal{T}$). A weak interpretation $\mathcal{I}$ *satisfies* an ABox $\mathcal{A}$ iff $\mathcal{I} \models_w \alpha$ for each assertion α in $\mathcal{A}$, where $\alpha \in \{C(a), R(a, b), a = b, a \neq b\}$. Such a weak interpretation is called a *QC weak model* of $\mathcal{A}$ (written $\mathcal{I} \models_w \mathcal{A}$).

The definition of strong interpretation is similar to weak interpretation. The main difference is that the definition for disjunction of concept in strong interpretation is more restricted.

Definition 4. *Let $\mathcal{I}$ be a pair $\mathcal{I} = (\Delta^{\mathcal{I}}, \cdot^{\mathcal{I}})$ with $\Delta^{\mathcal{I}}$ as domain, where $\cdot^{\mathcal{I}}$ is a function assigning elements of $\Delta^{\mathcal{I}}$ to individuals, and subsets of $\Delta^{\mathcal{I}} \times \Delta^{\mathcal{I}}$ to concepts. $\mathcal{I}$ is a* *strong interpretation in* $QC\ \mathcal{ALCNQ}$ *if the conditions in Table 2, except conjunction concept and disjunction concept, are satisfied and* $\mathcal{I}$ *satisfies as follows:*

- *conjunction of concept* $C_1 \sqcap C_2$: $(C_1 \sqcap C_2)$ *under strong interpretation* $\mathcal{I}$ *is defined as* $\langle +C_1 \cap +C_2, (-C_1 \cup -C_2) \cap (-C_1 \cup \overline{+C_2}) \cap (\overline{+C_1} \cup -C_2) \rangle$, *if* $C_i^{\mathcal{I}} = \langle +C_i, -C_i \rangle$ *for* $i = 1, 2$;
- *disjunction of concept* $C_1 \sqcup C_2$: $(C_1 \sqcup C_2)$ *under strong interpretation* $\mathcal{I}$ *is defined as* $\langle (+C_1 \cup +C_2) \cap (\overline{-C_1} \cup +C_2) \cap (+C_1 \cup \overline{-C_2}), -C_1 \cap -C_2 \rangle$, *if* $C_i^{\mathcal{I}} = \langle +C_i, -C_i \rangle$ *for* $i = 1, 2$.

Remark 2. In Definition 4, the disjunction of concept under strong interpretation is defined in order to hold three inference rules and the conjunction of concept under strong interpretation is defined in order to hold De Morgan law with disjunction of concept.

Definition 5. *Let* $\models_s$ *be a satisfiability relation called* $strong\ satisfaction$. *For a strong interpretation* $\mathcal{I}$, *we define* $\models_s$ *as follows:*

- $\mathcal{I} \models_s C(a)$ *iff* $a^{\mathcal{I}} \in +C$ *where* $C^{\mathcal{I}} = \langle +C, -C \rangle$;
- $\mathcal{I} \models_s R(a, b)$ *iff* $(a^{\mathcal{I}}, b^{\mathcal{I}}) \in +R$ *where* $R^{\mathcal{I}} = \langle +R, -R \rangle$;
- $\mathcal{I} \models_s C_1 \sqsubseteq C_2$ *iff* $\overline{-C_1} \subseteq +C_2$, $+C_1 \subseteq +C_2$ *and* $-C_2 \subseteq -C_1$, *for* $i = 1, 2$, $C_i^{\mathcal{I}} = \langle +C_i, -C_i \rangle$;
- $\mathcal{I} \models_s a = b$ *iff* $a^{\mathcal{I}} = b^{\mathcal{I}}$;
- $\mathcal{I} \models_s a \neq b$ *iff* $a^{\mathcal{I}} \neq b^{\mathcal{I}}$;

where C, C_1, C_2 *are concepts,* R *is a role and* a *is an individual.*

A strong interpretation $\mathcal{I}$ $satisfies$ a terminology $\mathcal{T}$ iff $\mathcal{I} \models_s C \sqsubseteq D$ for each GCI $C \sqsubseteq D$ in $\mathcal{T}$. Such a strong interpretation is called a $QC\ strong\ model$ of $\mathcal{T}$ (written $\mathcal{I} \models_s \mathcal{T}$). A strong interpretation $\mathcal{I}$ $satisfies$ an ABox $\mathcal{A}$ iff $\mathcal{I} \models_s \alpha$ for each assertion α in $\mathcal{A}$, where $\alpha \in \{C(a), R(a, b), a = b, a \neq b\}$. Such a strong interpretation is called a $QC\ strong\ model$ of $\mathcal{A}$ (written $\mathcal{I} \models_s \mathcal{A}$).

By Definition 4 and Definition 5, we conclude several properties in the following theorems.

Theorem 1. *Let* $L_1 \sqcup \cdots \sqcup L_n$ *be a clause and* a *be an individual.* $\mathcal{I} \models_s L_1 \sqcup \cdots \sqcup L_n(a)$ *iff for some* $L_i \in Lit(L_1 \sqcup \cdots \sqcup L_n)$, $a^{\mathcal{I}} \in +L_i$ *and* $a^{\mathcal{I}} \notin -L_i$; *or for all* $L_i \in Lit(L_1 \sqcup \cdots \sqcup L_n)$, $a^{\mathcal{I}} \in +L_i$ *and* $a^{\mathcal{I}} \in -L_i$; *where* $L_i^{\mathcal{I}} = \langle +L_i, -L_i \rangle$.

Theorem 2. *Let* C *be a concept and* R *be a role. For any weak (or strong) interpretation* $\mathcal{I}$ *defined in Definition 2 (or Definition 4), we have*

1. $(\neg(\leq nR.C))^{\mathcal{I}} = (> nR.C)^{\mathcal{I}}$ *and* $(\neg(\geq nR.C))^{\mathcal{I}} = (< nR.C)^{\mathcal{I}}$;
2. $(\exists R.C)^{\mathcal{I}} = (\geq 1R.C)^{\mathcal{I}}$ *and* $(\forall R.C)^{\mathcal{I}} = (< 1R.\neg C)^{\mathcal{I}}$.

Theorem 3. *Let* $\mathcal{I}$ *and* α *be an interpretation and a formula in* $\mathcal{ALCNQ}$ *respectively.*

$$\text{If } \mathcal{I} \models_s \alpha \text{ then } \mathcal{I} \models_w \alpha.$$

Theorem 3 shows that a strong model is a weak model but not vice versa.

Example 2. *Given an ABox* $\mathcal{A} = \{C(a), \neg C(a)\}$. *We assume that there exists an interpretation* $\mathcal{I}$ *such that* $a^{\mathcal{I}} \in +C$ *and* $a^{\mathcal{I}} \in -C$ *where* $C^{\mathcal{I}} = \langle +C, -C \rangle$. *Clearly,* $\mathcal{I} \models_w C \sqcup D(a)$ *while* $\mathcal{I} \not\models_s C \sqcup D(a)$.

Definition 6. *Given a set of formula* $\mathcal{K}$ *and an axiom* α *in* $\mathcal{ALCNQ}$, $\mathcal{K}$ *QC entails* α, *denoted by* $\mathcal{K} \models_Q \alpha$, *iff for every interpretation* $\mathcal{I}$, *if* $\mathcal{I}$ *is a strong model of every formula of* $\mathcal{K}$ *then* $\mathcal{I}$ *is a weak model of* α. *In this case,* $\models_Q$ *is a QC entailment (relation) between* $\mathcal{K}$ *and* α.

That $\models_Q$ is non-trivializable in the sense that when an ontology $\mathcal{O}$ is classically inconsistent, it is not the case that any axiom is entailed by $\mathcal{O}$ in QC $\mathcal{ALCNQ}$.

Example 3. *Given an ABox* $\mathcal{A} = \{B(a), \neg B(a)\}$ *and an atomic concept* A *in QC* $\mathcal{ALCNQ}$. *So* $\{B(a), \neg B(a)\}$ *is classically inconsistent. However it is not the case that* $\mathcal{A} \models_Q C(a)$ *holds, since there exists an interpretation* $\mathcal{I}$ *such that* $a^{\mathcal{I}} \in +B$ *and* $a^{\mathcal{I}} \in -B$ *where* $B^{\mathcal{I}} = \langle +B, -B \rangle$. *So* $\mathcal{I} \models_s B \sqcap \neg B(a)$, *but* $\mathcal{I} \not\models_w A(a)$ *since* $A(a)$ *does not occur in* $\mathcal{A}$.

Example 4. *Let an ABox* $\mathcal{A} = \{B \sqcup C(a), \neg B(a)\}$. *For all interpretations* $\mathcal{I}$ *such that* ($a^{\mathcal{I}} \in +B$ *or* $a^{\mathcal{I}} \in +C$) *and* $a^{\mathcal{I}} \in -B$, *if* $\mathcal{I} \models_s B \sqcup C(a)$ *and* $\mathcal{I} \models_s \neg B(a)$ *then* $\mathcal{I} \models_w C(a)$. *Hence,* $\mathcal{A} \models_Q B \sqcup C(a)$, $\mathcal{A} \models_Q \neg B(a)$, *and* $\mathcal{A} \models_Q C(a)$.

Example 5. *Suppose an ABox* $\mathcal{A}$ *is empty. Now consider the classical* $\top = A \sqcup \neg A$. *Here* $\mathcal{A} \models_Q A \sqcup \neg A(a)$ *does not hold. Since* $\mathcal{A}$ *strongly satisfies every formula in* $\mathcal{A}$, *but* $\mathcal{A}$ *does not weakly satisfy* $A \sqcup \neg A(a)$.

Example 5 shows that QC description logic is weaker than description logic in ability of reasoning.

Definition 7. *A complement of a concept* C *is defined by* $\overline{C}^{\mathcal{I}} = \langle \overline{+C}, \overline{-C} \rangle$ *if* $C^{\mathcal{I}} = \langle +C, -C \rangle$ *where* $\mathcal{I}$ *is a (strong or weak) interpretation of* C, *denoted by* $\overline{C}$. *We call* $\overline{C}$ *is the complement of* C.

Remark 3. The intuition behind complement of concepts is to reverse both the information of being true and of being false. And any $\mathcal{ALCNQ}$ ABox doesn't contain complement concepts.

In QC $\mathcal{ALCNQ}$, the set of signed formulae of $\mathcal{L}$ is denoted $\mathcal{L}^*$ and is defined as $\mathcal{L} \cup \{\overline{C} | C \in \mathcal{L}\}$. In the following, we discuss the QC tableau algorithm based on the language $\mathcal{L}^*$.

Theorem 4. *For concept name* A, *concepts* C, D *in* N_C, *individuals* a, b *in* N_I *and a role* R *in* N_R *in QC* $\mathcal{ALCNQ}$.

(1) $\mathcal{I} \models_x \overline{A}(a)$ *iff* $\mathcal{I} \not\models_x A(a)$
(2) $\mathcal{I} \models_x \overline{\neg A}(a)$ *iff* $\mathcal{I} \models_x \neg\overline{A}(a)$
(3) $\mathcal{I} \models_x \overline{\overline{A}}(a)$ *iff* $\mathcal{I} \models_x A(a)$
(4) $\mathcal{I} \models_w \overline{C} \sqcup \overline{D}(a)$ *iff* $\mathcal{I} \models_w \overline{C \sqcap D}(a)$
(5) $\mathcal{I} \models_w \overline{C} \sqcap \overline{D}(a)$ *iff* $\mathcal{I} \models_w \overline{C \sqcup D}(a)$
(6) $\mathcal{I} \models_x \overline{\top}(a)$ *iff* $\mathcal{I} \models_x \bot(a)$
(7) $\mathcal{I} \models_x \overline{(\forall R.C)}(a)$ *iff* $\mathcal{I} \models_x \exists R.\overline{C}(a)$
(8) $\mathcal{I} \models_x \overline{(\exists R.C)}(a)$ *iff* $\mathcal{I} \models_x \forall R.\overline{C}(a)$
(9) $\mathcal{I} \models_x \overline{(\geq nR)}(a)$ *iff* $\mathcal{I} \models_x \leq n-1R(a)$
(10) $\mathcal{I} \models_x \overline{(\leq nR)}(a)$ *iff* $\mathcal{I} \models_x \geq n+1R(a)$
(11) $\mathcal{I} \models_x \overline{(\geq nR.C)}(a)$ *iff* $\mathcal{I} \models_x \leq n-1R.C(a)$
(12) $\mathcal{I} \models_x \overline{(\leq nR.C)}(a)$ *iff* $\mathcal{I} \models_x \geq n+1R.C(a)$

Here $\models_x$ *is a place-holder for both* $\models_w$ *and* $\models_s$.

Theorem 5. *Let $\mathcal{I}$ be an interpretation and let C, D be concepts.*

1. $\mathcal{I} \models_s C \sqsubseteq D$ *iff for any individual* $a \in N_I$, $\mathcal{I} \models_s \neg C \sqcup D(a)$.
2. $\mathcal{I} \models_w C \sqsubseteq D$ *iff for any individual* $a \in N_I$, $\mathcal{I} \models_w \overline{C} \sqcup D(a)$.

In QC $\mathcal{ALCNQ}$, a $QC\ ABox$ ($QC\ TBox$) is a set whose axioms (concept inclusions) are formulae in $\mathcal{L}^*$. A concept C is $QC\ satisfiable$ w.r.t. a QC ABox $\mathcal{A}$ if there exists an individual a such that $\mathcal{A} \models_Q C(a)$, and $QC\ unsatisfiable$ w.r.t. $\mathcal{A}$ otherwise. A concept C is $QC\ satisfiable$ w.r.t. a QC ABox $\mathcal{A}$ and a QC TBox $\mathcal{T}$ if there exists an individual a such that $\mathcal{T} \cup \mathcal{A} \models_Q C(a)$, and $QC\ unsatisfiable$ w.r.t. $\mathcal{T}$ and $\mathcal{A}$ otherwise. A QC ABox $\mathcal{A}$ is QC $consistent$ if each concept occurring in $\mathcal{A}$ is QC satisfiable w.r.t. $\mathcal{A}$, and $QC\ inconsistent$ otherwise. A QC ABox $\mathcal{A}$ is $QC\ consistent$ w.r.t. a QC TBox $\mathcal{T}$ if each concept occurring in $\mathcal{A}$ is QC satisfiable w.r.t. $\mathcal{T}$ and $\mathcal{A}$, and $QC\ inconsistent$ w.r.t. $\mathcal{T}$ otherwise. A set of QC ABoxes $\mathcal{S}$ is $QC\ consistent$ iff there is a QC satisfiable QC ABox in $\mathcal{S}$, $QC\ inconsistent$ otherwise. A set of QC ABoxes $\mathcal{S}$ is $QC\ satisfiable$ iff there is a QC satisfiable QC ABox in $\mathcal{S}$, $QC\ unsatisfiable$ otherwise.

A concept C is in $QC\ NNF$, i.e., iff complement only occurs over concept names. By Theorem 4, for any concept $C \in \mathcal{L}^*$, $\overline{C}$ can be transformed into its QC NNF. For a concept, we define $clos(C)$ as the smallest set that contains C and is closed under sub-concepts, its NNF and QC NNF of its complement. We denote $clos(\mathcal{A}) = \bigcup_{C \in \mathcal{A}} clos(C)$. Clearly, the size of $clos(\mathcal{A})$ is polynomial in the size of $\mathcal{A}$.

Theorem 6. *Let $\mathcal{O} = (\mathcal{T}, \mathcal{A})$ be an $\mathcal{ALCNQ}$ ontology, a be a individual and ι be a new individual not occurring in $\mathcal{O}$. Then the following hold.*

1. $\mathcal{A} \models_Q C(a)$ *iff* $\mathcal{A} \cup \{\overline{C}(a)\}$ *is QC unsatisfiable.*
2. $\mathcal{A} \models_Q C \sqsubseteq D$ *iff* $\mathcal{A} \cup \{C \sqcap \overline{D}(\iota)\}$ *is QC unsatisfiable.*

4 QC Tableau Algorithm for $\mathcal{ALCNQ}$

In this section, we develop a tableau algorithm called by $QC\ tableau\ algorithm$ for reasoning with inconsistency in QC $\mathcal{ALCNQ}$. $R_{\mathcal{A}}$ denotes the set of roles occurring in $\mathcal{A}$, and $U_{\mathcal{A}}$ denotes the set of individuals occurring in $\mathcal{A}$.

4.1 QC Tableau

Definition 8. *Given a QC ABox $\mathcal{A}$, $T = (S, \mathcal{L}, \mathcal{E}, \mathcal{J})$ is a QC tableau for $\mathcal{A}$ iff*

- *S is a non-empty set;*
- *$\mathcal{L} : S \to 2^{clos(\mathcal{A})}$ maps each element in S to a set of concepts;*
- *$\mathcal{E} : R_{\mathcal{A}} \to 2^{S \times S}$ maps each role to a set of pairs of elements in S;*
- *$\mathcal{J} : U_{\mathcal{A}} \to S$ maps individuals occurring in $\mathcal{A}$ to elements in S.*

Furthermore, for all $s, t \in S$, $C, C_1, C_2 \in clos(\mathcal{A})$ and T satisfies:

(P1) if $C \in \mathcal{L}(s)$, then $\overline{C} \notin \mathcal{L}(s)$,
(P2) if $C_1 \sqcap C_2 \in \mathcal{L}(s)$, then $C_1 \in \mathcal{L}(s)$ and $C_2 \in \mathcal{L}(s)$,

(P3) if $C_1 \sqcup C_2 \in \mathcal{L}(s)$*, then*
  (a) if $\sim C_i \in \mathcal{L}(s)$ *for some* $(i \in \{1, 2\})$*, then* $\otimes(C_1 \sqcup C_2, C_i) \in \mathcal{L}(s)$*,*
  (b) else $C_1 \in \mathcal{L}(s)$ *or* $C_2 \in \mathcal{L}(s)$*,*
(P4) if $\forall R.C \in \mathcal{L}(s)$ *and* $\langle s, t\rangle \in \mathcal{E}(R)$*, then* $C \in \mathcal{L}(t)$*,*
(P5) if $\exists R.C \in \mathcal{L}(s)$*, then there is some* $t \in S$ *such that* $\langle s, t\rangle \in \mathcal{E}(R)$ *and* $C \in \mathcal{L}(t)$*,*
(P6) if $\leq nR.C \in \mathcal{L}(s)$*, then* $\sharp R^T(s, C) \leq n$*,*
(P7) if $\geq nR.C \in \mathcal{L}(s)$*, then* $\sharp R^T(s, C) \geq n$*,*
(P8) if $(\bowtie nR.C) \in \mathcal{L}(s)$ *and* $\langle s, t\rangle \in \mathcal{E}(R)$ *then* $C \in \mathcal{L}(t)$ *or* $\overline{C} \in \mathcal{L}(t)$*,*
(P9) if $a : C \in \mathcal{A}$*, then* $C \in \mathcal{L}(\mathcal{J}(a))$*,*
(P10) if $(a, b) : R \in \mathcal{A}$*, then* $\langle \mathcal{J}(a), \mathcal{J}(b)\rangle \in \mathcal{E}(R)$*,*
(P11) if $a \neq b \in \mathcal{A}$*, then* $\mathcal{J}(a) \neq \mathcal{J}(b)$*,*

where $\bowtie$ *is a place-holder for both* $\leq$ *and* $\geq$*, and* $R^T(s, C) = \{t \in S \mid \langle s, t\rangle \in \mathcal{E}(R)$ *and* $C \in \mathcal{L}(t)\}$*.*

Theorem 7. *A QC ABox* $\mathcal{A}$ *is QC satisfiable iff there is a QC tableau for* $\mathcal{A}$*.*

4.2 QC Tableau Algorithm

In this subsection, we develop a QC tableau algorithm based on QC transformation rules which are obtained by modifying the expansion rules in [18] in order to obtain a QC tableau for a QC ABox $\mathcal{A}$. In the following, we borrow the reasoning technique from Horrocks [18,19] to implement paraconsistent reasoning under QC semantics.

In QC $\mathcal{ALCNQ}$, a completion forest $\mathcal{F}$ for a QC ABox $\mathcal{A}$ is a collection of trees whose distinguished root nodes are possibly connected by edges in an arbitrary way. Moreover, each node x is labeled with a set $\mathcal{L}(x) \subseteq clos(\mathcal{A})$ and each edge $\langle x, y\rangle$ is labeled with a set $\mathcal{L}(\langle x, y\rangle) \in R$ of (possibly inverse) roles occurring in $\mathcal{A}$. Finally, completion forests come with an explicit inequality relation $\neq$ on nodes and an explicit equality relation $=$ which are implicitly assumed to be symmetric.

If nodes x and y are *connected* by an edge $\langle x, y\rangle$ with $R \in \mathcal{L}(\langle x, y\rangle)$, then y is called an *R-successor* of x and x is called an *R-predecessor* of y. If y is an R-successor of x, then y is called an *R-neighbor* of x. A node y is a *successor* (resp. *predecessor* or *neighbor*) of y if it is an R-successor (resp. R-predecessor or R-neighbor) of y for some role R. Finally, *ancestor* is the transitive closure of predecessor.

For a role R, a concept C and a node x in $\mathcal{F}$ we define $R^{\mathcal{F}}(x, C)$ by $R^{\mathcal{F}}(x, C) := \{y \mid y$ is R-neighbor of x and $C \in \mathcal{L}(y)\}$. A node is *blocked* iff it is not a root node and it is either directly or indirectly blocked. A node x is *directly blocked* iff none of its ancestors are blocked, and it has ancestors x', y and y' such that

- y is not a root node;
- x is a successor of x' and y is a successor of y';
- $\mathcal{L}(x) = \mathcal{L}(y)$ and $\mathcal{L}(x') = \mathcal{L}(y')$;
- $\mathcal{L}(\langle x', x\rangle) = \mathcal{L}(\langle y', y\rangle)$.

In QC $\mathcal{ALCNQ}$, given a QC ABox $\mathcal{A}$ and a role R, the algorithm initializes a completion forest $\mathcal{F}_{\mathcal{A}}$ consisting only of root nodes. More precisely, $\mathcal{F}_{\mathcal{A}}$ contains a root node x_0^i for each individual $a^i \in U_{\mathcal{A}}$, and an edge $\langle x_0^i, x_0^j\rangle$ if $\mathcal{A}$ contains an assertion $(a_i, a_j) : R$ for some R. The labels of these nodes and edges and the relations $\neq$ and $=$ are initialized as follows:

(1) $\mathcal{L}(x_i^0) := \{C \mid a_i : C \in \mathcal{A}\}$, and
(2) $\mathcal{L}(\langle x_0^i, x_0^j \rangle) := \{R \mid (a_i, a_j) : R \in \mathcal{A}\}, x_0^i \neq x_0^j$ iff $a_i \neq a_j \in \mathcal{A}$, and the $=$ -relation is initialized to be empty.

$\mathcal{F}_{\mathcal{A}}$ is then expanded by repeatedly applying the rules from Table 3.

Definition 9. *Let $\mathcal{O} = (\mathcal{T}, \mathcal{A})$ be an ontology, C_1, C_2, C concepts, R a role and x, y, z individuals in QC $\mathcal{ALCNQ}$. We define QC $transformation\ rules$ in Table 3.*

For a node x, $\mathcal{L}(x)$ is said to contain a $clash$ if $\{A, \overline{A}\} \in \mathcal{L}(x)$ for some concept name $A \in N_C$, or if there is some concept $\leq nR.C \in \mathcal{L}(x)$ and x has $n+1$ R-neighbors $y_0, \ldots, y_n$ with $C \in \mathcal{L}(y_i)$ and $y_i \neq y_j$ for all $0 \leq i < j \leq n$. A completion forest is $clash\text{-}free$ if none of its nodes contains a clash, and it is complete if no QC transformation rule from Table 3 can be applied to it. For a QC ABox $\mathcal{A}$, the algorithm starts with the completion forest $\mathcal{F}_{\mathcal{A}}$. It applies the QC transformation rules in Table 3, stopping when a clash occurs, and answers " $\mathcal{A}$ is QC satisfiable " iff the QC transformation rules can be applied in such a way that they yield a complete and clash-free completion forest; and answering " $\mathcal{A}$ is QC unsatisfiable " otherwise.

Theorem 8. *Let $\mathcal{A}$ be a QC ABox in QC $\mathcal{ALCNQ}$.*

1. *The QC tableau algorithm terminates when started for $\mathcal{A}$.*
2. *If the QC transformation rules can be applied to $\mathcal{A}$ such that they yield a complete and clash-free completion forest, then $\mathcal{A}$ has a QC tableau.*
3. *If $\mathcal{A}$ has a QC tableau, then the QC transformation rules can be applied to $\mathcal{A}$ such that they yield a complete and clash-free completion forest.*
4. *The QC tableau algorithm is a decision procedure for the consistency of QC ABoxes and the satisfiability and subsumption of concepts with respect to terminologies.*

Theorem 9. *Satisfiability of QC ABoxes is PSPACE-complete in $\mathcal{ALCNQ}$.*

5 Related Work

In this section, we compare *QC* description logic with four-valued description logic [7,9,20] which is a paraconsistent logic by integrating description logic with Belanp's four-valued logic [12]. The similarities and differences between QC logic and four-valued logic still exist between *QC* DLs and four-valued DLs.

In QC DLs, the weak semantics is analogous to the four-valued semantics in four-valued DLs. In fact, inconsistent or incomplete can be tolerated under two semantics and the contradiction can be taken as a part of ontologies. Below we give a presentation of the definition 4-model of four-valued DL which was proposed in [7,9,20].

Definition 10 (Ma [9]). *Let C, D be concepts and a be an individual in four-valued DLs. An interpretation $\mathcal{I}$ is a 4-model of an axiom $C(a)$ (written by $\mathcal{I} \models_4 C(a)$) iff $a^{\mathcal{I}} \in proj^+(C^{\mathcal{I}})$; $\mathcal{I}$ is a 4-model of a material inclusion $C \mapsto D$ (written by $\mathcal{I} \models_4 C \mapsto D$) iff $\overline{proj^-(C^{\mathcal{I}})} \subseteq proj^+(D^{\mathcal{I}})$; $\mathcal{I}$ is a 4-model of an internal inclusion $C \sqsubset D$ (written by $\mathcal{I} \models_4 C \sqsubset D$) iff $proj^+(C^{\mathcal{I}}) \subseteq proj^+(D^{\mathcal{I}})$ and $\mathcal{I}$ is a 4-model of a strong inclusion $C \rightarrow D$ (written by $\mathcal{I} \models_4 C \rightarrow D$) iff $proj^+(C^{\mathcal{I}}) \subseteq proj^+(D^{\mathcal{I}})$ and $proj^-(D^{\mathcal{I}}) \subseteq proj^-(C^{\mathcal{I}})$.*

Table 3. QC Transformation Rules in QC $\mathcal{ALCNQ}$

1. The $\rightarrow_{\sqcap}$-rule Condition: $C_1 \sqcap C_2 \in \mathcal{L}(x)$, x is not indirectly blocked, and $\{C_1, C_2\} \not\subseteq \mathcal{L}(x)$. Action: $\mathcal{L}(x) := \mathcal{L}(x) \cup \{C_1(x), C_2(x)\}$.
2. The $\rightarrow_{\sqcup}$-rule Condition: $C_1 \sqcup C_2 \in \mathcal{L}(x)$, x is not indirectly blocked, and $\{C_1, C_2, \sim C_1, \sim C_2\} \cap \mathcal{L}(x) = \emptyset$. Action: $\mathcal{L}(x) := \mathcal{L}(x) \cup \{E\}$ for some $E \in \{C_1, C_2\}$.
3.The $\rightarrow_{QC}$-rule Condition: $C_1 \sqcup C_2 \in \mathcal{L}(x)$, x is not indirectly blocked, and $\sim C_i \in \mathcal{L}(x)$(for some $i \in \{1, 2\}$). Action: $\mathcal{L}(x) := \mathcal{L}(x) \cup \{\otimes(C_1 \sqcup C_2, C_i)\}$.
4. The $\rightarrow_{\exists}$-rule Condition: $\exists R.C \in \mathcal{L}(x)$, x is not blocked, and x has no R-neighbor y with $C \in \mathcal{L}(y)$. Action: create a new node y with $\mathcal{L}(\langle x, y\rangle) := \{R\}$ and $\mathcal{L}(y) := \{C\}$.
5. The $\rightarrow_{\forall}$-rule Condition: (1) $\forall R.C \in \mathcal{L}(x)$, x is not indirectly blocked, and (2) there is an R-neighbor y of x with $C \in \mathcal{L}(y)$. Action: $\mathcal{L}(y) := \mathcal{L}(y) \cup \{C\}$.
6. The choose-rule Condition: $(\bowtie nR.C) \in \mathcal{L}(x)$, x is not indirectly blocked, and there is an R-neighbor y of x with $\{C, \overline{C}\} \cap \mathcal{L}(y) = \emptyset$. Action: $\mathcal{L}(y) := \mathcal{L}(y) \cup \{E\}$ for some $E \in \{C, \overline{C}\}$.
7. The $\rightarrow_{\geq}$-rule Condition: (1)$\geq nR.C \in \mathcal{L}(x)$, x is not blocked, and (2) there are no n R-neighbors $y_1, \ldots, y_n$ such that $C \in \mathcal{L}(y_i)$ and $y_i \neq y_j$ for $1 \leq i < j \leq n$. Action: create n new nodes $y_1, \ldots, y_n$ with $\mathcal{L}(\langle x, y_i\rangle) := \{R\}$, $\mathcal{L}(y_i) := \{C\}$, and $y_i \neq y_j$.
8. The $\rightarrow_{\leq}$-rule Condition: (1) $\leq nR.C \in \mathcal{L}(x)$, x is not indirectly blocked, and $\sharp R^{\mathcal{F}}(x, C) > n$, (2) there are R-neighbors y, z of x with not $y \neq z$, y is neither a root node nor an ancestor of z, and $C \in \mathcal{L}(y) \cap \mathcal{L}(z)$. Action: (1)$\mathcal{L}(z) := \mathcal{L}(z) \cup \mathcal{L}(y)$; and (2)if z is an ancestor of x then $\mathcal{L}(\langle z, x\rangle) := \mathcal{L}(\langle z, x\rangle)$ else $\mathcal{L}(\langle x, z\rangle) := \mathcal{L}(\langle x, z\rangle) \cup \mathcal{L}(\langle x, y\rangle)$ (3) $\mathcal{L}(\langle x, y\rangle) := \emptyset$; (4) set $u \neq z$ for all u with $u \neq y$.
9. The $\rightarrow_{\leq_r}$-rule Condition: (1)$\leq nR.C \in \mathcal{L}(x)$, and $\sharp R^{\mathcal{F}}(x, C) > n$ and (2) there are two R-neighbors y, z of x which are both root nodes, $C \in \mathcal{L}(y) \cap \mathcal{L}(z)$, and not $y \neq z$. Action: (1) $\mathcal{L}(z) := \mathcal{L}(z) \cup \mathcal{L}(y)$; (2) for all edges $\langle y, w\rangle$: (a) if the edge $\langle z, w\rangle$ does not exist, create it with $\mathcal{L}(\langle z, w\rangle) := \emptyset$ and (b) $\mathcal{L}(\langle z, w\rangle) := \mathcal{L}(\langle z, w\rangle) \cup \mathcal{L}(\langle y, w\rangle)$; (3) for all edges $\langle w, y\rangle$: (a) if the edge $\langle w, z\rangle$ does not exist, create it with $\mathcal{L}(\langle w, z\rangle) := \emptyset$ and (b) $\mathcal{L}(\langle w, z\rangle) := \mathcal{L}(\langle w, z\rangle) \cup \mathcal{L}(\langle w, y\rangle)$; (4) set $\mathcal{L}(y) := \emptyset$ and remove all edges to/from y; (5) set $u \neq z$ for all u with $u \neq y$ and set $y \neq z$.

Definition 11 (Ma [9]). *Let $\mathcal{O}$ be an ontology and α be an axiom in four-valued DL. $\mathcal{O}$ entails α under four semantics, denoted by $\mathcal{O} \models_4 \alpha$, iff for any model $\mathcal{I}$ of $\mathcal{O}$,[2] $\mathcal{I}$ is a 4-model of α.*

The follow property shows that there exists a close relationship between weak models and 4-models in description logic.

Theorem 10. *Let C, D be concepts, a be an individual and $\mathcal{I}$ be an interpretation in description logic.*

$$1.\ \mathcal{I} \models_w C(a) \text{ iff } \mathcal{I} \models_4 C(a);\quad 2.\ \mathcal{I} \models_w C \sqsubseteq D \text{ iff } \mathcal{I} \models_4 C \sqsubset D.$$

The follow property shows that strong models is stronger than 4-models in description logic.

Theorem 11. *Let C, D be concepts, a be an individual and $\mathcal{I}$ be an interpretation in description logic.*

1. $\mathcal{I} \models_4 C(a)$ *if* $\mathcal{I} \models_s C(a)$;
2. $\mathcal{I} \models_4 C \propto D$, *where* $\propto \in \{\mapsto, \sqsubset, \rightarrow\}$ *if* $\mathcal{I} \models_s C \sqsubseteq D$.

Theorem 12. *Let $\mathcal{O}$ be an ontology and α be a query in $\mathcal{ALC}$. If $\mathcal{O} \models_4 \alpha$ then $\mathcal{O} \models_Q \alpha$.*

Theorem 12 shows that QC entailment is stronger than four-valued entailment because QC entailment holds modus ponens, modus tollens, hypothetical syllogism and disjunctive syllogism which four-valued entailment does hold.

6 Conclusions

In this paper, we present QC OWL to handle inconsistency with holding three inference rules and intuitive equivalences. For the aim of paraconsistent reasoning, QC weak semantics is similar to four-valued semantics for OWL. In order to hold three inference rules, QC strong semantics is introduced by restricting disjunction of concepts in description logic. Compared with four-valued OWL, we redefine concept inclusion (or subsumption) in QC OWL to follow intuitive equivalences. For this purpose, concept inclusion under QC weak semantics is defined by internal inclusion of four-valued OWL and concept inclusion under QC strong semantics is defined by hybrid of three inclusions (material inclusion, internal inclusion and strong inclusion) of four-valued OWL. In order to find suitable for implementation of paraconsistent reasoning in QC OWL, we propose the QC tableau algorithm with technique of blocking for checking satisfiability of QC $\mathcal{ALCNQ}$ ABox. Technique of blocking is introduced to hold the terminability of QC tableau algorithm. Moreover, we conclude the terminability, soundness and completeness of the algorithm. The complexity of QC tableau algorithm is not more than the complexity of classical tableau algorithm for an $\mathcal{ALCNQ}$ ABox. Complement of

[2] An interpretation is a 4-models of an ontology iff it satisfies each assertion and each inclusion axiom in the ontology [9].

concept is introduced to QC semantics in order to build reasoning system of QC OWL by not rebuilding whole new reasoning system but borrowing from classical reasoning systems of OWL. It is quite easy to show that the QC tableau algorithm, which is presented in this paper, is obtained by modifying classical reasoning systems of OWL [18]. In the future, we will further study the similar paraconsistent approach for more complex DLs such as OWL Lite and OWL DL.

References

1. Bell, D.A., Qi, G., Liu, W.: Approaches to inconsistency handling in description-logic based ontologies. In: Meersman, R., Tari, Z., Herrero, P. (eds.) OTM-WS 2007, Part II. LNCS, vol. 4806, pp. 1303–1311. Springer, Heidelberg (2007)
2. Besnard, P., del Cerro, L.F., Gabbay, D.M., Hunter, A.: Logical handling of inconsistent and default information. In: Uncertainty Management in Information Systems, pp. 325–342 (1996)
3. Huang, Z., van Harmelen, F., ten Teije, A.: Reasoning with inconsistent ontologies. In: Proceedings of the Nineteenth International Joint Conference on Artificial Intelligence, IJCAI 2005, Edinburgh, Scotland, UK, July 30-August 5, 2005, pp. 454–459. Professional Book Center (2005)
4. Patel-Schneider, P.F., Hayes, P., Horrocks, I.: Owl web ontology language semantics and abstract syntax section 5. rdf-compatible model-theoretic semantics. Technical report, W3C (December 2004)
5. Schlobach, S., Cornet, R.: Non-standard reasoning services for the debugging of description logic terminologies. In: Proceedings of the Eighteenth International Joint Conference on Artificial Intelligence, IJCAI 2003, Acapulco, Mexico, pp. 355–362. Morgan Kaufmann, San Francisco (2003)
6. Haase, P., van Harmelen, F., Huang, Z., Stuckenschmidt, H., Sure, Y.: A framework for handling inconsistency in changing ontologies. In: Gil, Y., Motta, E., Benjamins, V.R., Musen, M.A. (eds.) ISWC 2005. LNCS, vol. 3729, pp. 353–367. Springer, Heidelberg (2005)
7. Straccia, U.: A sequent calculus for reasoning in four-valued description logics. In: Galmiche, D. (ed.) TABLEAUX 1997. LNCS, vol. 1227, pp. 343–357. Springer, Heidelberg (1997)
8. Ma, Y., Hitzler, P., Lin, Z.: Algorithms for paraconsistent reasoning with owl. In: Franconi, E., Kifer, M., May, W. (eds.) ESWC 2007. LNCS, vol. 4519, pp. 399–413. Springer, Heidelberg (2007)
9. Ma, Y., Hitzler, P., Lin, Z.: Paraconsistent resolution for four-valued description logics. In: Proceedings of the 2007 International Workshop on Description Logics (DL 2007), Brixen-Bressanone, near Bozen-Bolzano, Italy, June 8-10, 2007, vol. 250. CEUR-WS.org (2007)
10. Ma, Y., Hitzler, P., Lin, Z.: Paraconsistent reasoning for expressive and tractable description logics. In: Proceedings of the 21st International Workshop on Description Logics (DL 2008), Dresden, Germany, May 13-16, 2008, CEUR-WS.org (2008)
11. Qi, G., Pan, J.Z., Ji, Q.: Extending description logics with uncertainty reasoning in possibilistic logic. In: Mellouli, K. (ed.) ECSQARU 2007. LNCS, vol. 4724, pp. 828–839. Springer, Heidelberg (2007)
12. Belnap, N.D.: A useful four-valued logic. Modern uses of multiple-valued logics, 7–73 (1977)
13. Ma, Y., Qi, G., Hitzler, P., Lin, Z.: Measuring inconsistency for description logics based on paraconsistent semantics. In: Proceedings of the 2007 International Workshop on Description Logics (DL 2007), Brixen-Bressanone, near Bozen-Bolzano, Italy, June 8-10, 2007. CEUR Workshop Proceedings, vol. 250, CEUR-WS.org (2007)

14. Zhang, X., Lin, Z.: Paraconsistent reasoning with quasi-classical semantic in alc. In: Calvanese, D., Lausen, G. (eds.) RR 2008. LNCS, vol. 5341, pp. 222–229. Springer, Heidelberg (2008)
15. Besnard, P., Hunter, A.: Quasi-classical logic: Non-trivializable classical reasoning from incosistent information. In: Froidevaux, C., Kohlas, J. (eds.) ECSQARU 1995. LNCS, vol. 946, pp. 44–51. Springer, Heidelberg (1995)
16. Hunter, A.: Reasoning with contradictory information using quasi-classical logic. J. Log. Comput. 10(5), 677–703 (2000)
17. Baader, F., Calvanese, D., McGuinness, D., Nardi, D., Patel-Schneider, P. (eds.): The Description Logic Handbook: Theory, Implementation, and Applications. Cambridge University Press, Cambridge (2003)
18. Horrocks, I., Sattler, U., Tobies, S.: Reasoning with individuals for the description logic shiq. CoRR cs.LO/0005017 (2000)
19. Horrocks, I., Sattler, U.: A tableaux decision procedure for shoiq. In: Kaelbling, L.P., Saffiotti, A. (eds.) IJCAI, pp. 448–453. Professional Book Center (2005)
20. Patel-Schneider, P.F.: A four-valued semantics for terminological logics. Artificial Intelligence 38, 319–351 (1989)

How to Trace and Revise Identities*

Julien Gaugaz, Jakub Zakrzewski, Gianluca Demartini, and Wolfgang Nejdl

L3S Research Center, Leibniz Universität Hannover
Appelstrasse 9a, 30167 Hannover, Germany
{gaugaz,zakrzewski,demartini,nejdl}@L3S.de

Abstract. The Entity Name System (ENS) is a service aiming at providing globally unique URIs for all kinds of real-world entities such as persons, locations and products, based on descriptions of such entities. Because entity descriptions available to the ENS for deciding on entity identity—Do two entity descriptions refer to the same real-world entity?—are changing over time, the system has to revise its past decisions: One entity has been given two different URIs or two entities have been attributed the same URI. The question we have to investigate in this context is then: How do we propagate entity decision revisions to the clients which make use of the URIs provided by the ENS?

In this paper we propose a solution which relies on labelling the IDs with additional history information. These labels allow clients to locally detect deprecated URIs they are using and also merge IDs referring to the same real-world entity without needing to consult the ENS. Making update requests to the ENS only for the IDs detected as deprecated considerably reduces the number of update requests, at the cost of a decrease in uniqueness quality. We investigate how much the number of update requests decreases using ID history labelling, as well as how this impacts the uniqueness of the IDs on the client. For the experiments we use both artificially generated entity revision histories as well as a real case study based on the revision history of the Dutch and Simple English Wikipedia.

1 Introduction

Given a description of an entity as, for example, persons, locations or products, the *Entity Name System* (ENS) [4] provides globally unique URIs for such real-world entities. Because entity descriptions available to the ENS for deciding whether two entity descriptions refer to the same real-world entity (i.e., entity identity) are changing over time, the system sometimes has to revise its past decisions: One real-world entity has been given two different URIs[1] (e.g., `http://dbpedia.org/resource/Tim_Berners-Lee` and `http://data.semanticweb.org/person/tim-berners-lee`) or two entities have been

* This work is partially supported by the FP7 EU Large-Scale Integrating Project OKKAM Enabling a Web of Entities (contract no. ICT-215032).

[1] It is possible to find such duplicates by using a semantic search engine such as, for example, `http://sindice.com/search`

L. Aroyo et al. (Eds.): ESWC 2009, LNCS 5554, pp. 414–428, 2009.

attributed the same URI. For example, `http://dbpedia.org/page/Paris` could be the URI assigned to the city in France but, in DBpedia, it currently "disambiguates" (that is, refers) to 57 different entities. This could happen when not enough information is available when assigning a URI to an entity. When other entities called Paris are discovered new URIs should be assigned and the information be propagated. Our main research question in this paper is how to propagate entity decision revisions to the clients, which make use of the URIs provided by the ENS.

Providing a solution to the problem of propagating entity decision revisions is a crucial aspect of Semantic Web when the goal is to enable the Web of Entities. When software applications exchange information about entities it is essential to uniquely resolve the entities so that all the agents in the Web refer to the same real world entity using the same name, that is, the same URI. As entities can change over time, URI can be used multiple times by different sources to refer to the same entity, it is important to provide a framework for handling such identity revisions.

A lot of work have been done in the area of data provenance [9,11,6] and lineage retrieval [3]. However, today new challenges are present: with the growing size of the Semantic Web and with the spreading of semantic-aware user applications managing identity, the Web has to face the problem of scalability and efficiency. In this paper we analyse how to optimise identity management on the Web. In this paper:

- We describe a model for entity identifiers evolution on the Web (Section 2).
- We propose different labelling schemes allowing to dramatically reduce the number of identifier update requests to the Entity Name Service the client needs to perform, while preserving a reasonable uniqueness quality for its identifiers (Section 4).
- We perform extensive experiments measuring the quality of identifier uniqueness. The experiments, using both artificial and real revision history from Wikipedia, also measure the drop in identifier update requests when using the proposed labelling schemes (Section 5).
- We conclude the paper and describe possible future work in Section 6.

2 Application Setting and Scenarios

2.1 Motivation

It is known that "the Semantic Web can become an open and scalable space for publishing knowledge (in the form of RDF data) only if there will be a reliable (and trustworthy) support for the reuse of URIs" [5]. The ENS is providing such an infrastructure to the Semantic Web users.

It is also clear that in the current Web there is a proliferation of different URIs for the same real world entity. It is then necessary to provide a way of notifying the Semantic Web users of changes in the URIs. For example, a person such as Tim B. Lee has many different URIs assigned by different sources: it is desirable for applications and/or people on the Web referring to him to use the same identifier.

In this paper we present methods for the propagation of Identity Decision Revisions where such decision is taken by the ENS. In the following section we describe the context of Identity Decision Revisions and introduce different approaches for propagating such information to the interested clients.

2.2 The Entity Name System

The main goal of the ENS is to enable the Web of Entities. This is accomplished by supporting the use of globally unique IDs for entities. It is important to clarify that the information about entities (i.e., attribute names and values) stored in the ENS serves solely the purpose of distinguishing entities between themselves. The ENS is defined in [4] as: "a service which stores and makes available for reuse URIs for any type of entity in a fully decentralised and open knowledge publication space." The main functionalities of the ENS are: search for the identifier of an entity, generation of entity identifiers, matching entities present in the repository with external ones, and ranking entities by similarity to a given one. Notice that the question of how the ENS actually disambiguates entity identities is out of scope of this paper and will therefore not be discussed here.

2.3 Setting

We now describe by way of an example the possible operations that can be applied to an entity. A knowledge-based system, at some point in time, might have a wrong representation of an entity: for example, initially, a system has the knowledge that a person is called John Doe, but, after having acquired more evidence, the system knows that another entity representation, with name John J. Doe, refers actually to the same entity. In this case the two entity representations as well as their identifiers need to be *merged*. Also, the entity itself can change over time (e.g. semantic evolution, evolution of documents [1]): for example, the attribute *affiliation* of a person can change over time. For these reasons, we need to define the possible changes that the entity identifiers might undergo. The following operations can be applied on an entity:

Creation. When an entity is first encountered, a new ID is generated.

Split. When the system discovers that the same entity representation describes two different real world entities, two new IDs have to be created. For example, the system knows that a person is called "Andrea Rossi" but, at some point, it discovers that there are actually a man and a woman with the same name, in the dataset.

Merge. When two entities are matched, and recognised to be the same, they are merged, and the same has to be done to their IDs.

In the following we present a real word example where the propagation of Identity Decision Revision (IDR) is important.

Wikipedia History. We can see Wikipedia as a living information repository about entities. Therefore, it is possible to use it as a scenario of IDR propagation. Entities in Wikipedia have their own pages which can grow over time. It is also true that creation, split, and merge operations appear in this collection. New pages are created, articles about similar concepts are merged together, and articles which grow too much are usually split in two (or more) different ones. It is then easy to see the need that other pages linking to a merged/split page have of knowing about this decision. Links have to be changed and in Wikipedia one approach is to use the disambiguation pages: pages about a general concept leading to all the alternative uses of that term.

2.4 Models for Propagating Identity Decision Revisions

When we want to propagate the information of IDRs there are two main possibilities of doing it: by pushing or by pulling such information.

In the *push* model the ENS server notifies all the clients of an IDR. In this scenario the ENS must be aware of all the clients using a given URI. It is easy to see that this solution is not feasible for all clients[2] on the Semantic Web It is impossible for the ENS to keep track of which client uses which ID. More, in this case there would certainly be a problem of network traffic overload: a lot of communication would be needed for propagating IDRs.

In the *pull* model a client has to ask the ENS about changes for a given ID. In this case it is a duty of the client to decide when it is time to request for an update of the ID. That is, the client has to understand when an ID is obsolete. For example, if the client receives an e-mail which contains the ID of an entity, it as to discover that it is a newer version of an ID it already has (e.g., as a result of a split operation on the ENS). In this case it is necessary to query the ENS for an up-to-date version the ID.

It is clear that the trade-off between the two models is set by how often the IDs change. If there are few IDRs, then the push approach is more efficient as all the effort is centralised.

3 Concept Definitions for Identity Revision Management

In this section we define the concepts and relative notation which are needed for describing the approaches presented in Section 4.

In this paper we assume that the ENS, named in the following as *server*, is one central machine only. It will be future work to consider the effects of a distributed ENS platform on the identity management.

Additionally, several applications using entity identifiers are present and are asynchronously submitting entity resolution requests to the server. These can be end-user application like, e.g., e-mail clients or batch processes as web crawlers. We refer to applications that submit requests to the server as *clients*. The server

[2] The users on the Web are estimated to be 1.5 billions in 2008. Source: `http://www.internetworldstats.com/stats.htm`

will respond to requests using the knowledge available in the server's *central repository*: a container of all the up-to-date entities and identifiers. When more than one ID is up-to-date for a given deprecated ID, the ENS has to use *entity matching* [3] techniques to disambiguate between those candidates. Moreover, each client has its own *local repository* containing only the identifiers known by the client, that is, all the entities encountered up to now.

Clients are exchanging information about entities and their IDs as well. This means that clients are encountering new IDs as they receive new information from other clients.

As presented in Section 2.3, split and merge operations are possible on entity identifiers. We define as *Identity Revision Graph* the directed acyclic graph representing the history of one or more entity identifiers. Using the part of such a graph that it knows, a client can detect deprecated identifiers with no need of sending a request to the server.

4 Limiting the Number of Update Requests

A quick computation shows that if 50,000 clients request updates to the server for 3,000 IDs only once a day, it represents more than 1,500 requests per second, which is already significant, and it would be interesting to be able to limit this number of request. We propose to do so by trying to update only the IDs which most need to, with the idea that, as long as one ID in a client's local repository of entities refer to only one entity in this repository, it does not need an update since local uniqueness would be preserved. This is true whether the ID in question has been deprecated (split or merged) on the ENS server.

In the following we expose different techniques allowing to limit the number of update requests while preserving local uniqueness.

4.1 Lineage Preserving ID Labelling (LPID)

Wu et al. proposed in [10] a labelling scheme for transitive closure computation on Directed Acyclic Graphs (DAGs). Computing a transitive closure in a graph is used for identifying all ancestors (or descendants) of a given node in the graph. We describe hereafter the Lite version of Prime Numbers Labelling Scheme for DAGs (PLSD) proposed by Wu et al.

Considering the entity IDs evolution as a DAG, we can use the PLSD labelling scheme described in [10] to locally detect if an ID has been deprecated. As in Fig. 1(a), if an ID A is deprecated by another ID B, then there exists a directed edge $A \rightarrow B$. As a consequence the *merge* and *split* operations mentioned in Section 2.3 are modelled as illustrated in Fig. 1(b) and 1(c) respectively.

We annotate each node of the DAG such built with a pair of natural numbers: the self-label, and the ancestors-label. The self-label is a unique prime number (i.e., from all the self-labels, a self-label appears once and only once), and the ancestors-label of a vertex is the product of its self-label with the least common

[3] Also known as record linkage, or deduplication.

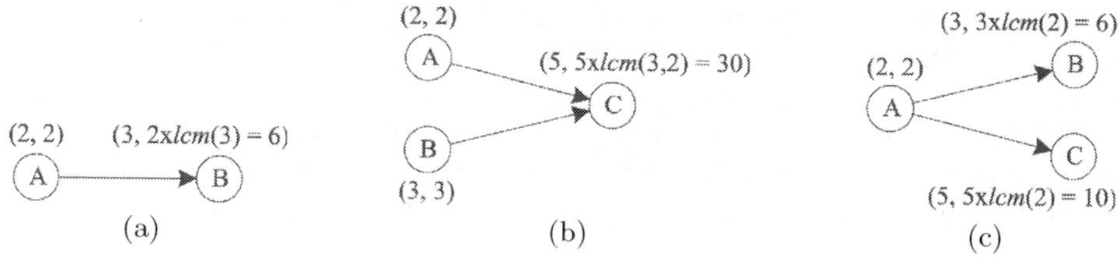

Fig. 1. (a) A is *deprecated* by B. (b) A and B are *merged* into C. (c) A is *split* into B and C.

multiple of the ancestors-labels of its ancestors (i.e., its in-component). Formally this gives:

Definition 1. *Let $G = (I, D)$ be a DAG with I a set of vertices representing IDs, and D a set of edges representing deprecation. We identify a vertex $i \in I$ by a pair $(i_{self}, i_{ancestors})$ with $i_{self}, i_{ancestors} \in \mathbb{N}^+$, where*

i) i_{self} is a prime number such that there exists a bijection between I and the set $I_{self} \equiv \{j_{self} | j \in I\}$
ii) $i_{ancestors} = i_{self} \cdot lcm(a^1_{self}, \ldots, a^n_{self})$ with $a^1, \ldots, a^n$ the ancestors (in-component) of i

We call $i = (i_{self}, i_{ancestors})$ the Lineage Preserving ID (LPID), i_{self} the self-label of i, and $i_{ancestors}$ its ancestors-label. □

Lemma 1 shows that, given two LPIDs as defined in Definition 1, the fact that the self-label of one LPID divides the ancestors-label of the other LPID implies that the former is deprecated.

Lemma 1. *Let $i = (i_{self}, i_{ancestors})$ and $j = (j_{self}, j_{ancestors})$ be two LPIDs. $j_{ancestors}/i_{self} \in \mathbb{N}$ implies that i is deprecated.* □

This approach is described in more detail in [7].

4.2 List Labelling

The LPID labelling uses an integer to encodes the ancestors' self-labels in an integer, product of those self-labels. This has the advantage to require only one division to find out whether an ID is an ancestor of another ID. On the other hand, it has the disadvantage to require the use of prime numbers as self-labels, which is costly to generate when the number of IDs is high. The obvious alternative is to use any natural number as a self-label, and store the ancestors simply in a list, separated by commas for example.

The difference with Definition 1, is that self-labels are ordinary natural numbers, and not restricted to prime numbers. And the definition of the ancestors-label is changed to:

ii) $i_{ancestors} = \{a^1_{self}, \ldots, a^n_{self}\}$ is set of the self-labels of $a^1, \ldots, a^n$, the ancestors (in-component) of i

Since this labelling is equivalent to the LPID labelling, Lemma 1 can be slightly modified for list labelling: and identifier a is an ancestor of b iff $a \in b_{ancestors}$.

The list labelling being equivalent to the LPID labelling in terms of selection of the identifiers to ask the ENS server for un update, the only difference between the two labellings is the storage space and and the selection performance.

4.3 Merge Epochs Labelling

The LPID and List Labellings allow to limit identifier update requests to some identifiers being in conflict. Once the identifiers to be updated have been selected, the client has no other choice to contact the ENS server to do so. This is because the client does not know whether there was a split or not between an identifier and its descendant. Consider Figure 2(b). The client has in its repository only identifier A, and he receives identifier C. Since A has been split, the client has to ask the ENS server whether to replace A with B or with C. However, if the IDR graph is the one in Figure 2(a) and if the client knew that A has been merged with another identifier into C, it could replace A with C directly, without contacting the ENS server. In this section, we introduce an additional label, the *merge-label*, allowing the client to know whether a split occurred between any identifier and any of its descendant.

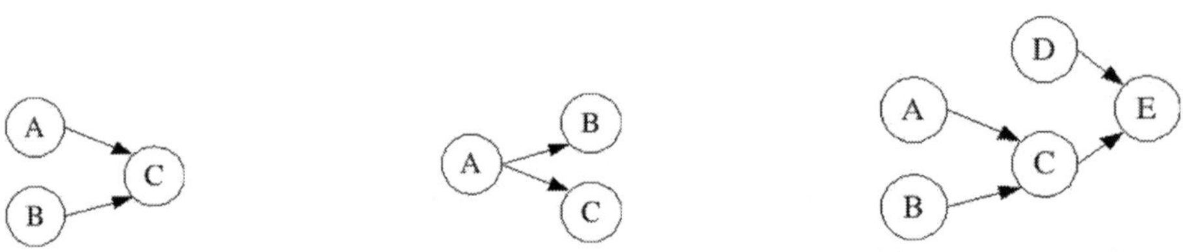

(a) A split is between A and C

(b) A merge is between A and C

(c) C has been deprecated

Fig. 2. IDR graphs. In (a) the client can replace A with C without contacting the ENS server. In (b), however, the client has to ask the ENS server to find out whether to replace A with B or C. And in (c) if the client uses the merge-label to replace A with C, it would use C which is deprecated. Whereas if the client contacted the ENS server, it would now use E which is the most up-to-date identifier for the entity considered.

As with the ancestors-label, recording which of the ancestors of an identifier can be replaced by it without asking the ENS server can be done using the lcm of the prime number self-labels of the ancestors, or simply the list of the ancestors. Thus, the Definitions 1 can be completed to include the merge-label as follows: the label for an identifier i is changed to $(i_{self}, i_{ancestors}, i_{merge})$, and the following definitions are added:

iii) $i_{merge} = lcm(m^1_{self}, \ldots, m^n_{self})$ with $m^k \in \{m^1, \ldots, m^s\} \equiv M$, the ancestors of i that are such that all identifiers between i (excluded) and m^k (included) have been merged with other identifiers.

Lemma 1 and its equivalent in Section 4.2 still hold for LPID Merge and List Merge labellings. For two LPID labellings $i = (i_{self}, i_{ancestors}, i_{merge})$ and $j = (j_{self}, j_{ancestors}, j_{merge})$: $i_{self}/j_{merge} \in \mathbb{N}$ implies that i can be replaced with j without risking to attribute an identifier to an entity which this identifier does not refer to. Similarly for List Merge labellings.

Drawbacks. As already mentioned, the merge labelling present the advantage to allow the client, in some cases, to replace an identifier with its descendant without contacting the ENS server. This comes at the price of two things: 1) the additional merge-label require more space to store the labelling of an identifier, and 2) the client might use deprecated identifiers, where if it contacted the ENS server, it would use the most up-to-date identifier. See Figure 2(c) for an illustration of the second point.

4.4 Baseline: No Labelling

As far as we know, since the ENS is a relatively new concept, the problem considered in this paper has not been approached before. For this reason, the baseline method which we use for comparing the above-presented techniques is the most straight forward one; that is, no labelling at all. In this case, when the client receives a description on an entity referred to by an identifier unknown to the client, the only way to make sure the new identifier does not conflict with another identifier in its repository is to simply ask the ENS server for each and every ID in its local repository.

This obviously would generate a lot of update requests to the ENS server, but it also has advantages over the labelling techniques we proposed. The first advantage is that a bigger part of the identifiers used by the client are up-to-date, since it more often asks for identifier updates. This implies that less identifiers in its repository refer to two or more entities in the world, and that less entities in its repository have more than one identifier. The second advantage is that since no labelling at all is used, the space requirement is less than when using lineage labelling.

5 Experiments

We conducted experiments aiming at comparing the different proposed labelling approaches to the baseline where no labelling is used. We compared the different approaches with respect to two main characteristics: network traffic and quality of the identifiers.

The network traffic is measured in terms of number of update requests a client addresses to the ENS server, and the size of the labelling metadata. The quality of the identifiers answers the question: how unique are the identifiers? Ideally the identifiers are globally unique, i.e. for each entity there exists exactly one identifier, and each identifier identifies exactly one entity. To measure this we used at the client level: the number of deprecated identifiers, the repository size, the number of entities per identifier, and the number of identifiers per entity.

5.1 Scenario

In our experiments, we considered the situation where a client receives semantic annotations describing different kind of entities. As the client encounters new identifiers, it first selects the identifiers in its repository which need update, according to the considered labelling scheme. The labelling scheme is one of the ones presented in Section 4: LPID, LPID Merge, List, List Merge, or the baseline without labelling. In each case it does its best to request update for all identifiers the scheme detects as deprecated. In the baseline case, since it does not have any information on which identifiers are outdated or not, it requests updates for all identifiers in its repository, plus the newly received identifiers. Once no more identifier is detected as outdated by the considered labelling scheme, it waits for a new arrival of identifiers, and repeats the process.

Before requesting for identifier updates, the merge labelling schemes perform the local replacements when applicable. Once the ENS receives an update request for identifier i, it returns one random[4] leaf identifier of the IDR graph which is a descendent of i. Howerver, for each identifier i to update, the returned up-to-date identifier is the same independently of the labelling scheme evaluated; this is important for the results to be comparable. The sets of identifiers added to the client are the same and in the same order for all evaluated labelling schemes as well.

5.2 Evaluation

In the remainder of this section we will use the following abbreviations for referring to the labelling schemes: LPID (Lineage Preserving IDentifier labelling), LPIDMerge (Lineage preserving identifier labelling with Merge labelling), List (List identifier lineage labelling), ListMerge (List identifier lineage labelling with Merge Labelling), and Classic for the Baseline without labelling.

Measuring Uniqueness. When identifiers are unique, it means there is a bijection between the set of identifiers and the set of entities. Informally this means there is exactly one identifier for each entity, and only one entity referred by each identifier. As we see unicity depends on the set of identifiers and the set of entities. We will talk about *local uniqueness* when considering the set of entities in a client's local repository and the set of identifiers related to those entities. And we will talk about *global uniqueness* when considering the set of entities in ENS server's repository and the set of identifiers related to those entities.

We will focus in this paper on global uniqueness. The reason is that we started to consider the problem of how to transmit identity decision revisions from the ENS server to the client, taking the point of view of the server, that is, global. We will then assume that each identifier in the client's repository refers to exactly

[4] In practice the ENS performs a matching operation, but since this is not the focus of this paper, we chose the identifier randomly, for the sake of simplicity.

one entity in the same repository, and devise a measure for how globally unique the identifiers in the client's repository are. This means we want to know how many entities in the world an identifier used in the client's repository refers to; and how many distinct identifiers there is in the client's repository for one entities in that repository. Those are the measures we will use in those experiments to compare the impact on uniqueness quality of the evaluated labelling schemes. We call them *number of entities per ID* and *number of IDs per entity.*

Measuring Network Traffic and ENS Server Workload. Identifier update requests makes use of two resources of concern in our scenario: 1) network bandwidth, and 2) ENS server workload. Each update request from a client needs to be sent over the network, including the identifiers to update, and if required the description of the entities whose identifier is to be updated. For each identifier update request the ENS server receives, if there is more than one leaf descendent, the server has to perform an entity matching process to find out which is the present up-to-date identifier for the requested entity.

In this paper we are not concerned with the details of the process the ENS server follows to return the up-to-date identifier. However we know that both network bandwidth consumption and ENS server workload are functions on the number of identifiers whose update is requested. This is what we will call in the remainder *number of update requests.* Note that, in practice, all identifiers needed to be updated might be sent to the server in one bulk update request. In this paper we assume that one update request concerns one identifier only.

In addition, impacting on the network bandwidth consumption, the *size of the lineage labelling metadata* will be reported.

Measuring Performance. As already mentioned, the reason why we introduced the identifier lineage labelling is to allow to reduce the number of request updates, while keeping a reasonable uniqueness quality. This is mainly a performance concern, and is therefore a central measure in this paper.

We report two durations. The first is the time it takes the client, using a given labelling scheme, to select which identifiers it needs to ask the ENS server for update. We call this the *selection time*, which is null for the baseline, since all identifiers are always updated. To be able to compare performance between the baseline and the labelling schemes, we thus need to consider also the total time it takes a client to update it's identifiers. This comprises the selection time, plus the transmission time (of the requests), plus the possible entity matching time of the ENS server, plus the transmission time back. We approximated the whole to be the selection time, plus some time constant per identifier update request. We chose arbitrarily 200 ms for this time constant, which is not an uncommon round trip time for an intercontinental internet connection. It makes it a low constant for our case, considered that it also includes the time the server needs to process the request. We call it *total matching time.*

Comparing the Labelling Schemes. Since most of the above proposed measures will yield different values for the different labelling schemes (and the baseline), we need some measure common to all of them to be able to compare them. For this we chose the the number of new identifiers added to the client.

5.3 Datasets

We experimented with 3 different IDR graphs. The two first based on the revisions history of the Dutch and the Simple English Wikipedia datasets, and the third one artificially generated. As introduced in Section 2.3, an IDR graph is defined by a sequence of three events: identifier creations, merges and splits.

Here is how we interpreted the Wikipedia revision history in terms of those three actions. When a Wikipedia page is created, it obviously corresponds to an identifier creation. The first revision of a Wikipedia page where a redirect appears is considered as a merge between the two identifiers corresponding to the Wikipedia page and its redirect target page. And the first revision of a page where a disambiguation tag appears is interpreted as a split for this page's identifier to the ones corresponding to its targets.

The artificial IDR graph was generated with the following probability distributions. At each iteration, we first choose what action to perform[5]: creation ($p = 0.77$), split ($p = 0.08$) and merge ($p = 0.15$). If split or merge was picked, the identifier(s) to which the action applies are picked at random (uniform distribution); and the number of identifiers into which the identifier is split, or the number of identifiers to merge follow a Zipf distribution with $\sigma = 2$, a maximum of 5, and a minimum of 2.

5.4 Experiment Results

We present in this section the results of our experiments. When significant difference is mentioned, this has been confirmed with a one-way ANOVA[6] with 5% of significance. For most of the graphs we present the results for each of for IDR graphs: artificially generated, simple English Wikipedia, and Dutch Wikipedia. For each we added to the client 100 times 500 (= 50,000) identifiers, and since the Dutch Wikipedia is the only having enough entities to allow it, we also tried 200 times 500 (= 100,000) identifiers for this IDR graph. Which gives us 4 different simulation for each scheme, which we note 'artificial', 'simplewiki', 'nlwiki100' and 'nlwiki200' respectively.

Identifiers Uniqueness. We show in Figure 3(a) the average number of entities per identifier at different steps (number of identifiers received by the client) in the simulation. The value for the classic scheme is one, as expected. For all datasets, the difference between using or not the merge-label happens to be non significant. The difference between the classic and the labelling schemes is visibly significant.

[5] Probabilities are estimated using statistics from Wikipedia articles operations.

[6] One-way ANalysis Of VAriance (ANOVA) is similar to the T-Test, but for more than two series of observations. See [8] for more details.

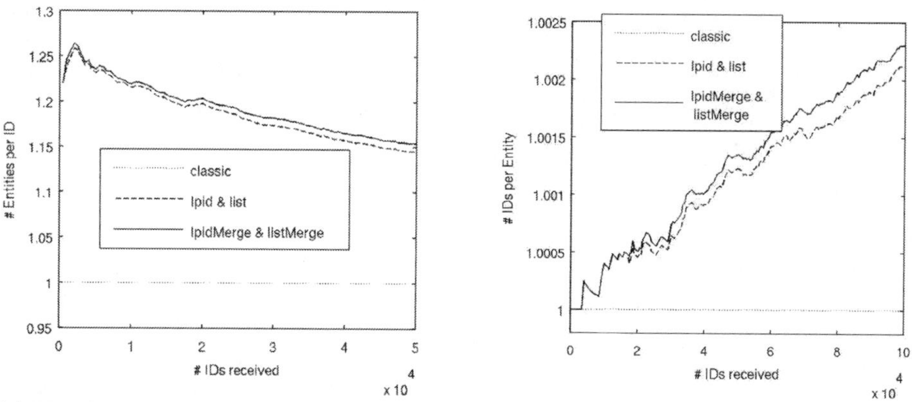

(a) Number of Entities per Identifier for the artificial IDR graph

(b) Number of Identifiers per Entity for Dutch Wikipedia IDR graph with 100,000 IDs received

Fig. 3. Uniqueness Quality Measurements

In Figure 3(b) we present the average number of identifiers per entity for each iteration step. Whereas for the classic scheme we have the expected values of one, it is significantly different from the others, but merge-labelling makes a significant difference only for the IDR graphs of Simple English and Dutch Wikipedia on 100,000 identifiers.

We note that the number of entities per identifier tends to decrease over time, which is a good sign in the sense that it converges to the ideal case of one entity existing on the ENS server per identifier in the client's repository. However the number of identifiers per entity tends to augment over time, and this more rapidly while using merge-labels as not using them. This is problematic, even though it is easier to detect duplicate entities in the client's repository than reverse an accidental merge of two distinct entities because they erroneously shared a same identifier. Note however that, even if the differences are significant, the values are still quite low, which indicates that the problem would occur rarely.

Network Traffic and ENS Server Workload. We present in Figure 4 the number of identifiers the client is requesting the ENS server for update for the baseline, as well as for the labelling schemes without merge-labels (LPID and List). The merge labelling schemes are similar to the ones without, except that, as expected, the number of updates is reduced from a factor 2 approximately. Because the values for the baseline are so huge compared to the labelling schemes, we present the plots per scheme instead of per dataset as for the other measures. In addition to the obvious significant explosion of updates of the classic scheme compared to using lineage labelling—approximately 50 to 500 more than with lineage labelling—the one-way ANOVA revealed that using merge-labels or not does not change significantly the number of update requests.

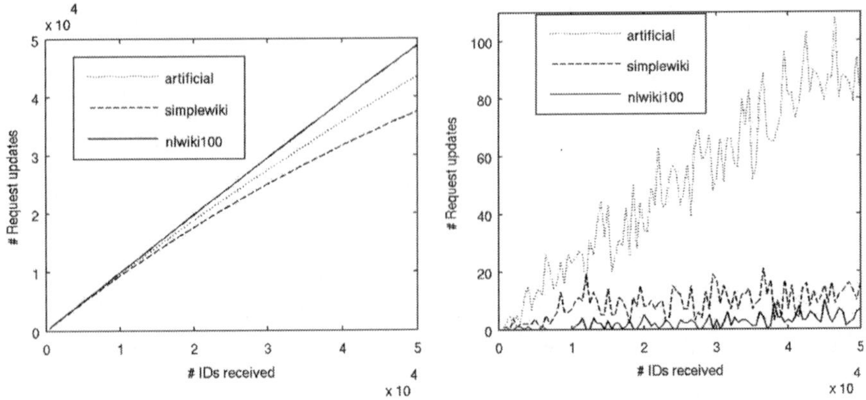

(a) Classic Scheme (b) LPID & List Schemes

Fig. 4. Number of ID update requests per iteration of 500 IDs added to the client

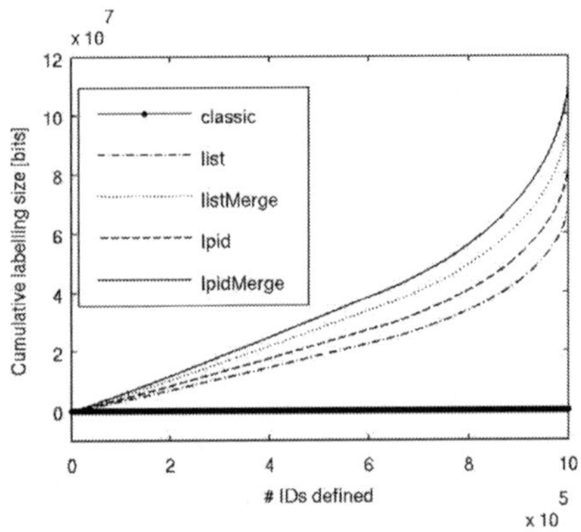

Fig. 5. Size of the labelling as the number of IDs grows on the Artificial IDR graph

Figure 5 present the cumulative size of the lineage labelling for all the identifier schemes on the artificial IDR graph. For example, the total size needed to store the lineage labelling for one million IDs using the LPIDMerge scheme is 13 MBytes (109,124,363 bits). Since the baseline is not using any labelling, its size is zero for all datasets. Additionally, we notice that the list labelling scheme uses less space than the LPID one, and naturally, adding merge-labels augments the size of the labelling. All differences are statistically significant, and similar for other IDR graphs.

More interestingly, the size of the lineage labelling schemes augments almost exponentially. In our case the biggest labelling is of 4,500 bits for one identifier in the artificial IDR graph, which has a bit more than one million identifiers. This is still reasonable, but this size always increases over time, and this will be even more pronounced as new identifiers will be created and new split and merge operations will occur.

Performance. Figure 6(a) shows the selection time. The main observation we can make regarding those figures, is that LPID selects the identifiers to be updated significantly faster than LPIDMerge, and the two List labellings. The classic is trivially different from the others since it is null, and all other labelling perform similarly. This analysis is confirmed by one-way ANOVA.

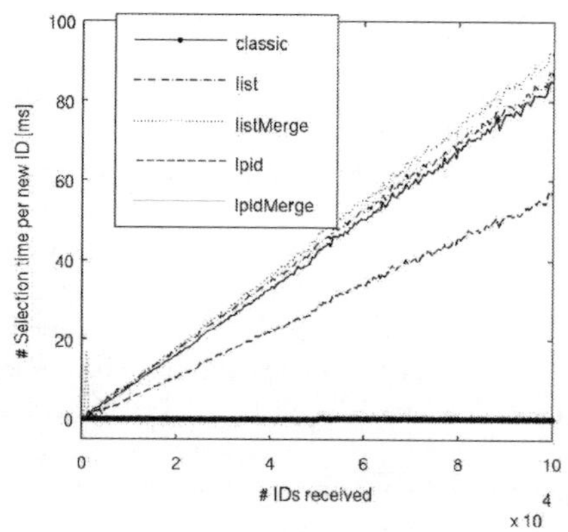

(a) Selection time per number of ID added to the client's repository on the Dutch Wikipedia IDR graph and 100,000 identifiers added to the client.

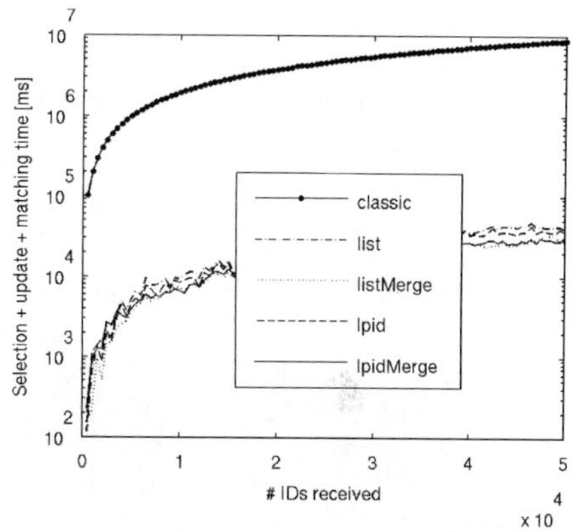

(b) Total matching time per number of identifier update request on the artificial IDR graph.

Fig. 6. Performance measurements

In Figure 6(b), we see the total matching times, i.e. the total time needed to update a clients repository, in average, for one identifier to update. It shows clearly that the classic scheme is much slower. This is mainly due to the very high number of identifier update requests it generates. Also, there is not a significant difference between the other schemes. This is confirmed by a one-way ANOVA.

6 Conclusions and Further Work

We proposed to use identifier lineage labelling, and experiments confirmed that they allow to reduce the number of identifier update requests to the ENS server, while keeping an acceptable quality of uniqueness at the level of client's local repository. We also proposed merge-labels to further reduce the number of update requests by allowing the clients to replace an identifier by one of its descendants when this does not decrease the local uniqueness quality of the identifiers on the client's repository, which was also confirmed by the experiments. In the future we need to address the exponential growth of the identifier labelling. To reduce the size of the labelling, we intend to investigate the use bloom filters [2] to summarise list of IDs. As well, avoiding to duplicate an ID in the merge-label and in the ancestors-label should also allow to reduce the size of merge-labelling, without solving the problem of exponential growth. The use of time-to-live (TTLs) associated to IDs indicating to the client when to request an update seems to be a promising approach as well.

References

1. Berberich, K., Bedathur, S.J., Neumann, T., Weikum, G.: A time machine for text search. In: SIGIR, pp. 519–526 (2007)
2. Bloom, B.H.: Space/time trade-offs in hash coding with allowable errors. Commun. ACM 13(7), 422–426 (1970)
3. Bose, R., Frew, J.: Lineage retrieval for scientific data processing: a survey. ACM Comput. Surv. 37(1), 1–28 (2005)
4. Bouquet, P., Stoermer, H., Bazzanella, B.: An Entity Name System ('ENS') for the Semantic Web. In: Bechhofer, S., Hauswirth, M., Hoffmann, J., Koubarakis, M. (eds.) ESWC 2008. LNCS, vol. 5021, pp. 258–272. Springer, Heidelberg (2008)
5. Bouquet, P., Stoermer, H., Niederée, C., Maña, A.: Entity name system: The backbone of an open and scalable web of data. In: ICSC, pp. 554–561. IEEE Computer Society Press, Los Alamitos (2008)
6. Cui, Y., Widom, J.: Lineage tracing for general data warehouse transformations. VLDB J. 12(1), 41–58 (2003)
7. Gaugaz, J., Demartini, G.: Entity identifiers for lineage preservation. In: IRSW, Tenerife, Spain (June 2008)
8. Miller Jr., R.G.: Beyond ANOVA: Basics of Applied Statistics. Texts in Statistical Science Series. Chapman & Hall/CRC, Boca Raton (1997)
9. Simmhan, Y.L., Plale, B., Gannon, D.: A survey of data provenance in e-science. SIGMOD Rec. 34(3), 31–36 (2005)
10. Wu, G., Zhang, K., Liu, C., Li, J.: Adapting Prime Number Labeling Scheme for Directed Acyclic Graphs. In: Li Lee, M., Tan, K.-L., Wuwongse, V. (eds.) DASFAA 2006. LNCS, vol. 3882, pp. 787–796. Springer, Heidelberg (2006)
11. Zhao, J., Goble, C.A., Stevens, R., Bechhofer, S.: Semantically linking and browsing provenance logs for e-science. In: Bouzeghoub, M., Goble, C.A., Kashyap, V., Spaccapietra, S. (eds.) ICSNW 2004. LNCS, vol. 3226, pp. 158–176. Springer, Heidelberg (2004)

Concept Search

Fausto Giunchiglia, Uladzimir Kharkevich, and Ilya Zaihrayeu

Department of Information Engineering and Computer Science
University of Trento, Italy
{fausto,kharkevi,ilya}@disi.unitn.it

Abstract. In this paper we present a novel approach, called Concept Search, which extends syntactic search, i.e., search based on the computation of string similarity between words, with semantic search, i.e., search based on the computation of semantic relations between concepts. The key idea of Concept Search is to operate on complex concepts and to maximally exploit the semantic information available, reducing to syntactic search only when necessary, i.e., when no semantic information is available. The experimental results show that Concept Search performs at least as well as syntactic search, improving the quality of results as a function of the amount of available semantics.

1 Introduction

Historically, there have been two major approaches to information retrieval that we call *syntactic search* and *semantic search*. Syntactic search use words or multi-word phrases as atomic elements in document and query representations. The search procedure is essentially based on the *syntactic matching* of document and query representations. Search engines, exploiting syntactic search, are known to suffer in general from low precision while being good at recall. Semantic search is based on fetching document and query representations through a semantic analysis of their contents using natural language processing techniques and, later, on retrieving documents by matching these semantic representations. The idea is that, differently from syntactic search, semantic search exploits the *meaning* of words, thus avoiding many of the well known problems of syntactic search, e.g., the problems of polysemy and synonymy. Semantics-based approaches, in general, allow to reach a higher precision but lower recall [11].

In this paper we propose a novel approach called *Concept Search* (*C-Search* in short) which extends syntactic search with semantics. The main idea is to keep the same machinery which has made syntactic search so successful, but to modify it so that, whenever possible, syntactic search is substituted by semantic search, thus improving the system performance. This is why we say that *C-Search* is *semantics enabled syntactic search.* Semantics can be enabled along different dimensions, on different levels, and to different extents forming a space of approaches lying between purely syntactic search and fully semantic search. We call this space the *semantic continuum.* In principle, *C-Search* can be tuned to work at any point in the semantic continuum taking advantage of semantics

L. Aroyo et al. (Eds.): ESWC 2009, LNCS 5554, pp. 429–444, 2009.

when and where possible. As a special case, when no semantic information is available, *C-Search* reduces to syntactic search, i.e., results produced by *C-Search* and syntactic search are the same.

The remainder of the paper is organized as follows. In Section 2, we first discuss information retrieval (IR) in general, and then focus on syntactic search. In this section, we provide a general model of IR which will be latter extended to semantic search. In Section 3, we introduce and describe the semantic continuum. In Section 4, we describe how *C-Search* is positioned within the semantic continuum. In Section 5, we show how *C-Search* can be efficiently implemented using inverted index technology. Section 6 presents experimental results. In Section 7, we discuss the state-of-the-art in semantic search and compare our approach with other related approaches. Section 8 concludes the paper.

2 Syntactic Search

The goal of an IR system is to map a natural language query q (in a query set Q), which specifies a certain user information needs, to a set of documents d in the document collection D which meet these needs, and to order these documents according to their relevance. *IR* can therefore be represented as a mapping function:

$$IR : Q \rightarrow D \tag{1}$$

In order to implement an IR System we need to decide (i) which models (*Model*) are used for document and query representation, for computing query answers and relevance ranking, (ii) which data structures (*Data Structure*) are used for indexing document representations in a way to allow for an efficient retrieval, (iii) what is an atomic element (*Term*) in document and query representations, and (iv) which matching techniques (*Match*) are used for matching document and query terms. Thus, an IR System can be abstractly modelled as the following 4-tuple:

$$IR\_System = \langle Model,\ Data\ Structure,\ Term,\ Match \rangle \tag{2}$$

The *bag of words model*, i.e., the model in which the ordering of words in a document is not considered, is the most widely used model for document representation. The *boolean model*, the *vector space model*, and the *probabilistic model* are the classical examples of models used for computing query answers and relevance ranking [13]. Conventional search engines rank query results using the cosine similarity from the vector space model with terms weighted by different variations of the *tf-idf* weight measure. Various index structures, such as the *signature file* and the *inverted index*, are used as *data structures* for efficient retrieval. Inverted index, which stores mappings from terms to their locations in documents, is the most popular solution [13]. Finally, in syntactic search, *Term* and *Match* are instantiated as follows:

- *Term* - a word or a multi-word phrase;
- *Match* - a syntactic matching of words or phrases.

In the simplest case, syntactic matching is implemented as search for equivalent words. These words are often stemmed. Furthermore, some systems perform approximate matching by searching for words with common prefixes or words within a certain edit distance with a given word.

There are several problems which may negatively affect the performance of syntactic search, as discussed below.

Polysemy. The same word may have multiple meanings and, therefore, query results, computed by a syntactic search engine, may contain documents where the query word is used in a meaning which is different from what the user had in mind. For instance, a document $D1$ (in Figure 1) which talks about *baby* in the sense of a very young mammal is irrelevant if the user looks for documents about *baby* in the sense of a human child (see query $Q1$ in Figure 1). An answer for query $Q1$, computed by a syntactic search engine, includes document $D1$, while the correct answer is the empty set.

Queries:

Q1: *Babies and dogs* Q2: *Paw print* Q3: *Computer table* Q4: *Carnivores*

Documents:

D1: *A small baby dog runs after a huge white cat.*

D2: *A laptop computer is on a coffee table.*

D3: *A little dog or a huge cat left a paw mark on a table.*

Fig. 1. Queries and a document collection

Synonymy. Two different words can express the same meaning in a given context, i.e., they can be synonyms. For instance, words *mark* and *print* are synonymous when used in the sense of a visible indication made on a surface, however, only documents using word *print* will be returned if the user query was exactly this word. An answer for query $Q2$ (in Figure 1), computed by a syntactic search engine, is the empty set, while the correct answer includes the document $D3$.

Complex concepts. Syntactic search engines fall short in taking into account complex concepts formed by natural language phrases and in discriminating among them. Consider, for instance, document $D2$ (in Figure 1). This document describes two concepts: *a laptop computer* and *a coffee table*. Query $Q3$ (in Figure 1) denotes concept *computer table* which is quite different from both concepts described in $D2$, whereas a syntactic search engine is likely to return $D2$ in response to $Q3$, because both words computer and table occur in this document. The correct answer is the empty set.

Related concepts. Syntactic search does not take into account concepts which are semantically related to the query concepts. For instance, a user looking for *carnivores* might not only be interested in documents which talk about carnivores but also in those which talk about the various kinds of carnivores such as *dogs* and *cats*. An answer for query $Q4$ (in Figure 1), computed by a syntactic search, is the empty set, while the correct answer might include documents $D1$ and $D3$, depending on user information needs and available semantic information.

3 The Semantic Continuum

In order to address the problems of syntactic search described in Section 2, we extend syntactic search with semantics. In the following, we identify three dimensions where semantics can improve syntactic search and represent these dimensions in the cartesian space shown in Figure 2.

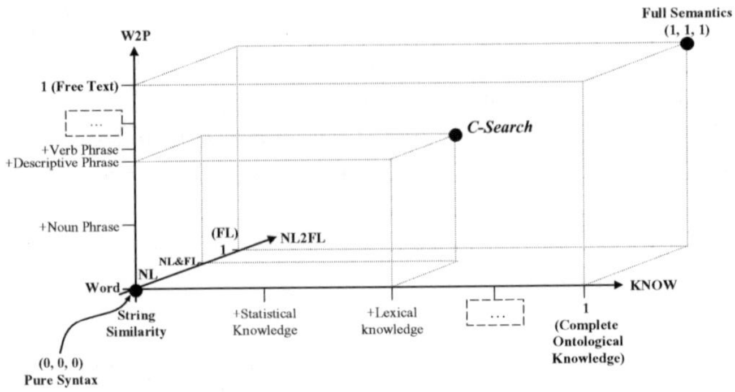

Fig. 2. Semantic Continuum

From natural language to formal language (NL2FL-axis in Figure 2). To solve the problems related to the ambiguity of natural language, namely, the problems of polysemy and synonymy, we need to move from words, expressed in a natural language, to concepts (word senses), expressed in an unambiguous formal language. An overview of existing approaches to sense based IR is presented in [17]. In the NL2FL-axis, 0 represents the situation where only words are used, while 1 represents the situation where only concepts are used. When we move from words to concepts, it is not always possible to find a concept which corresponds to a given word. The main reason for this problem is the lack of background knowledge [8], i.e., a concept corresponding to a given word may not exist in the lexical database. To address this problem, indexing and retrieval in the continuum are performed by using both syntactic and semantic information, i.e., a word itself is used as a concept, when its denoted concept is not known.

From words to phrases (W2P-axis in Figure 2). To solve the problem related to complex concepts, we need to analyze natural language phrases, which denote these concepts. It is well known that in natural language concepts are expressed mostly as noun phrases [19]. An example of a noun phrase parsing algorithm and its application to document indexing and retrieval is described in [23]. There are approaches in which the conceptual content of noun phrases is also analyzed (see e.g. [1]). In general, concepts can be expressed as more complex phrases (e.g., verb phrases) and possibly as a free text. In the W2P-axis, 0 represents the situation where only single words are used, while 1 represents the situation where complex concepts, extracted from a free text, are used.

From string similarity to semantic similarity (KNOW-axis in Figure 2). The problem with related concepts can be solved by incorporating knowledge about term relatedness. For instance, it can be statistical knowledge about word co-occurrence (see e.g. [6]), lexical knowledge about synonyms and related words (see e.g. [14]), or ontological knowledge about classes, individuals, and their relationships (see e.g. [15]). In the KNOW-axis, 0 represents the situation where only string similarity is used during the term matching process, while 1 represents the situation where complete ontological knowledge is used during term matching.

The three-dimensional space contained in the cube (see Figure 2) represents the *semantic continuum* where the origin (0,0,0) is a purely syntactic search, the point with coordinates (1,1,1) is a fully semantic search, and all points in between represent search approaches in which semantics is enabled to different extents. *C-Search* can be positioned anywhere in the semantic continuum with syntactic search being its base case, and semantic search being the optimal solution, at the moment beyond the available technology.

4 Concept Search

C-Search is implemented according to the model of Equations 1 and 2 from Section 2. In our proposed solution, *C-Search* reuses retrieval models (*Model*) and data structures (*Data Structure*) of syntactic search with the only difference in that now words (W) are substituted with concepts (C) and syntactic matching of words ($WMatch$) is extended to semantic matching of concepts ($SMatch$). This idea is schematically represented in the equation below:

$$\boxed{\textit{Syntatic Search} \xrightarrow{Term(W\ \rightarrow\ C),\ Match(WMatch\ \rightarrow\ SMatch)} \textit{C-Search}}$$

Let us consider in details how the words in W are converted into the complex concepts in C and also how the semantic matching $SMatch$ is implemented.

4.1 From Words to Complex Concepts ($W \rightarrow C$)

Searching documents, in *C-Search*, is implemented using complex concepts expressed in a propositional Description Logic (DL) language (i.e., a DL language without roles). Complex concepts are computed by analyzing meaning of the words and phrases.

Single words are converted into atomic concepts uniquely identified as *lemma-sn*, where *lemma* is the lemma of the word, and *sn* is the sense number in WordNet. For instance, the word *dog* used in the sense of a domestic dog, which is the first sense in WordNet, is converted into the atomic concept *dog*-1. The conversion of words into concepts is performed as follows. First, we look up and enumerate all meanings of the word in WordNet. Next, we perform word sense filtering, i.e., we discard word senses which are not relevant in the given context. In order to do this, we follow the approach presented in [22], which exploits POS tagging information and WordNet lexical database for disambiguation of words in short noun phrases. Differently from [22] we do not use the disambiguation technique which leaves only the most probable sense of the word, because of its low accuracy. If more than one sense is left after the word sense filtering step then we keep all these senses. If no senses from WordNet are found then *lemma* is used as the identifier for the atomic concept. In this case, *C-Search* is reduced to syntactic search.

Complex concepts are computed by extracting phrases and by analyzing their meaning. Noun phrases are translated into the logical conjunction of atomic concepts corresponding to the words. For instance, the noun phrase *A little dog* represents a concept, whose extension is the set of all dogs of a small size. It is translated into the concept $little\text{-}4 \sqcap dog\text{-}1$.

Concepts in natural language can be described ambiguously. For instance, the phrase *A little dog or a huge cat* represents a concept which encodes the fact that it is unknown whether the *only* animal described in the document is *a little dog* or *a huge cat*. In order to support complex concepts which encode uncertainty (partial information) coming from the coordination conjunction *OR* in natural language, we introduce the notion of *descriptive phrase*. We define descriptive phrase as a set of noun phrases, representing alternative concepts, connected by *OR*:

$$descriptive\_phrase ::= noun\_phrase \ \{OR \ noun\_phrase\} \quad (3)$$

Descriptive phrases are translated into logical disjunction of formulas corresponding to the noun phrases. For instance, phrase *A little dog or a huge cat* is translated into concept $(little\text{-}4 \sqcap dog\text{-}1) \sqcup (huge\text{-}1 \sqcap cat\text{-}1)$. To locate descriptive phrases we, first, follow a standard NLP pipeline to locate noun phrases, i.e., we perform sentence detection, tokenization, part-of-speech (POS) tagging, and noun phrase chunking. The first step allows as to locate noun phrases. As a second step, that we call descriptive phrase chunking, we locate descriptive phrases satisfying Formula 3. Our implementation is based on GATE [4].

Queries usually are short phrases (i.e., 1-3 words) and, as shown in [22], standard NLP technology, primarily designed to be applied on full-fledged sentences, is not effective enough in this application scenario. An atomic concept, in a query, can be computed incorrectly, because of the selection of a wrong part-of-speech tag. In order to address this problem, for *short queries*, we use a POS-tagger which is specifically trained on short phrases [22]. On the other hand, for *long queries* (i.e., 4 words or more), we use the standard NLP technology.

Even if atomic concepts are computed correctly, complex concepts can be erroneously computed. One of the reasons is that a complex concept can be represented as a sequence of words without following the grammar for noun phrases. For instance, the query *cat huge* is converted into two atomic concepts $cat\text{-}1$ and $huge\text{-}1$, while the correct concept might be $cat\text{-}1 \sqcap huge\text{-}1$. Another reason is that a query describing more than one concept, without properly separating them, can be recognized as a single complex concept. For instance, the query *dog cat* is converted into the concept $dog\text{-}1 \sqcap cat\text{-}1$, while the user might be actually looking for a document describing both animals, i.e., $dog\text{-}1$ and $cat\text{-}1$. The examples described above show that, in general, it is unknown how atomic concepts $A_1^q, \ldots, A_n^q$, extracted from short queries, should be combined in order to build complex query concepts. To represent this uncertainty we use the following query

$$(A_1^q \text{ AND } \ldots \text{ AND } A_n^q) \text{ AND } (A_1^q \sqcup \cdots \sqcup A_n^q) \quad (4)$$

where the first part (A_1^q AND ... AND A_n^q) encodes the fact that it is *known* that the query answer should contain documents which are relevant to all the atomic concepts in the query. The second part, i.e., the complex concept $A_1^q \sqcup \cdots \sqcup A_n^q$, can be equivalently rewritten as $(A_1^q) \sqcup (A_2^q) \sqcup \cdots \sqcup (A_1^q \sqcap A_2^q) \sqcup \cdots \sqcup (A_1^q \sqcap \cdots \sqcap A_n^q)$ and, therefore, encodes the fact that it is *unknown* which complex concept (e.g., $A_1^q \sqcap A_2^q$, or $A_1^q \sqcap \cdots \sqcap A_n^q$) should be actually described.

4.2 From Word to Concept Matching ($WMatch \rightarrow SMatch$)

In *C-Search*, we allow the search of documents describing concepts which are semantically related to query concepts. We assume that, when a user is searching for a concept, she is also interested in more specific concepts.[1] For example, the extension of concept $(little\text{-}4 \sqcap dog\text{-}1) \sqcup (huge\text{-}1 \sqcap cat\text{-}1)$ is a subset of the extension of concept *carnivore*-1. Therefore, documents describing the former concept should be returned as answers to the query describing the latter concept. Formally a query answer $A(C^q, \mathcal{T})$ is defined as follows:

$$A(C^q, \mathcal{T}) = \{d \mid \exists C^d \in d, \text{ s.t. } \mathcal{T} \models C^d \sqsubseteq C^q\} \quad (5)$$

where C^q is a complex query concept extracted from the query q, C^d is a complex document concept extracted from the document d, and $\mathcal{T}$ is a terminological knowledge base (the background knowledge) which is used in order to check if C^d is more specific then C^q. Equation 5 states that the answer to a query concept C^q is the set of all documents d, such that, there exists concept C^d in d which is more specific than the query concept C^q.

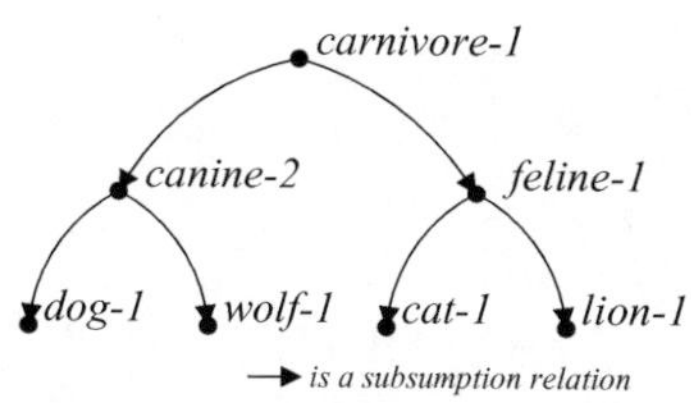

Fig. 3. Example of terminological knowledge base $\mathcal{T}_{WN}$

During query processing we need to compute $A(C^q, \mathcal{T})$ for every query concept C^q in the query. One approach is to sequentially iterate through each concept C^d, compare it to the query concept C^q using semantic matching [9], and collect those concepts for which semantic matching returns *more specific* ($\sqsubseteq$). However, this approach may become prohibitory expensive as there may be thousands and millions of concepts described in documents. In order to allow for the efficient computation of $A(C^q, \mathcal{T})$, we propose an approach described below.

Let us assume, as it is the case in the current implementation, that $\mathcal{T}$ consists of the terminological knowledge base $\mathcal{T}_{WN}$ generated from WordNet and

[1] This could be easily generalized to any set of semantically related concepts. The impact of this choice onto the system performance is part of the future work.

extended by words (represented as concepts) for which no senses in WordNet are found. One small fragment of $\mathcal{T}_{WN}$ is represented in Figure 3. $\mathcal{T}_{WN}$ can be thought of as an acyclic graph, where links represent subsumption axioms in the form $A_i \sqsubseteq A_j$, with A_i and A_j atomic concepts.

Concepts C^d and C^q, are created by translating descriptive phrases into propositional DL formulas (see Section 4.1 for details). The resulting concepts are disjunctions ($\sqcup$) of conjunctions ($\sqcap$) of atomic concepts (A) without negation, i.e., $C^d \equiv \sqcup \sqcap A^d$ and $C^q \equiv \sqcup \sqcap A^q$. For example, a possible document concept and query concept are:

$$C^d \equiv (little\text{-}4 \sqcap dog\text{-}1) \sqcup (huge\text{-}1 \sqcap cat\text{-}1)$$
$$C^q \equiv (small\text{-}4 \sqcap canine\text{-}2) \sqcup (large\text{-}1 \sqcap feline\text{-}1)$$

By substituting C^d with $\sqcup\sqcap A^d$, C^q with $\sqcup\sqcap A^q$, and $\mathcal{T}$ with $\mathcal{T}_{WN}$ in Equation 5, we obtain:

$$A(\sqcup \sqcap A^q, \mathcal{T}_{WN}) = \{d \mid \exists(\sqcup \sqcap A^d) \in d, \text{ s.t. } \mathcal{T}_{WN} \models \sqcup \sqcap A^d \sqsubseteq \sqcup \sqcap A^q\} \quad (6)$$

Let by $\mathbf{C}_{\sqcup\sqcap A^q}$ denote the set of all complex document concepts $\sqcup \sqcap A^d$ (in all documents d), which are equivalent to or more specific than $\sqcup \sqcap A^q$, in formulas

$$\mathbf{C}_{\sqcup\sqcap A^q} = \{\sqcup \sqcap A^d \mid \mathcal{T}_{WN} \models \sqcup \sqcap A^d \sqsubseteq \sqcup \sqcap A^q\} \quad (7)$$

Then Equation 6 can be rewritten as follows:

$$A(\sqcup \sqcap A^q, \mathcal{T}_{WN}) = \{d \mid \exists(\sqcup \sqcap A^d) \in d, \text{ s.t. } (\sqcup \sqcap A^d) \in \mathbf{C}_{\sqcup\sqcap A^q}\} \quad (8)$$

In order to compute set $\mathbf{C}_{\sqcup\sqcap A^q}$, as defined in Equation 7, we need to solve the following subsumption problem

$$\mathcal{T}_{WN} \models \sqcup \sqcap A^d \sqsubseteq \sqcup \sqcap A^q \quad (9)$$

Given that $\mathcal{T}_{WN}$ consists only of subsumption axioms between atomic concepts, and that concepts $\sqcup \sqcap A^d$ and $\sqcup \sqcap A^q$ do not contain negations, the problem in Equation 9 can be reduced to the set of subsumption problems

$$\mathcal{T}_{WN} \models \sqcap A^d \sqsubseteq A^q \quad (10)$$

This problem reduction is obtained by applying the following three equations:[2]

$$\mathcal{T}_{WN} \models \sqcup \sqcap A^d \sqsubseteq \sqcup \sqcap A^q \text{ iff for all } \sqcap A^d \text{ in } \sqcup \sqcap A^d, \mathcal{T}_{WN} \models \sqcap A^d \sqsubseteq \sqcup \sqcap A^q \quad (11)$$

$$\mathcal{T}_{WN} \models \sqcap A^d \sqsubseteq \sqcup \sqcap A^q \text{ iff there exists } \sqcap A^q \text{ in } \sqcup \sqcap A^q, \mathcal{T}_{WN} \models \sqcap A^d \sqsubseteq \sqcap A^q \quad (12)$$

$$\mathcal{T}_{WN} \models \sqcap A^d \sqsubseteq \sqcap A^q \text{ iff for all } A^q \text{ in } \sqcap A^q, \mathcal{T}_{WN} \models \sqcap A^d \sqsubseteq A^q \quad (13)$$

[2] Note, that in general, Equation 12 cannot be applied. One such case is when negation is allowed, a counterexample is $\models A_i \sqsubseteq A_j \sqcup \neg A_j$. A second case is when $\mathcal{T}$ contains axioms in the form $A_i \sqsubseteq A_j \sqcup A_k$; consider, e.g., $\mathcal{T} = \{A_i \sqsubseteq A_j \sqcup A_k\} \models A_i \sqsubseteq A_j \sqcup A_k$.

Notice that the second part of each equation is the same as the first part of the following equation, and that the first part of Equation 11 and the last part of Equation 13 are exactly Equations 9 and 10. This proves that the above problem reduction is correct.

If by $\mathbf{C}_C^{\sqcap}$ we denote a set of all conjunctive components $\sqcap A^d$, which are equivalent to or more specific than concept C, i.e.,

$$\mathbf{C}_C^{\sqcap} = \{\sqcap A^d \mid \mathcal{T}_{WN} \models \sqcap A^d \sqsubseteq C\}, \text{ where } C \in \{A^q, \sqcap A^q, \sqcup\sqcap A^q\} \tag{14}$$

Given Equations 11, 12, and 13, the query answer $A(C^q, \mathcal{T}_{WN})$, as defined in 8, can be computed by using the algorithm in the figure below. The five phases

Algorithm 1. Calculate $A(\sqcup\sqcap A^q, \mathcal{T}_{WN})$

1: $\mathbf{C}_{\sqcup\sqcap A^q}^{\sqcap} \leftarrow \emptyset$
2: **for all** $\sqcap A^q$ in $\sqcup\sqcap A^q$ **do**
3: $\quad i \leftarrow 0$
4: $\quad \mathbf{C}_{\sqcap A^q}^{\sqcap} \leftarrow \emptyset$
5: $\quad$ **for all** A^q in $\sqcap A^q$ **do**
6: $\quad\quad \mathbf{C}_{A^q}^{\sqcap} \leftarrow \{\sqcap A^d \mid \mathcal{T}_{WN} \models \sqcap A^d \sqsubseteq A^q\}$ | Phase 1
7: $\quad\quad$ **if** i = 0 **then**
8: $\quad\quad\quad \mathbf{C}_{\sqcap A^q}^{\sqcap} \leftarrow \mathbf{C}_{A^q}^{\sqcap}$
9: $\quad\quad$ **else**
10: $\quad\quad\quad \mathbf{C}_{\sqcap A^q}^{\sqcap} \leftarrow \mathbf{C}_{\sqcap A^q}^{\sqcap} \cap \mathbf{C}_{A^q}^{\sqcap}$
11: $\quad\quad$ **end if**
12: $\quad\quad i \leftarrow i + 1$
13: $\quad$ **end for** | Phase 2 (lines 5–13)
14: $\quad \mathbf{C}_{\sqcup\sqcap A^q}^{\sqcap} \leftarrow \mathbf{C}_{\sqcup\sqcap A^q}^{\sqcap} \cup \mathbf{C}_{\sqcap A^q}^{\sqcap}$
15: **end for** | Phase 3 (lines 2–15)
16: $\mathbf{C}_{\sqcup\sqcap A^q} \leftarrow \{\sqcup\sqcap A^d \mid$ all $\sqcap A^d$ in $\sqcup\sqcap A^d$ belong to $\mathbf{C}_{\sqcup\sqcap A^q}^{\sqcap}\}$ | Phase 4
17: $A \leftarrow \{d \mid$ there exists $\sqcup\sqcap A^d$ in d, s.t., $\sqcup\sqcap A^d$ belongs to $\mathbf{C}_{\sqcup\sqcap A^q}\}$ | Phase 5

of the algorithm can be explained as follows (the next section reports how these phases are actually implemented and their output on a running example):

Phase 1: (line 6) We compute $\mathbf{C}_{A^q}^{\sqcap}$, i.e., the set of all $\sqcap A^d$, such that, $\sqcap A^d \sqsubseteq A^q$.

Phase 2: (lines 5-13) We compute $\mathbf{C}_{\sqcap A^q}^{\sqcap}$, i.e., a set of all $\sqcap A^d$, such that, $\sqcap A^d \sqsubseteq \sqcap A^q$. As it follows from Equation 13, $\sqcap A^d \in \mathbf{C}_{\sqcap A^q}^{\sqcap}$ only if for every A^q in $\sqcap A^q$, $\sqcap A^d \in \mathbf{C}_{A^q}^{\sqcap}$. To compute $\mathbf{C}_{\sqcap A^q}^{\sqcap}$, we intersect sets $\mathbf{C}_{A^q}^{\sqcap}$ for all A^q in $\sqcap A^q$.

Phase 3: (lines 2-15) We compute the set $\mathbf{C}_{\sqcup\sqcap A^q}^{\sqcap}$, i.e., the set of all $\sqcap A^d$, such that, $\sqcap A^d \sqsubseteq \sqcup\sqcap A^q$. As it follows from Equation 12, $\sqcap A^d \in \mathbf{C}_{\sqcup\sqcap A^q}^{\sqcap}$ if $\sqcap A^d \in \mathbf{C}_{\sqcap A^q}^{\sqcap}$ at least for one $\sqcap A^q$ in $\sqcup\sqcap A^q$. To compute the set $\mathbf{C}_{\sqcup\sqcap A^q}^{\sqcap}$, we take the union of all the sets $\mathbf{C}_{\sqcap A^q}^{\sqcap}$ for all $\sqcap A^q$ in $\sqcup\sqcap A^q$.

Phase 4: (line 16) We compute the set $\mathbf{C}_{\sqcup\sqcap A^q}$, i.e., the set of all complex document concepts $\sqcup\sqcap A^d$, such that, $\sqcup\sqcap A^d \sqsubseteq \sqcup\sqcap A^q$. As it follows from Equation 11, $\sqcup\sqcap A^d \in \mathbf{C}_{\sqcup\sqcap A^q}$ only if all the conjunctive components $\sqcap A^d$ in $\sqcup\sqcap A^d$ belong to $\mathbf{C}_{\sqcup\sqcap A^q}^{\sqcap}$. To compute the set $\mathbf{C}_{\sqcup\sqcap A^q}$, we collect all $\sqcup\sqcap A^d$ which consist only from conjunctive components $\sqcap A^d$ in $\mathbf{C}_{\sqcup\sqcap A^q}^{\sqcap}$.

Phase 5: (line 17) We compute $A(\sqcup\sqcap A^q, \mathcal{T}_{WN})$ as defined in Equation 8, i.e., by collecting all the documents which contain concepts from $\mathbf{C}_{\sqcup\sqcap A^q}$.

5 Concept Search via Inverted Indexes

In *C-Search*, every document is represented as an enumerated sequence of conjunctive components $\sqcap A^d$ possibly connected by symbol "$\sqcup$". For example, in Figure 4 we show the sequences of $\sqcap A^d$ extracted from documents in Figure 1. Rectangles in Figure 4 represent either conjunctive components $\sqcap A^d$ or the disjunction symbol "$\sqcup$", a number in a square at the left side of a rectangle represents the *position* of the rectangle in the whole sequence. Note, that symbol "$\sqcup$" is used to specify that conjunctive components $\sqcap A^d$ connected by this symbol form a single disjunctive concept $\sqcup\sqcap A^d$, namely:

$$\sqcup\sqcap A^d ::= (\sqcap A^d)\{(\text{“}\sqcup\text{”})(\sqcap A^d)\} \tag{15}$$

For example, the first three positions in the sequence for document $D3$ in Figure 4 represent the concept $(little\text{-}4 \sqcap dog\text{-}1) \sqcup (huge\text{-}1 \sqcap cat\text{-}1)$.

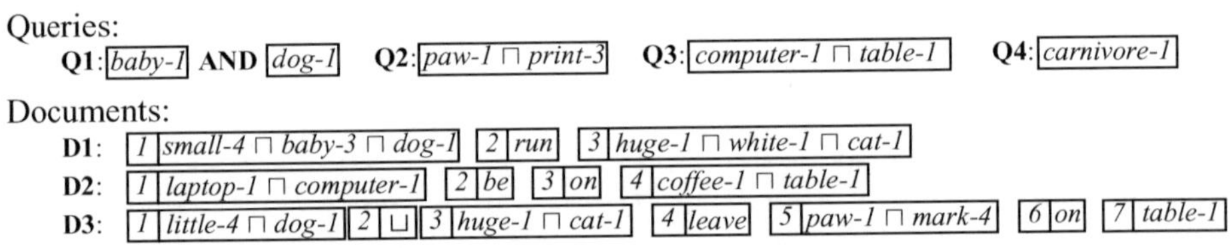

Fig. 4. Document and Query Representations

The resulting document representations (see Figure 4) are indexed using a positional inverted index. In a positional inverted index, as used in syntactic search, there are two parts: the *dictionary*, i.e., a set of terms (t) used for indexing; and a set of posting lists P(t). A posting list P(t) is a list of all postings for term t:

$$P(t) = [\langle d, freq, [position]\rangle]$$

where $\langle d, freq, [position]\rangle$ is a posting consisting of a document d associated with term t, the frequency $freq$ of t in d, and a list $[position]$ of positions of t in d.

In *C-Search*, we adopt a positional inverted index to index conjunctive components $\sqcap A^d$ by all more general or equivalent atomic concepts from $\mathcal{T}_{WN}$[3]. For example, in Figure 5 we show a fragment of the positional inverted index created by using the document representations in Figure 4. The inverted index

[3] In [7], we showed how searching for complex concepts can be implemented by indexing documents directly by these concepts. The problem of this method is that the size of the inverted index vocabulary, in the worst case, is exponential with respect to size of terminology $\mathcal{T}$. In the current paper, we propose an approach in which we index complex concepts by atomic concepts using the positional information in the inverted index. The size of the vocabulary in this case is the same as the size of $\mathcal{T}$.

Dictionary (t)	Posting lists (P(t))
$\sqcup$	$[\langle D3, 1, [2]\rangle]$
baby-3	$[\langle D1, 1, [1]\rangle]$
canine-2	$[\langle D1, 1, [1]\rangle; \langle D3, 1, [1]\rangle]$
carnivore-1	$[\langle D1, 2, [1, 3]\rangle; \langle D3, 2, [1, 3]\rangle]$
computer-1	$[\langle D2, 1, [1]\rangle]$
feline-1	$[\langle D1, 1, [3]\rangle; \langle D3, 1, [3]\rangle]$
leave	$[\langle D3, 1, [4]\rangle]$
little-4	$[\langle D1, 1, [1]\rangle; \langle D3, 1, [1]\rangle]$

Fig. 5. Positional Inverted Index

dictionary, in *C-Search*, consists of atomic concepts from $\mathcal{T}_{WN}$ (e.g., concepts *baby*-3 and *canine*-2 in Figure 5), and symbol "$\sqcup$" (e.g., the first term in Figure 5). The posting list $P(A)$ for an atomic concept A stores the positions of conjunctive components $\sqcap A^d$, such that, $\sqcap A^d \sqsubseteq A$. For instance, P(*canine-2*) $= [\langle D1, 1, [1]\rangle; \langle D3, 1, [1]\rangle]$, which means that at first position in documents $D1$ and $D3$ there are conjunctive components (i.e., *small*-4 $\sqcap$ *baby*-3 $\sqcap$ *dog*-1 and *little*-4 $\sqcap$ *dog*-1) which are more specific than *canine-2*. The posting list $P(\sqcup)$ stores the positions of the symbol "$\sqcup$".

Now, let us see how Algorithm 1 in Section 4.2 can be implemented by using the positional information of conjunctive components $\sqcap A^d$ stored in the inverted index. Notice that below instead of conjunctive components themselves we work only with their positions.

Phase 1: Positions of conjunctive components $\sqcap A^d$ in the set $\mathbf{C}^{\sqcap}_{A^q}$ (line 6 in Algorithm 1) are computed by fetching the posting list $P(A^q)$ for an atomic concept A^q. For instance (see Figure 5),

$$\mathbf{C}^{\sqcap}_{little\text{-}4} = [\langle D1, 1, [1]\rangle; \langle D3, 1, [1]\rangle]$$
$$\mathbf{C}^{\sqcap}_{carnivore\text{-}1} = [\langle D1, 2, [1, 3]\rangle; \langle D3, 2, [1, 3]\rangle]$$

Phase 2: The intersection of the sets of conjunctive components (line 10 in Algorithm 1) is implemented by the intersection of corresponding posting lists. For instance,

$$\mathbf{C}^{\sqcap}_{little\text{-}4 \sqcap carnivore\text{-}1} = [\langle D1, 1, [1]\rangle; \langle D3, 1, [1]\rangle]$$

Phase 3: The union of the sets of conjunctive components (line 14 in Algorithm 1) is implemented by uniting corresponding posting lists. For instance,

$$\mathbf{C}^{\sqcap}_{canine\text{-}2 \sqcup feline\text{-}1} = [\langle D1, 2, [1, 3]\rangle; \langle D3, 2, [1, 3]\rangle]$$

Phase 4: Every concept in set $\mathbf{C}_{\sqcup\sqcap A^q}$ (line 16 in Algorithm 1) should consists only from the conjunctive components in $\mathbf{C}^{\sqcap}_{\sqcup\sqcap A^q}$. In order to find the positions of such concepts, we take the union of the posting lists for $\mathbf{C}^{\sqcap}_{\sqcup\sqcap A^q}$ with the posting list for the symbol "$\sqcup$". Then we filter out all the positions which does not comply with the pattern defined in Equation 15. For instance, for complex query concept $canine\text{-}2 \sqcup feline\text{-}1$, we will find the following complex document concepts:

$\langle D1, 1, [1]\rangle \Rightarrow$ $\boxed{1 \mid small\text{-}4 \sqcap baby\text{-}3 \sqcap dog\text{-}1}$ $\langle D1, 1, [3]\rangle \Rightarrow$ $\boxed{3 \mid huge\text{-}1 \sqcap white\text{-}1 \sqcap cat\text{-}1}$

$\langle D3, 1, [1, 2, 3]\rangle \Rightarrow$ $\boxed{1 \mid little\text{-}4 \sqcap dog\text{-}1}$ $\boxed{2 \mid \sqcup}$ $\boxed{3 \mid huge\text{-}1 \sqcap cat\text{-}1}$

Phase 5: The query answer (line 17 in Algorithm 1) is computed by collecting the documents from all the postings. For instance,

$$A(canine\text{-}2 \sqcup feline\text{-}1, T_{WN}) = \{D1, D3\}$$

If n is a number of atomic concepts A^q in the query concept $\sqcup \sqcap A^q$, then to compute $A(C^q, T_{WN})$ it takes n posting list merges (i.e., intersections and unions). Note that, in a positional inverted index, the same number of posting list merges is required to process a phrase query consisting of $n + 1$ words [13].

In *C-Search*, query concepts C^q can be combined into more complex queries q by using the boolean operators AND and NOT. Query answer $A(q, T_{WN})$ in this case is computed by recursively applying the following rules:

$$A(q_i \text{ AND } q_j, T_{WN}) = A(q_i, T_{WN}) \cap A(q_j, T_{WN})$$
$$A(q_i \text{ NOT } q_j, T_{WN}) = A(q_i, T_{WN}) \;/\; A(q_j, T_{WN})$$

For instance, the query answer for query *baby*-1 AND *dog*-1 (in Figure 4) is computed as follows: $A(baby\text{-}1 \text{ AND } dog\text{-}1, T_{WN}) = A(baby\text{-}1, T_{WN}) \cap A(dog\text{-}1, T_{WN}) = \emptyset \cap \{D1, D3\} = \emptyset$.

Relevance of documents, in *C-Search*, is computed by adopting the cosine similarity from the vector space model. Concepts are weighted by *tf-idf* weight measure modified to take into account the semantic relatedness of concepts.

6 Evaluation

In order to evaluate our approach, we built two IR systems. The first system is an instantiation of syntactic search and is build on top of Lucene[4], an open source IR toolkit used in many search applications[5]. Standard tokenization and English Snowball stemmer were used for document and query preprocessing. The AND operator was used as a default boolean operator in a query. The second system is a semantics enabled version of Lucene, implemented following the methodology described in Sections 4 and 5.

As a data-set for our experiments, we used the TREC ad-hoc document collection[6] (disks 4 and 5 minus the Congressional Record documents) and three query sets: TREC6 (topics 301-350), TREC7 (topics 351-400) and TREC8 (topics 401-450). Only the title for each topic was used as a query. The whole data-set

[4] `http://lucene.apache.org/java/docs/index.html`

[5] `http://wiki.apache.org/lucene-java/PoweredBy`

[6] `http://trec.nist.gov/data/test_coll.html`

Table 1. Evaluation results

	MAP	P@5	P@10	P@15
TREC6 (301-350)				
Lucene	0.1361	0.3200	0.2960	0,2573
C-Search(Lucene)	0.1711(+25.7%)	0.3920(+22.5%)	0.3480(+17.6%)	0.3000(+16.6%)
TREC7 (351-400)				
Lucene	0.1138	0.3560	0.3280	0.3000
C-Search(Lucene)	0.1375(+20.8%)	0.4200(+18.0%)	0.3680(+12.2%)	0.3427(+14.2%)
TREC8 (401-450)				
Lucene	0.1689	0.4320	0.4000	0.3573
C-Search(Lucene)	0.2070(+22.6%)	0.4760(+10.2%)	0.4280(+7.0%)	0.4013(+12.3%)

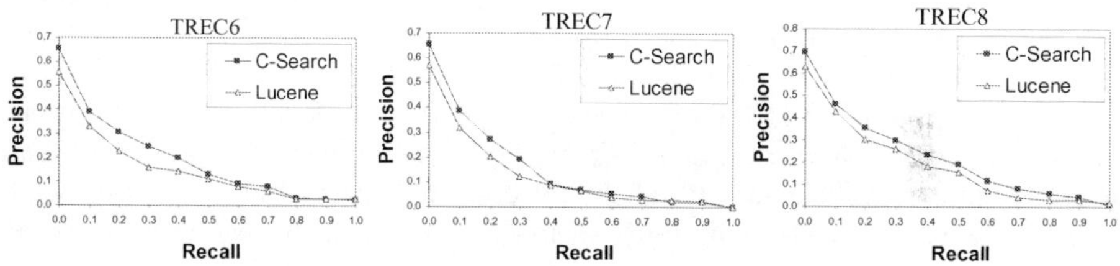

Fig. 6. Recall-Precision Graphs

consists of 523,822 documents and 150 queries. In the evaluation we used the standard IR measures and in particular the mean average precision (MAP) and precision at K (P@K), where K was set to 5, 10, and 15. The average precision for a query is the mean of the precision obtained after each relevant document is retrieved (using 0 as the precision for not retrieved documents which are relevant). MAP is the mean of the average precisions for all the queries in the test collection. P@K is the percentage of relevant documents among the top K ranked documents. MAP is used to evaluate the overall accuracy of IR system, while P@K is used to evaluate the utility of IR system for users who only see the top K results.

First, in Table 1, we report the evaluation results for the two systems and further, in figure 6, we provide recall-precision graphs, i.e., we plot precision as a function of recall, for these systems.

The experiments show that, on TREC ad-hoc data sets, *C-Search* performs better than purely syntactic search, which supports the underlying assumption of our approach. In particular, from Table 1 we observe that *C-Search* improves precision P@K for all K in all three TREC data sets. This is coherent with the intuition that semantics improve on precision. Notice that it means that we are able to show users more relevant documents at the top of the list. From Figure 6 we observe that the recall-precision graphs for *C-Search* are above those for Lucene, which means that the improvement in precision achieved by *C-Search* does not decrease recall.

7 Related Work

The fact that the syntactic nature of classical IR leads to problems with precision was recognized in the IR community long ago (e.g., see [20]). There were two major approaches to addressing this problem. The first is based on natural language processing and machine learning techniques in which (noun) phrases in a document corpus are identified and organized in a subsumption hierarchy which is then used to improve the precision of retrieval (e.g., see [21]). The second is based on using a linguistic database which is used to associate words in a document corpus with atomic lexical concepts in this database and then to index these documents with the associated concepts (e.g., see [18]). Our approach is different from both these approaches. In fact, the former approach is still essentially syntactic (and semantics is only implicitly derived with no guarantee of correctness), while in the latter approach only atomic concepts are indexed, whereas *C-Search* allows for indexing of complex concepts and explicitly takes into account possible relations between them. More importantly, our approach *extends* syntactic search and does not replace it as in the latter approach and, therefore, supports the continuum from purely syntactic to fully semantic search.

In the Semantic Web community, semantic search is primarily seen as the task of querying an RDF graph based on the mapping of terms appearing in the input natural language query to the elements of the graph. An analysis of existing semantic search systems is provided in [10]. Our approach is fundamentally different because, similarly to classical IR, the input query is mapped to document contents and not to elements of a knowledge representation structure. Document retrieval approaches developed in the context of the Semantic Web are surveyed in [12]. The matching of document and query representations, in these approaches, is based on query expansion (e.g., see [3]), graph traversal (e.g., see [16]), and RDF reasoning (e.g., see [5]). Differently from these approaches, *C-Search* is based on the semantic matching of *complex concepts*, where semantic matching is implemented by using positional inverted index. Hybrid Search [2] is similar to our approach in that it integrates syntactic search with semantic search. Differently from us, in [2], semantic search is implemented on metadata and is totally separated from syntactic search, implemented on keywords. Instead, in our approach semantic search is seamlessly integrated into syntactic search by substitution words with concepts and at the same time keeping the underlying machinery of syntactic search.

8 Conclusion

In this paper we have presented an approach where syntactic search is extended with a semantic layer. The proposed approach performs as good as syntactic search while allowing for an improvement whenever semantics are available and can be exploited. The reported experimental results provide a proof of concept of the quality of proposed solution.

We are still at the beginning and a lot of work still needs to be done. Future work includes:

– the development of more accurate document relevance metrics based on both syntactic and semantic similarity of query and document descriptions;
– the integration of more accurate algorithms for concept identification;
– providing support for queries in which concepts can be associated with a semantic scope such as equivalence, more/less general, disjoint;
– improving the performance of the system. Even if the system is reasonably fast in the sense that it provides answers in much less than a second, namely in a time which is reasonable for the user to wait, it is slower than Lucene and we believe that there is a lot of scope for improving its efficiency.

References

1. Andreasen, T., Anker Jensen, P., Fischer Nilsson, J., Paggio, P., Pedersen, B.S., Thomsen, H.E.: Content-based text querying with ontological descriptors. Data & Know. Eng. 48(2), 199–219 (2004)
2. Bhagdev, R., Chapman, S., Ciravegna, F., Lanfranchi, V., Petrelli, D.: Hybrid search: Effectively combining keywords and semantic searches. In: Bechhofer, S., Hauswirth, M., Hoffmann, J., Koubarakis, M. (eds.) ESWC 2008. LNCS, vol. 5021, pp. 554–568. Springer, Heidelberg (2008)
3. Celino, I., Della Valle, E., Cerizza, D., Turati, A.: Squiggle: a semantic search engine for indexing and retrieval of multimedia content. In: SEMPS (2006)
4. Cunningham, H., Maynard, D., Bontcheva, K., Tablan, V.: GATE: A framework and graphical development environment for robust NLP tools and applications. In: 40th Anniversary Meeting of the Association for Computational Linguistics (2002)
5. Davies, J., Weeks, R.: QuizRDF: Search technology for the semantic web. In: HICSS (2004)
6. Deerwester, S.C., Dumais, S.T., Landauer, T.K., Furnas, G.W., Harshman, R.A.: Indexing by latent semantic analysis. Journal of the American Society of Information Science 41(6), 391–407 (1990)
7. Giunchiglia, F., Kharkevich, U., Zaihrayeu, I.: Concept search: Semantics enabled syntactic search. In: SemSearch 2008 workshop at ESWC (2008)
8. Giunchiglia, F., Shvaiko, P., Yatskevich, M.: Discovering missing background knowledge in ontology matching. In: Proc. of ECAI (2006)
9. Giunchiglia, F., Yatskevich, M., Shvaiko, P.: Semantic matching: Algorithms and implementation. Journal on Data Semantics (JoDS) 9, 1–38 (2007)
10. Hildebrand, M., van Ossenbruggen, J., Hardman, L.: An analysis of search-based user interaction on the semantic web. Technical Report INS-E0706, CWI (2007)
11. Magnini, B., Speranza, M., Girardi, C.: A semantic-based approach to interoperability of classification hierarchies: evaluation of linguistic techniques. In: COLING 2004 (2004)
12. Mangold, C.: A survey and classification of semantic search approaches. Int. J. Metadata Semantics and Ontology 2(1), 23–34 (2007)
13. Manning, C., Raghavan, P., Schütze, H.: Introduction to Information Retrieval. Cambridge University Press, Cambridge (2008)
14. Moldovan, D.I., Mihalcea, R.: Using wordnet and lexical operators to improve internet searches. IEEE Internet Computing 4(1), 34–43 (2000)
15. Nagypál, G.: Improving information retrieval effectiveness by using domain knowledge stored in ontologies. In: Meersman, R., Tari, Z., Herrero, P. (eds.) OTM-WS 2005. LNCS, vol. 3762, pp. 780–789. Springer, Heidelberg (2005)

16. Rocha, C., Schwabe, D., de Aragao, M.: A hybrid approach for searching in the semantic web. In: 13th International World Wide Web Conference (2004)
17. Sanderson, M.: Retrieving with good sense. Inf. Retr. 2(1), 49–69 (2000)
18. Schutze, H., Pedersen, J.O.: Information retrieval based on word senses. In: 4th Annual Symposium on Document Analysis and Information Retrieval (1995)
19. Sowa, J.F.: Conceptual Structures: Information Processing in Mind and Machine. Addison-Wesley, Reading (1984)
20. Stokoe, C., Oakes, M.P., Tait, J.: Word sense disambiguation in information retrieval revisited, pp. 159–166 (2003)
21. Woods, W.A.: Conceptual indexing: A better way to organize knowledge. Technical Report TR-97-61, Sun Microsystems Laboratories, USA (1997)
22. Zaihrayeu, I., Sun, L., Giunchiglia, F., Pan, W., Ju, Q., Chi, M., Huang, X.: From web directories to ontologies: Natural language processing challenges. In: Aberer, K., Choi, K.-S., Noy, N., Allemang, D., Lee, K.-I., Nixon, L., Golbeck, J., Mika, P., Maynard, D., Mizoguchi, R., Schreiber, G., Cudré-Mauroux, P. (eds.) ASWC 2007 and ISWC 2007. LNCS, vol. 4825, pp. 623–636. Springer, Heidelberg (2007)
23. Zhai, C.: Fast statistical parsing of noun phrases for document indexing. In: Fifth Conference on Applied Natural Language Processing, pp. 312–319 (1997)

Semantic Wiki Search

Peter Haase[1], Daniel Herzig[1,⋆], Mark Musen[2], and Thanh Tran[1]

[1] Institute AIFB, Universität Karlsruhe (TH), Germany
{pha,dahe,dtr}@aifb.uni-karlsruhe.de
[2] Stanford Center for Biomedical Informatics Research (BMIR)
Stanford University, USA
musen@stanford.edu

Abstract. Semantic wikis extend wiki platforms with the ability to represent structured information in a machine-processable way. On top of the structured information in the wiki, novel ways to search, browse, and present the wiki content become possible. However, while powerful query languages offer new opportunities for semantic search, the syntax of formal query languages is not adequate for end users. In this work we present an approach to semantic search that combines the expressiveness and capabilities of structured queries with the simplicity of keyword interfaces and faceted search. Users articulate their information need in keywords, which are translated into structured, conjunctive queries. This translation may result in multiple possible interpretations of the information need, which can then be selected and further refined by the user via facets. We have implemented this approach to semantic search as an extension to Semantic MediaWiki. The results of a user study in the SMW-based community portal *semanticweb.org* show the efficiency and effectiveness of the approach as well as its ease of use.

1 Introduction

The availability of structured information on the Semantic Web enables new opportunities for information access. Search is no longer limited to matching keywords against documents, but instead complex information needs can be expressed in a structured way, with precise and structured answers as results [1,2,3].

A prominent application area are semantic wikis. Wikis in general have become a popular tool for collaborative creation and exchange of content on the Web. The content typically is maintained in wiki pages with plain text and occasional markup elements. Semantic wikis aim at extending wiki platforms with the ability to represent structured information in a machine-processable way. For example, Semantic MediaWiki (SMW) [4] is an extension of the well-known MediaWiki, which also powers Wikipedia. The main element of this extension is the notion of Typed Links, which enable users to generate structured information with a well-defined semantics. On top of the structured information in the wiki, novel ways to reuse, browse, and search the wiki content become possible.

For example, SMW already supports the so called ASK language[1] to perform structured queries. However, while such powerful query languages offer new opportunities

⋆ Work done while visiting student at BMIR, Stanford University.
[1] http://semantic-mediawiki.org/wiki/Help:Semantic_search

L. Aroyo et al. (Eds.): ESWC 2009, LNCS 5554, pp. 445–460, 2009.

for semantic search, users typically are not willing to use the syntax of formal query languages. What would be desired is a search interface that combines the expressiveness and capabilities of structured queries with the simplicity known from keyword interfaces and faceted browsing, which are much easier to handle for lay end users.

In this work we present such an approach to semantic search: Users articulate their information need using keyword queries, which are translated by the system into structured queries representing possible interpretations of the information need. The user is able to select the query that best matches his information need and to further refine it via facets. In doing so, the user does not need any knowledge about formal query languages nor any schema of the data. We extend our previous work [5] on semantic search based on a translation of keyword-based queries into structured queries against graph-structured data, thus enabling expressive queries while retaining a simple end user interface. We apply this approach to semantic search in the Semantic MediaWiki. We have deployed our extension for semantic search on several installations of Semantic MediaWiki. One of them is the wiki-based portal of the Semantic Web community – *semanticweb.org*, which gathers (semi-)structured information about news, people, events, etc. relevant for the Semantic Web community. Within *semanticweb.org* we have performed a user study that shows the efficiency and effectiveness of our approach.

The rest of the paper is organized as follows. In Section 2 we present on overview of our approach to semantic search from a process perspective. We describe the technical details of this process as well as the implementation in Section 3. We report on the results of our user evaluation study in Section 4. After a discussion of related work in Section 5 we conclude in Section 6.

2 Semantic Search – A Process View

In this section we describe the search process underlying our approach to semantic search. This process comprises three main steps:

1. *Articulation of information need:* Following the paradigm of keyword search, users express their information need using keyword queries.
2. *Query interpretation using keyword translation:* The information need expressed in keywords is interpreted via a translation into conjunctive queries. The user selects the query that best matches his information need.
3. *Result presentation and query refinement:* The answers to the selected query are presented to the user. The user can further refine the query using faceted search.

This process is based on a paradigm that differs from traditional search, where users issue a query, obtain a (set of) results, and – if the results do not fulfill the information need – start over with issuing a new query.

First, we introduce an additional step, in which different possible interpretations of the user information need are computed and presented to the user. Instead of presenting the results directly, which might actually belong to many distinct queries (representing different information needs), we allow the user to select the correct interpretation.

Second, we allow the user to refine the interpreted query to match the exact information need. The act of refinement is thus much more precise and direct: Instead of having to issue a new query, the user can directly refine the possible interpretations.

In the following, we briefly discuss some important aspects of these individual steps.

Articulation of the information need. In our approach, users articulate their information needs using keyword queries. We believe this is the most adequate form, as keyword interfaces have been widely adopted, and users are familiar with them both due to their simplicity and their presence in today's systems. For searching, the users do not need to know about the query syntax, the schema and even the labels of the data elements. Ideally, they can use their own words to express their information needs.

Query interpretation using keyword translation. This step is concerned with the translation of the user queries into system queries, i.e. structured conjunctive queries. Here we interpret keywords as elements of structured queries. The keyword search process contains an additional step, namely the presentation of structured queries. We consider this step as beneficial because these queries can be seen as descriptions, and can thus facilitate the comprehension of the answers. Also, refinement can be made more precisely on the structured query than on the keyword query. The details of the keyword translation will be described in Section 3.2. Briefly summarized, we follow a graph exploration to identify query graphs, i.e. substructures that connect all keyword elements. According to a query ranking scheme, the computed query graphs are sorted and presented to the user for *selection*. Here, the queries are not presented using a formal syntax, but using an intuitive, graph-based representation. Additionally, to better understand the query, we also present snippets of the query results. In certain cases, these snippets may already be the answers that fulfill the information need and the final step may not be needed.

Result presentation and refinement. Query results – which in the general case are sets of tuples satisfying the conjunctive query – are presented to the user in a structured, tabular form. Further, the user can refine the query following the paradigm of faceted search. Faceted search (often also referred to as faceted browsing) is a paradigm for exploring data through navigating along its hierarchical or multidimensional metadata or structure. The main characteristic of this paradigm is that it allows users to narrow and expand the data of interest according to their information need in an interactive way. The user actions of adding or removing facets are transparently converted to operations on the conjunctive query. The refined queries are immediately evaluated and the new results presented without the user having to explicitly issue a new query.

Refinement may be needed for several reasons. First of all, the computed interpretations may not exactly match the information need. Second, the user may also start out with an ill-defined information need, not knowing what exactly he is searching for. For these cases, the system can be used for faceted browsing to explore the knowledge base. The details of the faceted search will be described in Section 3.3.

3 Semantic Search in SMW

In this section we describe how the process of semantic search is realized and supported in our extension to Semantic MediaWiki. We first present the underlying data and query model within SMW, and then discuss the aspects of query interpretation, result presentation and query refinement.

3.1 Data and Query Model

Our search operates on the structured part of the data in Semantic MediaWiki. We briefly discuss the main primitives of structuring data in SMW along with a formal semantic interpretation of the wiki's structure in terms of OWL DL. The primary structural mechanism of MediaWiki is the organization of content within wiki pages. Every page can be assigned to one or many categories, which can be organised hierarchically. SMW introduces ways of adding further structure to MediaWiki by means of *annotating* textual content of the wiki: Properties are used to express binary relationships between one entity (as represented by a wiki page) and some other such entity or data value. The formal semantics of structured data in SMW is given via a mapping to the OWL ontology language [4].

Most elements are represented in OWL in a straight-forward manner, using the obvious mapping from wiki pages to OWL entities: normal article pages correspond to individuals, categories correspond to OWL classes, and SMW properties correspond to OWL properties. More precisely, SMW properties with article pages in their range are mapped to object properties, and SMW properties with datatypes in their range are mapped to datatype properties. Accordingly, property values can be individuals or data values (typed literals). Most annotations thus are directly mapped to simple assertions in OWL, similar to RDF triples. Finally, containment of pages in MediaWiki's categories is represented as class membership in OWL. MediaWiki supports the hierarchical organization of categories, and SMW can be configured to interpret this as an OWL class hierarchy. Moreover, SMW introduces a special property "subproperty of" that can be used for property hierarchies.

For querying, we distinguish between the notion of a *user query* and a *system query*. The user expresses his information needs in keyword queries.

Example 1. As a running example, we consider an information need of a user who is searching for upcoming conferences with their paper deadline, and as he likes to go to Greece, he is in particular interested in conferences in this country. He might express his information need using the keywords: "conferences Greece deadline".

The system queries are *conjunctive queries*. Formally, a *conjunctive query* is an expression of the form $(x_1, \ldots, x_k).\exists x_{k+1}, \ldots x_m.A_1 \wedge \ldots \wedge A_r$, where $x_1, \ldots, x_k$ are called *distinguished variables*, $x_{k+1}, \ldots, x_m$ are *undistinguished variables* and $A_1, \ldots, A_r$ are *query atoms*. These atoms are of the form $P(v_1, v_2)$, where P is called *predicate*, v_1, v_2 are variables or, otherwise, are called *constants*.

Conjunctive queries have high practical relevance because they are capable of expressing the large class of relational queries. The vast majority of query languages for many data models used in practice fall into this fragment, including large parts of SPARQL. Conjunctive queries can be used to address typical information needs frequently occurring when searching in a semantic wiki. We can distinguish the following categories of queries (corresponding to special cases of conjunctive queries):

1. *Entity Queries* are queries for certain entities (individuals in OWL), where the results correspond to a wiki page. An example information might be searching for the page of a person, e.g. "Rudi Studer".

2. *Fact Queries* are queries for concrete properties of particular objects (Example: "Rudi Studer phone"), however, the result does not correspond directly to a page, but to one (or more) statements on a page.
3. *General Conjunctive Queries* are queries that allow to retrieve n-ary tuple sets as results. An example could be the query "members AIFB phone" which asks for all members of the AIFB institute along with their phone numbers.[2]

In terms of the results, entity queries are similar to the existing keyword search supported in MediaWiki, which always returns single pages. The latter two types of queries go beyond the current search capabilities of MediaWiki.

3.2 Query Interpretation and Processing

The goal of this step is to interpret the keywords of the user query in terms of a structured, conjunctive query. Typically, due to the ambiguity inherent in keyword queries, such an interpretation is not unique. Therefore, we rely on a top-k procedure to generate candidate interpretation and allow the user to select the interpretation that best matches his information need. We build on our previous work [5] on translating keyword queries into structured queries based on a graph-exploration technique. For this purpose, we consider the knowledge base as a graph structure. (Intuitively, this graph corresponds to the RDF representation of an OWL knowledge base).

The computation of query graphs as interpretations of the user keywords involves three tasks: 1) construction of the query search space 2) top-k query graph exploration and 3) query graph ranking. Specific concepts and algorithms for these tasks have been introduced in [5].

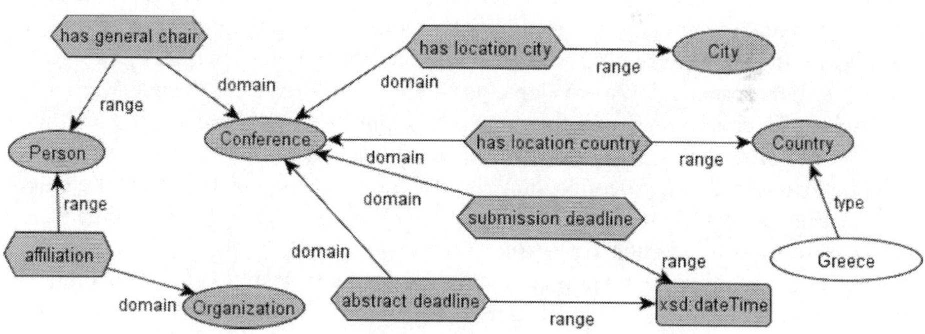

Fig. 1. Query space: schema graph (grey) augmented with keyword-matching elements (white)

Construction of the Query Search Space. The query search space contains all the elements that are necessary for the computation of possible interpretations. In our previous work [5], we have shown that keyword search is most efficient when the exploration for possible interpretations is performed on an "augmented" schema graph, instead of using the entire data graph (c.f. [6,7]). The schema graph can be trivially obtained from the

[2] Please note that this particular query could be represented using a tree-shaped conjunctive query, but in general the queries do not need to be tree-shaped.

class and property definitions (corresponding to the category and property definitions in the wiki). From experience we know that in semantic wikis the schema is typically incomplete. For example, often the domains and ranges of properties are not fully specified. Therefore, we additionally apply techniques for computing schema graphs automatically. In particular, a schema graph is derived from the data using the aggregation rules as described in [5].

The schema graph is augmented with elements that match the user keywords (to explore the query constants). We rely on a *keyword index* to map keywords to elements of the knowledge base, which in our approach might be classes, properties, individuals, and data values. IR concepts are adopted to support an imprecise matching that incorporates syntactic and semantic similarities. As a result, the user does not need to know the exact labels of the data elements when doing keyword search. Each element finally returned is associated with *a score* measuring the degree of matching, which is later used for ranking possible interpretations. The schema graph and the keyword elements are combined to obtain the query space.

Example 2. Figure 1 illustrates the query space constructed for our example keyword query. It consists of (the fragment of) a schema graph (nodes in grey). Keyword elements not covered by the schema graph are added. In particular, the keyword matching element *Greece* is added to *Country*, which it instantiates.

Exploration of Top-K Query Graphs. Given the query space, the remaining task is to search for the minimal query graphs in this space. Informally, a query graph is a matching subgraph of the augmented schema graph, such that for every keyword of the user query it contains at least one representative keyword matching element, and (2) the graph is connected, i.e. there exists a path from every graph element to every other graph element. A matching query graph is minimal if there exists no other query graph with a lower score. In [5], we have proposed a top-k procedure to find such query graphs. This procedure starts from the keyword elements and iteratively explores the query space for all distinct paths beginning from these elements. During this procedure, the path with the highest score so far is selected for further exploration. At some point, an element might be discovered to be a connecting element, i.e. there is a path from that element to at least one keyword element, for every keyword in the user query. These paths are merged to form a query graph. The explored graphs are added to the candidate list. The process continues until the upper bound score for the query graphs yet to be explored is lower than the score of the k-ranked query graph in the candidate list.

Example 3. An example query graph that can be found through exploration along these paths is shown in Figure 2. An alternative interpretation could have resulted in a query graph with *submission deadline* instead of *abstract deadline* as connecting element.

Scoring Query Graphs. During top-k computation, elements with highest scores are chosen for further exploration. Thus, the quality of the computed top-k query graphs depends largely on the scoring function. Factors that are considered in the element scoring function include the matching score (as obtained from the keyword index for the matching elements) and the importance (computed offline for elements of every schema graph). Since the cost of a path monotonically increases with the number of

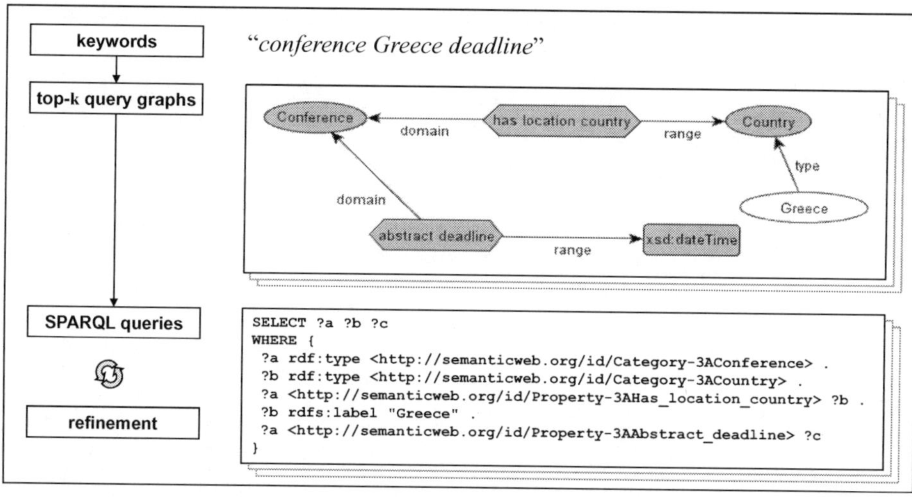

Fig. 2. Keyword translation

constituent elements, the length is also implicitly captured by this cost function. For the details of the scoring function, we refer to [5].

Query translation results in a list of top-k query graphs. Finally, the query graphs are translated to conjunctive queries in SPARQL query syntax. (See bottom of Figure 2 for an example.) Basically, edges are mapped to predicates, whereas vertices are mapped to variables and constants of the conjunctive query. (See [5] for a detailed description.)

3.3 Result Presentation and Query Refinement by Faceted Search

Since our search aims at lay end users, the interpretations in the form of conjunctive queries have to be presented to the user in an comprehensible and intuitively understandable way, so that the user can select the interpretation that fits best to his information need. A conjunctive query in SPARQL itself is far from being intuitively understandable, especially for end users. Therefore, we developed a concept for visualizing conjunctive queries suited for wiki-environments. We choose a table layout as the major pattern with a nested but minimal graph structure. The layout table is divided horizontally into three sections from top to bottom: A head-row, a relation-row and a preview section. The relation row spans the entire table, whereas the head row and the preview section are vertically divided into columns. There is one column for each distinguished variable of the query. The column labels denote the entity types (classes) and the arrows in between them represent the relationships (properties). In case there is no type information available, either the name of a property having the corresponding variable as the range is displayed in the column head, or if it occurs only as the domain of a property the placeholder "Something" is shown. All entities and relations tool have tips, which show the wiki page of the entity to help the user understand what entity is shown. To help the user grasp the meaning of the interpretation, the preview section contains a snippet of the query's results. The result tuples shown here are bindings to distinguished variables and are individuals of the classes denoted by the respective columns or data values.

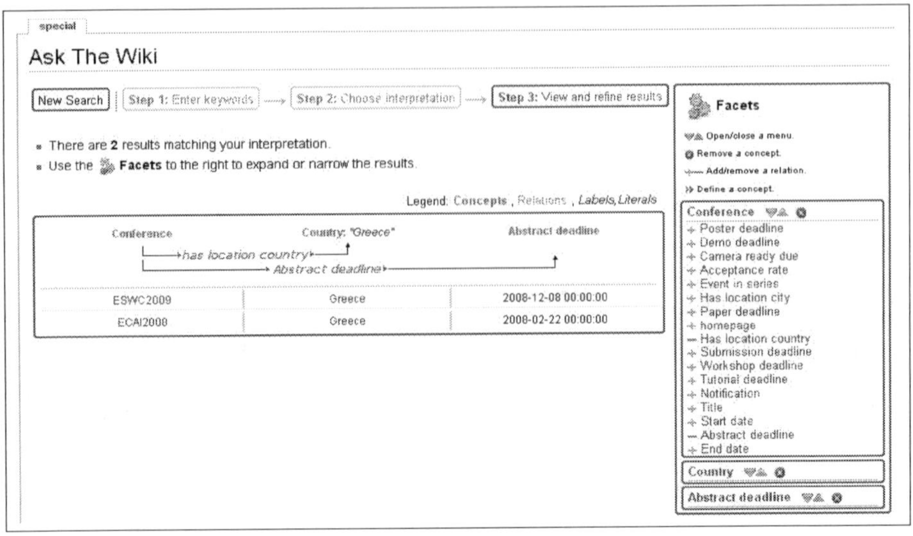

Fig. 3. Result presentation and query refinement. The selected query along with the query results are shown in the center. On the right side is the menu providing the faceted search capabilities.

This concept is sufficiently general for presenting answers conjunctive queries, in particular the three mentioned types of queries (cf. section 3.1).

1. *Entity Queries:* One single column for the entities in question.
2. *Fact Queries:* Two columns, one for a defined class and one for a data value/class.
3. *General Conjunctive Queries:* Several columns and relationships between them.

Figure 3 shows the visualization of the query from the example in Figure 2.

When an interpretation is chosen, the preview section is expanded and shows all matching entity tuples. In addition, the *facets menu* is loaded on the right side of the screen (Figure 3). The facets menu offers the faceted search capabilities. The facets menu has a sub menu for each distinguished variable of the query presenting the possible facets. Facets are properties or attributes of entities. The query modifications are performed by adding or removing facets to the interpretation. For example, in Figure 3 the user could drill down and refined the search result by adding the property "submission deadline" to (class) "Conference" in the interpretation. The new interpretation is immediately evaluated and the result presented to the user. We present only properties to the user, which would lead to a non-empty result when added to the interpretation. In case the range concept of an object property is not defined, a sub menu in the same style would offer the user to choose one from a list of valid concepts, which would narrow the search result accordingly. Instead of a adding a property, the user could also refine the search result by removing the property "has located country", he would then get all conferences and their abstract deadlines without the constraint that they are located in Greece. Furthermore, the user could also remove a variable and all properties bound to it, by pressing the "x" button in the head line of the corresponding sub menu. Using these modifications the user can fluently *browse* through the structured data of the semantic wiki by adapting the SPARQL query to his information need. All refinements are operations, namely adding or removing triple patterns to/from the SPARQL query.

3.4 Implementation

Our implementation – called *Ask The Wiki* – consists of two major parts, a Semantic MediaWiki (SMW) extension and a back-end, which we describe in the following. *Ask The Wiki* is integrated into the Semantic MediaWiki as a “Special page extension”. This extension provides the user interface and the faceted search methods via SPARQL modification. It is implemented in PHP and uses AJAX for the user interaction. Although, we precompute the schema, because it is often not fully specified in the wiki, the facets are retrieved on the fly for each query individually via AJAX from the back-end. All user selections, query modifications, and process states are kept on the client side.

The back-end is realized as a Java Servlet running on a Tomcat server. The servlet provides the keyword translation and top-k query construction, as well as the query evaluation. Before the keyword translation and query construction are available online, an offline preprocessing step is required. This step comprises computing the schema graph and indexing the OWL data export of SMW using special schema and keyword indexes, which are realized using Lucene. For storing the data and processing the SPARQL queries, in principle any triple store exposing a SPARQL endpoint can be used. We use Sesame with its native store and enabled inferencing.

The implementation is compatible with SMW 1.2 (and above) and Tomcat 5.5 (and above) running on Java 6. We have already deployed the system on several Semantic MediaWikis. In the following section, we report on evaluation experiments in one of the installations of *Ask the Wiki*.

4 Evaluation

The goal of the evaluation was to assess the potential and ability of our approach to semantic search in a real-life application. Since the search should make the potential of the underlying semantic technologies available to end users, we performed a user study to evaluate the system in terms of effectiveness and efficiency, as well as user satisfaction and usability.

4.1 Data Set and Evaluation Setting

We performed the evaluation within the community portal *semanticweb.org*, a wiki-based platform serving the Semantic Web community. The wiki contains information about the Semantic Web, in particular (but not limited to) events, publications, tools, and people. Its OWL data is available in the RDF/XML format and comprises a total of 55,365 triples, with 657 classes, 948 properties, 27,778 property instances, and 11275 individuals (as of Dec 4 2008). The data has been created by the users of the wiki over the last three years. Since the nature of a wiki is to provide unconstrained user editing, the data does not follow a predefined vocabulary or strict schema. The dataset is available at `http://semanticweb.org/RDF`. The search interface is publicly accessible at `http://semanticweb.org/wiki/Special:ATWSpecialSearch`.

The participants of the user study were 14 volunteers from the four different organizations active in the Semantic Web community. We chose a task based user evaluation with each task representing an information need that could typically occur when using the portal. Afterwards we asked the participants to answer a multiple choice questionnaire about their experience. The questions concerned their technical background (experience with other search engines, familiarity with certain technologies, etc.) and the experience and satisfaction with certain aspects of our search process. Additionally, the participants could give free text comments.

Each participant got five tasks and had up to three minutes to solve each task. The participants could give up before, if they felt that they could not solve the task. The participant received very limited information about the search interface upfront, namely that the search consists of three steps and that the search will not return a list of links like the common web search engines, but an interpretation of their keywords, also that they have to choose an interpretation, if necessary and that they could modify the interpretation in the third step. However, there was no walk-through introduction or information how to perform these steps.

The tasks were constructed so that all aspects of the systems functionality were covered, with different levels of difficulty. In particular, we created tasks that required queries that fall into the categories introduced before (i.e. entity queries such as *Find the page describing the AIFB institute*, fact queries such as *When is the paper deadline for the ASWC2008*, general conjunctive queries such as *Find the capitals of countries in Europe and the population of these cities*). Actions taken by the participants and system responses were logged. In particular, we logged the users' steps and keyword inputs and the system responses and measured how often users could solve the task, how much time it took them to choose an interpretation and how well the system performed in terms of keyword translation, query construction, and query evaluation. The evaluation was performed based on both the analysis of the log files as well as the questionnaire.

It should be noted that standard IR evaluation measures such as precision and recall do not directly apply to our approach to search, as the query results are always correct (sound and complete) with respect to the computed conjunctive query and the given dataset. Instead, a potential cause of imprecision lies in the translation of the keyword queries. In the evaluation, we thus put particular emphasis on this step.

4.2 Evaluation Results

We now report on the evaluation results, discussing first the overall effectiveness and efficiency, and then in more detail specific aspects of the individual steps of our search process. Due to space constraints, we present a much condensed form. A more detailed analysis with the complete results can be found in our Technical Report [8].

Overall Effectiveness and Efficiency. To measure the overall effectiveness, we have analyzed the ratio of tasks that have been successfully completed. To assess the efficiency, we measured the number of keyword queries that the user had to issue in order to complete the task. The results for these measures are shown in Figure 4. The results are aggregated over all queries (125 in total), grouped by the type of query that was needed to obtain the result. For the simple tasks (entity queries), the success rate was

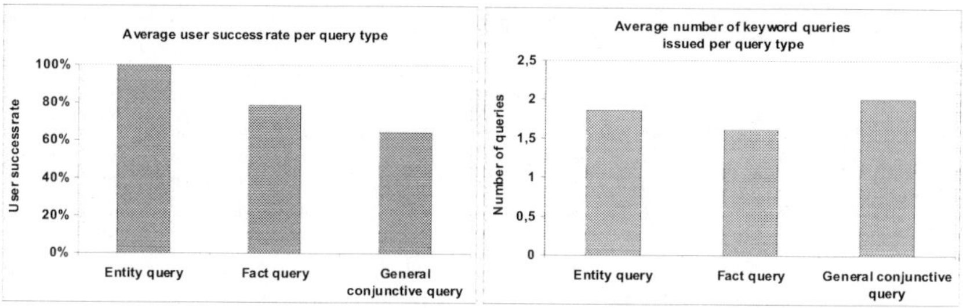

Fig. 4. Efficiency and Effectiveness of the Search

100%, the more complex tasks result in lower success rates: 79% for the fact queries and 64% for the general conjunctive queries. A typical reason why particular tasks were not completed was that the matching of the keywords against the available data was unsuccessful and no interpretations could be generated: The gap between the keywords and the underlying data was too large in certain cases. The more complex the queries (both in terms of structure and number of keywords), the larger was the effect (see discussion of keyword translation below).

Overall, 6 out of 14 participants were able to fulfill all tasks, 12 of the 14 were able to fulfill 60% or more. The other two users quickly gave up after the first or second query stating that they found the system too complicated. We find the overall success rate rather encouraging, considering that the participants used the system without detailed usage instructions and without knowing the schema of the underlying data.

For the efficiency, we see that on average the users needed to issue between 1.6 and 2 keyword queries to fulfill a task, depending on the query type. This number is surprisingly low, considering the structure and complexity of the generated queries and results. Expectedly, the value is larger for the more complex, general conjunctive queries (2 keyword queries per task) than for the simple types of queries.

We now discuss specific aspects of the three steps involved in our search process. In the questionnaire, we asked various questions related to the individual steps. The responses to these questions are shown in Figure 5.

Articulation of the information needs. The first question asked how difficult the users found it to express the information need in keywords. As expected, the users found it rather easy to do so, as all of them were familiar with keyword-based search interfaces.

Query interpretation by keyword translation. The next aspect we analyzed was the quality of the translation of the keyword queries to structured queries. First, we analyzed the robustness of the keyword matching: 88% of the keyword queries could be translated into structured queries. As mentioned above, the main reason why certain keyword matches failed was a gap between the keywords and the underlying data. As the users were not aware of the underlying schema at all, in some case no matching to the underlying data could be performed. However, after a more detailed analysis of the failures, we were able to improve the keyword matching algorithm after the user study.

Second, we measured the quality of the rankings of the possible interpretations. For this, we analyzed, which interpretation was selected by the user as the correct interpretation. We adopted a standard IR metric called Reciprocal Rank (RR) defined as

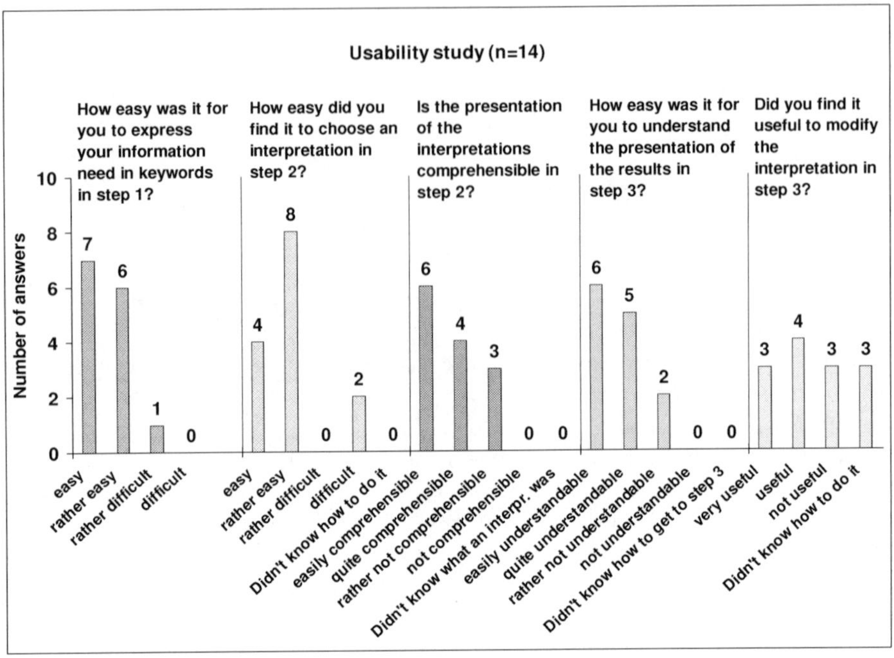

Fig. 5. Results of the usability study

$RR = \frac{1}{r}$, where r is the rank of the correct query. The mean RR for all queries was 0.84. For the successfully completed tasks, the users selected the top-ranked interpretation in 76% of the queries. The results indicate that the intended interpretation in most cases was ranked correctly, i.e. at first position.

Overall, the users found the representation of interpretations easily comprehendible, and it was easy for them to choose the right interpretation. Yet a few users had difficulties (see questions regarding step 2 in Figure 5), one reason being that in some cases the interpretations were so similar that the users could not easily tell the difference.

Result Presentation and Query Refinement. Finally, we analyzed the user satisfaction with the result presentation and query refinement. The majority of the users found the presentation of the results understandable. However, only seven users made use of the faceted search to refine a query. This corresponds to the question about how useful the modification was to the users: Seven participants found it *very useful* or *useful* to modify the interpretations, whereas three participants stated that they *did not know how to do it*. This suggests that it is useful, if the users know how to perform refinements. Unlike keyword search, faceted search is not a commonly used paradigm yet. As a consequence, effective use requires more detailed instructions, which were (deliberately) not given in our setting. Interestingly, the use of the faceted search was particularly effective for the more complex tasks. On average, 29.6% of the successfully completed tasks involved a refinement. For the most complex tasks involving general conjunctive queries, 38.9% of the successfully completed tasks involved refinements. We thus have reasons to believe that the overall success rate would have been much higher, if all users had known how to effectively utilize the faceted search.

Response Times. The evaluation of the system performance in terms of response times was not a primary goal of our study, as this aspect already had already been extensively analyzed in our prior work [5]: We showed that the query translation and query processing can be handled in near real-time even with considerably larger data sets. Still, for completeness we measured the time to translate keyword queries as well as the time for query answering. The keyword translation was performed on average in 132 ms, while the query answering (evaluation of the conjunctive query) on average took 31 ms.

5 Related Work

Our research relates to work from the three major areas, namely (1) search in wikis, (2) semantic search based on keyword translation, (3) faceted search.

Search in wikis. For wikis as knowledge sharing systems it is an important question how the knowledge in the wiki can be accessed, searched, and queried. MediaWiki, like most wiki systems, comes with a standard full-text search, which supports users to find pages in the wiki based on keyword matches. However, no structural information is exploited and only pages are regarded as search results. Also other semantic wiki systems, like AceWiki[3] or IkeWiki [4], which are powerful tools for editing and presenting structured data, do not take advantage of the structured data for search other than the standard full-text keyword search. Making the knowledge of Wikipedia accessible for structured queries is the goal of dbpedia[5], which extracts structured data from Wikipedia (and other datasets) and provides public access to it. However, users still need to know the syntax of formal query languages to search its content, which is not suitable for lay end users even though a graphical query builder OpenLink QBE [6] supporting the construction of queries is available.

Semantic search based on keyword translation. [3] provides a review of different semantic search tools and focuses on different modes of user interaction. Compared with other modes of interaction (form-based, view-based, or natural language), the advantages of keyword based lie in its simplicity and the familiarity most users already have with it. The problem of keyword queries on structured data has been studied from two different directions: 1) computing answers directly through exploration of substructures on the data graph [6,7] and 2) computing queries through exploration of a query space [5]. It has been shown in [5] that keyword translation operates on a much smaller query space, and is thus efficient. Besides, the structured queries presented to the user help in understanding the data (answer) and allow for more precise query refinement. We follow the second direction to keyword search and adapt it to wiki-based environments. [9] gives an analysis of existing semantic search systems, identifies different types of

[3] `http://attempto.ifi.uzh.ch/acewiki`
[4] `http://ikewiki.salzburgresearch.at`
[5] `http://dbpedia.org`
[6] `http://dbpedia.openlinksw.com:8890/isparql`

common features and also points out the lack of both the evaluation of search algorithms on publicly available real world datasets and the evaluation of user interfaces for semantic search. We addressed the first in our previous work [5] and the latter with the user evaluation presented in Section 4.

Faceted Search. Faceted Search (or Faceted Browsing) is increasingly used in search applications, and many websites already feature some sort of faceted search to improve the precision of their website search results. A crucial aspect of faceted search is the design of a user interface, which offers these capabilities in an intuitive way. This has been studied by [10,11] and applied in systems like Flamenco[7], Exhibit[8] or Parallax[9]. In a Semantic MediaWiki context, this paradigm has been applied by Semantic Drill Down[10] for browsing from top to bottom along the wiki's categories. Another cornerstone of faceted search is the question what is actually used as facets and if they are hierarchical or multidimensional, which obviously depends on the data corpus and its structure. Flamenco and Exhibit require a predefined set of properties, which each data item has to have and then allows browsing along the values of these properties. We use parts of the schema as facets, and therefore keep a domain independent and multidimensional faceted search. By multidimensional, we mean that vertical browsing is possible too. Although we precompute the schema, which is an indirect preparation of the facets, our approach determines the facets on the fly and provides those leading to non empty result for the each interpretation.

6 Conclusions and Future Work

We have presented an approach to search in semantic wikis, in which search is considered as a process with active user involvement: Information needs are articulated using keyword queries, which are translated into possible interpretation in terms of conjunctive queries. Users can select the interpretation and further refine it using faceted search. We have implemented this approach in an extension to Semantic MediaWiki and evaluated it in one of the most prominent installation of SMW, *semanticweb.org*.

The advantage of the approach lies in the combination of the expressiveness of structured query languages with the simplicity of keyword search and faceted search: While the user does not need to know the schema and structure of the underlying data, complex information needs can be answered that with prior technology would require either manual gathering of information (potentially distributed over a large number of pages), or the use of the formal syntax of structured query languages.

The results of our user study show that the search process is indeed adequate for end user: The participants of the study were able to use the search interface immediately without detailed instructions. Expressing the information needs in keywords, understanding and selecting interpretations in terms of conjunctive queries, and understanding the results caused little or no difficulties. However, the faceted search, while

[7] `http://flamenco.berkeley.edu`

[8] `http://simile.mit.edu/exhibit`

[9] `http://mqlx.com/~david/parallax/`

[10] `http://semantic-mediawiki.org/wiki/Help:SMW_extensions#Semantic_Drilldown`

extremely useful for refining queries to match the exact information need, was found to be too difficult by some users. Overall, it seems that people are used to the simplicity of the Google search process, and even if added value is provided, it is hard to depart from that. In this regard, there are several possibilities how the process can be simplified in certain cases: The second step in many cases already fulfills the information need: Namely, in the cases of queries for single entities and facts, the result snippets shown for the possible interpretations already contain the complete answers. Further, as our analysis of the query ranking shows, the top-ranked interpretation matches the intended user information need in most cases. Therefore, it seems reasonable to (optionally) skip the second step and show alternative interpretations only on request.

There are several further directions for future work. Currently, our search operates on the structured content of the wiki available in OWL. In this regard, we plan to extend our work to support a combination of keyword search over structured *and* unstructured content, building on our existing work on hybrid search [12]. Further, as Semantic MediaWiki is extended to support larger fragments of OWL, we intend to investigate how this can be supported and exploited for semantic search. Finally, we plan to do studies with installations of our search system in semantic wikis in other domains, where users have a less technical background.

Acknowledgements. Research reported in this paper was partially supported by the EU in the IST projects NeOn (IST-2006-027595), `http://www.neon-project.org/` and X-Media (IST-2006-026978), `www.x-media-project.org`

References

1. Guha, R., McCool, R., Miller, E.: Semantic search. In: WWW 2003: Proceedings of the 12th international conference on World Wide Web, pp. 700–709. ACM, New York (2003)
2. Mangold, C.: A survey and classification of semantic search approaches. International Journal of Metadata, Semantics and Ontologies 2(1), 23–34 (2007)
3. Uren, V.S., Lei, Y., Lopez, V., Liu, H., Motta, E., Giordanino, M.: The usability of semantic search tools: a review. Knowledge Eng. Review 22(4), 361–377 (2007)
4. Krötzsch, M., Vrandecic, D., Völkel, M., Haller, H., Studer, R.: Semantic wikipedia. Journal of Web Semantics 5(4), 251–261 (2007)
5. Tran, D.T., Wang, H., Rudolph, S., Cimiano, P.: Top-k exploration of query graph candidates for efficient keyword search on rdf. In: Proceedings of the 25th International Conference on Data Engineering (ICDE 2009), Shanghai, China (May 2009)
6. He, H., Wang, H., Yang, J., Yu, P.S.: BLINKS: ranked keyword searches on graphs. In: Chan, C.Y., Ooi, B.C., Zhou, A. (eds.) Proc. of the 2007 SIGMOD Int. Conf. on Management of data, pp. 305–316. ACM, New York (2007)
7. Kacholia, V., Pandit, S., Chakrabarti, S., Sudarshan, S., Desai, R., Karambelkar, H.: Bidirectional expansion for keyword search on graph databases. In: Proc. of the 31st Int. Conf. on Very Large Data Bases (VLDB), pp. 505–516. ACM, New York (2005)
8. Haase, P., Herzig, D., Musen, M., Tran, T.: Semantic wiki search. Technical report, University of Karlsruhe (2008), `http://www.aifb.uni-karlsruhe.de/WBS/pha/publications/semwikisearch.pdf`

9. Hildebrand, M., van Ossenbruggen, J., Hardman, L.: An analysis of search-based user interaction on the semantic web. Technical report, CWI, Amsterdam, The Netherlands (2007)
10. Hearst, M., Elliott, A., English, J., Sinha, R., Swearingen, K., Yee, K.P.: Finding the flow in web site search. Communications of the ACM 45(9), 42–49 (2002)
11. Hearst, M.: Design recommendations for hierarchical faceted search interfaces. In: ACM SIGIR Workshop on Faceted Search (August 2006)
12. Wang, H., Tran, T., Liu, C.: Ce2: towards a large scale hybrid search engine with integrated ranking support. In: CIKM, pp. 1323–1324. ACM, New York (2008)

Applying Semantic Social Graphs to Disambiguate Identity References

Matthew Rowe

OAK Group, Department of Computer Science, University of Sheffield, Regent Court, 211 Portobello Street, S1 4DP Sheffield, United Kingdom
m.rowe@dcs.shef.ac.uk

Abstract. Person disambiguation monitors web appearances of a person by disambiguating information belonging to different people sharing the same name. In this paper we extend person disambiguation to incorporate the abstract notion of identity. This extension utilises semantic web technologies to represent the identity of the person to be found and the web resources to be disambiguated as semantic graphs. Our approach extracts a complete semantic social graph from distributed Web 2.0 services. Web resources containing possible person references are converted into semantic graphs describing available identity features. We disambiguate these web resources to identify correct identity references by performing random walks through the graph space, measuring the distances between the social graph and web resource graphs, and clustering similar web resources. We present a new distance measure called "Optimum Transitions" and evaluate the accuracy of our approach using the information retrieval measure f-measure.

1 Introduction

In the modern web the need to automatically find information about a given individual has many applications such as monitoring web appearances for lateral surveillance, assessing the risk of identity theft, and protecting online visibility [3]. Malicious web practices have motivated average web users with a web presence to monitor their online visibility, and more importantly what aspects of their identity are visible in the public domain. The recent rise in the size of the social web, and particularly in the uptake of social networking sites has given members of the public their first searchable web presence from regular web entry points as sites such as Google[1] begin to index their online profiles. Therefore the need to monitor web appearances has grown, reflected in the rise in online monitoring services such Garlik Datapatrol[2].

One of the main challenges associated with monitoring web appearances is person disambiguation. For example if we are looking for information about a certain person there may be several people all sharing the same name. Therefore

[1] http://www.google.com
[2] http://www.garlik.com/

L. Aroyo et al. (Eds.): ESWC 2009, LNCS 5554, pp. 461–475, 2009.

a web page containing a person's name must be assessed to deduce whether the page contains a reference to the person we are interested in. The grounding for this problem is not uncommon in society where several people share the same name, we refer to such people as namesakes [14].

State of the art graph-based person disambiguation approaches rely on graph structures to measure distances between web resources, represented as sub graphs, according to available paths and edge weightings. These measures are then used to cluster resources where each cluster is representative of a different namesake. In this paper we present an complimentary approach to existing graph-based person disambiguation techniques using a three-stage procedure: We begin by generating individual semantic social graphs from dispersed Web 2.0 services, and link these graphs together to form a complete semantic social graph containing identity and social network information for a given person. Secondly we find possible web resources referencing a given person using a combination of crawling and searching both the web and the semantic web, and convert each resource into a semantic graph describing identity information found within that resource. Thirdly we integrate the social graph and resource graphs into a global graph space, merging instances of semantic concepts and replacing edges in the graph space. We cluster the web resource graphs closest to the social graph, thereby producing two cluster sets: Correct identity references, and ambiguous identity references.

Our intuition is that a given person will co-occur on a web page with people from his/her social network. Thus the focus of our approach is to utilise existing social information sources when disambiguating possible identity referencing web resources. When measuring distances between the social graph and web resource graphs we use two measures; commute time and optimum transitions. We evaluate the accuracy of our approach to clustering web resources using these measures against a baseline latent semantic analysis technique using f-measure.

The paper is structured as follows: Section 2 describes the overview of our approach, and details relevant requirements. Section 3 presents our methodology for social graph generation and linkage. Section 4 describes our methodology for web resource graph generation, and section 5 describes the techniques we have implemented for identity disambiguation in web resources. Section 6 describes our evaluation method; the data set we used, the chosen accuracy measures, and the results and discussions. Section 7 presents the state of the art associated with the three stages of our approach. We conclude the paper in section 8 by discussing the conclusions we have found from our work and future work.

2 Overview of the Approach

The identity disambiguation approach presented in this paper bootstraps an existing graph-based disambiguation process by incorporating a social graph. We believe that the use of the initial social graph when deriving measures among the nodes in the graph space provides a focussed approach. Current approaches rely on comparing every node with every other node in the graph space to derive distance measures, instead we only derive distances between the social graph and the web resource graphs. Due to the rise in usage of Web 2.0 services such as social networking and blogging platforms, web users build descriptions of their social identity in dispersed information spaces. By harnessing and collating this decentralised information a complete integrated social graph is compiled, providing useful information to correctly match web resources containing identity references. In order to focus our approach to identity disambiguation we have defined several requirements that must be fulfilled:

- Provide machine-readable semantic information from semi-structured Web 2.0 services.
- Handle web resources lacking machine-readable semantics.
- Bootstrap the identity disambiguation process with minimal user input.
- Handle information on a large scale.
- Disambiguate people, not just names.
- Provide accurate identity disambiguation results.

In our approach we use a semantic graph structure. We define a semantic graph as a graph containing nodes represented by classes from an ontology, and edges as the relations and properties between those nodes. The Resource Description Framework (RDF) is well suited to describing graph structures based on the structure of a triple, subject-object-predicate expression, being comparable to an edge in a graph linking two nodes. The advantages of using a semantic graph structure allows inferences to be made about the graph structure, aids with graph integration and node merging, and allows graph formalisation adhering to a formal standard. Figure 1 provides an overview of our approach to identity disambiguation, we divide the approach into three stages:

1. **Social Graph Generation.** Handles the extraction of multiple social graphs from different Web 2.0 services as semi-structured XML. The social graphs are then converted to RDF and linked together into a complete social graph.
2. **Resource Graph Generation.** Responsible for finding possible identity referencing web resources by crawling and searching the web. The resources are each converted into a semantic resource graph as RDF.
3. **Identity Disambiguation.** Integrates the web resource graphs and the social graph into the graph space. Random walks are performed through the graph space to derive distance measures within the graph. These measures are then used to cluster the web resources most similar to the social graph.

3 Social Graph Generation

3.1 Extracting Social Graphs

The rise in usage of Web 2.0 sites and services has encouraged average web users to sign up to multiple dispersed services spreading their identity and their attributed social graph throughout the web. The majority of such services contain social information hidden within a 'walled garden'. Accessing such information

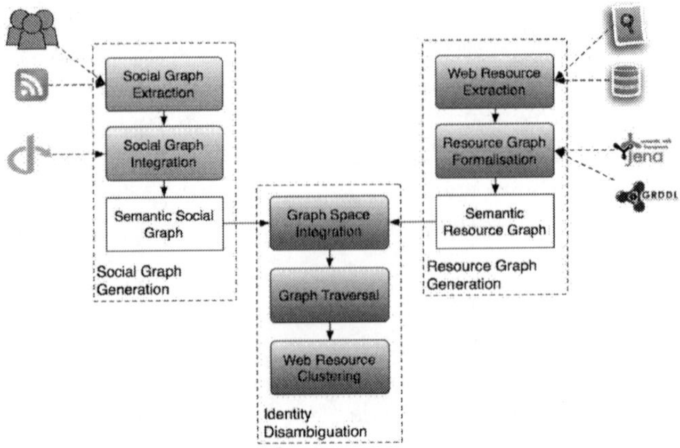

Fig. 1. Architecture of applying Semantic Social Graphs to Identity Disambiguation

is restricted to authorised API calls which - when validated - produce an XML response according to the service's XML schema.

Therefore the task of exporting this information is trivial in that in commonly involves mapping an existing XML schema produced from the API to existing semantic ontologies. We reuse an approach from our previous work [23] to produce RDF from the social networking site Facebook[3] by mapping the identity and social network infromation from the XML schema to concepts from the FOAF ontology [6], and geographical locations to the Geo [7] ontology using the top ranked URI for the location obtained from the GeoNames[4] web service. The selection of the FOAF ontology was a natural choice due to its ability to capture knowledge related to identity features and express social connections between instances of people. Figure 2 shows an example of the graph structure produced from a Facebook account. This social graph captures knowledge concerning the identity details of the person, and the social connections that exist with members of his/her social network described using the `foaf:knows` relation to connect two instances of `foaf:Person`. We replicate this process for social networking sites adopting Google's OpenSocial[5] standards such as MySpace[6], Orkut[7] and Hi5[8]. Available social information from accessing these services is converted into RDF acccording to the FOAF and Geo ontologies.

Many blogging platforms allow access to social information through unauthorised API calls, and return a response according to either the Atom [20] or RSS [28] syndication feed specifications depending on the platform. To generate a semantic graph we convert this information into RDF according to the FOAF

[3] http://www.facebook.com
[4] http://www.geonames.org
[5] http://code.google.com/apis/opensocial
[6] http://www.myspace.com
[7] http://www.orkut.com
[8] http://www.hi5.com

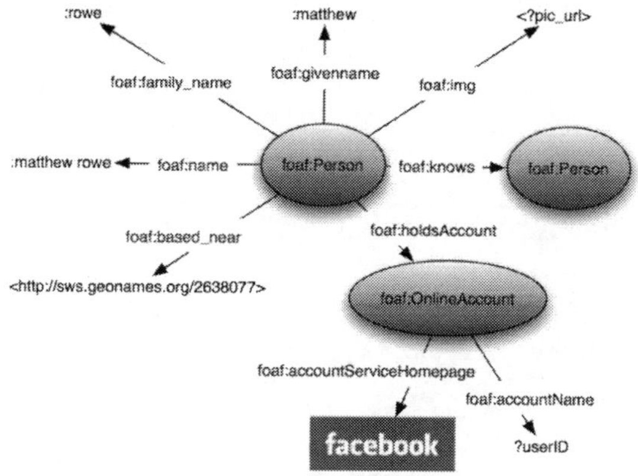

Fig. 2. RDF generated from Facebook according to the FOAF ontology

specification mapping the relevant concepts to capture all of the available social information. In the case of the Atom specification we convert the `atom:Author` concept to an instance of `foaf:Person`, the associated properties are also converted: `atom:name` and `atom:email` are converted to `foaf:name` and `foaf:mbox` respectively.

3.2 Linking Social Graphs

At this stage in the approach we have several social graphs each owned by the same person, expressed as a single unique instance of the `foaf:Person` class in each graph, we refer to this instance as the graph owner. We must link each of the graphs together and merge the structures. The problem we face is the lack of a unique URI that can be assigned to the graph owner, however we refrain from using the `foaf:name` property due to the ambiguity of person names. Therefore we assign an OpenID[9] resource using the `foaf:openid` relation to the graph owner in each of the social graphs. As `foaf:openid` is expressed as an inverse functional property of an instance of `foaf:Person` only one OpenID can be attributed to a given person. We also assume the converse to be true in that one person is restricted to only one OpenID to enforce social graph linkage.

Information in each sub graph is now linked together as figure 3 demonstrates, where all instances of the same `foaf:Person` class are now linked using the `owl:sameAs` property. These instances are then merged together to form one instance of `foaf:Person` for the owner of the complete social graph. We then merge the remaining social network content by comparing instances of `foaf:Person` from separate social networks for matches. We compare identity properties of each `foaf:Person` instance using the Smith-Waterman-Gotoh distance [8] for string literals, and look for matches when comparing identifiers such as `foaf:mbox` and `foaf:homepage`. If a match occurs then we merge the

[9] http://www.openid.net

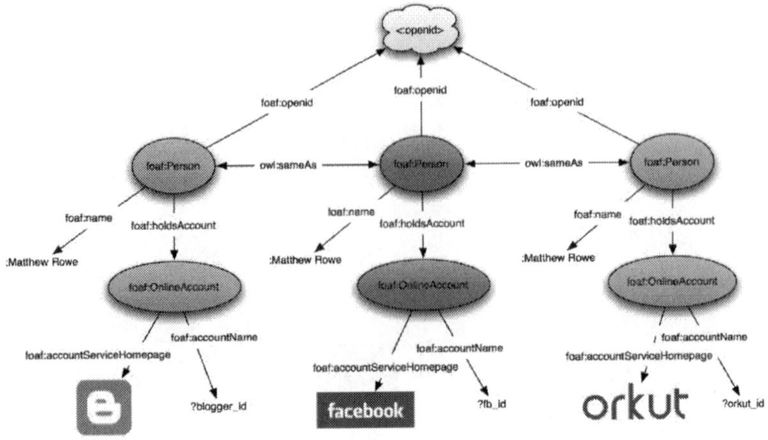

Fig. 3. Linked data across multiple social subgraphs

instances, otherwise we keep them separate in the complete social network. Our technique for linking social graphs is discussed in detail in [24].

3.3 Retrieving a Complete Social Graph

The complete social graph is stored within a triple store for exportation and linking of new triples. This allows SPARQL queries to be used in order to retrieve the complete social graph attributed to a given person using their OpenID. If we are to store social graphs for thousands of people in the same triple store, it is essential to have a URI for each person, hence the selection of OpenID when identiying the graph owners.

4 Resource Graph Generation

4.1 Gathering Resources

Our approach gathers resources using two methods; crawling and searching. We crawl web pages defined by the `foaf:homepage` property in the social graph, and known identity resources such as public telephone directories[10] and listings pages[11]. The crawler runs until a predefined number of web resources have been retrieved, the resources are then added to the resource list for graph conversion.

We search the web using search engines as entry points to generate a list of web resources for a given person using his/her name. URLs from the first 50 results are collected from Google and Yahoo[12] and are added to the resource list for graph conversion. The semantic web is searched using the semantic web

[10] http://www.thephonebook.bt.com
[11] http://www.192.com
[12] http://www.yahoo.com

search engines Watson[13] and Swoogle[14] to gather a list of resources containing semantic metadata using the person's name. Following this we then have at least 100 resources ready for graph generation for one identity.

4.2 Extracting Resource Graphs

A semantic graph structure is generated for each web resource using available identity information. Different techniques are used to perform the semantic graph generation depending on the available metadata within each resource. If the web resource is a web page containing lightweight semantics such as Microformats [15], eRDF [22], and RDFa [1], then the page is parsed using an XSL transformation [9] with GRDDL [10]. The advantage of this method is its adaptability to allow new lightweight semantic specifications to be adopted by simply including the accompanying XSL transformation. The extracted information in XML is converted to RDF using the FOAF, and Geo ontologies in a similar manner to the social graph generation obtaining a URI for the geographical location using the GeoNames service. We overcame the problem of differing usages of lightweight semantics (figure 4) by converting identity information into the same semantic format.

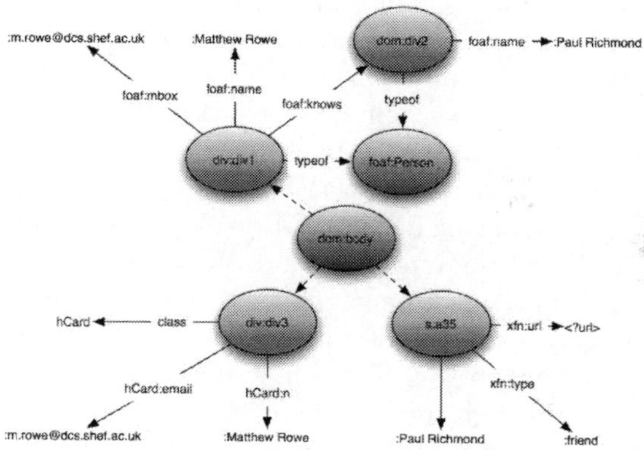

Fig. 4. Differing usage of lightweight semantics in web resources

Ontologies returned when searching the semantic web already contain machine-readable metadata therefore we parse the ontologies using Jena [17]. Any content described using the FOAF specification found within the ontology is extracted into a graph describing an instance of the `foaf:Person` class, attributed identity properties and social connections using the `foaf:knows` relation. We retrieve URIs described using the `foaf:seeAlso` property within the ontology and add these to the crawler, assuming that this linked data is likely to contain a reference to the

[13] http://watson.kmi.open.ac.uk/WatsonWUI
[14] http://swoogle.umbc.edu

person we are searching for. The derived resource graph can then be integrated into the graph space.

From our work we found that the majority of web resources contained no machine-readable semantics, instead the content is standard HTML. In such cases we use regular expressions formed from a person name gazetteer to match person names within the web resources. When a match occurs we create an instance of the `foaf:Person` class and assign the `foaf:name` property to express the found person name in the resource graph. When a person name match occurs, we match identity information such as email, homepage and date of birth using regular expressions surrounding the matched name through a window function of 50 words either side of the name. We use this window function on the assumption that any identity information relating to the given name would occur within the near proximity of the matched name.

5 Identity Disambiguation

5.1 Integration into the Graph Space

The graph space is initially populated with the social graph of the user, and iteratively with each web resource graph using an indexing paradigm. At a low-level this involves the comparison of semantic concepts and string literals in the case of semantic properties such as `foaf:name` to provide node deduplication. We use the Smith-Waterman-Gotoh string metric to compare string literal nodes, if the derived measure is above our assigned threshold we integrate the web resource graph into the graph space by replacing the internal node with the comparable node in the global graph. We also do the same for geographical semantic concepts by comparing location URIs. We perform this process until all web resource graphs have been integrated into the global graph space. Semantics play an important role in the merging of nodes within the graph space by limiting node comparisons to only nodes of the same semantic type. As more nodes are merged within the graph space the knowledge structure becomes enriched, providing more paths through the graph space.

In order to derive distance measures between nodes within the graph space we apply an edge weighting function based on the edge label. As each edge in the graph is described using a semantic property or relation it is possible to adjust the weighting model based on the preference of identity features. For example, the `foaf:homepage` property from the social graph acts as a URI therefore the edges labelled with the `foaf:homepage` property are given less weight to reduce the traversal cost. We normalise the weightings for semantic edge types in the graph using a standard weighting distribution.

5.2 Graph Traversal

Once the graph space has been fully integrated we derive the strongly connected component within the graph starting from the identity node, and prune the graph space of web resources containing no identity references. We define the identity node as the `foaf:Person` concept in the social graph representing the social graph owner. We use Tarjan's algorithm [26] to derive the component by performing a depth first search and returning all connected nodes and edges from the identity node and therefore the strongly connected component.

Using this pruned graph we now traverse the graph space using random walks [25] starting from the identity node. We begin by deriving the adjacency matrix A to compute the local similarity of adjacent nodes in the graph as described in [13] by deriving the cost of moving from each node i to each adjacent node j. Let $L = (l_1, \cdots, l_n)$ denote the set of distinct semantic edge types (properties and relations) in the graph space, such that the function $w(l_k)$ returns a weight based on the semantic edge type, therefore we populate A using:

$$a_{ij} = \begin{cases} \sum l_k \in L \frac{w(l_k)}{|(i,.) \in E : l(i,.) = l_k|}, & (i,j) \in E \\ 0, & otherwise \end{cases}$$

We compute the diagonal degree matrix using $D = \sum kA_{ik}$. A random walk through the graph is derived from a Markov chain [18] describing the probability of moving to node j given that the random walker is currently in node i given that t steps have been traversed through the graph space. Transition probabilities are defined as $P = D^{-1}A$ for one step, and $P^t = \left(D^{-1}A\right)^t$ for t steps. We now have a matrix which when consulted returns the probability of traversing from one node to another in t steps. At this stage we use two different techniques to derive distance measures between the identity node and the web resources in the graph space:

Commute Time. This measure derives the commute time taken to set off from the identity node, reach a web resource and then return again. We set the transition probability matrix for 2 steps due to the graph structure being relatively short in length between nodes. We use the formula described in [25] to derive the cost of moving to node k and returning back given that we are currently in state i:

$$m\left(k|i\right) = \begin{cases} 1 + \sum_{j=1}^{n} p_{ij} m(k|j), & i \neq k \\ 0 \end{cases}$$

This iterates through the graph cumulatively recording the cost of the path from the identity node to each web resource. Agglomerative clustering [27] is then used to cluster the most similar web resource nodes with the identity node based on the bottom-up clustering principle being well suited to comparison against a single social graph. The end product is 2 cluster sets: Correct identity references and ambiguous identity references (those resources that cannot be confirmed as referring to the identity we are interested in or not).

Optimum Transitions. The second measure derives the optimum number of steps needed to traverse from the identity node to web resource nodes within the graph space by incremening t until the transition probability has peaked. Given that we wish to traverse from node i to node j, we derive the optimum value for t using $max(P_{ij}^t)$. We then use agglomerative clustering to cluster the most similar web resource nodes with the identity node using the same method as the Commute Time clusters, we similarly derive 2 sets of clusters; correct identity references and ambiguous identity references.

6 Evaluation

6.1 Data Set

We evaluate our approach using a data set compiled from members of the Department of Computer Science[15] at the University of Sheffield. We selected members by analysing results from the search engines Google and Yahoo using each person's name, and manually identifying different entities referenced by the same name. If a large number of namesakes appeared in the results then the person was included for the evaluation. We compiled the data set by listing social web accounts, and referent web resources for each participant as described in section 4. This data set was then used to generate the complete social graph and web resource graphs, and compile the graph space.

6.2 Selection of a Baseline Technique

We compared our approach against a standard baseline latent semantic analysis method using hierarchical clustering as presented in [16]. In this paper we wish to demonstrate the effectiveness of implementing a semantic graph-based approach in comparison with standard feature comparison techniques that do not use graph structures. Therefore the selection of latent semantic analysis provides a comparable implementation due to its feature based methodology, clustering resources by calculating similarity measures based on features in each resource [16]. We perform the comparison between the baseline method and our approach by comparing the derived clusters. Our approach clusters the social graph with the most similar web resource graphs, whereas the baseline method produces a number of clusters each relating to a different namesake. Therefore we set the cluster containing the person we are searching for as the correct identity referents and merge all other clusters to form the ambiguous identity referents. This allows comparison of the cluster sets between our approach and the baseline using the following measures:

6.3 Accuracy Measures

We evaluate both our approach and the baseline approach using the information retrieval metric f-measure to assess the minimisation of the number of false positives and false negatives. We evaluate the accuracy of the correct identity referents cluster where f-measure is well suited to deciding the accuracy of information retrieved. We define p as the identity we are looking for, s_p as the set

[15] http://www.shef.ac.uk/dcs

of web resources that correctly reference the identity p, c_c as the cluster containing correct identity references, and c_a as the cluster containing ambiguous identity references. We define precision as $P(c_c) = |s_p \cap c_c| / |c_c|$ and recall as $R(c_c) = |s_p \cap c_c| / |s_p|$, and therefore f-measure as the harmonic mean of the precision and recall:

$$F(c_c) = \frac{2P(c_c)R(c_c)}{P(c_c) + R(c_c)}$$

6.4 Results and Discussion

Table 1 shows the results from the evaluation. We have included the column "Namesake Count" to highlight the ambiguity of the people we are searching for by counting how many different namesakes appeared in the search results. We achieve an average F-Measure of 0.76 for commute time and an average F-Measure of 0.78 for optimum transitions. In each instance both semantic graph-based methods outperform the baseline. The f-measures indicate that we achieve a healthy combination of true positives and true negatives when disambiguating web resources in comparison to the baseline technique.

Table 1. Evaluation results

Name	# of Namesakes	Commute Time	Optimum Transitions	Baseline
Matthew Rowe	22	0.88	**0.9**	0.85
Jonathan Butters	12	**0.72**	**0.72**	0.6
Rodrigo Carvalho	16	**0.76**	**0.76**	0.39
Sam Chapman	23	**0.6**	0.52	0.33
Neil Ireson	10	**0.88**	**0.88**	0.75
Paul Richmond	18	0.47	**0.87**	0.83
Christopher Brewster	16	**0.97**	0.92	0.83
Mark Greenwood	19	**0.81**	0.7	0.23
Joao Magalhaes	16	**0.81**	0.78	0.47
Diana Maynard	11	0.71	**0.72**	0.45
Overall		0.76	**0.78**	0.573

The results shows that there is little difference between each distance measure used to cluster resources, in certain cases the commute time outperforms optimum transitions and in others the converse. Overall optimum transitions produce a more consistent measure with less fluctuation than commute time. For instance, when searching for web resources relating to the person "Paul Richmond", commute time is outperformed by both the baseline and optimum transitions. In this instance this was due to the large number of paths available through the graph space from the social graph to each of the web resource graphs, optimum transitions works better when dealing with a large selection of possible paths. With the other participants the paths varied less through the graph space, and therefore commute time functioned better.

7 Related Work

7.1 Social Graph Generation

Generating and 'semantifying' content from Web 2.0 services has been addressed in [21] by extracting information from the photo sharing social networking site Flickr[16] and converting the API response into RDF annotated using both the FOAF and SIOC [5] ontologies. A similar approach[17] provides an exporter of RDF according to the FOAF ontology from the microblogging site Twitter[18]. Our previous work in [23] provides an approach to export RDF according to the FOAF ontology from the social networking site Facebook that we used within this paper. The problem faced following the extraction of individual semantic social graphs from dispersed Web 2.0 services is data linkage where the use of a person name is too ambiguous, therefore a URI to link the graphs is a necessity. The inclusion of the `foaf:openid` property in the latest FOAF specification was useful for our work in two ways: Firstly, Web 2.0 services never allow the extraction of email addresses as they are deemed private, thus rendering the `foaf:mbox` and `foaf:mbox_sha1sum` properties unusable. And secondly, this provides a single URI for retrieving the social graph from a triple store, and linking additional triples.

7.2 Resource Graph Generation

Lightweight specifications such as Microformats, RDFa (Resource Description Framework in Attributes) and eRDF (Embedded RDF) allow "lowercase" semantic metadata to be included within web pages using XHTML. Using an XSL transformation (XSLT) with GRDDL to extract available metadata allows semantic graphs to be generated from web pages. Work by [2] provides a mechanism to extract RDF from XML based content (including HTML) by merging XQuery and SPARQL to create a query language called XSPARQL therefore allowing semantic metadata to be extracted from HTML without the need for XSLT. Our approach complements the state of the art by using an XSLT transformation where possible to extract lightweight semantics with GRDDL and convert this into an RDF graph. The most commonly used specification for describing identity information is FOAF, such information available on the semantic web has been mined in [19] using a two-part approach to find ontologies containing FOAF content, extracting relationships from within the ontologies and deriving bond strengths using name co-occurrence from web queries. The semantic web is crawled for ontologies annotated with FOAF in work by [11], FOAF content is extracted and aggregated with other FOAF files. This allows assertions to be made about the people that were discovered during the crawl using the extracted semantic information. We have implemented both searching and crawling mechanisms for ontology gathering from the semantic web.

---

[16] http://www.flickr.com

[17] http://tools.opiumfield.com/twitter/mattroweshow/rdf

[18] http://www.twitter.com

7.3 Identity Disambiguation

Work by [16] uses an unsupervised method to perform person disambiguation by searching for web pages using a person name, and clustering web pages accordring to the community struture in each page. A similar approach is presented in [27] where named entities are extracted from within web pages, along with biographical information such as date of birth and email. These features are then used to cluster similar pages as belonging to different namesakes. Person name ambiguity is addressed in [12] by aligning a namesake with a relevant concept using a maximum entropy model, and clustering similar web resources based on the concept.

A graph-based approach is described in [4] utilising the link structure of web pages to cluster similar pages as belonging to different people working under the intuition that "Web pages of a group of acquaintances are likely to be interconnected". Link structures are modelled in a graph space, and web pages are clustered based on the distances between them. A similar approach is described in [27] by extracting names and hyperlinks from within web pages as features. Unlike [4] the graph model is extended, similar to our work, by setting nodes in the graph space as web pages, and named entities and hyperlinks found within those pages. Correlation clustering is then used to cluster similar web resources based on the strengths of the connections between them. A model of the graph space is populated in [13] by extracting various web page features, such as named entities, hyperlinks, and lexical tokens. Each feature is modelled as a node in the graph space, and edges are labelled between the nodes. Random walks are then performed through the graph space to derive distance measures according to the commute time between nodes. We address the disambiguation problem with respect to current state of the art methods by providing a semantic interpretation of the graph space using available ontologies to describe the nodes and edges. Our approach to identity disambiguation differs to the state of the art by clustering similar web resources with respect to the social graph rather than creating individual clusters for each namesake.

8 Conclusions

Our approach produced a semantic social graph from semi-structured XML describing the identity of the account holder and their social network, therefore fulfilling the requirement for producing machine-readable semantic information by generating RDF. The majority of web resources that were analysed for identity information contained no lightweight semantics. We believe that this is due to the slow uptake by web developers in annotating web content using lightweight semantics. Nevertheless, our approach was able to extract identity information from web resources using a gazetteer with regular expressions and convert this information into a semantic format.

Our work bootstraps an existing identity disambiguation technique minimising the user input by only involving the user in the social graph generation stage.

The bootstrapping focuses the derivation of distances within the graph space between certain nodes, by only measuring from the social graph. This requires less computations than the current state of the art by focussing the approach on one identity, rather than trying to produce a cluster containing web resources for each namesake, thus allowing more data to be handled.

We extended person disambiguation to perform identity disambiguation by modeling identity features captured throughout the social graph generation and resource graph generation stages of our approach. Preference is given to certain identity properties by weighting the semantic edge types in the graph space. For example, disambiguation is heavily influenced if the `foaf:mbox` (email address) property from the social graph is present in a web resource due to an email address being a URI. The results from our evaluation demonstrate the effectiveness of our approach of using a graph-based disambiguation in comparison with the baseline latent semantic analysis technique. The results also show that our optimum transitions measure for deriving distances between nodes in the graph space outperforms the commute time measure. We also experienced a more efficient process requiring less time to compute the distances, we plan to document this improved efficiency in comparison to commute time in our future work.

References

1. Adida, B., Birbeck, M.: RDFa Primer: Bridging the Human and Data Webs. World Wide Web Consortium (2008), http://www.w3.org/TR/xhtml-rdfa-primer/
2. Akhtar, W., Kopecky, J., Krennwallner, T., Polleres, A.: XSPARQL: Travelling between the XML and RDF Worlds – and Avoiding the XSLT Pilgrimage. In: Bechhofer, S., Hauswirth, M., Hoffmann, J., Koubarakis, M. (eds.) ESWC 2008. LNCS, vol. 5021, pp. 432–447. Springer, Heidelberg (2008)
3. Andrejevic, M.: The Discipline of Watching: Detection, Risk, and Lateral Surveillance. Critical Studies in Media Communication 23(5), 392–407 (2006)
4. Bekkerman, R., McCallum, A.: Disambiguating Web Appearances of People in a Social Network. In: Proc. 14th international conference on World Wide Web, Chiba, Japan, pp. 463–470 (2005)
5. Breslin, J., Harth, A., Bojars, U., Decker, S.: Towards Semantically Interlinked Online Communities. In: Gómez-Pérez, A., Euzenat, J. (eds.) ESWC 2005. LNCS, vol. 3532, pp. 500–514. Springer, Heidelberg (2005)
6. Brickley, D., Miller, L.: FOAF Vocabulary Specification (OpenID Edition). Creative Commons (2007), http://xmlns.com/foaf/spec/
7. Brickley, D.: Basic Geo Vocabulary. World Wide Web Consortium (2006), http://www.w3.org/2003/01/geo
8. Chapman, S., Norton, B., Ciravegna, F.: Armadillo: Integrating Knowledge for the Semantic Web. In: Proc. Dagstuhl Seminar in Machine Learning for the Semantic Web, Dagstuhl, Germany (2005)
9. Clark, J.: XSL Transformations. World Wide Web Consortium (1999), http://www.w3.org/TR/xslt
10. Connolly, D.: Gleaning Resource Descriptions from Dialects of Language. World Wide Web Consortium (2007), http://www.w3.org/TR/grddl/
11. Finin, T., Ding, L., Zhou, L., Joshi, A.: Social Networking on the Semantic Web. The Learning Organisation 1(5), 418–435 (2005)

12. Fleischman, M., Hovy, E.: Multi-Document Person Name Resolution. In: Proc. Workshop on Reference Resolution and its Applications: ACL 2004, Barcelona (2004)
13. Iria, J., Xia, L., Zhang, Z.: WIT: Web People Search Disambiguation using Random Walks. In: Proc. of the 4th International Workshop on Semantic Evaluations (Semeval 2007), Prague, Czech Republic (2007)
14. Kalashnikov, D., Chen, Z., Mehrotra, S., Nuray, R.: Web People Search via Connection Analysis. IEEE Transactions on Knowledge and Data Engineering 20(11), 1550–1565 (2008)
15. Khare, K.: Microformats: the next (small) thing on the Semantic Web? IEEE Internet Computing 10, 68–75 (2006)
16. Malin, B.: Unsupervised Name Disambiguation via Social Network Similarity. In: Proc. Workshop on Link Analysis, Counterterrorism, and Security, SIAM International Conference on Data Mining, Newport Beach, CA (2005)
17. McBride, B.: Jena: a Semantic Web toolkit. IEEE Internet Computing 6, 55–59 (2002)
18. Meyn, S., Tweedie, R.: Markov chains and stochastic stability. Springer, London (1993)
19. Mika, P.: Bootstrapping the FOAF-Web: An Experiment in Social Network Mining. In: Proc. Workshop on Friend of a Friend, Social Networking and the Semantic Web, Galway, Ireland (2004)
20. Nottingham, M.: Atom Syndication Specification. The Internet Society (2005), `http://www.atomenabled.org/developers/syndication/atom-format-spec.php`
21. Passant, A.: RDF Export of Flickr Profiles with FOAF and SIOC (2007), `http://apassant.net/blog/2007/12/18/`
22. RDF in HTML. Talis (2006), `http://research.talis.com/2005/erdf/`
23. Rowe, M., Ciravegna, F.: Getting to Me - Exporting Semantic Social Network Information from Facebook. In: Proc. Social Data on the Web Workshop, ISWC 2008, Karlsruhe, Germany (2008)
24. Rowe, M.: Interlinking distributed Social Graphs. In: Proc. Linked Data on the Web Workshop, WWW 2009, Madrid Spain (2009)
25. Saerens, M., Fouss, F., Yen, L., Dupont, P.: The principal components analysis of a graph, and its relationships to spectral clustering. In: Boulicaut, J.-F., Esposito, F., Giannotti, F., Pedreschi, D. (eds.) ECML 2004. LNCS, vol. 3201, pp. 371–383. Springer, Heidelberg (2004)
26. Tarjan, R.: Depth-first search and linear graph algorithms. SIAM Journal on Computing 1(2), 146–160 (1972)
27. Wan, X., Gao, J., Li, M., Ding, B.: Person resolution in Person Search Results: WebHawk. In: Proc. of the 14th ACM international conference on Information and knowledge management, pp. 163–170 (2005)
28. Winer, D.: RSS 2.0 Specification. Creative Commons (2007), `http://www.rssboard.org/rss-specification`

Middleware for Automated Implementation of Security Protocols

Béla Genge and Piroska Haller

"Petru Maior" University of Târgu Mureş, Department of Electrical Engineering,
N. Iorga Str., No. 1, (540088) Târgu Mureş, Romania
{bgenge,phaller}@engineering.upm.ro
http://www.upm.ro

Abstract. We propose a middleware for automated implementation of security protocols for Web services. The proposed middleware consists of two main layers: the communication layer and the service layer. The communication layer is built on the SOAP layer and ensures the implementation of security and service protocols. The service layer provides the discovery of services and the authorization of client applications. In order to provide automated access to the platform services we propose a novel specification of security protocols, consisting of a sequential component, implemented as a WSDL-S specification, and an ontology component, implemented as an OWL specification. Specifications are generated using a set of rules, where information related to the implementation of properties such as cryptographic algorithms or key sizes, are provided by the user. The applicability of the proposed middleware is validated by implementing a video surveillance system.

Keywords: Middleware, Web services, security protocols, automated execution, ontologies.

1 Introduction

In order to ensure security properties such as confidentiality, integrity or availability, Web services use technologies such as the Security Assertions Markup Language [19] (i.e. SAML) or WS-Security [20], providing a unifying solution for the authentication and authorization issues. The security tokens defined by WS-Security have been extended with additional ones and a set of new primitives in WS-Trust [21] allowing inter-domain authentication and authorization. The primitives defined by WS-Trust correspond to security protocols, denoting "communication protocols dedicated to achieving security goals" (C.J.F. Cremers and S. Mauw) [1]. The security protocols defined by WS-Trust consist of request-response messages with flexible message components.

Despite it's flexibility, WS-Trust does not define the operations that must be executed for each message that is constructed or processed. By defining these operations, services can execute new protocols without relying on predefined protocols.

L. Aroyo et al. (Eds.): ESWC 2009, LNCS 5554, pp. 476–490, 2009.

Based on these observations we propose a middleware for automated implementation of security protocols in Web services. The proposed middleware consists of two layers: the service layer and the communication layer. The *service layer* consists of several services providing the discovery of services and the authorization of client applications. We define four types of services: name services, specification services, authorization services and resource services. The *communication layer* is built on the SOAP layer and ensures the implementation of security protocols and the implementation of service protocols. By using these protocols we provide a secure communication channel for the service protocols implemented above.

Each system implementing the proposed platform defines a single name service, a single specification service, a single authorization service and multiple resource services. Services are accessed dynamically by using service protocol and security protocol specifications. For the automated execution of security protocol specifications we propose a semantic security protocol model (SSPM). The SSPM has two components: a sequential model and an ontology model. The first component is implemented as a WSDL-S [4] specification while the second component is implemented as an OWL [15] specification. The role of the WSDL-S implementation is to describe the message sequences and directions that must be executed by protocol participants. The role of the OWL implementation is to provide semantic information such as the construction, processing and implementation of cryptographic operations (e.g. encryption algorithm, encryption mode, key). The SSPM is constructed from a given security protocol model (SPM) and it must maintain the protocol's security properties. For this we propose several generating rules and algorithms that generate the SSPM.

The paper is structured as follows. In section 2 we present the architecture of the proposed middleware. We continue with the construction of the specification model in section 3. Based on the proposed middleware, in section 4 we exemplify the construction of specifications and we present a Web service-based video surveillance system, where video capturing resources can be accessed by automatically executing security protocols. We relate our work to others in section 5. We end with a conclusion and future work in section 6.

2 Middleware Architecture

2.1 Service Oriented Architecture

Client applications are able to access resources by first locating them, followed by the download of the service and security protocol specifications and by the execution of an authentication sequence. The services provided by the platform must provide a way to publish, locate and automatically access resource services. In addition, in order to face the challenges of rapidly changing protocol specifications, the protocol implementations must provide flexible and extensible components.

Based on these requirements, we define four types of services: name services, specification services, authorization services and resource services. Name services

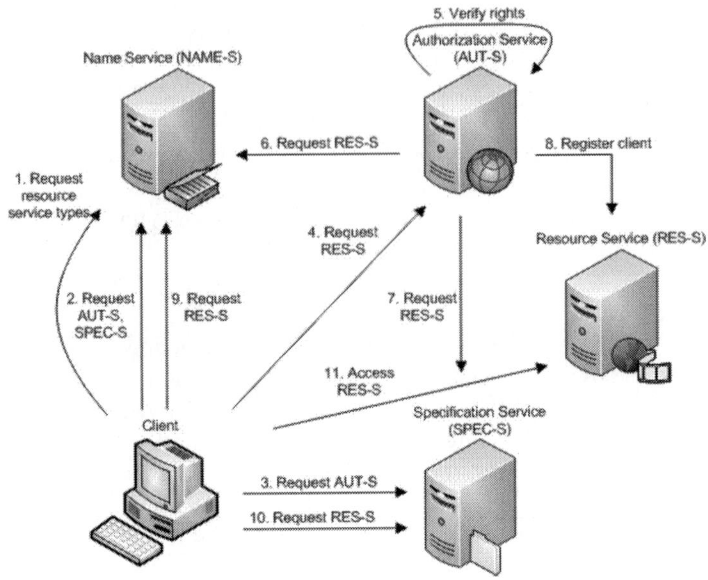

Fig. 1. Accessing services by client applications

(NAME-S) are implemented through UDDI [22] registries and are used to register, identify or locate existing services. Specifications are stored and managed by specification services (SPEC-S). Specifications are implemented using Web service technologies such as SAWSDL [12], WSDL-S [4] and OWL [15]. Authorization services (AUT-S) implement the verification mechanisms of client credentials and provide accessing mechanisms to the requested resources. Finally, the resource services (RES-S) implement a set of capabilities provided for client applications.

Accessing resources by a client application is done in several steps, as shown in figure 1. First, the client must establish the set of services implemented by the system (step 1). In order to access a resource, the client must be authenticated and its rights must be verified. This is done by accessing the AUT-S service, for which the specifications must be first downloaded. The client requests the location of AUT-S and SPEC-S (step 2) from NAME-S, followed by the request of the specifications for AUT-S (step 3). The request for accessing RES-S is sent to AUT-S (step 4), containing the user credentials. These are verified (step 5) and a security token is generated by AUT-S that is sent to RES-S (steps 6, 7, 8). The client receives the generated token and sends it to RES-S (steps 9, 10, 11), after which it is able to access the capabilities provided by the resource.

2.2 Software Architecture

The architecture of the software stack is given in figure 2. We identified two main layers: the communication layer and the service layer, given in figure 2.

The *communication layer* provides the implementation of the service and security protocols needed to access service capabilities. It is built on existing

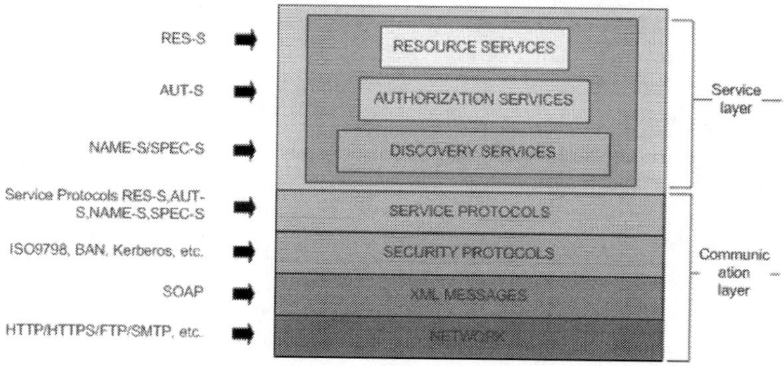

Fig. 2. Software stack

network and XML message-based protocols. Security protocols are implemented using extensions of the WS-Security standard, provided in our previous work [14]. The extensions consist of XML constructions for user names and binary keys required by key exchange and authentication protocols. The automated execution of security protocols is based on specifications developed using existing Web service technologies such as WSDL-S [4] and OWL [15], presented in the following sections. Service protocols are described by SAWSDL [12] specifications and are specific to each service. Name services define a set of messages for interrogating service registry, while specification services define messages for downloading specifications. Authorization services define messages for requesting access to services and to send security tokens. Resource services provide two types of specifications, for each external entity accessing them. The AUT-S specifications provide messages for setting user security tokens, while the client specifications provide client access messages.

The *service layer* provides the implementation of service capabilities. The capabilities of NAME-S, SPEC-S and AUT-S have already been discussed before. The capabilities of RES-S are specific to each implementation, ranging from video image capabilities to data storage capabilities. In order to implement new resource services, the only components that require change are the capabilities and the service protocol specifications. The security properties of communications with new resources are ensured by the underlying communication layer that remains unchanged.

3 Constructing Security Protocol Specifications

In order to provide an automated implementation of security protocols we need to construct a detailed specification model that includes sufficient information for protocol participants to generate and process messages. Specification models are generated from protocol models that contain a limited number of information. Based on the protocol model, we provide several rules and algorithms to generate specification models. However, this process is not entirely automated. The user must select parameters of cryptographic algorithms such as generated key and random number sizes.

3.1 Protocol Model

Protocol participants communicate by exchanging *terms* constructed from elements belonging to the following basic sets: P, denoting the set of role names; N, denoting the set of random numbers or *nonces* (i.e. "number once used"); K, denoting the set of cryptographic keys; C, denoting the set of certificates and M, denoting the set of user-defined message components.

In order for the protocol model to capture the message component types found in security protocol implementations [19], [20] we specialize the basic sets with the following subsets:

- $\mathsf{P}_{DN} \subseteq \mathsf{P}$, denoting the set of distinguished names; $\mathsf{P}_{UD} \subseteq \mathsf{P}$, denoting the set of user-domain names; $\mathsf{P}_{IP} \subseteq \mathsf{P}$, denoting the set of user-ip names; $\mathsf{P}_U = \{\mathsf{P} \setminus \{\mathsf{P}_{DN} \cup \mathsf{P}_{UD} \cup \mathsf{P}_{IP}\}\}$, denoting the set of names that do not belong to the previous subsets;
- N_T, denoting the set of timestamps; N_{DH}, denoting the set of random numbers specific to the Diffie-Hellman key exchange; $\mathsf{N}_A = \{\mathsf{N} \setminus \{\mathsf{N}_{DH} \cup \mathsf{N}_T\}\}$, denoting the set of random numbers;
- $\mathsf{K}_S \subseteq \mathsf{K}$, denoting the set of symmetric keys; $\mathsf{K}_{DH} \subseteq \mathsf{K}$, denoting the set of keys generated from a Diffie-Hellman key exchange; $\mathsf{K}_{PUB} \subseteq \mathsf{K}$, denoting the set of public keys; $\mathsf{K}_{PRV} \subseteq \mathsf{K}$, denoting the set of private keys;

To denote the encryption type used to create cryptographic terms, we define the following *function names*:

$$\begin{aligned} FuncName ::= &\ sk && (symmetric\,function) \\ &|\ pk && (asymmetric\,function) \\ &|\ h && (hash\,function) \\ &|\ hmac && (keyed\,hash\,function) \end{aligned}$$

The encryption and decryption process makes use of cryptographic keys. Decrypting an encrypted term is only possible if participants are in the possession of the decryption key pair. In case of symmetric cryptography, the decryption key is the same as the encryption key. In case of asymmetric cryptography, there is a public-private key pair. Determining the corresponding key pair is done using the function $\_^{-1} : \mathsf{K} \rightarrow \mathsf{K}$.

The above-defined basic sets and function names are used in the definition of *terms*, where we also introduce constructors for pairing and encryption:

$$\mathsf{T} ::= .\ |\ \mathsf{P}\ |\ \mathsf{N}\ |\ \mathsf{K}\ |\ \mathsf{C}\ |\ \mathsf{M}\ |\ (\mathsf{T},\mathsf{T})\ |\ \{\mathsf{T}\}_{FuncName(\mathsf{T})},$$

where the '.' symbol is used to denote an empty term.

Having defined the terms exchanged by participants, we can proceed with the definition of a *node* and a *participant chain*. To capture the sending and receiving of terms, the definition of nodes uses *signed terms*. The occurrence of a term with a positive sign denotes transmission, while the occurrence of a term with a negative sign denotes reception.

Definition 1. *A* node *is any transmission or reception of a term denoted as* $\langle\sigma, t\rangle$, *with* $t \in \mathsf{T}$ *and* σ *one of the symbols* $+, -$. *A node is written as* $-t$ *or* $+t$. *We use* $(\pm\mathsf{T})$ *to denote a set of nodes. Let* $n \in (\pm\mathsf{T})$, *then we define the function* $sign(n)$ *to map the sign and the function* $term(n)$ *to map the term corresponding to a given node.*

Definition 2. *A* participant chain *is a sequence of nodes. We use* $(\pm\mathsf{T})^*$ *to denote the set of finite sequences of nodes and* $\langle\pm t_1, \pm t_2, \ldots, \pm t_i\rangle$ *to denote an element of* $(\pm\mathsf{T})^*$.

In order to define a participant model we also need to define the preconditions that must be met such that a participant is able to execute a given protocol. In addition, we also need to define the effects resulting from a participant executing a protocol.

Preconditions and effects are defined using predicates applied on terms: $CON\_TERM : \mathsf{T}$, denoting a generated term; $CON\_PARTAUTH : \mathsf{T}$, denoting participant authentication; $CON\_CONF : \mathsf{T}$, denoting the confidentiality of a given term); $CON\_INTEG : \mathsf{T}$, denoting the integrity of a given term; $CON\_NONREP : \mathsf{T}$, denoting the non-repudiation property for a given term; $CON\_KEYEX : \mathsf{T}$, denoting a key exchange protocol.

The set of precondition-effect predicates is denoted by $\mathsf{PR\_CC}$ and the set of precondition-effect predicate subsets is denoted by $\mathsf{PR\_CC}^*$. The types attached to each protocol term are modeled using the following predicates: $TYPE\_DN : \mathsf{T}$ to denote distinguished names, $TYPE\_UD : \mathsf{T}$ to denote user-domain names, $TYPE\_NT : \mathsf{T}$ to denote timestamps, $TYPE\_NDH : \mathsf{T}$ to denote Diffie-Hellman random numbers, $TYPE\_NA : \mathsf{T}$ to denote other random numbers, $TYPE\_NDH : \mathsf{T} \times \mathsf{T} \times \mathsf{T} \times \mathsf{P} \times \mathsf{P}$ to denote Diffie-Hellman symmetric keys, $TYPE\_KSYM : \mathsf{T} \times \mathsf{P} \times \mathsf{P}$ to denote symmetric keys, $TYPE\_KPUB : \mathsf{T} \times \mathsf{P}$ to denote public keys, $TYPE\_KPRV : \mathsf{T} \times \mathsf{P}$ to denote private keys, and $TYPE\_CERT : \mathsf{T} \times \mathsf{P}$ do denote certificate terms.

The set of type predicates is denoted by $\mathsf{PR\_TYPE}$ and the set of type predicate subsets is denoted by $\mathsf{PR\_TYPE}^*$. Based on the defined sets and predicates we are now ready to define the participant and protocol models.

Definition 3. *A* participant model *is a tuple* $\langle prec, eff, type, gen, part, chain\rangle$, *where* $prec \in \mathsf{PR\_CC}^*$ *is a set of precondition predicates,* $eff \in \mathsf{PR\_CC}^*$ *is a set of effect predicates,* $type \in \mathsf{PR\_TYPE}$ *is a set of type predicates,* $gen \in \mathsf{T}^*$ *is a set of generated terms,* $part \in P$ *is a participant name and* $chain \in (\pm\mathsf{T})^*$ *is a participant chain. We use the* MPART *symbol to denote the set of all participant models.*

Definition 4. *A* security protocol model *is a collection of participant models such that for each positive node* n_1 *there is exactly one negative node* n_2 *with* $term(n_1) = term(n_2)$. *We use the* MPROT *symbol to denote the set of all security protocol models.*

3.2 Semantic Security Protocol Model

In this section we propose a new semantic security protocol model (SSPM) based on which we construct security protocol specifications that can be automatically executed by protocol participants. Protocols are given using their SPM model described in the previous section. Based on this model we generate the corresponding SSPM that has two components: the sequential model (SEQM) and the ontology model (ONTM). The first component is implemented as a WSDL-S specification while the second component is implemented as an OWL specification. In the remaining of this section we provide a description of each component and we provide a set of rules to generate SSPM from a given SPM.

We use the symbol URI to denote the set of *Uniform Resource Identifiers*, CONC to denote the set of all concepts and CONC^* to denote the set of subsets with elements from CONC.

Definition 5. *An* annotation *is a pair* $\langle uri, c\rangle$, *where* $uri \in \mathsf{URI}$ *and* $c \in \mathsf{CONC}$. *The set corresponding to a* SSPM *is denoted by* ANNOT *and the set of subsets with elements from* ANNOT *is denoted by* ANNOT^*.

By consulting the WSDL-S specification we define a message as a pair consisting of the message direction and an annotation.

Definition 6. *A* message *is a pair* $\langle d, a\rangle$, *where* $d \in \{in, out\}$ *and* $a \in \mathsf{ANNOT}$. *We define* MSG *to denote a set of messages and* MSG^* *to denote the set of subsets with elements from* MSG.

Next, we define the sequential model as a collection of preconditions, effects and messages, based on the previous definitions.

Definition 7. *A* sequential model *is a triplet* $\langle s\_prec, s\_eff, s\_msg\rangle$, *where* $s\_prec \in \mathsf{ANNOT}^*$ *is a set of preconditions,* $s\_eff \in \mathsf{ANNOT}^*$ *is a set of effects and* $s\_msg \in \mathsf{MSG}^*$ *is a set of messages.*

The ontology model follows the description of OWL.

Definition 8. *An* ontology model *is a triplet* $\langle conc, propr, inst\rangle$, *where* $conc \in \mathsf{CONC}$ *is a set of concepts,* $propr \in \mathsf{PROPR}$ *is a set of properties and* $inst \in \mathsf{INST}$ *is a set of instances. An element from propr is a pair* $\langle\alpha, \beta\rangle$, *where* α *is a unique id and* β *is a syntactic construction denoting the property name.*

Let $pr_1 = \langle\alpha_1, \beta_1\rangle$ *and* $pr_2 = \langle\alpha_2, \beta_2\rangle$. *Then* $pr_1 = pr_2$ *iff* $\alpha_1 = \alpha_2$ *and* $\beta_1 = \beta_2$. *We define the function* $(\_)_{\mathrm{id}}$ *to map the* α *component and the function* $(\_)_{\mathrm{nm}}$ *to map the* β *component of a given property.*

We use PROPR *to denote the set of all properties and* INST *to denote the set of all instances. We use* PROPR^* *to denote the set of all subsets with elements from* PROPR *and* INST^* *to denote the set of all subsets with elements from* INST.

In order to handle the previously defined ontology model we define the function $(\_)_{\mathrm{d}} : \mathsf{PROPR} \rightarrow \mathsf{CONC}$ to map the domain concept of a given property, $(\_)_{\mathrm{c}} : \mathsf{PROPR} \rightarrow \mathsf{CONC}$ to map the category concept of a given property, $(\_,\_)_{\mathrm{ci}} : \mathsf{CONC} \times \mathsf{PROPR} \rightarrow \mathsf{INST}$ to map the instance corresponding to a domain concept and property, $(\_)^{s}_{e} : \mathsf{CONC} \rightarrow \mathsf{CONC}^{*}$ to map the set of concepts for which the given concept is parent, $(\_)_{p} : \mathsf{CONC} \rightarrow \mathsf{PROPR}^{*}$ to map the set of properties for which the given concept is domain.

3.3 Generating the Semantic Security Protocol Model

In order to generate the SSPM for a given SPM, we start with a core ontology model (OM) (figure 3) that contains concepts found in classical security protocols. The core OM was constructed by consulting security protocols found in open libraries such as SPORE [18] or the library published by John Clark [5].

The core ontology is constructed from 7 sub-ontologies. The sub-ontologies that must be extended with new concepts for each SSPM are denoted in figure 3 by interrupted lines, while the permanent sub-ontologies are denoted by continuous lines.

The *SecurityProperty* sub-ontology contains concepts such as *Authentication*, *Confidentiality* or *Session_key_exchange*. The *TermType* sub-ontology includes concepts related to term types used in security protocol messages such as *SymmetricKey*, *PublicKey* or *ParticipantName*. Concepts related to cryptographic specifications such as encryption algorithms or encryption modes are found in the sub-ontology *CryptoSpec*. In order to model modules needed to extract keys, names or certificates we use the *LoadingModule* sub-ontology. The *ParticipantRole* sub-ontology defines concepts modeling roles handled by protocol participants such as *Initiator*, *Respondent* and *Third_Party*.

The *Knowledge* sub-ontology contains 5 concepts: *PreviousTerm*, *AccessedModule*, *InitialTerm*, *GeneratedTerm* and *DiscoveredTerm*. Each concept defines a class of terms specific to security protocols: terms from previous executions, modules, initial terms, generated terms and discovered terms.

The last sub-ontology is *CommunicationTerm*, which defines two concepts: *SentTerm* and *ReceivedTerm*. This sub-ontology is extended for each SEM-S with concepts that are sent or received. For each concept, functional properties are defined denoting the operations performed on the terms corresponding to concepts. The concepts used to extend the core ontology are specific to each protocol, however, the defined properties are applied on all constructions. From these properties we mention: *hasKey*, *isStored*, *isVerified*.

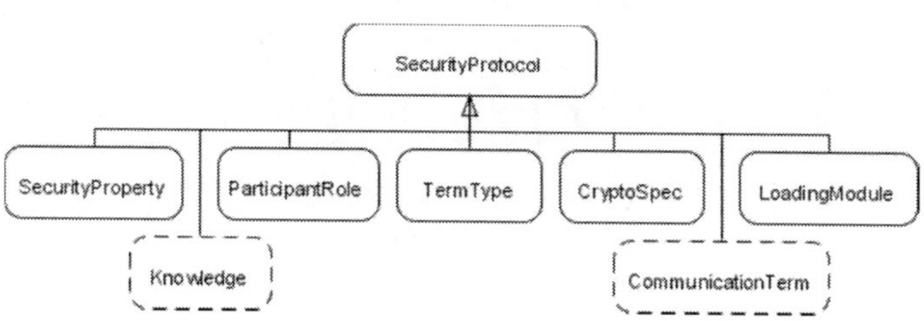

Fig. 3. Core ontology of SSPM

Next, we define a set of rules and algorithms to generate the SSPM for a given SPM. The developed rules use the $\_ \leftarrow_{\mathrm{r}} \_$ operator to denote set reunion and the $\_ \leftarrow_{\mathrm{a}} \_$ operator to denote a value transfer.

The first two rules generate the predicate concepts corresponding to preconditions *prec* from a SPM, where the function $gc : \mathsf{T} \rightarrow \mathsf{CONC}$ is used to generate the concept corresponding to a given term and the function $gcc : \mathsf{PR\_CC} \rightarrow \mathsf{CONC}$ is used to generate the concept corresponding to a given precondition predicate:

$$\frac{pr \in prec \;\; pr = CON\_TERM(t)}{c \leftarrow_{\mathrm{a}} gc(t) \;\; s\_prec \leftarrow_{\mathrm{r}} \{\langle uri, c\rangle\} \;\; (InitialTerm)^{\mathrm{s}}_{\mathrm{e}} \leftarrow_{\mathrm{r}} \{c\}} \text{pr\_term},$$

$$\frac{pr \in prec \;\; pr \neq CON\_TERM(t)}{s\_prec \leftarrow_{\mathrm{r}} \{\langle uri, gcc(pr)\rangle, \langle uri, gc(t)\rangle\}} \text{pr\_propr}.$$

The rules generating the effects have a similar structure because of the *eff* set. For each positive or negative node there is a corresponding concept in the *SentTerm* and *ReceivedTerm* sub-ontologies, generated by the following rules:

$$\frac{n \in chain \;\; sign(n) = +}{c \leftarrow_{\mathrm{a}} gtx(term(n)) \;\; s\_msg \leftarrow_{\mathrm{r}} \{\langle out, c\rangle\} \;\; (SentTerm)^{\mathrm{s}}_{\mathrm{e}} \leftarrow_{\mathrm{r}} \{c\}} \text{msg\_tx},$$

$$\frac{n \in chain \;\; sign(n) = -}{c \leftarrow_{\mathrm{a}} grx(term(n)) \;\; s\_msg \leftarrow_{\mathrm{r}} \{\langle in, c\rangle\} \;\; (ReceivedTerm)^{\mathrm{s}}_{\mathrm{e}} \leftarrow_{\mathrm{r}} \{c\}} \text{msg\_rx}.$$

The concatenated terms corresponding to each transmitted or received term are modeled using similar rules. For each sent term the SSPM must provide the construction operations and for each received term the SSPM must provide processing operations. Sub-concepts of *SentTerm* are connected to sub-concepts of *Knowledge* through the *isExtracted* property, generated according to the following rule, where we used the function $\mathsf{PR\_CC}^* \rightarrow \mathsf{ID}$ to generate a new property id:

$$\frac{c \in (SentTerm)^{\mathrm{s}}_{\mathrm{e}}}{p \leftarrow_{\mathrm{a}} \langle gid(propr), isExtracted\rangle \;\; (c)_{\mathrm{p}} \leftarrow_{\mathrm{r}} \{p\} \;\; (p)_{\mathrm{c}} \in (Knowledge)^{\mathrm{s}}_{\mathrm{e}}} \text{con\_extr}.$$

Processing of received terms is done according to the type of the given term and to the knowledge available to the user. The modeled operations introduce constraints on the type and location of knowledge through the following rules, where we used the $E\_SYM : \mathsf{CONC}$ predicate to denote symmetric encryption:

$$\frac{c \in (ReceivedTerm)^{\mathrm{s}}_{\mathrm{e}} \;\; p \in (c)_{\mathrm{p}} \;\; (p)_{\mathrm{nm}} = isDecrypted}{c' \leftarrow_{\mathrm{a}} (p)_{\mathrm{c}} \;\; E\_SYM(c') \vee E\_SYM(c') \;\; (c') \in (DiscoveredTerm)^{\mathrm{s}}_{\mathrm{e}}} \text{con\_decr},$$

$$\frac{c \in (ReceivedTerm)^{\mathrm{s}}_{\mathrm{e}} \;\; p \in (c)_{\mathrm{p}} \;\; (p)_{\mathrm{nm}} = isStored}{c' \leftarrow_{\mathrm{a}} (p)_{\mathrm{c}} \;\; (c') \in (DiscoveredTerm)^{\mathrm{s}}_{\mathrm{e}}} \text{con\_stored},$$

$$\frac{c \in (ReceivedTerm)^{\mathrm{s}}_{\mathrm{e}} \;\; p \in (c)_{\mathrm{p}} \;\; (p)_{\mathrm{nm}} = isVerified}{c' \leftarrow_{\mathrm{a}} (p)_{\mathrm{c}} \;\; (c') \in \{(DiscoveredTerm)^{\mathrm{s}}_{\mathrm{e}} \setminus (AccessedModule)^{\mathrm{s}}_{\mathrm{e}}\}} \text{con\_verif}.$$

In the *Knowledge* sub-ontology, each concept has an *isOfType* property attached based on which participants can decide on the operations to execute. For each type, additional properties are defined such as the *hasSymmAlg* or *hasKey* properties for symmetric encrypted terms. The rules based on which these properties are generated are specific to each type. For example, the following rules define the algorithm type and key for an encrypted term that must be processed or constructed:

$$\frac{c \in (Knowledge)^{\mathrm{s}}_{\mathrm{e}} \quad E\_SYM(c)}{p \leftarrow_{\mathrm{a}} \langle gid(propr), hasSymmAlg\rangle \quad (c)_{\mathrm{p}} \leftarrow_{\mathrm{r}} \{p\} \quad (c,p)_{\mathrm{ci}} \in (Symmetric)^{\mathrm{s}}_{\mathrm{e}}}\,\mathrm{sim\_alg},$$

$$\frac{c \in (Knowledge)^{\mathrm{s}}_{\mathrm{e}} \quad E\_SYM(c)}{p \leftarrow_{\mathrm{a}} \langle gid(propr), hasKey\rangle \quad (c)_{\mathrm{p}} \leftarrow_{\mathrm{r}} \{p\} \quad (p)_{\mathrm{c}} \in (Knowledge)^{\mathrm{s}}_{\mathrm{e}}}\,\mathrm{sim\_key}.$$

The rules presented above are executed by algorithms. For example, modeling positive nodes in SSPM is done through the use of algorithm 1. Here, the set of knowledge *KNOW* corresponding to each executing participant grows with the construction and reception of each new term. We used the function $mpart : \mathsf{T} \rightarrow \mathsf{T}^*$ to map the set of concatenated terms and the keyword "Exec" to denote the execution of sub-algorithms.

Algorithm 1. Model positive and negative nodes

```
Require: n ∈ (±T), sign(n) = +
  for all t ∈ mpart(term(n)) do
    Let c = gc(t)
    Let p ⇐ @con_extr(c)
    if t ∈ KNOW then
      (p)_c ←_a c
    else if t = {t'}_f(k) then
      (GeneratedTerm)^s_e ←_r {c}
      Exec ModelEncryptedGenerated(t)
    else if t ∈ gen then
      (GeneratedTerm)^s_e ←_r {c}
      Exec ModelPlainGenerated(t)
    else
      (DiscoveredTerm)^s_e ←_r {c}
      Exec ModelDiscoveredLoaded(t)
    end if
    KNOW ←_r t
  end for
```

4 Experimental Results

In this section we exemplify the construction of a SSPM from a given SPM and provide a few experimental result from implementing several generated SSPM.

4.1 Constructing the SSPM for the "BAN" Protocol

In order to provide an example for constructing an SSPM for a given SPM, we use the well-known "BAN Concrete Secure Andrew RPC" protocol [18]. This is a two-party protocol providing a session key exchange using symmetric cryptography. The protocol assumes that participants are already in the possession of a long-term key K_{ab}.

Because of space considerations, we only provide the construction of the SSPM for the A participant. Based on this, the construction of the SSPM for the second participant is straight-forward.

The precondition set $prec_A$ for participant A is $prec_A = \{CON\_TERM(A), CON\_TERM(B), CON\_TERM(K_{ab})\}$ and the effect set eff_A for the same participant is $eff_A = \{CON\_KEYEX(K_{ab})\}$. The set $type_A = \{TYPE\_UD\ (A), TYPE\_UD(B), TYPE\_KSYM(A, B, K_{ab}), TYPE\_KSYM(A, B, K), TYPE\_NA\ (N_a), TYPE\_NA(N_b)\}$ defines the type corresponding to each term and the set $gen_A = \{N_a\}$ defines the terms generated by participant A. The participant name is $part_A = A$ and the participant chain is $chain_A = \langle +(A, N_a), -\{N_a, K, B\}_{sk(K_{ab})}, +\{N_a\}_{sk(K)}, -N_b\rangle$.

By applying the rules and algorithms described in the previous sections we generate the SSPM model. Due to space considerations, instead of describing the actual SSPM we describe the implementation of the model. The sequential model is implemented as a WSDL-S specification, while the ontology model is implemented as a OWL specification.

Part of the resulted WSDL-S specification is given in figure 4 and part of the graphical representation of the OWL specification is given in figure 5.

```
...
<xsd:element name="Msg1Request">
  <xsd:complexType>
    <xsd:sequence>
      <xsd:element name="Term1" type="xsd:base64Binary"
                   wssem:modelReference=".../SecProt.owl#SentTerm1">
    </xsd:sequence>
  </xsd:complexType>
</xsd:element>
...
<wsdl:operation name="Msg1">
  <wsdl:output message="tns:Msg1Request"/>
</wsdl:operation>
<wssem:effect name="SessionKeyExchange"
              wssem:modelReference=".../SecProt.owl#SessionKey"/>
...
```

Fig. 4. Part of the sequential model's implementation

4.2 Case Study

We have generated several security protocol specifications corresponding to protocols such as ISO9798 and Kerberos V5. These specifications were stored by the SPEC-S and were downloaded and automatically executed by client test applications and the services provided by the platform.

Based on these specifications and the proposed middleware, we implemented a video surveillance system. The basic requirements of these systems include real time image transfer, low bandwidth consumption and access control procedures [13]. To these, we add another requirement: automated security protocol execution in heterogeneous environments.

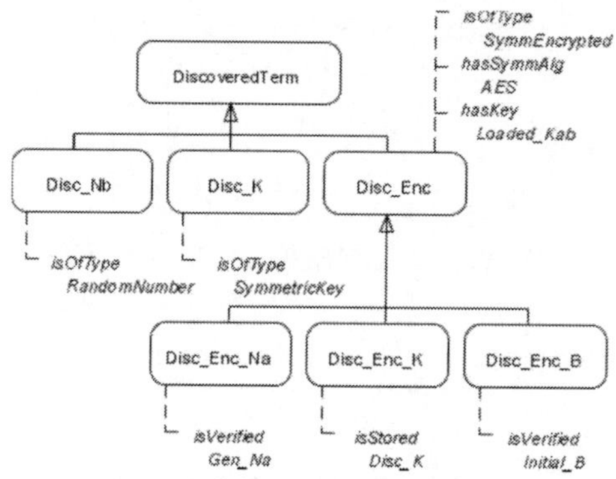

Fig. 5. Part of the ontology model's implementation

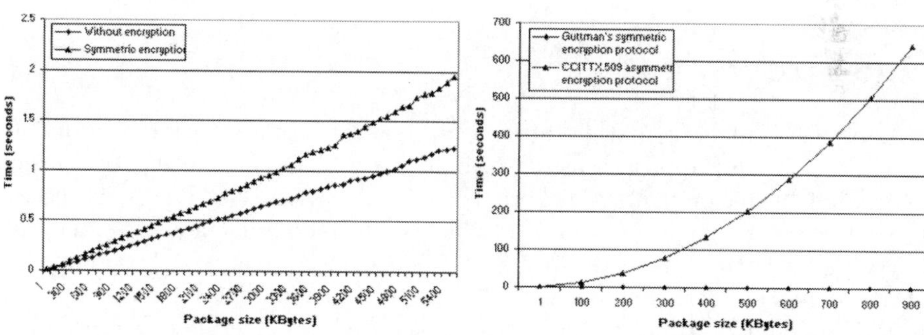

(a) (b)

Fig. 6. Performance of the communication layer: (a) Symmetrical encryption (b) Asymmetrical encryption

Our middleware satisfies all of the above formulated requirements. In figure 6 we have illustrated the performance of the communication layer. In the first case, we illustrated the performance overhead of using symmetric key-based protocols against the case when no cryptography is used. In the second case we illustrated the performance overhead of using security protocols based on asymmetric cryptography.

In order to provide automated access to resources we used asymmetric algorithms for key exchange protocols and symmetric algorithms for data transfer protocols. The keys exchanged using the first protocol types were used by the second protocols to encode data. All security protocol implementations were done using SOAP protocol header, according to the WS-Security standard.

Based on our experimental results, the time needed for a client application to access a service resource is around 520ms, as shown in figure 7, if the specifications are not cached. In case caching mechanisms are used, the time reduces to around 375ms. In both cases, if the specifications for the requested resources are

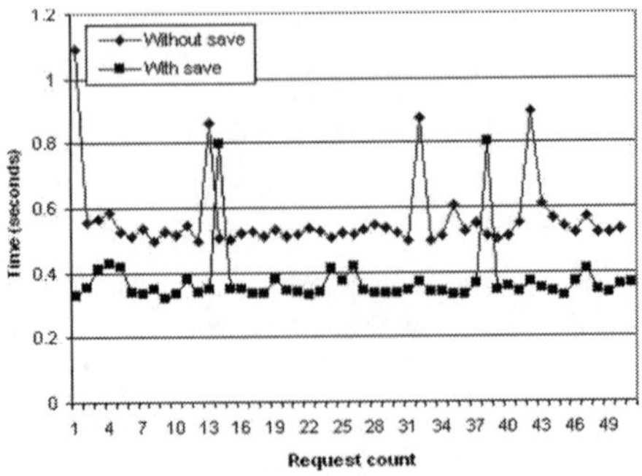

Fig. 7. Accessing system resources

not cached by AUT-S, the accessing time is much higher, as illustrated by the spikes from figure 7.

5 Related Work

An approach that aims at the automatic implementation of security protocols is given in [2]. This approach uses a formal description as a specification which is executed by participants. The proposed specification does not make use of Web service technologies, because of which inter-operability and extendability of systems executing the given specifications becomes a real issue.

Abdullah and Menasc propose in [3] a specification that is constructed as an XML document from which code is generated. The resulted code is then compiled and executed by participants. Because of this aspect, our proposal is more dynamic in the sense that applications can download and execute new protocols based on the developed specifications automatically, without having to stop program execution.

The authors from [8] propose a security ontology for resource annotation. The proposed ontology defines concepts for security and authorization, for cryptographic algorithms and for credentials. This proposal was designed to be used in the process of security protocol description and selection based on several criteria. In contrast, our ontologies, have a more detailed construction. For example, the ontology from [8] defines a collection of cryptographic algorithms, however, it does not define the algorithm mode, which is an implementation-specific information.

There have been several other security ontologies proposed [9], [10] which can be used to complete our core ontology with additional concepts and properties, for generating more complex protocol models.

6 Conclusion and Future Work

We developed a middleware for the automated execution of security protocols. The proposed middleware makes use of specifications generated from a semantic security protocol model. The sequential component of the proposed model is implemented as a WSDL-S specification while the ontology component is implemented as an OWL specification.

Constructing the SSPM model is not a trivial task and can induce new flaws in correct protocols that can lead to attacks. In order to ensure a correct construction process, we developed several generating rules and algorithms that map each component from the input protocol model to a component in the SSPM model. The resulted specifications were automatically executed by client applications and services implemented in a proposed video surveillance system. The experimental results have shown that the proposed middleware can be used in a wide variety of applications ranging from multimedia to eCommerce.

The proposed specification model encapsulates cryptographic properties that must be met by services and clients executing them. As future work we intend to develop a negotiation mechanism of such properties between RES-S and AUT-S as well as between client applications and AUT-S. These mechanisms can be implemented using the proposed communication layer enhanced with negotiation protocols.

References

1. Cremers, C.J.F., Mauw, S.: Checking secrecy by means of partial order reduction. In: Leue, S., Systä, T.J. (eds.) Scenarios: Models, Transformations and Tools. LNCS, vol. 3466. Springer, Heidelberg (2005)
2. Mengual, L., Barcia, N., Jimenez, E., Menasalvas, E., Setien, J., Yaguez, J.: Automatic implementation system of security protocols based on formal description techniques. In: Proceedings of the Seventh International Symposium on Computers and Communications, pp. 355–401 (2002)
3. Abdullah, I., Menascé, D.: Protocol specification and automatic implementation using XML and CBSE. In: IASTED conference on Communications, Internet and Information Technology (2003)
4. Akkiraju, R., Farrell, J., Miller, J., Nagarajan, M., Schmidt, M., Verma, K.: Web Service Semantics - WSDL-S. A joint UGA-IBM Technical Note (2005)
5. Clark, J., Jacob, J.: A Survey of Authentication Protocol Literature: Version 1.0. York University (1997)
6. Gavin, L.: Some new attacks upon security protocols. In: Proceedings of the 9th CSFW, pp. 162–169. IEEE Computer Society Press, Los Alamitos (1996)
7. Cremers, C.J.F.: Compositionality of Security Protocols: A Research Agenda. Electr. Notes Theor. Comput. Sci. 142, 99–110 (2006)
8. Kim, A., Luo, J., Kang, M.: Security ontology for annotating resources. In: Meersman, R., Tari, Z. (eds.) OTM 2005. LNCS, vol. 3761, pp. 1483–1499. Springer, Heidelberg (2005)
9. Blanco, C., Lasheras, J., Valencia-Garcia, R., Fernandez-Medina, E., Toval, A., Piattini, M.: A systematic review and comparison of security ontologies. In: Proc. of the Third International Conference on Availability, Reliability and Security, pp. 813–820 (2008)
10. Denkera, G., Kagal, L., Finin, T.: Security in the semantic web using owl. Information Security Technical Report 1(10), 51–58 (2005)

11. Gong, L.: Fail-Stop Protocols: An Approach to Designing Secure Protocols. In: Proceedings of the 5th IFIP Conference on Dependable Computing and Fault-Tolerant Systems, pp. 44–55 (1995)
12. Martin, D., Paolucci, M., Wagner, M.: Toward Semantic Annotations of Web Services: OWL-S from the SAWSDL Perspective. In: OWL-S: Experiences and Directions - workshop at 4th European Semantic Web Conf. (2007)
13. Ostheimer, D., Lemay, S., Ghazal, M., Mayisela, D., Amer, A., Dagba, P.: A Modular Distributed Video Surveillance System Over IP. In: Electrical and Computer Engineering Canadian Conference, pp. 518–521 (2006)
14. Genge, B., Haller, P.: Extending WS-Security to Implement Security Protocols for Web Services. In: International Conference on Recent Achievements in Mechatronics, Automation, Computer Science and Robotics, Targu Mures, Romania (to appear, 2009)
15. World Wide Web Consortium, OWL Web Ontology Language Reference, W3C Recommendation (2004)
16. Gutmann, P.: Cryptlib Encryption Toolkit, `http://www.cs.auckland.ac.nz/~pgut001/cryptlib/index.html`
17. OpenSSL Project, version 0.9.8h, `http://www.openssl.org/`
18. Laboratoire Specification et Verification, Security Protocol Open Repository, `http://www.lsv.ens-cachan.fr/spore`
19. Organization for the Advancement of Structured Information Standards, SAML V2.0 OASIS Standard Specification (2007), `http://saml.xml.org/`
20. Organization for the Advancement of Structured Information Standards, OASIS Web Services Security (WSS) (2006), `http://saml.xml.org/`
21. Organization for the Advancement of Structured Information Standards, WS-Trust (2007), `http://docs.oasis-open.org/ws-sx/ws-trust/200512/ws-trust-1.3-os.html`
22. Organization for the Advancement of Structured Information Standards, UDDI (2004), `http://www.uddi.org/pubs/uddi_v3.htm`

Can RDB2RDF Tools Feasibily Expose Large Science Archives for Data Integration?

Alasdair J.G. Gray[1], Norman Gray[2,3], and Iadh Ounis[1]

[1] Computing Science, University of Glasgow, Glasgow, UK
[2] Physics and Astronomy, University of Leicester, Leicester, UK
[3] Physics and Astronomy, University of Glasgow, Glasgow, UK

Abstract. Many science archive centres publish very large volumes of image, simulation, and experiment data. In order to integrate and analyse the available data, scientists need to be able to (i) identify and locate all the data relevant to their work; (ii) understand the multiple heterogeneous data models in which the data is published; and (iii) interpret and process the data they retrieve. RDF has been shown to be a generally successful framework within which to perform such data integration work. It can be equally successful in the context of scientific data, if it is demonstrably practical to expose that data as RDF.

In this paper we investigate the capabilities of RDF to enable the integration of scientific data sources. Specifically, we discuss the suitability of SPARQL for expressing scientific queries, and the performance of several triple stores and RDB2RDF tools for executing queries over a moderately sized sample of a large astronomical data set. We found that more research and improvements are required into SPARQL and RDB2RDF tools to efficiently expose existing science archives for data integration.

1 Introduction

Due to the ease with which information can be published on the Internet, scientists potentially have access to more data than at any time in history. The Internet makes it viable for researchers to publish the results of their experiments for others to use as text or binary files, XML documents, or relational databases. There are also numerous archive centres which gather information from various sources and make it available from one place, often as a query endpoint to a relational database due to the size and structure of the data.

For example, consider the situation in astronomy of the Virtual Observatory (VO), which is being developed by the many national and international projects that form the International Virtual Observatory Alliance (IVOA)[1]. One of the goals of the VO is to allow an astronomer to seamlessly access any piece of astronomical data, regardless of whether it has been archived in one of the many archive centres (e.g. ROE[2], ESAC[3]), generated by a simulation

[1] `http://www.ivoa.net` accessed 8 December 2008.
[2] `http://www.roe.ac.uk/ifa/wfau/` accessed 10 December 2008.
[3] `http://www.esa.int/SPECIALS/ESAC/` accessed 10 December 2008.

L. Aroyo et al. (Eds.): ESWC 2009, LNCS 5554, pp. 491–505, 2009.

(e.g. VIRGO[4], MODEST[5]), or published by an individual as the result of some analysis (e.g. NVO publish your own data interface[6]).

However, for astronomy, as with any other domain, there is no single data model of the world that is suitable for storing and querying all the available data—each data archive, simulation, etc., uses its own, tailor-made model that is a compromise between the storage and querying requirements for the particular data. Therefore, in order for an astronomer to be able to access, extract, and process the data relevant to their needs, they must be able to: (i) locate those data sources which potentially contain information relevant to their research; (ii) understand the data model of each relevant data source in order to compose a query to extract the required data; and (iii) interpret and process the result data with either bespoke or general purpose visualisation and analysis software.

The first of these problems is commonly addressed by using a registry service which stores details of the available resources. Users query the registry service by describing their information need which is *matched* to find those resources that are potentially relevant [7]. For example, in the case of the VO an XML Schema and access protocols are used to discover resources [15]. These registry services return a list of potentially relevant data sources, which the user must then understand in order to extract the data relevant to their research.

One solution to the second problem is to follow a traditional data integration approach where the data sources are queried through a global data model [9]. However, generally there is not a common data model that is capable of accessing all of the available data with the granularity required by all users. One of the benefits often stated for RDF is the ease with which data can be integrated from distributed RDF sources. This, in association with a peer-to-peer data integration approach—whereby the data models of the sources are exposed as RDF and linked to each other by alignment mappings [6]—should allow a user to query using a data model with which they are familiar and exploit the mappings in order to retrieve the data that is relevant to their research.

The third problem is reliant on a viable solution to the second. As such, there is currently limited software support within the VO for the analysis of data from multiple sources.

A first step to enable the peer-to-peer data integration approach is the ability to expose existing science archives as *virtual* RDF graphs. RDB2RDF tools provide such functionality for relational databases [13]. In this work, we consider the feasibility of using these tools to query existing relational databases as SPARQL endpoints, in the context of an astronomical data archive. Specifically, we investigate whether: (i) SPARQL can be used to express the types of query required in scientific applications, and (ii) the performance of RDF triple stores and RDB2RDF tools can meet the requirements of the large data sets encountered in the scientific domain.

---

[4] `http://www.virgo.dur.ac.uk/` accessed 10 December 2008.

[5] `http://manybody.org/` accessed 10 December 2008.

[6] `http://www.us-vo.org/publish.cfm` accessed 10 December 2008.

Section 2 presents the issues of integrating distributed scientific data and states how RDF can overcome these. Section 3 describes a typical scientific data set, its structure and the types of queries posed over it. The section then discusses the capabilities and limitations of SPARQL. In Section 4 we describe the performance experiments and discuss the results. Section 5 presents related work and our conclusions are given in Section 6.

2 Integrating Scientific Data

The aim of a data integration system is to seamlessly allow a user to access multiple autonomous data resources as if they were a single, coherent data source. This is generally achieved by (i) agreeing on a mediated *global* schema, (ii) providing a wrapper for each data source, and (iii) generating mappings that relate the source schema to the global schema (see [9] for an overview). However, there are a few well known problems with creating a data integration system.

First, there must be agreement on a single global schema. Within the astronomy setting of the VO, such a global schema does not exist, although over the last few years there have been attempts to create such a global schema [12]. Reaching this limited consensus has proven to be a time consuming process, and the resulting model still does not achieve the goal of a full global model: it falls short of being expressive enough to capture the existing data source models.

Second, insisting that all data sources map their data to a single global model leads to a loss of data precision. This is because the mappings must translate between the granularity of the data source and the global model, which generally limits the available data.

To evade these problems, we offer a more "peer-to-peer" approach. In such a system, the data resources would publish their schema in some common modelling language and produce mappings to one or more different models in the system. These mappings might be

Direct: if there is a close association between two (probably specialised) data models then equivalence mappings can be used, thus maintaining the level of expressible data and reducing the loss in data precision;

Indirect: using some widely understood data model which may be at a different level of granularity to link terms through a super-class relationship;

Multi-stepped: using some chain of mappings, although this risks some loss of fidelity.

An advantage of this approach is that the user can use a data model they are familiar with to express their information needs as a query. The system is able to convert the query into a set of queries over the linked data sources using the web of mappings that exist between the various data sources. Providing a suitable direct mapping exists, the user does not suffer from data granularity problems.

The peer-to-peer data integration approach outlined requires a common modelling language for expressing all of the available data source models. Since we consider that data sources can be published either as files, XML documents, or relational databases, we propose using RDF as the common data modelling language and leveraging the Semantic Web reasoning and mapping techniques available in RDFS and some limited OWL concepts to generate the mappings between the data models. It remains an open question whether such mappings are flexible enough, we do not cover this. A pre-condition for this form of data integration system is a wrapper for exposing an existing data source as an RDF model over which SPARQL queries can be posed.

Two approaches are possible. The first would be to replicate the data contained in a relational source into an RDF triple store such as Jena [14] or Sesame [4], aka Extract-Transform-Load (ETL) or an RDF dump. However, this would suffer from all the usual problems of replicating data, e.g. maintaining multiple copies of the data in different data models and data freshness. The second would be to use a wrapper that can expose the existing relational database schema as an RDF model and perform *on-the-fly* or *query-driven* conversions to translate SPARQL queries over the exposed RDF model into SQL queries over the relational schema. D2RQ [2] and SquirrelRDF [18] are two prototype RDB2RDF systems capable of performing these on-the-fly translations. This paper considers the viability of these data integration wrapper approaches in the context of scientific data.

3 Scientific Data Set

Various studies compare the performance of the storage and querying capabilities of RDF triple stores and RDB2RDF tools in non-scientific applications [11,19,3,17]. While these benchmarks have considered data sets with a significantly large number of triples (up to 100 million triples [3]), they have not considered data sets with similar structures to those of a science archive, nor have they considered queries using complex selection conditions, commonly found in scientific applications. In order to enable the integration framework outlined in Section 2, it is important to investigate the viability of using RDB2RDF tools to expose existing science archives. We conducted performance tests using real science queries over an existing archive.

3.1 The SuperCOSMOS Science Archive

The SuperCOSMOS Science Archive (SSA[7]) [10] contains data extracted from scans of photographic Schmidt plates, e.g. the position of objects, their luminosity, etc. The archive is about 4 terabytes in size and contains data on nearly 6.4 billion object detections covering the whole sky in three wavebands. This data is stored in a relational database on a Microsoft SQLServer 2000 DBMS located at the Royal Observatory, Edinburgh. It was chosen as the data set for the performance measures for several reasons.

First, it is a typical example of an astronomical data archive in that the information is stored in a relational database and contains a vast amount of data.

[7] `http://surveys.roe.ac.uk/ssa/` accessed 9 December 2008.

Table 1. Details of the relations contained in the Personal SuperCOSMOS Science Archive, providing the number of attributes in each relation and the number of rows

Relation	No. Attributes	No. Rows
CrossneighboursEDR	5	129,474
CurrIsoCal	5	3,535
Detection	46	169,558
Field	9	2,695
FieldSystem	3	4
HtmTable	6	263,200
HtmTableId	2	10

Relation	No. Attributes	No. Rows
IsoCalVersion	4	3
JunkDetection	7	6,125
Neighbours	3	585,560
Plate	152	3,535
PrevIsoCal	6	3,535
Source	57	90,947
Survey	24	9

The data is accessible for SQL query from a web page and also through various VO data access protocols. Second, the schema of the database was designed so that it could answer 20 scientifically significant queries which are representative of the types of query that would be expressed to the production system, *cf.* Gray *et al.* [8]. These queries, which are provided both as English statements and SQL, can be used to test the expressive power of SPARQL, and as the basis for the performance measures to test the retrieval speed of the various query engines. Finally, a subset of the data in the SSA is available for download, as the Personal SuperCOSMOS Science Archive (PSSA). The PSSA contains the data relating to the area of the sky given by ($184 < RA/deg < 186$, $-1.25 < Dec/deg < 1.25$) which is roughly 0.1% of the data contained in the full SSA. This provides a suitably sized dataset for running performance measures, without becoming a trivial toy example, and is comparable in size to other benchmark tests conducted over RDF triple stores [19].

3.2 Schema

The schema of the PSSA consists of 14 relations, the names of the relations together with the number of attributes in each relation and the number of data records are given in Table 1. The relations contain details of the photographic plates, the detection of objects on the plates, and the survey in which the photographic plate was captured. Of note, are the large number of attributes in the Detection, Plate, Source, and Survey relations. The large number of attributes are required to capture the parameters used when making the measurements, and is a distinguishing characteristic of scientific data sets not found in the existing benchmark datasets.

The schema of the PSSA also contains two views, ReliableGalaxies and ReliableStars. Both views select a subset of data from the Source relation by applying selection criteria that define ranges for certain attributes which when satisfied mean that the object can be reliably identified as a galaxy or a star respectively.

3.3 Example Scientific Queries

The design of the schema of the SSA was guided by 20 scientifically significant queries. The queries involve complex selection criteria for describing the required data as well as joins, including some that involve an external data source (the

```
Provide the positions and magnitudes of stars for which the magnitudes from
the two R band surveys differ by more than 3 magnitudes.

SELECT TOP 30 ra, dec, sCorMagB, sCorMagR1, sCorMagR2, sCorMagI
FROM   Source
WHERE  ellipR1 < 0.33 AND qualR1 < 2048 AND
       (prfstatR1 > -5 AND prfstatR1 < +5) AND
       ellipR2 < 0.33 AND qualR2 < 2048 AND
       (prfstatR2 > -5 AND prfstatR2 < +5) AND
       ABS (sCorMagR1-sCorMagR2) > 3 AND
       scorMagR1 > 0.0 AND sCorMagR2 > 0.0
ORDER BY ra,dec
```

Fig. 1. Query 2 from the 20 example SSA queries expressed as an English statement and as an SQL query over the SSA schema

Table 2. Features used in the 20 example queries

Feature	Contained in Query
Absolute value	2, 12, 15, 17, 19
Arithmetic in body	1, 2, 3, 4, 5, 7, 9, 12, 13, 15, 16, 17, 18, 19, 20
Arithmetic in head	7, 8, 9, 12, 13
Between	3, 5, 8, 12, 13, 15, 18, 19, 20
Count	7, 8, 9, 18
External data join	16, 17, 19
Group By	7, 8, 9
Join	12, 13, 15, 16, 17, 19
Log	4, 16
Negated sub-query	18, 20

Feature	Contained in Query
Ordering	1, 2, 3, 4, 5, 6, 7, 8, 10, 11, 12, 13, 14, 15, 16, 17, 19, 20
Power	4, 9, 16
Rounding	7
Self-join	14, 15
Server function	10, 11
Square root	4, 16
Trigonometry	8, 12
Type casting	7, 8, 12
Views	1, 3, 4, 5, 7, 8, 9, 10, 12, 15, 16, 17, 18, 19

Sloan Digital Sky Survey [1]). As an example of the types of constructs used in the queries, a selection query is shown in Fig. 1.

An analysis of the SQL features used in the 20 example queries is given in Table 2. The analysis shows a significant number of the queries involve arithmetic, mathematical, and trigonometric functions. This is likely to be, in part, a result of the selected test domain, astronomy. However, similar scientific domains will have similar features in their queries, and aggregate queries are a common SQL query form across a wide variety of domains.

Note that queries 10 and 11 involve a server defined function. This function identifies nearby objects by calculating their distance on the celestial sphere. Since the server function extends the capabilities of the query engine, for simplicity we decided to omit the queries for the purpose of the performance tests.

3.4 SPARQL Capabilities

Fig. 2 provides query 2 of the 20 SSA queries as a SPARQL query. Since there is no separate schema for an RDF graph, each of the returned variables must be explicitly bound in the query. Also, some of the syntactic shorthands found in SQL such as `between` and `abs` are missing from the SPARQL language, although their functionality can be provided using suitable `filter` conditions. The `filter` operations allow for basic comparisons of attributes. They can express numeric range expressions, e.g. $?attr > 35$; simple arithmetic operations, e.g. $?attr_1 * attr_2$;

```
PREFIX vocab: <http://surveys.roe.ac.uk/pssa/vocab/>
PREFIX rdf: <http://www.w3.org/1999/02/22-rdf-syntax-ns#>
PREFIX xsd: <http://www.w3.org/2001/XMLSchema#>
PREFIX rdfs: <http://www.w3.org/2000/01/rdf-schema#>

SELECT ?ra, ?dec, ?sCorMagB, ?sCorMagR1, ?sCorMagR2, ?sCorMagI
WHERE {
        ?source a vocab:source ; vocab:source_ra ?ra ; vocab:source_decl ?decl ;
                vocab:source_sCorMagB ?sCorMagB ; vocab:source_ellipR1 ?ellipR1 ;
                vocab:source_sCorMagR1 ?sCorMagR1 ; vocab:source_ellipR2 ?ellipR2 ;
                vocab:source_sCorMagR2 ?sCorMagR2 ; vocab:source_qualR1 ?qualR1 ;
                vocab:source_sCorMagI ?sCorMagI ; vocab:source_qualR2 ?qualR2 ;
                vocab:source_prfStatR1 ?prfStatR1 ; vocab:source_prfStatR2 ?prfStatR2 .
        FILTER (?ellipR1 < 0.33)
        FILTER (?qualR1 < 2048)
        FILTER (?prfStatR1 > -5 && ?prfStatR1 < 5)
        FILTER (?ellipR2 < 0.33)
        FILTER (?qualR2 < 2048)
        FILTER (?prfStatR2 > -5 && ?prfStatR2 < 5)
        FILTER (?sCorMagR1 > 0.0 && ?sCorMagR2 > 0.0 &&
                ((?sCorMagR1 - ?sCorMagR2 > 3) || (?sCorMagR1 - ?sCorMagR2 < -3)))
```

Fig. 2. Query 2 (see Fig. 1) from the 20 example SSA queries expressed as a SPARQL query over a suitable RDF version of the SSA

and string matching can be performed, e.g. regex(?*attr*, " $\wedge$ *Star*"). However, filter operations do not have the capabilities to perform mathematical functions such as logarithms, nor are they able to compute trigonometric functions.

There are also no group by or aggregate operators, which would also be an issue for a larger number of domain applications, e.g. for expressing business logic [16]. While it is possible to extend the functionality of SPARQL by using external functions to compute operators like aggregates and mathematical functions, this would have an impact on the performance of the query engines since the data must be transferred to the external function. When considering the size of the data sets found in scientific domains—several terabytes—this is not a feasible option due to the network delays of transporting the data.

SPARQL does not meet the requirements for a scientific query language. Of the 18 queries, which do not require a server function, only 9 are expressible in SPARQL: queries 1, 2, 3, 5, 6, 14, 15, 17, and 19. This is due to the lack of aggregate functions and to the large number of mathematical and trigonometric functions required in scientific queries.

4 Performance Test

The experimental hypothesis considered is

> RDF can be used to integrate data from existing scientific archives.

To test the hypothesis, we conducted experiments to investigate the feasibility of exposing an existing science archive as RDF. Specifically, we investigated whether

1. Large quantities of scientific data can be stored and queried as RDF;
2. SPARQL queries posed over an RDF model can be converted into queries over a currently deployed relational database.

To investigate these issues, we used the PSSA and the example queries used in its design.

Note that in Section 3.4 we identified that 9 of the example queries were expressible in SPARQL. For the experiments we were only able to use five of these queries; 1, 2, 3, 5, and 6. Queries 14 and 15 were not used as their execution did not complete with the RDB2RDF tools; this was most likely due to the self-join in these queries. Queries 17 and 19 were not used as they rely on additional data stored in the Sloan Digital Sky Survey [1].

4.1 Compared Systems

We compared the performance of five query engines: a relational database management system to provide a benchmark performance, two RDB2RDF systems, and two RDF triple stores which used a precomputed RDF version of the PSSA data set[8]. Details of the systems, the specific version used and any configuration required are outlined below.

MySQL v5.1.25 is an open source relational database management system. As a base case for the experiment we used the time to execute the SQL versions of the queries over the relational PSSA. The MySQL database was used in its distribution state, i.e. no changes were made to the configuration. The PSSA data was loaded using the downloadable scripts. In particular, note that no indexes were created other than the automatically created primary key indexes. The database was accessed using the MySQL JDBC connector v3.1.12.

D2RQ v0.5.2 [2] is a tool for mapping existing relational data to an RDF model. It provides mechanisms for automatically exposing a relational database as an RDF view, and a variety of mechanisms for accessing the relational data through the RDF model, including offering a SPARQL query endpoint. D2RQ relies on Jena v2.5.1 for providing the RDF model abstraction and access mechanisms. The PSSA database was hosted in the MySQL database used for the base case. The script for generating an RDF model of the relational data was used. The output of the script was edited so that the relations without primary keys, e.g. CrossneighboursEDR, could generate a unique URI to identify each row.

SquirrelRDF v0.1 [18] is a tool which allows an existing relational database to be queried using SPARQL. This is achieved by automatically generating a mapping file that exposes the relational schema as an RDF view. As with the D2RQ mappings, the SquirrelRDF automatically generated model needed to be edited for some of the relations so that each row of the relation could be identified with a unique URI. SquirrelRDF relies on Jena v2.4 for providing the RDF model abstraction and access mechanisms. The PSSA database was hosted in the MySQL database used for the base case.

[8] The RDF version of the PSSA was generated using D2RQ.

Jena v2.5.6 [14] provides a Java API for storing and accessing RDF data. To provide persistent storage for RDF data, Jena relies on a database system. However, there are a multitude of ways to configure Jena to work with a database system. After some initial experimentation, we opted for SDB[9] v1.1 as the mechanism to interface between Jena and the MySQL database system. Jena using SDB was configured to use the recommended settings, i.e. layout 2 with the InnoDB database engine. The RDF version of the PSSA was loaded using a Jena model and a chunk size of 200,000. Note that Jena with SDB uses a database system to store RDF triples. It did not access the relational version of the PSSA.

Sesame v2.1.3 [4] also provides a Java API for storing and accessing RDF data. Sesame provides a variety of mechanisms for storing RDF data: in a "MemoryStore" where small temporary RDF stores reside in memory data structures, in a "NativeStore" which stores data on disc using indexed data structures, or using a database system. We used the recommended NativeStore for the experiments.

Note that MySQL, Jena, and Sesame are mature software products whereas SDB is semi-mature, and D2RQ and SquirrelRDF are prototype systems.

4.2 Experiment Design

To compare the query engines of the systems, a Java application was used to measure the time taken to execute a query and provide the result set. Each run took two configuration parameters, the system to be tested and the query to be executed, and consisted of three parts. The *setup* phase installed and configured the desired test system, loaded the appropriate format of the PSSA, ran a warm-up query, and created an executable version of the test query. The *experiment* phase measured the time taken to execute the query by getting the current time in milliseconds before and after the query execution step in the Java application. Finally, the *checking* phase ensured that the result set was of the expected size and computed the difference between the start and end timestamps.

4.3 Experimental Environment

The performance measures were conducted on a cluster consisting of 8 quad core Intel Xeon 2.4GHz 64bit nodes. Each node has 4GB of RAM and 100GB of disc space. Java version 1.6.0_06 for the 64 bit architecture was used. Each repetition of the experiment was executed on a node in isolation. A cluster was used to allow separate repetitions to take place in parallel. This was essential due to the time taken to load in the data set, see Section 4.5 for more details.

4.4 Results

Table 3 presents the performance results for the five example SSA queries, expressible and executable as SPARQL queries, against each of the systems tested. The results presented are an average of ten runs for each query, together with their standard deviation.

[9] `http://jena.hpl.hp.com/wiki/SDB` accessed 19 September 2008.

Table 3. Average query execution times over ten runs for the five SSA queries

	MySQL (ms)		D2RQ (ms)		SquirrelRDF (ms)		Jena (ms)		Sesame (ms)	
	Mean	SD	Mean	SD	Mean	SD	Mean	SD	Mean	SD
Query 1	34	0.5	352	17.2	613	13.5	3,450	775.5	39	24.2
Query 2	38	1.0	5,339	36.5	21,492	259.7	485,932	169,800.6	83	12.0
Query 3	33	1.2	2,733	34.4	837	11.2	7,229	2,549.8	69	26.4
Query 5	34	2.5	4,090	43.0	1,307	10.0	17,793	8,849.5	65	13.4
Query 6	1	0.4	7,468	224.5	19,984	87.8	372,561	60,204.8	56	32.7

The MySQL database accessed through JDBC markedly outperformed all of the other systems on all of the queries, except for Sesame's performance on query 1, where the MySQL system still performed better. Sesame markedly outperformed all of the other SPARQL query engines. D2RQ outperformed SquirrelRDF on queries 1, 2, and 6.

All of the systems performed well with query 1, and for the four SPARQL query engines this was their best performing query. This is due to query 1 containing one selection condition over the view ReliableGalaxies (see below for more details about the effects of the views), although the condition does involve a multiplication which is why it is not the best performing query for MySQL.

With the exception of D2RQ, all the systems had their worst performance with query 2, and it was still one of the worst queries for D2RQ. This is probably a result of the number of variables returned, a total of 6, and the fact that the entire Source relation must be considered. There is no clear reason why D2RQ should perform worst with query 6 instead.

Query 6 is the simplest of the queries: it does not involving any arithmetic operations in either the head or the body of the query, nor any views. This explains the dramatic difference in speed in the MySQL execution of the query when compared to the other queries.

The SPARQL query engines perform well on the queries involving views: queries 1, 3, and 5. This is because both the RDB2RDF systems and the RDF triple stores benefit from the relational views[10]. The RDB2RDF systems benefit because they can treat the view as if it is a relation in the schema, and thus the underlying MySQL database system computes the view conditions for them. The triple stores benefit because the view has been stored as triples in the RDF data and can be directly referenced in the SPARQL queries, rather than being computed within the SPARQL query execution. This means that the queries involving the views do not need to consider so much data. The ReliableGalaxies view of the Source relation contains 1,242 rows while the ReliableStars view of the Source relation contains 5,123. Indeed, the size of these views accounts for the performance difference between queries 1 and 3, and query 5 which involve ReliableGalaxies and ReliableStars respectively, when compared to queries 2 and 6, which involve the Source relation containing 90,947 rows.

To investigate the effects of the relational views, we ran the experiments for queries 1, 3, and 5, over D2RQ and Sesame without using the relational views in the SPARQL queries, i.e. we converted the view definitions into the SPARQL

[10] The views were computed and represented as triples in the RDF version of the PSSA.

Table 4. Average query execution times over ten runs when the views are not used

	D2RQ (ms)		Sesame (ms)	
	Mean	SD	Mean	SD
Query 1	39,374	1,651.2	88	23.9
Query 3	40,021	1,803.1	122	73.7
Query 5	153,392	5,739.7	128	55.1

query. The results are displayed in Table 4. As expected, both systems took longer to execute the queries: a doubling in the execution time for the Sesame query engine, and at least 20 times slower for the D2RQ query engine. The scale of increase for the D2RQ query execution, demonstrates (i) the effects of query optimisation that the relational database management system is able to perform, and (ii) the translated query provided by D2RQ cannot be optimised by the DBMS (see Section 4.5 for more details).

4.5 Discussion

The experiments were conducted to investigate two issues relating to whether RDF makes a suitable data model for integrating scientific data. We will now discuss how well these objectives were achieved and the limitations of our study.

RDF as a Data Model for Scientific Data. The query performance of the RDF triple stores, when no precomputed version of the views was involved, were at least twice as slow as the performance of the relational database system. This is largely due to query optimisation research for SPARQL still being in its infancy.

A second issue is the time taken to load the data. Although this was not formally measured in this study, the MySQL database typically loaded in 8-10 seconds while Sesame took 2-3 hours and Jena was in the region of 30-35 hours. The difference in bulk loading mechanisms is very striking from a user perspective. For MySQL, batch files containing tabbed separated data were used. For Jena, a "chunk" size needed to be set to alter the number of triples read in for each stage of processing. This was a memory management issue, although varying the chunk size appeared to have little effect since most of the time was spent indexing the data. The Sesame system just read in the file of triples and managed itself. Again, indexing the data took most of the time. However, the loading time is a one-off penalty for which some delay is acceptable.

When dealing with domains containing large quantities of structured data, such as scientific applications, the RDF triple stores are currently unable to compete with the performance of relational database systems. This is because currently RDF triple stores are unable to exploit the structure of the data.

Mapping SPARQL queries to Relational Data. Both of the RDB2RDF systems performed poorly in the experiments, with query times being at least ten times slower than the database system, and at least four times slower than Sesame. However, the long execution times of the RDB2RDF systems is down to the way in which queries are translated—as shown by the log files—which prevents exploitation of the database system's capabilities to process large amounts of structured data and to optimise its query processing.

In both systems, a SPARQL query is translated into multiple queries to the database in order to retrieve the rows of the required relation. Additionally, the generated SQL queries involve a join for each bound variable in the SPARQL query. This often results in a large number of self-joins in the generated query since multiple attributes from the same relation need to be retrieved. This affects the performance of the underlying database system since for each self-join it must replicate the relation and the query optimiser cannot recognise that the query involves attributes from the same row in the relation. Once the RDB2RDF systems had retrieved all of the rows from the relations involved in the query, they then proceeded to perform comparisons to check the query conditions, no comparisons were "pushed-down" to the SQL query.

Limitations of the Study. The experiments focused on one scientific data archive. This meant that the data was structured, and was biased towards numeric data. This played to the advantages of the relational data model, but it is typical of the data required in certain application areas.

Only select-project queries were considered in the experiments, they did not involve any joins, although they did involve foreign key references. If queries 14 or 15 were to be executed, then a suitable index would be required to perform the self-join in a reasonable time period. This would benefit both the relational database and the RDB2RDF systems.

Both of the RDB2RDF tools considered relied upon Jena for providing their RDF modelling capabilities. The experiments showed that Jena did not perform well, and there are similar patterns in the execution times in these systems to the Jena execution times: the systems performed well on queries 1 and 3, moderately on query 5, and perform badly on query 2 and 6. However, this is most likely due to the use of views in the queries since Jena is not used for querying and storing data in the RDB2RDF tools.

There is a another class of RDB2RDF systems that offer a SPARQL query endpoint, but these have not been considered in this study. These are existing relational database management systems that have been extended to accept SPARQL queries, e.g. OpenLink's Virtuoso system [5]. These database system expose an RDF model of their relational data, and accept SPARQL queries over that model. The SPARQL queries are processed in the same way as a SQL query, i.e. they are parsed directly into a set of executable operations over the internal data structures of the database system. Thus, they can exploit the query optimiser of the database system. However, these systems were not considered in this study since existing legacy science archives are hosted on deployed DBMS without support for RDF and would require external tools to support SPARQL queries.

Summary. Current RDB2RDF systems are not capable of providing the query execution performance required to implement a scientific data integration system based on the RDF model. However, the considerable difference in the query execution times of D2RQ when the relational views were used, compared to when the relational views were not used, highlights the improvements that the underlying database system can provide when it is able to optimise the query execution for the underlying views. It is likely that with more work on query translation, suitable mechanisms for translating queries could be developed. These mechanisms should focus on exploiting the underlying database system's capabilities to optimise queries and process large quantities of structure data, e.g. pushing the selection conditions to the underlying database system.

5 Related Work

Several benchmark datasets for testing the performance of RDF triple stores have been developed and there is a community effort to catalogue these[11]. Generally, the benchmarks have considered models which contain far fewer attributes per relation than the SSA schema, and have only tried to simulate the real world. The focus of these benchmarks has been on the size of the data and query throughput rather than the complexity of the queries. Below we detail those benchmarks of most relevance to the current study.

The SP2B SPARQL performance benchmark [17] and the Berlin SPARQL Benchmark (BSBM) [3] both aim to test the SPARQL query engines of RDF triple stores. The SP2B uses bibliographic data from DBLP[12] as its test data set, while the BSBM benchmark considers eCommerce as its subject area. Both benchmarks allow for the creation of arbitrary sized data sets, although the number of attributes for any given class is lower than the numbers found in the SSA. Additionally, the queries used in the benchmarks are more simplistic than those found in the scientific domain, with few or only simple selection conditions.

The THALIA testbed is specifically aimed at benchmarking RDB2RDF tools [11]. The data set represents the university course catalogues of 45 computer science departments around the world. This limits the overall size of the data. While a variety of challenging queries have been considered, e.g. exploiting structural heterogeneity in the data, the complexity of the selection conditions are more limited than the example queries for the SSA, with a filter containing at most one condition and a maximum of two filter conditions in their most complex query.

Svihla and Jelinek [19] ran experiments to investigate the performance of their METAmorphoses processor in comparison to RDF triple stores, and RDB2RDF tools. Their experiments were conducted over a copy of the DBLP bibliographic database of comparable size to the PSSA. The experiments considered carefully constructed queries with increasing complexity. However, the most complex query considered consisted of a graph pattern with 4 attributes and one filter expression. This is far simpler than the queries found in the scientific queries considered for our experiments. The METAmorphoses processor was not included in our experiments because it does not provide a SPARQL query point.

[11] `http://esw.w3.org/topic/RdfStoreBenchmarking` accessed 10 December 2008.
[12] `http://www.informatik.uni-trier.de/~ley/db/` accessed 10 December 2008.

6 Conclusions

This paper has shown

> It is *currently unfeasible* to use RDB2RDF tools to expose large science archives for data integration.

This is due to two main issues which have been demonstrated in this paper.

The first problem highlighted in this study is the inability to express scientific queries in SPARQL—only 9 of the 18 standard queries could be expressed. Currently SPARQL only has support for basic arithmetic operators. However, in scientific domains such as astronomy there is a requirement for more advanced mathematical operations such as trigonometric functions. It is likely that as SPARQL matures and is used in more domains such functions will be added as standard features.

The second problem investigated in this study was the performance of current RDB2RDF tools—which enable existing relational databases to be exposed as RDF. The experiments showed that RDB2RDF tools were markedly slower than accessing the data using either RDF triple stores or the underlying relational model. A large amount of the poor performance of the RDB2RDF systems can be accounted for by (i) the translated queries not exploiting the strengths of the relational model or the query optimiser, and (ii) the selection conditions being computed in the RDB2RDF systems rather than being pushed down to the database.

This study has shown that more research and development on appropriate query translation and optimisation techniques for RDB2RDF is required. With such work, it should become possible to use RDB2RDF tools to expose existing legacy databases as RDF models which can then be used within a data integration framework.

Acknowledgements

This research has been funded by the EPSRC grant number EP/E01142X/1.

This research has made use of data obtained from the SuperCOSMOS Science Archive, prepared and hosted by the Wide Field Astronomy Unit, Institute for Astronomy, University of Edinburgh, which is funded by the UK Particle Physics and Astronomy Research Council.

References

1. Adelman-McCarthy, J.K., et al.: The fourth data release of the sloan digital sky survey. ApJSup 162(1), 38–48 (2006)
2. Bizer, C., Cyganiak, R.: D2RQ – lessons learned. In: W3C Workshop on RDF Access to Relational Databases, Cambridge, MA, USA (October 2007)
3. Bizer, C., Schultz, A.: Berlin SPARQL benchmark. Technical report, Free University of Berlin (September 17, 2008)

4. Broekstra, J., Kampman, A., van Harmelen, F.: Sesame: A generic architecture for storing and querying RDF and RDF schema. In: Horrocks, I., Hendler, J. (eds.) ISWC 2002. LNCS, vol. 2342, pp. 54–68. Springer, Heidelberg (2002)
5. Erling, O., Mikhailov, I.: RDF support in the Virtuoso DBMS. In: CSSW 2007, Leipzig, Germany, pp. 59–68 (September 2007)
6. Euzenat, J., Shvaiko, P.: Ontology matching, 1st edn. Springer, Heidelberg (2007)
7. Galanis, L., Wang, Y., Jeffery, S.R., Dewitt, D.J.: Locating data sources in large distributed systems. In: VLDB 2003, pp. 874–885, Berlin, Germany (September 2003)
8. Gray, J., Szalay, A.S., Thakar, A., et al.: Data mining the SDSS skyserver database. The Computing Research Repository (CoRR) (February 2002)
9. Halevy, A., Rajaraman, A., Ordille, J.: Data integration: The teenage years. In: VLDB 32, Seoul, Korea, pp. 9–16 (September 2006)
10. Hambly, N., Read, M., Mann, B., et al.: The SuperCOSMOS science archive. In: ADASS XIII, San Francisco,CA, USA, pp. 137–140 (2003)
11. Hammer, J., Stonebraker, M., Topsakal, O.: THALIA: Test harness for the assessment of legacy information integration approaches. In: ICDE 2005, Tokyo, Japan, pp. 485–486 (April 2005)
12. Louys, M., Richards, A., Bonnarel, F., et al.: Data model for astronomical dataset characterisation. Recommendation, IVOA (November 8, 2007)
13. Malhotra, A.: Progress report from the RDB2RDF XG. In: Poster and Demo Session, ISWC 2008, Karlsruhe, Germany (October 2008)
14. McBride, B.: Jena: A semantic web toolkit. IEEE Internet Computing 6(6), 55–59 (2002)
15. Plante, R., Benson, K., Graham, M., et al.: VOResource: An XML encoding schema for resource metadata. Recommendation, IVOA (February 22, 2008)
16. Pöss, M., Floyd, C.: New TPC benchmarks for decision support and web commerce. SIGMOD Record 29(4), 64–71 (2000)
17. Schmidt, M., Hornung, T., Lausen, G., Pinkel, C.: SP2Bench: A SPARQL performance benchmark. Technical report, Albert-Ludwigs-Universität Freiburg (April 28, 2008)
18. Seaborne, A., Steer, D., Williams, S.: SQL-RDF. In: W3C Workshop on RDF Access to Relational Databases, Cambridge, MA, USA (October 2007)
19. Svihla, M., Jelinek, I.: Benchmarking RDF production tools. In: Wagner, R., Revell, N., Pernul, G. (eds.) DEXA 2007. LNCS, vol. 4653, pp. 700–709. Springer, Heidelberg (2007)

A Flexible API and Editor for SKOS

Simon Jupp, Sean Bechhofer, and Robert Stevens

School of Computer Science, University of Manchester, UK
first.last@manchester.ac.uk

Abstract. We present a programmatic interface (SKOS API) and editor for working with the Simple Knowledge Organisation System SKOS. The SKOS API has been designed to work with SKOS models at a high level of abstraction to aid developers of applications that use SKOS. We describe a SKOS editor (SKOSEd) that is built on the Protege 4 framework using the OWL and SKOS API. As well as exploring the benets of the principled extensibility afforded by this approach, we also explore the limitations placed upon SKOS by restricting SKOSEd to OWL-DL.

1 Introduction

We present an editor for The Simple Knowledge Organisation System (SKOS) called SKOSEd[1] and an associated SKOS API[2]. SKOSEd uses the SKOS ontology as a schema to drive the API, editor and the development of SKOS vocabularies. In doing so, SKOSEd takes advantage of existing OWL 2 DL infra-structure and gains the benefits of a *principled* means of extending SKOS vocabularies through committing to an OWL 2 DL version of SKOS.

SKOS is a developing specification for representing and publishing classification schemes, thesauri, taxonomies and subject heading systems on the web. It is currently under development as part of the W3C Semantic Web Deployment Group (SWDWG)[3] and Editor's Working Drafts of the SKOS Reference[4] and SKOS Primer[5] are available. In order to facilitate adoption and use of SKOS there is now a need for tools that can support its use in the form of APIs, editors, browsers and validators.

There has been a long tradition in the library and information sciences in using thesaurus-like artefacts to index and manage large volumes of data, such as inventories or cataloging systems [10], [12], [9]. For the purposes of this paper we will refer to these artefacts as Knowledge Organisation Systems (KOS). Applications of KOS include indexing; search and information retrieval; text processing (e.g. text mining); and document collection navigation.

[1] http://code.google.com/p/skoseditor/
[2] https://sourceforge.net/projects/skosapi/
[3] http://www.w3.org/2006/07/SWD/
[4] http://www.w3.org/TR/skos-reference/
[5] http://www.w3.org/TR/skos-primer

L. Aroyo et al. (Eds.): ESWC 2009, LNCS 5554, pp. 506–520, 2009.

These KOS can span multiple domains and vary considerably, as they are often designed for a particular application or institution. Despite this, they often share many features in common, especially in their content type and structure. As we move towards a Semantic Web, providing structure to content on the web, a lot can be gained from exploiting existing KOS artefacts and technology. Technologies such as RDF and OWL [6] are seen as key elements for building a Semantic Web. The SKOS model is built in accordance with these technologies and has a serialisation to the Resource Description Framework (RDF). In general, KOS differ significantly from formal ontologies, as represented using OWL, as they do not contain detailed intensional descriptions of concepts [3], and often contain relationship between concepts where the semantics are weakly defined or ambiguous [4].

The SKOS model can be used to structure and represent any knowledge artefact that contains statements about concepts and the relationships between them. The shared features of these KOS are primarily in the form of a lexical resource along with some semantic relationships between each resource. The semantic relationships between resources are typified by synonym, hypernym, hyponym, antonym, broader, narrower, and related.

SKOS provides a data model that can be used to express these kinds of relationships between resources and is designed to be extensible and modular. Central to SKOS is the core vocabulary deemed sufficient to represent most of the common features found in concept schemes. A *concept* can be considered any unit of cognitive thought. *Lexical labels* allow the association of lexical forms (preferred labels, alternative lables and so on) with each concept. *Semantic relations* capture relationships between concepts including hierarchical broader-narrower relationships and general associative relationships.

A goal of SKOS is to provide flexibility—demands and expectations for what a KOS should provide vary across communities and different aspects of its requirements will be reflected in how different tools support its use. There are a community of users who wish to use SKOS to represent a complete KOS, which could then in turn be used as a reference system for providing annotations. SKOS also provides vocabulary for metadata on the web—for example, its use as a meta-data vocabulary for annotating OWL ontologies is one area where SKOS is seen as of potential use [8].

Standardised representations are key to interoperation, but additional support is needed in order to build applications that make use of these languages. In particular, language APIs, frameworks or toolkits facilitate the construction of applications and services, allowing developers to work at a higher level of abstraction, and providing some insulation from issues relating to serialization or parsing of data structures. For example, frameworks like Jena[6] or the OWL API[7] [2] provide code libraries for working with RDF[8] or OWL[9].

[6] `http://jena.sourceforge.net`
[7] `http://owlapi.sourceforge.net`
[8] `http://www.w3.org/RDF`
[9] `http://www.w3.org/2004/OWL/`

SKOS is defined as an RDF vocabulary, and vanilla RDF infrastructure could be used to support SKOS in applications. As with earlier work on OWL [2], we, however, argue that appropriate abstractions facilitate application development. In this paper we describe work towards such abstraction in an API for SKOS.

The flexibility of SKOS is, in part, given through its specification and description in OWL Full (See section 2) . This then allows further specialisation of the vocabulary in a *principled* manner—for example through the introduction of sub-properties of the SKOS semantic relations. Through the use of existing OWL 2 DL infrastructure, we are able to provide a flexible collection of classes for the manipulation of SKOS models. As an illustration of the use of the API, we present a prototype SKOS editor – SKOSEd – and discuss plans for a larger suite of tools providing SKOS validation and support for conversion between SKOS and OWL. Our approach brings a number of benefits:

- Support for extension of the basic SKOS vocabulary through the definition of subproperties;
- Basic reasoning about inverse relationships;
- Characteristics of properties can be asserted (for example functionality or transitivity), with corresponding effects on the results of queries;
- We have access to additional expressivity. For example the use of property chains in order to capture additional semantics, such as the interaction between locative and partitive properties.

In all the above cases, the use of the underlying OWL 2 DL infrastructure provides easy access to reasoning services that support the use and extension of the vocabulary.

2 SKOS and OWL

Standardised representations such as RDF and OWL are key to the Semantic Web vision. RDF provides a common data model and syntax for metadata representation on the web. RDFS and OWL provide schema languages for modeling data on the web. OWL can be used for modeling domain knowledge and has a formal semantics for building expressive and powerful knowledge artefacts. The style of knowledge artefact found in most KOS often lack a precise formal semantics that can make direct conversion into an OWL TBox inappropriate [13]. KOS are really just facts asserted about things (concepts) – an OWL ABox (a collection of axioms about individuals) is thus more appropriate. We can, however, use an OWL TBox as a meta-vocabulary and a data model for KOS and use OWL inference to aid construction, maintenance and use.

A combination of RDF and OWL are used to define the SKOS data model. The data model is itself an OWL ontology, any thesaurus represented in SKOS is then an instantiation of this ontology. The SKOS ontology provides a minimal vocabulary for representing a KOS, this vocabulary can be extended with specialisations of the SKOS constructs. A useful type of extension to SKOS may be extending the basic *Semantic Relationships* – `broader`, `narrower`, and `related`

– that are asserted between Concepts. The broader and narrower relationships provide some hierarchical structure to a KOS, however, it can be desirable to express different kinds of broader/narrower relationships, for example, extensions to broader such as broaderGeneric and broaderPartitive, to specify a class subsumption and partitive relationship between Concepts respectively. The ability to extend SKOS gives KOS authors the flexibility to more accurately represent their own knowledge artefact, whilst keeping it defined within the SKOS model for interoperation with other SKOS aware applications on the web. The ability to extend the SKOS data model in terms of OWL is a key motivator for using existing OWL infrastructure as the basis for the SKOS API and SKOSEd.

The SKOS data model is described using OWL Full, the most expressive OWL species. OWL Full has few constraints on the use of the vocabulary, providing a useful and flexible language for describing data models. The existing SKOS data model is designed with flexibility in mind, and uses elements of OWL Full where necessary. While this approach has benefits for SKOS as an expressive language for describing a KOS, it does have implications for applications and tools that wish to support SKOS. This is mainly down to the fact that SKOS is inherently coupled to the RDF syntax, it does not yet have a complete data model that is abstract from this syntax. Any dedicated SKOS editor would need to provide support for not only SKOS, but also the OWL Full and RDF infrastructure. Current support for OWL Full editors is limited and the interfaces to these application are not necessarily familiar to users of conventional KOS software. By defining a restricted, yet valid, SKOS model that is abstract from the RDF syntax we are able to provide support in the form of an API and editor for SKOS that is more akin to standard KOS editing software.

We have taken inspiration from the ontology community where the use of a restricted sub language of OWL Full, called OWL 2 DL, has been used to define an abstract data model for OWL ontologies that also benefits from being based on a Description Logic (DL). The OWL 2 DL sublanguage describes a model that can be abstracted from any concrete syntax, such as RDF, and has been exploited in tools for working with ontologies, such as the OWL API[2,5] and Protégé 4[10]. The majority of constructs defined in the SKOS model can be expressed using OWL 2 DL, the work presented here aims to exploit tools currently available for working with OWL 2 DL ontologies, and build a similar infrastructure for working with SKOS, or even combinations of both OWL and SKOS.

One of the main defining features of the OWL 2 DL sub language is the ability to use DL reasoners to compute entailments and check consistency [11]. Although the expressivity of SKOS is only a small fraction of that which is capable of use by current reasoners, we can take advantage of tools that exists to infer any entailments that can be made using the SKOS semantics. The kinds of entailment that can be computed within SKOS are relatively simple in comparison to what is capable with OWL 2 DL. They are, however, useful and should still be

[10] http://protege.stanford.edu

Asserted facts

```
a rdf:type skos:Concept
b rdf:type skos:Concept
ex:has-part rdf:type owl:ObjectProperty
ex:part-of rdf:type owl:ObjectProperty
ex:has-part rdfs:subPropertyOf skos:narrower
ex:has-part owl:inverseOf ex:part-of
a ex:has-part b
```

Inferred facts

```
ex:part-of rdfs:subPropertyOf skos:broader
b ex:part-of a
```

Fig. 1. Example of inferences that follow from the SKOS semantics

supported by the API and SKOSEd. Consider the example shown in figure 1, `Concept a` is asserted to be related to `Concept b` via a `has-part` relationship. The `part-of` relationship is a sub-property of the SKOS `narrower` property. We can use the fact that SKOS `broader` and `narrower` properties are defined as `inverse` properties in SKOS, to query for all the concepts that are related to `Concept b` via `part-of`. Although this is not asserted in our vocabulary, we can use the semantics of SKOS to infer that `Concept b` is related to `Concept a` via `part-of`.

There is great potential for combining OWL and SKOS in bridging the gap between domain modeling in OWL and language models built in SKOS. Building a knowledge base that contains an expressive OWL ontology combined with a model of the language in SKOS is potentially powerful and accommodates various axes for query and navigation of the information space, keeping SKOS in OWL 2 DL is likely to make this more feasible.

3 API Design

Many KOS authors interested in publishing data on the web may prefer not to be exposed to the underlying technical details such as RDF and OWL. We suggest that a SKOS data model that provides an appropriate abstraction from the technical issues of concrete representations (such as RDF) is of benefit. This approach proved successful in designing an API for OWL [2] and subsequent tools such as an OWL Validator [1] and editor, Protégé [11]. Such tools are of benefit to the growing community of OWL users, and it is hoped that building a suite of tools for working with SKOS will have similar benefits.

The first step in designing the API and building a subsequent editor on top of that API is to define an abstract model for SKOS.

The key issues addressed in this paper are about providing access to the underlying data model to support extensions, and to have built in support for

[11] `http://protege.stanford.edu/`

interpreting the SKOS semantics, and how this requirement can be achieved using OWL. Many tools exist for working with RDF, including APIs[12] and graphical editors[13], however, these often operate at too low a level, exposing the underlying RDF structures which may be unfamiliar or unintuitive to KOS developers. Tools exist for building KOS, yet few have direct support for SKOS or provide any programmatic interface to the model. With these considerations in mind, our approach was to exploit current OWL tools and build an API and editor on top of these.

Protégé is the most widely used open source ontology editor available to the OWL community. The latest version of Protégé, Protégé 4, is built on top of the OWL API [5]. The OWL API is a new implementation of the original Wonder Web API [2] and has full support for the emerging OWL 2 DL standard[14]. The API provides interfaces to an abstract representation of the OWL constructs for working with OWL ontologies. It comes with an in-memory implementation and supports a wide range of parsing and rendering services in addition to access to DL reasoners. Given our commitment to providing access to the underlying OWL model from within the API, and our desire to use some kind of inference engine to handle SKOS entailments, we chose to build an extension of the OWL API to support SKOS. In addition to the advantages the OWL API provides in its own right, its relationship to the Protégé interface provides us with an existing platform upon which to build a SKOS editor.

The SKOS schema is OWL Full – primarily due to the existence of subproperty axioms concerning annotation properties[15]. As a result of this, our API and editor will not support arbitrary SKOS/RDF vocabularies. All vocabularies produced using the tool will be valid SKOS, but there may be cases where arbitrary vocabularies produced outside of the tools may not be processable. We are taking a pragmatic approach here, and hope that the advantages we gain through the reuse of the OWL API and Protégé framework outweigh requirements to support arbitrary SKOS/RDF vocabularies.

It is likely that the OWL 2 DL specifications will address many of the issues that result in SKOS being OWL Full (for example through the support for punning and rich annotations). As the OWL API is committed to supporting OWL 2 DL, we believe that this will then enable us to support and handle the vast majority of SKOS vocabularies.

Our choice to use the OWL API as the basis for the SKOS API is driven by our requirements to have access to the OWL data model that will support the ability to move between pure SKOS and a combination of SKOS and OWL. The SKOS API provides a set of interfaces for working with SKOS along with an implementation that uses the OWL API. Separation of the interface from

[12] `http://jena.sourceforge.net/`

[13] `http://www.altova.com/products/semanticworks/semantic_web_rdf_owl_editor.html`

[14] `http://www.w3.org/2007/OWL`

[15] Based on the latest Editor's Draft revision of the SKOS Reference, `http://www.w3.org/2006/07/SWD/SKOS/reference/20081001/`, revision 1.45 as of 15/12/200

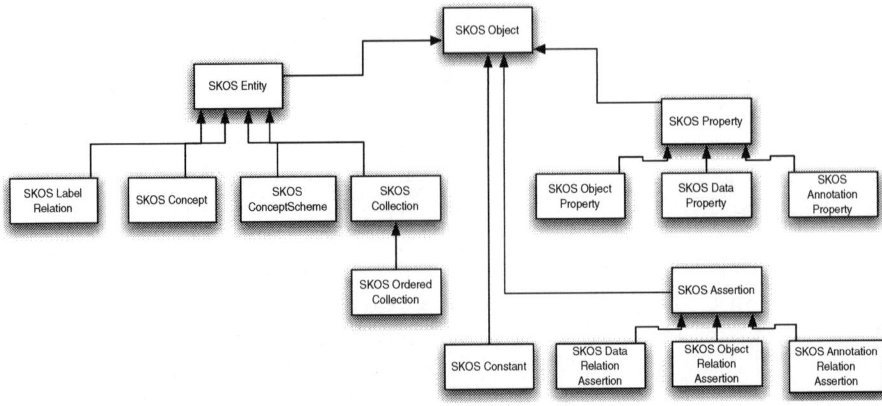

Fig. 2. Abstraction of the SKOS data model showing the base object we have in the SKOS API, these have further specialisations (see API documentation for full details)

implementation leaves the API open for others to provide their own backend implementation. Figure 2 is an object diagram showing the basic SKOS data model represented in our SKOS interface.

The interface provides a set of common objects for building and querying a SKOS model. Most of these are convenience methods for working with the types of object found in SKOS. These interfaces alone could be implemented with any of the popular APIs for working with RDF; we chose to use the OWL API and have extended only the interfaces needed to work with SKOS. The implementation is essentially a wrapper around the OWL API, along with a set of convenience methods for working with SKOS. Access to the underlying OWL API objects that are being manipulated is always available via the *SKOStoOWL-Converter* utility class, which may be desirable when exploiting the ability to extend the SKOS model. We also reuse the practice adopted by the OWL API to separate asserted and inferred models. A DL reasoner can be used to provide the inferred SKOS model for more sophisticated querying.

4 API Functionality

Much of the key functionality of the SKOS API is provided *via* the use of the OWL API.

Parsing and Rendering. We take advantage of the OWL API serialisation and parsing mechanisms which support handling of multiple concrete syntaxes for SKOS. For the most part RDF/XML will be the default and the most common.

Change control. Manipulation of a SKOS via the addition and removal of objects and assertions. This includes a mechanism for tracking changes that are made to a SKOS.

Extension to the model. The OWL API provides interfaces for manipulating the underlying OWL ontology and SKOS model, our implementation provides support for using the OWL API objects with the SKOS API.

Inference. We can exploit the power of DL reasoners to build an inferred model of a SKOS data set. The API provides support for working solely with an asserted model, or querying an inferred model.

Overall control and manipulation of SKOS is handled by a *SKOSManager* class. The *SKOSManager* is responsible for loading, creating and saving SKOS data sets. A *SKOSManager* can contain multiple SKOS data sets. The SKOS data set contains all the assertion about a particular KOS, this may include multiple concept schemes and sets of concepts that have some relationships asserted on them. We propose a set of SKOS objects in the API to represent those types of object contained in the data set.

We have strived to keep the API simple and ensure its usage is consistent. The SKOS API is structured as a hierarchical set of SKOS Objects. Four main types of objects are used in combination to build the SKOS model. These are: (i) *SKOSEntity*, (ii) *SKOSRelationship*, (iii) *SKOSAssertion* and (iv) *SKOSLiteral*. These objects are further extended to some more specific types. All objects are accessed *via* an object factory – *SKOSFactory*. Assertions are built using SKOS objects and added to data sets using the SKOS Manager.

SKOSEntity. A SKOS entity represents the major types of objects found in a SKOS, this includes `Concept`, `ConceptScheme`, `Collection` and `LabelRelation`. As the SKOS reference is finalised, more entities may be added. We also plan to provide support for SKOS extensions proposed by the SWDWG. These Classes have convenience methods for accessing the kinds of SKOS data asserted on these entities, for example, a *SKOSConcept* has methods for accessing related concepts and documentation properties such as preferred label and definitions.

SKOSProperty. The *SKOSProperty* class is extended to represent the major types of properties found in SKOS. For example, *SKOBroaderProperty*, essentially an OWL Object Property and the SKOS label properties that are OWL Data Properties. We also have the SKOS document properties that can be used as either OWL object, data or annotation properties. Despite the OWL full implication of this use of property punning, we are able to handle this within the OWL API.

SKOSLiteral. The *SKOSLiteral* class represents our RDF literals, these can be typed, such as String, or untyped, with the untyped values having support for XML language tags. These are usually used as the filler of a relationship along a data or annotation property.

SKOSAssertion. *SKOSAssertion* is used to represent relationships between SKOS Objects. These are further divided into SKOS object, SKOS data and SKOS annotation property assertion. An example of a SKOS object assertion might be to make a broader relationship between two concepts. Each SKOS change is enacted with respect to a particular SKOS data set *via* the manager. The use of explicit change objects supports the management of changes being applied to a particular dataset (as discussed in [2]). This is particularly useful in the context of editing, as it facilitates the support of undo and redo actions.

SKOSReasoner. *SKOSReasoner* is an interface to query an inferred representation of the SKOS dataset. This class can take any DL reasoner that implements the *OWLReasoner* interface from the OWL API. There are currently three publicly available reasoners that implement the *OWLReasoner* class that can be used to classify SKOS dataset; Pellet[16], Fact++[17] and HermiT[18]. Whilst all three reasoners are more than capable of processing the simple entailments that can be inferred using SKOS semantics, all three systems vary considerably in terms of memory consumption and speed.

5 SKOS API Performance

In order to demonstrate the capabilities of the SKOS API we ran some simple tests with a set of publicly available thesauri. Scalability is still a big issue with Semantic Web technologies and as thesauri have a tendency to be rather large, it is important that SKOS tools are scalable. We chose five thesauri that have been converted into a SKOS representation, we chose these based on their public availability, size and the fact that they use more than SKOS's basic expressivity. We measured the time it takes for the SKOS API to read the RDF files into memory and the amount of memory consumed. These test were performed on an Apple Mac Pro with two 2.2 GHz Dual-Core Intel processor, the Java virtual machine was limited to 2GB of addressable memory.

- GTT[19] thesaurus is used to index and retrieve books from the Dutch National Library
- AGROVOC is a multilingual structured thesaurus of all subject fields in Agriculture, Forestry, Fisheries, Food security and related domains.
- The Library of Congress Subject Headings (LCSH) comprise a thesaurus of subject headings, maintained by the United States Library of Congress, for use in bibliographic records.
- Medical Subject Headings (MeSH) is a large controlled vocabulary for the purpose of indexing journal articles and books in the life sciences.
- GEMET is a compilation of several multilingual thesauri that provides a terminology for the environment.

---

[16] http://pellet.owldl.com

[17] http://owl.man.ac.uk/factplusplus/

[18] http://hermit-reasoner.com/

[19] GTT is a Dutch abbreviation for GOO keyword thesaurus, GOO referring to the so-called Joint Subject Indexing system used by many libraries throughout the Netherlands.

Table 1. SKOS API load times and memory consumption

SKOS data set	Number of SKOS Entities	Number of triples	Time to load (seconds)	Memory consumption (MB)
GEMET	5245	350,357	5	119
MeSH	24,765	615,403	11	215
GTT	65,306	514,619	13	174
AGROVOC	28,440	848,348	24	483
LCSH	266,858	2,441,515	67	845

The SKOS API successfully loaded all our test sets (see Table 1), however, it is worth noting that the parsing and creation of the objects are mostly being handled by the OWL API implementation that sits under the SKOS API. These results show that developing tools for manipulating the largest SKOS dataset using the SKOS API would be no problem on modern desktop machines.

6 SKOSEd

SKOSEd is a plugin for the OWL ontology editor Protégé 4. It aims to provide support for working with SKOS. The decision to use the Protégé environment as a platform for SKOSEd follows on from our decisions and justifications for using the OWL API as a basis for the SKOS API. Protégé is primarily an editor for OWL ontologies and, as mentioned previously, the SKOS model is described in terms of an OWL ontology. Consequently, a standard installation of Protégé is already a capable SKOS editor, however, the interface is not necessarily suitable or intuative to KOS authors. SKOSEd provides a suite of *Views*, that expose the structure of SKOS, along with support for editing and authoring.

There are a number of existing tools for working with KOS – Willpower Information [20] lists over 40 thesaurus editors alone. Many existing applications use proprietary data exchange formats, lack a programmable interface, and have little or no support for SKOS. One open source editor that does have support for SKOS is ThManager [7]. ThManager is a stand alone application for working with SKOS, and has good support for managing thesaurus meta-data. Two of our requirements for a SKOS editor are currently not met by ThManager: One being the ability to extend the SKOS model and the second being able to exploit the RDFS and OWL semantics. By building SKOSEd on top of the Protégé environment we essentially have this functionality for free.

It is still too early in SKOS development to gather a complete set of user requirements that would satisfy all SKOS communities. The development of SKOSEd is, at least in part, an exercise in gathering these requirements. The SKOS use case and requirements documents [21] describes a wide range of requirements from which we can draw ideas about features a SKOS editor should offer.

[20] `http://www.willpowerinfo.co.uk/thessoft.htm`

[21] `http://www.w3.org/TR/skos-ucr/`

These range from simply viewing and editing the core SKOS constructs to more advanced tasks such as manipulating the underlying OWL model and handling inference. SKOSEd is a powerful SKOS editor and offers more functionality than most users new to SKOS will require. The flexible nature of the Protégé architecture should enable development of more lightweight interfaces that essentially hide the more advanced features from new users. We have defined a basic list of functional requirements for SKOSEd that we believe should be the minimum set of features required for any SKOS authoring tool:

- Ability to create, edit and delete SKOS entities such as Concepts, Concept Schemes etc...
- Assert SKOS relationships between the SKOS entities, for example `skos:inScheme`, `skos:hasTopConcept`.
- Visualise the broader/narrower hierarchy in the form of a navigable tree.
- Assert and edit labelling properties such as `skos:prefLabel`, along with support for multilingual labeling.
- Support for meta-data such as SKOS documentation properties and other standard vocabularies like Dublic Core.
- Provide alternate renderings in the editor, especially for multilingual thesauri.
- Access to the SKOS data model and support for its extention.
- Query interface that supports SKOS entailments.
- Change tracking, the ability to undo/redo actions and track those changes.
- Ability to parse and render a range of SKOS serialisations, including RDF/XML.

In order to work with SKOS in the Protégé environment we needed a version of the SKOS ontology that was restricted to the OWL 2 DL species. A cut-down OWL 2 DL version of the SKOS schema, has been defined[22]. This scheme is imported into Protégé *via* the standard Protégé import mechanisms.

Protégé 4 features a set of views, wizards and menu actions that are useful when working with ontologies. Similarly there will be a set of views and actions that are commonly needed for working with thesaurus like artefacts represented in SKOS. The SKOSEd plugin for Protégé 4 is exactly this, a suite of views and tools for working specifically with SKOS. SKOSEd is designed to use the same Protégé API and editor kit that the existing Protégé 4 interface uses, this ensures that all the SKOS views and actions integrate seamlessly with all the default Protégé tools. This flexible plugin framework offered by Protégé 4 is of additional benefit to 3rd party developers who want to extend SKOSEd or add additional tools that would interact with SKOSEd.

One of the core requirements for SKOSEd is a view that displays the hierarchical relationships between Concepts often found in KOS (See Figure 3). SKOSEd provides a Concept hierarchy view that displays the structure of the hierarchy and allows for its manipulation. For any concept selected in the hierarchy a set of other views are available that display asserted information on the concepts.

[22] `http://www.w3.org/2006/07/SWD/SKOS/reference/20081001/skos-dl.rdf`

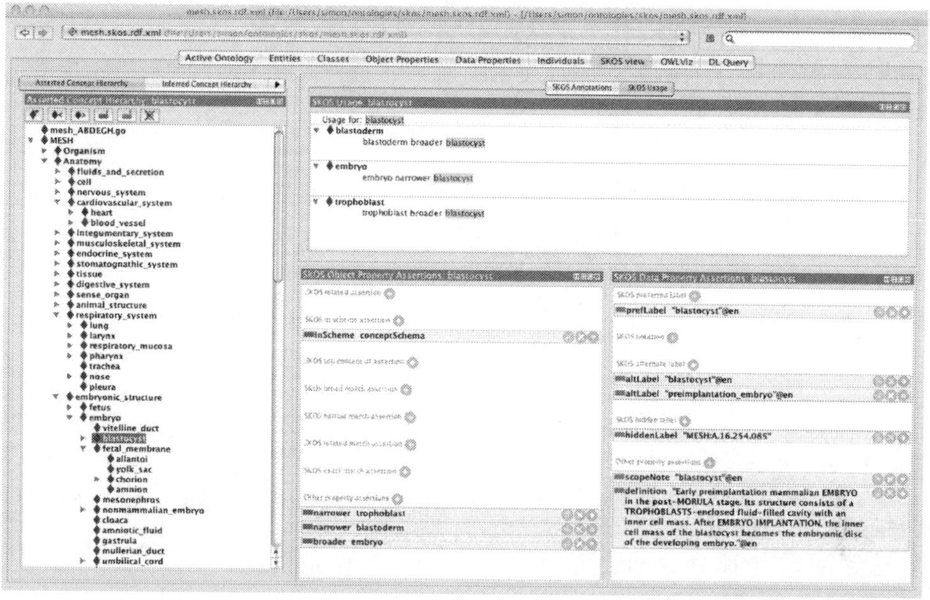

Fig. 3. A screenshot of SKOSEd plugin in Protégé, loaded with a SKOS representation of MeSH. The left hand panel shows the Concept hierarchy, the right hand panel shows asserted information about the selected Concept.

Whilst Protégé by default can display any asserted data, object or annotation properties asserted on any individual, SKOSEd does this in a more intuitive fashion. Rather than having a long list of assertions, SKOSEd provides a separate window for the different types of information asserted on SKOS concepts. Each of these windows has a set of convenience buttons for manipulating SKOS data along with context driven action depending on the type of SKOS assertion being made. This set of views include a view for the semantic relationships, such as broader, narrower and related; a view for adding labeling informations; a view for adding SKOS notes and a default panel for adding any other arbitrary type of relationship that is not part of the SKOS core schema.

In addition to views, SKOSEd provides menu items and wizards to simplify some common tasks when creating and editing SKOS. One of these is a wizard to create collections. Collections in SKOS are a meaningful grouping of Concepts. To create a collection in the default Protégé OWL interface the user would have to create a new instance on the Collection class and assert which concepts are members of this collection using the `skos:member` property. SKOSEd provides a Collection wizard that allows the user to simply select some Concepts from a list and choose the "Create Collection" action in the menu. Another useful feature is the support for multilingual KOS, Figure 4 shows a portion of AGROVOC that has been rendered along the `skos:prefLabel` property in Chinese. SKOSEd allows a user to specify an order of preference for properties and language to render.

Fig. 4. SKOSEd supports multilingual rendering. The figure shows the asserted Concept hierarchy of AGROVOC with the concepts rendered along the preferred label in Chinese.

SKOSEd currently follows the Protégé 4 practice of separating the asserted and inferred models. We have chosen to adopt the same practice with the existing SKOSEd views. For every asserted view available there is a read-only inferred view that shows inferred statements after classification. These inferred views are particularly useful when looking at SKOS datasets that have missing information. For example, we can exploit the inverse nature of the SKOS broader/narrower semantic relationships to display any Concepts that are Narrower than a selected Concept when only broader relationships have been asserted in the SKOS source file. The inferred views are also useful when extensions have been made to the underlying data model. In this way users can create specialisations of the SKOS schema that more accurately reflect their own data model. Whilst the asserted views will not be able to interpret these extensions, the inferred views will allow them to view their data with respect to the extended SKOS data model.

It will take time to gain feedback from users to ascertain how practical this notion of asserted and inferred views is to SKOSEd users. We take the view that users new to SKOS will not care much for these features initially, the functionality offered by the basic asserted views should suffice for most SKOS development. More advanced SKOS users, who have a better understanding of the underlying OWL data model, will benefit from the inferred views and will also benefit from having access to all the default OWL machinery offered by Protégé 4. The ability to edit and work on inferred models of ontologies, and now SKOS, has long been a problem for tool developers. This is due to the fact that existing reasoners must completely re-classify an ontology after a single axiom change in order to compute any inconsistencies or new entailments. In the past this has not been practical for Ontology developers, especially when the ontologies are large and expressive. Advancements in incremental reasoning are almost at a stage where asserted views in ontology editors like Protégé 4 may not be necessary, allowing users to work solely on an inferred model. A re-evaluation of the SKOSEd interface will be likely should these tools become available.

7 Discussion

In this paper, we presented an API for working with SKOS, along with a graphical editor SKOSEd, implemented using the API. We discussed the requirements for such tools and our justifications for reusing existing OWL infrastructure to provide support when working with SKOS. We highlighted two key issues that we believe are desirable when working with SKOS; 1/ the ability to extend the SKOS model via an API or editor; and 2/ the exploitation of DL reasoners to compute entailments for richer querying of data represented in SKOS. The first can benefit different communities in allowing the specialisation of SKOS to better represent their data. Additional properties that are extensions of SKOS can provide multiple axes on which to query their knowledge. Access to the underlying data model helps to support this effectively in tools. By reusing an existing API (the OWL API) and ontology editing software (Protégé 4), we were able to rapidly develop early tool support for the emerging SKOS specification.

SKOS is well placed to bridge the gap between the social tagging culture that has already gained popularity on the web, and the kind of rich metadata and structured knowledge provided by ontologies, that is needed to realise a Semantic Web. SKOSEd aims to provide the community with a scalable application for working and managing SKOS data. By adopting the Protégé framework we hope that the existing plugin will be extended with additional plugins provided by the community, to improve the way we work with not only SKOS, but combinations of both the SKOS and OWL languages. The SKOS API provides a flexible set of interfaces for working with SKOS, along with a concrete implementation using the OWL API. The API is independent from the editor and additional tools are planned, including a re-implementation of the interface to provide some persistent storage mechanism, such as a database backend, or extensions to support alternate querying interfaces, such as SPARQL. Additional plans include the use of the SKOS API as a basis for SKOS validation services. SKOS has a mixture of both semantic constraints that are represented in the schema, along with additional constraints defined in the SKOS reference documentation. For example, constraints on the use of the `skos:prefLabel` property state that every concept can only have one preferred label in a given language. This type of consistency checking can easily be done at the API level and made accessible to users of SKOSEd.

SKOSEd is available as open source[23]. We encourage the reader to download and use the application and welcome further comments or suggestions for enhancements.

Acknowledgements. We would like to acknowledge Matthew Horridge for his initial work on the SKOSEd plugin and valuable expertise on the OWL API and Protégé 4. SJ is supported by the Sealife project (IST-2006-027269).

[23] `http://code.google.com/p/skoseditor/`

References

1. Bechhofer, S., Carroll, J.J.: Parsing OWL DL: Trees or Triples? In: Proceedings of the World Wide Web Conference, WWW 2004, pp. 266–275. ACM Press, New York (2004)
2. Bechhofer, S., Volz, R., Lord, P.: Cooking the Semantic Web with the OWL API. In: Fensel, D., Sycara, K.P., Mylopoulos, J. (eds.) ISWC 2003. LNCS, vol. 2870, pp. 659–675. Springer, Heidelberg (2003)
3. Bechhofer, S., Yesilada, Y., Stevens, R., Jupp, S., Horan, B.: Using Ontologies and Vocabularies for Dynamic Linking. IEEE Internet Computing 12(3), 32–39 (in press, 2008)
4. Hepp, M., de Bruijn, J.: Gentax: A generic methodology for deriving owl and rdf-s ontologies from hierarchical classifications, thesauri, and inconsistent taxonomies. In: Franconi, E., Kifer, M., May, W. (eds.) ESWC 2007. LNCS, vol. 4519, pp. 129–144. Springer, Heidelberg (2007)
5. Horridge, M., Bechhofer, S., Noppens, O.: Igniting the OWL 1.1 Touch Paper: The OWL API. In: Proceedings of OWLEd 2007: Third International Workshop on OWL Experiences and Directions (2007)
6. Horrocks, I., Patel-Schneider, P.F., van Harmelen, F.: From $\mathcal{SHIQ}$ and RDF to OWL: The making of a web ontology language. Journal of Web Semantics 1(1), 7–26 (2003)
7. Lacasta, J., Nogueras-Iso, J., López-Pellicer, F.J., Muro-Medrano, P.R., Zarazaga-Soria, F.J.: Thmanager: An open source tool for creating and visualizing skos. Information Technology and Libraries (ITAL) 26(3), 39–51 (2007)
8. Miles, A.: SKOS: requirements for standardization. In: DCMI 2006: Proceedings of the 2006 international conference on Dublin Core and Metadata Applications, pp. 55–64. Dublin Core Metadata Initiative (2006)
9. National Library of Medicine. Medical subject headings: main headings, subheadings, and cross references used in the index medicus and the national library of medicine catalog, 1st edn. (1960)
10. Peterson, T.: Introduction to the Art and Architecture Thesaurus. Oxford University Press, Oxford (1994)
11. Sirin, E., Parsia, B., Cuenca Grau, B., Kalyanpur, A., Katz, Y.: Pellet: A practical owl-dl reasoner. J. Web Sem. 5(2), 51–53 (2007)
12. Soergel, D., Lauser, B., Liang, A., Fisseha, F., Keizer, J., Katz, S.: Reengineering thesauri for new applications: the AGROVOC example. Journal of Digital Information 4(4) (March 2004)
13. Soualmia, L.F., Goldbreich, C., Darmoni, S.J.: Representing the mesh in owl: Towards a semi-automatic migration. In: First International Workshop on Formal Biomedical Knowledge Representation (KR-MED 2004) (June 2004)

An Ontology of Resources: Solving the Identity Crisis

Harry Halpin[1] and Valentina Presutti[2]

[1] School of Informatics, University of Edinburgh, 2 Buccleuch Place, EH8 9LW Edinburgh, Scotland, UK

[2] Semantic Technology Laboratory, ISTC-CNR, Via Nomentana 56 00161, Rome, Italy

H.Halpin@ed.ac.uk, valentina.presutti@cnr.it

Abstract. The primary goal of the Semantic Web is to use URIs as a universal space to name anything, expanding from using URIs for webpages to URIs for "real objects and imaginary concepts," as phrased by Berners-Lee. This distinction has often been tied to the distinction between information resources, like webpages and multimedia files, and non-information resources, which are everything from real people to abstract concepts like 'the integers.' Furthermore, the W3C has recommended not to use the same URI for information resources and non-information resources, and several communities like the Linked Data initiative are deploying this principle. The definition put forward by the W3C, that non-information resources are things whose "essential nature is information" is a difficult distinction at best. For example, would the text of Moby Dick be an information resource? While this problem could safely be ignored up until recently, with the rise of Linked Data and projects like OKKAM, it appears that this problem should be modelled formally. An ontology called IRW (Identity and Reference on the Web) of various types of resources and their relationships, both for the hypertext Web and the Semantic Web, is presented. It builds upon Information Object Lite (an extension of DOLCE Ultra Lite for describing information objects) and IRE (an earlier ontology of and aligns with other work in this area. This ontology can be used as a tool to make the Semantic Web more self-describing and to allow inference to be used to test for membership in various classes of resources.

Keywords: Web architecture, resource, ontology, linked data.

1 Introduction

The Semantic Web constitutes a truly new scenario for knowledge representation that may have a strong impact on existing models due to its use of URIs as a mechanism of identification of resources. The key feature of the Semantic

L. Aroyo et al. (Eds.): ESWC 2009, LNCS 5554, pp. 521–534, 2009.

Web is not its use of knowledge representation technologies like ontologies and inference per se, but the introduction of these technologies to operate over the Web, as defined as a space of URIs. On the Semantic Web, we can use URIs in order to name anything, which is in contrast to the hypertext Web, which primarily used URIs just to access webpages. Due to this bifurcation, there are now several types of resources on the Web, and the W3C has prescribed different behaviors for their identification and access. For this reason it is of key importance to make it clear what kinds of resources can be identified on the Semantic Web, and how to distinguish between them. Yet, there is still debate within Web circles as to what the definition of a 'resource' is, what kinds of resources there are, and how these relate to the Semantic Web and to the pre-Semantic Web hypertext Web. We model these terms and the debates around them using a lightweight formal ontology in OWL-DL, which we call *IRW*, for 'Identity of Resources on the Web.' IRW is meant to be an helpful formal tool for resolving conflicting arguments about identity and URIs, and as a consequence, it provides a supporting vocabulary for implementing practical solutions in a variety different scenarios.

One of the most difficult tasks of Web architecture is stating the definitions of terms in natural language, the obvious first step to formalization. First, what is a resource? The W3C TAG state in their *Architecture of the Web* that 'resource' is used in a general sense for whatever might be identified by a URI [19]. Previously, a resource was thought of as strictly network-accessible objects such as webpages, since the term 'resource' is defined by Fielding in the first HTTP RFC as "a network data object or service, identified by a URI" [1]. However, Berners-Lee broadened the concept of resource in IETF RFC 2396, stating that "a resource can be anything that has identity. Familiar examples include an electronic document, an image, a service (e.g., 'today's weather report for Los Angeles'), and a collection of other resources. Not all resources are network 'retrievable'; e.g., human beings, corporations, and bound books in a library can also be considered resources" [4].

One distinction that has been upheld by Hayes and others is the distinction between reference and access [17]. Making an analogy between URIs and names, *access* means "that the name provides a causal pathway to the thing, perhaps mediated by the Web" while *reference* means that "the name is being used to mention the thing," which may or may not coincide with access [17]. Some thing is then 'Web-accessible' if it can accessed via the use of HTTP. This use of the term 'resource' for both referring to non-Web accessible things and for naming Web-accessible things is continued in URI RFC 3986, the current IETF RFC, which states that "this specification does not limit the scope of what might be a resource ... likewise, abstract concepts can be resources, such as the operators and operands of a mathematical equation, the types of a relationship (e.g., 'parent' or 'employee'), or numeric values (e.g., zero, one, and infinity)" [5]. It seems that *very different* kinds of things are being described by the notion of a resource.

[1] `http://tools.ietf.org/html/draft-fielding-http-spec-00`

The most obvious distinction is between a resource that could *in principle* be Web-accessible, like a webpage, and a resource that is not in principle Web-accessible, like the Eiffel Tower itself. This distinction is given by the W3C TAG as the distinction between an information resource and something that may not be an information resource [19]. The W3C TAG then define an *information resource* as something "whose essential characteristics can be conveyed in a message," which is a controversial definition [19]. As noted by the Linked Data tutorial, this implies there is another kind of resource, *non-information resources*, for things that are not possibly Web-accessible, like a URI whose primary purpose is to refer to the Eiffel Tower [7]. Furthermore, one can distinguish 'Web resources' (a subset of information resources) that *are* usually Web-accessible, such as web-pages, from things that simply carry information, like the text of Moby Dick, regardless of whether it is on the Web or not. Again, let us emphasize that some find these distinctions very intuitive, while others do not. Lastly, in order to distinguish URIs for non-accessible things on the Semantic Web (the 'Cool URIs for the Semantic Web') from the normal use of URIs on the hypertext Web, we call the former *Semantic Web URIs* [26]. In Web architecture circles, what are typically called 'webpages' are called instead 'representations' of a resource, since multimedia and other not-hypertext pages can be retrieved via the Web [19]. In order to distinguish the use of the word 'representation' in Web architecture circles from its normal usage, the word *Web Representation* is used in this paper to designate a more encompassing notion of web-page as any set of bits that is 'coming down the wire' on the Web.

2 The Identity Crisis

What does a URI identify? For multiple agents to exchange knowledge representations on the Semantic Web, they must be able to identify what *precise* kinds of resource a URI identifies if they are to make any non-trivial inferences. For example, how can an agent determine whether or not a URI for the Eiffel Tower is used to refer to the Eiffel Tower in Paris? If so, does the aforementioned URI access a Web-accessible representation? This cluster of questions has been dubbed the *Identity Crisis* of the Semantic Web.

2.1 Two Viewpoints

There are at least two distinct positions on this question. The first position, the *direct reference position*, holds that on the Web a URI identifies whatever it was intended by the owner to identify. The owner of the URI should be able to unambiguously declare and communicate the meaning of any URI, including a Semantic Web URI. In this position, the referent is generally considered to be some individual unambiguous *single* thing, like the Eiffel Tower or the concept of unicorns. The URI is given its referent by fiat by the owner or creator of the URI, who then communicates this via some causal chain, in a similar manner to Kripke's causal theory of proper names [21]. This viewpoint is the one generally held by many Web architects, like Berners-Lee, who imagine it holds not just for the Semantic Web, but the entire Web. Of course this seems true for the hypertext Web, since a URI naturally identifies a web-page by virtue of providing access to the web-page.

The second position, the *logicist position*, holds that for the Semantic Web, a URI refers to whatever model(s) – including actual things – that satisfy the formal semantics of the Semantic Web. Adherents of this position hold that the referent of a URI is almost always ambiguous, as many different models can satisfy an interpretation of a RDF graph. This position has been championed extensively against Berners-Lee by Hayes, as Hayes believed that the direct reference position "doesn't make sense, that it isn't true, and that it could not possibly be true" as it contradicts the standard interpretation of Tarski-style formal semantics [16]. A URI has no identity in and of itself, but only in the context of its use in a graph or, in a minor variation argued for by Parsia and Patel-Schneider, the explicit use of `owl:imports` [23]. This position is generally held by those who claim that the Semantic Web is entirely distinct from the hypertext Web. Others like Booth [9] hold a hybrid position that an agent should be able to access logical descriptions from the URI itself so that whatever things satisfy these accessible logical descriptions are the referents.

2.2 The TAG's Resolution

The TAG officially resolved *httpRange-14* by saying that the `303 See Other` HTTP header can serve to disambiguate between information resources and possible non-information resources. The official resolution to Identity Crisis by the TAG is given below as [13]:

- If an HTTP resource responds to a GET request with a `2xx` response, then the resource identified by that URI is an information resource;
- If an HTTP resource responds to a GET request with a `303 (See Other)` response, then the resource identified by that URI could be any
- If an HTTP resource responds to a GET request with a `4xx` response, then the nature of the resource is unknown.

One concrete example would be an agent is trying to access a URI that refers to the Eiffel Tower itself, `http://dbpedia.org/resource/Eiffel_Tower`. Upon attempting to access that resource with a HTTP GET request on a URI, since the Eiffel Tower itself is not an information resource, no Web representations are directly available. Instead, the agent gets a `303 See Other` that in turn redirects them to an information resource that hosts Web representations about the Eiffel Tower, such as `http://dbpedia.org/page/Eiffel_Tower`. When this URI returns the 200 status code in response to an HTTP GET request, the agent can infer that `http://dbpedia.org/page/Eiffel_Tower/` is actually an information resource. The Semantic Web URI used to refer to the Eiffel Tower itself, `http://dbpedia.org/resource/Eiffel_Tower`, could be any kind of resource and so *could* be a non-information resource [13]. This example is illustrated in Figure 1, using terms from the IRW ontology introduced in Section 5.

An alternative to the 303 redirection is the *hash convention*, in which one uses the fragment identifier of a URI to get redirection 'for free' with smaller RDF vocabularies. If one wanted a Semantic Web URI that referred to the Eiffel Tower itself without the hassle of a 303 redirection, one would use the URI `http://www.tour-eiffel.fr/#it` to refer to the Eiffel Tower itself. Since browsers either dispose of or treat the fragment identifier as a fragment of a hypertext document or some other Web representation, if an agent tries to access via HTTP GET a Semantic Web URI that uses the hash convention, the server will not return a `404 Not Found` status code, but instead will resolve to the URI before the hash, `http://www.tour-eiffel.fr`, which can then be an Web resource capable of returning Web representations, which is called an 'associated description' in the Linked Data community [7]. In this way, Semantic Web inference engines can keep the Semantic Web URI that refers to the Eiffel Tower and an associated description about the Eiffel Tower separate by taking advantage of the predefined behaviour in web browsers. However, practically the 303 redirection of the W3C TAG and the hash convention leave the question of whether a resource is an information resource or non-information resource indeterminate, since there is nothing to prevent 303 redirection from being used to redirect from one information resource to another information resource, and the hash convention is dependent on media types, being more often used for named parts in the document in HTML instead of as a shortcut for distinguishing non-information resources and their associated descriptions.

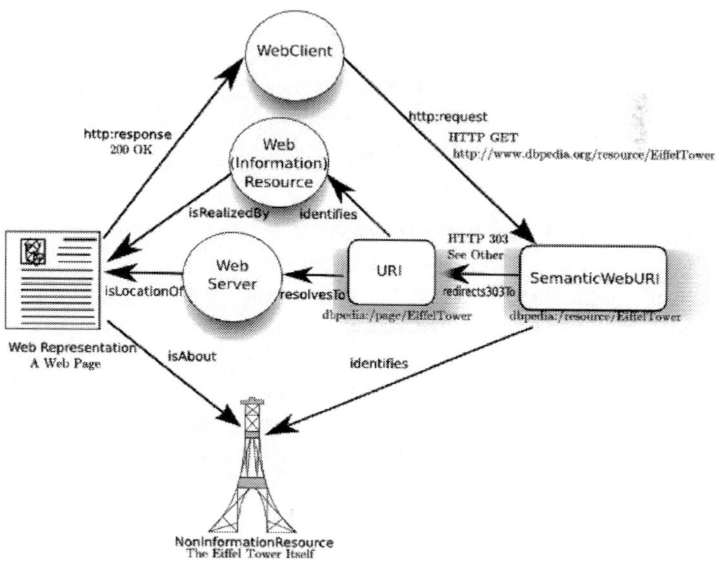

Fig. 1. 303 Redirection for Semantic Web URIs

3 Related Work

There has been some related work in this area. Mogul has suggested that there are fundamental disagreements about what precisely the difference between an HTTP entity and a "representation of a resource" are, and that this leads to widespread problems with caching implementations in HTTP [22]. David Boorh has proposed an informal categorisation of what can be identified by a URI, noticing the confusion between 'naming' and 'identifying' and even 'describing' [8]. Hayes has long attempted to elucidate the fundamental difference between the use of resources to access webpages and the use of a URI to refer to some non-Web accessible thing [17]. Furthermore, the use of URIs to refer to physical entities and the subsequent clarification of the direct reference position has led to the OKKAM project, a project to build a catalogue of 'entity' URIs that is supposed to directly refer to physical entities [10]. This general line of thinking has led to a number of workshops at conferences such as the World Wide Web Conference and the European Semantic Web Conference devoted to this topic [11, 12].

Within the W3C, there is an informal activity of the W3C TAG called the 'Architecture of the Semantic Web' (AWWSW) that has for over a year attempted to decipher Web architecture formally, in part prompting by the need to model HTTP in RDF directly in order for HTTP transactions to be validated via EARL, the RDF-based *Evaluation and Report Language* used by the W3C to validate new W3C standards and describe test-cases [1, 20]. Yet, HTTP in RDF currently does not model the notion of 'resource' except with a misuse of `rdf:Alt`, so it must be corrected by integrating an ontology of resources like IRW. While both EARL and the AWWSW are attempting a much more detailed and low-level description of HTTP transactions than we attempt, the lightweight IRW ontology described in this paper should allow specifications like HTTP in RDF to directly address the notion of a 'resource.'

4 The Use of a Formal Ontology

The primary use of a formal ontology in the context of Web architecture and the arguments around the Identity Crisis is to allow those involved in debates to model formally their positions using a common ontology as a starting point, and so clarify disagreements, as shown in Section 7. To this aim, IRW can be discussed, reviewed, and comment on the `ontologydesignpatterns.org` wiki[2]. To serve the aim of elucidating arguments, additional modules of IRW have been developed and are briefly introduces in Section 5.

There have been previous attempts to model at least a subset of the notions outlined beforehand in a formal ontology, but all lack coverage of some crucial concepts. For example, while the ontology given by RDF Schema touches upon the vocabulary of resources via its term `rdfs:Resource`, it does not cover the distinction between information and non-information resources. The IRE

[2] `http://ontologydesignpatterns.org/wiki/Submissions:IRW`

(Identifiers, Resources, and Entities), based on Dolce Ultra Lite (DUL),[3] a light version of the widely-known DOLCE foundational ontology and its extension for describing information objects[4] (IOL, described in [14]), attempted to model some of these concepts earlier [25]. However, many aspects were not included in IRE, such as the distinctions between resources and their Web representations, or the concept of accessing a web-page via a web server, that are crucial to the efforts within the W3C, while many of the distinctions drawn by DUL+IOL were found to be too 'heavy-weight' for these communities [15]. In response to these concerns, the IRE ontology has been evolved into the IRW ontology.

5 The IRW Ontology

The prefix `irw:` is for the namespace `http://purl.org/NET/irw/` of the IRW ontology. The stable version of the ontology can also be accessed via its PURL. The latest version of the IRW ontology may be accessed at:

`http://ontologydesignpatterns.org/ont/web/irw.owl`

The prefix `rdfs:` is used for the RDF(S) namespace

`http://www.w3.org/2000/01/rdf-schema#`

Note `ir:` is

`http://www.ontologydesignpatterns.org/cp/owl/informationrealization.owl`

While the IRW ontology in full can not explicated due to lack of space, the primary classes and properties are given in Figure 2. The IRW-related elements needed for the example of 303 redirection are given in Figure 1. The IRW ontology starts with `irw:Resource`. While this class expresses the same intuition as `rdfs:Resource`, we have defined it because this version of IRW is within OWL-DL expressivity. In OWL Full, this class is equivalent to `rdfs:Resource`. Now, we move to modelling the debates around the Identity Crisis.

Identification and reference. The notion of a URI is modeled as a class, `irw:URI` that has exactly one value for the datatype property `irw:hasURI` allowing to specify its value. Modelling URIs as a class allows us to talk about different kinds of URIs, such as IRIs (Internationalized Resource Identifiers) and Semantic Web URIs. According to the *direct reference position*, URIs identify exactly one resource. This is modeled in IRW by the functional property `irw:identifies`, having range `irw:Resource` (and inverse property, `irw:isIdentifiedBy`). Of course, those that disagree with this viewpoint may not use `irw:identifies`, and so it is given sub-properties `irw:accesses` and `irw:refersTo`. The viewpoint of the *logicist position*, namely that a URI can refer to more than one resource, is modeled by the object property `irw:refersTo` (and inverse property,

[3] `http://www.loa-cnr.it/ontologies/DUL.owl`

[4] `http://www.loa-cnr.it/ontologies/IOLite.owl`

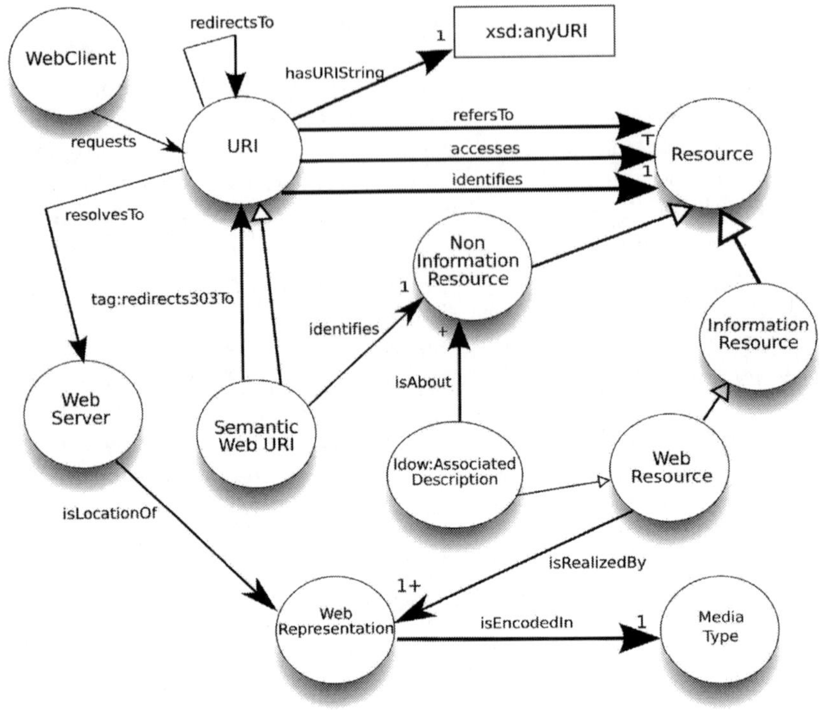

Fig. 2. The IRW ontology illustrated as a graph. Rounded nodes are classes, while rectangular ones are datatypes. Arcs ending with an empty triangle are `rdfs:subClassOf` relationships. Arcs ending with a filled triangle are either object properties or datatype properties depending of the range node. Arcs' direction indicates the domain and range of the property. A '1' associated to a property means it is functional, a 'T' means it is transitive, '1+' means 'at least one'. Prefixes are indicated only if different from `irw:`.

`irw:isReferencedBy`). One condition on this property is that the object of reference should be "immediately causally disconnected" from its subject [17]. This is important, as reference is the relationship to both URIs for non-information resources like the Eiffel Tower or integers, but also applies to the relationship of an information resource to some non-information resource, like the relationship of Tim Berner-Lee's homepage to Berners-Lee himself.

Access and redirection. Distinct from reference is the `irw:accesses` relationship, which is a causal connection to the thing identified. This is modelled again as a relationship between URIs and resources, although it is transitive, unlike `irws:refersTo`. If one can access a and a accesses b then a accesses c (via b). Although a wide notion, access allows us to model the typical HTTP request-response Web transactions between a Web client and a server. A URI may also have a `irw:redirectsTo` property, a sub-property of `irw:accesses`, that we can use to model HTTP redirection. However, since redirection can be used between just information resources that have nothing to do with the Semantic Web, their domain and range say nothing about the type of resource. In order to model explicitly the redirection solution to the "Identity Crisis" by the W3C TAG, two distinct sub-properties of this have been added in a TAG-specific module of IRW[5] that contains `tag:redirects303To` property and a `tag:redirectsHashTo` property. Obviously, `tag:redirects303To` models the TAG's 'solution' to *httpRange-14* while `tag:redirectsHashTo` represents the hash convention.

Types of resources. Having defined reference and redirection, we can now categorize resources. There are two main disjoint sub-classes of `irw:Resource`. The first subclass is given as `irw:InformationResource`, which is an information object, such as a musical composition, a text, a word, or a picture. An information object is an object defined at a level of abstraction, independently from how it is concretely realized. So an `irw:InformationResource` expresses the same intuition and is an equivalent class to the DUL+IOL information object [14]. This means an information resource has, via the `ir:realizes` property (with inverse `ir:isRealizedBy`), at least one `ir:InformationRealization`, a concrete *realization.* This term is again imported from DUL+IOL [14]. So an information resource's "essential characteristics can be conveyed in a single message" implies that everything from a bound book to an HTTP message can be a realization of an information resource [19]. Furthermore, the property `irw:isAbout` (and inverse property, `irw:isTopicOf`) expresses the relationship of an information resource to a resource or resources the information is 'about.' Examples of this are descriptions of a resource using natural language or depictions of a resource using images. Information resources also can, but not necessarily, be identified (either accessed or referred to) with a URI. In this manner, the text of Moby Dick can be an information resource since it could be conveyed as a single message in English, and can be realized by both a particular book or a webpage containing that text.

Note `irw:NonInformationResource` complements `irw:InformationResource` from which it is disjoint with. Such class represents things that can not themselves – for whatever reason – be realized as a single digitally encoded message. A number of different kinds of things may be `irw:NonInformationResources`. Since this concept is the cause of much confusion and debate, it is detailed with three disjoint sub-classes. These kinds of IRW distinctions are not normative, as there are other possible plausible, more detailed modeling choices. Our aim here is to communicate the intuition behind the concepts of information and non-information resources without entering the philosophical debate about top-level ontologies. IRW contains three sub-classes of `irw:NonInformationResources:`[6] `irw:PhysicalEntityResource`, is a resource that is 'touchable' like physical

[5] `http://www.ontologydesignpatterns.org/ont/web/tag2irw.owl` associated with prefix `tag:`.

[6] Note that the three classes does not constitute an exhaustive partition.

people, artifacts, places, bodies, chemical substances, biological entities; `irw:ConceptualResource`, which refer to resources that are created in a social process that can not be completely realized digitally, such as legal entities, political entities, social relations, as well as the concept of horse and imaginary objects like unicorns; and finally `irw:AbstractResource`, which refers to abstract combinatorial spaces that cannot be located in space-time such as formal entities like functions or the integers as well as more mundane resources like the infinite set of names that constitute the resource identified by URIs themselves.

A sub-class of `irw:InformationResource` is `irw:WebResource`, which is an information resource identified by at least one URI and realized by at least one `irw:WebRepresentation`, so that a Web resource is just an information resource that is realized by at least one accessible Web representation like a webpage. `irw:WebRepresentation` is a sub-class of `irw:InformationRealization` with constraints added to make the cardinality of `ir:isRealizedBy` and `irw:isIdentifiedBy` both at least 1. In this way IRW can distinguish between a resource for the text of 'Moby Dick' in general and a webpage about 'Moby Dick.'

Hypertext Web transactions. The typical hypertext Web transaction can be modelled by IRW. We begin with `irw:WebClient`, which is some client in the context of the Web that can have a `irw:requests` relationship to a URI (note that `irw:requests` serves as an hook to the alignment of IRW with *HTTP in RDF* [20]), as exemplified by a typical `HTTP GET` request). The `irw:requests` property is a sub-property of `irw:access`. A `irw:WebClient` then `irw:requests` a `irw:URI`. We also introduce the class `irw:WebServer`, which has a `irw:isResolutionOf` property that relates a URI to a concrete Web server (inverse property `irw:resolvesTo`). This `irw:resolvesTo` property is currently implemented by mapping a URI to an IP address or addresses. So each `irw:WebServer` is the resolution of at least one `irw:URI`. Additionally, a `irw:WebServer` has a `irw:isLocationOf` property with at least one `irw:WebRepresentation` (inverse property, `locatedOn`), indicating the Web server concretely can respond to an HTTP request with a particular Web Representation.

Semantic Web (Linked Data) transactions. The typical Semantic Web transaction is also modeled. A new sub-class of `irw:URI`, `SemanticWebURI` is given, where the Semantic Web URI has a constraint that it must have at least one `irw:redirects` property.

In the Linked Data Initiative, another important kind of resource is "associated descriptions," which is just an Web resource that can be accessed via redirection from a Semantic Web URI [7]. For example, in DBPedia[7] the resource `dbpedia:/resource/Eiffel_Tower` redirects to some RDF/XML at `dbpedia:/`

[7] Prefix `dbpedia:` is used for the namespace `http://dpedia.org`

data/Eiffel_Tower, and to an HTML page at dbpedia:/page/Eiffel_Tower depending on the requested media type [3]. This Linked Data typical scenario can be generalized: a irw:WebClient irw:requests a irw:SemanticWebURI x and the request is redirected (e.g. via hash or 303 redirection) to another URI, where this second URI identifies an ldow:AssociatedDescription,[8] which has one irw:isAbout property to a non-information resource. We model ldow:AssociatedDescription as a subclass of irw:WebResource.

6 Resolving the Identity Crisis

One purpose of this ontology is to describe, in formal detail, the exact nature of the conflicts between the various sides of the *I*dentity Crisis debate. The main conflict between the *logicist* position of Hayes and the *direct reference* position of Berners-Lee can then be cast as an argument over three IRW properties. Berners-Lee's slogan that "URIs identify one thing" is modelled by having the irw:identifies property be *functional*, i.e. a URI can only identify one resource. Furthermore, he would also hold that a irw:SemanticWebURI irw:refersTo exactly one irw:NonInformationResource.

Hayes's response would be that irw:identifies should be eliminated and there can be no constraints whatsoever on irw:refersTo and thus no constraints on the usage of URIs for referring to things on the Semantic Web, while typical hypertext Web transactions can be modelled functionally with irw:accesses. Although IRW models Berners-Lee's more general notion of identification via irw:identifies, IRW also captures Hayes's perspective with the properties irw:refersTo and irw:accesses. Lastly, the criticisms of redirection modeled with irw:redirectsTo has mainly to do with the fact that the domain can only be simply irw:URI rather than irw:SemanticWebURI, which we also explicitly model. Thus, there is no way to ever definitely be sure that a URI is a Semantic Web URI and so one can never be sure that a URI identifies a non-information resource. We show how IRW can solve this problem in Section 8.

7 Aligning IRW to Other Ontologies

In this section, we present a number of suggested alignments, as given in Table 1. The alignments are to the three primary other ontologies, the *RDF in HTTP* ontology [20], and the IRE ontology as well as an ontology for HTTP used by the Tabulator Browser [25, 6]. The namespaces for ont is http://www.w3.org/2007/ont/http. IRE, due to its modular construction and re-use of terms from DUL+IOL patterns, uses many namespaces, but they can be found at http://www.ontologydesignpatterns.org/cpont/ire.owl. The http namespace is http://www.w3.org/2006/http#.

[8] Typical Linked Data terminology is represented in a specific module of IRW represented here by the prefix ldow: referring to the namespace http://ontologydesignpatterns.org/ont/web/ldow2irw.owl

Table 1. Mapping of IRW to Other Ontologies

Class or Property	Alignments
irw:WebRepresentation	owl:equivalentClass http:Message owl:equivalentClass ont:ResponseMessage rdfs:subClassOf ire:InformationRealization rdfs:subClassOf ir:InformationRealization
http:Content	rdfs:subClassOf ir:InformationRealization
http:MessageHeader	rdfs:subClassOf ir:InformationRealization
irw:InformationResource	owl:equivalentClass ir:InformationObject
irw:SemanticWebURI	ire:SemanticWebURI
irw:identifies	ire:isExactProxyFor
irw:isAbout	ire:about

8 The Self-describing Semantic Web

The main application of this ontology is to clarify debates over Web architecture, in particular, the 'Identity Crisis' debate. However, the IRW ontology can also solve the problem noted earlier that currently it is impossible to describe whether or not some resource describes some non-Web accessible thing, such that there is no "definition, description, some other kind of indication of what the identifier is intended to identify" [24]. Solving this can be done on via adding IRW statements to associated descriptions accessible via Semantic Web URIs. There would be a number of advantages if webpages that have RDF content could distinguish themselves as such, in the same way that HTML 'valid' documents are currently validated by W3C Validators. This can be done by embedding a IRW statement in RDF/XML documents, RDF returned from SPARQL endpoints, and RDFa or GRDDL statement in XHTML or XML documents [2]. Ideally, this would be in conjunction with some sort of graphical logo to distinguish the page as 'Semantic Web Enabled,' much as current web-pages can be marked up with a logo for 'XHTML 1.0 Valid.' This is useful because detecting RDF 'in the wild' on the Web, such as embedded RDFa, can be difficult for humans. The main problem is that an `irw:NonInformationResource` has no Web representation to embed such a statement in. Take for example the Semantic Web URI created by Pat Hayes for himself: `www.ihmc.us/users/phayes/PatHayes.html`. While originally a stand-alone web-page, currently Hayes has the URI use 303 redirection to `http://www.ihmc.us/users/phayes/PatHayesAbout.html`.[9] This latter web-page could easily use a combination of RDFa as IRW to mark itself up as a representation of a non-information resource by including the statement that `phayes:PatHayes.html rdf:type irw:NonInformationResource` and adding `phayes:PatHayes.html tag:redirects303To phayes:PatHayesAbout.html`. The IRW ontology can help explicitly model and make available to the rest of the Semantic Web the often subterranean details of Web architecture.

[9] Let `phayes:` stand for `http://www.ihmc.us/users/phayes`

9 Conclusion and Future Work

Overall, the IRW ontology is a beginning, yet it should serve as foundational contribution of modelling Web architecture, the "Dark Side of Semantic Web" that Hendler believes may give the Semantic Web a crucial advantage over previous efforts in knowledge representation [18]. What is surprising is that it has taken so long for this problem to be solved in a relatively straightforward manner. However, as the problem space is highly contentious and the various documents that describe this problem are spread throughout many informal and semi-formal notes and standards (and arguments over W3C lists), systematising the problem and modelling it formally was perhaps more difficult than would be expected. Such problems that are closest the Semantic Web are often the most difficult to model objectively, such as meta-modelling in general. Future work needs to be done to standardise IRW or a descendant thereof through the W3C, which will doubtless result in refinements to IRW, and to encourage its use within the Linked Data community in the context of various validators, debuggers, and search engines. By developing a consistent vocabulary for describing the identity of resources in IRW, the first step has been taken.

Acknowledgements

We would like to thank Aldo Gangemi for his insightful comments. Harry Halpin was partially supported by a Microsoft 'Beyond Search' award. Valentina Presutti was partially supported by 'Interactive Knowledge Stack' EU FP7 Project.

Bibliography

[1] Abou-Zahra, S.: Evaluation and Report Language (EARL) 1.0 Schema. W3C Working Draft, W3C (2007), http://www.w3.org/TR/EARL10-Schema/
[2] Adida, B., Birbeck, M., McCarron, S., Pemberton, S.: RDFa in XHTML: Syntax and Processing. W3C Recommendation, W3C (2008), http://www.w3.org/TR/rdfa-syntax/
[3] Auer, S., Bizer, C., Lehmann, J., Kobilarov, G., Cyganiak, R., Ives, Z.: DBpedia: A nucleus for a web of open data. In: Aberer, K., Choi, K.-S., Noy, N., Allemang, D., Lee, K.-I., Nixon, L., Golbeck, J., Mika, P., Maynard, D., Mizoguchi, R., Schreiber, G., Cudré-Mauroux, P. (eds.) ASWC 2007 and ISWC 2007. LNCS, vol. 4825, pp. 722–735. Springer, Heidelberg (2007)
[4] Berners-Lee, T., Fielding, R., Masinter, L.: IETF RFC 2396 Uniform Resource Identifier (URI): Generic Syntax (1998), http://www.ietf.org/rfc/rfc2396.txt (last accessed on September 15, 2008)
[5] Berners-Lee, T., Fielding, R., Masinter, L.: IETF RFC 3986 Uniform Resource Identifier (URI): Generic Syntax (January 2005), http://www.ietf.org/rfc/rfc3986.txt (last accessed on April 2, 2008)
[6] Berners-Lee, T., Hollenbach, J., Lu, K., Presbrey, J., Prud'hommeaux, E., Schraefel, M.: Tabulator Redux: Browsing and Writing Linked Data. In: Proceedings of the WWW 2007 Workshop on Linked Data on the Web (2008)
[7] Bizer, C., Cygniak, R., Heath, T.: How to publish Linked Data on the Web (2007), http://www4.wiwiss.fu-berlin.de/bizer/pub/LinkedDataTutorial/ (last accessed on May 28, 2008)

[8] Booth, D.: URIs and the myth of resource identity. In: Proceedings of Identity, Reference, and the Web Workshop at the WWW Conference (2006), `http://www.ibiblio.org/hhalpin/irw2006/dbooth.pdf`
[9] Booth, D.: URIs declaration versus use. In: Proceedings of Identity, Reference, and the Semantic Web Workshop at the European Semantic Web Conference (2008)
[10] Bouquet, P., Stoermer, H., Giacomuzzi, D.: OKKAM: Enabling a Web of Entities. In: i3: Identity, Identifiers, Identification. Proceedings of the WWW 2007 Workshop on Entity-Centric Approaches to Information and Knowledge Management on the Web, CEUR Workshop Proceedings, Banff, Canada, May 8 (2007), ISSN 1613-0073, `http://CEUR-WS.org/Vol-249/submission_150.pdf`
[11] Bouquet, P., Stoermer, H., Tummarello, G., Halpin, H. (eds.): Proceedings of the WWW 2007 Workshop I^3: Identity, Identifiers, Identification, Entity-Centric Approaches to Information and Knowledge Management on the Web, Banff, Canada, May 8, 2007. CEUR Workshop Proceedings, CEUR-WS.org (2007)
[12] Bouquet, P., Stoermer, H., Tummarello, G., Halpin, H. (eds.): Proceedings of the ESWC 2008 Workshop on Identity, Reference, and the Web, Tenerife, Spain, June 1, 2008. CEUR Workshop Proceedings (2008)
[13] Connolly, D.: A pragmatic theory of reference for the web. In: Proceedings of Identity, Reference, and the Web Workshop at the WWW Conference (2006), `http://www.ibiblio.org/hhalpin/irw2006/dconnolly2006.pdf` (last accessed November 22, 2008)
[14] Gangemi, A.: Norms and plans as unification criteria for social collectives. Journal of Autonomous Agents and Multi-Agent Systems 16(3) (2008)
[15] Gangemi, A., Guarino, N., Masolo, C., Oltramari, R., Schneider, L.: Sweetening ontologies with DOLCE. In: Gómez-Pérez, A., Benjamins, V.R. (eds.) EKAW 2002. LNCS, vol. 2473, pp. 166–181. Springer, Heidelberg (2002)
[16] Hayes, P.: Message to www-rdf-comments@w3.org (2003), `http://lists.w3.org/Archives/Public/www-tag/2003Jul/0198.html`
[17] Hayes, P., Halpin, H.: In defense of ambiguity. International Journal of Semantic Web and Information Systems 4(2), 1–18 (2008)
[18] Hendler, J.: The Dark Side of the Semantic Web. IEEE Intelligent Systems 22(1), 2–4 (2007)
[19] Jacobs, I., Walsh, N.: Architecture of the World Wide Web. Technical report, W3C (2004), `http://www.w3.org/TR/webarch/` (last accessed October 12, 2008)
[20] Koch, J., Velasco, C.A., Abou-Zahra, S.: HTTP Vocabulary in RDF. W3C Working Draft, W3C (2008), `http://www.w3.org/TR/EARL10-Schema/`
[21] Kripke, S.: Naming and Necessity. Harvard University Press, Cambridge (1972)
[22] Mogul, J.: Clarifying the fundamentals of HTTP. In: Proceedings of the 11th International World Wide Web Conference, pp. 444–457 (2002)
[23] Parsia, B., Patel-Schneider, P.F.: Meaning and the Semantic Web. In: Proceedings of Identity, Reference, and the Web Workshop at the WWW Conference (2006), `http://www.ibiblio.org/hhalpin/irw2006/bparsia2006.pdf`
[24] Pepper, S.: The case for published subjects. In: Proceedigs Identity, Reference, and the Web Workshop at the WWW Conference (2006), `http://www.ibiblio.org/hhalpin/irw2006/spepper2.pdf`
[25] Presutti, V., Gangemi, A.: Identity of resources and entities on the web. International Journal of Semantic Web and Information Systems 4(2), 49–72 (2008)
[26] Sauerman, L., Cygniak, R.: Cool URIs for the Semantic Web. W3C Note, W3C (2008), `http://www.w3.org/TR/cooluris/`

Mining Semantic Descriptions of Bioinformatics Web Resources from the Literature

Hammad Afzal, Robert Stevens, and Goran Nenadic

School of Computer Science, University of Manchester, Manchester, UK
{Hammad.Afzal@postgrad.,R.Stevens@,G.Nenadic@}manchester.ac.uk

Abstract. A number of projects (myGrid, BioMOBY, etc.) have recently been initiated in order to organise emerging bioinformatics Web Services and provide their semantic descriptions. They typically rely on manual curation efforts. In this paper we focus on a semi-automated approach to mine semantic descriptions from the bioinformatics literature. The method combines terminological processing and dependency parsing of journal articles, and applies information extraction techniques to profile Web services using informative textual passages, related ontological annotations and service descriptors. Service descriptors are terminological phrases reflecting related concepts (e.g. tasks, approaches, data) and/or specific roles (e.g. input/output parameters, etc.) of the associated resource classes (e.g. algorithms, databases, etc.). They can be used to facilitate subsequent manual description of services, but also for providing a semantic synopsis of a service that can be used to locate related services. We present a case-study involving full text articles from the BMC Bioinformatics journal. We illustrate the potential of natural language processing not only for mining descriptions of known services, but also for discovering new services that have been described in the literature.

1 Introduction

The bioinformatics domain has recently witnessed a number of tools and data sources available on the Web that can be used to retrieve or carry out on-line data analysis. For example, as indicated by the BioCatalogue[1] project, there are around 200 Web servers that provide Web Service interfaces that are becoming both an essential and a critical part of bioinformatics research. These resources need to be organised and their functionalities semantically described to make them accessible by both bioinformaticians and search/discovery engines. A number of projects[2] (e.g. myGrid, BioMOBY, BioCatalogue, myExperiment etc.) have been initiated to comprehensively catalogue these resources and provide their semantic descriptions (see Table 1 for an example). Most of the cataloguing frameworks, however, rely on manual annotation that has resulted in a backlog of non-described services, reducing the chance of their discovery and use in the community [1]. In order to deal with the huge number of resources

---

[1] http://www.biocatalogue.org

[2] For details about these projects see: http://www.mygrid.org.uk/, http://www.biomoby.org/, and http://www.myexperiment.org/

L. Aroyo et al. (Eds.): ESWC 2009, LNCS 5554, pp. 535–549, 2009.

Table 1. A partial service description of service Emma as described for Feta [2]. The input and output of operations are specified in terms of parameters that have name, description and semantic type. Operation tasks and semantic types of parameters are linked to the myGrid ontology [3].

Service name:		Emma	
Description:		Performs a multiple alignment of nucleic acid or protein sequences using ClustalW program.	
Operation	**Name:**	Emma	
	Description:	Performs a multiple alignment of nucleic acid or protein sequences using ClustalW program.	
	Task	www.mygrid.org.uk/ontology/**multiple_local_aligning**	
	Input	**Parameter**	Name: **sequence_usa**
			Description: The Uniform Sequence Address, or USA, is a standard way of specifying a sequence to be read into a program in EMBOSS. …
			SemanticType: www.mygrid.org.uk/mygrid-moby-service#**simpleParameter**
	Output	**Parameter**	Name: **outseq**
			Description: Returns a multiple sequence alignment report
			SemanticType: http://www.mygrid.org.uk/ontology# **multiple_sequence_alignment_report**

(tools, databases, Web services etc.) that need semantic description, (semi)automated approaches are needed.

A number of resources (along with their typical use-cases) have been described and presented in various textual documents (scientific articles, blogs, documentation, user manuals, etc.). In this paper we focus on the extraction of functional descriptions of bioinformatics tools and Web services from scientific, full-text articles. In our approach, semantic descriptions include informative textual passages, related ontological annotations and service descriptors. Service descriptors are terminological phrases reflecting related concepts (e.g. tasks, roles, approaches) typically used with specific resource classes (e.g. algorithms, databases, etc.). These descriptions can be used not only to facilitate subsequent manual description of services, but also for providing a semantic synopsis of a service that can be used to locate related services. We present a case-study involving a subset of full text articles from the BMC Bioinformatics journal, and discuss the manually evaluated results.

2 Background and Related Work

Automatic semantic annotation of Web services is a relatively new area, involving description of functionality, input and output parameters, etc. A number of approaches have been based on invoking services automatically, or on using corresponding Web

Service Definition Language (WSDL) files and existing descriptions of similar services or workflows. For example, Carman and Knoblock presented an automatic approach to learn definitions and semantic descriptions of online information resources by invoking them and comparing the output they produce with that of known sources of information [4]. On the basis of this comparison, they used the metadata associated with known sources to add annotations to unknown sources. The method was evaluated on 25 services from five domains (geospatial, financial, weather, hotels and cars) – the precision ranged between 56% and 91%. Lerman et al. presented a meta-data based classification algorithm that used WSDL files to perform semantic labelling of inputs and outputs of Web services [5]. The semantic types were organised in the form of a domain ontology. The method was evaluated on 313 WSDL files from different domains, with an overall F-measure of approximately 80%. Previously, Hess and Kushmerick annotated Web services using machine learning to classify metadata used to describe those services [6]. Information about the services given in WSDL files was used in this work. Finally, Belhajjame et al. performed automatic annotation of parameters using workflow annotations, by inferring from their links to other (annotated) operation parameters within existing workflows [7].

These methods rely on existing annotations and classification methods in order to infer descriptions of new Web services. In general, the applied techniques demonstrated reasonable performance only in cases of highly focused domains with a large number of training examples (cf. [4, 6, 7]). Finally, neither relies on textual sources (journals articles, application notes etc).

Our work uses techniques from natural language processing, in particular terminological processing, phrase structure identification (e.g. noun and verb phrases) and predicate-argument structures (PAS) to extract service descriptions. For example, a PAS can comprise a verb along with its arguments, e.g. subject and object(s). The Stanford parser[3], for instance, identifies dependencies in the form *dep(abc, xyz)* where *dep* is the dependency type (e.g. nominal subject (nsub), direct object (dobj), etc.), and *abc* and *xyz* are arguments. For example, dependency statement *nsubj (applied, BLAST)* denotes *BLAST* as the nominal subject of *applied*. Wattarujeekrit et al. argued the significance of using PAS-based extraction compared to regular expressions applied on shallow parsed sentences, in particular for event extraction in the biomedical domain [8]. Similarly, Tateisi et al. demonstrated the need and benefits of recognising predicate-argument structures for improving the performance of information extraction systems in biology [9]. Our method also uses predicate-argument structures, but additionally incorporates the phrase-structure recognition and intensive terminological processing in the process of service profiling.

3 Methodology

Our method combines terminological processing and dependency parsing of documents, and applies information extraction techniques to profile Web tools and services. It is focused around the concepts of (1) *semantic classes* (SC) associated with

---

[3] More details on the Stanford parser: http://nlp.stanford.edu/software/lex-parser.shtml; a full list of typed dependencies: http://nlp.stanford.edu/software/dependencies_manual.pdf

bioinformatics resources of interest (e.g. algorithms, applications, etc.), and (2) *semantic descriptors* that represent semantic roles of related SC instances. Descriptors are used to profile a given resource and/or to link it to a domain ontology (e.g. frequent descriptors are *gene expression*, *phylogenetic tree*, *microarray experiment*, *hierarchical clustering*, *amino acid sequences*, *motif*, etc.).

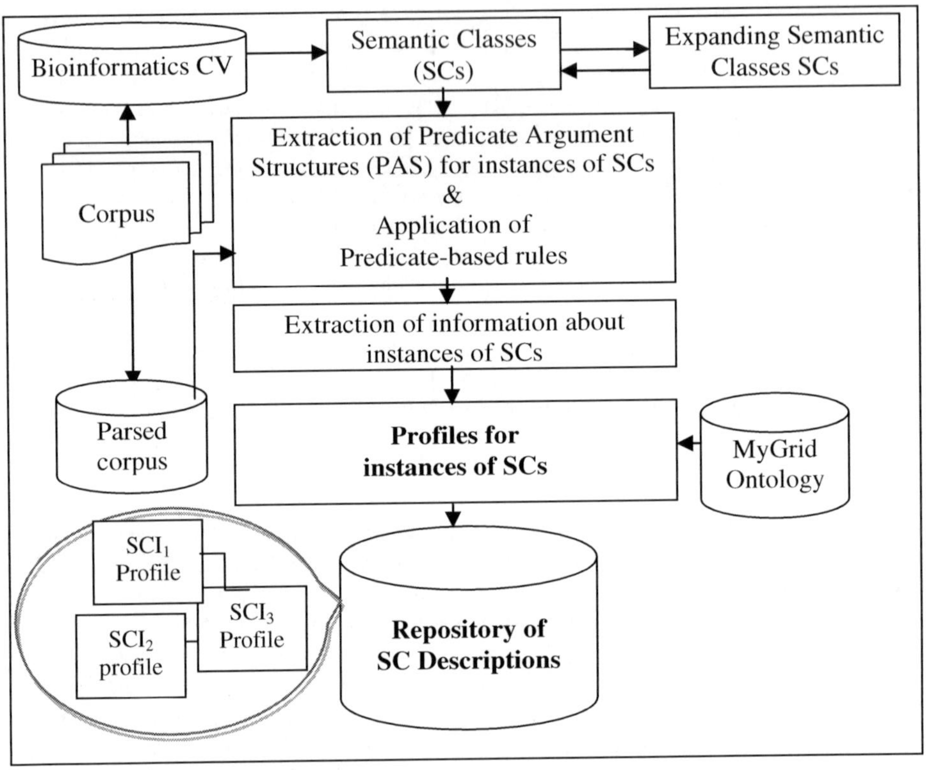

Fig. 1. System Architecture

The method has 3 steps: in the first step, the instances of the semantic classes are identified in the literature in sentences that potentially contain their descriptions. In the second step, semantic descriptors are extracted from the sentences using string matching and a rule based approach, and are potentially mapped to a domain ontology that describes service functionalities. Finally, in the third step, the most relevant sentences are selected to complete the semantic profile of a service. All semantic profiles are collected into a repository that eventually can be used by human curators to provide annotations. The system architecture is presented in Figure 1.

3.1 Semantic Classes

Semantic classes represent the major concepts used in the description of bioinformatics services and tools. The instances of these classes are both extraction targets and can be the key components of semantic descriptions of other services. These concepts were engineered from the myGrid bioinformatics ontology, and include *algorithm, application, data, data resource* and *task* concepts (see example instances in Table 2). As opposed to other classes whose instances are expressed using noun phrases, tasks are typically represented by verbs or verbal phrases portraying functions related to other classes. We therefore followed two methods to identify SCs: for the first four classes we compiled dictionaries with instances, while for the task class we used information extraction rules (see Section 3.3). In the following text we refer to the non-task classes as SCs for simplicity, while tasks will be considered separately.

Table 2. Examples of semantic classes and their instances

Semantic class	Example instances
Algorithm	SigCalc *algorithm*, CHAOS local *alignment*, SNP *analysis*, KEGG Genome-based *approach*, GeneMark *method*, K-fold cross validation *procedure*
Application	PreBIND Searcher *program*, Apollo2Go *Web Service*, FLIP *application*, Apollo Genome Annotation curation *tool*, GenePix *software*, Pegasys *system*
Data	GeneBank *record*, Genome Microbial CoDing *sequences*, Drug Data *report*
Data resource	PIR Protein Information *Resource*, BIND *database*, TIGR *dataset*, BioMOBY Public Code *repository*

We used iterative terminological processing to collect and classify instances belonging to the SCs. In the first step, we extracted a number of instances by identifying typical terminological heads of a given SC (e.g. Smith-Waterman *algorithm* is an algorithm). The set of key terminological heads was obtained by exploring the first level of the myGrid ontology. We then examined a list of 100,000 bioinformatics terms previously collected in the process of building a controlled vocabulary[4] (CV) of bioinformatics [10]. We selected terms and associated them with the corresponding SC if their terminological head belonged to the following set:

Heads(application) = {*application, tool, service, software, system, program*}
Heads(algorithm) = {*algorithm, method, approach, procedure, analysis, alignment*}
Heads(data) = {*record, report, sequence*}
Heads(data resource) = {*resource, database, dataset, repository*}

In the second step, we collected instances that did not have these discriminative terminological heads, but appeared in specific contexts with other instances from a given SC. For example, we collected all instances that co-appeared in coordination and apposition expressions with other instances using patterns similar to those used by Hearst and Shütze [11].

In the final step, we identified instances that demonstrated behaviour similar to the instances identified in the previous step. Similar behaviour was modelled by analysing dependency structures that involved nsubj, dobj and nsubjpass (nominal subject in

[4] Available at: http://gnode1.mib.man.ac.uk/bioinf/CV

passive) dependencies. The motivation here was to firstly identify verbs with which known SC instances appeared in the capacity of nominal subject, direct object and passive nominal subject, and then – using these verbs – extract terms appearing in a similar context. For each SC, a list of frequent and discriminative verbs was automatically compiled (different frequency thresholds were set for different SCs). Newly identified instances were grouped with the corresponding SCs based on a majority vote.

3.2 Semantic Service Descriptors

Service descriptors are terminological phrases used in existing descriptions to refer to the related concepts and specific roles (e.g. input/output parameters, etc.) corresponding to the semantic classes. The reason behind collecting these terms was to identify terms that have been previously used in the description of Web services and tools. The hypothesis was that their presence in the "free" literature would indicate useful sentences for annotating SC instances.

We have used two sources to build a dictionary of service descriptors. The first resource was the list of terms collected from the bioinformatics ontology used in the MyGrid project [3]. This list contains 443 terms describing concepts in *informatics* (the key concepts of data, data structures, databases and metadata); *bioinformatics* (domain-specific data sources e.g. model organism sequencing databases, and domain-specific algorithms for searching and analyzing data e.g. the sequence alignment algorithm); *molecular biology* (higher level concepts used to describe bioinformatics data types, used as inputs and outputs in services e.g. protein sequence, nucleic acid sequence); and *tasks* (generic tasks a service operation can perform e.g. retrieving, displaying and aligning). The second resource contained the terms (recognised by the TerMine[5] service) and frequent noun phrases obtained from the existing descriptions of bioinformatics Web services available in Feta[6] and other WS providers' websites[7].

3.3 Extraction of Functional Service Descriptions

Our methodology for the extraction of functional descriptions (tasks) was predicate-centric, i.e. organised around verbs that appeared with the instances of the SCs. We applied a set of extraction patterns (see Appendix 1) on sentences that were parsed by the Stanford parser to obtain predicate-argument structures and phrase structures (separately). Here, our hypothesis was that by keeping one of the arguments to be an SC instance, the verb and other argument would provide a clue about the functionality of that instance.

We employed two parsing methodologies to extract the arguments and potential functions: one by integrating the dependency parse and phrase-structure parse, and the other using only the phrase structure parse. In the first case, we derived one of the arguments (either Arg-1 or Arg-2) using *nsubj, dobj or nsubjpass* dependencies.

[5] http://www.nactem.ac.uk/software/termine/

[6] http://www.mygrid.org.uk/feta/mygrid/descriptions/

[7] See http://www.mygrid.org.uk/wiki/Mygrid/BiologicalWebServices

These SC instances were the recognised terms that have been pre-marked in sentences. Finding an exact mention (i.e. scope) of the other argument, however, was not as straightforward. In addition, we were also interested in the sub-clause appearing with the PAS which might contain useful descriptive information for the SC instance. For this purpose, we integrated the phrase-structure parse with the associated verb phrase containing the predicate and sometimes, sub-clause information as well.

```
<NP>
    <NNP>Term</NNP>
</NP>
<VP>
    <VBZ>generates</VBZ>
    <NP>
        <NP>
            <JJ>similarity/identity</JJ>
            <NNS>matrices</NNS>
        </NP>
        <PP>
            <IN>for</IN>
            <NP>
                <NN>DNA</NN>
                <CC>or</CC>
                <NN>protein</NN>
                <NNS>sequences</NNS>
            </NP>
        </PP>
    </NP>
</VP>
```

```
nsubj(generates-2, Term-1)

amod(matrices-4, similarity/identity-3)

dobj(generates-2, matrices-4)

nn(sequences-9, DNA-6)

conj_or(DNA-6, protein-8)

prep_for(matrices-4, sequences-9)

prepc_without(generates-2, needing-11)

dobj(needing-11, pre-alignment-12)

det(data-15, the-14)

prep_of(pre-alignment-12, data-15)
```

Fig. 2. Integrating parsing results: part of the phrase-structure parsed tree (left) and dependency parsed tree (right) for the sentence "Matrix Global Alignment Tool MatGAT generates similarity/identity matrices for DNA or protein sequences"

Figure 2 provides an example and illustrates the approach. Here, the term *Matrix Global Alignment Tool MatGAT* was recognised as a tool instance, and was replaced by a Term in the original sentence that was then dependency parsed. The dependency parse identified that the term was a subject (nsubj) and *matrices* was a direct object (dobj) of the verb *generates*. In order to get the full dobj argument, we used the phrase structure parse to identify the full verb phrase (<VP>) involving the verb in question (*generate*), and then integrated the two. In this case, we derived that *Matrix Global Alignment Tool MatGAT* uses *DNA or protein sequence* to generate *similarity/identity matrices*.

We compiled a list of 108 verbs appearing with the SC instances along with their PASs as described above. The verbs providing similar semantic information or indicating the presence of similar functional "content" were manually grouped in clusters (e.g. the verbs related to inputs and/or outputs; see Table 3 for examples). For each of the groups, we devised a predicate-centric set of patterns for the extraction of functional service descriptions. For example, *App % accepts % Input* specifies a common pattern that provides information about input for a given application. The patterns created for the most frequent verbs are given in Appendix 1.

Table 3. Examples of verbs grouped according to the type of functional "content" they typically provide

Verbs	Typical functional content
applied, access [*to*], *achieve, align, allow, based, developed, implemented, present, provide, used*	general function description, task specification
accept, applied, create, provide, query, retrieve, starts with, take [*input*], *used*	inputs, outputs
outperform, perform [*better/worse*], *compare*	comparison, similar tasks
implemented [*in/using*]	technique, programming language
contained, constructed, generated	composition, subtasks

3.4 Semantic Profiling of Bioinformatics Resources

The ultimate aim was to provide a separate profile for each instance of semantic classes of interest, resulting in a repository of descriptions that could be used for annotation. Each profile contained two conceptual parts: related key-terms and textual descriptions. The related key-terms included: associated myGrid ontology concepts, service descriptors and other related terms (including other SC instances and input/output parameters), all identified by co-occurrence within the same sentence with a given SC instance. Textual descriptions contained the extracted structured information presenting functional content (obtained by applying the rules on predicate-argument structures) and associated sentences. Figure 3 gives a summary of semantic profile of an instance, while Appendix 2 provides a complete example (more examples are available in the supplementary materials[8]).

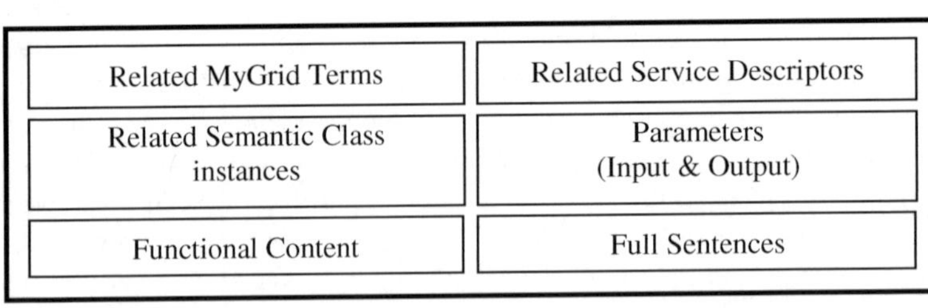

Fig. 3. Semantic profile of a semantic class instance

4 Experiments and Results

We performed two experiments to evaluate the methodology presented above. We first examined to what extent we could reconstruct the existing bioinformatics service descriptions. In the second experiment, we examined how useful the extracted semantic service profiles were for the manual curation process. Both experiments have been done in the context of the MyGrid project [12].

[8] Available at: http://gnode1.mib.man.ac.uk/bioinf/descriptions

The semantic descriptors were collected from 471 descriptions from the Feta repository and 450 descriptions retrieved from various service providing websites. These descriptors were collected by applying automatic term recognition (TerMine was used) to these descriptions. Literature data have been compiled from 2120 full text open-access articles comprising the entire publication output of BMC Bioinformatics published before March 2008. It was obvious that this task requires full-text articles as resource descriptions are not likely to appear in abstracts (unless a publication introduces a resource).

Table 4 gives the number of service instance mentions that have been identified in the corpus using terminological head comparisons and coordination co-occurrences (Section 3.1). Table 5 presents the number of functional descriptions collected for each of the four semantic classes (Section 3.3). For each service instance, a complete profile was generated (Figure 3).

Table 4. The number of instances of the semantic class in the BMC Bioinformatics corpus

Semantic Class	# of instances identified using terminological head comparison	# of instances identified using coordination co-occurrences	Total # of instances
Algorithm	5658	64	5722
Application	1862	43	1905
Data	1424	18	1442
Data Resource	2307	34	2341

Table 5. The total number of descriptions compiled for each SC

	Algorithm	Application	Data	Data resource
General description, function	1222	316	468	226
Parameters (input/output)	122	50	90	38
Comparison (similar instances)	229	37	149	35
Implementation environment	10	5	1	3

The extracted description profiles were evaluated manually by a MyGrid bioinformatics service curator, who evaluated the following three components: the quality of the associated myGrid ontological terms, the quality of service descriptors and the quality of textual descriptions. The descriptions were assessed by their capability to be used for semantic description of a given bioinformatics service. Each component was scored as follows:

– 0: completely irrelevant (e.g. sentence *The HeatMapper tool has already proven to be very useful in several studies* is irrelevant for semantic description of HeatMapper).

- 1: partially useful description (e.g. sentence *To compare Kalign to other MSA programs, the following test sets were used* is partially useful as it specified that Kalign was a multiple-sequence alignment algorithm).
- 2: contains information useful for annotation (e.g. sentence *To add a new species to the COG system, the annotated protein sequences from the respective genome were compared to the proteins in the COG database by using the BLAST program and assigned to pre-existing COGs by using the COGNITOR program* explains the functionality of COGNITOR).

In the first experiment, we randomly selected five well-known bioinformatics services that were already manually curated and then evaluated their generated profiles against the existing "gold standard" descriptions. As an example, consider a manual description of the ClustalW service[9]:

> *ClustalW is a fully automatic program for global multiple alignment of DNA and protein sequences. The alignment is progressive and considers the sequence redundancy. Trees can also be calculated from multiple alignments.*

The following description was extracted from the literature (for full results see the supplementary materials):

- **MyGrid Terms:** *phylogenetic tree, Clustalw format*;
- **service descriptors:** *ClustalW program, DNA sequence, phylogenetic tree, protein sequence, Phylogenetic Analysis, Classification tree, homology, …*
- **descriptive sentences** (selection)**:** (1) *Finally, a phylogenetic tree for the SARS-CoV isolates has been produced using the CLUSTALW program, showing high compatibility with former qualitative classification.* (2) *We also used the CLUSTALW program for multialignment as a control process, as well as for phylogenetic investigations.*

Overall (for the five existing services examined), the extracted sentences were deemed as fully useful (all textual service descriptions were scored as 2). While, semantic descriptors were considered as more than partially useful (the average score of 1.33), the extracted myGrid ontology terms were only partially useful (the average score was 1) – the main reason was that ontological terms did not appear frequently in the corpus, and thus were too sparse to be useful (see also discussion).

In the second experiment we considered five randomly chosen bioinformatics services that have not yet been manually annotated. For these services there were generally fewer documents containing mentions of the service. Still, the results were only slightly worse: the textual descriptions were deemed highly useful (the average score of 1.67), while semantic descriptors were uniformly given a score of 1. Interestingly, there were no MyGrid terms extracted for any of the services examined.

[9] This description is from the EBI's repository.

5 Discussion

The main aim of our work was to provide a technology to facilitate semantic description of bioinformatics services for both service curators and for automated annotation understandable in the context of the Semantic Web. The results of the application of our methodology suggested that text mining techniques could produce accurate and efficient descriptions of bioinformatics services. The quality of the extracted data was measured from the curator's perspective only (i.e. how useful extracted descriptions are for their task). As expected, sentences were most useful as these provided a broader context for the curator to annotate a service. Key service descriptors were only partially useful for curation, but – on the other hand – these are more likely to be appropriate for providing a semantic synopsis of a service that could be used to represent the service in the Semantic Web context and/or locate related services automatically (by a search agent). Although there are limitations for the accurate capturing of predicate-argument structures, translating an extracted PAS into a related Resource Description Framework (RDF) statement would make the description a direct component for the Semantic Web representation of services.

As expected, there were obvious limitations in retrieving related ontological terms from the literature. As previously shown for other ontologies, conceptual descriptions are not likely to appear in text in the form in which they appear in an ontology [13]. Therefore, additional string comparison methods (including similarity distance, n-grams, variation, etc.) would need to be employed to improve the coverage [14].

Despite the number of mentions identified, the recognition of tool and service instances by using only lexical and syntactical rules proved to be difficult. As we used terminological heads to identify instances, in some cases references to generic classes were also selected (e.g. *clustering algorithm*). Therefore, context-driven rule based patterns (including PAS) would be needed to recognise them more accurately and differentiate between mentions of generic concepts and specific instances.

Phrase structure parses enabled us to extract the information in a more descriptive form rather than just by key entities. This information could be used to describe specific aspects of services/tools and thus have a potential to bridge the gap between co-occurrence based service descriptors and full sentence extraction. Still, providing and integrating phrase structures with dependency parsing was challenging, since there were numerous verbs used to communicate similar information and thus this step was not as productive as expected. In addition, some of the semantic classes were similar to each other in their "textual" behaviour (e.g. both Algorithms and Applications perform a specific functionality on a particular input, generating a specific output), and therefore the extraction patterns would need to take into account both the semantic classes and verbs involved, as well as their specific roles for the given class.

6 Conclusion

In this paper we presented a literature mining approach to automatically extract information that semantically described bioinformatics Web services and tools. The work had two aims: reducing the amount of effort invested by domain experts in manual curation, and providing semantic synopses of services that could be used in the context of the Semantic Web.

The suggested methodology was based on the concept of major semantic classes associated to bioinformatics resources (e.g. applications, algorithms, data, data resources) that were compiled from the myGrid bioinformatics ontology. The instances of these classes were collected from the bioinformatics literature along with related sentences. Sentence filtering was followed by the extraction of semantic descriptors that referred to frequent semantic roles of tools and services in a particular context. The sentences were also dependency parsed and integrated with phrase structures separately recognised. The integrated dependency structures were used to identify specific functional content. A semantic profile of each service/tool was generated by combining descriptive sentences, semantic descriptors, links to other related tools, and semantic labels linked to the myGrid ontology. The presented case-study involving a subset of full text articles from BMC Bioinformatics illustrated the potential of natural language processing not only for mining descriptions of known services, but also for discovering and describing new services that have been mentioned in the literature.

To the best of our knowledge, this is the first attempt to mine semantic service description from the domain literature. Although the initial results are promising, there is room for improvements which would include a context-based recognition approach to service/tools mentions, improving specificity and coverage of PAS-based information extraction, and inferring semantic descriptions from textually related services. Finally, the utility of the proposed approach will be further tested within the BioCatalogue project, which is a registry that provides documentation, location and semantic annotations for Web Services for Life Sciences.

Acknowledgements

We are grateful to Franck Tanoh for the evaluation of the extracted service descriptions. This work was partially supported by the UK Biotechnology and Biological Science Research Council (BBSRC) via the projects "Mining Term Associations from Literature to Support Knowledge Discovery in Biology" and "BioCatalogue". We are thankful to the National University of Sciences & Technology Pakistan's Higher Education Program for additional support.

References

1. Cannata, N., Merelli, E., Altman, R.B.: Time to Organize the Bioinformatics Resourceome. PLoS Computational Biology 1, e76 (2005)
2. Lord, P., Alper, P., Wroe, C., Goble, C.: Feta: A Light-Weight Architecture for User Oriented Semantic Service Discovery. In: Gómez-Pérez, A., Euzenat, J. (eds.) ESWC 2005. LNCS, vol. 3532, pp. 17–31. Springer, Heidelberg (2005)
3. Wolstencroft, K., Alper, P., Hull, D., Wroe, C., Lord, P.W., Stevens, R.D., Goble, C.A.: The myGrid Ontology: Bioinformatics Service Discovery. International Journal of Bioinformatics Research and Applications 3, 326–340 (2007)

4. Carman, M.J., Knoblock, C.A.: Learning Semantic Descriptions of Web Information Sources. In: International Joint Conferences on Artificial Intelligence, Hyderabad, pp. 2695–2701 (2006)
5. Lerman, K., Plangrasopchok, A., Knoblock, C.A.: Automatically Labeling the Inputs and Outputs of Web Services. In: Proc. of AAAI 2006, Boston, MA, USA, pp. 149–181 (2006)
6. Hess, A., Kushmerick, N.: Learning to Attach Semantic Metadata to Web Services. In: Proc. 2nd International Semantic Web Conference, Sanibel Island, Florida, USA (2003)
7. Belhajjame, K., Embury, S.M., Paton, N.W., Stevens, R., Goble, C.A.: Automatic Annotation of Web Services based on Workflow Definitions. In: Cruz, I., Decker, S., Allemang, D., Preist, C., Schwabe, D., Mika, P., Uschold, M., Aroyo, L.M. (eds.) ISWC 2006. LNCS, vol. 4273, pp. 116–129. Springer, Heidelberg (2006)
8. Wattarujeekrit, T., Shah, P., Collier, N.: PASBio: predicate-argument structures for event extraction in molecular biology. BMC Bioinformatics 5, 155 (2004)
9. Tateisi, Y., Ohta, T., Tsujii, J.: Annotation of Predicate-argument Structure on Molecular Biology Text. In: Workshop on the 1st International Joint Conference on Natural Language Processing (IJCNLP 2004) (2004)
10. Afzal, H., Stevens, R., Nenadic, G.: Towards Semantic Annotation of Bioinformatics Services: Building a Controlled Vocabulary. In: Proc. of the Third International Symposium on Semantic Mining in Biomedicine, Turku, Finland, pp. 5–12 (2008)
11. Hearst, M.A., Schutze, H.: Customizing a lexicon to better suit a computational task. In: Corpus processing for lexical acquisition, pp. 77–96. MIT Press, Cambridge (1996)
12. Oinn, T., Li, P., Kell, D.B., Goble, C., Goderis, A., Greenwood, M., Hull, D., Stevens, R., Turi, D., Zhao, J.: Taverna/myGrid: aligning a workflow system with the life sciences community. In: Dennis, B., Ian, G., Taylor, J., Deelman, E., Shields, M. (eds.) Workflows for e-Science: scientific workflows for Grids, pp. 300–319. Springer, Guildford (2007)
13. Spasic, I., Ananiadou, S., McNaught, J., Kumar, A.: Text mining and ontologies in biomedicine: Making sense of raw text. Briefings in Bioinformatics 6, 239–251 (2005)
14. Rebholz-Schuhmann, D., Arregui, M., Gaudan, M., Kirsch, H., Jimeno, A.: Text processing through Web services: Calling Whatizit. Bioinformatics 24(2), 296–298 (2007)

Appendix 1: A Collection of Predicates, Patterns and Their Usage for Information Extraction

In these patterns, <SC_I> represents an instance of any SC, whereas <Algo>, <Data>, <App>, <Data Resource> depict instances of the corresponding SCs.

Predicate	**Pattern**
Accept	<App/Algo> % accepts % <Input>
Access	<App> provides % access to % <Data Resource>
achieve	<App/Algo> % achieves % <Performance-quality>
Align	<App/Algo> % aligns the % <Data>.
Allow	<SC_I> allows user to <Description>
Applicable	<App/Algo> % are/is applicable to % <Problem/Input>
Applied	< App >% applied % <Algo> % to <NP: Input(s)>/ <VP: Function>
Available	< App > is available at/on <Forum/Website>
Based	<Algo > is based on <Concept>
Called	Gives specific name of a generalized concept
Compare	<SC_I>compares<SC_I>
Constructed	<Data/Data Resource> constructed from/using <composition>
Constructed	<Data/Data Resource/Output> constructed by/using < App/Algo>
Contain	<Database > contains [information] <composition>
Create	<SC_I> used <SC_I> to create <Output>
Developed	< App /Algo> developed <Algo> that/which/to/for <Function>
Generated	<Data/Data Resource/Output> generated by < App /Algo>
Generated	< Data/Data Resource/Output> generated using <App/Algo>
Implemented	<App> implemented <Algo> to <Function>
Implemented	<Algo> implemented using/in/as <tech>/<language>/<part-of>
Outperform	< App/Algo> outperforms <App/Algo>
Outperform	<SC_I> outperforms <SC_I>
Perform	<SC_I> performs better than < SC_I>
Present	present <Algo> that/which<Function>
Querie (s/d)	<SC_I> queries/queries <Data Resource (Database)>
Retrieve	<SC_I> retrieves/retrieved <Output/Result>
Take	<App/Algo> takes input <Input>
Used	<SC_I> % is used to % <Function>

Appendix 2: Extracted Service Descriptions for the GeneClass Algorithm (Example)

Related service descriptors

Descriptor	Frequency of co-occurrence
GeneClass algorithm	5
Motif data	4
Reliable predictive model	2
genome-wide protein-DNA binding data	2
Differential gene expression	3
Transcriptional gene regulation	2

Descriptive sentences

1. *We also show how to incorporate genome-wide protein-DNA binding data from ChIP chip experiments into the GeneClass algorithm, and we use an improved noise model for gene expression data.* [PMCID:1810316]
2. *Target set: We extend the original GeneClass algorithm to use all target genes for which both motif and expression data is available.* [PMCID:1810316]
3. *Motif data versus CHIP chip data: In order to study different aspects of target gene regulation we use different sets of motifs and parents with the GeneClass algorithm.* [PMCID:1810316]
4. *The GeneClass algorithm for predicting differential gene expression starts with a candidate set of motifs; representing known or putative regulatory element sequence patterns and a candidate set of regulators or parentSS.* [PMCID:1810316]

Predicate-argument structures and associated functional content

Predicate	Subject	Object	Type
show	*We*	*how to incorporate genome-wide protein-DNA binding data from ChIP chip experiments into the GeneClass algorithm*	Functional Description
predict	*The GeneClass algorithm*	*differential gene expression starts with a candidate set of motifs x003bc*	Input/ Output
extend	*We*	*the original GeneClass algorithm to use all target genes for which both motif and expression data is available*	Functional Description

Hybrid Adaptive Web Service Selection with SAWSDL-MX and WSDL-Analyzer

Matthias Klusch, Patrick Kapahnke, and Ingo Zinnikus

German Research Center for Artificial Intelligence
Stuhlsatzenhausweg 3, Saarbrücken, Germany
{klusch,patrick.kapahnke,ingo.zinnikus}@dfki.de

Abstract. In this paper, we present an adaptive, hybrid semantic matchmaker for SAWSDL services, called SAWSDL-MX2. It determines three kinds of semantic service similarity with a given service request, that are logic-based, text-based and structural similarity. In particular, the degree of structural service similarity is computed by the WSDL-Analyzer tool [12] by means of XMLS tree edit distance measurement, string-based and lexical comparison of the respective XML-based WSDL services. SAWSDL-MX2 then learns the optimal aggregation of these different matching degrees over a subset of a test collection SAWSDL-TC1 based on a binary support vector machine-based classifier. Finally, we compare the retrieval performance of SAWSDL-MX2 with a non-adaptive matchmaker variant SAWSDL-MX1 [1] and the straight forward combination of its logic-based only variant SAWSDL-M0 with WSDL-Analyzer.

1 Introduction

As a W3C recommendation dated August 28, 2007, the SAWSDL[1] specification proposes mechanisms to enrich Web services described in WSDL[2] (Web Service Description Language) with semantic annotations. Among others, one goal of these additional descriptions is to support intelligent agents in automated service selection, a task which is hard to accomplish using pure syntactic information of service profiles based mainly on XML-Schema definitions. Typical application scenarios that require or benefit from a service matchmaking component include for example negotiation and coalition forming among agents and automated or assisted service composition. The first hybrid semantic service matchmaker SAWSDL-MX1 for SAWSDL we proposed in [1] adopted the ideas of our hybrid matchmakers OWLS-MX and WSMO-MX (see [5,3]) for OWL-S, respectively, WSML.

However, SAWSDL-MX1 focused on semantic annotations of the signature but not on the XML structure of the Web service as a whole. This is taken into

[1] http://www.w3.org/TR/sawsdl/
[2] http://www.w3.org/TR/wsdl/,
http://www.w3.org/TR/wsdl20/

L. Aroyo et al. (Eds.): ESWC 2009, LNCS 5554, pp. 550–564, 2009.

account by the WSDL-Analyzer tool presented in [12] by means of measuring the XML tree edit distances between given pair of services through XML type compatibility, token-based text and lexical similarity measurements. Besides, SAWSDL-MX1 combines logic-based and text-similarity based matching in a fixed manner: It applies five logical matching filters and ranks service offers that share the same logical matching degree with respect to a given request according to their text similarity value. The hybrid variant SAWSDL-M0+WA does the same as SAWSDL-MX1 except that its ranking of services with the same logical matching degree is according to their structural similarity value as computed by the WSDL-Analyzer. Finally, the adaptive hybrid variant SAWSDL-MX2 computes three kinds of semantic matching, logical, text and structural similarity-based. In addition, it learns the optimal aggregation of these different types of semantic matching to decide on the semantic relevance of a service to a given request.

One major question concerns the practical applicability of these different matchmakers in general, not restricted to some given domain-specific and/or very small-sized scenario, by means of their retrieval performance over a given initial test collection, SAWSDL-TC1, that consists of more than 900 SAWSDL services from different application domains. The results of our experiments shows that all hybrid matchmaker variants outperform the single matching type variants (logic-based or text or structural only) in terms of precision, while the performance of all three hybrid variants do not significantly differ with respect to SAWSDL-TC1.

The remainder of the paper is structured as follows. After a brief introduction to SAWSDL in section 2, the matching approach of the non-adaptive matchmaker SAWSDL-MX1 is recapitulated in section 3. Section 4 presents the structural matching of WSDL services performed by the WSDL-Analyzer tool. The adaptive aggregation of matching results based on a binary SVM-based classifier by our new adaptive matchmaker SAWSDL-MX2 is described in section 5. Results of our experimental evaluation over the given test collection SAWSDL-TC1 in terms of recall, (macro-)averaged precision and average query response time are shown in section 6. We comment on related work in section 7, and conclude in section 8.

2 Service Descriptions in SAWSDL

SAWSDL is designed as an extension of WSDL enabling service providers to enrich their service descriptions with additional semantic information. For this purpose, the notions of *model reference* and *schema mapping* have been introduced in terms of XML attributes (tags) that can be added to already existing WSDL service description elements as depicted in figure 1.

Semantic annotation of WSDL services. More precisely, the following extensions are used for semantic annotations of WSDL services:

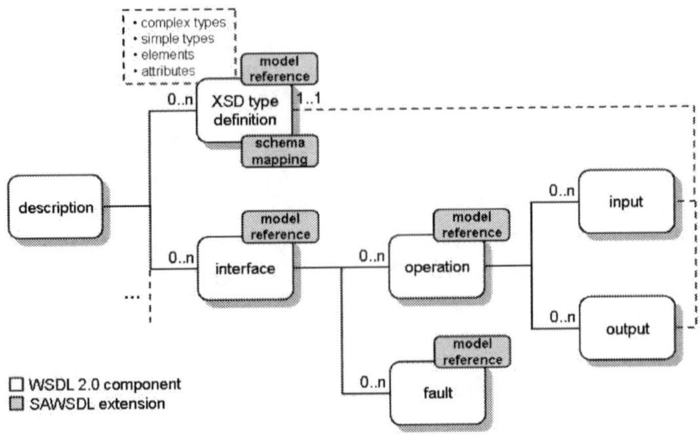

Fig. 1. SAWSDL extensions of WSDL interface components

- *modelReference*: A *modelReference* points to one ore more concepts with equally intended meaning expressed in an arbitrary semantic representation language. They are allowed to be defined for every WSDL and XML Schema element, though the SAWSDL specification defines their occurrence only in WSDL interfaces, operations, faults as well as XML Schema elements, complex types, simple types and attributes. The purpose of a model reference is mainly to support automated service discovery.
- *liftingSchemaMapping*: Schema mappings are intended to support automated service execution by providing rules specifying the correspondences between semantic annotation concepts defined in a given ontology (the "upper" level) to the XML Schema representation of data actually required to invoke the Web service (the "lower" level), and vice versa. A *liftingSchemaMapping* describes the transformation from the "lower" level in XML Schema up to the ontology language used for semantic annotation.
- *loweringSchemaMapping*: The attribute *loweringSchemaMapping* describes the transformation from the "upper" level of a given ontology to the "lower" level in XML Schema.

However, the current specification of SAWSDL model references imposes quite some problems for semantic service matchmaking as follows.

No uniform, formal ontology language. Unlike OWL-S or WSML, the specification of SAWSDL does not restrict the developer to any uniform, formal ontology langage like OWL or WSMO. As a result, any mean of automated semantic service selection has to cope with the semantic interoperability problems of heterogeneous domain ontologies and ontology languages. While this problem could be resolved in some cases by means of syntactic and semantic transformations - such as for OWL-DL and WSML-DL - it remains hard in general.

Multiple references to different ontologies. The same holds especially for references to different kinds of ontologies like plain or structured text files, annotated image archive, or logic theories. In fact, SAWSDL allows multiple references to different kinds of ontologies for annotating even the same service description element. How shall any semantic service matchmaker know how to process them to understand the semantics of that single element? Are its annotations meant to be complementary or equivalent? If complementary, how to aggregate them, if equivalent, which one to select best for further processing? This opens up a wide range of pragmatic solutions for SAWSDL service matching.

Top-level vs bottom-level annotations. According to the SAWSDL specification, semantic annotation by means of so-called *top-level annotation* and *bottom-level annotation* shall be considered both independent from each other and applicable at the same time. While *top-level annotation* refers to the annotation of a complex type or element definition of a message parameter by means of a model reference as a whole, any *bottom-level annotation* focuses only on a single (atomic) XML element. Unfortunately, it remains unclear how to evaluate a matching *between* top-level and low-level annotated parameters, or which one to prefer if both levels of annotation are available for a complex service description element. In addition, element and type definition specifying a message component can be annotated at the same time.

Pragmatic assumptions for SAWSDL service matching. Regarding the above mentioned problems of SAWSDL service matching, the following pragmatic assumptions were made for using our SAWSDL-MX matchmaker variants SAWSDL-M0+WA, SAWSDL-MX1, and SAWSDL-MX2:

- References to formal ontologies in description logics only. The current implementation of SAWSDL-MX performs reasoning on logic-based annotations in OWL-DL[3] but is not restricted to it: It supports other description logics (DL) if they are translated into the standard DIG 1.1[4] interface representation format.
- Only top-level semantic annotations of service parameters are considered for service matching. However, any direct top-level annotation of a WSDL message part has priority over the top-level annotation of the respectively referenced (and likewisely annotated) XML-Schema element or type.
- In case of multiple annotations of a single element at the same level, one of them is selected uniformly at random. Only semantic annotations of service (IO) parameters are considered, but not annotations of entire operations or interfaces. However, the proposed matching variants could easily be adopted for this purpose.

In the following sections, we describe each of these different SAWSDL matchmaker variants and results of their comparative experimental evaluation of performance.

[3] http://www.w3.org/2004/OWL/
[4] http://dig.sourceforge.net/

3 SAWSDL-MX: Logic and Text Similarity-Based Signature Matching

In this section, we describe the hybrid service signature matching performed by both SAWSDL-MX matchmaker variants, the non-adaptive SAWSDL-MX1 and the adaptive SAWSDL-MX2. The logic-based only variant of SAWSDL-MX is called SAWSDL-M0. Service requests and offers are presumed to be formulated in SAWSDL, each comprising one or multiple operations with semantically annotated signatures.

3.1 Hybrid Service Interface Matching

For each pair of service offer O and service request R, the matchmaker determines their semantic similarity by evaluating every combination of their operations in terms of either logic-based only (SAWSDL-M0) or text similarity-based only operation matching, or both (SAWSDL-MX1, SAWSDL-MX2). The process of logic-based and text similarity-based (service) operation matching is described in more detail below.

To determine an injective mapping between service offer and request operations that is optimal regarding their matching degrees, SAWSDL-MX applies bipartite operation graph matching. Nodes in the graph represent the operations and the weighted edges are built from possible one-to-one assignments with their weights derived from the computed degree of (logical/text/hybrid) operation match. If there exists such a mapping, then it is guaranteed that there exists an operation of the service offer for every requested operation, disregarding the quality of their matching at this point.

For example, consider the service request and service offer given in figure 2. Every request operation RO_i (with $i \in \{1,2\}$) is compared to every advertisement operation O_j (with $j \in \{1,2,3\}$) with respect to logic-based filters defined in the next section. In this example, RO_1 exactly matches with O_1, but fails for O_2 and O_3. O_3 is a weaker plug-in match for RO_2 (the subsumed-by match of RO_2 with O_2 is even weaker than a plug-in match). The best (max) assignment of matching operations is $\{\langle RO_1, O_1\rangle, \langle RO_2, O_3\rangle\}$.

One conservative (min-max) option of determining the matching degree between service offer and request based on their pairwise operation matchings is to assume the worst result of the best operation matchings. In other words, we guarantee a fixed *lower* bound of similarity for *every* requested operation - which is what SAWSDL-MX is doing. In the example shown in figure 2, the service offer is considered a *plug-in* match for the request. Other not yet implemented possibilities would be to merge the operation matching results based on their average syntactic similarity values and to provide more detailed feedback to the user on the operation matchings involved.

3.2 Logic-Based Operation Matching

The logic-based operation matching by SAWSDL-MX bases on the successive application of the following four filters of increasing degree of relaxation to a

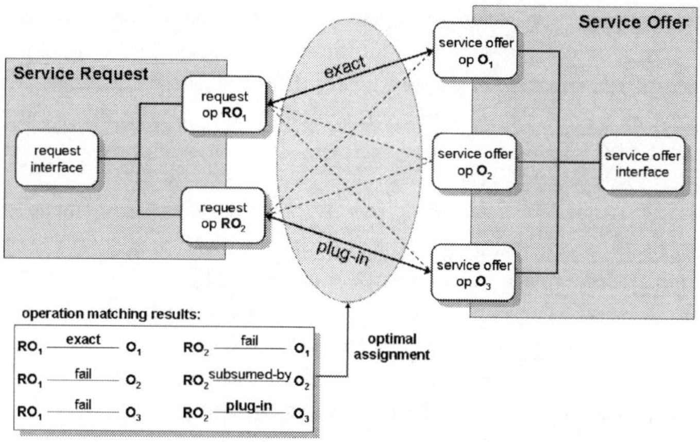

Fig. 2. Interface level matching of SAWSDL-MX

given pair service offer operation O_O and service request O_R: *Exact*, *Plug-in*, *Subsumes* and *Subsumed-by*. These filters have been originally developed for the matchmaker OWLS-MX but extended with bipartite concept graph matching to ensure an injective mapping between I/O concepts of service offer and request, whenever possible. As an overview to description logic and DL reasoning, we refer to [9]. The sets $lgc(c)$ and $lsc(c)$ contain the least generic concepts of c (direct parent) and the least specific concepts of c (direct child), respectively.

Exact match: Service operation O_O *exactly* matches service operation $O_R \Leftrightarrow$ ($\exists$ injective assignment $M_{in} : \forall m \in M_{in} : m_1 \in in(O_O) \wedge m_2 \in in(O_R) \wedge m_1 \equiv m_2$) $\wedge$ ($\exists$ injective assignment $M_{out} : \forall m \in M_{out} : m_1 \in out(O_R) \wedge m_2 \in out(O_O) \wedge m_1 \equiv m_2$). There exist a one-to-one mapping of perfectly matching inputs as well as perfectly matching outputs. Assuming that an operation fullfills a requesters need if every input can be satisfied and every requested output is provided, the assignments only require to be injective (but not bijective), thus additional available information not required for service invocation and additional provided outputs not explicitly requested are tolerated.

Plug-in match: Service operation O_O *plugs into* service operation $O_R \Leftrightarrow$ ($\exists$ injective assignment $M_{in} : \forall m \in M_{in} : m_1 \in in(O_O) \wedge m_2 \in in(O_R) \wedge m_1 \sqsupseteq m_2$) $\wedge$ ($\exists$ injective assignment $M_{out} : \forall m \in M_{out} : m_1 \in out(O_R) \wedge m_2 \in out(O_O) \wedge m_2 \in lsc(m_1)$). The filter relaxes the constraints of the *exact* matching filter by additionally allowing input concepts of the service offer to be arbitrarily more general than those of the service request, and advertisement output concepts to be direct child concepts of the queried ones.

Subsumes match: Service operation O_O *subsumes* service operation $O_R \Leftrightarrow$ ($\exists$ injective assignment $M_{in} : \forall m \in M_{in} : m_1 \in in(O_O) \wedge m_2 \in in(O_R) \wedge m_1 \sqsupseteq m_2$) $\wedge$ ($\exists$ injective assignment $M_{out} : \forall m \in M_{out} : m_1 \in out(O_R) \wedge m_2 \in$

$out(O_O) \wedge m_1 \sqsupseteq m_2$). This filter further relaxes constraints by allowing service offer outputs to be arbitrarily more specific than the request outputs (as opposed to the *plug-in* filter, where they have to be direct children). Thus, a *plug-in* can be seen as special case of a *subsumes* match resulting in a more fine-grained view at the overall service ranking.

Subsumed-by match: Service operation O_O is *subsumed by* service operation $O_R \Leftrightarrow$ ($\exists$ injective assignment $M_{in} : \forall m \in M_{in} : m_1 \in in(O_O) \wedge m_2 \in in(O_R) \wedge m_1 \sqsupseteq m_2$) $\wedge$ ($\exists$ injective assignment $M_{out} : \forall m \in M_{out} : m_1 \in out(O_R) \wedge m_2 \in out(O_O) \wedge m_2 \in lgc(m_1)$). The idea of the *subsumed-by* matching filter is to determine the service offers that the requester is able to provide with all required inputs and at the same time deliver outputs that are at least closely related to the requested outputs in terms of the inferred concept classification.

The matching degree of *fail* is true if and only if none of the matching filters defined above succeed. As a result, services are ranked according to their matching degree in the following decreasing order: *exact* > *plug-in* > *subsumes* > *subsumed-by* > *fail.*

SAWSDL-M0. The logic-based only matchmaker variant SAWSDL-M0 applies the above logical matching filters only, and ranks service offers that share the same logical matching degree with respect to a given request uniformly at random.

3.3 Text Similarity-Based Operation Matching

The hybrid variants of SAWSDL-MX also perform complementary text similarity-based matching by means of classical token-based text similarity measures *Loss-of-Information*, *Extended Jaccard*, *Cosine* and *Jensen-Shannon* as implemented, for example, in SimPack[5]. For this purpose, the signatures of both request and offer are considered as text such that the degree of semantic similarity is measured in terms of their averaged text similarity.

More concrete, each semantic service signature is transformed into a pair of weighted keyword vectors for input, respectively, output - according to the classical vector space model of information retrieval. For this purpose, each input concept is logically unfolded in the shared ontology (as defined for standard tableaux reasoning algorithms) and concatenated with all others to a complex logical expression containing only primitive components and logical operators. This expression is treated as mere text string being processed to a TFIDF weighted keyword vector; the same is done with service output concepts. The TFIDF term weighting values are computed over two distinct text indices depending on whether service inputs or outputs are compared.

SAWSDL-MX1. The hybrid semantic service matchmaker SAWSDL-MX1 applies the logical matching filters mentioned above and ranks service offers that

[5] http://www.ifi.uzh.ch/ddis/research/semweb/simpack/

share the same logical matching degree with respect to a given request according to their text similarity value.

4 WSDL-Analyzer: Structural WSDL Matching

The WSDL-Analyzer (WA) tool introduced in [12] performs a strucrual only matching of WSDL 1.1 services. In fact, it ignores the semantic annotations of SAWSDL descriptions and treats a SAWSDL description as a mere WSDL description. The WA tool detects similarities and differences between WSDL files and can be used to find a list of non-logic-based semantically relevant Web services. Since its similarity algorithm produces a mapping between WSDL service descriptions, the tool can also be used for supporting mediation between services.

More concrete, the WA tool exploits various types of schema information such as element names, datatypes and structural properties, and characteristics of data instances, as well as background knowledge from dictionaries and thesauri. The similarity algorithm calculates the similarity between the XML structures of a requested and a candidate services, respects the structural information of complex datatypes and is flexible enough to allow for relaxed matching as well as matching between parameters that come in different orders in service parameter lists.

The comparison of two WSDL services is a multi-step process. It starts off with (1) the comparison of the operation sets offered by the services, which is based on (2) the comparison of the structures of the operations input and output messages, which, in turn, is based on (3) the comparison of the datatypes of the objects communicated by these messages. The recursive structural matching of two XML-based WSDL service descriptions is performed as follows.

A WSDL description is represented as a labelled tree where leaf nodes are the basic built-in datatypes provided by the XML schema specification[6]. Let $L = \{l_1, l_2, ..., l_n\}$ be a set of labels.

A **labelled tree** $T = (N, E, root(T), \varphi)$ is an acyclic, connected graph with:

- $N = \{n_1, n_2, ..., n_n\}$ is a set of nodes.
- $E \subset N \times N$ is a set of edges.
- $root(T)$ the root of the tree.
- $\varphi : N \to L$ is a function which assigns a label to each node with basic datatypes $D \subset L$.

The process of calculating the similarity of two trees T_1 and T_2 starts with the roots $root(T_1)$ and $root(T_2)$, and traverses these trees recursively:

For $a \in N_{T_1}$ and $b \in N_{T_2}$ do compute,

$$sim(a,b) = \begin{cases} \omega_{name} * sim_{name}(\varphi(a), \varphi(b)) \\ +\omega_{struct} * max(\oplus_{i,j}(sim(n_i, m_j))), & \varphi(a), \varphi(b) \notin D \\ sim_{type}(\varphi(a), \varphi(b)), & \varphi(a), \varphi(b) \in D \end{cases} .$$

[6] http://www.w3c.org/TR/xmlschema-2/

where $(a, n_i) \in E_{T_1}$, $(b, m_j) \in E_{T_2}$ and $\oplus_{i,j}(n_i, m_j)$ denotes the sum of pairs $sim(n_i, m_j)$ for $1 \leq i \leq card(n)$ and $1 \leq j \leq card(m)$ such that each n_i and m_j occur at most once in the sum. If $card(n) \neq card(m)$, some of the nodes cannot be matched. Weights ω_{name} and ω_{struct} are used to either increase or decrease the effect of element (label) name or structural similarity. Computation of type similarity sim_{type} bases on a given type compatibility table which assigns a value to each combination of basic data types. The similarity of labels sim_{name} can be calculated with different measures such as string edit distance, substring containment or WordNet[7] similarity (semantic proximity). In order to improve the mapping results, we used substring matching and WordNet. Experiments showed that especially in rather standardized areas the results are better than with pure data type mapping.

SAWSDL-M0+WA. The hybrid semantic service matchmaker SAWSDL-M0+WA applies the logical matching filters only, and ranks service offers that share the same logical matching degree with respect to a given request according to their degree of structural similarity as computed by the WSDL-Analyzer.

5 SAWSDL-MX2: Adaptive Matching Aggregation

Inspired by [2,4], we developed an adaptive, hybrid semantic service matchmaker SAWSDL-MX2. This matchmaker (a) separately computes three different kinds of semantic service matching degrees, that are logical, text and structural matching (each of them as described in previous sections), and then (b) learns over a given traing set how to best aggregate them to decide on the semantic relevance (ranking) of a service to a given request. For the latter purpose, SAWSDL-MX2 exploits a Support Vector Machine (SVM) to learn a binary classification function, which is characterized by a hyperplane in a given feature space. This classification function evaluates for any given pair of service offer and request, their binary relevance class membership, i.e. *relevant* or *irrelevant*, for the matching problem at hand. For result ranking, the distances of training samples to this plane are computed, and then taken as reference similarity values for deciding on the relevance of unknown services.

More concrete, let $X = \{0,1\}^5 \times [0,1] \times [0,1]$ feature space with feature vectors $x_i = (f_1, f_2, f_3, f_4, f_5, f_6, f_7)$ feature vectors where each feature f_1 to f_5 represents the logic-based matching result for a service query/offer pair including *fail*, feature f_6 stands for the text-based similarity value, and f_7 for the structural similarity value computed by the WSDL-Analyzer. For example, the feature vector $(0,0,1,0,0,0.6,0.7)$ represents the hybrid semantic matching result computed by SAWSDL-MX as follows: A logical *subsumes* match with text similarity of 0.6 and structural similarity of 0.7. The y_i values of the training samples equal -1 for an irrelevant service offer given a query, and 1 for the relevant case. As usual, relevance sets in the test collection are subjectively defined

[7] http://wordnet.princeton.edu/

by domain experts. Eventually, the input to the SVM of SAWSDL-MX2 is a set of training examples $\{(x_1, y_1), \ldots, (x_m, y_m)\}$ with $x_i \in X$ and $y_i \in \{-1, 1\}$.

The result of running a SVM on such input is a hyperplane, possibly in a higher dimensional space, which separates training examples in the feature space as good as possible while the distance of the nearest points of each category is maximized to avoid biased categorization. This is expressed in the following optimization problem:

$$\text{minimize } w,b,\zeta\text{: } \frac{1}{2}w^T w + C\sum_{i=1}^{N} \zeta_i$$

$$\text{subject to } \forall 1 \leq i \leq N : y_i(w^T \phi(x_i) + b) \geq 1 - \zeta_i, \zeta_i \geq 0,$$

where w and b define the optimal hyperplane according to the previously mentioned characteristics. The error term $C\sum_{i=1}^{N} \zeta_i$ is introduced to allow for outliers in a non-linear separable training set, where the error penalty parameter C must be specified beforehand. ϕ is a predefined function which maps features into a higher dimensional space. There exists also a dual problem description, which utilizes *Lagrange multipliers* to express the hyperplane as linear combination of support vectors. This form allows for the introduction of a *kernel* function K, which implicitly defines the original function ϕ of the primal problem. For our experiments, the *radial basis function* (RBF) has been used as kernel. It is controlled by a second parameter γ and is defined as follows: $K(x_i, x_j) = e^{-\gamma\|x_i - x_j\|^2}$. To find a good parameter setting (C, γ), the n-fold cross validation and grid-search approach proposed in [13] has been conducted. For more details on SVM's in general and on the dual problem solving, we refer the interested reader for example to [14].

Implementation of SAWSDL-MX2. SAWSDL-MX2 has been fully implemented in Java using the sawsdl4j[8] API (handling SAWSDL for WSDL 1.1) and the OWL API[9] for access to SAWSDL and OWL files, the DIG 1.1 as standard interface to handle SHOIQ knowledge base queries, and the Pellet[10] reasoner as inference engine for logic-based matchmaking. As SVM implementation, we used libSVM[11].

6 Evaluation of Performance

For the adaptive integration of SAWSDL-MX and WSDL Analyzer by SAWSDL-MX2, a retrieval performance evaluation based on the well known measures recall and precision has been conducted. To prove the statistical significance of different matching variants, we applied the *Friedman Test*. In the following, we focus

[8] http://knoesis.wright.edu/opensource/sawsdl4j/
[9] http://owlapi.sourceforge.net/
[10] http://pellet.owldl.com/
[11] http://www.csie.ntu.edu.tw/ cjlin/libsvm/

on the comparative evaluation of the three SAWSDL-MX variants described in previous sections: the non-adaptive SAWSDL-M0+WA, SAWSDL-MX1 and the adaptive SAWSDL-MX2. For more detailed results on SAWSDL-MX1 alone, we refer to [1].

6.1 Evaluation Setup

The experimental evaluation of service retrieval performance is based on the first SAWSDL test collection SAWSDL-TC1. It is semi-automatically derived from OWLS-TC 2.2[12] using the OWLS2WSDL[13] tool, as there is currently no other standard test collection for SAWSDL available. OWLS2WSDL transforms OWL-S service descriptions (and concept definitions relevant for parameter description) to WSDL through syntactic transformation. Top-level annotations taken from the original OWL-S descriptions have been added for XML Schema type definitions used to describe message inputs and output. The collection consists of around 900 Web services covering different application domains: education, medical care, food, travel, communication, economy and weaponry. It also includes a set of queries and binary relevance sets subjectively specified by domain experts. As one result, each service in SAWSDL-TC1 contains only a single interface with one operation. All automatically derived model references point to OWL ontologies. Therefore, this test collection can only be seen as a first attempt towards a commonly agreed testing environment for SAWSDL service discovery and our evaluation has to be considered as preliminary. The performance tests have been conducted on a machine with Windows 2000, Java 6, 1,7 GHz CPU and 2 GB RAM using the SME^2 tool[14] for automated evaluation.

6.2 Performance Tests

For retrieval performance evaluation, we measured precision and recall: $Prec = \frac{|A \cap B|}{|B|}$ and $Rec = \frac{|A \cap B|}{|A|}$, where A is the set of all relevant documents, and B the set of all retrieved documents for a request. Further, we measured the macro-averaged precision: $Prec_i = \frac{1}{|Q|} \cdot \sum_{q \in Q} \max\{P_o | R_o \geq Rec_i \wedge (R_o, P_o) \in O_q\}$, where O_q denotes the set of observed pairs of recall/precision values for query q when scanning the ranked services in the answer set for q stepwise for true positives in the relevance sets of the test collection. For evaluation, the answer sets are the sets of all services registered at the matchmaker which are ranked with respect to their (totally ordered) matching degree. In other words, we computed the mean of precision values for answer sets returned by the matchmaker for all queries in the test collection at standard recall levels Rec_i $(0 \leq i < \lambda)$. Ceiling interpolation is used to estimate precision values that are not observed in the answer sets for some queries at these levels; that is, if for some query there is no precision value at some recall level (due to the ranking of

[12] http://projects.semwebcentral.org/projects/owls-tc/
[13] http://projects.semwebcentral.org/projects/owls2wsdl/
[14] http://projects.semwebcentral.org/projects/sme2/

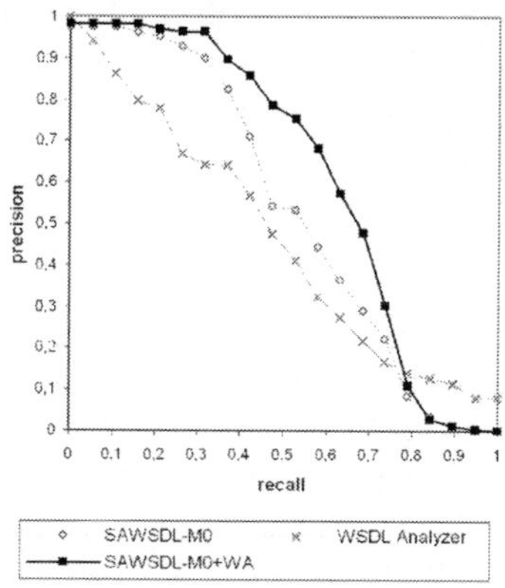

(a) integration vs. basic strategies

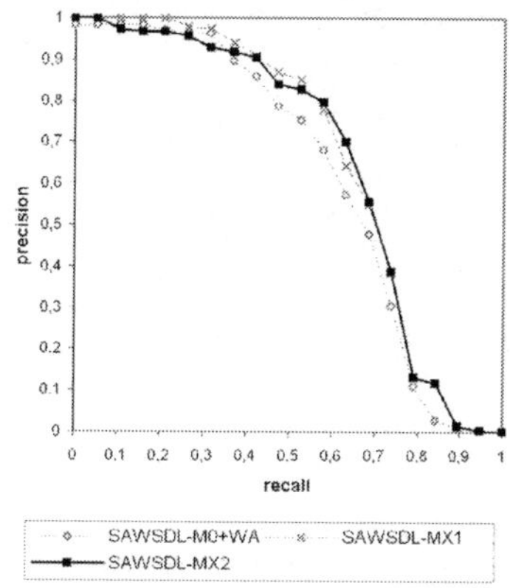

(b) different integration variants

Fig. 3. Performance of matching variants

services in the returned answer set by the matchmaker) the maximum precision of the following recall levels is assumed for this value. The number of recall levels from 0 to 1 (in equidistant steps $\frac{n}{\lambda}, n = 1 \ldots \lambda$) we used for our experiments is $\lambda = 20$. The *Average Precision* (AP) measure produces a single-valued rating of a matchmaker for a single query result: $AP = \frac{1}{|R|} \sum_{r=1}^{|L|} isrel(r) \frac{count(r)}{r}$, where R is the set of relevant items previously defined by domain experts for the examined query, L the ranking of returned items for that query, $isrel(r) = 1$ if the item at rank r is relevant and 0 otherwise and $count(r)$ the number of relevant items found in the ranking when scanning top-down, i.e. $count(r) = \sum_{i=1}^{r} isrel(i)$. The AP measure is independent from the way and size of ranking.

As a first experiment, we compared the retrieval performance of SAWSDL-M0+WA to that of both approaches applied solely. This experiment was conducted mainly to check, whether even such a simple hybrid combination of logic-based and non-logic-based semantic matching as in SAWSDL-M0+WA can improve upon the performance of each of both (SAWSDL-M0 and Text-IR) individually. As shown in figure 3(a), the combination of both performs best at almost every recall level except towards full recall. This is in perfect line with our experimental results on SAWSDL-MX1 reported in [1]. As we already pointed out there, ontologies currently found in the Web are merely inclusion hierarchies or taxonomies rarely making use of elaborated logical concept definitions for service annotation, which still dampens the benefit of any logic-based semantic matching approach.

To compare the performance of the adaptive hybrid matchmaker SAWSDL-MX2 (logic, text, structural similarity) with that of the non-adaptive variants SAWSDL-M0+WA (logic and structural similarity) and SAWSDL-MX1 (logic and text similarity), we conducted a second evaluation experiment. As shown figure in 3(b), the adaptive SAWSDL-MX2 performs better than SAWSDL-M0 (logic-based only) but is as good as the non-adaptive variant SAWSDL-MX1 utilizing logic-based matching and extended Jaccard text similarity-based matching.

Table 1. Average precision and query response time (in seconds) of SAWSDL-MX matchmaker variants

	SAWSDL-M0+WA	SAWSDL-MX1	SAWSDL-MX2
AP	0.61	0.66	0.65
AQRT	15.48s	8.17s	18.8s

This is mainly to the fact, that text similarity computation as described in section 3.3 is closely related to structural matching when applied to mere is-a ontologies (inclusion hierarchies, taxonomies). In fact, for the given test collection, where SAWSDL files have been semi-automatically derived from OWL-S and the XML Schema parameters origin from OWL concept definitions, the WSDL-Analyzer (WA) indirectly performs both structural and text similarity-based concept matching which makes it partly redundant to SAWSDL-MX2 in such cases. Nevertheless, for the general case, the adaptive approach of SAWSDL-MX2 enables an easy and well-defined integration of arbitrary matching mechanisms to improve result rankings.

Table 1 summarizes the averaged precision (AP) and query response time of the discussed SAWSDL matchmakers. We emphasize that these results strongly depend on the test collection used. In summary, the non-adaptive SAWSDL-MX1 (logic + text) still performs best even over the adaptive variant SAWSDL-MX2 which exhibits a longer response time in average due to its comparatively most complex matching (logic + text + structural).

6.3 Statistical Significance Tests

The differences in the performance evaluation results can be shown to be statistically significant or insignificant by means of the so-called Friedman test. This is a non-parametric test for simultaneously analyzing ranked result sets of at least two different (service matching) methods and has been shown in [11] to be a vital explanatory component of a comparative retrieval performance evaluation.

We are using the Friedman Test variant proposed in [10] as $F_N = \frac{MSR}{MSE}$, where MSR is the mean-squared difference between the different matching variants and MSE the mean-squared error. The resulting value can be compared to the F-distribution with $m-1$ and $(n-1)(m-1)$ degrees of freedom, where n is the number of queries anf m the number of tested matching variants. The resulting *p-value* indicates, if there is a significant difference between the variants which one can not interpret as being an implication of the *null hypothesis*, i.e. that variations of the matchmaker rankings per query are insignificant. As a threshold value for p, we rely on $\alpha = 0.05$, which is frequently used for tests like this. To produce the rankings for the test, averaged AP values have been used.

The test resulted in a value of $p = 0.026$ for SAWSDL-M0+WA compared to the WSDL-Analyzer, and $p = 0.0028$ compared to SAWSDL-M0. Both values are below the threshold α which means that the recall/precision results are a significant improvement at 5% level. However, the test returned a value of $p = 0.331$ for the second evaluation experiment which means that none of the evaluated matchmaker variants performs significantly better than any of the others regarding the used test collection. As already meantioned before, this is mainly due to the fact, that the additional structural comparison implicitly performs partly redundant matching, as the semi-automatically derived test collection services mainly differ with respect to the XML Schema parameter definitions derived from the original OWL concepts used in the original test collection OWLS-TC 2.1.

7 Related Work

To the best of our knowledge, there exist only very few implemented semantic service discovery systems for SAWSDL. In FUSION [7], any SAWSDL service description is classified at the time of its publishing and then mapped to UDDI to allow for fast lookups. In case of unknown semantic service requests reasoning has to be done at query time. In contrast to SAWSDL-MX, the FUSION discovery relies on one infered logical matching degree only, Like SAWSDL-MX, FUSION is strictly bound to OWL-DL. Lumina [8] developed in the METEOR-S project[15] follows a similar approach based on mapping of WSDL-S (and later on SAWSDL respectively) to UDDI but performs syntactic (structural and text similarity-based) service matching only. Finally, the URBE matchmaker by Plebani[16] performs non-logic-based matching in terms of text similarity and structural comparisons. Unfortunately, no public information is available on this matchmaker yet. For a survey of semantic service matchmakers in general, we refer the interested reader to [6].

8 Conclusion

We discussed three different hybrid SAWSDL service matchmakers all of which outperforming the individual types of matching they combine for detecting the semantic relevance of a service with interfaces with multiple operations to a given request. The combination with structural service matching by the WSDL-Analyzer tool turned out to be of benefit compared to logic-based only matching but not with respect to logical and text similarity-based matching. In addition, applied to the given test collection SAWSDL-TC1, the adaptive hybrid approach combining all three types of matching did not outperform the non-adaptive variant in [1] yet. We are currently working on a new hybrid matchmaker variant SAWSDL-MX3 that also supports different knowledge representation formalisms. SAWSDL-MX1 and SAWSDL-TC1 are publicly available at *semwebcentral.org*.

References

1. Klusch, M., Kapahnke, P.: Semantic Web Service Selection with SAWSDL-MX. In: Proceedings of the 2nd Workshop on Service Matchmaking and Resource Retrieval in the Semantic Web (SMR2) (2008)

[15] http://lsdis.cs.uga.edu/projects/meteor-s/
[16] http://www-ags.dfki.uni-sb.de/ klusch/s3/html/2008.html

2. Joachims, T., Radlinski, F.: Search Engines that Learn from Implicit Feedback. Computer 40(8), 34–40 (2007)
3. Kaufer, F., Klusch, M.: WSMO-MX: A Logic Programming Based Hybrid Service Matchmaker. In: Proceedings of the 4th IEEE European Conference on Web Services (ECOWS 2006). IEEE CS Press, Zurich (2006)
4. Kiefer, C., Bernstein, A.: The Creation and Evaluation of iSPARQL Strategies for Matchmaking. In: Bechhofer, S., Hauswirth, M., Hoffmann, J., Koubarakis, M. (eds.) ESWC 2008. LNCS, vol. 5021, pp. 463–477. Springer, Heidelberg (2008)
5. Klusch, M., Fries, B., Sycara, K.: Automated Semantic Web Service Discovery with OWLS-MX. In: Proceedings of 5th International Conference on Autonomous Agents and Multi-Agent Systems (AAMAS), Hakodate, Japan. ACM Press, New York (2006)
6. Klusch, M.: Semantic Web Service Coordination. In: Schumacher, M., Helin, H., Schuldt, H. (eds.) CASCOM - Intelligent Service Coordination in the Semantic Web, ch. 4. Birkhäuser Verlag/ Springer (2008)
7. Kourtesis, D., Paraskakis, I.: Combining SAWSDL, OWL-DL and UDDI for Semantically Enhanced Web Service Discovery. In: Bechhofer, S., Hauswirth, M., Hoffmann, J., Koubarakis, M. (eds.) ESWC 2008. LNCS, vol. 5021, pp. 614–628. Springer, Heidelberg (2008)
8. Li, K., Verma, K., Mulye, R., Rabbani, R., Miller, J.A., Sheth, A.P.: Semantic Web Processes and Their Applications. In: Cardoso, J., Sheth, A. (eds.) Designing Semantic Web Processes: The WSDL-S Approach. Springer, Heidelberg (submitted)
9. Baader, F., Calvanese, D., McGuinness, D.L., Nardi, D., Patel-Schneider, P.F.: The Description Logic Handbook: Theory, Implementation, and Applications. Cambridge University Press, Cambridge (2003)
10. Iman, R.L., Davenport, J.M.: Approximations of the critical region of the friedman statistic. Communications in Statistics A9, 571 (1980)
11. Hull, D.: Using statistical testing in the evaluation of retrieval experiments. In: Proceedings of the 16th annual international ACM SIGIR conference on Research and development in information retrieval, pp. 329–338 (1993)
12. Zinnikus, I., Rupp, H.J., Fischer, K.: Detecting Similarities between Web Service Interfaces: The WSDL Analyzer. In: Second International Workshop on Web Services and Interoperability (WSI 2006), Pre-conference Workshop of Conference on Interoperability for Enterprise Software and Applications, I-ESA 2006, March 20-21, Bordeaux (2006)
13. Hsu, C.-W., Chang, C.-C., Lin, C.-J.: A Practical Guide to Support Vector Classification (2007)
14. Boser, B.E., Guyon, I.M., Vapnik, V.N.: A training algorithm for optimal margin classifiers. In: Proceedings of the fifth annual workshop on Computational learning theory Pittsburgh, Pennsylvania, United States, pp. 144–152 (1992)

Enhancing Service Selection by Semantic QoS

Henar Muñoz Frutos[1], Ioannis Kotsiopoulos[2], Luis Miguel Vaquero Gonzalez[1], and Luis Rodero Merino[1]

[1] Telefónica Research & Development, C/ Emilio Vargas, 6
28043 Madrid, Spain
[2] School of Computer Science, University of Manchester, Kilburn Building
Oxford Road, Manchester, M13 9PL, United Kingdom
henar@tid.es, ioannis@cs.man.ac.uk, lmvg@tid.es, rodero@tid.es

Abstract. The increasing number of functionally similar services requires the existence of a non-functional properties selection process based on the Quality of Service (QoS). Thus, in this article, authors focus on the provision of a QoS model, an architecture and an implementation which enhance the selection process by the annotation of Service Level Agreement (SLA) templates with semantic QoS metrics. This QoS model is composed by a specification for annotating SLA templates files, a QoS conceptual model formed as a QoS ontology and selection algorithm. This approach, which is backward compatible, provides interoperability among customer-providers and a lightweight alternative. Finally, its applicability and benefits are shown by using examples of Infrastructure services.

Keywords: Service Selection, QoS, semantic annotations, SA-SLA, ontology.

1 Introduction

The marketplace model, where customers and providers interact for buying and selling services, is evolving to incorporate infrastructure services such as storage or processing ones [1], which can be accessible through the Internet [2] boosting the number of available services [3]. In this setting of increasing number of functionally similar services, the discovery process to meet user requirements while ensuring the QoS [32] can be seriously hampered. Therefore, there is a need for a service selection process based on specified customer preferences or restrictions which are considered as non-functional properties [31] [32] such as price, reputation and reliability. Some of the non-functional properties can involve QoS attributes which may impact the quality of the service offered by a Web Service [15] [32]. Moreover, this QoS specification can be contained inside the Service Level Agreement (SLA) Templates (TSLA) files which are used to achieve an agreement between customer and provider about the terms required in the service provision [18][20]. Thus, a service specification offered by a provider can be materialized as a capability (functional description) provided in the service description and a set of QoS constraints (non-functional descriptions) in the associated TSLA files.

L. Aroyo et al. (Eds.): ESWC 2009, LNCS 5554, pp. 565–577, 2009.

In this highly dynamic scenario, providers and customers may use different terminologies or even different languages, to describe the conditions and requisites [21] (like *Performance, Throughput or Response Time*). This heterogeneity in data and interfaces may turn the selection process almost impossible when using the conventional approaches. For instance, UDDIe [16] and the QoS Broker developed by Serhani et al [17] extend the Universal Description, Discovery and Integration (UDDI) structure to include QoS attributes in the service description. However, they model QoS metrics syntactically, which is poor and not extensible [15], and cannot solve the heterogeneity problem. In this matter, it is required a mutual agreement on the service(s) functional and non-functional properties [21] [22] [23] to develop a method for automatically matching the offers and the requests [22].

In order to increase the interoperability among customers-providers, some works introduce semantic technologies in the definition of QoS metrics. Wang et al [27], Tsesmetzis et al [30], Maximilien et al [33] and Ren et al [32] present different conceptual models formed as ontologies so as to extend service descriptions. However, these solutions are not backward compatible, and demand providers to change their service offer descriptions. In addition, Oldham et al [22] and Chaari et al [23] propose the definition of QoS metrics inside standards to manage non-functional aspects of WS-Agreement [18] and WS-Policy [19] respectively, so that, they do not modify current service descriptions. Although, they introduce semantic annotations inside the standards, they extend their standards XML schemas, being alternatives not backward compatible.

Thus, in this paper we try to tackle the detected inflexibility of current QoS metrics specification approaches providing a backward compatible approach. We enrich the expression capabilities of current QoS frameworks by developing an agreed conceptual model captured in the form of ontologies for improving the interoperability among parties and a semantic specification for annotating SLA files. Thus, the contribution is related to the provision of a QoS model composed by i) a semantic specification for annotating SLA files and ii) a QoS ontology which contains also non-functional properties for Infrastructure services. Moreover, a reference architecture is designed to implement this QoS model and an implementation to test its applicability and benefits.

The remaining of this document is structured as follows. Section 2 presents the related work; Section 3 shows the proposed QoS model including the semantic annotations and the ontology. This framework is mapped to a reference architecture (Section 4) and validated with experiments in Section 5. Finally, Section 6 summarizes the obtained results and presents the future work.

2 Related Work

Work in the area of Grid/Web Service (GWS) QoS typically involves syntactic aspects. Serhani et al [17] presents an architecture based on QoS certification approach employing an extended UDDI for supporting Web Service selection using QoS attributes. Moreover, UDDIe [16] extends the structure to include QoS attributes in the context of Grid computing. In addition, other initiatives enable standardized QoS specification for Web Service defined on top of existing protocols and standards for the establishment of an agreement between two parties. There are two main standards related to SLA, which use QoS metrics to define the terms agreed: i) Web Services

Agreement Specification (WS-Agreement) [18], which extends current Web Services standards to allow the establishment of an agreement and ii) Web Service Level Agreement Language (WSLA) [20] which provides a framework for monitoring and evaluating SLA. Moreover, some work has been carried out towards the combination of both standards with the result of the TrustCoM SLA specification [21]. Moreover, before signing the contract, TSLA files are using to negotiate the QoS metrics [21] in order to achieve an agreement between customer and provider about the terms required in the service provision. Thus, the provider describes the service offers in terms of QoS metrics in the TSLA.

In spite of helping guaranteeing QoS levels, these initiatives do not support open environments [15], where flexible QoS is a must to enhance the interaction and understanding between providers and consumers. The reason is because GWS definition languages are mainly XML-based [4], which lacks the semantic expression power for its flexible definition [7] [8] in an evolving market place were many consumers and providers often express their preferences using different terminologies. In fact, machines are not able to process some information, which is only comprehensible by humans [9]. As a result, services are rigid and cannot be adapted to changes without human intervention [7]. Ontologies can overcome the heterogeneity problem providing some automatic support [9], since they allow data semantics to be expressed in a formal and more expressive language understandable by machines [7].

Similar to our approach, other works in the area of GWS QoS employ semantic technologies to increase interoperability. Wang et al [27], Tsesmetzis et al [30], Maximilien et al [33] and Ren et al [32] present different models for discovery services with QoS constrains by using ontologies as well as some algorithms for matchmaking purposes. However, as they extend current service descriptions, these solutions are not backward compatible, and demand providers to change their service offer descriptions. Moreover, the entire approach can lack in efficiency and performance, since all customers' requests are translated into rules and ontologies increasing the time in providing the answer. On the other hand, Tian et al [24] offer a specification of several QoS aspects and a mapping of QoS requirements regarding network performance onto the actual transport technology at runtime. However, this work is focused towards network parameters monitoring and not business requirements.

Finally, the approach discussed here is similar to the contribution presented by Oldham et al [22], which enriches WS-Agreement with semantic annotations. This approach specifies the QoS metrics inside the standards instead of the service description. However, the solution provided is limited due to:

- It annotates TSLA by using semantic QoS metrics extending the current WS-Agreement schema. This approach fails since it is not backward compatible, so that, previous components, which do not understand the new specification, cannot work with it.
- It transforms all customer requests into ontology and rule descriptions, which makes the process be so slowly, since it implies an updating of the ontology each time a request is done.

In order to improve interoperability between customer and provider and have a backward compatible solution, which allows for working in open environments, our approach provides a QoS model, which is explained in next Section.

3 Selection of Services: The QoS Model

This work proposes a QoS model to allow for service selection based on QoS metrics, which are specified inside TSLA files, where providers describe their offers in terms of non-functional properties. Moreover, in order to increase the interoperability among customers and providers, we are going to introduce semantic technologies in terms of semantic annotations of the QoS metrics included in the TSLA. The main drivers behind this approach are: i) the necessity for a backward compatible approach, so that, components non aware of semantic annotation can continue working with SLA files (there will be annotated and not-annotated SLA files) and ii) a lightweight approach to perform any operation (selection, negotiation…) at runtime [21]. In order to implement this approach, a QoS model is defined, which is composed by i) a specification for introducing semantic QoS annotations in SLA files, ii) a conceptual model for QoS metrics formed as ontologies to be shared between customers and providers and iii) a QoS selection algorithm. For the selection algorithm, we took the work done in Wang et al [27], since it tries to solve similar problems to ours. In this work, we skip QoS metrics collection as part of the QoS model, since we assume that providers have already obtained and measured the QoS metrics values to be included in the TSLA files.

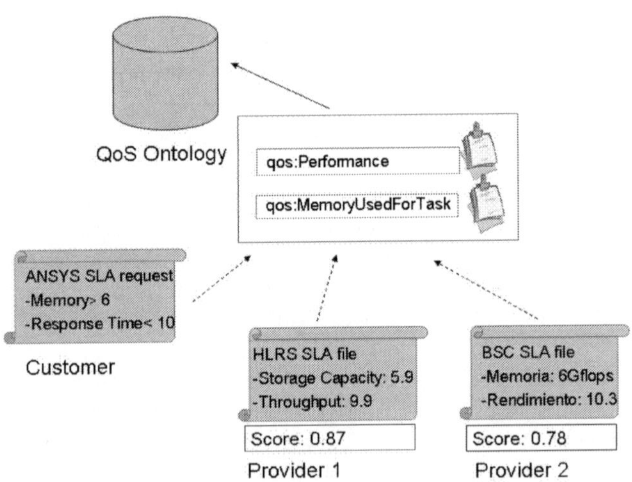

Fig. 1. A QoS Model example

Thus, in order to analyze the QoS model, we use an example obtained from the Virtual Scenario of the BREIN project [34] which is represented in Figure 1. There, an engineering company (ANSYS) tries to outsource the computing infrastructure required for software simulations in order to reduce costs. In this matter, due to the existence of several infrastructure providers (BSC, HLRS…), the framework should be able to select the best provider in each moment which satisfies the customer's requests. In the example, the customer asks for a computing service together with some QoS parameters like "*memory higher than 6*" and "*response time lower than 10*" to guarantee the service to the final clients. In real business, clients and providers can use different terminologies, or even different languages, to describe their requests or offers. This situation is shown in the example, where the customer and the providers are using different terminologies and even languages (e.g. *Response time* vs. *Throughput* vs. *Rendimiento (Performance)*). Thus, the selection process cannot return any answer in case a keyword search is carried out.

In addition, Figure 1 represents the previous example in the specification of the QoS model. By using the QoS model, the SLA parameters in the TSLA files are annotated by QoS terms which are linked to concepts from the QoS ontology, extending the common conceptual model. In addition, the TSLA files obtain a score indicating the customer' satisfaction (in term of customer' conditions) by the QoS selection algorithm.

3.1 Specification of Semantic Annotations for SLA

Our work provides a specification for annotating TSLA files called Semantic Annotation for Service Level Agreement (SA-SLA). SA-SLA is based on the Semantic Annotations for Web Service Description Language (SA-WSDL) [5][6], which has become the dominant approach in the area of Semantic Web Services [20]. It provides a standard description format extending SLA template files such as WS-Agreement, TrustCoM templates or WSLA. Thus, it allows for annotating SLA descriptions with pointers to semantic concepts from more expressive ontologies, coded in formal languages as Web Ontology Language (OWL) [11] or Web Service Modelling Language (WSML) [13], so that SLA-aware components can interpret the content and automate the tasks. Concretely, in this paper contribution, we are linking SLAParameters and Metrics in TSLA with semantic descriptions by using the QoS concepts ontology.

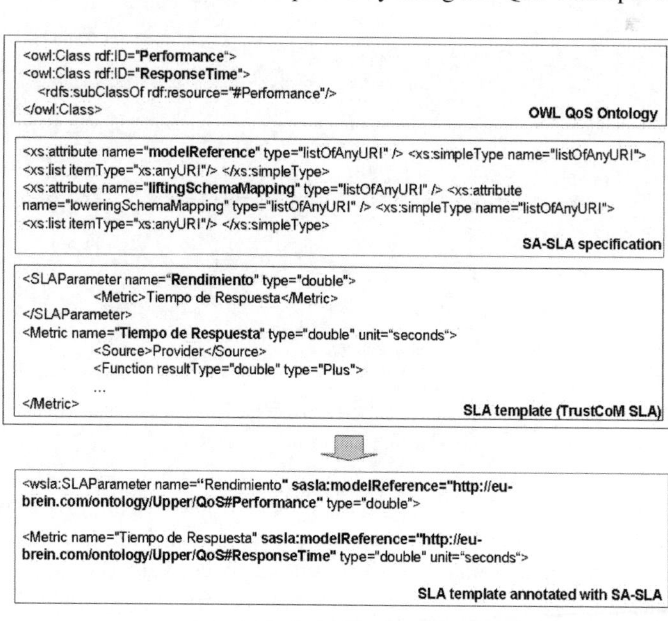

Fig. 2. TSLA example annotated with SA-SLA specification

The SA-SLA specification is constituted mainly by: i) model reference, which is an association between a SLA schema and a concept in some semantic model, and ii) schema mappings for specifying mappings between semantic data and XML [6]. Figure 2 represents an example of how an SLA template according to TrustCoM SLA specification can be annotated by some concepts (*Performance* and *ResponseTime*) from the BREIN QoS ontology following the SA-SLA specification, mainly by using the model reference element.

3.2 QoS Ontology

Since services can be provided by third parties and invoked dynamically over the Internet, their QoS can vary greatly. Therefore, it is important to have a framework capturing the QoS provided by the supplier and for the QoS required by the customer, and ultimately the match between the two when discovering the service best match the required QoS [29].

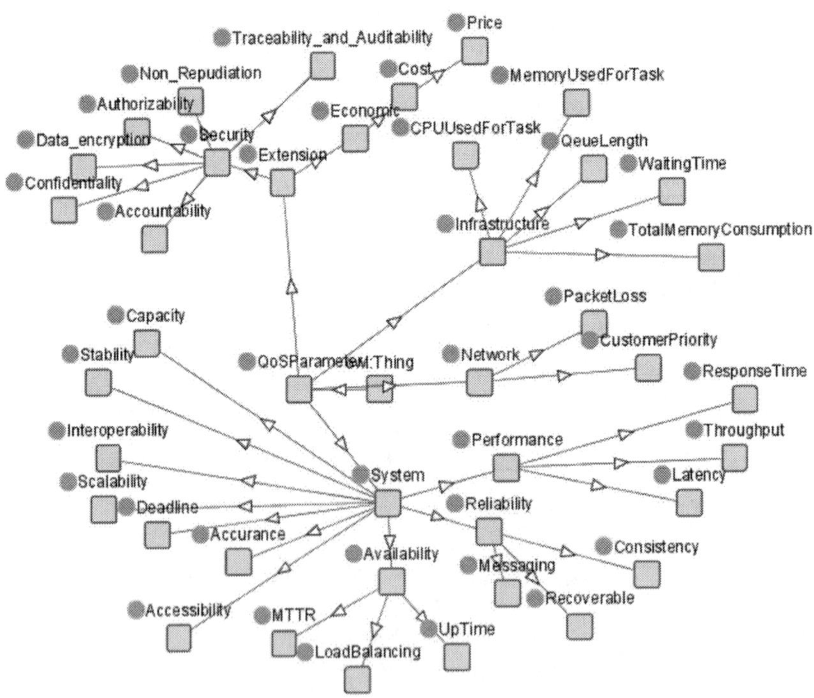

Fig. 3. QoS ontology

Thus, BREIN provides a QoS conceptual model, which is shared by customers and providers, and formed as an OWL QoS ontology to provide a common understanding of QoS parameters and their semantics. The ontology consists of basic concepts to define QoS parameters and the relationship between them required for describing non-functional properties for services included Infrastructure ones. Basically, the QoS ontology is based on the QoS conceptual model provided by Wang et al [27], Lee et al. [28], Shupping et al [29], Tsesmetzis et al [30], Maximilien et al [33] and Ren et al. [32], and it will be extended in the future work to relate business QoS metrics like availability to infrastructure ones.

The QoS ontology defines a generic set of QoS parameters, which is extensible, and can be customized to include domain-specific QoS parameters, as Figure 3 depicts. In order to conceptualize this domain, we organize them into QoS categories, which are grouped into different types: i) system, which describes non-functional properties in systems which are encapsulated as services, ii) network, which involves the typical QoS requirements in the network, iii) infrastructure, related to requirements coming back infrastructure services or task to be executed in the resources, and iv) some extension concepts focused on security and economical issues.

Moreover, according to the example shown in Figure 1, a method to work with different terminologies of the same concept in the ontology is required (e. g. "Response Time" vs. "Tiempo de Respuesta" with respect to qos:ResponseTime concept). Thus, the QoS model uses the Simple Knowledge Organization System (SKOS) [10], which provides a standard way to represent knowledge organisation systems using the Resource Description Framework (RDF) [14], so that, it allows for linking different representations to a QoS concept. In this matter, concepts in the QoS ontology can be mapped into different representations by SKOS files. One example of this specification can be seen on Figure 4, where the QoS concept ResponseTime is mapped into several different representations corresponding to different languages.

```
<rdf:Description rdf:about="http:// brein.com/Upper/QoS#ResponseTime">
<skos:inScheme rdf:resource="http://eu-brein.com/thesaurus/QoSmetrics"/>
     <rdf:type
rdf:resource="http://www.w3.org/2004/02/skos/core#Concept"/>
     <skos:prefLabel xml:lang="en">Response Time</skos:prefLabel>
     <skos:altLabel xml:lang="es">Tiempo de Respuesta</skos:altLabel>
/rdf:Description>
```

Fig. 4. SKOS representation example

4 Implementation

The previous QoS model needs to have a framework which implements its specification, conceptual model and selection algorithm. Figure 5 shows an architecture, whose main two modules have been implemented inside the BREIN project. Mainly, it is composed by:

- Semantic Enhanced Service Selector (SESS): This component allows for parsing of TSLA annotated files offered by providers and matched the providers' offers with the customer' parameters. It will interact with the SLA Translator (SLAT) which supports SESS to understand the semantic annotations. Moreover, the selector module implements the algorithm proposed. As a result, the best service, which satisfies customer's request, is returned.
- SLA translator: It provides the semantic support for SLA aware components, concretely by the translation of providers' metrics into predefined

ones, solving the heterogeneity problem commented above. It is composed by an ontology manager and query engine to manage ontologies. Moreover, it has an SKOS module to take into account different representations of the same concept through the SKOS standard.

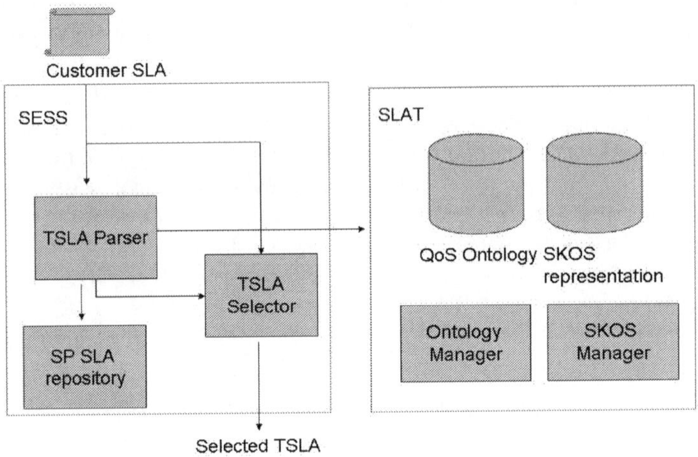

Fig. 5. The framework which implements the QoS model

These two main components are composed by a set of functional modules:

- TSLA Parser: It parses the TSLA files, which define the providers' offers, in order to obtain the QoS metrics required by the customer and their values. Thus, it performs two kind of matchmakings depending on the metrics (annotated or not): i) syntactic, which involves a keyword search and ii) semantic, which interact with SLAT in order to matchmake the semantic QoS metrics with customer' requests.
- Service Provider (SP) SLA repository: It stores the TSLA templates associated to the different services candidates in the selection process.
- TSLA Selector: It applies a selection algorithm taken from the work done by Wang et al [27]. This algorithm normalizes the matrix constituted by QoS metrics (Q_s) to map all variables to a common range [0, 1] by scaling the value ranges with the maximum and minimum values of each quality metric [27]. Finally, it sums the normalized metrics value in order to obtain the TSLA score. Thus, the selector supplies each provider's score and chooses the one which has the highest score.
- Ontology Manager: It allows for managing the QoS ontology, by using Jena as ontology manager and Pellet as reasoner working with OWL ontologies.
- SKOS Manager: It allows for managing the SKOS representation of the QoS concepts. It also uses Jena to work with RDF representations.

5 Experimentation

The scenario chosen for demonstration has been explained in Section 3 and shown in Figure 1. Thus, for experimentation proposes, we built an early simulation testbed composed by a set of TSLA files (6 files concretely) based on the TrustCoM specification, which are associated to different services with similar functionality (processing capacity), so that, a selection process is required. These TSLA files have different QoS metrics to characterize the service offers. We are supposing the usage of different terminologies in the metric definition (as performance vs. response time or memory vs. storage capacity) as well as different languages (English, Spanish, and French), which is the problem found in the open environments. Moreover, some TSLA are semantically annotated by QoS concepts from the QoS ontology. Figure 6 represents this testbed, which includes a customer' request asking for 4 QoS metrics and different providers' TSLA described by using some of these metrics, which can be semantically annotated.

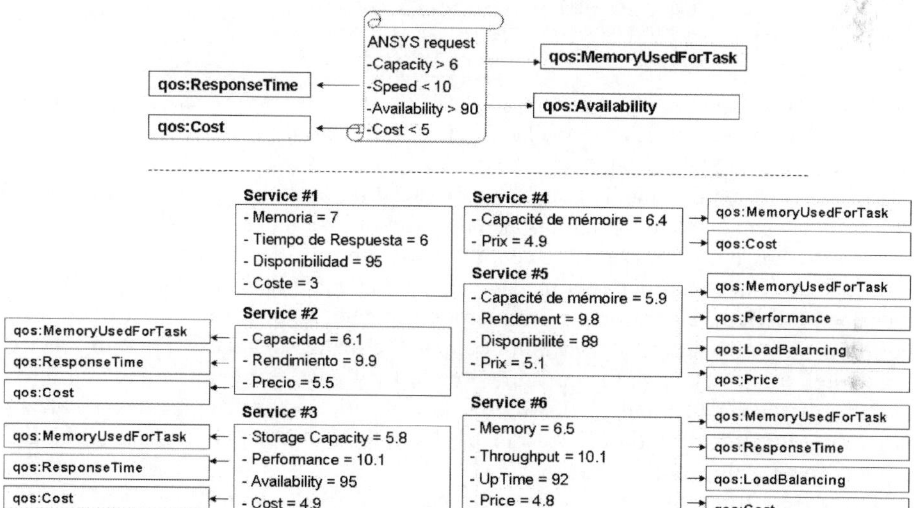

Fig. 6. SLA templates testbed

Figure 7 shows a representation of the results obtained by testing the implementation of the QoS model. Concretely, it presents the total score of each TSLA which is formed as the sum of the normalized value of the QoS metrics in the TSLA involved in the testbed. As a result of the experimentation, we see that service #1 although having all metrics required by the customer and the best values in the testbed, its score is 0, since it lacks of semantic annotations and the terminology used in the metrics definition is different from the customer one. Thus, as the selector module is doing a syntactic search (due to the lack of semantic annotations), there are not results in the matchmaking process. On the other hand, despite the fact that service #6 does not satisfy all customer' restrictions and their values are worst than service #1, it has the best score, since all its metrics have been annotated.

In addition, we can observe that service #5, in spite of using different annotations than the customer (e.g. qos:ResponseTime vs. qos:Performance or qos:Cost vs. qos:Price), its metrics are matchmaked, since they are using related concepts belonging to QoS ontology. Furthermore, we demonstrate that it is possible to work with non-annotated metrics, only if they are defined by the customer' terminology as the Availability metric in service #3.

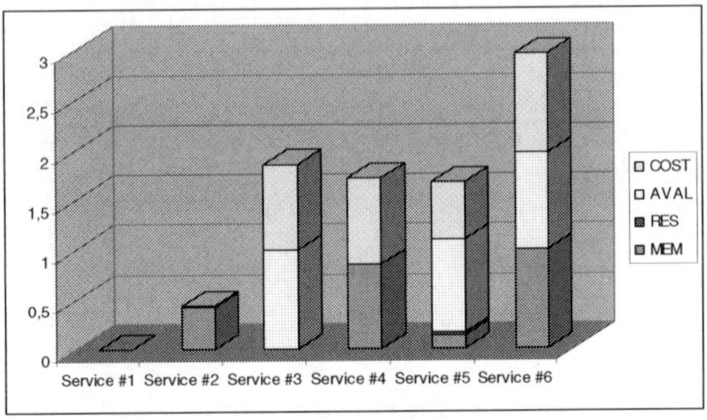

Fig. 7. Experimentation results from the implementation (where MEM= Memory User For Task, RES=Response Time, AVA= Availability, COST=Price)

However, we see that the TSLA scores obtained do not fully satisfy the customer' request. For instance, there is not any penalization in case the TSLA does not include metrics asked by the customer. Service #4 has the third position in the ranking despite the fact it does not consider two metrics (Response Time and Availability). On the other hand, service #2 considers most metrics annotated and its score is low. In fact, due to the way of scaling by maximum values, the metric Response Time in service #2 although it is so close to the customer one (10.1 vs. 10), for being a maximum, its normalized value is 0. Finally, although a normalization in metrics values is carried out by the selection algorithm [27] to map all variables to a common range [0,1], the output of this result is not scaled into [0,1].

Thus, these results show that in an open market, where new customers and providers can enter, mechanisms to improve several aspects such as interoperability, performance and easy of use are key aspects. In this matter, improving semantic interoperability during the selection phase implies that more providers are able to be introduced in the market. Moreover, by allowing the users to define their own terms for TSLA descriptions and reducing the task of selection process from customer's requirements by automatic support, we achieve decreasing the entrance barrier of new actors in the market. However, the algorithm used is not the best one due to the limitation analysed by the examples.

6 Conclusions and Future Work

The increasing number of functionally similar services requires the existence of a non-functional properties selection process. This work has focused on the service selection based on QoS metrics used inside TSLA files. Moreover, it has introduced semantic technologies for annotating QoS metrics to increase the interoperability between customers and providers, which can use their own and different terminologies. Thus, our contribution has provided a QoS model composed by i) SA-SLA specification and ii) a QoS ontology. Mainly, the formalization of the QoS conceptualization model in terms of ontology increases the interoperability as the experimentation has shown. In addition, the SA-SLA specification makes the QoS model be backward compatible and a lightweight approach, which allow for continuing working with previous implementations. Furthermore, our work provides a reference architecture together its development which implements the QoS model proposed. From our experimentation, it is possible to see the advantages of using semantics in this open environment.

As future work, we will make the QoS ontology more expressive and improve the relationships between QoS metrics mapping the business requirements to infrastructure ones. Moreover, the QoS ontology will be part of a conceptualization model for business Grid, in order to include more non-functional properties in the selection process (also related to the resources which the Infrastructure services virtualize). In addition, we plan to continue working in the QoS model in order to extend the algorithm we took to adjust/extend it to our requirements considering the feedback we obtained from the experimentation. Basically, the idea is to change the normalization method for scaling metrics, to include penalization in case not all the customer' conditions have included and to modify the algorithm so that output is scaled also between [0,1]. Finally, current implementation will be extended to take into account previous changes. Moreover, we will try to apply some Peer To Peer technologies in order to provide scalability to our solution required when the number of entities improves.

Acknowledgments

This work has been supported by the BREIN project and has been partly funded by the European Commission's IST activity of the 6th Framework Programme. This paper expresses the opinions of the authors and not necessarily those of the European Commission. The European Commission is not liable for any use that may be made of the information contained in this paper.

References

1. Brunner, R., Freitag, F., Navarro, L.: Towards the development of a decentralized market information system: Requirements and architecture. In: IEEE International Symposium on Parallel and Distributed Processing IPDPS 2008, pp. 1–7 (2008)
2. Shaw, M.J.: Building an E-Business from Enterprise Systems. Information Systems Frontiers 2(1), 7–17 (2000)
3. Cusumano, M.A.: The changing software business: Moving from products to services. Computer 41(1), 20–27 (2008)
4. Papazoglou, M.P., Heuvel, W.J.: Service oriented architectures: approaches, technologies and research issues. The VLDB Journal 16(3), 389–415 (2007)

5. Kopecký, J., Vitvar, T., Bournez, C., Farrell, J.: SAWSDL: Semantic Annotations for WSDL and XML Schema. IEEE Internet Computing 11(6), 60–67 (2007)
6. Farrell, J., Lausen, H.: Semantic Annotations for WSDL and XML Schema, W3C. Recommendation (August 28, 2007), http://www.w3.org/TR/sawsdl/
7. Gruber, T.R.: Towards Principles for the Design of Ontologies Used for Knowledge Sharing. In: Guarino, N., Poli, R. (eds.) Formal Ontology in Conceptual Analysis and Knowledge Representation, Deventer, The Netherlands. Kluwer Academic Publishers, Dordrecht (1993)
8. Fensel, D., Lausen, H., Polleres, A., de Bruijn, J., Stollberg, M., Roman, D., Domingue, J.: Enabling Semantic Web Services: The Web Service Modeling Ontology. Springer, New York (2006)
9. Krummenacher, R., Hepp, M., Polleres, A., Bussler, C., Fensel, D.: Www or what is wrong with Web Services. In: ECOWS 2005: Proceedings of the Third European Conference on Web Services, Washington, DC, USA. IEEE Computer Society, Los Alamitos (2005)
10. The OWL Services Coalition. OWL-S 1.1 beta release (July 2004), http://www.dalm.org/services/owl-s/
11. OWL Working Group, Web Ontology Language (OWL) (2004), http://www.w3.org/2004/OWL/
12. W3C Semantic Web Deployment Working Group, SKOS Simple Knowledge Organization System Reference, http://www.w3.org/TR/skos-reference
13. ESSI WSML working group, Web Service Modeling Language, http://www.wsmo.org/wsml/
14. RDF Core Working Group, Resource Description Framework (RDF): Concepts and Abstract Syntax, W3C recommendation, http://www.w3.org/TR/rdf-concepts
15. Kritikos, K., Plexousakis, D.: A semantic qos-based Web Service discovery algorithm for over-constrained demands. In: NWESP 2007: Proceedings of the Third International Conference on Next Generation Web Services Practices, Washington, DC, USA, pp. 49–54. IEEE Computer Society, Los Alamitos (2007)
16. ShaikhAli, A., Rana, O.F., Al-Ali, R., Walker, D.W.: UDDIe: An Extended Registry for Web Services. In: Proceedings of the 2003 Symposium on Applications and the internet Workshops (Saint 2003 Workshops), January 27 - 31, 2003, p. 85. IEEE Computer Society, Washington (2003)
17. Serhani, M.A., Dssouli, R., Hafid, A., Sahraoui, H.: A QoS Broker Based Architecture for Efficient Web Services Selection. In: Proceedings of the IEEE international Conference on Web Services, ICWS, July 11 - 15, 2005, pp. 113–120. IEEE Computer Society, Washington (2005)
18. Grid Resource Allocation Agreement Protocol WG (GRAAP-WG), Web Services Agreement Specification (WS-Agreement), Version 2005/09 (September 2005), http://www.ggf.org/Public_Comment_Docs/Documents/Oct-2005/WS-AgreementSpecificationDraft050920.pdf
19. W3C, Web Service Policy, http://www.w3.org/TR/ws-policy/
20. Web Service Level Agreement (WSLA) Language Specification, ver. 1.0 (January 2003), http://www.research.ibm.com/wsla/WSLASpecV1-20030128.pdf
21. Kotsiopoulos, I., Soler Juber, S., Tenschert, A., Benedicto Cirujeda, J., Koller, B.: Using Semantic technologies to improve Negotiation of Service Level Agreements, eChallenges 2008 (2008)

22. Oldham, N., Verma, K., Sheth, A., Hakimpour, F.: Semantic WS-agreement partner selection. In: Proceedings of the 15th international Conference on World Wide Web, WWW 2006, Edinburgh, Scotland, May 23-26, 2006. ACM, New York (2006)
23. Chaari, S., Badr, Y., Biennier, F.: Enhancing Web Service selection by QoS-based ontology and WS-policy. In: Proceedings of the 2008 ACM Symposium on Applied Computing, SAC 2008, Fortaleza, Ceara, Brazil, March 16 - 20, 2008, pp. 2426–2431. ACM, New York (2008)
24. Tian, M., Gramm, A., Ritter, H., Schiller, J.: Efficient Selection and Monitoring of QoS-Aware Web Services with the WS-QoS Framework. In: Proceedings of the 2004 IEEE/WIC/ACM international Conference on Web intelligence, Web Intelligence, September 20 - 24, 2004, pp. 152–158. IEEE Computer Society, Washington (2004)
25. Somasundaram, T.S., Balachandar, R.A., Kandasamy, V., Buyya, R., Raman, R., Mohanram, N., Varun, S.: Semantic Description and Discovery of Grid Services using WSDL-S and QoS based Matchmaking Algorithm. In: Advanced Computing and Communications, 2006. ADCOM 2006, December 20-23, pp. 113–116 (2006)
26. Li, H., Du, X., Tian, X.: Towards Semantic Web Services Discovery with QoS Support Using Specific Ontologies. In: Proceedings of the Third international Conference on Semantics, Knowledge and Grid, SKG, October 29 - 31, 2007, pp. 358–361. IEEE Computer Society, Washington (2007)
27. Wang, X., Vitvar, T., Kerrigan, M., Toma, I.: A QoS-aware selection model for semantic Web Services. In: Dan, A., Lamersdorf, W. (eds.) ICSOC 2006. LNCS, vol. 4294, pp. 390–401. Springer, Heidelberg (2006)
28. Lee, K., Jeon, J., Lee, W., Jeong, S., Park, S.: QoS for Web Services: Requirements and Possible Approaches, W3C Working Group Note 25 (2003), `http://www.w3c.or.kr/kr-office/TR/2003/ws-qos/`
29. Ran, S.: A Model for Web Services Discovery With QoS. ACM SIGecom Exchanges 4(1) (March 2003)
30. Tsesmetzis, D.T., Roussaki, I.G., Papaioannou, I.V., Anagnostou, M.E.: QoS awareness support in Web-Service semantics. In: Proceedings of the Advanced int'L Conference on Telecommunications and int'L Conference on internet and Web Applications and Services, AICT-ICIW, February 19 - 25, 2006, p. 128. IEEE Computer Society, Washington (2006)
31. Garcia, D.Z., de Toledo, M.B.: Semantics-enriched QoS policies for Web Service interactions. In: Proceedings of the 12th Brazilian Symposium on Multimedia and the Web, WebMedia 2006, Natal, Rio Grande do Norte, Brazil, November 19 - 22, 2006, vol. 192, pp. 35–44. ACM, New York (2006)
32. Ren, K., Chen, J., Chen, T., Song, J., Xiao, N.: Grid-Based Semantic Web Service Discovery Model with QoS Constraints. In: Third International Conference on Semantics, Knowledge and Grid, October 29-31, 2007, pp. 479–482 (2007)
33. Maximilien, E.M., Singh, M.P.: A Framework and Ontology for Dynamic Web Services Selection. IEEE Internet Computing 8(5), 84–93 (2004)
34. The BREIN Project, Web site (December 9, 2008), `http://www.eu-brein.com/`

Towards an Agent Based Approach for Verification of OWL-S Process Models

Alessio Lomuscio and Monika Solanki

Department of Computing, Imperial College London, UK
{a.lomuscio,m.solanki}@imperial.ac.uk

Abstract. In this paper we investigate the transformation of OWL-S process models to ISPL - the system description language for MCMAS, a symbolic model checker for multi agent systems. We take the view that services can be considered as agents and service compositions as multi agent systems. We illustrate how atomic and composite processes in OWL-S can be encoded into ISPL using the proposed transformation rules for a restricted set of data types. As an illustrative example, we use an extended version of the BravoAir process model. We formalise certain interesting properties of the example in temporal-epistemic logic and present results from their verification using MCMAS.

1 Introduction

The verification of web service behaviour and interaction protocols is now an integral aspect of several frameworks providing service oriented solutions to the IT industry. The Increasing complexities that arise during service composition, make offline verification as model checking [4] crucial in successfully implementing and using services. Model checkers typically use specialised formats for the specification of behaviour, different from those commonly used for describing services. Examples of such system description languages include Promela, used with the checker SPIN [9] and NuSMV, used with the checker NuSMV [3]. However in the web service domain, WSBPEL [16], WSCDL [15] and OWL-S [17], are some popular and widely used standards for describing service behaviour, their composition and interaction protocols.

The languages above work at different levels of abstraction. In order to verify services, an important first step is to investigate how the input language to the model checker can be adapted to encode a suitable abstraction of the service behaviour, which has been described using one of the above standards. Figure 1 illustrates the general architecture of a verification framework for services. As highlighted, a crucial component is the "compiler" that takes as input the service specification, and generates a suitably abstracted model/program, encoded in the system description language for the checker. The model and the desired properties to be verified are fed to the checker. By performing a systematic exploration of the complete set of states that can be generated during an interaction between a service and its clients, the model checker is able to verify desirable properties of the composition. Generating a well abstracted model is thus crucial to the verification of services. However, developing a tool that generates such

L. Aroyo et al. (Eds.): ESWC 2009, LNCS 5554, pp. 578–592, 2009.

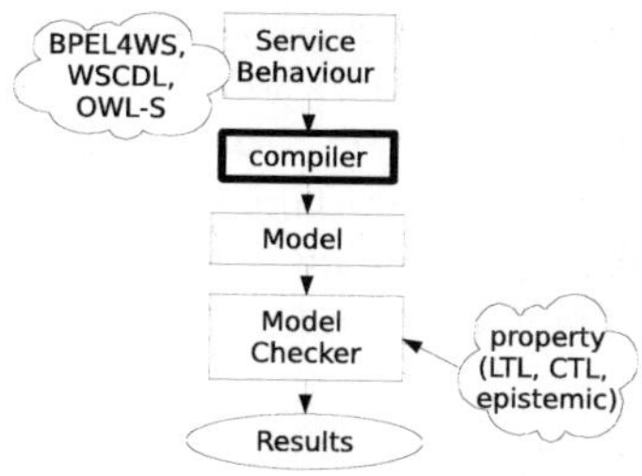

Fig. 1. General architecture for service verification

a model is non trivial. Mapping rules between the languages are required to be established before any automated translation can be undertaken. The rules provide the basis and rationale for development of a compiler providing (semi)automatic compilations from one abstraction to the other.

In this paper, we explore the generation of transformation rules from the process model of OWL-S, a well-established language for the description of web services on the semantic web, to ISPL (Interpreted Systems Programming Language) in view of verifying the results using MCMAS [13]. We are interested in using MCMAS, because it enables the user to verify rich specifications. MCMAS supports not only temporal logic, but also epistemic and deontic modalities. We take the view that a web service can be modelled as an "agent" [6]. Keeping this in perspective, a composition of web services can be viewed as a multi agent system [19]. An OWL-S process model specifies the composition and interaction between agents/services and their clients. Control constructs similar to those found in programming languages can be used to compose services. Earlier work [14,1] on using MCMAS for services has focused on exploiting its verification capabilities. In this paper we provide the transformation rules and a compiler implementation of the rules, that combined with MCMAS, could aid in automated verification of services.

The paper is structured as follows: Section 2 outlines our running example for the paper, a flight booking and managing service, which is an extended version of the BravoAir process from the OWL-S suite of examples. In Section 3 we provide a brief overview of OWL-S, ISPL and MCMAS. Section 4 discusses the mapping rules from OWL-S to ISPL and presents our implementation of the compiler. We present a brief account of the analysis and verification for the case study in Section 5. Finally, we conclude in Section 6.

2 Case Study

As a running example, we use an extended version of the BravoAirProcess model from the OWL-S suite of examples. BravoAir functions as a flight booking agent. It allows a client to perform several tasks such as searching, selecting and booking flights. Bookings can be made as individual or as groups. The top level process, `BravoAir`, is a composite process. It is composed of a *sequence* of processes. Components of the sequence are `GetDesiredFlightDetails` and `SearchAvailableFlight`, and a composite process, BookFlight. We extend the BookFlight process as a *sequence* whose components are LogIn, followed by a choice between Individual Bookng and GroupBooking and finally ConfirmReservation. Group bookings can be done for a group of more than 10 people and the discount offered is 10% of the total booking fee. However when a group booking is cancelled, the cancellation fee is 15% rather than 10% for individual bookings. We elaborate on the GroupBooking process in Section 4.

We further extend the top level process with a composite process ManageBooking, to be executed in *sequence* with BookFlight. ManageBooking is composed as a *choice* between the atomic process ChooseSeats and the composite processes, ChangeBooking and CancelBooking. If a booking is cancelled the amount which is charged for cancellation depends on whether the booking was made at an individual or group level. ChangeBooking is composed of a *split+join* whose components are the three atomic processes ChargeCard- for economy bookings, AllocateNew Booking- for club and business class bookings and SendUnAvailability- when a change of booking is not possible. The outcome from a choice between the first two processes is composed in *sequence* with SendConfirmation. ChargeCard is also invoked when a booking is cancelled. Figure 2 illustrates the various control constructs and processes used in the composition between BravoAir and a potential client. The composition can be viewed as a multi-agent system where individual processes are abstracted as agents. Within the above settings certain interesting properties of the composition can be verified. We enumerate some of these below and formalise them in temporal-epistemic logic in Section 5.

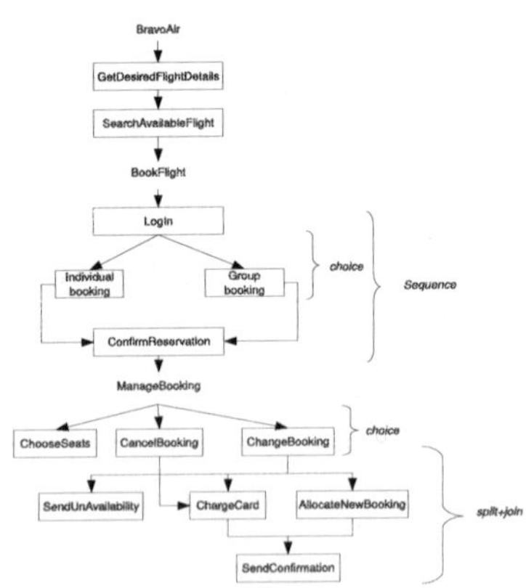

Fig. 2. An extended version of the BravoAir process

- if there is a request confirmation of reservation, the ConfirmReservation agent knows that the booking has been successful and payment has been made.
- If the number of people is more than 10, group booking will always be successful.
- Whenever a booking change is requested, it will always be confirmed.
- If a card is not charged when a booking is changed, the ChangeBooking agent knows that the reference is a business class booking.

- if a confirmation is received, the Customer agent knows that his booking was changed.
- If a card is charged after a booking has been made, it always implies that the booking has been cancelled.

3 Preliminaries

OWL-S [17] is an OWL-based Web service ontology which supplies Web service providers with a core set of constructs for describing the properties and capabilities of their Web services. It defines an upper ontology for services with *Service Profile*, *Service Model* and *Service Grounding* models. The ServiceProfile provides the information needed for an agent to discover a service, while the ServiceModel and ServiceGrounding, taken together, provide enough information for an agent to make use of a service, once found. Due to space restrictions we do not present any further details of the language here and refer the interested reader to [17].

ISPL (Interpreted System Programming Language) is based on the formal semantics of interpreted systems [7] and is the input language for the model checker MCMAS [13,12]. The syntax of ISPL includes the following

- The definition of agents describes the local behaviour of every agent in terms of states, actions, protocols and evolution function. Each agent has a set of local variables. The states of an agent, each of which contains a valuation of its local variables can be further partitioned into two disjoint sets: a non-empty set of allowed ("green") states and a set of disallowed ("red") states. Currently, ISPL allows three types of variables: Boolean, enumeration and bounded integer.
- The global evaluation function of the system defines atomic propositions held over global states which are a combinations of local states of agents defined in the model.
- The local initial state for each agent in the system.
- Specification to be checked defined as formulae in temporal, epistemic and deontic logic and fairness formuale.

MCMAS is a specialised model checker for the verification of multi-agent systems. It builds on symbolic model checking via OBDDs as its underlying technique, and supports CTL, epistemic and deontic logic. The current version of MCMAS has the following features:

- Support for variables of the following types: Boolean, enumeration and bounded integer. Arithmetic operations can be performed on bounded integers.
- Counterexample/witness generation for quick and efficient display of traces falsifying/satisfying properties.
- Support for fairness constraints. This is useful in eliminating unrealistic behaviours.
- Support for interactive execution mode. This allows users to step through the execution of their model.
- A graphical interface provided as an Eclipse plug-in which includes a graphical editor with syntax recognition, a graphical simulator, and a graphical analyser for counterexamples.

4 Encoding OWL-S Processes as ISPL Models

In an OWL-S process model, inputs and outputs are process parameters that have concrete datatypes. The current version of ISPL provides support for variables of types *bounded integers*, *boolean* and *enum*. Mapping between ontologies and these types can be done as discussed in [11]. Most existing model checkers including MCMAS, are not equipped to support OWL object types as also highlighted in [2]. We therefore abstract from defining object types for the transformation presented in this paper.

Conditions explicitly occur in OWL-S models as `Precondition`, and as part of the `Result` and `if` statement. In ISPL conditions are defined as formulae when specifying the protocol and evolution functions. In this section we first propose the following step-by-step methodology for transforming an atomic process to an ISPL program. We then show how composite processes can be transformed. These rules also facilitate the generation of a semi-automatic compiler from OWL-S to ISPL.

4.1 Encoding Atomic Processes

Agent. For every atomic process in OWL-S, we define an agent, qualified as `ProcessName` in ISPL. Recall, that the definition of an agent in ISPL includes: local variables, red states, actions, protocol functions and evolution functions.

Variables and Local States. The local states of an agent in ISPL are defined in terms of valuation of the local variables. We define the set of local variables for an agent by transforming the ontological inputs and outputs in the process model, to variables with the same identifiers and datatypes in the ISPL model. Bounds for integer variables are interactively assigned keeping the domain and context of the process model in perspective. For an atomic process, we identify two kinds of states: (1) An "Input" or initial state and (2) several "Result" states depending on the number of results defined for the process.

Let V_I denote the set of integer variables, V_B the set of Boolean variables and V_E the set of variables of type enum. We define the set of their valuations as Val_I, Val_B and Val_E respectively. The set of local variables for an agent is therefore $V = V_I \cup V_B \cup V_E$. The set of local states, L_{lstate}, of an agent can now be defined as

$$L_{lstate} : (V_I \rightarrow Val_I) \cup (V_B \rightarrow Val_B) \cup (V_E \rightarrow Val_E).$$

We enumerate L_{lstate} for an agent as follows -

- the initial state (l_0) where local variables are assigned initial values. We denote the set of variables at the initial state as $V_0 \subseteq V$.
- the set of states L_{result}, where each $l \in L_{result}$ corresponds to a non deterministic `Result` state, defined for the process. For example, a credit card validating service may produce two results: *ValidationSuccess* with boolean output $validated$ as $true$, and $ValidationnFailed$ with boolean output $validated$ as $false$. The set of variables at each of the result state is denoted as $V_i \subseteq V, i = 1 \ldots \mid L_{result} \mid$. The valuations for the variables are computed as per the evolution function described below.

- finally, l_f, a failure state which is reached when the preconditions for the process evaluates to $false$. We denote the set of variables at this state as $V_f \subseteq V$. Valuations for the variables are again computed as per the evolution function.

Red States. They are reached when an agent performs an undesirable action. This feature of ISPL is most useful while encoding faults and recovery in complex systems. The red states of an agent are represented by a Boolean formula, f^{red}, over its local variables.

Actions. The internal actions taken by a service cause a transition from the input state to one of the several result states. Actions for the agent are enumerated as follows:

- the null action ϵ,
- the set of internal actions, $A_{int} = \{a_i | i = 1 \dots n\}$, the agent takes at the input state to reach one of the several result states.
- the internal action a_f taken when the precondition fails, to reach state l_f.
- the set of actions $A_{send} = \{s_i | i = 1 \dots n\}$. The agent takes an action $s \in A_{send}$ at each $l \in l_{result}$ respectively to send the corresponding results to the client.
- the action s_f which the agent takes to send the precondition failure message at l_f.
- It follows that the total number of actions is:

$$N_{actions} = \mid A_{int} \cup \{a_f\} \mid + \mid A_{send} \cup \{s_f\} \mid + \mid \{\epsilon\} \mid$$

For simplicity we assume, $\mid A_{int} \cup \{a_f\} \mid = \mid A_{send} \cup \{s_f\} \mid$, and simplify the above to,

$$N_{actions} = 2\times \mid A_{int} \cup \{a_f\} \mid +1$$

Protocols. Protocols for the agent can be enumerated as follows:

$$\begin{aligned} f^{pre} &: \{a_i | i = 1 \dots \mid A_{int} \mid\} \\ !f^{pre} &: a_f \\ f_i^{res} &: \{s_i | i = 1 \dots \mid A_{send} \mid\} \cup s_f \end{aligned}$$

where f^{pre} (precondition), f_i^{res}, $i = 1 \dots \mid A_{send} \mid$ (condition in results) and $!f^{pre}$ are Boolean formulae over the set of local variables at the input state, result states[1] and the failure state respectively. Note that ISPL and MCMAS allow non determinism in the specification of protocols.

At execution time an agent at l_0 takes an action, $a \in A_{int}$ if f^{pre}, i.e., the precondition holds and action a_f if $!f^{pre}$ holds. This causes a transition to one of the result states $l \in L_{result} \cup l_f$, where the conditionals from the results, f_i^{res}, $i = 1 \dots \mid A_{send} \mid$ are required to hold. At l, the agent take an action, $s \in A_{send} \cup \{s_f\}$.

[1] Note that for simplicity we do not consider the case, when due to some internal failure of the service, the result conditions do not hold, but this may well be possible and additional transitions would have to be defined to consider such scenarios.

Evolutions (Transitions). The evolution function determines how local states evolve based on the agent's current local state and a set of actions. An evolution consists of a set of assignments of local variables in V and an enabling condition which is a Boolean formula, over local variables and actions of all agents.

l_0 if f_i^{res} and ProcessName.Action $= s_i$ or ProcessName.Action $= s_f, i = 1 \ldots \mid A_{send} \mid$
l_i if f^{pre} and ProcessName.Action $= a_i, i = 1 \ldots \mid A_{int} \mid$
l_f if $!f^{pre}$ and ProcessName.Action $= a_f$

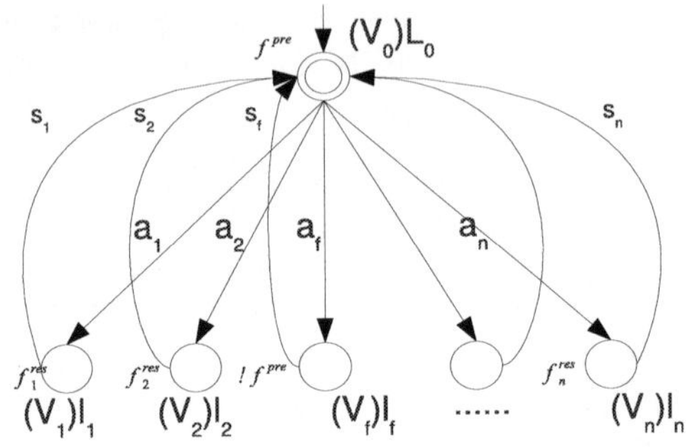

Fig. 3. Mapping between an OWL-S atomic process and ISPL

We have implemented a semi automatic compiler for the transformation, by extending the CMU OWL-S API [5]. Given an OWL-S process model, the compiler extracts the agent name, inputs and outputs from the process model as ISPL variables and enumerates the actions for the agent. Currently the definition of the red states, protocol and evolution function are interactively given, but we hope to automate the process in future versions of the paper.

The psuedocode of an algorithm, which we implemented as part of our compiler for compiling an atomic process to ISPL is presented as algorithm 1 towards the end of the paper.

4.2 ISPL Encoding of the Atomic Process: GroupBooking

Figure 4 illustrates an atomic process "Group Booking " from the case study presented in Section 2. The process takes $noOfPeople$, $flightdetails$, $carddetails$ and $loggedInStatus$ as inputs. In order to perform a group booking, the preconditions on the process are that the payment card details must be provided, the number of people in a group must be atleast 10 and the booking client must be a logged in. It returns as output, a $successMsg$ message, an $invalidCardMsg$ or an $invalidNumMsg$ message depending on the conditions $isBookingSuccessful$, $isValidCard$ and $isValidNumberOfPeople$ being $true$ or $false$. For a successful booking it also returns the discounted booking cost.

We specify the "GroupBooking" agent using the presentation syntax of OWL-S along with its corresponding ISPL code in Table 1. The inputs and outputs are mapped

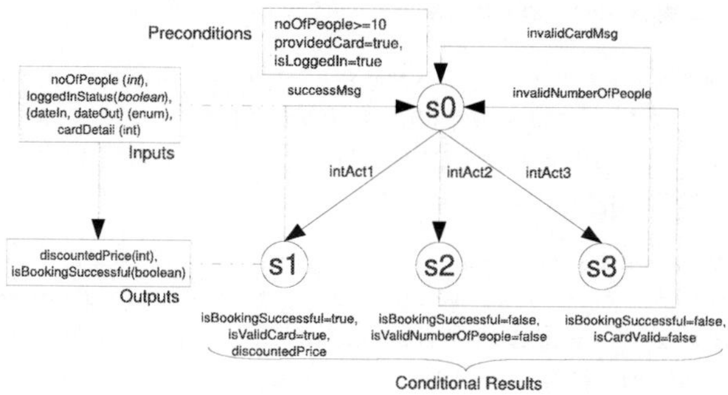

Fig. 4. Atomic Process: Group Booking

Table 1. The Group Booking Atomic Process

OWL-S atomic process

```
define atomic process Group
(inputs:(
noOfPeople - xsd:integer
 cardDetails - xsd: string
dout - xsd:date
din - xsd:date
),
exists:(
loggedInStatus - xsd:boolean
),
preconditions :( (loggedInStatus)
& (noOfPeople>=10)
& providedCard(cardDetails),
outputs:( invalidCardMsg - xsd:string
  successMsg - xsd:string
 invalidNumMsg -xsd:string
 discountedPrice - xsd:String
 )
 results :(
 isValidCard & noOfPeople>=10)|-> output(isBookingSuccessful -
xsd:boolean, discountedPrice - xsd:integer),
isValidCard
|-> output(invalidCardMsg-xsd:string),
isValidNumberOfPeople
|-> output(invalidNumMsg),
)
```

ISPL Agent

```
Agent GroupBooking
Vars:
noOfPeople:1..20;
isValidNumberOfPeople:boolean;
loggedIn:boolean;
providedCard:boolean;
isValidCard:boolean;
isBookingSuccessful:boolean;
price:1000..200000;
discountedPrice:100..20000;
dates:{dout, din};
successMsgSent:boolean;
cardFailureMsgSent:boolean;
numberFailureMsgSent:boolean;
end Vars
RedStates:
end RedStates
Actions = {intAct1, intAct2, intAct3, intAct4,
     invalidCardMsg, invalidNumMsg, successMsg, nothing};
Protocol:
loggedIn=true and providedCard=true and noOfPeople >=10 :
        {intAct1, intAct3};
loggedIn=true and noOfPeople <10 : {intAct2};
isValidCard=false:{invalidCardMsg};
isValidNumberOfPeople=false:{invalidNumMsg};
isBookingSuccessful=true:{successMsg};
end Protocol
Evolution:
isBookingSuccessful=true and isValidCard=true and
        discountedPrice=price -(price * 1/10) if
loggedIn=true and providedCard=true and noOfPeople>=10 and
         GroupBooking.Action=intAct1;
isBookingSuccessful=false and isValidCard=false if
         providedCard=true and
         GroupBooking.Action=intAct3;
isBookingSuccessful=false if noOfPeople<=10 and
         GroupBooking.Action=intAct2;
successMsgSent=true if isBookingSuccessful=true and
         GroupBooking.Action=successMsg;
cardFailureMsgSent=true if isBookingSuccessful=false and
         GroupBooking.Action=invalidCardMsg;
end Evolution
end Agent
```

as "Vars" in ISPL. Actions are interactively enumerated in accordance with the operation names specified in the grounding model defined for the process. Preconditions such as $loggedIn$, $providedCard$ and $noOfPeople \leq 10$ are specified as Boolean formula on the LHS of the protocol function. For the precondition, $loggedIn = true$ and $providedCard = true$ and $noOfPeople \geq 10$ the internal actions $intAct1$ or $intAct3$ would be non deterministically chosen by MCMAS as specified in the protocol function. The conditional part of results is specified as Boolean formula on the LHS of the "if" in the evolution function.

4.3 Encoding Composite Processes

A composite process may use one of several control constructs such as *sequence*, *if-then-else*, found in programming languages. In what follows, we discuss the modelling of composite processes, for some of the control constructs. Transformation to other constructs follows intuitively from those presented below.

Sequence. The *sequence* specifies a list of processes to be executed in a certain order. The modelling of OWL-S sequence requires explicit synchronisation. In ISPL, the definition of evolution for an agent encodes this synchronisation. We illustrate sequential composition through an example of processes composed in sequence.

Consider the process `BookFlight`, from the BravoAir model, which is a sequential composition of three atomic processes, `Login`, `GroupBooking` and `Confirm Reservation`. After receiving the result of a successful booking from the `Group Booking` process, the client invokes the `ConfirmReservation` process with inputs *isbookingSuccessful=true* and *confirmFlight=true*. The precondition for the execution of `ConfirmReservation` is *isbookingSuccessful=true*. Note that this was also the result condition of the `GroupBooking` process. The `ConfirmReservation` process returns a single result as a complex message consisting of a *reservationID* and *seatNumber*. The processes are synchronised for these inputs on the final state of `GroupBooking` and the initial state of `ConfirmResearvation`. It may also be the case that the client provides all the inputs for both the processes in the initial state of the `GroupBooking` process. In such a scenario the `GroupBooking` process invokes the `ConfirmResearvation` process at its final state using those inputs. We encode both the atomic processes in ISPL using the approach outlined in Section 4.1. We then define synchronisation between the processes for the sequential composition as outlined above. Figure 5 illustrates the composition. A typical evolution function for the "GroupBooking" agent would now be:

```
Evolution:
isBookingSuccessful=true and isValidCard=true and discountedPrice=price -(price * 1/10) if
loggedIn=true and providedCard=true and noOfPeople>=10 and GroupBooking.Action=act1;
isBookingSuccessful=false and isValidCard=false if
 providedCard=true and GroupBooking.Action=act3;
isBookingSuccessful=false if
noOfPeople<=10 and GroupBooking.Action=act2;
sucessMsgSent=true if
isBookingSuccessful=true and GroupBooking.Action=successMsg and
ConfirmResearvation.Action=recBookingSuccessMsg;
cardFailureMsgSent=true if
isBookingSuccessful=false and GroupBooking.Action=invalidCardMsg;
numberFailureMsgSent=true if
isBookingSuccessful=false and GroupBooking.Action=invalidNumMsg;
end Evolution
```

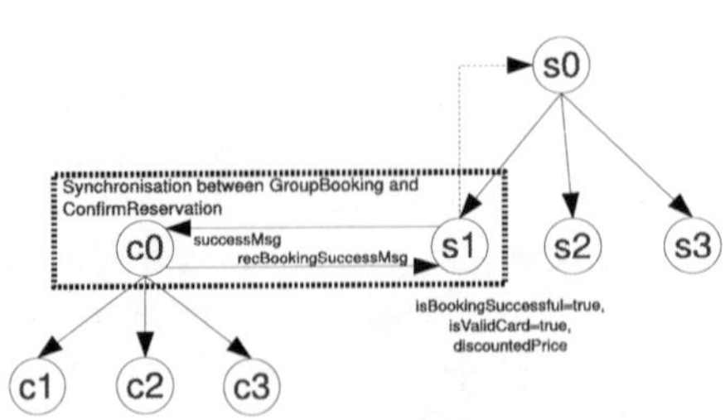

Fig. 5. Sequential composition of GroupBooking and ConfirmReservation

Split, Split+Join. OWL-S provides two types of constructs for concurrent execution: *split* and *split+join*. The components of a split process are a set of processes to be executed concurrently. Split completes when all its component processes have been scheduled for execution whereas split+join completes when all of its component processes have been completed. In both these types of constructs there is a parent process that spawns off the component processes. Split is encoded as illustrated in Figure 6.

The parent process is agent P which spawns three child processes A, B, C at the initial state. Synchronisation between the processes is defined at the initial states of the parent and child processes. Note that in ISPL the parent and child processes are encoded as agents. The number of child agents to be defined can be extracted automatically from the OWL-S definition of composite process.

Similar to split, split+join is encoded as illustrated in Figure 7. In addition to the synchronisation on the initial states of the child processes, the parent process is also synchronised on their final states.

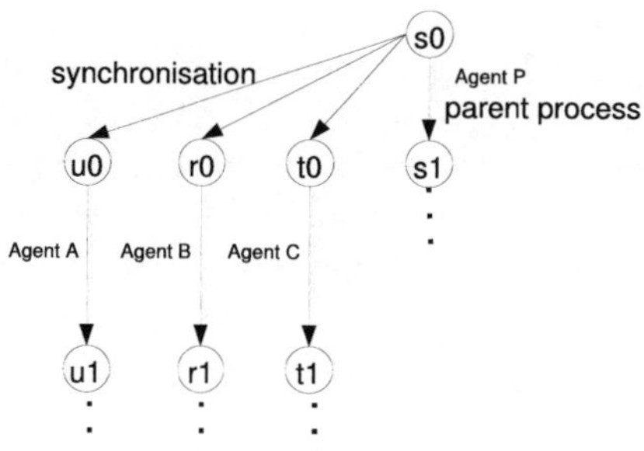

Fig. 6. Encoding split

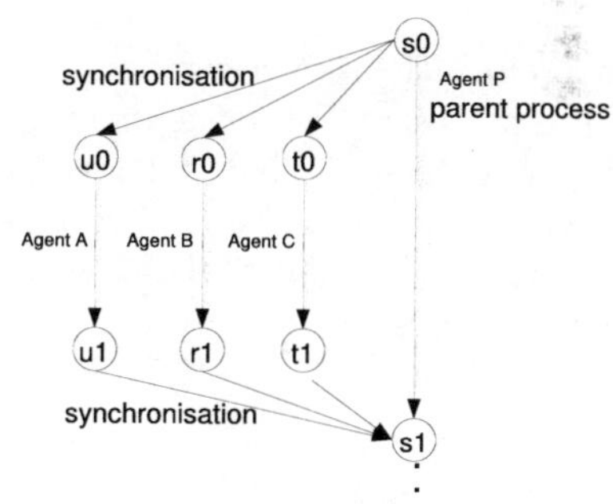

Fig. 7. Encoding split+join

Consider, our extended model for BravoAir. The parent process `ChangeBooking` is composed of two child atomic processes, `ChargeCard` and `AllocateNewBooking` using *split+join*. The processes are synchronised at the states indicated in Figure 8. `ChangeBooking` sends an input message to `ChargeCard` which includes the client's details and the extra payment to be charged. As output `ChargeCard` returns transaction details. Concurrently, `ChangeBooking` sends the client's details, original booking and the requested new booking to `AllocateNewBooking`. The process returns new booking for the client. All messages are encoded as propositions as discussed in section 4.1.

Choice. The *choice* construct allows the invoking process to choose one process non deterministically, from a set of processes. Once the choice is made, the composition essentially resolves to a sequential composition between the invoking process and the chosen process. The parent process as well as the set of processes are encoded as agents in ISPL, as illustrated in Figure 9.

In our extended example for BravoAir, `ManageBooking` is a process, composed as a choice between `ChooseSeats` and the `ChangeBooking` and `CancelBooking`. Due to space restrictions, we do not discuss this in the paper.

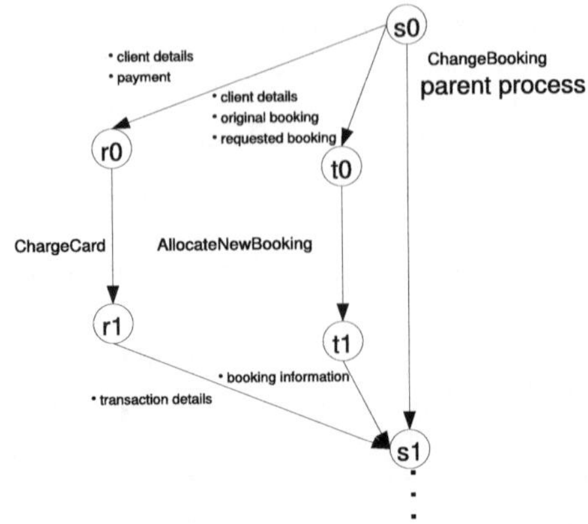

Fig. 8. The ChangeBooking process

If-then-else. The *if-then-else* construct allows to conditionally choose a process from a pair of processes. We define three agents, corresponding to the invoking process and the pair of processes. Since the condition to be checked is a boolean formulae, similar to checking preconditions, we define two states, one where the condition holds and the other where it does not. At these two states we define the synchronisation between the invoking process and the processes in the pair respectively as illustrated in Figure 10.

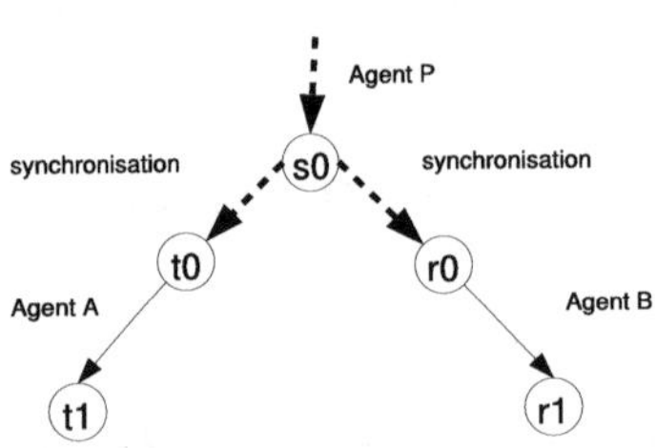

Fig. 9. Encoding choice

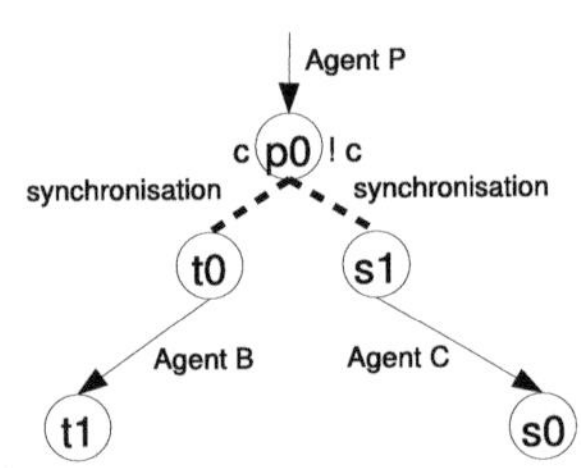

Fig. 10. Encoding if-then-else

Iterate, Repeat-while, Repeat-Until. The *iterate* construct allows the unconditional repeated invocation of a process. Two agents are defined corresponding to the invoking and the iterating process. *Repeat-while* and *Repeat-Until* allow repeated and conditional invocation of a process Synchronisation between the processes is illustrated in Figure 11 and 12.

5 Analysis and Verification

In this section we show the results of verification of some interesting properties of the BravoAir process. Properties are specified as CTL and epistemic formulae[2].

[2] For further details on how such properties could be specified and verified using MCMAS, the interested reader is referred to [1].

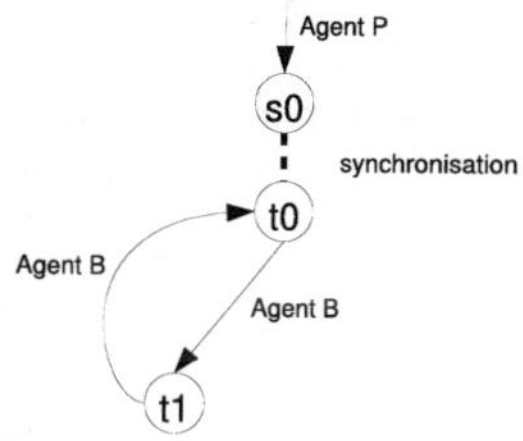

Fig. 11. Encoding iterate

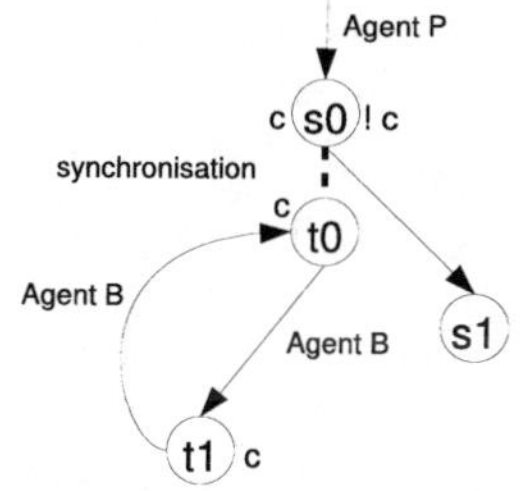

Fig. 12. Encoding repeat-while-repeat-until

- if there is a request for flight booking confirmation, the ConfirmReservation agent (CR) knows that the customer (C) is an authorised customer.

$$EF((confirmBookingRequest) \rightarrow K_{CR}(authorisedCustomer))$$

- If the number of people is more than 10, group booking will always be successful.

$$AF((noOfPeople > 10) \rightarrow EF(isBookingSuccessful))$$

Intuitively the property does not hold because if the card details provided are not valid, the booking will not be successful.
- Whenever a booking change is requested, it will eventually be confirmed.

$$EF((bookingChangeRequest) \rightarrow EF(sendConfirmation))$$

Intuitively the property does not hold because if there are no alternative bookings available, the change will not be confirmed.
- If a card is not charged when a booking is changed, the `ChangeBooking` agent CB knows that the reference is a business class booking.

$$EF(bookingChanged \wedge \neg cardCharged \rightarrow K_{CB}(businessBooking))$$

- if a confirmation is received, the Customer agent(C) knows that his booking was changed.

$$EF(receivedConfirmation \rightarrow K_C(bookingChanged))$$

We encoded the scenario and the specification above in ISPL and verified it using MCMAS. Our system was running on Linux Ubuntu 8.10 (kernel 2.6.27) on Intel Core 2 Duo T5500 1.66GHz with 2GB memory. We encoded 20 ISPL agents by using 120 BDD variables: 43 BDD variables for local states (the same number of BDD variables are constructed for the transition relation) and 34 for local actions. The total number of global states is approximately 10^5. it took about 41 seconds with 34 MB memory space for MCMAS to verify 15 properties. The verification results were in accordance with what expected.

6 Conclusions

Although extensive research has been done on the automated verification of web service composition using BPEL4WS, work on verification of OWL-S process models is relatively scant. Approaches closely related to our work are [2] and [10]. In [2] the mapping rules are defined for Promela to be used with SPIN, an explicit model checker. In [10], the rules are defined for a C-like specification language to be used with BLAST [8]. The limitation in both cases is that only LTL (SPIN) and LTL-like (BLAST) properties can be verified. Using our approach, it is possible to verify LTL, CTL, epistemic and deontic properties with MCMAS. For example, for the case study in Section 2 the composition can be viewed as a multi-agent system where individual processes are abstracted as agents.

In this paper we have proposed mapping rules from the process model of OWL-S to ISPL. We believe the rules are sound as the semantics of ISPL are based on standard kripke semantics and it has been shown that OWL-S processes can be encoded as transition systems [18]. We have shown the mapping for atomic processes and for certain control constructs used for composing them. Our approach provides the first steps necessary to compile automatically OWL-S process models to ISPL. We have developed a compiler that implements the proposed mapping rules. The compiler generates ISPL agents for atomic processes and processes composed in sequence. We are now in the process of enhancing the compiler for other control constructs such as *choice* and *if-then-else*.

Acknowledgements. The research described in this paper is partly supported by the European Commission Framework 6 funded project CONTRACT (IST Project Number 034418).

References

1. Lomuscio, A., Qu, H., Solanki, M.: Towards verifying compliance in agent-based web service compositions. In: Proceedings of The Seventh International Joint Conference on Autonomous Agents and Multi-agent systems (AAMAS 2008). ACM Press, New York (2008)
2. Ankolekar, A., Paolucci, M., Sycara, K.P.: Towards a formal verification of owl-s process models. In: Gil, Y., Motta, E., Benjamins, V.R., Musen, M.A. (eds.) ISWC 2005. LNCS, vol. 3729, pp. 37–51. Springer, Heidelberg (2005)
3. Cimatti, A., Clarke, E., Giunchiglia, F., Roveri, M.: NuSMV: A new symbolic model verifier. In: Halbwachs, N., Peled, D.A. (eds.) CAV 1999. LNCS, vol. 1633, pp. 495–499. Springer, Heidelberg (1999)
4. Clarke, E.M., Grumberg, O., Peled, D.A.: Model Checking. MIT Press, Cambridge (1999)
5. CMU: An API for OWL-S. CMU OWL-S API, `http://www.daml.ri.cmu.edu/owlsapi/index.html`
6. Booth, D., Haas, H., McCabe, F., Newcomer, E., Champion, M., Ferris, C., Orchard, D.: Web service architecture. W3c working group note (February 11, 2004), `http://www.w3.org/TR/ws-arch/`
7. Fagin, R., Halpern, J.Y., Moses, Y., Vardi, M.Y.: Reasoning about Knowledge. MIT Press, Cambridge (1995)

8. Henzinger, T.A., Jhala, R., Majumdar, R., Sutre, G.: Software verification with BLAST. In: Ball, T., Rajamani, S.K. (eds.) SPIN 2003. LNCS, vol. 2648, pp. 235–239. Springer, Heidelberg (2003)
9. Holzmann, G.J.: SPIN Model Checker, The: Primer and Reference Manual. Addison Wesley Professional, Reading (2003)
10. Huang, H., Tsai, W., Paul, R., Chen, Y.: Automated model checking and testing for composite web services. In: Proc. of ISORC 2005, pp. 300–307. IEEE Computer Society, Los Alamitos (2005)
11. Kalyanpur, A., Jimnez, D.: Automatic mapping of owl ontologies into java. In: Proceedings of Software Engeering and Knowledge Engeering (SEKE 2004) (2004)
12. Lomuscio, A., Qu, H., Raimondi, F.: Mcmas 0.9 alpha (2008), `http://sourceforge.net/projects/ist-contract/`
13. Lomuscio, A., Raimondi, F.: MCMAS: A model checker for multi-agent systems. In: Hermanns, H., Palsberg, J. (eds.) TACAS 2006. LNCS, vol. 3920, pp. 450–454. Springer, Heidelberg (2006)
14. Lomuscio, A., Qu, H., Sergot, M.J., Solanki, M.: Verifying temporal and epistemic properties of web service compositions. In: Krämer, B.J., Lin, K.-J., Narasimhan, P. (eds.) ICSOC 2007. LNCS, vol. 4749, pp. 456–461. Springer, Heidelberg (2007)
15. Kavantzas, N., Burdett, D., Ritzinger, G., Fletcher, T., Lafon, Y.: Web Services Choreography Description Language Version 1.0: W3C Working Draft (December 17, 2004)
16. OASIS Web service Business Process Execution Language (WSBPEL) TC. Web service Business Process Execution Language Version 2.0 (2007)
17. The OWL-S Coalition. OWL-S 1.1 Release (2004), `http://www.daml.org/services/owl-s/1.0/`
18. Traverso, P., Pistore, M.: Automated composition of semantic web services into executable processes. In: McIlraith, S.A., Plexousakis, D., van Harmelen, F. (eds.) ISWC 2004. LNCS, vol. 3298, pp. 380–394. Springer, Heidelberg (2004)
19. Wooldridge, M.: An introduction to multi-agent systems. John Wiley, England (2002)

Transformation Algorithm from OWL-S to ISPL

Algorithm 1. Psuedocode: OWL-S - ISPL mapping

read `ProcessName.owl` {parse the atomic process}
print Agent `ProcessName` {define the agent}
print Vars: {begin the extraction and definition of local variables}
for all hasInput in ProcessName.owl **do**
    print InputID: datatype;
end for
for all hasOutput in ProcessName.owl **do**
    print OutputID: datatype;
    print end Vars {end of variable definition}
end for
$N_{hasResult} \Leftarrow$ count(`<process:hasResult>`){count the number of `<process:hasResult>` elements}
$N_{actions} \Leftarrow 2 \times N_{hasResult}$+2 {determine the number of actions for the process}
print RedStates:{begin the definition the red states}
f^{red} {define the Boolean formula for the red states}
print end RedStates {end definition of red states}
print Actions ={ϵ, a_f {begin enumeration of actions}
for all i such that $1 \leq i \leq \mid A_{int} \mid$ **do**
    print a_i, {enumeration of actions taken at states l_0}
end for
for all i such that $1 \leq i \leq \mid A_{send} \mid$ **do**
    print s_i, {enumeration of actions taken at states $\{l_i \mid i = 1 \ldots \mid A_{send} \mid\}$}
end for
print s_f} {end enumeration of actions}
print Protocol: {begin enumeration of the protocols}
printf^{pre} : {define the Boolean formula for the precondition}
for all i such that $1 \leq i \leq \mid A_{int} \mid$ **do**
    print $\{a_i \mid i = 1 \ldots \mid A_{int} \mid\}$ {enumerate the actions}
end for
print;{end of line}
print!f^{pre} : a_f; {define the protocol for precondition failure}
for all i such that $1 \leq i \leq \mid A_{send} \mid$ **do**
    print f_i^{res} : $\{s_i\}$
    print;{end of line}
end for
print end protocol {end enumeration of the protocol}
print Evolution: {begin enumeration of the evolutions}
print l_0 if (
for all i such that $1 \leq i \leq \mid A_{send} \mid$ **do**
    print (l_i and ProcessName.Action=s_i)
end for
print;{end of line}
for all i such that $1 \leq i \leq \mid A_{int} \mid$ **do**
    print l_i if (l_0 and ProcessName.Action=a_i);
end for
print l_f if (Lstate =l_0 and ProcessName.Action=a_f);
end Evolution; {end enumeration of evolutions}
end Agent {end agent definition}

Leveraging Semantic Web Service Descriptions for Validation by Automated Functional Testing

Ervin Ramollari[1], Dimitrios Kourtesis[1], Dimitris Dranidis[2], and Anthony J.H. Simons[3]

[1] South East European Research Centre (SEERC),
Research Centre of the University of Sheffield and CITY College
Mitropoleos 17, 54624, Thessaloniki, Greece
erramollari@seerc.org, dkourtesis@seerc.org
[2] Computer Science Department, CITY College,
Affiliated Institution of the University of Sheffield
Tsimiski 13, 54624 Thessaloniki, Greece
dranidis@city.academic.gr
[3] Department of Computer Science, University of Sheffield
Regent Court, 211 Portobello Street, Sheffield S1 4DP, UK
a.simons@dcs.shef.ac.uk

Abstract. Recent years have seen the utilisation of Semantic Web Service descriptions for automating a wide range of service-related activities, with a primary focus on service discovery, composition, execution and mediation. An important area which so far has received less attention is service validation, whereby advertised services are proven to conform to required behavioural specifications. This paper proposes a method for validation of service-oriented systems through automated functional testing. The method leverages ontology-based and rule-based descriptions of service inputs, outputs, preconditions and effects (IOPE) for constructing a stateful EFSM specification. The specification is subsequently utilised for functional testing and validation using the proven Stream X-machine (SXM) testing methodology. Complete functional test sets are generated automatically at an abstract level and are then applied to concrete Web services, using test drivers created from the Web service descriptions. The testing method comes with completeness guarantees and provides a strong method for validating the behaviour of Web services.

Keywords: Semantic Web Services, Web service testing, Service Validation.

1 Introduction

The vision in Semantic Web Services (SWS) research [13], [14] is to bring formal semantics into the Web services realm such that service characteristics can be explicated in an unambiguous, computer-interpretable manner that allows for automating a broad range of service design-time and run-time activities. Recent years have seen the development of numerous SWS frameworks that aim to address this goal. The most prominent ones have been OWL-S [15], WSMO [4], and WSDL-S [1]. The latter

L. Aroyo et al. (Eds.): ESWC 2009, LNCS 5554, pp. 593–607, 2009.

served as the basis for the recently ratified W3C Recommendation of SAWSDL [7] which is currently the only standard in the area. Research around SWS frameworks has mostly focused on the development of methods and tools for enabling automated Web service discovery, composition, and execution [13], [14], while the areas of service testing and validation have remained largely unexplored.

The ability to validate that a Web service implementation conforms to given functional/behavioural specifications through functional testing is instrumental in engineering dependable service-based systems. However, the process of Web service testing is currently highly error-prone and laborious, as most of the activities, and especially the construction of test cases, need to be carried out manually. In this paper we show that in the presence of service descriptions with formal semantics, the process of generating test cases and performing validation through testing can be automated to a great extent. In particular, we propose a method for deriving a formal specification of a Web service's behaviour as a Stream X-machine (SXM) [9] model, by transforming ontology-based and rule-based descriptions of the service's inputs, outputs, preconditions, and effects (IOPE). The derived SXM model is utilised for automatically generating abstract test cases, which are executed against the deployed Web service to perform validation.

An important aspect of our approach is the strength of the validation method, which stems from the completeness guarantees that the SXM testing method can provide. The SXM testing method is guaranteed to generate a complete, finite set of test cases able to reveal all inconsistencies among an SXM specification and an implementation under test. Another important aspect is the adoption and promotion of standards. The semantic descriptions of Web service IOPE from which the SXM model is derived, are encoded using an existing and a forthcoming standard from the Semantic Web technology stack. OWL [16] is employed for specifying the service's data model, and RIF-PRD (Rule Interchange Format – Production Rule Dialect) [21] is employed for specifying the service's functionality. As a forthcoming standard for rule interchange, RIF [22] can assist in overcoming the impedance mismatch among today's heterogeneous SWS frameworks, and can enable our approach to remain generic and even reusable with different frameworks like WSMO or OWL-S.

The rest of this paper is organised as follows. Section 2 presents an overview of related works in the area of Web service testing, validation and verification that involves the use of semantic service descriptions. Section 3 discusses how stateful Web service behaviour can be represented in terms of IOPE, and how Semantic Web technologies such as OWL and RIF can be employed for this purpose, with the help of an example Web service. Section 4 provides an overview of the Stream X-machine modelling formalism which we adopt for modelling Web service behaviour and generating test cases in an automated way. Section 5 provides a detailed step-by-step description of the method of deriving an SXM model from the IOPE-based service description. Section 6 discusses the overall testing approach, the generation of test cases from the SXM model, and the way in which tests can be executed against a Web service. Finally, section 7 concludes the paper with a summary of the key points and an outlook to future research objectives.

2 Related Work

Methods and tools for Web service testing, verification and validation are essential for engineering and managing dependable service-based software applications, and are receiving increasing attention. In this section we consider related research works that address the validation of Web services through model-based testing, where a model is used to generate a set of test cases.

Heckel and Mariani [8] employ graph transformation rules to model the behaviour of individual operations of stateful Web services in terms of preconditions and updates on the world state. The graph transformation rules are used to generate test cases using domain partitioning and data-flow criteria. The method is used in a registry-based testing approach to validate conformance of advertised services.

Bertolino et al [2] also propose a registry-based testing approach where the provider augments the WSDL document with behavioural descriptions in a UML 2.0 Protocol State Machine (PSM) diagram, which is translated to a Symbolic Transition System (STS). The registry uses the STS model to generate a set of test cases, which are run on the Web service under test for validating behavioural conformance.

Wang et al [23] use the OWL-S description of a composite service process as a starting point to derive an equivalent Petri-net model through an informal transformation algorithm. The Petri-net model is then used to generate test cases and perform model-checking.

Narayanan and McIlraith [17] also describe a manual method to construct a Petri net from descriptions in DAML-S (the precursor of OWL-S). They propose using this model for a number of activities, including testing. However, no testing approach and test-case generation algorithm is described, since the authors assume that a test suite already exists and the model serves as a test oracle to predict the expected outputs.

Sinha and Paradkar [19] utilise IOPE-based descriptions of service operations using SWRL (Semantic Web Rule Language) rules and OWL ontologies to derive an extended finite state machine (EFSM) model. The authors propose various manual or automated techniques for generating test cases based on the EFSM model, with varying adequacy criteria, coverage and completeness. The generated EFSM consists of a single state and lacks information on explicit sequencing of operations (or conversation protocol) necessary to guide state-based testing.

Keum et al [11] provide a high-level, manual algorithm for constructing a multi-state EFSM model from plain WSDL specifications, with additional information provided by the user, who fills a WSDL analysis template. The described algorithm mainly focuses on the derivation of states, while little or no detail is given for obtaining the transitions and other EFSM elements.

In this paper we propose a method that utilises IOPE-based semantic descriptions of Web service operations that are encoded using RIF-PRD to generate a multi-state Stream X-machine model. In addition to the states and transitions that define a finite state machine, other constructs which are specific to SXMs, such as memory and processing functions, are also derived. The SXM testing strategy [9] allows for the derivation of a test set with completeness guarantees, making it possible to prove functional conformance of an implementation to the SXM model. The automated generation of test cases and their execution on a Web service under test are supported by an existing suite of tools.

3 IOPE-Based Descriptions of Web Service Functionality

A popular approach for describing functional properties of stateful and conversational Web services is in terms of service inputs, outputs, preconditions and effects (IOPE) [20]. IO and PE descriptions are means to represent two different aspects of a service: (i) the information transformation that it produces through the inputs it consumes and the outputs it generates, and (ii) the state-wise conditions that need to hold before service operations can be invoked (preconditions) as well as the state changes that will be applied after the invocation (effects).

Descriptions of inputs and outputs allow service providers to explicate the semantics of the data vocabulary that the service operates on, independently of the service's behaviour. This type of specification can be provided for any kind of service, ranging from simple information-providing services to complex transactional services, and enables high-precision service discovery through semantic matchmaking, as well as run-time data mediation through semantic adapters. On the other hand, descriptions of preconditions and effects are only useful for Web services whose behaviour depends on the state of the world, and whose execution may cause changes to it. Examples of stateful Web services with side-effects could be an atomic ticket reservation service that consists of a single WSDL operation, or a bank account management service that comprises multiple WSDL operations. Our approach is focused on behavioural validation for services of the latter type, i.e. conversational Web services that maintain information about internal state between invocations of operations, and enforce a specific interaction protocol (choreography) with the service consumer.

Variations of the IOPE model are employed within all of the existing SWS frameworks. OWL-S process descriptions are based on an Input, Output, Precondition and Result (IOPR) model. A Result is associated to some Condition specifying the premises under which the result can occur, the corresponding Output, and the Effects to be produced. In descriptions of WSMO Goals and Capabilities a more fine-grained notion of Pre-conditions, Assumptions, Post-conditions, and Effects is employed to define state-wise constraints. In WSDL-S and SAWSDL, preconditions and effects can be associated with a Web service operation via a modelReference annotation on a WSDL portType, in the same way in which inputs and outputs are specified via modelReference annotations on WSDL operations or XSD elements.

In all cases, the representation of IOPE requires a combination of ontology language for encoding descriptions of inputs and outputs, and rule language for encoding preconditions and effects as logic-based expressions. Each of the previously mentioned SWS frameworks adopts a different approach to achieve this. OWL-S requires the representation of IO using OWL and allows encoding of logical expressions (treated as literals) for PE in a variety of rule languages, although it appears that only SWRL-like expressions have actually been supported in implemented OWL-S tools. In WSMO, Preconditions, Postconditions, Assumptions and Effects involve the specification of axioms as First Order Logic formulae that can be encoded in WSML dialects such as WSML-Rule or WSML-Flight. In WSDL-S and SAWSDL the format of the logical expressions is left to be defined by the modellers, and their semantic representation language of choice.

The kind of rules and rule language to be used for representing preconditions and effects depends on the intended semantics of the Web service's effects. We can generally distinguish among the use of deductive (inference) rules which have logical, non-monotonic semantics, and the use of reactive rules that carry operational, non-monotonic semantics. Reactive rules can be Event-Condition-Action (ECA) rules, which allow modelling changes to the state of the world as a reaction to an event, or Condition-Action (CA) rules, also known as production rules, which allow modelling state changes without reference to a particular event. A significant difference among deductive and reactive rules is found in the conclusion part of the rule. Conclusions of deductive rules contain only logical statements that should be true, whereas conclusions of reactive rules contain actions can add, delete, or modify facts in the knowledge base, and have other side-effects [21].

Our approach in this paper utilises production rules for specifying the behaviour of stateful and conversational Web services, and the rule language that we adopt is the Rule Interchange Format - Production Rule Dialect (RIF-PRD) [21]. RIF [22] is a forthcoming standard for the representation and exchange of rules that is currently under development by the W3C RIF Working Group, and is put forward as the next generation rule language for the Semantic Web. At the time of this writing RIF-PRD is still work in progress, but expected to become a W3C recommendation soon.

An important reason for using RIF-PRD is that the semantics of PRD rules can be specified as a labelled transition system [18], where states correspond to possible states of working memory and transitions correspond to actions taken by the rules to assert/retract terms. The fact that PRD has operational semantics makes the definition of an equivalence relationship between PRD and SXM specifications possible, and thus enables to prove the correctness of the transformation. Moreover, RIF-PRD can serve as a standard interface language between existing SWS formalisms and tools for derivation of SXM models and generation of test cases, thus bringing added value to existing SWS frameworks by allowing semantic descriptions to be reused for the purposes of testing and validation.

3.1 Example Description of a Stateful Web Service

To provide an example of a stateful Web service whose functionality is modelled using the approach described above, consider the case of a service that allows performing elementary operations on a bank account. For simplicity, let us assume that the service interface comprises five operations: (i) open, (ii) deposit, (iii) withdraw, (iv) getBalance, and (v) close. When an account is created it is initialised as inactive and therefore needs to be set to active (opened) before any transaction can be performed. The deposit of an amount will result in increasing the balance of the account as appropriate, while the withdrawal of an amount can take place only if the amount does not exceed the balance, and will result in reducing the balance accordingly. A successful deposit or withdrawal will also result in having the updated balance returned to the client as part of the invocation response message. Finally, an account can be closed only if its balance is zero, and once closed cannot be re-activated. The listing below provides an encoding of preconditions and effects for this example Web service, using RIF-PRD presentation syntax.

```
Prefix(func http://www.w3.org/2007/rif-builtin-function#)
Prefix(pred http://www.w3.org/2007/rif-builtin-predicate#)


(* wsdl:operation open *)
Forall     ?account ?status ?balance ?request (
And      (  ?account#Account
            ?account[hasStatus->?status]
            ?account[hasBalance->?balance]
            ?request#OpenRequest )
If       (  External (pred:string-equal(?status "INITIAL") )
Then Do  (  Retract (?account[hasStatus->?status])
            Assert (?account[hasStatus->"ACTIVE"])
            Retract (?account[hasBalance->?balance])
            Assert (?account[hasBalance->0])
            Retract (?request)
            (?response New(?response#OpenResponse))
            Assert (?response[hasMessage->"Account opened"]) ) )

(* wsdl:operation close *)
Forall     ?account ?status ?balance ?request (
And      (  ?account#Account
            ?account[hasStatus->?status]
            ?account[hasBalance->?balance]
            ?request#CloseRequest )
If And   (  External (pred:string-equal(?status "ACTIVE")
            External (pred:numeric-equal(?balance 0) )
Then Do  (  Retract (?account[hasStatus->?status])
            Assert (?account[hasStatus->"CLOSED"])
            Retract (?request)
            (?response New(?response#CloseResponse))
            Assert (?response[hasMessage->"Account closed"]) ) )

(* wsdl:operation getBalance *)
Forall     ?account ?status ?balance ?request (
And      (  ?account#Account
            ?account[hasStatus->?status]
            ?account[hasBalance->?balance]
            ?request#GetBalanceRequest )
If       (  External (pred:string-equal(?status "ACTIVE") )
Then Do  (  Retract (?request)
            (?response New(?response#GetBalanceResponse))
            Assert (?response[hasAmount->?balance) ) )

(* wsdl:operation deposit *)
Forall     ?account ?status ?balance ?request ?depositAmount (
And      (  ?account#Account
            ?account[hasStatus->?status]
            ?account[hasBalance->?balance]
            ?request#DepositRequest
            ?request[hasAmount->?depositAmount] )
If And   (  External (pred:string-equal(?status "ACTIVE")
            External (pred:numeric-greater-than(?depositAmount 0) )
Then Do  (  Retract (?account[hasBalance->?balance])
            Assert (?account[hasBalance->External(func:numeric-add(?balance
            ?depositAmount))])
            Retract (?request)
            (?response New(?response#DepositResponse))
            (?newBalance ?account[hasBalance->?newBalance])
            Assert (?response[hasAmount->?newBalance) ) )

(* wsdl:operation withdraw *)
Forall     ?account ?status ?balance ?request ?withdrawAmount (
And      (  ?account#Account
            ?account[hasStatus->?status]
            ?account[hasBalance->?balance]
            ?request#WithdrawRequest
            ?request[hasAmount->?withdrawAmount] )
If And   (  External (pred:string-equal(?status "ACTIVE")
```

```
          External (pred:numeric-greater-than-or-equal(?balance
          ?withdrawAmount) )
Then Do ( Retract (?account[hasBalance->?balance])
          Assert(?account[hasBalance->External(func:numeric-subtract(?balance
          ?withdrawAmount))])
          Retract (?request)
          (?response New(?response#WithdrawResponse))
          (?newBalance ?account[hasBalance->?newBalance])
          Assert (?response[hasAmount->?newBalance) ) )
```

In addition to the rule-based description of service behaviour that is encoded in RIF-PRD, we assume the existence of an ontology-based description of the service's data model, encoded in OWL. The ontology is to be used for two purposes: i) specifying the structure of the service's inputs, outputs and internal state-related variables which are not visible on its interface, and ii) specifying the set of facts that hold when the service is found at its initial state (e.g. after original deployment or resetting). This information is utilised for the derivation of the SXM model, as described in section 5. The following table lists the OWL entities that the ontology would need to comprise:

Table 1. Ontology entities

Classes (11)	Account, OpenRequest, OpenResponse, DepositRequest, DepositResponse, GetBalanceRequest, GetBalanceResponse, WithdrawRequest, WithdrawResponse, CloseRequest, CloseResponse
Datatype properties (4)	hasStatus (domain: Account, range: string enumeration {"INITIAL", "CLOSED", "FALSE"}), hasBalance (domain: Account, range: nonNegativeInteger), hasAmount (domain: {GetBalanceResponse, DepositRequest, DepositResponse, WithdrawRequest, WithdrawResponse}, range: nonNegativeInteger) hasMessage (domain: {OpenResponse, CloseResponse}, range: string)
Individuals (1)	acc1 (asserted Account individual, hasStatus: "INITIAL", hasBalance: 0)

Note that the RIF-PRD rules are not meant to refer to the ontology classes and properties in a direct way. Rather, they are meant to refer to in-memory representations of these entities which are created when the transformation procedure is initiated by reading-in the ontology contents. This hybrid integration approach has been chosen since RIF-PRD does not support importing of OWL ontologies in RIF-PRD documents and using ontology entities within rules, as is the case with RIF-BLD (Basic Logic Dialect). The reason for this is the incompatible semantics of OWL and RIF-PRD; OWL has monotonic model theoretic semantics, while production rules have non-monotonic operational semantics (due to the actions performed).

To complete the description of the example Web service, we also need a way to associate each of the production rules with the corresponding WSDL operation whose behaviour is meant to be described. This association can be specified by placing SAWSDL modelReference annotations on the corresponding wsdl:operation elements, and adopting a convention for identifying and referencing different rules within a RIF XML document (e.g. through XPointer fragment identifiers). Semantic annotations on the WSDL document can also serve for specifying the indirect linkage among the rule-based description of service behaviour and the ontology-based description of the service data model, which needs to be known for the SXM derivation. A complete description of an SAWSDL-based approach for linking the WSDL, RIF, and OWL artefacts is however beyond the scope of this paper.

4 Stream X-Machines and Web Service Modelling

Stream X-machines (SXMs) are a computational model capable of representing both the data and the control of a system. SXMs are special instances of the X-machines introduced in 1974 by Samuel Eilenberg [6]. They employ a diagrammatic approach of modelling control flow by extending the expressive power of finite state machines. In contrast to finite state machines, SXMs are capable of modelling non-trivial data structures by employing a memory attached to the state machine. Moreover, transitions between states are not labelled with simple input symbols but with processing functions. Processing functions receive input symbols and read memory values, and produce output symbols while modifying memory values.

Apart from being formal as well as proven to possess the computational power of Turing machines [9], SXMs offer a highly effective testing method for verifying the conformance of a system's implementation against a specification. The SXM testing method, which is a generalization of the W-method [3], is guaranteed to generate a complete, finite set of test cases that can reveal all inconsistencies among an SXM specification and an implementation under test [10]. More details about the derivation of the test sequences can be found in [5].

Parallels can be drawn among a stateful Web service and a Stream X-machine, since they both accept inputs and produce outputs, while performing specific actions and moving from one internal state to another. SXM inputs and outputs correspond to SOAP request and response messages. SXM processing functions correspond to Web service operation invocations in specific contexts (an operation invocation may map to more than one processing functions because of the potentially different input parameters and the different state). In [5] the process of modelling and test generation for a stateful Web service are demonstrated in more detail, while in [12] we present an approach that integrates SXM-based modelling and test generation for extending a service registry with functional testing and behavioural verification capabilities.

5 Derivation of an SXM Model from an IOPE Description

This section describes the steps of the transformation of IOPE-based descriptions of Web service behaviour to SXM models. The example of the Account Web service is used throughout, for illustration purposes.

Identifying State Variables. All properties p_i of frame atomic formulas `t[`p_1`->`v_1 ... p_n`->`v_n`]` that appear in the action part of all rules are identified as state variables. Excluded are the frames associated to the reserved objects `?request` and `?response`, which represent inputs and outputs, respectively.

In the account service example two state variables are identified: `hasStatus` and `hasBalance`.

Partition Analysis of State Variables. The domains of the state variables are determined by consulting their respective *ranges* in the *datatype properties* section of the OWL ontology. Then, the domain of each state variable p_i is partitioned based on the *preconditions* of the production rules. For each variable v_i bound to a property p_i, which is a state variable, all formulas in the precondition of the rules are evaluated.

In the account example, the domains of the state variables are:

`hasStatus ::= {INITIAL, ACTIVE, CLOSED}` and `hasBalance::= 0…∞`.

The preconditions may restrict the state variables to actual values or ranges of values. In the rules of the account service example, the variables `?status` and `?balance` bound to the state variables `hasStatus` and `hasBalance` are evaluated in the following conditions:

```
hasStatus:
    • ?status = INITIAL
    • ?status = ACTIVE
hasBalance:
    • ?balance = 0
    • ?balance ≥ ?withdrawAmount AND ?withdrawAmount > 0
```

By application of naive arithmetic rules on the last condition listed above, it can be deduced that `?balance>0`. It should be noted that mathematical deductions like this can be hard to automate, and would require manual intervention. After examining the domains of the state variables and inferring the complements, the final partitions are:

```
hasStatus: {INITIAL}, {ACTIVE}, {CLOSED}
hasBalance: {0}, {x | x>0}
```

Identifying Preliminary States. The preliminary state space is defined as the product of the state variable partitions. The initial state q_0 is determined from the initial values of the respective state variables, which are specified in an individual in the OWL ontology.

In our example there are six preliminary states, which are labelled as `INITIAL_0`, `INITIAL_>0`, `ACTIVE_0`, `ACTIVE_>0`, `CLOSED_0`, and `CLOSED_>0`.

In the account specification the initial value for `hasStatus is INITIAL`, and for `hasBalance` is zero. Therefore state `INITIAL_0` is the initial state.

Determining Inputs and Outputs. An input (output) is defined for each service operation request (response) specified in the RIF rules. Inputs appear at the bindings and outputs at the action parts of the rules.

The inputs in the example are:

```
OpenRequest()
GetBalanceRequest()
DepositRequest(depositAmount)
WithdrawRequest(withdrawAmount)
CloseRequest()
```

The outputs are:

```
OpenResponse()
GetBalanceResponse(balance)
DepositResponse(newBalance)
WithdrawResponse(newBalance)
CloseResponse()
```

Determining Transition Pre-States. An *input* is *accepted* at a state (pre-state) iff the preconditions are satisfied at the pre-state. In that case the input triggers a *transition* of the Stream X-machine from the *pre-state* to another state (the *post-state*). For each input (service operation invocation) and each state, it is determined whether the input is accepted at that state.

For example, the preconditions of the rule for input `OpenRequest` are `(hasStatus=INITIAL)`, so they are satisfied at states `INITIAL_0` and `INITIAL_>0`. On the other hand, input `WithdrawRequest` is a different case. By applying the previously mentioned mathematical deduction we can infer the following:

```
hasBalance≥withdrawAmount AND withdrawAmount>0 → hasBalance>0
```

Therefore, the input `WithdrawRequest` is only accepted at pre-state `ACTIVE_>0`. The inputs and the pre-states at which they trigger a transition are as follows:

```
OpenRequest: {INITIAL_0, INITIAL_>0}
GetBalanceRequest: {ACTIVE_0, ACTIVE_>0}
DepositRequest: {ACTIVE_0, ACTIVE_>0}
WithdrawRequest: {ACTIVE_>0}
CloseRequest: {ACTIVE_0}
```

State Merging. Preliminary states in which the same set of inputs is accepted are merged, since these states cannot be distinguished.

In our example, in both states `INITIAL_0` and `INITIAL_>0` the set of inputs `{OpenRequest}` is accepted. Therefore, those two states are merged into one. The resulting states are:

```
INITIAL: (hasStatus=INITIAL, hasBalance=*)
ACTIVE_0: (hasStatus=ACTIVE, hasBalance=0)
ACTIVE_>0: (hasStatus=ACTIVE, hasBalance>0)
CLOSED:   (hasStatus=CLOSED, hasBalance=*)
```

where * denotes any value.

Determining Transition Post-States. For each state-input pair identified in a previous step, we determine the possible transition destinations, or post-states, by applying the effects of the invoked operation on the pre-state. Isolated states (there does not exist any sequence of transitions starting from q_0 and ending in that state) are removed from the model.

In our example, applying these effects potentially involves reasoning with standard operations, such as the mathematical operations of *addition* and *subtraction*. Accepting input `OpenRequest` at pre-state `INITIAL: (hasStatus=INITIAL)` and applying the effects `(hasStatus ← ACTIVE` and `hasBalance ← 0)` results in a post-state where `hasStatus=ACTIVE` and `hasBalance=0`. This post-state matches exactly with the state `ACTIVE_0`. On the other hand, applying the effects of accepting the input `WithdrawRequest`, and appealing to mathematical rules for subtraction, there are two possible post-states `(ACTIVE_0` and `ACTIVE_>0)`, depending on the amount withdrawn. The two different transitions start from the same state, so they are labelled by different processing functions, for instance, `WithdrawRequest1` and `WithdrawRequest2`. Using the same approach for the rest of the state-input pairs, the resulting transitions are as in the following table.

Table 2. Transition pre-states and post-states

Input	Pre-State	Effects	Post-State
OpenRequest	INITIAL	hasStatus ← ACTIVE hasBalance ← 0	ACTIVE_0
GetBalanceRequest	ACTIVE_0	-	ACTIVE_0
	ACTIVE_>0	-	ACTIVE_>0
DepositRequest	ACTIVE_0	hasBalance ← hasBalance+depositAmount	ACTIVE_>0
	ACTIVE_>0	hasBalance ← hasBalance+depositAmount	ACTIVE_>0
WithdrawRequest	ACTIVE_>0	hasBalance ← hasBalance-withdrawAmount	ACTIVE_0
	ACTIVE_>0	hasBalance ← hasBalance-withdrawAmount	ACTIVE_>0
CloseRequest	ACTIVE	hasStatus ← CLOSED	CLOSED

The final states and transitions define the *associated finite automaton* of the Stream X-machine, as illustrated in Fig. 1.

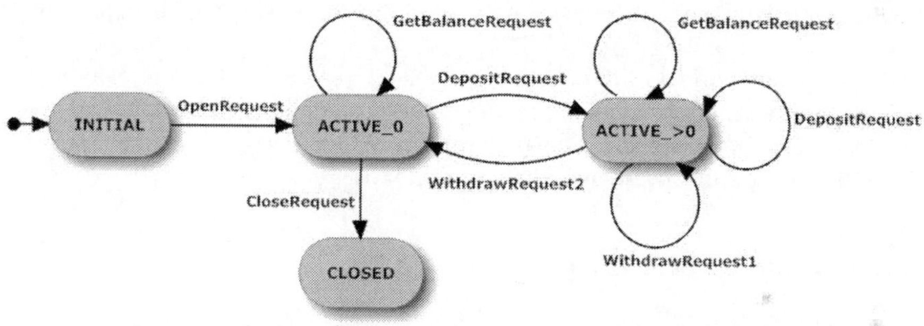

Fig. 1. State diagram of the SXM model that is the result of the transformation

Determining Memory. The memory consists of all the state variables, except for those whose all possible values were covered by the partitions identified during partition analysis.

In the account service example, the enumerated state variable `hasStatus` was partitioned into all its possible values. Therefore, only state variable `hasBalance` is put in the memory. The initial memory m_0 is the initial value of `hasBalance`, i.e. m_0 = (0).

Determining Guard Conditions for Processing Functions. For each state-input pair where the input triggers a unique transition, the guard conditions are the same as the corresponding rule preconditions; any predicates already satisfied in the pre-state are omitted.

For instance, when attempting to accept input `OpenRequest` at state `INITIAL`, all preconditions are already satisfied at the pre-state, so the guard condition is empty.

For each state-input pair where the input triggers more than one transition, the preconditions have to be split into mutually disjoint guard conditions, one for each processing function. For each transition, starting from its post-state, the effects are "undone" (i.e. the updates of the effects are inversed) and it is derived what should hold at the pre-state.

This is the case with input `WithdrawRequest(withdrawAmount)`. The only predicates which are not satisfied in the pre-state are `(hasBalance≥withdrawAmount AND withdrawAmount>0)`. In the post-state of `WithdrawRequest1`, `hasBalance>0`, while in the post-state of `WithdrawRequest2`, `hasBalance=0`. The effect of accepting the input `WithdrawRequest` is: `newBalance = hasBalance - withdrawAmount`. Since the value of updated balance in the post-state of `WithdrawRequest1` is greater than zero, then:

```
(hasBalance - withdrawAmount)>0 → hasBalance>withdrawAmount
```

Thus, the guard condition of `WithdrawRequest1` is `(hasBalance>withdrawAmount AND withdrawAmount>0)`. Similarly it can be deduced that the guard condition of `WithdrawRequest2` is `(hasBalance=withdrawAmount AND withdrawAmount>0)`.

Determining Memory Updates for Processing Functions. The memory updates of each processing function consist only of the effects which update the *memory variables*.

For example, the effect of the rule for `CloseRequest` does not involve any memory variables, thus the memory update function is empty. Similarly, for input `DepositRequest` the memory update function is `hasBalance ← hasBalance+depositAmount`, which is the only action affecting memory variable `hasBalance`.

6 Generation of Test Cases

The main benefit of obtaining a Stream X-machine model of the Web service is the ability to automatically generate complete functional test sets. This formalism is associated with a complete testing strategy which under certain assumptions [9] is proven to find all faults in the implementation. In addition, the SXM model can be animated and thus serve as a test oracle for predicting the service's outputs, which are compared with the real outputs returned by the service implementation.

Test generation starts by applying the W-method [3] on the associated *finite automaton* of the SXM, where processing functions are considered as simple inputs. As a result, the test set *X* for the associated finite automaton consists of sequences of processing functions and is given by the formula:

$$X = S(\Phi^{k+1} \cup \Phi^{k} \cup \ldots \cup \Phi \cup \{\epsilon\})W,$$

where *W* is a *characterization set*, *S* a *state cover* of the associated finite automaton, and *k* is an estimate of maximum path length between redundant states in the implementation. A characterization set is a set of sequences of processing functions for which any two distinct states of the machine are *distinguishable* and a state cover is a set of sequences of processing functions such that all states are reachable from the initial state. The next step is to convert the sequences of processing functions to sequences of inputs. This is achieved by the fundamental test function as described in [9].

The above test generation procedure is supported by a suite of tools that we have developed in Java. The test generation tool takes as input an XML representation of the SXM model and automatically produces sequences of abstract input and expected output pairs. Another tool converts the abstract test cases to JUnit test cases. In order to execute the generated JUnit test cases on the Web service under test, a Java client stub is required as an adapter. We use the open source WSDL2Java API provided by Apache Axis2 to automatically generate a client stub, which can invoke Web service operations when its Java methods are called.

In order to illustrate the generation of test cases for the account example, we have implemented and deployed a Web service that behaves according to the rule-based description provided in section 3. We have further created an SXM model based on the prescribed behaviour, generated test cases, and executed them on the deployed Web service. The table below lists the number of generated test sequences for different values of k, and the time taken to execute them. It should be noted that in all cases the time required for test generation was negligible (less than one second).

Table 3. Test sequences and execution times for different testing configurations

Value of k	No. of test sequences	Execution time
$k = 0$	25	3.9 s
$k = 1$	72	11.5 s
$k = 2$	230	45.2 s

7 Conclusions and Outlook

Despite the significant amount of research interest in the area of Semantic Web Service technologies, and the numerous results already contributed towards automating service discovery, composition and mediation, the possibility of leveraging SWS descriptions for automated generation of test cases and behavioural validation has remained largely unexplored. In this paper we propose the use of Stream X-machines as a powerful behavioural modelling formalism for Web services, which can facilitate automated generation of test cases with completeness guarantees. The test cases can be used for performing validation of Web services through automated functional testing (in a black-box manner). We have described a method for deriving an SXM model from an IOPE-based description of a Web service's functionality, and provided an example of encoding a stateful Web service's IOPE using RIF-PRD and OWL, to subsequently serve as input to the transformation.

The particular strengths of the presented approach can be summarised in three points. Firstly, a significant advantage of SXMs compared to other formalisms, is in the strength of their associated testing method, which can be used for proving functional equivalence among a specification and an implementation under test. This brings a significant advantage over other approaches which are not based on a formal theory of testing, or can only provide guarantees for weaker notions of equivalence. Secondly, the method of transforming an IOPE-based description of a Web service that has been encoded in RIF-PRD and OWL, into an SXM specification, makes our proposed approach generic and framework-independent. Given a service description compliant with some SWS framework, and a method of transforming such a description to one that is based on our presented modelling conventions, the derivation of an SXM model and generation of test cases will be possible. This is in contrast to other approaches for test case generation which are bound to specific SWS frameworks (e.g. OWL-S). Thirdly, the tasks of test case generation and test execution are readily supported by existing tools.

As future work we intend to investigate certain aspects of the SXM derivation method in greater detail, especially those steps that appear to be most challenging to automate, such as the partitioning of domains for state variables. We will further focus on the formalisation of the algorithm, considering complexity and decidability properties, and providing a proof of equivalence among the original rule-based specification and the derived SXM model. We will finally work towards the implementation of tools that automate the transformation, as well as experimental validation and collection of empirical evidence concerning the effectiveness of the overall approach.

References

1. Akkiraju, R., Farrell, J., Miller, J., Nagarajan, M., Schmidt, M.T., Sheth, A., Verma, K.: Web Service Semantics - WSDL-S. W3C Member Submission (2005)
2. Bertolino, A., Frantzen, I., Polini, A., Tretmans, J.: Audition of Web Services for Testing Conformance to Open Specified Protocols. In: Reussner, R., Stafford, J.A., Szyperski, C. (eds.) Architecting Systems with Trustworthy Components. LNCS, vol. 3938, pp. 1–25. Springer, Heidelberg (2006)
3. Chow, T.S.: Testing Software Design Modelled by Finite State Machines. IEEE Transactions on Software Engineering 4, 178–187 (1978)
4. de Bruijn, J.d., Bussler, C., Domingue, J., Fensel, D., Hepp, M., Keller, U., Kifer, M., Konig-Ries, B., Kopecky, J., Lara, R., Lausen, H., Oren, E., Polleres, A., Roman, D., Scicluna, J., Stollberg, M.: Web Service Modeling Ontology (WSMO). W3C Member Submission (2005)
5. Dranidis, D., Kourtesis, D., Ramollari, E.: Formal Verification of Web Service Behavioural Conformance through Testing. Annals of Mathematics, Computing & Teleinformatics (AMCT) 1(5), 36–43 (2007)
6. Eilenberg, S.: Automata, Languages and Machines, vol. A. Academic Press, New York (1974)
7. Farrell, J., Lausen, H. (eds.): Semantic Annotations for WSDL and XML Schema. W3C Recommendation (2007)
8. Heckel, R., Mariani, L.: Automatic Conformance Testing of Web Services. In: Cerioli, M. (ed.) FASE 2005. LNCS, vol. 3442, pp. 34–48. Springer, Heidelberg (2005)
9. Holcombe, M., Ipate, F.: Correct Systems: Building Business Process Solutions. Springer, Berlin (1998)
10. Ipate, F., Holcombe, M.: An Integration Testing Method that is Proved to Find All Faults. International Journal of Computer Mathematics 63, 159–178 (1997)
11. Keum, C., Kang, S., Ko, I.Y.: Generating Test Cases for Web Services using Extended Finite State Machine. In: Uyar, M.Ü., Duale, A.Y., Fecko, M.A. (eds.) TestCom 2006. LNCS, vol. 3964, pp. 103–117. Springer, Heidelberg (2006)

12. Kourtesis, D., Ramollari, E., Dranidis, D., Paraskakis, I.: Discovery and Selection of Certified Web Services through Registry-Based Testing and Verification. In: Camarinha-Matos, L., Pickard, W. (eds.) Pervasive Collaborative Networks, IFIP, vol. 283, pp. 473–482. Springer, Boston (2008)
13. Martin, D., Domingue, J., Brodie, M.L., Leymann, F.: Semantic Web Services, Part 1. IEEE Intelligent Systems 22(5), 12–17 (2007)
14. Martin, D., Domingue, J., Sheth, A., Battle, S., Sycara, K., Fensel, D.: Semantic Web Services, Part 2. IEEE Intelligent Systems 22(6), 8–15 (2007)
15. Martin, D., Burstein, M., Hobbs, J., Lassila, O., McDermott, D., McIlraith, S., Narayanan, S., Paolucci, M., Parsia, B., Payne, T., Sirin, E., Srinivasan, N., Sycara, K.: OWL-S: Semantic Markup for Web Services. W3C Member Submission (2004)
16. McGuinness, D.L., van Harmelen, F.: OWL Web Ontology Language Overview, W3C Recommendation (2004)
17. Narayanan, S., McIlraith, S.A.: Simulation, Verification and Automated Composition of Web Services. In: Proceedings of the 11th International Conference on the World Wide Web, pp. 77–88 (2002)
18. Plotkin, G.D.: A Structural Approach to Operational Semantics. Journal of Logic and Algebraic Programming 60-61, 17–139 (2004)
19. Sinha, A., Paradkar, A.: Model-based Functional Conformance Testing of Web Services Operating on Persistent Data. In: Proceedings of Workshop on Testing, Analysis and Verification of Web Services and Applications (TAV-WEB 2006), pp. 17–22 (2006)
20. Urbieta, A., Azketa, E., Gomez, I., Parra, J., Arana, N.: Analysis of Effects- and Preconditions-Based Service Representation in Ubiquitous Computing Environments. In: Proceedings of the 2008 IEEE International Conference on Semantic Computing, pp. 378–385 (2008)
21. W3C RIF Production Rule Dialect (RIF-PRD), http://www.w3.org/TR/rif-prd/
22. W3C Rule Interchange Format (RIF) Working Group, http://www.w3.org/2005/rules/
23. Wang, Y., Bai, X., Li, J., Huang, R.: Ontology-Based Test Case Generation for Testing Web Services. In: Proceedings of Eighth International Symposium on Autonomous Decentralized Systems (ISADS 2007), pp. 43–50 (2007)

Neighborhood-Based Tag Prediction*

Adriana Budura[1], Sebastian Michel[1], Philippe Cudré-Mauroux[2], and Karl Aberer[1]

[1] Ecole Polytechnique Fédérale de Lausanne (EPFL), Switzerland
adriana.budura@epfl.ch, sebastian.michel@epfl.ch, karl.aberer@epfl.ch
[2] MIT, USA
pcm@csail.mit.edu

Abstract. We consider the problem of tag prediction in collaborative tagging systems where users share and annotate resources on the Web. We put forward HAMLET, a novel approach to automatically propagate tags along the edges of a graph which relates similar documents. We identify the core principles underlying tag propagation for which we derive suitable scoring models combined in one overall ranking formula. Leveraging these scores, we present an efficient top-k tag selection algorithm that infers additional tags by carefully inspecting neighbors in the document graph. Experiments using real-world data demonstrate the viability of our approach in large-scale environments where tags are scarce.

1 Introduction

The past few years have witnessed the rise of Web 2.0 applications promoting the sharing of documents through online communities. Documents shared by users in such applications vary largely, from scientific publications on *CiteU-Like* (*http://www.citeulike.org*) to images on *Flickr* (*http://www.flickr.com*) or bookmarks on *del.icio.us* (*http://del.icio.us*). One standard practice in all these scenarios is to rely on user-provided metadata to foster search capabilities. *Tags* are short textual annotations used to describe documents and represent such user-generated metadata in Web 2.0 applications. Tags are essential in resolving user queries targeting shared documents, yet require human attention to be generated. Users typically tag a small fraction of the shared documents only, leaving most of the other documents with incomplete metadata. This lack of metadata seriously impairs search, as documents without proper annotations are typically much harder to retrieve than correctly annotated documents.

Several attempts have already been made to foster the creation of additional tags in such a context. A number of successful projects, such as the E.S.P. game (*http://www.espgame.org*), are based on incentives compelling the end-users to create additional tags. Automatic tag creation techniques can produce good

* The work presented in this paper was partially supported by the National Competence Center in Research on Mobile Information and Communication Systems (NCCR-MICS), a center supported by the Swiss National Science Foundation under grant number 5005-67322.

L. Aroyo et al. (Eds.): ESWC 2009, LNCS 5554, pp. 608–622, 2009.

results for specific tag categories; they fail however in general, as automating the labeling of arbitrary documents is an inherently challenging research problem.

In the following, we present a tag inference technique to assign tags to previously un-tagged documents or to extend the set of tags for already tagged documents by *propagating* existing tags from one document to related documents. We put forward HAMLET (*Harvesting Adjacent Metadata in Large-ScalE Tagging Systems)*, a suite of principles, scoring models and algorithms for metadata propagation. HAMLET is based on the assumption that similar documents share similar tags, therefore we take advantage of a graph relating similar documents, and suppose that a tag attached to one particular document in the graph might be appropriate for the neighboring documents as well. Our approach is triggered by the observation that in our considered scenarios links very often stay in the topic of the source document. As this has been shown for Web pages, it is even more visible in citation of scientific papers (where authors refer to related work).

We apply our principles on real data extracted from two popular Web portals: CiteULike and del.icio.us. CiteULike is an application where users can create online libraries of academic papers, tag papers, and search for scientific articles while del.icio.us provides similar functionalities for organizing bookmarks.

1.1 Computational Model

Our model is based on three concepts: *documents*, *tags*, and *document neighborhoods*. A document $d \in \mathcal{D}$ is uniquely identifiable and represents in our scenarios scientific publications or Web pages. The documents are created and shared by users who interact with the tagging system. A *tag* $t \in \mathcal{T}$ represents a user-provided, unstructured piece of metadata attached to a document. We write $t \mapsto d$ to indicate that a particular tag t is attached to a document d and $T(d)$ to denote the bag of tags $\{t \mid t \mapsto d\}$ attached to a given document d.

The document graph is generated by the authors of the documents by references (citations) to other publications in the case of scientific publications or by linking via html links in the case of Web pages. To propagate tags from one document to the others, we take advantage of document graphs organizing related documents together. We write $d_i \rightarrow d_j$ to indicate that the author of a document has decided to relate document d_i to document d_j through a link in the graph. The *neighborhood* of a document d is denoted $N(d)$ and contains all documents which are reachable from d by traversing the edges of the graph. We will devise a top-k algorithm to dynamically explore the neighborhood of a particular document in order to infer new tags which are relevant for the considered document.

1.2 Contributions and Outline

Our contributions can be summarized as follows:

- We identify document graphs explicitly created by end-users in two real world scenarios (citation graph for scientific publications and Web graph) as an instrument for tag inference;
- We present a scoring model to assess the relevance of tags propagated along edges of the document graph;

- We show how to cast our tag propagation problem into a top-k selection problem and adapt state-of-the-art top-k algorithms to our context;
- We experimentally evaluate the performance of our approach by analyzing the propagation of tags for documents from two real world tagging portals.

We start below by reviewing related work in Section 2. We highlight the peculiarities of our setting, present the principles underpinning tag propagation and introduce our new scoring model in Section 3. In Section 4 we describe how to cast our tag propagation problem into a top-k aggregation problem, and present a new algorithm tailored to our needs. We discuss experimental results in Section 5 before concluding in Section 6.

2 Related Work

Recently the sphere of social annotations has gained increasing attention. The term *folksonomy* has been widely used to denote the process of collaboratively generating unstructured annotations for documents. Efforts in this area have concentrated on formalizing generic collaborative tagging systems. The work by Mika [17] considers a tri-partite model of instances, users, and concepts and mines the relationships among these three entities. [8,16,7] have concentrated on understanding the tagging process and the resulting social annotations by examining tag distributions, constructing tag-tag correlation networks and extracting statistics of users' tagging behavior. [14] investigate the overlap between anchor-text and tags for a Web page. [9] present an in-depth analysis of the del.icio.us tagging portal and conclude that the data contained in tagging portals represents valuable information which could enhance the performance of search engines. A similar conclusion is drawn by Yanbe et al. [24] who also investigate the possibility of a hybrid search approach, based both on PageRank and on social annotations. Bao et al. [2] introduce two algorithms which integrate information extracted from the tags into the search process. In [19] Schenkel et al. propose a new method to leverage social tags for improved content search. While their method also incorporates users in the query processing, it does not consider tag propagation using document-to-document similarities. Their method is orthogonal to our method as we focus on tag retrieval which is an important ingredient for meaningful document retrieval in social networks.

A different body of work deals with the problem of extracting semantic information from unstructured annotations. A probabilistic model has been proposed in [23] in order to derive the emergent semantics which characterize the collaborative tagging process. The authors also describe a framework for discovering semantically related tags based on the relations between the three entities: tags, users and resources. Rattenbury et al. [18] extract events and place semantics from Flickr tags based on geographical and temporal metadata attached to images. Schmitz et al. [20] address a similar problem and propose a solution based on a subsumption model for deriving an ontology from social annotations. While the above mentioned approaches are similar to our work, they are actually orthogonal since the authors do not directly address the problem of inferring tags for resources.

Addressing the problem of inferring new annotations for resources, the approach in [3] generates personal annotations based on the content of the documents or on data from the personal desktop. Our method enhances the annotation set of documents by solely relying on their context and uses content-based information only to optimize the results.

Sigurbjörnsson et al. [21] address a similar problem of automatic tag suggestion in the Flickr portal by using a co-occurrence measure for assessing tag relatedness. As opposed to our approach, the authors do not consider the local context of a resource and rely exclusively on global statistics.

The work by Jäschke et al. [13] considers a tripartite model of users, resources, and tags and employ their Folkrank [11] algorithm, a PageRank like, iterative algorithm, to assign tags to resources. While the general idea is similar to ours, it requires knowledge of the global graph.

The same problem is tackled by Heymann et al. [10] where the authors predict tags for Web pages in del.icio.us based on the anchortext, Web page content and on the surrounding hosts. The authors predict tags from the set of 100 most frequent tags found in del.icio.us by training a classifier which runs a binary classification task for each tag. The classifier is trained on a set of very popular bookmarks. As opposed to this method our approach does not rely on classification. In addition, we do not restrict ourselves to a predefined set of tags which are predicted. By doing so we could expect a gain in precision; however, this would drastically restrict our general approach to only a small number of tags which can be inferred.

Tag prediction is treated as a classification problem by Song et al. [22] who propose a real-time automatic tag recommendation method. The authors start by building document clusters offline and then assign a new document to those clusters with some probability in an online process. Again, our method is different since we do not consider any learning algorithms.

Our idea of using the context of a document (defined by its neighborhood in the graph) is furthermore supported by the observations in, e.g., [4], showing that in a Web graph most outgoing hyperlinks stay in the same topic as the originating page.

Existing work in the area of graph based classification [15,1] is also potentially relevant to our approach. [1] is in particular related to our approach since it considers content based node similarities as edge weights in the citation graph which supports our initiative of using links to relate documents. However, these classification algorithms are inherently limited to only a few classes, usually rely on global computations, or require several iterations, which makes these approaches hardly applicable in our envisioned scenario.

3 Scoring Model

In the following we introduce the main characteristics of our setting, discuss generic principles underpinning our tag propagation framework and show how these principles can be applied to create a scoring model for assessing tag relevance w.r.t. a document.

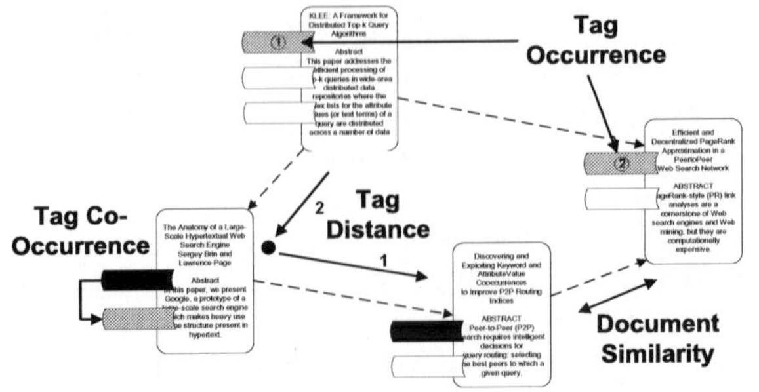

Fig. 1. The four principles underpinning our notion of tag propagation

Our setting has the following characteristics:

Scarce Information: tagging documents requires human attention, which is one of the scarcest resources today. Realistically, only a small subset of users will make the effort of tagging a given shared document and most probably the number of tags attached will be small too.

Document Organization: the document collections we analyze are not flat, as documents are organized through user-defined links which are in our case html links or citations. Hence, in addition to the notion of document collection classically defined in information retrieval, we always take into consideration document neighborhoods defined by following series of links from one given document.

3.1 Principles

Our tag propagation algorithm relies on four key principles (cf. Figure 1) which help us determine whether a tag should be propagated towards a given document or not, i.e., to which extent the tag is relevant to the given document.

i. the frequency with which a tag appears in a given document neighborhood (*tag occurrence*);
ii. the frequency with which a pair of tags is simultaneously attached to a document (*tag co-occurrence*);
iii. the distance between a potential tag and a given document in the resource graph (*tag distance*);
iv. the similarity between the documents the tags are attached to (*document similarity*).

In the following we will present the different measures which we use to capture these principles and which will be later combined into an overall scoring function.

3.2 Measures of Tag Relevance

Tag Occurrence. The importance of a term in a document is usually considered as increasing proportionally with the term frequency, i.e., with the number of times the term appears in the document. In our setting where most tags are missing, however, tags rarely occur more than once for a given document, proving the measure pointless most of the time. Instead, we define the notion of tag occurrence as the number of times a tag t' appears in a certain neighborhood of a document d, $N(d)$: $occ_N(t') = \sum_{d' \in N(d)} \mathbb{1}_{d'}(t')$, where $\mathbb{1}_{d'}(t')$ is an indicator function equal to one if $t' \mapsto d'$, to zero otherwise. Hence, the tag occurrence can be seen as capturing the popularity of a tag in a neighborhood. Note that we consider only a boolean decision, i.e., a tag is attached to a document or not. We do not consider how often a tag is assigned to a document as we aim at searching for support of tags in the entire neighborhood and do not want to push the influence of tags that occur often in some documents but rarely in the neighborhood.

Tag Co-Occurrence. The tags attached to a document are usually not independent. Some tags regularly appear together as they describe correlated facets of a given document (e.g., *Macintosh* and *Apple*). Therefore we consider measures for tag co-occurrence like the Dice or Jaccard coefficients or conditional probabilities as one ingredient of assessing tag relatedness. Using such measures we can assess the potential relevance $co\_occ(t', t)$ of a tag t' for a document that already contains a tag t. As documents typically contain more than one tag, we extend this notion for sets of tags $T(d)$ attached to a document d: $co\_occ(t', T(d)) = \sum_{t \in T(d)} co\_occ(t', t)$.

As one particular instantiation, we consider in our work the conditional probability that a document contains tag t' given that it already contains tag t: $P(t' \mapsto d \mid t \mapsto d)$. We treat all initial tags individually, i.e., we do not consider an aggregated score for an observed tag $t' \in N(d)$ to all possible subsets of the initial set of tags attached to d. In the case when the set of initial tags is empty we synthetically introduce a tag t_s for the considered document that co-occurs once with t' in the whole collection.

Document Similarity. Another ingredient of our scoring model is the degree of similarity between the documents to which the tags are attached. We employ a standard vector space model and use the following cosine similarity metric:

$$sim(d, d') = \frac{\sum_i w_d[i] * w_{d'}[i]}{\sqrt{\sum_i w_d[i]^2} * \sqrt{\sum_i w_{d'}[i]^2}},$$

where w_d is the vector representation of document d and $w_d[i]$ its i-th entry, i.e. reflecting the importance of term $term_i$ for d. There are multiple ways to quantify this importance, such as the relative term frequency $rtf = tf/tf_{max}$ which we have chosen in this work, where tf is the term frequency which is defined as the number of occurrences for a particular term in a particular document. tf_{max} denotes the largest tf value for one document. In order to assess the relatedness of documents which are *several* hops away we consider the product of similarities along the path between the observed document and the initial document:

$$\Pi_{i=0}^{|path(d,d')|-1} sim(d_i, d_{i+1})$$

Tag Distance. We assume that documents which are closer to each other in the graph have more related tags. Thus, an important metric is the distance from from one document d to another document d'. We define this distance as the shortest path (in terms of the number of graph edges): $dist(d, d') = path(d, d')$. We will use this distance as a dampening factor reducing the score of those tags that are located further away in the graph.

Due to the top-k fashion of our algorithm that explores the neighbors in a greedy way, we cannot (as we will see in Section 4) use the exact shortest path length. Instead, we compute the path with the smallest "cost" and use its length to approximate the minimal distance between the documents. The "cost" of a path is defined in this case in terms of 1 minus the similarity score of the path as described above.

3.3 Combined Scoring Function

We combine the four principles described above and produce a scoring function, used to assess the degree of relevance of a tag t' (observed in the neighborhood of d) for a document d.

$$score(t', d) = \sum_{d' \in N(d) \wedge t' \mapsto d'}^{m} S * O$$

The final score of tag t' observed up to m times in the neighborhood of the initial document d represents the sum of partial scores for each of its single occurrences. The parameter m controls the number of tag occurrences to be considered in the final aggregation. As we will see in the following section, m is required to compute the upper and lower bound scores to enable tag pruning in the top-k retrieval and therefore influences the size of the neighborhood which will be explored.

The partial score for each observation of t' is computed as a product of two factors. The first factor, coined S, consists of the aggregated similarities along the path between a document d' and the initial document d, i.e., principle (iv), dampened by the distance between the documents. The rationale behind this dampening stems from the fact that the relevance of a tag decreases with its distance to the initial document, which is one of our fundamental assumptions, i.e., principle (iii).

$$S = \frac{\Pi_{i=0}^{|path(d,d')|-1} sim(d_i, d_{i+1})}{log(1 + dist(d, d'))}$$

The second factor, denoted O, expresses the relatedness of a particular tag t' to the set of tags obtained from the initial document $T(d)$, which reflects the second principle (ii).

$$O = log(1 + co\_occ(t', T(d)))$$

In the overall scoring function, the sum aggregates the various occurrences of tag t' in the neighborhood of d, $N(d)$, following principle (i). Overall, the term (i.e., tag) specific influence is weighted by the objective importance of the term. This importance is in our case modeled by the distance from the point of observation to the initial document.

4 Top-k Tag Selection

Our algorithm starts from an initial document for which new tags should be inferred and traverses its neighborhood while computing scores for each observed tag. In the following, we show how tag propagation can be cast into a top-k selection problem that finds the most relevant tags for a given document in the graph while minimizing the number of documents visited.

As of today, the most efficient top-k query processing algorithms are based on the family of threshold algorithms (TA) (cf., e.g., [5]). In particular, the NRA (or TA-sorted) algorithm [6] only uses sequential (sorted) data access and hence is particularly interesting for our graph-based top-k problem. Our top-k tag selection algorithm is based on the concepts developed in this context, but adapts this family of algorithms to fit our unique context.

A common feature of these threshold algorithms is that they consider monotone aggregation functions, on which we focus here as well. We employ summation as the aggregation function, as it reflects the paradigm of considering the tag occurrences in a neighborhood.

Our computational model is the following: We consider the tag propagation along multiple edges in the graph towards an initial document d as a top-k aggregation problem where every document $d' \in N(d)$ represents one particular index list consisting of a series of $(tag, score)$ pairs. These $(tag, score)$ pairs are sorted in descending order w.r.t. to their score and express the relevance of a particular tag to the initial document d. Each edge in the document graph leads from one index list to other index lists (the actual number of index lists it leads to depends on the out-degree). The algorithm starts at the document for which we want to infer new tags and traverses the graph while computing scores for each observed tags t'.

Algorithm 1 depicts our nested top-k graph traversal and top-k aggregation algorithm. The algorithm performs a greedy traversal of the graph as follows: In the beginning, the algorithm inserts all neighbors of the initial document into a priority queue (Q_{docs}). Then, it chooses the neighbor with the highest document similarity (line 8) and aggregates the $(tag, score)$ pairs of the selected document incorporating the document similarity as a weight as mentioned above. Subsequently, the neighbors of the current document are inserted in the priority queue. Index lists are thus accessed on-demand depending on their expected contribution to the final result.

The algorithm maintains a list of potential candidate tags (Q_{cand}) and a list of the current top-k results ($topk$). Following the standard concepts of top-k query processing [6], we define $worstscore(t')$ (line 11) as the lower bound score for tag t' by considering those scores that have already been reported for t'. Similarly, $bestscore(t')$ denotes the upper bound score for tag t', that is the

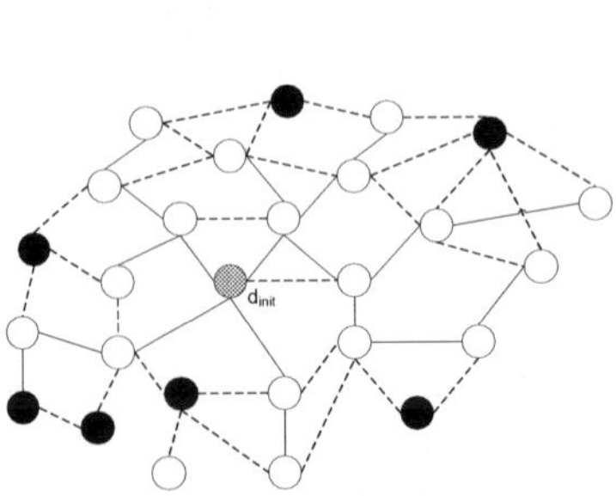

Fig. 2. An example of a particular state during the top-k aggregation. The thick lines indicate the edges of the spanning tree. All nodes touching these edges have been visited so far. The black circles denote the current border line (candidate nodes to be visited in the next step) which is responsible for the score upper bounds of the tags.

```
getTopKTags(Document d_init)
1  PriorityQueue<document, score> Q_docs = ∅
2  PriorityQueue<tag, score> Q_cand = ∅
3  Array(T, float) topk = ∅
4  Q_docs.enqueue(firstlevelneighbors(d_init))
5  s_max = 1
6  min-k = 0
7  while (Q_cand.size > 0 and m * s_max > min-k){
8     d_current = Q_docs.dequeue()
9     top_border = Q_docs.peek().score
10    for t' ∈ d_current{
11       t'.worstscore+ = score(t, d_current)
12       t'.seen + +
13       if t'.worstscore > min-k then {
14          topk.update(t')
15          min-k = topk[k]
16       }
17       t'.bestscore = ... (see below) ...
18       if t'.bestscore ≤ min-k then Q_cand.delete(t')
19    }
20    Q_docs.enqueue(firstlevelneighbors(d_current))
21    Q_cand.updateBestScore(m, s_max)
22    s_max = top_border * min(co_occ)
23 }
24 return topk
```

Algorithm 1. Top-K Algorithm

score considering all previously seen scores plus the largest possible scores for t' coming from unexplored regions of the graph. We denote the *worstscore* of the tag currently at rank-k as $min\text{-}k$. A tag t' with $bestscore(t') < min\text{-}k$ can be safely dismissed since it has no chance of getting into the final top-k results (line 18).

The main difficulty in our scenario stems from the fact that the number of documents in the neighborhood is unknown. Considering summation as the aggregation function is then problematic since it does not allow for pruning candidate tags during the top-k query processing, as the expected score mass is potentially unbounded. In our algorithm, the document queue describes, at any point in time, the borderline to the currently unexplored graph regions (cf. the solid black circles in Figure 2). Since we cannot look beyond that line, we are not aware of the tags and documents occurring there. However, we know the score of the best path (or of the multiple best paths) to one (or more) of the documents at the border line, coined top_{border}. The *bestscore* (upper bound score) of a tag t' observed so far m' times with *worstscore* (lower bound score) is based on the most promising document on the borderline and on the co-occurrence value of the tag w.r.t. the initial document. A tag that has been seen m' out of m possible observations can at most gain $m - m'$ times the current maximum score, which is similar to existing work, solely based on the tag scores. More formally we can write:

$$bestscore = worstscore + (m - m')\left(\frac{top_{border}}{log(1 + dist(d_{init}, d_{border}))} * log(1 + co\_occ)\right).$$

5 Experiments

5.1 Experimental Setup

We experimentally confirm the validity of our approach using two datasets which we have crawled in 2007 from two popular tagging portals. All algorithms have been implemented in Java 1.6. The simulation ran on a 2x2.33 GHz Quad-Core Intel Xeon CPU with 8GB RAM. The data is stored in an Oracle 11g database.

We consider the following real-world datasets:

CiteULike: This dataset consists of documents and tags from CiteULike, a tagging portal for academic publication. We take advantage of the bibliographic information, i.e., citations, to define the document graph: we add an edge $d_i \rightarrow d_j$ whenever d_i cites d_j. We automate the citation extraction by relying on the citation information available on CiteSeer. Our dataset consists of approximately 540,000 papers annotated with 195,000 distinct tags. From this set of papers we have randomly crawled a total of 2200 pdf documents which we indexed using *Lucene* (*http://lucene.apache.org*).

Del.icio.us: We have obtained the del.icio.us crawl from [19]. This dataset consists of approximately 120,000 html pages, annotated by users with 59,143 distinct tags. We have built the document graph by adding an edge $d_i \rightarrow d_j$ whenever a page d_i has a hyperlink pointing to another page d_j. The Web pages were crawled and indexed using *Lucene*.

We tested our approach having three specific metrics in mind:

- *Precision:* We considered the "precision at k" of the tag propagation, which is defined as follows: $precision@k = \#relevant\_tags/k$, where the relevance of the tag is either taken from a human expert evaluation or automatically computed. In order to have an assessment as fair as possible, we disregard all inferred tags that were among the initial tags of the document, i.e., we concentrate on k *new* tags systematically.
- *NDCG:* In addition to the precision we also report on the NDCG (Normalized Discounted Cumulative Gain) [12]. This measure takes into consideration the ranking of the results in addition to their relevance.
- *Number of documents visited:* To assess the practical viability of our algorithm, we decided to measure the number of nodes visited in the document graph during the tag inference process. This was preferred over CPU or processing time in order to get easily reproducible results.

Note that we do not report on recall since (i) we focus on top-k retrieval, and (ii) the number of relevant tags for a document is unknown and potentially very large.

In order to compute the precision, we consider two different kinds of relevance assessments: human expert evaluation and automatic evaluation. In the first case we ask a group of experts to mark the newly inferred tags as relevant or non-relevant. This measure is referred to as *absolute precision*. Secondly, we make use of existing tags attached to documents in order to run an automatic precision evaluation. We "hide" parts of the already assigned sets of tags, then try to infer new tags with our algorithm. We use the "hidden" tags to assess the relevance of the inferred ones, i.e., an inferred tag is considered relevant only if it occurred in the set of initial (but hidden) tags. This is actually a *relative precision* measure, since it is reported w.r.t. to the set of already assigned tags. Naturally, the relative precision values tend to be smaller, as potentially relevant tags are deemed irrelevant because they did not appear in the initial set of tags. However this approach allows us to run experiments on a larger scale which is difficult to achieve when relying on manual relevance assessments.

As an illustrative example, consider for instance the paper entitled "Routing Indices For Peer-to-Peer Systems". The set of initial tags contained the tags {*"index"*, *"search"*, *"20050923"*, *"p2p"*, *"p2p-search"*, *"20050923"*, *"peer-to-p"*, *"semantic"*, *"survey"*, *"unstructured-network"*} which our algorithms extended for $k = 5$ with the additional tags {*"network"*, *"search"*, *"distributed"*, *"security"*, *"system"*}, yielding a precision of 80%.

Algorithms under Comparison. Our experiments start by randomly selecting a set of initial documents (academic papers or bookmarked Web pages). For each of these initial documents we run our algorithm to infer the top-k most relevant tags which are new for the initial document (i.e., not in the set of already assigned tags). We always report on the average values of precision@k, NDCG@k and visited neighbors for the set of initial documents.

In the following we present the set of algorithms on whose performance we will later report. Besides our approach, we include a set of baselines which implement different measures of tag relevance for a given document.

Document Similarity Baseline: Our first baseline ignores the document graph and only considers the document similarity as an indicator for tag relevance. For each initial document d we compute a ranked list of the most similar documents $d' \in \mathcal{D}$ from the dataset. As a measure for document similarity we employed the cosine similarity $sim(d, d')$ presented in Section 3.1. In order to build a set of the most relevant k tags, we traverse the ranked list starting with the most similar document and copy its tags $T(d')$ into our top-k list. We stop when we have copied k new tags and compute the corresponding precision@k. In case the last document selected delivers more than k tags we randomly select among them.

Tag Co-Occurrence Baseline: Our second baseline is build upon the tag co-occurrence as a measure of relevance of a tag t' to a document d. For each of the initial documents d we build a ranked list of tags which have the highest co-occurrence scores $co\_occ(t', T(d))$ w.r.t. the initial set of tags $T(d)$ of the document. These scores are computed as described in Section 3.1. The precision@k is computed by selecting the first k tags from this list.

Tag Overlap Baseline: Our last baseline considers those documents which have the highest number of tags in common with the initial document as good candidates for tag propagation. Therefore we rank all documents in our dataset $d' \in \mathcal{D}$ according to the size of the intersection between their sets of tags $T(d')$ and the set of tags $T(d)$ of the initial document d: $I = T(d) \cap T(d')$. We then compose our top-k list by copying the remaining tags from the set $T(d') \setminus I$.

HAMLET: This is our ranking and graph traversal algorithm which combines the measures of tag relevance and the top-k ranking algorithm presented in the previous sections. For each of the initial documents d we explore the neighborhood $N(d)$ in a greedy best-first manner by carefully selecting only those neighbors d' which have a high chance of delivering potentially relevant tags for the initial document according to our principles of tag relevance. Our algorithm stops when there are no more possible top-k tags to be discovered.

5.2 Experimental Results

We start by presenting the results obtained for the *citeulike* benchmark where we considered 30 initial documents and collected manual assessments of tag relevance from 20 colleagues in order to evaluate the precision@k. Table 1 reports on the precision@k and NDCG@k values and on the number of visited neighbors for HAMLET. Our precision@k is promising in particular for $k < 10$, where we achieve 73% for $k = 3$ and 65% for $k = 5$. Intuitively, the precision values decrease with a higher k.

The parameter m, which controls the number of occurrences that are considered for a tag, influences the number of documents that are visited during run-time, e.g., for $k = 3$ our algorithm visits only 41 neighbors when $m = 3$ and 293 for $m = 10$. This is expected since m directly influences the upper bound scores of observed tags which are essential for the candidate pruning and hence the stopping of the algorithm. With m large, a higher number of tags is kept in the candidate queue as their *bestscore* might still be larger than the score of the current tag at rank-k (*min-k*). However, the influence of m on the precision@k is almost negligible which supports the initial hypothesis that the most promising

Table 1. Precision@k, NDCG@k and number of visited neighbors for the **citeulike** benchmark (30 initial documents) with manual precision evaluation

m	k	Precision@k	NDCG	Neighbors
3	3	0.73	0.74	41
3	5	0.65	0.68	93
3	7	0.55	0.60	74
3	10	0.51	0.58	217
3	15	0.44	0.53	234
5	3	0.70	0.72	124
5	5	0.65	0.69	153
5	7	0.57	0.63	247
5	10	0.52	0.59	328
5	15	0.45	0.53	386
7	3	0.72	0.74	243
7	5	0.65	0.69	257
7	7	0.57	0.64	356
7	10	0.51	0.59	421
7	15	0.46	0.54	486
10	3	0.74	0.76	293
10	5	0.65	0.69	385
10	7	0.57	0.64	468
10	10	0.51	0.59	517
10	15	0.46	0.54	543

Table 2. Relative Precision@k, relative NDCG@k and number of visited neighbors for the **del.icio.us** benchmark (140 initial documents) with automatic precision evaluation

m	k	Rel. Prec.@k	NDCG	Neighbors
3	3	0.50	0.51	1.65
3	5	0.42	0.46	1.65
3	7	0.36	0.42	1.67
3	10	0.30	0.37	1.67
3	15	0.23	0.31	1.67
5	3	0.50	0.51	1.67
5	5	0.42	0.46	1.65
5	7	0.36	0.42	1.67
5	10	0.30	0.37	1.67
5	15	0.23	0.31	1.67
7	3	0.50	0.51	1.67
7	5	0.42	0.46	1.65
7	7	0.36	0.42	1.67
7	10	0.30	0.37	1.67
7	15	0.23	0.31	1.67
10	3	0.50	0.51	1.67
10	5	0.42	0.46	1.66
10	7	0.36	0.42	1.67
10	10	0.30	0.37	1.67
10	15	0.23	0.31	1.67

Table 3. Precision@k for the three baselines for the **citeulike** benchmark (30 initial documents) with manual precision evaluation. Summarized HAMLET performance for ease of comparison.

k	CoOcc	DocDoc	Overlap	HAMLET
3	0.53	0.20	0.14	0.70 - 0.74
5	0.39	0.20	0.16	0.65
7	0.31	0.20	0.16	0.55 - 0.57
10	0.25	0.17	0.15	0.51 - 0.52
15	0.22	0.17	0.14	0.44 - 0.46

Table 4. NDCG@k for the three baselines for the **citeulike** benchmark (30 initial documents) with manual precision evaluation. Summarized HAMLET performance for ease of comparison.

k	CoOcc	DocDoc	Overlap	HAMLET
3	0.54	0.19	0.15	0.74 - 0.76
5	0.45	0.19	0.16	0.69 - 0.68
7	0.39	0.19	0.16	0.60 - 0.64
10	0.35	0.17	0.16	0.58 - 0.59
15	0.31	0.17	0.15	0.53 - 0.54

tags are actually found in the close vicinity of a document and that exploring further regions does not contribute substantially to the top-k result.

The results of the three baseline algorithms for the *citeulike* experiment are presented in Table 3 (precision) and Table 4 (NDCG). As expected, the precision values of the individual baselines are much lower than those achieved by our combined scoring scheme. This is mainly due to the fact that without any links between the documents we loose the context information captured by document graph and therefore the quality of the chosen tags, based only on global statistics, drops considerably. Still, the co-occurrence baseline achieves a precision of 53% which supports the general claim that the co-occurrence score of two tags is closely related to their semantics.

For the *del.icio.us* dataset we took advantage of the fact that popular bookmarks are annotated with a large set of tags and we implemented the above described automatic method for evaluating our precision. We selected bookmarks which were annotated with more than 20 and less than 40 tags and for each of them we randomly chose 4 tags (corresponding to the average number of tags per document in the whole dataset) as an input for our algorithm and disregarded the rest. We then ran our algorithm for each initial document and its 4 initial tags and computed the relative precision@k w.r.t. the hidden tags. We expect the real precision values to be higher, since the set of relevant tags for a document is potentially much bigger than the tags which have already been assigned by users. Table 2 reports on the results of our algorithm on the *del.icio.us* benchmark consisting of 140 initial documents and for which we automatically assessed the relative precision@k and NDCG@k. This second set of results shows the performance of our algorithm (e.g., 50% precision for $k = 3$) when ran on a larger scale. As described above this is a measure of relative precision as we use as the ground truth the set of initial tags already attached to a document. This set of tags presumably represents only a small portion of all the relevant tags for that document. Another observation is that the number of visited neighbors is much lower and remains constant as compared to the *citeulike* experiment. This is caused by the sparsely annotated general Web graph, i.e., documents have often only a few neighbors which are also tagged in *del.icio.us*. Therefore, in this the setting parameter m looses its importance since the annotated neighborhood of a document is small. However, our reported relative precision proves that our approach works well in both settings and does not depend significantly on the properties of the considered datasets.

Table 5. Relative Precision@k for the three baselines for the **del.icio.us** benchmark (140 initial documents) with automatic precision evaluation. Summarized HAMLET performance for ease of comparison.

k	CoOcc	DocDoc	Overlap	HAMLET
3	0.04	0.006	0.13	0.50
5	0.03	0.005	0.13	0.42
7	0.02	0.003	0.13	0.36
10	0.02	0.004	0.12	0.30
15	0.02	0.004	0.11	0.23

Table 6. Relative NDCG@k for the three baselines for the **del.icio.us** benchmark (140 initial documents) with automatic precision evaluation. Summarized HAMLET performance for ease of comparison.

k	CoOcc	DocDoc	Overlap	HAMLET
3	0.05	0.007	0.13	0.51
5	0.04	0.006	0.13	0.46
7	0.04	0.005	0.13	0.42
10	0.03	0.005	0.13	0.37
15	0.03	0.005	0.12	0.31

The results of the baseline algorithms for the *del.icio.us* benchmark can be observed in Table 5 (precision) and Table 6 (NDCG). Similarly to the previous experiment the relative precision@k values are much lower than those achieved by our algorithm. The baseline precisions are very low in this case and this is mainly due to the heterogeneity of the *del.icio.us* dataset. While *citeulike* represents a more restrictive domain of academic publications, our *del.icio.us* dataset represents a small subset of the Web, which means that the global measures will give far less accurate results. Our algorithm achieves higher precision values by limiting the scope of the tag inference to the neighborhood of a document and thus filtering out tags which are not directly related to that particular context.

In general, we can consider our method as accurate, given the inherent complexity of the task at hand. By taking advantage of a user-defined document graph and devising specific scoring and pruning methods, we reach a precision of 73% for the newly inferred tags when we use manual relevance assessments and 50% relative precision in the automatic case. Furthermore, our approach operates without any domain-specific knowledge, and with a number of computational operations sufficiently low to accommodate large settings.

6 Conclusions

Retrieving information from Web 2.0 applications opens up new challenges, as well as many exciting opportunities. We have put forward HAMLET, an approach for metadata propagation which is based on a socially-driven graph relating similar documents. The techniques we have developed throughout this work can potentially be extended beyond our showcase scenario focusing on academic papers and html bookmarks. As concrete examples, we believe it would be reasonable to extend our algorithm to images, songs, or movies explored online by communities of users. In addition, our approach could be used to rank existing tags of a document based on their support from the neighborhood.

References

1. Angelova, R., Weikum, G.: Graph-based text classification: learn from your neighbors. In: SIGIR (2006)
2. Bao, S., Xue, G., Wu, X., Yu, Y., Fei, B., Su, Z.: Optimizing web search using social annotations. In: WWW (2007)

3. Chirita, P.A., Costache, S., Nejdl, W., Handschuh, S.: P-tag: large scale automatic generation of personalized annotation tags for the web. In: WWW (2007)
4. Davison, B.D.: Topical locality in the web. In: SIGIR (2000)
5. Fagin, R.: Combining fuzzy information from multiple systems. J. Comput. Syst. Sci. 58(1) (1999)
6. Fagin, R., Lotem, A., Naor, M.: Optimal aggregation algorithms for middleware. J. Comput. Syst. Sci. 66(4) (2003)
7. Golder, S., Huberman, B.A.: Usage patterns of collaborative tagging systems. Journal of Information Science 32(2) (2006)
8. Halpin, H., Robu, V., Shepherd, H.: The complex dynamics of collaborative tagging. In: WWW (2007)
9. Heymann, P., Koutrika, G., Garcia-Molina, H.: Can social bookmarking improve web search? In: WSDM (2008)
10. Heymann, P., Ramage, D., Garcia-Molina, H.: Social tag prediction. In: SIGIR (2008)
11. Hotho, A., Jäschke, R., Schmitz, C., Stumme, G.: Information retrieval in folksonomies: Search and ranking. In: Sure, Y., Domingue, J. (eds.) ESWC 2006. LNCS, vol. 4011, pp. 411–426. Springer, Heidelberg (2006)
12. Järvelin, K., Kekäläinen, J.: Cumulated gain-based evaluation of IR techniques. ACM Trans. Inf. Syst. 20(4) (2002)
13. Jäschke, R., Marinho, L.B., Hotho, A., Schmidt-Thieme, L., Stumme, G.: Tag recommendations in folksonomies. In: Kok, J.N., Koronacki, J., Lopez de Mantaras, R., Matwin, S., Mladenič, D., Skowron, A. (eds.) PKDD 2007. LNCS, vol. 4702, pp. 506–514. Springer, Heidelberg (2007)
14. Kinsella, S., Budura, A., Skobeltsyn, G., Michel, S., Breslin, J.G., Aberer, K.: From Web 1.0 to Web 2.0 and back - how did your grandma use to tag? In: WIDM (2008)
15. Kleinberg, J.M., Tardos, É.: Approximation algorithms for classification problems with pairwise relationships: Metric labeling and markov random fields. In: FOCS (1999)
16. Marlow, C., Naaman, M., Boyd, D., Davis, M.: Ht06, tagging paper, taxonomy, flickr, academic article, to read. In: HYPERTEXT (2006)
17. Mika, P.: Ontologies Are Us: A Unified Model of Social Networks and Semantics. In: Gil, Y., Motta, E., Benjamins, V.R., Musen, M.A. (eds.) ISWC 2005. LNCS, vol. 3729, pp. 522–536. Springer, Heidelberg (2005)
18. Rattenbury, T., Good, N., Naaman, M.: Towards automatic extraction of event and place semantics from flickr tags. In: SIGIR (2007)
19. Schenkel, R., Crecelius, T., Kacimi, M., Michel, S., Neumann, T., Parreira, J.X., Weikum, G.: Efficient top-k querying over social-tagging networks. In: SIGIR (2008)
20. Schmitz, P.: Inducing ontology from flickr tags. In: Workshop on Collaborative Tagging at WWW (2006)
21. Sigurbjörnsson, B., van Zwol, R.: Flickr tag recommendation based on collective knowledge. In: WWW (2008)
22. Song, Y., Zhuang, Z., Li, H., Zhao, Q., Li, J., Lee, W., Giles, C.: Real-time automatic tag recommendation. In: SIGIR (2008)
23. Wu, X., Zhang, L., Yu, Y.: Exploring social annotations for the semantic web. In: WWW (2006)
24. Yanbe, Y., Jatowt, A., Nakamura, S., Tanaka, K.: Towards improving web search by utilizing social bookmarks. In: Baresi, L., Fraternali, P., Houben, G.-J. (eds.) ICWE 2007. LNCS, vol. 4607, pp. 343–357. Springer, Heidelberg (2007)

User Evaluation Study of a Tagging Approach to Semantic Mapping

Colm Conroy, Rob Brennan, Declan O'Sullivan, and Dave Lewis

Knowledge and Data Engineering Group,
Trinity College Dublin
{coconroy,rob.brennan,declan.osullivan,dave.lewis}@cs.tcd.ie

Abstract. A key aspect of semantic interoperability is the semantic mapping process itself. Traditionally, semantic mapping processes conducted by knowledge engineers have been proposed to bridge this gap. However, knowledge engineers alone are unlikely to cope with the ever increasing amount of mapping work required, especially as mappings themselves begin to be specialised for different contexts. One solution is to develop new mapping processes that enable users to participate in the mapping process themselves. In this paper we present an evaluation study of our user-driven tagging approach to the semantic mapping process. In our approach, users actively participate in generating mappings by categorising automatically generated candidate matches presented in natural language over a long time period. In the evaluation study three groups of users generated mappings between their personal ontologies and a sports ontology describing sports news content from RSS feeds. The mapping process was embedded within the users' work environment as a Firefox browser extension. The study is discussed, focusing on whether the mapping process is unintrusive, engaging and simplified for the user. The evaluation results were promising and indicate that people with various levels of expertise could become active in the semantic mapping process.

1 Introduction

The development of user-friendly ontology authoring approaches have been predicted to "lead to a step change in the deployment of the Semantic Web" [1]. However any technology that produces a proliferation of ontologies will lead to an avalanche of semantic interoperability problems unless we simultaneously empower users to resolve semantic mapping tasks themselves. Semantic mapping has been identified as one of the most time-consuming data management problems [2], with high requirements for user input [2] where fully automated solutions are impossible [3]. Thus a scalable solution to wide-scale semantic interoperability cannot rely on using knowledge engineers to create mappings which bridge the semantic gap between the increasingly personalised user experience [4] and domain knowledge models. New user-centric semantic mapping approaches, processes and tools are required, with a consequent emphasis on usability evaluation rather than matching algorithm optimisation. Such a mapping approach, based on casual web users categorising automatically

L. Aroyo et al. (Eds.): ESWC 2009, LNCS 5554, pp. 623–637, 2009.

generated candidate matches presented in natural language over a long time period, is described here. Unique aspects of our approach include the combination of: (a) natural language-based presentation; (b) integration within the normal desktop work environment as a web-browser extension; (c) the use of Web 2.0-style tags as our match categorisation mechanism used to generate mappings from candidate matches; (d) the freedom users have to express in their own terms the relationship of candidate matches; (e) the segmentation of the mapping process into small sessions which allow individual tasks to be completed quickly and unobtrusively while the users can see the feedback effects of their mapping actions on the trial semantic interoperability application.

Key factors for evaluation of the success of our approach are: (a) the *simplification* of the semantic mapping process so that non-knowledge engineers can accurately generate meaningful mappings in domains of interest to them; (b) creation of a sufficiently *engaging* mapping process that will encourage users to continue to participate over time since our mapping approach splits the normally long mapping task into a large number of small sessions; (c) *unintrusive* integration of the mapping activity with a user's everyday tasks so that over time the mapping process does not become distracting or annoying for the user. This paper presents an experimental study that examines the usability of our prototype user-driven tagging semantic mapping application in terms of whether the users found it provided an *unintrusive*, *engaging* and *simplified* mapping process. The remainder of the paper is organised as follows: Section 2 discusses related work, Section 3 presents the tag mapping tool, Section 4 give the details of the experimental study carried out, and Section 5 offers some conclusions and future work.

2 Related Work

Even though the performance of automatic matching tools [5] is steadily improving there is still a need for users to be involved in the mapping process for the foreseeable future [6]. The number of user trials on mapping tools has been small [7] and those that have taken place have tended to focus purely on the mapping effectiveness and have not addressed usability or visualization issues. User evaluations for the PROMPT and Chimera mapping tools have been performed in the medical domain [8] and results showed that each user found merging quite difficult, and in particular performing any non-automated procedures was highlighted as extremely tough.

Although the goal of mapping tools is to map between different information sources, the existing tools focus primarily on the calculation of candidate matches – they *do not aim to provide the assistance and support for the user during the entire mapping process*. Some recent work, however, has focused on improving usability. For example, Robertson et al. [9] presents better ways to display mappings to users using standard visualisation techniques and demonstrates their effectiveness with a user study. Bernstein et al. have presented on an incremental interactive schema matching technique [10] which displays suggested candidate matches for a selected schema element to the user rather than for the global set of elements. The history of the user's mapping actions is also taken as a heuristic in selecting candidate matches for each schema element. Falconer and Storey [2] have presented a theoretical framework for cognitive support in mapping tools that gives design guidelines based on state of the art of cognitive support research and results gathered from an observation case study. The paper gives a list of tool requirements and framework principles which they have use to develop a tool called CoGZ. Mocan et al. [11] have proposed a formal model for ontology mapping creation which provides the use of connected perspectives to deemphasize non relevant aspects of the ontologies being matched while displaying the relevant entities for each match. In contrast to the previous works, our research is focused on allowing users of various different levels of expertise, not just technically aware users who can grasp the meaning of a graph type structure, to participate in the mapping process.

Shvaiko and colleagues [12] have provided a semantic mapping system which provides proofs and explanations of mappings. The proofs are displayed using short, high level explanations in natural language without any technical detail. These are designed to be intuitive and understandable by ordinary users. The results of an experimental study have been promising and shown the potential to scale to the semantic web. Kaufmann and Bernstein [13] have conducted an evaluation study on the usability of natural language interfaces (NLIs) and concluded that they are indeed useful for casual end-users. Recently a number of NLIs for authoring ontologies have emerged which are tasked with allowing domain experts to participate in the ontology engineering cycle, e.g. [1]. Soantag [14] has proposed an approach for a dialogue-based interaction mapping process where medical domain experts will validate mappings via verbal communication.

Increasingly techniques from the semantic desktop research community will enable the creation of a personal ontology on behalf of a user [15]. Some of the techniques used are reverse engineering techniques, re-use of ontologies, and conceptualization of folksonomies. There has been significant focus on modelling the users own interest based on the tags they used in folksonomies [16]. Results have been promising and showed a far richer interest profile can be generated when multiple folksonomies are combined. There have been various studies focusing on analysing user incentives and motivations behind tagging. In particular [17] found that the motivations tend to be personally-focused and [18] that the user-generated tags are consistent with the content they are attached to. In contrast to these approaches, the work described in this paper uses tagging as a way to categorise the algorithmically derived candidate match results for relationships between concepts within two different ontologies.

3 Tag Mapping Tool Design

The tag mapping tool supports the creation of personalised mappings between users' sporting interests and a sport news ontology used to represent and store sports news stories from RSS feeds[1]. These mappings are used to deliver personal sporting news to the users by filtering the sport news relevant to the users sporting interests. The users are required to progressively validate candidate matches via a tagging interface. The design of the tag mapping tool allows the mapping cycle to occur over an extended duration within the user's environment via a mapping process which has been

[1] The ontologies & matching file can be found at http://www.cs.tcd.ie/~coconroy/tagmatch/

described in detail in previous work [19, 20]. The main steps, each performed iteratively, include: (a) Determine when to present mapping tasks (section 1 aspect e); (b) Determine what mapping tasks to present; (c) Present the mapping task; (d) Perform the mapping interaction; and (e) Evaluation of the user response. The tool is split into two parts: a Firefox extension and a mapping server. Fig. 1 shows the architecture of the tool. A Firefox extension was picked as the implementation framework since this allows the mapping process to occur within an environment the user will use every day, i.e. browsing (section 1 aspect b). Graphical message alerts via a browser status bar icon is displayed as a prompt for the user if there are either any mapping tasks to perform or if any new RSS information is available. The RSS information is displayed via a sidebar in the browser window and also in message alerts whenever new RSS information is available. In this prototype the only mapping task the user is prompted to perform is to *tag a matching result*. Future extensions of the prototype could add additional mapping tasks.

The mapping server is used to execute the mapping logic of the application and also stores the ontologies and mappings. The same sport ontology formed each user's own personal ontology and candidate matches were derived to the sport news ontology with Align [21] (18 missed matches were appended). To personalise their ontology users choose their specific interests within the general sport domain when registering, e.g. football, tennis, and so on. These interests link to a concept within their own ontology and is used to locate the RSS items of interest for the user via mappings constructed with the sport news ontology. Although every user has a set of 'concepts interests' in the ontology which corresponds to their sport interest, each user is still required to categorise all of the candidate matches between the ontologies. In this initial prototype the ontologies remain stable, further work will investigate the case of evolving ontologies.

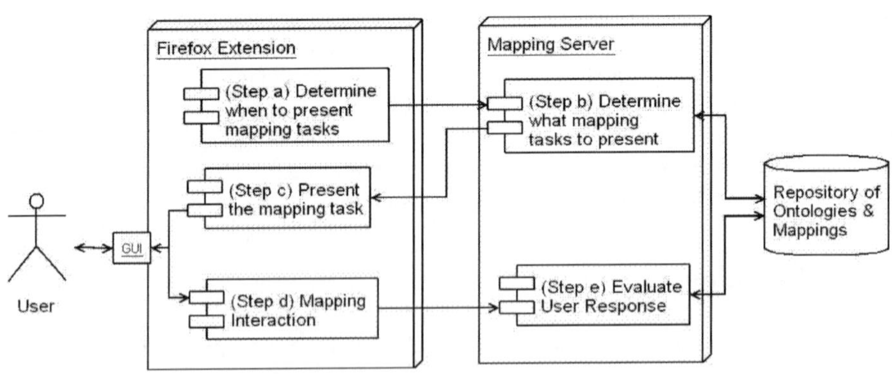

Fig. 1. Mapping architecture of tagging mapping tool

The extension prompts the user to undertake only three mapping sessions (each with 3 mapping tasks) every time the browser is opened, step (a). The first session is prompted five minutes after the browser is opened while the other sessions are prompted hourly afterwards until all three sessions have been prompted. When a mapping session is prompted the mapping server determines which mapping tasks are to be asked, step (b), according to both the accuracy of the candidate matches (via matching tool) and the number of RSS news item instances available via the matching. The ontological information is visually displayed, step (c), to the user in natural language, see Fig. 2. The natural language used represents the ontological information using bullet points to outline the structure of the concept (section 1 aspect a), i.e. child, parent, property. This choice of presentation interface was informed by a previous experiment [22] which established that natural language mapping representations were useful for end-users. The user is then required to tag the candidate match, step (d), to gather the user's feedback on the suitability of the match (section 1 aspect c). The user may input the tags on offer from the suggested list or new tags they can create (section 1 aspect d). Note that the user may use multiple tags. The list of tags entered by the user is then evaluated, step (e), on the mapping server which will categorise the candidate match based on the tags entered. The categories used in our approach are: *Equivalent*, *Equivalent Sometimes*, *Different*, and *Unknown*. The equivalent and equivalent sometimes categorises are used to categorise candidate matches which are valid mappings. Equivalent sometimes mappings are also used where the target concept is more general than the source concept. The different category is used to categorise matches in which the mappings is not valid. The unknown category is used to categorise candidate matches which were only tagged with new tags the user created. In the current prototype there is no support for re-categorising matches in the unknown category but in future work we plan to re-classify these matches via monitoring (see section 5). The focus of the current work is to evaluate the effectiveness of the basic tagging approach. The evaluation of this experiment is discussed in section 4.2.

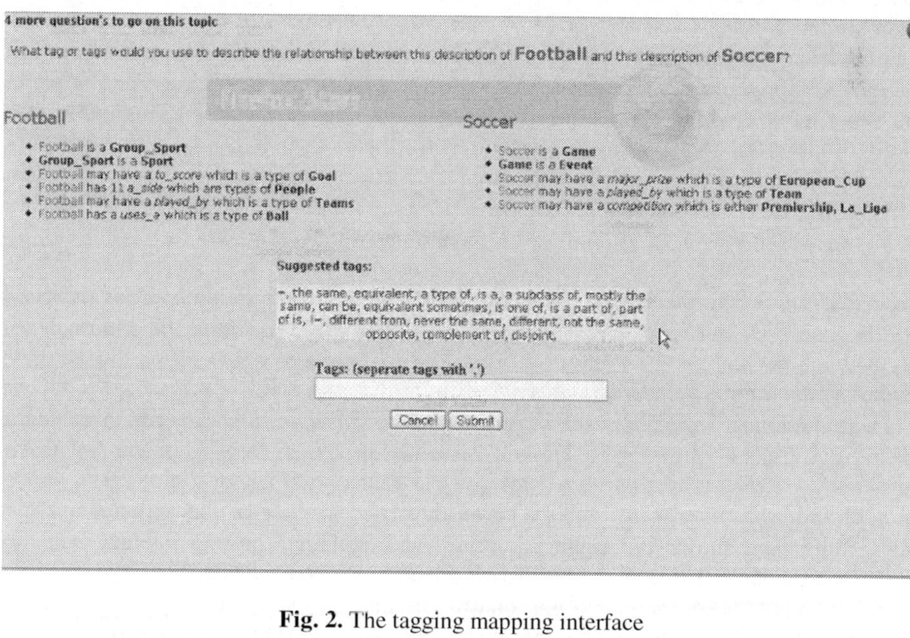

Fig. 2. The tagging mapping interface

4 Experimental Study

To assess the effectiveness of our tag mapping tool, we conducted an experimental study addressing three groups of questions: (1) **Unintrusive** – How interfering for the user is the tool? Is the tool disruptive for the user? (2) **Engaging** – Is the user participating in the mapping process over time? Are users answering mapping tasks regularly? (3) **Simplified** – Are the tasks confusing? Does the user understand the information? What is the quality of the mappings?

4.1 Experimental Design

Scenario. The experiment involved the participants using the Firefox extension of the tag mapping tool within their browser on their own personal computer over a two week period. At the start of the experiment each participant was asked to select their personal interests from a list of sport concepts. During the experiment each participant was asked to answer the mapping session prompts when convenient and to look at any RSS feed items they found interesting. The total number of mappings required to answer by the user during the course of the experiment was 32. The reason for the low number of mappings is because we anticipate that a user's interests will be relatively shallow and that mapping tasks will be done over time in small sessions as their interests vary.

Participants. Three different groups of users were used to examine the impact of the experiment upon each of the groups: *Ontologically aware* users who have ontology work experience (volunteers from our research group), *Technology aware* users who have database/UML modelling experience but no ontology experience (volunteers from students) and *Non-Technology Aware* users who have basic computer experience with no database/UML modelling or ontology experience (volunteers from the university sports centre). Four volunteers from each of the three groups took part in the experiment, making for a total of 12 participants. Note that the low numbers for each group will not give definitive outcomes for the experiment. Instead the numbers used were kept low so we could gain more detailed feedback and analysis from each user through interviewing techniques, with a view to learning from this experiment to assist the design for the large scale experiment that we have planned for 2009.

Procedure. Our sample size satisfies the requirements for reliable SUS evaluations [23] but we used our own questionnaires so we that could examine the approach in-depth. Each participant was required to use the tool over a two week period on their own personal computer. It included several steps.

The participants were first asked to fill in a *pre-study questionnaire* which included several questions about their background so we could assess their technical skills. An *introduction to the tag mapping tool* was given to each participant describing how to use the tool and what they were required to do during the course of the experiment. The participants were shown the tagging interface and explained how to interact with the tool. Each participant was also required to select their personal interests from a list of sport concepts to form their 'concept interests'. After the introduction was finished the participants were asked to *use the tool* over a two week period. While the participants were using the tool all their actions with the tool were logged through Ajax calls to the mapping server.

The participants were required to answer a *mid experiment questionnaire*[2] to give their initial thoughts of the tool which were focused on the usability of the tool, e.g. 'Is the questions and supplied information confusing'. A four point scale was taken to allow users to express different levels of satisfaction. After the experiment, participants filled out a *post-study questionnaire* to evaluate their final thoughts. Again the questionnaire focused on the usability of the tool with a four point scale. The same questions were asked from the mid experiment questionnaire and other questions were added, e.g. 'how abrupt was the timing of the question prompts'.

Finally an *interview* was conducted with each participant to give their general impression and reaction to the tool. The survey answers were taken into account during each interview. Some of the questions asked were: 'how simple was the mapping task', 'how engaging was the mapping task' and 'how interruptive were the mapping tasks'.

Data Collected. The following data was collected: (a) *Mid experiment Questionnaire* – used to examine the initial usability and usefulness of the tool, (b) *Post Questionnaire* – used to examine the long-term usability and usefulness of the tool, (c) *Interviews* – used to give the in depth account of the participants experience with the tool, (d) *Log of users actions* – used to track the user's behaviour with the tool, (e) *Resultant Mappings* – the quality of the mappings were analysed to give the precision for each group. The number of participants used (12) in the experiment was relatively small but the number of events was much larger (2,538 total giving approx. 21 events per day for every user) giving a reasonable data to analyse and which allows us to draw some indications and conclusions, especially for the event-centric metrics.

Metrics. To evaluate our approach to semantic mapping we separate the evaluation into three key areas already stated before: **unintrusive**, **engaging** and **simplified.** By *unintrusive* we mean that the mapping process should not disrupt the user from their daily work and that the process ought to blend into the background while they work. A couple of specific (there are others) quantitive metrics and qualitative feedback we defined for unintrusive tasks were,

- Percentage of mapping sessions entered (Fig. 3): an unintrusive task will have a high rate of response to task prompts.
- How interruptive was the mapping task (Interviews): an unintrusive task will not be perceived as interruptive by users.

By *engaging* we mean that the user will enter and finish mapping sessions regularly over time seeing benefits with personalised RSS feed sports news through mappings accomplished. A couple of specific (there are others) quantitive metrics and qualitative feedback we defined for unintrusive tasks were

- Percentage of mappings done (Fig. 4): an engaging task will have a high rate of mappings done.
- How engaging were the mapping tasks (Interviews): an engaging task will lead to users answering that they felt engaged to complete the task.

---

[2] All Questionnaires can be found at http://www.cs.tcd.ie/~coconroy/tagmatch/

By *simplified* we mean that the user will be able to achieve rich mappings in a time efficient way that is not confusing to the user. A couple of specific (there are others) quantitive metrics and qualitative feedback we defined for simplified tasks were:

- Mappings accuracy w.r.t. gold standard (Fig. 6): a simplified task will assist the users in making valid mapping decisions such that they have a high rate of generation of correct categorised mappings.
- How simple was the mapping task (Interviews): a simplified task will lead to users answering that they found the task straightforward to achieve.
- Mean mapping session time (Fig. 4): a simplified task will be quick to complete.
- What was the quality of user response: The quality of tags should be high.

4.2 Evaluation

In this section we present the results gathered from the experiment. The graphs give the logged event data for the overall browsing hours. The browsing hours were also separated in half, in order to analyse the results per half.

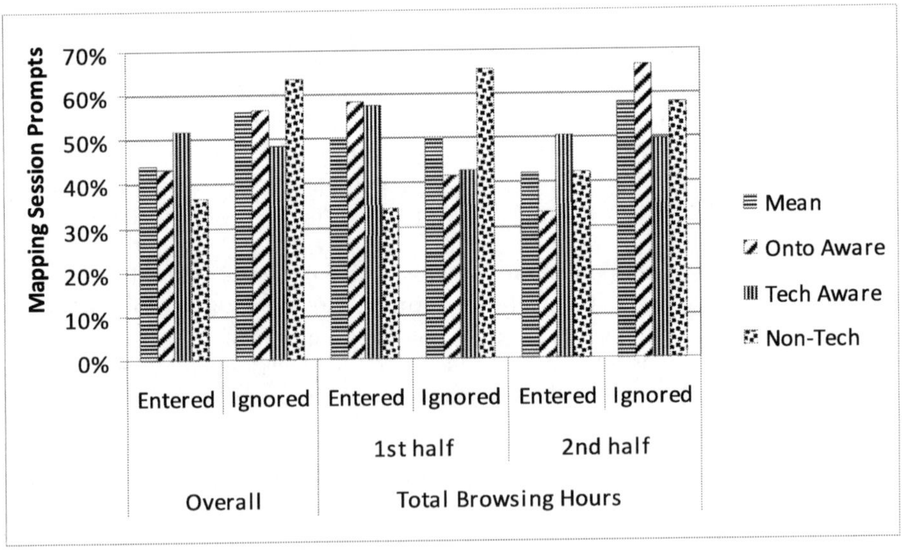

Fig. 3. Mapping Session Prompts

4.2.1 Unintrusive

To assess whether users were using the mapping application we tracked how many mapping session prompts they clicked on or ignored, which is an indication of how annoying the tool is to users, i.e. intrusive. These results, in Fig. 3, give an indication of the participation rate of each group and to a lesser extent the perceived utility of the task for users within each group. The results show on average that almost half of the prompts asked were entered (approx 45%). On the other hand there is a slight increase (<10%) in the number of prompts ignored in the second half. However, the entered rate is still quite high in the second half (approx 42%) which indirectly demonstrate that users did not find the mapping sessions in general to be overly inconvenient in their daily work, otherwise we would expect to have seen a greater drop in the number of sessions entered in the second half of the experiment.

Fig. 4 gives the results for the mean time taken for each user group to achieve a mapping and indicates that the mapping sessions are not time-consuming with a mapping task taking on average 38 seconds (session approx 2 minutes). The mean task time reduced slightly (by 2 seconds) over the second half, however with the non-technical aware group the time doubles. This is an alarming increase but looking at the number of mappings done by non-technical users (Fig. 4) might suggest that this increase in part could be attributed to the low number of mappings achieved by the users. The standard deviation is approximately 14 seconds. In the survey taken a 73% majority felt that the application blended into the background while they were undertaking their daily work while all the users felt the application was not disruptive to their daily work rising from 92% in the mid-week survey. Another 93% of respondents did not find the mapping session prompts to be disruptive.

When looking at participation of the groups individually in entering mapping sessions (Fig. 3), it shows that the ontology aware group 'ignored rate' increases sharply (approx 25%) in the second half which is in contrast to the non-technical users who have a reduction in the number of prompts ignored (approx 8%). When asked in the interviews why they were ignoring the mapping session prompts, users answered that a lot of the prompts were displayed when they were busy with a daily task which would lead to the user ignoring them. Some of the comments given by users were: 'The mapping tasks were not overly interruptive just when I was working it was, it didn't bother me at any other time' and also 'When I am doing something and I did not want to be interrupted it was annoying but when I am just browsing the web it is fine. Just when I am concentrating it is not'. The feedback given suggests that users would prefer mapping tasks to be asked in a more suitable context.

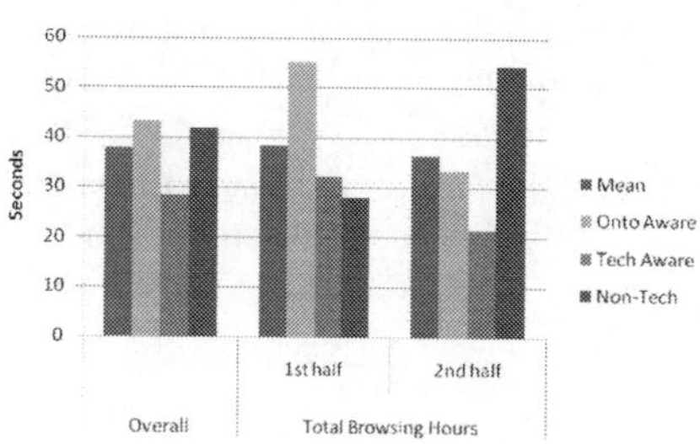

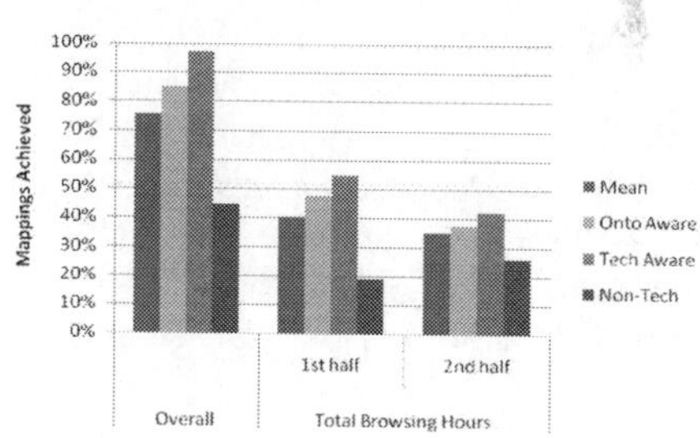

Fig. 4. Mean Time per Mapping (left) & Percentage of Mappings Achieved (right)

4.2.2 Engaging

From Fig. 4 we can see that the overall number of mappings done is quite high at 75% done. However looking at the results in depth we can see a minor drop in the percentage done in the second week but even more worrying is that the non-technical group only achieved 45% of the mappings. The low percentage of mappings done by non-technical users can be attributed to the low number of hours browsing by the users over the course of the experiment, average of 7 ½ hours. Contrasting this with the non-technical users, average was 17 hours, and ontology aware user, average 13 ½ hours, shows the non-technical group had the browser open for less than 50% of the time of technical users and less 60% of the ontology aware users. When taking the number of mappings done per browsing hour it shows that 6% are done by non-technical, 5.71% are done by technical and 6.30% are done by ontology aware. This suggests that the non-technical users were quite competitive when compared to the two other groups when taking browsing statistics into account.

For every time a user responded to a mapping session prompt, we tracked the users' actions which allows us to calculate the number of mapping sessions finished (Fin), abandoned (Abd) and not started (Non). A "not started" mapping session is one in which no more user events are recorded, this could be due to system unreliability or if a user simply ignores the session after it starts. These results will give us an indication if users are finding the mapping task too complicated or too time-consuming. In the results, see Fig. 5, 8% of the sessions were abandoned which is quite low and could possibly be explained by the user suddenly getting busy, e.g. they just got an email or they got a message on internet messenger. More alarming and interesting is that 30% of the mapping session entered were not even started, i.e. the user exited immediately without finishing a mapping task. During interviews conducted with the participants when asked about this they answered that they exited the mapping task when it was asking them something they were not interested in, i.e. if a user had not included football as an interest then they were perplexed as to why it was being asked as a mapping task and chose not to answer. A comment given was 'I was engaged with the tasks when it was something I was interested in but when it was something I was not interested in I did not want to do it'. Also a couple of the users mentioned that the submit and close mapping session buttons were placed too close together on the screen and that they accidental pressed the close button when submitting tags for a mapping.

In the survey taken only 25% of the users found that answering the mapping tasks tailored the sports news. Instead a massive 42% found it did nothing at all and 33% said it only slightly changed. With 75% of the users not seeing the sport news getting more tailored it shows that most users are not seeing the benefits that the mappings should be bringing. A suggested reason as to why some users did not see benefits may be due to the important mappings, the mappings which affect the users interests the most, being miss-categorised through the wrong choice of tags or undetermined tags being used. With 59% of the mappings correct (Fig. 6) with respect to the gold standard this gives our reason some supporting evidence.

4.2.3 Simplified

In the interviews we asked how simple they found the mapping task. The general response from each group was that they found the mapping tasks were very straightforward and not difficult at all. Some of the comments given were 'it was very simple, I have many options and the choice was obvious' and 'it was pretty simple just click on tags. I understood the information it was pretty straightforward'. The participants also mentioned they were confused by some of the candidate matches generated, e.g. Basket and Motor Racing, which were very poor candidates for mappings. The users did not like being asked to categorise these types of mappings. Finally everybody mentioned in their interviews that they liked the tagging approach as it gave more choices than just 'yes' or 'no' and also allow them to be expressive.

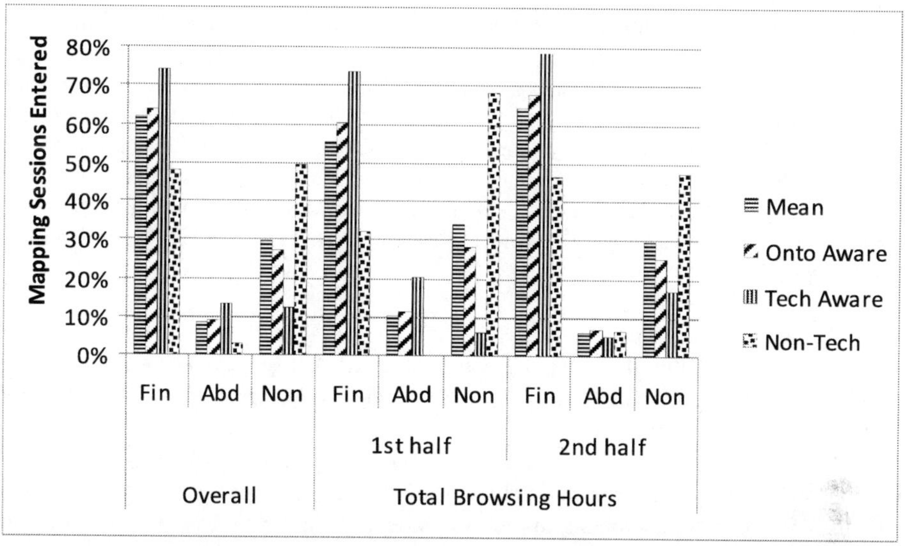

Fig. 5. Mapping Sessions Entered. The mapping sessions entered are separated into the sessions finished (Fin), abandoned (Abd) and not started (Non).

Table 1 shows the distribution of the three most popular suggested tags for each category. This gives an impression of what suggested tags the users like to use the most. The results show that the term 'same' is a frequently used tag in the most popular suggested tags for each category. In the equivalent sometimes category 'mostly the same' was the fourth rank. Also the terms 'equivalent' and 'different' are popularly used by the users. These terms have the highest frequency of use and so might be expected to have the most well-defined semantics for the user. However, some of these 'terms' were the tags used for incorrectly categorised mappings (29% of the incorrect mappings). But looking at the incorrect mappings made this had more to do with the mapping being unclear to the user rather than the tag, i.e. 'Football' and 'American Football' might be viewed as the same concept without reading the description or given some context.

Table 1. The list of suggested tags used by users: Ranking and percentage for the top three tags in each category are given. The tag categories are *equivalent*, *equivalent sometimes* and *different*. The suggested tags accounted for 81% of the overall number of tags used.

Category / Rank	Equivalent		Equivalent Sometimes		Different	
	Tag	**% used**	**Tag**	**% used**	**Tag**	**% used**
1	The same	40%	Part of is	32%	Different from	28%
2	Equivalent	30%	Equivalent sometimes	24%	Not the same	22%
3	=	22%	Is a part of	23%	Never the same	16%

After analysing both the new tags added by users and the matches that the new tags were applied to, the set of new tags could be categorised as: *Miss-spelt tags* - tags which are spelt incorrectly; *Meaning of a suggested tag* – tags which have the same meaning as a suggested tag; *Tag giving semantic meaning* – tags which correspond to a property; *Alert to bad mapping* – tags users have used to indicate that the mapping is wrong; *Other* – tags which indicate a general correspondence. In future work we hope to categorise these newly added tags by users through various techniques including monitoring the mapping usage and the history of other users, with similar interests, mapping actions.

4.2.4 Quality of Mappings

Fig. 6 shows the results of the mapping accuracy for each group. The results showed that overall 59% of the mappings made were deemed to be accurate. Looking through each group shows the ontology aware group achieved the highest accuracy with 74% while both the technical and non-technical groups had an accuracy of 51%. In addition the results show that the inconclusive mappings were higher than the incorrect mappings for both the technical and non-technical groups which is in contrast to ontology aware group. Upon examination of logs this is because the technical and non-technical groups are creating more new tags than the ontology aware group. There is a significant error rate for both the non-technical group at 21% and the ontology aware group at 19%. Examining all of the incorrect mappings showed that 73% had a relationship correspondence, e.g. Golf and Golf Course, which were categorised as Equivalent or Equivalent Sometimes. The other 27% were matching's in which the two matched concepts had a common parent concept, e.g. Golf and Soccer, which were mostly categorised as Equivalent. Examining all of the tags used in the incorrect mappings showed:

1. 59% were tags which, in the interviews, the participants mentioned they thought the tags had different meanings that what they were supposed to be, e.g. 'part of' was meant to be used to represent the union set theory relation but instead it was used as a way to express a concept to property correspondence.
2. 20% were an 'equivalent sometimes' tag for matches which in fact had a correspondence, i.e. 'a subclass of' used for the match 'Basket' and 'Basketball'.
3. 12% were tags which suggest the user misunderstood the descriptions of the concepts being matched, e.g. 'the same' used for 'Football' and 'American Football'.
4. 9% were an 'equivalence' tag for matches which in fact had a correspondence, e.g. 'the same' on the match 'Wimbledon' and 'Wimbledon Trophy'.

The majority of these points indicate users made a wrong choice of tag which could be due to certain tags being unclear, e.g. 'part of', while in other cases it may be due to users being unsure of what tag to use. It should also be noted that some users may be not inclined to add their own tags and just preferred to click on a suggested tag that seemed appropriate to them. In contrast to the other points, the third point suggests that some users were just tagging based on the concept names and not the description this could be due to the descriptions being unclear or users being not inclined to read the description. These are important issues for mapping interface design and suggest that only standard tags that should have a clear meaning should be displayed at first to users and that each of the tags created by a user will have a personal meaning to them. It would also be beneficial to track the use of the mappings to make sure they are categorised correctly to see if users are using suggested tags in a personalised way, i.e. evaluate if the mappings are showing benefits to the user by monitoring the use of the mappings.

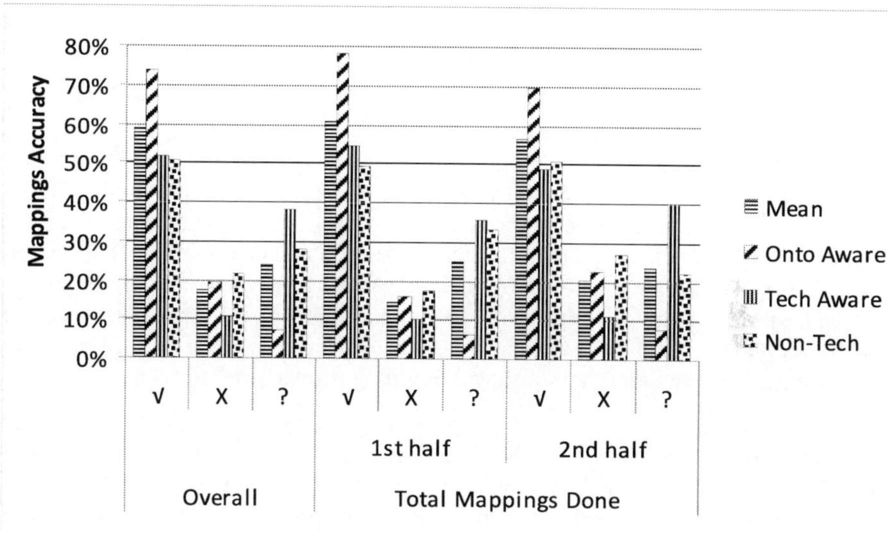

Fig. 6. Mapping Precision. The mappings are separated into the categories which are correct (√), incorrect (X) and inconclusive (?).

5 Conclusions and Future Work

The main goal of this experiment was to assess the effect of a tagging approach to semantic mapping in the user's own work environment. We evaluated the approach under three main usability areas: **unintrusive**, **engaging** and **simplified**. The analysis of the results points to several issues. First, we found that the tag mapping tool does not hamper the daily work of users and indeed it blends into the background, although when the user is busy they find it annoying and distracting whenever a mapping session prompt appears. We envisage that this issue would be resolved if the mapping tasks were context-sensitive to what the user is doing. Second, if users are asked mapping tasks that are not related to their interests or if the benefits from the mappings are unclear then the users would become less engaged in the mapping process. This observation shows it is important to keep the mapping tasks related to the users interests and to also keep showing the benefits of the mapping process to maintain the user interest. Third, users can achieve rich mappings through the use of tagging although some of the suggested tags were unclear to some users. The tagging mapping task is simple and straightforward enough for users but there is still a need to make the information given clearer. Although we have shown the mapping process can be implemented within the casual web users work environment with a tagging approach there is still need for improvement. The authors' have the following suggestions:

- It is important that users do not see the mapping process as a distraction or interfering in their work. The mapping session must be asked in **context** which is suitable and agreeable to the user.
- To make sure the users are not confused by the mapping tasks, it is vital that the mapping task and supporting ontological information are represented in an **easily identifiable and clear way** to the users.
- The **personalisation of a user tag set** may improve the quality of mappings.
- It is necessary to only present **quality matches** for the user to tag.
- To solve the problem of miss-categorised mappings we propose actively **evolving the mappings.** The system should assess whether a mapping, over both time and context, is bringing benefits to the user, and if not the mapping should evolve. Types of evolving include changing the mapping category, making it more specific, getting subsumed or subsuming mappings, and generating child mappings.
- Gather **support from other users** in the mapping process. For example a user may be not inclined to achieve any mapping task so it would be beneficial if other users with similar interest could help if they are willing (directly or indirectly).

In conclusion, we believe our results are promising in that our experimental analysis indicates that users with varied levels of expertise can become active in the mapping process over time in their own area of interest. However, there is still a need to refine our approach to help engage the user more in the mapping process. In our next experiment we will focus primarily on improving the context for mapping tasks and evolution of mappings.

Acknowledgments. This research is partly funded by the Irish Higher Education Authority under the PRTLI 4 NEMBES project.

References

1. Dimitrova, V., Denaux, R., Hart, G., Dolbear, C., Holt, I., Cohn, A.G.: Involving Domain Experts in Authoring OWL Ontologies. In: Sheth, A., Staab, S., Dean, M., Paolucci, M., Maynard, D., Finin, T., Thirunarayan, K. (eds.) ISWC 2008. LNCS, vol. 5318, pp. 1–16. Springer, Heidelberg (2008)
2. Falconer, S.M., Storey, M.: A Cognitive Support Framework for Ontology Mapping. In: Aberer, K., Choi, K.-S., Noy, N., Allemang, D., Lee, K.-I., Nixon, L., Golbeck, J., Mika, P., Maynard, D., Mizoguchi, R., Schreiber, G., Cudré-Mauroux, P. (eds.) ASWC 2007 and ISWC 2007. LNCS, vol. 4825, pp. 114–127. Springer, Heidelberg (2007)
3. Noy, N.: Semantic Integration: A Survey of Ontology-Based Approaches. ACM SIGMOD Record 33(4), 65–70 (2004)
4. Huhns, M., Stephens, L.: Personal Ontologies. IEEE Internet Computing 3(5), 85–87 (1999)
5. Euzenat, J., Shvaiko, P.: Ontology Matching. Springer, Heidelberg (2007)
6. Shvaiko, P., Euzenat, J.: Ten Challenges for Ontology Matching. In: Meersman, R., Tari, Z. (eds.) ODBASE 2008. LNCS, vol. 5332, pp. 1164–1182. Springer, Heidelberg (2008)

7. Jameson, A.: Usability and the Semantic Web. In: Sure, Y., Domingue, J. (eds.) ESWC 2006. LNCS, vol. 4011, p. 3. Springer, Heidelberg (2006)
8. Lambrix, P., Edberg, A.: Evaluation of ontology merging tools in bioinformatics. In: Pacific Symposium on Biocomputing 2003, pp. 589–600. World Scientific, Singapore (2003)
9. Robertson, G.G., Czerwinski, M.P., Churchill, J.E.: Visualization of mappings between schemas. In: CHI 2005: Proceedings of the SIGCHI conference on human factors in computing systems, pp. 431–439. ACM Press, New York (2005)
10. Bernstein, P.A., Melnik, S., Churchill, J.E.: Incremental Schema Matching. In: VLDB, pp. 1167–1170 (2006)
11. Mocan, A., Cimpian, E., Kerrigan, M.: Formal Model for Ontology Mapping Creation. In: Cruz, I., Decker, S., Allemang, D., Preist, C., Schwabe, D., Mika, P., Uschold, M., Aroyo, L.M. (eds.) ISWC 2006. LNCS, vol. 4273, pp. 459–472. Springer, Heidelberg (2006)
12. Shvaiko, P., Giunchiglia, F., da Silva, P.P., McGuinness, D.L.: Web Explanations for Semantic Heterogeneity Discovery. In: Gómez-Pérez, A., Euzenat, J. (eds.) ESWC 2005. LNCS, vol. 3532, pp. 303–317. Springer, Heidelberg (2005)
13. Kaufmann, E., Bernstein, A.: How Useful are Natural Language Interfaces to the Semantic Web for Casual End-users? In: Aberer, K., Choi, K.-S., Noy, N., Allemang, D., Lee, K.-I., Nixon, L., Golbeck, J., Mika, P., Maynard, D., Mizoguchi, R., Schreiber, G., Cudré-Mauroux, P. (eds.) ASWC 2007 and ISWC 2007. LNCS, vol. 4825, pp. 281–294. Springer, Heidelberg (2007)
14. Soantag, D.: Towards Dialogue-Based Interactive Semantic Mediation in the Medical Domain. In: 3rd International Workshop on Ontology Matching (OM 2008) at ISWC 2008 (2008)
15. Groza, T., Handschuh, S., Moeller, K., Grimnes, G., Sauermann, L., Minack, E., Mesnage, C., Jazayeri, M., Reif, G., Gudjonsdottir, R.: The NEPOMUK Project – On the way to the Social Semantic Desktop. In: Pellegrini, T., Schaffert, S. (eds.) Proceedings of I-Semantics 2007, JUCS, pp. 201–211 (2007)
16. Szonszor, M., Alani, H., Contador, I., O'Hara, K., Shadbolt, N.: Semantic Modelling of User Interest Based on Cross-Folksonomy Analysis. In: Sheth, A., Staab, S., Dean, M., Paolucci, M., Maynard, D., Finin, T., Thirunarayan, K. (eds.) ISWC 2008. LNCS, vol. 5318, pp. 632–648. Springer, Heidelberg (2008)
17. Golder, S.A., Huberman, B.A.: Usage patterns of collaborative tagging systems. J. Information Science 32, 198–208 (2006)
18. Li, X., Guo, L., Zhao, Y.E.: Tag-based social interest discovery. In: Proc. 19th Int. World Wide Web Conf. (WWW), Beijing, China (2008)
19. Conroy, C., O'Sullivan, D., Lewis, D.: Ontology Mapping Through Tagging. In: International Workshop on Ontology Alignment and Visualisation, CISIS 2008, pp. 886–891 (2008) ISBN 0-7695-3109-1
20. Conroy, C.: Towards Semantic Mapping for Casual Web Users. In: Sheth, A., Staab, S., Dean, M., Paolucci, M., Maynard, D., Finin, T., Thirunarayan, K. (eds.) ISWC 2008. LNCS, vol. 5318, pp. 907–913. Springer, Heidelberg (2008)
21. Euzenat, J.: An API for ontology alignment. In: McIlraith, S.A., Plexousakis, D., van Harmelen, F. (eds.) ISWC 2004. LNCS, vol. 3298, pp. 698–712. Springer, Heidelberg (2004)
22. Conroy, C., O'Sullivan, D., Lewis, D.: A Tagging Approach to Ontology Mapping. In: 2nd International Workshop on Ontology Mapping (OM 2007) at ISWC 2007 (2007)
23. Tullis, T.S., Stetson, J.N.: A comparison of questionnaires for assessing website usability. In: Usability Professionals' Association Conference, Minneapolis, Minnesota (June 2004)

Fuzzy Annotation of Web Data Tables Driven by a Domain Ontology

Gaëlle Hignette, Patrice Buche, Juliette Dibie-Barthélemy, and Ollivier Haemmerlé

INRA/AgroParisTech Unité Mét@risk - Université de Toulouse le Mirail, 16 rue Claude Bernard, F-75231 Paris Cedex 5, France
{hignette,buche,dibie}@agroparistech.fr, ollivier.haemmerle@univ-tlse2.fr

Abstract. We propose an automatic system for annotating accurately data tables extracted from the web. This system is designed to provide additional data to an existing querying system called MIEL, which relies on a common vocabulary used to query local relational databases. We will use the same vocabulary, translated into an OWL ontology, to annotate the tables. Our annotation system is unsupervised. It uses only the knowledge defined in the ontology to automatically annotate the entire content of tables, using an aggregation approach: first annotate cells, then columns, then relations between those columns. The annotations are fuzzy: instead of linking an element of the table with a precise concept of the ontology, the elements of the table are annotated with several concepts, associated with their relevance degree. Our annotation process has been validated experimentally on scientific domains (microbial risk in food, chemical risk in food) and a technical domain (aeronautics).

1 Introduction

The web is usually modeled as a set of unstructured documents interconnected by a hyperlink graph. However, those documents contain a significative amount of relational data stored in tables. Those tables can be seen as small relational databases even if they lack the explicit metadata associated with a database. Those databases are of course very interesting potential external sources to feed the data warehouse of a company, dedicated to a given domain of application. In general, those data warehouses are already composed of local relational databases fed with internal data belonging to the company. The aim of the company, retrieving external data from the Web, is to complement local data and/or to compare external data to local ones. In order to realize this objective, it is necessary to integrate external data into local data. One possibility is to index external data using the vocabulary already used to index local data. This vocabulary can be easily built from explicit metadata associated with relational local databases, which correspond to the relational schemas of the databases and their attributes with their associated domains. This approach has been experimentally tested on three different domains (microbial risk in food, chemical risk in food and aeronautics): three OWL ontologies have been created within a

L. Aroyo et al. (Eds.): ESWC 2009, LNCS 5554, pp. 638–653, 2009.

couple of hours thanks to preexisting information retrieved from local databases and a very simple tool which translates automatically csv files containing the metadata into an OWL ontology.

In this paper, we present a method to annotate automatically and accurately Web data tables with a domain ontology expressed in OWL. This method relies on heuristics which uses the relations signatures, the symbolic and numeric attributes names, associated domains and constraints (set of symbolic possible values for the symbolic attributes, interval of numeric values and units for the numeric attributes) of the ontology and can be compared with works in schema matching (see [1] for a review). Our original contribution in this paper compared to such works is threefold. Firstly, we instanciate accurately the recognized relations of the ontology for each row of the Web data table. By accurate, we mean that each cell of the row which corresponds to a symbolic (or numeric) attribute is associated with a set of symbolic similar values (or with an interval of possible numeric values) of the ontology. Thanks to this annotation process, it is possible, using the MIEL querying system (see [2]) and the ontology, to query simultaneously the local relational databases and the annotated Web tables thanks to adapted wrappers. Secondly, the heuristics used by our annotation method are based on a set of criteria which can be easily extended. Thirdly, in our approach, thanks to the ontology easily built from preexisting metadata, we don't need to use matching learning methods which imply to learn manually the classifiers on a part of the corpus of documents (see [3] by example). As we will present in this paper, we obtain experimental results which are similar to machine learning methods to identify the type of symbolic columns. This method has been applied with success in the three following different application fields: microbial risk in food, chemical risk in food and aeronautics.

Recent propositions in the Semantic Web community propose to extract, filter, annotate and query Web data tables (see [4,5,6]), but they have not been designed with the same objectives. TableSeer (see [4]) for instance permits to extract a set of predefined metadata (caption, cell content, geographical position of the table in the HTML page, ...) from Web tables, but it does not compare the schema of the Web tables to preexisting schemas defined in a ontology. We can also cite WebTables (see [5,6]) which proposes a system to identify relational tables in the huge amount of tables included in HTML documents, index them to query and rank them. Nevertheless, the WebTables querying language is only composed of a set of key-words which are compared to the attribute names of the Web tables. The row content of the Web tables is not used in the querying process which is only based on global co-occurences frequencies statistics of attribute names. Our approach can be compared to the construction of frames from tables described in [7] but they use a generic ontology and create new relations according to the table signature, whereas we want to recognize predefined relations in an ontology specific to the target domain. In [8], relations from an ontology are instanciated using various HTML structures including tables. However, they only identify binary concept-role relations between instances that are assumed to be already annotated (manually or using another information extraction system). Our work differs as we focus on the recognition of n-ary relations and we propose a step-by-step algorithm including the recognition of element types. From this point of view, the work presented in [9] is nearer to ours, as they transform tables of different structures into a common relational database schema with n-ary relations. However, their approach depend upon the manual definition of an "extraction ontology" that defines extraction rules for each set of objects, giving all synonyms for an attribute name or defining the context of apparition of a string to extract. Our work differs as the ontology is built independently from the annotation process.

Our annotation process is divided into several steps: first, find documents on the web that are relevant to the application domain by regularly crawling sites that are specialized in the domain of interest (for example, JIFSAN[1] or EFSA[2] for risk in food purposes) or even by accessing the "hidden web" via subscription using alerting systems such as RSS feeds (for example on a site like Web Of Knowledge[3]); second, extract the tables from the documents; third annotate the tables; and finally, query the tables using an extended version of the MIEL system (see [2]). The three first steps are implemented in the @Web software, developed using the standards of the WebContent platform[4]. The Web data tables are stored in an XML document using the predefined data model of the WebContent platform which permits to exchange data between web services available on the platform. The annotations generated by the Web service are associated with the tuples of the Web data table and expressed in RDF.

This paper will focus on the third step of the @Web software: the automatic annotation of Web data tables. The structure and OWL representation of our ontology is presented in section 2. We then describe the annotation process: section 3 presents how we recognize the semantic relations represented by a table, and section 4 shows how we create the actual annotations for the table.

2 The Ontology

In the current MIEL system, when a user ask a query on a database, he/she can define the selection criteria according to a predefined vocabulary. We have brought together this vocabulary and the domain information available within the databases schemas to create an ontology. The given examples come from the ontology of the food microbiology domain, but the structure we propose is generic and can be applied to many other domains such as chemical risk in food or aeronautics.

2.1 The Ontology Structure

The structure of our ontology relies on the structure of database schemas and is organised into numeric types, symbolic types and relations.

---

1 Joint Institute for Food Safety and Applied Nutrition, http://www.jifsan.umd.edu
2 European Food Safety Authority, http://www.efsa.europa.eu
3 http://www.isiwebofknowledge.com/
4 http://www.webcontent.fr

Symbolic types are described by a type name, a list of synonyms for the type name and a taxonomy of possible values. In our ontology on the food microbiology domain, the type *Food Product* is associated with a taxonomy of more than 500 food products, the type *Microorganism* with a taxonomy of more than 150 microorganisms and the type *Response* is a flat list of the possible responses of a microorganism to a treatment: growth, absence of growth or death.

Numeric types are described by a type name, a list of synonyms for the type name and the set of units in which the type can be expressed and eventually a numerical range. Our ontology on food microbiology contains 18 numeric types, for example *Temperature* which can be expressed in °C or °F and has no numerical range, or *pH* which has no unit and is restricted to the range $[0, 14]$.

Relations are described by the name of the relation and its signature. The signature of a relation is divided into one result type (the range of the relation) and several access types (the domain of the relation). Our ontology on food microbiology contains 16 relations. For example, in the relation *Growth kinetics*, the result type is *Microorganism concentration* and the access types are *Microorganism*, *Food product*, *Temperature* and *Time*.

The names of types and relations, as well as the possible values of a symbolic type defined in its taxonomy, are called terms. Those terms will be used to annotate Web tables extracted from the web.

2.2 The Ontology in OWL Format

The OWL representation of the ontology is divided in two parts. First, the definition of the structure of the ontology is domain independent: what is a symbolic type, a numeric type or a relation and what are the properties of these different objects. Then, the second part of the ontology presents the definition of the actual types and relations of the domain.

We have separated in the OWL representation of the ontology the concepts (i.e. the types, the relations and the elements of the taxonomies) and the vocabulary (i.e. the actual terms with their words). Any concept in the ontology can be linked to its corresponding term via a property *AssociatedTerm*. The taxonomy of a symbolic type is viewed as a hierarchy of subclasses from a generic OWL class *Taxonomy*: the symbolic type is associated to the root of its hierarchy via the property *HasForTaxonomy*. Each concept in the hierarchy (except the taxonomy root) is associated with its corresponding term. Relations in the ontology are n-ary. They are thus represented as advised by [10], a relation being an OWL class associated to the types of its signature via the properties *AssociatedKey* (for the access types) and *AssociatedResult* (for the result type). The complete OWL definition of our ontology is available for download at `http://metarisk.inapg.inra.fr/UserFiles/File/Gaelle/onto.owl`

3 Recognising Relations Represented in a Table

Given the ontology and given a table extracted from the web, we want to find which semantic relations described in the ontology are represented in the table.

For that purpose, we use an aggregation approach, first looking at the contents of the cells, then computing the type of the columns and finally comparing the signature of the table (the column types) with the signatures of the relations in the ontology. We first distinguish between symbolic and numeric columns, using some of the knowledge described in the ontology (mainly the units): for better description of this step, please refer to [11] which is a preliminary version of this work. We present here how we find the type of symbolic then numeric columns, and finally the relations represented in the table.

3.1 Finding the Type of a Symbolic Column

When a column contains symbolic data, we want to find which symbolic type of the ontology corresponds to the contents of the column. For that purpose, we compute a score of each symbolic type of the ontology for the column. This score is a combination of the score of the type for the column according to the column contents and the score of the type for the column according to the column title. Both these scores use the notion of term similarity: let a and b be two terms, represented as weighted vectors $(a_1, \ldots, a_n)$ and $(b_1, \ldots, b_n)$, the coordinate axis corresponding to the lemmatised words, the coordinate values corresponding to the weight of the word in the term (0 if the word is not part of the term); the term similarity between a and b is the cosine similarity [12] between the two vectors.

$$sim(a,b) = \frac{\sum_{i=1}^{n} a_i b_i}{\sqrt{\sum_{i=1}^{n} a_i^2 \times \sum_{i=1}^{n} b_i^2}} \tag{1}$$

Each word, of terms from the web and in the ontology, is given a weight of 1 (except for stopwords such as articles or prepositions, as well as words that contain only one letter, which are given a weight of 0). The score of a symbolic type t for a column col according to the column title, noted $score_{title}(t, col)$, is the term similarity between the type name and the column title.

To compute the score of a symbolic type for the column according to the column contents, we first explore each cell of the column (excluding the title). For each cell of the column, we compute the score of each symbolic type for the cell as the sum of the term similarities between the content of the cell and the terms in the taxonomy of the symbolic type. Let $cell$ be the term in a cell of the column and t be a symbolic type, with $taxo(t)$ the set of terms in the taxonomy of possible values associated with t. Then the score of type t for the cell $cell$ is:

$$score(t, cell) = \sum_{x \in taxo(t)} sim(cell, x) \tag{2}$$

For each cell of the column, we compute the proportional advantage of the type having the best score: let $best$ be the type with the best score and $secondBest$ be the type with the second best score. We compute the proportional advantage of $best$ for the cell $cell$:

$$advantage(best, cell) = \frac{score(best, cell) - score(secondBest, cell)}{score(best, cell)} \tag{3}$$

The cell is affected the type *best* if its proportional advantage is higher than a specified threshold (for our experiments, this threshold is set to 10%). Otherwise, the cell is considered as having an unknown type.

Once all cells in the column are affected a type, we compute the score of a type for the column according to the column contents as the proportion of cells in the column having this type. Let *col* be a symbolic column with n_{col} the number of cells in the column (excluding the column title), t a symbolic type with $n(t, col)$ the number of cells in the column having the type t, then

$$score_{contents}(t, col) = \frac{n(t, col)}{n_{col}} \quad (4)$$

The score of a symbolic type t for the column *col* is then computed as a combination of the score from the title and the score from the contents:

$$score_{final}(t, col) = 1 - (1 - score_{title}(t, col))(1 - score_{contents}(t, col)) \quad (5)$$

The type of the column is the type having the best final score for the column, assuming that its proportional advantage (see equation 3, replacing the score for the cell by the final score for the column) is greater than a specified threshold (in our experiments, this threshold was set to 10%). Otherwise the column is considered as having an unknown type.

We have experimented our process on 81 columns containing symbolic data, extracted from tables that appeared in publications on food microbiology. Those columns were manually classified in one of the three symbolic types of our ontology, or *Other* if it corresponded to data not modeled in the ontology. Then we ran our process to automatically classify the columns according to their symbolic type. The columns that were of unknown type at the end of our type recognition process were considered as belonging to the *Other* category. In order to assess the quality of our results, we wanted to compare our method with a machine learning classifier. To our knowledge, there is no classifier that is dedicated to the classification of symbolic data using a domain ontology. We have compared our process to the SMO classifier [13]: as it is an optimized version of the well-known SVM, it allows comparing our results with a machine learning method. It is thus a comparison between two alternatives: SMO uses no domain knowledge *but* uses learning, while our method is based on domain knowledge *but* has no learning phase. For the SMO classifier, we used the following pre-treatment: each distinct lemmatized word present in a column results in an attribute; the value of this attribute for a given column is the frequency of the word in the column. The SMO classifier was evaluated using a leave-one-out cross-validation, with default parameters of the Weka[5] implementation. Results of this experiment are given in Table 1.

Over the three symbolic types of the ontology, we obtain a precision of 89% and a recall of 81% with our annotation method, while the SMO classifier gives a 84% precision and a 90% recall. Our method, which uses domain knowledge but

[5] http://www.cs.waikato.ac.nz/ml/weka

Table 1. Classification results on 81 symbolic columns

manual \ computed	our method using the ontology				SMO			
	Food	Micro.	Resp.	Other	Food	Micro.	Resp.	Other
Food	34	0	0	12	45	0	0	1
Micro.	0	16	0	0	4	12	0	0
Response	0	0	1	0	0	0	0	1
Other	3	3	0	12	7	0	0	11

no learning phase, gives similar results to the learning classifier, with a slightly better precision and lower recall. This is mainly because we have biased our annotation technique towards precision: whenever we do not know for sure the type of a column, it is considered as unknown.

Our process has been also experimented on two other corpora: chemical risk in food whose ontology contains 3 symbolic types and aeronautics whose ontology contains 4 symbolic types. Over the 22 columns of the chemical risk in food corpus containing symbolic data, we obtain a precision of 100% and a recall of 100% with our annotation method. Over the 46 columns of the aeronautics corpus containing symbolic data, we obtain a precision of 82% and a recall of 100% with our annotation method.

3.2 Finding the Type of a Numeric Column

As it has been done for the symbolic columns, we find the type of a numeric column by computing the score of each numeric type in the ontology for the column, this score being a combination of the score of the type for the column according to the units in the column and the score of the type for the column according to the column title.

The score of a numeric type t for a column *col* according to the column title, noted $score_{title}(t, col)$ is the term similarity between the type name and the column title.

To compute the numeric type for the column according to the units in the column, we first compute the score of the numeric type for each unit that is present in the column: a numeric type has a score for each unit, depending on the number of numeric types that can be expressed in this unit. Let u be a unit and T_u the set of numeric types for which this unit has been declared in the ontology, the score of a type t for the unit u is $score(t, u) = 0$ if $t \notin T_u$, and $score(t, u) = \frac{1}{|T_u|}$ if $t \in T_u$.

The score of a numeric type for the column is computed as the maximum of the scores of the type for each unit present in the column. If no unit from the ontology was identified in the column, then the column is considered as presenting the unit "no unit", which is treated as a normal unit in the ontology.

The final score of a numeric type t for the column *col* is a combination of the score of t for *col* according to the title of the column ($score_{title}(t, col)$), and the score of t for *col* according to the units in the column ($score_{unit}(t, col)$). However, types are filtered according to the numeric values presented in the column:

- if the column contains a numeric value outside the range of values for the type t, then $score_{final}(t, col) = 0$.
- if all values in the column are compatible with the range of type t, then

$$score_{final}(t, col) = 1 - (1 - score_{title}(t, col))(1 - score_{unit}(t, col)) \quad (6)$$

The type of the column is then the type having the best final score for the column, assuming that its proportional advantage is greater than a specified threshold (in our experiments, this threshold was set to 10%). Otherwise the column is considered as having an unknown type.

We have experimented our method of numeric column annotation on 261 columns containing numeric data extracted from tables found in publications on food microbiology. The columns were manually annotated with the 18 numeric types of our domain ontology, and we compared the manual annotation with the types computed by our method. On the 261 columns, 243 were correctly annotated, 9 were considered as unknown and 9 were annotated with a wrong type (i.e. 96% precision, 93% recall). By comparison, when considering the score from title as the final score with no use of the units defined in the ontology (as was done in [14]), precision was 96% but recall was only 83%: the use of the units defined in the ontology allows better recall with no more errors.

Our process has been also experimented on two other corpora: chemical risk in food whose ontology contains 5 numeric types and aeronautics whose ontology contains 29 numeric types. Over the 65 columns of the chemical risk in food corpus containing numeric data, we obtain a precision of 100% and a recall of 85% with our annotation method. Over the 114 columns of the aeronautics corpus containing numeric data, we obtain a precision of 96% and a recall of 86% with our annotation method.

3.3 Finding the Relations Represented in the Table

Once all columns of the table are assigned a type, we want to find the semantic relations that are represented in the table. For that purpose, we compute a score of each relation of the ontology for the table. This score is a combination of a score of the relation for the table according to the table title and a score of the relation for the table according to the table signature.

The score of a relation for the table according to the table title is computed as the term similarity between the table title and the relation name.

The score of a relation r for the table *tab* according to the table signature is computed as follows:

- if the result type of the relation r was not recognized as a type of a column of the table, then $score_{signature}(r, tab) = 0$
- else, the score of the relation for the table is the proportion of types in its signature that were recognized in the table columns. Let $Sign(r)$ be the set

of types in the signature of relation r and $Sign_{tab}$ the set of types that were recognized for the table columns, then

$$score_{signature}(r, tab) = \frac{|\ Sign(r) \cap Sign_{tab}\ |}{|\ Sign(r)\ |} \tag{7}$$

Then the final score of a relation r for the table tab is computed as:

$$score_{final}(r, tab) = 1 - (1 - score_{title}(r, tab))(1 - score_{signature}(r, tab)) \tag{8}$$

When the scores of all relations of the ontology have been computed for the table, we choose the relation(s) to keep for the table. A table can represent several relations at a time: for example, if a table gives the pH and the water activity of a food product, we will consider it as two separate relations: *food pH* and *food water activity*. We define the notion of concurrent relations: two relations are called concurrent if they have the same result type. If a relation has a non-zero score for the table and has no concurrent relation, this relation is considered as represented in the table. If there are several concurrent relations with non-zero scores for the table, then we only keep the one with the highest score for the annotation of the table. If several concurrent relations have the same highest score, we keep them all.

We have experimented our method on 60 tables extracted from publications on food microbiology. Those tables were manually annotated with the 16 relations of the ontology: one table was typically annotated with 1 to 5 relations, which gives a total of 123 relations (the manually annotated tables are available at `http://metarisk.inapg.inra.fr/UserFiles/File/Gaelle/tables.zip`).

We ran the different steps of our relation recognition system without validating the intermediate steps, i.e. even columns that were wrongly recognized were further annotated and used for the relation annotation. Over the 123 relations in the manual annotation, 119 were correctly recognized in our process, 4 were not recognized and there were 30 relations that were kept for the tables while they should not have been recognized. This gives a precision of 80% for a recall of 97%. The relation recognition is biased towards recall, as we accept partially recognized relations (only the result type is mandatory): it often happens that we recognize two different binary relations for one result type (for example *Food property: pH* and *Growth parameter: pH*) — one of the two relations is the one represented in the table, the other is false.

Our process has been also experimented on two other corpora: chemical risk in food whose ontology contains 4 relations and aeronautics whose ontology contains 26 relations. Over the 34 relations in the manual annotation of the chemical risk in food corpus, 27 were correctly recognized in our process, 7 were not recognized and there were 2 relations that were kept for the tables while they should not have been recognized. This gives a precision of 93% for a recall of 79% with our annotation method. Over the 113 relations in the manual annotation of the aeronautics corpus, 100 were correctly recognized in our process, 13 were not recognized and there were 2 relations that were kept for the tables while they should not have been recognized. This gives a precision of 98% for a recall of 88% with our annotation method.

4 Instanciating Relations and Annotating the Tables

To annotate the tables, we annotate each row of the table with an instance of each relation that was recognized for the table. These instances of relations are linked to the instances of numeric and symbolic types of their signature, corresponding to the data available in the table row. The annotations we generate are fuzzy: they allow one to take into account the imprecision of the initial data in the table (for example an interval for a numeric type) as well as the problems of matching between the vocabulary of the table and the vocabulary of the ontology, and the uncertainty of the recognition of relations. We first present briefly the theory of fuzzy sets that we use for our annotations, then we present how we instanciate numeric types, symbolic types and relations.

4.1 Fuzzy Sets

We use the definition of fuzzy sets given by [15,16]. The notion of fuzzy set is an extension of classical subsets. In the classical case, elements of a reference set X that have some properties belong to a subset A, and elements that do not have these properties belong to the complementary subset of A in X. In a fuzzy set, elements can belong partially to the fuzzy set, with a membership degree comprised between 0 (element which is not part of the fuzzy set) and 1 (element which is completely part of the fuzzy set). The membership degree of an element $x \in X$ for the fuzzy set A is noted $\mu_A(x)$. When X is a continuous domain, we talk about a continuous fuzzy set; if X is discrete, we talk about a discrete fuzzy set.

The support of a fuzzy set A defined on a reference set X is the set (in the classic definition) of elements $x \in X$ so that $\mu_A(x) > 0$. The kernel of a fuzzy set A defined on a reference set X is the set (in the classic definition) of elements $x \in X$ so that $\mu_A(x) = 1$. A trapezoid fuzzy set TFS is a special continuous fuzzy set that is described only by its support $sup = [min_{sup}, max_{sup}]$ and its kernel $ker = [min_{ker}, max_{ker}]$. The membership degree of a numeric value x in the reference set, is then defined as follows:

- if $x \leq min_{sup}$ or $x \geq max_{sup}$ then $\mu_{TFS}(x) = 0$;
- if $min_{ker} \leq x \leq max_{ker}$ then $\mu_{TFS}(x) = 1$;
- if $min_{sup} \leq x \leq min_{ker}$ then $\mu_{TFS}(x) = \frac{x - min_{sup}}{min_{ker} - min_{sup}}$;
- if $max_{ker} \leq x \leq max_{sup}$ then $\mu_{TFS}(x) = \frac{x - max_{ker}}{max_{ker} - max_{sup}}$.

There are several semantics for fuzzy sets, defined in [17]:

- preferences: the elements with the higher membership degrees are the preferred elements. This is used in the MIEL querying system for the user to define query preferences;
- uncertainty or imprecision: there exists a "true" value, but we do not know it. The higher is the membership degree of a value x, the more probable is x to be the "true" value. This is used in our annotations to represent the imprecision of the original data in the tables (in the instanciation of numeric types);

– similarity: a new object is represented by its similarity with known objects. The higher is the membership degree of a known object x, the more it is similar to the new object. This is used in our annotations to represent the similarity between a term from the web and terms from the ontology.

4.2 Instanciating Numeric Types

When instanciating a numeric type t in the signature of a relation that has been recognised for the table, there are three possibilities:

1. There is one column in the table (thus one cell in the row to annotate) that has been recognized as having the numeric type t. In this case, the values in the cell are used to instanciate the type: it can be an isolated value, an enumeration of isolated values, an interval or a mean with a standard error. Intervals and mean with standard error are recognized using specific patterns; if those patterns are not recognized, then all numeric values in the cell are considered as isolated.
2. There are several columns in the table that have been recognized as having the numeric type t. In this case, we have to find the relations between the columns: it is done by looking for keywords in the column title. A column can represent a minimum value, a maximum value or an optimum value (comprised between the minimum and maximum); it can also represent a mean value or a standard error.
3. There is no column in the table that was recognized as having the type t. If the numeric type t has a defined unit, we search for occurences of a numeric value followed by this unit, in the title of the table or in the titles of columns: those occurences are then considered as isolated values.

An instance of a numeric type is represented as a fuzzy set, the reference set being the value range defined in the ontology for the numeric type. The fuzzy set used for annotation is a union of trapezoid fuzzy sets. The trapezoid fuzzy sets used for the instanciation of a numeric type are constructed as follows:

– when recognizing an isolated value x in the table, we construct a trapezoid fuzzy set with $sup = ker = [x, x]$;
– when recognizing an interval $[a, b]$ in the table, either in one cell or when a is the value in a cell recognized as minimum and b is the value in a cell recognized as maximum with no cell recognized as optimum, we construct a trapezoid fuzzy set with $sup = ker = [a, b]$;
– when recognizing a cell as minimum, its minimum numeric value being min, a cell as maximum, its maximum numeric value being max and a cell as optimum, its values being comprised in the minimum interval $[a, b]$, we construct a trapezoid fuzzy set with $sup = [min, max]$ and $ker = [a, b]$;
– when recognizing a mean m and a standard error e, we construct a trapezoid fuzzy set with $sup = [m - e, m + e]$ and $ker = [m, m]$.

Once all trapezoid fuzzy sets are created, the annotation is constructed as the union of all those sets (for example, there can be a union of several isolated values).

Example 1. Table 2 is annotated with the relation of the ontology *Growth parameter: pH*, with the access type *Microorganism* and the result type *pH*. This result type is instanciated, for the first row of data in the table, as a unique trapezoid fuzzy set with $sup = [5, 8.8]$ and $ker = [6, 7]$.

Table 2. Excerpt of a table: Ecologic values for some foodborne bacterial pathogens (extracted from [18])

Species	pH Min	pH Opt.	pH Max
Bacillus cereus	5	6–7	8.8

Below is the RDF representation of the fuzzy set used for the instanciation of type *pH* in the first data row of Table 2.

```
<onto:CFS rdf:about="Row:2/GrowthParameterPH/PH/CFS">
   <onto:IsComposedOf rdf:resource="Row:2/GrowthParameterPH/PH/CFS/TFS:1"/>
</onto:CFS>
<onto:TrapezoidFuzzySet rdf:about="Row:2/GrowthParameterPH/PH/CFS/TFS:1">
   <onto:HasForMinSupport >5</onto:HasForMinSupport >
   <onto:HasForMaxSupport >8.8</onto:HasForMaxSupport >
   <onto:HasForMinKernel >6</onto:HasForMinKernel >
   <onto:HasForMaxKernel >7</onto:HasForMaxKernel >
</onto:TrapezoidFuzzySet>
```

To evaluate our annotation system, we have instanciated the 119 relations that were correctly recognized in the 60 tables used for the experiment on relation recognition. The instanciation of numeric values was analyzed for the first data row of each table: we assume that the structure is homogeneous inside a table, so the instanciation of the first row is representative of what happens for the whole table. On the 119 relations, there were 2 errors on the extraction of numeric values (one was an error of type recognition, one was an error of numeric value recognition). For 5 tables (corresponding to 13 relations), the numeric type *Temperature* was not instanciated because its value was not present in the table but in the textual environment of the table in the original publication. There also were 3 errors in interval reconstruction (values were considered as isolated while it was an interval) and one error in the construction of a minimum/optimum/maximum trapezoid fuzzy set (values were considered as isolated). For all 100 remaining relations, all numeric values were correctly instanciated.

4.3 Instanciating Symbolic Types

To instanciate a symbolic type t in the signature of a relation that has been recognized for the table, we construct a fuzzy set. The reference set of this fuzzy set is the set of all terms in the taxonomy of the type. The membership degree of a term x in the annotation fuzzy set is the term similarity between x and the term in the cell that has been recognized as having type t. In our 60 tables, it did not happen that several columns in a table were recognized as having the same type t, however, would that happen, we would construct a union of fuzzy sets (one fuzzy set for each column).

Below is the RDF representation of the fuzzy set used for the instanciation of type *Microorganism* in the first data row of Table 2 (we only represent the elements of the taxonomy for which the membership degree is not equal to 0).

```
<onto:DFS rdf:about="Row:2/GrowthParameterPH/Microorganism/DFS">
  <onto:HasForElement rdf:resource="Row:2/GrowthParameterPH/Microorganism/DFS
       /elt:1"/>
  <onto:HasForElement rdf:resource="Row:2/GrowthParameterPH/Microorganism/DFS
       /elt:2"/>
</onto:DFS>
<onto:BacillusCereus rdf:about="Row:2/GrowthParameterPH/Microorganism/DFS/elt
     :1">
  <onto:HasForMembershipDegree>1.0</onto:HasForMembershipDegree>
</onto:BacillusCereus >
<onto:Bacillus rdf:about="Row:2/GrowthParameterPH/Microorganism/DFS/elt:2">
  <onto:HasForMembershipDegree>0.7</onto:HasForMembershipDegree>
</onto:Bacillus>
```

The quality of the annotation has been evaluated on 185 instances of food products. For those food products, the "best match" in the ontology (i.e. the term in the ontology that was the nearest to the meaning of the term in the table) was manually defined. The evaluation is done by looking at the position of the "best match" in the automatic annotation, by order of descending membership degree. The position is evaluated at worse, i.e. if there are several terms in the ontology having the same membership degree in the fuzzy set used for the annotation, the "best match" is always considered as being in last position. This evaluation at worse is due to the need for manual validation of the annotations: if we present a user with the 5 terms having the best membership degree for him to choose the best, we want to be sure that the "best match" will be among those 5. On the 185 terms from the web, 78% had a "best match" for which the term similarity with the term from the web was not null. 46% of the terms from the web had their "best match" in first position in the computed annotation, while 66% had their "best match" among the five best positions. This validates the approach of keeping a fuzzy set for instanciating the symbolic types, instead of keeping only the concept in the taxonomy having the best term similarity with the term in the table.

4.4 Representing the Relations in the Table Annotation

Once all types of the signature of a relation have been instanciated for a row of the table, we can annotate the row with the relation: we create an instance of the relation, which is related to a degree of certainty (the score that has been computed during the relation recognition phase) via the property *HasForScore*, and is related to the instances of numeric and symbolic types of its signature that were created for the row.

Below is the annotation of the first data row of Table 2, given the DFS and CFS that were already defined during the intanciation of symbolic and numeric types.

```
<onto:GrowthParameterPH rdf:about="Row:2/GrowthParameterPH" >
   <onto:HasForScore>1.0</onto:HasForScore>
   <onto:AssociatedKey rdf:resource="Row:2/GrowthParameterPH/Microorganism"
       />
   <onto:AssociatedResult rdf:resource="Row:2/GrowthParameterPH/PH" />
</onto:GrowthParameterPH>
<onto:Microorganism rdf:about="Row:2/GrowthParameterPH/Microorganism">
  <onto:IsConstructedFrom rdf:resource="Row:2/Text:1"/>
  <onto:IsAnnotatedBy>
    <onto:DFS rdf:resource="Row:2/GrowthParameterPH/Microorganism/DFS"/>
  </onto:IsAnnotatedBy>
</onto:Microorganism>
<onto:PH rdf:about="Row:2/GrowthParameterPH/PH">
  <onto:IsConstructedFrom rdf:resource="Row:2/Text:2"/>
  <onto:IsConstructedFrom rdf:resource="Row:2/Text:3"/>
  <onto:IsConstructedFrom rdf:resource="Row:2/Text:4"/>
  <onto:IsAnnotatedBy rdf:resource="Row:2/GrowthParameterPH/PH/CFS" />
</onto:PH>
```

5 Conclusion and Perspectives

In this paper we have described a complete annotation system to annotate data tables extracted from the web. It is implemented in the @Web software which has been developed using the standards of the WebContent platform. In @Web, the workflow is divided into three steps: first, find documents on the web that are relevant to the application domain; then, extract the tables from the documents; finally annotate the tables. @Web proposes very efficient and powerful tools which permit to transform semi-automatically the structure of a Web data table before its automatic annotation using a domain ontology. To the best of our knowledge, @Web is the only software which permits first to deal efficiently with the heterogeneity and the complexity of the table structures available on the Web and second to annotate accurately a Web table with a domain ontology. This annotation entirely relies on the domain knowledge given in the ontology, with no learning phase. After segregating between numeric and symbolic columns, we recognize the types of the columns using both the column title and the contents of the column. Then, we recognize the relations represented in the table: the use of both the table title and the table signature helps finding the relations with a very good recall and an acceptable precision. The relations are then instanciated for each row of the table: the instanciation of numeric types gives very good results, the instanciation of symbolic types being more difficult due to language variation from one author to another. Annotations associated with each row are expressed in a fuzzy extension of RDF representing either imprecises data or the semantic distance between Web data tables and the ontology. This fuzzy representation allows the use of approximate reasoning techniques in the querying step of the annotated Web data tables using the ontology (see [2]).

In the very next future, we want to explore two ideas. The first one consists in enhancing the performance of the annotation system using machine learning techniques on the knowledge of the ontology but without manual training on a subset of the corpus. The second perspective will be to exploit the annotated Web data tables in three different applications. The Web data tables annotated by our method will be incorporated in two applications in the field of microbial and chemical risk in food. A third application in economic watch in the field of aeronautics will be also realized with EADS.

Acknowledgments

Financial support from the French National Research Agency (ANR) for the project WebContent in the framework of the National Network for Software Technology (RNTL) is gratefully acknowledged.

References

1. Rahm, E., Bernstein, P.A.: A survey of approaches to automatic schema matching. VLDB J. 10(4), 334–350 (2001)
2. Buche, P., Dibie-Barthélemy, J., Hignette, G.: Flexible querying of fuzzy rdf annotations using fuzzy conceptual graphs. In: Eklund, P., Haemmerlé, O. (eds.) ICCS 2008. LNCS, vol. 5113, pp. 133–146. Springer, Heidelberg (2008)
3. Doan, A., Domingos, P., Halevy, A.Y.: Learning to match the schemas of data sources: A multistrategy approach. Machine Learning 50(3), 279–301 (2003)
4. Liu, Y., Bai, K., Mitra, P., Giles, C.L.: Tableseer: automatic table metadata extraction and searching in digital libraries. In: JCDL, pp. 91–100. ACM Press, New York (2007)
5. Cafarella, M.J., Halevy, A.Y., Zhang, Y., Wang, D.Z., Wu, E.: Uncovering the relational web. In: WebDB (2008)
6. Cafarella, M.J., Halevy, A.Y., Wang, D.Z., Wu, E., Zhang, Y.: Webtables: exploring the power of tables on the web. PVLDB 1(1), 538–549 (2008)
7. Pivk, A., Cimiano, P., Sure, Y.: From tables to frames. In: McIlraith, S.A., Plexousakis, D., van Harmelen, F. (eds.) ISWC 2004. LNCS, vol. 3298, pp. 166–181. Springer, Heidelberg (2004)
8. Tenier, S., Toussaint, Y., Napoli, A., Polanco, X.: Instantiation of relations for semantic annotation. In: Int. Conf. on Web Intelligence, pp. 463–472 (2006)
9. Embley, D.W., Tao, C., Liddle, S.W.: Automatically extracting ontologically specified data from HTML tables of unknown structure. In: Spaccapietra, S., March, S.T., Kambayashi, Y. (eds.) ER 2002. LNCS, vol. 2503, pp. 322–337. Springer, Heidelberg (2002)
10. Noy, N., Rector, A., Hayes, P., Welty, C.: Defining n-ary relations on the semantic web. W3C working group note (2006), `http://www.w3.org/TR/swbp-n-aryRelations`
11. Hignette, G., Buche, P., Dibie-Barthélemy, J., Haemmerlé, O.: An ontology-driven annotation of data tables. In: WISE Workshops 2007. Web Data Integration and Management for Life Sciences., Nancy, France, pp. 29–40 (December 2007)
12. Van Rijsbergen, C.J.: Information Retrieval, 2nd edn. Dept. of Computer Science, University of Glasgow (1979)
13. Platt, J.C.: Fast training of support vector machines using sequential minimal optimization, pp. 185–208. MIT Press, Cambridge (1999)
14. Gagliardi, H., Haemmerlé, O., Pernelle, N., Saïs, F.: An automatic ontology-based approach to enrich tables semantically. In: AAAI Context and Ontologies Workshop (2005)

15. Zadeh, L.: Fuzzy sets. Information and control 8, 338–353 (1965)
16. Zadeh, L.: Fuzzy sets as a basis for a theory of possibility. Fuzzy Sets and Systems 1, 3–28 (1978)
17. Dubois, D., Prade, H.: The three semantics of fuzzy sets. Fuzzy Sets and Systems 90(2), 141–150 (1997)
18. Cliver, D.O., Hajmeer, M.N., Jay-Russell, M.: Foodborne infections and intoxications. School of Veterinary Medicine, University of California (2004)

An Integrated Approach to Extracting Ontological Structures from Folksonomies

Huairen Lin, Joseph Davis, and Ying Zhou

School of Information Technologies, The University of Sydney, Australia
{lin,jdavis,zhouy}@it.usyd.edu.au

Abstract. Collaborative tagging systems have recently emerged as one of the rapidly growing web 2.0 applications. The informal social classification structure in these systems, also known as folksonomy, provides a convenient way to annotate resources by allowing users to use any keyword or tag that they find relevant. In turn, the flat and non-hierarchical structure with unsupervised vocabularies leads to low search precision and poor resource navigation and retrieval. This drawback has created the need for ontological structures which provide shared vocabularies and semantic relations for translating and integrating the different sources. In this paper, we propose an integrated approach for extracting ontological structure from folksonomies that exploits the power of low support association rule mining supplemented by an upper ontology such as WordNet.

Keywords: Ontological structure, Folksonomy, Collaborative tagging system.

1 Introduction

Organization of digital resources has been a major challenge on the World Wide Web. Recently, collaborative tagging systems (CTS) have emerged as a mechanism for users to organize and share such resources. CTS allow users to use any keywords or tags relevant to the content to annotate their favorite resources on the web. For example, Flickr[1], an online photo management and sharing application launched in Feb. 2004, counts more than 40 million monthly visitors and is ranked as one of the world's top 30 websites. As of November 2007, 2 billion photos are stored on the site [1]. Another example is Citeulike[2], a free online bibliography manager allowing users to gather, organize, and share scholarly papers. It has become popular especially among researchers and other academic users.

This kind of informal social classification in CTS where users use their own language or terminology to describe and classify the content was first recognized as folksonomy by Vander [2]. When a user tags an online resource, s/he is creating an informal taxonomy. These tags are aggregated to help find the information they represent. With bottom-up, user-driven and freely chosen vocabularies, folksonomies stand

---

1 http://www.flickr.com
2 http://www.citeulike.org

L. Aroyo et al. (Eds.): ESWC 2009, LNCS 5554, pp. 654–668, 2009.

in contrast to taxonomies which use controlled terms. The relationships among the terms are typically contributed by domain experts in order to classify resources.

As the amount of resources annotated using folksonomies has increased significantly, exploration and retrieval of the annotated resources pose challenges. The major problem with folksonomies is that the tags used to describe the content can be idiosyncratic and not understood by many users. Most tags are chosen based on individual users' own experience and linguistic styles and preferences. Furthermore, the concept and internal structure are not explicit to the machine or to other systems even though the tags may be meaningful and coherent to the user who created them [3]. Folksonomies tend to include all kinds of tags ranging from standard dictionary words and compound expressions created by individual users such as "evolutionary- genomics" to jargon and nonsense words.

Various solutions have been proposed to improve the quality of queries based on folksonomy. One stream of research has attempted to refine the query result using meaningful knowledge derived from the folksonomy itself. Clustering and tag clouds are widely used approaches. Clustering techniques group the search result into several subsets and recommend related resources based on selected tags. However, clustering techniques rely heavily on statistical association or co-occurrence of tags. The effectiveness of this approach can be limited as the relations derived are not based on meaning. Tag cloud is a somewhat rough approach to organizing tags. It shows a subset of frequently used tags in sizes relative to their frequencies [4]. It is easy for the user to see the "hot" keywords. However, tag cloud normally contains very general terms such as "computer" or "picture" and do not show any semantic relation between the tags. Another stream of research takes an existing upper ontology as the base structure and uses it to facilitate organizing query results. Although the need for relevant ontological structures to support CTS systems is well understood, the upper ontology may not be well matched with the tags in the folksonomy[5]. For example, WordNet is widely used as an upper ontology because it describes very general concepts across all domains. However, methods heavily dependent on WordNet frequently get poor results for accuracy due to the fact that many of the tags from collaborative tagging system do not exist in WordNet.

In this paper, we propose an integrated approach to extract ontological structures from folksonomies. Our approach combines the knowledge extracted from folksonomies using data mining techniques with the relevant terms from an existing upper-level ontology. Specifically, low support association rule mining is used to analyze a large subset of a folksonomy. Knowledge is expressed in the form of new relationships and domain vocabularies. Standard tags in the vocabulary are mapped to WordNet to get semantic relations. Jargon tags and user defined compounds are then incorporated into the hierarchy based on domain knowledge extracted from folksonomy. Thus, the hidden knowledge embedded in the folksonomies is transformed into formalized knowledge in the form of ontological structures.

In particular, this paper answers the following questions:

- How to extract shared vocabularies from large data sets?
- How to find the semantic relations for these shared vocabularies?
- How to handle the non-standard tags in the folksonomies?

The rest of the paper is organized as follows: In section 2, related research is reviewed. In section 3, we present our integrated approach and discuss the detailed steps. Then in section 4, experimental results are presented and evaluated. Finally, we discuss our conclusion and future work.

2 Related Research

One of the main strengths of folksonomies and CTS is that they directly reflect the users' vocabularies, their choices in terminology, and subjective meaning [6]. Despite the fact that people assign one or more freely chosen tags to each of the resources and these tags are based on their own knowledge or professional background, there is a common basis of understanding of the tags used to communicate with each other [7]. Thus, a folksonomy has its uses and has the potential to be a weak knowledge base. Since users add new contextual dimension during collaborative tagging, most of the tags or keywords annotated by users tend to be more highly correlated than keywords automatically extracted by machine, such as Term Extraction Web Service from Yahoo[3]. Thus, folksonomies and CTS can be seen as potential sources of semantic information to support ontology evolution [8].

Approaches such as clustering and related tags do not make the hierarchical relations explicit between tags. As a result, it is difficult for a user to find related resources with broader or narrower tags during navigation which may better represent the user's current interests and help a user who has limited subjective knowledge. By representing folksonomy as a tripartite network of users, tags and objects, semantics such as relation between broader/narrower tags has been unveiled through a process of graph transformation using social network analysis [9]. Aiming to converting a large corpus of tags from folksonomy into a navigable hierarchical taxonomy of tags, an algorithm using graph centrality has been proposed. Cosine similarity between tags has been used to measure the distance from one tag to another and organize them into a hierarchical tree by starting with a single "root" node representing the top of the tree, and adding other tags to the tree in decreasing order of distance [10].

Association rule mining has also been adopted to analyze and structure folksonomies. Since folksonomies provide a three-dimensional dataset (user, tag, and resources), Schmitz proposed a conceptual level notation to reduce the three-dimensional folksonomy to a two-dimensional formal context and apply association rule mining. The output of association rule mining on a folksonomy data set are association rules like $A \rightarrow B$, which shows that users assigning the tag A to some resources often tend to also assign the tag B to them [11]. Association rule based approach has been extended in [12] to mine structural features of taxonomies by pruning the less important relations between tags.

Significant research progress in the field of semantic techniques has offered promising prospects for extracting semantic structures and relations from the folksonomies. To further discover the relationships within tags in clusters, several existing ontology resources can be used as references, such as WordNet (despite its limitations) and other semantic web resources. Ontology mapping and matching techniques are commonly applied to identify relationships between tags, between tags and lexical resources, and between tags and elements in an existing ontology. For example, by

[3] http://developer.yahoo.com/search/content/V1/termExtraction.html

mapping "apple and fruit" in a food ontology, we can find the relation that "apple" is a subclass of "fruit" [13] [14]. WordNet has been successfully applied in many applications as a reliable upper ontology. An et al [15] presented an approach to automatically build a domain ontology by interweaving sub-taxonomies of WordNet with information extracted from deep web pages. In this research, concept and relations from WordNet have been used to bridge the concept gap and tie together ontology fragments into a single ontology. Laniado et al [16] illustrated an approach to integrating WordNet noun hierarchy in the *related tags* panel of del.icio.us (a collaborative tagging system which allows users to tag, manage and share Web pages). By mapping *related tags* to WordNet and getting the related terms, the tags and terms will be organized into a navigation tree according to a semantic criterion.

The ontological structure extracted from folksonomies can be useful in many areas of CTS, such as providing multi-dimensional views, cataloguing and indexing, query translation and tagging suggestion. It can be used to organize the search result into different dimensions like topic, date, location, etc. For each dimension, relevant resources can be organized in a hierarchical structure. Second, ontological structures provide an expressive way to catalog and index large digital resources. While a query needs a pre-specified keyword for information retrieval, the ontological structures give users a quick understanding of the subjective knowledge and let them directly browse for further information. Third, with query translation based on an ontological structure, we can enhance the precision and recall by matching the query keywords and the potential results at the level of semantics. For example, the keyword in the query can be replaced by their approximations and related instances will be returned to user. Finally, suggesting relevant ontological classes to the user will not only improve the tagging experience but increase classification quality [7, 17, 18].

In summary, folksonomies have their own shared vocabularies and relations, which can be extracted as an ontological structure and used to improve the exploration and retrieval of digital resources. Although several approaches have been proposed to bring structure to folksonomies, they do not come without limitations. These include the inability to decide the rules generated by association rule mining as to which term is more general or narrow, and tags that cannot be found in the upper ontologies.

3 System Architecture

In folksonomies, natural language has been used to annotate resources and to recall resources. As the result of non-controlled human language, the vocabularies used in folksonomy are shaped into following four word-formations:

- Standard tags, which can be found in traditional dictionaries, e.g. "genomics".
- Compound tags, a non-standard expression, part of them can be found in dictionary, e.g. "evolutionary-genomics".
- Jargon tags, another non-standard expression frequently used to quickly express user's ideas, e.g. "scientometrics", "folksonomy", "CSCW".
- Other nonsense tags, such as misspelling tags.

In this section, we present our integrated bottom-up and top-down architecture that aims to extract ontological structures from folksonomies based on the four word-formations. A visual representation of the entire extraction architecture has been presented in Fig. 1. The system proceeds as follows:

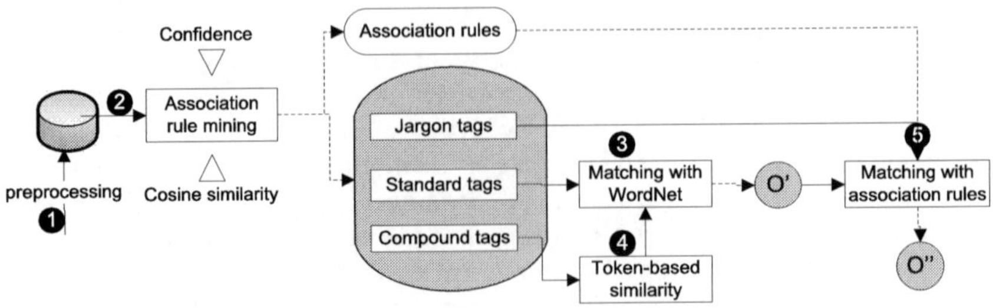

Fig. 1. The extraction process

1) In the data pre-processing phase, resources with only one tag or tagged by languages other than English are excluded. However, we should be very careful in this step not to delete jargon and compound tags. Thus methods like traditional dictionary filtering are not appropriate in our approach.
2) Low support association rule mining is applied to generate association rules representing relations between correlated tags. In brief, it has three sub tasks:
   - Discovering shared vocabularies or essential tags, where a tag should have certain relation with other tags, which is the basis for ontological structure.
   - Extracting the association rules between jargon and standard tags. Those association rules are treated as ontology matchers to incorporate jargon into ontological structure.
   - Retrieving associated terms and excluding non-relevant terms during the mapping and matching with WordNet.
3) WordNet is implemented as an upper ontology providing the semantic is-a relation, which is called as hypernyms in WordNet. After the standard tags have been connected to each other via semantic relations from WordNet, they were organized into a hierarchical structure.
4) A series of similarity filters are employed to interpret the compound tags before matching them with WordNet.
5) Jargon tags are incorporated into the previously built ontological structure by matching tags using association rules and similarity coefficient.

In following subsections, we discuss in detail each of the steps.

3.1 Low Support Association Rules Mining

Association rule mining technique is applied to the dataset to discover possible pair wise associations between tags. A priori association rule mining algorithm has been proposed to solve the "supermarket basket" problem and to discover interesting relations between items. If 90% of the transactions that include *butter* and *bread* also include *milk,* the relation is showed *as* $\{butter, bread\} \rightarrow milk$ with a confidence value of 0.9 [19, 20]. Such analysis is based on the past transaction data consisting of a set of transaction D = $(d_1, d_2 ... d_k)$ and a set of items, I = $(i_1, i_2 ... i_k)$. In our approach, given a dataset from CTS, where every resource was annotated with a set of tags by several online users, the resources set corresponds to transactions D and the set of tags correspond to items I.

The aim of association rule mining in CTS is to generate associations in the form $t_a \rightarrow t_c$ between tags t_a and t_c that have support and confidence above certain thresholds, called minimum support and minimum confidence. Support of a rule is simply computed as the percent of the resources containing the tag pair. Confidence is computed as the ratio of the number of resources containing both t_a and t_c and the number of resources containing t_a only. While the confidence threshold reflects the strength of the rule, support threshold measures the coverage.

As folksonomy is collectively built by various users, the tags in folksonomies usually follow a Zipf distribution. Except for a few general tags, majority of the tags do not occur very frequently in the dataset. Traditional association rule mining algorithms normally set a relatively high support and confidence threshold to find common and strong rules. However, this is not the case for folksonomies. Setting a relatively high support threshold is likely to miss important associations among tags in the long tail of the Zipf distribution[21]. Hence we adopt a very low support threshold to include tags that do not occur very frequently in our analysis. Lower support may inadvertently bring lots of noise in the rule set. To offset this effect, we introduce cosine similarity[3, 22] to filter out possible noise.

To apply this measure, we first convert datasets from folksonomies into a metric space V. Given a pair of tag (x, y), tag x is expressed as a vector $\vec{x}$ in this space, where each dimension corresponds to a resource and value indicating whether or not a tag appears in a resource [23]. This tag-resources model can be converted into a 0/1 matrix because the times of a tag appears in a resource should be 0 or 1.

Table 1 shows such 0/1 matrix for tag(x, y), where each column represents a resource and each row represents a tag x or y, intersection (row, column) = 1 if a specific tag appear in the resource. If not, the value is 0. The traditional cosine similarity between (x, y) can be measured as Eq. (1). Considering the occurrence value is only 1 and 0 in folksonomies, the Eq. (1) can be simplified as Eq. (2) where the capital letter X and Y correspond to the set of resources having tags x or y.

Table 1. 0/1 matrix view

$$\begin{bmatrix} & t_1 & t_2 & t_3 & t_4 & \cdots & \cdots & \cdots & t_i \\ x & 1 & 0 & 1 & 0 & \cdots & 0 & 0 & 0 \\ y & 0 & 1 & 1 & 1 & \cdots & 0 & 0 & 0 \end{bmatrix}$$

$$\cos(x, y) = \frac{\sum_{i \in |v|} \vec{x}_i * \vec{y}_i}{\sqrt{\sum_{i \in |v|} \vec{x}_i^2} * \sqrt{\sum_{i \in |v|} \vec{y}_i^2}} \tag{1}$$

$$\cos(x, y) = \frac{|X \cap Y|}{\sqrt{|X| * |Y|}} \tag{2}$$

Compared to the support measure, cosine similarity measure not only provides a correlation value between two tags, but it also enable us to prune the rule set because it does not calculate the resources that contain neither x nor y. The cosine similarity measure also helps to exclude high confidence but poorly correlated rules.

Considering the above mentioned specialty in our approach, Apriori, the earliest and highly efficient algorithm to mine association rules, does not fit our purpose well. We modify it and develop a simplified version of Apriori algorithm, LApriori: We only calculate the relationship between tag pairs; both antecedent and consequent can only have one tag; additional cosine similarity threshold is set to offset the noise caused by low support and to compare the relevance between tags.

Algorithm 1. LApriori to discover association rules mining in folksonomies

```
L1 = count frequent 1-item sets;
for every resource r do
        for each pair of tags {t_a , t_c} in r do
                if t_a ⊂ L1 and t_c ⊂ L1 then
                        increase support of {t_a , t_c} by 1;
                end if
        end for
end for

for every frequent 2-item set {x,y} do
        cos(x,y)= Support(x,y)/sqrt(support(x)*support(y));
        if cos(x,y)> min_sim  then return x->y;
        end if
end for
```

3.2 Standard Tags: Ontology Matching with WordNet

We use WordNet as the upper ontology and compute each semantic relation between tags in terms of hypernym relation from WordNet. A term that is more generic or more abstract than a given term is considered as a hypernym. For example, in table 2, the term wine has the following upper hypernyms: alcohol, beverage, drink, red, etc.

Table 2. A hypernym example of "wine" from WordNet 3.0

Sense 1:	Sense 2:
wine, vino	wine
=> alcohol, alcoholic beverage, intoxicant,	=>red
=> beverage, drink, drinkable, potable	=>color
=> food, nutrient	

Possible semantic relations between them are described as *more general* ($\supseteq$), *less general* ($\subseteq$), *equivalence* (=) [24]. x $\supseteq$ y, if x is a hypernym of y. For example alcohol is a hypernym of wine, then we can say that alcohol is *more general* than wine, or wine *is-a* kind of alcohol, alcohol $\supseteq$ wine.

In folksonomies, we added another two definitions: *essential tags*, and *candidate hypernyms*.

Essential tags are all distinct tags existing in association rules filtered by predefined thresholds.

Candidate hypernyms: Given a tag, only the hypernyms that exist in its related tags are regarded as *candidate hypernyms*. For example, if *beverage* and *food* are two hypernyms for *wine* and also related to *wine* through association rules, then *beverage* and *food* are *candidate hypernyms* for wine. On the other hand, although *alcohol* and *red* are also hypernyms for wine, we do not consider them as *candidate hypernyms* because they have no relationship with *wine* in the generated association rules.

We only use hypernyms both existing in WordNet and association rules, because those hypernym terms not related to certain tags in folksonomies do not reflect the subjective knowledge well.

Based on the above mentioned considerations, we design the following Folk2Onto algorithm to find *more general* term for each *essential tag*.

Algorithm 2. Folk2Onto to find *more general* term for each *essential tag*

for each tag t_k in *essential tags* **do**
U_k = the *more general* term for t_k , set U_k = null;
S_k = get all tags related to t_k from association rules;
W_k = get all hypernyms for t_k from WordNet;
candidate hypernyms $\{h_1 \dots h_n \dots\} = S_k \cap W_k$;
    for each h_n **do**
        if U_k is null
            $U_k = h_n$;
        else if U_k is not null **and** h_n is a hyperhym of U_k
            break ;
        else if U_k is not null **and** U_k is a hyperhym of h_n
            $U_k = h_n$;
        end if
    end for
end for

For example, given a set of tags {food, beverage, wine, milk}, following semantic relations or ontological structure were generated as Fig 2:

beverage $\supseteq$ wine,
beverage $\supseteq$ milk,
food $\supseteq$ beverage,

Fig. 2. An ontological structure for "wine"

Beside *hypernyms*, WordNet also provides semantic relations such as *meronyms, synonyms, and antonyms which* can potentially be helpful in our approach.

3.3 Compound Tags: Token-Based Similarity Matching

Compound tags are non-standard terms and thus cannot be processed by WordNet without transformation. Here we adopt a series of filters provided by Jawbone[25] to analyze the compound tags. If they match certain defined criteria, the compound tags will be reserved and represented by its base term for more general parent finding.

In detail, the following term filters are applied to check whether the compound tag has a particular relationship to another term existing in WordNet:

1) EndWithFilter operates by splitting the compound into independent token of standard terms. The last one is used to represent the whole compound. For instance, "collaborative_tagging" is represented by tagging.
2) StartsWithFilter operates in a similar way as EndWithFilter except that the first token is used to represent the whole word. We apply this filter after the EndWithFilter because the first part of a compound is usually a definitive term while the last part is usually a subject which reflects the main meaning of the compound tag.

Note that we do not replace or transform the compound to the standard term, but only use them as interpreters for semantic relation discovery.

3.4 Jargon Tags: Combination of Association Rules and Similarity Ranking

In this step, jargon tags are incorporated to the previously built ontological structure with a matcher using graph centrality in a similarity graph of tags[10]. Although jargon tags are also non-standard and cannot be recognized, the association rules show their relations with other common tags. Considering each jargon and its related standard tags as separate subset in vector spaces, the tag similarity graph for each subset is a subgraph where each tag is represented by a vertex and the cosine similarity measures the distance between them.

The incorporation considers each jargon tag as the central node of a subgraph. Then it adds each related standard tags in the subgraph. Based on the matcher between this jargon and its related standard tag, the jargon tag is incorporated to the structure. If there is more than one standard tag associated with the jargon tag, the tag with the highest cosine similarity index will have the priority. Association rules involving jargons usually have the jargon as the antecedent. Thus, the jargon tag will be considered as a child of its consequent in the rule. This incorporation repeats until all jargon tags have been connected with their related standard tags in the structure.

For example, a jargon tag "folksonomy" is associated with four standard tags, "tagging, plurality, social, ontology". Ranking by cosine similarity, the rule "folksonomy $\rightarrow$ tagging" was selected. Based on this match, folksonomy was incorporated to the ontological structure as a child of tagging.

4 Experimental Evaluation

The experiment was based on collections from two separate CTS, citeulike.org and flickr.com. The collection from Citeulike was crawled from using several keywords, including "science", "philosophy", "research". We got 30,769 rows of data, where each row represents a paper and a set of tags described by online users. Another dataset from Flickr was collected with flickr API, which consists of a set of callable methods for users, photos, photosets and other uniquely identifiable objects. We crawled the data using a narrow keyword "fruit" and collected 18,555 rows of data.

Pre-processing operations were performed to clean up the datasets. For dataset from Flickr, we only kept one record for each user because many users batch upload multiple photos with same tags. These repetitive tags will give us biased support count in association rules mining step and thus were excluded. Other common cleanup methods were applied to remove the tags called "no-tag" (a system generated tag for empty tag). We also removed objects with only one tag.

The descriptive statistics on the datasets after pre-processing is shown in Table 3:

Table 3. Statistics of collections used in experiment

	Collection	
	Citeulike	**Flickr**
Resources	30,769	18,555
After cleaning	25,937	6,462
Distinct tags	26,709	16,832
Users	4,068	6,462
Seed keywords	science, philosophy, research	fruit

4.1 Low Support Association Rules

There are three parameters, minimum support, confidence, and cosine similarity (denoted as *minsup*, *minconf*, and *mincos*) need to determine in our approach. We counted the number of *essential tags* with different *minsup* threshold and observe that most of the *essential tags* do not occur frequently (see Table 4). Moreover, the investigation of the initial association rule set reveals some interesting patterns of the cosine similarity. The value of similarity between pair of synonyms or sub-classes under same upper class tends to be high, sometimes close to 1. On the other hand the similarity value between a sub-class tag and its parent tag or upper class tag tends to be low. For instance, *food* is the parent of *beverage* in WordNet, the cosine similarity between *food* and *beverage* is low because *food* is a general term and it is associated with many other tags in the data set.

Thus the *mincos* was set to a relatively low value 0.2 to preserve the relations between upper and sub class tags and the relations between subclass tags. The *minsup* was set a very low value, 0.02% to include low-occurrence tags and reflect their relations. As usual, minconf was set to a 0.8, a relatively high value.

We observe that in total 152,372 rules are generated from citeulike at 0.02% minsup. These rules are significantly reduced to 24,025 by 0.2 cosine similarity and 0.8 confidence thresholds. Approximately 4,000 essential tags have been found. Table 5 also demonstrates the necessity of very low support threshold. In both these experiments, support 0.02% retains relations between around 4,000 essential tags. It only keeps relations between 300 essential tags if we increase the support threshold to 0.18%, a low support in traditional associational rule mining.

Table 4. Distribution of essential tags

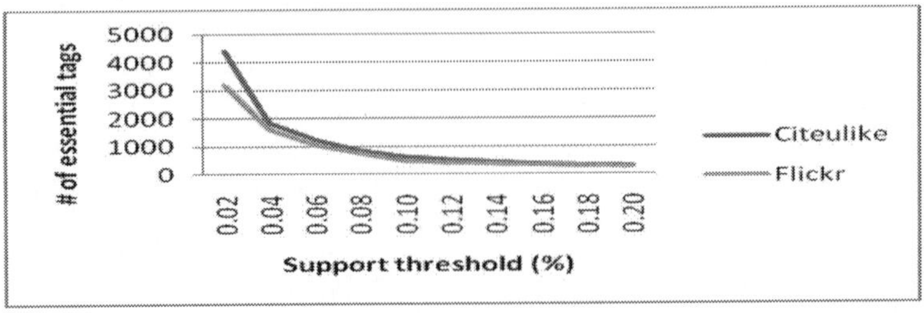

Table 5. Rules generated with 0.02% support, 80% confidence

Rules			Support	Confidence	Cosine	Accept?
folksonomy	→	tagging	1.59%	0.82	0.722	Y
macroeconomics	→	economics	0.09%	0.96	0.2671	Y
cyber-ethnography	→	ethnography	0.06%	1.00	0.2872	Y
asc	→	collaboratior	0.03%	1.00	0.172	N
final	→	social	0.04%	0.90	0.1679	N
seeking	→	information	0.03%	0.85	0.1605	N

Table 5 shows the effect of the three thresholds. It contains 6 randomly selected low support rules generated at support threshold 0.02% and confidence threshold 0.8. Low support value helps to preserve rarely occurred pairs while cosine similarity acts as a guard to exclude rules consisting of tag pairs not highly related. For example, the relation between *macroeconomics* and *economics* was revealed under low support threshold. On the other hand, although the confidence for the rule *final* →*social* is higher than 0.8, it was excluded because cosine similarity is less than *mincos*. If we set *minsup* higher than 0.18% or *mincos* higher than 0.3, both second and third rules will not be discovered and will not be included in the final ontological structure.

4.2 Ontological Structures and Evaluation

In this section, we present and evaluate the results. Taking the task-based evaluation approaches [26], we measure how far the extracted ontological structure will help to influence and improve the results of certain tasks including multi-dimensional view and cataloguing and indexing.

Multi-dimensional View: The result retrieved with the "fruit" keyword was successfully organized into several dimensions in our approach. In these concept dimensions, "*produce, plant, food,* and *color*" contain most sub-classes. In Fig 3, it shows that detailed sub-classes are organized into multiple level ontological structures.

We provide a subjective evaluation of our ontological structure (Fig 3) by comparing it to an ontology (Fig 4) extracted from sei.cmu.edu and cluster results from flickr.com (Fig 5).

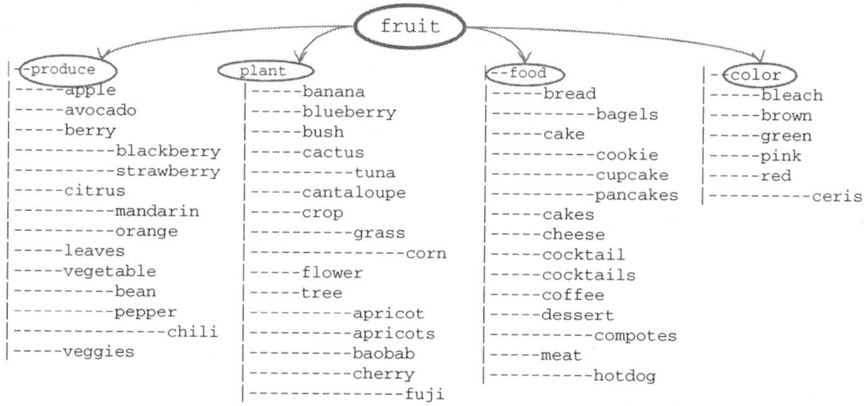

Fig. 3. A fragment output of "fruit" ontological structure, extracted from Flickr dataset

We observe that the terms from our result have a reasonable coverage and similar structure with the ontology in Fig 4. Such as the name and color of fruits like "berry, citrus, and red". The result also shows that our method produces more specific terms and additional levels than the ontology in Fig 4. For example, the "citrus" includes sub-classes, "orange" and "mandarin". However, our result does not provide enough property information for each term, such as flavor and seedless of strawberry in the ontology in Fig 4. The reason is that we currently only consider the hypernym relation from WordNet.

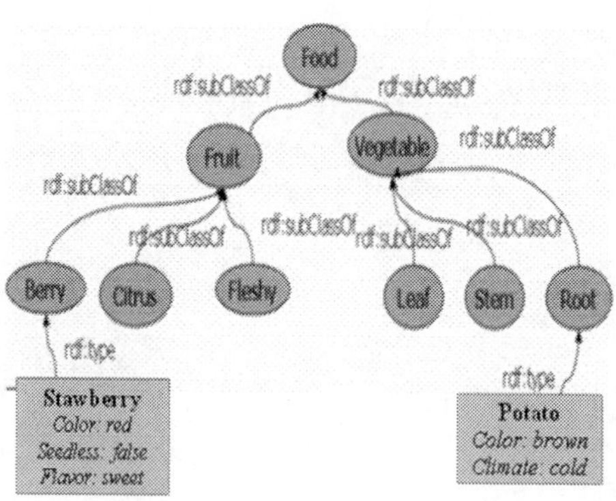

Fig. 4. An ontology of food[4]

Fig. 5. Fruit clusters[5] from Flickr

After that, we compared our result with clusters of query result using keyword "fruit" from flickr.com (Fig 5). There are three main clusters in the screenshot. First is "red, food, etc." Second is "yellow, banana, etc." The third is "nature, tree, plant, etc." Although we can see that the third cluster are mainly about nature and is right to be separated from other clusters, there is no significant difference between first and

[4] http://www.sei.cmu.edu/isis/guide/technologies/owl-s.htm
[5] http://www.flickr.com/photos/tags/fruit/clusters/

second clusters since most of them are names of fruits. Furthermore, tags like "food, yellow, red" are not distinguished correctly, and are mixed in the two clusters.

On the other hand, the terms related to fruit are clearly classified into four dimensions in our results in Fig 3. Furthermore, our structure provides detailed sub-classes in each dimension, for example, the "berry" is placed under the "produce" dimension and could be further navigated into "blackberry and strawberry".

Although terms like "fleshy" are missing in our structure, it is actually rarely used by users. By checking the photos annotated with "fleshy" in Flickr website, we found that only 0.00003% photos were tagged by "fleshy.

In short, the extracted ontological structure reflects the fruit domain knowledge well and organizes the related resources into several navigable dimensions.

Cataloguing and Indexing: Fig 6 illustrates a fragment of the experiment result in science domain from Citeulike. The related terms are organized into a 5-level ontological structure, which gives users an overview of knowledge in science domain. It also provides a conceptual framework for cataloguing and indexing the resources. For example, from general to specific, anthropology and biology are organized under science catalogue. Then, biology is further divided into genetics and neurobiology. The number beside each term shows how many papers are contained in the corresponding catalogue.

```
|-science  (762)
|------anthropology  (111)
|------------ethnography (128)
|------biology  (256)
|------------genetics  (154)
|------------------evolutionary-genomics (41)
|------------------evolutionary-proteomics (22)
|------------------genomics (250)
|------------------proteomics (127)
|------------neurobiology (41)
|------------------neuroscience (199)
|------------------------neurophysiology (24)
|------------sociobiology (26)
|------system_biology (6)
|------------sysbio (74)
|------cryptography (25)
|------economics (259)
|------------macroeconomics (21)
|------informatics (141)
|------ip (54)
|------mathematics (163)
|------------geometry (78)
|------------statistics (456)
|------medicine (105)
|------------toxicology (12)
|------------biomedicine (11)
```

Fig. 6. A fragment of ontological structure in science domain

We evaluated the catalogues manually and observe that compound and jargon terms have been appropriately incorporated at the correct hierarchical level, such as, "evolutionary-genomics", "evolutionary-proteomics" and "sociobiology" (see fig 6). In total, 1540 terms were incorporated into the ontological structure. Among those terms, 35.65% of them were standard terms and more than 64% were non-standard terms, including 36.17% compound and 28.18% jargon terms.

5 Conclusion and Future Work

In this paper, we have proposed an integrated bottom-up and top-down approach to extract ontological structures from collaborative tagging systems. Through the investigation into four kinds of word-formations (standard tags, jargon tags, compound tags, and nonsense tags) in folksonomies, our approach has produced promising initial results using two datasets from Flickr and Citeulike.

Though WordNet as an upper ontology resource contains a sufficiently wide range of common words, it does not cover special domain vocabulary and cannot reflect usage change. In CTS, many of the tags are in the form of jargon and compound terms. Mapping terms with WordNet ontology is obviously not enough to find the relationships among them. Thus, additional consideration was given to incorporate there terms into ontological structures by matching tags using association rule mining and token-based similarity. Rather than the clustering technique, association rule mining is a unsupervised data mining method to find interesting association between data sets. In this paper, we applied the association rules to find semantically related tags, which is the basis for further ontology building. Furthermore, we simplified the a priori algorithm to find 2-item set rules and introduced a new cosine coefficient, which significantly improved the efficiency in low support mining.

We observe that the ontological structures obtained could be enriched and deepened using larger tag datasets, other semantic relations provided by WordNet, and more specialized semantic lexical resources such as thesauri and subject-specific dictionaries. Additional work to represent the extracted ontologies in the web using RDF data format and SPARQL query language[6] will enable the integration with other web services, such as a collaborative ontology evolvement environment to reflect the knowledge and usage changes in CTS.

References

1. Auchard, E.: Flickr to map the world's latest photo hotspots. Reuters.com (2007)
2. Vander, T.: Folksonomy Coinage and Definition (2007), `http://www.vanderwal.net/folksonomy.html`
3. Euzenat, J., Shvaiko, P.: Ontology matching. Springer, Heidelberg (2007)
4. Wikipedia: Tag cloud. Wikipedia, The Free Encyclopedia (2009)
5. Suchanek, F.M., Vojnovic, M., Gunawardena, D.: Social tags: Meaning and Suggestions. In: ACM Conference on Information and Knowledge Management (CIKM 2008), pp. 223–232. ACM, Napa (2008)
6. Mathes, A.: Folksonomies-Cooperative Classification and Communication Through Shared Metadata. Computer Mediated Communication, LIS590CMC (2004)
7. Stuckenschmidt, H., Harmelen, F.V.: Information Sharing on the Semantic Web. Springer, Heidelberg (2005)
8. Al-Khalifa, S., Davis, C.: Measuring the Semantic Value of Folksonomies. Innovations in Information Technology (2006)
9. Mika, P.: Ontologies are us: A unified model of social networks and semantics. Web Semantics 5, 5–15 (2007)

[6] http://www.w3.org/TR/rdf-sparql-query/

10. Heymann, P., Garcia-Molina, H.: Collaborative Creation of Communal Hierarchical Taxonomies in Social Tagging Systems. Stanford InfoLab Technical Report (2006)
11. Schmitz, C., Hotho, A., Jäschke, R., Stumme, G.: Mining Association Rules in Folksonomies. In: Proceedings of the 10th IFCS Conference, Studies in Classification, Data Analysis, and Knowledge Organization (2006)
12. Schwarzkopf, E., Heckmann, D., Dengler, D., Kroner, A.: Mining the Structure of Tag Spaces for User Modeling. In: Workshop on Data Mining for User Modeling (ICUM 2007) (2007)
13. Specia, L., Motta, E.: Integrating Folksonomies with the Semantic Web. In: Franconi, E., Kifer, M., May, W. (eds.) ESWC 2007. LNCS, vol. 4519, pp. 624–639. Springer, Heidelberg (2007)
14. Damme, C.V., Hepp, M., Siorpaes, K.: FolksOntology: An Integrated Approach for Turning Folksonomies into Ontologies. In: Proc. of the ESWC Workshop Bridging the Gap between Semantic Web and Web (2007)
15. An, Y.J., Geller, J., Wu, Y.-T., Chun, S.A.: Automatic Generation of Ontology from the Deep Web. Database and Expert Systems IEEE 2007 (2007)
16. Laniado, D., Eynard, D., Colombetti, M.: Using WordNet to turn a folksonomy into a hierarchy of concepts. In: Proc. of 4th Italian Semantic Web Workshop, Italy (2007)
17. Schreiber, A.T.G., Dubbeldam, B., Wielemaker, J., Wielinga, B.: Ontology-Based Photo Annotation. IEEE Intelligent Systems (2001)
18. Schmitz, P.: Inducing Ontology from Flickr Tags. In: Proceedings of the Collaborative Web Tagging Workshop (WWW 2006), Edinburgh, UK (2006)
19. Agrawal, R., Imielinski, T., Swami, A.: Mining Association Rules between Sets of Items in Large Databases. In: Proceedings of the 1993 ACM SIGMOD international conference on Management of data (1993)
20. Agrawal, R., Srikant, R.: Fast Algorithms for Mining Association Rules in Large Databases. In: Proceedings of the 20th International Conference on Very Large Data Bases (1994)
21. Cohen, E., Datar, M., Fujiwara, S., Gionis, A., Indyk, P., Motwani, R., Ullman, J., Yang, C.: Finding interesting associations without support pruning. Transactions on Knowledge and Data Engineering 13, 64–78 (2001)
22. Spertus, E., Sahami, M., Buyukkokten, O.: Evaluating similarity measures: a large-scale study in the orkut social network. In: Proceedings of the 11th ACM SIGKDD intl. conference on Knowledge Discovery in Data mining, Chicago, USA (2005)
23. Salton, G., McGill, M.J.: Introduction to Modern Information Retrieval. McGraw-Hill, Inc., New York (1986)
24. Giunchiglia, F., Shvaiko, P., Yatskevich, M.: S-Match: an Algorithm and an Implementation of Semantic Matching. In: Bussler, C.J., Davies, J., Fensel, D., Studer, R. (eds.) ESWS 2004. LNCS, vol. 3053, pp. 61–75. Springer, Heidelberg (2004)
25. Wallace, M.: Jawbone (2007), `http://mfwallace.googlepages.com/jawbone.html`
26. Dellschaft, K., Staab, S.: Strategies for the evaluation of ontology learning. In: Buitelaar, P., Cimiano, P. (eds.) Ontology learning and population: Bridging the gap between text and knowledge. IOS Press, Amsterdam (2008)

Reducing Ambiguity in Tagging Systems with Folksonomy Search Expansion

Jeff Z. Pan, Stuart Taylor, and Edward Thomas

Dept. of Computing Science
Univ. of Aberdeen
Aberdeen, UK

Abstract. Search facilities are vital for folksonomy (or social tagging mechanism) based systems. Although these systems allow great malleability and adaptability, they also surfer from problems, such as ambiguity in the meaning of tags, flat organisation of tags and some degree of unstabilising factor on consensus about which tags best describe some certain Web resources. It has been argued that folksonomy structure can be enhanced by ontologies; however, as suggested by Hotho et al., a key question remains open: how to exploit the benefits of ontologies without bothering untrained users with its rigidity. In this paper, we propose an approach to address the problem of ambiguity in tagging systems by expanding folksonomy search with ontologies, which are completely transparent to users. Preliminary implementations and evaluations on the efficiency and the usefulness of such expansions are very promising.

1 Introduction

Web 2.0 applications are emerging providers of a new type of Web resources. On the one hand, more and more Web resources become available in popular folksonomy (or social tagging mechanism) based systems, such as Flickr[1], YouTube[2] and del.icio.us[3]. For example, in April 2008 alone, there are about 2.5 million photos geotagged[4] in Flickr. On the other hand, mashup applications combine Web resources from more than one folksonomy based system into various integrated experience [11]. These two observations indicate that search facilities are vital *both* within folksonomy based systems *and* across folksonomy based systems.

There are advantages and disadvantages of the tagging approach [9]. The folksonomy based systems allow great malleability and adaptability, and the simplicity of adding and using tags allows them to be more accessible to users than well defined (more complex) classification systems. However, these systems also surfer from problems [17], such as ambiguity in the meaning of tags and flat organisation of tags (see Section 3.1).

For example, in the results from a search on YouTube for "Focus" (the Dutch progressive rock band), only 3 of the top 20 items in the list are relevant to that band. The

[1] `http://www.flickr.com/`
[2] `http://www.youtube.com/`
[3] `http://del.icio.us/`
[4] Geotagging is the art of adding location information to things like a photo, according to Flickr.

L. Aroyo et al. (Eds.): ESWC 2009, LNCS 5554, pp. 669–683, 2009.

remaining 17 search results relate to alternative meanings of the word "focus". This simple example illustrates that ambiguity in the meaning of tags is a significant problem for domain specific folksonomy searches. Furthermore, for each of the 3 relevant results, the tags list for each result contained few other relevant tags than "Focus" and the name of the song in the video. In every case, information on at least of: the type of music, the album, the band members and other relevant information was omitted from the tag list.

Halpin et al. expressed a major concern of the folksonomy approach, i.e., whether or not in the folksonomy based system consensus about which tags best describing some certain Web resources can be stabilised, so that such best tags are used most often with the folksonomy based systems [9]. There could be a serious problem if no coherent categorisation scheme can emerge at all from coolaborative tagging. This is a crucial point because ranking techniques employed by major Web search engines today are, to a large extent, influenced by people's opinion of Web resources [10]. Although Halpin et al. concluded that tagging distributions *tend to* stabilise into power law distributions, the process of stabilisation (w.r.t. certain contexts) depends on the number of active users and to some extent on the temporal duration of the tagging process, according to their empirical study. Concrete predictions are not (yet) available on the number of active users and duration of time needed to complete the stabilisation process. The situation of searching across multiple folksonomy based system could be even worse, since different styles and environments provided by different systems could make it harder to reach stabilised consensus. Some might even argue that it is impossible to reach such Web scale stabilised consensus, as consensus from different systems might be conflicting with each other.

There has been significant interests on combining Semantic Web and Web 2.0 [3,6,12]. It has been suggested [10,17,1,4] that Semantic Web technologies could help improve the internal structure of folksonomies. Passant proposed to encourage users to link tags with concepts and individuals in ontologies [17]. However, as suggested by Hotho et al. [10], a key question remains open: how to exploit the benefits of ontology without bothering untrained users with its rigidity.

In this paper, we propose a novel approach to using ontology entailments to expand folksonomy search, without bothering untrained users to use ontologies directly. More precisely, the paper makes the following major contributions:

1. Our approach allows Web application developers to exploit the benefits of ontology without bothering untrained users with its rigidity. *Ontology transparency* is achieved by (i) automatic associating keywords with classes, properties and individuals in domain ontologies (Section 4.1) and (ii) relating the search keyword(s) with keywords based on the structure of domain ontologies (Section 4.2).
2. We propose a general framework of ontology-based folksonomy search expansion methods for reducing ambiguity (Section 4), based on some use cases and observations on search expansions (Section 3). For example, a search for "focus" could be expanded to include (some) album titles, songs titles, live performance venues, as well as keywords related to the domain in general. Hence, more meaningful results are returned to users than making use of the tagging systems directly. An interesting feature of the proposed methods is that all the required ontology entailments

are included in OWL2 QL semantic approximations [15] of given domain ontologies (Section 2). This suggests that theoretically the proposed methods can be very efficient.

3. We report the implementations in our ontology infrastructure of the proposed automatic keyword-ontology association and the proposed ontology-based folksonomy search expansion methods (Section 5). Preliminary evaluations indicate that these expansion methods are efficient even for some very large ontologies (Section 6).
4. We present a semantic mashup application, called MusicMash2, to demonstrate how the above ontology infrastructure could enhance current Web 2.0 application and experience (Section 5.2). MusicMash2 provides music domain-specific search features to retrieve data from folksonomy based systems, offering search by genre, artist, release and song, etc. We show the usefulness of our proposed methods by comparing MusicMash2 against using some folksonomy based system directly. To the best of our knowledge, this is the first published application which shows both (i) that expressive ontology languages like OWL DL can be useful for Web applications and (ii) that querying service for OWL DL can be scalable enough to support such applications.

2 Preliminary

2.1 Folksonomies

Informally speaking, a folksonomy describes users, resources and tags in such a way that users assign tags to resources [10]. Here we follow Mika's [13] and Hotho et al.'s [10] formal model of folksonomies.

Formally, a folksonomy is a tuple $\mathbf{F} = \langle \mathcal{U}, \mathcal{T}, \mathcal{R}, A, \Theta \rangle$, where $\mathcal{U}$, $\mathcal{T}$, $\mathcal{R}$ are finite sets of users, tags and Web resources, resp., $A \subseteq \mathcal{U} \times \mathcal{T} \times \mathcal{R}$ is called tag assignments, Θ is a set of tag relations; i.e., for each member $\prec$ of Θ, $\prec \subseteq \mathcal{U} \times \mathcal{T} \times \mathcal{T}$ is a tag relation. As Mika [13] indicated, tag relations are often thesaurus relations, such as those specified in the SKOS specification[5]. In this paper, we consider tag relations such as 'narrower' $\prec_n$, 'broader' $\prec_b$ and 'related' $\prec_r$. For example, the following assignment $\langle u1, \text{"}cat\text{"}, r1 \rangle$ says user u1 uses "cat" to tag the Web resource r1; the following tag relation $\langle \text{"}cat\text{"} \prec_n^{u1} \text{"}animal\text{"} \rangle$ says that the tag "cat" is narrower than the tag "animal" for user u1. There are some properties of the above three tag relations:

1. If $\langle tag1 \prec_n^u tag2 \rangle$, then $\langle tag2 \prec_b^u tag1 \rangle$, and vice versa.
2. If $\langle tag1 \prec_r^u tag2 \rangle$, then $\langle tag2 \prec_r^u tag1 \rangle$.

It should also be noted that here we simply present a formal model of tag relations. How to extract such relations from folksonomy systems goes beyond the scope of the paper.

2.2 Ontologies and Their Semantic Approximations

As a 'representation of a shared conceptualisation' of a specific domain [7,21], ontology is regarded as a key component of the Semantic Web. Practically, ontologies are used to

[5] http://www.w3.org/TR/swbp-skos-core-spec/

capture the intended meaning of a set of shared and common vocabulary of a specific domain, including important concepts (such as Artist, Album and Song), properties (such as hasAlbum and hasSong) and individuals (objects). Recently, W3C standardised the OWL Web Ontology Language [2] in its Semantic Web Activity.

It has been argued that the Web needs lightweight ontologies [13]. Recently, Calvanese et al proposed the OWL2 QL (also called DL-Lite) [5] ontology language which can express most features in UML class diagrams but still has a low reasoning overhead (worst case polynomial time, compared to worst case exponential time in the case of most widely used Description Logics-based ontology languages). A OWL2 QL ontology ($\mathcal{O}$) is a set of axioms of the following forms:

1. concept inclusion axioms: $B \sqsubseteq C$ where B is a basic concept $B := \mathsf{A} \mid \exists R \mid \exists R^-$ and C is a general concept $C := B \mid \neg B \mid C1 \sqcap C2$ (where A denotes an named concept and R denotes a named property);
2. property inclusion axioms: $R1 \sqsubseteq R2$, where $R1, R2$ are named properties;
3. individual axioms: $A(\mathsf{a}), R(\mathsf{a}, \mathsf{b})$ where a and b are named individuals.

Straccia later proposed fuzzy DL-Lite [19]. Pan et al. [14] propose a large family of fuzzy query languages and corresponding query answering algorithms for fuzzy OWL2 QL. The TrOWL[6] ontology reasoning infrastructure consists an ontology repository, where users can submit and share ontologies and some scalable ontology query engines, including Quill for OWL2 QL and a fuzzy OWL2 QL, supporting a large family of fuzzy queries proposed by Pan et al. [14]. Preliminary evaluations [14] have shown that Quill operates efficiently and scalably, being able to handle millions of individuals.

Pan and Thomas [15] propose a general approach to semantically approximating of ontologies, i.e., to approximate a source ontology in an expressive ontology langauge (such as OWL DL) into its (lowest) upper bound in a lightweight ontology language (such as OWL2 QL). This process allows conjunctive querying against OWL DL ontologies, after some careful query rewriting, to be performed within a database, while still giving soundness guaranteed results for all queries, and sound and complete (w.r.t. the source ontology) results for database-style queries. Preliminary evaluation [15] shows that the semantic approximation based OWL DL query engine in TrOWL can handle millions of individuals.

3 Motivations and Use Cases

3.1 Limitations of Folksonomy Searching

Let us first revisit the formal model of folksonomy to illustrate some limitations of tagging from the aspect of search. Given a folksonomy based system $\mathcal{S}$ that uses the folksonomy $\mathbf{F} = \langle \mathcal{U}, \mathcal{T}, \mathcal{R}, A, \Theta \rangle$, there are three main limitations of tagging regarding search in $\mathcal{S}$ [17]:

– Tag variation (synonymy). Consider the following two assignments: $\langle u1, \text{“}cat\text{”}, r1 \rangle$, $\langle u2, \text{“}kitty\text{”}, r2 \rangle$ (users $u1$ and $u2$ use “cat” and “kitty” to tag two different resources $r1$ and $r2$, resp.). Even if $u1$ and $u2$ have the same meaning in mind for the two different tags, the system $\mathcal{S}$ is not able to relate these tags in a general sense.

[6] http://trowl.eu/

- Tag ambiguity (polysemy). Consider the following two assignments: $\langle u1, \text{“}ma\text{”}, r3\rangle$, $\langle u2, \text{“}ma\text{”}, r4\rangle$ (users $u1$ and $u2$ use “ma” to tag two different resources $r3$ and $r4$). Even if $u1$ and $u2$ use “ma” to express different ideas, the system $\mathcal{S}$ would still return both $r3$ and $r4$ when another user (who might have a totally different idea of “ma”) searches for “ma”.
- Flat organisation of tags. Consider the following two assignments: $\langle u1, \text{“}cat\text{”}, r5\rangle$, $\langle u2, \text{“}animal\text{”}, r6\rangle$ (users $u1$ and $u2$ use “cat” and “animal” tag two different resources $r5$ and $r6$, resp.). When another user searches for “animal”, the system $\mathcal{S}$ is able to return $r6$, but not $r5$.

3.2 Why Can Ontologies Help

Here are two observations of the above three limitations:

1. The issues of tag variation and flat organisation of tags indicate that the tag relations in Θ are not useful enough, because these relations are with respect to some user (or user group) only, rather than in the system level (all users).
2. The issue of ambiguity indicates we need to be able to associate tags with some contexts, so as to reduce the ambiguity.

Observation 2 clearly suggests ontology could be helpful, as they contain domains-specific knowledge. Observation 1 suggests that ontological structures should (at least) be used in the system level. Now the question is: *How can ontologies help?* In the next sub-section, we will illustrate this with some use cases.

3.3 Use Cases

This section outlines some prominent examples of ambiguity in tagging systems and how our Folksonomy Search Expansion approach could be applied in each case. The overall goal in each use case is to increase the precision of the search results. These use cases assume that an appropriate domain ontology is available in order to identify the correct individuals for expansion. The **expansion(s)** identified in each use case are shown in Table 1.

Music Videos

- *Example Search Term(s):* “Focus”
- *Description:* A user wants to search YouTube for videos relating to a specific music artist. $Focus$ is an ambiguous search term on YouTube and does not commonly relate to the artist Focus. However, YouTube users typically tag videos with both the artist name and song title. By experimentation, it can be shown that using the combination of artist name and song title as the search terms is sufficient to increase precision significantly (see Section 6.1).
- *Example Expansion:* The expansion concatenates the original search term $Focus$ with the title of each song created by Focus, resulting in a set of searches.
- *Example Output:* { “Focus Hocus Pocus”, “Focus Silvia”, ... }.
- **Expansion:** Individual Property Value Expansion (IPVE) with the property specified as $createdSong$ (see Table 1).

Table 1. Outline of Expansion Methods (where i represents individuals related to the search keyword(s))

Name	Description
ICE	The search terms are expanded by concatenating keywords relating to any class C, such that $C(i)$ entailed by the target ontology.
IPE	The search terms are expanded by concatenating keywords relating to any property P, such that the relation $P(i, \_)$ entailed by the target ontology.
IPVE(P)	The search terms are expanded by concatenating keywords relating to the individual q, using the specified property P such that the relation $P(i, q)$ entailed by the target ontology.

Music Images

- *Example Search Term(s):* "Metallica"
- *Description:* A user wants to search Flickr for images relating to a specific music artist. However, the search term $Metallica$ returns many unrelated results [7]. Adding further search terms increases the number of relevant results returned. However, we cannot reuse the same expansion from the previous use case, since users do not typically tag images with particular song names. In this example, the names of band members and music-related keywords (e.g., Live, Concert, Music etc.) are added to increase the search precision. Band member names would typically be found in the ABox of the ontology and music-related keyword would typically be found in the TBox.
- *Example Expansion (Band Members):* The expansion concatenates the original search term $Metallica$ with the name of each band member of Metallica.
- *Example Output (Band Members):* { "Metallica James Hetfield", "Metallica Lars Ulrich", ... }.
- **Expansion (Band Members):** Individual Property Value Expansion (IPVE) with the property specified as $hasBandMember$.
- *Example Expansion (Music-related Keywords I):* The expansion concatenates the original search term $Metallica$ with keywords related to the artist class in the ontology. Note that this particular expansion depends on the quality of annotations in the target ontology. In this example we assume that $Metallica$ is an instance of classes such as $LiveBand$, $MusicArtist$ etc.
- *Example Output (Music-related Keywords I):* { "Metallica Live", "Metallica Concert", "Metallica Music", ... }.
- **Expansion (Music-related Keywords I):** Individual Class Expansion (ICE)
- *Example Expansion (Music-related Keywords II):* Depending on the structure of the target ontology, the expansion can also be performed by concatenating the original search term $Metallica$ with the keywords related to the individual's properties. For example, the property $performedLiveAt$ contains the keywords 'performed' and 'live'.
- *Example Output (Music-related Keywords II):* { "Metallica Live", "Metallica Performed", ... }.
- **Expansion (Music-related Keywords II):** Individual Property Expansion (IPE).

[7] Flickr's tag search returned 3 out of 25 relevant results (12-Dec-2008).

Film Clips

- *Example Search Term(s):* "The Wicker Man"
- *Description:* A user wants to search YouTube for videos clips related to the 1973 film The Wicker Man. However, $The\ Wicker\ Man$ could refer to the Iron Maiden song The Wicker Man or the 2006 remake of the original film. Adding some context to the original search terms can help disambiguate between the various different meanings of $The\ Wicker\ Man$. In this example, including the names of associated people such as the director, writer and staring actors increases the relevance of the search results.
- *Example Expansion (associated people):* The expansion concatenates the original search terms $The\ Wicker\ Man$ with the names of staring actors, writer and director, resulting in a set of searches.
- *Example Output:* {"The Wicker Man Edward Woodward", "The Wicker Man Christopher Lee", "The Wicker Man Anthony Shaffer", ...}.
- **Expansion:** Individual Property Value Expansion (IPVE) with the property specified as $hasAssociatedPerson$, such that in the target ontology $hasAssociated\ Person$ is a super-property of the $hasActor$, $hasDirector$ and $hasWriter$ properties.

Table 1 outlines the three general Folksonomy Search Expansion methods motivated by the use cases. Each expansion method firstly identifies individuals i that relate to the original search keyword(s) provided by the user. The output from each expansion method is a set of expanded search terms constructed by concatenating each result of the expansion to the original search term. In the next section we will formally define the proposed expansions.

4 Folksonomy Search Expansion

The section presents the details of our Folksonomy Search Expansion approach to address the problem of ambiguity in tagging systems. Firstly we show how keywords are associated with ontologies. We then present a general framework of ontology-based folksonomy search expansion methods based on the use cases in Section 3.3.

4.1 Automatic Keyword-Ontology Association

Keywords appear in annotations and datatype properties of ontologies. These keywords can be associated with concepts, properties and individuals (in ontologies), with scores n based on ranking factors[8]. These scores in turn can be used to calculate the $tf{\cdot}idf$ [18] for each keyword, and normalise them using a sigmoid function such as the one shown in (1) to a degree between 0 and 1.

$$w(n) = \frac{2}{1.2^{-n} + 1} - 1 \tag{1}$$

Formally, for each ontological element i (concept, property, object) in a given ontology $\mathcal{O}$, there exists a set of associated keywords $\mathsf{K}(i)$ and each keyword $k \in \mathsf{K}(i)$ associates to i with a degree $d(i,k) \in (0,1]$. Similarly, each keyword k is associated with a set of classes $\mathsf{C}(k)$, properties $\mathsf{P}(k)$ and individuals $\mathsf{O}(k)$.

[8] http://www.seomoz.org/article/search-ranking-factors

4.2 Ontology-Based Folksonomy Search Expansions

Informally, the main ideas of ontology-based folksonomy search expansion is to make use of the (implicit or explicit) structure of ontologies to relate keywords with other keywords. Accordingly, a search expansion $\mathsf{E}(k, expansionMethod)$ is a function that takes a tag (keyword) k and the expansion method as inputs and returns a set of keywords. Now we formally define the search expansions identified in Section 3.3.

Individual Property Value Expansion (IPVE) returns the set of associated keywords of all individuals that relate (via a given property P) to all the individuals in $\mathsf{O}(k)$.

$$\mathsf{E}(k, IPVE(P)) = \bigcup_{\forall i \in \mathsf{O}(k)} \; \bigcup_{\forall q: \mathcal{O} \models P(i,q)} \mathsf{K}(q).$$

Individual Class Expansion (ICE) returns the set of associated keywords of all classes of all the individuals in $\mathsf{O}(k)$.

$$\mathsf{E}(k, ICE) = \bigcup_{\forall i \in \mathsf{O}(k)} \; \bigcup_{\forall C: \mathcal{O} \models C(i)} \mathsf{K}(C).$$

Individual Property Expansion (IPE) returns the set of associated keywords of all properties that relate to all the individuals in $\mathsf{O}(k)$.

$$\mathsf{E}(k, IPE) = \bigcup_{\forall i \in \mathsf{O}(k)} \; \bigcup_{\forall P: \mathcal{O} \models P(i,\_)} \mathsf{K}(P).$$

As we consider entailments in the given ontology $\mathcal{O}$, the computational cost could be expensive, since entailment problem in OWL DL is NEXPTIME-complete. It is important to point out that, all the required ontology entailments in the above expansion methods are included in the DL-Lite semantic approximation of a given OWL DL ontology. This suggests that theoretically, by making use of semantic approximations, the proposed methods can be scalable. This is confirmed by our evaluation presented in Section 6. Furthermore, if the set of tags returned by search expansions is too large, one could apply some strategies to keep it small, such as by setting a threshold for the association degrees. A typical scenario where our approach could be used is in domain-specific mashup application. In such a system the developer should tailor the exact expansion used to their own ontology, application domain and target tagging system.

5 Implementations

This section describes how the proposed approach is implemented in our ontology infrastructure and illustrates how to make use of the ontology infrastructure in a semantic mashup application.

5.1 Ontology Infrastructure

The proposed automatic keyword-ontology association and the proposed ontology-based folksonomy search expansion methods are implemented in TrOWL the ontology reasoning infrastructure (which is made use of by ONTOSEARCH2 [16,20] the ontological search engine for entailment-plus-keyword search) and Taggr the ontology-based folksonomy search optimiser, respectively.

A useful feature of the TrOWL repository is that it automatically associates keywords (in values of annotation properties as well as implicit metadata in target ontologies) with concepts, properties and individuals in the ontologies. Default annotation properties include the `rdfs:label`, `rdfs:comment`, `rdfs:seeAlso`; we also define the `dc:title`, `dc:description`, and `foaf:name` properties (from Dublin Core[9] and FOAF[10]) as annotation properties. Implicit metadata is drawn from the namespace and ID of each artifact in the ontology. When users upload their ontologies, TrOWL computes the semantic approximation of the ontologies as described in [15] and computes keyword-ontology association as presented earlier in the paper.

Taggr provides an ontology-enabled common interface for folksonomy based systems[11]. It stores a basic ontology (which we refer to as the "tagging database") in TrOWL, capturing the relationships between users, tags and resources in the folksonomy based systems it supports. It provides the functionality of gathering resources and their related tags from the systems that it supports, and then populate them to its tagging database from time to time. Furthermore, Taggr allows users to specify which search expansion method(s) and which reference ontology(-ies) to use for the expansion. The extended searches are evaluated against its tagging database. As the ontological constraints needed for the proposed search expansion require only the expressive power of OWL2 QL, Taggr can make use of the semantic approximation(s) of the reference OWL DL ontology(-ies) for all the entailment checking. Due to the logical properties of semantic approximation, TrOWL can provide sound and complete results for all the needed entailment checking. As mentioned in Section 4, all the proposed search expansion methods are available for application developers to choose, which allows search expansions to be tailored to the nature of tagging in different domains and types of resources that are to be retrieved.

Fig. 1. MusicMash2 Alpha: http://www.musicmash.org/

[9] http://dublincore.org/

[10] http://www.foaf-project.org/

[11] Taggr currently supports YouTube and Flickr.

5.2 Case Study: MusicMash2

MusicMash2 is a semantic mashup application (see Figure 1), which is intended to integrate music-related content from various folksonomy based tagging systems and music meta-data Web services.

As discussed in Section 3.1, search both within and across folksonomy based systems is an open problem. A naive approach to folksonomy search, such as those provided by most tagging systems[12], results in unacceptable precision in domain specific searches (see Section 6.1). MusicMash2 addresses this problem by making use of the Folksonomy Search Expansion methods provided by Taggr. More specifically, since the music content of interest to MusicMash2 can be naturally expressed as a populated domain ontology. Currently MusicMash2 addresses the problem of ambiguity when search for music videos related to music artists, albums and tracks. Instance Property Value Expansion (IPVE) is chosen as the search expansion method for MusicMash2. For example, when searching for videos relating to a specific artist, MusicMash2 will call Taggr to make the following expansion, $\mathsf{E}(artistName, IPVE(\text{foaf:made}))$, i.e., to search with some artist name ($artistName$) for everything related to it via the foaf:made property. The goal of this expansion is to reduce the ambiguity of the original search term $artistName$ by providing a set of $artistName, songTitle$ keyword sets.

To use the IPVE search expansion method, developers of MusicMash2 need to have a domain ontology on music, populated with data needed by MusicMash2. Firstly, the Music Ontology[13] is chosen here as it provides the main classes and properties for describing music on the Web. Secondly, to populate the Music Ontology, MusicMash2 makes use of Web services which provide needed information (each in their own proprietary format), such as, MusicBrainz[14], Last.fm[15] and DBpedia[16]. MusicMash2 maps the data from each music meta-data Web service to the standardised Music Ontology format and submits the populated ontology to ONTOSEARCH2. Such populated ontology is exploited by the IPVE search expansion method. Since it is stored in ONTOSEARCH2, the populated ontology can also be reused by users of ONTOSEARCH2.

6 Evaluation

In this section, we present our evaluations (on usefulness and scalability) of Individual Property Value Expansion (IPVE) to address the problem of ambiguity illustrated by the Music Video use case in Section 3.3. We have used our case study application, MusicMash2, to provide a basis to refine the implementation of this particular expansion method. For this reason, we believe that an evaluation of IPVE provides a good practical illustration of how a Folksonomy Search Expansion method can perform in a real-world application.

[12] YouTube Developer API: `http://www.youtube.com/dev/`
[13] `http://www.musicontology.com/`
[14] `http://www.musicbrainz.org/`
[15] `http://www.last.fm/`
[16] `http://dbpedia.org/`

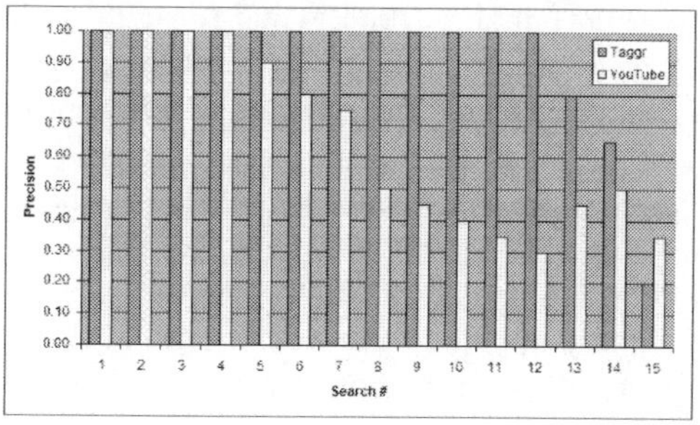

Fig. 2. Result set precision for *top k* search

6.1 Usefulness of Search Expansions

Based on our case study (see Section 5.2), we evaluated the usefulness of our approach by performing identical *top k* searches for music artists with YouTube and Taggr using the IPVE search expansion. We provide this evaluation to show how our system deals with ambiguity in the music domain, which we feel is a typical application of our Folksonomy Search Expansion approach. The set of 15 search terms (see Table 2) used in the evaluation have been chosen to evaluate how well Taggr with IPVE search expansion handles ambiguity in the music domain when compared with the standard YouTube search facilities. We have also included some less ambiguous artist names such as Radiohead and Metallica for comparison.

First of all, we performed *top k* searches using each system to retrieve the top 20 results. We then classified each result as either relevant or irrelevant with respect to the meaning of the original search. In the case of this evaluation, a result was classified as relevant if it featured the specified artist. The precision of each result set (see Figure 2) was calculated using the function (2); where n_{tp} is the number of relevant results and n_{fp} is the number of irrelevant results in the result set.

$$P(n_{tp}, n_{fp}) = \frac{n_{tp}}{n_{tp} + n_{fp}} \tag{2}$$

To prepare for this evaluation, Taggr's tagging database was populated with the top 500 results from YouTube for each of the search terms in Table 2. These 500 results typically included both music and non-music related content. Taggr IPVE made use of a populated Music Ontology for search expansion, with the role specified as foaf:made. More precisely, the expansion $\mathsf{E}(artistName, IPVE(\text{foaf:made}))$ was used, where $artistName$ is the full name of the artist used as the search term.

Fig. 2 shows that in all but one search, Taggr outperformed YouTube in terms of precision. YouTube returned good results for unambiguous artist names but poor results for ambiguous names. In searches #1 - 4 both YouTube and Taggr return results with perfect precision with these unambiguous artist names. With the more ambiguous artist names in searches #5 - 12 YouTube preforms worse, while Taggr retains perfect precision. In searches #13 and #14, Taggr loses some precision for the highly ambiguous artist names "Yes" and "Kiss". In the final search, #15, Taggr performs slightly worse than YouTube, by 15%. Our initial reaction is that for highly ambiguous artist names and song titles, the artist name plus song title combination used by IPVE is less effective than YouTube's built in search facilities. A simple approach to addressing this issue would be to attempt to rank the results by examining the complete set of tags for a resource and checking how many are actually matched by the keywords generated in the search expansion.

This evaluation is intended to show how our approach performs in terms of reducing ambiguity in a real-word system. We have chosen the IPVE search expansion to illustrate the approach, since this particular expansion method seems to be most widely applicable (see Section 3.3).

Table 2. Artist names used in the evaluation

Search #	Search Term	Search #	Search Term
1	The Beatles	9	Thursday
2	Porcupine Tree	10	Pelican
3	Radiohead	11	Focus
4	Metallica	12	Isis
5	Eels	13	Yes
6	Doves	14	Kiss
7	Finch	15	Jet
8	Strung Out		

6.2 Scalability of Search Expansions

First of all, it should be noted that the scalability of this application depends on the performance of TrOWL. Several evaluations [15,14] of TrOWL have been made using the Lehigh University Benchmark [8], and these have shown that TrOWL is scalable for large ontologies, containing of millions of individuals.

We have evaluated the performance of IPVE in Taggr by using two different sized datasets. More specifically, Taggr was used to perform the following search expansion $\mathsf{E}(artistName, IPVE(\text{foaf:made}))$, i.e., search with some artist name $(artistName)$ for everything related to it via the foaf:made property. The evaluation was carried out using a typical desktop computer (see Table 3).

This evaluation relies on a populated Music Ontology being present in ONTOSEARCH2. MusicMash2 was used to prepare the data for the evaluation. A list of artists was created by selecting the most popular artists from the Audioscrobbler web services (using the example user profile from the API documentation[17]). The list was then expanded by including the top 50 similar artists for each of the artists. MusicMash2 then performed a batch process to retrieve information relating to each artist (including other artists) from MusicBrainz[18], DBpedia[19] and Audioscrobbler[20] web services. This information was used to populate the two versions of the Music Ontology.

[17] http://ws.audioscrobbler.com/1.0/user/RJ/topartists.xml

[18] http://www.musicbrainz.org/

[19] http://www.dbpedia.org/

[20] http://ws.audioscrobbler.com/

Table 3. Specifications of test computer

Operating System	Windows XP Pro
CPU	Intel Pentium 4 3.0GHz Dual Core
RAM	2Gb
Hard Disk	80Gb SATA
JVM	Sun JVM version 1.6
Database	PostgreSQL 8.2

Table 4. Individual Property Value Expansion (IPVE) Performance

$N_{Artists}$	N_{Albums}	N_{Songs}	N_{Axioms}	T_{IPVE} (ms)
1303	2,531	30,182	$\sim$1,000,000	155
5521	11,336	136,040	$\sim$4,500,000	2036

Table 4 shows the performance of the system working with a small dataset of 1,300 artists (30,000 total individuals) and a larger dataset of 5,500 artists (136,000 total individuals). The larger dataset results in an ontology containing over 4,500,000 ground axioms. It is clear to see that further evaluations with larger datasets will need to be carried out to determine how this approach scales with extremely large datasets. Given the time (about 2 seconds for the larger dataset, see the T_{IPVE} column in Table 4) used for the IPVE search expansion, we believe that with a dedicated database server and further optimisations of our prototype implementation, this application will scale effectively over ontologies containing billions of axioms.

7 Discussion and Outlook

How to apply Semantic Web technologies to improve folksonomy based systems and social networks has been a pressing issue for the Semantic Web community [13]. In this paper, we have addressed the problem of ambiguity in tagging systems with a general framework of Folksonomy Search Expansion methods, based on ontology entailments, showing that different expansion methods can be useful in potentially different situations. In particular, we use a semantic mashup application, i.e. MusicMash2, to demonstrate the use of the IPVE search expansion method and evaluate its usefulness and scalability. The results of our preliminary evaluation indicate that the proposed technique can handle very large ontologies and allows folksonomies to be searched more effectively, returning more relevant results for folksonomy search than the keyword based approach.

By using our ontology infrastructure to expand folksonomy search transparently for users, we have solved the open problem of how to exploit the benefits of ontology to improve folksonomy search, without bothering untrained users with its rigidity. This means that users, such as those of MusicMash2, can exploit the benefits of ontology enhanced folksonomy search without the knowledge of existing of neither the ontology infrastructure nor the related ontologies. Interestingly, for all the Folksonomy Search Expansion methods, the required ontology entailments are within the expressive power of OWL2 QL, which allows us to make use of OWL2 QL semantic approximations [15] of given domain ontologies. This means that, although the domain ontologies can very complicated, the Folksonomy Search Expansion methods only require the OWL2 QL approximations of such ontologies. It has been argued that the Web needs lightweight ontologies [13]. Our work suggests that the Web needs lightweight representation/approximations of (potentially complicated but still useful) ontologies.

While this paper addresses the problem of ambiguity in tagging systems, we believe that our Folksonomy Search Expansion approach can be extended to address the lack of structure and tag variation limitations. For example, users searching for a high level concept such as *animal* may find that resources in tagging systems are not frequently tagged with such a high level concept. To address the problem of search across tagging systems, different expansions could be used for each target tagging system. For example, in Section 3.3 we use different expansions to search YouTube and Flickr for resources related to music artists. We are currently working on implementing and evaluating these extensions to our approach and expect that they should have similar benefits to our ambiguity approach presented in this paper.

References

1. Ankolekar, A., Krotzsch, M., Tran, T., Vrandecic, D.: The two cultures: mashing up web 2.0 and the semantic web. In: Proc. of the 16th international World Wide Web conference (WWW 2007), pp. 825–834 (2007)
2. Bechhofer, S., van Harmelen, F., Hendler, J., Horrocks, I., McGuinness, D.H.L., Patel-Schneider, P.F., Stein, L.A. (eds.): OWL Web Ontology Language Reference, Feburary (2004), `http://www.w3.org/TR/owl-ref/`
3. Benjamins, R.V., Davies, J., Baeza-Yates, R., Mika, P., Zaragoza, H., Greaves, M., Gomez-Perez, J.M., Contreras, J., Domingue, J., Fensel, D.: Near-term prospects for semantic technologies. IEEE Intelligent Systems 23(1), 76–88 (2008)
4. Bhogal, J., Macfarlane, A., Smith, P.: A review of ontology based query expansion. Information Processing & Management (4), 866–886 (2007)
5. Calvanese, D., De Giacomo, G., Lenzerini, M., Rosati, R., Vetere, G.: DL-Lite: Practical Reasoning for Rich DLs. In: Proc. of the DL 2004 Workshop (2004)
6. Greaves, M.: Semantic web 2.0. IEEE Intelligent Systems 22(2), 94–96 (2007)
7. Gruber, T.R.: Towards Principles for the Design of Ontologies Used for Knowledge Sharing. In: Guarino, N., Poli, R. (eds.) Formal Ontology in Conceptual Analysis and Knowledge Representation, Deventer, The Netherlands. Kluwer Academic Publishers, Dordrecht (1993)
8. Guo, Y., Pan, Z., Heflin, J.: LUBM: A Benchmark for OWL Knowledge Base Systems. Journal of Web Semantics 3(2), 158–182 (2005)
9. Halpin, H., Robu, V.n., Shepherd, H.: The complex dynamics of collaborative tagging. In: Proc. of the 16th International World Wide Web Conference (WWW 2007), pp. 211–220 (2007)
10. Hotho, A., Jäschke, R., Schmitz, C., Stumme, G.: Information retrieval in folksonomies: Search and ranking. In: Sure, Y., Domingue, J. (eds.) ESWC 2006. LNCS, vol. 4011, pp. 411–426. Springer, Heidelberg (2006)
11. Jackson, C., Wang, H.J.: Subspace: secure cross-domain communication for web mashups. In: Prec. of 16th International World Wide Web Conference (WWW 2007), pp. 611–620 (2007)
12. Lassila, O., Hendler, J.: Embracing "web 3.0". IEEE Internet Computing 11(3), 90–93 (2007)

13. Mika, P.: Ontologies are us: A unified model of social networks and semantics. In: Gil, Y., Motta, E., Benjamins, V.R., Musen, M.A. (eds.) ISWC 2005. LNCS, vol. 3729, pp. 522–536. Springer, Heidelberg (2005)
14. Pan, J.Z., Stamou, G., Stoilos, G., Taylor, S., Thomas, E.: Scalable Querying Services over Fuzzy Ontologies. In: Proc. of the 17th International World Wide Web Conference (WWW 2008) (2008)
15. Pan, J.Z., Thomas, E.: Approximating OWL-DL Ontologies. In: Proceedings of the 22nd Conference on Artificial Intelligence (AAAI 2007), pp. 1434–1439 (2007)
16. Pan, J.Z., Thomas, E., Sleeman, D.: ONTOSEARCH2: Searching and Querying Web Ontologies. In: Proc. of WWW/Internet 2006, pp. 211–218 (2006)
17. Passant, A.: Using Ontologies to Strengthen Folksonomies and Enrich Information Retrieval in Weblogs. In: Proc. of 2007 International Conference on Weblogs and Social Media (ICWSM 2007) (2007)
18. Salton, G., McGill, M.J.: Introduction to modern information retrieval. McGraw-Hill, New York (1983)
19. Straccia, U.: Answering vague queries in fuzzy DL-Lite. In: Proc. of the 11th International Conference on Information Processing and Management of Uncertainty in Knowledge-Based Systems (IPMU 2006), pp. 2238–2245 (2006)
20. Thomas, E., Pan, J.Z., Sleeman, D.: ONTOSEARCH2: Searching Ontologies Semantically. In: Proc. of OWL: Experiences and Directions (OWL-ED 2007) (2007)
21. Uschold, M., Gruninger, M.: Ontologies: Principles, Methods and Applications. The Knowledge Engineering Review (1996)

Ontology-Based Service Discovery Front-End Interface for GloServ

Knarig Arabshian[1], Christian Dickmann[2], and Henning Schulzrinne[3,⋆]

[1] Alcatel-Lucent Bell Labs
[2] VMWare
[3] Dept of Computer Science, Columbia University

Abstract. This paper describes an ontology-based service discovery front-end interface for GloServ. GloServ is a service discovery engine, which is an ontology-based distributed service discovery system that allows sophisticated querying of services. The working implementation of the front-end interface demonstrates how GloServ can be used for different types of web service discovery. The front-end generates a search form from the service class ontology. It also allows multiple services to be queried for in a single search by generating cascaded forms for combined service queries. It then converts the input to a GloServ query and displays the results to the user in a coherent manner. The use cases that are demonstrated with this implementation are service discovery for location-based services, tagged services and collaborative search with other users.

Keywords: User interface, service discovery, ontologies, OWL, CAN, peer-to-peer.

1 Introduction

This paper describes an ontology-based service discovery front-end user interface to GloServ [12] [13] [11]. GloServ is an ontology-based global service discovery system. It uses the Web Ontology Language Description Logic (OWL DL) [5] to classify services in an ontology and map knowledge obtained by the ontology onto a hybrid hierarchical peer-to-peer network. It operates in wide as well as local area networks and supports a large range of services that are aggregated and classified in ontologies. A partial list of these services include: events-based, physical location-based, communication, e-commerce or web services.

GloServ provides a generic back-end service discovery framework for different front-end systems to interact with it. The contribution of this paper is a working implementation of a web-based front-end which demonstrates how GloServ can be used for different types of web service discovery. The use cases that are demonstrated with this implementation are service discovery for location-based services, tagged services and collaborative search with other users.

⋆ This work was developed at the Dept of Computer Science, Columbia University.

L. Aroyo et al. (Eds.): ESWC 2009, LNCS 5554, pp. 684–696, 2009.

In order to build a web-based front-end to GloServ, the following problems need to be addressed: 1) downloading the service class ontologies from the GloServ back-end; 2) generating a search form from the service class ontology files; 3) allowing multiple services to be queried for in a single search by generating cascaded forms for combined service queries; 4) converting the user input to a GloServ query; 5) displaying the results to the user in a coherent manner.

Below, we describe the solution of the design and implementation of the web-based front-end system. We begin by giving an overview of the back-end GloServ service discovery system in Section 2. In Section 3 we motivate the need for a web-based front-end by describing a few use cases. Sections 4 and 5 describe the front-end system and its implementation, respectively. Finally we discuss related work in service discovery in Section 6 and conclude in Section 7.

2 Overview of GloServ

This section gives a brief overview of GloServ, the back-end service discovery architecture. For further details on the design, implementation and evaluation of GloServ as well as related work to service discover, we encourage the reader to refer to [12] [13] [11].

One of the main components of GloServ is the service classification ontology. At a high-level, services are classified in a pure hierarchy where each service class is disjoint from the other. Hence, service instances will only be classified within one of the branches. At the lower levels of the ontology, classes may have relationships with other classes and a pure hierarchy is not maintained. The upper hierarchical ontology which defines high-level services is mapped onto a hierarchical network and the low-level ontologies are mapped to a peer-to-peer network.

Another component of GloServ is the back-end hybrid hierarchical peer-to-peer service discovery network. The high-level hierarchical ontology maps to a physical hierarchical network. Each service class ontology maps to the Content Addressable Network (CAN) [16] which is a peer-to-peer network architecture. Since GloServ achieves load distribution, fast query and update processing time, while maintaining reliability, we have elected this as the underlying service discovery system.

GloServers maintain three types of information: a service classification ontology, a thesaurus ontology and, if part of a peer-to-peer network, a CAN lookup table. The high-level service classification ontology is not prone to frequent changes and thus can be distributed and cached across the GloServ hierarchical network. Each high-level service will have a set of properties that are inherited by all of its children. As the subclasses are constructed, the properties become specific to the particular service type. The thesaurus ontology maps synonymous terms of each service to the actual service term within the system. Figure 1 gives an overview of the GloServ architecture generated from the *Restaurant* ontology and how queries are routed to the correct servers.

At the lower levels, maintaining a purely hierarchical ontology structure becomes difficult as classes tend to overlap. Thus, in order to efficiently distribute service instances according to similar content, servers that hold information on

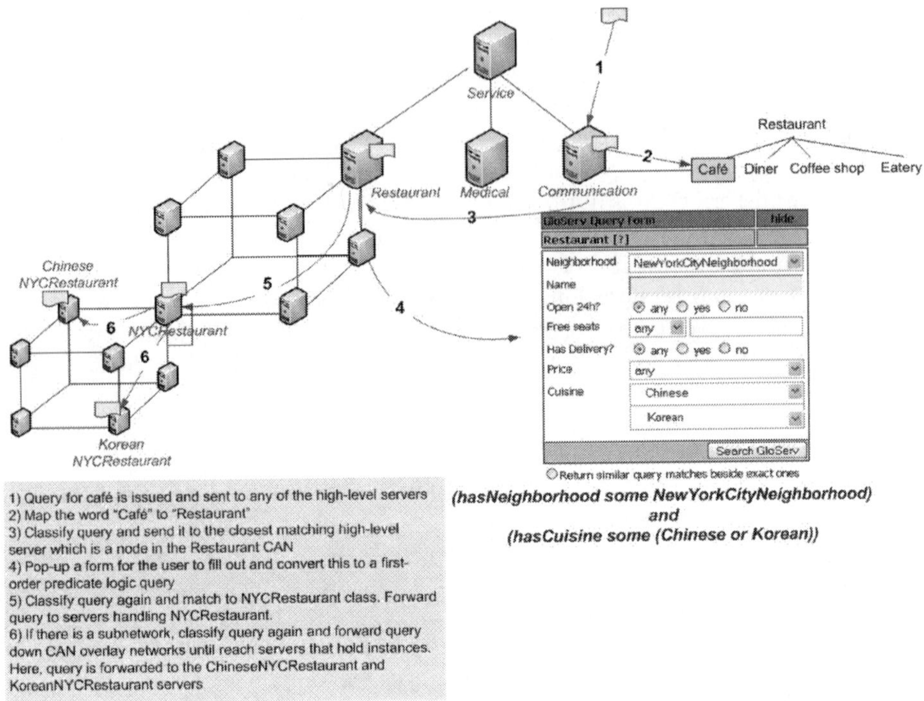

Fig. 1. Finding servers in GloServ

similar classes are distributed in a peer-to-peer network. The CAN peer-to-peer architecture distributes classes with similar content and is generated as a network of n-level overlays, where n is the number of subclasses nested within the main class. The first CAN overlay is a d-dimensional network which has the first level of subclasses of the *Restaurant* class. The number of dimensions is determined by the maximum number of nodes which can be added into the CAN. This is estimated to be $(log_2 n)/2$ to ensure that the number of query hops are $O(log_2 n)$. Services are represented as instances of the service classes and reside in the CAN servers.

A user initially contacts a GloServ user agent and enters a service name. The initial GloServer is found after following the steps outlined in Figure 1. Since each hierarchical node handles a class which is disjoint from its siblings, the query is routed down only one branch reducing the query hops considerably. Once the correct GloServer is contacted, the user agent obtains the ontology pertaining to that service class. The interface to the user can either be human-centric or automated, depending on the implementation. In either case, a query is formed and sent to the GloServer. The query is a first order predicate logic statement that contains restrictions on various properties such as: (*hasLocation* **some** `NYC`) and (*hasCuisine* **some** (`Korean` **or** `Chinese`)). The server handling that service class creates a class with this query restriction and classifies it in its ontology. Since the subclasses of the *Restaurant* class are restricted by location, the query class gets classified as a subclass of the *NYCRestaurant* class. The query is then forwarded to the nodes that handle *NYCRestaurant* classes. When a node is found, the query class is classified again. Since the *NYCRestaurant* class has subclasses that have cuisine restrictions, when the reasoner classifies the query class, it becomes a subclass of *NYCRestaurant* and a superclass of *KoreanNYCRestaurant* and *ChineseNYCRestaurant* classes. The classification indicates that the query must be routed to the servers handling Chinese and Korean restaurants. In order to route the query within a CAN, the query needs to reduce to a dimension and key. The dimension and key values assigned to each of these classes are used during the CAN network generation to convert the ontology class to a $< dimension, key >$ pair.

GloServ allows ontology-based querying with key word matching. Since the service ontology may not capture all parts of a particular service, we allow each ontology to have a keyWords property which service providers can populate with keyword terms that describe the specifics of their service. The ontology-based query is executed and the remaining results are refined by matching the keywords entered. Gloserv also allows a combination of queries across different service classes. Thus, one can search for services with a common property, such as location, in a single search. An example of this is querying for a restaurant and a nearby movie theater.

3 Use Cases

The motivation to create a web-based front-end to GloServ is to demonstrate that GloServ can be easily plugged into a Web Service Architecture [8] and perform interesting service discovery searches. The Web Service Architecture may have different types of service discovery systems such as centralized registries as in UDDI [10], text-based search engines as in Google or peer-to-peer architectures such as GloServ. Using GloServ for web service discovery shows that service discovery can be done using a richer description language as well as within a distributed architecture. We demonstrate that a web service front-end can be implemented which discovers services using GloServ. However, we have not built a front-end which interacts with the service providers directly using SOAP and HTTP as this is not applicable to the service discovery phase and is beyond the scope of this paper. Below, we describe three use cases that a web-based front-end to GloServ can be used for: location-based service discovery, searching for tagged services and collaboratively searching for services.

3.1 Location-Based Services

Consider location-based services such as restaurants, theaters, and traffic. Imagine a couple is planning an evening out in New York City and would like to find a seafood restaurant near the waterfront in Manhattan that also has a theater nearby playing an action movie. In order to find the quickest route to the restaurant, the closest restaurant which has the least congested route needs to be found. It should also be located near a movie theater which is featuring action movies with showtimes that are some time after dinner is over.

To perform the above search, location data must be integrated across the different domains. First, seafood restaurants are searched for in New York City which have an additional attribute of being located near a waterfront. The location of these restaurants are then fed into a traffic service network which returns the best route to the restaurant. The locations of these restaurants are chosen and then fed into the theater service network along with attributes for movie and showtimes. The results retrieved will be all the restaurants and nearby theaters which are located in places that offer the least congested route.

Currently, in order to perform such a search, a user must issue multiple queries to restaurant, theater and traffic sites in order to determine the best restaurant and theater to go to. Since data for each of these services is represented in different formats, there is no way to integrate the data while composing services. The use of ontologies would greatly aid in performing these types of complicated queries. Furthermore, if we take into account all types of location-based services, global distribution of data becomes necessary. Thus, we have chosen to use GloServ as the underlying service discovery system since it provides an ontology-based service search in addition to global distribution of services.

In this use case, three service classes are queried: *Restaurant*, *Theater*, and *Traffic Reports*. These classes all share a common property, namely, location. Since an ontology is useful in mapping semantically equivalent concepts to one another a *Location* ontology can be defined and shared across all the different service domains. The front-end downloads ontologies for the multiple service classes by providing links to the service forms. It then exhibits three cascading forms that appear to the user, one for each service class with the shared location field appearing once. The results are displayed in a Google map.

3.2 Service Tagging

Another example of a use case for a web-based front-end is searching for services and their tags. Currently, tagging is becoming a common phenomenon in various web services such as Flickr [3] and de.licio.us [2]. We define a *tag* as an attribute of a service that is not part of the service description itself. For example, many users like to read and give feedback pertaining to services they will use or have used. This review system can be deemed as a *Tag* service class which has any number of different rating services such as Zagat [9] for restaurants, Better Business Bureau for businesses, or regular user reviews. Given this type of service composition, one can search for any service which has a particular rating and give feedback for this service as well. A *Tag* service class is defined with its own service classification ontology and inserted into the GloServ network.

The relationships between GloServ service classes and their corresponding tag services can be determined using an ontology mapper. For example, tags from the *Zagat* and *NY Times* classes match the *Restaurant* service class for restaurant ratings and the *NY Times* class also matches the *Theater* class for movie ratings.

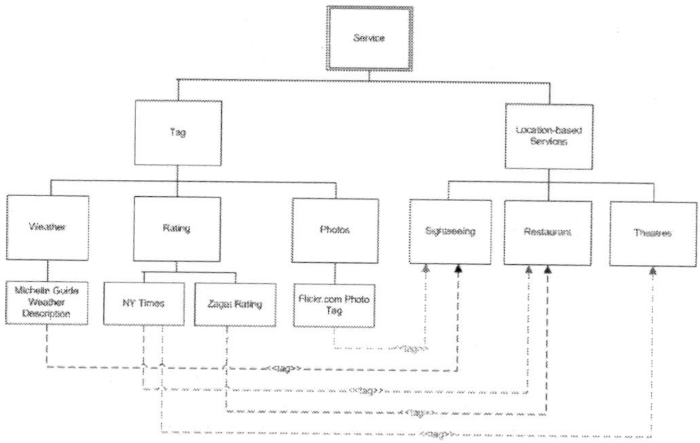

Fig. 2. Relationship ontology of the Tag class and the Location-based Services class

Flickr and *Weather* tag classes match the *SightSeeing* service class because each of these can tag various travel sites such as travel brochures or tourist spots. The front-end establishes the relationship table of a given service class and its corresponding tagging service by using an ontology mapper.

Figure 2 shows the relationships between *Tag* services and *Location-based Services*. Front-end web services handling sightseeing or restaurant services know about the tagging providers that contribute tags to their services via this relationship ontology. In most cases, these tags will be contributed by third party entities.

Searching for restaurants and theaters with specific reviews becomes simple with the front-end we have designed. The user queries for *Restaurant* and *Rating* by choosing these service links and entering data into the restaurant and rating forms. Users can also add tags to a service by providing the service URN and the reviews in the rating form.

3.3 Collaborative Search

An interesting application of GloServ is letting users collaborate in a single search. Imagine two mobile users who decide on a last minute dinner meeting. They would like to perform the search together but one of them is sitting in a meeting and is unable to talk. Thus, her friend invites her to collaborate with her on a service discovery search through GloServ. The invitation is sent via email or text message which includes a link to the GloServ collaborative interface. As *user A* clicks on her preferences, the values change on *user B*'s front-end as well, by synchronizing the interfaces. The values entered by both users is converted to a regular GloServ query and issued to GloServ.

4 Web-Based Front-End

The front-end is a web server that runs Apache and PHP [6] (version 5.1.x). It allows users to register and query for location-based services. Currently, the service classes that are supported are for *Restaurant*, *Theater*, *Weather* and *Tagging* services. The interface provides links for each service class. When the user clicks on a service class name, a form is generated for that service class. The query results are displayed in a Google map as well as in a list. The overall GloServ front-end can be seen in Figure 3. This section describes how the front-end operates in order to perform location-based service discovery using GloServ.

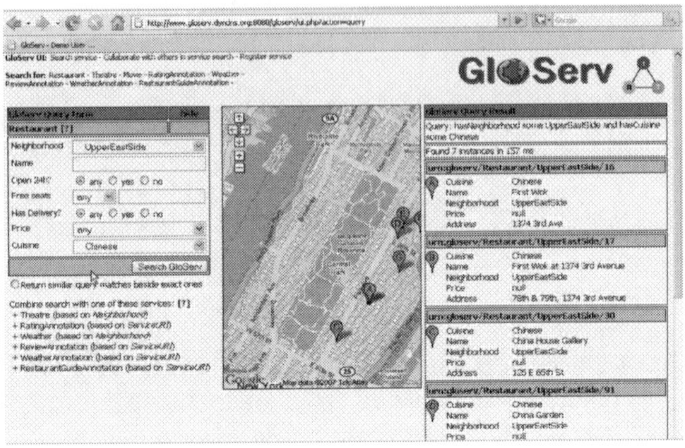

Fig. 3. GloServ Front-End

4.1 Generating a Search Form

The first step that a front-end needs to accomplish is to download the correct service ontology files from the GloServ back-end. Depending on what the web service is, it contacts the GloServ back-end by submitting the service class name. This query is routed through the GloServ hierarchy as described in Section 2 and the ontology is returned to the front-end server. The front-end caches these ontology files as well as the hostnames for the CAN super nodes of the service classes and refreshes these periodically. In this way, the query does not have to route through the hierarchical servers but can go directly to the CAN level.

The front-end parses the ontology and displays each property as a field in the form. Each property is annotated with a label property which is used for the graphical user interface. The ontology parser parses out each property's label and displays it as the field's label in the form. The form field is determined by the property's range. If the range is an object property, all the classes in the range are parsed and displayed as a drop-down menu. For datatype properties, the field is a regular text box. For example, the hasCuisine property has annotation properties label.us=`Cuisine` and label.de=`Kueche`. These two labels signify how to display the hasCuisine field in the English and German languages. The form field is a drop down menu of the *Cuisine* class and displays all the subclasses of

Fig. 4. Ontology Form

Cuisine. Similarly, the hasName datatype property is labeled `Name` and its field is a text box. Figure 4 illustrates this concept.

4.2 Cascading Forms for Combined Service Queries

We have built a front-end that supports combined querying of services. In order to accomplish this, the user interface needs to be able to display more than one service form at a time. The front-end server downloads a number of ontologies for each service class. In order to allow multiple services to be searched for in a single query, the relationship between these service classes needs to be established. This is accomplished by passing the ontologies through an ontology mapper which establishes the relationship between the ontologies. We implement a simple ontology mapping tool by storing the relationship between services in a relationship table. This also includes the corresponding matching properties.

The front-end interface displays the combined service classes as links. When users click on these links, cascading forms are displayed. The matching property or properties are only displayed once and inserted as a shared property in the combined query. Figure 5 below shows a combination of *Restaurant* and *Theater* service forms displayed with one *Neighborhood* field which is the common property.

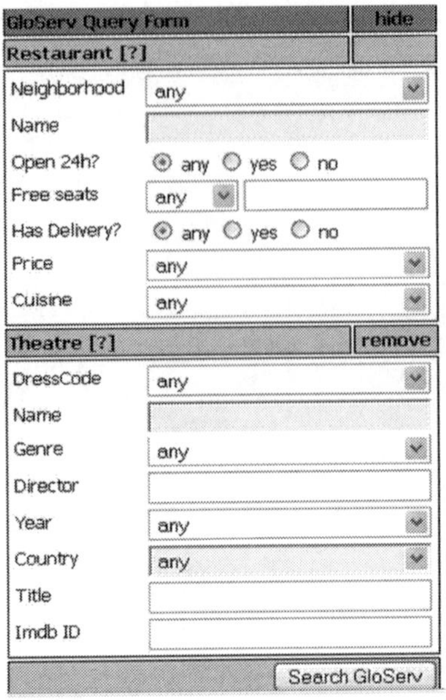

Fig. 5. Cascading Service Forms

4.3 Creating GloServ Queries

Single service ontology queries in GloServ are a conjunction of all the property values. Although more complicated queries can be formed (ie. disjunctive queries, cardinality assignments, and set equivalences), a conjunctive query suffices for the purpose of the demonstrative use cases. Besides creating simple ontology queries, the front-end also checks to see if there are combined queries or keywords entered and creates the query accordingly. The information entered in by the user is converted to a GloServ query. For example, hasCuisine **some** `Italian`, signifies that the object property hasCuisine is assigned to `Italian`.

For combined queries, the front-end creates a *primary* query out of the first service class and *nested* queries from subsequent service classes. In the example above, the *Restaurant* service query will be the primary one and the *Theater* query, the nested one. If there were more services cascaded in the form then these would be nested within each other.

For the collaborative user search, a single query is formed given inputs from multiple users. The interface is synchronized across user group. Every time a user enters a value for a property, her collaborators see an updated form with the property value the user entered. Once every user inputs a value for a given property, the query that is constructed is a conjunction of each property value entered by the users.

The query is put in an XML message which indicates what type of query it is (service registration or query). It is then passed onto the front-end server which constructs the appropriate GloServ message and issues the query to the GloServ backend.

4.4 Displaying Results

For single service query results, the results are displayed in a list where each instance and its properties are displayed. In combined queries, results are displayed using an instance tree that shows the relationship between each service instance. An example of this is the combined query of *Restaurant* and *Theater*, where the tree is formed such that instances of the *Restaurant* class are parents of matching *Theater*.

For location-based services, the results are also displayed in a Google map. Each service instance is labeled with a pin on the Google map. For combined service queries, the map shows the primary query tagged in red and the nested query tagged in green. When clicking on either the red or green tags, the user can choose to show only corresponding services. This grays out all the services

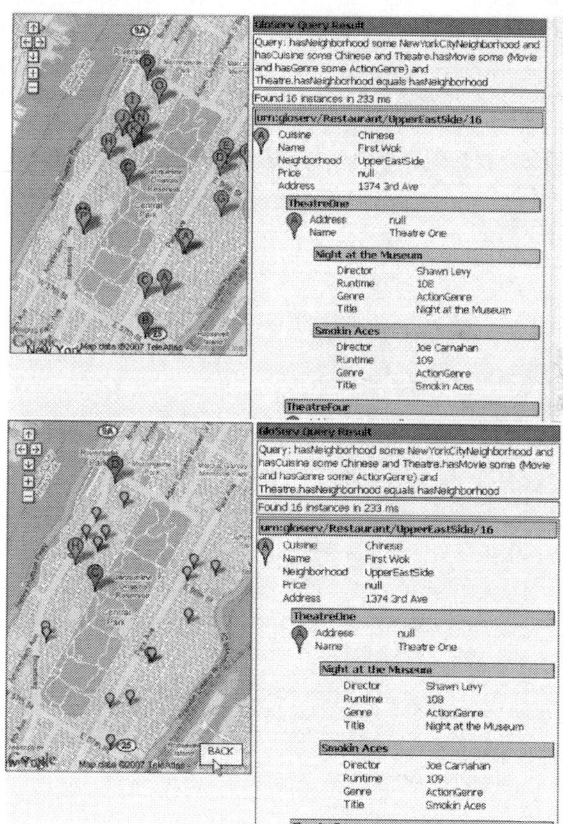

Fig. 6. Google map results for combined service queries

except for those services which match the current one that has been clicked. In this way, users can easily navigate through the map and can determine what the matching services are. Figure 6 shows how the query results are displayed for a combined restaurant and movie search.

5 Implementation

The front-end implementation is done mostly in PHP. For the collaborative interface, functionality in JavaScript [4] has also been defined. The front-end uses a Smarty template engine to generate HTML code for PHP [7]. Smarty facilitates a manageable way to separate application logic and content from its presentation. The collaborative search interface requires user view synchronization and a central data storage. The Asynchronous JavaScript and XML (AJAX) [1] framework is used to communicate with the client side and web server. AJAX provides a framework for creating efficient and interactive web applications.

GloServ nodes communicate with each other using a form of property-value-pair encoding. The implementation separates the message format and encoding from the logic. Figure 7 below shows how the messages are exchanged between the user's browser, the front-end server, the Smarty template engine and the GloServ back-end interface. It is assumed that the front-end server already has the ontology file in the cache.

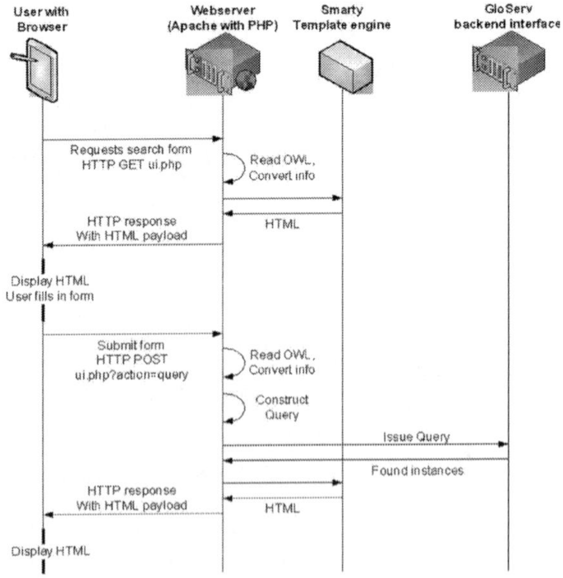

Fig. 7. Query Code Walk

6 Related Work

There are a few service discovery protocols in use today. Most service discovery mechanisms are localized and use attribute-value pairs for service descriptions. Below we describe each of these and compare them to GloServ.

SLP [14] and Jini [15] are both similar in that they have agents that manage services, users and directories of services. Agents advertise each others' presence to each other using either multicast or unicast. In SLP, service registration and queries are broadcast to the directory agents or directly between the service and user agents depending on if the directory agents are present. In Jini, however, a client downloads the service proxy and invokes through Java RMI in order to access the service through a discovery process. Service descriptions in SLP are done in simple attribute-value pairs whereas Jini matches interfaces. SLP is mainly used in local area networks. Jini can span to a larger enterprise networks. The Universal Description, Discovery and Integration (UDDI) [10] specification is used to build discovery services on the Internet. UDDI provides a consistent publishing interface and allows programmatic discovery of services. Services are described in XML and published using a Publisher's API. Consumers access services by using the Programmer's API built on top of SOAP. Services in UDDI are stored in a centralized business registry. The main drawback of UDDI is that it has a centralized architecture and does not span to a global area.

Since these architectures are either centralized or scale to a local level, GloServ provides the best architecture for service composition in a globally distributed network. Also, because of its use of ontologies, a friendlier user interface can be implemented which allows intelligent querying of services.

7 Conclusion

GloServ provides ontology service descriptions as well as a framework for different service classes to be aggregated in a single network. Due to these attributes, we have built a web-based front-end to demonstrate the interesting use cases for GloServ. These include searching for a combination of location-based services, tagged services as well as collaborating with other users on a single search.

References

1. Asp.net ajax, `http://www.asp.net/ajax/`
2. del.icio.us, `http://del.icio.us/`
3. Flickr, `http://www.flickr.com`
4. Javascript, `http://www.javascript.com/`
5. Owl: Web ontology language, `http://www.w3.org/2004/OWL/`
6. PHPs hypertext processor, `http://www.php.net/`
7. Smarty template engine, `http://www.smarty.net`
8. Web services architecture, `http://www.w3.org/TR/ws-arch/`
9. Zagat survey, `http://www.zagat.com/`

10. UDDI technical white paper. white paper, uddi (universal description, discovery and integration). Technical report, OASIS (September 2000)
11. Arabshian, K., Dickmann, C., Schulzrinne, H.: Service composition in an ontology-based global service discovery system. Technical report, Columbia University, New York, NY (September 2007)
12. Arabshian, K., Schulzrinne, H.: An ontology-based hierarchical peer-to-peer global service discovery system. Journal of Ubiquitous Computing and Intelligence (JUCI) 2, 133–144 (December)
13. Arabshian, K., Schulzrinne, H.: Combining ontology queries with key word search in service discovery. In: ACM/IFIP/USENIX 8th International Middleware Conference, Newport Beach, California (2007)
14. Guttman, E., Perkins, C., Veizades, J., Day, M.: Service location protocol, version 2. RFC 2608, Internet Engineering Task Force (June 1999)
15. Sun Microsystems. Jini architectural overview. Technical report (1999)
16. Ratnasamy, S., Francis, P., Handley, M., Karp, R., Shenker, S.: A scalable content addressable network. In: Proceedings of ACM SIGCOMM 2001, San Diego, CA (2001)

A Resource List Management Tool for Undergraduate Students Based on Linked Open Data Principles

Chris Clarke

Talis Group Limited,
Knights Court, Birmingham Business Park, B37 7YB
chris.clarke@talis.com

Abstract. Resource List Management tools help educators create and publish lists of resources relevant to students undertaking a particular module, course or assignment. Traditional approaches to online delivery of these lists have been limited by lack of interoperability and integration with other campus systems, poor take-up by instructors and brittle linking strategies which break as institutions shift suppliers for e-content from year to year. In this paper we present a Resource List Management tool that uses Semantic Web technologies and Linked Data principles to overcome these limitations and improve the interoperability of data contained within such systems.

1 Introduction

In this paper we present a Resource List Management system for university students based on Semantic Web technologies [1] and Linked Data principles [2], and in active use at a number of UK universities.

Resource Lists are collections of text books, journal articles, Web pages and/or audio visual content intended to be companions for students to degree courses, modules or assignments. These lists, defined by course instructors, were traditionally provided to the student as a paper handout comprising a list of citations, the intention being for the student to visit the campus library or bookshop to obtain the listed resources.

Over the past 10 to 15 years, these materials have increasingly become available over the Web, with publishers and content owners moving to an e-delivery model for both monographs and serials. Instructors have been making increasing use of blogging, podcasting and multimedia learning objects in their syllabuses. Resource List Management (RLM) tools emerged over this period, moving paper lists to online electronic signposting solutions.

While this transition has created many benefits by improving access to published lists, current RLM solutions have a number of limitations, particularly regarding interoperability of and wider access to the data they contain. In Section 2 we will discuss these limitations in more detail. Section 3 presents the system from the user perspective and outlines its usage context at the pilot sites (University of Plymouth and University of Sussex). In Section 4 we describe our technical approach based on the Semantic Web technology stack [1] and Linked Data principles [2], and detail many of the design decisions taken when developing the system. We conclude the paper in Section 5 with final observations and directions for future work.

L. Aroyo et al. (Eds.): ESWC 2009, LNCS 5554, pp. 697–707, 2009.

2 Limitations of Traditional RLM Tools

2.1 Limited Interoperability of Resource List Data

Historically, the interoperability of data between publishing platforms, the University library catalogue, the University Virtual Learning Environment (VLE) and RLM tools themselves has been poor. RLM tools have often been no more than electronic replicas of the paper based list, storing details about the lists and resources themselves within proprietary database schemas, accessible only by a HTML or other user-orientated interface, with no open APIs available for integration.

Assuming that a new solution can provide better integration with other systems and the wider web, the issue of licensing data provided by APIs is oft-overlooked. We recognise the importance of being specific in how you wish others to use data in section 4.4.

2.2 Few Opportunities for Collaboration or Personal Contributions

RLM tools have not traditionally encouraged, or even allowed for, contributions from the majority of the users of the system (the students). Instead the emphasis has been on instructors creating a static list that is then used in read-only mode by the students. There has been little attempt to allow students to add valuable annotations to lists. Examples would include communicating whether they have found an item useful in attaining their learning goals; annotating resources with study notes; or forming new collections of resources for a particular essay or group exercise.

Not only do these restrictions contradict the notion of the Web as a read/write medium, recently popularised by trends often labelled as Web2.0 [3], but it also contradicts the view that learning could or should be approached as a collaborative process [4].

From the perspective of instructors, the task of discovering new or revised content to update existing or create new lists has been largely unaddressed by existing systems beyond offering a simple bookmarking system or search interface over the local university library catalogue. With hundreds if not thousands of instructors worldwide facing the task of creating, for example, an introductory physics resource list there is great potential for a more collaborative solution.

2.3 Complexity in Access Mechanisms

While the transition to electronic resources has improved the access landscape to some extent, it has also led to the situation where a single journal subscription can be fulfilled by numerous different publisher or aggregator platforms. This is known as the appropriate copy problem [5]. As an institution shifts subscription agents from year to year, the landscape of access rights a university has purchased becomes increasingly complex. Links to e-content harvested by the instructor into the RLM to may function this year, but expire or move to a different platform the next.

Institutional link resolver solutions, such as SFX from ExLibris[1], go some way to solving these issues. However, in our experience, such library-biased solutions often represent a poor experience for the student and are potentially forgotten about by the instructor when assembling lists. Instead of searching the resolver and harvest-

[1] http://www.exlibris.co.il/category/SFXOverview

ing the resource's resolver page, many instructors naturally harvest direct from the source web page to which the resolver points.

2.4 Stock Management

Traditional RLM tools have overcome a number of the limitations of paper-based resource lists, most notably in the area of stock management. Web-based lists that link to the resources in the university's library catalogue can be used to give advance warning to library acquisition staff that particular resources will be in high demand during certain periods and that stock levels may need to be increased.

However, existing systems are primitive and brittle in their approach, relying on catalogue database identifiers to link to resources, and fail to identify identical or equivalent items available in the catalogue such as duplicate copies or subsequent editions of a work. This results in sub-optimal algorithms for predicting both usage and availability for a resource.

3 The Talis Resource List Management System

In partnership with 15 UK and Eire universities (the focus group), we embarked on a project to replace Talis' existing RLM tool and seek to address some of these issues through the use of Semantic Web technologies.

The system, known as Talis Aspire[2], was launched at the University of Plymouth, one of the existing focus group partners, in September 2008, initially with just 1000 students. Throughout the autumn semester, Plymouth have been increasing adoption of the system with the aim of giving access to all 22000 students in 2009. During January 2009, the University of Sussex joined the pilot, increasing access to around 40000 users in total.

Before we discuss our technical approach we will give an overview of the application from the perspective of our two primary user groups – students and instructors.

3.1 The Instructor's User Experience

The lists themselves can be constructed using sections (see figure 1), as defined in the Resource List ontology, to group resources according to topics, time periods, or importance. For example an instructor may group items into a time-based section (e.g. "Lecture 1") or use a topic-based (e.g. "Accounting for Intangibles") arrangement. This, along with any class notes the instructors add, both add extra context for the student and enhance the semantics of the list, and indicate intended uses for a group of resources. This makes the lists much more semantically-rich, and provides high quality contextual data for the future construction of recommendation services (see section 5).

The resources themselves are gathered into the system by a process known as harvesting. Using a bookmarklet, and in a similar pattern to sites such as del.icio.us[3],

---

[2] http://www.talis.com/aspire

[3] http://del.icio.us

instructors can harvest resources from a multitude of web sites, including the library catalogue. These are placed in a holding area known as "My Library" where they can be further annotated, tagged or commented upon. Later, when the instructor is making changes to an existing list, or creating a new list, the library appears on the interface and resources can be dragged and dropped onto the appropriate place on the list.

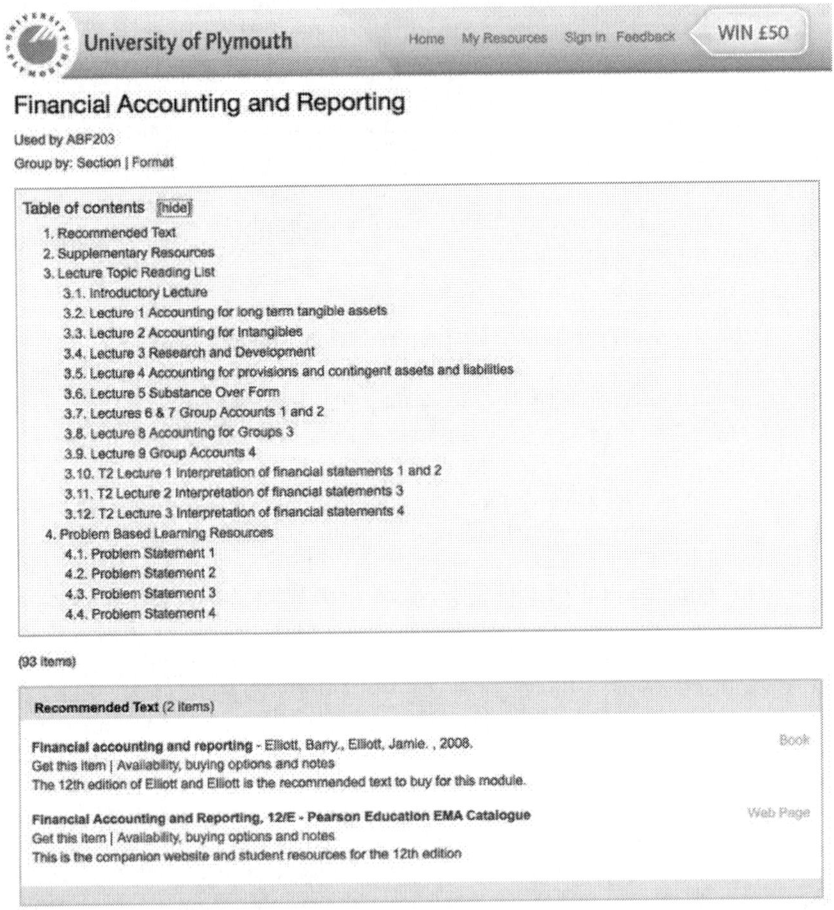

Fig. 1. Outline view of the list, with the table of contents in open mode. The blue box represents a section containing two item references.

The difference between the system and most bookmarking services is that the system attempts to identify the resource the page is describing, rather than just recording the location of the page itself. A series of recognisers are applied to the page content to determine the particulars of the item being harvested. Recognisers use a variety of techniques, including querying of other sources, based on facts gleaned from the page content, or evaluating microformat-style markup such as COinS[4] or embedded RDF to pull metadata from the page. Only if a resource cannot be identified by automatic means is the user prompted to supply metadata and classify the type of resource being harvested.

3.2 The Student's User Experience

The student can interact with the resources within the lists in a number of ways. For physical items, the system will pull in stock availability from the local library catalogue, as well as offer a number of buying options. However, there are a number of functions designed not only to improve the experience for the student, but also increase the richness of the data (see figure 2).

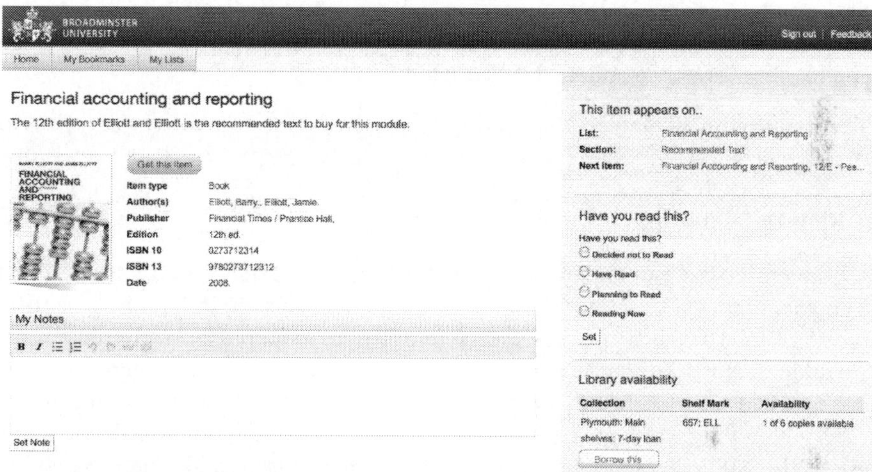

Fig. 2. The student experience (individual item view) showing the harvested metadata, personal study notes, read status and library availability. This screenshot is from the demonstration system, depicting the fictitious Broadminster University.

For example, using the Resource List ontology [6], the system allows the student to annotate each item on a list, augmenting the instructor's class notes with their own, and allowing to categorize each of the resources by intention (e.g. "Planning to read", "Reading now"). The student can then use the tool to manage their library of resources, typically running into the thousands, much more effectively. These intentions are stored as RDF triples in the application backend (invisibly to the user) and could be used as the basis for other applications. For example, a book loan service could be created, connecting level 3 students who own a core text with incoming level 2 students that are seeking to obtain it. Additionally, these intentions will be used in a forthcoming version of the system to enhance feedback given to instructors as to the popularity of the resources, by overlaying the raw usage statistics gathered from web server logs with the intentions of the students in the class.

---

[4] http://ocoins.info/

4 Technical Approach

As mentioned above, the system is built on Semantic Web technologies (RDF, SPARQL, RDFa) and Linked Data principles[5], which set out a recipe for linking and publishing data on the web via de-referenceable URIs. It is our view that this approach affords the following advantages in the context of this application:

- Unify the descriptions of resources using existing ontologies
- Allow these descriptions to be merged and matched with other bibliographic repositories
- Use ontologies designed for appropriation to describe the relationships between these resources and how they fit into the institutional hierarchy
- Improve the interoperability of the data using linked open data principles
- Encourage semantic enrichment of the data by students and instructors to enable context-aware recommendation functionality

The application is built in PHP, utilising the ARC[6]- and Moriarty[7] libraries, and stores its data natively in RDF.

4.1 RDF Data Storage

The system uses the Talis Platform [7] to store, index and query across its RDF data. The application makes use of a group of distinct stores to separate graphs with different concerns and levels of privacy (see figure 3). For example, publicly linkable and discoverable data is stored in a different store from private user content or details, such as preferences, passwords or private study notes. Data from these stores can be merged within the system at run time when required. Using this approach encourages re-use of the public data by other applications. Full SPARQL and search capabilities can be offered externally without exposing sensitive or private data.

As previously described, instructors harvest metadata about resources into the system using an intelligent bookmarklet. Once obtained, the metadata about these resources is stored in the instructor's library as RDF using the Bibliographic Ontology [8] increasing the interoperability of the harvested data with other systems and workflows. Bibliographic Ontology is already in use by a number of other groups, including Zotero[8] a tool to help collect research sources, and the LIBRIS project[9] at the National Library of Sweden which is a linked open data online library catalogue.

The system employs a mechanism of trusted sources to improve the quality of harvested metadata. Rather than rely solely on the metadata provided by the source web page as bookmarked by the user, the system scans the source for identifiers such as DOIs[10] or ISBNs. These are then used to query trusted sources such as the

[5] http://linkeddata.org/
[6] http://arc.semsol.org/
[7] http://n2.talis.com/wiki/Moriarty
[8] http://www.zotero.org
[9] http://libris.kb.se/?language=en
[10] http://www.doi.org/

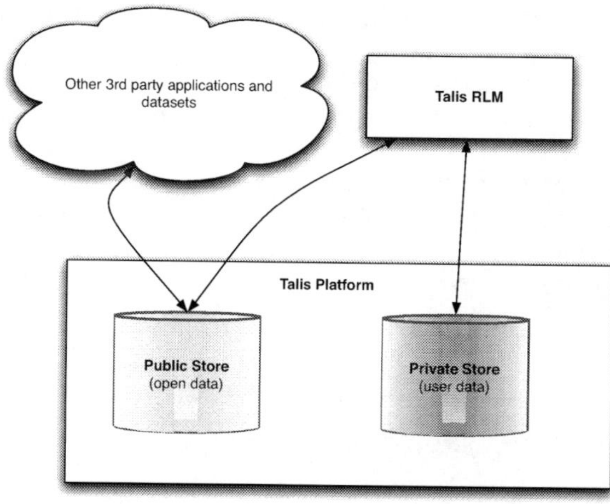

Fig. 3. Partitioning of data stores, with publicly linkable data in the store on the left, and private data, only available to the application, in the store on the right

institution's library catalogue, or the publisher's record for a journal article. A logical extension to this in the near future is to use a common identifier to mint a URI according to a known algorithm to produce a de-referenceable URI in the linked open data cloud, and retrieve the metadata from there. It would be easy to mint a URI referencing a resource in the RDF Book Mashup[11] from an ISBN, for example, and follow the URI to discover metadata about the resource.

By storing metadata about the resource being described, rather than the page describing it, more resilient strategies can be employed to ensure content links do not break if the library decides in future to change supplier. Much of this more sophisticated logic is simply invoked by the user by clicking the "Get this item" button (see figure 1). The philosophy of this feature is to get the user as close to the resource as possible. To improve the user experience, content such as eBooks or journal articles that are available on-line are in-lined in the page without linking to an external location, in the style of YouTube[12] and the Google Books[13] services.

The interface to build or edit lists (see figure 4) uses a WYSIWYG metaphor implemented in Javascript to allow the user to drag and drop resources and edit data quickly, without the need to round trip back to the server on completion of each operation. Behind the scenes the actions of moving, adding, grouping or editing resources create RDFa [9] markup on the page. When the user has finished editing, they hit a save button which serialises the RDFa model in the page into an RDF/XML model which is submitted back to the server to be persisted to disk.

Adoption of this approach has significantly reduced complexity in the persistence logic of the application. The user effectively builds the RDF representation of the list

[11] http://www4.wiwiss.fu-berlin.de/bizer/bookmashup/
[12] http://www.youtube.com
[13] http://books.google.com

themselves on the client, and submits the new model along with the model as it appeared before changes were applied to the server at the end of the editing session. The system simply compares the two models to calculate a delta of changes, and these are submitted to the appropriate store using the Changeset protocol [10].

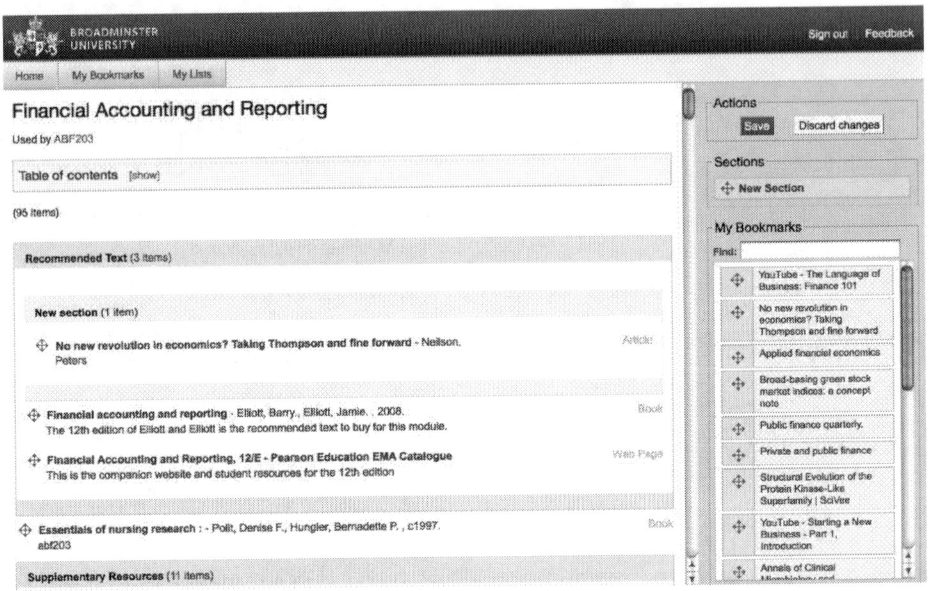

Fig. 4. The list editing. Harvested references can be dragged from the right and dropped onto the list on the left. References already on the list can be dragged and re-ordered.

4.2 Vocabularies and Ontologies Used

As well as re-using existing ontologies such as FOAF [11], SKOS [12], Biblio Ontology [8] and Dublin Core [13], we developed and published two new ontologies as part of the project. The Resource List Ontology [6] underpins the semantics of the relationships between resources and their intended uses. The Academic Institution Internal Structure ontology (AIISO) [14] describes the courses, modules, departments and schools that make up an institution, which was required to enable instructors to link lists to the relevant module or course, to enable students to find lists.

4.3 Interoperability of Data

As well as standardising the description of resources by using ontologies in use by other groups, the RLM tool encourages re-use and linking of data to other external sets by adhering to linked open data principles [15]. This means that not only did the design of the data have to fit the immediate use cases of the user, but also unknown use cases beyond the system boundary. This is referred to as designing for appropriation.

Unlike many web applications, the system is architected to use a noun-based URL structure to describe entities, rather than a verb-based approach to describe operations (see table 1). This works with the mechanisms that underpin the web, rather than against them, and focuses on describing resources via dereferencable URIs rather than describing operations [16]. This approach lends itself well to publishing linked open data, with URLs representing entities rather than actions, and the system uses content negotiation approaches [13] to allow each URL in the system to be served via a HTTP GET request as XHTML (for humans) and RDF/XML or JSON for software agents.

Table 1. Noun-based vs. verb-based URL Structure

URL Structure	
Noun-based	http://lists.lib.plymouth.ac.uk/lists/abf203
Verb-based	http://example.org/view?courseCode=ABF203

In the linked data world, agents can follow the dereferencable URIs in one document to another if they wish to augment the description of a particular resource. This can rapidly become unwieldy if documents contain large quantities of links - agents may need to make a large amount of additional HTTP calls for even basic use cases. In line with designing for appropriation, the system makes a reasonable estimate of shape of data that a user agent might expect when requesting a description of a particular class of resource. For example, a GET issued against the URI of a list will also return details of the items in the list.

In general, using a linked data approach makes it very easy for system integrators to either simply link to the system's human-consumable interpretation of the user interface (via the dereferencable URI), or create their own interpretation within their respective VLE or institutional portal environments.

Other services on the web can link to resources described within the system to augment their own descriptions, and mine rich relationships to other related topics or resources.

4.4 Licensing of Data

An oft-overlooked aspect of publishing data on the web is the inclusion of specific license terms on what rights the owner of the data set is willing either to exert or waive [17]. As the linked open data web expands, the importance of attaching specific licensing terms will become as important as it has in the open source software world. Knowing that you are free to incorporate a dataset within your own will become as important and knowing that you cannot. The Open Data Commons Public Domain Dedication and Licence[14] is available for publishers of data to explicitly gift data to the commons according to community norms.

Talis is consulting with customers of the RLM tool to make sure they understand the implications of publishing data on the open web, and encourages them to do so

[14] http://www.opendatacommons.org/odc-public-domain-dedication-and-licence/

with explicit license terms. Part of the implementation process involves the selection of an appropriate strategy according to the specific customer's requirements. To encourage the widespread re-use and linking of data, Talis is offering savings on product subscriptions for those customers who select open licensing strategies for their data.

5 Conclusions and Future Work

By using existing ontologies and publishing the data within the RLM as linked open data, the tool not only unifies the description of learning resources, but it also unlocks silos of semantically rich descriptions of inter-related resources and the links and relationships between them, as described by domain experts (the instructors). This enables both more effective linking strategies and better interoperability of data between enterprise systems compared to existing systems.

The Talis RLM tool encourages user-enriched descriptions of the data it manages by increasing the opportunities for users to annotate existing data. The aggregate of these enrichments form the basis for explicit feedback to the list creators.

Having unified the descriptions of lists, added semantic richness to the data and encouraged mass user annotations, future development will focus on the issues around context-aware recommendation systems. The first iteration of such a system will be implemented at the point at which the instructor harvests items into the system. Because the system recognises resources not just web links, it can make well informed suggestions based on context rather than keywords. For example, the system may suggest an alternative, newer edition of a book available from library stock, rather than the older copy the instructor is harvesting from Amazon.

This richness in description, allowing the identification of similar or equivalent items, can also be applied to the problem of identifying patterns in stock usage. Combined with the collective intentions of the user base, gleaned from annotations made by students as to their likely use of the item, customers can predict and plan an optimal stock strategy more accurately than in a system that relies on proprietary identifiers alone.

At the time of writing, Talis are continuing the process of rolling the system out to further focus group partners, and will continue to implement the system across the existing RLM customer base of 1/3rd of UK Universities (representing over 800,000 students and over 20,000 instructors) through 2009. Published as linked open data, this will represent a significant corpus of discoverable data about how instructors use resources for teaching.

References

1. World-Wide Web Consortium: Semantic Web Layer Cake (2007), http://www.w3.org/2007/03/layerCake.svg (accessed December 12, 2008)
2. Berners-Lee, T.: Linked Data (2006), http://www.w3.org/DesignIssues/LinkedData.html
3. O'Reilly, T.: What Is Web 2.0 (2005), http://www.oreillynet.com/pub/a/oreilly/tim/news/2005/09/30/what-is-web-20.html (accessed December 12, 2008)

4. Bruffee, K.: Collaborative Learning: Higher Education, Interdependence, and the Authority of Knowledge. Johns Hopkins University Press (1993)
5. Breeding, M.: The Many Facets of Managing Electronic Resources. In: Proc. Computer in Libraries Conference (2004)
6. Shabir, N., Styles, R.: Resource List Ontology, http://vocab.org/resourcelist/schema-20080519.html (accessed December 15, 2008)
7. Leavesley, J. and Davis, I.: Talis Platform: Harnessing Sophisticated Mass Collaboration on a Global Scale, http://www.talis.com/platform/resources/assets/harnessing_sophisticated_mass.pdf (accessed December 12, 2008)
8. Darkus, B.: Bibliographic Ontology, http://purl.org/ontology/bibo/ (accessed December 15, 2008)
9. Adida, B., Birbeck, M.: RDFa Primer, http://www.w3.org/TR/xhtml-rdfa-primer/ (accessed December 15, 2008)
10. Talis Group Limited: Changeset Protocol, http://n2.talis.com/wiki/Changeset_Protocol (accessed December 15, 2008)
11. Brickley, D., Miller, L.: FOAF Ontology, http://xmlns.com/foaf/spec/ (accessed December 16, 2008)
12. Miles, A., Brickley, D.: SKOS Core Guide, http://www.w3.org/TR/2005/WD-swbp-skos-core-guide-20050510/ (accessed December 16, 2008)
13. Dublin Core Metadata Initiative: XML Schemas to support the Guidelines for implementing Dublin Core in XML, http://dublincore.org/schemas/xmls/ (accessed December 16, 2008)
14. Shabir, N., Styles, R.: Academic Institution Internal Structure Ontology, http://vocab.org/aiiso/schema-20080514 (accessed December 15, 2008)
15. Bizer, C., Cyganiak, R., Heath, T.: How to Publish Linked Data on the Web (2007), http://sites.wiwiss.fu-berlin.de/suhl/bizer/pub/LinkedDataTutorial/
16. Styles, N.: Pages, Screens, MVC and Not Getting It, http://dynamicorange.com/2008/10/03/pages-screens-mvc-and-not-getting-it/ (accessed December 12, 2008)
17. Miller, P., Styles, R., Heath, T.: Open Data Commons, a License for Open Data. In: Proceedings of the 1st International Workshop on Linked Data on the Web (LDOW), 17th International World Wide Web Conference, Beijing, China (April 2008)

SCOVO: Using Statistics on the Web of Data

Michael Hausenblas[1], Wolfgang Halb[2], Yves Raimond[3], Lee Feigenbaum[4], and Danny Ayers[5]

[1] DERI, National University of Ireland, Galway
IDA Business Park, Lower Dangan, Galway, Ireland
michael.hausenblas@deri.org
[2] Institute of Information Systems & Information Management,
JOANNEUM RESEARCH, Steyrergasse 17, 8010 Graz, Austria
wolfgang.halb@joanneum.at
[3] BBC Audio & Music interactive
London, United Kingdom
yves.raimond@bbc.co.uk
[4] Cambridge Semantics,
PO Box 425003, Cambridge, MA 02142, United States
lee@cambridgesemantics.com
[5] Talis Ltd,
Knights Court, Birmingham Business Park, B37 7YB, United Kingdom
danny.ayers@talis.com

Abstract. Statistical data is present everywhere—from governmental bodies to economics, from life-science to industry. With the rise of the Web of Data, the need for sharing, accessing, and using this data has entered a new stage. In order to enable proprietary, closed-world formats, to enter the Web of Data, we propose a framework for modelling and publishing statistical data. To illustrate the usefulness of our approach we demonstrate its application in real-world statistical datasets.

1 Motivation

Statistical data is present everywhere—from governmental bodies to economics, from life-science to industry. With the rise of the Web of Data, the need for sharing, accessing, and using this data has entered a new stage. Available technologies are either not compatible with the Semantic Web or are too complex to be useful for a range of use cases. We have identified the need for a more general and flexible solution to the problem of modelling and publishing statistics on the Web.

Our motivation stems from ongoing work in three distinct efforts, namely riese ("RDFizing and Interlinking the EuroStat Data Set Effort") [11,12], U.S. Census Bureau's annual Statistical Abstract of the United States, and making the UN data accessible on the Web of Data. In riese, we provide statistical data about European citizens. One of our main use cases of riese was in the context of an advertising analysis application [23] allowing for example a market researcher to better and faster understand a certain market or product. Further, in the U.S. Census Bureau's annual Statistical Abstract of the United States case, one of the authors and representatives of the U.S. Environmental Protection Agency

L. Aroyo et al. (Eds.): ESWC 2009, LNCS 5554, pp. 708–722, 2009.

were exploiting the semantics implicit in spreadsheets published yearly since 1878 by the U.S. Census Bureau. This data corpus is a comprehensive collection of social, political, and economic statistics compiled from information from over 250 agencies. The goal here was to enable a fast and efficient publishing of the statistics on the Web. Currently, this means making MS Excel documents and PDF documents available for download from the Web.

The paper is structured as follows: Firstly, we review related efforts in section 2. Then, in section 3 we discuss issues with representing statistical data and derive requirements for a generic modelling. The core of the work is presented in section 4—where we propose a modelling framework for statistical data—and section 5 in which two reference implementations are discussed. Finally, we conclude our work and outline future steps in section 6.

2 Related Work

Representing statistical data has a long tradition, hence a plethora of proposals and solutions exists [3]—mainly driven by governmental and international institutions dealing with high volumes of data. In the 1990's the U.S. Bureau of the Census has developed a statistical metadata content standard, which allows to describe all aspects of survey design, processing, analysis, and data sets [15]. Later, the "United Nations Economic Commission for Europe" (UN-ECE) has developed guidelines [27] covering search, navigation, interpretation, and post-processing of statistical data in their realm. A standard proposed by the International Organization for Standardization (ISO) is the "Statistical data and metadata exchange" (SDMX) [14]. More recently, the OECD has released a report on the management of statistical metadata at the OECD [18]. Another related effort is the Data Documentation Initiative (DDI)[1], which aims at establishing an XML-based standard for the content, presentation, transport, and preservation of documentation for datasets in the social and behavioural sciences. As we have already pointed out in [11], there are known attempts concerning the modelling and use of statistical data on the Web of Data [4,10,24,25]. However, unlike earlier attempts such as [16,19], we aim at a light-weight solution enabling a quick uptake and wide deployment.

The **Web of Data** is understood as the part of the Web where the linked data principles are applied. The basic idea of linked data was outlined by Sir Tim Berners-Lee [5]. The Linking Open Data (LOD) community project[2] is an open, collaborative effort applying the linked data principles. It aims at bootstrapping the Web of Data by publishing datasets in RDF on the Web and creating large numbers of links between these datasets [6].

We finally highlight the modelling issue with n-ary relations on the Web of Data. The data model of the Web of Data is RDF, hence modelling n-ary relations is a non-trivial task. In 2006 the W3C Semantic Web Best Practices and Deployment Working Group has published a note dealing with this issue [17]; we will use this as a base for our framework.

[1] `http://www.ddialliance.org/`

[2] `http://linkeddata.org/`

3 Requirements and Issues

3.1 Issues with Modelling Statistical Data

Independent of the original format of the data (such as a table in an Excel sheet, etc.), the issues discussed in the following need to be addressed properly to ensure a lossless representation of the statistical data.

Handling of Multiple Dimensions. It is very often the case that a data item has several dimensions. Roughly, two types of dimensions can be identified, (i) generic dimensions, such as location and time, and (ii) domain specific dimensions. For example we may be interested in the reliability of flights (domain specific) from Cambridge, MA, US to London, UK (both are locations) between 2003 and 2007 (time period). It is crucial that a vocabulary aiming at representing statistics is capable of denoting such dimensions and allows to attach as many as needed to a single data item.

Reusability and Uptake. Statistics are no ends in themselves; rather they are **about** something—be it money, flights, death rates or the consumption of YouTube videos. It is therefore essential being able to reuse existing information; both on the schema as on the instance level. Related to reusability is the issue of community uptake: Most statistical metadata formats are rather complex, yielding a small deployment.

Structural vs. Domain Semantics. Two kinds of semantics come into mind when modelling statistics:

- structural semantics, stemming from how statistics are presented, such as grouping into time-periods, primary dimensions, etc.;
- domain semantics, stemming from the domain the statistic is about (money, airports, etc.).

Performance and Scalability Issues. As discussed elsewhere [13] performance and scalability issues may arise from the way data is encoded and served.

3.2 Requirements

Based on the issues listed above we state the following requirements for our framework:

1. The framework *must* be directly usable on the Web of Data. This implies for example that a vocabulary used in the framework must be expressed in RDF. This requirement addresses the issue of structural and domain semantics, as well as ensuring reusability;
2. The framework *must* be extensible both on the schema level and the instance level, enabling reusability;
3. The framework *should* be light-weight, addressing uptake, and performance and scalability issues.

It has to be noted that the first requirement does not stem from the issues discussed earlier—it was rather introduced to benefit from the large deployment base of domain vocabularies[3] as well as available tools and systems[4].

4 Statistical Modelling Framework

Driven by the requirements we propose a modelling and publishing framework for statistics on the Web of Data consisting of:

- a core vocabulary for representing statistical data
- a "workflow" to create the statistical data

The framework is depicted at a glance in Fig. 1. The lower half is the generic part, defined in this work, the upper part is the application-specific part, depending on the technologies used to implement the framework.

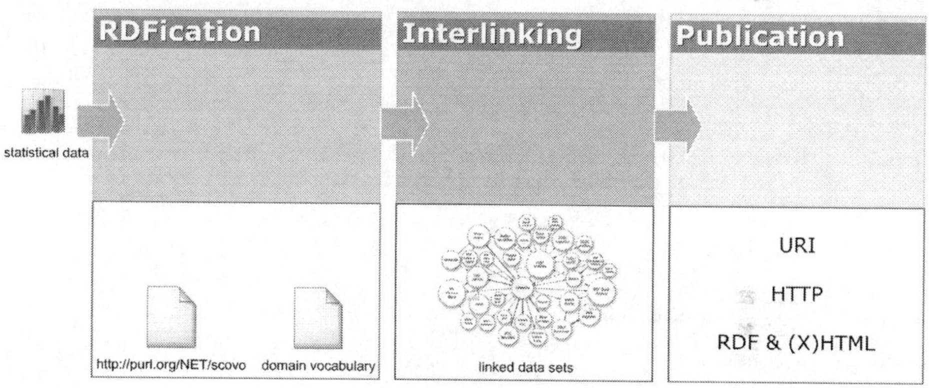

Fig. 1. The Statistical Modelling Framework

4.1 Statistical Core Vocabulary (SCOVO)

One of the main contributions of our work at hand is the Statistical Core Vocabulary (SCOVO)[5]. SCOVO defines three basic concepts:

- a dataset, representing the container of some data, such as a table holding some data in its cells;
- a data item, representing a single piece of data (e.g. a cell in a table);
- a dimension, representing some kind of unit of a single piece of data (for example a time period, location, etc.)

[3] `http://schemacache.test.talis.com/`
[4] `http://esw.w3.org/topic/SemanticWebTools`
[5] `http://purl.org/NET/scovo`

A statistical dataset in SCOVO is represented by the class `Dataset`; it is a SKOS concept [22] in order to allow hooking into a categorisation scheme. A statistical data item `Item` belongs to a dataset (cf. inverse properties `dataset` and `datasetOf`). An `Item` is subsuming the `Event` concept, as defined in the Event ontology[6]. The Event ontology essentially adopts the view from Allen and Fergusson [2]:

> [..] events are primarily linguistic or cognitive in nature. That is, the world does not really contain events. Rather, events are the way by which agents classify certain useful and relevant patterns of change.

An event is then defined in this ontology as the way by which cognitive agents classify arbitrary time/space regions. Our `Item` concept is subsuming this `Event` concept—a statistical item is a particular classification of a time/space region. Dimensions of a statistical item are factors of the corresponding events, attached through the `dimension` property, pointing to an instance of the SCOVO `Dimension` class.

This model is easily extensible by defining new factors and agents pertaining to the actual statistical data. For example, we can relate to a statistical data item the institutional body responsible of it as well as the methodology used. A `Dimension` can have a minimum (and respectively a maximum) range value, captured through the `min` and `max` properties.

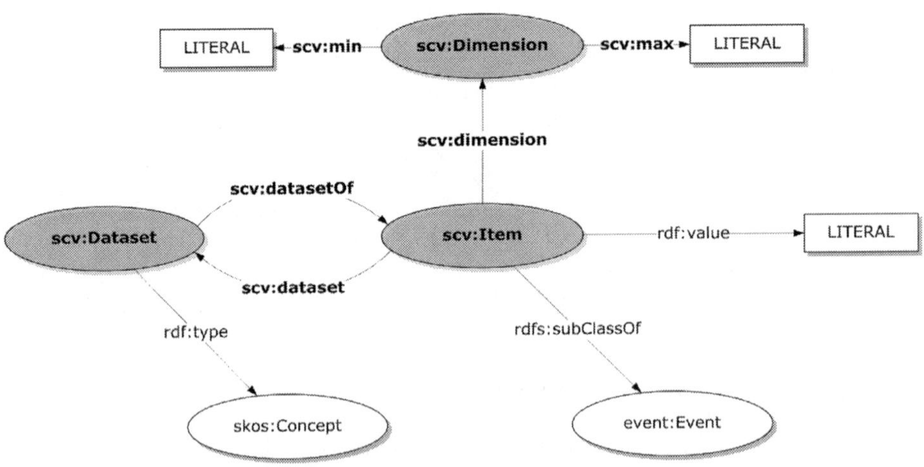

Fig. 2. The Statistical Core Vocabulary (SCOVO)

The Statistical Core Vocabulary (depicted in Fig. 2) is currently defined in RDF-Schema. It is possible to express SCOVO in OWL-DL, if advanced reasoning is of necessity. Although we have depicted the range of both `:min` and `:max` in Fig. 2 being of literal value, we emphasize that in the RDF-Schema the ranges have not been specified in order to allow an extension for domain-specific purposes. Hence, this can be seen as a kind of recommendation for the default case.

[6] `http://purl.org/NET/c4dm/event.owl`

Example. To demonstrate the usage of SCOVO, let us assume we want to model airline on-time arrivals and departures. The input in our example is the "Table 1047. On-Time Flight Arrivals and Departures at Major U.S. Airports: 2006"[7] (cf. Fig. 3) from the US Census data set. Every airport, for each time period has an on-time arrival percentage and an on-time departure percentage.

Table 1047. On-Time Flight Arrivals and Departures at Major U.S. Airports: 2006

[See Notes]

Airport	On-time arrivals (percent) 2006				On-time departures (percent) 2006			
	1st quarter	2d quarter	3d quarter	4th quarter	1st quarter	2d quarter	3d quarter	4th quarter
Total major airports	77,8	76,7	75,6	73,7	79,8	78,5	77,7	76,8
Atlanta, Hartsfield	73,9	75,5	68,0	70,4	76,0	74,3	66,2	70,3
Boston, Logan International	75,6	66,8	71,9	72,8	80,5	74,8	76,5	77,9
Baltimore/Washington International	82,6	77,7	79,0	80,8	80,3	75,9	78,2	80,3
Charlotte, Douglas	81,2	76,1	74,3	73,8	81,6	75,8	76,5	75,7
Cincinnati, Greater Cincinnati	86,4	84,7	82,0	78,9	87,6	87,0	84,3	78,4
Washington, Reagan National	80,6	76,4	74,9	73,4	84,9	81,7	81,0	80,2
Denver International	77,5	81,7	80,4	75,0	75,3	80,3	79,7	76,6
Dallas-Fort Worth International	79,6	73,0	79,8	76,8	77,5	74,2	76,7	75,1
Detroit, Metro Wayne County	80,6	79,6	76,1	69,4	79,6	79,6	78,2	73,4
Newark International	63,3	63,0	64,5	59,3	74,6	71,8	70,1	71,1
Fort Lauderdale-Hollywood International	80,2	79,3	75,8	74,7	80,4	81,3	81,7	79,0
Washington/Dulles	78,7	75,7	73,6	74,6	77,9	74,3	71,9	75,2
Houston, George Bush	77,4	76,0	80,8	76,5	80,5	77,0	81,9	79,7
New York, JFK International	72,7	73,7	67,3	65,2	77,6	81,1	71,9	71,2
Las Vegas, McCarran International	75,2	77,6	77,3	75,7	74,1	75,5	76,1	75,3
Los Angeles International	76,4	78,7	78,3	75,3	79,7	81,9	81,0	79,2
New York, La Guardia	66,2	64,5	64,9	61,0	76,5	74,4	74,6	73,5

Fig. 3. On-Time Flight Arrivals and Departures at Major U.S. Airports: 2006

In listing 1 an excerpt[8] of the modelling of the airline on-time arrivals and departures is shown[9]. We note that the example has `http://example.org/on-time-flight#` as its base along with the prefix `ex:` (line 1). Lines 4 to 16 define domain-specific entities, such as a `ex:TimePeriod`, an `ex:Airport`, etc. From line 18 to 22 the one and only dataset (`ex:ontime-flights`) is defined, corresponding to the entire Excel table used as an input. Further, from line 24 on an exemplary data item is defined, stating that the on-time arrival of the "Atlanta, Hartsfield" airport in the first quarter of 2006 was round 74%. This corresponds to the highlighted cell in Fig. 3.

The SPARQL query from listing 2 can be used to explore high-performing airports. With "high-performing" we define in this context airports with on-time arrivals or departures higher than 85%. Lines 2 to 6 provide the generic pattern for an `Item`. Note how in line 7 and 8 the dimensions are constrained. Line 9 of listing 2 basically expresses "give me all kinds of on-timeness", and finally line 10 implements the "high-performance" filter criteria.

The query result—depicted in Fig. 4—shows the list of high-performing airports along with the time period, starting with the best airport in terms of "on-timeness". We note that the complete example, including the exemplary queries in an executable form, is available at `http://purl.org/NET/scovo`

[7] `http://www.census.gov/compendia/statab/tables/08s1047.xls`

[8] The full example in RDF/XML is available at `http://sw.joanneum.at/scovo/otf-example-full.rdf`

[9] Note, that in all of our examples the well-known prefixes such as rdf:, rdfs:, etc. have been omitted due to readability reasons.

```
@prefix ex: <http://example.org/on-time-flight#> .
@prefix scv: <http://purl.org/NET/scovo#> .

ex:TimePeriod rdfs:subClassOf scv:Dimension; dc:title "time period" .

ex:Q12006 rdf:type ex:TimePeriod; dc:title "2006 Q1";
          scv:min "2006-01-01"^^xsd:date ;
          scv:max "2006-03-31"^^xsd:date .

ex:OnTime rdfs:subClassOf scv:Dimension; dc:title "on-time ..." .

ex:ota rdf:type ex:OnTime; dc:title "on-time arrivals" .

ex:Airport rdfs:subClassOf scv:Dimension; dc:title "airport" .

ex:AtlantaHartsfield rdf:type ex:Airport; dc:title "Atlanta, ..." .

ex:ontime-flights rdf:type scv:Dataset ;
                  dc:title "On-time Flight Arrivals ..." ;
                  scv:datasetOf ex:atl-arr-2006q1 .

ex:AtlantaHartsfield-ota-2006-q1 rdf:type scv:Item ;
                                 rdf:value 74 ;
                                 scv:dataset ex:ontime-flights ;
                                 scv:dimension ex:Q12006 ;
                                 scv:dimension ex:ota ;
                                 scv:dimension ex:AtlantaHartsfield .
```

Listing 1. Modelling flight on-time arrival statistics

```
SELECT ?airport_name ?percent_ontime ?period ?ontime_type WHERE {
  ?item rdf:type scv:Item ;
        scv:dimension ?airport ;
        scv:dimension ?time_period ;
        scv:dimension ?ontime ;
        rdf:value ?percent_ontime .
  ?airport rdf:type ex:Airport; dc:title ?airport_name .
  ?time_period rdf:type ex:TimePeriod; dc:title ?period .
  ?ontime rdf:type ex:OnTime; dc:title ?ontime_type .
  FILTER (?percent_ontime > 85)
} ORDER BY DESC (?percent_ontime)
```

Listing 2. SPARQL query for high-performing airports

4.2 Workflow—Good Practice Rules

In this section we discuss the overall workflow as shown in Fig. 1. Based on our findings from publishing real-world statistical datasets, the following should be seen as strong advises, helping to avoid failings and to enable a quick adoption.

RDFication. In the very first step, the data needs to be converted into an RDF-based form. This is equally true for the schema level as for the instance level. The schema level (e.g. XSD, etc.) is a typical starting point which is followed by the conversion of the actual data in a second step. While creating and populating the ontology with instances several issues arise.

URI Design. It has to be ensured that every entity has a URI assigned, which is usually referred to as "URI minting"; cf. [7] for a more detailed discussion on URI

airport_name	percent_ontime	period	ontime_type
Cincinnati, Greater Cincinnati	88	2006 Q1	on-time departures
Salt Lake City International	88	2006 Q2	on-time departures
Cincinnati, Greater Cincinnati	87	2006 Q2	on-time departures
Salt Lake City International	87	2006 Q3	on-time departures
Salt Lake City International	86	2006 Q2	on-time arrivals
Portland International	86	2006 Q3	on-time departures
Cincinnati, Greater Cincinnati	86	2006 Q1	on-time arrivals
Portland International	86	2006 Q2	on-time departures

Fig. 4. Results for high-performing airports

design. For example `http://dbpedia.org/resource/Airport` has been minted by DBpedia to represent the concept of an airport.

More specifically we recommend using HTTP URIs in order to be compliant with the linked data principles: When dereferencing the aforementioned URI for "Airport" it yields

```
curl -I http://dbpedia.org/resource/Airport
HTTP/1.1 303 See Other
Server: Virtuoso/05.00.3028 (Solaris) x86_64-sun-solaris2.10-64  VDB
Content-Type: text/html; charset=UTF-8
Date: Thu, 08 May 2008 10:32:29 GMT
Location: http://dbpedia.org/page/Airport
```

basically telling us that the "concept URI" redirects to an information resource at `http://dbpedia.org/page/Airport`, see also [21].

Domain Ontologies. As already mentioned, statistics are always about a certain domain. In order to use domain vocabularies together with SCOVO, several "hooks" can be used:

- Subclassing the SCOVO-`Dimension` class. In most cases it is sufficient to use this technique to incorporate domain-specific concepts, for example

  ```
  ex:Airport rdfs:subClassOf scv:Dimension
  ex:AtlantaHartsfield rdf:type ex:Airport
  ```

  from the example in listing 1;
- Use the built-in support for `event:Event` and `skos:Concept`. The latter is of particular help if an existing taxonomy or thesaurus is used as a base. The earlier can be used to capture more information pertaining to the creation of a particular statistical item;
- Defining sub-properties of using SCOVO-`min` and `max`. Whenever the need arises to more explicitly declare what kind of range is intended, this technique can be used (e.g. an `xsd:date`).

Interlinking. Classes and instances of the domain vocabulary *should* be interlinked to existing LOD entities. The rationale behind is that any dataset can be enriched through this at low costs. For example, to connect the airports to the LOD datasets, one could use the query from listing 3 to find according targets in DBpedia (note that this query can be executed at `http://dbpedia.org/snorql/`).

```
SELECT ?airports_state ?airport
WHERE {
 ?airports_state skos:broader
 <http://dbpedia.org/resource/Category:Airports_in_the_United_States> .
 ?airport skos:subject ?airports_state ;
          <http://dbpedia.org/property/name> ?name .
 FILTER regex(?name, "Atlanta", "i")
}
```

Listing 3. Interlinking airports to DBpedia

The result of the query from listing 3 may subsequently be used to enrich our example, that is adding for example the triple

```
ex:AtlantaHartsfield owl:sameAs
<http://dbpedia.org/resource/Hartsfield-Jackson_Atlanta_International_
Airport>
```

in order to express that the two URIs are actually identifying the same thing. With this interlinking we have significantly broadened the possibilities for querying our dataset; for example we could issue a location-based query with the geo-data from DBpedia or could further follow down the path to other LOD datasets containing even more information related to the Hartsfield airport.

Publication. When publishing the dataset, one needs to make choices on the formats to be used for the data. While certain circumstances may require the usage of specialized and/or proprietary formats such as PDF or the SPSS file format, there are four basic technologies that we (unsurprisingly) see central to our setup: URIs, HTTP, RDF and (X)HTML; every publishing system on the Web of Data *should* use these as primary technologies. We have discussed URI minting above. Regarding HTTP—beside its basic transport function—we encourage people to use light-weight REST interfaces. One particular issue, however, is how to deploy the metadata. Several options exist, we list some widely used in the following:

- use an RDF standalone format such as RDF/XML, N3, etc. along with 303 redirects or links such as described in [21];
- use XHTML+RDFa[10] for both humans and machines (see also [11]);
- SPARQL-endpoints and RDF dumps [20,13].

Note that in practice very often the approaches listed above are used in combination. For example offering an RDF dump (in N-Triples) for semantic search engines such as Sindice [26] along a SPARQL-endpoint for cross-site query is a typical pattern.

To allow semantic search engines to efficiently and effectively process the dataset it is advisable to use proper announcement mechanisms such as the semantic crawler sitemap extension protocol [8].

[10] `http://www.w3.org/TR/rdfa-syntax/`

4.3 Comparison with Other Approaches

The following table presents a comparison between three different approaches for modelling statistics in RDF. The comparison is based on [11] and highlights some differences between the modelling from the D2R Server for Eurostat[12], the 2000 U.S. Census Data [25], and SCOVO itself.

The most distinguishing feature of SCOVO is the ability to express complex statistics over time while still keeping the structural complexity very low. Both other approaches are not capable of representing historical data and only provide statistics for one point-in-time. From the table below we conclude further that SCOVO seems to be the best combination of flexibility and usability, allowing to recreate the data-table structures with a reasonable degree of fidelity in another environment (that is, on the Web). Additionally, in our understanding SCOVO is more aligned with the linked data principles, compared to D2R Eurostat and 2000 U.S. Census.

	D2R Eurostat	**2000 U.S. Census**	**Scovo**
Expressivity	Simple, limited	Complex	Complex
Modelling of time	Point-in-time*	Point-in-time	Over time
Historic data	No*	No	Yes
Easy table (re-)generation	No	Yes	Yes
Access to actual statistical data	Using individual predicates	Using individual predicates	Value attached to items
Location of classifying features	Concatenated in the name of the predicate	Concatenated in the name of the predicate and a chain of predicates	Attached to items
Structural complexity	Flat model	Complex, related via individual predicates; sometimes inconsistent	Relatively flat model; all information attached to item; datasets as container
Use of blank nodes	No	Yes	No
Different predicates	Many	Many	Few
Knowledge needed for query	Predicate names of desired dimensions	Predicate names of desired dimensions and nesting chain	Name or URI of desired dimensions
What is modelled?	Real-world, ignoring statistical artefacts such as time, table, sub-tables, etc.	Statistical domain in question	Statistics in general
Use of deref.-able URIS	Very limited	Limited	Each statistical item has an explicit URI

* The use of different named graphs for different points in time is planned in D2R.

[11] http://www.thefigtrees.net/lee/blog/2008/03/modeling_statistics_in_rdf_a_s.html

[12] http://www4.wiwiss.fu-berlin.de/eurostat/

5 Usages in the Wild

5.1 Eurostat Data—riese

In riese[13], the "RDFizing and Interlinking the EuroStat Data Set Effort" [11,12] we have RDFized, interlinked, and published the Eurostat dataset on the Web. First of all the dimensions and dataset hierarchies defined in Eurostat get translated to RDF and interlinked. The actual data is translated to RDF on-the-fly from the raw Eurostat tables. Both human users and machines can access the data from the same location thanks to embedding RDF on the human-readable pages with XHTML+RDFa. Additional access methods (as described below) are available as well.

RDFication. The translation to RDF is performed using SWI-Prolog. The SWI-Prolog Semantic Web Library provides an infrastructure for reading, querying and storing Semantic Web documents; additionally the Prolog-2-RDF (p2r) modules[14] and individually defined mappings are used for translating the input data to RDF. The input data consists of a table of contents in HTML defining the overall structure of the datasets, dictionary files for resolving the approximately 80,000 different data codes used, and the actual data itself in tab-separated values format. The Eurostat data consists of more than 4,000 datasets containing roughly 350 million data values/items.

In riese HTTP URIs are used, which are compliant with the linked data principles. For instance the currency dimension "Euro" has been minted with the "concept URI" of `http://riese.joanneum.at/dimension/currency/eur` which can be dereferenced for accessing the information resource that describes the concept. Accordingly datasets and individual items have an URI assigned for unambiguous identification of all resources.

Several domain ontologies are being re-used in riese. Geographical dimensions such as countries, etc. for instance make use of the Geonames ontology as they are subclasses of `geonames:Feature` which allows more expressive descriptions. It is hence easily possible to express further classifications such as the class of the geographical dimension (e.g. country, administrative division, etc.). Furthermore, the riese schema re-uses the Event ontology in the same way as described above.

Interlinking. The automated interlinking of country descriptions between riese and Geonames for instance is done using the ISO-3166 alpha2 country codes which are available in both datasets assuring that exactly the same resource is addressed. In the practical implementation this means that a search using the country code is performed in both datasets. According to the nomenclature used by Eurostat it is also possible to identify only country descriptions in the source dataset. The result from the target dataset is restricted to return only countries as well. Finally all matches are being interlinked using `owl:sameAs`. In this case it is possible to create exact matching high-quality interlinks. The generation of interlinks to other LOD datasets such as DBpedia, CIA Factbook, and Wikicompany follows a similar approach.

Publication. All data on riese is accessible for both humans and machines (Semantic Web agents) equally. An Apache 2 Server with a set of PHP scripts is used to render the pages in XHTML+RDFa. A dump of the entire data is also available in RDF/XML which should be used by Semantic Web agents like indexers that want to crawl the

[13] `http://riese.joanneum.at`

[14] `http://moustaki.org/p2r/`

entire content. Providing a simple file download enables the indexers to easily acquire all data and reduces the server load. Furthermore a SPARQL endpoint using SWI-Prolog and p2r is provided. It operates on the raw input data and maps the data on-the-fly. Experiments have shown that for average queries this approach offers acceptable response times at reasonable server cost.

5.2 In Other Vocabularies

SCOVO is used in voiD, the "Vocabulary of Interlinked Datasets" [1] to express information about the number of triples, resources and so forth. Using SCOVO in voiD allows a simple and extendable description of statistical information, however, a shortcoming has been identified: as `scovo:Items` are grouped into `scovo:Datasets`, there is an implicit assumption that all items in such a dataset share the same dimensions. This yields to complex SPARQL expressions, as it will often require a verbose check to make sure that an item has only certain dimensions and no others. An examplary usage of SCOVO in voiD is given below in listing 4. Further, we have gathered that SCOVO is used in the RDFStats framework[15], see Fig. 5 (kudos to Andreas Langegger for the screen shot), that generates statistics for datasets behind SPARQL-endpoints and RDF documents.

```
:DBpedia a void:Dataset ;
         void:statItem [
           scovo:dimension void:numberOfTriples ;
           rdf:value 212576239 ;
         ] ;
}
```

Listing 4. Usage of SCOVO in voiD

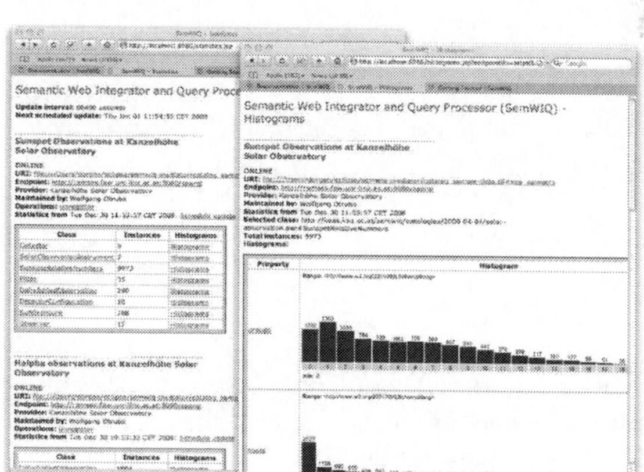

Fig. 5. SCOVO driving the statistics in RDFStats

15 `http://semwiq.faw.uni-linz.ac.at/node/9`

6 Conclusion and Future Work

We have proposed a vocabulary, SCOVO, and discussed good practice guidelines for publishing statistical data on the Web in this paper. The framework aims at supporting people to publish their statistics on the Web of Data in an effective and efficient manner. The framework includes good practice rules stemming from our experience in publishing linked datasets. Finally, we have demonstrated the implementation of our framework and discussed practical issues with it.

However, there are limitations we are aware of. We advocate simplicity, hence there exist edge cases where it is hard to find an appropriate semantic modelling. Take for example our `scovo:Dataset` concept. In the current representation it is not explicitly defined how the overall range of the dataset is expressed. This may yield performance problems when one determines to figure out the overall range of a dataset. It is for sure possible to concatenate single dimensions used on the `scovo:Item`-level—for example concluding from the range of the four quarters `ex:Q12006` to `ex:Q42006` that the dataset actually is referring to the year 2006. Additionally, from the application of SCOVO in voiD we have learned that there is a demand for aggregates. Hence, we plan to add support for data aggregation in a future version of the SCOVO schema.

The United Nations provide a range of statistics about various domains[16]. We plan to publish this available data using the Talis Platform [9]. Due to the high volume of the data, scalability issues are at the centre of the entire design. For example, one part of the UN data set—the Commodity Trade Statistics Database (COMTRADE)—alone provides commodity trade data for all available countries and areas since 1962, containing almost 1.1 billion records. Further, our ongoing work focuses on broadening the deployment base available[17], making converters (from and to SCOVO) available, and extending the framework itself. One very important issue is what we call "statistical-presentation fidelity". When the data is present in a table with a certain layout, it turns out to be advantageous to not only repurpose and link the data, but also reuse the data table in the author's intended form.

Acknowledgements

The research reported in this paper was partially supported by the "Understanding Advertising" (UAd) project[18], funded by the Austrian FIT-IT Programme and the ICT-2007.1.2 ROMULUS project[19], partially funded under the 7th Framework Programme of the European Commission.

References

1. Alexander, K., Cyganiak, R., Hausenblas, M., Zhao, J.: Describing Linked Datasets - On the Design and Usage of voiD, the 'Vocabulary of Interlinked Datasets'. In: WWW 2009 Workshop: Linked Data on the Web (LDOW 2009), Madrid, Spain (2009)

16 http://unstats.un.org/unsd/databases.htm

17 by for example publishing as RDF the OECD data at http://www.sourceoecd.org/database/OECDStat

18 http://www.sembase.at/index.php/UAd

19 http://www.ict-romulus.eu/

2. Allen, J.F., Fergusson, G.: Actions and events in interval temporal logic. Technical Report TR521, University of Rochester, Computer Science Department (1994), http://citeseer.ist.psu.edu/allen94actions.html (last accessed February 2008)
3. Antoch, J.: Environment for statistical computing. Computer Science Review 2(2), 113–122 (2008)
4. Assini, P.: NESSTAR: A Semantic Web Application for Statistical Data and Metadata. In: International Workshop Real World RDF and Semantic Web Applications, 11th International World Wide Web Conference (WWW 2002) (2002)
5. Berners-Lee, T.: Linked Data (2007), http://www.w3.org/DesignIssues/LinkedData.html
6. Bizer, C., Heath, T., Ayers, D., Raimond, Y.: Interlinking Open Data on the Web (Poster). In: 4th European Semantic Web Conference (ESWC 2007), pp. 802–815 (2007)
7. Booth, D.: URI Declaration Versus Use (2008), http://dbooth.org/2007/uri-decl/
8. Cyganiak, R., Delbru, R., Tummarello, G.: Semantic Web Crawling: A Sitemap Extension (2007), http://sw.deri.org/2007/07/sitemapextension
9. Davis, I.: The Talis Platform (2005), http://www.talis.com/applications/downloads/white_papers/TalisPlatform.pdf
10. Grossenbacher, A.: Semantic Web: Basics, RDF, DC and the description of a statistical site (2007), http://tinyurl.com/2d5gta
11. Halb, W., Raimond, Y., Hausenblas, M.: Building Linked Data For Both Humans and Machines. In: WWW 2008 Workshop: Linked Data on the Web (LDOW 2008), Beijing, China (2008)
12. Hausenblas, M., Halb, W., Raimond, Y.: Scripting User Contributed Interlinking. In: 4th Workshop on Scripting for the Semantic Web (SFSW 2008), Tenerife, Spain (2008)
13. Hausenblas, M., Slany, W., Ayers, D.: A Performance and Scalability Metric for Virtual RDF Graphs. In: 3rd Workshop on Scripting for the Semantic Web (SFSW 2007), Innsbruck, Austria (2007)
14. ISO. Statistical data and metadata exchange (SDMX). Standard No. ISO/TS 17369:2005 (2005)
15. Lestina, G.J., LaPlant, W.P., Gillman, D.W., Appel, M.V.: Technical Development of the Proposed Statistical Metadata Standard. Report, Bureau of the Census (1996)
16. McClean, S., Grossmann, W., Froeschl, K.: Towards Metadata-Guided Distributed Statistical Data Processing. In: Proc. of New Techniques and Technologies for Statistics (NTTS) (1998)
17. Noy, N., Rector, A.: Defining N-ary Relations on the Semantic Web. W3C Working Group Note, W3C Semantic Web Best Practices and Deployment Working Group (2006)
18. OECD. Management of Statistical Metadata at the OECD. Report, Organisation for Economic Co-operation and Development, OECD (2006)
19. Papageorgiou, H., Pentaris, F., Theodorou, E., Vardaki, M., Petrakos, M.: Modeling statistical metadata. In: Thirteenth International Conference on Scientific and Statistical Database Management, 2001. SSDBM 2001. Proceedings, pp. 25–35 (2001)
20. Prud'hommeaux, E., Seaborne, A.: SPARQL Query Language for RDF. W3c working draft 4 october 2006, W3C RDF Data Access Working Group (2006)

21. Sauermann, L., Cyganiak, R.: Cool URIs for the Semantic Web. W3C Interest Group Note, W3C Semantic Web Education and Outreach Interest Group (2008)
22. Semantic Web Deployment Working Group. SKOS Simple Knowledge Organization System Reference. W3C Working Draft, Semantic Web Deployment Working Group (2008)
23. Softic, S., Hausenblas, M.: Towards Opinion Mining Through Tracing Discussions on the Web. In: Social Data on the Web (SDoW 2008) Workshop at the 7th International Semantic Web Conference, Karlsruhe, Germany (2008)
24. Stuckenschmidt, H., van Harmelen, F.: Information Sharing on the Semantic Web. Springer, Heidelberg (2005)
25. Tauberer, J.: The 2000 U.S. Census: 1 Billion RDF Triples (2007), `http://www.rdfabout.com/demo/census/`
26. Tummarello, G., Delbru, R., Oren, E.: Sindice. com: Weaving the Open Linked Data. In: Aberer, K., Choi, K.-S., Noy, N., Allemang, D., Lee, K.-I., Nixon, L., Golbeck, J., Mika, P., Maynard, D., Mizoguchi, R., Schreiber, G., Cudré-Mauroux, P. (eds.) ASWC 2007 and ISWC 2007. LNCS, vol. 4825, pp. 552–565. Springer, Heidelberg (2007)
27. UN. Guidelines for Statistical Metadata on the Internet. Report, United Nations Economic Commission for Europe (UNECE) (2000)

Media Meets Semantic Web – How the BBC Uses DBpedia and Linked Data to Make Connections

Georgi Kobilarov[2], Tom Scott[1], Yves Raimond[1], Silver Oliver[1], Chris Sizemore[1], Michael Smethurst[1], Christian Bizer[2], and Robert Lee[3]

[1] British Broadcasting Corporation, London, UK
firstname.lastname@bbc.co.uk
[2] Freie Universität Berlin, Berlin, Germany
firstname.lastname@fu-berlin.de
[3] Rattle Research, Sheffield, UK
robl@rattlecentral.com

Abstract. In this paper, we describe how the BBC is working to integrate data and linking documents across BBC domains by using Semantic Web technology, in particular Linked Data, MusicBrainz and DBpedia. We cover the work of BBC Programmes and BBC Music building Linked Data sites for all music and programmes related brands, and we describe existing projects, ongoing development, and further research we are doing in a joint collaboration between the BBC, Freie Universität Berlin and Rattle Research in order to use DBpedia as the controlled vocabulary and semantic backbone for the whole BBC.

1 Introduction

The Linking Open Data project[1] became one of the main showcases for successful community-driven adoption of Semantic Web technologies in the last year. It aims at developing best practices to opening up the "data gardens" on the Web, interlinking open data sets on the Web and enabling web developers to make use of that rich source of information. But the data made available in that process, the practices and technologies developed, are not only useful for open web data, they also provide benefits to end users and the enterprises at large.

In this paper, we describe how Linked Data technologies [1][2] were applied within the BBC, one of the world's largest broadcasters, and how DBpedia [3], often considered as the linking hub of the Linking Open Data project, and MusicBrainz are used in that process as both an interlinking vocabulary and a data provider. The paper is structured as follows: Frst, we describe the former status quo of multiple (largely unconnected) data sources and categorization systems within the BBC and describe the problems and the requirements for a transition from those unconnected data sources to a more interlinked ecosystems of data.

[1] `http://esw.w3.org/topic/SweoIG/TaskForces/CommunityProjects/LinkingOpenData`

L. Aroyo et al. (Eds.): ESWC 2009, LNCS 5554, pp. 723–737, 2009.

We then describe how we publish programmes data, how we link concepts with DBpedia and through that documents to each other and how that manifests in actual benefits for the users of the BBC main website at www.bbc.co.uk.

1.1 Problem Definition

The BBC publishes large amounts of content online – both text and audio & video. Historically this has focused largely on supporting broadcast brands and a series of domain specific microsites (e.g. food, gardening etc.) to the exclusion of wider integration with the rest of bbc.co.uk, let alone the rest of the web; albeit with some notable exceptions such as *news online*. That is, the focus has been on providing separate, largely stand alone, microsites designed to be accessed via HTML on the desktop.

This means that the BBC has been able to support the primary use cases within the context of the individual microsite but not those use cases that span programme brands or domains. For example, we can tell you who presents Top Gear, but not what else those people have presented. By developing self-contained microsites the BBC has produced some very popular services but it has also failed to reach its full potential because it hasn't been able to join up all of its resources. By failing to link up the content (on both a data and a user experience level) the content published has never been able to be greater than the sum of its parts. As a user it is very difficult to find everything the BBC has published about any given subject, nor can you easily navigate across BBC domains following a particular semantic thread. For example, you can't yet navigate from a page about a musician to a page with all the programmes that have played that artist.

Furthermore, while BBC Backstage[2] has made great progress in making BBC data available for third party developers we realized that there was much more that could be achieved if machine representations were made available in the same way as the HTML views.

As with many large publishers of Web content, the BBC has divided its different services by domain, e.g. food, music, news etc. Each service is maintained by a different team; as have the programme support sites, which have historically been commissioned independently of each other. This has made it difficult to coordinate interlinking between services and programme sites.

The development of bbc.co.uk/music/[beta] sought to address the interlinking of services around the music domain - music played on programmes, at events or in sessions. Likewise bbc.co.uk/programmes aims to provide a central programme support service which the rest of the business can use.

However, neither music nor programmes solved the problem of cross domain linking, nor do they address the issue of disambiguation between multiple controlled vocabularies - the fact that Madonna is both an artist (in MusicBrainz), and an actor and a person (in Wikipedia). And possibly more significantly nor does it address the other domains the BBC has an interest in - news, food, books,

[2] http://backstage.bbc.co.uk

sport, natural history etc. To address these issues the BBC, Freie Universität Berlin and Rattle Research are investigating using DBpedia to provide a common "controlled" vocabulary and equivalency service, which in turn is used to add "topic badges" to existing, legacy web pages.

1.2 Objectives

As we've seen in the previous section, there are a large number of data sources within the BBC that are either unconnected or only cross-connected for specific use cases. This scenario is not an uncommon one in large enterprises. While - in theory - when designing a data infrastructure it should be possible to design a system that connects and interlinks all domains – the reality however is almost always different. Organic growth of organizations and supporting systems leads to a diverse ecosystem and trying to simply start again from scratch is clearly an untenable option.

Enterprises the size of the BBC have invested significant resources, over a number of years, into their existing web offerings and redesigning and redeploying all of them at once is impossible not only from a deployment point of view, but also from an organizational one. Therefore an approach needs to be taken that results in better connections and interlinking of existing systems – providing a soft transition and reducing the impact on existing systems – while at the same time, where possible, developing new services to maximize the interlinking of domains. The Semantic Web, especially Linked Data technologies, can offer such a soft transition.

Our objectives then are fourfold:

1. to develop a new service that supports the branding of our Radio stations, TV channels and programmes (bbc.co.uk/programmes) while at the same time ensuring that users and third party developers were able to traverse the graph of BBC data though to other data sources elsewhere in the Linked Data cloud – following contextually related links across all BBC content or content elsewhere on the web.
2. to develop a new music offering (bbc.co.uk/music/beta) that builds on existing open web standards and is fully integrated with the emerging programme support service.
3. to retrofit simple navigational elements (i.e. Topic Badges and term extraction) to existing, legacy pages, and new pages built with legacy systems to support contextual, semantic navigation.
4. to provide a common set of web scale identifiers to help classify all BBC online content (and external URLs) and to help create equivalency between multiple vocabularies.

In meeting these objectives it is our hope that bbc.co.uk becomes a more coherent, useful place by providing contextual, semantic links connecting content across different domains and providing meaningful navigation paths for our users.

So the main part of this paper will first describe how BBC Programmes and Music were developed in order to publish Linked Data, both for use internal and for external developers. We will then describe legacy systems, in particular a content categorization system called CIS and how we interlinked concepts from CIS with DBpedia as the controlled vocabulary for integrating the different BBC domains. Then we turn to documents, which still are the main part of the content published by the BBC, and how we interlink those through the work of using DBpedia as vocabulary for concepts. And finally we will present how our users benefit from better, more coherent, user journeys across www.bbc.co.uk.

2 Publishing and Linking BBC Programmes and Music

The previous piecemeal approach to programme support afforded an opportunity to develop a new service bbc.co.uk/programmes (which launched in Summer 2007) with the objective of providing one URI per programme - for every programme the BBC broadcasts - allowing other teams within the BBC to incorporate those pages into new and existing programme support sites, TV Channel and Radio Station sites.

Likewise there was an opportunity to redevelop the music site along the same principles - but in this case a URI for every artist (and eventually a URI for every track) the BBC plays. However, although there are commonalities between programmes and music there are also differences in the development approach that are worth noting.

BBC programmes is underpinned with a proprietary database – PIPs – which seeks to contain the definitive record of all public facing BBC programme metadata. The public representation of this data is, in part, exposed at bbc.co.uk/iplayer and also at bbc.co.uk/programme. The later representation is published in accordance with the programmes ontology[3].

Unlike BBC programmes the BBC does not "own" the music domain and as such it is important that the BBC adopts existing web identifiers to ease the development effort for both BBC software engineers and third party engineers. MusicBrainz, the community maintained music metadata service, provides such web identifiers.

However, artists also "exist" in other domains – for example Madonna is a singer and an actress (hence is also part of a movies domain) and also a person. The BBC therefore needs a mechanism to create equivalence between two or more identifiers from different domains.

Both BBC Music and Programmes provide persistent web identifiers and a set of corresponding representations for BBC programmes and music artists, bringing BBC data to the Semantic Web. These services were designed according to the Linked Data principles coined in [1]:

- Web identifiers are used to denote things: entities within the scope of the Programmes Ontology (brands, series, episodes, versions, services and broadcasts) and of the Music Ontology [4] (music artists);
- These web identifiers have multiple representations, including:

[3] http://purl.org/ontology/po/

- An XHTML representation, designed for human interaction;
- An RDF representation, exposing structured BBC data;
- These representations hold links to further web identifiers, allowing to discover more structured data. For example, the representation of an artist in BBC Music holds an *owl:sameAs* link to the corresponding artist in DBpedia.

We were also convinced that the value in programme websites is not in the implicit metadata of the domain model but rather in the way this domain model overlapped and intersected with other domains. As ever, the links are more important than the nodes because that's where the context lives:

programmes:segment <features> music:track,
programmes:segment <features> food:recipe etc.

In this way we could weave new "user journeys" into and out of /programmes, into and out of bbc.co.uk. But to achieve that, we needed to bring more of our legacy content and systems into the picture.

3 Cross-Linking Legacy Content and Legacy Systems

We have seen how the BBC programmes domain has successfully modeled programme specific relationships. This, though progress, represents a comparatively small amount of the BBC content and it is desirable to link to further BBC domains. One way of doing this was through an *about*-relationship between programmes and people, places and subjects. This data was created with a legacy auto-categorization system called CIS.

CIS holds a hierarchy of terms in five main top-level classes: proper names, subjects, brands, time periods and places. The system was originally used for annotating regional news stories in a content management system. This meant that the vocabulary was primarily focused around British regional content. This included many local bands, people and events. In addition to this at its core is a more useful general subject vocabulary and list of locations that would be of use to the annotation of BBC programming. Based on this more useful parts of its vocabulary (around about 40% of the total concepts) and the expertise already existing in the BBC it was chosen as an initial solution for annotating BBC programmes.

CIS was put in place as an automated system to categorize programmes based on their textual description. For example, the programme synopsis "a look ahead to the Beijing Olympics, including a preview of British boxing hopes" would be categorized with "Beijing" (place), "British" and "Boxing" (subjects). The term "Beijing" can then become the link between this programme and other programmes about Bejing. Addtionally this could be used to link to BBC news stories (if news use similar identifiers) related to Beijing. This approach could help the interlinking of different BBC services, whilst keeping developing them independently. The key to this is ensuring there are mappings between the various vocabularies in use.

However, even though this approach highlights the need for a common vocabulary, shared across the different BBC services and acting as a set of links between them, a single categorization system is difficult to maintain, and it is difficult to cover every single entity that might be of interest. Moreover, no relations relating terms together are available within CIS. For example, it would be impossible using a CIS-based framework to access the relationship between Beijing and the Beijing Olympics. In order to provide a satisfying user experience, we need richer relationships between these different terms. Furthermore CIS terms will only ever be an internal identifier and so will never help automating the linking of BBC resources to non-BBC data.

In addition, the objects identified with /programmes and /music are also to be found within other domains. Ideally we need a mechanism by which we map between equivalent terms.

So we decided to look for a common set of web identifiers for the BBC. The DBpedia project[4], which is bringing information extracted from Wikipedia to the Semantic Web, had at that time already developed into the de-facto interlinking hub for the Linking Open Data project, so it was the obvious choice to rely on DBpedia identifiers and this way at the same time joining the Linked Data Web. And while DBpedia doesn't only provide Linked Data URIs for a broad range of concepts, but also structured data about those concepts and their relationships, that data can be used to power (semi-)automatic interlinking algorithms, serve as a source of data to be used and displayed in BBC applications, and provide the rich relationships between terms we need. So DBpedia becomes the controlled vocabulary to connect our BBC domains such as Music, News, Topics and Programmes as shown in Figure 1. Although considering Wikipedia as a "controlled" vocabulary is arguable due to changes being made to Wikipedia article URIs, Hepp et al. [5] note that these changes occur less often than expected, and DBpedia can provide mechanisms to capture those changes through redirect resolution.

BBC domains contain both concepts and structured data as well as documents and other content. In order to achieve our cross-domain linking of concepts and content, we will in the next section describe our attempts to first interlink concepts and then content about concepts with DBpedia.

4 Linking CIS Concepts to DBpedia

In order to quickly bridge our existing CIS content with the DBpedia-centric ecosystem, we developed a system to automatically interlink CIS concepts with DBpedia. In the following sections, we will describe our approach, give an overview of the algorithm we have implemented and provide a preliminary evaluation of the interlinking results.

4.1 Interlinking Approach

CIS contains 150,000 terms overall, covering four different domains: BBC brands, locations, people and general subjects. Each domain has its one term hierarchy categorizing the terms expressed in SKOS [9].

[4] http://dbpedia.org

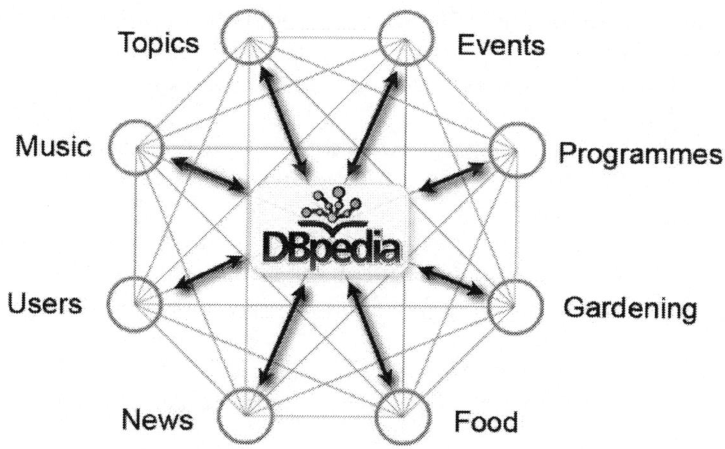

Fig. 1. Linking BBC Domains

For the interlinking only the CIS taxonomy itself could be used, for various reasons there was no access to documents tagged with CIS terms. So the interlinking algorithm had to work only with the information we could get from CIS, including labels and categorization information. The main challenge was to find the most probable matches based on label lookup of CIS terms in DBpedia and to disambiguate those matches using classification information.

This problem is a well known one in the research area of ontology matching [6][7]. However, we needed to developed a solution that is tailored to our very specific use case, taking into account the very limited information on CIS terms that were available for the matching implementation, as well as some of the specific characteristics of DBpedia.

The core idea of our approach can be described as similarity clustering based on the "context" of concepts. While the term "apple" is in itself ambiguous, given the context of the terms "microsoft" and "google", the meaning of "apple" referring to Apple Inc. becomes clear. This assumption forms the basis for our CIS interlinking. We use the available classification information in CIS to build similarity clusters (concepts in the same category or parenthesis-based class) that help us disambiguate the meaning of a given term. In DBpedia, we can then calculate a similarity metric for the tuples of meanings based on the node distances in the DBpedia categorization and classification graph.

So the algorithm we developed is divided into two parts: DBpedia label lookup and context-based result disambiguation. The overall automated interlinking approach is therefore similar to the one described in [8], but differs in the way the context identification is performed and the results are ranked.

4.2 DBpedia Label Lookup

The core of the linking process is a weighted label lookup [10]. DBpedia contains 2.5 million concepts, and most input strings from CIS match to several concepts:

"shakespeare" matches to over 50 DBpedia resources. In order to find the most likely matches for a given term, the system uses a weighted label lookup, using Wikipedia inter-article-links as weight indicator. PageRank works based on the assumption that the more important a web page is, the more hyperlinks on the web will point to that page. We found out that in Wikipedia the number of inter-article-links pointing to an article can be similarly treated as indicator for the articles' overall relevance. The Wikipedia article about William Shakespeare has over 6000 links pointing to it, while the article about Nicholas Shakespeare only has 18. In order to handle synonyms and abbreviations (such as "EU" referring to the European Union) we included Wikipedia redirects into the index, calculating their weight as `links(redirect)*log2(weight(article))`.

DBpedia Lookup builds on a custom Lucene index which combines our relevance metric with Lucene's string-similarity-based ranking, and is available as web-service[5].

4.3 Context-Based Disambiguation

In order to disambiguate the possible matches, we identify similarity contexts of CIS terms by clustering matches and finding according contexts in DBpedia. We used the CIS categorization hierarchy and parenthesis texts as clusters. The concept "Mary (1985 sitcom)" falls into the clusters "television" (the category), "1985" and "sitcom" (parenthesis texts). The algorithm creates these clusters for all CIS terms and identify matching DBpedia categories, classes and templates for each cluster based on the multiple possible DBpedia matches.

Those identified contexts are then used to disambiguate matches for every CIS concept. In addition, we added restrictions based on the CIS domain: stemming is used for subjects, not for people's names, locations must use a geo-coordinate template, and for brands the most relevant templates in DBpedia were identified manually.

With the previous example of "Mary (1985 sitcom)" falling into the clusters together with other sitcoms and television shows, we were able to reject the top ranked label-based result "Mary (Holy Mother)" for the search term "Mary" due to its DBpedia class, reject "Something about Mary" and "Mary Tylor Moore Show" and accept the match `dbpedia:Mary_(1985_TV_series)` based on the DBpedia category "1980s American television series" (see figures 2 and 3).

4.4 Evaluation

The goal of this first automated interlinking approach was to only create links with high confidence values (i.e. reducing the number of false positives or false links) while accepting that we will have missed many potentially correct links (i.e. having many false negatives). Table 1 shows our interlinking results.

One reason why we have only interlinked 20% - 30% of the brands, locations and names datasets is that many concepts don't have their own Wikipedia article. TV Episodes for example are often merged into one single list-article, which

[5] http://lookup.dbpedia.org/api/search.asmx

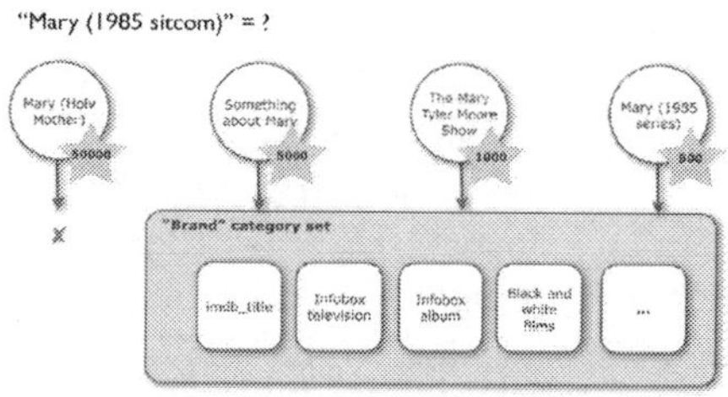

Fig. 2. Context Identification

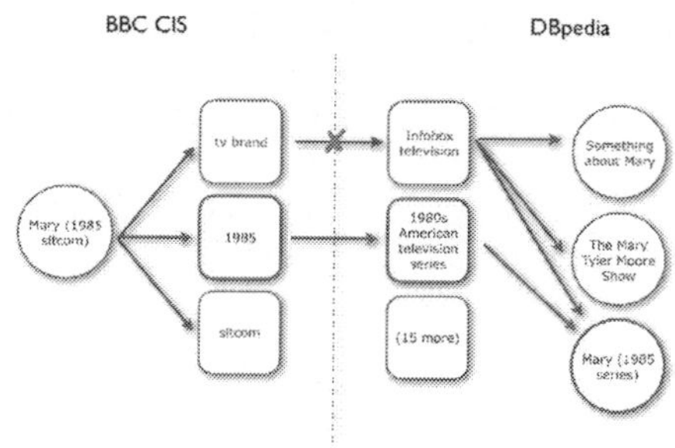

Fig. 3. Context-based Disambiguation

Table 1. CIS / DBpedia Interlinking Results

	Total	Linked	Precision (test set)	Recall (test set)
Brand	6,630	1,267 (19%)	86%	41%
Location	55,943	11,316 (20%)	99%	77%
Name	73,442	22,341 (30%)	92%	67%
Subject	11,231	6,822 (61%)	92%	75%

would be inappropriate to use as URI for every episode. And many people don't exist in Wikipedia due to low notability. We manually created a test set of 540 CIS-DBpedia links for important concepts of each domain (100 brands, 150 locations, 130 names, 160 subjects), in order to check the precision and recall of our algorithm. While we have included CIS concepts with no according DBpedia concept into the test sets to test for false positives, their percentage in the test sets does not represent their real distribution in the main data sets. We want to also note that this work has been done before the release of DBpedia 3.2, which provides a much cleaner dataset and a hierarchical ontology. We are convinced that we would now be able to achieve much less false negative results, but haven't evaluated that yet.

5 Linking Documents to Concepts

In the previous sections we focused on how we have published and interlinked existing structured data and concepts. Now we turn to documents. The BBC has got a lot of textual documents, such as editorial web sites and news articles. Interlinking the CIS vocabulary with DBpedia brings a part of the available documents into our linked data ecosystem, but there are much more. In this section, we describe how we use a named entity extraction system called Muddy Boots to process BBC News articles, recognize entities in those articles, and use DBpedia identifier for those entities in order to bring BBC News into the picture. Muddy Boots also provides us with the text processing capabilities we need for our semi-automated Content Link Tool, described later.

5.1 Named Entity Recognition: Muddy Boots

During the past year, Named Entity Recognition (NER) became one of the main focus of so called Semantic Web startups and companies. Products such as OpenCalais[6], Twine[7], and Zemanta[8] built on NER in order to extract concepts (mostly people, locations and companies) from textual input. OpenCalais and Zemanta provide APIs to enable bloggers and businesses to use their NER systems. But those systems did use their own entity identifiers, so we looked for a system that actually reuses existing web identifiers, i.e. Wikipedia/DBpedia URIs. (OpenCalais and Zemanta recently announced to link their entity IDs to DBpedia URIs.)

The BBC worked with Rattle Research to create a system that used DBpedia URIs as a controlled vocabulary to identify entities [11] [12] in their news archive, called Muddy Boots. The Muddy Boots system parses the story body from a given BBC news URI and then uses a NER system in combination with the Yahoo Term Extraction API[9] to extract the main entities from the story. These entities are just text and have no semantic meaning or classification attached to them at this point. They are matched to possible resources in DBpedia using a simple fuzzy logic algorithm that compares DBpedia resource titles to the extracted entities. In the event of multiple DBpedia resources being identified, the system allocates each of these to the extracted term as a possible valid match.

The system follows any DBpedia `redirects` and `disambiguates` predicate and compiles a list of possible DBpedia resources for each term and ranks each identified DBpedia resource, using a contextual disambiguation. For a story about 'Apple Inc', there will be terms such as 'Steve Jobs', 'Iphone' and 'Macbook'. By using the complete list of extracted terms from the story and applying a similarity algorithm across the resources identified to rank them in terms of their similarity to a document consisting of just the extracted terms, the system uses the highest ranking resource as the disambiguation for the extracted term.

This allows Muddy Boots to create a mapping between a list of extracted terms to their probable counterparts in DBpedia. The final step is to identify the resources that correspond to 'people' and 'companies', by examining the predicates for each resource and creating a scoring algorithm that assigns points to a categorisation (either 'people' or 'company') based on the predicates that are present, such as 'birthdate' or 'birthplace' for a person. The categorisation with the highest score is assigned to the resource and an overall 'confidence' metric is assigned using a weighted algorithm that combines results from the similarity metric, the categorisation score and the ambiguity of any given term.

The Muddy Boots prototype produced a system that uses DBpedia as a controlled vocabulary to unambiguously identify the main actors in a piece of content. The system has been further developed to utilise the new DBpedia ontology and only attempts to categorise resources that do not have a classification

[6] http://www.opencalais.com/
[7] http://www.twine.com
[8] http://www.zemanta.com
[9] http://developer.yahoo.com/search/content/V1/termExtraction.html

within this. This allows the system to identify an increased number of entities in a wider variety of categories. The system currently powers the Content Link Tool resource recommendation system, described in the next section, and is also available as a set of publically available web services (http://www.muddy.it).

5.2 Content Link Tool

No matter how good auto-categorisation methods are there will always be the need for editorial intervention. This meant the BBC would need a tool with which to add or remove DBpedia identifiers from any given BBC URL. The DBpedia concepts and associated URIs could be then stored centrally to be made available for any service that required BBC URLs about a DBpedia concept.

Previous experience with semi-automated tagging had taught us that the user interface of any annotation tool was critical to the creation of high quality metadata. The tool would also need to incorporate high quality automated suggestions to make the annotation process as painless as possible. The Muddy Boots system was chosen as the source of automatic suggestions based on the DBpedia dataset. In addition users could search for terms that had not been suggested. The search box incorporates auto-completion of keywords based on DBpedia resource titles and synonyms using the DBpedia Lookup web service for auto-completion, and displayed abstracts make it easy for BBC editors to choose from the suggested concepts. Selected terms are simply added to the list of concepts for that URL. Figure 4 shows a screenshot of the Content Link Tool.

The cultural aspect of editorial annotation was also considered and lessons have been learnt from the past. Any added or removed concepts will be immediately reflected in the related links on the content page. This means there is a user facing editorial aspect to the annotation, concepts end up as links on the page as opposed to hidden in the HTML. Another change has been the move away from the language of tagging to the language of linking, where as editors will use the 'Content Link Tool' as opposed to the 'Content Tag Tool'. These are attempts to ensure content creators take annotation seriously and see it as an integral part of the content creation process.

6 Using Concept Links to Create User Journeys: Topic Pages and Navigation Badges

The lack of linked structured data discussed in the beginning made it difficult to present a coherent BBC website. Scalable cross site navigation on the whole is dependent on the interlinking of well modeled data via agreed identifiers. Obviously this is more difficult with the unstructured content, which, in the case of the BBC, is the majority of it.

One solution to this problem is the creation of aggregation pages of unstructured and structured content. These pages pull together the modeled world of BBC programmes (CIS identifiers mapped to DBpedia) and the unstructured world of BBC News articles. These types of aggregation pages (topic pages)

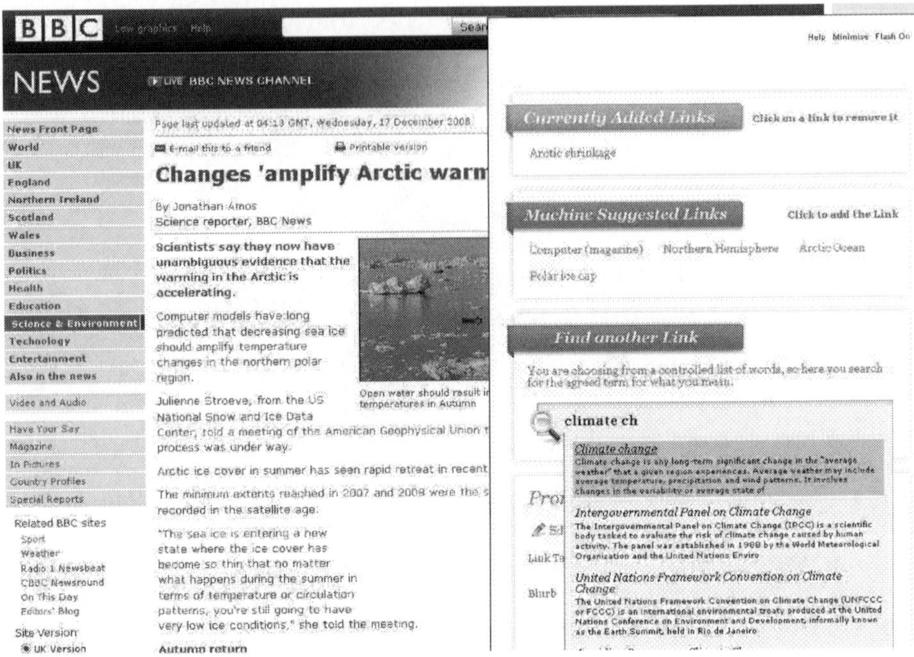

Fig. 4. Tagging and Linking BBC News Articles

have become popular with sites like the New York Times and CNN primarily because of their ability to focus search engines on a given keyword. For the BBC aggregation pages have the additional benefit as navigational nodes to facilitate journeys across unlinked content domains.

When deciding on a vocabulary to drive the BBC aggregation pages a decision was made to use DBpedia. The DBpedia vocabulary offered a number of advantages including:

- Interoperability with other BBC domain specific data such as MusicBrainz.
- It is cheaper to maintain than existing internally maintained vocabularies.
- DBpedia has additional data available, for example: short descriptions, temporal and geo-location data to enrich aggregation pages.
- DBpedia offers rich associative and hierarchical relationships.
- DBpedia's descriptive text can be used as training material for auto-categorisation systems.

Using DBpedia as our vocabulary meant we could join the domain modeled structured data that are using Linking Open Data identifiers (like MusicBrainz) with the unstructured content aggregated using auto-categorisation systems based on the DBpedia dataset (in this case an implementation of Autonomy). In addition any machine annotation can be editorially managed using the Content Link Tool.

These aggregation pages allow a user to go from a given person, place or subject to any BBC content area. But this is only half the story because once a user has entered an area of BBC content there are few links through to other related content. An example might be going from the 'carrots' aggregation page to a particular recipe (BBC Food) but then not being able to go to a related health story (BBC Health). What is needed is a link on the content page back to the related aggregation page linking up the user journey. Providing this link is the role of the navigation badge.

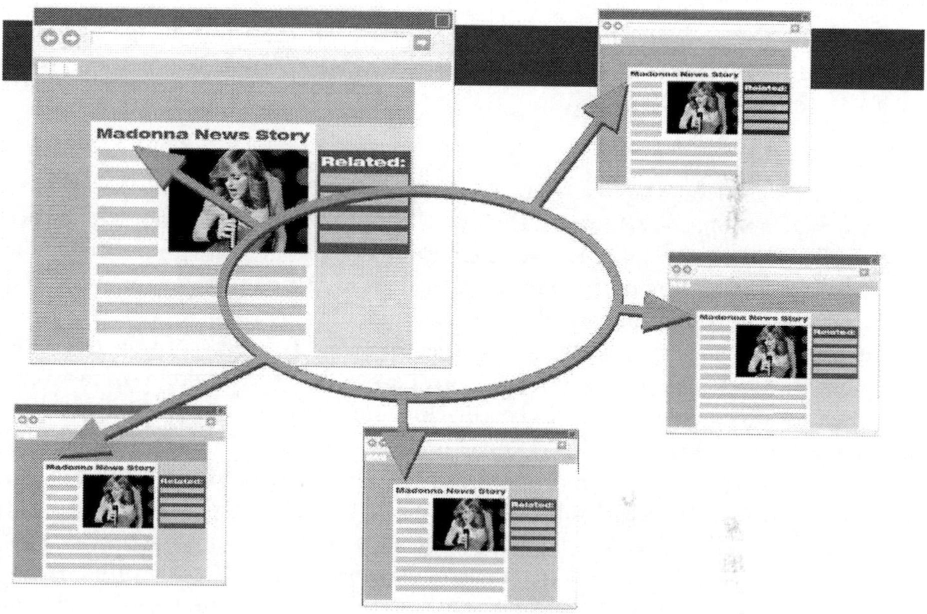

Fig. 5. Linking Documents via Navigation Badges

A navigation badge is dynamically created based on entity extraction of the content on the page. The identified concepts are then compared to existing aggregation pages and where an aggregation page exists it is linked to. This is done using Muddy Boots via the Content Link Tool. In addition to the automatic application of links to aggregation pages editorial staff will be able to add and remove links as appropriate. The navigation badge 'knows' which navigation page to link to because both services are based on DBpedia identifiers. Figure 5 illustrates how different stories about Madonna are interlinked via navigation badges, in the figure shown as "related"-boxes.

The result is a coherent and scalable user journey via a combination of aggregation pages and navigation badges. It is DBpedia in this case that is providing the semantic backbone with which to create the nodes on these journeys and ensure all services are using the same identifiers to express a given concept.

7 Conclusion and Outlook

In this paper, we described how Linked Data technologies were applied by the BBC. We mentioned two BBC services, BBC Music and BBC Programmes, publishing data in two different domains. We then described a categorisation system, CIS, allowing us to interlink data items in these different services because of the successful interlinking of CIS categories with corresponding DBpedia web identifiers, in order to get access to richer relationships between concepts. Finally, we demonstrated how these links between data items can benefit our user facing web sites, through topic pages and navigation badges.

There are three main next development steps we want to outline, as those need to be addressed to continue our path of better connecting BBC content with the help of DBpedia:

DBpedia Live. The DBpedia dataset is currently updated only every 2-3 months based on the releases of Wikipedia database dumps. In order to make DBpedia a real-time RDF view of Wikipedia, the DBpedia framework will be adjusted to support live extraction and updates of the dataset. Live extraction will provide instant feedback for users adding information and make that information available for use immediately. BBC News editors can at the time they write an article about an event create according DBpedia concepts (such as a previously unknown person being involved in some event) and instantly access them via the mechanisms described in this paper.

More domains. Along the same lines as programmes and music we are looking to add additional domains, specifically, events (e.g. music festivals) and natural history (species, habitats) content. This will allow the BBC to more accurately reflect the nature of events such as the Proms - cross linking with those programmes covering the event and those artists featured at the event. The work to create and publish a nature history domain at bbc.co.uk will allow the cross linking of programmes, news stories etc. through natural history concepts.

Equivalency Engine. Finally, our work linking CIS and DBpedia was just a starting point. There are more ID systems and concept stores in use at the BBC, and we want to build a generic equivalency engine based on the methods described above (and similar to link discovery tools being developed in the meantime[13][14]) in order to interlink those remaining systems. We think that such an equivalency engine would not only help the BBC interlink their internal datasets with each other and with DBpedia, but would also be beneficial for outside developers, enterprises and the larger web.

In conclusion we believe that the BBC and their users can largely benefit from the better connected ecosystem of content we are creating, and we hope that more content providers will join us on that road of interlinking content with the Linking Open Data project and in particular with DBpedia and MusicBrainz to create more meaningful navigation paths not only within websites but across the whole web.

Acknowledgments

We are grateful to all the people who made this work possible, in particular to Patrick Sinclair, Nicholas Humfrey, Derek Harvie, Matthew Wood, Frances McNamara, Andrew Shearer and Sophie Walpole from the BBC and Christian Becker from Freie Universität Berlin.

References

1. Berners-Lee, T.: Design Issues: Linked Data (2006), http://www.w3.org/DesignIssues/LinkedData.html
2. Bizer, C., Cyganiak, R., Heath, T.: How to publish Linked Data on the Web (2007), http://www4.wiwiss.fu-berlin.de/bizer/pub/LinkedDataTutorial/
3. Auer, S., Bizer, C., Kobilarov, G., Lehmann, J., Cyganiak, R., Ives, Z.: DBpedia - A Nucles for a Web of Data. In: Aberer, K., Choi, K.-S., Noy, N., Allemang, D., Lee, K.-I., Nixon, L., Golbeck, J., Mika, P., Maynard, D., Mizoguchi, R., Schreiber, G., Cudré-Mauroux, P. (eds.) ISWC 2007. LNCS, vol. 4825, pp. 722–735. Springer, Heidelberg (2007)
4. Giasson, F., Raimond, Y.: Music Ontology Specification (2007), http://musicontology.com
5. Hepp, M., Bachlechner, D., Siorpaes, K.: Harvesting Wiki Consensus - Using Wikipedia Entries as Ontology Elements. In: Proceedings of the Workshop on Semantic Wikis at the ESWC 2006 (2006)
6. Euzenat, J., Shvaiko, P.: Ontology Matching. Springer, Heidelberg (2007)
7. Shvaiko, P., Euzenat, J.: A survey of schema-based matching approaches. Journal on Data Semantics IV (2005)
8. Raimond, Y., Sutton, C., Sandler, M.: Automatic Interlinking of Music Datasets on the Semantic Web. In: Proceedings of the Linked Data on the Web (LDOW 2008) Workshop at 17th International World Wide Web Confernce (2008)
9. Miles, A., Matthews, B., Wilson, M., Brickley, D.: SKOS Core: Simple Knowledge Organisation for the Web. In: Proceedings of the International Conference on Dublin Core 2005 (2005)
10. Zaragoza, H., et al.: Ranking Very Many Typed Entities on Wikipedia. In: Proceedings of the Sixteenth ACM International Conference on Information and Knowledge Management (2007)
11. Dakka, W., Cucerzan, S.: Augmenting Wikipedia with Named Entity Tags. In: Proceedings of the 3rd International Joint Conference on Natural Language Processing (2008)
12. Bunescu, R., Pasca, M.: Using Encyclopedic Knowledge for Named Entity Disambiguation. In: Proceedings of the 11th Conference of the EACL (2006)
13. Volz, J., Bizer, C., Gaedke, M., Kobilarov, G.: Silk – A Link Discovery Framework for the Web of Data. In: Proceeding of the 2nd Workshop about Linked Data on the Web (2009)
14. Hassanzadeh, O., et al.: A Declarative Framework for Semantic Link Discovery over Relational Data. Poster at 18th World Wide Web Conference (2009)

Creating Digital Resources from Legacy Documents: An Experience Report from the Biosystematics Domain

Guido Sautter[1], Klemens Böhm[1], Donat Agosti[2], and Christiana Klingenberg[3]

[1] Universität Karlsruhe (TH), Am Fasanengarten 5, 76128 Karlsruhe
{sautter,boehm}@ipd.uka.de
[2] Am. Mus. of Nat. Hist., Central Park West at 79th, New York, NY 10024-5192
[3] Staatliches Museum für Naturkunde, Erbprinzenstr. 13, 76133 Karlsruhe

Abstract. Digitized legacy document marked up with XML can be used in many ways, e.g., to generate RDF statements about the world described. A prerequisite for doing so is that the document markup is of sufficient quality. Since fully automated markup-generation methods cannot ensure this, manual corrections and cleaning are indispensable. In this paper, we report on our experiences from a digitization and markup project for a large corpus of legacy documents from the biosystematics domain, with a focus on the use of modern tools. The markup created covers both document structure and semantic details. In contrast to previous markup projects reported on in literature, our corpus consists of large publications that comprise many different semantic units, and the documents contain OCR noise and layout artifacts. A core insight is that digitization and automated markup on the one hand and manual cleaning and correction on the other hand should be tightly interleaved, and that tools supporting this integration yield a significant improvement.

Keywords: Semantic XML Markup, Digital Resources, RDF Generation.

1 Introduction

Documents marked up according to specific XML schemas are important resources in the sciences. They can be stored in databases or can be used for RDF generation, facilitating highly specific search, data mining, and machine reasoning. A prerequisite is that the markup is accurate. Fully automated digitization and markup-generation methods cannot ensure this. This makes (manual) correction indispensable.

The objective of our work has been to make all publications on the ant fauna of Madagascar available as machine-readable resources, marked up as TaxonX documents [4]. The project started in January 2007, and two biologists have worked on it for more than a year. Marking up documents has consumed most of their time, and the goal of this paper is to report on the experiences we have gained with this work. There are 119 documents in total, comprising about 2,500 pages with roughly 1,000,000 words. The most important parts are the more than 4,000 taxonomic treatments, i.e., atomic semantic data units within the documents. The 119 documents are from the last 250 years, with a broad variety in layout and style. The documents are written in

L. Aroyo et al. (Eds.): ESWC 2009, LNCS 5554, pp. 738–752, 2009.

five different languages: English (32 documents), French (54), Italian (5), German (16), and Latin (12). Consequently, reliable rules for automated markup are hard to find/formulate, and automated markup generation requires considerable manual corrections. That high diversity is one reason why our project has been more ambitious than previous ones. Our original idea had been to run automatic digitization and markup tools over the corpus and then correct the result manually. However, we realized soon that this approach is too undifferentiated and not efficient. The reason is the complex structure of the markup and of its generation process, which consists of many steps. Some steps depend on others, i.e., the results of some steps serve as input for later ones. The markup process and correction should be intertwined, i.e., markup generated should typically be corrected right away, before other steps make use of it. Otherwise, errors propagate and add up. This means that we had to deal with the markup process much more extensively than anticipated originally.

The Madagascar corpus has a number of characteristics that have added to the difficulty of markup generation: (1) In the printed originals, the logical reading order of the text is not always identical to the physical one. To improve the search opportunities, it has been necessary to restructure the text to its logical reading order. In particular, individual treatments should not be interleaved with each other or with other parts of a document. (2) The text needs to be free of artifacts originating from print layout, like page numbers. Inclusions like captions and footnotes have to be moved out of paragraphs if they interrupted the text flow otherwise, and hyphenated words need to be rejoined. (3) The individual treatments must be marked up and become accessible individually. (4) To facilitate specific queries and RDF generation, domain-specific details (taxonomic names and locations in this current context) need to be marked up. (5) 'Important' abbreviations, e.g., those within taxonomic names must be resolved to their full meaning. This is because their meaning is clear in the context of the publication as a whole, but not in a treatment looked at in isolation.

Other legacy documents whose digitization would be valuable have similar characteristics. In biomedicine, for instance, historical reports on diseases are valuable to reconstruct the spread of epidemics. Digitization of such documents has to deal with Issues 1, 2, 4, and 5. Historical newspapers and books are important resources as well. Projects creating machine-readable resources from legacy documents will have to deal with most the above Issues 1-5. While various projects aim at digitizing printed information (e.g., the Google Libraries Project), they do not always go beyond automated OCR and spend relatively little effort on error correction. The markup we have added to the Madagascar corpus is hardly within their scope. Further, previous corpus-creation projects which bear some similarity with ours have not faced the same complexity. While projects like [13] have shown that NLP (natural language processing) in combination with subsequent manual correction is viable, their setting has been simpler. A characteristic of [13] and others is that it has dealt with small pieces of text that are individual semantic units (in particular, MEDLINE abstracts [16]) that are clean, i.e., the markup is accurate, and annotations only refer to words within sentences. We in turn deal with large publications comprising many treatments. As an effect of digitization, the documents also contain a lot of optical character recognition (OCR) errors and layout artifacts, which need to be cleaned up.

To support interleaving automated markup generation with manual correction, we have developed the GoldenGATE Editor [18]. It seamlessly integrates both automated markup components and custom dialogs that support users in correcting the results of automated steps. Further, while users can fully edit textual content, XML editing works on element level. This is helpful in two ways: The user does not have to deal with XML syntax details, i.e., on the character level, and the editor can keep the document well-formed. In this way, it combines the functionality of commercial XML editors like Altova XMLSpy [2] with that of NLP frameworks like GATE [9].

Many insights we have gained with our work are not domain-specific, but are of general interest when creating machine-readable resources from legacy documents: (1) Appropriate ordering of the individual steps may reduce the effort significantly. For instance, instead of first identifying sections and then their titles in a top-down fashion, the reverse order yields higher accuracy. This is because the titles help to identify where sections start. (2) To assist users in correcting auto-generated markup, it is worthwhile to create specialized document views that only provide the required information and options, and render work more intuitive. (3) The effort for manual corrections depends on characteristics of the components that generate the markup, especially when extracting document details: While a user has to search for false negatives by sifting through the whole document, he can sort out false positives rather easily using, for instance, a list view of all extracted details. Hence, it is advantageous to optimize for recall, even at the cost of some precision. (4) A plug-in based architecture like that of the GoldenGATE editor is useful: It facilitates quick deployment of enhanced automated components that follow new insights gained during the work.

Paper outline: Section 2 discusses related work. Section 3 describes the document corpus, Section 4 the software used. Section 5 describes the markup process. Section 6 reviews some statistics and features an outlook. Section 7 concludes.

2 Related Work

Annotated corpora have been created in different domains, to provide input data for document analysis. Projects like the Penn Treebank Project [13] show that NLP reduces the effort. According to publications [5, 8, 12, 13], most of these projects work in two steps: First, NLP tools process the texts. Second, domain experts manually correct the output of the first step, up to a 98% level of inter-annotator agreement in the case of [13]. The annotations[1] cover low levels in the document structure, i.e., part-of-speech and sentence structure [13], word sense [8], named entities [5], and domain-specific terms [12]. However, we are not aware of any studies on how to mark up higher levels of the structure, like sections, treatments, subsections, and paragraphs. This is different from corpus annotation in two ways: First, corpus annotation usually works on small pieces of text that form single semantic units, e.g., the MEDLINE abstracts [16] used for the GENIA corpus [12], and annotations are usually restricted to words and phrases (comparable to the taxon names and locations in our documents, but without the semantic appreciation that comes, for instance, with geo-referencing). In contrast, the documents we work with consist of several (up to 100) treatments, and the markup covers both them and their subsections. This results

[1] Annotations usually refer to small pieces of text, usually individual words, with no specific representation. As opposed to this, XML markup can also express document structure.

in a much more complex procedure. Second, as opposed to uniform and clean documents like MEDLINE abstracts, our documents are diverse, e.g., different print layouts used over time. Further, they contain layout artifacts like page titles and noise, notably OCR errors, which require extensive cleanup.

3 The Documents

In this section, we describe the documents and the XML markup we have created.

The **Biosystematic Documents** we have worked with are scientific publications from the biosystematics domain. They categorize, describe and discuss living organisms and relationships between them. They have a typical structure: (1) The body contains so-called **treatments**. A treatment is a section referring to one **taxon**. It consists of subsections, each covering one aspect of the taxon (to be explained below and in Section 5.1). A taxon is a node in the tree of life at any level, e.g., a genus or a species. (2) In its **nomenclature** subsection, each treatment provides the reference to the taxon it deals with, usually right at its beginning. (3) Each treatment may comprise a **morphological description** of its taxon and **locations** where specimens have been collected. In addition, treatments may also include subsections covering other aspects. Only markup covering treatment boundaries and structure, taxon names, and locations and morphological descriptions, respectively, allows for creating useful RDF statements from the document content.

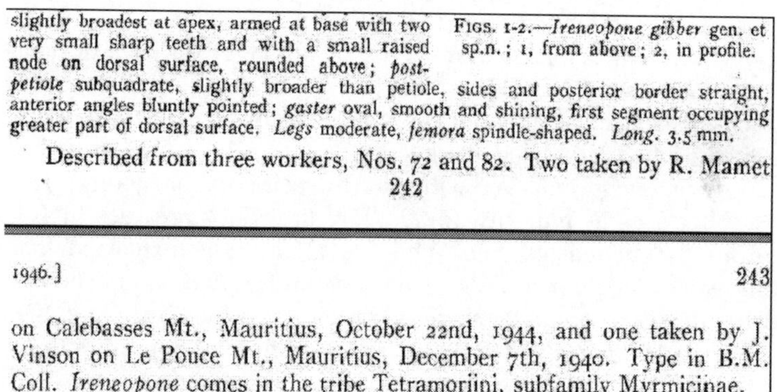
slightly broadest at apex, armed at base with two very small sharp teeth and with a small raised node on dorsal surface, rounded above; *post-petiole* subquadrate, slightly broader than petiole, sides and posterior border straight, anterior angles bluntly pointed; *gaster* oval, smooth and shining, first segment occupying greater part of dorsal surface. *Legs* moderate, *femora* spindle-shaped. *Long.* 3.5 mm.

FIGS. 1-2.—*Ireneopone gibber* gen. et sp.n.; 1, from above; 2, in profile.

Described from three workers, Nos. 72 and 82. Two taken by R. Mamet

242

1946.] 243

on Calebasses Mt., Mauritius, October 22nd, 1944, and one taken by J. Vinson on Le Pouce Mt., Mauritius, December 7th, 1940. Type in B.M. Coll. *Ireneopone* comes in the tribe Tetramoriini, subfamily Myrmicinae.

Fig. 1. Page break in the middle of a paragraph

The documents from the last 250 years show a wide range of writing styles, layouts, and levels of print quality. With many documents, the logical structure (sections, etc.) is not in line with the layout structure. The print layout is also responsible for artifacts in the document text after digitization. E.g., there are page numbers, page titles, footnotes, or copyright notices. Figure 1 gives an example: A page border (the gray bar), page numbers and a page title interrupt the paragraph "*Described from three workers ... the tribe Tetramoriini, subfamily Myrmicinae.*". In addition, there may be hyphenations from the print layout, which are in the way of NLP.

For the **Output XML Documents** to yield maximum benefit, they should be marked up according to TaxonX [4], a dedicated XML schema for biosystematics. It provides markup for several levels of document structure, like sections or treatments, and for specifics like taxon names or locations, together with normalizations, like geo-coordinates. TaxonX imports the DarwinCore [7] schema – a de-facto standard in biosystematics – for the normalized data. TaxonX markup facilitates many usages of the documents. A reasoner can compute the dispersal of a taxon (i.e., all locations where it occurs), the fauna of a location (i.e., all taxa that occur at a given location), or the taxon a specimen belongs to, based on observable morphological features.

4 Tools Used

For the initial Optical Character Recognition process (OCR), we used Versions 8 and 9 of **ABBYY FineReader** [1], a commercial scanning and OCR tool, with a specific setup: Since our source documents are written in different languages, sometimes within one document, we use the multiple languages mode. The program then considers more than one language for the spell check. We have integrated two domain-specific dictionaries we have built ourselves, containing parts of known taxon names and technical terms occurring in morphological descriptions. For text recognition, ABBYY FineReader can deploy user patterns, i.e., trained configurations for a given layout style or font. They help processing old documents with uncommon layout where the built-in type recognition would fail otherwise. We have created training files for 15 journals that contain some of the articles.

To support users correcting markup as far as possible, we have designed and implemented the **GoldenGATE Editor**[2] [18]. It tightly integrates automated markup generation and assistance features for subsequent manual correction. To facilitate deployment of tools for automated markup generation, the editor provides extension interfaces for their integration. We will refer to such components integrated through one of these interfaces as plug-ins. In addition, the editor provides high-level assistance functions for manual editing: While users can edit textual content as usual, XML editing works on element level. This is helpful in two ways: The user does not have to deal with XML syntax details (i.e., on the character level), and the editor can always keep the document well-formed. This is helpful for users who are domain experts like biologists rather than XML professionals. A laboratory experiment [19] has shown that the combination of assisted manual editing and NLP-based markup generation reduces the markup effort significantly, compared to commercial XML editors. In the experiment, the users worked with small documents with markup according to a simple XML schema. This paper in turn will show that the functionality described in this paragraph and in [18] is even more useful in a real-world setting.

As an example of a plug-in we have developed for the GoldenGATE editor, Figure 2 shows the **Annotation Selector**. It displays XML elements as a list to provide a view of some of the document content. An XPath expression selects the elements to display. The user can edit and remove each element. This view is helpful to correct auto-generated detail markup, in particular to sort out false positives.

---

[2] Available for free download and use at http://idaho.ipd.uka.de/GoldenGATE/

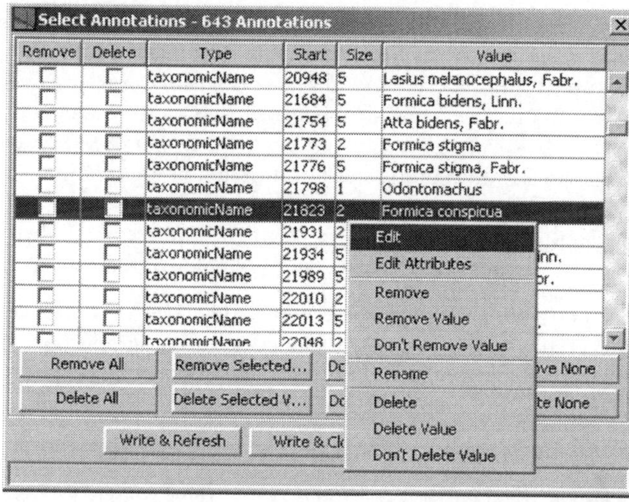

Fig. 2. List view of XML elements for correcting detail markup

For a detailed description of the GoldenGATE Editor please see [18] or the documentation at http://idaho.ipd.uka.de/GoldenGATE/.

5 Document Markup

In this section, we describe the automated steps to mark up the documents, and how we have interleaved them with manual correction to prevent error propagation. Many steps are generic, i.e., are necessary whenever creating digital resources from digitized legacy documents, e.g., artifact removal and normalization. Others are specific to our project to some degree, but similar steps are likely to be part of projects in other domains as well. We also discuss how arranging the steps in an appropriate order can minimize errors and thus the correction effort.

5.1 The Markup Steps

We now describe the digitization and markup steps, grouped as follows: OCR & normalization; structural markup; and detail markup. This is not the actual order of the steps, but eases the presentation. We will mention various automated markup-generation tools in this section. Most of them are plug-ins to the GoldenGATE editor. Unless stated otherwise, we have implemented them ourselves. Paragraphs labeled *INSIGHTS* contain some generalizations which others might find helpful.

5.1.1 OCR Artifact Removal and Normalization

The step for OCR, error correction, and normalization is not specific to our project. It is an endemic part of digitizing and marking up any corpus of legacy documents.

Scanning and OCR. ABBYY FineReader scans and transforms the printed legacy documents to character data via OCR. This step produces an HTML document, the starting point for further processing with GoldenGATE. We use the following output options of ABBYY FineReader: (a) *Remove all formatting*. The text quality in older publications sometimes is so bad that fonts and style may wrongly change several times within a word. (b) *Simple (compatible with all browsers).* (c) *Keep line breaks.* Line breaks have to be retained initially, because they help to detect and re-join hyphenated words. (d) *Use solid line as page break (keep page breaks).* Likewise, we initially keep page breaks, to identify certain artifacts like page titles, page numbers, and footnotes, which usually are next to page breaks. (e) *Do not save with images.* Since the XML documents are intended as XML resources, we do not keep images. However, the URLs of images can be stored as attributes of their respective captions.

Layout-Artifact Detection. To prevent errors in later steps, we have to make sure that the structure of the text is intact. In printed text, paragraphs can be broken, for two reasons: (1) They can run over page breaks, and page number and page title interrupt the text flow. There can also be footnotes prior to the page break. (2) Paragraphs may have figures or tables with captions embedded, which also interrupt the text flow. To recognize and remove these insertions, a dedicated tool first marks up the pages, from one page break to the next one. Other tools then search for page titles, page numbers etc. at the top and at the bottom of the pages. One of them turns the page numbers into attributes of the paragraphs of the page to facilitate page-level citation.

Paragraph-Border Correction. Paragraph borders are sometimes faulty in OCR output, due to varying indentation and spacing schemes. I.e., a paragraph may be split in the middle, or two paragraphs are marked up as one. Respective checks can be automated to some degree, e.g., based on capitalization or the punctuation. But these checks can only point the user to likely errors, since the evidence tends to be too incomplete and unreliable for automated correction. Before removing recognized artifacts, like page titles and footnotes, we manually check which paragraphs are artifacts and which are not. A specialized document view supports this, displaying the paragraphs of one page at a time for better overview.

Paragraph Normalization. Once the paragraph boundaries are in place, a respective plug-in removes line breaks from within the paragraphs and re-joins hyphenated words, e.g., 'fau-<line break>na' becomes 'fauna'. It also removes the page markup, which is not needed any longer.

MODS Referencing. To ease citation and provide a unique identifier for the XML document, we import respective metadata [14], using the MODS data format[3]. A plug-in obtains the metadata from a web service[4] and inserts it into the document.

Structural Normalization. The next step is to remove the layout artifacts. Deleting page titles is unproblematic because they do not contain useful information. Footnotes and captions are harder to deal with because they represent content. Thus, if they do not interrupt the text flow of another paragraph, they stay in place. Otherwise, a dedicated plug-in moves them to the position behind the paragraph which they interrupt. This changes the structure of the text, but not significantly. Further, footnotes are usually printed right before a page break, independent of the exact position in the text

---

[3] Metadata Object Description Schema

[4] http://atbi.biosci.ohio-state.edu:210/hymenoptera/hym_utilities.format_ref?style=MODS&id=<documentIdNumber>

referencing them. Captions in turn stand next to the figures they belong to, and the position of the latter typically depends on layout considerations. Thus, moving footnotes and captions out of paragraphs is unproblematic.

INSIGHTS. *(a) Since correcting OCR errors causes considerable effort, OCR quality should be favored over full automation and throughput if markup generation is an objective of the project. For mass digitization in turn, where markup generation might only be an option later on, we recommend providing a link from the OCR-ed documents to the original page images. These images have turned out to be helpful for later corrections. (b) Even though marking up the layout structure of the printed document creates additional artifacts, having this markup temporarily is helpful for cleaning. In particular, marked up pages are reliable evidence for finding other artifacts, e.g., page title or footnotes.*

5.1.2 Structural Markup

The markup we create introduces two levels of document structure, namely the treatments and their inner structure.

Treatment Markup. Once the document is structurally clean, a respective plug-in marks up the treatments. Like sections in other documents, a treatment consists of one or several paragraphs. I.e., the plug-in uses rules to group the paragraphs into treatments and sort out the ones that do not belong to any treatment. To simplify correction of the treatment boundaries, we again use a specialized document view: The user can visually check if the paragraphs are correctly grouped into treatments and can correct errors.

Table 1. Types of subsections in treatments

Type	Primary Content
Description	Morphological description of a taxon
Materials Examined	Locations where the specimens described in Description have been collected
Biology	Behavior of taxon, interactions with environment, nesting preferences
Diagnosis	Explanation of differences of taxon to related taxons, plus identifying features.
Discussion	Explanation why these differences justify making them a taxon of their own
Distribution	Summary and distribution pattern of a taxon (e.g., 'throughout southern Africa'), based on material presented in Materials Examined
Etymology	Etymology (origin) of the taxon name

Structure of Treatments. A treatment usually consists of several subsections: The nomenclature subsection always exists and provides the name of the taxon the treatment refers to. Other subsection types do not always exist and need to be marked up if they do (cf. Table 1). As the treatments, their subsections also consist of one or several paragraphs each. A rule-based plug-in groups the paragraphs of each treatment into subsections. To make manual corrections as simple as possible, we use the same approach as for the treatment boundaries: In a dedicated document view, the user can visually correct the grouping of the paragraphs into subsections.

INSIGHTS. *Elements of coarser structure, e.g., treatments, typically are groups of paragraphs. It turns out that marking up the paragraphs before the coarser elements is favorable over a top-down approach. This is because, given the paragraphs, one can mark up the coarser elements by grouping paragraphs relatively easily. Respective document views have turned out to be very helpful to correct the groupings.*

5.1.3 Detail Markup

The markup we create comprises two types of semantic details, namely taxon names and locations. Both are normalized, to facilitate search, data mining, and creating RDF statements. In addition, we add identifier numbers to the former and geographical coordinates to the latter. While the detail markup generated is specific to our context, the general problem of marking up important details is not: In the context of historical medical documents, for instance, geographic details are essential to compute the spread of an epidemic, as are bacteria names.

Taxon-Name Markup. The taxon names are the most important details in the documents. To detect and mark them up automatically, we use a dedicated algorithm [17], implemented in a plug-in. It is both rule- and dictionary-based, using regular expression patterns for the rules. In addition, it extends the lexica dynamically as it finds new taxonomic names in the documents. Further, it can be tuned towards precision or recall. A list view of the taxonomic names avoids sifting through the document for corrections. Since this view only helps with false positives, we tuned the component towards maximized recall to avoid false negatives.

Since older publications in particular show a variety of layout styles, including capitalization and punctuation schemes for taxonomic names, we had to adjust the rules frequently in the beginning. To simplify adaptation to new styles, we ended up implementing a specialized editor for the regular expression patterns of the rules.

The next step is to extract the normalized form of the taxon names, to facilitate exact searching and simplify RDF generation. In particular, this means adding the 'meaningful' parts of the name as attributes to the XML element marking up the name. The meaningful parts are the ones that carry taxonomic information, as opposed to those clarifying name authority or labeling meaningful parts. For example, in the taxonomic name *Drosophila melanogaster* Meigen, *Drosophila* is the genus, *melanogaster* the species, and Meigen the author of the species name *melanogaster* (the author of the genus name *Drosophila* is Fallén, which is not given here). For search and data mining purposes, we are only interested in the meaningful parts *Drosophila* and *melanogaster*, i.e., the parts that specify the position of a taxon in the tree of life. We add these parts as *genus* and *species* attributes to the XML element marking up the taxon name. We do not make an attribute of Meigen.

A plug-in parses the meaningful parts from the taxon names as they occur in the document. Since some parts of taxon names may be abbreviated or omitted completely, the next step is to fill in these gaps, based on other taxon names in the document. *Drosophila melanogaster*, for instance, may be referred to as *D. melanogaster* later in the same document. In the document as a whole, the meaning of the abbreviation *D.* is clear, but this may not hold for a treatment in isolation. We therefore resolve the abbreviation to *Drosophila*. Similarly, co-reference analysis in any other domain may resolve pronouns and last names to a full person name mentioned earlier in a document. So even without much context information, it is clear to whom a sentence refers to. To correct a normalization result, we use a dedicated dialog that allows for selecting the full form of an abbreviated part from a respective list. In addition, users can change the meaningful parts of the taxon names (i.e., the parts of the name that carry taxonomic semantics, see example above). A parser then automatically adjusts the meaning of the other parts to comply with the correction. Throughout our work, we have kept improving this dialog based on feedback from the users. The improvements include displaying the page number where the taxon name occurs in the document, and displaying the context of the taxon name in the document, about 10 words to the left and right. Both measures simplify deciding which part of the taxon name has which meaning. In its current status, the dialog reduces the effort for the normalization of taxon names by 95% according to our estimations; see Section 6.

After the meaning of all taxon names is clear (i.e., there are no unresolved abbreviations), the next step links these names to an existing biosystematics database. In particular, it obtains their Life Science Identifiers (LSIDs, [3]) from the Hymenoptera Name Server [11]. LSIDs allow assembling very precise RDF statements on a specific taxon. In addition, they facilitate integrating the treatments into mashups, i.e., brief compositions of all data available on a given taxon from various sources. If a taxon name is not yet in the database of that name server, i.e., there exists no LSID for it, we upload it to the server and obtain a newly generated LSID in return.

Location Markup. Another important piece of information in the documents is the locations. We have implemented a combination of named entity recognition techniques [6, 15] to mark them up automatically. Cleaning location markup is similar to the one of taxon names, i.e., using the list view to sort out false positives.

To give the locations a unified unambiguous representation, a respective plug-in adds attributes with their geographical longitude and latitude. It obtains this information from the GeoNames web service [10]. For ambiguous location names, the user can select the correct coordinates from a respective dialog.

INSIGHTS. *(a) Marking up important details is a common step when creating digital resources from legacy documents, since the details are essential to extract RDF statements. The high diversity of legacy literature, however, poses new challenges: For instance, to reduce the correction effort, we frequently had to adjust the rules for taxonomic name extraction. The situation is similar when patterns are used for detail extraction, due to the numerous variants of spelling, capitalization and punctuation in legacy literature, which are impossible to foresee. The pattern editor has turned out to be helpful, and we would also expect this for other projects. (b) An important optimization goal should be maximizing recall, even at the cost of (some) precision, as opposed to f-score or accuracy, which are relevant criteria in NLP research. Namely, sorting out false positives from a list view of marked up details has turned out to cause much less effort than searching the document for false negatives.*

5.2 Using Generic Markup

Instead of creating TaxonX markup directly, the preprocessors produce and work with generic 'out-of-schema' XML elements. Only the last preprocessor transforms the documents to TaxonX using XSLT. We have chosen this approach for several reasons:

(1) **Temporary Markup.** Some markup steps require XML elements that are temporary. The detection of layout artifacts, for instance, makes use of markup identifying the pages of the printed document. Since the pages are not part of the final markup, TaxonX does not provide an element for them. Thus, we have to use 'out-of-schema' XML elements, and the working documents would not be valid TaxonX anyhow. (2) **Target-Schema Evolution.** In order to react to various unforeseeable special cases we encountered in the project, e.g., whole treatments written in captions or footnotes, we have proposed significant refinements and modifications of the TaxonX schema itself. Generic markup allows dealing with those cases right away (i.e., without having to initiate schema modifications and to wait for the outcome) without encountering validity problems. Generic markup is also helpful with cases which the target schema cannot represent, e.g., a nesting of structural markup elements that is not valid. (3) **Storing Additional Information.** When marking up the documents, we obtain additional useful information, e.g., normalized forms of the taxon names and geo-coordinates of the locations. TaxonX stores such extra data as textual content under the tax:xmldata element, as imposed by DarwinCore [7]. Thus, creating the target markup right away would mix up artifact elements with document text. This would be in the way of further markup steps, especially NLP-based ones. To avoid this, we use generic markup that allows storing the extra data as attributes. (4) **Reusability of Automation Components.** Implementing specialized document views to correct the results of the individual steps has been a significant effort, though – in the long term – this effort is much lower than doing the corrections manually in a generic view. To facilitate reuse of these views, it is helpful not to develop them for a specific XML schema. (5) **Integration of Third Party Components.** Using existing NLP components for automated markup generation becomes harder when producing markup that must conform to a schema. This is because then, the output of the component has to be transformed right away, to comply with the schema.

Most of these advantages also hold for other projects which mark up legacy documents. The only item specific to our project at first sight is the representation of the taxon names in TaxonX. However, final markup generated by an intermediate step may be in the way of subsequent processing steps. Thus, the distinction between "working markup" and final markup should be helpful in other domains as well.

INSIGHTS. *Working with generic markup and transforming it into the target schema only at the end is advantageous: Compared to schema-bound markup, one is more flexible regarding the deployment of NLP components, the explicit representation of extra information, and structural variations it can represent.*

5.3 Order of Markup Steps

Performing the markup steps in top-down order, i.e., first structure, then details, results in unnecessary effort. In particular, details are reliable evidence to avoid errors when marking up the document structure. This leads to an alternative order of the steps (see Figure 3). We first mark up the details, and then use these details as evidence when generating the structural markup.

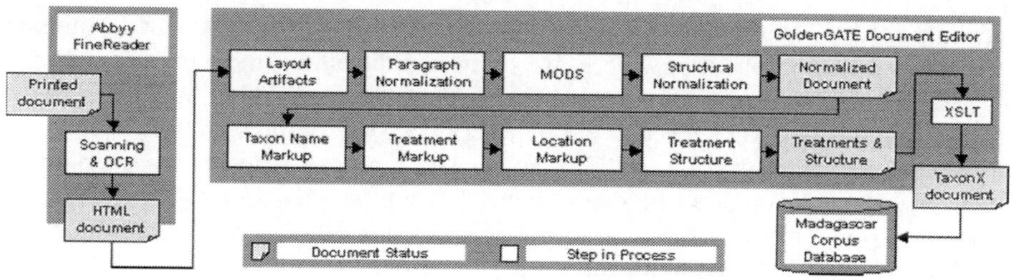

Fig. 3. The markup & cleaning process as we use it

Treatment Markup. We have observed that almost every treatment starts with its nomenclature subsection. This subsection usually contains one rather short paragraph and starts with a taxon name. This means that the taxon names reliably identify nomenclature paragraphs. Based on these paragraphs, a respective plug-in generates the treatment markup, assuming that a treatment starts at the start of a nomenclature paragraph and ends immediately before the next nomenclature paragraph. However, this assumption does not always hold: There can be document text between treatments, and there usually is text after the last treatment, namely acknowledgements and the references sections. Thus, we cannot eliminate manual correction altogether, but we can reduce the effort considerably.

Internal Structure of Treatments. We observe that certain details identify some subsections of treatments rather clearly: (1) A series of taxon names identifies a discussion or diagnosis subsection. (2) The presence of location names indicates a distribution or 'Materials Examined' subsection. (3) The presence of the names of body parts (observable morphological features of specimens) indicates a description, diagnosis, or discussion subsection. – Further, we have observed that 'discussion' subsections are much more frequent than 'diagnosis' subsections, and 'Materials Examined' subsections are frequent than subsections with plain distribution data. With these observations, we have been able to formulate more reliable rules for the component that groups the paragraphs of a treatment into subsections.

INSIGHTS. *(a) When several markup steps need to be done, yielding markup of different granularities, a good ordering of the steps can facilitate a higher degree of automation and prevent errors, compared to the top-down approach. (b) Our approach for treatment markup is applicable in other settings as well. In general, details are reliable evidence to identify the document structure. For instance, once section titles are marked up, one can mark up sections automatically, with few errors. (c) Classification rules similar to the ones used here for the inner structure of treatments can avoid errors when marking up the document structure: For instance, in scientific publications regardless of the domain, the presence of person (author) names, titles, and years is a reliable hint that the section is a References section.*

6 Statistics

One of the biologists of our project has marked up about 100 documents; Figure 4 graphs the average working time per page. When our work started, both the GoldenGATE editor and the markup generation plug-ins were less sophisticated than they currently are (cf. Subsection 5); marking up one document page took over 12 minutes. Over time, based on our experiences, we improved the editor, the markup-generation plug-ins, and the specialized document views. The average working time decreased to about 2 minutes per page, a reduction by over 80%. The outliers are documents with large pages or with very poor OCR quality. The biologist who did the markup accounts about 3 to 4 minutes of the 10 minute speedup per page to training effects. The remaining 6 to 7 minutes result from the refinement of the tools. Unfortunately, we could not validate these numbers more rigidly without actually disturbing the users in their work. This is because this project took place under regular working conditions, with real output, not as a laboratory experiment. The user, according to her own assessment, was familiar with the tools after about 10 to 15 documents, i.e., about 100 pages. The average working time per page was still around 9 to 10 minutes at that point. Thus, this might serve as further evidence that the remaining speedup is largely due to technical improvements.

Table 2 graphs the amount of time users spent with correcting the markup from the different steps. Correcting the document structure takes considerable time. This is because the OCR errors do not follow any regularity, at least according to our

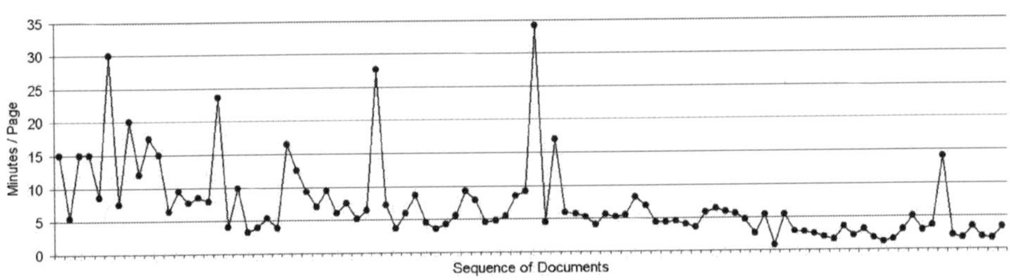

Fig. 4. Average working time per page

Table 2. Time per step (in minutes) for 10 pages

Markup Step	Time (With Automation)	Time (Manually)
Scanning & OCR	**38:00**	
Layout Artifact Detection	3:16	6:25
Paragraph Normalization	0:10	2:39
Get MODS Metadata	0:43	
Taxon Name Markup	1:41	8:21
Taxon Name Normalization	0:45	20:10
LSID Referencing	3:53	14:40
Treatment Markup	1:02	2:32
Location Markup	1:40	2:11
Treatment Structure Markup	3:20	17:30
Structural Normalization	0:05	2:10
Markup Total	**16:35**	**77:21**
Total	**54:35**	**115:12**

perception. Consequently, users have to clean up every page. Getting the LSIDs for the taxon names takes long due to the web-service lookups. Cleaning the inner structure of treatments takes rather long as well. This is because the respective classification technique still yields many errors. Despite this, the effort with the respective document view (3:20) is still less than 20% of what it would be without it (17:30).

Outlook. Our work on the Madagascar corpus[5] is now finished. While this corpus allows biosystematicists to experiment with new methods (like machine reasoning and data mining) and showcases the benefit of converting legacy documents into digital resources[6], it comprises only a small fraction (< 1%) of the available literature. We plan to investigate how to perform this conversion more efficiently: (a) To improve document normalization, we plan to analyze the errors the automated components have produced on the Madagascar corpus more systematically. To further improve the markup of treatment subsections, we plan to refine the respective rules, e.g., by multiple levels of automated classification, and by more extensive use of keywords. (b) Despite the high level of assistance and the low error rate achieved in this project, a small group of domain experts cannot take on 100,000 or more pages of legacy documents. We plan to investigate how to distribute the work more efficiently and get more individuals involved.

7 Conclusions

Converting legacy documents into digital resources for semantic web applications is a complex task. Running the complete markup procedure and correcting the final result has turned out to be inefficient. This is because the generation of complex markup involves many steps, and some steps use the output of previous ones as input. Automated processing is rarely 100% accurate, and dependencies between the steps result in error propagation and thus induce high correction effort. Interleaving markup generation and manual correction prevents error from propagating. To support the users, we have developed the GoldenGATE Editor. It tightly integrates automated markup-generation components with support for manual correction, i.e., integrates the basic editing functionality found in conventional XML editors with the markup generation facilities of NLP frameworks. In this paper, we have reported on experiences gained from converting a large corpus of digitized legacy literature into semantic digital resources. The work took place under regular working conditions, not as a laboratory experiment. Biologists from Australia, Brazil, Germany, Taiwan, and the USA have expressed interest in our tools. On the other hand, while the domain currently is systematic biology, our insights should be of interest to other fields as well.

---

[5] Corpus available at http://antbase.org/databases/madagascar.htm

[6] You can try out search functionality based on marked up semantic details at http://plazi.org:8080/GgSRS/search - type "Probolomyrmex tani" in the search field labeled "Name" for a start. The "GoogleMaps" link in the result shows one of the possible applications of detail markup: It plots a map from marked and geo-referenced locations, visualizing the distribution of the ant species *Probolomyrmex tani*.

References

1. Abbyy FineReader, http://www.abbyy.com/finereader_ocr/
2. Altova GmbH, http://www.altova.com
3. Brazma, A., Krestyaninova, M., Sarkans, U.: Standards for systems biology. Nature Reviews Genetics 7, 593–605 (2006)
4. Catapano, T., et al.: TaxonX: A Lightweight and Flexible XML Schema for Mark-up of Taxonomic Treatments. In: Proceedings of Annual Meeting of Taxonomic Data Working Group 2006, St. Louis, MO, USA (2006)
5. Chinchor, N.: MUC-7 Named Entity Task definition. In: Proceedings of Message Understanding Conference, Washington, DC, USA (1997)
6. Cucerzan, S., Yarowsky, D.: Language Independent Named Entity Recognition Combining Morphological and Contextual Evidence. In: Proceedings of the Joint SIGDAT Conference on EMNLP and VLC, University of Maryland, College Park, MD, USA (1999)
7. DarwinCore2, http://darwincore.calacademy.org/
8. Fellbaum, C. (ed.): WordNet: An Electronic Lexical Database. MIT Press, Cambridge (1998)
9. General Architecture for Text Engineering, http://gate.ac.uk
10. GeoNames, http://www.geonames.org
11. Johnson, N.F.: The Hymenoptera Name Server, http://atbi.biosci.ohiostate.edu:210/hymenoptera/nomenclator
12. Kim, J.-D., Ohta, T., Tateisi, Y., Tsujii, J.: GENIA corpus – a semantically annotated corpus for bio-text-mining. Bioinformatics, i180–i182 (2003)
13. Marcus, M.P., Santorini, B., Marcinkiewicz, M.A.: Building a Large Annotated Corpus of English: The Penn Treebank. Computational Linguistics 19(2), 313–330 (1994)
14. Metadata Object Description Schema, http://www.loc.gov/standards/mods/
15. Mikheev, A., Moens, M., Grover, C.: Named Entity Recognition without Gazetteers. In: Proceedings of Annual Meeting of European Association Computational Linguistics, Bergen, Norway (1999)
16. National Library of Medicine: MEDLINE, http://www.nlm.nih.gov/pubs/factsheets/medline.html
17. Sautter, G., Böhm, K., Agosti, D.: A combining approach to find all taxon names (FAT). Biodiversity Informatics 3 (2006), https://journals.ku.edu/index.php/jbi/index
18. Sautter, G., Agosti, D., Böhm, K.: Semi-automated XML Markup of Biosystematics Legacy Literature with the GoldenGATE Editor. In: Proceedings of Pacific Symposium on Bio-computing, Weilea, HI, USA (2007)
19. Sautter, G., Böhm, K., Padberg, F., Tichy, W.: Empirical Evaluation of Semi-Automated XML Annotation of Text Documents with the GoldenGATE Editor. In: Proceedings of European Conference on Research and Advances in Digital Libraries, Budapest, Hungary (2007)

Collaborative Ocean Resource Interoperability: Multi-use of Ocean Data on the Semantic Web

Feng (Barry) Tao, Jon Campbell, Maureen Pagnani, and Gwyn Griffiths

National Oceanography Centre Southampton, UK
{bt,joc,gxg,mred}@noc.soton.ac.uk

Abstract. Earth Observations (EO) collect various characteristics of the objective environment using sensors which often have different measuring, spatial and temporal coverage. Making individual observational data interoperable becomes equally important when viewed in the context of its expensive and time-consuming EO operations. Interoperability will improve reusability of existing observations in both the broader context, and with other observations. As a demonstration of the potential offered by semantic web technology, we have used the National Oceanography Centre Southampton's Ferrybox project (where suites of environmental sensors installed on commercial ships collect near real time data) to set up an ontology based reference model of a Collaborative Ocean, where relevant oceanographic resources, such as sensors and observations, can be semantically annotated by their stakeholders to produce RDF format metadata to facilitate data/resource interoperability in a distributed environment. We have also demonstrated an infrastructure where common semantic management activities are supported, including ontology management, semantic annotation, storage, and reuse (navigating, inference and query). Once the method and infrastructure are adopted by other related oceanographic projects to describe their resources and move their metadata onto the semantic web, it would be possible to see better interoperability within the Collaborative Ocean initiative to facilitate multiuse of ocean data, as well as making more EO data available on the semantic web.

Keywords: Semantic Web, data/resource interoperability, oceanographic earth observation, Ferrybox, Multiuse of ocean data.

1 Introduction

Advances in Web and sensor technologies have made available more and more environmental observations, both remote and in-situ, accessible in real time and as archived datasets. For example, the Ferrybox project [1] at the National Oceanography Centre Southampton (NOCS) has been in operation for six years, making oceanographic observations from various "Ships of Opportunity". One of the Ferryboxes is installed in the engine room of the *Pride of Bilbao* passenger ferry that sails between the south of England and Northern Spain. The Ferrybox maintains a closed system where near-to-surface sea water is pumped through a set of sensors to carry out observations of various environmental parameters such as temperature, salinity, fluorescence (as an indicator of the quantity of chlorophyll), turbidity, dissolved oxygen and carbon dioxide. Since 2002 a subset of the data collected form the Pride of Bilbao has been automatically telemetered to computers at NOCS every 10 minutes. These data are used to generate near real-time, public web display of ship track and oceanographic measurement information [7].

L. Aroyo et al. (Eds.): ESWC 2009, LNCS 5554, pp. 753–767, 2009.

Datasets are available for downloading and quality-controlled data are eventually archived at British Oceanographic Data Centre (BODC) to allow future data analysis requirement by environmental scientists to understand the change of our natural environment. The Ferrybox project observations are by nature, dense in time coverage but restricted to only a small spatial coverage for a set of limited measurements; Coarse but more global-scale measurements can be obtained through remote sensing observations, from regional aircrafts to earth-orbiting satellites. However, the lack of standardizations between these observations has made it difficult to collate measurements from multiple providers in order to automate intelligent processes, such as measurement cross-validation, data/services mash-up, and personalised knowledge representation based on intelligent processing of distributed measurement data. In other words, the lack of standards and semantics will "isolate important data streams and intensify the existing problem of too much data and not enough knowledge"[2].

On the other hand, as a multi-disciplinary subject, ocean science often involves collaborative work from different stakeholders and roles such as sensor development researchers, oceanographic instrument deployment engineers, data management, and environmental scientists, each of whom provides, as well as consumes data/services that are shared within the domain. Within the context of Oceans 2025, which is a strategic marine science programme funded by the UK Natural Environment Research Council (NERC), bringing together marine researchers to increase people's knowledge of the marine environment so that they are better able to protect it for future generations [13]. For example, in the Ferrybox project, engineers need to collaborate with the sensor providers in order to select and integrate sensors into the Ferrybox Instruments, which are then deployed on platforms (e.g. Ships of Opportunity in this case). Observations are sampled and brought back regularly by telecommunication though satellite and on-site visits to collect the physical memory storage. Real time monitoring is used to monitor the Ferrybox status. Calibrations for sensors are inevitably needed as the measurements drift due to bio-fouling and other reasons, hence their calibration history should be recorded to allow easy management for future reuse. Datasets of the observations are then archived by data managers with various metadata and descriptions attached so that domain scientists can investigate a particular environmental phenomenon, e.g. a bloom of plankton and its relationship with temperature and salinity [8]. All these activities through out the workflow have access to different yet related resources which can be annotated against a domain reference model to reduce ambiguity (every resource has a unique formal identifier) and improve connectivity (resources are linked to other resources through defined relations). This eventually will support sharing earth observation data and knowledge.

1.1 Related Work and Motivation

The Semantic Sensor Web (SSW) is a framework for providing well defined meaning using the semantic web technology in W3C that realises the idea of "having data on the Web defined and linked in a way that it can be used by machines not just for display purposes, but for automation, integration, and reuse of data across various applications" [4]. The underpinning modelling language for the semantic web is the RDF (Resource Description Framework), which is a directed-graph model using XML syntax that can also be expressed in many other forms, e.g. Notation3 (N3). The

RDF data model describes the real world by making statements in the form of *triples* (subject-predicate-object) where every item is either a literal or a URI (Uniform Resource Identifier) prefixed with a name space. RDF is an open system in that the namespaces can be unlimited, each one corresponding to an ontology that defines a formal conceptual representation within a domain - concepts and relationships in particular. The simple RDF data model allows computers to store, exchange, and to some extent reason (think!) using information from across the Web, providing opportunities for computer-assisted information handling with more assurance and certainty.

Semantic Web based knowledge management has been previously introduced and successfully applied in engineering design search optimization [14] and e-learning [15]. Both domains are single core inter-disciplinary, i.e. between computer science and one more area. The same application in environmental domain is multi-core inter-disciplinary area where the domain is more diversified to include for example oceanography, electronics, chemistry, biology, etc. Therefore setting up the infrastructure and gluing/reusing existing domain ontology are more important.

With these objectives in mind, we have created a Collaborative Ocean Group (COG) in the context of Marine Metadata Interoperability (MMI) [5] to demonstrate this proof of concept for multi-use of ocean data. In order to coordinate oceanographic activities and facilitate resource sharing and data interoperability, we describe in this article a semantic web infrastructure that helps different roles to semantically annotate their resources so that they can be managed as linked-data which are W3C compatible to facilitate semantic web processing.

2 Ontology Based Metadata Modeling

Metadata has been widely used in environmental data management systems to improve both the accessibility and the quality of the observational information. However, interoperability between these systems is often restricted due to the lack of semantics and closeness of their metadata reference models. In our work, we aim to improve the situation by adopting the Semantic Web approach. For example, the concepts of namespace and URI in the Semantic Web make it possible to describe resources in an open and globally accessible manner. Further more, recent developments in the Semantic Web community have made available more tools and APIs that can handle data on the semantic web for parsing, querying and inference.

2.1 An Ontology Based Reference Model for Collaborative Ocean

Ontology is a formal representation of a set of commonly agreed concepts and defined relationships between them. It provides shared but controlled vocabularies, makes domain assumptions explicit and can be used to describe resources in a domain in a way that all domain users can collaborate with each other (Figure 1 and Figure 2). In short, it is an accepted consensus of domain conceptualisation based on which semantics can be added to enrich domain resources.

RDF and OWL has been the main model for expressing ontology. The RDF data model is based on the idea of making statements about web resources in the form of subject-predicate-object triples, where each entity in the triple is either a literal or a RDF resource with a URI (Uniform Resource Identifier). A set of triples using RDF resources can be viewed as a directed RDF graph. OWL (Web ontology language)

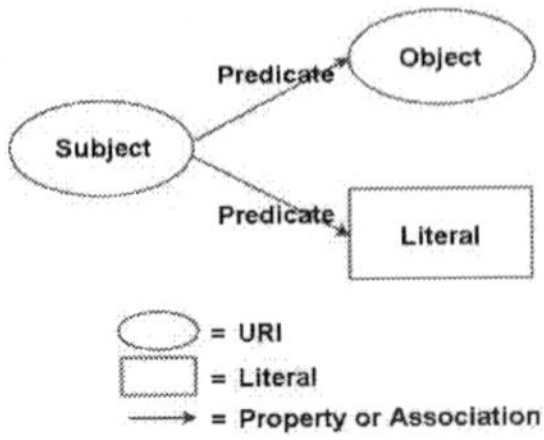

Fig. 1. RDF direct graph model

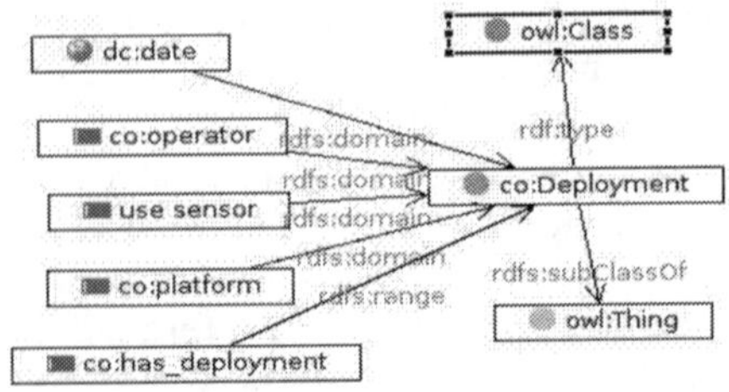

Fig. 2. An example in the Ferrybox project

extends the RDF Schema with additional vocabulary, e.g. on property restrictions (*allValuesFrom*) and restricted cardinality (*minCardinality*), etc.

The ontologies used can be categorized into three types according to their levels of generic description to the domain

Upper level and meta ontology

There are many generic conceptualisations that are domain independent and therefore can be used to provide cross-domain resource description.

DC - Dublin Core is a cross-domain XML/RDF based metadata standard designed to describe general media on the web to make them easier to find. It is divided into two levels - Simple Dublin Core (Title, Creator, Date, Subject, etc) and Qualified Dublin Core (Provenance and Rightsholders, etc). We use the DC to describe generic attributes of Ferrybox resources such as datasets collected and put online.

W3C RDF – defined internal vocabulary for RDF, such as *rdf:resource* and *rdf:datatype*, etc.

W3C RDFS - this is the root vocabulary and syntax specification originally defined in the RDF model. We use rdfs:seeAlso to maintain linkages between our parameters and BODC controlled vocabularies.

SKOS - noted as Simple Knowledge Organization Schema, it is designed to express generic relationship before concepts, such as *skos:broader*, *skos:related*, etc. BODC uses SKOS to express relationship among their control vocabularies.

Middle Level ontology

These ontologies are less generic than the upper level ontology but more generic than the overall domain ontology.

A common feature of EO data is its geospatial properties. Almost all EO measurements are related to location. We adopted the widely use W3C GEO vocabulary for this modelling requirement.

Another feature is the inclusion of social network related properties which are often required to describe the different roles in oceanography in such a way as to allow them to communicate and collaborate with each other in order to carry out EO tasks. We adopted the FOAF ontology that is originally proposed in W3C to describe social network related properties.

Domain ontology

Domain ontology is specific to the domain application and is normally non-transferring for other different domains. We build the CO ontology by investigating the domain.

CO - Collaborative Ocean ontology

This is the main specific reference model ontology designed to provide domain conceptualisation needed for describing domain related resource annotations. The ontology has the namespace "*co*" and focuses on concept taxonomy and how they are connected to other concepts through defined properties (Figure 3).

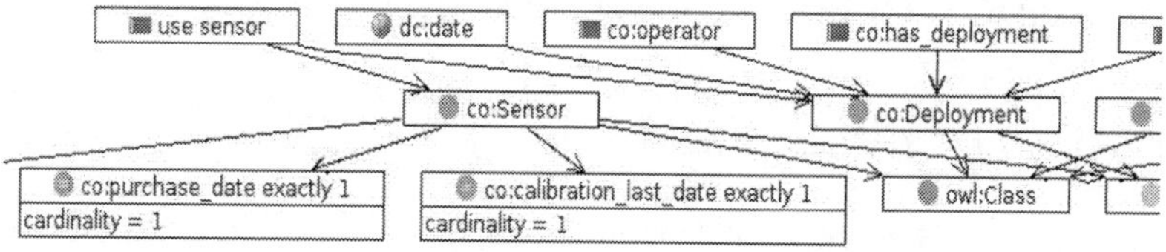

Fig. 3. A section of the CO ontology

Some of the important concepts are Sensor, Platform, Deployment, Datasets, etc. RDFS and OWL has been used to explicitly describe concept taxonomy, relationship and constraints in term of other define properties.

2.2 Interoperability with Existing Reference Model

Interoperability with other reference models such as MMI device ontology and BODC vocabulary is maintained by crossed linkage and references. One example is shown below where we have reused BODC measurement parameter vocabulary by adding SKOS annotations. This would provide the semantic linkage between two reference models when navigating on the semantic web from one to the other.

In parallel with the development of the Collaborative Ocean ontology, we have also participated in the MMI Device Ontology Working Group, with an aim to contribute our experience and requirement for sensor annotation metadata and to eventually reuse

```
<co:ChemicalParameter
rdf:about="http://www.soton.ac.uk/
~bt/ontology/ocean/collaberativeoce
an_ins.rdf#salinity">
<rdfs:seeAlso
rdf:datatype="http://www.w3.org/20
01/XMLSchema#string"
>http://vocab.ndg.nerc.ac.uk/ter
m/P021/25/PSAL</rdfs:seeAlso>
<co:unit
rdf:datatype="http://www.w3.org/20
01/XMLSchema#string"
>\u2030</co:unit>
<co:unit
xml:lang="en">ppt</co:unit>
</co:ChemicalParameter>
```

* **RDF snippet that links to BODC vocabulary**

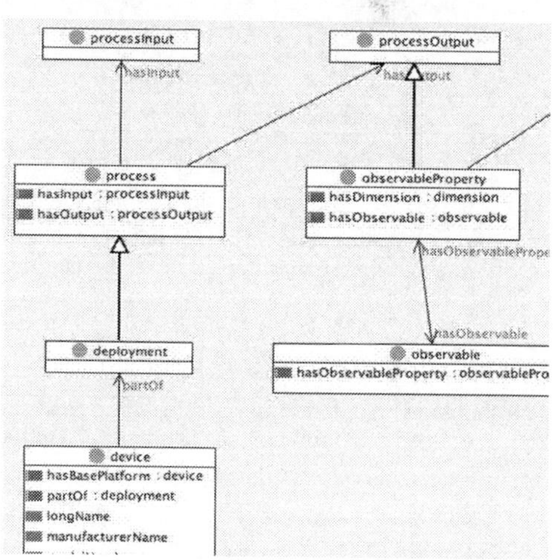

Fig. 4. MMI Device ontology

their device ontology in the CO ontology. This will provide end users with more choices to carry out their sensor device semantic annotation and better enable interoperability in a broader EO context. Figure 4 is a snapshot of the MMI device ontology concept diagram in development.

3 A Semantic Web Infrastructure for the Collaborative Ocean

In order to efficiently describe, store and reuse the oceanographic EO operational resource and measurement data on the semantic web, we have set up a layered infrastructure as shown in Figure 5. An advantage of this structure is to allow distributed management of the semantic web information at different granularities, and to expose a different level of technical details to different roles through customized interfaces and toolkits.

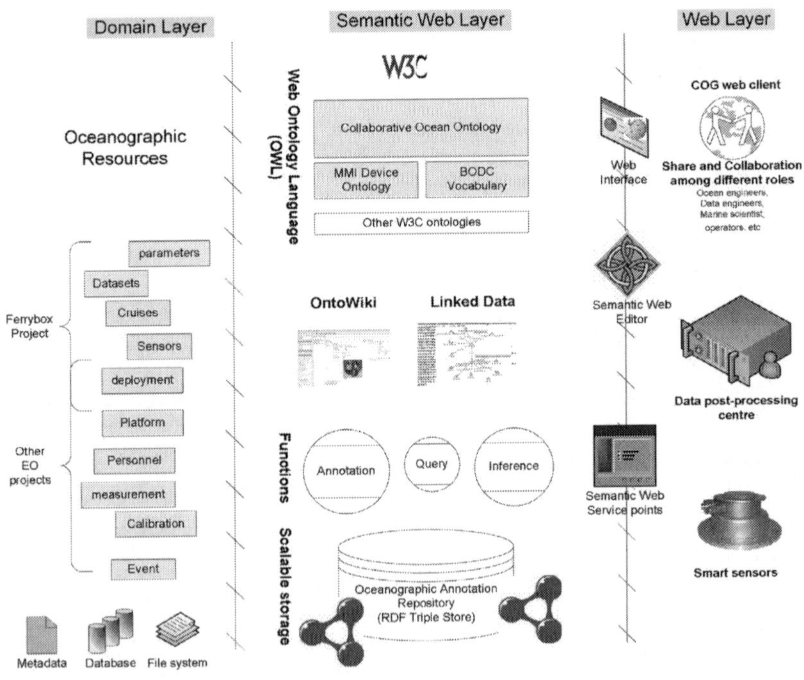

Fig. 5. Layered diagram of the semantic web infrastructure

Domain layer

Domain resources such as measurement datasets, calibration reports, and sensor profiles as well as existing metadata are stored in this layer, in the form of flat files or database records. The Collaborative Ocean Ontology and relevant controlled vocabulary are modelled and adopted based on study of the resources in this layer as well as best practice directly from experts operating the Ferrybox project. We use not only existing metadata such as Dublin Core and FOAF but also related domain vocabulary such as MMI device ontology and BODC vocabulary. The CO ontology serves as the domain reference model that provides important concepts and assumptions of the domain resource conceptualisation to allow proper semantic annotation of the Ferrybox resources and their publishing on the semantic web. Collaborative Ocean ontology has been built using Collaborative Protégé [9], which is an extension of the existing Protégé system that supports collaborative ontology editing. The Ontology is further maintained in the Topbraid Editing environment [10] to exploit its advantage in managing multiple ontology namespaces and its RDF triple representation of the ontology as well as the semantic annotations simulation. The advantage of Topbraid composer over protégé is its capability in managing multiple external namespaces and accessibility to the openRDF triple store.

During the Knowledge Acquisition (KA) process, we have adopted an agile knowledge modelling method that encourages domain experts to participate in the KA as an iterative process. A Wiki page has been set up to assist recording sensor profile text as well as key description of resources available in the Ferrybox project. A Protégé server has been setup to drive a collaborative protégé client to collect concepts and resource taxonomy from various domain experts in the group for a very initial version of the CO ontology. The RDF data is then fed into an OntoWiki [11] web based knowledge acquisition system as an information map where intuitive authoring of semantic RDF content as well as simulation of the semantic annotation process is made possible to enhance understanding of the domain. The OntoWiki adopted an agile and adaptive knowledge engineering strategy that is similar to a wiki-style paradigm of "making it easy to correct mistakes, rather than making it hard to make them" [12]. The Semantic Wiki is another similar tool that integrates the Semantic Web into a Wiki system. Interested users can find a recent review at [18].

Semantic Web layer

This is the layer where semantic web based functionalities such as domain resource annotation, browsing, query and inference are made available as services though integrating existing semantic web APIs with the ontologies and domain resources in the domain layer. The layer is driven by the CO ontology from the domain layer. We use Jena Semantic Web API for local semantic processing and remote access to a Sesame triple store, which is a persistent storage of all the semantic annotations in RDF triple format. The HP Jena Semanitc Web API is so far the most comprehensive java API to manipulate RDF models while the sesame openRDF delivers a better semantic web framework. Several methods have been used to generate and manage these RDF triples, namely the Collaborative Protégé, OntoWiki, TopBraid and a COG web client.

The Topbraid composer is a flexible and versatile semantic web editing tool that can manage the semantic web triples in different formats such as illustrated in Figure 6, where relevant RDF triples can be collected through standard SPARQL query to the Sesame triple store and presented as a directed linked graph as well as XML based RDF files.

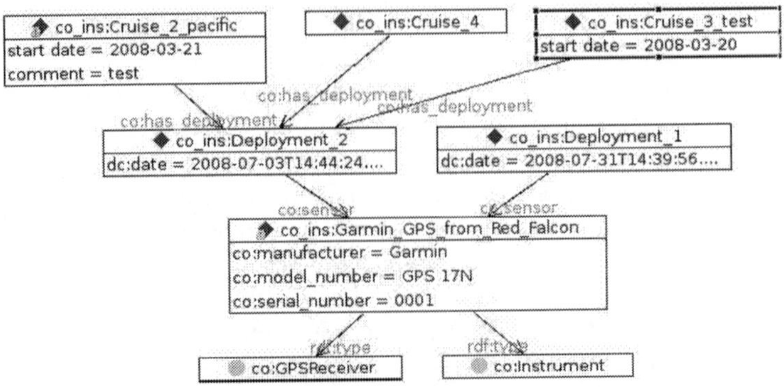

Fig. 6. RDF graph representation of the triples

Similar semantic web annotation and query functionalities are also interfaced to a COG web client (at the Web layer) using Java Server Face technology to allow users with less semantic web experience to contribute and reuse semantic web metadata on oceanographic observational resources.

Web layer
This is where applications are deployed to interface end users with the functionalities offered by the semantic web layer in the context of the domain. We have implemented a COG web client that allows manual and semi-automated semantic management of oceanographic resource share and collaboration using semantic web functionalities offered in the semantic web layer of the system. More intelligent, semantic web driven operations, such as data post-processing and smart sensor activities can be also supported in the future.

4 Technical Implementation and How the System Works

The system components/services have been deployed onto multiple nodes of different operating systems (Windows XP and Linux) to demonstrate a distributed Service Oriented Architecture (SOA). As shown in Figure 7, we have chosen to use Java and PHP as the main development languages due to the fact that most semantic web APIs (such as Jena and Sesame) are written in Java and its portability in different systems. PHP is another popular Web language especially suitable for large scale distributed systems across the Web. We have used Apache and Tomcat to run the web server and provide a Java servlet container where various services are deployed, such as an ontology driven Java Facelet web interface service that invokes semantic web functionalities such as query and inferences to interact with the semantic information in the RDF triple store. The triple store is a Sesame openRDF repository package deployed in another node across the network, where OntoWiki is also deployed to offer an alternative access to the semantic information.

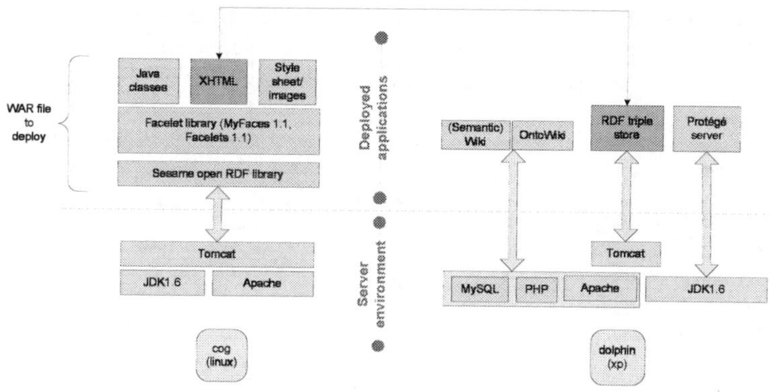

Fig. 7. Technical architecture of the system

Connecting to the system

A Sesame server address is required to specify the location where all the semantic annotations are stored as RDF triples. Multiple repositories can co-exist in the same server to store the RDF triples.

The user can also indicate whether inference is enabled. The inference is at the RDFS level to include inferred new triples according to RDF schematic reasoning rules. More interesting domain rules can be designed to produce inferred triples based on domain business logics.

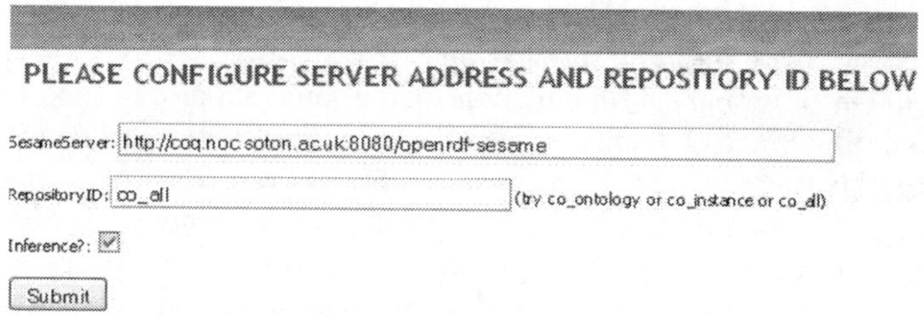

Fig. 8. Connecting the RDF triple store

Dynamic ontology driven semantic annotation of domain resources

Most initial semantic annotations are entered through other channels by our domain experts using Protégé and Topbraid Semantic web toolkits. Once they are stored as RDF triples in the repository, other interested users will be able to reuse part of these triples to generate further semantic annotations in a web interface. Meanwhile, reasoners can be used to generate further RDF statements that represent inference closure results of existing triples.

The web based semantic annotation page is dynamically generated by querying the CO ontology using SPARQL, therefore it is adaptive during the maturing the CO ontology. SPARQL is the querying language standard for collecting data from the semantic web. It is similar to the SQL for database, but with added semantic web features based on the RDF model and multi-source RDF query across the Web. Below is a SPARQL query for defined properties of the domain class *co:ChemicalSensor*, used to dynamically generate the semantic annotation forms as shown in Figure 9.

```
SELECT
DISTINCT ?property
WHERE { {?property
rdfs:domain ?class_X}
{co:ChemicalSensor rdfs:
subClassOf ?class_X }
UNION {?property
rdfs:domain
co:ChemicalSensor} }
```

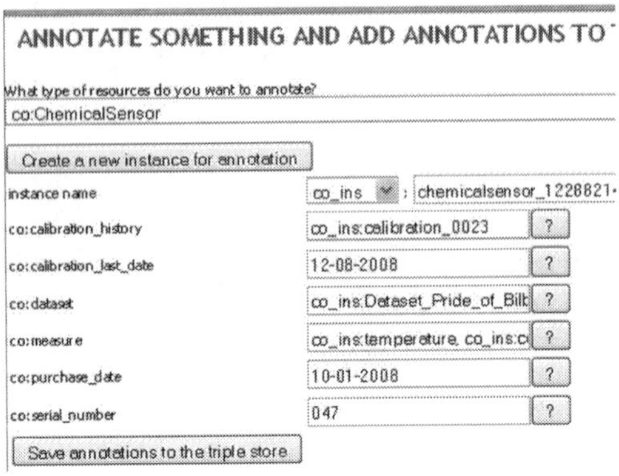

Fig. 9. Ontology driven semantic annotation web form generation

With the ontology driven annotation form, users can annotate datasets by providing values for a set of properties as defined in the ontology. These annotations will be converted as RDF triples and stored in the semantic web triple store so that other users can access through further query or browsing.

Semantic search, browsing and navigation

Semantic search, browsing and navigation of the annotations can also be achieved through submitting SPARQL semantic queries to the triple store and rendering results as linked data [19] on the web. Users will be able to view a list of semantically enriched instances of a particular class. They can click on any interesting element of the semantic annotations, for which the system will generate a new query accordingly, and get back another set of matched triples for rendering on the web. As all the resources can be annotated by their stakeholders at any time, these changes will be reflected in other semantic web management activities that reuse these triples.

An example of the semantic search and navigation is illustrated in Figure 10. According to the semantic annotations, the Ferrybox deployments are associated with the "*co:Cruise*" concept where additional properties such as "*co:aboutSea*", "*co:startPort*" and "*co:dataset*" can be assigned with values during the semantic annotation process. The semantic query retrieves a list of annotated cruise instances on which users can further browse their semantics and navigate to other linked semantic annotations.

Similarly, queries to dataset annotations can help users to discover their linkages to other annotated resources in the reference model, such as measurement parameters, original dataset producers and visualization plots generated, etc. Figure 11.

A notable feature is the resolvable namespace prefixes that point to corresponding published ontologies that define the underlying semantics.

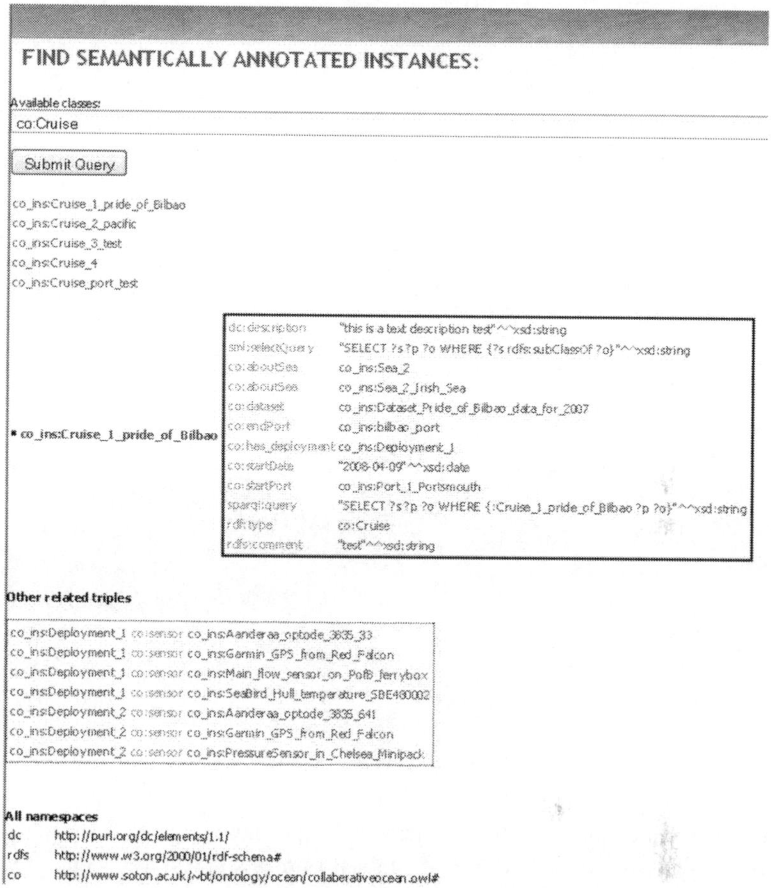

Fig. 10. Semantic browsing of Cruises

Mashing up with OGC service

Since the establishment of the Open Geospatial Consortium (OGC) in 1994, a lot of standards and services have been developed to coordinate the geospatial industry all over the world. One of the most relevant initiatives in the OGC is Sensor Web Enablement (SWE), which aims to enable real-time integration of heterogeneous sensor observations into the Web infrastructure. The main specifications include Sensor Model Language (SensorML), Sensor Observation Service (SOS), Sensor Planning Service (SPS), etc [6]. Various XML schemas are defined in SWE to accommodate XML files used in service in/output, e.g. in SOS, SensorML and Transducer Markup Language (TML) are used in *describeSensor* method response. The *getObservation* method has an input XML schema that uses Geographical Markup Language (GML) to specify time-spatial coverage of the observations. Observation offerings in SOS also provide non-overlapping observations constrained by a number of parameters such as geographical region and phenomena being sensed. This allows easy integration of spatially referenced data to layers in Web Map Service (WMS), another important OGC specification.

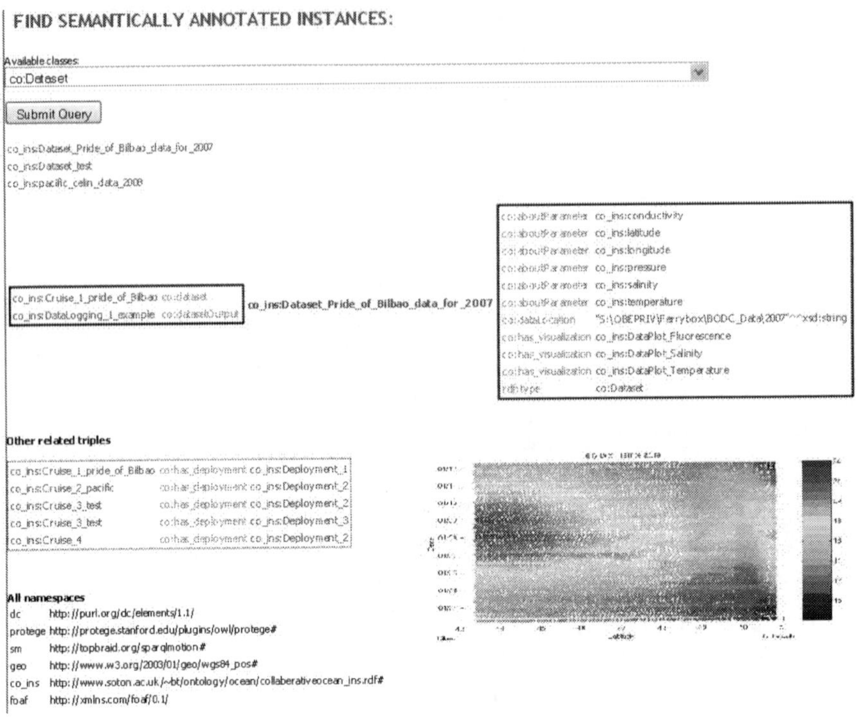

Fig. 11. Semantic browsing of datasets

In the Ferrybox context, we have done some experiments to demonstrate that some resource annotations, such as those of the sea ports, which have been semantically annotated using geospatial metadata, can be queried and their results mashed up with third-party WMS services such as Google map to visualize their geospatial presence. This demonstrates interoperability potential between RDF triples and OGC standard. We believe that as more measurement data are annotated with geospatial references, the mashing up will provide more interesting integration between the W3C semantic web and the OGC services.

5 Discussion and Future Work

The Semantic Web technology has moved from a pure research topic to more and more applications in different domains [15] [16] thanks to the growing availability of tools and implementation of W3C specifications. The bottle-neck that prevents abundant publishing of semantic content still lies in the semantic annotation. This is because users tend to hesitate in carrying out the somewhat tedious semantic annotation tasks without being able to see immediate benefit of the input. We believe that this problem can be relieved through two approaches - distributing the task load and batch data conversion.

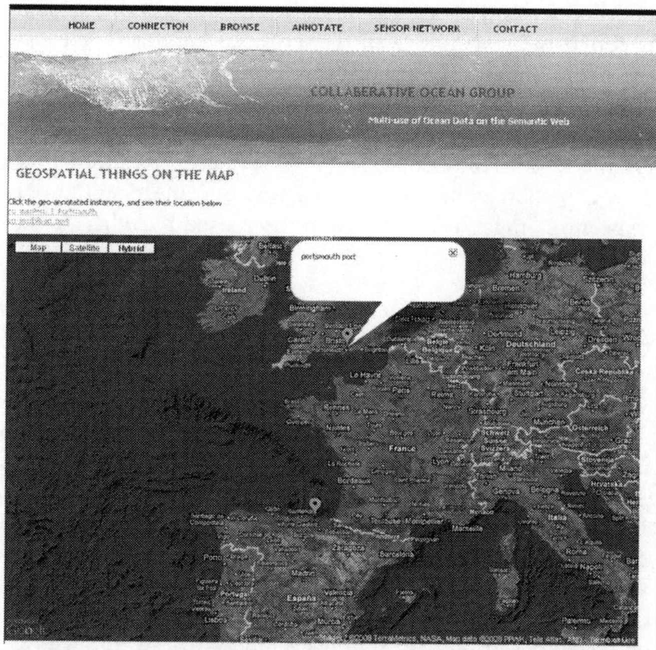

Fig. 12. Query and plotting of spatial annotations on the Google map

Since 2002 the data collected from the Pride of Bilbao Ferrybox every 10 minutes has been stored in a database which is used to generate annual measurement graphs shown on the web. However, users or software agents reading the web site have no access to formal descriptions of other related resources that actually collect the measurement datasets. It would be interesting to access such semantic information, e.g. sensor profiles, their calibration and deployment history, from the semantic annotation repository, as they are annotated by their stakeholders in the RDF triple store. This could also help the user to find other similar EO data for comparison by matchmaking and reasoning their semantic annotations on certain aspects of ontological reference model. We also plan to extend the system to a broader context in EO to incorporate more resource annotations into the system.

As can be seen in this paper, there is a clear distinction between ontology and semantic annotation, both of which can be represented in RDF. The ontology, which contains higher level schema-like domain knowledge, has more restricted written access once it reaches a finalized version and is published at a URL. It then serves as a set of formal reference points for definition of all necessary semantics used in the semantic annotations. The semantic annotations are however stored in the triple store which allows more flexible management including modifying, appending, and removing of the RDF triples. This is called the Semantic Web approach. One the other side, for a more open system with less control on the domain schema, ontology can be viewed as having the same level as semantic annotations. That means both of them are stored in the RDF triple store without real distinctions. In this way the ontology can also be modified by end users just as they can do with the semantic annotation triples. This is called the semantic web (lower case) approach. We use the first approach in our work but plan to try the second approach as well for comparison.

With the initial COG infrastructure set up at NOCS, it is now possible to semantically enrich oceanographic resources with ontology based metadata, either defined locally or externally. We have encouraged engineers at NOCS to contribute semantic annotations of their resources in a distributed way and the resulting RDF triples are stored in a persistent storage so that they can be reused by other engineers and data management applications in a later phase.

6 Conclusion

In this paper we presented our work on Collaborative Ocean Initiative using the Ferrybox project as a context. The aim is to facilitate resource and data interoperability in oceanographic Earth Observation (EO) to allow multi use of the ocean data. We investigated various technologies and methods, in particular the semantic web approach. By following a semantic web based knowledge life cycle, we have carried out activities on ontology modelling, semantic web annotation of domain resources and their storage as RDF triples and publishing/reuse on the. A semantic web infrastructure has been set up to accommodate these activities. Based on domain knowledge acquired, an ontology based domain reference model has been developed to provide domain conceptualisation and assumptions in the Ferrybox project. Through using different Semantic Web tools and interfaces, we have semantically annotated operational resources such as sensor and platform, as well as observation datasets in the Ferrybox project on the ontology based domain reference model. The ontology and semantic annotations are stored as OWL and RDF triples in a semantic web triple store where they can be further reused to facilitate dynamic ontology driven semantic annotation, linked data browsing, query and data integration.

We have learned through these semantic web activities that the data interoperability problem in EO can be improved by adding a semantic web layer where the ontology based domain reference model is used to enrich EO resources with standard metadata shared within the community.

Acknowledgements

This study is supported by the UK Natural Environment Research Council (NERC) through the Oceans 2025 core programmes of the National Oceanography Centre, Southampton.

References

1. Hydes, D.J., Campbell, J.M., Dunning, J.: Systematic Oceanographic Data Collected by FerryBox, Sea Technology (Feburary 2004)
2. Semantic Sensor Web at Wikipedia, http://en.wikipedia.org/wiki/Semantic_Sensor_Web2

3. Sheth, A., Henson, C.: Semantics & Services - Semantic Sensor Web. IEEE Computer Society, Los Alamitos (2008); 1089-7801/08, IEEE IEEE INTERNET COMPUTING, Kno.e.sis Center, Wright State University
4. The World Wide Web Consortium (W3C), http://www.w3.org
5. MMI-COG, http://marinemetadata.org/community/teams/cog
6. OpenGIS® Sensor Observation Service Implementation Specification, OGC 06-009r1, 2006-02-13
7. Ferrybox real time monitoring, http://www.noc.soton.ac.uk/ops/ferrybox_index.php
8. Hydes, D.J., Hartman, M.C., Bargeron, C.P., Campbell, J.M., Cure, M.S., Woolf, D.K.: A study of gas exchange during the transition from deep winter mixing to spring bloom in the Bay of Biscay measured by continuous observation from a ship of opportunity. Journal of Operational Oceanography 1(2), 41–59 (2008)
9. Tudorache, T., Noy, N.F., Tu, S.W., Musen, M.A.: Supporting collaborative ontology development in Protege. In: Sheth, A., et al. (eds.) ISWC 2008. LNCS, vol. 5318, pp. 17–32. Springer, Heidelberg (2008)
10. TopBraid Composer, http://www.topquadrant.com/topbraid/composer/index.html
11. Auer, S., Dietzold, S., Riechert, T.: OntoWiki – A Tool for Social, Semantic Collaboration. In: Cruz, I., Decker, S., Allemang, D., Preist, C., Schwabe, D., Mika, P., Uschold, M., Aroyo, L.M. (eds.) ISWC 2006. LNCS, vol. 4273, pp. 736–749. Springer, Heidelberg (2006)
12. Leuf, B., Cunningham, W.: The Wiki Way: Collaboration and Sharing on the Internet. Addison-Wesley Professional, Reading (2001)
13. Ocean2025, http://www.oceans2025.org/
14. Tao, F., Shadbolt, N., Chen, L., Xu, F., Cox, S.: Semantic Web based Content Enrichment and Knowledge Reuse in e-Science. In: Meersman, R., Tari, Z. (eds.) OTM 2004. LNCS, vol. 3290, pp. 654–669. Springer, Heidelberg (2004)
15. Tao, F., Millard, D., Zalfan, M., Chen, L., Davis, H.: Knowledge based Learning Experience Management on the Semantic Web. In: IADIS International Conference of e-Learning, Lisbon, Portugal, July 6-8 (2007)
16. Lewis, M., Cameron, D., Xie, S., Arpinar, B.: ES3N: A Semantic Approach to Data Management in Sensor Networks. In: Semantic Sensor network workshop, the 5th International Semantic Web Conference, ISWC 2006, November 5-9, Athens, Georgia, USA (2006)
17. Berners-Lee, T., Hendler, J., Lassila, O.: The Semantic Web. Scientific American, 34–43 (May 2001)
18. Fowler, D., Guan, T., Tao, F., Wills, G., Crowder, R., Shadbolt, N.: Comparison of cfd wikis. Technical Report at School of ECS University of Southampton (2008)
19. W3C Linked Data, http://www.w3.org/DesignIssues/LinkedData.html

ONKI SKOS Server for Publishing and Utilizing SKOS Vocabularies and Ontologies as Services

Jouni Tuominen, Matias Frosterus, Kim Viljanen, and Eero Hyvönen

Semantic Computing Research Group (SeCo)
Helsinki University of Technology (TKK) and University of Helsinki
firstname.lastname@tkk.fi
http://www.seco.tkk.fi/

Abstract. Vocabularies are the building blocks of the Semantic Web providing shared terminological resources for content indexing, information retrieval, data exchange, and content integration. Most semantic web applications in practical use are based on lightweight ontologies and, more recently, on the Simple Knowledge Organization System (SKOS) data model being standardized by W3C. Easy and cost-efficient publication, integration, and utilization methods of vocabulary services are therefore highly important for the proliferation of the Semantic Web. This paper presents the ONKI SKOS Server for these tasks. Using ONKI SKOS, a SKOS vocabulary or a lightweight ontology can be published on the web as ready-to-use services in a matter of minutes. The services include not only a browser for human usage, but also Web Service and AJAX interfaces for concept finding, selecting and transporting resources from the ONKI SKOS Server to connected systems. Code generation services for AJAX and Web Service APIs are provided automatically, too. ONKI SKOS services are also used for semantic query expansion in information retrieval tasks. The idea of publishing ontologies as services is analogous to Google Maps. In our case, however, vocabulary services are provided and mashed-up in applications. ONKI SKOS was published in the beginning of 2008 and is to our knowledge the first generic SKOS server of its kind. The system has been used to publish and utilize some 60 vocabularies and ontologies in the National Finnish Ontology Service ONKI www.yso.fi.

1 Introduction

Thesauri and other controlled vocabularies are used primarily for improving information retrieval [1]. This is accomplished by using concepts or terms of a thesaurus in content indexing, content searching or in both of them, thus simplifying the matching of query terms and the indexed resources (e.g. documents) compared to using natural language. For users, such as content indexers and searchers, to be able to use thesauri, publishing and finding methods for them as well as methods for integrating them with applications [2] are needed [3]. Thesauri are of great benefit for the Semantic Web[1], enabling semantically disambiguated data exchange and integration of data from different sources, though

[1] http://www.w3.org/2001/sw/

L. Aroyo et al. (Eds.): ESWC 2009, LNCS 5554, pp. 768–780, 2009.

not to the same extent as ontologies [4] where the semantics of concepts are defined in more refined and "machine understandable" ways.

Publishing and utilizing thesauri is a laborious task because the representation formats of thesauri and features they provide differ from each other. When utilizing thesauri one has to know how to locate a given thesaurus, and also be familiar with the software the thesaurus is published with. A thesaurus can even be published as a plain text file or even worse, as a paper document, with no proper support for utilizing it. In such a case the users have to implement applications for processing the thesaurus in order to exploit it. Therefore, standard ways for expressing and publishing thesauri would greatly facilitate the publishing and utilizing processes of thesauri.

The Simple Knowledge Organization System (SKOS)[2] [5] being developed within W3C is a data model and a syntax for expressing concept schemes such as thesauri. SKOS provides a standard way for creating vocabularies and migrating existing vocabularies to the Semantic Web. SKOS solves the problem of diverse, non-interoperable thesaurus representation formats by offering a standard convention for presentation. Existing thesauri can be transformed into SKOS format via conversion processes. When a thesaurus is expressed as a SKOS vocabulary, it can be published as a RDF file on the web, allowing the vocabulary users to fetch the files and process them in a uniform way. However, this does not solve the problem of users having to implement their own applications for processing vocabularies.

This paper argues that there are many common, sharable tasks in such vocabulary-aware applications related to e.g. term/concept finding, browsing, selecting, fetching, and query expansion. Lots of work and costs can be saved by implementing such functionalities in standard ways [6] and by providing them for production use as ready-to-use services without having to reimplement the functionalities separately for each local application case. In this way, the use patterns of utilizing vocabularies in the user interface can be harmonized, which makes systems easier to learn and use.

Ontology servers [7,8] have been proposed for publishing ontologies and vocabularies on the Semantic Web. Such servers are used for managing ontologies and offering users access to them. For accessing SKOS vocabularies, there is the Web Service based SKOS API[3] developed in the SWAD-Europe project, and the terminology service by Tudhope et al. [9]. There is also ongoing research by the Networked Knowledge Organization Systems/Services (NKOS) community[4] that focuses on enabling knowledge organization systems as networked interactive information services to support the description and retrieval of diverse information resources through the Internet. However, general tools for providing out-of-the-box support for utilizing SKOS vocabularies in, e.g., content indexing, without needing to implement application specific user interfaces for end users do not exist.

[2] `http://www.w3.org/TR/skos-reference/`

[3] `http://www.w3.org/2001/sw/Europe/reports/thes/skosapi.html`

[4] `http://nkos.slis.kent.edu/`

This paper fills this gap by presenting the ONKI SKOS Server for publishing and utilizing thesauri and lightweight ontologies. In the following, ways of presenting thesauri on the Semantic Web are first overviewed. After this, the approach and implementation of the ONKI SKOS Server is presented followed by descriptions of use cases of its services. In conclusion, contributions of the work are summarized and directions for further research outlined.

2 Presenting Thesauri on the Semantic Web

W3C's SKOS data model provides a vocabulary for expressing the basic structure and contents of concept schemes, such as thesauri, classification schemes and taxonomies. The concept schemes are expressed as RDF graphs by using classes and properties defined in the SKOS specification, thus making thesauri compatible with the Semantic Web. SKOS is capable of representing vocabularies which loosely conform to the influential ISO 2788 thesaurus standard [6].

Although semantically richer RDFS/OWL ontologies enable more extensive ways to perform logical inferencing than SKOS vocabularies, in several cases thesauri represented with SKOS are sufficient. In our opinion, the first and the most obvious benefit of using Semantic Web ontologies/vocabularies in content indexing is their ability to disambiguate concept references in a universal way. This is achieved by using persistent URIs as the identification mechanism and is a tremendous advantage when compared with the traditional idea of controlled vocabularies where plain concept labels are used as identifiers. In such a case, identification problems can be encountered e.g. with polysemous and homonymous labels, when dealing with multi-lingual resources, and with resources whose labels may change as time goes by. For example, human names are transliterated in many ways in different languages, may change due to marriage, a person may have nick names, and so on. In such cases the labels of concepts are not a permanent identification method, and the references to the concepts may become ambiguous or invalid.

URIs provide not only an identification mechanism, but also means for accessing the concept definitions and thesauri, when using the http URI Scheme[5]. Such URIs can act as URLs, and with a proper server configuration provide the users additional information about the referred concept[6]. In addition to these general RDF characteristics, SKOS provides a way for expressing relations between concepts suitable for the needs of thesauri, thus providing conceptual context for concepts.

In short, using a common representation model such as SKOS for thesauri greatly reduces the cost of (a) sharing thesauri, (b) using different thesauri in conjunction within one application, and (c) development of standard software to process them [6].

[5] `http://www.iana.org/assignments/uri-schemes.html`

[6] `http://www.w3.org/TR/swbp-vocab-pub/`

3 Utilizing Thesauri with ONKI SKOS Services

ONKI SKOS is an ontology server implementation for publishing and utilizing thesauri and lightweight ontologies. It conforms to the general ONKI vision and API, and is thus usable via ONKI ontology services as easily integrable user interface components and APIs [2].

The Semantic Web applications typically use ontologies which are either straightforward conversions of well-established thesauri, application-specific vocabularies or semantically richer ontologies, that can be presented and accessed in similar ways to thesauri [3,10]. Since SKOS defines a suitable model for expressing thesauri, it was chosen as the primary data model supported by the ONKI SKOS Server.

ONKI SKOS can be used to browse, search and visualize any vocabulary conforming to the SKOS specification, and also RDFS/OWL ontologies with additional configuration. ONKI SKOS does simple reasoning, e.g., transitive closure over class and part-of hierarchies. The implementation has been piloted using various thesauri and ontologies from diverse domains. Piloted ontologies include ontologies for cultural heritage (Art & Architecture Thesaurus AAT[7], 28,000 concepts; Iconclass[8], 27,000 concepts), health ontologies (Medical Subject Headings MeSH[9], 24,000 concepts), business ontologies (United Nations Standard Products and Services Code UNSPSC[10], 21,000 concepts), geographical instance ontologies (The Getty Thesaurus of Geographic Names TGN[11], 143,000 concepts), upper ontologies (General Finnish Upper Ontology YSO [3], 21,000 concepts), governmental ontologies (Finnish Governmental Thesaurus VNAS, 6,300 concepts) and natural science taxonomies (Ontology of Birds of the World AVIO, 11,000 concepts; Ontology of Mammals of the World MAMO, 5,000 concepts). However, not all of these ontologies are freely accessible due to licensing issues.

When utilizing thesauri represented as SKOS vocabularies and published on the ONKI SKOS server, several benefits are gained. Firstly, SKOS provides a universal way of expressing thesauri. Thus processing different thesauri can be done in the same way, eliminating the use of thesaurus specific processing rules in applications or separate converters between various formats. Secondly, ONKI SKOS provides access to all published thesauri in the same way, so content indexers and end-users do not have to use thesaurus specific implementations of thesaurus browsers and other tools developed by different parties, which is the predominant way. Also, one of the goals of the ONKI ontology services is that all the essential ontologies/thesauri can be found at the same location, thus eliminating the need to search for other thesaurus sources.

The typical way to use thesaurus specific publishing systems in content indexing and searching is either by using their browser user interface for finding

[7] `http://www.getty.edu/research/conducting_research/vocabularies/aat/`
[8] `http://www.iconclass.nl/`
[9] `http://www.nlm.nih.gov/mesh/meshhome.html`
[10] `http://www.unspsc.org/`
[11] `http://www.getty.edu/research/conducting_research/vocabularies/tgn/`

desired concepts and then copying and pasting the concept label to the used indexing system[12], or by using APIs for accessing and querying the thesaurus [9]. Both methods have some drawbacks. The first method introduces rather uncomfortable task of constant switching between two applications and the clumsy copy-paste procedure. The second method leaves the implementation job of the user interface entirely to the parties utilizing the thesaurus.

While ONKI SKOS supports both the aforementioned thesauri utilizing methods, in addition, as part of the ONKI ontology services, it provides a lightweight web widget for integrating general thesauri accessing functionalities into web based applications on the user interface level. The ONKI Selector widget depicted in Figure 1 can be used to search and browse thesauri, fetch URI references and labels of desired concepts and storing them in a concept collector. Similar ideas have been proposed by Hildebrand et al. [11] for providing search widget for general RDF repositories, and by Vizine-Goetz et al. [12] for providing widget for accessing thesauri through the side bar of the Internet Explorer web browser.

When the desired concepts have been selected with the ONKI Selector they can be stored into, e.g., the database of the application by using an HTML form. Either the URIs or the labels of the concepts can be transferred into the application providing support for the Semantic Web and legacy applications. For browsing the context of the concepts in thesauri, the ONKI SKOS Browser can be opened by pressing a button. Desired concepts can be fetched from the browser to the application by pressing the "Fetch Concept" button. Thus, there is no need for copy-paste procedures or user interface implementation projects.

For information retrieval use cases, ONKI SKOS provides support for finding suitable query terms and expanding them by using ontological inference, e.g., based on the concept subsumption, partonymy or associative relations between concepts. In legacy systems, which do not support URIs, the labels of the concepts can be used as query terms. In this context, ONKI SKOS supports multilingual search. If a thesaurus contains, e.g., English and Finnish labels for the concepts, the relevant query concepts can be searched in English, while in the actual information retrieval task the used query terms are in Finnish. This is useful when the end-user does not understand the language of the terms used in indexing the content of the underlying system.

The ONKI SKOS Browser (see Figure 2) is the graphical user interface of ONKI SKOS. It consists of three main components: 1) *semantic autocompletion concept search*, 2) *concept hierarchy* and 3) *concept properties*. When typing text to the search field, a query is performed to match the concepts' labels. The result list shows the matching concepts, which can be selected for further examination.

When a concept is selected, its concept hierarchy is visualized as a tree structure. The ONKI SKOS Browser supports multi-inheritance of the concepts (i.e. a concept can have multiple parents). Whenever a multi-inheritance structure is met, a new branch is formed to the tree. This leads to cloning of nodes, i.e.

[12] This is the way the Finnish General Thesaurus YSA has been used previously via the VESA Web Thesaurus Service, `http://vesa.lib.helsinki.fi/`

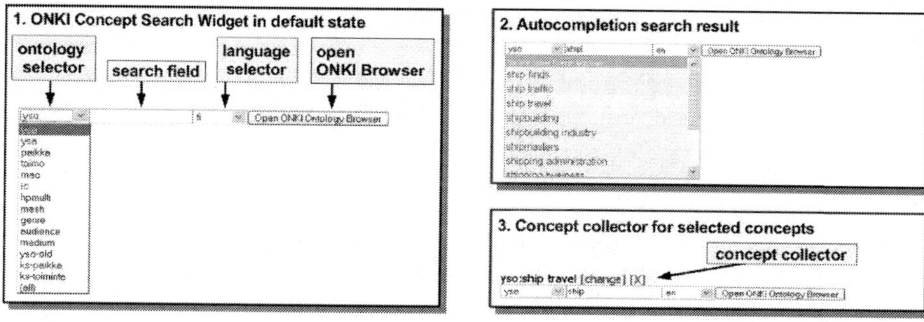

Fig. 1. ONKI Selector for concept finding

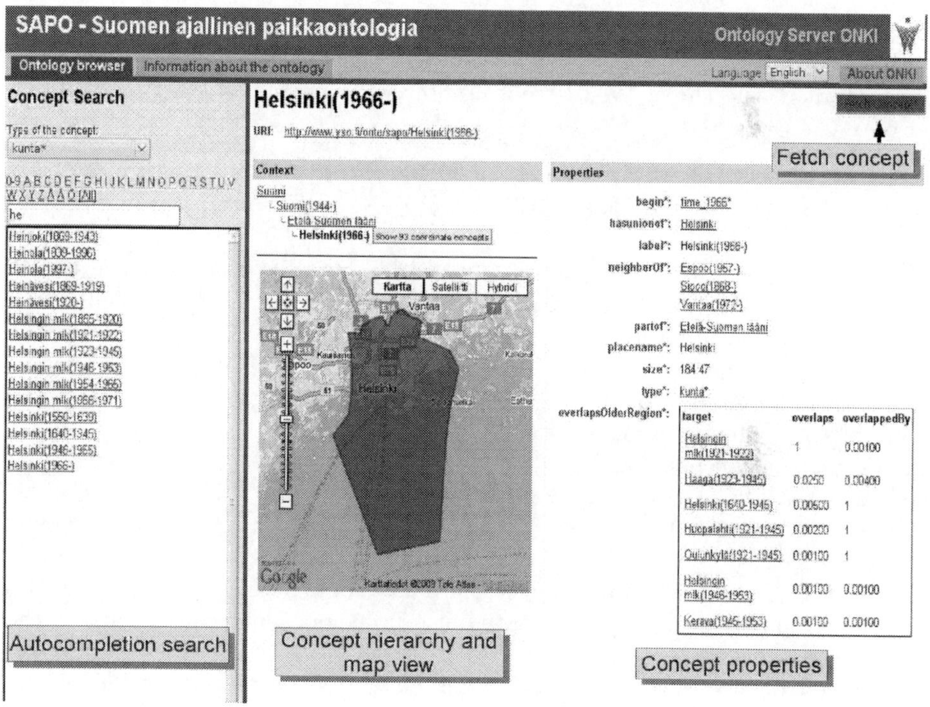

Fig. 2. The ONKI SKOS Browser

a concept can appear multiple times in the hierarchy tree. Next to the concept hierarchy tree, the properties of the selected concept are shown in the user interface. Mappings and other relations between concepts act as links allowing user to browse thesauri. In context of geographical ontologies, the place instances can be visualized on a Google Maps[13] view, as depicted in the Figure 2, if the instances contain coordinate information.

[13] http://maps.google.com/

The National Finnish Ontology Service ONKI[14] provides documentation and tools for software architects developing ontology-based applications and integrating ONKI services into them [13]. With ONKI Selector Builder a developer can set the desired configuration properties for the ONKI Selector and generate the needed JavaScript code to integrate the ONKI Selector into an application. Also, a hands-on demonstration page for integrating the ONKI Selector is available. The page contains a HTML form, into which the user can type HTML/JavaScript code and see how the resulting page will look like in the end-user's web browser.

The Web Service and AJAX interfaces of the ONKI SKOS server can be used for querying for concepts by label matching, performing query expansion, getting label for a given URI and for querying for supported languages and type URIs of a thesaurus. For testing Web Service methods, a demonstration page is provided at the ONKI service. One can try out the methods of the ONKI API and examine the resulting SOAP request and response messages.

ONKI SKOS is implemented as a Java Servlet using the Jena Semantic Web Framework[15], the Direct Web Remoting library[16] and the Lucene[17] text search engine.

4 Configuring ONKI SKOS with SKOS Structures

ONKI SKOS supports straightforward loading of SKOS vocabularies with minimal configuration needs. For using other data models than SKOS, various configuration properties are specified to enable ONKI SKOS to process the thesauri/ontologies as desired. The configurable properties include the properties used in hierarchy generation, the properties used to label the concepts, the concept to be shown in the default view and the default concept type used in restricting the concept search.

When the ONKI SKOS Browser is accessed with no URL parameters, information related to the concept configured to be shown as default is shown. Usually this resource is the root resource of the vocabulary, if the vocabulary forms a full-blown tree hierarchy with one single root. In SKOS concept schemes the root resource is the resource representing the concept scheme itself, i.e. the instance of *skos:ConceptScheme*.

The concept hierarchy of a concept is generated by traversing the configured properties. In SKOS these properties are *skos:narrower* and *skos:broader* and they are used to express the hierarchical relations between concepts. Hierarchical relations between the root resource representing the concept scheme and the top concepts of the concept scheme are defined with the property *skos:hasTopConcept*.

Labels of concepts are needed in visualizing search results, concept hierarchies, and related concepts in the concept property view. In SKOS the labels are

[14] `http://www.yso.fi/`
[15] `http://jena.sourceforge.net/`
[16] `http://directwebremoting.org/`
[17] `http://lucene.apache.org/java`

expressed with the property *skos:prefLabel*. The label is of the same language as the currently selected user interface language, if such a label exists. Otherwise any label is used.

The semantic autocompletion search of ONKI SKOS works by searching for concepts whose labels match the search string. To support this, the labels of the concepts are indexed. The indexed properties can be configured. In SKOS these properties are *skos:prefLabel*, *skos:altLabel* and *skos:hiddenLabel*. When the user searches, e.g., with the search term "cat", all concepts which have one of the aforementioned properties with values starting with the string "cat" are shown in the search results. The autocompletion search also supports wildcards, so a search with a string "*cat" returns the concepts which have the string "cat" as any part of their label.

If a SKOS vocabulary contains concepts that are instances of subclasses of skos:Concept, the search can be limited to certain types of concepts only. To accomplish this, the types of the concepts (which are expressed with the property *rdf:type*) are indexed. It is also possible to limit the search to a certain subtree of the concept hierarchy by restricting the search to subconcepts of a specific concept. To support this, also the parents of concepts are indexed.

Many thesauri include structures for representing categories of concepts. To support category-based concept search in the ONKI SKOS Browser, another search field is provided. When a category is selected from the category search view, the concept search is restricted to the concepts belonging to the selected category. SKOS includes a concept collection structure, *skos:Collection*, which can be used for expressing such categories. However, *skos:Collection* is often used for slightly different purposes, namely for node labels[18]. Thus, instances of *skos:Collection* are not used for category-based concept search by default.

5 Converting Thesauri to SKOS—Case YSA

Publishing a thesaurus in the ONKI SKOS server is straightforward. To load a SKOS vocabulary into the server, only the location path of the RDF file of the vocabulary needs to be configured manually. After rebooting the ONKI SKOS, the RDF file is loaded, indexed and made accessible for users. ONKI SKOS provides the developers of thesauri a simple way to publish their thesauri.

There exists quite amount of well-established keyword lists, thesauri and other non-RDF controlled vocabularies which have been used in traditional approaches in harmonizing content indexing. In order to reuse the effort already invested developing these resources by publishing these vocabularies in ONKI SKOS server, conversion processes need to be developed. This idea has also been suggested by van Assem et al. [6] and Summers et al. [14]. We have implemented transformation scripts for, e.g., MARCXML format[19], XML dumps from SQL databases and proprietary XML schemas.

---

[18] A construct for displaying grouping concepts in systematic displays of thesauri. They are not actual concepts, and thus they should not be used for indexing. An example node label is "milk by source animal".

[19] `http://www.loc.gov/standards/marcxml/`

An example of the SKOS transformation and publishing process is the case of YSA, the Finnish General Thesaurus[20]. YSA is developed by the National Library of Finland and is the most widely used controlled vocabulary in Finland. YSA is published in the VESA Web Thesaurus Service, where it receives over 12 million user hits per year. YSA is simultaneously published in the ONKI SKOS, currently in a pilot phase. During this year, the operation of the VESA service will be ended, and the ONKI SKOS becomes the official publishing channel of YSA.

To begin the SKOS transformation process, YSA is exported into MARCXML format from the thesaurus management system of the National Library of Finland. The resulting constantly up-to-date file is published at the web server of the National Library of Finland, from where it is fetched via OAI-PMH protocol[21] to our server. This process is automated and the new version of the XML file is fetched daily. Instead of fetching the full file every day, the incremental update feature of the OAI-PMH could be used, but we have not tested it. After fetching a new version of the file, the transformation process depicted in Figure 3 is started by loading the MARCXML file (ysa.xml). The Java-based converter first creates the necessary structure and namespaces for the SKOS model utilizing Jena Semantic Web Framework. Next, the relations in YSA are translated into their respective SKOS counterparts, which is depicted in Figure 4.

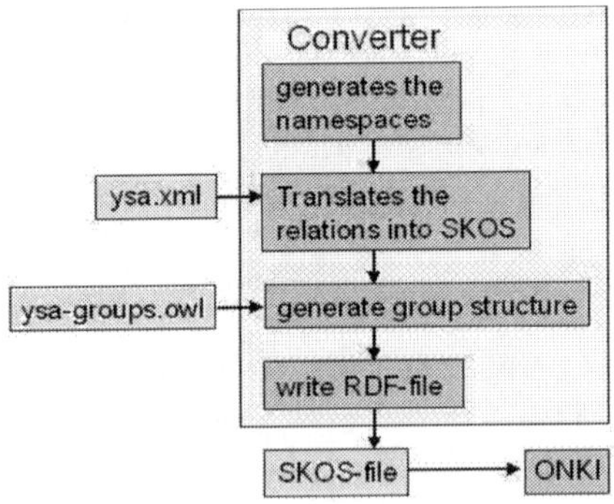

Fig. 3. The SKOS transformation process of YSA

A URI for the new concept entry is created through the unique ID in the source file. The preferred and alternative labels are converted straightforwardly from one syntax to another. Similarly the type and scheme definitions are added to the SKOS model. Since the relations in the MARCXML refer not to the identifiers but rather to the labels, the source file is searched for an entry that has the given label and then its ID is recorded for the SKOS relation.

MARCXML record for the concept "pistols" can be seen in the left part of Figure 4. The ID of the concept is denoted by the *controlfield* with *tag* attribute's

[20] `http://www.nationallibrary.fi/libraries/thesauri/ysa.html`

[21] `http://www.openarchives.org/OAI/openarchivesprotocol.html`

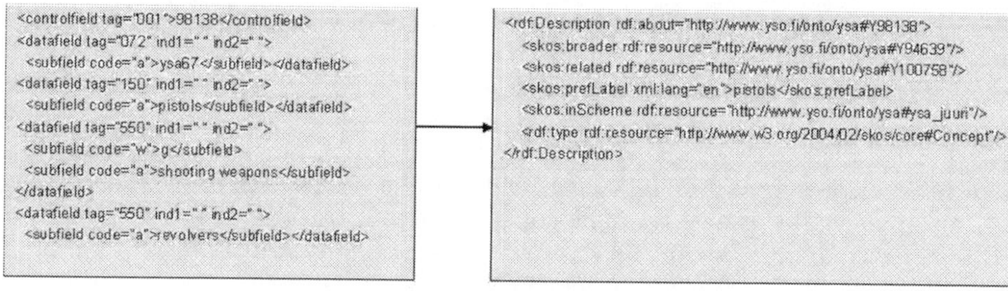

Fig. 4. An example of the SKOS transformation of YSA

value "001" ("98138"). This ID value becomes the local name of the URI of the corresponding SKOS concept. The category in which the concept belongs to is depicted with the *datafield* with *tag* "072" ("ysa67"). The preferred label is the value provided in the *datafield* with *tag* value "150" ("pistols"). Hierarchical and associative relations are marked with *datafield* with *tag* value "550", while the *subfield*'s attribute *code* specifies which one of the relations is in question.

Once the SKOS transformation is ready, the converter fetches the labels for the concept categories from a separate file (ysa-groups.owl) - these labels are not included in the MARCXML file. Finally, a RDF file is written and imported into ONKI SKOS.

6 Use Cases of ONKI SKOS

ONKI SKOS has been tested in several content indexing and information retrieval pilot applications.

6.1 Content Indexing

The browser-based annotation editor SAHA[22] [15] has been used for creating semantic content for semantic portals. The two major applications demonstrating Semantic Web technologies are HealthFinland[23] [16] in the health promotion context, and CultureSampo[24] [17] in the cultural heritage context. Both systems use the SAHA editor with the services provided by ONKI SKOS and the National Finnish Ontology Service ONKI. ONKI Selector has been integrated into SAHA for finding suitable concepts in content indexing.

The web laboratory Owela[25] of the VTT Technical Research Centre of Finland has implemented a service for collecting and sharing text and image clips from the Web[26]. In the service one can organize the clips into folders and tag them with

[22] `http://demo.seco.tkk.fi/saha/?lang=en`
[23] `http://demo.seco.tkk.fi/tervesuomi/`
[24] `http://www.kulttuurisampo.fi/`
[25] `http://owela.vtt.fi/`
[26] `http://owela.vtt.fi/tilkut`

different categories. The ONKI Selector is used for tagging the clips keywords from vocabularies published in the ONKI service.

6.2 Information Retrieval

Kantapuu.fi[27] is a web user interface for browsing and searching for collections of Finnish museums of forestry. The collection items are annotated with Finnish terms from General Finnish Thesaurus YSA, Thesaurus for Museum Domain MASA and Agriforest Thesaurus. Kantapuu.fi search page is a web form into which query strings are typed as free text. The query strings can be placed into specific fields, e.g. “keywords”, “place of use” or “time of use”. We have created a demonstration page containing a Kantapuu.fi's search form with integrated ONKI Selectors which can be used for selecting query terms to be used in the Kantapuu.fi search[28]. The ONKI Selector is used for finding terms from vocabularies of the ONKI service. The used vocabularies are the same as those used in the annotation process of the items, or actually their ontologized versions. To find suitable query terms user can utilize the autocompletion search functionality or the ONKI Browser. Thus, the user does not need to be familiar with the vocabularies used in the annotations of the items, as in the case of free text search. The ONKI Selector performs query expansion based on the selected query terms. So, for example a query term “animals” would return items annotated with term “cats”. When the desired query terms are selected, the actual search to the Kantapuu.fi system can be executed. The users of the pilot system have given positive feedback especially on the multilingual search possibility.

The ONKI Selector Widget has also been integrated into the Viikki Science Library[29] reference database system eViikki[30]. eViikki is a search interface for the library's collections, which consist of scientific literature on agriforestry. The literature is annotated with terms from General Finnish Thesaurus YSA and Agriforest vocabulary. The ONKI Selector is used for populating the “keywords” field of the search form of eViikki. The fetched concept labels are used in the information retrieval task. Query expansion is not performed currently.

7 Discussion

The main contribution of this paper is to show, both in principle and as a deployed implemented service in use, how thesauri can be published and utilized easily on the Semantic Web, emphasizing the benefits of the use of W3C's SKOS data model as a uniform vocabulary representation framework. The ONKI SKOS server is presented as a proof of concept for a cost-efficient thesauri utilization method. By using ONKI SKOS, general thesauri accessing functionalities can be easily integrated into applications without the need for users to reimplement their own user interfaces for this.

[27] http://www.kantapuu.fi/
[28] http://www.yso.fi/lusto
[29] http://www.tiedekirjasto.helsinki.fi/english/
[30] http://www-db.helsinki.fi/eviikki/eviikkihaku.html

The utilization of the SKOS structures in an ontology server was described in context of the ONKI SKOS server. The case of the Finnish General Thesaurus acts as one example on how an existing, widely used thesaurus can be converted into the SKOS format and be published on the ONKI SKOS server. At the moment some 60 international and national vocabularies, taxonomies, and lightweight ontologies have been published online using ONKI SKOS and the number is increasing. Finally, in order to demonstrate end-user benefits of ONKI SKOS services, we provided some real world use cases of ONKI SKOS in content indexing and information retrieval.

Future work for ONKI SKOS includes creating a more extensive Web Service interface for supporting, e.g., querying for properties of a given concept and for discovering concepts which are related to a given concept. The starting point for this API will be the SKOS API. Also, version control techniques and other support for vocabulary management (e.g. protocol for communicating vocabulary changes to users) would be needed to support the development phase of vocabularies.

Acknowledgements

We thank Ville Komulainen for his work on the original ONKI server. This work is a part of the National Semantic Web Ontology project in Finland[31] (FinnONTO) and its follow-up project Semantic Web 2.0[32] (FinnONTO 2.0, 2008-2010), funded mainly by the National Technology and Innovation Agency (Tekes) and a consortium of 38 private, public and non-governmental organizations.

References

1. Aitchison, J., Gilchrist, A., Bawden, D.: Thesaurus Construction and Use: A Practical Manual, 4th edn. Europa Publications (2000)
2. Viljanen, K., Tuominen, J., Hyvönen, E.: Publishing and using ontologies as mashup services. In: Proceedings of the 4th Workshop on Scripting for the Semantic Web (SFSW 2008), Tenerife, Spain, June 1-5 (2008)
3. Hyvönen, E., Viljanen, K., Tuominen, J., Seppälä, K.: Building a national semantic web ontology and ontology service infrastructure—the FinnONTO approach. In: Bechhofer, S., Hauswirth, M., Hoffmann, J., Koubarakis, M. (eds.) ESWC 2008. LNCS, vol. 5021, pp. 95–109. Springer, Heidelberg (2008)
4. Staab, S., Studer, R. (eds.): Handbook on ontologies. Springer, Heidelberg (2004)
5. Miles, A., Matthews, B., Wilson, M., Brickley, D.: SKOS Core: Simple knowledge organisation for the web. In: Proceedings of the International Conference on Dublin Core and Metadata Applications (DC 2005), Madrid, Spain, September 12-15 (2005)

[31] http://www.seco.tkk.fi/projects/finnonto/

[32] http://www.seco.tkk.fi/projects/sw20/

6. van Assem, M., Malaisé, V., Miles, A., Schreiber, G.: A method to convert thesauri to SKOS. In: Sure, Y., Domingue, J. (eds.) ESWC 2006. LNCS, vol. 4011, pp. 95–109. Springer, Heidelberg (2006)
7. Ding, Y., Fensel, D.: Ontology library systems: The key to successful ontology reuse. In: Proceedings of SWWS 2001, The first Semantic Web Working Symposium, Stanford University, USA, pp. 93–112 (August 2001)
8. Ahmad, M.N., Colomb, R.M.: Managing ontologies: a comparative study of ontology servers. In: Proceedings of the eighteenth Conference on Australasian Database (ADC 2007), Ballarat, Victoria, Australia, January 30–February 2, pp. 13–22. Australian Computer Society, Inc (2007)
9. Tudhope, D., Binding, C.: Towards terminology service: experiences with a pilot web service thesaurus browser. In: Proceedings of the International Conference on Dublin Core and Metadata Applications (DC 2005), Madrid, Spain, September 12-15, 2005, pp. 269–273 (2005)
10. van Assem, M., Menken, M.R., Schreiber, G., Wielemaker, J., Wielinga, B.: A method for converting thesauri to RDF/OWL. In: McIlraith, S.A., Plexousakis, D., van Harmelen, F. (eds.) ISWC 2004. LNCS, vol. 3298, pp. 17–31. Springer, Heidelberg (2004)
11. Hildebrand, M., van Ossenbruggen, J., Amin, A., Aroyo, L., Wielemaker, J., Hardman, L.: The design space of a configurable autocompletion component. Technical Report INS-E0708, Centrum voor Wiskunde en Informatica (CWI), Amsterdam (2007)
12. Vizine-Goetz, D., Childress, E., Houghton, A.: Web services for genre vocabularies. In: Proceedings of the International Conference on Dublin Core and Metadata Applications (DC 2005), Madrid, Spain, September 12-15 (2005)
13. Viljanen, K., Tuominen, J., Hyvönen, E.: Ontology libraries for production use: The Finnish ontology library service ONKI. In: Aroyo, L., et al. (eds.) ESWC 2009. LNCS, vol. 5554, pp. 781–795. Springer, Heidelberg (2009)
14. Summers, E., Isaac, A., Redding, C., Krec, D.: LCSH, SKOS and Linked Data. In: Proceedings of the International Conference on Dublin Core and Metadata Applications (DC 2008), Berlin, Germany, September 22-26 (2008)
15. Valkeapää, O., Alm, O., Hyvönen, E.: Efficient content creation on the semantic web using metadata schemas with domain ontology services (system description). In: Franconi, E., Kifer, M., May, W. (eds.) ESWC 2007. LNCS, vol. 4519, pp. 819–828. Springer, Heidelberg (2007)
16. Hyvönen, E., Viljanen, K., Suominen, O.: HealthFinland—Finnish health information on the semantic web. In: Aberer, K., Choi, K.-S., Noy, N., Allemang, D., Lee, K.-I., Nixon, L., Golbeck, J., Mika, P., Maynard, D., Mizoguchi, R., Schreiber, G., Cudré-Mauroux, P. (eds.) ASWC 2007 and ISWC 2007. LNCS, vol. 4825, pp. 778–791. Springer, Heidelberg (2007)
17. Hyvönen, E., Ruotsalo, T., Häggström, T., Salminen, M., Junnila, M., Virkkilä, M., Haaramo, M., Mäkelä, E., Kauppinen, T., Viljanen, K.: Culturesampo—Finnish culture on the semantic web: The vision and first results. In: Robering, K. (ed.) Information Technology for the Virtual Museum. LIT Verlag, Berlin (2007)

Ontology Libraries for Production Use: The Finnish Ontology Library Service ONKI

Kim Viljanen, Jouni Tuominen, and Eero Hyvönen

Semantic Computing Research Group (SeCo)
Helsinki University of Technology (TKK) and University of Helsinki
firstname.lastname@tkk.fi
http://www.seco.tkk.fi/

Abstract. This paper discusses problems of creating and using ontology library services in production use. One approach to a solution is presented with an online implementation—the Finnish Ontology Library Service ONKI— that is in pilot use on a national level in Finland. ONKI contributes to previous research on ontology libraries in many ways: First, mashup and web service support with various tools is provided for cost-efficient utilization of ontologies in indexing and search applications. Second, services covering the different phases of the ontology life cycle are provided. Third, the services are provided and used in real world applications on a national scale. Fourth, the ontology framework is being developed by a collaborative effort by organizations representing different application domains, such as health, culture, and business.

1 Introduction

The Semantic Web[1] is based on ontologies [1,2,3]. With the help of ontologies, the content and services on the Web can be described with metadata in an explicit, machine "understandable" way which enables, for example, interoperability on a semantic level and intelligent semantic searching and browsing of heterogeneous distributed content in semantic portals [4,5,6]. Utilizing ontologies, including thesauri and other vocabularies, in new and existing applications requires efficient tools for finding, managing, searching and browsing ontologies. *Ontology library systems*[2] offer functions for managing, adapting and standardizing groups of ontologies, for indexing content with ontologies, and for utilizing ontologies in applications [6,7,8,9].

This paper discusses the requirements for ontology library systems from a practical viewpoint, and presents an approach for building such a service. As a concrete result of the research and a case study, the national Finnish Ontology Library Service ONKI[3] is presented. ONKI is a major objective of the National

[1] http://www.w3.org/2001/sw

[2] Ontology library systems are referred to in the literature with terms "ontology servers" and "ontology services", too. We use the term "ontology library systems" to encompass all these meanings.

[3] http://www.yso.fi/

L. Aroyo et al. (Eds.): ESWC 2009, LNCS 5554, pp. 781–795, 2009.

Semantic Web Ontology project FinnONTO (2003–2010)[4] which aims at developing a Semantic Web ontology infrastructure on a national level in Finland [6]. The consortium behind the initiative represents a wide spectrum of functions of the society, including libraries, health organizations, cultural institutions, government, media, and education.

In the following, we first set requirements for an ontology library system in terms of services at different phases of ontology development and usage. After this, the ONKI system and its services are presented along the same phases, and the implementation is discussed. The system has been used in several case applications that are briefly surveyed next. In conclusion, related work is discussed and contributions of ONKI summarized.

2 Requirements for an Ontology Library Service

Requirements for ontology library services can be identified by analyzing the life cycle of an ontology. Based on literature [8,7] and our own work, we identify following parts to be typical in a life cycle of an ontology.

1) Designing the ontologies. The first step in the life cycle of an ontology is to design the structure and modelling principles of the ontology based on analysis of the subject domain and by identifying the business and application problems the ontology is intended to solve. The foundational classes, properties and instances of the ontology are created. Sometimes the ontology may also be based on existing vocabularies such as a thesaurus, which are then "ontologised". The main actors of this phase are workgroups and individual ontologists. Ontology libraries should provide support e.g. for collaborative ontology editing, reuse and alignment [10]. *2) Populating the ontologies.* Ontologies may consist of huge amounts of instances such as people, organisations and places. Populating and maintaining the information can either be a one-time effort or constantly continuing process even after the ontology has been published. Populating may be e.g. a community-based distributed effort or based on utilizing existing registries and sources as input. Ontology libraries should support such content collection and updating processes. *3) Publishing the ontologies.* When an ontology has been created, methods for publishing and promoting it are needed to ensure that the ontology is actively used to achieve the benefits of creating the ontology in the first place. In addition to provide such publishing and promoting mechanisms, the ontology library should also have mechanisms - both manual and automatic - for ensuring the quality of the ontologies to be published. The main actors of this phase are the ontology owners. *4) Finding, comparing and commiting to ontologies.* When considering using ontologies for some purpose, the finding of suitable ontologies require support from the ontology library. Typical users of this phase are information architects. *5) Ontology based semantic application creation.* Learning, evaluating and implementating ontology services to applications require functionalities from the ontology library service for making this

[4] Our work is funded by the National Funding Agency for Technology and Innovation (Tekes) and a consortium of 38 companies and public organizations.
http://www.seco.tkk.fi/projects/finnonto/

process as fluent and easy as possible - and to support the wide usage of ontologies in applications [11,12]. Typical users of this phase are software architects. *6) Ontology based semantic content creation.* Ontologies are mostly used for describing and indexing content semantically. Ontology libraries should support the work of content indexers by providing efficient tools and services for e.g. browsing ontologies, finding and fetching concepts for annotating purposes [9] or automatically indexing documents [13]. *7) Ontology-based end-user applications.* Ontological content search, semantic browsing, semantic portals are examples of typical ways to provide the end-user with benefits from using ontologies in an applications (see e.g. [4,5,6,14]). Ontology libraries should support creating such end-user applications by providing services for the application builders. Ontology libraries may also provide services directly to the end-users such as the possibility to learn about some domain with the help of ontologies.

3 Finnish Ontology Library Service ONKI

The Finnish Ontology Library Service ONKI is a pilot system for addressing the requirements of an ontology library service on a national scale, but with the special focus on ontology publishing and using them in content indexing, and information retrieval through both user and application interfaces [14,6]. ONKI contains currently over 40 ontologies from various domain areas (see Table 1). Most of the ontologies are freely available to anybody to test and use in their applications.

3.1 Designing the Ontologies, Populating the Ontologies

Ontologies developed within the ONKI framework are mostly created with the Protégé[5] editor. Version management of the ontology files is done with Subversion[6]. In many cases, the ontologies are based on an existing thesaurus or other content which have first been transformed with an custom-made program to OWL and then been refined ontologically by the ontologist using Protégé and by aligning the ontologies with the Finnish Upper Ontology YSO [6]. Populating the ontologies have been done either with Protégé by the ontologist(s) or collaboratively with the browser-based annotation editor SAHA [15]. For many ontologies, populating have been done with custom-made programs.

Quality of the ontologies is controlled using three ways: gate keeping, quality requirements and training. Gate keeping is practised by selecting only trusted participants in the ontology creation and publishing which include both companies, governmental and non-governmental organizations. Typically the main author of any single ontology in ONKI is the leading authority in Finland of the respective domain. Quality requirements enforced in ONKI cover ontology presentation and ontology creation process issues. The ontology should be presented using some RDF-based ontology language, such as SKOS, OWL or RDF

[5] http://protege.stanford.edu
[6] http://subversion.tigris.org/

Table 1. A selection of ontologies currently available in ONKI (The amount of concepts consists classes and/or instances, depending on the ontology.)

Ontology	Concepts	Format	Public?
Upper and Holistic Ontologies			
Holistic Collaborative Finnish Ontology KOKO	ca. 30,000	OWL	yes
General Finnish Upper Ontology YSO	20,649	OWL	yes
General Finnish Thesaurus YSA	26,633	SKOS	yes
Wordnet	ca. 230,000	SKOS	yes
Cultural Ontologies			
Ontology for Museum Domain MAO	6,775	OWL	yes
Ontology of Applied Arts TAO	29,940	OWL	yes
Finnish Ontology of Photography VALO	22,596	OWL	yes
Ontology for music MUSO	21,650	OWL	yes
Art and Iconography classification Iconclass	26,636	SKOS	yes
Kaunokki thesaurus for fictive literature	4,373	SKOS	yes
Music thesaurus MUSA	931	SKOS	yes
Art & Architecture Thesaurus AAT	27,992	OWL	no
Agriforest and Natural Science Ontologies			
Agriforest Ontology AFO	26,612	OWL	yes
Ontology of Birds AVIO	11,161	SKOS	no
Ontology of Mammals MAMO	6,059	SKOS	no
Health Ontologies			
Medical Subject Headings MeSH	24,355	SKOS	yes
European Multilingual Thesaurus on Health Promotion HPMULTI	1,271	SKOS	yes
Business Ontologies, Governmental Ontologies			
Seafaring thesaurus MESA	1,448	SKOS	yes
United Nations Standard Products and Services Code UNSPSC	20,794	SKOS	no
Finnish Governmental Thesaurus VNAS	6,342	SKOS	yes
Instance Ontologies			
Finnish Geo-ontology SUO	ca. 800,000	OWL	yes
Finnish Time-Location Ontology SAPO	1,102	OWL	yes
Getty Thesaurus of Geographic Names TGN (exluding USA)	142,990	OWL	no
Getty Union List of Artist Names ULAN	ca. 100,000	OWL	no

Schema. The consistency of the ontologies is checked including syntax checking (valid RDF) and conceptual checking manually by the ontologists. One of the most important ways to enforce the quality is that the ontologies in ONKI library are created using a common development process and modelling idea which are supervised by the core YSO developer team [6]. This promotes using compatible development processes in all other ontologies also and thus provides a more compatible collection of ontologies as a result. If possible, ontologies are aligned with a common upper ontology, the Finnish Upper Ontology YSO. This alignment to YSO adds value to YSO, the ontology at hand and the ONKI Library as a whole, because each additional alignment adds new possibilities to find concept relations. To spread good practices and knowledge about the modelling methods used in YSO and other relevant ontologies, training is provided to ontology developers. To enforce the reuse of ontologies, the license of the published ontologies should allow publishing, using and redeveloping the ontology as freely as possible. The default license used for ONKI ontologies is the Creative Commons license[7].

3.2 Publishing the Ontologies

Publishing an ontology in ONKI typically contains the following phases: First, the ontology to be published is added to the Subversion repository and the needed configurations for the ontology are created [16]. Second, a URI normalization for the ontology to be published is done where the original URIs are transformed to persistent numeric URIs (PURIs). Instead of (typically) human readable URIs we propose that URIs should not contain any reference to human languages to avoid unnecessary needs for changing the concept URIs e.g. when translating the ontology to some other language[8]. For example, instead using the URI "myonto:semanticweb" we propose using the URI "myonto:p12345". Third, if the ontology is published part of the KOKO ontology, the automatic updating of KOKO takes place. Finally, the ontology is added to ONKI and made available via different services such as human user interfaces and machine APIs.

If the ontology is maintained in some external system it can be published using ONKI by establishing a publishing pipeline from the external system to ONKI. This method has been used e.g. in publishing the General Finnish Thesaurus YSA, maintained by the National Library of Finland [16]. The thesaurus is fetched each night from the National Library's server using the MARCXML format[9]. The content is then transformed to SKOS and finally published in ONKI. ONKI provides also an upload functionality "Your ONKI" for publishing SKOS (or other) ontologies in the library. When an ontology has been uploaded, it is moderated by the server administration and if the content is suitable, it will be added to the library. ONKI quality requirements presented earlier are recommended also for Your ONKI submissions.

3.3 Ontology Discovery, Ontology Library Service Evaluation

To support finding, evaluating and choosing an ontology for specific purposes each ontology is described with metadata including title, description, classification, version information and available access methods. Depicted in Figure 1 is the main user interface with the list of available ontologies, which also shows the available access methods for each ontology. Access methods are described later, but include e.g. the possibility to browse and search the ontologies. The ontologies can be described and documented in a wiki, part of the ONKI system.

The list of ontologies is available in RDF for machine usage. When allowed by the publishing license, current and previous versions of ontologies are available for downloading.

---

[7] http://creativecommons.org/

[8] The idea of stable URIs is also discussed in http://www.w3.org/Provider/Style/URI

[9] http://www.loc.gov/standards/marcxml/

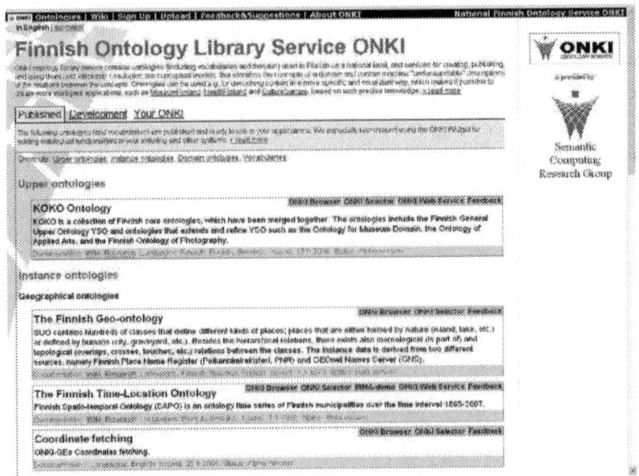

Fig. 1. List of ontologies in ONKI. Each ontology contains the links to the available access methods such as the ontology-specific browser.

3.4 Ontology-Based Semantic Content Creation

For users creating semantic content, e.g. by describing resources by using ontological concepts, ONKI service provides the ONKI Selector, depicted in Figure 2 [9]. With ONKI Selector content creators can find suitable concepts for their annotation tasks. When the ONKI Selector is integrated into a HTML input field, the field turns into a semantic autocompletion search interface. When typing search string into the input field, the matching concepts are returned as a hit list. Desired concepts can be selected from the list and added to the content creation application. Depending on the use case, concept's URI, label or both of them can be fetched to the application.

In combination with the ONKI Selector, domain-specific ONKI browsers can be used to browse the ontologies when searching for suitable concepts. The browsers have a "Fetch Concept" button which returns the selected concept into the content creation application. ONKI SKOS Browser[16] is an ontology

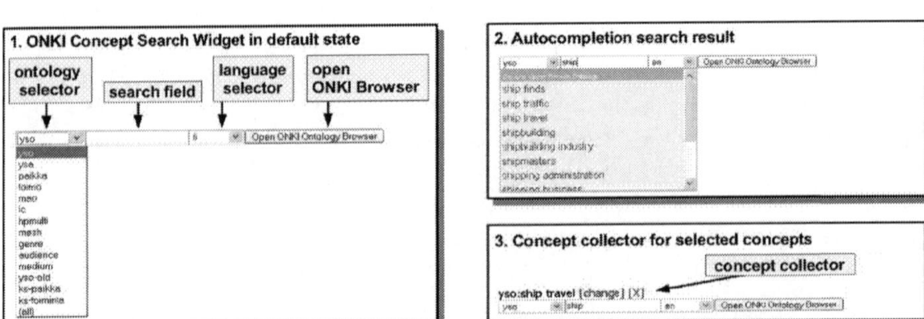

Fig. 2. ONKI Selector

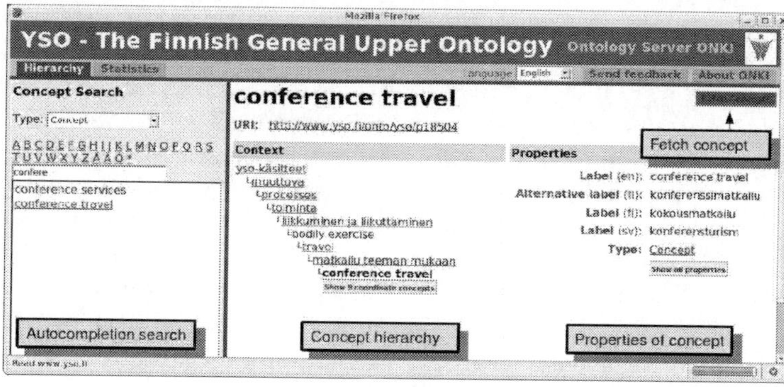

Fig. 3. ONKI SKOS Browser

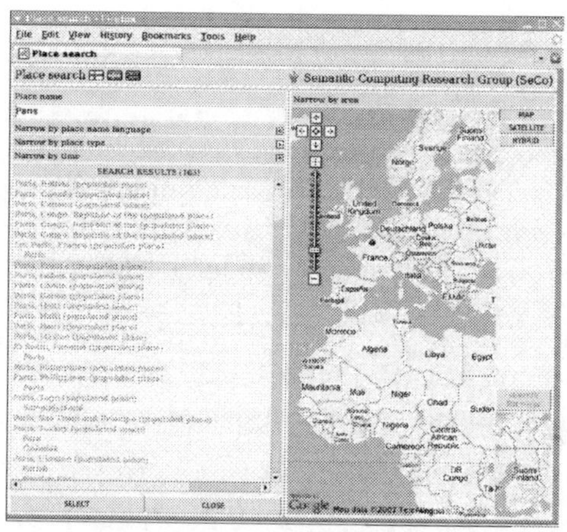

Fig. 4. ONKI Geo Browser

browser for thesaurus-like class ontologies. It supports visualizing and browsing of vocabularies conforming to SKOS recommendation, and also RDF(S) and OWL ontologies with additional configuration. ONKI SKOS Browser consists of three main components: 1) *concept search with semantic autocompletion*, 2) *concept hierarchy* and 3) *concept properties*, as depicted in Figure 3. ONKI Geo Browser [17] is used for accessing geographical instance data with a map interface, as depicted in Figure 4. It provides unambiguous place identifiers (URIs) and coordinates for arbitrary points or polygons to be used in content annotation. ONKI People [18] is used for browsing and searching ontologies of persons, organizations, and similar instance registries.

3.5 Ontology-Based End-User Applications

For ontology-based end-user applications ONKI service provides means for finding ontological concepts and using them, e.g., in information retrieval tasks.

Compared to a simple free text search field, the ONKI Selector aids user to find query concepts with autocompletion search and ontology browsers. The ONKI Selector is useful even if the application is not ontology-based. In that case the labels of the concepts can be used as query terms.

To increase the recall of the information retrieval tasks, ONKI Selector performs query expansion by ontological inference. The properties used for performing the query expansion can be configured separately for each ontology. In class ontologies, a concept is typically expanded to its subconcepts. The query expansion could also be based on partonomy, associative relationships or other relations between concepts. In geographical instance data, a place instance is expanded to places that have historically had overlapping regions with the place. Other possible query expansion methods include partonomy, places with shared regional borders etc.

3.6 Ontology-Based Semantic Application Creation

ONKI services may be integrated into semantic applications at the user interface level by using the ready-to-use user interface component ONKI Selector, domain-specific ONKI Browsers, and by using application programming interfaces (API). ONKI supports the software developer in using ONKI services with the help of helper applications such as the ONKI Selector Builder, depicted in Figure 5, which helps the developer to generate the JavaScript code needed for integrating ONKI Selector into web-based applications. When the desired configuration properties have been set in the ONKI Selector Builder, the resulting JavaScript code can be copied into the application being developed.

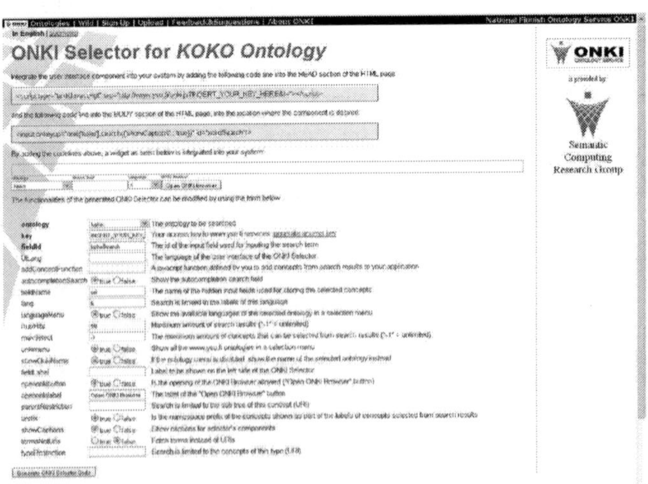

Fig. 5. ONKI Selector builder

ONKI API provides methods for accessing ontologies, e.g., for searching for concepts, getting metadata of an ontology and performing ontology-based query expansion. ONKI API is implemented as Web Service (SOAP) and JavaScript interface. ONKI API contains the following methods:

- *search(query, lang, maxHits, type, parent)* - for searching for the ontological concepts. Returns a list of matching concepts.
- *expandQuery(URI, lang, maxHits, type)* - for querying for the query expansion for a concept. Returns a list of concepts.
- *getLabel(URI, lang)* - for fetching a label for a given concept URI in a given language.
- *getAvailableLanguages()* - for querying for the supported languages of an ontology. Returns a list of language codes.
- *getAvailableTypeUris()* - for querying for the concept types (rdf:type relations) existing in the ontology. Returns a list of URIs.

Software developers can also utilize the RDF files of the ontologies published in the ONKI service. All concept and instance URIs are designed so that they function also as URLs. When the URI of a concept is accessed with a web browser, the relevant view is opened in the ONKI browser. This means that the URI itself acts as a functional link when added to a HTML page. In accordance to W3C[10], if the URI is accessed with an RDF aware system, the machine readable RDF presentation of the content is returned instead of the ONKI browser's HTML presentation.

4 Implementation and Usage Statistics

The ONKI service is constituted of a loosely coupled set of independent applications such as the ONKI SKOS and ONKI Geo servers which are combined by using a lightweight facade service made with an Apache web server, Apache rewrite rules and PHP scripts. Back-end ONKI applications conform also to the ONKI API described in section 3.6. Technologies used for implementing the various back-end applications include Java, Semantic Web Framework Jena[11], MySQL database, Lucene index[12], Direct Web Remoting DWR for AJAX functionalities[13], Varnish HTTP accelerator[14] and shell scripts and Subversion version management system. The ontologies are presented internally in various RDF formats, typically in SKOS or OWL. With the help of ontology-specific configurations, the ontologies are served to the user in a uniform way.

The ONKI is running as a pilot service publicly available on the web. It was officially launched in September 2008[15]. During year 2008 ONKI had ca. 36,000

[10] http://www.w3.org/TR/swbp-vocab-pub/
[11] http://jena.sourceforge.net
[12] http://lucene.apache.org
[13] http://www.directwebremoting.org
[14] http://varnish.projects.linpro.no
[15] http://www.youtube.com/watch?v=qG2YhK17ifs

Table 2. ONKI usage statistics for November 2008

Service	Hits
Human interfaces	
ONKI-SKOS Browser	89,346
ONKI Selector Widget	54,384
Persistent URI redirects	4,415
ONKI Selector builder	1,103
Web Service builder	1,403
ONKI-IRMA	495
ONKI-Geo Browser	203
ONKI-Geo coordinates Browser	168
Application interfaces	
Web service calls	87,388
Javascript calls	18,816
Total	257,721

unique visitors and ca. 104,000 visits. 91 organizations outside the research group have registered an access key for using the JavaScript and web service interfaces which of 25 have actually implemented a test application using the components. For an overview, table 2 presents the usage statistics of the main ONKI functionalities for a representative month[16]. The most popular ontologies are the Medical Subject Headings, the Finnish Upper Ontology YSO, the General Finnish Thesaurus YSA and the Ontology for Museum Domain MAO where each ontology got over ten thousand hits during the month.

5 Case Applications Using ONKI

5.1 Content Creation: HealthFinland, CultureSampo and Tilkut

HealthFinland and CultureSampo are two major pilot applications of the FinnONTO project [6]. They demonstrate the usage of Semantic Web technologies in the contexts of health promotion and cultural heritage. Both systems uses ONKI as the ontology server for indexing content especially with the the browser-based annotation editor SAHA[17] [15]. For example (Figure 6), the Finnish General Ontology YSO [6] has been added as a ONKI Selector component to SAHA for finding and fetching annotation concepts.

The web laboratory Owela[18] of the VTT Technical Research Centre of Finland has implemented a service for collecting and sharing text and image clips from the Web[19]. In the service one can organize the clips into folders and tag them with different categories. The ONKI Selector is used for tagging the clips.

[16] The hits represents user activities of the system. Outliers caused e.g. by web crawlers have been removed as far as possible.
[17] http://www.seco.tkk.fi/services/saha/
[18] http://owela.vtt.fi/
[19] http://owela.vtt.fi/tilkut

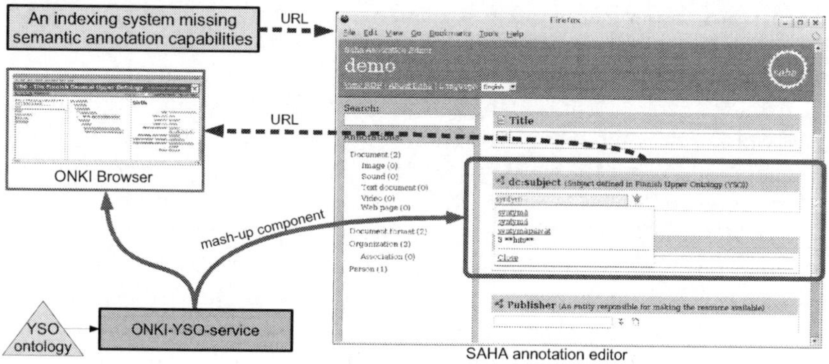

Fig. 6. The Finnish General Ontology connected to SAHA

5.2 Content Search: Kantapuu.fi and eViikki

Kantapuu.fi[20] is a web user interface for browsing and searching for collections of Finnish museums of forestry. The collection items are annotated with terms from General Finnish Thesaurus YSA, Thesaurus for Museum Domain MASA and Agriforest Thesaurus. Kantapuu.fi search page is a web form into which query strings are typed as free text. The query strings can be placed into specific fields, e.g. "keywords", "place of use" or "time of use". We have created a demonstration page containing a Kantapuu.fi's search form with integrated ONKI Selectors which can be used for selecting query terms to be used in the Kantapuu.fi search[21]. The ONKI Selector is used for finding terms from vocabularies of the ONKI service. The used vocabularies are the same as those used in the annotation process of the items, or actually their ontologized versions. To find suitable query terms user can utilize the autocompletion search functionality or the ONKI Browser. Thus, the user does not need to be familiar with the vocabularies used in the annotations of the items, as in the case of free text search. The ONKI Selector performs query expansion based on the selected query terms. So, for example a query term "animals" would return items annotated with term "cats". When the desired query terms are selected, the actual search to the Kantapuu.fi system can be executed.

The ONKI Selector Widget has also been integrated into the Viikki Science Library[22] reference database system eViikki[23]. eViikki is a search interface for the library's collections, which consist of scientific literature on agriforestry. The ONKI Selector is used for populating the "keywords" field of the search form of eViikki. The fetched concept labels are used in the information retrieval task. Query expansion is not performed currently.

[20] http://www.kantapuu.fi/
[21] http://www.yso.fi/lusto
[22] http://www.tiedekirjasto.helsinki.fi/english/
[23] http://www-db.helsinki.fi/eviikki/eviikkihaku.html

6 Related Work

Based on reviews on ontology library systems [7,8], the main focus in existing systems tends to be in supporting ontology development and not the runtime usage of ontologies such as indexing and ontology-based end-user applications. Although ONKI provides support for the whole ontology life cycle, a major contribution of ONKI is the support for indexing and other runtime needs.

The DAML Ontology Library[24] is a classic implementation of an ontology library. The ontologies can be accessed via different categories, such as metadata describing the ontologies such as keywords, Open Directory category[25] and submitting organization. Also information obtained from the ontologies such as names of classes and properties can be used for finding relevant ontologies. The main method for using the ontologies is to download them. In comparison, ONKI provides application support for e.g. adding ontologies as mash-up and web services to applications.

The Ontology Library Service ONKI provides methods for finding ontologies and concepts amongst the ontologies published in the centralized service, whereas Swoogle [19] and Watson [20] act as global Semantic Web search engines. They crawl the web and index the RDF files they find. Such search engines can be useful when searching for suitable ontologies to use in applications, providing an overview of ontologies of some domain published on the web. The ONKI Service is based on a different approach. It aims to be a community-based service that gathers together the users of ontologies providing them ready-to-use ontological functionalities which can be integrated into semantic applications.

Dameron et al. proposes that ontology services should be provided as Ontology Web Services (OWS) which could be used in applications for automatically find and use ontologies [12]. In ONKI we support the idea of providing application interfaces to the ontology library, but extend the idea to a higher abstraction level by providing also ready-to-use user interface components to avoid duplicated work by re-implementing user interface and visualization functionalities.

Faviki[26] is a semantic bookmarking service which uses Wikipedia's[27] term identifiers for tagging web pages. In comparison, ONKI is focusing on publishing ontologies and to support the creation of ontology-based indexing, content search and other applications.

Freebase[28] is a data repository on the web that aggregates information from many sources and provides a single topic and identifier for each logical entity, e.g. a person. One goal of ONKI is also to provide (optimally) single, shared identifiers for ontological concepts which can be used to aggregate distributed content repositories. Freebase is based on a bottom-up approach based on existing information e.g. in Wikipedia whereas ONKI ontologies are (typically) based on top-down analysis of a domain and its relevant concepts.

---

[24] http://www.daml.org/ontologies/

[25] http://www.dmoz.org

[26] http://www.faviki.com

[27] With the help of DBPedia, http://dpbedia.org

[28] http://www.freebase.com

7 Discussion

This paper discussed the requirements of an ontology library system to support the different phases of an ontology life cycle and related user needs for creation, publishing, maintaining and using ontologies. The Finnish national ontology library service ONKI addresses all phases of the ontology life cycle and contributes especially in providing support for 1) collaborative ontology publishing, 2) content indexing, and 3) information searching. The ontology services can be used in external legacy and other applications as ready-to-use functionalities. The new idea here is to support mash-up usage of ontologies in a way similar to Google Maps and other similar services. Our approach of providing an integrable autocompletion widget for external systems is the same as in [21].

The ONKI system supports syntactically, structurally and semantically heterogeneous content. RDF-based content representations such as RDF Schema, SKOS and OWL can be easily published by the ONKI SKOS server. SKOS generators for especially thesauri presentation formats such as MARCXML, various database schemas and text files has been implemented. Also multilingual content is supported. ONKI has been built for and tested with real world data consisting of ontologies, well-known thesauri and registries. The geographical ontology SUO contains over 800,000 places in Finland, which is published using the ONKI Geo server. The typical size of ontologies and thesauri published using the ONKI SKOS server is tens of thousands concept, e.g., YSO contains 20,600 concepts. For testing the scalability of ONKI SKOS, the Wordnet with 230,000 concepts has been successfully presented with the system. The ONKI services have been tested in various ways: the ONKI Selector as part of the SAHA editor in creating content for e.g. HealthFinland and CultureSampo and by external parties in their indexing and search applications. The ONKI Browser has been tested by expert users from e.g. the National Library of Finland.

To conclude, this full scale national ontology library service ONKI is novel and has not been done before. There are currently thousands of individual users and hundreds of organizations from different domains testing the ONKI system and using it in pilot applications. The National Library's commitment to ONKI means that a substantial part of public and private organizations in Finland begin to use ONKI – and most promisingly also the KOKO ontology – for indexing and search, but also for publishing and accessing their ontologies and thesauri – and to join the Semantic Web.

Future work include continuing observing how the ontology development, publishing and using community continues to build up around ONKI. We are most interested in seeing what kind of new applications will emerge based on ONKI and how well the concept of interlinked ontologies works in practice. DBPedia could be an interesting repository to be published in ONKI. Research topics include developing further methods for supporting community-based ontology development, managing changes in ontologies and utilizing change history and change propagation in e.g. inference and searching.

References

1. Gruber, T.R.: A translation approach to portable ontology spesifications. Knowledge Acquisition 5(2), 199–220 (1993)
2. Staab, S., Studer, R.: Handbook on ontologies. Springer, Heidelberg (2004)
3. Fensel, D.: Ontologies: Silver bullet for knowledge management and electronic commerce, 2nd edn. Springer, Heidelberg (2004)
4. Reynolds, D., Shabajee, P., Cayzer, S.: Semantic Information Portals. In: Proceedings of the 13th International World Wide Web Conference on Alternate track papers & posters. ACM Press, New York (2004)
5. Hyvönen, E., Mäkelä, E., Salminen, M., Valo, A., Viljanen, K., Saarela, S., Junnila, M., Kettula, S.: MuseumFinland—Finnish museums on the semantic web. Journal of Web Semantics 3(2), 224–241 (2005)
6. Hyvönen, E., Viljanen, K., Tuominen, J., Seppälä, K.: Building a national semantic web ontology and ontology service infrastructure—the finnonto approach. In: Bechhofer, S., Hauswirth, M., Hoffmann, J., Koubarakis, M. (eds.) ESWC 2008. LNCS, vol. 5021, pp. 95–109. Springer, Heidelberg (2008)
7. Ding, Y., Fensel, D.: Ontology library systems: The key for successful ontology reuse. In: Cruz, I.F., Decker, S., Euzenat, J., McGuinness, D.L. (eds.) Proceedings of SWWS 2001, The first Semantic Web Working Symposium, Stanford University, California, USA, July 30 - August 1, 2001, pp. 93–112 (2001)
8. Ahmad, M.N., Colomb, R.M.: Managing ontologies: a comparative study of ontology servers. In: ADC 2007: Proceedings of the eighteenth conference on Australasian database, pp. 13–22. Australian Computer Society, Inc., Australia (2007)
9. Viljanen, K., Tuominen, J., Hyvönen, E.: Publishing and using ontologies as mashup services. In: Bechhofer, S., Hauswirth, M., Hoffmann, J., Koubarakis, M. (eds.) ESWC 2008. LNCS, vol. 5021. Springer, Heidelberg (2008)
10. Euzenat, J., Shvaiko, P.: Ontology matching. Springer, Heidelberg (2007)
11. Mäkelä, E., Viljanen, K., Alm, O., Tuominen, J., Valkeapää, O., Kauppinen, T., Kurki, J., Sinkkilä, R., Känsälä, T., Lindroos, R., Suominen, O., Ruotsalo, T., Hyvänen, E.: Enabling the semantic web with ready-to-use web widgets. In: Proceedings of the First Industrial Results of Semantic Technologies Workshop, ISWC 2007, November 11 (2007)
12. Dameron, O., Noy, N.F., Knublauch, H., Musen, M.A.: Accessing and manipulating ontologies using web services. In: McIlraith, S.A., Plexousakis, D., van Harmelen, F. (eds.) ISWC 2004. LNCS, vol. 3298. Springer, Heidelberg (2004)
13. Vehviläinen, A., Hyvänen, E., Alm, O.: A semi-automatic semantic annotation and authoring tool for a library help desk service. In: Emerging Technologies for Semantic Work Environments: Techniques, Methods, and Applications. IGI Group, Hershey (2008)
14. Viljanen, K., Tuominen, J., Känsälä, T., Hyvönen, E.: Distributed semantic content creation and publication for cultural heritage legacy systems. In: Proceedings of the 2008 IEEE International Conference on Distibuted Human-Machine Systems, Athens, Greece. IEEE Press, Los Alamitos (2008)
15. Valkeapää, O., Hyvönen, E., Alm, O.: A framework for ontology-based adaptable content creation on the semantic web. J. of Universal Computer Science (2007)
16. Tuominen, J., Frosterus, M., Viljanen, K., Hyvönen, E.: ONKI SKOS server for publishing and utilizing skos vocabularies and ontologies as services. In: Aroyo, L., et al. (eds.) ESWC 2009. LNCS, vol. 5554, pp. 768–780. Springer, Heidelberg (2009)

17. Kauppinen, T., Henriksson, R., Sinkkilä, R., Lindroos, R., Väätäinen, J., Hyvönen, E.: Ontology-based disambiguation of spatiotemporal locations. In: Proceedings of the 1st international workshop on Identity and Reference on the Semantic Web (IRSW2008), Tenerife, Spain, June 1-5 (2008)
18. Kurki, J.: Finding people and organizations on the semantic web. In: AI and Machine Consciousness - Proceedings of the 13th Finnish Artificial Intelligence Conference STeP, August 20-22 (2008)
19. Ding, L., Finin, T., Joshi, A., Pan, R., Cost, R.S., Peng, Y., Reddivari, P., Doshi, V., Sachs, J.: Swoogle: a search and metadata engine for the semantic web. In: CIKM 2004: Proceedings of the thirteenth ACM international conference on Information and knowledge management, pp. 652–659. ACM Press, New York (2004)
20. d'Aquin, M., Baldassarre, C., Gridinoc, L., Sabou, M., Angeletou, S., Motta, E.: Watson: Supporting next generation semantic web applications. In: Proceedings of IADIS International Conference on WWW/Internet (ICWI 2007), Vila Real, Portugal (2007)
21. Hildebrand, M., van Ossenbruggen, J., Amin, A., Aroyo, L., Wielemaker, J., Hardman, L.: The design space of a configurable autocompletion component. Technical Report INS-E0708, CWI, Amsterdam (November 2007)

SAscha: Supporting the Italian Public Cooperation System with a Rich Internet Application for Semantic Web Services

Alessandro Adamou

Semantic Technology Lab, CNR-ISTC

Abstract. We present SAscha, a web browser-based application for authoring semantic Web Service descriptions by the SAWSDL standard. Even though it was conceived in a specific real domain scenario, i.e. the Italian framework for service interoperability, it is in fact a general-purpose SAWSDL tool that showcases a singular feature set including, but not limited to, simplicity of use, availability as an infrastructural service and XML Schema support. Although upper-level ontologies are expected to be stored in an ad-hoc repository, SAscha allows for OWL ontologies being stored anywhere locally or on the web.[1]

1 Background

The Public Cooperation System (abbreviated SPCoop) is the framework for Public Administration interoperability that the Italian government has been promoting since its inception in 2007. It is built on top of a large ad-hoc connectivity infrastructure called Public Connectivity System (SPC) and is a Service Oriented Architecture based on Semantic Web Services, which are exposed by each public organization and assembled as e-Government workflows. The adoption of semantic technologies in the scope of this architecture is aimed at facilitating service discovery, composability and reuse as well as reducing startup costs for new organizations intending to join the system [1].

SPCoop conforms to the WS-I Basic Profile 1.0 [2], hence specifies that service descriptions be exposed as sets of formal specifications (called Service Agreements), which include XML documents written in WSDL (Web Service Description Language) version 1.1, a format that the newer WSDL 2.0 is only partially retro-compatible with. Concept-level service specifications are integrated with the SAWSDL extension for Semantic Web Services [3], allowing services to reference concepts describing their components and the information conveyed [4]. SAWSDL represents the state of the art in terms of semantic Web Service definition and exposure, as it allows immediate validation by matching service element annotations with related semantic entities. In SPCoop these entities may

[1] This work is the result of a combined effort of the Italian National Agency for Digital Administration (CNIPA), the University of Rome "La Sapienza" - Department of Computer Science, and the National Research Council - Institute of Cognitive Sciences and Technologies (CNR-ISTC).

L. Aroyo et al. (Eds.): ESWC 2009, LNCS 5554, pp. 796–800, 2009.

include, but are not limited to elements from a shared knowledge repository called Schema and Ontology Catalog.

The need for a simple, richly interactive tool for authoring SAWSDL model references on existing Service Agreement components, as is SASCHA, is due to a number of different factors. For one thing, the application suite developed specifically for supporting SPCoop administration or participation therein relies upon third-party tools for annotating Service Agreements. These tools, which we will look into in the following section, do not fully cover the application of SAWSDL extensions to WSDL 1.1 documents and stand-alone XML Schemas. Also, it is not possible to presume any degree of knowledgeability from domain experts in each organization, with respect to knowledge modelling and ontology engineering. This is due to two reasons: first, semantic annotation of services in SPCoop occurs when such services are published, which may happen anytime after the domain knowledge has been formalized and registered; then, multi-provider services may be managed by a neutral third party, which may have had no part in authoring domain knowledge. Finally, by following a Software as a Service approach, these functionalities could be provided as remote infrastructural services available anytime, anywhere, thus favoring the adoption of the Web application paradigm. Originally presented in [5], the AJAX application SASCHA was developed to address these issues.

2 Relation with Existing Work

SASCHA makes an extensive use of tried-and-true technologies which encompass several key components of semantic web applications, from the user interface to core functionalities. Most of the GUI was written using GWT-Ext, an open-source Google Web Toolkit wrap of the Ext widget library, which is already being employed for the *OwlSight* ontology browser[2] as well as the *WebProtege* editor[3]. Although additional parsing for XML syntax highlighting is performed on the client, all the business logic (namely serialization and parsing of WSDL and XSD, ontology loading and reasoning) is implemented server-side and made available as GWT servlets. The OWL API is employed for ontology management, and is backed by *Pellet* for additional reasoning tasks. Interpretation of WSDL 1.1 documents relies upon the *WSDL4J* library, to which SAWSDL model reference support is provided through a modified version of the *SAWSDL4J* library. Because their object model JWSDL does not offer direct support for stand-alone XML Schemas, a small plugin (SAXSD4J) was also developed ad-hoc.

Having conceived SASCHA in response to standards and features which fall short of coverage among existing Semantic Web Service tools, an evaluation of similarities and differences with the state of the art is appropriate. For comparison, we have focused on three Semantic Web Services tools, which all happen to be based on the Eclipse framework: the LSDIS *Radiant*[4] plugin, *WSMO Studio*[5]

[2] http://pellet.owldl.com/ontology-browser

[3] http://protegewiki.stanford.edu/index.php/WebProtege

[4] http://lsdis.cs.uga.edu/projects/meteor-s/downloads/index.php?page=1

[5] http://www.wsmostudio.org

and *Semantic Tools for Web Services*[6] by IBM alphaWorks. These tools were tested with reference to the W3C SAWSDL implementation report[7], against evaluation criteria including support for SAWSDL and version 1.1 of the WSDL standard, the ability to annotate non-embedded XML Schemas and the availability of a common view for representing and interacting with annotation source and target - what is called "Shared UI Perspective" in the following table.

Table 1. Comparison of semantic Web Service annotation tools

	SAWSDL Support	**WSDL 1.1 Support**	**Stand-alone XSD Support**	**Shared UI Perspective**
Radiant	yes	partial	yes	yes
WSMO Studio	yes	partial	no	no
Semantic Tools for WS	no	yes	no	N/A
SAscha	yes	yes	yes	yes

The Semantic Tools for Web Services were developed with support for the WSDL-S specification, which preceded the SAWSDL standard. As for the remaining candidates, Radiant has shown to be the only one to support stand-alone XML Schemas and to offer simultaneous display for ontologies and annotation targets, which can prove helpful for inexperienced users. However, it still shows minor interface quirks and limited WSDL support, which only encompasses common elements between versions 1.1 and 2.0 of the standard. On the other hand, WSMO Studio provides additional support for the `attrExtensions` extensibility element, which is required for annotating operations in WSDL 1.1. It does, however, allow blank `attrExtensions` elements, which is generally not a desirable feature.

3 Demonstration Scenario

In the demonstration session, visitors will learn how a WSDL description and an XML Schema can be annotated in SAscha with references to entities from several sources. They will also learn how an annotation can be edited or deleted, as well as alternate paths to perform these actions across different resource views. A localized user interface in English will be provided for the purpose.

The example task is to load an annotated resource and replace its most generic model references with new ones pointing at appropriate, more specific entities (such as subclasses or instances of originally referenced classes). The demonstration will consist of the following steps:

1. Loading an annotation target (WSDL or XSD), at least one suitable ontology and their dependencies into SAscha.

[6] http://www.alphaworks.ibm.com/tech/wssem

[7] http://www.w3.org/2002/ws/sawsdl/CR

2. Introducing visitors to the three available display methods - i.e. tree, stack and source code view - which are used for WSDL and XSD documents as well as OWL ontologies.
3. Removing existing model references that must be replaced. Technically, this is not stricly necessary, but it is useful for showing the context menus that are invoked on tree nodes representing one model reference each.
4. Performing SAWSDL annotations through drag-and-drop actions across both tree and stack views.
5. Verifying the impact of previous actions on the source code view, which is updated in real-time on every change in the model reference set.

When visitors are introduced to the SAscha user interface, the functions of the lower tabbed panel, which is used for diagnostics, ontology status and dependency listing as well as entity slots, will also be displayed. An example of this setup is shown in Figure 1.

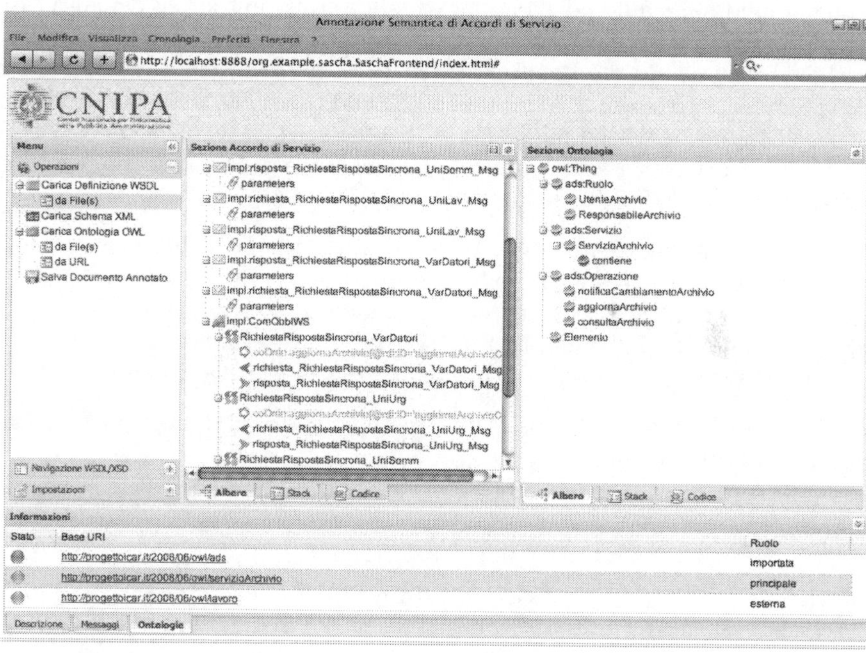

Fig. 1. Opening a WSDL document and a set of ontologies in SAscha

At least one WSDL document and one XML Schema will be loaded as active items at different times. If these documents reference any other XML Schemas by means of `import`, `include` and `redefine` elements, the referenced schemas will be brought in as well. As for ontologies, they may or may not be bound by OWL `imports` declarations. Users have the option to instruct the reasoner so that it will infer a class hierarchy from the whole set of ontologies the system is fed, rather than only the active one, which is always the last in the upload queue. Failure to load ontology dependencies is not fatal, as users are able to operate with the available set despite being warned about missing imports.

When annotating a single element, multiple semantic entities can be selected at a time if needed. If an entity belongs to a namespace that has not been bound to any prefix on the target document (which is going to occur at least once), a dialog will pop up to request a prefix for shorthand references. This action is optional, as the full entity URI will be used as a model reference if no prefix is entered.

4 Final Remarks and Future Work

Our demonstration will outline the features of a platform-independent, easy-to-use tool that provides a straightforward interface and several alternate paths for adding semantic value to Web Services descriptions and XML Schemas, and sets the ground for a full-fledged application for easily authoring Semantic Web Service descriptions from scratch.

Although tightening the bond with SPCoop, by releasing modules for supporting Service Agreement navigation and Schema and Ontology Catalog querying, is still top-priority, the migration of SAscha to a complete SAWSDL environment is part of the development plan. Some key development points are listed as follows:

1. Decision support by mapping WSDL and XSD schema definitions to mediator ontology patterns which can be best-matched against the active ontology;
2. Coverage of SAWSDL schema mappings applied to XSD elements;
3. Support for WSDL 2.0 and non-normative semantic annotations;
4. Natural language annotations through `wsdl:documentation` nodes.

At the time of writing, SAscha is undergoing testing for its first public release, possibly as LGPL software.

References

[1] Baldoni, R., Fuligni, S., Mecella, M., Tortorelli, F.: The Italian e-Government Service-Oriented Architecture. Strategic Vision and Technical Solutions (2007)

[2] Ballinger, K., Ehnebuske, D., Gudgin, M., Nottingham, M., Yendluri, P.: Basic Profile Version 1.0, The Web Services-Interoperability Organization (2004), http://www.ws-i.org/Profiles/BasicProfile-1.0-2004-04-16.html

[3] Farrell, J., Lausen, H.: Semantic Annotations for WSDL and XML Schema, W3C Recommendation (August 28, 2007), http://www.w3.org/TR/sawsdl

[4] Sheth, A., Verma, K.: Semantically Annotating the Web Services. IEEE Internet Computing 11(2), 83–85 (2007)

[5] Adamou, A.: SAscha: a RIA approach for supporting Semantic Web Services in the Italian interoperability framework. In: Proceedings of the 5th Workshop on Semantic Web Applications and Perspectives (SWAP 2008) (2008)

Folksonomy Enrichment and Search

Sofia Angeletou, Marta Sabou, and Enrico Motta

Knowledge Media Institute (KMi)
The Open University, Milton Keynes, United Kingdom
{S.Angeletou,R.M.Sabou,E.Motta}@open.ac.uk

Abstract. The Semantic Web community has expressed its interest on how the Semantic Web technology can be applied more efficiently in a manner that supports real world applications. Additionally, the popularity of social tagging systems has demonstrated a clear need for organisation and more flexible ways of querying the user contributed content. This work presents FLOR, a folksonomy enrichment algorithm, which exploits a variety of knowledge sources to apply structure on the user tagspaces. In addition, a query mechanism is presented demonstrating how the enriched folksonomies structures can be interrogated by transforming the user keyword queries on folksonomies to formal queries on semantic structures. The first prototype of the FLOR enrichment algorithm and a first instance of the query mechanism have been implemented and a demonstration is available online[1].

1 Introduction

The interaction between Web2.0 and Semantic Web has been given special attention by pioneers of both fields [2,4,5]. On the one hand, Semantic technology is expected to solve folksonomies problems such as *lack of structure, ambiguity, synonymy, basic level variation, syntactical variation* that may impede folksonomy search either by returning few results but mainly, by not supporting meaningful or representative result presentation. On the other hand, the socially derived tagspaces and their emerging semantics are claimed to potentially provide valuable information for overcoming certain impediments of the Semantic Web such as slow evolution of ontologies. The efforts to combine folksonomies and the Semantic Web have followed two main lines of research, statistical methods based on co-occurrence and explicit application of semantics on tags.

FLOR's novelty mainly lies on its entirely automated nature. FLOR selects relevant knowledge on the fly without pre-selection and user interference and automatically enriches the tags with all the appropriate semantic information available. In addition to the FLOR enrichment algorithm, we present a first approach on interrogating the semantically enriched folksonomies to test the Semantic Web usefulness in this context.

[1] `http://flor.kmi.open.ac.uk`

L. Aroyo et al. (Eds.): ESWC 2009, LNCS 5554, pp. 801–805, 2009.

2 Technique

In the following we describe the FLOR enrichment algorithm that leads to semantically structured folksonomies and the query mechanism to translate the user queries into semantic queries on these structures.

2.1 FLOR Folksonomy Enrichment

FLOR is an algorithm that automatically transforms tagspaces into semantic structures by making use of multiple knowledge sources. FLOR enrichment is performed in four phases as depicted in Fig. 2. An earlier version of the algorithm is explained in more detail in [1].

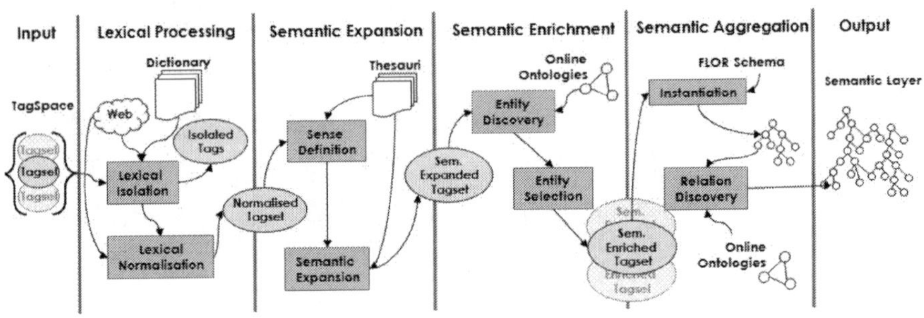

Fig. 1. FLOR Phases

PHASE 1: Lexical Processing. The first step of this phase, **Lexical Isolation)**, uses a set of heuristics to identify the tags not to be further processed by FLOR. Currently we isolate tags shorter than three characters, non-English tags and tags containing numbers or special characters. The **Lexical Normalisation** step produces a number of lexical representations for the tag, aiming to bridge the naming conventions used in folksonomies, ontologies and other knowledge resources.

PHASE 2: Sense Definition and Semantic Expansion. In the first step of Phase 2, **Sense Definition and Disambiguation**, sense disambiguation is performed by utilising WordNet based similarity techniques as explained in [1]. Another issue this phase tries to address is the Semantic Web sparseness. Expanding the tag with synonyms and lexical variations increases the possibility of finding the correct Semantic Web Entity in Phase 3. We perform **Semantic Expansion** for each tag, using WordNet.

PHASE 3: Semantic Enrichment. The third phase of FLOR identifies the Semantic Web Entities (SWEs) that are relevant for each tag taking as input the results of Phases 1 and 2. These SWEs are selected during the **Entity Discovery** step by querying a gateway to online ontologies [3]. Finally in the **Entity Selection** step we filter the SWEs in order to identify the ones that correspond to the relevant tags.

PHASE 4: Semantic Aggregation. The final phase returns the output of FLOR, a semantic structure with the definitions of the concepts corresponding to each tag and the relations of all the tags within the input tagspace. This is done by incrementally aggregating the enrichments of individual tagsets. The first step of Phase 4, the **Instantiation**, is executed after a tagset has been enriched and is the step where each individually enriched tagset is integrated into a final schema with which we represent the concepts of *Resource* `isTaggedWith` *Tag* `hasDefinition` *Sense* `isEnrichedWith` *Semantic (Web) Entity*. Finally, **Relation Discovery** is the last step of FLOR and returns the semantic layer linking together the tags. The actual relation discovery happens for the senses of the tags; among tags of a specific tagset but also cross-tagset, among tags of the overall tagspace (see Fig. 1). The Scarlet relation discovery algorithm ([6]) will be applied here. This step adds the statement: *Semantic (Web) Entity* `isRelatedto` *Semantic (Web) Entity* to the schema. The property `isRelatedto` can be any type of Object Property.

2.2 Querying Mechanism

At the moment the query mechanism supports only single keyword queries. As a next step we plan to support multiple keyword queries which will require a more complex query mechanism. The current demonstration performs search on a subset of Flickr resources (photos). The three different possible result sets for each query are Tags, Synonyms - Lexicals and Related.

The **Tags (A)** result set contains all the photos that have been tagged with the query keyword explicitly and show the same results as retrieved by a tag query in Flickr. The retrieval of this set is straightforward.

The **Synonyms - Lexicals (B)** set of results contains the photos that are tagged with synonyms or lexical variations of the query keyword. The notions of Synonym and Lexical are supported by our schema as Datatype Properties of a Sense. For example, *Sense_(Leaf)* `hasSynonym` "Foliage" and `hasLexical` "Leaves". The process of selecting this result set first identifies the senses that have the query keyword as a value in the `hasSynonym` or `hasLexical` properties. The mapping then to the tags and resources is straightforward.

The **Related (C)** set contains photos that are tagged with a sense that relates to keyword in the following ways.

- **SubClass**, tagged with subclasses of the query keyword.
- **SuperClass**, tagged with superclasses of the query keyword
- **Sibling**, tagged with senses sharing the same superclasses with the keyword
- **Generic Relation**, tagged with senses that relate with the keyword with all other possible relations (excluding disjointness) for example, meronymy.

More specifically, the process of selecting the set Related (C) first maps the keyword query to a sense in the same manner as in Synonyms - Lexicals (B). Then, the related sense is discovered according to the above cases and the inverse process is followed to identify tags and photos. This means that we retrieve all the photos that are tagged with synonyms or lexicals of the previously discovered related sense.

3 Implementation

The first prototype has been implemented and evaluated on a dataset from Flickr. Currently we have partly implemented Phase 4 of FLOR (Semantic Aggregation) and only the SubClass case of Related (C) from the query mechanism. With regards to the data, we selected a Flickr subset, from the group **Plant [directory]**[2] (5.943 members and 63.454 photos on 24-07-2008). We then randomly selected 12233 photos with a total of 89446 tags. For the first evaluation of this setup 11 users asked the system queries related to plants (45 queries in total) and retrieved **36% additional results to the ones that are returned without the FLOR enrichment with a mean precision of 94%.**

4 Demonstration Plan

This demo will focus on showcasing the functionality of querying on a Flickr dataset enriched by FLOR as described. The enrichment algorithm will be explained and demonstrated for each case of user query and how the FLOR results are obtained in detail. The basic part of the demonstration will be performed on the Web application implementing the setup described in Section 3.

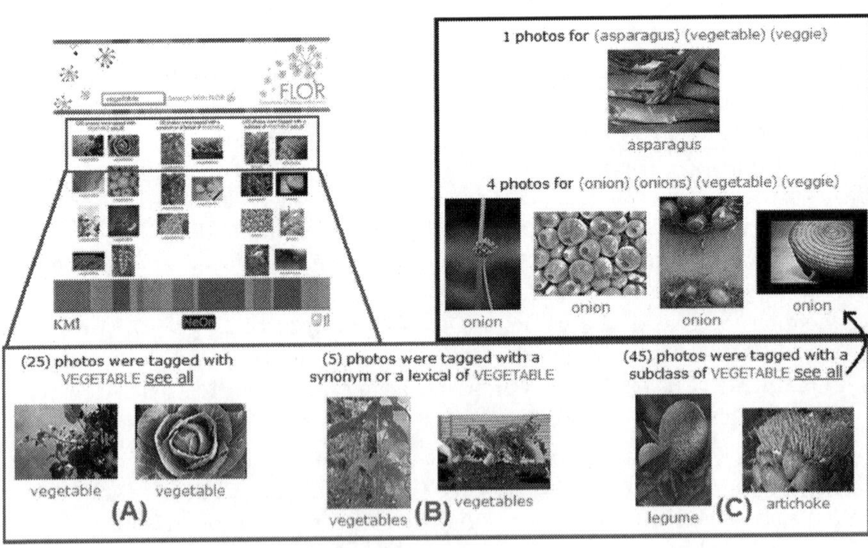

Fig. 2. Demonstration Screens

The initial page of the application consists of a search box. The first result page is broken down into three columns as shown in the left part of Fig.2. Column (A) presents the photos that are explicitly tagged with the query keyword, e.g.,

[2] `http://www.flickr.com/groups/plantdirectory/`

vegetable and the results are exactly the same as the ones returned with the current tag-based search of folksonomies. Column(B) presents the results that are tagged with synonyms or different lexical representations of the query term e.g., veggies, vegetables. Finally, column (C) presents the photos that are tagged with subclasses of the query keyword e.g., legume, artichoke.

To avoid visual clutter, a maximum of ten photos are presented in each column. If a column contains more than ten responses, these can be accessed by clicking on "**see all**". For the first two columns this will simply lead to a "bag" display of photos. In the case of column (C), the "**see all**" page categorises the photos under each subclass of the query keyword as demonstrated in the top-right part of Fig.2.

Under each photo we present the tag which was mapped against the query term. For example, in column (C) of Fig. 2 the tags legume and artichoke were found as subclasses of vegetable. By clicking on a photo the user obtains a larger view of it with all its associated tags.

We expect the visitor to learn how the integration of Semantic Web with Web2.0 is realised and experience a real world application that is useful to the casual web user and user-friendly.

Acknowledgements

This work was funded by the NeOn project sponsored under EC grant number IST-FF6-027595.

References

1. Angeletou, S., Sabou, M., Motta, E.: Semantically enriching folksonomies with FLOR. In: Proc. of the 5th ESWC: CISWeb, Tenerife, Spain (2008)
2. Benjamins, R., Davies, J., Baeza-Yates, R., Mika, P., Zaragoza, H., Greaves, M., Gomez-Perez, J., Contreras, J., Domingue, J., Fensel, D.: Near-term prospects for semantic technologies. IEEE Intelligent Systems 23, 76–88 (2008)
3. d'Aquin, M., Sabou, M., Dzbor, M., Baldassarre, C., Gridinoc, L., Angeletou, S., Motta, E.: Watson: A gateway for the semantic web. In: 4th ESWC, Innsbruck, Austria (2007)
4. Greaves, M.: Semantic Web 2.0. IEEE Intelligent Systems 22(2), 94–96 (2007)
5. Lassila, O., Hendler, J.: Embracing "Web 3.0". IEEE Internet Computing 11(3), 90–93 (2007)
6. Sabou, M., d'Aquin, M., Motta, E.: Exploring the semantic web as background knowledge for ontology matching. Journal of Data Semantics (2008)

K-Tools: Towards Semantic Knowledge Management

Sam Chapman[1], Vitaveska Lanfranchi[2], and Ravish Bhagdev[1,2]

[1] Knowledge Now Limited, 217 Portobello, Sheffield, S1 4DP, United Kingdom
N.Surname@k-now.co.uk
[2] Department of Computer Science, Regent Court, 211 Portobello, Sheffield, S1 4DP, United Kingdom
N.Surname@dcs.shef.ac.uk

Abstract. This paper details the use of Semantic Web tools for supporting networked knowledge acquisition, search and sharing in large distributed organisations. The demonstration will showcase from a user perspective an application developed to aid knowledge management in large organisations, detailing the problems and technical solutions employed.

Keywords: Semantic Web, Search, Forms, Knowledge Management, Business Intelligence, K-Now, K-Tools, Knowledge Capture, Enterprise.

1 Introduction

The amount of information that large organisations need to collect and store in digital archives and repositories is growing exponentially. Traditional knowledge management and sharing practices fail to address this data explosion: centralised fixed capture methodologies struggle to cope with dynamic, varied and specialised knowledge (i.e. the long tail of an organisations knowledge). Although such knowledge is highly variable it is also essential to specific parts of organisations that want to share and collaborate using it in conjunction with existing centralised knowledge. Many organisations attempt to capture this growing knowledge in shareable digital formats to encourage reuse. Semantic wikis [1] are sometimes utilised but they require training and time to use, lack integration with pre-existing knowledge representations and do not encourage knowledge integration and discovery across an organisation.

We propose a general knowledge lifecycle management methodology where Semantic Web Technologies are adopted to improve efficacy and efficiency in preserving and sharing organisational memory [2]. We therefore introduce a new framework for knowledge management, K-Tools, which uses Semantic Web Technologies to empower distributed networked representation, capturing, retrieval and sharing of knowledge within large enterprises. K-Tools is described in detail (including technologies and implementation choices), followed by the demonstration plan.

2 K-Tools

K-Tools is a Semantic Knowledge Management Framework that targets the whole knowledge lifecycle in an easy and usable way, thus maximising the potential adoption inside an organisation. K-Tools is composed of two main technologies:

L. Aroyo et al. (Eds.): ESWC 2009, LNCS 5554, pp. 806–810, 2009.

1) **K-Forms** supports knowledge acquisition and representation (the definition of the knowledge structures, e.g. an ontology) as well as capture (the creation of instances);
2) **K-Search** enables flexible searching and sharing of information and knowledge across differing views, perspectives and documents.

Both tools use the concept of forms as the user interaction paradigm. Form-based solutions are very familiar to users being very common in knowledge management and in every-day life therefore they are very easy to understand and use.

2.1 Key Technologies Adopted in K-Tools

K-Forms supports knowledge acquisition allowing easy creation and deployment of web forms that can instantly be used to capture knowledge at its generation time.

Forms are divided into sections and fields. Sections having further subsections and fields, e.g. the section designed to hold data about a person will be a concept; a person will have a series of properties e.g. name, address and date of birth. Sections are presented as sub-forms to be filled. Fields are typed (e.g. text field, integer field, text area, checkbox, option list, etc.). Fields represent meta-properties of the document (e.g. author, date, etc.) or its content (e.g. an issue to be reported). Simple fields will require just the inputting of a value at filling time (a number, a string, a text, a date, etc.). When a field is defined, the user selects a field type and must input appropriate properties for the specific field (e.g. size, validation constraints, default value, help text, etc.). If parts of an existing form are imported, the system automatically imports any associated constraint (e.g. on data types).

When a form is created (see Figure 1), K-Forms translates a template into an OWL Ontology with the form being represented as an OWL class (*<FormName> Class*) with its own NameSpace to avoid name conflicts. Form sections are represented as OWL classes (*<ConceptName> Class*) which can have subsections (related classes) or individual fields (properties). Relations between concepts are represented as OWL relations between classes. Adding relations between parents (section or entire form) and the contained subsections (object type properties) enables reuse of entire sections. Relational tables can be represented as advanced sections. The domain of some relations may be the overarching *<FormName> Class*. When concepts are introduced at the top level, a relation is formally created domain *<FormName> Class* and range *<ConceptName> Class*. Individual fields (such as text box, text area, check box) can be added as a property of each section, subsection, or directly in form classes; they are represented as OWL properties. Some of them (as *FormCompilationDate*) are properties of the overarching *<FormName> Class*. Restrictions can be set for the possible values of the individual fields using xml datatype schema (for example declaring a type to be a positive integer). Again individual fields can be reused between forms and classes. When concepts are reused across forms, owl:equivalentClass is created. This is due to the fact that uses may reuse the concept but change non conceptual details regarding its visualisation.

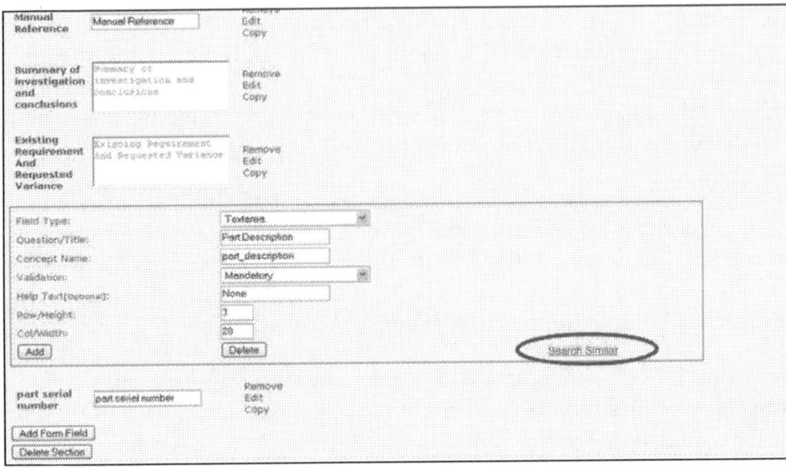

Fig. 1. K-Forms showing the generation of a new form

When (parts of) a form are reused (see Figure 1), an explicit mapping between (parts of) the two underlying ontologies is established. This means that instead of creating new concepts into the form namespace, the new ontology will use the concept in the other form's namespace. If the concept is modified, then a link between the two concepts is maintained by automatically creating a meta-class capturing the common parts. This may happen either at the single concept level or at the whole form level. This creates automatically a network of interconnected ontologies.

Knowledge acquisition using the new form is then realised by transforming all input data into RDF statements related to the form ontology, which can then be searched using standard query languages such as SPARQL and SERQL. This allows all captured knowledge to be automatically linked across an organisation for instantaneous reuse. K-Forms is described in detail in [3].

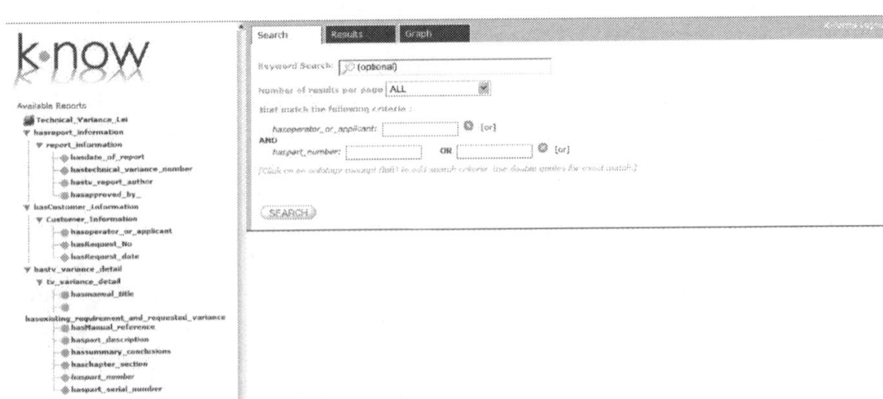

Fig. 2. The Ontology (left) produced by K-Forms after a new form is created can be instantly viewed in K-Search for searching newly generated knowledge

After knowledge has been stored, it can be searched and explored using K-Search (see Figure 2). Going beyond keyword approaches, K-Search enables Hybrid Search (HS) [4] over multiple repositories at once, from the perspective of a document type, a departmental view, a personal view or from a full company ontology. In HS, users can combine within the same query:

(i) ontology-based search;
(ii) keyword-based search and
(iii) keyword-in-context based search.

Ensuring consistency in the framework, K-Search has been designed following the same form-based interaction paradigm, providing an interface that allows users to flexibly query conditions on both structured knowledge and unbounded keywords within the same query.

The query output of K-Search is a set of ranked documents displayed as a list on the mid-right panel of the interface; each item in the list is identified by the values in the metadata that satisfy the ontology-based search. Moreover, K-Search provides ways of inspecting the results of the query as bi-dimensional graphs (pie or bar chart) according to two elements e.g. issue and component); items in the graph (e.g. pie chart slices) are clickable to focus on the associated subset of data (e.g. associated documents or data). The retrieved triples can be exported if needed (in RDF, CSV or XLS formats) for external analysis or automated processing. K-Search is described in details in [4].

3 System Demonstration

The demonstration focuses on showing how K-Tools fit the dynamic needs of large organisations, collecting knowledge across multiple sites/departments with various changing focuses. For demonstration purposes, K-Tools will utilise data from pre-existing industrial applications and then show in real-time how pre-existing forms can be modified or reused and how newly created content can be queried in a flexible way alongside pre-existing knowledge.

The system will be demonstrated from a user perspective, showing through K-Tools interfaces how a single user within a large organisation can create forms, distribute them, fill them and search the available knowledge. The underlying technology will be explained at each stage (although normally hidden in such a way as to allow any user within an organisation to use the system with minimal training).

Firstly from the point of capturing new knowledge relevant to an individuals task, e.g. a simple form design process whereby a user can create/reuse/modify fields or sections of forms previously entered and then use them as they wish. This wide flexibility helps addressing the long tail of organisational knowledge allowing large companies to capture specific knowledge across departments/groups/users. It will be shown how a user can easily create, modify and reuse form components by selecting some simple properties (e.g. type, name, validation constraints, default value, help text, etc.) – the underlying ontology(-es) will then be shown to better showcase the semantic mapping processes happening in background.

For the knowledge capturing stage, during the demonstration it will be possible to fill in created forms and then analyse the RDF triples that have been automatically generated by the system.

Knowledge sharing and reuse will be demonstrated through K-Search, showing how the previously created/modified forms and the collected information are immediately available for exploration by keyword, semantic and hybrid queries in a single interface. Most importantly this demonstration will highlight real world cases where semantics alone cannot solve a problem and show how integration with legacy data/systems aids in organisational adoption. The hybrid search modality will be showcased to explain how K-Search bridges the gaps with pre-existing unstructured knowledge and how it is possible to search information that is not fully represented by the available ontologies.

Acknowledgments. The work was supported by 1) IPAS, a project jointly funded by the UK DTI (Ref. TP/2/IC/6/I/10292) and Rolls-Royce plc, 2) X-Media, an Integrated Project on large scale knowledge management across media, funded by the European Commission as part of the IST programme (IST-FP6-026978), (www.x-media-project.org) and 3) Knowledge Now Ltd.

References

1. Krötzsch, M., Vrandecic, D., Völkel, M., Haller, H., Studer, R.: Semantic Wikipedia. Journal of Web Semantics 5, 251–261 (2007)
2. Stern, E.W.: Organizational memory: Review of concepts and recommendations for management. International Journal of Information Management, 17–32 (1995)
3. Bhagdev, R., Chakravarthy, A., Chapman, S., Ciravegna, F., Lanfranchi, V.: Creating and Using Organisational Semantic Webs in Large Networked Organisations. In: Sheth, A., et al. (eds.) ISWC 2008. LNCS, vol. 5318, pp. 723–736. Springer, Heidelberg (2008)
4. Bhagdev, R., Chapman, S., Ciravegna, F., Lanfranchi, V., Petrelli, D.: Hybrid Search: Effectively Combining Keywords and Semantic Searches. In: Bechhofer, S., Hauswirth, M., Hoffmann, J., Koubarakis, M. (eds.) ESWC 2008. LNCS, vol. 5021, pp. 554–568. Springer, Heidelberg (2008)

The XMediaBox: Sensemaking through the Use of Knowledge Lenses

Aba-Sah Dadzie[1], José Iria[1], Daniela Petrelli[2], and Lei Xia[1]

[1] Dept. of Computer Science
[2] Dept. of Information Studies
The University of Sheffield
{a.dadzie,j.iria,d.petrelli,l.xia}@sheffield.ac.uk

Abstract. Sensemaking is the process of analysing complex situations in order to make informed decisions. Semantic Web technology can be effectively used to create new sensemaking systems that focus on concepts and knowledge instead of documents. We demonstrate how this is achieved using information extraction to acquire knowledge and create a semantic repository that can then be semantically searched. A domain ontology is used to support the creation of an analysis tree; the semantic visualisation enables knowledge discovery, a core aspect of sensemaking.

Keywords: Semantic web, Knowledge acquisition, Semantic visualisation, Sensemaking.

1 Background

Sensemaking is the process of analysing complex and uncertain situations in order to take informed decisions [1]. It requires knowledge workers to find relevant information and formulate hypotheses based on the knowledge contained, test the hypotheses and revise the effectiveness of the decisions adopted. While some systems support only document manipulation in sensemaking (e.g., [2]), the X-Media project [3] uses Semantic Web (SW) technology and ontology-based visualisation to enable abstraction from the low level of multiple instances and more easily compose a conceptual model of a situation. Knowledge acquisition and retrieval, and visualisation are basic components of semantic sensemaking.

Knowledge acquisition (KA) transforms information into knowledge via classification and extraction. While information is heterogeneous and dispersed, coming from many sources, e.g., text documents, images, tables, etc., knowledge is homogeneous and instantiates concepts and relations captured in domain ontologies. In *knowledge retrieval*, users search knowledge repositories to find those missing elements that would improve their understanding of the current situation. *Knowledge visualisation* supports visual thinking by explicitly laying out the retrieved knowledge, information and personal reflections.

We demonstrate the steps we follow during analysis using the XMediaBox, developed as part of X-Media, using a non-restricted dataset. The XMediaBox has also been evaluated with experts in a realistic, knowledge-intensive setting: aerospace engineers working on gas turbine engines [4].

L. Aroyo et al. (Eds.): ESWC 2009, LNCS 5554, pp. 811–815, 2009.

2 Semantic Web for Sensemaking

We make use of SW technologies in three ways, by: i) *capturing* knowledge from archive documents via automated KA from multimedia; ii) *searching* archive information via hybrid search (semantic and keyword-based); and iii) *visualising* knowledge-in-context to support visual thinking.

2.1 Capturing Knowledge

A wealth of information is "hidden" within unstructured text, image, and mixed-media content. Information Extraction (IE) from text produces a semantic representation of the information (entities and relations) expressed in the text in natural language. Analogously, Content Based Image Retrieval systems produce a semantic representation of image content. Our applications represent information using ontologies; in particular, we use the Core Ontology of Multimedia (COMM) [5] to represent the knowledge captured from multimedia documents.

We use state-of-the-art algorithms from the IE and IR (information retrieval) literature in addition to rule- and machine learning-based approaches (Saxon [6], T-Rex [7]) to extract knowledge from single and multiple media documents. We perform entity recognition and relation extraction from text and region-of-interest classification from images to populate the knowledge repository ([8] details our approach). Fig. 1 shows an example of the semi-structured, multimedia documents analysed in X-Media, with (form) fields mapping to concepts in the domain ontology and images whose content may be described in the text.

2.2 Searching Knowledge

The KA step creates a knowledge base (KB) that the user can search using specific tools. We will demonstrate a hybrid search engine that combines ontology- and keyword-based searching to increase the effectiveness of retrieval [9]. The knowledge retrieved and associated documents may be inspected by the user, who decides relevance to the current analysis. Adding the knowledge retrieved into the ongoing sensemaking scenario enriches the metadata attached to it, and as a result, the knowledge captured in the KB. This also aids data categorisation, leading to more accurate retrieval across different media and document types.

2.3 Visualising Knowledge

Information visualisation provides effective support for intuitive analysis [10] and understanding of ontology structure [11] and ontology-based information. In X-Media we go a step further and provide multiple ways of visualising knowledge in order to support better understanding of complex scenarios. We have developed *Knowledge Lenses*, alternative views on the underlying data that support different levels of abstraction. Each knowledge lens has a different focus: the interconnections between knowledge in documents, human understanding (via notes), or the knowledge in the semantic repository. The semantics behind the data enable a smooth transfer from document to knowledge to personal notes.

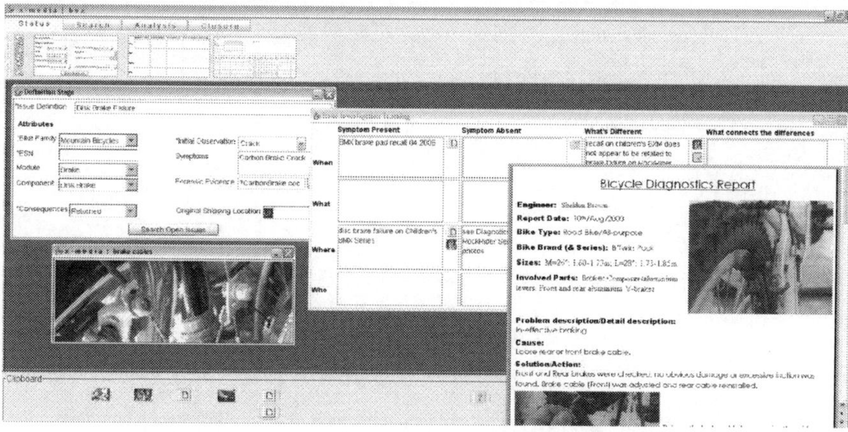

Fig. 1. A definition form in the XMediaBox (top, left) maps form fields to relevant concepts in the domain ontology. The interactive table (top, right) contains users' thoughts linked to evidence. The panel at the bottom is used to bookmark and share knowledge objects. Two examples of the data referenced within the system are shown.

3 Sensemaking via Knowledge Lenses

The sensemaking process within the XMediaBox is articulated in several steps, each providing a different view on the current situation and related information. Each step corresponds to a tab on the user interface (see the top of Fig 1: *status*, *search*, *analysis* and *closure*). The sensemaking process may span several months; however it always begins by defining the purpose of the investigation. A definition form contextualises the issue using selected domain ontology concepts. The information input by the user initialises a search of legacy corpora and other pertinent repositories to identify information relevant to the current case. The documents retrieved may be inspected and, if found to be relevant, bookmarked for future use, for e.g., as evidence to support a hypothesis.

As the sensemaking progresses evidence is collected and users' understanding develops. The status of the investigation is recorded using *interactive tables* (see Fig. 1), contextualising the implicit knowledge embedded in the documents using free text and semantic annotation via ontologies. The tables summarise the knowledge exploration, supporting sharing with other knowledge workers.

While search is an activity carried out only when the user needs to collect new evidence, analysis, which is the core of sensemaking, is sustained throughout the process. User understanding is built up during the systematic analysis of evidence and counter-evidence, and on the evaluation of which hypotheses explain the phenomena observed. The XMediaBox captures this rationalisation process in the *interactive analysis tree* (see top, Fig. 2) that helps users formulate new and analyse existing hypotheses. The system supports users in building the analysis tree from suggestions in the domain ontology, focusing on the relations *hasCause* (with inverse *hasEffect*), and *subpart* and *isA* relations between physical objects.

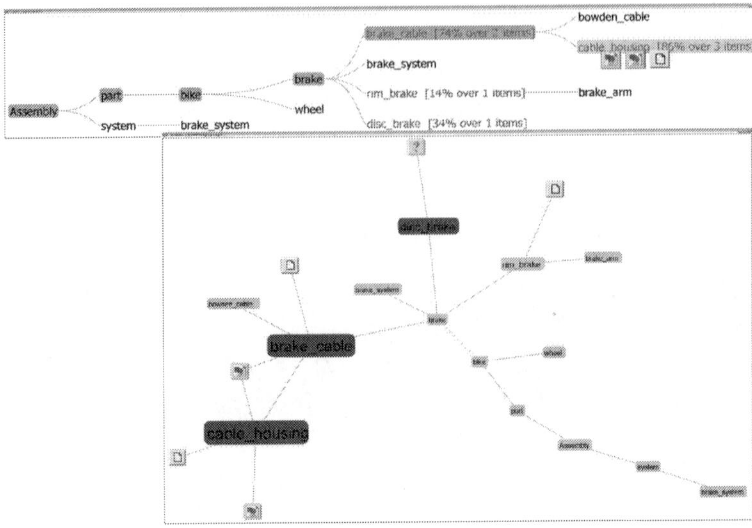

Fig. 2. Two knowledge lenses based on a domain ontology. The *knowledge tree* (top) provides structured visual analysis, and the *knowledge graph* (bottom) clusters hypotheses with evidence in common, simultaneously highlighting outliers. Attaching evidence to nodes tags them with the corresponding hypotheses, enriching the metadata and leading to more effective knowledge creation, categorisation, retrieval and reuse. [The modules were built using the prefuse visualisation toolkit (`http://prefuse.org`)].

Relevant documents, snippets of evidence and comments retrieved from the KB or created within the system may be attached to each (cause/hypothesis) node in the tree during the sensemaking process, creating an individual perspective on the hypotheses being investigated and the resulting knowledge space. The knowledge lens thus created supports pattern spotting, helping users determine which branch in the investigation contains the more valid hypothesis and which others need further investigation. Attaching the multimedia document in Fig. 2 to the node *brake_cable* indicates that it contains evidence that the hypothesis "failure of the brake cable" may be the root cause of the issue. Extracting a snippet that links the concept *component* to *disc_brake* with the relation *removed* (with a record of its provenance), is equivalent to manually annotating the document content. Both actions update the metadata in the knowledge repository; subsequent searches on the ontology concepts will retrieve this document.

While the *knowledge tree* provides a structured view of the knowledge space for the current analysis, the *knowledge graph* provides a more complete overview of the relationships between the hypotheses (concepts of interest) and the knowledge and annotated evidence. The knowledge graph serves as an alternative, less restricted view on the knowledge space, showing interconnections between evidence that may not be easily identified otherwise. It further supports the discovery of existing, but not yet identified knowledge during exploratory analysis, by providing a means for browsing the contents of the knowledge repository.

A prototype of the XMediaBox has been evaluated by 12 aerospace engineers during a simulation of a sensemaking activity using a grounded, domain-specific scenario [4]. Documents fed into the knowledge capture phase used multiple formats: text only, images only, or multimedia – both text and images. The documents formed part of a very large company archive, processed using the KA techniques summarised. Guided by ontology-based suggestions, the engineers extended an analysis tree, populated the tree with the factual evidence retrieved, and recorded their understanding of the knowledge space created. The evaluation showed that Semantic Web technology effectively supports different levels of abstraction: table summaries, tree analysis and exploratory graphs, shifting the sensemaking activity from document manipulation and management to a higher conceptual level, that of knowledge management and discovery.

Acknowledgements

The research described in this paper is funded by the X-Media project, sponsored by the European Commission as part of the Information Society Technologies (IST) programme under EC grant number ISTFP6-026978. Thanks also to Fabio Ciravegna for the invaluable discussions and contributions to the paper.

References

1. Klein, G., Moon, B., Hoffman, R.: Making sense of sensemaking 1: Alternative perspectives. IEEE Intelligent Systems, 70–73 (2006)
2. Wright, W., Schroh, D., Proulx, P., Skaburskis, A., Cort, B.: The Sandbox for analysis: concepts and evaluation. In: Proceedings of the SIGCHI conference on Human Factors in Computing Systems, pp. 801–810 (2006)
3. The X-Media Project, `http://www.x-media-project.org`
4. Dadzie, A.-S., Lanfranchi, V., Petrelli, D.: Seeing is believing: Linking data with knowledge. Information Visualization (in press)
5. Arndt, R., Troncy, R., Staab, S., Hardman, L., Vacura, M.: Comm: Designing a well-founded multimedia ontology for the web. In: Aberer, K., Choi, K.-S., Noy, N., Allemang, D., Lee, K.-I., Nixon, L., Golbeck, J., Mika, P., Maynard, D., Mizoguchi, R., Schreiber, G., Cudré-Mauroux, P. (eds.) ASWC 2007 and ISWC 2007. LNCS, vol. 4825, pp. 30–43. Springer, Heidelberg (2007)
6. Greenwood, M., Iria, J., Ciravegna, F.: Saxon: An extensible multimedia annotator. In: 6th International Conference on Language Resources and Evaluation (2008)
7. Iria, J.: T-Rex: A flexible relation extraction framework. In: 8th Annual Colloquium for the UK Special Interest Group for Computational Linguistics (2005)
8. Dadzie, A.S., Bhagdev, R., Chakravarthy, A., Chapman, S., Iria, J., Lanfranchi, V., Magalhães, J., Petrelli, D., Ciravegna, F.: Applying semantic web technologies to knowledge sharing in aerospace engineering. Intelligent Manufacturing (2008)
9. Bhagdev, R., Chapman, S., Ciravegna, F., Lanfranchi, V., Petrelli, D.: Hybrid search: Effectively combining keywords and semantic searches. In: Bechhofer, S., Hauswirth, M., Hoffmann, J., Koubarakis, M. (eds.) ESWC 2008. LNCS, vol. 5021, pp. 554–568. Springer, Heidelberg (2008)
10. Keim, D., Andrienko, G., Fekete, J.D., Görg, C., Kohlhammer, J., Melançon, G.: Visual analytics: Definition, process, and challenges. In: Kerren, A., Stasko, J.T., Fekete, J.D., North, C. (eds.) Information Visualization: Human-Centered Issues in Visual Representation, Interaction, pp. 154–175. Springer, Heidelberg (2008)
11. Geroimenko, V., Chen, C. (eds.): Visualizing the Semantic Web: XML-based Internet and Information Visualization, 2nd edn. Springer, Heidelberg (2006)

Controlled Natural Language for Semantic Annotation

Brian Davis[1], Pradeep Varma[1], Siegfried Handschuh[1], Laura Dragan[1], and Hamish Cunningham[2]

[1] Digital Enterprise Research Institute, National University of Ireland, Galway
{brian.davis,pradeep.varma,siegfried.handschuh,laura.dragan}@deri.org
[2] Sheffield NLP Group, University of Sheffield
hamish@dcs.shef.ac.uk

Abstract. *Sovereign* is a novel collection of resources for authoring(1), annotating(2) and accessing(3) knowledge on the Social Semantic Desktop. With respect to (2), the Sovereign Semantic Annotator allows the non-expert user in a novel way to semi-automatically author and annotate meeting minutes and status reports *simultaneously* using Controlled Natural Language(CNL). The metadata is captured as knowledge on the Social Semantic Desktop for later aggregation and access. The annotator is based on Controlled Language for Information Extraction (CLIE) technology. Furthermore, it is available as a plugin for a semantic note-taking application for the Social Semantic Desktop. We intend to present a working prototype of the Sovereign annotator for the Social Semantic Desktop at ESWC.

1 Introduction and Background

Semantic technologies are difficult to access for non-expert users wishing to author(1), annotate(2) and/or access(3) knowledge. Our research investigates how Human Language Technology (HLT) interfaces; specifically Controlled Natural Languages(CNL)[1] and applied Natural Language Generation(NLG) can provide a user friendly means for non-expert users and organisations seeking to exploit Semantic Web technologies. *Sovereign*[2] is a novel collection of resources for authoring(1), annotating(2) and accessing(3) knowledge on the Social Semantic Desktop. This demonstration focuses on the annotation resource(2) in Sovereign, however we will also discuss briefly resources (1) and (3) since all three are interrelated.

With respect to **authoring(1)** - specifically ontology authoring, few ontology editing tools are aimed at non expert users, in particular users wishing to

[1] CNLs are "subsets of natural language whose grammars and dictionaries have been restricted in order to reduce or eliminate both ambiguity and complexity". See http://www.ics.mq.edu.au/~rolfs/controlled-natural-languages/

[2] Semantic annOtation VErbal REsources for ExtractIon and Generation of kNowledge.

L. Aroyo et al. (Eds.): ESWC 2009, LNCS 5554, pp. 816–820, 2009.

create simple structures without delving into the intricacies of knowledge representation languages. The CLIE (Controlled Language for Information Extraction) approach however allows users to create and edit ontologies quite simply by using a restricted version of the English language. This 'controlled natural language' is based on an open vocabulary and a restricted set of grammatical constructs. Sentences written in this language unambiguously map into a number of knowledge representation formats including OWL and RDF-S. Previous work has used CLIE to generate ontologies from the input CNL called CLOnE (Controlled Language for Ontology Engineering)[1]. The reverse of the process involves the generation of CLOnE from an existing ontology using NLG, specifically shallow NLG. The NL generator and the authoring process both combine to form a RoundTrip Ontology Authoring (ROA) environment: one can start with an existing or empty ontology, create an ontology using CLIE, reproduce CLOnE using the NL generator, modify or edit the text as required and subsequently parse the text back into the ontology using the CLIE environment. The process can be repeated as necessary until the required result is obtained.

With regard to **knowledge access(3)**, NLG can act as a human-readable window into the otherwise formally structured database. Rather than summarising the contents of an ontology for the purposes of modification or quality assessment as is specified above in the context of ROA(1), the use case of NLG is to provide a user friendly means of presenting semantically annotated knowledge captured within the Social Semantic Desktop in a human readable form and furthermore at the appropriate level of detail to the non-expert user, based on his/her specific queries.

1.1 Semantic Annotation

Concerning **annotation(2)**, richly interlinked, machine-understandable data constitute the basis for the Semantic Web, and by extension the Social Semantic Desktop[2]. Manual semantic annotation is a complex and arduous task both time-consuming and costly often requiring specialist annotators. (Semi)-automatic annotation tools attempt to ease this process by detecting instances of classes within text and relationships between classes, however their usage often requires knowledge of Natural Language Processing(NLP) and/or formal ontological descriptions. This challenges researchers to develop user-friendly annotation environments within the knowledge acquisition process. CNLs offer an incentive to the novice user to annotate, while simultaneously authoring, his/her respective documents, in a user-friendly manner as well as shielding him/her from the underlying complex knowledge representation formalisms. A natural overlap exists between tools used for both ontology creation and semantic annotation. However, there is a subtle difference between both processes. "Semantic annotation is described as a process, as well as the outcome of the process. Hence it describes i) the process of addition of semantic data or metadata to the content given an agreed ontology and ii) it describes the semantic data or metadata itself as a result of this process"[3]. Of particular importance here is the notion of the addition or association of semantic data or metadata to *content*.

2 A Use Case for Controlled Natural Language for Semantic Annotation

In our scenario Nepomuk-KDE[3] - the KDE instance of the Social Semantic Desktop serves as the platform on which to build our tools and conduct our experiments. Firstly, however with respect to **annotation(2)**, CNLs cannot offer a panacea for semi-automatic annotation since it is unrealistic to expect users to annotate every textual resource using CNL, however there are certain use-cases where CNLs can offer an attractive alternative as a means for semi-automatic semantic annotation, particular in contexts, where controlled vocabulary or terminology is implicit such as health care patient records or business vocabulary. Our use case focuses on administrative tasks such taking minutes during a project team meeting and weekly status reports. Very often such note taking tasks can be repetitive and boring. In our scenario the user is a member of a research group which is part of an integrated EU research project. Based on pre-defined templates, the user *simultaneously authors and annotates* his/her meeting minutes or status reports in CNL, using a semantic note taking tool - SemNotes[4], which is an application available for Nepomuk-KDE. The newly created metadata is thus available for immediate use, inclusive of both querying and aggregation, whereby the retrieved RDF triples can be passed to a **Natural Language Generator(3)** to produce tailored textual reports and summaries. In addition, wrt **authoring(1)**, ROA supports the ontology authoring process to create a common vocabulary on which to base annotation and knowledge capture and subsequent text generation. The authoring, annotation and NLG functionalities, are made available as plugins to SemNotes. These three plugins are collectively called **Sovereign** - *Semantic annOtation VErbal REsources for ExtractIon and Generation of kNowledge.*

3 Implementation

As mentioned earlier, the Sovereign CNL annotator is available as a plugin to SemNotes. Furthermore, the CNL is realised within a semantic note. The CNL is anchored to existing semi-structured data such as a *AgendaTitle*, *Scribe* or *ActionItem* based on a predefined meeting minutes or status report template. The annotator is based on CLIE. The CNL itself is very similar to the CLOnE language, with some modifications. The annotator architecture contains a standard GATE pipeline[5](see Figure 1) consisting of the following language and processing resources: The GATE English tokeniser, the Hepple part-of-speech tagger,

[3] http://nepomuk.kde.org/
[4] http://smile.deri.ie/projects/semn
[5] General Architecture for Text Engineering, See http://gate.ac.uk/

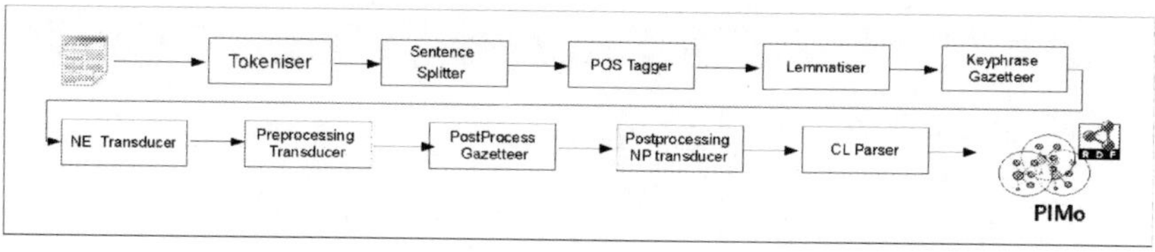

Fig. 1. The CNL Semantic Annotator pipeline

a morphological analyser, a gazetteer list component for recognising useful key-phrases, such as structured elements from the templates and reserved phrases in the controlled language. Any sentences preceded by a `Comment:` element are considered candidates for controlled language parsing. Any remaining tokens from the CNL sentence which are not recognised as reserved CNL key-phrases are used as names to generate ontological objects(See Figure 2). This is followed by a standard Named Entity(NE) transducer in order to recognise useful NEs, a `preprocessing` JAPE[6] finite state transducer(FST) for identifying quoted strings, chunking Noun Phrases(NPs) and additional preprocessing. A second gazetteer list look up is applied to identify trigger phrases associated with NEs which intersect with quoted and unquoted NP annotation spans. Additional feature values are then added to the NP chunks to indicate the appropriate class to link an NP chunk as an instance to. The last FST parses the CNL from the text and generates the metadata. The current tool is bootstrapped via the Nepomuk Core Ontologies[7] and currently the application creates/populates a meeting minutes/status report ontology, which references the users Personal Information Model Ontology(PIMO) [8] via the GATE Ontology API. We have also modified the code to write directly to Nepomuk KDE RDF store.

4 The Demo

Our demonstration will provide the audience with an opportunity to obtain live, hands-on experience using the **Sovereign CNL Annotator**. The demo will consist of a running Nepomuk KDE installation as a fictitious user (one of the personas from the NEPOMUK project[9]). The demo will be supported by a dataset consisting of prepopulated PIMO instances of the fictitious user amongst others as well as a dataset based on a corpus of **real-world** status-reports and meeting-minutes which has been collected over the three year duration of the Nepomuk Project. A demonstration of the user authoring/annotating meeting minutes and a status report based on the screenshot in Figure 2 will be provided in order to showcase what can achieved with the CNL annotator. We will also

[6] Java Annotations Pattern Engine.
[7] http://www.semanticdesktop.org/ontologies/
[8] http://www.semanticdesktop.org/ontologies/2007/11/01/pimo/
[9] http://nepomuk.semanticdesktop.org/

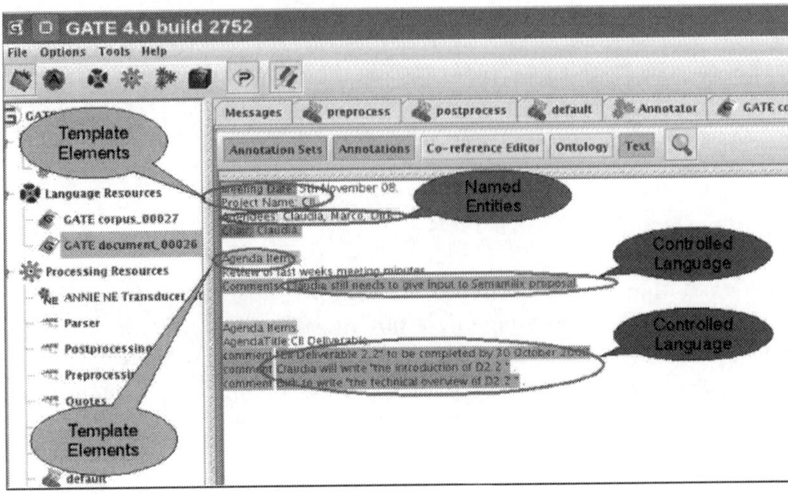

Fig. 2. CNL Annotator visualised in GATE

provide a small CNL reference guide. Hence visitors will be invited to write CNL as well as experiment with and test the annotator as they see fit.

Acknowledgements

The work presented in this paper was supported (in part) by the Líon project supported by Science Foundation Ireland under Grant No. SFI /02/CE1/I131 and (in part) by the European project NEPOMUK No FP6-027705.

References

1. Davis, B., Iqbal, A.A., Funk, A., Tablan, V., Bontcheva, K., Cunningham, H., Handschuh, S.: Roundtrip ontology authoring. In: Sheth, A.P., Staab, S., Dean, M., Paolucci, M., Maynard, D., Finin, T.W., Thirunarayan, K. (eds.) ISWC 2008. LNCS, vol. 5318, pp. 50–65. Springer, Heidelberg (2008)
2. Decker, S.: The social semantic desktop: Next generation collaboration infrastructure. Information Services and Use 26(2) (2006)
3. Handschuh, S.: Creating Ontology-based Metadata by Annotation for the Semantic Web. PhD thesis (2005)

Multilingual and Localization Support for Ontologies

Mauricio Espinoza, Asunción Gómez-Pérez, and Elena Montiel-Ponsoda

UPM, Laboratorio de Inteligencia Artificial, 28660 Boadilla del Monte, Spain
{jespinoza,asun,emontiel}@fi.upm.es

Abstract. This demo proposal aims at providing support for the localization of ontologies, and as a result at obtaining multilingual ontologies. We briefly present an advanced version of LabelTranslator, our system to localize ontology terms in different natural languages. The current version of the system differs from previous works reported in [1,2] in that it relies on a modular approach to store the linguistic information associated to ontology terms. Additionally, it uses a synchronization method to maintain the conceptual and linguistic information updated.

Keywords: Ontology localization, Multilingual ontologies.

1 Introduction

Multilinguality in ontologies has become an impending need for institutions worldwide with valuable linguistic resources in different natural languages. Since most ontologies are developed in one language, obtaining multilingual ontologies implies to localize or adapt them to a concrete language and culture community, as defined in [5,6]. For this end, we designed and implemented LabelTranslator [1,2], a system that suggests translations of ontology labels among English, Spanish and German, with the purpose of building a multilingual ontology.

The current trend in the integration of multilinguality in ontologies suggests the suitability of keeping ontology knowledge and linguistic (multilingual) knowledge separated and independent. However, the first version of LabelTranslator followed a non-modular approach, in which multilingual information was embedded in the ontology by means of the RDFS/OWL predicates. This approach has important limitations related to the restricted amount of linguistic information that can be attached to ontology concepts. Furthermore, multilingual information is limited to strings without information about senses in their respective languages, nor about provenance of the information, which makes concept localization to different natural languages quite difficult.

In this paper we describe the features and design aspects of the second version of the LabelTranslator system. The current version of our tool supports the Linguistic Information Repository (LIR) [3,4] model, which follows the current trend in the integration of linguistic information in ontologies. The LIR model consists of a set of linguistic classes, whose nature accounts for the localization of ontology terms in a certain language.

L. Aroyo et al. (Eds.): ESWC 2009, LNCS 5554, pp. 821–825, 2009.

The rest of this paper is structured as follows. In Section 2 we briefly describe the LIR model designed for the representation of multilingual information in ontologies. Then, a comprehensive solution to the problems of managing the conceptual knowledge and the linguistic knowledge by means of synchronization techniques can be found in Section 3.

2 Ontology Enrichment by Means of the LIR Model

The first version of our tool used some fields of the NeOn ToolKit[1] for storing the multilingual information generated by the system. The link we established between the terms in the ontology and its associated translations was characterized by simple references between concepts and labels (as offered by the standard owl:comment and rdfs:label properties). However, for the second version of LabelTranslator we decided to go for a modular approach, in which the conceptualization is kept apart from the multilingual information.

With this purpose in mind, an ontology of linguistic and terminological concepts has been developed within the NeOn project in order to capture relevant linguistic information that linked to domain ontologies enables their localization. The LIR is a portable model that can be associated to any element of an OWL ontology by means of an OWL meta-ontology. The main classes that compose the LIR (Lexicalization, Sense, Definition, UsageContext, Note and Source) are organized around the LexicalEntry class, which is related to any ontology element (by means of the hasLexicalEntry relation). The set of LIR concepts enables a complete description in natural language of the ontology element they are associated to. Additionally, by means of typical lexical relations within the same language (e.g. hasSynonym) and across languages (hasTranslation), the LIR organizes linguistic information within and across languages with the aim of enabling ontology localization. The LIR provides a selection of ISO standard classes for describing linguistic and terminological knowledge, ensuring in this way its eventual extension to cover other types of linguistic data (e.g. morphological or syntactic description), if required.

When the ontology user creates/imports a new OWL ontology in Neon, the LabelTranslator system automatically builds an empty linguistic model associated to the ontology under consideration. This model is used by our system to store the linguistic information associated with each ontology term. Figure 1 shows the *Linguistic Information* page associated to the sample ontology term *FAO*. LabelTranslator fills in runtime the majority of fields of the linguistic page, according to the information obtained by the system in the translation process. A more detailed description of the translation process can be found in [1].

Initially, the linguistic information page shows five sections that correspond to the lexical entries of the selected ontology element (*FAO* in our example) and the associated information of each lexical entry: *lexicalizations*, *lexical entry senses*, *usage contexts*, *sources*, and *notes*. In the example given in figure1 we illustrate also part of the information supported by the LIR model. Thus, three

[1] http://www.neon-toolkit.org/

lexical entries (LexicalEntry-1,LexicalEntry-2, and LexicalEntry-3) are associated with the same concept (FAO:Class), which means that they are all terms that identify one and the same concept. Two lexical entries (LexicalEntry-1 and LexicalEntry-2) belong to the same Language (English), whereas the third lexical entry (LexicalEntry-3) belongs to Spanish. The two English lexical entries are considered synonyms, and translations of the Spanish lexical entry. This information is shown in the field *Lexical Entry Relationships*.

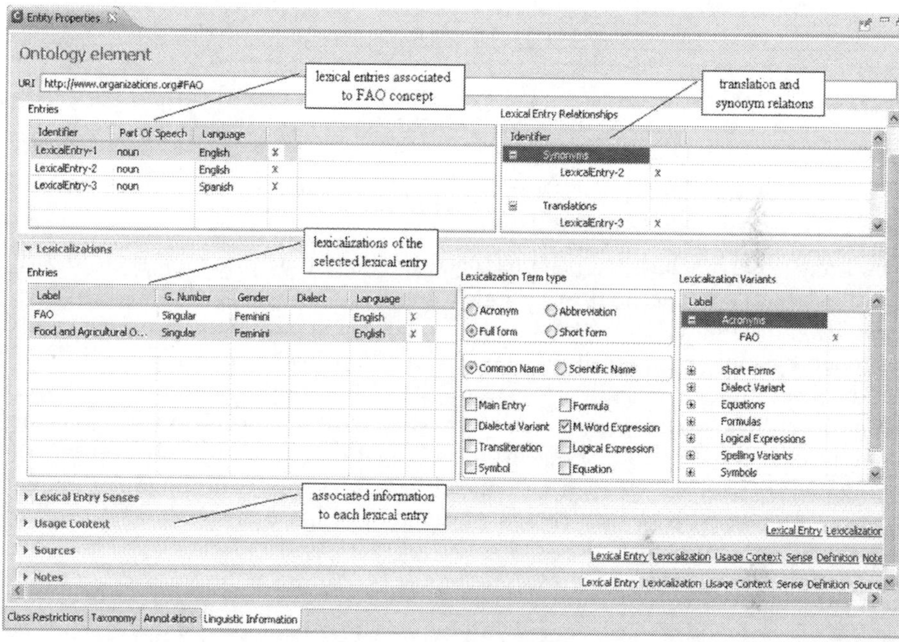

Fig. 1. Linguistic Information page with data of the concept FAO

Of course, every time that the user chooses a new entry, the interface automatically displays the information correlated in the different sections. Thus, in our example LexicalEntry-1 includes two lexicalizations whose labels are FAO and Food and Agriculture Organization, respectively. Food and Agriculture Organization has like acronym FAO, and, moreover, it is considered a *common name* (in opposition to scientific name) and a *multi-word expression*. This information is shown in the fields *Lexicalization Term Type* and *Lexicalization Variants*.

3 Localization Management

In this section we present a comprehensive solution to the problems of managing changing on the ontology model (OM) and propagating these changes to linguistic model (LM). In order to keep both models synchronized we first need

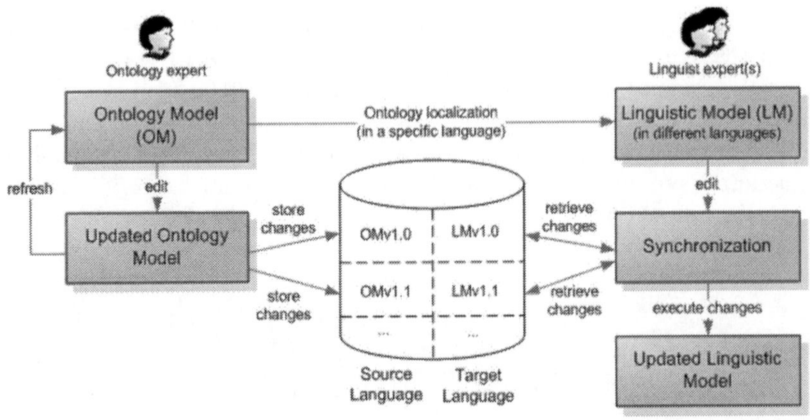

Fig. 2. Synchronization of ontology and linguistic model

to find out exactly what has been changed in the ontology model, then find the equivalent places in the linguistic model, and only then start the updating.

Addition of new terms in the ontology, or deletion of an existing term can be controlled by some mechanism of change synchronization. In the NeOn ToolKit, an advanced change tracking, based on Resource Delta[2], is able to capture changes even when ontological terms have changed their position within the ontology model. By adopting this feature, our system can accurately identify the minimal set of changes needed to adjust the structure of the linguistic model, a critical first step to ensure that a matching change is made in the localized ontology.

Simplifying localization management using synchronization. LabelTranslator provides a model where sets of ontology terms and linguistic information associated (in different languages) are separately stored. Therefore, it would be very difficult for a person to update/review all the linguistic definitions and information associated with a particular concept. We believe that this process will be done by different people at different times and in different countries. Thus, the maintenance cycle for each language will often be separated. Figure 2 illustrates the localization management used in our system to synchronize the conceptual and linguistic information. In the following we analyze the process in more detail, describing the actions performed by each actor of our scenario.

- ***Ontology expert.*** (S)he is responsible for editing the changes in the ontology model. All the changes executed in each user session are stored in a repository as a new version. The types of changes that our system can manage are the following: changes of the label content (e.g., ontology label rename) and ontology structure changes (e.g., delete or add operations). For each case, LabelTranslator stores the type of operation executed and

[2] A resource delta represents changes in the state of a resource tree between two discrete points in time.

its additional information (e.g., the name of the renamed label). This information is used in our system to synchronize the conceptual and linguistic information.

- ***Linguist expert(s).*** The linguist expert in a specific target language is responsible for performing the localization process. Notice that this process always uses the last version of an ontology. When the linguist needs to update the linguistic model (LM), our system tries to synchronize both models, performing the following actions: (1) obtaining the current version of the LM to be updated, (2) extracting the last version of the changes in the ontology model (OM) from which the last localization was taken (normally the one with the same number as the LM), (3) performing all the actions of the file of changes in the LM, and (4) updating the LM version in the repository.

Acknowledgements

This work is supported by the European Commission's Sixth Framework Program under the project name: Lyfecycle support for networked ontologies (NeOn) (FP6-027595), and the National Project "GeoBuddies" (TSI2007-65677C02).

References

1. Espinoza, M., Gómez-Pérez, A., Mena, E.: Enriching an ontology with multilingual information. In: Bechhofer, S., Hauswirth, M., Hoffmann, J., Koubarakis, M. (eds.) ESWC 2008. LNCS, vol. 5021, pp. 333–347. Springer, Heidelberg (2008)
2. Espinoza, M., Gómez-Pérez, A., Mena, E.: Labeltranslator - automatically localizing an ontology. In: Bechhofer, S., Hauswirth, M., Hoffmann, J., Koubarakis, M. (eds.) ESWC 2008. LNCS, vol. 5021, pp. 792–796. Springer, Heidelberg (2008)
3. Montiel-Ponsoda, E., Aguado, G., Gómez-Pérez, A., Peters, W.: Modelling multilinguality in ontologies. In: Coling 2008: Companion volume - Posters and Demonstrations, Manchester, UK (2008)
4. Montiel-Ponsoda, E., Peters, W., Aguado, G., Espinoza, M., Gómez-Pérez, A., Sini, M.: Multilingual and localization support for ontologies. In: NeOn Project Deliverable 2.4.2 (August 2008)
5. Suárez-Figueroa, M.C., Gómez-Pérez, A.: First attempt towards a standard glossary of ontology engineering terminology. In: Proc. of 8th International Conference on Terminology and Knowledge Engineering (TKE 2008) (2008)
6. Suárez-Figueroa, M.C., Gómez-Pérez, A.: Towards a glossary of activities in the ontology engineering field. In: Proc. of 6th International Conference on Language Resources and Evaluation (LREC 2008) (2008)

WSMX 1.0: A Further Step toward a Complete Semantic Execution Environment

Federico Michele Facca, Srdjan Komazec, and Ioan Toma

Semantic Technology Institute (STI) – Innsbruck
ICT Technologiepark, Technikerstrasse 21a, 6020 Innsbruck, Austria
firstname.lastname@sti2.at

Abstract. The Web Service Execution Environment (WSMX) project is a continuously ongoing effort that aims at delivering a middleware covering all the Semantic Web Services life cycle. WSMX represents the reference implementation of the Semantically enabled Service Oriented Architecture (SESA) [1]. In this demonstration we aim to present the latest achievements that include: Web Service monitoring, Web Service ranking and Web Service grounding.

1 Introduction

The Web Service Execution Environment (WSMX) [1] is an open-source development and execution environment which implements a significant part of the SESA facilities. It is a platform characterized by strong component decoupling, goal-driven Web service usage and direct support for mediation facilities. It is a reference implementation of the Web Services Modeling Ontology (WSMO) [2] which is a conceptual model used to describe various aspects related to Semantic Web Services (SWS). WSMO is based on four major fundamental elements: Ontologies, Web services, Goals and Mediators. WSMX can achieve the goal of the user by dynamically discovering and selecting a matching Web service (where both Web services and goals are described in terms of ontologies) relying on data and process mediation facilities in various phases of the SWS life cycle. In this demo we show some of the most interesting innovations introduced in WSMX 1.0:

- monitoring for Semantic Web Services and the SESA environment will enable fault handling and quality of service monitoring;
- ranking engine to rank discovered services according to user preferences;
- service grounding engine based on W3C SAWSDL recommendation[1].

This paper is structured as follows: Section 2 presents the overall architecture of WSMX. In Section 3 we present our latest achievements toward Web Services and SESA monitoring. Section 4 presents the ranking component. In Section 5 we give an overview of our Web Service grounding solution. Finally in Section 6 we draw conclusions and describe the demonstration plan.

[1] http://www.w3.org/TR/sawsdl/

L. Aroyo et al. (Eds.): ESWC 2009, LNCS 5554, pp. 826–830, 2009.

2 WSMX Architecture

In order to provide for the complete Semantic Web Service life cycle support, the WSMX platform consists of several components with clear functional responsibilities [1], as depicted in Figure 1. The *WSMX Manager Core* component is the central part of WSMX, managing all other components and interactions between them. The Core is responsible for a number of the tasks such as the business logic of the system, the events engine, the internal workflow engine, and the loading of distributed components. The *Resource Manager* offers an interface for WSMX persistent storage facilities for WSMO artifacts. The *Discovery* component is concerned with finding Web service descriptions that match the goal specified by the service requester. In case there is more than one service capable of fulfilling the requester's goal, the *Ranking* component ranks them based on non-functional properties like price, guaranteed response time, etc. The *Data Mediator* component has the role of reconciling the data heterogeneity problems that can appear during discovery, composition, selection or invocation of Web Services based on mapping rules retrieved from the resource manager. Mapping rules have been created in advance at design time. The *Communication Manager* component is responsible for dealing with the various protocols for sending and receiving messages to and from WSMX. It represents the entry point to the system for external entities that want to consume WSMX functionality. The *Choreography* component defines the communication patterns in terms of messages exchanged in order to interact with the Web service. The *Grounding* component enables lifting from SOAP messages to WSMO ontologies and lowering from WSMO ontologies to SOAP messages. The *Invoker* component handles communication between WSMX and external Web services.

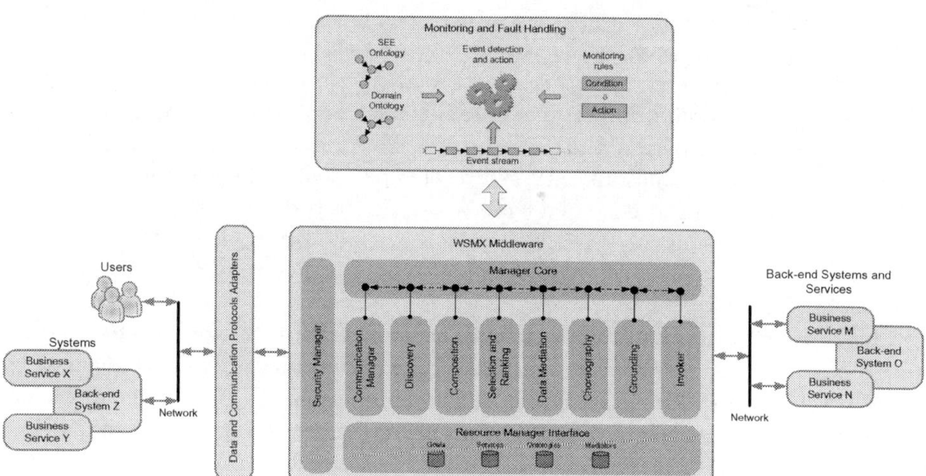

Fig. 1. WSMX architecture

3 Web Service Monitoring

Semantic Web Services monitoring is a relevant research issue. In particular, Vaculin et al. [3] presented a solution based on OWL-S as the foundational SWS ontology. The authors are covering only choreography execution while our solution tends to tackle the overall SWS lifecycle. Additionally, Pedrinaci et al. [4] introduced a set of ontologies for business process analysis and gave outlines for the semantic process mining and monitoring in the form of sequence which consist of 5 steps: observe, evaluate, detect, diagnose, and resolve.

Figure 1 gives a brief architectural overview of the WSMX monitoring and fault-handling subsystem. As suggested by the figure, all the WSMX middleware components are appropriately instrumented, thus enabling detection and extraction of low-level events and generation of the semantically-enhanced event descriptions. These descriptions are defined in the context of the domain ontology which formalizes notions of the problem domain, but also the SEE ontology which specifies structure and relationships inside WSMX. Based on the event streams produced by the monitored objects, as well as observational goals expressed in the ECA compliant rule-based monitoring language (inspired by the work of Behrends et al. [5] and built on top of the WSML-Rule), the solution will employ event-detection and reasoning techniques in order to ensure the reactiveness of the system in a timely fashion. Depending on the ECA rules definition the subsystem is capable of dealing with a variety of issues such as fault-handling, verification of service compositions and monitoring and enforcement of the Service Level Agreements.

4 Web Service Ranking

Web Service Ranking is the process which generates an ordered list of services out of the candidate services set according to ranking criteria specified by the user. The candidate services set is constructed by an a priori discovery process. Various non-functional properties can be specified by the user and obtained from the goal description as ranking criteria. On the service side the requested non-functional property values are either directly specified in the service description or are provided (computed or collected) by a monitoring tool (see Section 3). Non-Functional Properties (NFPs) specified in goal and service descriptions are expressed using the WSML language, by means of logical rules using terms from NFP ontologies [6].

The core of the ranking process is the evaluation of the logical rules used to model NFPs of services. Additional data is required during the rule evaluation process. This data represents mainly user preferences and includes: (1) which NFPs user is interested in, (2) the level of importance of each of these NFPs, (3) how the list of services should be ordered (i.e. ascending or descending) and (4) concrete instances extracted from the goal description.

In a nutshell the multi-criteria ranking algorithm works as follows. First a set of tuples containing nonfunctional properties and their associated importance is extracted out of the goal description. If no importance is specified the default value is consider to be 0.5 which specifies a moderate interest in the nonfunctional property. The importance is a numeric value ranging from 0 to 1, where 1 encodes the fact that the user is extremely interested in the nonfunctional property and 0 encodes the fact that the nonfunctional property is not of interest. Later on, instance data from the goal is extracted and a knowledge base is created. The last step in extracting relevant information for the ranking process is to identify how the results should be ordered i.e. ascending or descending. Once the preprocessing steps are done, each service is checked to see if the requested nonfunctional properties specified in the goal are available in the service description. In case of a positive answer the corresponding WSML logic rules are extracted and evaluated using a reasoning engine. A quadruple structure is built containing the service, the nonfunctional property, the computed value and its importance. An aggregated score is computed for each service by summing the normalized values of NFPs weighted by importance values. Finally the score values are sorted according to the ordering sense extracted from the goal and the final list of services is returned.

5 Web Service Grounding

Grounding is a fundamental step in the Semantic Web Service life cycle that enables the communication between the semantic layer and and the back-end services. Several ongoing initiatives like SAWSDL [7], and microWSMO attend to tackle this problem. In WSMX, to be able to provide a flexible solution to the grounding problem, we designed and implemented an extensible grounding framework. The framework, according to the different types of service to be invoked (e.g. WSDL, RESTful, ...), is capable of selecting the correct grounding engine. The grounding engine is determined by evaluating the grounding endpoints in the WSMO Web Service descriptions. The current version, since WSMX platform concentrates mainly on WSDL services, includes only a grounding engine based on SAWSDL. To enable decoupling between remote services and WSMX, we implemented a mechanism to replace remote WSDL descriptions with local SAWSDL ones during the execution. This enables the use of the SAWSDL approach even if remote services are not annotated by creating a local annotated copy of them. The engine interprets the SAWSDL descriptions and, according to the XML Schema element to be sent or received, performs the required lowering (WMSO→XML) or lifting (XML→WSMO). Current lifting and lowering solution are based on XSLT scripts or on Java code (this allows for seamless integration of grounding code used in the previous WSMX releases). The introduction of a specific grounding component is a relevant step toward the evolution of WSMX in order to make it more flexible and dynamically configurable. Furthermore it easily enables the publication of WSMO goals as standard Web Services to enable easier B2B integration through the platform.

6 Demonstration Plan and Conclusion

In this paper we presented a set of new functionalities introduced in the new WSMX release. The new functionalities will be demonstrated at ESWC. In particular the demonstration is based on a scenario for the COIN project, which is dealing with automatic composition of product retailer services and shipment services: a company building aircraft needs to integrate components for the cockpit consisting of the various parts such as displays, sensors and actuators, data buses and processors. The company has specific preferences regarding the purchasing and shipping phase so as to minimize the overall price and to minimize the shipping time. The scenario aims to show how the new integrated components empower WSMX.

The demo will present to the visitor a Web interface to create a goal and associated ranking preferences. After the submission of the goal the visitor will be able to follow the behavior of the platform internals through the console. For a better understanding the demo will be supported by an animated architecture diagram.

Acknowledgments

The work presented in this paper is partially founded by FP7 COIN EU Project (216256). We would like to thank Jacek Kopecký for his contributions to the evolution of the grounding component.

References

1. Fensel, D., Kerrigan, M., Zaremba, M. (eds.): Implementing Semantic Web Services: The SESA Framework, 1st edn. Springer, Heidelberg (2008)
2. Fensel, D., Lausen, H., Polleres, A., de Bruijn, J., Stollberg, M., Roman, D., Domingue, J.: Enabling Semantic Web Services: The Web Service Modeling Ontology. Springer, Heidelberg (2006)
3. Vaculin, R., Sycara, K.: Specifying and Monitoring Composite Events for Semantic Web Services. In: Proceedings of the 5th IEEE European Conference on Web Services (November 2007)
4. Pedrinaci, C., Domingue, J.: Towards an Ontology for Process Monitoring and Mining. In: Proceedings of the Workshop on Semantic Business Process and Product Lifecycle Management (SBPM 2007), vol. 251, CEUR-WS (2007)
5. Behrends, E., Fritzen, O., May, W., Schenk, F.: Combining ECA Rules with Process Algebras for the Semantic Web. In: RULEML 2006: Proceedings of the Second International Conference on Rules and Rule Markup Languages for the Semantic Web, Washington, DC, USA, pp. 29–38. IEEE Computer Society, Los Alamitos (2006)
6. Toma, I., Roman, D., Fensel, D., Sapkota, B., Gomez, J.M.: A multi-criteria service ranking approach based on non-functional properties rules evaluation. In: Krämer, B.J., Lin, K.-J., Narasimhan, P. (eds.) ICSOC 2007. LNCS, vol. 4749, pp. 435–441. Springer, Heidelberg (2007)
7. Kopecký, J., Vitvar, T., Bournez, C., Farrell, J.: Sawsdl: Semantic annotations for wsdl and xml schema. IEEE Internet Computing 11(6), 60–67 (2007)

MoKi: The Enterprise Modelling Wiki

Chiara Ghidini[1], Barbara Kump[2], Stefanie Lindstaedt[2,3], Nahid Mahbub[1], Viktoria Pammer[2,3], Marco Rospocher[1], and Luciano Serafini[1]

[1] FBK-irst. Via Sommarive 18 Povo, 38050, Trento, Italy
[2] Knowledge Management Institute, TU Graz. Inffeldgasse 21a, 8010 Graz, Austria
[3] Know-Center. Inffeldgasse 21a, 8010 Graz, Austria

Abstract. Enterprise modelling focuses on the construction of a structured description, the so-called *enterprise model*, which represents aspects relevant to the activity of an enterprise. Although it has become clearer recently that enterprise modelling is a collaborative activity, involving a large number of people, most of the enterprise modelling tools still only support very limited degrees of collaboration. Within this contribution we describe a tool for enterprise modelling, called MoKi (MOdelling wiKI), which supports agile collaboration between all different actors involved in the enterprise modelling activities. MoKi is based on a Semantic Wiki and enables actors with different expertise to develop an enterprise model not only using structural (formal) descriptions but also adopting more informal and semi-formal descriptions of knowledge.

1 Research Background: Enterprise Modelling

According to [1], "an *enterprise model* is a computational representation of the structure, activities, processes, information, resources, people, behavior, goals, and constraints of a business, government, or other enterprise". We refer to *enterprise modelling* as the process of creating an enterprise model. Often, an enterprise model focuses on the description of two specific aspects of an enterprise: (i) its processes and activities, and / or (ii) the business domain within which the enterprise operates. Additionally, other aspects of an enterprise, like goals, human resources, organisational structure and roles etc. may also be important assets to be described in an enterprise model. This is due to the central role that enterprise models are playing in the development of a large number of applications, including Internet and (Semantic) Web based applications.

Building an enterprise model requires a number of skills. These skills span from *knowing* the different aspects that have to be described in the models, to having the ability of *encoding* such knowledge into formal statements, to having the ability of *integrating* different aspects, such as structure, activities, processes, information, resources, people, behavior, goals, and constraints into a uniform and coherent vision. Given the complexity of enterprise modelling, it is unrealistic to assume that all the actors possess all the skills above. For instance different actors may have very different expertise in encoding some content into formal languages, or may know only of specific aspects of an enterprise. For this

L. Aroyo et al. (Eds.): ESWC 2009, LNCS 5554, pp. 831–835, 2009.

reason, enterprise modelling is inherently a *collaborative* activity, performed by different actors with different know-how and technical skills. To support these different actors we foresee a system in which content can be represented at different degrees of formality. This will allow domain experts to create, review and modify models at a rather informal/human intelligible level and knowledge engineers to check the quality of the formal definitions and their correspondence with the informal parts they intend to represent. In order not to increase the overhead of human work, translation between different levels of formality should be as much automatised as possible. To support a coherent development and integration of the different components of the enterprise model we also foresee a tool in which all the relevant aspects of an enterprise can be modelled in a collaborative and cooperative manner, thus exploiting the synergy of "having to think the same thing out only once". To meet this vision we are developing MoKi (**Mo**delling Wi**ki**), a collaborative tool that (i) supports access to the enterprise model at different levels of formality (informal, semi-formal and formal), (ii) supports integrated modelling of several aspects of an enterprise, and (iii) ensures a coherent development of the formal part. In the current implementation of MoKi, we focus on an enterprise model describing the domain, the processes and the competencies of an enterprise. The choice of these aspects was originally motivated by the EU-project APOSDLE described in Section 3.

2 Key Technology: Semantic Wiki

MoKi is a tool based on Semantic MediaWiki (SMW) [2]. It extends SMW offering specific support for enterprise modelling. The choice of developing MoKi on top of a semantic wiki was made for several reasons. First, wikis are a state of the art collaborative tool. Second, users are quite familiar with wikis and editing of wiki pages, due to the growing popularity of wiki-based web sites. Third, it provides a uniform tool for the (informal) specification of the different components of an enterprise model (domain, processes, and competencies in our case). Fourth, the natural language descriptions inserted in a semantic wiki can be structured according to predefined templates, with the help of semantic constructs like attributes and relations. As a consequence, the informal descriptions in natural language contain enough structure to be automatically translated in formal models, thus allowing the reuse of informal descriptions for automatic ontology creation. Finally only a web-browser is required on the end user side to use the system.

Model elements are associated to MoKi pages which contain informal but structured information about the element itself. The typical page contains (i) an informal description of the element considered in natural language (images or drawings can be used as well), and (ii) a structured part, where the element is described by means of triples of the form *(subject, relation, object)*, with the element itself playing the role of the subject. This natural language based, but also structured, description provides a natural bridge between formal and informal representation of knowledge. The user fills a page via forms, so he/she does not

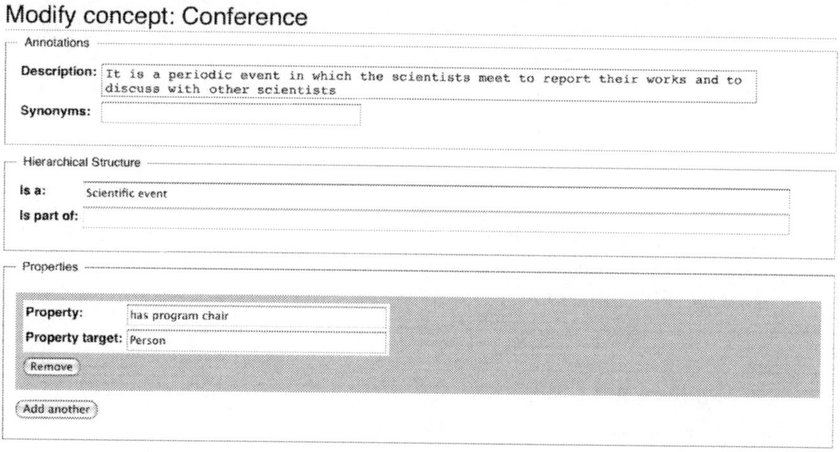

Fig. 1. An example of MoKi page

need to know any particular syntax or language to participate in the creation of the enterprise model. All the actors involved in the modelling activities can also interact with each others and exchange further ideas and comments using the *discussion* SMW's built-in functionality. An example of a MoKi page is shown in Figure 1. To support modelling, MoKi provides groups of functionalities, described below, that can be accessed via a wiki's style menu bar.[1]

Import. We provide two types of import functionalities. First, functionalities to import pre-existing structured knowledge. Currently it is possible to upload OWL domain ontologies, and to import list of domain concepts or tasks organized in a simple textual structure where indentation is used to indicate a hierarchy. Depending on the user's choice, the hierarchy is interpreted as taxonomy or partonomy. The import functionality takes care of creating the indicated concepts as well as the properties denoting the desired hierarchy. Second, text analysis functionalities such as the extraction of relevant terms from digital resources and the clustering of such terms according to their relatedness are included to support the utilization of available unstructured knowledge relevant for the modelling activity. The latter functionalities are provided by the KnowMiner [3], an advanced text analysis tool developed by the Know-Center.

Model Management. This set of functionalities supplies the basic functionality each modelling tool necessarily provides: creating, editing and deleting model elements. Depending on the type of element, pre-defined templates are loaded when it is created or edited. Such templates contain for instance properties for specifying a taxonomy or partonomy, or a sequence in the case of tasks.

Visualization. These functionalities allow to produce different types of graphical overviews of the models: they help the actors to deal with the global picture

[1] In what follows, we indicate with "model element" any basic component of the models (e.g. a concept, a relation, a task, a competency and so on).

on the models and not only with the single model elements. For example, with the help of the IsA Browser and the PartOf Browser, the user can easily visualize and rearrange via drag 'n' drop the taxonomy and partonomy of concepts in the domain model.

Export. These functionalities support the fully automatic export of knowledge of the enterprise model into standard knowledge representation languages. At the moment the supported formalism is OWL for the whole enterprise model. On-going work is devoted to the addition of formal languages especially for task/process specification, where a number of standards already exist.

MoKi is built to facilitate the plugging-in of new or existing state-of-the-art tools as we have done with KnowMiner. We plan to use this feature to add further functionalities in the future (e.g. tools providing knowledge elicitation techniques like card-sorting or laddering). Concerning the informa/semi-formal/formal structuring of a page we currently focus on the semi-formal part, which also contains some informal description. A demo version of MoKi can be tried out on-line at `moki.fbk.eu`. A detailed description of the current version of MoKi is contained in the MoKi manual, available at the MoKi web site.

3 Use Cases and Application

MoKi has been successfully applied within the EU-project APOSDLE (See `www.aposdle.org`). Within this context, MoKi has been used to develop enterprise models in the following domains: environmental consultancy, electromagnetism simulation, innovation and knowledge management, requirements engineering, and statistical data analysis. Some of the experiences of an early usage of the system are described in [4]. In addition MoKi is used in applications that go beyond typical enterprise modelling: the representation of medical guidelines encoded in the ASBRU language (as part of the OncoCure project [5]), and the collection of data for the Personal Health Record of the Province of Trento, Italy (as part of the TreC project, see `trec.fbk.eu`). In each of these cases, MoKi has been applied in the context of a collaborative paradigm as described in [6].

4 System Demonstration

In the live demonstration we will present the main MoKi features that can be used to develop an enterprise model. We will first show how pages describing entities of the enterprise model look like, and we will describe the basic editing functionalities (add, edit, delete, rename). Then, we will demonstrate the import functionalities, starting from the domain model import. We will show that it is possible to create and pre-fill (annotations, class subclass/partof relations) pages in MoKi by importing an already available ontology. Then we will show that if there is an already available list of domain concepts (tasks) organized via plain-text indentation according to a taxonomy or partonomy (task/subtask decomposition), it it is possible to load pre-filled pages in MoKi preserving this kind of hierarchical structure. We will also present the term extraction functionality by uploading a digital resource (a pdf or MS Word document) to the MoKi and retrieving a list of candidate concepts, from which we will choose some to add as entities in our enterprise model. As regards the visualisation functionalities, we will demonstrate how to get different overviews of the models: we will show (i) how to display (and edit) in a tree-view the taxonomy or partonomy hierarchy of the element of the domain model, and (ii) how to display in a tabular-view the list of all elements (and their main properties) in the models. Finally, we will demonstrate that the enterprise model (its components) designed in MoKi can be exported into an OWL formal model, and later refined with any OWL ontology editor.

5 Conclusions

In this demonstration we have presented a new collaborative tool, called MoKi, to support the creation of integrated enterprise models. MoKi supports collaboration in the modelling activities among people with different skills and levels of expertise. We are currently improving the tool in several directions, which span from including support for the modelling of competencies, improving support for processes modelling and processes visualization, and improving the preservation of the alignment between the content described in MoKi pages and the formal models created.

Acknowledgements. We thank all the people involved in the modelling activities within the APOSDLE project for their useful suggestions and feedback. APOSDLE is partially funded under the FP6 of the EU Commission within the IST work program (FP6-IST-2004-027023). The Know-Center is funded by the Austrian Competence Center program Kplus under the auspices of the Austrian Ministry of Transport, Innovation and Technology and by the State of Styria.

References

1. Fox, M.S., Grüninger, M.: Enterprise modeling. AI Magazine 19(3), 109–121 (1998)
2. Krotzsch, M., Vrandecic, D., Volkel, M.: Wikipedia and the semantic web - the missing links. In: Proc. of the 1st Int. Wikimedia Conference, Wikimania 2005 (2005)
3. Pammer, V., Scheir, P., Lindstaedt, S.: Two protégé plug-ins for supporting document-based ontology engineering and ontological annotation at document-level. In: 10th International Protégé Conference, Budapest, Hungary, July 15-18 (2007)
4. Christl, C., Ghidini, C., Guss, J., Pammer, V., Rospocher, M., Lindstaedt, S., Scheir, P., Serafini, L.: Deploying semantic web technologies for work integrated learning in industry. a comparison: Sme vs. large sized company. In: Sheth, A., et al. (eds.) ISWC 2008. LNCS, vol. 5318, pp. 709–722. Springer, Heidelberg (2008)
5. Eccher, C., Ferro, A., Seyfang, A., Rospocher, M., Miksch, S.: Modeling clinical protocols using semantic MediaWiki: the case of the Oncocure project. In: ECAI workshop on Knowledge Management for Healthcare Processes, K4HelP (2008)
6. Rospocher, M., Ghidini, C., Serafini, L., Kump, B., Pammer, V., Lindstaedt, S.N., Faatz, A., Ley, T.: Collaborative enterprise integrated modelling. In: SWAP. CEUR Workshop Proceedings, vol. 426 (2008)

The Personal Knowledge Workbench of the NEPOMUK Semantic Desktop

Gunnar Aastrand Grimnes, Leo Sauermann, and Ansgar Bernardi

DFKI GmbH, Kaiserslautern, Germany
gunnar.grimnes@dfki.de, leo.sauermann@dfki.de, ansgar.bernardi@dfki.de

Abstract. In this paper we describe some of the personal information management features found in the PSEW prototype of the NEPOMUK Semantic Desktop project. Focussing on importing and adding new knowledge to the system as well as navigating, browsing, and searching the existing knowledge we show how a user can use NEPOMUK to integrate her information across application boundaries and create and discover new structures in her information space.

1 The NEPOMUK Semantic Desktop

The paper presents parts of the outcome of the now completed NEPOMUK project[1]. NEPOMUK aimed to create a specification for a Social Semantic Desktop and a prototype implementation of this specification. This paper will only describe the details around the prototype implementation, for information on the Semantic Desktop specifications such as the ontologies, the architecture blueprint and service descriptions we refer the reader to the project deliverables available from the project website as well as [1,2].

The NEPOMUK project resulted in two parallele implementations of the Semantic Desktop, one written in C++ which is tightly integrated with the KDE4 desktop environment, and a cross platform version written in Java. This paper deals only with the Java version. Furthermore, we focus only the Personal Information Management (PIM) components of the NEPOMUK Semantic Desktop, that is, only the components used on a single desktop machine. NEPOMUK also includes social features allowing various kinds of distributed search and information sharing, but they will not be covered here.

The NEPOMUK approach to personal information management aims to overcome the limitation of having a separate applications for each type of data, such as email, photo-manager, publication-managers, etc. Instead all the information is integrated into one information-space and the user is free to relate and annotate information elements across application boundaries. The central hub for working with this information-space is called the NEPOMUK Personal Semantic Workbench (PSEW) and a typical view is show in Figure 1. Here we see a range of different views on the user's data, and a *person* concept, "Claudia",

[1] http://nepomuk.semanticdesktop.org

L. Aroyo et al. (Eds.): ESWC 2009, LNCS 5554, pp. 836–840, 2009.

being editted. The central idea for personal information management in NEPOMUK is *semantic annotation*, ranging from simple *tagging* to more sophisticated ontology based use of typed relations, class hierarchies, etc. Although semantically rich relations are possible in the framework, we have found the general *hasTag* relation to be at a level of complexity the users can easily understand and work with, and is also enough for most applications [3]. In NEPOMUK we go beyond the simple text tags used by web2.0 sites such as del.icio.us or Flickr, and each tag is a *PIMO concept*; that is, tags belong to one of the classes in the Personal Information Modelling Ontology (PIMO). PIMO is the NEPOMUK way of letting the user express her world-model on the Semantic Desktop by extending and sub-classing the starting PIMO which provides general classes such as *Document*, *Person*, *Company*, etc. Users are free to get as involved as they wish, if they are happy with a flat list of tags, all tags can be left as the default *pimo:Topic*, or they can create a more elaborate structuring scheme. Automated services that will automatically assign the correct type to known cities, people and countries are also available.

From a technical point of view NEPOMUK works by adding all desktop knowledge into a central RDF store, paired with a Lucene index for full-text queries of the content of RDF literals [4]. We have defined a set of ontologies for representing desktop items, known collectively as NIE (Nepomuk Information Elements)[2]; NIE includes ontologies for contacts (Nepomuk Contact Ontology – NCO), messages (Messaging Ontology – NMO). The PIMO ontology is used to organise and interrelate this knowledge. A set of core information handling services, such as PIMO manipulation, data-crawling, etc., are available through HTTP based interfaces and a Eclipse RCP GUI is built in top of these. Additional plugins for existing applications are also built using the HTTP interfaces.

2 Adding Knowledge

When starting the use a new knowledge management tool, bootstrapping is an important problem. To the end-user, the value of the Semantic Desktop increases as more data is added, allowing richer interrelations and increased chances of finding interesting patterns.

In NEPOMUK the Aperture framework[3] is used for importing existing data into the Semantic Desktop. Aperture can crawl information from a wide range of sources, such as file systems, email servers, web2.0 web-sites and existing desktop applications such as MS Outlook or Mozilla Thunderbird. From the crawled information sources Aperture can extract meta-data as well as the textual contents of documents; most common document formats are supported, such MS Office documents, PDFs, HTML, JPEGs, PNGs, MP3s, other multi-media files, and many more.

When NEPOMUK is first installed the user has a choice of what data-sources to setup and crawl, common data-sources such as files and email are automati-

[2] http://www.semanticdesktop.org/ontologies/

[3] http://aperture.sourceforge.net

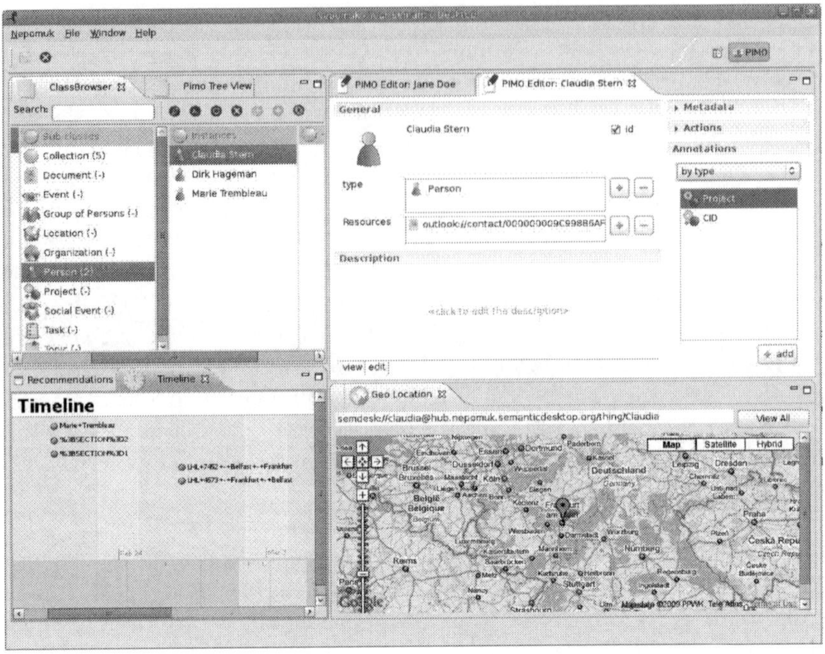

Fig. 1. The NEPOMUK Personal Semantic Workbench

cally detected by the system. Crawling is done regularly and iteratively, meaning changes to the original data will soon be picked up by NEPOMUK.

Once NEPOMUK has been setup and existing information has been crawled, NEPOMUK offers several ways to add new information directly into the Semantic Desktop data-cloud. Firstly, NEPOMUK includes a *DropBox*, a special folder that is monitored for changes, when a new file is downloaded from the web it can be saved in this folder. NEPOMUK will detect the file and analyse the content a pop-up a dialog with suggested tags, as shown in Figure 2. Here the file *prototypearchitecture.pdf* has been dropped, and the tags *Marco Andriotti* and *Technical Workpackage Meeting* has been proposed, the user himself has added a new tag: *cidprototype*. Each tag in NEPOMUK can be associated with a folder on disk, when a tag with an associated folder is assigned in the DropBox, the file is automatically moved to the appropriate location as well as tagged with the correct concept in one quick operation.

For content that is not downloaded into the DropBox NEPOMUK offers a bookmarklet for easy tagging of web-pages, as well as plugins for Mozilla Thunderbird and MS Outlook for tagging of emails.

3 Information Enrichment and Annotation

Importing the knowledge into a common RDF store only solves part of the information integration problem, as the data still forms separate islands for each

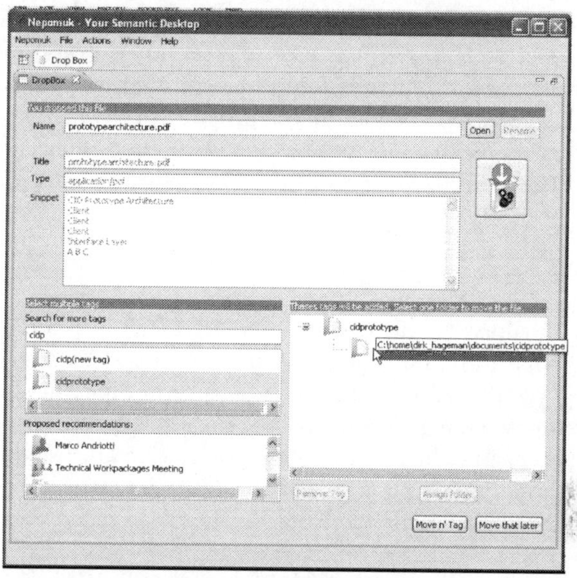

Fig. 2. The NEPOMUK DropBox

data-source. Creating the relations that for instance links the author of a PDF on disk to the sender of an email from Thunderbird is an important part of NEPOMUK and we call the process Local Data Alignment (LDA). Space does not allow us to go into details here, but NEPOMUK uses several heuristics for doing named entity merging; assigning the right class to known cities and countries; using the semantically enabled encyclopedia DBPedia[4] for finding additional information about concepts; and graph mining approaches for finding related resources. For crawled data, the data-alignment happens automatically. For manually created concepts the relations proposed by LDA are shows as recommendations in the user-interface.

4 Browsing and Searching

One of the key benefits of using Semantic Web technology to structure desktop information is that one is no longer tied to a single scheme of categorisation, i.e. in a traditional file-system a file resides in one folder, and one folder only. If one has chosen to arrange ones files by *project*, there is no way to later go back and view them by *year*. On the Semantic Desktop we can offer a wide range of views on the same data, for example, a timeline view that shows what concepts have been active in a certain period; a map view that shows the geographical location of concepts; a facetted browser; a tree-based view of the personal ontology, etc. For example, Figure 1 shows the hierarchical class-browser for browsing concepts based on their class, a timeline view, a map view, as well as a concept being editted.

[4] http://dbpedia.org

For the cases when browsing is not sufficient, NEPOMUK also offers a powerful search perspective, allowing the user to search her entire information space based either simple keyword queries as well as a more powerful structured query builder [5]. The structured query builder provides a visual interface for creating SPARQL queries by drag-dropping, and also offers heuristics for ontology relaxation when the exact ontology of the data queried is not known.

5 Summary

We have briefly outlined the features the NEPOMUK Semantic Desktop offers for personal information management, focussing on the process of adding information into the system and later browsing and re-finding this information. The NEPOMUK system we have discussed is open source, and the system, as well as the test-data and scripts that will be used for the demo, are all available from the project website.

References

1. Bernardi, A., Decker, S., van Elst, L., Grimnes, G., Groza, T., Handschuh, S., Jazayeri, M., Mesnage, C., Moeller, K., Reif, G., Sintek, M., Sauermann, L.: The Social Semantic Desktop: A New Paradigm Towards Deploying the Semantic Web on the Desktop. IGI Global (October 2008)
2. Groza, T., Handschuh, S., Moeller, K., Grimnes, G., Sauermann, L., Minack, E., Mesnage, C., Jazayeri, M., Reif, G., Gudjonsdottir, R.: The nepomuk project – on the way to the social semantic desktop. In: Proceedings of the Third International Conference on Semantic Technologies (I-SEMANTICS 2007), Graz, Austria (2007)
3. Sauermann, L., Heim, D.: Evaluating long-term use of the gnowsis semantic desktop for pim. In: Sheth, A., et al. (eds.) ISWC 2008. LNCS, vol. 5318, pp. 467–482. Springer, Heidelberg (2008)
4. Minack, E., Sauermann, L., Grimnes, G., Fluit, C., Broekstra, J.: The sesame lucene sail: Rdf queries with full-text search. Technical Report 2008-1, NEPOMUK Consortium (February 2008)
5. Kawase, R., Minack, E., Nejdl, W., Araäjo, S., Schwabe, D.: Incremental end-user query construction for the semantic desktop. In: International Conference on Web Information Systems and Technologies (WEBIST), Lisboa, Protugal (2009)

Utilizing Semantics in the Production of iTV Shows

Georg Güntner, Dietmar Glachs, and Rupert Westenthaler

Salzburg Research Forschungsgesellschaft m.b.H.
Jakob Haringer Straße 5/III, A-5020 Salzburg, Austria
{georg.guentner,dietmar.glachs,
rupert.westenthaler}@salzburgresearch.at

Abstract. This paper gives an overview of the semantic aspects of an advanced, semantics-based broadcasting production support system designed to enable the creation of interactive multi-channel television shows. The "Intelligent Media Framework" forms the middleware of this system and was developed in the context of the European integrated project LIVE. The envisaged "intelligence" is based on formal, machine understandable descriptions of the content and the events. We demonstrate the successful usage of the system by a broadcasting corporation in a field trial with several hundred end consumers, conducted during the Olympic Games in Beijing, August 2008.

1 Introduction

The European integrated project "LIVE Staging of Media Events" (LIVE, www.live-ist.org) aims at the creation of novel intelligent content production methods and tools for broadcasters to stage live media events such as the Olympic Games 2008 in Beijing. In the terminology of the project, "staging live media events" is a notion for the real-time creation of non-linear multi-channel video shows, whose format is adaptive according to the interests of the consumers. While the development of the envisaged interactive content formats forms a research strand of its own [1] the concept also imposes several conceptual and technical challenges from a software engineering and semantic modelling point of view. Fig. 1 depicts the environment the LIVE production support system needs to work in to support the preparation and production of shows as envisioned above.

Staging of media events in the novel TV format comprises all knowledge of an event e.g. the schedule and participants; knowledge of all related information about the event including archived content and - foremost - a detailed knowledge of how the staged event is to be presented to the targeted consumer audience. Typically such information is collected during the preparation phase of a show. However, the LIVE production support system addresses not only the preparation, but also the real time staging of media events where archival knowledge related to the event must be combined with real-time information of what is actually happening during the event. Furthermore, the actual production is influenced by accurate, real-time feedback from the viewing audience. An overview of the preparation and staging workflow is also shown in Fig. 1 which shows the activities of the professional users in their dedicated user spaces.

L. Aroyo et al. (Eds.): ESWC 2009, LNCS 5554, pp. 841–845, 2009.

The remainder of this paper is structured as follows: Section 2 provides details of the LIVE production support system used for this demonstration; Section 3 outlines how the Intelligent Media Framework (IMF) uses semantic technologies to support knowledge and information integration in the LIVE staging domain.

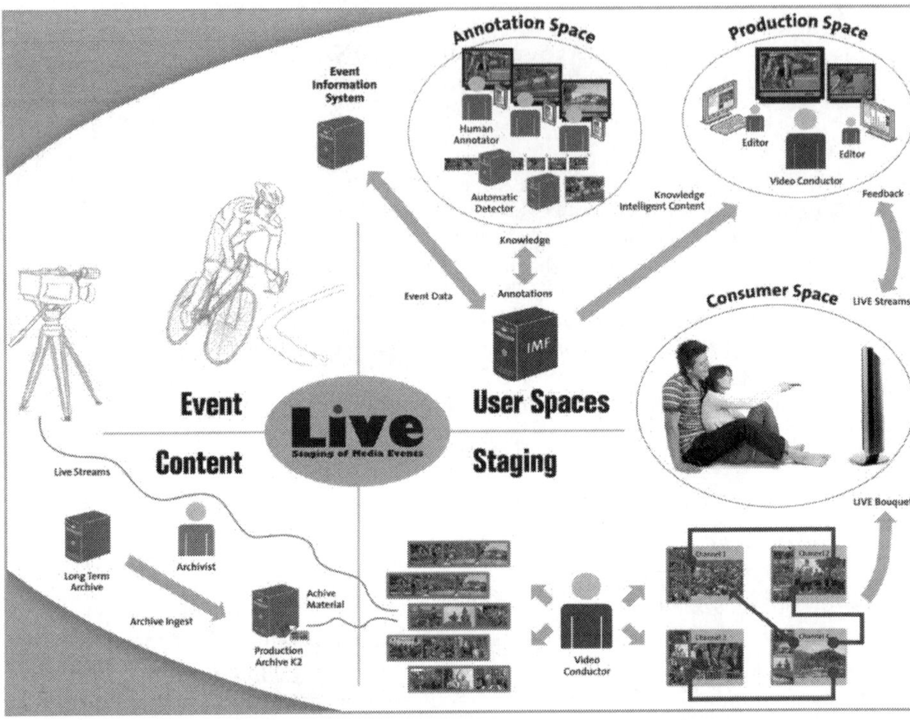

Fig. 1. Overview on the LIVE staging environment

2 Showcasing the LIVE Production Support System

Our demonstrator is based on the LIVE production support system used in a field trial during the Summer Olympic Games in August 2008. Over 500 households participated in the iTV trial and gave explicit (e.g. voting) and implicit (e.g. channel switching) feedback to the production team. At the production site in Vienna at ORF, the Austrian Broadcasting Corporation, annotators manually annotated up to twelve live video feeds from the Olympic Games. ATOS origin, the supplier for the Olympic Sports Information System provided event information in real time. This resulted in about 12.000 messages per hour that were sent to the Intelligent Media Framework (IMF). The IMF had to interpret the messages, align them to the knowledge model, maintain the semantic index and propagate a filtered set of messages to the consuming systems (e.g. the recommender system). A dedicated application called "Staging Console" displayed a set of messages filtered according to their relevance for the production process (e.g. priority, schedule).

The demonstrator outlines the complete information flow by starting with the captured sport event and by simulating the information created by the Olympic Sport Information System. The "Staging Console" as shown in Fig. 2, visualizes the scenery by outlining the observed live feeds (1), the scheduled events presented along a time line (2) and as a schedule of upcoming events (3). It further shows manually and automatically generated messages to the production team (4). Optionally, details of each message can be retrieved in a dedicated panel (5).

Besides the "Staging Console", the "Recommender Application" (see [3]) is part of the LIVE system and allows the production team to easily retrieve video clips from the production archive according to the context of the show. The "Feedback Panel" shows the viewers' channel switching behaviour (implicit feedback) and polling results (explicit user feedback). In the field trial the feedback was collected via the back channel of the set-top boxes of the used IPTV infrastructure (aon.tv[1]).

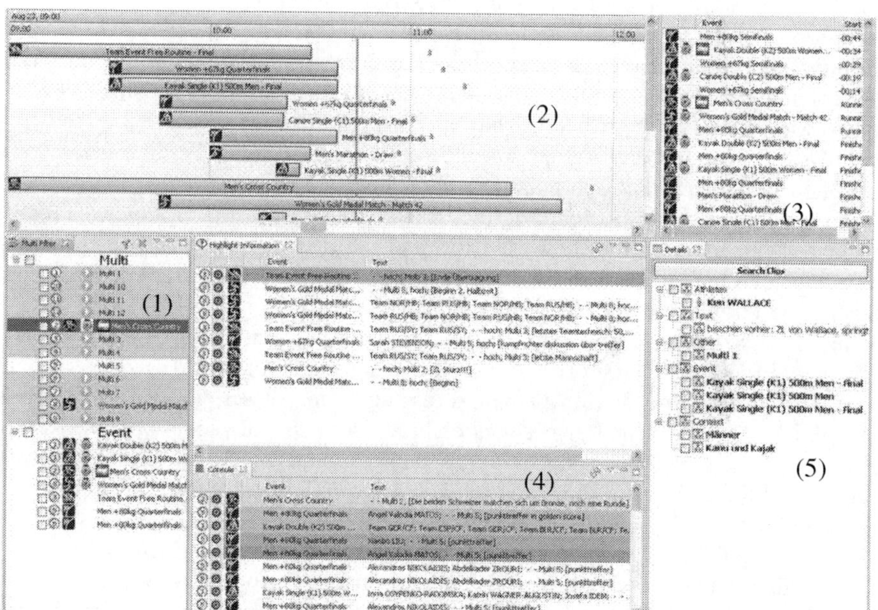

Fig. 2. LIVE Staging Console

The demonstration is a time shifted replay based on data gathered during the Olympic field trial. The demonstration shows the usage of the different tools in the workflow starting with the live capture of video streams, showing the semantic annotation and the bundling of streams into "shows" by the so called "video conductor" who can also react to consumer feedback and change the broadcast accordingly.

[1] aon.tv – see http://www.aon.tv/ (in German) – last visited 06.03.2009.

3 Intelligent Media Framework

The Intelligent Media Framework (IMF) implements the knowledge based middleware component of the LIVE production support system. This section gives first an overview of the knowledge model used and then describes the principal architecture of the LIVE production support system.

The IMF is based on a knowledge structure divided into three sub-models: the Term Model, The Event Domain Model and the Intelligent Content Model. The Term Model forms the foundation of this knowledge structure and is based on functionalities defined in the Simple Knowledge Organization System (SKOS). It supports the features of SKOS, but in addition it defines a base typology for terms, the "Term Type Hierarchy". This typology aims to simplify the structuring of the controlled vocabulary and provides additional semantics to the other knowledge models of the IMF. All terms in the controlled vocabularies are aligned to the Term Type Hierarchy. Details of the entire knowledge model can be found in [2].

On top of the Term Model is the Event Domain Model. This model supports the definition of formal statements describing the current state (annotating) or a future state (planning) of events. Such statements are used to describe the staged event as well as the staging process itself. They are instantiations of concepts defined in the Event Domain Model (e.g. Agent, Activities, Events, Settings). Terms and part of the controlled vocabulary are used to parameterize such statements.

The Intelligent Content Model defines knowledge-based content objects called Intelligent Media Assets (IMAs) which manage the relation between content and the knowledge. To achieve this, an IMA consists of two main parts: (i) Content annotations define knowledge about features of the essence such as storage location(s), encoding, access rights or publication rights. Dedicated elements of existing standards such as MPEG-7 and NewsML are used for such annotations. (ii) Subject matter annotations define information items expressing formal descriptions of what the essence is about. Subject matter descriptions include a spatial-, temporal-, topic based (includes terms as well as instances of the event domain model) and free text classification. In addition they hold information about the creator (professional user/software agent and user role) to support collaborative filtering.

The Intelligent Media Framework provides the knowledge infrastructure for the LIVE production support system and is based on a multi-tier service-oriented architecture in which all interfaces to external components are modeled as Web Services. The service interfaces are structured along the three parts of the knowledge model and provide access to the controlled vocabulary, the event domain model and the intelligent media assets. In addition to these pull-type services, the IMF uses a messaging system to support the real-time aspect of the staging process. Asynchronous messaging is used to process information from internal and external sources (automatic annotators, human annotator and information sent by the event information system). The IMF aligns the information to the already available knowledge of the event as well as the show and propagates the resulting knowledge by publishing it to subscribing applications of the LIVE production support system. This information enrichment uses statements as defined by the Event Domain model to formally describe the current state of the event as well as the staged show.

The IMF also maintains a semantic index over the whole knowledge base. This semantic index uses the relations defines in the Event Domain- and the Term Model to provide tagging-like classifications for all resources. Real-time changes to the knowledge structure are reflected in the semantic index.

During the Olympic Trial the IMF processed around 20.000 terms, around 100.000 semantic relations between the terms; 120.000 event participations and about 100.000 statements describing specific situations in the various events. Around 20.000 Intelligent Media Assets were created during the Olympic Field Trial.

4 Conclusion

In the field trial, content-based information was combined with event information in real-time, using a common knowledge model. The combination of content management systems with event information systems was successfully tested. The LIVE production support system demonstrated how a knowledge based system can enable knowledge workers to create the envisioned LIVE iTV format: Editors and directors at the production site at ORF, the Austrian Broadcasting Corporation, confirm that it would not be feasible to create such non-linear multi-channel video shows by using traditional production methods.

A survey by GfK Austria showed that the quality of the new format was rated with 1,5 (out of 5 with "1" expressing highest satisfaction) by the viewers. 82% of them expressed a high interest in a continuation of the interactive multi-channel format.

The LIVE approach differs from related research primarily with respect to the real time nature of the production methods (see [4] for a comparable approach in the area of non-linear, yet non real time narrative structures).

Acknowledgments

The LIVE project ("Live Staging of Media Events") is an integrated project partially funded by the European Commission within the 6th Framework of the IST under grant number FP6-27312. All statements in this work reflect the personal ideas and opinions of the authors and not necessarily the opinions of the European Commission.

References

1. Grünvogel, S.M., Wages, R., Bürger, T., Zaletelj, J.: A Novel System for Interactive Live TV. In: Ma, L., Rauterberg, M., Nakatsu, R. (eds.) ICEC 2007. LNCS, vol. 4740, pp. 193–204. Springer, Heidelberg (2007)
2. Güntner, G., Bürger, T., Westenthaler, R., Glachs, D.: Basic Specification of the Intelligent Media Framework: Public Synopsis. LIVE Deliverable D7.4 (2007), `http://www.ist-live.org`
3. Zaletelj, J.: Recommender System for the Multi-channel TV Production. In: Tscheligi, M., Obrist, M., Lugmayr, A. (eds.) EuroITV 2008. LNCS, vol. 5066, pp. 102–106. Springer, Heidelberg (2008)
4. Zsombori, V., Ursu, M., Wyver, J., Kegel, I., Williams, D.: ShapeShifting Documentary: A Golden Age. In: Tscheligi, M., Obrist, M., Lugmayr, A. (eds.) EuroITV 2008. LNCS, vol. 5066, pp. 40–50. Springer, Heidelberg (2008)

Knowledge Applications for Life Events: How the Dutch Government Informs the Public about Rights and Duties in the Netherlands

Ronald Heller and Freek van Teeseling

Be Informed, Linie 620, 7325 DZ Apeldoorn, Netherlands
{R.Heller,F.vanTeeseling}@beinformed.nl

Abstract. The Dutch government has several ontology driven applications on the internet that inform the public about rights and duties that are relevant for certain life events. Life events such as "New to Holland", "The death of a close relative" and "Working and studying" are examples of events in life where you want to be informed about your rights and duties. In the demonstration we show how we make use of an ontology to determine a profile of the user and how we give him or her the information that is tailored to his/her specific situation.

Keywords: Profile, ontology, life events, government, knowledge-models, customer oriented.

1 Background and Application Context

Citizens demand better support and service from the government than ever before. The world has become more and more complex and citizens demand that governmental organizations adapt to this ever-changing world more quickly. Agility is no longer just wanted or needed, it is crucial. Citizens see the government as one big organization from their viewpoint, although it is actually a huge complex of tasks and responsibilities, divided over many different smaller organizations. This results in the fact that many means are used to inform citizens about all kind of laws and regulations, e.g. the internet, brochures, commercials etc. The main problem is that citizens are unable to receive a complete overview of all the relevant information. This is why they ask the government for one clear answer with regards to which regulation they must comply to. They don't care which organization is involved. Politicians in the Netherlands have acknowledged this and strive for a more agile and efficient government. In the specific case of the Netherlands, this means that over 1300 governmental organizations are asked to change their way of serving and informing the public. At the moment most information and services are offered to the customers from a supplier point of view. The Dutch government intends to change this into a citizen-centric approach. Dutch governmental organizations are now working together towards a better customer interaction. They identified so called life events that are important for citizens. Examples of such life events are the death of a close relative [7] or the immigration into the Netherlands [4]. The important factor should be the situation and information status of the

L. Aroyo et al. (Eds.): ESWC 2009, LNCS 5554, pp. 846–850, 2009.

citizen, not the product or service of the individual governmental organization. Several governmental organizations are now working together to be able to inform a citizen when needed and with exactly the information needed.

2 Demonstration

The demonstration covers one life event: New to Holland [5]. It is an application giving background information for people who want to migrate to the Netherlands. Twelve organizations were involved in the development of this application, such as the Immigration office, organizations for taxes, grants and more. When you want to come to the Netherlands as a student you will need other information compared to when you want to work or live or even come to the Netherlands as an "Au pair". The demo will show how the personal situation of the customer is identified. An ontology is used for this profiling, and from this the ontology is used to decide on the relevant information. A WYSIWYG technique is used to create and store the knowledge models. There is no separation between model and implementation. A visual graph with subject-relation-object triples can be seen in the screenshot in figure 2. The customization to a specific application as New to Holland is essentially done by setting up an appropriate metamodel and the definition of patterns and guidelines necessary for the application [2].

In the case of live events this means that the governmental organizations are able to change the ontology by modeling the triples themselves using the graphical modeling environment. The tool supports an easy way of specifying which piece of information is relevant for certain citizens using strong typed relations between concepts.

The demonstration will show the application, including the ontologies it is based on. In the demonstration we will also discuss the process of development, aspects of maintenance using ontologies and the reuse of models over ontologies and applications.

3 Profiling Using Ontologies

In order to be able to make consistent knowledge models that can be managed and maintained by the information owners (business users), we defined a metamodel based on the Be informed methodology [2].

3.1 The Metamodel

The metamodel for the life event is far from complex. It consists of three layers: profile, advice and descriptions and annotations.

First we identify the profile of the customer, and following this it is determined which advices should be offered. The relevance of each piece of advice is dependent on the situation of the customer. One of the most important factors is the reason of his of her stay in the Netherlands (work, study, living). Other factors are also important, like the age of the customer and the nationality. The relevant advices are then linked to pieces of information to be shown with that advice (Description and annotations). Adding this layer also enables us to provide the information multi-lingual. In this case Dutch and English. This metamodel is used for all "Life Event"Applications.

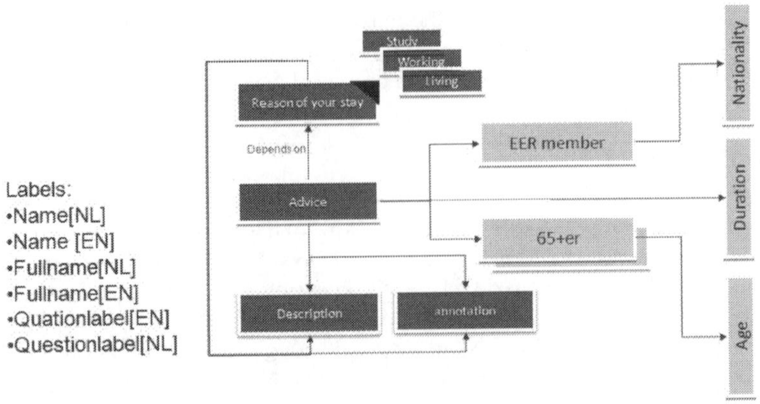

Fig. 1. The metamodel

3.2 Modeling the Advice

The involved organizations came up with roughly a hundred different pieces of information that they thought were of any interest to the public. All that had to be done was stating when these pieces where relevant, using the metamodel as mentioned above. This metamodel allows to select among a set of goals and purposes for people to immigrate to the Netherlands (working, studying, asylum, working as an "Au pair" etc.) in interaction with the user and dynamically. The metamodel in essence specifies which are the relevant questions to ask to decide dynamically on the purpose for the immigration and proceed accordingly. This resulted in knowledge models like shown in figure 2.

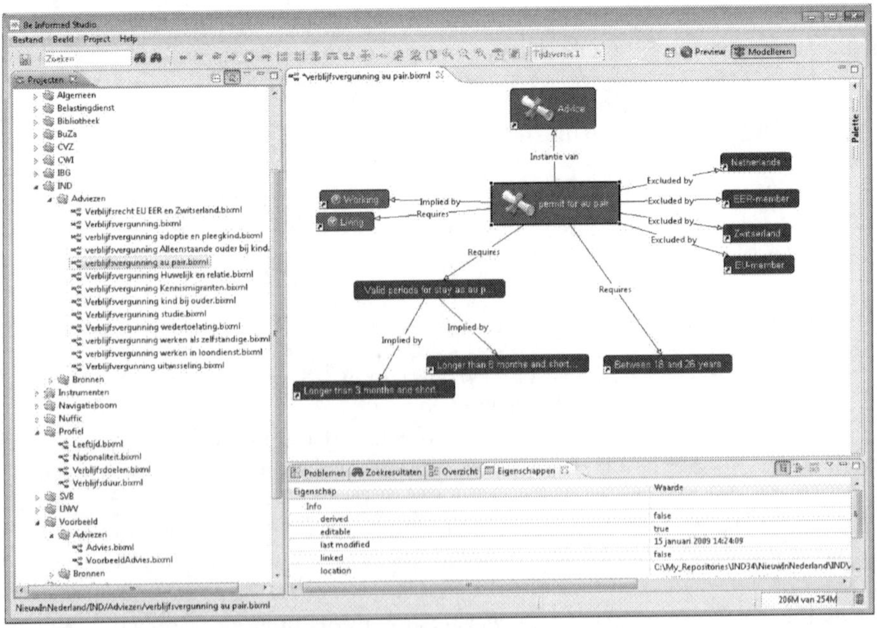

Fig. 2. Knowledge model example

In the example "Permit for Au Pair" is a concept of type Advice, triggered by purposes combined with length of stay, nationality and age. Reuse of knowledge is key to knowledge management, so knowledge about nationalities is reused where needed. In the model above the deducted concept of EU-membership is important for inferring the au pair advice, and is reused in different deductions.

4 Life Events on the Internet

On the web one can currently find several life event applications. Go to http://www.overheid.nl [5] to find the most up to date list. The demonstration will mainly cover the New to Holland application.

4.1 New to Holland

In the application a personal checklist is the actual profile that we use to infer which pieces of information are relevant for the customer. The left screenshot shows the start interface for the users. The right screenshot shows the inferred result. Again this is rather a simple screen where users can read relevant information, that applies to that specific profile. The application uses the power and possibilities of an ontology which remains in the background and opaque to the user. It does however offer easy use in managing and maintaining the knowledge involved using the Studio environment from figure 2.

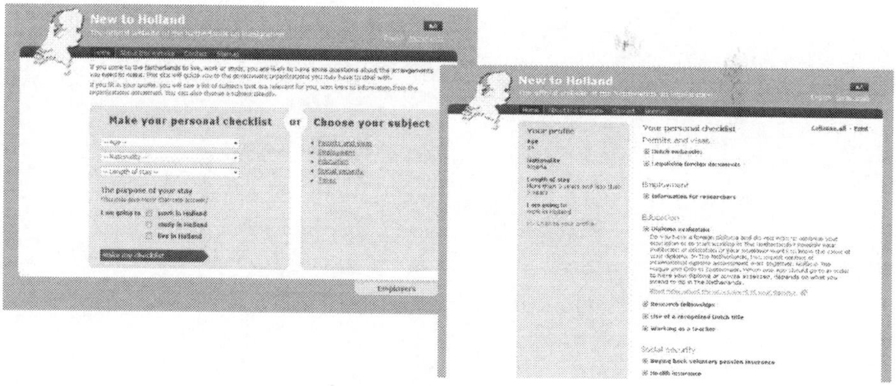

Fig. 3. New to Holland website

References

1. Verbeek, J., Rensen, G.: A Citizen Centric Government Enabled by Semantic Technologies. In: Semantic Technology Conference, May 18-22, San Jose, CA (2008)
2. Van Teeseling, F., Heller, R.: Patterns in Metamodelling. Be Informed White paper (2008)
3. http://www.newtoholland.nl

4. http://www.arbeidsmigratie.nl
5. http://www.overheid.nl
6. http://onderwijsenbijverdienen.overheid.nl
7. http://www.vertreknaarhetbuitenland.overheid.nl
8. http://www.overlijden-overheid.nl

CultureSampo: A National Publication System of Cultural Heritage on the Semantic Web 2.0

Eero Hyvönen, Eetu Mäkelä, Tomi Kauppinen, Olli Alm, Jussi Kurki, Tuukka Ruotsalo, Katri Seppälä, Joeli Takala, Kimmo Puputti, Heini Kuittinen, Kim Viljanen, Jouni Tuominen, Tuomas Palonen, Matias Frosterus, Reetta Sinkkilä, Panu Paakkarinen, Joonas Laitio, and Katariina Nyberg

Semantic Computing Research Group (SeCo)
Helsinki University of Technology (TKK) and University of Helsinki
first.last@tkk.fi
http://www.seco.tkk.fi/

Abstract. CULTURESAMPO is an application demonstration of a national level publication system of cultural heritage contents on the Web, based on ideas and technologies of the Semantic (Web and) Web 2.0. On the semantic side, the system presents new solutions to interoperability problems of dealing with multiple ontologies of different domains, and to problems of integrating multiple metadata schemas and cross-domain content into a homogeneous semantic portal. A novelty of the system is to use semantic models based on events and narrative process descriptions for modeling and visualizing cultural phenomena, and for semantic recommendations. On the Web 2.0 side, CULTURESAMPO proposes and demonstrates a content creation process for collaborative, distributed ontology and content development including different memory organizations and citizens. The system provides the cultural heritage contents to end-users in a new way through multiple (nine) thematic perspectives, based on semantic visualizations. Furthermore, CULTURESAMPO services are available for external web-applications to use through semantic AJAX widgets.

1 A Basis for Cultural Heritage on the Semantic Web

In our view, a cross-domain semantic cultural heritage portal [1] should be built on three pillars: First we need a cross-domain content infrastructure of ontologies, metadata standards, and related services, that is developed and maintained on a global level through collaborative local efforts. Second, the process of producing ontologically harmonized metadata should be organized in a collaborative fashion, where distributed content producers create semantically correct annotations cost-efficiently through centralized services. Third, the contents should be made available to human end-users and machines thought intelligent search, browsing, and visualization techniques. For machines, easy to use mash-up APIs and web services should be available. In this way, the collaboratively aggregated, semantically enriched knowledge base can be exposed and reused easily as services in other portals and applications in the same vein as Google Ads or Maps[1].

[1] http://code.google.com/apis/maps/

L. Aroyo et al. (Eds.): ESWC 2009, LNCS 5554, pp. 851–856, 2009.

CULTURESAMPO[2] is an operational online demonstration on a national Finnish level of implementing of such a semantic portal. The system consists of a national ontology infrastructure [2], content creation process, and an online portal [3]. In this paper, the main parts of CULTURESAMPO are first shortly overviewed. After this contibutions of the work are summarized w.r.t. related work.

2 Collaborative Ontology Infrastructure

An integral part of CULTURESAMPO are the ontologies and services of the FinnONTO infrastructure [2]. The general idea here is to extend the generic, logic based W3C recommendations with domain specific ontologies in different domains. Most of the FinnONTO ontologies were developed by transforming nationally used thesauri into light weight ontologies in a semi-automatic process. In the FinnONTO model, the ontologies are developed in a distributed fashion by collaborating expert groups of different fields, and are mapped together to form a large national ontology called KOKO encompassing all domains. At the moment, KOKO includes an upper ontology YSO (20 600 concepts), a museum ontology MAO (6800 concepts), an agriforestry ontology AFO (5500 concepts), an applied art ontology TAO (2600 concepts) and a photography ontology VALO (1900 concepts). In addition, CULTURESAMPO also utilizes a geographical registry of 800 000 places in Finland, a spatiotemporal ontology of Finnish counties 1865–2007 [4], an ontology of persons and organizations, and ontologized international systems, such as the Iconclass[3], and the Union List of Artists Names (ULAN)[4].

3 Distributed, Cross-Domain Content Creation

The contents of CULTURESAMPO come from 22 museums, archives, and libraries, most of which produce their contents independently from each other using heterogeneous cataloging systems and practices. The system also contains external web contents: all Wikipedia articles (in English and Finnish) that have coordinate information, as well as related photographs from the Panoramio service can be found on CULTURESAMPO's map views.

CULTURESAMPO also has an internal Web 2.0 commenting facility by which citizens can contribute new knowledge to individual content items, e.g. identify persons in an old photograph of a museum collection. In these ways, citizens are able to contribute to the "national semantic memory". Furthermore, interactive content production based on the SAHA editor [5] has been used in the system by participating organizations.

The system contains cultural objects of 67 different content types: artifacts, paintings, drawings, sculpture, abstract art, novels, comics, web pages, three types of folklore, five types of folk music, photos, aerial photos, persons, organizations, biographies, historical events, skills, videos, buildings, archeological sites, and others. The metadata

[2] The portal prototype is publicly in use at http://www.kulttuurisampo.fi/

[3] http://www.iconclass.nl/

[4] http://www.getty.edu/research/conducting_research/vocabularies/ulan/

is represented using 204 different properties and 253 additional properties used in ontologies. The content types harvested from the collections of the participating memory organizations are represented using 18 different metadata schemas. The aggregated knowledge base contains 134,000 cultural objects and 285,000 additional cultural resources (persons, places etc.), described by 2.7 million triples. The content is enriched using reasoning, that generates 8,7 million additional RDF triples in the knowledge base.

From a semantic modeling viewpoint, a research focus of our work has been event- and process-based annotations used in artificial intelligence and knowledge representation [6]. Events have been used for modeling cultural processes and narrative stories; a simplified version of the model [7] is in use.

An example of a cultural process is "Production of Ceramics" created by experts at the University of Applied Arts in Helsinki. It illustrates and explains the composition of different work phases when manufacturing ceramics. Our process model has also been applied to videos. A case example available on the portal describes how the shoe maker Onni Wirlander manufactured a pair of traditional leather boots. The video can be viewed directly starting from different points of interest. When watching the video, the recommendation system creates dynamically, for each subsequence separately, recommendation links to materials of interest in the portal, such as tools related to the sequence.

Content creation in CULTURESAMPO is supported by a number of generic tools, such as the National Ontology Library Service ONKI[5], the metadata editor SAHA, information extraction tool POKA[6], and the semantic content validator VERA[7].

4 The Semantic Web 2.0 Portal CULTURESAMPO

The portal is an end-user application for both 1) human users and 2) machines, and is multilingual (Finnish, Swedish, and English). A major novelty of the portal is to provide its contents through nine "thematic perspectives":

Map views. There are 4 map views [4] available using Google Maps: One for viewing all objects and filtering them in terms of their relation to the places; one for finding old Finnish counties with digitized limits; one for viewing historical maps layered semi-transparently over modern Google Maps ; one for finding near-by objects of interest. It is also possible to find Wikipedia articles and Panoramio photos by using the map interfaces.

Relational search. This view is a demonstration of relational search [8], where the idea is not to search for objects but associative relation chains between objects. We used the ULAN registry of 120,000 artists and organizations with 390,000 names. The user types in two names, using semantic autocompletion, and CULTURESAMPO tells how the persons or organizations are related with each other by the social network based on

[5] http://www.yso.fi/

[6] http://www.seco.tkk.fi/tools/poka/

[7] http://www.seco.tkk.fi/services/vera/

some 50 different social roles (e.g., parent-of, teacher-of, patron-of etc.). The underlying social RDF/OWL network can also be browsed by a graphical network browser.

Search and organize. In this view, a new kind of generalized version of the faceted-search paradigm has been developed, called domain centric search [9]. This advanced search and analysis facility is used for filtering contents and for visualising the results on facets, maps or a timeline. For example, it is possible to find out how the beard fashion, as depicted in art and photographs, has changed in the 19th century in Finland.

Collections view. Here the contents can be accessed based on an organizational view. Each participating organization has an automatically generated home page in the system with links to subcollections and the actual collection items.

Finnish history view. This view is based on an ontology representing events in the Finnish history. These events are of interest of their own, but are also used to create semantic recommendations to other CULTURESAMPO contents, e.g. to biographies of persons participating in the events.

Skills and processes. This view is used for finding cultural procedural descriptions in the system, i.e. semantic models of skills, processes, and documentations on videos.

Biographies. In this view, biographies of the National Biography[8] are used to access CULTURESAMPO contents. When reading biographies, related contents are shown based on the concepts extracted from the text using the information extraction tool POKA.

Semantic Kalevala. This view [10] contains a semantically annotated version of the national epic of Finland, Kalevala, that is related in many ways to Finnish art and culture. The epic is based on old Finnish folklore; thousands of runes and pieces of folk music are available in the portal, too. When reading Kalevala, annotations related to its subsequences can be viewed to help reading, and semantic recommendations to related materials, such as art, artifacts etc. in CULTURESAMPO are automatically produced.

Karelia view. This view contains Wikipedia articles about the Karelia area in Finland that has been influential to Finnish culture. Like in the biographies view, the POKA system is used for extracting concepts from the texts (here web pages) and for generating semantic recommendations.

In addition to humans, the system can be used by machines via an AJAX interface. In this way, the collaboratively aggregated and semantically enriched national CULTURESAMPO knowledge base can be used by other portals and systems on the web, in the same spirit as Google Ads or Maps.

5 Discussion and Conclusions

The vision and implementation of CULTURESAMPO goes beyond current semantic web portals for cultural heritage [1], such as MuseumFinland [11], MultimediaN 2006 [12],

[8] http://www.kansallisbiografia.fi/

and CHIP Demonstrator 2007 [13]: The CULTURESAMPO system 1) is highly cross-domain with lots of content types and metadata schemas (usually only one schema such as Dublin Core or VRA is used), 2) it makes use of sophisticated semantic annotation models including events and processes, 3) it uses new kind of semantic search and recommendation techniques, 4) it has exceptionally versatile selection of semantic visualizations available (different map views, timelines, graphs, process visualization, semantic video viewing), 5) it is based on a large nation wide collaboratively maintained infrastructure of ontologies and ontology services, 6) it includes a model of and tools for collaborative semantic content creation, and 7) the services are available for machines, too.

Acknowledgements. This work is a part of the national FinnONTO[9] research project 2003–2007, 2008–2010, funded mainly by Tekes and a consortium of 38 companies and public organizations, partly by the EU FP7 SmartMuseum project2008–2010, and by the Finnish Cultural Foundation (2008–2010).

References

1. Hyvönen, E.: Semantic portals for cultural heritage. In: Handbook on Ontologies, 2nd edn. Springer, Heidelberg (forthcoming, 2009)
2. Hyvönen, E., Viljanen, K., Tuominen, J., Seppälä, K.: Building a national semantic web ontology and ontology service infrastructure—the FinnONTO approach. In: Bechhofer, S., Hauswirth, M., Hoffmann, J., Koubarakis, M. (eds.) ESWC 2008. LNCS, vol. 5021, pp. 95–109. Springer, Heidelberg (2008)
3. Hyvönen, E., Mäkelä, E., Kauppinen, T., Alm, O., Kurki, J., Ruotsalo, T., Seppälä, K., Takala, J., Puputti, K., Kuittinen, H., Viljanen, K., Tuominen, J., Palonen, T., Frosterus, M., Sinkkilä, R., Paakkarinen, P., Laitio, J., Nyberg, K.: CultureSampo—Finnish culture on the Semantic Web 2.0. Thematic perspectives for the end-user. In: Museums and the Web 2009 Proceedings, Archives & Museum Informatics, Toronto (2009), `http://www.archimuse.com/mw2009/papers/hyvonen/hyvonen.html`
4. Kauppinen, T., Väätänen, J., Hyvönen, E.: Creating and using geospatial ontology time series in a semantic cultural heritage portal. In: Bechhofer, S., Hauswirth, M., Hoffmann, J., Koubarakis, M. (eds.) ESWC 2008. LNCS, vol. 5021, pp. 110–123. Springer, Heidelberg (2008)
5. Valkeapää, O., Alm, O., Hyvönen, E.: Efficient content creation on the semantic web using metadata schemas with domain ontology services. In: Franconi, E., Kifer, M., May, W. (eds.) ESWC 2007. LNCS, vol. 4519, pp. 819–828. Springer, Heidelberg (2007)
6. Sowa, J.: Knowledge Representation. Logical, Philosophical, and Computational Foundations. Brooks/Cole (2000)
7. Junnila, M., Hyvönen, E., Salminen, M.: Describing and linking cultural semantic content by using situations and actions. In: Robering, K. (ed.) Information Technology for the Virtual Museum. LIT Verlag (2008)
8. Kurki, J., Hyvönen, E.: Relational semantic search: Searching social paths on the semantic web. In: Poster Proc. of the ISWC + ASWC 2007, Busan, Korea (2007)
9. Mäkelä, E., Suominen, O., Hyvönen, E.: Automatic exhibition generation based on semantic cultural content. In: Proc. of the Cultural Heritage on the Semantic Web Workshop at ISWC + ASWC 2007 (2007)

[9] http://www.seco.tkk.fi/projects/finnonto/

10. Hyvönen, E., Takala, J., Alm, O., Ruotsalo, T., Mäkelä, E.: Semantic Kalevala—accessing cultural contents through semantically annotated stories. In: Proc. of the Cultural Heritage on the Semantic Web Workshop at ISWC + ASWC 2007, Busan, Korea (2007)
11. Hyvönen, E., Mäkela, E., Salminen, M., Valo, A., Viljanen, K., Saarela, S., Junnila, M., Kettula, S.: MuseumFinland—Finnish museums on the semantic web. Journal of Web Semantics 3(2), 224–241 (2005)
12. Schreiber, G., Amin, A., van Assem, M., de Boer, V., Hardman, L., Hildebrand, M., Hollink, L., Huang, Z., van Kersen, J., de Niet, M., Omelayenko, B., van Ossenbruggen, J., Siebes, R., Taekema, J., Wielemaker, J., Wielinga, B.J.: MultimediaN e-culture demonstrator. In: Cruz, I., Decker, S., Allemang, D., Preist, C., Schwabe, D., Mika, P., Uschold, M., Aroyo, L.M. (eds.) ISWC 2006. LNCS, vol. 4273, pp. 951–958. Springer, Heidelberg (2006)
13. Wang, Y., Stash, N., Aroyo, L., Gorgels, P., Rutledge, L., Schreiber, G.: Recommendations based on semantically-enriched museum collection. Journal of Web Semantics 6(4), 43–50 (2008)

A Rule System for Querying Persistent RDFS Data*

Giovambattista Ianni[1], Thomas Krennwallner[2], Alessandra Martello[1], and Axel Polleres[3]

[1] Dipartimento di Matematica, Università della Calabria, I-87036 Rende (CS), Italy
{ianni,a.martello}@mat.unical.it
[2] Institut für Informationssysteme, Technische Universität Wien, A-1040 Vienna, Austria
tkren@kr.tuwien.ac.at
[3] Digital Enterprise Research Institute, National University of Ireland, Galway
axel.polleres@deri.org

Abstract. We present GiaBATA, a system for storing, aggregating, and querying Semantic Web data, based on declarative logic programming technology, namely on the dlvhex system, which allows us to implement a fully SPARQL compliant semantics, and on DLV^{DB}, which extends the DLV system with persistent storage capabilities. Compared with off-the-shelf RDF stores and SPARQL engines, we offer more flexible support for rule-based RDFS and other higher entailment regimes by enabling custom reasoning via rules, and the possibility to choose the reference ontology on a per query basis. Due to the declarative approach, GiaBATA gains the possibility to apply well-known logic-level optimization features of logic programming (LP) and deductive database systems. Moreover, our architecture allows for extensions of SPARQL by non-standard features such as aggregates, custom built-ins, or arbitrary rulesets. With the resulting system we provide a flexible toolbox that embeds Semantic Web data and ontologies in a fully declarative LP environment.

1 Introduction

Thanks to W3C standards such as RDF(S), OWL, and recently RIF, – see `www.w3.org/2001/sw/` for an overview –, an increasing bulk of machine-readable data and knowledge encoded in rules and ontologies are becoming available on the Web. This vast amount of information uses rather lightweight ontologies in contrast to the full power of expressive description logic reasoning, which imposes new demands on systems to deal with Semantic Web data. A Semantic Web system should be able to *handle and evaluate arbitrarily complex queries on large amounts of RDF instance data* that are not manageable in memory. Thus, current RDF stores (see Sec. 2) provide persistent storage facilities and typically support SPARQL [1,2]. On top of SPARQL, data processing requires *ontological inferences as well as custom, possibly nonmonotonic rules*, which are crucial, e.g., for modelling mappings between vocabularies. Moreover, it should be possible to *query combinations of different data graphs, ontologies,*

* This work has been partially supported by the Italian Research Ministry (MIUR) project Interlink II04CG8AGG, the Austrian Science Fund (FWF) project P20841, by Science Foundation Ireland (SFI) project Líon (SFI/02/CE1/I131), and by the EU FP6 project inContext (IST-034718).

L. Aroyo et al. (Eds.): ESWC 2009, LNCS 5554, pp. 857–862, 2009.

```
@prefix bob: <http://bob.org/foaf.rdf#> .
bob:me foaf:name "Bob"; foaf:knows _:a .
_:a foaf:name "Alice" ;
    rdfs:seeAlso <http://alice.org/foaf.rdf> .
```

(a) Graph `bob.org/foaf.rdf`

```
@prefix alice: <http://alice.org/foaf.rdf#> .
alice:me a foaf:Person; foaf:name "Alice" ;
    foaf:knows [ foaf:name "Bob"] ;
    foaf:knows [ foaf:name "Charles"] .
```

(b) Graph `alice.org/foaf.rdf`

```
@prefix foaf: <http://xmlns.com/foaf/0.1/> .
...
foaf:knows rdfs:range foaf:Person.    myOnt:isFriendOf rdfs:subPropertyOf foaf:knows.
...
```

(c) Graph `example.org/myOnt.rdfs`, an extension of the FOAF ontology

Fig. 1. Two RDF graphs, and an extract of the FOAF ontology

and rules dynamically harvested from the Web. In this respect, most available RDF databases only offer limited support for dynamic inference, custom reasoning via rules, or querying upon datasets of choice.

We propose an architecture aiming at closing this gap by providing the GiaBATA system which caters for knowledge-intensive applications on top of Semantic Web data by combining (a) SPARQL querying on varying datasets and varying schemes, (b) rule-based RDFS and (limited) OWL inference, and (c) a persistent storage system based on deductive database technology. Based on systems described in Sec. 2, our proposed RDF store fulfils the above goals and is easily extensible by features such as using arbitrary aggregates and built-in predicates as well as adding custom rules to SPARQL [3].

Motivating Example. Take the two data graphs about our well-known friends Alice, Bob, and Charles in Fig. 1(a)+(b). Both use vocabulary defined in an extended FOAF ontology (`http://xmlns.com/foaf/spec/`); the relevant part is in Fig. 1(c).

For instance, a SPARQL query could be used to extract names of persons mentioned in graphs that belong to friends of Bob. We assume that Bob provides links to the graphs associated to the persons he knows using `rdfs:seeAlso` statements:

```
select ?N from <http://example.org/myOnt.rdfs> from <http://bob.org/foaf.rdf>
          from named <http://alice.org/foaf.rdf>                                (1)
where { bob:me myOnt:isFriendOf ?X . ?X foaf:seeAlso ?G .
        graph ?G { ?P a foaf:Person; foaf:name ?N } }
```

The *dataset* $DS = (G,N)$ of a SPARQL query is defined by the *default graph* G obtained from the RDF merge [4] (denoted $\uplus$) of all graphs mentioned in **from** clauses plus a set of named graphs $N = \{(g_1, G_1), \ldots, (g_k, G_k)\}$, i.e., pairs of IRIs given in **from named** clauses and their corresponding graphs. In our example, the dataset is being specified explicitly by a set of **from** and **from named** clauses, i.e., the dataset of query (1) would be $DS_1 = (\, G_{\texttt{myOnt}} \uplus G_{\texttt{bob.org}}, \{(\texttt{<http://alice.org/>}, G_{\texttt{alice.org}})\})$.

Since neither of the three source graphs contain a triple matching the pattern `bob:me myOnt:isFriendOf ?X` in the **where** clause, most existing RDF stores with SPARQL support would return an empty result on (1). This is because typically SPARQL engines only take simple RDF entailment into account when answering queries, which boils down to basic graph pattern matching. When taking ontological inference by the statements of the `myOnt` ontology into account, however, one would expect to obtain three matchings for the variable ?N: {?N/"Alice", ?N/"Bob", ?N/"Charles"}.

In order to obtain this expected answer, SPARQL's basic graph pattern matching needs to be extended, see [1, Section 12.6]. In principle, this means that the graph pattern in the **where** clause needs to be matched against a graph enlarged by applying inferences prescribed by the given entailment regime.[1] SPARQL extensions providing entailment regimes other than simple RDF entailment are still an open research problem.[2]

Approximations of entailment regimes such as RDFS and OWL [5,6,7], however, can be modeled by finite Datalog rulesets that cover sound but not necessarily complete RDFS and OWL inferences. For instance, the RDF Semantics document contains an informative set of entailment rules, a subset of which is implemented by most available RDF stores. However, these rules could be dynamically evaluated upon query time, or inferences could be materialized at graph loading time, producing a persistent extended graph which queries are issued on. Materialization upon loading time has drawbacks though: when the dataset includes a number of different graphs, it is not clear upfront (i) which ontology should be taken into account for which data graph, and (ii) to which graph the inferred triples "belong," which in turn complicates querying named graphs.

As for (i), assume that a user agent wants to issue another query on graph `bob.org` with the original FOAF ontology instead of our modified version within the default graph. Since the property `myOnt:isFriendOf` is stated to be a subproperty of `foaf:knows`, the triple bob:me `myOnt:isFriendOf` _ : a, which could be inferred from $G_{\texttt{myOnt}} \uplus G_{\texttt{bob.org}}$, would most probably contribute in an undesired way to such a different query. Current RDF stores however prevent to parametrize inference with an ontology of choice, since typically inferences are computed upon loading time *once and for all*.

As for (ii), queries upon datasets including named graphs are even more problematic. Even if inference was supported properly, the answer to (1) over dataset DS_1 is just $\{$?N/`"Alice"`$\}$. The `myOnt` ontology is only merged into the default graph, but not into the named graph. Thus, there is no way to infer that `"Bob"` and `"Charles"` are actually names of `foaf:Persons` within the named graph $G_{\texttt{alice.org}}$. Indeed, SPARQL does not allowed to explicitly specify a dataset to be merged with named graphs.

In order to overcome both point (i) and (ii) above, the current nonintuitive semantics of SPARQL with RDFS can be mitigated by slightly extending SPARQL's syntax. We give the possibility of specifying ontologies which are used for the entire dataset, including named graphs, by means of the new directive **`using ontology`**. For instance, to specify the dataset $DS_2 = (G_{\texttt{myOnt}} \uplus G_{\texttt{bob.org}}, \{(\texttt{<http://alice.org/>}, G_{\texttt{myOnt}} \uplus G_{\texttt{alice.org}})\})$, we allow the following syntax replacing the first line of query (1):

```
select ?N using ontology <http://example.org/myOnt.rdfs> ...
```

Hence, the **`using ontology`** construct allows for coupling the given dataset with the proper class/property hierarchy of the `myOnt` data schema. The possibility of query dependent and dynamic ontology inference shows the effectiveness of this approach.

2 System Description and Evaluation

GiaBATA is based on a combination of the DLV^{DB} [8] and dlvhex [9] systems, and caters for persistent storage of both data and ontology graphs. The former system is a variant of the DLV system with built-in database support. The latter is a solver for the so called HEX-programs [9]. dlvhex features an extensible plugin system which

[1] The entailment rules for getting the expected results for (1) are RDFS3 and RDFS7 from [4].
[2] cf. `http://www.polleres.net/sparqltutorial/`, Unit 5b, for details

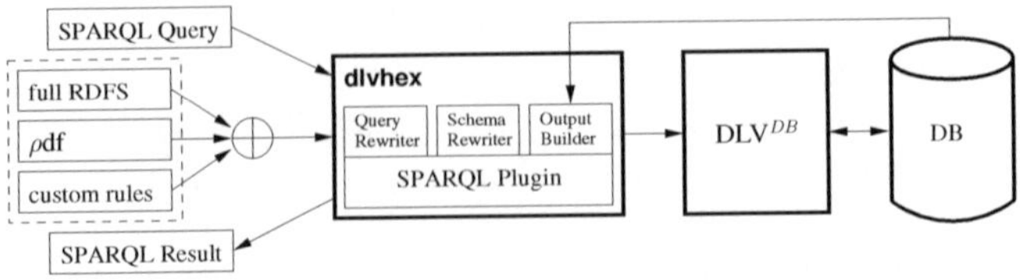

Fig. 2. The GiaBATA system

we used for developing a rewriter-plugin able to translate SPARQL queries to HEX-programs. Our prototype supports the standard SPARQL specifications extended with the syntax outlined above, arbitrary entailment regimes (currently, we support RDFS), and the addition of custom rules. We provide a SPARQL-protocol compliant Web service interface. Fig. 2 shows a high-level view of our system architecture. It consists of four main components: (i) a SPARQL to datalog rewriter as part of a plugin for dlvhex (see [10,11]), (ii) a schema rewriter which manipulates the rewritten datalog rules, adds auxiliary definitions in order to match the underlying database schema, and implements the chosen semantics by adding rules to the input of the module in (iii) the DLV^{DB} datalog engine, which rewrites the input program to native SQL queries, and thereby accesses triples persistently stored in (iv) a DBMS of choice storing the RDF data according to a particular storage schema. This architecture provides a fully declarative implementation of SPARQL and brings support for further (deductive) database and LP optimization and extensibility.

By and large, the chosen database schema for the triples exploits a main table storing quadruples composed of subject, predicate, object, and the source graph of each triple. This representation has been improved [12] by additional relations mapping URIs/literal string to integer values, which avoids string matching in favour of integer comparison.

We have investigated the persistent storage facilities as well as SPARQL features of some of the state-of-the-art RDF stores, namely, Sesame RDF database 2.0, Owlim v3.0b8, AllegroGraph 3.0.1, and Jena/Joseki 3.2.[3] All systems support persistent storage, either based on RDBMS or on other dedicated backends, and deal with RDFS inferencing and partial or non-standard OWL fragments. While a persistence strategy is provided, both reasoning and query evaluation are usually performed in main memory, adopting a materialization strategy. In our tests we focused on querying under ρDF inference regime [5], requiring the capability to match patterns against named graphs. The investigation revealed that combining RDFS and named graph queries lead to unexpected behavior. While inference is properly handled as long as the query ranges over the whole dataset, RDFS reasoning fails in case of queries using explicit default or named graphs. That leaves dynamic named graph queries with inference infeasible in current systems.

3 Related Work and Outlook of the Demo

There are earlier approaches to integrate ontological inference in a logic programming environment. As opposed to SWI Prolog's Semantic Web library [13] which also offers

[3] `http://www.openrdf.org/`, `http://www.ontotext.com/owlim/`, `http://jena.hpl.hp.com/`, and `http://agraph.franz.com/`

SPARQL support, we support Answer Set Programming as underlying LP paradigm, instead of Prolog. In contrast to OntoDLV [14], which supports proprietary ontology and query languages, the focus of GiaBATA is on compliance with Web standards such as SPARQL, RDF(S), OWL, and RIF. We currently work on the efficiency of FILTER expressions by rewriting them to SQL queries over the underlying database, see [15,16]. Moreover, one of the next steps includes to support arbitrary database schemes. Applying optimization techniques from both the DB and LP area should boost evaluation performance. Extensions of SPARQL by aggregates and custom built-ins presented in earlier works [11] carry over to our persistent storage version with minor modifications. We are going to add support for extended graphs, that is SPARQL "views" as defined also in [11].

The live discussion will provide a better insight into GiaBATA showing use cases that cover the key features of the system in accessing, storing and querying Semantic Web data. For features explanation we will refer to largely adopted benchmark suites,[4] while ad hoc showcases will help to understand the advantages of our proposed **`using ontology`** extension. Finally, the users will get an overview of GiaBATA's capabilities and will learn how to interact with it.

References

1. Prud'hommeaux, E., Seaborne A. (ed.): SPARQL Query Language for RDF, W3C Recommendation (January 2008)
2. Pérez, J., Arenas, M., Gutierrez, C.: Semantics and complexity of SPARQL. In: Cruz, I., Decker, S., Allemang, D., Preist, C., Schwabe, D., Mika, P., Uschold, M., Aroyo, L.M. (eds.) ISWC 2006. LNCS, vol. 4273, pp. 30–43. Springer, Heidelberg (2006)
3. Polleres, A., Scharffe, F., Schindlauer, R.: SPARQL++ for mapping between RDF vocabularies. In: Meersman, R., Tari, Z. (eds.) OTM 2007, Part I. LNCS, vol. 4803, pp. 878–896. Springer, Heidelberg (2007)
4. Hayes, P.: RDF semantics, W3C Recommendation (February 2004)
5. Muñoz, S., Pérez, J., Gutiérrez, C.: Minimal deductive systems for RDF. In: Franconi, E., Kifer, M., May, W. (eds.) ESWC 2007. LNCS, vol. 4519, pp. 53–67. Springer, Heidelberg (2007)
6. ter Horst, H.J.: Combining RDF and part of OWL with rules: Semantics, decidability, complexity. In: Gil, Y., Motta, E., Benjamins, V.R., Musen, M.A. (eds.) ISWC 2005. LNCS, vol. 3729, pp. 668–684. Springer, Heidelberg (2005)
7. Ianni, G., Martello, A., Panetta, C., Terracina, G.: Efficiently querying RDF(S) ontologies with Answer Set Programming. J. Logic Computation (August 2008)
8. Terracina, G., Leone, N., Lio, V., Panetta, C.: Experimenting with recursive queries in database and logic programming systems. Theor. Pract. Log Prog. 8(2), 129–165 (2008)
9. Eiter, T., Ianni, G., Schindlauer, R., Tompits, H.: Effective Integration of Declarative Rules with External Evaluations for Semantic Web Reasoning. In: Sure, Y., Domingue, J. (eds.) ESWC 2006. LNCS, vol. 4011, pp. 273–287. Springer, Heidelberg (2006)
10. Polleres, A.: From SPARQL to rules (and back). In: WWW 2007, pp. 787–796. ACM Press, New York (2007)
11. Polleres, A., Schindlauer, R.: dlvhex-sparql: A SPARQL-compliant query engine based on dlvhex. In: ALPSWS 2007, CEUR WS Proceedings, vol. 287, pp. 3–12 (September 2007)

[4] LUBM Benchmark available at `http://swat.cse.lehigh.edu/projects/lubm/`

12. Abadi, D.J., Marcus, A., Madden, S., Hollenbach, K.J.: Scalable Semantic Web Data Management Using Vertical Partitioning. In: VLDB 2007, pp. 411–422. ACM Press, New York (2007)
13. Wielemaker, J., Hildebrand, M., van Ossenbruggen, J.: Prolog as the Fundament for Applications on the Semantic Web. In: ALPSWS 2007 (2007)
14. Ricca, F., Gallucci, L., Schindlauer, R., Dellarmi, T., Grasso, G., Leone, N.: OntoDLV: An ASP-based System for Enterprise Ontologies. J. Logic Computation (August 2008)
15. Lu, J., Cao, F., Ma, L., Yu, Y., Pan, Y.: An Effective SPARQL Support over Relational Databases. In: SWDB-ODBIS 2007, co-located with VLDB 2007, pp. 57–76 (2007)
16. Theoharis, Y., Christophides, V., Karvounarakis, G.: Benchmarking Database Representations of RDF/S Stores. In: Gil, Y., Motta, E., Benjamins, V.R., Musen, M.A. (eds.) ISWC 2005. LNCS, vol. 3729, pp. 685–701. Springer, Heidelberg (2005)

RaDON — Repair and Diagnosis in Ontology Networks*

Qiu Ji, Peter Haase, Guilin Qi, Pascal Hitzler, and Steffen Stadtmüller

Institute AIFB
Universität Karlsruhe (TH), Germany
{qiji,pha,gqi,phi}@aifb.uni-karlsruhe.de,
steffen.stadtmueller@stud.kit.edu

Abstract. One of the major challenges in managing networked and dynamic ontologies is to handle inconsistencies in single ontologies, and inconsistencies introduced by integrating multiple distributed ontologies. Our RaDON system provides functionalities to repair and diagnose ontology networks by extending the capabilities of existing reasoners. The system integrates several new debugging and repairing algorithms, such as a relevance-directed algorithm to meet the various needs of the users.

1 Introduction

Next generation semantic applications are characterized by a large number of ontologies, some of them constantly evolving. As the complexity of semantic applications increases, more and more knowledge is embedded in applications, typically drawn from a wide variety of sources. This new generation of applications thus likely relies on ontologies embedded in a network of other ontologies. Ontologies and metadata have to be kept up to date when application environments and users' needs change. One of the major challenges in managing these networked and dynamic ontologies is to handle potential inconsistencies in single ontologies, and inconsistencies introduced by integrating multiple distributed ontologies.

For inconsistency handling in single, centralized ontologies, several approaches are known, see the survey in [4,3]. There are mainly two ways to deal with inconsistent ontologies [5]. One way is to simply live with the inconsistency and to apply a non-standard reasoning method to obtain meaningful answers. The second way to deal with logical contradictions is to resolve logical modeling errors whenever a logical problem is encountered [16,15,12].

There is relatively little work done on handling inconsistency in networked ontologies. For example, in [1], the authors deal with the problem of inconsistency in DDL by removing some bridge rules which are responsible for the inconsistency. However, there is no tool available to diagnose and repair inconsistencies in networked ontologies.

To meet the above mentioned needs, we develop the RaDON system to repair and diagnose not only single ontologies but networked ones, where several new debugging algorithms, such as a relevance-directed algorithm and a paraconsistency-based algorithm, have been integrated into RaDON to meet the various needs of the users.

* This work is partially supported by the EU in the IST project NeOn (IST-2006-027595).

L. Aroyo et al. (Eds.): ESWC 2009, LNCS 5554, pp. 863–867, 2009.

2 The RaDON System

RaDON[1] has been developed to deal with inconsistency[2] and incoherence[3] for ontology networks. It supports OWL-DL and is implemented in Java as a plug-in for the NeOn toolkit[4], which is an extensible ontology engineering environment to handle multiple networked ontologies. The RaDON plugin has already been applied in the FAO[5] case study in the context of diagnosing and repairing automatically learned ontologies. Results of these applications have been reported in NeOn Deliverable D1.2.2 [13].

In this section, we first introduce the functionalities of the RaDON system and then we present the process to debug and repair ontology networks. Afterwards, we describe the two new algorithms respectively: relevance-directed algorithm and paraconsistency-based algorithm for debugging ontologies.

2.1 Functionalities of RaDON

RaDON provides a set of techniques for dealing with inconsistency and incoherence in ontologies. In particular, RaDON supports novel strategies and consistency models for distributed and networked environments.

RaDON extends the capabilities of existing reasoners with the functionalities to deal with inconsistency and incoherence. Specifically, the functionalities provided by RaDON include: (1) debugging an incoherent or inconsistent ontology to explain why a concept is unsatisfiable or why the ontology is inconsistent, (2) repairing an ontology automatically by computing all possible explanations w.r.t. all unsatisfiable concepts if the ontology is incoherence, or w.r.t. the inconsistent ontology if it is inconsistent, (3) repairing an ontology manually based on the debugging results. For the manual repair, the user can choose the axioms to be removed for restoring the coherence or consistency. (4) coping with inconsistency based on a paraconsistency-based algorithm.

2.2 Process to Debug and Repair Ontology Networks

The plug-in can be used to diagnose and repair not only a single ontology, but also multiple ontologies that are networked. In particular, we consider ontologies that are networked via *mappings*. Mappings essentially are correspondences between the elements of two different ontologies, in the most simple case in the form of subclassOf or equivalentClasses axioms (cf. [2] for a definition of the networked ontology model). The mapping assertions may additionally be annotated with confidence values.

RaDON takes as input an ontology network consisting of ontologies and mappings between the ontologies. For our definition of inconsistency in the ontology network, we follow a global model semantics: An ontology network is inconsistent (resp. incoherent) if the ontology obtained by merging the ontologies and their mappings is inconsistent

[1] http://radon.ontoware.org/demo.htm

[2] An ontology is inconsistent iff it has no model.

[3] An ontology is incoherent iff it contains at least one unsatisfiable concept in its terminology. A concept is unsatisfiable if it is mapped to the empty set in all models of the ontology.

[4] http://www.neon-toolkit.org/

[5] http://www.fao.org/

(resp. incoherent). For the repair process we focus on repairing the mappings, assuming the ontologies are already individually coherent and consistent. Consequently only axioms in the mappings could be removed to resolve incoherence or inconsistency.

If the merged ontology is incoherent, the system computes all the unsatisfiable concepts. When an unsatisfiable concept is selected, all the minimal unsatisfiability-preserving subsets (MUPS)[6] w.r.t. this concept are computed. Similarly, inconsistencies in the ontology network are debugged: For inconsistencies, all the minimal inconsistent subsets (MIS)[7] w.r.t. the ontology are computed.

After obtaining the debugging results, we can repair the ontology automatically according to the proposed solution by our system. Also, the user can repair the ontology by manually choosing the axioms to be removed. To help the user to make a decision, our system provides the confidence values or scores for each axiom which could be removed to resolve incoherence or inconsistency. See [13] for more details.

2.3 Relevance-Directed Algorithm

A key problem of diagnosing and repairing ontology networks is to compute all or some MUPS for an unsatisfiable concept efficiently. The RaDON system provides various strategies to compute some or all MUPS based on the relevance-directed algorithm [7]. This algorithm incrementally selects sub-ontologies using a *selection function* and finds a set of MUPS from these sub-ontologies for an unsatisfiable concept. Our algorithm is adapted from the algorithm given in [8] which is based on Reiter's Hitting Set Tree (HST) algorithm [14].

Specifically, the relevance-directed algorithm provides the following strategies when computing MUPS for an unsatisfiable concept:

- Compute one MUPS;
- Compute all MUPS and all hitting sets[8] for the set of all MUPS;
- Compute some (not all) MUPS and some hitting sets for the set of all MUPS.

Therefore, the users could choose different strategies according to the ontology and their purpose. For example, if the testing ontology is relatively small, they could try to compute all MUPS. Also, if the users only intend to resolve the incoherence with some but not all solutions, we can choose the third strategy.

2.4 Paraconsistency-Based Algorithm

In RaDON we also provide the functionality to provide inconsistency-tolerant reasoning. Specifically, an algorithm for paraconsistent reasoning given in [10] has been integrated into RaDON which helps the user to cope with the inconsistency. This algorithm

[6] Let C be a named concept which is unsatisfiable in an ontology $\mathcal{O}$. A set $\mathcal{M} \subseteq \mathcal{O}$ is a *minimal unsatisfiability-preserving subset (MUPS)* [16] of $\mathcal{O}$ if C is unsatisfiable in $\mathcal{M}$, and C is satisfiable in every subset $\mathcal{M}' \subset \mathcal{M}$.

[7] Given an inconsistent ontology $\mathcal{O}$, a set $\mathcal{M} \subseteq \mathcal{O}$ is a *minimal inconsistent subset (MIS)* w.r.t. $\mathcal{O}$ if $\mathcal{M}$ is inconsistent and every subset $\mathcal{M}' \subset \mathcal{M}$ is consistent.

[8] Given a set $S = \{\mathcal{M}_1, ..., \mathcal{M}_n\}$ of MUPS of an ontology $\mathcal{O}$ for an unsatisfiable concept, a *hitting set* T for S is a subset of $\mathcal{O}$ such that $\mathcal{M}_i \cap T \neq \emptyset$ for all $1 \leq i \leq n$.

considers inconsistent ontologies to be based on a four-valued semantics. Such ontologies are translated to new consistent ontologies, in a way such that classical two-valued reasoning tasks with the translated ontologies result in the same conclusions as paraconsistent four-valued reasoning. The user can decide whether he wants to translate all inconsistent ontologies automatically or select the ontologies, that should be translated. Further information on the theory can be found in [10].

3 Related Work

To our best of knowledge, the most closely related tools are Swoop, Protégé, PION and DION.

Swoop [9] provides a user interface for computing explanations of an unsatisfiable concept and for repairing an ontology using several kinds of ranking methods. Our plug-in differs from theirs in the following main aspects. First of all, Swoop only considers a single ontology while RaDON can deal with not only single ontologies but ontology networks. Secondly, RaDON can repair both incoherence and inconsistency. Thirdly, the internal repair algorithm by using confidence values is different (more details can be found in [13]). Finally, we integrated our plug-in into the NeOn toolkit which provides a flexible framework to deal with ontology networks.

As for Protégé [11] (here, we use Protégé 4.0 alpha), it is able to compute explanations and provide a manual way to repair an ontology. The differences mentioned above for Swoop can apply to Protégé as well. Furthermore, no functionality to repair an ontology automatically is implemented in Protégé.

PION [6] and DION [17] have been developed in the SEKT project[9] to deal with inconsistency. PION is an inconsistency-tolerant reasoner which can return meaningful answers when querying an inconsistent ontology. It tries to answer a query by selecting a consistent sub-ontology from the inconsistent ontology. DION provides debugging support to compute MUPS or MIPS. But DION does not consider repairing inconsistent or incoherent ontologies. Furthermore, DION it cannot be applied to deal with very expressive ontologies, such as OWL DL ontologies.

4 Conclusion and Future Work

In our demonstration, we present the RaDON system, which provides support to diagnose and repair ontologies automatically or manually. In particular, RaDON supports novel strategies to handle incoherence and inconsistency for ontology networks. To repair an ontology network in which the ontologies are related via mappings, we focus on the repair of mappings between ontologies, assuming the individual ontologies are locally coherent and consistent and are more important and reliable than the mapping. In the demonstration, the process to debug and repair ontology networks will be shown with practical examples. Besides, for various needs of the users different algorithms to deal with incoherence and inconsistency will be demonstrated.

As future work, we intend to support ontology networks in which ontologies are networked in different way, e.g. via dependency and extension relationships in modular ontologies.

[9] http://www.sekt-project.com/project

References

1. Christian Meilicke, A.T., Stuckenschmidt, H.: Repairing ontology mappings. In: Proceeding of 22nd National Conference on Artificial Intelligence (AAAI), pp. 1408–1413 (2007)
2. Haase, P., Brockmans, S., Palma, R., Euzenat, J., d'Aquin, M.: D1.1.2 updated version of the networked ontology model. NeOn Project Deliverable D1.1.2, Universität Karlsruhe, TH (2007)
3. Haase, P., Qi, G.: An analysis of approaches to resolving inconsistencies in DL-based ontologies. In: Proceeding of International Workshop on Ontology Dynamics (IWOD) (2007)
4. Haase, P., van Harmelen, F., Huang, Z., Stuckenschmidt, H., Sure, Y.: A framework for handling inconsistency in changing ontologies. In: Gil, Y., Motta, E., Benjamins, V.R., Musen, M.A. (eds.) ISWC 2005. LNCS, vol. 3729, pp. 353–367. Springer, Heidelberg (2005)
5. Huang, Z., van Harmelen, F., ten Teije, A.: Reasoning with inconsistent ontologies. In: Proceeding of Nineteenth International Joint Conference on Artificial Intelligence (IJCAI), pp. 254–259 (2005)
6. Huang, Z., van Harmelen, F., ten Teije, A., Groot, P., Visser, C.: D3.4.1 reasoning with inconsistent ontologies: a general framework. SEKT Project Deliverable D3.4.1, Vrije Universiteit Amsterdam (2004)
7. Ji, Q., Qi, G., Haase, P.: A relevance-based algorithm for finding justifications of DL entailments. Technical report, University of Karlsruhe (2008), `http://www.aifb.uni-karlsruhe.de/WBS/gqi/papers/RelAlg.pdf`
8. Kalyanpur, A., Parsia, B., Horridge, M., Sirin, E.: Finding all justifications of OWL DL entailments. In: Aberer, K., Choi, K.-S., Noy, N., Allemang, D., Lee, K.-I., Nixon, L., Golbeck, J., Mika, P., Maynard, D., Mizoguchi, R., Schreiber, G., Cudré-Mauroux, P. (eds.) ASWC 2007 and ISWC 2007. LNCS, vol. 4825, pp. 267–280. Springer, Heidelberg (2007)
9. Kalyanpur, A., Parsia, B., Sirin, E., Grau, B.C., Hendler, J.A.: Swoop: A web ontology editing browser. Journal of Web Semantics 4(2), 144–153 (2006)
10. Ma, Y., Hitzler, P., Lin, Z.: Algorithms for paraconsistent reasoning with OWL. In: Franconi, E., Kifer, M., May, W. (eds.) ESWC 2007. LNCS, vol. 4519, pp. 399–413. Springer, Heidelberg (2007)
11. Noy, N.F., Sintek, M., Decker, S., Crubézy, M., Fergerson, R.W., Musen, M.A.: Creating semantic web contents with protégé-2000. IEEE Intelligent Systems 16(2), 60–71 (2001)
12. Parsia, B., Sirin, E., Kalyanpur, A.: Debugging OWL ontologies. In: Proceeding of 14th International World Wide Web Conference (WWW), pp. 633–640 (2005)
13. Qi, G., Haase, P., Ji, Q., Völker, J.: D1.2.2 consistency models for networked ontologies evaluation. NeOn Project Deliverable D1.2.2, Universität Karlsruhe (2007)
14. Reiter, R.: A theory of diagnosis from first principles. Artificial Intelligence 32(1), 57–95 (1987)
15. Schlobach, S.: Diagnosing terminologies. In: Proceeding of 20th National Conference on Artificial Intelligence (AAAI), pp. 670–675 (2005)
16. Schlobach, S., Cornet, R.: Non-standard reasoning services for the debugging of description logic terminologies. In: Proceeding of 18th International Joint Conferences on Artificial Intelligence (IJCAI), pp. 355–362 (2003)
17. Schlobach, S., Huang, Z.: D3.6.1 inconsistent ontology diagnosis: Framework and prototype. SEKT Project Deliverable D3.6.1, Vrije Universiteit Amsterdam (2005)

Supporting the Reuse of Global Unique Identifiers for Individuals in OWL/RDF Knowledge Bases*

Xin Liu[1], Heiko Stoermer[1], Paolo Bouquet[1], and Shu Wang[2]

[1] University of Trento
Department of Information Science and Engineering,
Trento, Italy
{liu,stoermer,bouquet}@disi.unitn.it
[2] Jilin University
Department of Computer Science and Technology,
Chang Chun, Ji Lin, China
wangshu83@gmail.com

Abstract. In the current state of Semantic Web knowledge representation, it is fairly common that individuals in ontologies have different identifiers even if they are meant to denote the same object in the real world. This makes data integration at the level of individuals quite difficult. In this demonstration, we show two plug-ins called Okkam4N and Okkam4P developed for the NeOn Toolkit and Protégé software respectively, which support the reuse of globally unique identifiers for individuals in OWL/RDF knowledge bases. The key idea is that users are given a chance to look for an existing URI in the publicly available Entity Name System for the corresponding individual in the local OWL/RDF knowledge bases. They are available and tested for the latest official release versions, the NeOn Toolkit 1.2 and Protégé 3.4.

1 Introduction

One of the major problems that have emerged in the Semantic Web(SW) effort is the problem of uniquely identifying entities [1], or "individuals" on OWL terms. Entities play a major role for the SW since they represent atomic objects of reference and reasoning. Nevertheless, we currently face the problem of uniquely identifying and referencing these entities, which prohibits us from making the next step towards the goal of the SW, that of reasoning about entities in a globally distributed knowledge base. The problem derives from the fact that different users, or systems, assign and use different identifiers for the same real-world entity. As a result, we cannot effectively reason about this entity, exactly because it is not consistently being assigned the same identifier across distributed data sources.

* This work is partially supported by the FP7 EU Large-scale Integrating Project **OKKAM – Enabling a Web of Entities** (contract no. ICT-215032). For more details, visit http://www.okkam.org.

L. Aroyo et al. (Eds.): ESWC 2009, LNCS 5554, pp. 868–872, 2009.

If we could assume that individuals in different ontologies which represent the same actual resources have the same URI, this kind of heterogeneous data problem would be far less significant. In this way, recognizing the same resource distributed in heterogeneous data sources would be only based on simply comparing their identifiers rather than depending on their data structures or content similarities for approximating identity. However, in the current infrastructure of the SW, there is no component to support content creators to reuse already existing URIs of individuals which have already been referred to in other applications or repositories.

The Okkam project[2] is dedicated to solving the issue of identifiers crisis in the SW. It proposes a web-scale infrastructure called the Entity Name System (*ENS*)[2] to support the reuse of existing URIs for any type of individuals (person, location, organization, event, etc) and provides a few APIs to applications to exchange information with the *ENS*. The principal mechanism of the *ENS* is to match the request provided by users or knowledge bases and the entity properties stored in the *ENS* repository in order to return a list of candidate individuals. This can enable knowledge creation applications to reuse globally unique identifiers for entities.

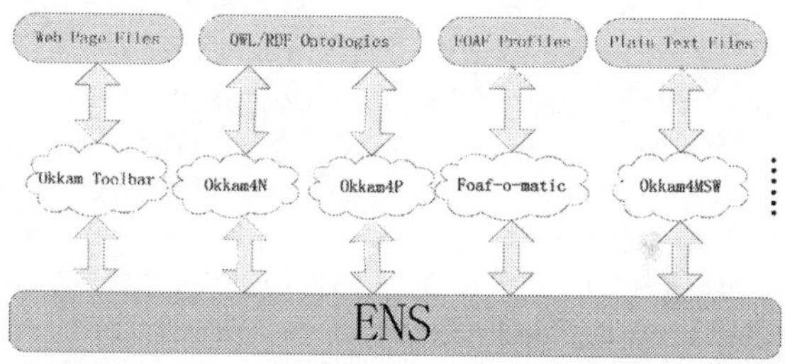

Fig. 1. Currently available *ENS*-Empowered Tools

But there are various types of content in the SW, html files, RDF documents, relation database management systems, plain text documents and so on. So different *ENS*-empowered tools[1] are required to aim at different types of knowledge bases. So far, a few client applications have already been developed to deal with entities in some given types of knowledge bases. E.g. *Foaf-o-matic*[3] is for FOAF profiles; *Okkam toolbar* is for website contents; *Okkam4MSW* is for Microsoft word; the two plug-ins *Okkam4P*[4] and *Okkam4N* we present here are for OWL/RDF ontologies. As illustrated in Fig.1.

In this demonstration, we show two plug-ins named Okkam4N and Okkam4P for the NeOn Toolkit[2] and Protégé software[3] respectively, as well as their primary

[1] *ENS*-empowered tools are standard end-user applications extended with functionalities which facilitate the creation of okkamized content through the use of *ENS*. All the currently available tools can be found at `http://www.okkam.org/test-tubes`

[2] `http://www.neon-toolkit.org`

[3] `http://protege.stanford.edu`

principles and benefits. With the two plug-ins, ontology editors can create ontologies which feature globally unique, shared identifiers for individuals through the use of the *ENS*. In this way, different knowledge bases can be easily aligned at the level of individuals.

2 Okkam4N and Okkam4P

2.1 Primary Principles

The Okkam4N uses the Eclipse[4] plug-in infrastructure to extend the NeOn Toolkit and the Okkam4P uses the plug-in framework developed at Stanford[5] to extend the functionality for Protégé-OWL editor. Both of them make use of an extension point which allows a plug-in to appear in the right-click popup menu of a selected individual and can be effortlessly hooked into the application architectures to implement the function. The information exchange with the *ENS* is achieved by invoking the web services provided by the Okkam project.

2.2 Benefits

In the files generated by the NeOn Toolkit or Protégé, the name of an individual relies on the automatic mechanism of the applictions or the input of human beings. These identifiers are difficult to be understood by other knowledge bases or applications because their scope is usually local. But after being enriched by *ENS*-empowered tools, the identifiers of various nodes which present the same real-world resource can be assigned to the same globally unique identifier stored in the *ENS*. So various knowledge bases can be linked together through the nodes which have the same identifier automatically by applications, consequently more useful information will be retrieved from the integrated one.

3 Demonstration

We will demonstrate the current versions of Okkam4N and Okkam4P. In order to show the plug-ins more clearly, we present the sequence diagram (Fig.2) to show the process of reusing a globally unique identifier for an individual.

We take the NeOn Toolkit with the Okkam4N plugin and an individual "Barack Obama" with some properties (firstname, lastname, nationality, birthdate, etc) as an example to show how to reuse a pre-existing identifier in local knowledge bases. First of all, users should create and edit the property values of the individual in local applications. Then, in order to reuse a pre-existing identifier for this individual, a request, which consists of the individual type and individual characterism description, will be sent to the *ENS*. The description is a collection of datatype property name-value pairs and used to distinguish one individual from all the others. Users can get each property from a new popup

[4] `http://www.eclipse.org`

[5] `http://www.stanford.edu/`

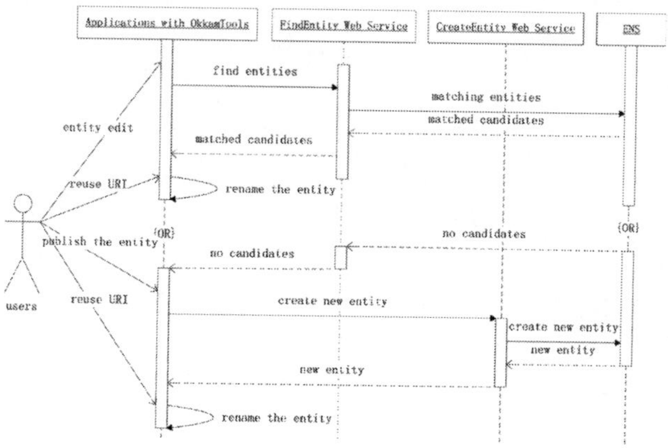

Fig. 2. Sequence diagram of reusing a pre-existing URI

dialog (Fig.3a) and assist to generate the request. After looking up the similar individuals, the *ENS* will return users a ranked list of matched candidates (Fig.3b), among which users can select one correct individual and reuse its identifier, thus causing the substitution of the locally generated identifier with the pre-existing one in the XML code of the ontology; otherwise if no matching individual is found or no matched candidates returned from the *ENS*, users can choose to publish this individual with its properties to the *ENS*, which will create a new globally unique identifier for it so that other applications or users could reuse it. In this case, the newly created identifier will be used in the XML code.

In this way, various ontology files mentioning "Barack Obama" are interlinked relying on the nodes which share the same identifier, thus generating a semantic environment of this individual (as showed in Fig.4).

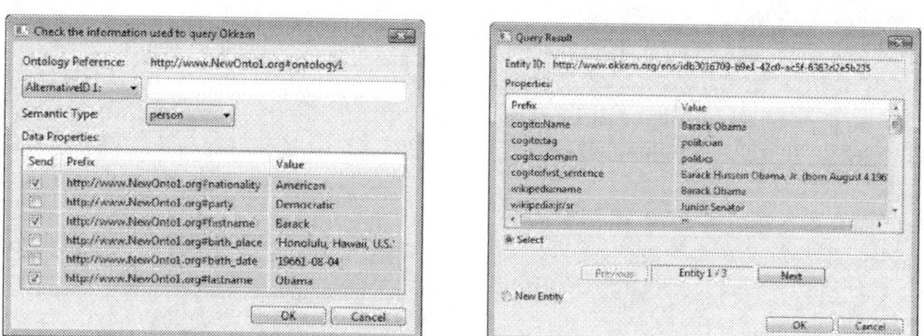

(a) The information of the specific individual

(b) The matched candidate list

Fig. 3. The screensnap of Okkam4N

Fig. 4. The semantic environment of Barack Obama

4 Conclusion

In this demonstration, we show two plug-ins named Okkam4N and Okkam4P, which can reuse the shared, globally unique identifiers for individuals issued by the *ENS* in the OWL/RDF knowledge bases created by the NeOn Toolkit or Protégé software.

With the effort of Okkam4N, Okkam4P and other *ENS*-empowered tools, more useful information would be retrieved from various types of knowledge bases. The diffusion of the use of *ENS*-empowered tools can help to address the difficulties during the process of data integration.

References

[1] Bouquet, P., Stoermer, H., Giacomuzzi, D.: OKKAM: Enabling a Web of Entities. In: i3: Identity, Identifiers, Identification. Proceedings of the WWW 2007 Workshop on Entity-Centric Approaches to Information and Knowledge Management on the Web, CEUR Workshop Proceedings, Banff, Canada, May 8 (2007) ISSN 1613-0073, `http://CEUR-WS.org/Vol-249/submission_150.pdf`

[2] Bouquet, P., Stoermer, H., Niederee, C., Mana, A.: Entity Name System: The Backbone of an Open and Scalable Web of Data. In: Proceedings of the IEEE International Conference on Semantic Computing, ICSC 2008, number CSS-ICSC 2008-4-28-25, pp. 554–561. IEEE Computer Society, Los Alamitos (2008)

[3] Bortoli, S., Stoermer, H., Bouquet, P.: Foaf-O-Matic - Solving the Identity Problem in the FOAF Network. In: Proceedings of the Fourth Italian Semantic Web Workshop (SWAP 2007), Bari, Italy, December 18-20 (2007), `http://CEUR-WS.org/Vol-314/43.pdf`

[4] Bouquet, P., Stoermer, H., Xin, L.: Okkam4P - A Protégé Plugin for Supporting the Re-use of Globally Unique Identifiers for Individuals in OWL/RDF Knowledge Bases. In: Proceedings of the Fourth Italian Semantic Web Workshop (SWAP 2007), Bari, Italy, December 18-20 (2007), `http://CEUR-WS.org/Vol-314/41.pdf`

Modeling and Enforcement of Business Policies on Process Models with Maestro

Ivan Markovic[1], Sukesh Jain[1], Mahmoud El-Gayyar[2], Armin B. Cremers[2], and Nenad Stojanovic[3]

[1] SAP Research, Karlsruhe, Germany
{ivan.markovic,su.jain}@sap.com
[2] Department of Computer Science III, University of Bonn, Germany
elgayyar@cs.uni-bonn.de, abc@iai.uni-bonn.de
[3] FZI, Karlsruhe, Germany
nenad.stojanovic@fzi.de

Abstract. Business policies and rules govern and guide the business processes of an organization. Enterprises usually only document their business policies and rules in natural language. This makes the procedure of determining which business policies and rules apply to a certain process and their enforcement on this process very costly and cumbersome. We present a tool that supports formal specification of policies and rules and their automated enforcement on process models. We explain the research background underlying the tool, and we overview what will be demonstrated at ESWC.

1 Introduction

Directives within an organization exist to govern and guide its business processes and they are categorized as business policies and business rules. According to OMG's Business Motivation Model (BMM) [1], a business policy is a high level directive that exists to control, guide and shape how an enterprise realizes its courses of action. Business policies define what is allowed/not allowed to be done, and may indicate how or pose constraints on how it should be done. Business rules on the other hand are derived from and used as a mechanism for implementing business policies.

Usually enterprises either do not capture directives formally or just document them in natural language. This makes the procedure of determining which directives apply to a certain process and their enforcement on this process very costly and cumbersome. In order to support explicit specification and automated enforcement of business policies and rules on business processes, we have devised an approach supported by semantic technologies and the process-oriented enterprise ontology framework [2]. In our approach we semantically model business policy and business rules which implement a policy. The approach is based on the notion of anti-patterns for process models [3], i.e. explicitly capturing the violating scenario and looking for its occurence in a process model, rather than checking for the correctness of a process model. Thus, we explicitly capture violation of

L. Aroyo et al. (Eds.): ESWC 2009, LNCS 5554, pp. 873–877, 2009.

a business policy in terms of formalized business rules which implement this policy and enforce them as constraints on process models. The approach is prototypically implemented in the form of extensions to *Maestro*, an SAP Research modeling tool. We explain our approach, tool support and various technologies used to realize this tool using the *Separation of Duty* (SoD) policy [4] as an example scenario for its wide application as a security mechanism in the financial management domain.

There have been several attempts to support specification of business rules and their enforcement. ARIS Design platform offers predefined "ARIS Semantic Checks" to ensure modeling standards are adhered by business process models [5]. For programming additional checks, the user must be familiar with the underlying database schema in which models are stored and also with the scripting language. This results in logic of checks being hard-coded and scattered. In [6], the rules are modeled as simple IF/THEN textual rules and lack visualization. Moreover, the work does not show how complex nature of processes, i.e. abstraction levels and other enterprise models, are handled.

2 Business Policy and Rule Specification

In order to support the visual modeling of business policies and rules, we have developed a new modeling perspective within *Maestro*, named *Maestro4BPR* (see Fig. 1).

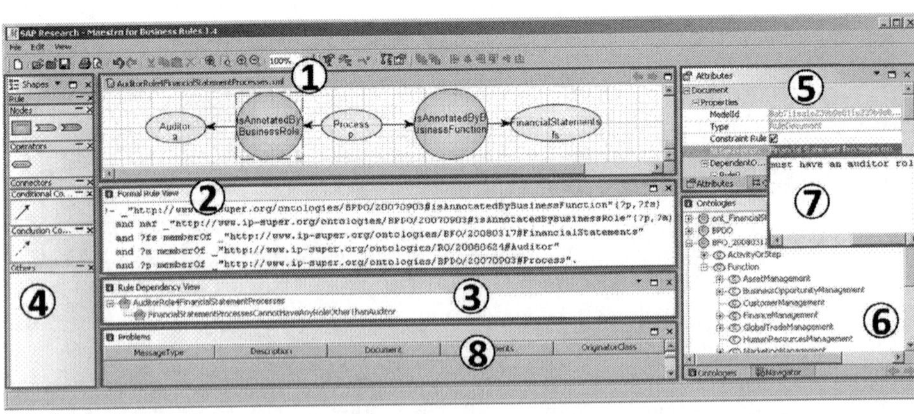

Fig. 1. Visual modeling of business rules using Maestro4BPR

The user interface of Maestro4BPR is built using the *Tensegrity Graph Framework*[1] which provides features for introducing new visual elements, adding new views, auto-layouting algorithms etc. The tool allows users to build business rules by dragging and dropping concepts, instances, relations and attributes

[1] http://www.tensegrity-software.com/

defined in any application ontology[2]. The *Ontologies* view (see no. 6 in Fig. 1) provides functionality of loading the required application ontologies for building a business rule. The ontologies are built using Flight variant of the *Web Service Modeling Language* (WSML) [7]. *WSMO4J*[3] provides the necessary Java API for ontologies described in WSML. The user does not need to have any prior knowledge of WSML language, as the rules are built by dragging elements defined in various application ontologies and additional elements provided in the *Shapes* view (see no. 4 in Fig. 1) to the *Modeling* view (see no. 1 in Fig. 1).

The business rule in Fig. 1 represents one of the violation scenarios of SoD policy. The oval and circle shapes (see no. 1 in Fig. 1) indicate respectively the concepts and relations between concepts defined in the application ontologies. The dotted square (see no. 1 in Fig. 1) around relation (circle) represents the *negation as failure* (naf) construct [7]. Business rule in Fig. 1 indicates that a violation occurs when any *process* '*p*' that is annotated with any instance of *FinancialStatements* business function '*fs*' and not (naf) annotated with any instance of *Auditor* role '*a*'. WSML Flight works with variables (*p*, *fs*, *a*) which are then checked with concrete instances for violation during enforcement phase. The concepts *FinancialStatements*, *Auditor and Process* are defined in the functional, organizational and process perspective in [2], respectively. In addition, users can specify natural language description (see no. 7 in Fig. 1) and other properties of the modeled rule in the *Attributes* view (see no. 5 in Fig. 1).

Visually modeled business rule is automatically transformed to WSML Flight logical expressions in the background (see no. 2 in Fig. 1), thus hiding the complexity of the formal language from the user. The tool provides possibility for users to specify other related business rules which need to be enforced along with the currently modeled one (see no. 3 in Fig. 1) in the *Rule Dependency* view. The related business rule in Fig. 1 (see no. 3) depicts violation scenario when *FinancialStatements* processes have other roles in addition to the *Auditor* role. It complements the first business rule in that it prevents the users from specifying other additional roles to *FinancialStatements* processes. This helps the users to organize related business rules. In addition, *Maestro4BPR* assists users by providing them feedback while modeling a business rule through the *Problems* view (see no. 8 in Fig. 1).

When modeling a business policy, the tool allows to model its context (applicability criteria) by dragging and dropping instances defined in application ontologies, e.g. assigning an organizational unit instance would mean that the modeled business policy is applicable to business processes involving this organizational unit. In addition, the user can define independent business rules and other business policies that together implement the currently modeled business policy. For the SoD scenario, in addition to the above mentioned business rule for *FinancialStatements* processes, similar business rules are required for *Accounting* processes to ensure that only the *Accountant* role is allowed to perform

[2] Application ontologies are domain-specific ontologies built using the process-oriented enterprise ontology framework [2].

[3] http://wsmo4j.sourceforge.net/

them and a business rule ensuring that no users defined in the organizational structure model have both roles i.e. Accountant and Auditor. These independent business rules together implement SoD business policy. Finally, business policies and rules are formalized as instances of *Business Policy & Rule Ontology* [2].

3 Policy Recommendation and Enforcement

Not all business policies are applicable for each business process, hence the user needs to be assisted in finding the relevant ones for a given process model. *Policy Recommender* application (see no. 1 in Fig. 2) addresses this issue using matchmaking technique to provide a ranked list of relevant policies for the currently modeled process. This application is an extension of the *Maestro4BPMN* modeling view used for designing processes in the BPMN[4] notation (see Fig. 2).

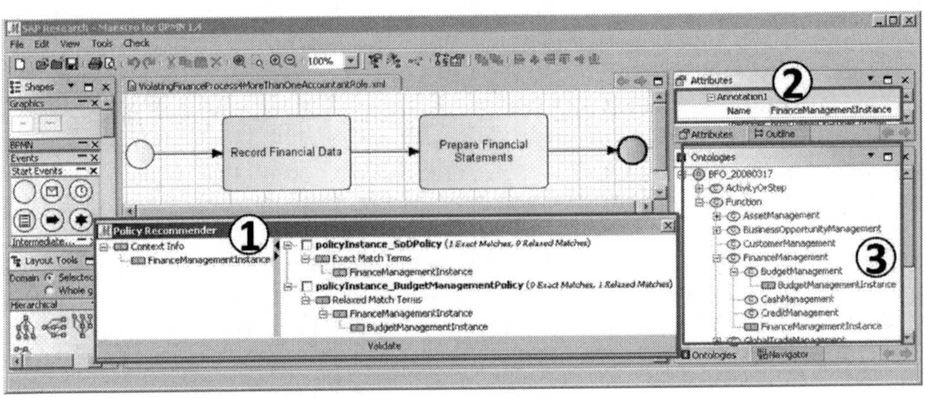

Fig. 2. Policy recommender application based on matchmaker

Matchmaker uses the context information of the currently modeled process (see no. 2 in Fig. 2), which is captured by annotating it with instances defined in the application ontologies. Each instance in the process context is looked for corresponding matching instances in the policy context. Ranking is done based on a number of instances in the process context with a corresponding matching instance in the policy context and the type of match. We distinguish two types of match, namely *Exact* and *Relax*. Exact match occurs if the same instance is used in process and policy contexts. Relaxed match occurs if an instance in policy context is either from the same concept, super concept, subconcept, or any ancestor concept in the concept hierarchy of an instance in process context (see no. 3 in Fig. 2 for concept hierarchy). Exact match is ranked higher compared to relaxed (see no. 1 in Fig. 2). Query mechanism provided by underlying formalization is used to perform the matchmaking.

[4] www.bpmn.org

Users can select from a list of business policies (see no. 1 in Fig. 2) for enforcement. The enforcement of all specified business policies against a process model is performed by considering them as knowledge base and checking for consistency of the knowledge base via logical reasoning using the *Integrated Rule Inference System* (IRIS) [8] inference engine. WSML2Reasoner[5] is used to transform ontologies in WSML to the underlying IRIS language. On violations, users are provided with natural language description of violated business rules in the *Problems* view (see no. 8 in Fig. 1).

4 Demo Scope

During the demonstration, we will first show how a user can assign the context information to an SoD policy. In the next step, the user will switch the perspective to *Maestro4BPR* in order to model the business rules which implement the policy. After each of the 3 implementing business rules are visually modeled, we show their translation to the underlying formal representation. Finally, for a given process model we search the relevant policies based on their context annotations, select the matched SoD policy and enforce it on the process.

5 Conclusion

We have presented extensions to *Maestro* which demonstrate tool support for modeling and enforcement of business policies and rules on process models. This enables the automation in ensuring compliance of organization's business processes with the established business policies and rules. We illutrate the tool support using an SoD business policy scenario.

References

1. `http://www.omg.org/spec/BMM/1.0/PDF` (2008)
2. Markovic, I., Hasibether, F., Jain, S., Stojanovic, N.: Process-oriented semantic business modeling. In: Wirtschaftsinformatik (2009)
3. Koehler, J., Vanhatalo, J.: Process anti-patterns: How to avoid the common traps of business process modeling. IBM WebSphere Dev. Tech. Journal 10(2), 4 (2007)
4. Congress of the United States: Public Company Accounting Reform and Investor Protection Act (Sarbanes-Oxley Act). Pub. l. no. 107-204, 116 stat. 745 edn. (2002)
5. Davis, R., Brabänder, E.: ARIS Design Platform. Springer, Heidelberg (2007)
6. Happel, H.J., Stojanovic, L.: Ontoprocess - a prototype for semantic business process verification using SWRL rules. In: Demosession of the 3rd (ESWC 2006) (June 2006)
7. de Bruijn, J., Lausen, H., Polleres, A., Fensel, D.: The Web Service Modeling Language WSML: An Overview. In: Sure, Y., Domingue, J. (eds.) ESWC 2006. LNCS, vol. 4011, pp. 590–604. Springer, Heidelberg (2006)
8. Bishop, B., Fischer, F.: IRIS — Integrated Rule Inference System. In: ARea Workshop (2008)

[5] http://tools.sti-innsbruck.at/wsml2reasoner/

A Reasoning-Based Support Tool for Ontology Mapping Evaluation

Christian Meilicke[1], Heiner Stuckenschmidt[1], and Ondřej Šváb-Zamazal[2]

[1] University of Mannheim
{christian,heiner}@informatik.uni-mannheim.de
[2] University of Economics, Prague
ondrej.zamazal@vse.cz

Abstract. In this paper we describe a web-based tool that supports the human in revising ontology alignments. Our tool uses logical reasoning as a basis for detecting conflicts in mappings and exploits these conflicts to propagate user decision. The proposed approach reduces the effort of the human expert and points to logical problems that are hard to find without support.

1 Motivation

The alignment of ontologies is a common problem on the semantic web as many tasks such as information integration or semantic search rely on integrated background knowledge. As manual ontology alignment is a difficult and time-consuming process, a variety of algorithms and systems for computing matches between elements of different ontologies have been developed [1]. Almost all of these methods rely on linguistic or structural heuristics for deciding whether to regard elements from different ontologies as equivalent or not. These heuristics are bound to fail in many situations resulting in erroneous mappings that need to be corrected manually. As it turns out, this manual correction is a very difficult task that requires a good understanding of and sufficient information about the ontologies to be aligned. Motivated by this experience gained from five years of ontology alignment benchmarking carried out in the context of the Ontology Alignment Evaluation Initiative (OAEI) [2] we developed a web-based tool that supports the human user when evaluating a mapping. The distinguishing feature of this tool is the use of logical reasoning as a basis for detecting conflicts in mappings and implications of evaluation decisions. As shown in previous work [3] this approach can significantly reduce the effort of a manual evaluation and as argued in [4] also provides the basis for improving the correctness of mappings. In the following we first briefly discuss the kind of reasoning performed by our system. We then describe our system and present a typical usage example. We close with a brief description of the systems used in the context of the OAEI and some discussion of future work.[1]

[1] A demo version applied to the use case of Section 4 can be found at http://webrum.uni-mannheim.de/math/lski/ontdebug/mapdebug/debugpages/

L. Aroyo et al. (Eds.): ESWC 2009, LNCS 5554, pp. 878–882, 2009.

2 Reasoning-Based Mapping Verification

Similar to Euzenat and Shvaiko [1] we define a mapping as a set of correspondences. A correspondence is a 4-tuple $\langle i\#E, j\#E', R, n\rangle$ where $i\#E$ and $j\#E'$ are concepts or properties of the matched ontologies $\mathcal{O}_i$ and $\mathcal{O}_j$, $R \in \{\sqsubseteq, \equiv, \sqsupseteq\}$ is a semantic relation, and n is a confidence value. In [3] we distinguished between two reasoning tasks useful for supporting manual mapping revision. On the one hand it can be decided whether a correspondence can be derived from a mapping, on the other hand it can be decided whether a mapping is incoherent.

We focused on the second aspect using a sound but complete reasoning technique that is based on two kinds of propagation patterns to detect incoherence conflicts between pairs of correspondences. We refer to these pattern as *subsumption propagation pattern* and *disjointness propagation pattern* both depicted in Figure 1. The subsumption propagation pattern works as follows. Suppose that a mapping $\mathcal{M}$ contains correspondences $\langle i\#A, j\#B, \sqsupseteq, n\rangle$ and $\langle i\#C, j\#D, \sqsubseteq, n'\rangle$, represented as arrows. Further suppose that $\mathcal{O}_i \models i\#A \sqsubseteq i\#C$. Due to the $\mathcal{M}$ it can be derived that $\mathcal{O}_i \cup \mathcal{M} \cup \mathcal{O}_j \models j\#B \sqsubseteq j\#D$ and also $\mathcal{O}_i \cup \mathcal{M} \cup \mathcal{O}_j \models j\#E \sqsubseteq j\#D$ for each subconcept $j\#E$ of $j\#B$. Suppose now that $j\#E$ and $j\#D$ are defined to be disjoint in $\mathcal{O}_j$. From this we can conclude that $j\#E$ has become an unsatisfiable concept due to the additional terminological axioms introduced by $\mathcal{M}$. Thus $\mathcal{M}$, respectively its subset $\{\langle i\#A, j\#B, \sqsupseteq, n\rangle, \langle i\#C, j\#D, \sqsubseteq, n'\rangle\}$ is incoherent. We have to remove at least one element to resolve the incoherence.

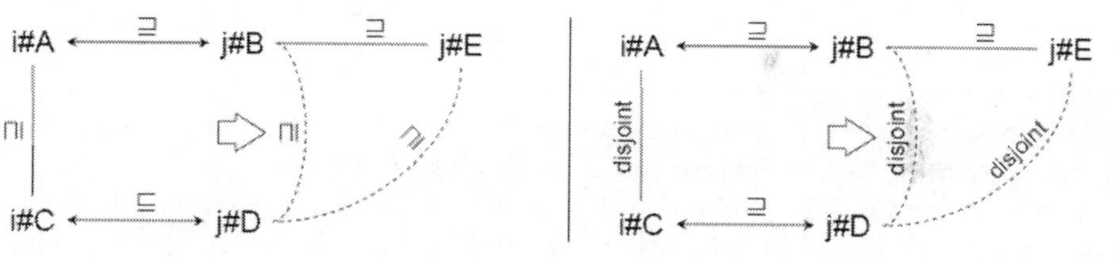

Fig. 1. Subsumption (on the left) and disjointness propagation pattern (on the right)

The disjointness propagation can be understood as the inverse of the subsumption propagation, thus we omit a detailed description. Once we classified both ontologies, the presence of these patterns has to be checked in both direction by a sequence of lookups for all possible pairs of correspondences. This process can be seen as a diagnosis of the incoherence. While we presented both patterns for subsumption correspondences, our approach is in the same way applicable for equivalence correspondences since such a correspondence can be decomposed into two subsumption correspondences. We apply the same approach to correspondences between properties in a similar way by exploiting their domain and range restrictions.

3 A Web Based Tool for Mapping Verification

When a mapping is loaded, the system first performs a diagnosis step according to the approach described in the previous section. In this step, conflicts between pairs of correspondences are identified. The system then presents a list of correspondences together with their confidence value as provided by the matching system. Depending on the status of the evaluation, each correspondence is labeled with one of the following symbols (see Figure 2 for an example):

The correspondence has been *manually accepted*. It is assumed that this correspondence is correct and any conflicting correspondences are incorrect.

The correspondence has *not yet been evaluated and does not conflict* with any accepted or not yet evaluated correspondence.

The correspondence *conflicts* with an accepted or not yet evaluated correspondence.

The correspondence has been *dismissed automatically* as it conflicts with a manually accepted correspondence.

The correspondence has been *manually dismissed* by the user. Any correspondence previously in conflict with this correspondence changes its status if there exists no other non-dismissed conflicting correspondence.

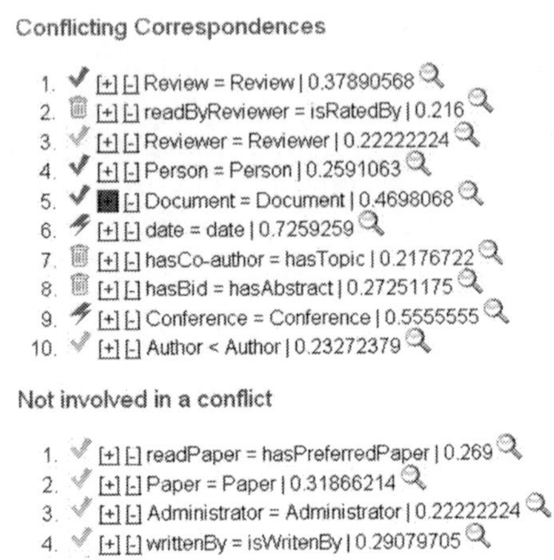

Fig. 2. User interface

In an ideal case, the evaluation is carried out until all correspondences are labeled with , or indicating that it has been confirmed directly or indirectly by the human expert to be either correct or incorrect. In the case of large mappings, this will not always be the case as the user will focus on resolving conflicts. In this case, correspondences labeled with and are assumed to be correct and correspondences labeled with , or are assumed to be incorrect. Figure 3 shows a typical example of how the evaluation tool is used. The data used in the example is the result of the DSSim matching system on the CMT and CRS ontology both of which are part of the OntoFarm dataset [5]. This matcher generates twelve correspondences eight of which are involved in a conflict. These correspondences are shown in Figure 3(a) as presented to the user at the beginning of the evaluation process.

In a first step the user might check correspondence *Review* = *review*. The tool supports his decision by showing the context of the matched elements when clicking on the lens symbol. In the case of concepts, the path to the top of the hierarchy is presented for both concepts. Figure 3(b) shows the corresponding context for the first correspondence in the set. Based on this additional information the user will decide that the correspondence is correct and mark it as

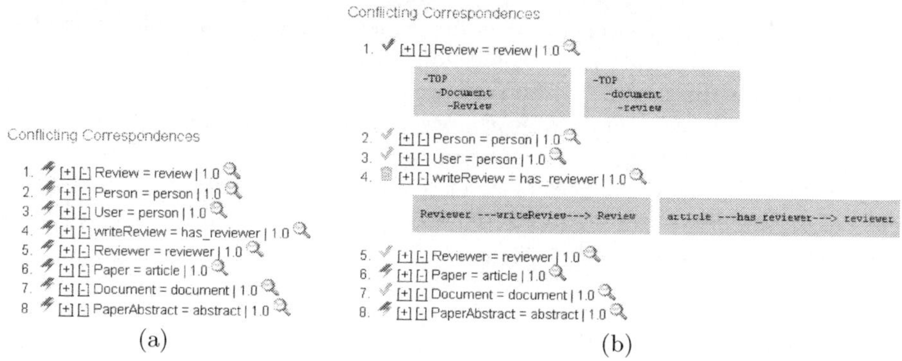

Fig. 3. Examples of an evaluation process

manually accepted by clicking on the (+) symbol changing its label from to . This single decision triggers a number of events. In particular, correspondence 4 (*writeReview* = *has_reviewer*) can now be decided to be incorrect as it conflicted with correspondence 1 and will be relabeled as . As Figure 3(b) shows assuming that correspondence 4 is incorrect resolves a number of other conflicts which leads to a relabeling of correspondences 2, 3, 5 and 7 from to . The fact that the rejection of correspondence 4 resolves many other conflicts is already an indication that the decision should be correct. In order to be sure that the correspondence is actually incorrect, the user can additionally inspect the context information provided by the system which is shown in Figure 3(b). Looking at the context clearly shows that the two relations cannot be the same as they connect completely different concepts ('Reviewer' and 'Review' in the first and 'article' and 'reviewer' in the second case). In a similar way, the evaluation can be continued until all correspondences are correctly labeled.

4 Application in the OAEI Challenge

We applied our tool in the context of the OAEI conference track. This track deals with a collection of ontologies in the domain of conference organization known as the *OntoFarm* collection [5]. Since 2006 the evaluation procedure of the conference track the submitted mappings had been manually evaluated over the last years. This procedure resulted in a large corpus of annotated correspondences which can be seen as a first step towards building reference mappings. Since an enormous number of correspondences had to be processed, it can be expected that not every correspondence has been annotated correctly. Thus, we used the annotated corpus as input to our tool, revised it, and generated precise reference mappings over five ontologies (i.e. 10 mappings).

Using our tool we could explore conflicts between correspondences and track which correspondences would be dismissed provided another correspondence is evaluated as correct. This information turned out to be very useful in the revision process as we already argued with respect to the example presented above. It would have required an extensive and time consuming inspection of the involved ontology to gain this information without support. We experienced that the best way to use the tool is to label the most certain correspondences at first. This resulted very often in automatically eliminating doubtful correspondences without the need for further investigation. Overall, we significantly reduced the input corpus of correspondences previously annotated as correct and could thus create a basis for highly precise reference mappings that will be used in future evaluations of the conference track.

5 Conclusions

There are different ontology editors that provide functionality for modeling and visualizing mappings such as Protege and OntoStudio. The support of these systems, however, is limited to some form of visualization of mapped elements and their context. Based on discussions with researchers in the area, we have identified a strong need for a more substantial support that we want to provide at least partially with our system. As the methods implemented in our system rely on logical reasoning, the system is particularly useful for evaluating mappings between richly axiomatized ontologies. However, as shown in previous work [6], it is also applicable to light-weight ontologies. Currently, the presentation of context information to the user is rather basic. In future work, we will enhance this feature. We plan to implement our tool as Protege-Plugin using Protege functionalities for visualizing the context of mapped elements.

References

1. Euzenat, J., Shvaiko, P.: Ontology Matching. Springer, Heidelberg (2007)
2. Caracciolo, C., Euzenat, J., Hollink, L., Ichise, R., Isaac, A., Malaise, V., Meilicke, C., Pane, J., Shvaiko, P., Stuckenschmidt, H., Svab-Zamazal, O., Svatek, V.: Results of the OAEI 2008. In: Proc. of the ISWC 2008 Workshop on Ontology Matching, Karlsruhe, Germany (2008)
3. Meilicke, C., Stuckenschmidt, H., Tamilin, A.: Supporting manual mapping revision using logical reasoning. In: Proc. of the 23rd AAAI Conference on Artificial Intelligence, Chicago, Illinois, US (2008)
4. Meilicke, C., Stuckenschmidt, H.: Incoherence as a basis for measuring the quality of ontology mappings. In: Proc. of the ISWC 2008 Workshop on Ontology Matching, Karlsruhe, Germany (2008)
5. Sváb, O., Svátek, V., Berka, P., Rak, D., Tomasek, P.: Ontofarm: Towards an experimental collection of parallel ontologies. In: Poster Proceedings of the International Semantic Web Conference 2005 (2005)
6. Meilicke, C., Völker, J., Stuckenschmidt, H.: Learning disjointness for debugging mappings between lightweight ontologies. In: Proc. of the 16th International Conference on Knowledge Engineering and Knowledge Management, Acitrezza, Italy (2008)

Semanta – Semantic Email in Action

Simon Scerri, Ioana Giurgiu, Brian Davis, and Siegfried Handschuh

Digitial Enterprise Research Institute,
National University of Ireland Galway, IDA Business Park, Galway, Ireland
{simon.scerri,ioana.giurgiu,brian.davis,
siegfried.handschuh}@deri.org

Abstract. Semanta is a system supporting Semantic Email, implemented as an add-in to two popular Mail User Agents, using existing email transport technology and integrated with the Social Semantic Desktop. It enables machines to support email users with correctly interpreting, handling and keeping track of action items within email messages, visualizing email workflows, and extracting tasks and appointments from email messages.

Keywords: Semantic Email, Email Workflows, Speech Act Theory.

1 Introduction

Recent surveys have shown that about 80% of knowledge workers prefer emails as a business communication tool[1]. However email has been known to present knowledge management problems for its users, who become overwhelmed with the tasks they need to perform and keep track of, eventually resulting in their failure to keep up with the ongoing email workflows whilst gaining a loss in productivity. We do not restrict email use, but support its management by providing structure and semantics to email's characteristic ad-hoc workflows. In supporting these workflows, we support the users. In our approach, the content of an email message can be summarized into a number of *Action Items* (e.g. *Meeting Request*, *Task Assignment*, *File Request* etc.). Once exchanged, every single action item can be seen as the start, or continuation of a separate workflow. We have presented the theoretical ideas behind the sMail (Semantic Email) Framework in detail in previous work [1], and demonstrated [2] how Semantic Email can perform an important role in the social fabric of the Social Semantic Desktop [3]. We will now demonstrate the usefulness of its implementation - Semanta, via a common email exchange scenario between two knowledge workers, after an insight into related work and the system's architecture.

2 Related Work

The notion of Semantic Email was originally introduced by Dowell et. al. [4] to refer to an email message consisting of a structured query coupled with corresponding

[1] http://www.highbeam.com/doc/1G1-100628406.html

L. Aroyo et al. (Eds.): ESWC 2009, LNCS 5554, pp. 883–887, 2009.

explanatory text, based on a number of Semantic Email Processes that represent commonly occurring workflows within email. The system, successfully implemented within Mangrove [5], provided templates which exposed structured knowledge about these scenarios to both humans and machines. Semanta is more flexible, since although we do provide templates, our approach is more oriented towards ad-hoc email processes, i.e. we support email workflows as they evolve, without restricting the user's actions whatsoever. We also apply Speech Act Theory [6] to provide more adequate representation of email action items. Although the same theory was applied to email beforehand – particularly to address the problems of email task management [7][8] and email classification [9][10][11]; it was never combined with semantic technologies. We represent action items by speech act instances from within the sMail Ontology[2]. Furthermore, by grounding Semanta within the Nepomuk[3] Social Semantic Desktop, semantic knowledge obtained from email is integrated with that on the semantic desktop so that other applications can re-use it.

3 Semanta

The following is a very brief overview of Semanta's architecture, which consists of four main components:-

- Text Analytics Service - provides semi-automatic annotation of email content by employing Ontology-Based Information techniques via the use of the GATE Framework [12] and the Jape Patterns Engine [13]
- Social Semantic Desktop - provides an RDF store that Semanta can read and write to, and the conceptual modelling required for Semantic Email data to be integrated in the Semantic Desktop knowledge base.
- Simplistic Graphical User Interface - abstracts theoretical models and semantics from the user beneath add-ins to popular Mail User Agents (Outlook, Thunderbird).
- Semantic Email Service – Exposes knowledge in the sMail Framework to Semanta and serves as the link between the GUI and the other components

4 Scenario

We highlight the support for ad-hoc email workflows provided by Semanta via a typical email exchange scenario[4] between two (more participants are supported) hypothetical colleagues - Claudia and Dirk. Claudia proceeds to write an email via Microsoft Outlook, her preferred email client. When she is done, Semanta attempts automatic annotation of action items in the content, and finds two action items:

1. "How is the review report going?" - an *Information Request* addressed to *Dirk*
2. "Can we discuss the review tomorrow afternoon?" - a *Meeting Request* between *Claudia* and *Dirk* addressed to *Dirk*

---

[2] http://ontologies.smile.deri.ie/smail#
[3] http://www.nepomuk.semanticdesktop.org/
[4] More demonstrations available at http://smile.deri.ie/projects/semanta/#Demonstrations

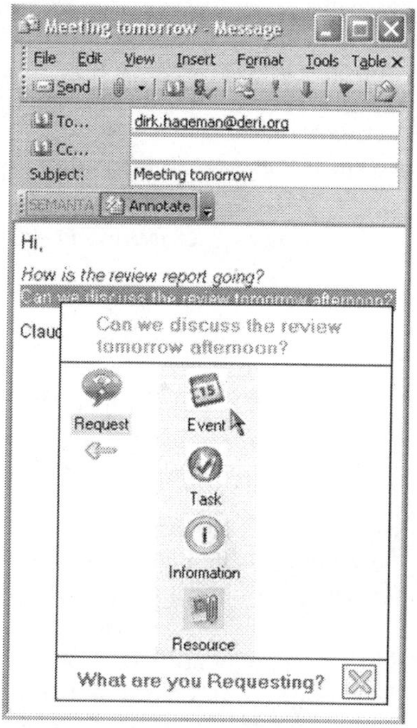

Fig. 1. Annotation Wizard

Although Claudia is happy with these annotations, she can change or create annotations via the Semanta Annotation Wizard. The latter assists the user with the intuitive, step-by-step construction or modification of semantic annotations representing action items (Fig. 1). When Claudia sends the email, metadata is invisib ly transported alongside the email content in the email headers. When Dirk receives the email, Semanta detects action items within and flags it appropriately in the inbox. After opening it, Dirk can react to the action items via a right mouse click. Depending on its type, the user is given a number of appropriate options. In Fig. 2, Dirk has reacted to the *Information Request* by providing the required information ("So far, so good"). As regards the *Meeting Request* apart from approving, ignoring, declining or amending it, Dirk can react in any other way via the 'Other..' option. In fact Dirk wants to question the purpose of the meeting. The annotation wizard is shown right away so that Dirk can choose an action (e.g. *Request Information*) and provide text for this action ("What is there to be discussed?"). This constitutes an additional action item in this ad-hoc email workflow. When Dirk closes the incoming email, its semantic representation on his desktop is updated, and a reply email is automatically sent back to Claudia.

Claudia replies to Dirk's question and although he acknowledges her reply, he forgets to reconsider her meeting request. This is a frequent email problem hindering collaborative work, but it can be addressed by Semanta's *Action Item Tracker*. Dirk uses the tracker to view all action items he received but still needs to process, via the 'Pending Incoming Item' view. One of these items represents Claudia's *Meeting Request* (Fig. 3). Action items are shown in the context of the workflow that they belong to. In the 'Context' pane on the right, Dirk can see that he reacted to this request by asking for more information regarding the event ("What is there to be discussed?"). He also sees that although Claudia replied to the latter request ("The way forward!"), he forgot to reconsider the meeting proposal. After making up his mind, he opens the containing email directly from the tracker and accepts the *Meeting Request*.

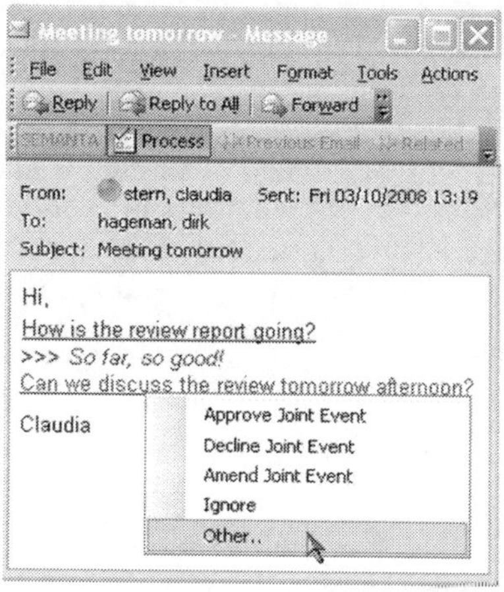

Fig. 2. Processing Incoming Email

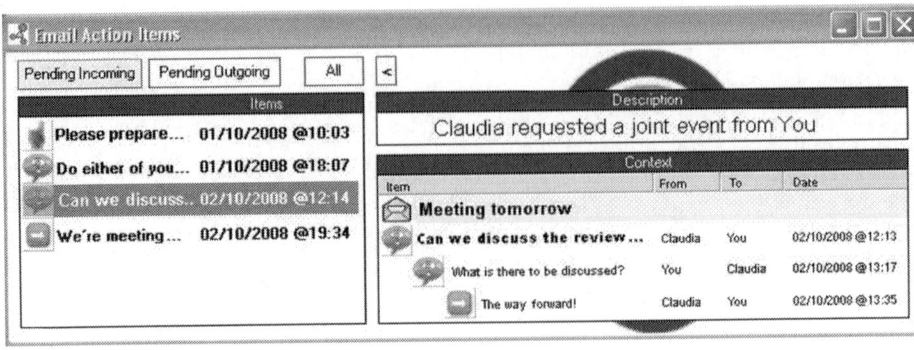

Fig. 3. Semanta's Action Item Tracker, visualizing email workflows

When Dirk accepts the meeting, Semanta detects that an event involving Dirk has been generated (Fig. 4). Dirk can either dismiss detected events/tasks, or add them to his Outlook Calendar/Tasklist. Dirk chooses to add the event to his calendar, whereby the semi-populated, native Outlook Appointment Item creation window is brought up. The subject of this appointment item is obtained from the text of the original *Meeting Request* followed by the text of Dirk's *Meeting Approval.* The meeting attendees for this appointment are known given the semantic annotation for the *Meeting Request* as shown in annotation 2 at the start of our scenario. Dirk's meeting approval also triggers an automatic email reply notifying Claudia that her meeting proposal has been approved. When Claudia reads this email, she is also assisted with the addition of this event to her calendar.

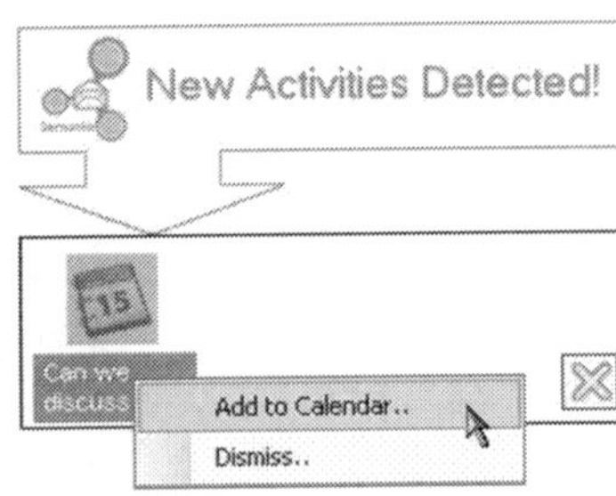

Fig. 4. Event Detection

The objects involved in this email exchange and their relationships are represented semantically using the Social Semantic Desktop ontologies[5], e.g. the emails (nmo:Email), the action items (smail:SpeechAct), the appointment item (ncal:Event), Claudia/Dirk (nco:Contact), etc. In Fig. 3 we saw how semantic relationships between action items in email workflows and their relationship to the messages used to carry out these workflows can be utilised for the user's benefit. Semanta provides three other useful functions – the 'Related Email' button directly loads the email that generated the appointment from within the appointment item window; the 'Related Activity' button points back to the generated appointment from within the email; and the 'Previous Email' button enables Dirk to traverse up the email thread.

5 Conclusion

We have demonstrated how Semanta - a fully-implemented system supporting Semantic Email, can assist the user with the execution of previously implicit ad-hoc

[5] http://www.semanticdesktop.org/ontologies/

email workflows. By exposing knowledge about email messages, contacts involved, action items within and the workflows they generate, resulting events and tasks, as well as their relationships to machines, the user can be assisted with better managing their email use. Furthermore, the integration of Semanta within the Social Semantic Desktop means that the aggregated semantic knowledge is immediately exposed to other desktop applications. The results of an initial evaluation [14] show that its support for semantic email processes is applicable to everyday email usage, and that it does not constrain the conventional ways in which people tend to use email.

Acknowledgments. The work presented in this paper was supported (in part) by the Lion project supported by Science Foundation Ireland under Grant No. SFI/02/CE1/I131 and (in part) by the European project NEPOMUK No FP6-027705.

References

1. Scerri, S., Mencke, M., Davis, B., Handschuh, S.: uating the Ontology underlying sMail – the Conceptual Framework for Semantic Email Communication. In: Proc. Language Resources and Evaluation 2008, Marrakech, Morocco (2008)
2. Scerri, S., Handschuh, S., Decker, S.: Semantic Email as a Communication Medium for the Social Semantic Desktop. In: Bechhofer, S., Hauswirth, M., Hoffmann, J., Koubarakis, M. (eds.) ESWC 2008. LNCS, vol. 5021, pp. 124–138. Springer, Heidelberg (2008)
3. Decker, S., Martin, F.: The Social Semantic Desktop. Technical Report (2004)
4. McDowell, L., Etzioni, O., Halevey, A., Levy, H.: Semantic Email. In: Proc. World Wide Web (2004)
5. McDowell, L., Etzioni, O., Gribble, S., Halevey, A., Levy, H., Pentney, W., Verma, D., Vlasseva, S.: Evolving the Semantic Web with Mangrove. UW Tech Report (2003)
6. Searle, J.: Speech Acts. Cambridge University Press, Cambridge (1969)
7. Khoussainov, R., Kushmerick, N.: Email task management: An iterative relational learning approach. In: Conference on Email and Anti-Spam, CEAS 2005 (2005)
8. Corston-Oliver, S., Ringger, E., Gamon, M., Campbell, R.: Task-focused summarization of email. In: Text Summarization Branches Out workshop, ACL 2004 (2004)
9. Khosravi, H., Wilks, Y.: Routing email automatically by purpose not topic. Natural Language Engineering 5, 237–250 (1999)
10. Goldstein, J., Sabin, R.E.: Using Speech Acts to Categorize Email and Identify Email Genres. In: System Sciences, HICSS 2006 (2006)
11. Carvalho, V., Cohen, W.: On the collective classification of email speech acts. In: SIGIR 2005, pp. 345–352 (2005)
12. Cunningham, H., Maynard, D., Bontcheva, K., Tablan, V.: GATE: A Framework and Graphical Development Environment for Robust NLP Tools and Applications. In: Proc. 40th Anniversary Meeting of the Association for Computational Linguistics (ACL 2002), Philadelphia (2002)
13. Cunningham, H., Maynard, D., Tablan, K.: JAPE: a Java Annotation Patterns Engine, 2nd edn. Technical report CS–00–10, University of Sheffield, Department of Computer Science (2000)
14. Scerri, S., Davis, B., Handschuh, S., Hauswirth, M.: Semanta – Semantic Email Technology for the Masses. In: Proc. ESWC 2009, Crete, Greece (2009)

KiWi – A Platform for Semantic Social Software (Demonstration)

Sebastian Schaffert, Julia Eder, Szaby Grünwald, Thomas Kurz, and Mihai Radulescu

Salzburg Research Forschungsgesellschaft
Jakob Haringer Str. 5/II, A-5020 Salzburg, Austria
firstname.lastname@salzburgresearch.at

Abstract. Semantic Wikis have demonstrated the power of combining Wikis with Semantic Web technology. The KiWi system goes beyond Semantic Wikis by providing a flexible and adaptable platform for building different kinds of Social Semantic Software, powered by Semantic Web technology. While the KiWi project itself is primarily focussed on the knowledge management domain, this demonstration shows how KiWi aspects like the Wiki Principles and Content Versatility can be used to build completely different kinds of Social Software applications. The first application we show is an "ordinary" Semantic Wiki system preloaded with content from a online news site. The second application called TagIT is a map-based system where locations and routes on a map can be "tagged" by users with textual descriptions, SKOS categories, and multimedia material. Both applicatons are built on top of the same KiWi platform and actually share the same content.

1 Motivation: From Semantic Wikis to KiWi

Semantic Wikis have been under investigation in the research community since 2005 (see e.g. [1,2]). Although there are many different systems with many different properties, a common trait between all Semantic Wikis is that they aim to combine ordinary Wiki content and technology with Semantic Web technologies, either in order to provide better wikis ("Semantic Web for Wikis") or to ease the creation of Semantic Web data ("Wikis for the Semantic Web") [3]. Many also see Semantic Wikis as the "Semantic Web in a Nutshell", because – like Wikis show similar traits to the Web as a whole – Semantic Wikis share many properties and also problems with the envisioned Semantic Web.

The EU-funded project KiWi ("Knowledge in a Wiki") takes the Semantic Wiki approach to the next level by providing a platform that allows to build many different kinds of Social Semantic Software, based on the conviction that most social software follows the "Wiki Principles". By "Wiki Principles", we mean that the term "Wiki" does not refer to technology alone. It is more a new philosophy of working with Web content influenced by ideas in the OpenSource community. These principles have revolutionised the way we work with content in the (Social) Web:

L. Aroyo et al. (Eds.): ESWC 2009, LNCS 5554, pp. 888–892, 2009.

– **Wikis allow anyone to edit:** The core principle is that there are no access restrictions or strict hierarchies on the content of a wiki. Anyone can easily contribute his or her own knowledge, his or her own ideas, and his or her own content.
– **Wikis are easy to use:** Anyone who is sufficiently familiar with the basic functionalities of word processing software (write, delete, save) has all the skills required to edit, correct and expand a wiki.
– **Wiki content is linkable:** By allowing users to create links between words and as such between concepts, wikis also allow for the creation of semantic relations, i.e. of meaning.
– **Wikis support versioning:** Never does information disappear on a wiki. If a page is edited, the previous version is still stored somewhere. This has an important psychological effect as it takes away the wiki writer's block: the fear that something might get lost through editing.
– **Wikis support all media:** Wikis are web-based. So whichever type of content you have, be it text, images, audio, spreadsheets, documents – anything that can be displayed in a web browser can be displayed in a wiki. And even if a file cannot be displayed in the browser itself, it can still be downloaded.

This same philosophy is underlying not only Wikis (technically-speaking), but also a large array of other Social Software applications: e.g., a weblog or social networking platform can be seen as just a different user interface (and different way of using), but is otherwise very similar concerning the underlying principles and also technology.

2 Content Versatility: Same Content, Different Views

Following this line of argumentation, it is close at hand to build generic platforms for developing different kinds of Social Software. And indeed, there are several products already available that aim to deliver a platform that allows to combine wikis, weblogs, social networking, etc. Such platforms are for example Clearspace Community[1] from Jive Software, Community Server[2] from telligent, and Liferay Social Office[3] from Liferay. However, all these systems only provide integration on the user interface level and still see wiki articles, blog posts, etc. as separate kinds of content that is visualised in a specific way.

KiWi follows a completely different approach which we call "Content Versatility". The underlying principle is that every piece of information is a combination of human-readable content and associated meta-data, and that the same piece of information can be presented to the user in many different forms, even in parallel: as a wiki page, as a blog post, as a comment to a blog, as a photo, or even in a bubble in a map-based application. The decision how the information is displayed is taken based on the context of the content and the user (e.g. metadata, user preferences, different applications).

[1] http://www.jivesoftware.com/products/clearspace-community
[2] http://communityserver.com/
[3] http://www.liferay.com/web/guest/products/social_office

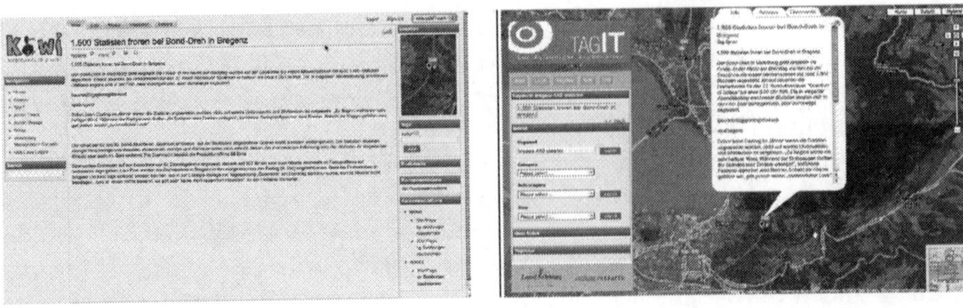

Fig. 1. KiWi Applications: the same content item in the KiWi Wiki on the left, and in TagIT on the right

In KiWi, we call the smallest piece of information a "content item". A content item is identified by a URI and consists of human-readable content in the form of XHTML and associated meta-data in the form of RDF. The KiWi core system provides means to store, update, search, and query content items, and offers automatic versioning of all updates (content and meta-data). Whereas core properties of a content item (like the content, the author, and the creation date) are persisted in a relational database, all other properties can be flexibly defined using RDF properties or relations. The URI of a content item is generated in such a way that it is possible to make a KiWi system part of the Linked Open Data cloud [4]. In the following, we present two KiWi applications that share the same content but have completely different ways of presenting it.

The KiWi Wiki. The primary and most generic interface to the KiWi system is a Semantic Wiki. The layout and functionality (Figure 1 left) of the KiWi Wiki is inspired by its predecessor IkeWiki[5]: the browser view is divided into three columns where the left column provides navigation functionality and access to tools, the center column contains the main (human-readable) wiki content, and the right column contains dynamic widgets that display additional information about the content item based on its metadata (e.g. a map or incoming and outgoing links). The center column by default displays the content of the content item and the user can switch to editing, to annotation, and to history by using tags displayed on top of the content. The KiWi Wiki interface can be used in the same way as IkeWiki, with some additional features.

TagIT. TagIT is an application originally developed in a separate project with the goal to create the "Youth Atlas of Salzburg", which has now been ported to the KiWi platform (Figure 1 on the right). In TagIT, users browse through a map (based on Google Maps) where they can "tag" locations with descriptions and provide interesting information about them, e.g. cafés, bars, sports parks, hiking tours, etc. Such tags can be associated with categories from a SKOS thesaurus and with additional multimedia data like photos or videos. Other users can then browse or search for interesting locations using the same interface. In addition to the web-based platform, TagIT also offers a mobile client that can run on a GPS-enabled mobile phone. When visiting an interesting location, users can then start the TagIT application, take a picture of the location, add a short description and directly upload the "tag" using UMTS. The tag is automatically geolocated and immediately available for other users.

Although quite different on the user interface and in the way it is used, TagIT actually closely follows the wiki principles: everyone can add and edit tags, the system is extremely easy to use, tags can be linked, tags are versioned, and different kinds of content are supported. On top of KiWi, tags are realised as content items and can thus be displayed in both the TagIT user interface and the previously described KiWi Wiki (in which case a small map widget is displayed in the right column showing the position).

The two user interfaces are currently implemented as separate applications on top of the KiWi core system. In further phases of the project, the KiWi system will also provide reasoning and personalisation support that will allow to flexibly decide which presentation of a content item to choose for a given user and context.

3 Demo Outline

The demonstration we aim to give at ESWC will first show the core features of the KiWi system, namely the use and features of the KiWi wiki and the TagIT system and their semantic properties. A particular focus will be on the demonstration of the "Content Versatility", as it is a fundamental concept of KiWi and demonstrates the power and flexibility of using Semantic Web technologies for Social Software applications.

For the purpose of demonstrating the KiWi system in a real-world scenario, we will preload the system with a selection of news articles with geoinformation from our partner, *Salzburger Nachrichten*. We will then show how such articles can be browsed, searched, and edited in the KiWi wiki and in the TagIT system, and we will demonstrate that both user interfaces are really just different views on the same content. To demonstrate the flexibility of the system, we will also show how the mobile client can be used to add new geo-located content to the system and browse the system for interesting locations. Depending on the state of the KiWi system at the time of the conference, we furthermore aim to show advanced features like personalisation, and we might be able to demonstrate the integration of additional Salzburger Nachrichten content beyond news articles.

4 Perspectives and Conclusion

KiWi is an ongoing research project of which we have only demonstrated the first results. Many more salient aspects are still to come. Particularly, KiWi will be extended with "enabling technologies" in the following areas:

- **Reasoning and Reason Maintenance:** In this area, the KiWi system will be extended with a custom, rule-based querying and reasoning component where advanced users will be able to add custom inference rules to the knowledge base; reasoning will be based on content as well as meta-data [6]. In addition, there will be a reason maintenance component that allows users to get explanations why a certain information has been inferred; reason maintenance is also beneficial for update performance and might even allow for different users having different rule sets.

- **Information Extraction:** The KiWi system will furthermore provide a component that semi-automatically extracts meta-data from the content that is stored in the knowledge base. Information Extraction will be performed in interaction with the user to minimise the number of errors.
- **Personalisation:** Based on the meta-data for a content item, a user model, and the rule-based reasoning, KiWi will also offer a personaliation component that allows to further customise the presentation of a content item. The personalisation component will further demonstrate the flexibility of the Content Versatility approach taken by KiWi.

TagIT is now further developed as part of the Salzburg NewMediaLab project "Future Content Platforms" together with our partner *Salzburger Nachrichten*. The goal is to integrate in the same interface not only wiki content and TagIT locations but also news articles, blog posts, and small advertisements from our partner. For the purpose of the demonstration, we have already imported 20.000 online news articles, but the system is designed to scale to hundreds of thousands.

Acknowledgement. This research has been partly funded by Salzburg New Media Lab and by the the European Commission within the 7th Framework Programme project KiWi - Knowledge in a Wiki (No. 211932). The latest KiWi source code is available as Open Source at the KiWi website `http://www.kiwi-project.eu`. The demonstration system is published from time to time at `http://showcase.kiwi-project.eu`

References

1. Völkel, M., Schaffert, S. (eds.): 1st Workshop "From Wiki to Semantics" (SemWiki 2006) – colocated with ESWC 2006, Budva, Montenegro (2006)
2. Lange, C., Schaffert, S., Skaf-Molli, H., Völkel, M. (eds.): 3rd Workshop "From Wiki to Semantics" (SemWiki 2008) – colocated with ESWC 2008, Tenerife, Spain (2008)
3. Schaffert, S.: Semantic Social Software: Semantically Enabled Social Software or Socially Enabled Semantic Web? In: Semantics 2006, Vienna, Austria (2006)
4. Bizer, C., Heath, T., Ayers, D., Raimond, Y.: Interlinking Open Data on the Web. In: Franconi, E., Kifer, M., May, W. (eds.) ESWC 2007. LNCS, vol. 4519. Springer, Heidelberg (2007)
5. Schaffert, S., Westenthaler, R., Gruber, A.: IkeWiki: A User-Friendly Semantic Wiki. In: Sure, Y., Domingue, J. (eds.) ESWC 2006. LNCS, vol. 4011. Springer, Heidelberg (2006)
6. Bry, F., Kotowski, J.: Towards reasoning and explanations for social tagging. In: Proceedings of 3rd International Workshop on Explanation-aware Computing, Patras, Greece, July 21–22 (2008)

Pattern-Based Annotation of HTML-Streams

Florian Schmedding, Max Schwaibold, and Kai Simon

Institute of Computer Science,
Albert-Ludwigs-University Freiburg, Germany
{schmeddi,schwaibo,ksimon}@informatik.uni-freiburg.de

Abstract. Web pages containing RDFa markup facilitate a broad range of new agents that improve their usability for human readers. Unfortunately, there still exist only few web sites featuring such annotations. In this paper, we demonstrate Atheris, a system that annotates structured web pages by means of our web data extraction tool ViPER. Atheris runs inside a web proxy service, making it transparently available. Our approach enables the web browser—the mostly used web agent—to operate intelligently on the displayed page by providing a semantic view over previously 'meaningless' data in order to support human readers.

1 Introduction

Many new applications are facilitated by *RDFa* [1], a recent W3C-recommendation. RDFa-aware agents are able to combine the visual representation of data with their meaning, thereby offering additional value to users (cf. [2] and Sec. 3). In this paper, we will demonstrate our approach to annotating structured web pages to enable the application of such agents. An example is shown in Fig. 1. It depicts a product list of an online pharmacy with information about medicine (price, producer, and additional information). Here, an agent could compare products from other pages with the current products depending on the user's actual interaction with the displayed page and his current focus of interest (w. r. t. the displayed content or the mouse pointer, for instance).

Image	Title	Producer	Price
	Aspirin 20 pcs. - pills article no. 1738533	Bayer	Normal price: ~~3.99 $~~ Your price: 2.99 $

Fig. 1. Product list excerpt from an online pharmacy

Atheris, our semi-automatic web data annotation tool, recognizes such pattern structures and annotates the contained data. By implementing it into a web proxy we make the annotation step invisible to the user and provide a semantic view on previously not machine-readable web sites.

L. Aroyo et al. (Eds.): ESWC 2009, LNCS 5554, pp. 893–897, 2009.

```
_:a a pp:pharmaProduct ;
    pp:productLink
      ex:shop/Aspirin.html ;
    pp:productName "Aspirin" ;
    pp:pzn "1738533" ;
    pp:manufacturer "Bayer" ;
    commerce:costs _:b2 .
_:b a commerce:Price ;
    commerce:amount "2.99" ;
    commerce:currency "$" .
```

Listing 1. RDF statements for the pharmacy example

```
<tr typeof="pp:pharmaProduct"
  xmlns:pp="http://www.example.org/pharma_product#"
  xmlns:commerce="http://purl.org/commerce#">
  <td><a href="/shop/Aspirin.html"><img
    src="/img/Aspirin.png" alt="Aspirin" /></a></td>
  <td><a href="/shop/Aspirin.html"
        property="pp:productName" rel="pp:productLink"
        >Aspirin</a><br />20 pcs. — pills<br />article
    no. <span property="pp:pzn">1738533</span></td>
  <td><span property="pp:manufacturer">Bayer</span></td>
  <td rel="commerce:costs"><table><tr><td>Normal price:
     </td><td>3.99 $</td></tr></table>
    <table typeof="commerce:Price"><tr><td>Your price:
    <span property="commerce:amount">2.99</span> 
    <span property="commerce:currency">$</span></td></tr>
    </table></td></tr>
```

Listing 2. XHTML snippet. RDFa inserted according to Listing 1 appears in **bold print**.

With RDFa, arbitrary RDF statements [3] can be embedded into XHTML at any position using the structure and content of the document itself to express them. We refer to the reference specification in [4] instead of explaining RDFa and show a reasonable RDF description (Listing 1) and its embedding (Listing 2) for the before-mentioned example.

1.1 Related Work

Research in the field of web data extraction and annotation has brought up a broad range of approaches and tools (cf. the surveys in [5] and [6]). Some of these systems are specialized in extracting relevant data from pages (e. g. Lixto [7], SparqPlug [8]), but w. r. t. RDFa-agents it is important to include the semantic annotations into the page. For instance, Piggy Bank [9] uses extractions and is thus unable to adapt the offered functionality to the user's current interest. Manual tools like Annotea[1] do not fit the requirements either although Annotea relates the annotations with corresponding page elements by XPointer expressions. MnM [10] or S-CREAM [11] are better suited. Both are able to add additional XML-tags around the relevant data but further knowledge is necessary to convert them into RDF. Similar to PowerMagpie[2], they are rather designed for continuous text than for structured content. Thresher [12] is most related to our approach yet uses a non-standardized formalism to relate terms and concepts within documents. Other systems like Greasemonkey[3] and Chickenfoot [13] are also adding further functionality to page elements identified by screen-scraping techniques. However, this is achieved by hard-coded scripts within the document.

In contrast, our system uses RDFa to annotate the identified concepts and relations between them. Hence, its output can be used by existing RDFa-agents and so is independent from the annotation process. Additionally, Atheris is available as proxy service to provide an easy access for web browsers.

[1] http://www.w3.org/2001/Annotea/

[2] http://powermagpie.open.ac.uk/

[3] http://greasemonkey.mozdev.org/

1.2 ViPER

Our wrapper extraction tool ViPER [14] automatically identifies and extracts different regularly structured content with high accuracy according to several visual features. We build Atheris upon it due to its fast and robust pattern matching algorithm. ViPER was refined into RexViPER [15] to also derive post-processing rules from the extracted and aligned data. The purpose of these rules is the automatic labeling of named entities, the splitting of strings, and the removal of text from data entities to simplify further rule assignments. Rules can be assigned manually or with a drag-and-drop tool for non-expert users as implemented in [16]. The generated extraction pattern matching the structure of the data records in our example is described by the (abbreviated) HTML tag sequence:

Position	0	1	2	3	4	5	6	7	8	9	10	11	12	13	14	...	37
Pattern	<TR>	<TD>	<A>	<IMG>	</A>	</TD>	<TD>	<A>	TEXT	</A>	 	TEXT	 	TEXT	</TD>	...	</TR>

The attached set of rules (depicted in Listing 3) consists of both automatically learned rules found with the aid of RexViPER and manually assigned RDFa annotation rules to enrich the appropriate data items with RDFa semantics. The first argument of a rule references a pattern element by its position. For instance, the rule `split(32, )` splits the content of the 32nd element according to hard space. Here, the content will be split into Your, price: 2.99, and $. The resulting substrings can be referenced using the sub-indices 32.0, 32.1 and 32.2. RDFa-specific rules are shown in the next section.

2 Atheris

To annotate pages, Atheris has to extend their structure to make the pattern's data 'visible' to agents instead of disregarding it. Thus, being an extraction tool, some modifications inside ViPER were necessary to facilitate annotations. The changes are summarized by three points. First, to insert the RDFa-markup correctly (mostly attributes and *span*-elements), we store all tag positions w. r. t. the HTML-stream. Second, we implement new rules besides adapting the existing ones to the stored tag positions. Third, to ensure the well-formedness of the output, changes to the handling of formatting elements like *b* and *i* are necessary.

Annotation Rules. Listing 3 represents the rules used to annotate the example page. Additionally, two rules are used to set up the namespaces in the *tr*-element. We illustrate the respective results with Listing 2. The rules `typeof` and `relation` insert *@typeof* into elements 0 and 29 respectively *@rel* into elements 7 and 18. This suffices to express triples 1, 2, 6, and 7 in Listing 1. The remaining statements follow from `property` which delimits the desired literals by *span*-tags (except for element 7) and inserts the associated predicate into the *@property*-attribute. Rules to declare datatypes or other constructs are not used here and therefore left out. We want to highlight the `join`-rule that combines separated text elements into a single literal: In the example, this could be used to attach elements 11 and 13 to a predicate like `pp:additionalInformation`.

```
typeof(0, pp:pharmaProduct)        relation(7, pp:productLink)
property(7, pp:productName)        property(13, pp:pzn)
remove(13, article no.)            property(16, pp:manufacturer)
relation(18, commerce:costs)       typeof(29, commerce:Price)
split(32,  )                  property(32.1, commerce:amount)
remove(32.1, price:)               property(32.2, commerce:currency)
```

Listing 3. Annotation rules

Two or More Patterns. When a web page contains more than one type of structured information, it is not known in advance which kind of data is going to be used by agents in the annotation scenario. It is therefore necessary to annotate all patterns before sending the page to the client. We enhanced the pattern recognition engine in ViPER to execute concurrent parallel pattern matching threads in spite of running Atheris for each pattern serially. This approach benefits from a better performance by avoiding multiple input parsing.

2.1 Proxy

To deliver the annotated pages easily to web browsers, we decided to implement an HTTP proxy service due to the transparent access. With the proxy, users also benefit from sharing the annotation patterns and rules. Incoming requests are classified as either annotable or others. Not annotable requests are handled by the proxy[4] itself while annotable requests are passed over to Atheris and processed according to the proxy's pattern and rule database.

Like ViPER, Atheris is working on the raw HTML stream in favor of a smaller latency for the clients. This streaming capability is the reason why it is possible to use chunked encoding for the annotation handling giving better response times and thus making the proxy as transparent as possible for clients.

3 Conclusion and Future Work

We have presented Atheris, our system to annotate web pages with RDFa-markup. Atheris relies on the robust pattern-matching technique of ViPER and introduces an abstraction layer between web pages and additional functionality for these pages (changing the appearance, adding information, etc.) by embedding meta data with RDFa. Due to the provided semantic view over previously not machine-readable data, the web browser can be turned into a semantic user agent operating on structured pages in a meaningful way that has been impossible before. Because of the clear separation between data and functionality it is also possible to change either of the components without side effects.

Despite its just mentioned advantages, an HTTP proxy is unable to annotate streams in secure channels (cf. also [13]). It has also limited resources and can quickly induce a bottleneck under heavy load. Therefore, and because the primary use of annotated pages would occur in combination with the rendered presentation, it is an option to implement the system directly inside a browser. If the annotation step could be performed directly inside the browser's parser,

[4] Subclass of Mentalis proxy, `http://www.mentalis.org/soft/projects/proxy`

this would even improve the performance. Of course, to keep the benefit of collaboration, the system should request the patterns and rules from a common store. To make it available to other RDFa-consuming systems it could be published as web service, too. We are also improving the annotation rules to capture more cases. For instance, the rule to annotate currencies should implement an automatic conversion from symbols to the required ISO-4217 literals (`$` to `USD` in our example) for `commerce:currency`. Additionally, we are currently working on a browser plugin that integrates RDFa into the DOM of web pages to identify the annotated parts of a page with SPARQL queries. This enables the plugin to show feedback from other agents just beside the annotations.

References

1. Adida, B., Birbeck, M.: RDFa Primer (2008), `http://www.w3.org/TR/2008/NOTE-xhtml-rdfa-primer-20081014/`
2. Adida, B., Hausenblas, M.: RDFa Use Cases (2007), `http://www.w3.org/TR/2007/WD-xhtml-rdfa-scenarios-20070330/`
3. Manola, F., Miller, E.: RDF Primer (2004), `http://www.w3.org/TR/2004/REC-rdf-primer-20040210/`
4. Adida, B., Birbeck, M., McCarron, S., Pemberton, S.: RDFa in XHTML: Syntax and Processing (2008), `http://www.w3.org/TR/2008/REC-rdfa-syntax-20081014`
5. Chang, C.H., Kayed, M., Girgis, M.R., Shaalan, K.F.: A Survey of Web Information Extraction Systems. IEEE TKDE 18(10), 1411–1428 (2006)
6. Uren, V., Cimiano, P., Iria, J., Handschuh, S., Vargas-Vera, M., Motta, E., Ciravegna, F.: Semantic Annotation for Knowledge Management: Requirements and a Survey of the State of the Art. Web Semantics (4), 14–28 (2006)
7. Baumgartner, R., Flesca, S., Gottlob, G.: Visual Web Information Extraction with Lixto. In: VLDB, San Francisco, CA, USA, pp. 119–128 (2001)
8. Coetzee, P., Heath, T., Motta, E.: SparqPlug: Generating Linked Data from Legacy HTML, SPARQL and the DOM. In: LDOW, Bejing, China (2008)
9. Huynh, D., Mazzocchi, S., Karger, D.: Piggy Bank: Experience the Semantic Web inside your Web Browser. Web Semant 5(1), 16–27 (2007)
10. Vargas-Vera, M., Motta, E., Domingue, J., Lanzoni, M., Stutt, A., Ciravegna, F.: MnM: Ontology Driven Semi-Automatic and Automatic Support for Semantic Markup. In: Gómez-Pérez, A., Benjamins, V.R. (eds.) EKAW 2002. LNCS, vol. 2473, pp. 379–391. Springer, Heidelberg (2002)
11. Handschuh, S., Staab, S., Ciravegna, F.: S-CREAM — Semi-automatic CREAtion of Metadata. In: Gómez-Pérez, A., Benjamins, V.R. (eds.) EKAW 2002. LNCS, vol. 2473, pp. 358–372. Springer, Heidelberg (2002)
12. Hogue, A., Karger, D.: Thresher: Automating the unwrapping of semantic content from the world wide web. In: WWW, Chiba, Japan, pp. 86–95 (2005)
13. Bolin, M., Webber, M., Rha, P., Wilson, T., Miller, R.C.: Automation and Customiztion of Rendered Web Pages. In: UIST, Seattle, WA, USA, pp. 163–172 (2005)
14. Simon, K., Lausen, G.: ViPER: Augmenting Automatic Information Extraction with Visual Perceptions. In: CIKM, Bremen, Germany, pp. 381–388 (2005)
15. Simon, K., Hornung, T., Lausen, G.: Learning Rules to Pre-process Web Data for Automatic Integration. In: RuleML, Athens, GA, USA, pp. 107–116 (2006)
16. Hornung, T., Simon, K., Lausen, G.: Mashing Up the DEEP Web - Research in Progress. In: WEBIST (2), Madeira, Portugal, pp. 58–66 (2008)

OntoComp: A Protégé Plugin for Completing OWL Ontologies

Barış Sertkaya*

Theoretical Computer Science, TU Dresden, Germany
sertkaya@tcs.inf.tu-dresden.de

Abstract. We describe OntoComp, a Protégé 4 plugin that supports ontology engineers in completing OWL ontologies. More precisely, OntoComp supports an ontology engineer in checking whether an ontology contains all the relevant information about the application domain, and in extending the ontology appropriately if this is not the case. It acquires complete knowledge about the application domain efficiently by asking successive questions to the ontology engineer. By using novel techniques from Formal Concept Analysis, it ensures that, on the one hand, the interaction with the ontology engineer is kept to a minimum, and, on the other hand, the resulting ontology is complete in a certain well-defined sense.

1 Introduction

Ontologies play a key role for the Semantic Web. Since the standardization of OWL as the ontology language for the Semantic Web, several ontology editors now support OWL [4,3], and ontologies written in OWL are employed in more and more semantic web applications in various domains. As the number and size of these ontologies grows, tools that support improving their quality become more important. The tools available until now mostly deal with detecting inconsistencies and inferring consequences, i.e., implicit knowledge that can be deduced from the knowledge explicitly represented in the ontology. There are also promising approaches that allow to pinpoint the reasons for inconsistencies and for certain unwanted consequences. These approaches address the quality dimension of *soundness* of an ontology, both within itself (consistency) and w.r.t. the intended application domain (no unwanted consequences). In our previous work [1,5], we have considered a different quality dimension, namely *completeness* of the knowledge in an ontology. We have provided a formally well-founded technique called *ontology completion*, that supports the ontology engineer in checking whether an ontology contains all the relevant information about the application domain, and in extending the ontology appropriately if this is not the case.

An OWL ontology typically consists of two parts, the terminological part (TBox), which defines concepts and also states additional constraints (so-called

* Supported by German Research Foundation (DFG) under grant BA 1122/12-1.

L. Aroyo et al. (Eds.): ESWC 2009, LNCS 5554, pp. 898–902, 2009.

general concept inclusions or GCIs) on the interpretation of these concepts, and the assertional part (ABox), which describes individuals and their relationship to each other and to concepts. Given an application domain and an OWL ontology describing it we can ask:

- Are all the relevant constraints that hold between concepts in the domain captured by the TBox?
- Are all the relevant individuals existing in the domain represented in the ABox?

Such questions cannot be answered by an automated tool alone. Clearly, to check whether a given relationship between concepts—which does not already follow from the TBox—holds in the domain, one needs to ask a domain expert, and the same is true for questions regarding the existence of individuals not described in the ABox. The role of the ontology completion tool here is to ensure that the expert is asked as few questions as possible; in particular, she should not be asked trivial questions, i.e., questions that could actually be answered based on the represented knowledge.

2 Motivating Example

As an example, of how ontology completion supports the ontology engineer in practice, consider the OWL ontology for human protein phosphatases that has been described and used in [7]. This ontology was developed based on information from peer-reviewed publications. The human protein phosphatase family has been well characterised experimentally, and detailed knowledge about different classes of such proteins is available. This knowledge is represented in the terminological part of the ontology. Moreover, a large set of human phosphatases has been identified and documented by expert biologists. These are described as individuals in the assertional part of the ontology. One can now ask whether the information about protein phosphatases contained in this ontology is complete. That is, are all the relationships that hold among the introduced classes of phosphatases captured by the constraints in the TBox, or are there relationships that hold in the domain, but do not follow from the TBox? Are all possible kinds of human protein phosphatases represented by individuals in the ABox, or are there phosphatases that have not yet been included in the ontology or even not yet been identified?

Clearly, these questions need to be answered by a biologist. In this example, answering a non-trivial question regarding human protein phosphatases may require the biologist to study the relevant literature, query existing protein databases, or even to carry out new experiments. Thus, the expert may be prompted to acquire new biological knowledge.

3 Ontology Completion

The key technologies lying under ontology completion are Description Logic reasoning, and the attribute exploration method developed in Formal Concept

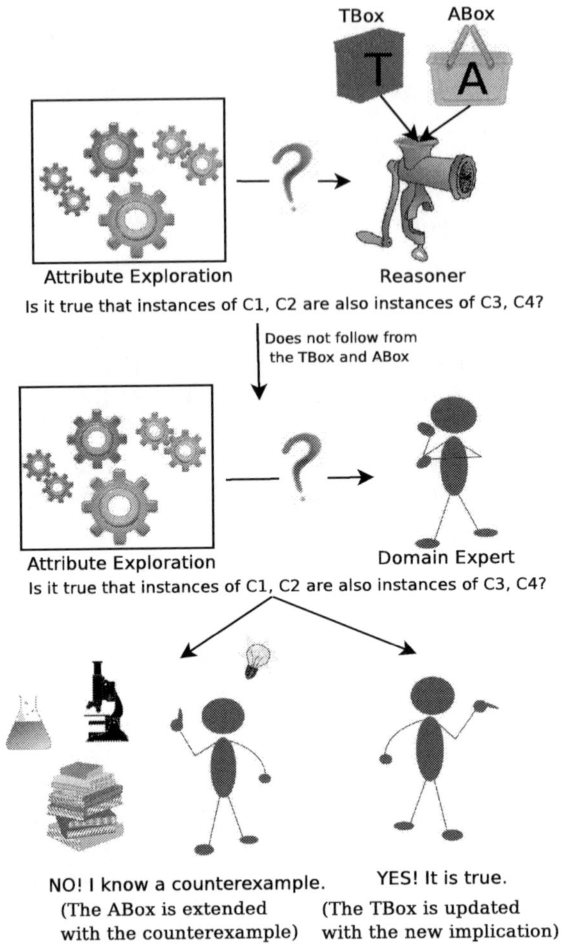

Fig. 1. Ontology completion process

Analysis (FCA) [2]. FCA is a field of applied mathematics that aims to formalize the notions of a concept and a conceptual hierarchy by means of mathematical tools. It is used for conceptual data analysis and knowledge processing. Attribute exploration is a knowledge acquisition method of FCA that is used to acquire complete knowledge about an application domain by asking successive questions to a domain expert. It asks the expert questions of the form "*is it true that instances of the classes* $C_{i1}, \ldots, C_{ik}$ *also instances of* $C_{j1}, \ldots, C_{jl}$*?*". When such a question is asked, the expert is expected to either confirm the question, in which case a new implication in the application domain has been discovered, or reject it. If the expert rejects such a question, she is expected to give a counterexample, i.e., an individual that is instances of the classes $C_{i1}, \ldots, C_{ik}$ but is not an instance of at least one of $C_{j1}, \ldots, C_{jl}$. This counterexample is then added to the ontology as a new individual, and the next question is asked.

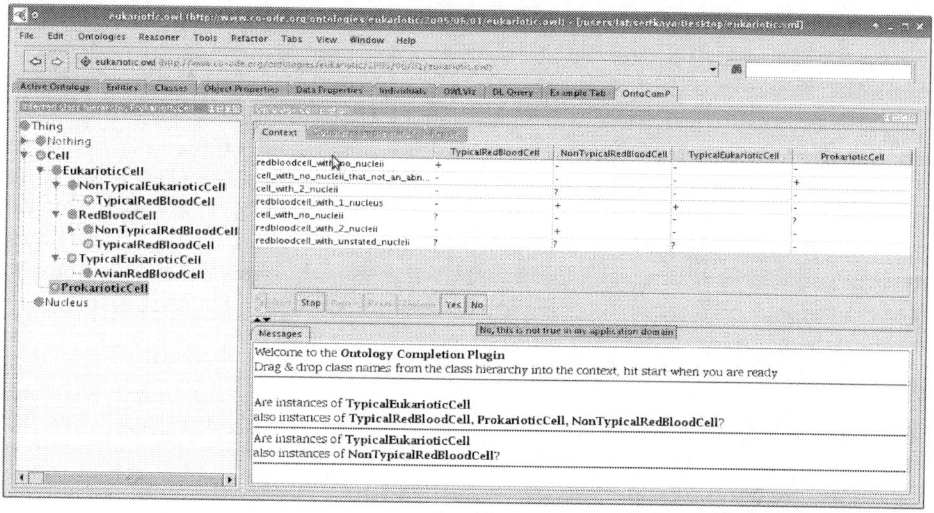

Fig. 2. OntoComP window during completion

What makes attribute exploration an attractive method for capturing expert knowledge is that it guarantees to make best use of the expert's answers, and to ask the minimum possible number of questions that suffices to acquire complete knowledge about the application domain.

4 OntoComP

Based on our results in [1,5], we have implemented an open-source ontology completion tool called ONTOCOMP[1], which stands for ONTOlogy COMpletion Plugin. It is written in the Java programming language as a plugin for the Protégé 4 ontology editor. It can be easily installed by just copying the `jar` file provided under the URL given below into the `plugins` directory of an existing PROTÉGÉ 4 installation. Then upon a new start, PROTÉGÉ will find the ONTOCOMP plugin and open a new tab for it.

In order to complete an ontology with ONTOCOMP, the user first classifies the ontology with a reasoner that is supported by PROTÉGÉ 4, namely FACT++ or PELLET. Once the ontology is classified, the ONTOCOMP tab displays the class hierarchy on the left. At this point the user can drag "interesting" class names, which she wants to have in the completion process, from this hierarchy and drop them into the `Context` tab on the right. Simultaneously, the instances of these classes will also be displayed in a table in the `Context` tab. In this table a `+` in row a and column C means that the individual a is an intance of the class C, a `-` means that a is an instance of the complement of C, and a `?` means that nothing is known about the membership of a in class C. When the user is done with selecting the relevant classes, she starts the completion by hitting

[1] Available under `http://ontocomp.googlecode.com`

the *Start* button and sees the first question in the `Messages` tab. If she confirms the question by hitting the *Yes* button, OntoComP comes up with the next question. If she rejects the question by hitting *No* button OntoComP opens the `Counterexample editor` tab where the user can generate a counterexample to the rejected question. The process continues until all questions have been answered by the user. During completion, at any time the user can suspend the completion process and see her answering history. If she thinks she has made an error at some point, she can repair it by undoing that particular erroneous answer and can continue completion. A similar FCA-based ontology refinement tool that is related to OntoComP, but mainly aimed at acquiring domain-range restrictions has been described in [6].

5 The Demo

During the demo, the audience will have the opportunity to see OntoComP in action on ontologies from real world application domains, which do not require specialized expert knowledge (so that the demo is still understandable for the whole audience). We are going to demonstrate in detail how to use OntoComP and how it supports the ontology engineer in completing an ontology. We are going to prepare scenarios where the ontology engineer makes errors during ontology completion and demonstrate how to repair them.

References

1. Baader, F., Ganter, B., Sertkaya, B., Sattler, U.: Completing Description Logic Knowledge Bases using Formal Concept Analysis. In: Proc. of the Twentieth International Joint Conference on Artificial Intelligence (IJCAI 2007). AAAI Press, Menlo Park (2007)
2. Ganter, B., Wille, R.: Formal Concept Analysis: Mathematical Foundations. Springer, Berlin (1999)
3. Kalyanpur, A., Parsia, B., Sirin, E., Grau, B.C., Hendler, J.A.: Swoop: A web ontology editing browser. Journal of Web Semantics 4(2), 144–153 (2006)
4. Knublauch, H., Fergerson, R.W., Noy, N.F., Musen, M.A.: The protégé OWL plugin: An open development environment for semantic web applications. In: McIlraith, S.A., Plexousakis, D., van Harmelen, F. (eds.) ISWC 2004. LNCS, vol. 3298, pp. 229–243. Springer, Heidelberg (2004)
5. Sertkaya, B.: Formal Concept Analysis Methods for Description Logics. Ph.D. dissertation, Institute of Theoretical Computer Science, TU Dresden, Germany (2007)
6. Völker, J., Rudolph, S.: Fostering web intelligence by semi-automatic OWL ontology refinement. In: Proc. of the 7th International Conference on Web Intelligence (WI 2008) (2008)
7. Wolstencroft, K., Brass, A., Horrocks, I., Lord, P.W., Sattler, U., Turi, D., Stevens, R.: A little semantic web goes a long way in biology. In: Gil, Y., Motta, E., Benjamins, V.R., Musen, M.A. (eds.) ISWC 2005. LNCS, vol. 3729, pp. 786–800. Springer, Heidelberg (2005)

Demo: HistoryViz – Visualizing Events and Relations Extracted from Wikipedia

Ruben Sipoš[1], Abhijit Bhole[2], Blaž Fortuna[1], Marko Grobelnik[1], and Dunja Mladenić[1]

[1] Jozef Stefan Institute, Jamova 39, 1000 Ljubljana, Slovenia
{ruben.sipos,blaz.fortuna,marko.grobelnik,dunja.mladenic}@ijs.si
[2] Indian Institute of Technology, Bombay, Mumbai - 400076, India
abhijit.bhole@cse.iitb.ac.in

Abstract. HistoryViz provides a new perspective on a certain kind of textual data, in particular the data available in the Wikipedia, where different entities are described and put in historical perspective. Instead of browsing through pages each describing a certain topic, we can look at the relations between entities and events connected with the selected entities. The presented solution implemented in HistoryViz provides user with a graphical interface allowing viewing events concerning the selected person on a timeline and viewing relations to other entities as a graph that can be dynamically expanded.

1 Introduction

The challenge we are facing with growing amount of electronic data on the Web is how to find and extract the knowledge we want form the vast amount of the available data. The most common way is searching using keywords or even just browsing using entity names and links. The main drawback of this approach, when we are looking for a broader overview of some topic and not only a specific fact, is the amount of human work required to extract and consolidate facts from a number of web pages found using a search engine.

Another approach to fact discovery and knowledge acquisition is using textual data that is already in (more or less) structured format. Instead of browsing through dozens of search engine results to find relevant facts about e.g., iron we can go to Wikipedia[1] article and find everything on a single page. Similarly, we could use RDF data from some provider. However, doing this we do not fully exploit the potential that data sources provide us with. If we used those data sources as a whole and not only the explicitly relevant bits, we could gain some additional knowledge. While this can be done manually, it is too time consuming because we have to manually combine a lot of small pieces into a larger picture. Because data sources we are considering are at least partially structured we can use text mining [3] and similar approaches to automatically extract data from multiple subsections and present it's summary to the user.

In this demo, we present HistoryViz[2], a web application allowing user to explore events (extracted from Wikipedia) connected with selected persons presented on a

[1] http://www.wikipedia.org/
[2] http://historyviz.ijs.si/

L. Aroyo et al. (Eds.): ESWC 2009, LNCS 5554, pp. 903–907, 2009.

timeline and to browse the network consisting of persons described on Wikipedia. HistoryViz uses data extracted using text mining and presents a new perspective on indirectly available data contained in Wikipedia that can be acquired automatically. Our main focus was on the entity network and timeline of events although using a similar approach one can also gain some other information.

1.1 Related Work

The demo presented in this paper is using some of the data extracted from Wikipedia using the approaches described in [1], where articles from Wikipedia were categorized (using binary linear SVM) into articles describing persons, places and organizations. They also show how to extract events, described in Wikipedia articles about persons, and their time frame and extract link data used for building the entity network and determining the importance of nodes. In our work we have extended the existing approaches by using and incorporating other data sources, such as Freebase[3] and performing some additional steps (e.g., associating keywords with events) in processing the data. Due to a space limitation of this paper, only the closely related work is discussed here, more can be found in [1].

1.2 Motivation

The advent of web sources providing semi-organized data has drastically improved user's experience when using the web to gain new knowledge. Still, if user wants some information that can be obtained by gathering facts from different articles or entries, there is usually a lot of manual work involved. Because the data is not strictly structured we can not use most of existing tools. However, we can use text mining on this data and provide new ways to view data to the user.

This demo shows how to use already existing data in Wikipedia to provide user with an interface displaying temporal data and social network based on data obtained by combining pieces of information gathered from many articles. For example, user writing a report on some important historical person might have to read through many articles to gather enough biographical data and discover relations with other persons. On the other hand, by using our interface as implemented in HistoryViz it is easy to see all important relations (presented as an entity network) and timeline of the events which happened in a life of the selected person.

Moreover, we wanted to show that it is possible to simplify certain user tasks and provide additional views on already existing data using currently available technologies. Many data sources are still not fully exploited because it would require too much manual effort and because the structure is too lax to allow use of current tools. Nevertheless, we can now take advantage of that still unused potential.

2 HistoryViz

HistoryViz is a web based application that allows the user to view events related to selected person's life on a timeline and to explore social network based on relations found in Wikipedia. There are numerous browsers which simplify exploring of

[3] http://www.freebase.com/

Wikipedia. However, they usually present hyperlinks between articles in a graphical manner. Most of them directly use links whereas we consider the whole entity network based on relations found by examining content of the articles. Moreover, we provide timeline view of data automatically collected from separate articles, while other solutions take already consolidated data or gather data from strictly structured sources such as RDF.

In timeline view shown on Figure 1 the user selects person by entering full or partial name into query box. System finds the corresponding article in Wikipedia describing that person. Events (extracted from Wikipedia) are then shown on the timeline, which consists of two parts. The top part shows bars representing the life from birth to death of the found persons. Also, the most closely related persons to the one user selected are shown. They are chosen by exploring local neighborhood in the entity network and using heuristics mostly based on PageRank [2] and cosine similarity of the link structure. This enables the user to easily explore correlation of events in lives of closely related persons.

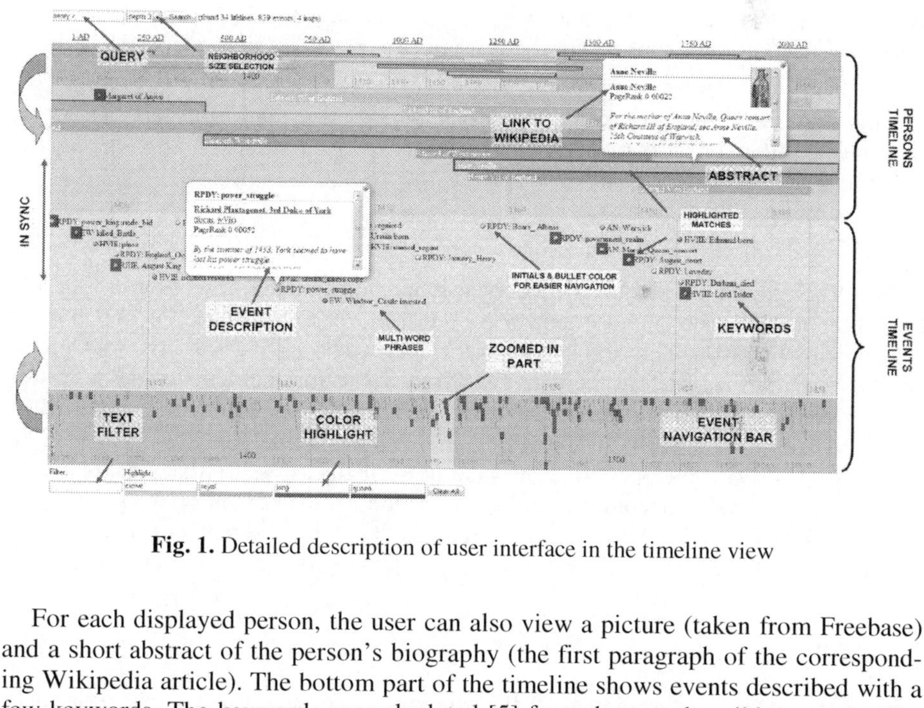

Fig. 1. Detailed description of user interface in the timeline view

For each displayed person, the user can also view a picture (taken from Freebase) and a short abstract of the person's biography (the first paragraph of the corresponding Wikipedia article). The bottom part of the timeline shows events described with a few keywords. The keywords are calculated [5] from the text describing events. The user can also read the event description after selecting the desired event.

The user can explore network of entities using a java applet (based on Prefuse[4] [3]) shown in Figure 2. The starting node is selected by searching for article corresponding to the user's query. The graph is dynamically expanded as the user traverses over it.

[4] http://prefuse.org/

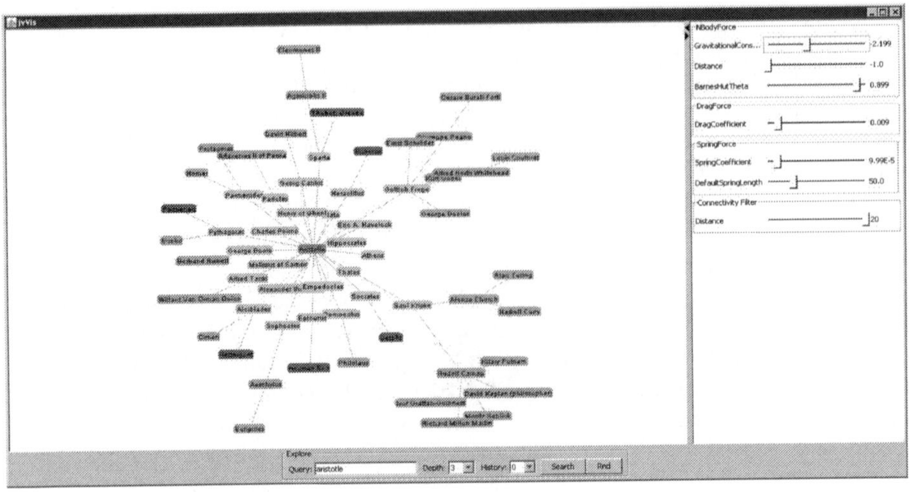

Fig. 2. Example of a part of the entity network using our java applet. The starting point was query "Aristotle". Colors represent entity categories (persons, places and organizations).

Both views are interconnected. The user can always switch from one to another. By doing this the currently displayed items in one view are transferred to the other to help user keep the current context.

2.1 Implementation

HistoryViz is implemented as client/server application. Server is written in C++ and has all the needed data stored in binary format to lower access time. The pipeline for data processing (from raw dumps to internal indexed binary format) is streamlined and therefore any updates to the data can be easily performed. Clients are Ajax web application displaying timeline and java applet for browsing the entity network. Both communicate with the server using HTTP and exchanging data in XML format. This architecture allows developers to easily add new clients with different functionalities and keeps a few gigabytes of data on the server side while exchanging only small portions of it with the client. In this demo we have 3.2GB on the server side and typically exchange 200-400kB with the client (most of it being textual descriptions of events).

3 Demo

During our demo, visitors will have the opportunity to try out our application, hear detailed description of all the implemented features and if interested see some parts of the application that are hidden from the end user (such as, intermediate data files, details about server implementation, contents of client/server communication and how we use data from other sources e.g., Freebase). We will provide sample tasks and show how our application simplifies them. Also, we will demonstrate how users can discover new and interesting facts using our application because the data is presented in a different way.

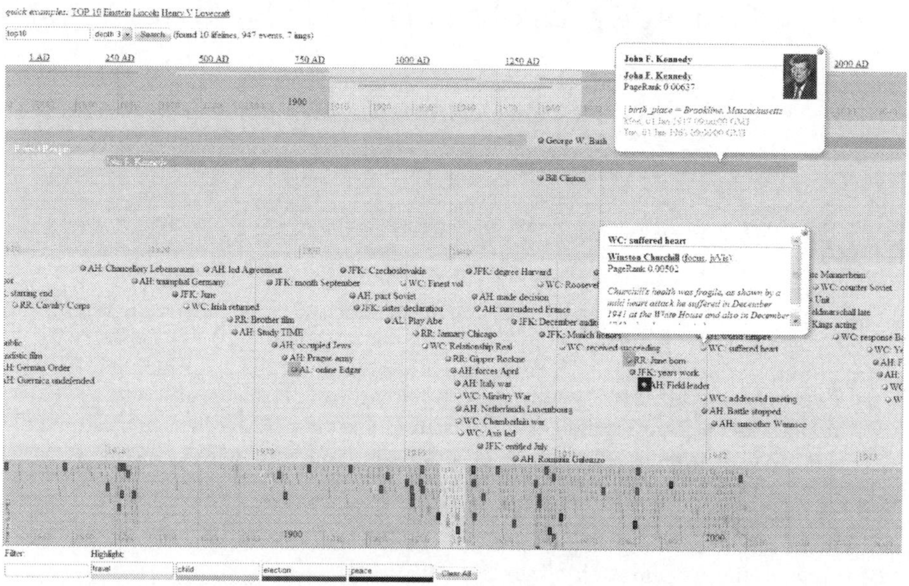

Fig. 3. Persons displayed on this timeline were selected as the most important nodes given by PageRank. This can give some additional insight into Wikipedia's content and world history.

References

1. Bhole, A., Fortuna, B., Grobelnik, M., Mladenic, D.: Extracting Named Entities and Relating Them over Time Based on Wikipedia. Informatica, Part 4 31, 463–468 (2007)
2. Page, L., Brin, S., Motwani, R., Winograd, T.: The PageRank citation ranking: Bringing order to the web. Stanford Digital Libraries Working Paper (1998)
3. Sebastiani, F.: Machine learning in automated text categorization. ACM Computing Surveys 34(1), 1–47 (2002)
4. Heer, J., Card, S.K., Landay, J.A.: Prefuse: a toolkit for interactive information visualization. In: Proceedings of the SIGCHI Conference on Human Factors in Computing Systems, Portland, Oregon, USA, April 2-7 (2005)
5. Grobelnik, M., Mladenic, D.: Text Mining Recipes. Springer, Heidelberg (2009), http://www.textmining.net

Ontology Evolution with Evolva

Fouad Zablith, Marta Sabou, Mathieu d'Aquin, and Enrico Motta

Knowledge Media Institute (KMi), The Open University
Walton Hall, Milton Keynes, MK7 6AA, United Kingdom
{f.zablith,r.m.sabou,m.daquin,e.motta}@open.ac.uk

Abstract. Ontology evolution is a painstaking and time-consuming process, especially in information rich and dynamic domains. While ontology evolution refers both to the adaptation of ontologies (e.g., through additions or updates possibly discovered from external data sources) and the management of these changes, no existing tools offer both functionalities. The Evolva framework aims to be a blueprint for a comprehensive ontology evolution tool that would cover both tasks. Additionally, Evolva proposes the use of background knowledge sources to reduce user involvement in the ontology adaptation step. This demo focuses on the initial, concrete implementation of our framework.

1 Introduction

Ontologies form the backbone of Semantic Web systems. In today's information rich environments, ontologies need to be kept up-to-date so that the dependent systems remain operational. We highlight two research approaches to this problem: The first considers ontology evolution as the management of user-performed changes [5,7,9,10], while the second primarily focuses on dynamically updating and learning ontologies [1,2,6]. We understand ontology evolution as the "timely adaptation of an ontology to the arisen changes and the consistent management of these changes" [4]. This definition reflects the need of combining the two aforementioned approaches for supporting both evolution tasks. Yet, to our knowledge, no practical and complete solutions exist that cover all stages of ontology evolution.

Our ontology evolution framework, Evolva, proposes a comprehensive approach to evolution by identifying tasks that need to be performed both during ontology adaptation and change management (Section 2). The Evolva tool (Section 3 and 4) is a partial, concrete implementation of the framework. One of its main novelties, which will be captured by the demo, is in pioneering the use of background knowledge to decrease user input [12].

2 Underlying Approach: The Evolva Framework

The Evolva framework (detailed in [11]) consists of five main components as shown in Figure 1 and described next.

Information Discovery. New information can be potentially discovered in external data sources. Contrasting the ontology with the content of such sources is

L. Aroyo et al. (Eds.): ESWC 2009, LNCS 5554, pp. 908–912, 2009.

a way of detecting new knowledge that should be reflected by the base ontology. Data sources can be either unstructured (e.g., text documents, folksonomy tagspaces) or structured (e.g., databases, ontologies). This component handles each data source differently: (1) Text documents are processed using information extraction, ontology learning or entity recognition techniques. (2) Other external ontologies are subject to translation for language compatibility with the base ontology, and (3) database content is translated into ontology languages.

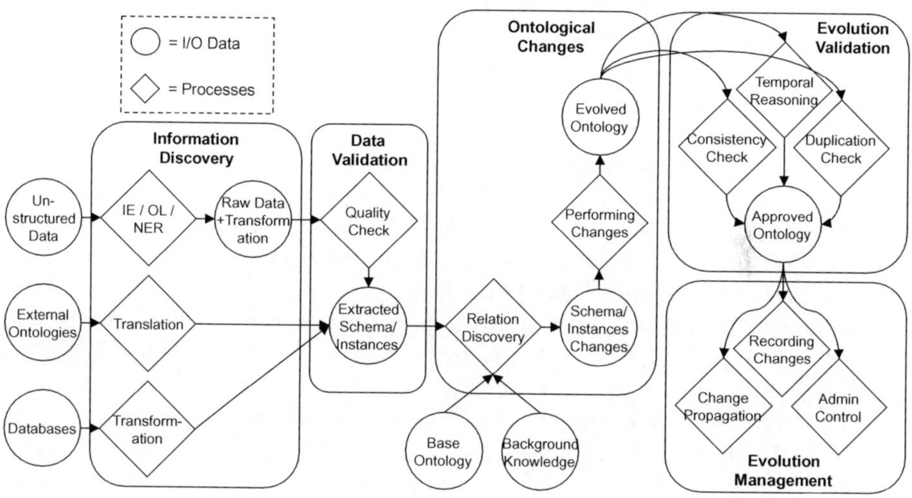

Fig. 1. Evolva Framework Overview

Data Validation. Discovered information is validated in this component. We rely on a set of heuristic rules such as the length of the extracted terms. This is especially needed for information discovered from text documents, as information extraction techniques are likely to introduce noise. For example, most of the less than two-letter terms extracted from a corpora are meaningless and should be discarded. In the case of structured data, this validation is less needed as the type of information is explicitly defined.

Ontological Changes. This component is in charge of establishing relations between the extracted terms and the concepts in the base ontology. These relations are identified by exploring a variety of background knowledge sources. Appropriate changes are then performed on the ontology.

Evolution Validation. Performing ontological changes could generate some problems such as conflicting statements, data duplication and time related inconsistencies. We deal with these problems at the level of the evolution validation component, formed of the consistency and duplication checks, in addition to temporal reasoning.

Evolution Management. Managing the evolution is about recording changes, and giving the ontology curator a degree of control over the evolution, as well as propagating changes to other dependent ontologies (e.g., in the case of an ontology network) or applications. User control will deal with tracking ontology changes, spotting and solving unresolved problems.

3 Implementation: The Evolva NeOn Toolkit Plugin

The Evolva NeOn toolkit plugin is a concrete implementation of our framework and will be the object of the proposed demonstrations (see screenshot in Figure 2). Currently, the plugin supports the first three steps depicted in Figure 1.

In the *information discovery* step Text2Onto is used to identify concepts in a corpus of text documents (selected in the "Data Sources" panel). *Data validation* relies on string similarity and term length measures. We use a customizable Jaro-based [3] calculation to compute the similarity of the extracted and existing concepts. Concepts with the similarity under a given threshold are considered new and are validated based on their length. The user can set the parameters of the validation techniques and also manually indicate which concepts should be considered for integration ("Data Validation").

Ontological changes are identified by finding links between validated terms and ontology concepts. We use two main sources of background knowledge: WordNet and online ontologies accessed through Scarlet[1]. Scarlet [8] is a Semantic Web based relation discovery engine. It automatically selects and explores online ontologies to discover relations between two given concepts. For example, when relating two concepts labeled *Researcher* and *AcademicStaff*, Scarlet 1) identifies (at run-time) online ontologies that can provide information about how these two concepts inter-relate and then 2) combines this information to infer their relation. [8] describes two increasingly sophisticated strategies to identify and to exploit online ontologies for relation discovery. Hereby, we rely on the first strategy that derives a relation between two concepts if this relation is defined (directly or through transitivity) within a single online ontology. Besides subsumption relations, Scarlet is also able to identify disjoint and named relations. As in the case of WordNet, the derived relations are accompanied by a path of inferences that lead to them ("Relation Discovery" panel). We performed an experiment on the potential use of background knowledge in our context, and we got a promising 77% precision[2]. A list of ontology changes are deduced from the relation discovery step and applied to the ontology ("Ontology Changes").

4 Demonstration Plan

The demonstration will focus on the various functionalities of the Evolva plugin (Figure 2) that support ontology evolution from textual documents. A particular emphasis will be on the use of background knowledge sources to automatically identify relations between terms and ontology concepts. This gives our tool a

[1] `http://scarlet.open.ac.uk`

[2] Additional details of the experiment can be found in [12]

cutting edge advantage in decreasing user input. It is a critical feature as this is where user input is usually most needed. The following functionalities will be demonstrated:

Ontology selection and loading in NeOn toolkit. At the left side of Figure 2, the user will see how to load and open an ontology in the NeOn toolkit, and start Evolva.

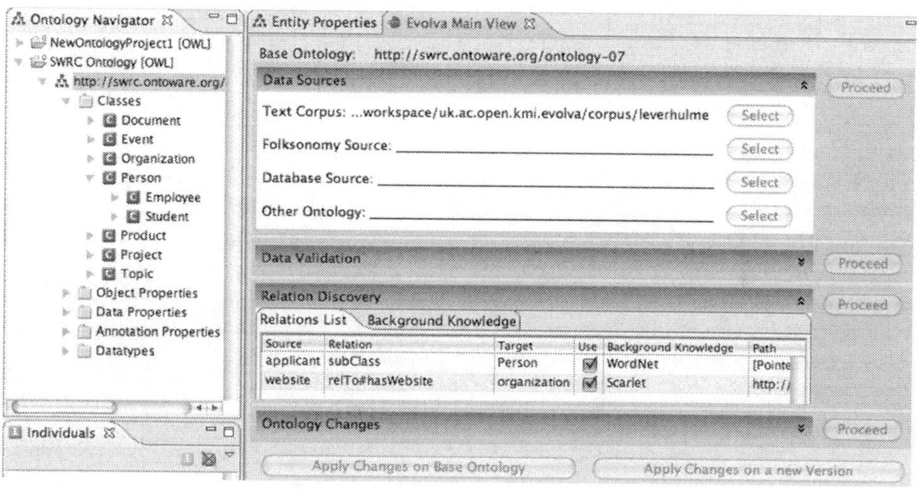

Fig. 2. Evolva Pilot System Screenshot

The selection of data sources. The user selects the text documents from where terms are extracted, then proceeds to the next step in the process.

Data validation and parameters settings. At this level the list of extracted terms are displayed to the user. The validation shows the terms that correspond to already existing ontology concepts. Further noise is removed based on the term length. This is performed by a customizable string length checker to exclude for example one or two letter terms. Parameters for controlling the automated validation include changing the minimum length of acceptable terms and controlling the similarity threshold. In addition to the automatic validation, the user is able to manually select which terms should be included in the following step.

Relation discovery between terms and ontology concepts. The user will experience how background knowledge helps identifying links between the validated terms and ontology concepts. For example, consider the case of the base ontology containing the concepts *Person* and *Organization*. From the text documents, Evolva discovers *Applicant* and *Website* as potential concepts to be added to the ontology. WordNet is able to link *Applicant* as *subClassOf Person*, and online ontologies link the new concept *Website* through a *hasWebsite* relation to *Organization*. The "Relation Discovery" panel shows these relations and allows their inspection. The complete relation path is displayed to help in verifying the relation's correctness. This panel also contains a set of customizable parameters such as: the background knowledge source to be used (WordNet, Scarlet or both); the maximum inference path for deriving a relation; the type of relations to be discovered.

Applying changes to the ontology. Based on the discovered relations, the user will be able to see the list of changes to be applied to the ontology. This includes the type of changes and their corresponding details. At the bottom of the screen, the user has the choice to apply the changes either on the base ontology or to create a new detached version. In the latter case, the new ontology version will appear in the left panel, having the current date and time appended to its URI.

Acknowledgements

This work was funded by the NeOn project sponsored under EC grant number IST-FF6-027595.

References

1. Alani, H., Harris, S., O'Neil, B.: Winnowing ontologies based on application use. In: Sure, Y., Domingue, J. (eds.) ESWC 2006. LNCS, vol. 4011, pp. 185–199. Springer, Heidelberg (2006)
2. Bloehdorn, S., Haase, P., Sure, Y., Voelker, J.: Ontology Evolution, pp. 51–70. John Wiley & Sons, Chichester (2006)
3. Cohen, W.W., Ravikumar, P., Fienberg, S.E.: A comparison of string distance metrics for name-matching tasks. In: Proceedings of the IJCAI 2003 Workshop on Information Integration on the Web (IIWeb 2003) (2003)
4. Haase, P., Stojanovic, L.: Consistent evolution of OWL ontologies. In: Gómez-Pérez, A., Euzenat, J. (eds.) ESWC 2005. LNCS, vol. 3532, pp. 182–197. Springer, Heidelberg (2005)
5. Klein, M.: Change Management for Distributed Ontologies. PhD thesis, Vrije Universiteit in Amsterdam (2004)
6. Novacek, V., Laera, L., Handschuh, S.: Semi-automatic integration of learned ontologies into a collaborative framework. In: IWOD (2007)
7. Noy, N.F., Chugh, A., Liu, W., Musen, M.A.: A framework for ontology evolution in collaborative environments. In: Cruz, I., Decker, S., Allemang, D., Preist, C., Schwabe, D., Mika, P., Uschold, M., Aroyo, L.M. (eds.) ISWC 2006. LNCS, vol. 4273, pp. 544–558. Springer, Heidelberg (2006)
8. Sabou, M., d'Aquin, M., Motta, E.: Exploring the Semantic Web as Background Knowledge for Ontology Matching. Journal on Data Semantics XI (2008)
9. Stojanovic, L.: Methods and Tools for Ontology Evolution. PhD thesis, FZI (2004)
10. Vrandecic, D., Pinto, H.S., Sure, Y., Tempich, C.: The DILIGENT knowledge processes. Journal of Knowledge Management 9, 85–96 (2005)
11. Zablith, F.: Dynamic ontology evolution. ISWC Doctoral Consortium (2008)
12. Zablith, F., Sabou, M., d'Aquin, M., Motta, E.: Using background knowledge for ontology evolution. In: IWOD Workshop at ISWC (2008)

Cupboard – A Place to Expose Your Ontologies to Applications and the Community*

Mathieu d'Aquin[1] and Holger Lewen[2]

[1] Knowledge Media Institute (KMi), The Open University, United Kingdom
m.daquin@open.ac.uk
[2] Institute AIFB, Universität Karlsruhe (TH), Germany
lewen@aifb.uni-karlsruhe.de

Abstract. In this demo, we present the Cupboard system for ontology publishing, sharing and reuse. This system is intended to support both ontology engineers and ontology users/practitioners. For the developers of ontologies, it offers a complete infrastructure to host their ontologies in online ontology spaces, providing mechanisms to describe, manage and effectively exploit these ontologies (through APIs). Furthermore, these ontologies are then exposed to the community, providing users with a complete, friendly environment to find, assess and reuse ontologies.

1 Introduction

In the recent years, there has been an ever-increasing amount of ontologies indexed by ontology search engines such as Watson[1]. Yet, even with so many ontologies now available to reuse, the tasks of finding, assessing and effectively making use of existing ontologies remain difficult. We argue that the main factor inhibiting ontology reuse relates to the lack of support for ontology practitioners to find, assess and exploit existing ontologies, and incidentally to the lack of support for ontology engineers to effectively deploy and expose their ontologies for use and reuse.

Ontology search engines provide a partial answer to this problem, as they support users in finding, selecting and, in some cases, using ontologies (see e.g., [1]). However, they still lack mechanisms for assessing ontologies, for providing rich ontology metadata, for providing alignments between ontologies, etc. In other terms, they only focus on providing the ontologies, without leaving ontology publishers and users any control or any possibility of interaction among themselves and with the ontologies. In addition, there have been several initiatives coming up with ontology repositories lately (see e.g., [2,3]). However, these systems are generally limited in their scope (e.g., considering only a particular domain or a particular type of ontologies) and leave only little control to the user over the ontology publication process.

* This work was partially supported by the European Commission in the IST project NeOn (IST-2006-027595).

[1] http://watson.kmi.open.ac.uk

L. Aroyo et al. (Eds.): ESWC 2009, LNCS 5554, pp. 913–918, 2009.

Apart from the Watson ontology search engine already mentioned, the NeOn project[2] is developing a number of technologies that, if put together and integrated in a broader system, can provide a more comprehensive answer to the aforementioned problem. More specifically, Oyster [4] is a peer-to-peer ontology sharing system which relies on rich metadata for ontologies, using the OMV format [5]. The alignment server [6] is an open system dedicated to the creation, management and evaluation of ontology alignments. In addition, a *topic specific open rating system* (TS-ORS) [7] for ontologies is being developed to allow users to review different properties of ontologies and to express trust in reviewers.

In this demo, we present Cupboard, an online system that relies on these components and integrates them to provide a complete and friendly environment for ontology publishing, sharing and reuse. One of the most important questions addressed in this development concerns the incentives provided for an ontology engineer to share his/her ontology and invest the time to properly add it to a repository. For many users, setting up their own server infrastructure, generating search indexes and setting up SPARQL-endpoints was quite a cumbersome task. Cupboard does not only allow users to add their ontologies in a personal space, but actually indexes them (using the Watson engine), provides mechanism to link them (using the Alignment Server), hosts them and exposes them through APIs and SPARQL, thus offering the infrastructure for engineers to actually deploy their ontologies with minimal efforts. Moreover, Cupboard is designed to be a community tool, helping ontology users and practitioners (including ontology developers) in finding and reusing ontologies, through the use of rich ontology metadata (thanks to Oyster and OMV) and to advanced ontology review mechanisms (using the TS-ORS system).

2 The Cupboard System

Cupboard is different from classical ontology repository systems as it does not provide one single space where ontologies are exposed. On the contrary, each user can create his/her own *ontology space*, containing the ontologies he/she has selected. In a sense, Cupboard can be seen more as a system to host ontology repositories, than as an ontology repository itself.

2.1 Overview

The main advantage in relying on this notion of ontology spaces is that each ontology space virtually implements a complete infrastructure for building semantic applications. Indeed, each uploaded ontology is automatically indexed using the Watson engine. Watson is designed to be an ontology search engine, but focuses on providing the necessary access mechanisms (through services, APIs, SPARQL, etc.) to enable semantic applications to dynamically exploit ontologies made available online (see [1]). In Cupboard, each ontology space

[2] `http://neon-project.org`

becomes like a little Watson, making it possible for developers to easily build applications exploiting the exposed ontologies.

However, for ontology and application developers, Cupboard's ontology spaces should do more than listing ontologies. To be used jointly in applications, ontologies have to be related with each other. Using the Alignment Server [6] as a core component, Cupboard allows users to populate their ontology spaces not only with ontologies, but also with alignments (mappings). Using the *Alignment Format* [6], alignments can be uploaded for a given (pair of) ontology(ies). The Alignment server then offers to Cupboard the necessary features to support the management, evaluation, and even production of alignments.

Uploading an ontology into Cupboard is not only a way to build applications, but also a way to share this ontology with the community for reuse. It is therefore important to provide efficient ways for users to find, assess and select ontologies in Cupboard's ontology spaces. The system naturally inherits the powerful search mechanisms implemented in the Watson engine. In addition, information for example on the provenance of the ontology or the methodology used to build it are essential for ontology selection. Cupboard relies on the Oyster system [4] to provide users with the possibility to enter rich metadata about ontologies (in the OMV format [5]) and to manage these metadata.

Finally, one of the most important issues hampering reuse concerns the evaluation of ontologies. In Cupboard, we intend to exploit the community feature of the system, so that users can provide precise reviews on ontologies (reviewing separately different characteristics, like reusability, complexity, coverage, etc.) The TS-ORS System [7], used to handle and manage these reviews, implements a mechanism to rank ontologies according to these evaluations. However, reviews are subjective evaluations, as users might have different views on ontologies. When searching for an ontology to reuse, not only the review information should be exploited to support the user, but also the information about how much he/she would trust a particular review or a particular reviewer. Such trust information is obtained through the classical "Did you find this review useful?" question, and handled by the TS-ORS system to compute trust information used to personalize ontology ranking.

Figure 1 shows different parts of the Web interface of Cupboard, in particular displaying an ontology space, uploading an ontology, inspecting and searching the content of a particular ontology, reviewing it, etc.

2.2 Architecture

While the different technologies employed and combined in the system have been described and implemented before, Cupboard provides homogeneous (user and programming) interfaces to these components and features. The main challenges during integration were defining the interfaces between components, optimizing the performance and generating an easy-to-use Web interface. However, this has been made easier by the fact that the integrated systems are designed to work as independent, remote components, resulting in a naturally distributed architecture for Cupboard.

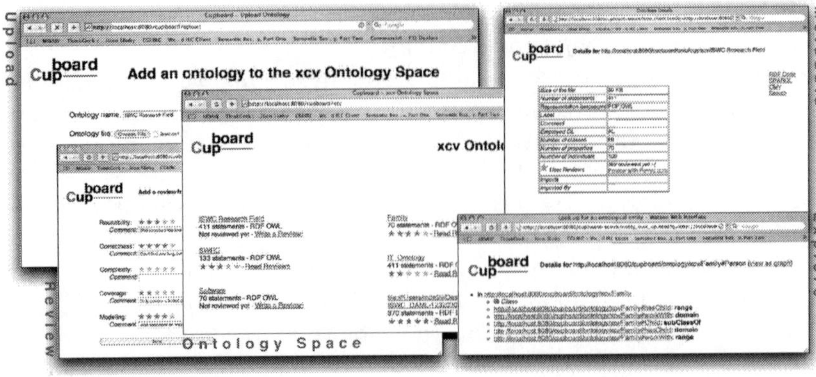

Fig. 1. Screenshots of some components of the Cupboard Web interface

As depicted in Figure 2, the central component of Cupboard is a specifically made component, called the *Cupboard core*, which is in charge of orchestrating the interaction between all the other components, external applications and users. It heavily relies on the Watson engine to provide basic functionalities such as storing the ontologies, indexing them for search, validating them and exposing them to applications. The TS-ORS open rating system, the Oyster ontology metadata sharing system and the Alignment Server each provide specific functionalities to the system, that are federated and exposed in an homogeneous way to the Cupboard Web interface and APIs, through the Cupboard core.

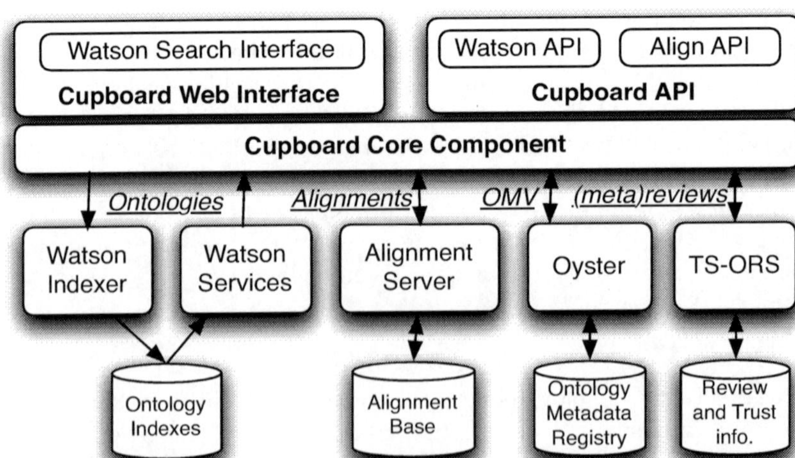

Fig. 2. Cupboard's architecture

3 Demonstration and Availability

In the demonstration, users will learn how to use Cupboard, to upload, expose and explore their ontologies, as well as to find, select, reuse and assess other users' ontologies. They will learn about the benefits of using Cupboard through a variety of concrete, real-life scenarios, supporting different user groups (ontology engineers, application developers, ontology publishers, etc.) in tackling different tasks (ontology sharing, reuse, application, evaluation, etc.).

At the time of writing, Cupboard is being tested by a selected group of users. It will be made openly available (at `http://cupboard.open.ac.uk`) by June 2009, allowing users to register and create their own ontology spaces. At a later stage, Cupboard will also be distributed as an open source system, to allow organizations to deploy their own Cupboard servers, in case for example intellectual property restrictions prevent them from sharing their ontologies online.

4 Conclusion

In this document, we presented Cupboard, an online ontology publishing and sharing environment that addresses two major problems faced by the Semantic Web community: the deployment of inter-linked ontologies for exploitation in applications, often requiring heavy infrastructures, and the reuse of existing ontologies where, besides search mechanisms, rich metadata and proper assessment of the ontologies are required. To tackle these problems, we propose a complete system, based on a number of well-established technologies, allowing ontology engineers to deploy their ontologies, providing the necessary infrastructures to support their exploitation, and ontology users in reusing available knowledge, providing essential, community-based functionalities to facilitate the search, selection and exploitation of the available ontologies. To the best of our knowledge, Cupboard is the first system to put together all these functionalities to create an essential infrastructure component for Semantic Web developers and more generally, a useful, shared and open environment for the ontology community.

References

1. d'Aquin, M., Motta, E., Sabou, M., Angeletou, S., Gridinoc, L., Lopez, V., Guidi, D.: Toward a new generation of semantic web applications. IEEE Intelligent Systems 23(3), 20–28 (2008)
2. Musen, M., Shah, N., Noy, N., Dai, B., Dorf, M., Griffith, N., Buntrock, J., Jonquet, C., Montegut, M., Rubin, D.: BioPortal: Ontologies and Data Resources with the Click of a Mouse. In: Proc. of the Annual Symposium proceedings/AMIA Symposium. AMIA Symposium, AMIA Annu. Symp. Proc., 1223 (2008)
3. Buitelaar, P., Eigner, T., Declerck, T.: OntoSelect: A Dynamic Ontology Library with Support for Ontology Selection. In: Proc. of the Demo Session at the International Semantic Web Conference, Hiroshima, Japan (2004)
4. Palma, R., Haase, P.: Oyster - sharing and re-using ontologies in a peer-to-peer community. In: Gil, Y., Motta, E., Benjamins, V.R., Musen, M.A. (eds.) ISWC 2005. LNCS, vol. 3729, pp. 1059–1062. Springer, Heidelberg (2005)

5. Hartmann, J., Palma, R., Sure, Y., del Carmen Suárez-Figueroa, M., Haase, P., Gómez-Pérez, A., Studer, R.: Ontology metadata vocabulary and applications. In: Meersman, R., Tari, Z., Herrero, P. (eds.) OTM-WS 2005. LNCS, vol. 3762, pp. 906–915. Springer, Heidelberg (2005)
6. Euzenat, J.: An API for ontology alignment. In: McIlraith, S.A., Plexousakis, D., van Harmelen, F. (eds.) ISWC 2004. LNCS, vol. 3298, pp. 698–712. Springer, Heidelberg (2004)
7. Lewen, H., Supekar, K., Noy, N.F., Musen, M.A.: Topic-Specific Trust and Open Rating Systems: An Approach for Ontology Evaluation. In: Proc. of the 4th International Workshop on Evaluation of Ontologies for the Web (EON 2006) at the 15th International World Wide Web Conference (WWW 2006) (2006)

Effects of Using a Research Context Ontology for Query Expansion

Carola Carstens

German Institute for International Educational Research, Information Center for Education
Schloßstraße 29, 60486 Frankfurt, Germany
carstens@dipf.de

Abstract. This thesis investigates the question whether and how ontologies such as the ones currently evolving in the Semantic Web can serve as knowledge structures for the generation of query expansion terms in information retrieval systems. This issue is examined using a specific example domain, namely educational research. Initial results support the already well-researched finding that query expansion can increase recall. Subsequent experiments will focus on comparing the effectiveness of thesaurus and ontology-based query expansion, on identifying ontological relationships which are especially useful for the generation of query expansion terms in the domain of educational research, and on evaluating the usefulness of different ontological relationships as expansion terms for different types of queries, as well as for different query expansion modes.

Keywords: Information Retrieval, Ontologies, Semantic Web, Query Expansion.

1 Research Interest

Following the Semantic Web vision [1], more and more ontologically organized Semantic Web data is currently being produced. With regard to the retrieval of this data, the Semantic Web has a strong focus on data retrieval, allowing to query the data by formal query languages. In addition to being directly queryable for certain ontology objects, ontological knowledge bases can also support information retrieval scenarios. Ontologically structured Semantic Web data constitutes a new abundance of corpus-independent knowledge that may be exploited for the implementation of query expansion mechanisms.

While the existing research on query expansion focuses on the use of thesauri, this research aims at investigating the use of ontologies for that purpose, their semantic relationships usually being more numerous than those typically comprised in thesauri. As research context related data is widely available on the Semantic Web, the use of a research context ontology for query expansion is examined. More specifically, the domain of educational research serves as an example application area.

The main research interest of this work is to investigate if ontologies such as the ones currently evolving in the Semantic Web are promising knowledge structures for

L. Aroyo et al. (Eds.): ESWC 2009, LNCS 5554, pp. 919–923, 2009.

the implementation of query expansion mechanisms in information retrieval systems. Furthermore, the following more-detailed research questions are addressed:

- Can ontologies generate added value in query expansion mechanisms, as compared to thesauri?
- Which ontological relationships are most useful as query expansion terms for the field of educational research?
- Which ontological query expansion terms are most suitable for which type of query terms (concept, project, person, organization queries)?
- Which ontological relationships are suitable for automatic query expansion; which for interactive query expansion?

2 State of the Art

In the Semantic Web, a large proportion of the data that is published is research context related data. This is for example illustrated by the datasets of the ISWC and ESWC conferences[1] and the data of the AIFB semantic portal of the University of Karlsruhe[2]. For the purpose of publishing such data, the vocabulary of evolving standard ontologies such as SWRC [2], FOAF[3] and SKOS[4] can be used.

While this semantically interpretable data can be easily retrieved with the help of formal query languages (data retrieval), the role that full-text search and information retrieval play on the Semantic Web is still an open research issue [3, 4]. One step towards the integration of data and information retrieval is the augmentation of traditional search engine results with structured Semantic Web data. For example, this is achieved in the Semantic Search application by Guha et al. [5] and is also supported by Yahoo!'s SearchMonkey initiative[5].

Another possible integration of ontologies and information retrieval is the use of ontologically structured data as background knowledge for the implementation of query expansion mechanisms in information retrieval systems. Much research has already been conducted on the use of corpus-independent thesauri for this purpose [6, 7, 8, 9]. This research has shown that thesaurus-based query expansion often induces an increase in recall, usually accompanied by a significant loss in precision.

Taking a more detailed look at the effect of certain thesaurus relationships on the effectiveness of query expansion, Greenberg determined that synonyms and narrower terms are well suited for automatic query expansion, because they "increased relative recall with a decline in precision that was not statistically significant" [6]. This finding was further reinforced in her follow-up study focusing on the differences between automatic query expansion and interactive query expansion [7]. A more recent study by Navigli and Velardi examined the use of expansion terms derived from WordNet [10], coming to the conclusion that the use of gloss words for query expansion achieved top scores for the precision@10 measure, outmatching query expansion by synsets and hyperonyms, for example.

[1] http://data.semanticweb.org/dumps/
[2] http://www.aifb.uni-karlsruhe.de/english
[3] http://www.foaf-project.org/
[4] http://www.w3.org/2004/02/skos/
[5] http://developer.yahoo.com/searchmonkey/

As in the aforementioned study, several other authors also refer to thesauri as ontologies in the context of query expansion, although these lightweight ontologies are mostly restricted to typical thesaurus relationships, namely synonyms, broader terms, narrower terms and related terms [11]. Query expansion based on ontologies with additional semantic relationships, by contrast, has yet to be studied extensively.

3 Research Design

In order to simulate the existence of Semantic Web data for the domain of educational research, a domain-specific ontological knowledge base is created in the first step. The research context ontology (rescon) schema is modeled around the basic concepts *person*, *organization*, *project* and *concept*. With the aim of representing commonly used Semantic Web relationships, the schema partly integrates vocabulary from the FOAF, SWRC and SKOS ontologies, as illustrated in figure 1. This schema is instantiated with data from several domain-specific, structured or semi-structured data sources, resulting in a total of 27,082 concepts, 5,526 persons, 7,665 organizations and 540 projects. As shown in figure 1, the objects are interrelated by semantic relationships such as *swrc:employs*, *skos:related* and *foaf:topic_interest*.

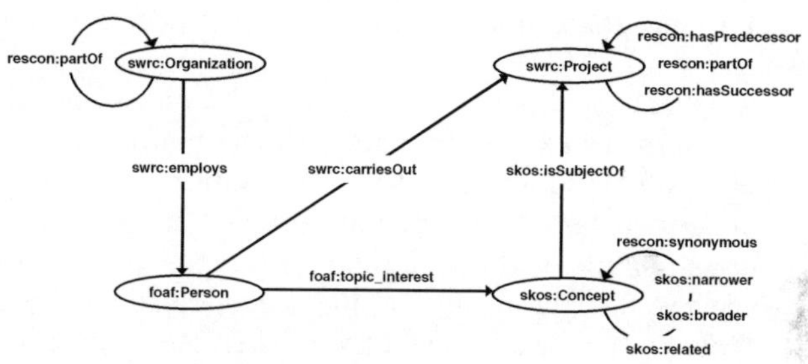

Fig. 1. Research context ontology schema

In the second step, a prototypical retrieval system based on Lucene[6] is implemented, incorporating both an automatic and an interactive mode for query expansion. Its configuration determines which ontology relationships are used for the generation of query expansion terms.

In the automatic query expansion mode, the expansion terms are added directly to each of the original query terms with the Boolean OR operator, before the query is sent to the Lucene index. This is known as the building-block strategy [12]. In the interactive mode, by contrast, the system generates expansion terms as suggestions that are presented to the user, who can then decide whether or not to make use of them to start a new search. Moreover, the user has the option to either broaden the original query or to narrow it down with the Boolean AND and OR operators.

The system's effectiveness is then measured in information retrieval experiments. This evaluation phase consists of two parts: the first part focuses on the automatic query expansion mode, while the second part investigates the interactive mode.

[6] http://lucene.apache.org/

In the automatic mode, different ontology relationships are used sequentially as query expansion terms to determine how they affect information retrieval effectiveness. This can be evaluated in terms of the standard measures recall and precision [13]. Depending on the type of query – be it a *concept*, a *person*, an *organization* or a *project* – the use of each semantic relationship as a query expansion term is analyzed sequentially. Accordingly, the concept query *Musikunterricht* (*music lessons*) can be expanded by the synonym *Musikerziehung* (*music education*), the related term *Tanzunterricht* (*dance lessons)*, the broader term *Unterricht* (*lessons)* and the narrower term *Privatmusikunterricht* (*private music lessons*).

Compared to the automatic mode, the effectiveness of interactive retrieval systems is more difficult to evaluate, because the system's effect is difficult to isolate from other variables, such as user influences [14:15]. To circumvent this problem, a pre-questionnaire is used to record data about the search experience and the topic-related knowledge of the participants. In a between-groups design, one group of users works with a baseline retrieval system while another group works with the ontology enhanced system. As in the final TREC Interactive Track experiments [14], users are given realistic tasks to accomplish, such as to identify documents that give evidence of the cooperation of two organizations, for example. Comparable topics are defined for each of the different types of queries. Finally, it is possible to analyze which group accomplishes the tasks most effectively based on the number of result documents gathered in a certain time frame and the number of steps necessary for collecting the result documents.

This research setup makes it possible to identify which ontological relations are most suitable for which query expansion mode, which relations are most effective as query expansion terms for the domain under consideration, which relations are most effective for certain query types, and if ontological relations can generate added value compared to thesaurus relations. Although the experiments apply to one specific domain, it can be expected that the results on the use of relationships from a general research context ontology may be at least partly transferable to other domains.

4 Preliminary Results and Outlook

At the current stage, a pre-test for the automatic mode has been conducted, which already has some implications for the design of further experiments. For this pre-test, an excerpt from the German Education Index[7], a bibliographic database for the domain of educational research, served as a test corpus. Its 215,948 documents were all in German and contained full-text abstracts. A Lucene index was built using the document fields *abstract*, *title* and *subtitle*. The intellectually assigned keywords of the documents served as relevancy judgments. As these keywords were restricted to terms of the type *concept*, only the relations *synonymous (SYN)*, *narrower (NT)*, *broader (BT)* and *related (RT)* have been used sequentially as query expansion terms in this experiment. The average effects of these relations on information retrieval effectiveness are listed in table 1[8]. On the whole, this experiment took into account 3,067 SYN, 1,404 NT, 1,625 BT and 1,013 RT expansion terms.

[7] http://www.fachportal-paedagogik.de/start_e.html

[8] Expansion terms that did not occur in the test corpus were excluded from this calculation.

Table 1. Average influence of expansion terms on retrieval effectiveness (rounded)

	SYN	NT	BT	RT
Average recall deviation from baseline query	+ 8 %	+ 1%	+ 15%	+ 5%
Average precision deviation from baseline query	- 5%	- 3%	- 38%	- 15%

These preliminary results are in line with the results attained by Greenberg [6] insofar as they show that query expansion with semantic relationships can induce an increase in recall. Nevertheless, Greenberg reports a much higher increase in recall, up to nearly one quarter. This difference may be due to the fact that certain expansion terms only scarcely occur in the test corpus, thereby not permitting the expanded query to detect a high number of additional relevant documents. Therefore, the corpus frequency of the expansion terms will be taken into account in further calculations.

These pre-test experiences will inspire the more comprehensive upcoming experiments. By applying further evaluation measures, such as the precision at different ranks, and by studying user influences in the interactive mode, they can be expected to produce even more meaningful results in practice.

References

1. Berners-Lee, T., Hendler, J., Lassila, O.: The Semantic Web. Scientific American (May 2001)
2. Sure, Y., Bloehdorn, S., Haase, P., Hartmann, J., Oberle, D.: The SWRC Ontology - Semantic Web for Research Communities. In: Bento, C., Cardoso, A., Dias, G. (eds.) EPIA 2005. LNCS, vol. 3803, pp. 218–231. Springer, Heidelberg (2005)
3. Scheir, P., Pammer, V., Lindstaedt, S.N.: Information Retrieval on the Semantic Web - Does it exist? In: LWA 2007, pp. 252–257. Martin-Luther-University Halle-Wittenberg (2007)
4. Finin, T., Mayfield, J., Joshi, A., Cost, R.S., Fink, C.: Information Retrieval and the Semantic Web. In: 38th Annual Hawaii International Conference on System Sciences (2005)
5. Guha, R., McCool, R., Miller, E.: Semantic Search. In: 12th International Conference on World Wide Web, pp. 700–709 (2003)
6. Greenberg, J.: Automatic Query Expansion via Lexical-Semantic Relationships. Journal of the American Society for Information Science and Technology 52(5), 402–415 (2001)
7. Greenberg, J.: Optimal Query Expansion (QE) Processing Methods with Semantically Encoded Structured Thesauri Terminology. Journal of the American Society for Information Science and Technology 52(6), 487–498 (2001)
8. Kristensen, J.: Expanding End-Users' Query Statements for Free Text Searching with a Search-Aid Thesaurus. Information Processing and Management 29(6), 733–744 (1993)
9. Voorhees, E.M.: Query Expansion using Lexical-Semantic Relations. In: Annual ACM Conference on Research and Development in Information Retrieval, pp. 61–69 (1994)
10. Navigli, R., Velardi, P.: An Analysis of Ontology-based Query Expansion strategies. In: 14th European Conference on Machine Learning, pp. 42–49 (2003)
11. ISO 2788: Documentation - Guidelines for the Establishment and Development of Monolingual Thesauri. International Organization for Standardization (1986)
12. Harter, S.P.: Search Strategies and Heuristics. In: Online Information Retrieval. Concepts, Principles and Techniques, pp. 170–204. Academic Press, Orlando (1986)
13. Salton, G., McGill, M.: Introduction to Modern Information Retrieval. McGraw-Hill, New York (1983)
14. Voorhees, E.M., Harman, D.K.: TREC: Experiment and Evaluation in Information Retrieval. MIT Press, Cambridge (2005)

Towards a Semantic Infrastructure for User Generated Mobile Services

Marcin Davies

Telecommunications Research Center Vienna (ftw.), Austria
davies@ftw.at

Abstract. This paper presents a research approach towards a semantic infrastructure for user generated mobile services. Building on the concept of semantic *microservices*, the aim of this work is to sufficiently lower the barrier for end-users in order to enable easy ad-hoc creation, customisation, and discovery of mobile services.

Keywords: Semantic web services, end-user development, user-centric service creation, microservices, mobile services, web 2.0.

1 Introduction and Problem Description

The current trend of active user participation and service creation on the Internet (Web 2.0, mashups, widgets, etc.) will certainly continue within mobile environments. The enabler technology is mostly available as mobile terminals have an enormous capacity in terms of processing power, interaction, and interfacing. Furthermore, high-speed data coverage as well as flat-fee data plans are increasing to enable an 'always-on' experience in the mobile world.

There is massive potential when mobile devices not only present data but provide valuable information and services to other users. Instant mobile services like traffic jam information, recommended places, or friend locations can be of benefit to a large audience and will pave the way to a new mobile internet empowering end-users to easily find and provide relevant, context-aware information.

Although early approaches towards end-user service creation on the web exist (e.g. Yahoo! Pipes [1]), they cannot be directly transferred to mobile environments (user interface issues, different architecture, etc.). Furthermore, they are too complex for a widespread use (i.e. requiring programming skills) and to facilitate spontaneous service creation on the go.

Driven by the main research question *whether and how much semantic technologies can lower the barrier for end-users in order to enable easy ad-hoc creation, customisation, and discovery of mobile services*, this thesis introduces the concept of *microservices*[1] — small, sharply focused, and semantically annotated services that can be flexibly created, adapted, and discovered by end-users. Leveraging this concept, this work will elaborate a semantic infrastructure as a step towards a broad adoption of mobile microservices.

[1] The term also refers to the fact that these services are running on small devices.

L. Aroyo et al. (Eds.): ESWC 2009, LNCS 5554, pp. 924–928, 2009.

2 Related Work and Areas of Contribution

This thesis project will be based on and contribute to the state of the art in three main areas, namely (1) Service Description and Modelling, (2) Knowledge Warehousing and Service Discovery, and (3) End-User Service Creation Paradigms.

2.1 Service Description and Modeling

Semantic Web Services (SWS) commonly operate on service descriptions encoded in OWL [2] or RDF [3] and service models such as OWL-S [4] or WSMO [5]. In addition, there exist more lightweight approaches such as SAWSDL [6] and SA-REST [7] that put a semantic layer on traditional, non-semantic service description methods such as WSDL [8]. However, none of these techniques are tailored to the needs of mobile services, which are fundamentally different in terms of e.g. life span, context dependency, or service availability. Moreover, existing toolkits for describing SWS (e.g. [9]) are directed towards professionals and are not feasible for end-users while on the go. Therefore, the thesis will explore user-friendly ways of mobile service description and annotation.

Microservice descriptions are interpreted by an execution environment (similar to a virtual machine) on a mobile device and can be shared and customised by other users to create new services. Since existing approaches such as WS-BPEL [10] and model-driven architectures and languages like Executable UML [11] are too much focused on big software systems, this work will present a lightweight microservice model building on the REST [12] architecture that allows for easy integration with existing Web standards and protocols.

2.2 Knowledge Warehousing and Service Discovery

A central part of this thesis will consider management and discovery of the previously discussed semantic microservice descriptions. Existing and widespread software like Jena [13] or Sesame [14] allows for arbitrary ontology management, but needs to be enriched by an additional layer supporting community-driven ontologies [15] and ontology matching [16] for microservices and their provided content, revision of knowledge (due to a possibly limited life span of microservices) and content- and context-aware recommendation mechanisms (to deliver meaningful results in mobile scenarios).

2.3 End-User Service Creation Paradigms and Methods

Current examples for end-user created software on the Web such as Microsoft Popfly [17] allow users to create new web applications by visual editing of functional blocks. However, these examples can only be regarded as a first step, both in terms of ease of use (still requiring significant knowledge) and functionality (mostly restricted to aggregation of information not provision of it). With respect to the mobile domain, current service creation approaches (e.g. [18,19]) are very limited (again in terms of functionality) or not tailored to mobile devices and usage scenarios (with regard to user interfaces, capabilities, and context information).

The research area of End-User Development [20] considers different approaches for end-users to create software (e.g. Programming by Example). However, these directions will need to be explored in the context of software services, open communities, and mobile usage patterns. In addition, current end-user development methods are still at an early stage and can benefit from semantic technologies (e.g. by using annotations or rules to describe the behavior of functional blocks).

3 Methods and Approach

The overall methodology for the thesis is shown in Fig. 1 and explained in detail below.

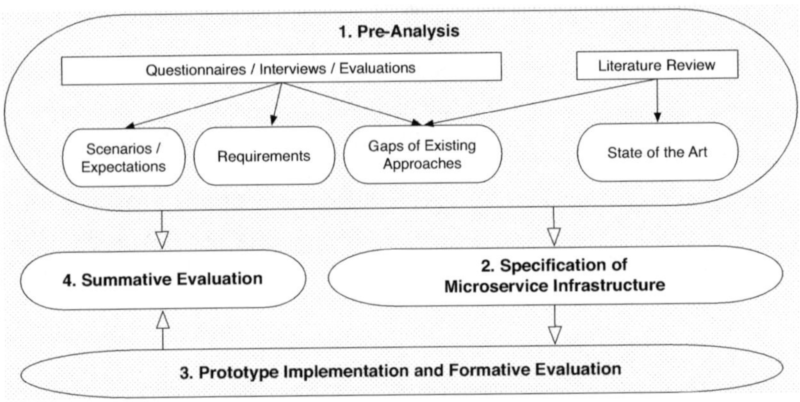

Fig. 1. Overall Methodology

1. *Pre-Analysis*
   In this first phase of the thesis, initial requirements, possible usage scenarios, and user expectations for microservices will be collected by interviews and web questionnaires. In addition, existing frameworks and tools (e.g. LabView [21], Yahoo! Pipes [1], Microsoft Popfly [17]) and small mockups for alternative approaches (e.g. question-based service creation) will be evaluated with regard to specific criteria, e.g. time needed, perceived complexity, and quality of experience (QoX). This pre-analysis should also help to explore gaps in existing solutions that semantic microservices might be able to fill.
2. *Specification of Microservice Infrastructure*
   Based on these results and, of course, review of existing work and the state of the art, a suitable semantic microservice infrastructure will be specified and described. This also includes semantic descriptions of the basic components needed to build microservices.
3. *Prototype Implementation and Formative Evaluation*
   The next step of the thesis will be a prototype implementation of the semantic microservice infrastructure. This framework will consist of server-side software to search and discover microservices (knowledge warehouse) as well as a mobile client that allows for creation, customisation, and execution

of microservices. By following an iterative design approach, the semantic components derived in early stages of the development will be evaluated in small-scale user studies[2] in order to gather early user feedback and to refine the prototype implementation.
4. *Summative Evaluation*
Finally, the validity of the developed infrastructure will be assessed in a final user study with regard to easier service creation, customisation, and discovery and by considering the criteria derived in the pre-analysis phase. In addition, the study will also show the applicability to different usage scenarios, e.g. tourism, gaming, recreation, or social networking.

4 Preliminary Results

Besides a first definition of scenarios, requirements, and a general system architecture for microservices (cf. [23]), the most relevant achievements so far are towards a specification for the microservice description language. As outlined in Sect. 2.1, microservice descriptions will be used during the whole microservice lifecycle (e.g. creation, modification, execution) and are therefore an essential part of a microservice infrastructure. Currently, the specification foresees the following language parts (more information can be found in [24]):

- *Microservice Profile:* provides a descriptive and functional definition of the microservice by an extensible set of metadata elements
- *Microservice Logic:* controls the data flow, capability and content handling, and all other operational aspects of a microservice
- *Microservice Rendering:* defines the service user interface and provides an abstract and device-independent layer for service presentation
- *Microservice Content:* describes the content that is manipulated or offered by a microservice

5 Conclusions and Future Work

This paper introduced the idea of microservices and described related work and areas of contribution. After outlining the planned approach, this work also presented some preliminary results. According to the methodology shown in Sect. 3, the next major steps are to complete the pre-analysis phase (already started) in order to receive a sound basis for the infrastructure specification (where work has been started too — cf. Sect. 4). Finally, this infrastructure will be prototypically implemented and evaluated in user studies.

Acknowledgements. This work has been supported by the European Commission, ICT FP7 Collaborative Project m:Ciudad [25]. The Telecommunications Research Center Vienna (ftw.) is supported by the Austrian government and the City of Vienna within the competence center program COMET. Thanks to Anna V. Zhdanova for extensive feedback and supervision of this work.

[2] If needed, 'Wizard-of-Oz' techniques [22] will be applied that simulate behavior that is actually not yet or fully implemented.

References

1. Yahoo! Pipes, http://pipes.yahoo.com
2. McGuinness, D.L., Van Harmelen, F.O.: Web Ontology Language Overview, W3C Recommendation (February 10, 2004)
3. Manola, F., Miller, E.: RDF Primer, W3C Recommendation (February 10, 2004)
4. Martin, D., et al.: OWL-S: Semantic Markup for Web Services, W3C Member Submission (November 22, 2004)
5. de Bruijn, J., et al.: Web Service Modeling Ontology (WSMO), W3C Member Submission (June 3, 2005)
6. Farrell, J., Lausen, H.: Semantic Annotations for WSDL and XML Schema, W3C Recommendation (August 28, 2007)
7. Lathem, J., Gomadam, K., Sheth, A.P.: SA-REST and (S)mashups: Adding Semantics to RESTful Services. In: International Conference on Semantic Computing (ICSC), pp. 469–476 (September 2007)
8. Christensen, E., Curbera, F., Meredith, G., Weerawarane, S.: Web Services Description Language (WSDL) 1.1, W3C Note (March 15, 2001)
9. Kerrigan, M., Mocan, A.: The Web Service Modeling Toolkit. In: Bechhofer, S., Hauswirth, M., Hoffmann, J., Koubarakis, M. (eds.) ESWC 2008. LNCS, vol. 5021, pp. 812–816. Springer, Heidelberg (2008)
10. Barreto, C.: WS-BPEL Version 2.0 Primer
11. Mellor, S.J., Balcer, M.J.: Executable UML: A Foundation for Model-driven Architecture. Addison-Wesley, Reading (2002)
12. Fielding, R.T.: Architectural Styles and the Design of Network-based Software Architectures. PhD thesis, University of California, Irvine (2000)
13. Jena: A Semantic Web Framework for Java, http://jena.sourceforge.net
14. SourceForge.net: Sesame, http://sourceforge.net/projects/sesame
15. Zhdanova, A.V.: Community-Driven Ontology Construction in Social Networking Portals. Web Intelligence and Agent Systems 6(1), 93–121 (2008)
16. Zhdanova, A.V., Shvaiko, P.: Community-Driven Ontology Matching. In: Sure, Y., Domingue, J. (eds.) ESWC 2006. LNCS, vol. 4011, pp. 34–49. Springer, Heidelberg (2006)
17. Microsoft Popfly, http://www.popfly.com
18. WidSets: Mobilize Your Web, http://www.widsets.com
19. Shin, Y., Yu, C., Chung, S., Kim, S.: End-User Driven Service Creation for Converged Service of Telecom and Internet. In: Fourth Advanced International Conference on Telecommunications (AICT), pp. 71–76 (June 2008)
20. Lieberman, H., Paternò, F., Wulf, V.: End-User Development. Springer, Heidelberg (2006)
21. National Instruments: LabVIEW, http://www.ni.com/labview
22. Dow, S., MacIntyre, B., Lee, J., Oezbek, C., Bolter, J.: Wizard of Oz Support throughout an Iterative Design Process. IEEE Pervasive Computing (January 2005)
23. Davies, M., et al.: m:Ciudad: An Infrastructure for Creation and Sharing of End User Generated Microservices. In: Poster and Demonstrations Proceedings of the 1st Future Internet Symposium (FIS), Vienna, Austria, pp. 24–26 (September 2008)
24. Carrez, F., et al.: D2.3: m:Ciudad U+ Service Description. Project Deliverable, m:Ciudad ICT FP7 Collaborative Project (January 2009)
25. Project m:Ciudad FP7, http://www.mciudad-fp7.org

Relational Databases as Semantic Web Endpoints

Matthias Hert

Department of Informatics, University of Zurich,
Binzmuehlestrasse 14, CH-8050 Zurich, Switzerland
hert@ifi.uzh.ch

Abstract. This proposal explores the promotion of existing relational databases to Semantic Web Endpoints. It presents the benefits of ontology-based read and write access to existing relational data as well as the need for specialized, scalable reasoning over that data. We introduce our approach for translating SPARQL/Update operations to SQL, describe how scalable reasoning can be realized by using the power of the database system, and outline two case studies for evaluating our approach.

1 Motivation

Relational Databases (RDBs) are used in most current enterprise environments to store and manage data. While RDBs are well suited to handle large amounts of data they were not designed to preserve the semantics of that data. This is especially relevant if data from different parties needs to be exchanged or integrated.

The Semantic Web provides a common framework that allows data to be shared and reused across application, enterprise, and community boundaries. Although invented for the Web, these technologies have proven to be useful in other domains as well. Recent research[1] has shown that an ontology-based access to relational data reduces the barriers for data exchange and integration. The expressive and formal semantics increases the value of the existing data and enables new applications on that data.

Ontology-based access to relational data can be achieved in two different ways: (1) All relational data is converted at once to a Semantic Web format, eliminating the need for the original database or (2) the relational database is preserved and RDF data is provided on demand through a mediator that translates between Semantic Web requests and SQL. The one-time conversion approach is often not feasible due to the following reasons: (1) *Scalability* of RDF storage systems is still a topic of research, while relational databases were highly optimized over the past thirty years; (2) The existence of *legacy applications* prohibits the elimination of the relational database; (3) Mirroring the data in an RDF store does not only increase the total data size but also leads to *consistency*

[1] E.g. the Linked Data project at http://linkeddata.org

L. Aroyo et al. (Eds.): ESWC 2009, LNCS 5554, pp. 929–933, 2009.

issues if the original data or the RDF mirror is updated. In contrast, the on demand provision has the drawback that it does not constitute a native Semantic Web data source and therefore leaves behind a conceptual gap between the two worlds of relational data and the Semantic Web.

The goal of this thesis is to promote relational databases to Semantic Web endpoints that enable ontology-based read/write access and offer scalable reasoning support. This is achieved with a wrapper in front of the database that mediates between the relational and the Semantic Web representations. Therefore, we profit from the strength of both technologies, the scalability in terms of response time and database size as well as the explicit, formal semantics that enable data exchange, integration, and reasoning.

2 State of the Art

Ontology-based Access to RDBs: The W3C developed SPARQL [1] as the standard query language for RDF which also provides a protocol [2] for remote read access to RDF data on so-called SPARQL endpoints. Relational.OWL [3] defines an ontology to represent relational structures (tables, attributes, keys, etc.) in RDF. This enables Semantic Web technologies to access the relational data, but the structure of the RDF data is limited to the database schema. RDQuery [4] enables SPARQL-based access to relational data by translating the queries to SQL via mappings expressed in Relational.OWL. D2R Server [5] wraps existing RDBs into SPARQL endpoints based on its own declarative mapping language D2R MAP [6] that maps database tables and attributes to corresponding classes and properties in an user-defined ontology. This allows to reuse existing vocabularies and avoids the exposure of the database schema to the ontology representation. OpenLink Virtuoso[2] is a commercial database system featuring RDF Views that allow querying the relational data with SPARQL. The World Wide Web Consortium (W3C) has recognized the importance of mapping relational data to the Semantic Web by starting the RDB2RDF incubator group (XG)[3] to investigate the need for standardization. The XG recommends that the W3C starts a working group to produce a standard RDB to RDF mapping language.

SPARQL/Update is a language to express updates to an RDF store that reuses the syntax of SPARQL. It supports operations to insert and delete RDF data in existing graphs as well as create and remove graphs in an RDF store [7]. To the best of our knowledge, none of the existing RDB to RDF mapping approaches support data manipulations via SPARQL/Update.

Reasoning Over Data in RDBs: Reasoning over large ABoxes as found in mapped RDBs pose a scalability limitation to current reasoners. OWLDB [8] aims to overcome this by offloading the reasoning task to the relational database system. The inference rules of OWL DL are translated to equivalent SQL queries that are executed to materialize and store the inference inside the database.

[2] `http://virtuoso.openlinksw.com`
[3] `http://esw.w3.org/topic/Rdb2RdfXG`

MASTRO-I [9] is a data integration framework that uses DL-Lite [10] ontologies as the global schema. The family of DL-Lite description logics were designed to enable a maximum of expressiveness while remaining within the LOGSPACE data complexity that allows to perform reasoning as standard SQL query evaluation.

3 Proposed Approach

Current approaches for mapping relational data to the Semantic Web as described in Section 2 are limited to read-only access. The relational data is immutable to the Semantic Web layer, which means modifications to the RDF data cannot be propagated back into the RDB. However, applications of Semantic Web technologies apart from the Web domain (e.g. in enterprise environments) would benefit from the possibility of persistent modifications to RDF data. Furthermore, mapping industry-size relational data to RDF generates a large amount of instance data which blasts current in-memory reasoning approaches.

In this PhD thesis, we strive for a promotion of existing relational databases to Semantic Web Endpoints, i.e., ontology-based read and write access to RDBs as well as support for scalable reasoning over the relational data. The use of relational databases as the storage layer should ideally be transparent to the developer of ontology-based applications, however due to the conceptual differences (relational versus triple-orientied representation of the data) this cannot be realized in general. Our goal is to achieve a maximum of transparency while providing support to the developer in the cases where the underlying storage system is exposed (e.g. if a SPARQL/Update insert operation does not contain enough data to create a legal tuple in a database table). This will be achieved through a mediator in front of existing RDBs that, based on mappings, translates requests, results, and errors between the relational representation and the Semantic Web. Ontology-based read access will be provided through a SPARQL endpoint and write access through a corresponding SPARQL/Update endpoint. This enables a clear separation between data and applications, independence from any specific Semantic Web framework, and the possibility to remotely access the data. As shown in [11], SPARQL and relational algebra have equivalent expressive power, which means SPARQL queries can be translated completely to SQL for efficient execution by the database engine. We will investigate how the SPARQL/Update language can be translated to the SQL data manipulation language (DML) and in particular how each of the SPARQL/Update operations is mapped to SQL DML with respect to our application scenario. We will build a prototype implementation that can be set up in front of an existing database to enable ontology-based access to the relational data.

Reasoners that need to materialize all data in main memory are not feasible in our application scenario as the instance data in the database exceeds the main memory size. In addition, the support for data updates renders approaches based on pre-computing and caching of the inference unsuitable as it would need to be validated or recomputed after every update. Therefore, efficient reasoning needs to be performed at query time to avoid scalability problems. The OWL 2 QL profile,[4] which is based on the DL-Lite description logic, was specifically designed for this kind of application as it allows to reduce reasoning to standard SQL query evaluation. We plan to integrate DL-Lite reasoning through query rewriting in our approach.

4 Evaluation Strategy

We plan to deploy our approach in two case studies. The first is derived from an industry project where we develop an ontology-based inventory management and verification system. Our industry partner employs relational databases to store data about individual parts and product configurations. This relational data is the foundation of many legacy applications that can neither be replaced nor adapted in the short-run, resulting in the absolute need to preserve the RDBs. The goal of the project is to establish a semantic layer on top of the existing systems that verifies new product configurations according to a set of rules. The layer is realized with Semantic Web technologies and implements our approach for read/write access to the data as well as for reasoning.

The second case study is motivated in the domain of software quality analysis where the application of Semantic Web technologies recently gained popularity. Software analysis builds on different data sets (source code repositories, bug tracking systems, etc.) typically stored in individual relational databases. Semantic Web technologies provide an ideal mean to integrate this data and reasoning is used for certain analyses. We plan to integrate our approach into the existing tool set as a semantic interface to the source data, enabling ontology-based and existing analysis tools to (co-)operate on the same data.

Furthermore, we will evaluate the performance and expressiveness of our approach with the Berlin SPARQL Benchmark (BSBM).[5] BSBM is currently focused on SPARQL queries, therefore we plan to develop a set of representative SPARQL/Update operations to cover all features of our approach.

4.1 Current State of Our Research

In a first step, we defined a simple mapping language to bridge the conceptual gap between the relational representation of the data and an ontology. Mappings defined in this language not only contain the correspondences between database and ontology terms but also all structural information of the database schema needed to support write operations (e.g. primary keys, foreign key relationships, etc.). Based on this mapping language, we implemented a first prototype that supports inserting and deleting RDF data through SPARQL/Update. The prototype exploits the information stored in the mapping to support even complex insert operations spanning multiple tables with foreign key relationships. We also investigated how the SPARQL/Update operations are mapped to semantically equivalent SQL DML queries and their required order of execution.

[4] `http://www.w3.org/TR/owl2-profiles/#OWL_2_QL`

[5] `http://www4.wiwiss.fu-berlin.de/bizer/BerlinSPARQLBenchmark/`

5 Conclusion

In this proposal, we outlined the benefits that Semantic Web technologies can have in application scenarios where data is stored in relational databases. We presented the need for full read and write access as well as support for scalable reasoning and the advantages the preserving of the RDB has over the one-time data conversion. We introduced our approach that will enable ontology-based read and write access to relational data through SPARQL+SPARQL/Update endpoints as well as provide support for scalable reasoning performed at query time to use the power of the database engine and to avoid problems occurring from data updates. We further described where we plan to apply our approach for evaluation purposes.

References

1. Prud'hommeaux, E., Seaborne, A.: SPARQL Query Language for RDF (2008), `http://www.w3.org/TR/rdf-sparql-query/`
2. Clark, K.G., Feigenbaum, L., Torres, E.: SPARQL Protocol for RDF (2008), `http://www.w3.org/TR/rdf-sparql-protocol/`
3. de Laborda, C.P., Conrad, S.: Relational.OWL - A Data and Schema Representation Format Based on OWL. In: Proceedings of the 2nd Asia-Pacific Conference on Conceptual Modelling (2005)
4. de Laborda, C.P., Zloch, M., Conrad, S.: RDQuery - Querying Relational Databases on-the-fly with RDF-QL. In: Proceedings of the 15th International Conference on Knowledge Engineering and Knowledge Management (2006)
5. Bizer, C., Cyganiak, R.: D2R Server - Publishing Releational Databases on the Semantic Web. In: Proceedings of the 5th International Semantic Web Conference (2006)
6. Bizer, C.: D2R MAP - A Database to RDF Mapping Language. In: Proceedings of the 12th International World Wide Web Conference (2003)
7. Seaborne, A., Manjunath, G., Bizer, C., Breslin, J., Das, S., Davis, I., Harris, S., Idehen, K., Corby, O., Kjernsmo, K., Nowack, B.: SPARQL Update - A language for updating RDF graphs (2008), `http://www.w3.org/Submission/2008/SUBM-SPARQL-Update-20080715/`
8. Auer, S., Ives, Z.: Integrating Ontologies and Relational Data. Technical Report MS-CIS-07-24, Department of Computer and Information Science, University of Pennsylvania (2007)
9. Calvanese, D., Giacomo, G.D., Lembo, D., Lenzerini, M., Poggi, A., Rosati, R.: MASTRO-I: Efficient Integration of Relational Data through DL Ontologies. In: Proceedings of the 20th International Workshop on Description Logics (2007)
10. Calvanese, D., Giacomo, G.D., Lembo, D., Lenzerini, M., Rosati, R.: Tractable Reasoning and Efficient Query Answering in Description Logic: The DL-Lite Family. Journal of Automated Reasoning (2007)
11. Angles, R., Gutierrez, C.: The Expressive Power of SPARQL. In: Sheth, A., et al. (eds.) ISWC 2008. LNCS, vol. 5318, pp. 114–129. Springer, Heidelberg (2008)

The Relevance of Reasoning and Alignment Incoherence in Ontology Matching

Christian Meilicke

KR& KM Research Group, University of Mannheim, Germany
christian@informatik.uni-mannheim.de

Abstract. Ontology matching has become an important field of research over the last years. Although many different approaches have been proposed, only few of them are committed to a well defined semantics. As a consequence, the possibilities of reasoning are not exploited to their full extent. A reasoning based approach will not only improve ontology matching, but will also be necessary to solve certain problems that hinder the progress of the whole field. We focus on the notion of alignment incoherence to understand the capabilities of reasoning in ontology matching.

1 Problem Statement

Ontology matching has been identified as key element towards realizing the vision of the semantic web by bridging the gap between different conceptualizations of similar or overlapping domains. The core problem is the detection of semantic relations between concepts, properties or instances of two ontologies. In addition, many important side issues have been tackled, e.g. collaborative ontology matching, matcher selection, versioning of alignments, etc. Nevertheless, the role of semantics and in particular the role of reasoning in the context of ontology matching has been neglected or dealt with in an imprecise manner. This can be explained by the fact that many ontologies typically used as testcases within the matcher community are hierarchies that do not contain expressive constructs such as disjointness or property restrictions. Some of these ontologies, for example, are not modeled by knowledge engineers but based on automatically converting poorly structured knowledge bases into OWL.[1] Thus, any reasoning based approach will not exploit its full potential on most accepted testcases. Another explanation can be found in the history of ontology matching which can be seen as a further development of matching database or XML schemas. Ontologies are in most cases interpreted within a DL based semantics, where we find clear definitions of notions like e.g. entailment, satisfiability and inconsistency. Contrary to this, a comparable model based semantics is often not

[1] In the Ontology Alignment Initiative [1], for example, we find a large deal of ontologies that are automatically generated OWL representations of Thesauri or Web directories or have been converted from formalisms like OBO.

L. Aroyo et al. (Eds.): ESWC 2009, LNCS 5554, pp. 934–938, 2009.

given in the context of database semantics or cannot be applied directly to the field of ontology matching.

In our work we are concerned with the role of semantics in the context of ontology matching. In particular, we focus on the problem of alignment incoherence and show that a reasoning based approach can be applied in different contexts to enrich and optimize ontology matching.

2 Related Work and Motivation

Due to the lack of space, an exhaustive overview on related work cannot be given. Instead, we focus on specific problems to clarify the need for a reasoning based approach.

Evaluation. The classical measures used to evaluate ontology matching results are precision and recall. They are based on comparing an alignment $\mathcal{A}$ against a reference alignment $\mathcal{R}$ on a syntactic level. Suppose, that there exists a correspondence c in $\mathcal{A}$ not included in $\mathcal{R}$ but derivable from $\mathcal{R}$. On a pure syntactic level c has to be counted as incorrect correspondence, which is obviously an unreasonable choice. In many accepted testcases the classical syntactic approach has no negative effects, since $\mathcal{A}$ and $\mathcal{R}$ are restricted to equivalences where the deductive closure coincides with the deductive reduction. The problem has been dealt with in [2] under the notion of semantic precision an recall. Although, the problem has been precisely described, there are still some issues that have to be solved, e.g. the fact that the deductive closure of an alignment is an infinite set. Computing semantic precision and recall will become even more important, when complex correspondences come to the fore.

Semantics. Lots of research is concerned with theoretical issues related to alignment semantics. In [3], for example, the authors formally investigate different alignment formalisms. Nevertheless, both lines of research - the theory of alignment semantics vs. the practically oriented task of ontology matching - are only loosely coupled. Most matching systems are not committed to a certain semantics. Neither do they ensure the coherence of generated alignments nor do they allow to generate the deductive closure with respect to a certain semantics. We find a counterpart in the procedure of the Ontology Alignment Initiative (OAEI) [1], where in none of the subtasks results are expected to agree on a certain semantics.

Incoherence. We already claimed that no matching system ensures the coherence of its results. One might raise the objection that system like e.g. ASMOV and Lily make both use of a debugging component (see section 10 in [1]), which is also referred to as semantic verification. To the best of our knowledge, these debugging mechanisms are pattern based, which means that the final alignment is checked for certain patterns that indicate erroneous combinations of correspondences. Non of these patterns is explicitly backed by a certain semantics and their motivation, although comprehensible, is only an ad hoc explanation. Notions like completeness and soundness are not applicable with respect to these patterns.[2]

3 Approach and Methodology

Euzenat and Shvaiko have listed the problem of reasoning with alignments as one of the ten outstanding challenges in ontology matching [5]. The previous section indicates that there exists no unique approach to cope with this challenge. We propose an approach centered around the notion of alignment incoherence. In particular, we are concerned with the following research questions.

(1) How can the incoherence of an alignment be measured and interpreted?
(2) How can reasoning about incoherence be applied to improve the results of a matcher?
(3) Can incoherence be used to support manual alignment revision?

First of all, our approach requires to define the notion of alignment incoherence. Given a semantics $\mathcal{S}$ and an alignment $\mathcal{A}$, we defined $\mathcal{A}$ as incoherent with respect to $\mathcal{S}$, whenever there exists a satisfiable class in one of the aligned ontologies that becomes unsatisfiable due to interpreting $\mathcal{A}$ in terms of $\mathcal{S}$. This definition abstracts from the specifics of $\mathcal{S}$ and is applicable for any semantics $\mathcal{S}$ that supports the notion of satisfiability. An answer to the first question requires to elaborate a theory of alignment incoherence and to apply the resulting measures in the evaluation process. For answering the second question we have to compare an alignment generated by a matching system against a subset of this alignment where all incoherences have been resolved automatically. The third issue requires to take at least two aspects into account. On the one hand we have to compare the resulting alignment against the results of an unsupported revision. On the other hand we have to analyze in how far the effort for the human in the loop can be decreased.

4 Results

The origin of our work, starting two years ago, can be found in a paper that describes how Distributed Description Logics (DDL) can be used to reason about ontology alignments [6]. Based on the proposed framework, we used DDL as alignment semantics in our first experiments. In [7] we focused on automatically repairing alignments generated by different state of the art matching systems. In particular, we applied a greedy approach to resolve alignment incoherences by removing the correspondence with lowest confidence from minimal sets of

[2] S-Match [4], on the contrary, employs sound and complete reasoning procedures. Nevertheless, the underlying semantic is restricted to propositional logic due to the fact that ontologies are interpreted as tree-like structures.

conflicting correspondences. This approach increased the precision of the results between 2% and 19% depending on the particular matching system. In [8] we used a similar approach but imposed this choice on the user. In particular, we observed that the effort of a reasoning supported revision is in average reduced to 40% compared to a complete manual revision.[3]

Since DDL is mainly motivated by its possibilities to reason in a distributed environment, we proposed another semantic motivated by the use case of merging ontologies. This semantics can be seen as the natural translation of correspondences into DL-axioms. We first defined this semantics in a study where we applied the approach to synthetic ontologies, experimenting with a decision procedure that aims to remove a minimum number of correspondences weighted via their confidences [10] instead of the greedy approach explained above. A comparison of several methods for choosing a coherent subset of an alignment and their application can be found in [11], where we observed minor improvements often based on a trade off between precision and recall. Our experiments indicate that a heuristic that aims to achieve a global optimum will result in better choices whenever we have alignments with a high number of conflicts.

Recently, we focused on the first research question and proposed four ways to measure the degree of incoherence of an alignment [12]. Even though our definitions are based on merging ontologies, they are directly applicable to any other alignment semantics that supports the notion of satisfiability. We first applied this approach to the submissions of the OAEI 2008 conference track, where we observed that in average about 15% of the correspondences in an alignment have to be removed for logical reasons.[4] In particular, it turned out that even systems with debugging components could not ensure the coherence of their alignments. Moreover, we proved that a certain way to measure the incoherence of an alignment $\mathcal{A}$ results in a strict upper bound for the precision of $\mathcal{A}$ which was an unexpected and important result of our research.

5 Conclusions and Future Work

We motivated our work by general consideration about the role of reasoning in ontology matching. While there are many open problems, we focused on a specific family of interrelated problem centered around the notion of alignment incoherence. In particular, we implemented a reasoning system that allows to reason about alignments to repair incoherent alignments and to measure the degree of incoherence. The implemented techniques, partially derived from the field of ontology debugging, require a full fledged reasoning approach that will become problematic for large ontologies. In future work we have to tackle this problem by optimizing our reasoning strategies. In addition, we extended our system to support subsumption correspondences and correspondences between

[3] A summary of both approaches and their results can be found in [9].

[4] Notice that this result clearly shows that the capabilities of reasoning in ontology matching are not yet exploited to their full extent. Detailed results are reported in [1].

properties. First results show that the property extension significantly increases the number of detectable coherence conflicts. Nevertheless, further experiments have to confirm our first exploratory studies.

We started our work with automatically debugging ontology alignments. It turned out that we increased the quality of an alignment in most cases. Nevertheless, we often observed a trade off between precision and recall. We assume that this is based on the fact that we worked on the final alignment generated by a matcher, e.g. extracted from a similarity matrix. The benefits of our reasoning component can be further increased if correspondences eliminated by the extraction method of the matching system, would also be available for the decision process of our component. To check this hypotheses we are currently implementing our own matching system where the input to the reasoning component is not restricted to a previously filtered set of correspondences.

References

1. Caracciolo, C., Euzenat, J., Hollink, L., Ichise, R., Isaac, A., Malaise, V., Meilicke, C., Pane, J., Shvaiko, P., Stuckenschmidt, H., Svab-Zamazal, O., Svatek, V.: Results of the OAEI 2008. In: Proc. of the ISWC 2008 Workshop on Ontology Matching, Karlsruhe, Germany (2008)
2. Euzenat, J.: Semantic precision and recall for ontology alignment evaluation. In: Proc. of the 20th International Joint Conference on Artificial Intelligence, Hyderabad, India (2007)
3. Serafini, L., Stuckenschmidt, H., Wache, H.: A formal investigation of mapping language for terminological knowledge. In: Proc. of the International Joint Conference on Artificial Intelligence, Athens, GA, USA (2005)
4. Giunchiglia, F., Yatskevich, M., Shvaiko, P.: Semantic matching: Algorithms and implementation. Journal on Data Semantics (2007)
5. Shvaiko, P., Euzenat, J.: Ten challenges for ontology matching. In: Proc. of the 7th international conference on ontologies, dataBases, and applications of semantics, Monterrey, Mexico (2008)
6. Stuckenschmidt, H., Serafini, L., Wache, H.: Reasoning about ontology mappings. In: Proc. of the ECAI 2006 Workshop on Contextual Representation and Reasoning (2006)
7. Meilicke, C., Stuckenschmidt, H., Tamilin, A.: Repairing ontology mappings. In: Proc. of the 22nd Conference on Artificial Intelligence, Vancouver, Canada (2007)
8. Meilicke, C., Stuckenschmidt, H., Tamilin, A.: Supporting manual mapping revision using logical reasoning. In: Proc. of the 23rd AAAI Conference on Artificial Intelligence, Chicago, Illinois, US (2008)
9. Meilicke, C., Stuckenschmidt, H., Tamilin, A.: Reasoning support for mapping revision. Journal of Logic and Computation (2008)
10. Meilicke, C., Stuckenschmidt, H.: Applying logical constraints to ontology matching. In: Proc. of the 30th German Conference on Artificial Intelligence, Osnabrück, Germany (2007)
11. Meilicke, C., Stuckenschmidt, H.: Analyzing mapping extraction approaches. In: Proc. of the ISWC 2007 Workshop on Ontology Matching, Busan, Korea (2007)
12. Meilicke, C., Stuckenschmidt, H.: Incoherence as a basis for measuring the quality of ontology mappings. In: Proc. of the ISWC 2008 Workshop on Ontology Matching, Karlsruhe, Germany (2008)

Towards a Semantic Service Broker for Business Grid

Henar Muñoz Frutos

Telefónica Research & Development
Valladolid, Spain
henar@tid.es

Abstract. The increasing number of infrastructure services requires the existence of mechanisms to discover and select services and resources, called service broker, based on customer requirements. This mechanism should improve the interoperability among customer-provider and be a backward compatible and light weight approach. The introduction of semantic annotations in service description (both functional and non-functional properties) as well as a conceptual model for business Grid can help to achieve them. Finally, the extension of the semantic Open Grid Service Architecture (S-OGSA) with the incorporation of the semantic service broker can incorporate the required capabilities to the Grid middleware.

Keywords: Service Broker, Business Grid, semantic annotations, ontology, Quality of Service.

1 Introduction

Current business Grid markets allow enterprises to collaborate in a dynamic environment in order to buy and sell resources (computation, storage, applications, etc ...). Moreover, these resources are virtualized as services called Grid services to reduce their complexity and be available through the Internet, boosting the number of available services. Due to the increase of the available services, some automatic mechanisms are required in order to simplify and reduce the human intervention in business.

The Web service technology provides an infrastructure for developing distributed systems and performing electronic business operations [1]. This technology implements the Service Oriented Computing (SOC) paradigm, which considers three main stakeholders: i) service producer, which offers, in our e, ii) service broker and iii) service consumer. In our case, the service producers, concretely infrastructure providers, offers resources as services, to service consumers which specify their requirements. In addition, the service broker, on behalf of the customer, discovers a set of available services for the service consumer with a certain level of the consumer satisfaction.

Thus, the service broker should locate services that fulfil a concrete user request normally by capabilities constrains. However with the increasing number

L. Aroyo et al. (Eds.): ESWC 2009, LNCS 5554, pp. 939–943, 2009.

of functional similar services in the Web [2], the discovery process is not able to find the best service to meet user requirements while ensuring the Quality of Service (QoS) [3]. Thus, a service selection process is required based on specified customer preferences or restrictions, which are considered as non-functional properties such as price, reputation and reliability, which are QoS attributes and may impact the quality of the service offered by a Web Service [3].

As part of the non-functional properties, some information related to the way of provisioning the service by the provider side can be useful. This is part of the Service-Level Agreement (SLA) templates, where besides service descriptions, it is defined information about the resources (hardware and software) to be provided, the quality of service (QoS) level to be maintained, etc [4]. Thus, a service specification offered by a provider can be materialized as a capability (functional description) provided and a set of non-functional constraints (QoS metrics, service provision terms) by using the TSLA files.

However, in this highly dynamic scenario, providers and customers may use different terminologies or even languages, to describe the conditions and requisites [5]. In this matter, it is required a mutual agreement on the service(s) functional and non-functional properties to develop a method for automatically matching the offers and the requests. In order to increase the interoperability among customers-providers, some works introduce semantic technologies in the definition of QoS metrics. Domingue et al [6] and Ren et al [3] present different conceptual models formed as ontologies so as to extend service descriptions. However, these solutions are not backward compatible, and demand providers to change their service offer descriptions. Moreover, they are not lightweight approaches, which increase the complexity to user in annotating services and the framework to implement such solutions.

In addition, considering lightweight approaches, the work introduced by Kopecky [7] defines the usage of semantic annotations for describing services based on the Web Service Modelling Ontology (WSMO). However, this approach lacks in the information provided by providers related to SLA. In addition, Oldham et al [8] propose the introduction of semantic annotations for QoS terms inside WS-Agreement [4], but extending the XML schemas, being an alternative not backward compatible.

Thus, our approach should be able to build a service broker with the following features: i) improve the interoperability among customer-providers, ii) be backward compatible and iii) be a lightweight approach.

2 Proposed Approach

The proposed approach tries to satisfy the main requirements commented above composed by a service broker model and an architecture which implements this model. In order to describe the non-functional properties (including the service provision) together service functional descriptions, we use TSLA from the WS-Agreement recommendation [4]. Moreover, in order to increase the interoperability among customers and providers, we are going to introduce semantic technologies in terms of semantic annotations both in service description and QoS metrics in the TSLA. Semantic annotations allows to link SLA standards terms with concepts belonging to a conceptual model as part as the document [9]. The main drivers behind this approach are: i) the necessity for a backward compatible approach, so that, components non aware of semantic annotation can continue working with service descriptions and ii) a lightweight approach to perform any operation (selection, negotiation ...) at runtime [5].

2.1 Service Broker Model

In order to introduce the semantic annotations, the service broker model is defined by i) a specification for introducing semantic annotations, ii) a conceptual model for service descriptions formed as ontologies and iii) a service selection algorithm. The *specification for Semantic Annotation for SLA (SA-SLA)* is based on the W3C recommendation Semantic Annotation for Web Service Description Language (SA-WSDL) [9], which allows WS-Agreement terms (as metrics, KPI, service descriptions) to be pointed to semantic concepts from more expressive ontologies. Regarding the *conceptual model*, the approach tries to define a generic set of concepts (functional and non-functional) in a business Grid scenario, which are extensible, and can be customized to include domain-specific concepts. Finally, the *service selection algorithm* will involve i) the discovery of service regarding functional properties and ii) the selection of services based on customer' requirements in terms of non-functional properties. This algorithm is going to allow for working with imprecise requirements in order to provide a way of ranking and incorporate more amount of services in the compete market.

2.2 Architecture

Corcho et al [10] defines a Reference Architecture for the introduction of semantic technologies to Grid, called semantic Open Grid Service Architecture (S-OGSA). Concretely, S-OGSA defines a model, the capabilities and the mechanisms for the Semantic Grid. Thus, we can extend current S-OGSA architecture with the Semantic Grid Service Broker, being a new Semantically Aware Grid Service (SAGS), which should work with the rest of S-OGSA architecture to fulfill their complete functionality.

The service broker should work with different modules: i) a Grid service discovery and ii) a non-functional properties selector, besides all modules corresponding to the semantic infrastructure. In the broker design, we intent to create a discovery pattern that will be based on the well-known principle of the "separation of concerns". This principle will be applied decoupling the functional properties from the non functional properties matching.

3 Methodology

Methodologically, we propose a quantitative experimentation, obtaining results from different experiments. Concretely, for the experimentation, we are going to take the Virtual Scenario of the BREIN project [11]. There, an engineering company (ANSYS) tries to outsource the computing infrastructure required for software simulations in order to reduce costs. Thus, due to the existence of several infrastructure providers (BSC, HLRS), the framework should be able to select the best provider in each moment which satisfy the customer's requests. Thus, for experimentation proposes, we are going to build a simulation testbed composed by a set of service descriptions and TSLA files, which are semantically annotated.

As the main three statement we want to demonstrate are: i)backward compatible, ii)light weigh and iii) improve interoperability, we have to design the experiments in order to show it. Thus, we will provide some annotated and non-annotated services with different used terminologies. Finally, the idea is to compare our approach with others like for instance WSMO, in order to analyze the advantages of our solution.

4 Results

In the implementation of the solution, we are following an agile project management approach, such as depicted by SCRUM and others originally developed for large scale IT development tasks. With this approach a set of goals considered relevant for the idea objectives is identified and implemented. Moreover, to have the full implementation some other external implementations or algorithms can be taken in order to be able to have early results. Table 1 can summarize the main goals defined and their current status.

Table 1. Goals in the semantic Grid service broker

	Goals	Status	Comment
1	SA-SLA specification	90%	To be submitted to GRAAP (OGF)
2	Ontologies	35%	QoS ontology, service ontology.
3	QoS Broker	80%	Extension according to ontologies
4	QoS selection algorithm	0%	Including imprecise constrains
5	Metrics Collection	0%	Obtain metric value for providers
6	Functional properties discovery	0%	Annotating WS-Agreement service description with SA-SLA

5 Conclusions and Future Work

This document has established the main pillars towards the realization of a semantic grid service broker for business Grid. It has started from the definition of the context, where mechanisms are required for the discovery and selection of Grid services considering non-functional properties. Moreover, the first analysis of the state of the art pointed out that the existing approaches are not usually focused on service Grid and do not solve the heterogeneity problem planned in a open market place. Finally, other approaches, which can satisfy these requirements, lack in a backwards compatible and light weight approach, with allows continuing working with current components. Our proposed approach relies on the introduction of semantic annotations in TSLA and ontologies to work towards these requirements. Currently, there are some developed results as the QoS broker, but we should continue working on the state of the art to guarantee that the idea is original, as well as some bullet points marked in the proposed approach.

References

1. Papazoglou, M.P., Heuvel, W.J.: Service oriented architectures: approaches, technologies and research issues. The VLDB Journal 16(3), 389–415 (2007)
2. D'Mello, D.A., Ananthanarayana, V.S., Thilagam, S.: A qos broker based architecture for dynamic web service selection. In: AMS 2008: Proceedings of the 2008 Second Asia International Conference on Modelling & Simulation (AMS), pp. 101–106. IEEE Computer Society, Washington (2008)
3. Ren, K., Chen, J., Chen, T., Song, J., Xiao, N.: Grid-based semantic web service discovery model with qos constraints. In: SKG 2007: Proceedings of the Third International Conference on Semantics, Knowledge and Grid, Washington, DC, USA, pp. 479–482. IEEE Computer Society, Los Alamitos (2007)
4. Grid Resource Allocation Agreement Protocol WG (GRAAP-WG): Web Services Agreement Specification (WS-Agreement),Version 2005/09 (2005)
5. Kotsiopoulos, I., Juber, I.S., Tenscbert, A., Cirujeda, J.B., Koller, B.: Using semantic technologies to improve negotiation of service level agreements (2008)
6. Domingue, J., Cabral, L., Galizia, S., Tanasescu, V., Gugliotta, A., Norton, B., Pedrinaci, C.: Irs-iii: A broker-based approach to semantic web services. Web Semant. 6(2), 109–132 (2008)
7. Kopecký, J., Vitvar, T.: Wsmo-lite: Lowering the semantic web services barrier with modular and light-weight annotations. In: ICSC 2008: Proceedings of the 2008 IEEE International Conference on Semantic Computing, Washington, DC, USA, pp. 238–244. IEEE Computer Society, Los Alamitos (2008)
8. Oldham, N., Verma, K., Sheth, A., Hakimpour, F.: Semantic ws-agreement partner selection. In: WWW 2006: Proceedings of the 15th international conference on World Wide Web, pp. 697–706. ACM, New York (2006)
9. Kopecký, J., Vitvar, T., Bournez, C., Farrell, J.: Sawsdl: Semantic annotations for wsdl and XML schema. IEEE Internet Computing 11(6), 60–67 (2007)
10. Corcho, O., Alper, P., Kotsiopoulos, I., Missier, P., Bechhofer, S., Goble, C.: An overview of s-ogsa: A reference semantic grid architecture. Web Semantics: Science, Services and Agents on the World Wide Web 4(2), 102–115 (2006)
11. The BREIN Consortium: The brein project (2008)

Evolva: A Comprehensive Approach to Ontology Evolution

Fouad Zablith*

Knowledge Media Institute (KMi), The Open University
Walton Hall, Milton Keynes, MK7 6AA, United Kingdom
f.zablith@open.ac.uk

Abstract. Ontology evolution is increasingly gaining momentum in the area of Semantic Web research. Current approaches target the evolution in terms of either content, or change management, without covering both aspects in the same framework. Moreover, they are slowed down as they heavily rely on user input. We tackle the aforementioned issues by proposing Evolva, a comprehensive ontology evolution framework, which handles a complete ontology evolution cycle, and makes use of background knowledge for decreasing user input.

1 Problem and Methodology

Ontologies form the basis of Semantic Web systems. As such, they need to be kept up-to-date for the dependent systems to remain usable. With the increase of complexity and changes occurring in the represented domains, ontology evolution becomes a painstaking and time-consuming process. Thus research has witnessed an increased interest in ontology evolution. We regard ontology evolution as the "timely adaptation of an ontology to the arisen changes and the consistent management of these changes" [10]. "Timely adaptation" suggests a quick adaptation, that can only be achieved by decreasing user involvement in the evolution process. However, most of current approaches heavily rely on user input. Moreover, the definition suggests that a successful evolution can only be achieved by having both "adaptation" and "change management". Yet, as we discuss in Section 4, no existing approach handles the two tasks in one framework. One set of approaches considers evolution as the management of changes performed by users for preserving consistency [11,13,16,17], while another set targets techniques for integrating new knowledge into the ontology [2,5,12,14], without an extensive handling of change and evolution management. Our research tackles the following two main questions:

Question 1: How to cover a complete ontology evolution cycle? The focus here is on providing both ontology adaptation and change management.

Question 2: How to decrease user involvement in ontology evolution? In order to be as agile as possible, especially in dynamic domains where information is

* Work funded by the NeOn project under EC grant IST-FF6-027595.

L. Aroyo et al. (Eds.): ESWC 2009, LNCS 5554, pp. 944–948, 2009.

abundantly generated, ontology evolution should be performed with the minimum user involvement.

Our methodology for resolving the above questions has three main phases: in the first phase we analyzed a concrete case within our KMi Semantic Web portal[1] that helped us determine the main required tasks for evolution. Then we propose an ontology evolution framework covering all previously identified tasks. In the second phase, we are conducting an experiment on using background knowledge, in addition to a pilot system implementation. The third phase will include the evaluation of our approach.

2 Proposed Approach

We propose a comprehensive ontology evolution framework, Evolva[2], covering a complete ontology evolution cycle including a) performing changes based on external data sources, and b) the management of these changes as shown in Figure 1. Such framework is designed to meet the requirements of our research Question 1.

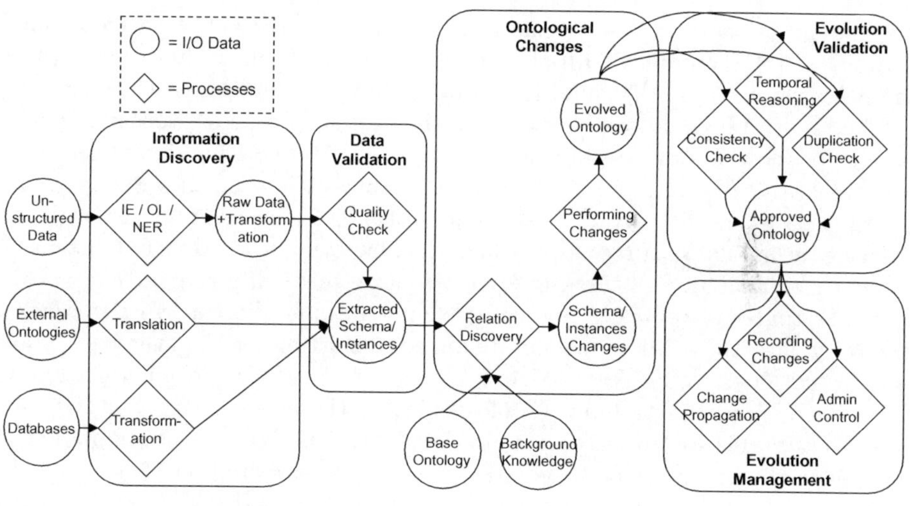

Fig. 1. Evolva Framework Architecture

We focus on identifying the need for evolution by contrasting the content of the ontology to external data sources. Such sources could be text documents, folksonomies, databases, or other ontologies. Each source requires a different method of content extraction handled by the *information discovery* component. The *data validation* component identifies new terms that are relevant for the

[1] http://semanticweb.kmi.open.ac.uk

[2] Detailed description of the framework can be found in [18].

ontology. It also checks the quality of content and filters out noise generated from the information discovery component. The validated information is passed to the *ontology changes* component in which background knowledge plays a core role in automating the integration of new information to the ontology. The evolution could generate conflicts and problems that are handled at the level of the *evolution validation* component. We plan to have in this component temporal reasoning for time related problems, coupled with duplication and consistency checks. Finally the validated ontology is passed to the *evolution management* component (bottom right part of Figure 1) where the user has control over the evolution, and changes are recorded and propagated to dependent ontologies.

With response to Question 2, Evolva uses background knowledge for decreasing, or even eliminating user input throughout the evolution. Background knowledge exists through various formats including lexical databases e.g. WordNet [7], online ontologies, and unstructured web documents. Our relation discovery step uses background knowledge to determine the relationship path of new knowledge to existing knowledge in the ontology.

3 Initial Results and Conclusions

By the end of the first year of the PhD, we have implemented part of Evolva within the NeOn ontology editor[3]. Our second major result is an experiment on the potential use of background knowledge as a replacement of user input.

The Evolva plugin allows users to run the evolution process step-by-step and to control the relevant parameters for each step. Figure 2 highlights the main features of the Evolva plugin. The left panel shows the evolved ontologies, and the right panel shows the various steps of Evolva. Using those, the user can specify the external data sources from which the new content is extracted. Then the content is validated, and the user has the option to modify the automatic validation settings, as well as to manually exclude unwanted terms. In the relation discovery part, the validated terms are linked to existing terms in the base ontology through background knowledge. For example, WordNet linked *Applicant* as *subClassOf Person*, and online ontologies linked the new concept *Website* with a *hasWebsite* relation to *Organization*. Finally, the user can choose to apply the changes to the base ontology itself, or create a new, updated ontology.

Our background knowledge experiment, described in [19], focused on using WordNet and online ontologies for relation discovery in the context of the KMi portal ontology. Online ontologies are accessed using Scarlet [15], a Semantic Web based relation discovery engine, that matches [6] terms to online ontological entities for resolving relations. The overall precision of around 77% shows that background knowledge can largely contribute to automating the integration of new knowledge into the ontology. This is where user input is traditionally most needed. We also found that precision can be further increased through introducing validation techniques like using the ontology itself as a relation validator, and by using filter mechanisms for excluding irrelevant terms.

[3] http://www.neon-toolkit.org

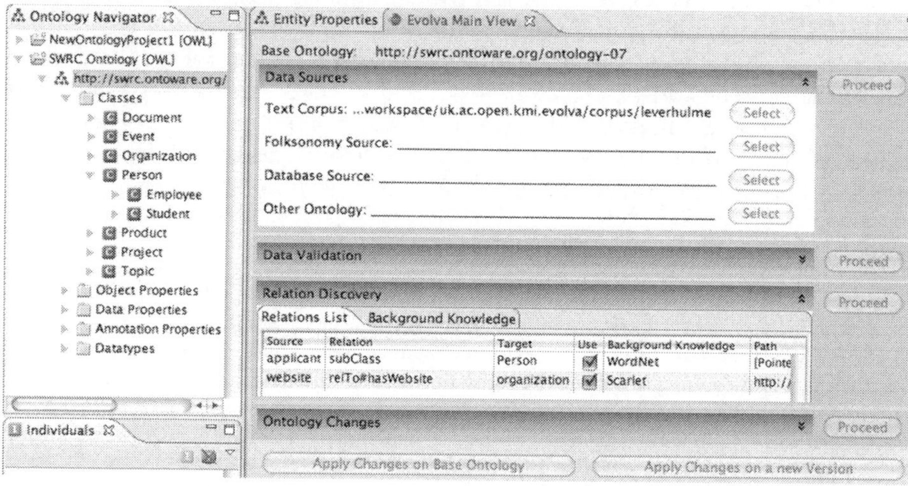

Fig. 2. Evolva Pilot System Screenshot

4 State of the Art

Research on ontology evolution is roughly divided in two groups: The first group considers ontology evolution as the process of managing changes performed by users [11,13,16,17]. In brief, the idea is to evolve an ontology as a closed entity by analyzing its content by e.g. identifying changes based on data mining techniques [16]. A key achievement, common in [11] and [16], is the representation of changes in an ontology of changes. However, the user here is the only source of new knowledge added to the ontology. The second group focuses on identifying new information to be added to the ontology [2,5,12,14]. This is done either through analyzing trends in user behavior [2,5], or by exploiting changes in external data sources such as databases or text documents [12,14]. However, the user still plays a major role throughout their process, and in this case, little attention is given to the management of evolution. The intention of Evolva is to cover both approaches.

Background knowledge has been successfully used in various tasks. For example, WordNet helps in solving word sense disambiguation issues [9] and ontology matching tasks [8]. Medical domain ontologies [3] and online ontologies [15] are employed in ontology matching. Folksonomy tagspaces [4] and ontology learning [1] are also using online ontologies as a background knowledge source. In our research, we make use of background knowledge in the field of ontology evolution.

5 Future Plans

Our plan for the second year of the PhD is to study automated techniques for relation validation, crucial to increase Evolva's precision. We also plan to fine-tune and extend our prototype to cover the remaining components. Evaluating the end research product forms an important part of our planned tasks and will be conducted during the third year. We plan to run tests in real domains such as the UN's Food and Agriculture Organization (FAO)[4] fisheries or agriculture. Such tests should give us firstly qualitative measures, which reflect the correctness of the content added to the ontology; and secondly, quantitative measures to analyze the performance of Evolva in terms of time, number of ontological entities added, and number of sources analyzed. Comparisons will be performed against other ontology building tools and against ontology designers.

References

1. Alani, H.: Position Paper: Ontology Construction from Online Ontologies. In: Proceedings of WWW (2006)
2. Alani, H., Harris, S., O'Neil, B.: Winnowing ontologies based on application use. In: Sure, Y., Domingue, J. (eds.) ESWC 2006. LNCS, vol. 4011, pp. 185–199. Springer, Heidelberg (2006)
3. Aleksovski, Z., Klein, M., Kate, W., Van Harmelen, F.: Matching Unstructured Vocabularies using a Background Ontology. In: Staab, S., Svátek, V. (eds.) EKAW 2006. LNCS(LNAI), vol. 4248, pp. 182–197. Springer, Heidelberg (2006)
4. Angeletou, S., Sabou, M., Motta, E.: Semantically enriching folksonomies with FLOR. CISWeb (2008)
5. Bloehdorn, S., Haase, P., Sure, Y., Voelker, J.: Ontology Evolution, pp. 51–70. John Wiley & Sons, Chichester (2006)
6. Euzenat, J., Shvaiko, P.: Ontology Matching. Springer, Heidelberg (2007)
7. Fellbaum, C.: Wordnet: An Electronic Lexical Database. MIT Press, Cambridge (1998)
8. Giunchiglia, F., Shvaiko, P., Yatskevich, M.: Discovering Missing Background Knowledge in Ontology Matching. In: Proc. of ECAI (2006)
9. Gracia, J., Lopez, V., d'Aquin, M., Sabou, M., Motta, E., Mena, E.: Solving semantic ambiguity to improve semantic web based ontology matching. In: Workshop on Ontology Matching at ISWC and ASWC (2007)
10. Haase, P., Stojanovic, L.: Consistent evolution of OWL ontologies. In: Gómez-Pérez, A., Euzenat, J. (eds.) ESWC 2005. LNCS, vol. 3532, pp. 182–197. Springer, Heidelberg (2005)
11. Klein, M.: Change Management for Distributed Ontologies. PhD thesis, Vrije Universiteit in Amsterdam (2004)
12. Novacek, V., Laera, L., Handschuh, S.: Semi-automatic integration of learned ontologies into a collaborative framework. In: IWOD (2007)
13. Noy, N.F., Chugh, A., Liu, W., Musen, M.A.: A framework for ontology evolution in collaborative environments. In: Cruz, I., Decker, S., Allemang, D., Preist, C., Schwabe, D., Mika, P., Uschold, M., Aroyo, L.M. (eds.) ISWC 2006. LNCS, vol. 4273, pp. 544–558. Springer, Heidelberg (2006)
14. Ottens, K., Glize, P.: A multi-agent system for building dynamic ontologies. In: Proceedings of AAMAS (2007)
15. Sabou, M., d'Aquin, M., Motta, E.: Exploring the Semantic Web as Background Knowledge for Ontology Matching. Journal on Data Semantics XI (2008)
16. Stojanovic, L.: Methods and Tools for Ontology Evolution. PhD thesis, FZI (2004)
17. Vrandecic, D., Pinto, H.S., Sure, Y., Tempich, C.: The DILIGENT knowledge processes. Journal of Knowledge Management 9, 85–96 (2005)
18. Zablith, F.: Dynamic ontology evolution. In: ISWC Doctoral Consortium (2008)
19. Zablith, F., Sabou, M., d'Aquin, M., Motta, E.: Using background knowledge for ontology evolution. In: IWOD (2008)

[4] http://www.fao.org

A Context-Aware Approach for Integrating Semantic Web Technologies onto Mobile Devices

Stefan Zander

University of Vienna
Department of Distributed and Multimedia Systems
Liebiggasse 4/3-4, A-1010 Vienna
stefan.zander@univie.ac.at

Abstract. Semantic Web technologies such as RDF are usually incorporated in the infrastructure of desktop and web applications and can currently not be entirely deployed on mobile devices. Therefore, the unique opportunities and novel features the Semantic Web offers are not amenable in mobile application scenarios, in which context and context-awareness is essential. In this thesis, we propose a context-sensitive Semantic Web framework for mobile devices by which contextual data can be processed semantically and integrated with data sets and services from other communities. Mobile devices augmented with those capabilities can operate autonomously and independent in different environments.

Keywords: Mobile computing, Semantic Web, context, context-awareness.

1 Introduction and Problem Description

Mobile phones and mobile computing in general became central parts in our everyday life. Additionally, new research areas such as Ambient Intelligence or pervasive and ubiquitous computing gain increasing interest. Their idea is to create smart, intelligent, and adaptive environments in which mobile devices can request and consume proliferated services and resources autonomously, where decision making processes are not only based on the user's presence or preferences, but also on the surrounding environment [1]. Central to this idea is the notion of context and context-awareness [2,3]. Ambient Intelligence environments must elaborate on the semantics and meaningfulness of user-related tasks and activities in order to provide appropriate services and valuable information.

Currently, Semantic Web based applications and services can hardly be deployed on mobile devices because of the lack of suitable implementations for RDF storage and management, although the necessity of Semantic Web technology and ontologies in particular for realizing a context-aware mobile computing infrastructure is broadly acknowledged [1,4]. In this thesis, we elaborate on the technical and conceptual background needed for integrating Semantic Web technology onto mobile devices and employ technologies such as RDF for the realization of a powerful, adaptable, and scalable context-aware infrastructure for harnessing contextual data that can be used for requesting ambient services and data sets from linked data repositories.

L. Aroyo et al. (Eds.): ESWC 2009, LNCS 5554, pp. 949–953, 2009.

2 Related Work and State of the Art

2.1 Mobile Semantic Web Frameworks

We have analyzed two XML parsers for mobile environments, *NanoXML for J2ME*[1] and *kXML*[2], as well as two mobile RDF frameworks, *Mobile RDF*[3] and *μJena*[4]. Our survey reveals that μJena is the most advanced framework providing ontology and inferencing support, although its API is currently in prototypical status and only allows for processing RDF data serialized in N-Triples format. None of the evaluated frameworks support queries on RDF data via SPARQL or other query languages. A serialization mechanism between RDF data and the internal storage mechanisms used by mobile devices could also not be found. Such mechanisms are needed since many mobile platforms do not use a file system for permanent data storage but provide platform-specific storage systems such as the Record Management System (RMS).

2.2 Context and Context-Awareness

The notion of context is characterized by "considerable confusion" according its nature, usage, and its role in interactive systems [5]. Technical disciplines primarily concentrate on sensorial factors such as location, time, activity etc., whereas social disciplines consider context as a dynamic, emergent phenomenon evolving in the course of interaction that cannot be represented by specific properties. Context needs an interdisciplinary and holistic research approach for understanding the interdependence between context and technology [6,5]. A context model should be developed according to its application domain since the generality of a context model in terms of expressiveness and powerfulness is inversely proportional to its practical applicability and usability [7]. Replacing the prevailing system-centric view of context-awareness by a user-centric view is suggested by [6], since context-awareness is an activity-driven process, centered around user activities. The advantages context provides to the Semantic Web are elaborated in [8] and [4], introducing an architecture for a context modeling and processing framework for pervasive applications ([4]) as well as a context mechanism for the Semantic Web to support data aggregation ([8]).

2.3 Related Projects

DBpedia Mobile[5], a location-aware mobile application, allows users to access information from the DBpedia project [9] about the physical environment surrounding them. Users are able to receive additional information by exploring links to other resources located in the Semantic Web. A similar approach is

[1] `http://sourceforge.net/projects/nanoxml-j2me/`
[2] `http://kxml.sourceforge.net/`
[3] `http://www.hedenus.de/rdf/index.html`
[4] `http://poseidon.elet.polimi.it/ca/?page_id=59`
[5] `http://wiki.dbpedia.org/DBpediaMobile`

taken by the *mSpace Mobile*[6] project in which access to location-based information according to the user's current situation is provided via a spatial browser. Considered contexts are time, space, and subject. The *IYOUIT*[7] project collects contextual information about certain aspects of the user's lifestyle such as visited places or people met and reflects this information on the Web. People are able to share their personal contexts within a community portal.

Although these projects make use of Semantic Web technologies such as RDF, the processing of contextual data is done on external servers/applications rather than on the device itself. Certainly, this dependency is useful for a variety of scenarios but embodies risks in case of connectivity problems, server breakdowns etc. If devices were able to process semantic data directly, they could operate autonomously and independent from external services and specific environments.

3 Proposed Approach

Our overall objective is to make Semantic Web technologies attractive to mobile computing by showing the unique opportunities and novel features these technologies can provide. Therefore, we have composed an application scenario demonstrating how a mobile device supports the user by understanding the situations and different contexts in which she operates:

The invitation to a project meeting John received via email mentions the topic, place, and participants. His mobile device semantically interprets this information and adds a calendar entry, the location's GPS coordinates, and information about the other meeting participants (e.g., by retrieving their FOAF profiles[8]) to John's personal dataspace. By combining this information the device could infer on the situation in which John might be at that time ("in meeting") and take appropriate actions when necessary. At the meeting, John's address book already contains the contact data of the other participants together with additional information about them (e.g., personal homepage, department or company website, recently published articles, etc.) and corresponding events, retrieved from exposed LOD[9] repositories.

The realization of this scenario requires mobile devices to be augmented with capabilities to process semantic data (RDF) directly using a scalable, lightweight context model based on ontological semantics expressed by means of OWL. This model serves as the basis for collecting and interpreting sensed data, which must be done on different levels [4]. Machine learning techniques guarantee that the context model continually adapts according to the user's intentions and tasks. This process is supported by feedback mechanisms implicitly and explicitly to minimize the misinterpretation of contextual data. Time and history have significant influence on the user's behavior and allow for inferencing on what the user has done in comparable situations to get a better understanding on current and future actions. Since the device is aware of John's current situation, it is able to

[6] `http://mspace.fm/projects/mobile`

[7] `http://www.iyouit.eu/portal/`

[8] `http://www.foaf-project.org/`

[9] `http://esw.w3.org/topic/SweoIG/TaskForces/CommunityProjects/LinkingOpenData`

operate accordingly and suggest appropriate actions (e.g., detect the importance of an incoming call; detect whether John will arrive at the meeting belatedly and notify the other participants before; record all the places one visited within a certain time frame and automatically interlinks them with related data from LOD repositories etc.). We believe that the computational model of the semantic desktop [10] also applies to mobile devices and their personal dataspaces.

4 Methodology and Evaluation

The application scenario introduced earlier serves as a basis for deducing requirements of the underlying Semantic Web framework. Since none of the evaluated frameworks provide the required functionalities a priori, we decided to build on the μJena framework and extend it as needed. To get an impression about the interdependence between context, context-awareness, and mobile technology we surveyed how context and context-awareness is defined and understood across different communities. Questions on how to align and synthesize them with semantic technologies under the restrictions mobile devices impose will be addressed and redefined in the course of the thesis.

Our Semantic Web framework aims to facilitate the deployment of context-sensitive mobile Semantic Web applications and will be iteratively enhanced and extended. We propose a multi-layer architecture where each layer is responsible for a certain group of context-related tasks (e.g., sensing, acquisition, aggregation, interpretation, inferencing, representation etc.). Questions on how to efficiently process and store triple-based data on devices with limited processing power, memory and power capacity are of concern too.

As proof of concept, the proposed application scenario will be developed as a Google Android[10] application, where the requirements deduced serve as the basis for a qualitative evaluation. Additionally, feedback questionaries will be developed and answered by industrial partners involved in the related project MobiSem[11] to evaluate user satisfactoriness according to the functionalities proposed. We also perform measurements with respect to efficiency in processing, storing, and inferencing semantic data as well as memory consumption on the basis of a quantitative evaluation.

5 Current Status and Results Achieved

Our conducted survey on existing mobile Semantic Web frameworks reveals that almost all evaluated frameworks are in very early stages, the μJena framework being the most advanced one. Reviewing the literature on context and context-awareness indicates that these terms are used very ambiguously across communities. Recent research into this topic claims for dynamic, user-centric context models to better reflect the characteristics and emergent nature. A prototypical

[10] `http://code.google.com/android`

[11] `http://www.mobisem.org`

Android application based on the μJena packages has been implemented allowing for retrieving N-Triple serialized RDF data that can be stored and processed directly on a mobile device. We are currently implementing a simple form of inferencing to reason about the RDF data stored and take specific contextual parameters for requesting RDF data from the *Linked Geo Data* project[12].

6 Conclusion

Semantic technologies are on the doorstep to become central parts of the applications operating on desktop computers as well as on the Web. We want to take this a step further by building a context-sensitive Semantic Web framework for integrating Semantic Web technologies into the applications operating on mobile devices. Our framework will enable the collection and interpretation of contextual data that can be used for requesting ambient services and allocating resources that have a specific relationship to the user's current situation.

References

1. Preuveneers, D., Van den Bergh, J., et al.: Towards an extensible context ontology for ambient intelligence. In: Markopoulos, P., Eggen, B., Aarts, E., Crowley, J.L. (eds.) EUSAI 2004. LNCS, vol. 3295, pp. 148–159. Springer, Heidelberg (2004)
2. Remagnino, P., Foresti, G.L.: Ambient intelligence: A new multidisciplinary paradigm. IEEE Transactions on Systems, Man and Cybernetics, Part A 35(1), 1–6 (2005)
3. Coutaz, J., Crowley, J.L., Dobson, S., Garlan, D.: Context is key. Commun. ACM 48(3), 49–53 (2005)
4. Euzenat, J., Pierson, J., Ramparany, F.: Dynamic context management for pervasive applications. Knowl. Eng. Rev. 23(1), 21–49 (2008)
5. Dourish, P.: What we talk about when we talk about context. Personal Ubiquitous Comput 8(1), 19–30 (2004)
6. Teo, H.-S.: An activity-driven model for context-awareness in mobile computing. In: MobileHCI 2008: Proceedings of the 10th international conference on Human computer interaction with mobile devices and services, pp. 545–546. ACM, New York (2008)
7. Bolchini, C., Curino, C., Quintarelli, E., Schreiber, F., Tanca, L.: A data-oriented survey of context models. SIGMOD Rec. 36(4), 19–26 (2007)
8. Guha, R., McCool, R., Fikes, R.: Contexts for the semantic web. In: McIlraith, S.A., Plexousakis, D., van Harmelen, F. (eds.) ISWC 2004. LNCS, vol. 3298, pp. 32–46. Springer, Heidelberg (2004)
9. Auer, S., Bizer, C., Kobilarov, G., Lehmann, J., Ives, Z.: Dbpedia: A nucleus for a web of open data. In: Aberer, K., Choi, K.-S., Noy, N., Allemang, D., Lee, K.-I., Nixon, L., Golbeck, J., Mika, P., Maynard, D., Mizoguchi, R., Schreiber, G., Cudré-Mauroux, P. (eds.) ASWC 2007 and ISWC 2007. LNCS, vol. 4825, pp. 722–735. Springer, Heidelberg (2007)
10. Heath, T., Motta, E., Dzbor, M.: Context as a foundation for a semantic desktop. In: Proc. of Semantic Desktop Workshop at the ISWC, Galway, Ireland, November 6, vol. 175 (2005)

[12] `http://aksw.org/Projects/LinkedGeoData`

Dealing with Inconsistencies in *DL-Lite* Ontologies*

Liping Zhou

Beijing Jiaotong University, Beijing, China
dearliping@gmail.com

1 The Research Problem

As a shared conceptualization of a particular domain, ontologies play an important role for the success of the Semantic Web. However it is often difficult to create an absolutely consistent ontology. Inconsistency can occur due to several reasons, such as modeling errors, migration or merging ontologies, and ontology evolution. So it is essential to study how to deal with inconsistent ontologies. Many approaches have been proposed to solve this problem. These approaches are mainly used for dealing with inconsistency in expressive Description Logics(DLs). In our work, we consider inconsistency handling in the *DL-Lite* family, which is a family of DLs that preserve tractable reasoning and are specifically tailored to deal with large amounts of data. Like other DLs, inconsistencies in *DL-Lite* can also easily occur because disjoint axioms are allowed.

2 The State of Art

There are two main ways to deal with inconsistent ontologies [5]. One way is to simply avoid the inconsistency and to apply a non-standard reasoning method to obtain meaningful answers. For example, Huang et al. in [5] have proposed a general framework for reasoning with inconsistent ontologies by using a selection function. In [7,8,12], four-valued logics have been applied to reason with inconsistent ontologies.

The other is to resolve logical modeling errors whenever a logical problem is encountered. For example, some researchers proposed methods to debug erroneous terminologies and have them repaired when inconsistencies are detected [17,18,13,16,11,4]. Some researchers considered resolving the logical inconsistencies during ontology evolution through some revision operators (see, for example, [15]).

When resolving inconsistencies, we always need some additional information to decide which solutions are better. Therefore, some researchers provided methods to measure the incoherence [14] or inconsistency [10,9,3] of ontologies. For example, some researchers provided methods for measuring inconsistency of axioms to identify which

* I would like to acknowledge my supervisor Houkuang Huang, my co-supervisor Youli Qu, my external supervisors Zhisheng Huang and Guilin Qi for their guidance, support and fruitful discussions. Our research is supported by the National Grand Fundamental Research 973 Program of China under grant (No.2007CB307100, No.2007CB307106) and by the Specialized Research Foundation of Doctoral Program of Higher Education of China under grant (No.20050004008).

L. Aroyo et al. (Eds.): ESWC 2009, LNCS 5554, pp. 954–958, 2009.

axioms need to be removed or modified to resolve an inconsistency (see [3,10]). Qi and Hunter in [14] provide a method for measuring the incoherence of an ontology.

The above approaches to dealing with inconsistency are usually based on expressive DLs which suffer from worst-case exponential time behavior of reasoning [1]. This may hinder their applications to very large real life ontologies.

As a family of tractable DLs, *DL-Lite* can keep all the reasoning tasks tractable, in particular, with polynomial time complexity with respect to the size of the knowledge base [2]. Recently, there are some researchers began to study the problem of dealing with inconsistent *DL-Lite* ontologies. For example, Lembo et al. in [6] provided an approach for consistent query answering over ontologies specified in *DL-Lite*. Ma et al. in [8] proposed four-valued semantics for *DL-Lite*.

3 Proposed Approach

In our work, we mainly focus on how to handle inconsistency in *DL-Lite*. We will explore the following approaches. The list of approaches is not exhausting and new approach may be considered. First, we will work on a method for debugging a *DL-Lite* ontology by exploring the specific feature of *DL-Lite*. Second, we will work on a method for consistent query answering in *DL-Lite* based on multi-valued semantics. Third, we will propose a novel method for measuring inconsistency in a *DL-Lite* ontology. The main expected results of our research are as follows:

1. To give a method for measuring the inconsistency degree of *DL-Lite* ontologies. The inconsistency measures can be used to evaluate the quality of the ontology, to rank axioms in an ontology and to give guidance to resolve inconsistency. We expect that the inconsistency degree of an ontology is a single value. Unlike the work given in [10], our work uses a single value as the inconsistency degree of an ontology. Arguably, this single value is easier to be used to deal with inconsistencies, we will find an algorithm to compute the inconsistency degree in polynomial time.
2. To find a debugging algorithm tailored for *DL-Lite* ontologies. We will explore the tracing techniques for consistency checking algorithm in *DL-Lite*.
3. To design algorithms for query answering over inconsistent *DL-Lite* ontologies. Based [6], we plan to find a solution for query answering over inconsistency *DL-Lite* ontologies based on multi-valued semantics.
4. To implement the proposed algorithm for dealing with inconsistencies in *DL-Lite* ontologies.

In this paper, we will elaborate the first task, i.e., measuring inconsistency in *DL-Lite* ontologies. Our work is based on a three-valued semantics. Compared with classical semantics, three-valued semantics allows for a third truth contradictory such that even an inconsistency ontology can has three-valued models. Since we only aim to analyze inconsistency, there is no need to adopt other multi-valued semantics, such as four-valued semantics which contains a fourth truth value for expressing incomplete knowledge.

4 Research Methodology and Results

4.1 Three-Valued Semantics

Inspired by the four-valued semantics [7,8], we give a three-valued semantics for *DL-Lite*. Based on the three-valued semantics, the truth values about concepts or roles have three values instead of the classical two. The three truth values are $true$, $false$ and $contradictory$, we use the symbols $t, f, \mathbb{B}$ respectively to denote them. For a given domain Δ, we give each concept or role an extended truth value $\langle P, N\rangle$, where P (resp., N) is the subset of Δ (for concept) or $\Delta \times \Delta$ (for role) that supports concept or role to be true (resp., false). We denote $proj^+(\langle P, N\rangle) = P$ and $proj^-(\langle P, N\rangle) = N$. Given a three-valued interpretation I, we say that I satisfies a concept inclusion axiom $B \sqsubseteq C$ (resp., a role inclusion axiom $R_1 \sqsubseteq R_2$) if $proj^+(B^I) \subseteq proj^+(C^I)$ (resp., $proj^+(R_1^I) \subseteq proj^+(R_2^I)$). I satisfies a function assertion (Funct Q) if $\forall x, y, z, (x, y) \in proj^+(Q^I) \wedge (x, z) \in proj^+(Q^I) \rightarrow (y, z) \in proj^+((=)^I)$. Furthermore, I satisfies an atomic concept assertion $C(a)$ (resp., an atomic role assertion $R(a, b)$) if $a^I \in proj^+(C^I)$ (resp., $(a^I, b^I) \in proj^+(R^I)$). A three-valued model of a *DL-Lite* ontology $\mathcal{K}$ is a three-valued interpretation I which satisfies each assertion and each axiom in $\mathcal{K}$. A *DL-Lite* ontology is three-valued satisfiable (unsatisfiable) if there exists (does not exist) such a model. Furthermore, any *DL-Lite* ontology has the finite model property under three-valued semantics.

4.2 Approach to Measuring Inconsistency

As we have discussed above, we want to compute a single inconsistency degree of a *DL-Lite* ontology. We start with studying on the conflicting atomic assertions. Considering the novel property of *DL-Lite* that a *DL-Lite* knowledge base $\mathcal{K} = \langle \mathcal{T}, \mathcal{A}\rangle$ is satisfiable iff *db*($\mathcal{A}$) is a model of $\langle cln(\mathcal{T}), \mathcal{A}\rangle)$ [2] and a contradiction on a *DL-Lite* KB may hold only if a membership assertion in the ABox contradicts a functionality assertion or a negative inclusions (NI) implied by the closure $cln(\mathcal{T})$ [2], we think that $db(\mathcal{A})$ and $\langle cln(\mathcal{T}), \mathcal{A}\rangle$ may be useful for measuring the inconsistency degree. We plan to do as follows:

To compute all membership assertions that maybe cause inconsistency. Consider a *DL-Lite* ontology $\mathcal{K} = \langle \mathcal{T}, \mathcal{A}\rangle$ without functional assertions, we find that all conflicting assertions in $\langle cln(\mathcal{T}), \mathcal{A}\rangle$ obtained through a three-valued preferred model with the domain $\Delta^{db(\mathcal{A})}$ are also in $\mathcal{A}$. Furthermore, we find that $\mathcal{K}$ will be consistent when we remove from $\mathcal{K}$ the conflict assertions of $\langle cln(\mathcal{T}), \mathcal{A}\rangle\rangle$.

Now we discuss functional assertions. Consider a *DL-Lite* ontology $\mathcal{K} = \langle \mathcal{T}, \mathcal{A}\rangle$, where $\mathcal{T}$ only contains functional assertions. If $\mathcal{K}$ is inconsistent, we can know that there are some membership assertions in $\mathcal{A}$ contradicting with some functional assertions. By computing preferred models of $\mathcal{K}$ with any domain, we find that the conflict sets obtained in any preferred model with respect to different domains are the same and assertions in conflict sets are all in the form $= (a, b)$ [1]. For example, for $\mathcal{K} = \{(FunctR), R(a, b), R(a, c)\}$, the conflict set is $\{= (b, c)\}$. We also note that the conflict set is only related with $\mathcal{A}$, so we have the following definition.

[1] We enforce the unique name assumption on constants [1].

Definition 1 (Inconsistency Degree). *Let $\mathcal{K} = \langle \mathcal{T}, \mathcal{A} \rangle$ be a DL-Lite$_\mathcal{R}$ or a DL-Lite$_\mathcal{F}$ ontology, and let I be a three-valued preferred model of $\langle cln(\mathcal{T}), \mathcal{A} \rangle$ w.r.t $\Delta^{db(\mathcal{A})}$. The inconsistency degree of DL-Lite$_\mathcal{R}$ ontology (resp., DL-Lite$_\mathcal{F}$ ontology), called $OntoInc$ ($\mathcal{K}$), is defined as: $OntoInc(\mathcal{K}) = \frac{|ConSet(I,\langle cln(\mathcal{T}),\mathcal{A}\rangle)|}{|GroundSet(\mathcal{K})|}$ (resp., $OntoInc(\mathcal{K}) = \frac{|ConSet(I,\langle cln(\mathcal{T}),\mathcal{A}\rangle)|+ConEqus(I,\langle cln(\mathcal{T}),\mathcal{A}\rangle)}{|GroundSet(\mathcal{K})|}$), where $ConSet(I, \langle cln(\mathcal{T}), \mathcal{A} \rangle)$ is the set of conflicting atomic individual assertions in $\langle cln(\mathcal{T}), \mathcal{A} \rangle$ regardless of functional assertions and $ConEqus(I, \langle cln(\mathcal{T}), \mathcal{A} \rangle)$ is the set of conflicting assertions obtained through functional assertions and $GroundSet(\mathcal{K})$ is the collection of all possible atomic individual assertions.*

From Definition 1, we can see that the inconsistency degree is only a single value. The next step is only to compute a preferred model of $\langle cln(\mathcal{T}), \mathcal{A} \rangle$ with the domain $\Delta^{db(\mathcal{A})}$ so as to obtain $ConSet(I, \langle cln(\mathcal{T}), \mathcal{A} \rangle)$ and $ConEqus(I, \langle cln(\mathcal{T}), \mathcal{A} \rangle)$.

4.3 Algorithm to Compute a Preferred Model of $\langle cln(\mathcal{T}), \mathcal{A} \rangle$

Using the property that $cln(\mathcal{T})$ only contains NIs and functional assertions and the definition of chase in [2], we can give a definition of chase which mainly adapt to $\langle cln(\mathcal{T}), \mathcal{A} \rangle$ which is finite. Then based on this chase, we give an algorithm to compute a three-valued model of $\langle cln(\mathcal{T}), \mathcal{A} \rangle$ which is a preferred model of $\langle cln(\mathcal{T}), \mathcal{A} \rangle$ with the domain $\Delta^{db(\mathcal{A})}$. So we can obtain $ConSet(I, \langle cln(\mathcal{T}), \mathcal{A} \rangle)$ and $ConEqus(I, \langle cln(\mathcal{T}), \mathcal{A} \rangle)$. We show that the complexity of the algorithm is in PTime. Finally, we can obtain the inconsistent degree of $\langle \mathcal{T}, \mathcal{A} \rangle$ based on Definition 1.

4.4 Evaluation

We plan to implement the algorithm for measuring the inconsistency degree proposed above and give a prototype. Meanwhile we will give some potential applications of our tool. For example, users can input some *DL-Lite* ontologies and obtain the inconsistency degree of each ontology; users can either resolve inconsistency if the degree is high (e.g. greater than 0.7 [10]) or ignore it.

5 Conclusion

Dealing with inconsistency in ontologies is an important topic in the Semantic Web. In our paper, we mainly discuss how to handle inconsistency in *DL-Lite* ontologies. We first give some proposals for our work, then we illustrate how to measure inconsistency in a *DL-Lite* ontology based on a three-valued semantics.

References

1. Baader, F., Calvanese, D., McGuinness, D., Nardi, D., Patel-Schneider, P.: Description logic terminology. In: Baader, F., et al. (eds.) The Description Logic Handbook: Theory, Implementation, and Applications. Cambridge University Press, Cambridge (2003)

2. Calvanese, D., De Giacomo, G., Lembo, D., Lenzerini, M., Rosati, R.: Tractable reasoning and efficient query answering in description logics: The *DL-Lite* family. J. Autom. Reasoning 39(3), 385–429 (2007)
3. Deng, X., Haarslev, V., Shiri, N.: Measuring inconsistencies in ontologies. In: Franconi, E., Kifer, M., May, W. (eds.) ESWC 2007. LNCS, vol. 4519, pp. 326–340. Springer, Heidelberg (2007)
4. Friedrich, G., Shchekotykhin, K.: A general diagnosis method for ontologies. In: Gil, Y., Motta, E., Benjamins, V.R., Musen, M.A. (eds.) ISWC 2005. LNCS, vol. 3729, pp. 232–246. Springer, Heidelberg (2005)
5. Huang, Z., van Harmelen, F., ten Teije, A.: Reasoning with inconsistent ontologies. In: Kaelbling, L.P., Saffiotti, A. (eds.) Proceedings of the 19th International Joint Conference on Artificial Intelligence, pp. 454–459 (2005)
6. Lembo, D., Ruzzi, M.: Consistent query answering over description logic ontologies. In: Marchiori, M., Pan, J.Z., de Sainte Marie, C. (eds.) RR 2007. LNCS, vol. 4524, pp. 194–208. Springer, Heidelberg (2007)
7. Ma, Y., Hitzler, P., Lin, Z.: Algorithms for paraconsistent reasoning with OWL. In: Franconi, E., Kifer, M., May, W. (eds.) ESWC 2007. LNCS, vol. 4519, pp. 399–413. Springer, Heidelberg (2007)
8. Ma, Y., Hitzler, P., Lin, Z.: Paraconsistent reasoning for expressive and tractable description logics. In: Baader, F., Lutz, C., Motik, B. (eds.) Description Logics (2008)
9. Ma, Y., Qi, G., Hitzler, P., Lin, Z.: An algorithm for computing inconsistency measurement by paraconsistent semantics. In: Mellouli, K. (ed.) ECSQARU 2007. LNCS, vol. 4724, pp. 91–102. Springer, Heidelberg (2007)
10. Ma, Y., Qi, G., Hitzler, P., Lin, Z.: Measuring inconsistency for description logics based on paraconsistent semantics. In: Mellouli, K. (ed.) ECSQARU 2007. LNCS, vol. 4724, pp. 30–41. Springer, Heidelberg (2007)
11. Parsia, B., Sirin, E., Kalyanpur, A.: Debugging OWL ontologies. In: Ellis, A., Hagino, T. (eds.) Proceedings of the 14th International World Wide Web Conference, pp. 633–640 (2005)
12. Patel-Schneider, P.F.: A four-valued semantics for terminological logics. Artif. Intell. (3), 319–351 (1989)
13. Qi, G., Haase, P., Huang, Z., Ji, Q., Pan, J.Z., Völker, J.: A kernel revision operator for terminologies - algorithms and evaluation. In: Sheth, A.P., et al. (eds.) ISWC 2008. LNCS, vol. 5318, pp. 419–434. Springer, Heidelberg (2008)
14. Qi, G., Hunter, A.: Measuring incoherence in description logic-based ontologies. In: Aberer, K., Choi, K.-S., Noy, N., Allemang, D., Lee, K.-I., Nixon, L., Golbeck, J., Mika, P., Maynard, D., Mizoguchi, R., Schreiber, G., Cudré-Mauroux, P. (eds.) ASWC 2007 and ISWC 2007. LNCS, vol. 4825, pp. 381–394. Springer, Heidelberg (2007)
15. Qi, G., Liu, W., Bell, D.A.: Knowledge base revision in description logics. In: Fisher, M., van der Hoek, W., Konev, B., Lisitsa, A. (eds.) JELIA 2006. LNCS, vol. 4160, pp. 386–398. Springer, Heidelberg (2006)
16. Schlobach, S.: Diagnosing terminologies. In: Veloso, M.M., Kambhampati, S. (eds.) Proceedings of the 20th National Conference on Artificial Intelligence, pp. 670–675 (2005)
17. Schlobach, S., Cornet, R.: Non-standard reasoning services for the debugging of description logic terminologies. In: Gottlob, G., Walsh, T. (eds.) Proceedings of the 17th International Joint Conference on Artificial Intelligence, pp. 355–362 (2003)
18. Schlobach, S., Huang, Z., Cornet, R., van Harmelen, F.: Debugging incoherent terminologies. J. Autom. Reasoning 39(3), 317–349 (2007)

Author Index

FSC
www.fsc.org
MIX
Papier | Fördert
gute Waldnutzung
FSC® C083411

Kolibri 360 GmbH
Ferdinand-Jühlke-Straße 7
99095 Erfurt, Deutschland
produktsicherheit@kolibri360.de